INTERNATIONAL TABLE OF ATOMIC WEIGHTS (1987)

Based on relative atomic mass of $^{12}C = 12$.

The following values apply to elements as they exist in materials of terrestrial origin and to certain artificial elements. Values in parentheses are the mass number of the isotope of longest half-life.

Name	Symbol	Atomic Number	Atomic Weight	Name	Symbol	Atomic Number	Atomic Weight	Name	Symbol	Atomic Number	Atomic Weight
Actinium[d,e]	Ac	89	(227)	Helium[g]	He	2	4.002602	Radium[d,e,g]	Ra	88	(226)
Aluminum[a]	Al	13	26.981539	Holmium[a,b]	Ho	67	164.93032	Radon[d,e]	Rn	86	(222)
Americium[d,e]	Am	95	(243)	Hydrogen[b,c,g]	H	1	1.00794	Rhenium	Re	75	186.207
Antimony (Stibium)	Sb	51	121.75	Indium	In	49	114.82	Rhodium[a]	Rh	45	102.90550
				Iodine[a]	I	53	126.90447	Rubidium[g]	Rb	37	85.4678
Argon[b,g]	Ar	18	39.948	Iridium	Ir	77	192.22	Ruthenium[g]	Ru	44	101.07
Arsenic[a]	As	33	74.92159	Iron	Fe	26	55.847	Samarium[g]	Sm	62	150.36
Astatine[d,e]	At	85	(210)	Krypton[c,g]	Kr	36	83.80	Scandium[a]	Sc	21	44.955910
Barium	Ba	56	137.327	Lanthanum[g]	La	57	138.9055	Selenium	Se	34	78.96
Berkelium[d,e]	Bk	97	(247)	Lawrencium[d,e]	Lr	103	(260)	Silicon[b]	Si	14	28.0855
Beryllium[a]	Be	4	9.012182	Lead[b,g]	Pb	82	207.2	Silver[g]	Ag	47	107.8682
Bismuth[a]	Bi	83	208.98037	Lithium[b,c,g]	Li	3	6.941	Sodium (Natrium)[a]	Na	11	22.989768
Boron[b,c,g]	B	5	10.811	Lutetium[g]	Lu	71	174.967				
Bromine	Br	35	79.904	Magnesium	Mg	12	24.3050	Strontium[b,g]	Sr	38	87.62
Cadmium	Cd	48	112.411	Manganese[a]	Mn	25	54.93805	Sulfur[b]	S	16	32.066
Calcium[g]	Ca	20	40.078	Mendelevium[d,e]	Md	101	(258)	Tantalum	Ta	73	180.9479
Californium[d,e]	Cf	98	(251)	Mercury	Hg	80	200.59	Technetium[d,e]	Tc	43	(98)
Carbon	C	6	12.011	Molybdenum	Mo	42	95.94	Tellurium[b]	Te	52	127.60
Cerium[b,g]	Ce	58	140.115	Neodymium[g]	Nd	60	144.24	Terbium[a]	Tb	65	158.92534
Cesium[a]	Cs	55	132.90543	Neon[c,g]	Ne	10	20.1797	Thallium	Tl	81	204.3833
Chlorine	Cl	17	35.4527	Neptunium[d,e]	Np	93	(237)	Thorium[b,f,g]	Th	90	232.0381
Chromium	Cr	24	51.9961	Nickel	Ni	28	58.69	Thulium[a,b]	Tm	69	168.93421
Cobalt[a]	Co	27	58.93320	Niobium[a]	Nb	41	92.90638	Tin[g]	Sn	50	118.710
Copper	Cu	29	63.546	Nitrogen[b,g]	N	7	14.00674	Titanium	Ti	22	47.88
Curium[b]	Cm	96	(247)	Nobelium[d,e]	No	102	(259)	Tungsten (Wolfram)	W	74	183.85
Dysprosium[e,g]	Dy	66	162.50	Osmium[g]	Os	76	190.2				
Einsteinium[d,e]	Es	99	(252)	Oxygen[b,g]	O	8	15.9994	Unnilquadium[d]	Unq	104	(261)
Erbium[g]	Er	68	167.26	Palladium[g]	Pd	46	106.42	Unnilpentium[d]	Unp	105	(262)
Europium[g]	Eu	63	151.965	Phosphorus[a]	P	15	30.973762	Unnilhexium[d]	Unh	106	(263)
Fermium[d,e]	Fm	100	(257)	Platinum	Pt	78	195.08	Unnilseptium[d]	Uns	107	(262)
Fluorine[a]	F	9	18.9984032	Plutonium[d,e]	Pu	94	(244)	Uranium[c,f,g]	U	92	238.0289
Francium[d,e]	Fr	87	(223)	Polonium[d,e]	Po	84	(209)	Vanadium	V	23	50.9415
Gadolinium[g]	Gd	64	157.25	Potassium (Kalium)	K	19	39.0983	Xenon[a,c,g]	Xe	54	131.29
Gallium	Ga	31	69.723					Ytterbium[g]	Yb	70	173.04
Germanium	Ge	32	72.61	Praseodymium[a]	Pr	59	140.90765	Yttrium[a]	Y	39	88.90585
Gold[a]	Au	79	196.96654	Promethium[d,e]	Pm	61	(145)	Zinc	Zn	30	65.39
Hafnium	Hf	72	178.49	Protactinium[f]	Pa	91	231.03588	Zirconium[g]	Zr	40	91.224

[a] Elements with only one stable nuclide.

[b] Element for which known variation in isotopic abundance in terrestrial samples limits the precision of the atomic weight given.

[c] Element for which users are cautioned against the possibility of large variations in atomic weight due to inadvertent or undisclosed artificial separation in commercially available materials.

[d] Element has no stable nuclides.

[e] Radioactive element that lacks a characteristic terrestrial isotopic composition.

[f] An element, without stable nuclide(s), exhibiting a range of characteristic terrestrial compositions of long-lived radionuclide(s) such that a meaningful atomic weight can be given.

[g] In some geological specimens this element has an anomalous isotopic composition, corresponding to an atomic weight significantly different from that given.

General Chemistry

Fourth Edition

Kenneth W. Whitten

University of Georgia, Athens

Kenneth D. Gailey

Late of University of Georgia, Athens

Raymond E. Davis

University of Texas, Austin

Saunders College Publishing

Harcourt Brace Jovanovich College Publishers

Fort Worth   Philadelphia   San Diego   New York
Orlando   Austin   San Antonio
Toronto   Montreal
London   Sydney   Tokyo

General Chemistry

Fourth Edition

Requests for permission to make copies of any part of the work should be mailed to Permissions Department, Harcourt Brace Jovanovich, Publishers, 8th Floor, Orlando, Florida 32887.

Text Typeface: Times Roman
Compositor: General Graphic Services, Inc.
Acquisitions Editor: John Vondeling
Developmental Editor: Richard Koreto
Managing Editor: Carol Field
Project Editor: Martha Brown
Copy Editor: Mary Patton
Manager of Art and Design: Carol Bleistine
Art Assistant: Caroline McGowan
Text Designer: Tracy Baldwin
Cover Designer: Lawrence R. Didona
Text Artwork: J&R Art Services, Inc.
Layout Artist: Dorothy Chattin
Director of EDP: Tim Frelick
Production Manager: Bob Butler
Marketing Manager: Marjorie Waldron

Cover Credit: Spray of glass fibre optics consisting of 2000 individual strands each measuring 60 microns thick. This type of cable is used for decorative lighting. Adam Hart-Davis/Science Photo Library © Photo Researchers, Inc.

Title page credit: W. H. Woodruff, Los Alamos National Laboratory.

Printed in the United States of America

GENERAL CHEMISTRY, Fourth Edition

ISBN 0-03-072373-6

Library of Congress Catalog Card Number: 91-050635

2345   039   987654321

To the memory of
Kenneth Durwood Gailey

To the Professor

AT&T Bell Laboratories

In revising GENERAL CHEMISTRY and GENERAL CHEMISTRY WITH QUALITATIVE ANALYSIS, we have incorporated many helpful suggestions that we received from professors who used the earlier editions. Facts, explanations, and concepts are presented in a direct and concise fashion. The text is easier to read, while the scientific rigor has been improved significantly. The full-color presentation includes more than 500 photographs of substances, reactions, procedures, and applications. In the artwork and in the textual material, the pedagogic use of color has been expanded.

The fourth edition of the text is larger than the third edition, but the size of the *basic text* has changed very little. New features that make the text more useful and more attractive to students include the following:

1. A **plan,** i.e., a brief outline of the principles and logic used to solve each illustrative example, is given before the solution.
2. Even **more illustrative examples** (308) are worked out in detail.
3. Approximately **2400 end-of-chapter exercises** are included (300 more than the 3rd edition); of these more than 1400 are new.
4. **Enrichment boxes** provide more insight into selected topics for better prepared students. These can be omitted with no loss of continuity.
5. **Chemistry in Use** essays, several by guest authors, relate chemistry to topics of current or historical interest.
6. A list of **objectives** is given at the beginning of each chapter.
7. A more **open format** makes reading easier.

We have exerted great effort to make our text more **flexible.** Some examples follow.

1. We have clearly delineated the parts of Chapter 15, **Thermodynamics,** that can be moved forward for those who wish to cover **thermochemistry** (Sections 15-1 through 15-10) after **stoichiometry** (Chapters 2 and 3).
2. Chapter 4, **Some Types of Chemical Reactions,** is based on the periodic table. This material has been thoroughly reorganized and rewritten to introduce chemical reactions just after stoichiometry. Reactions are classified into the following classes: (a) precipitation, (b) acid–base, (c) displacement, and (d) oxidation–reduction reactions. There is no loss in continuity when **thermochemistry** is covered before Chapter 4. Chapter

4 can be moved to several positions later in text, e.g., after **structure and bonding,** for those who prefer this order.

3. Some professors prefer to discuss **gases** (Chapter 12) after **stoichiometry.** Chapter 12 can be moved into that position with no difficulty.

4. Chapters 5, **(The Structure of Atoms)**, 6 **(Chemical Periodicity)** and 7 **(Chemical Bonding and Inorganic Nomenclature)** provide comprehensive coverage of these key topics.

5. As in earlier editions, **Molecular Structure and Covalent Bonding Theories** (Chapter 8) includes parallel comprehensive VSEPR and VB descriptions of simple molecules. This approach has been widely accepted. However, some professors prefer to present separate descriptions of covalent bonding. The chapter has been carefully *organized into numbered subdivisions* to accommodate these professors; detailed suggestions are included at the beginning of the chapter.

6. Chapter 9 **(Molecular Orbitals in Chemical Bonding)** is a "stand alone chapter" that may be omitted or moved with no loss in continuity.

7. Chapter 10 **(Reactions in Aqueous Solutions I: Acids, Bases, and Salts)** and Chapter 11 **(Reactions in Aqueous Solutions II: Calculations)** include (a) comprehensive discussions of acid–base and redox reactions in aqueous solutions, and (b) solution stoichiometry calculations for acid–base and redox reactions.

We have used color extensively to make the text easier to read and comprehend. A detailed description of our pedagogical use of color is given on page xiv. The result of all these changes is an improved clarity, accuracy, and simplicity of expression throughout the text.

We have kept in mind that chemistry is an experimental science and have emphasized the important role of theory in science. We have presented many of the classical experiments followed by interpretations and explanations of these milestones in the development of scientific thought.

We have defined each new term as accurately as possible and illustrated its meaning as early as practical. We begin each chapter at a very fundamental level and then progress through a series of carefully graded steps to a reasonable level of sophistication. *Numerous* illustrative examples are provided throughout the text and keyed to end-of-chapter exercises (EOC). The first examples in each section are quite simple, the last considerably more complex. The unit–factor method has been emphasized where appropriate.

We have used a blend of SI and the more traditional metric units, because many students are planning careers in areas in which SI units are not yet widely used. The health-care fields, the biological sciences, home economics, and agriculture are typical examples. We have used the joule rather than the calorie in nearly all energy calculations.

We have included throughout the text some interesting historical notes. Marginal notes have been used to point out historical facts, to provide additional bits of information, to further emphasize important points, to relate information to ideas developed earlier, and to note the "relevancy" of various discussions.

We welcome suggestions for improvements in future editions.

Organization

There are thirty-two chapters in GENERAL CHEMISTRY, and GENERAL CHEMISTRY WITH QUALITATIVE ANALYSIS includes eight additional chapters.

We present stoichiometry (**Chapters 2** and **3**) before atomic structure and bonding (**Chapters 5–9**) to establish a sound foundation for a laboratory program as early as possible. However, these chapters are as nearly self-contained as possible to provide flexibility for those who wish to cover structure and bonding before stoichiometry.

Because much of chemistry involves chemical reactions, we have introduced chemical reactions in a simplified, systematic way early in the text (**Chapter 4**). A logical, orderly introduction to formula unit, total ionic, and net ionic equations is included so that this information can be used throughout the remainder of the text. There are many references to this material in later chapters. Solubility rules are presented in this chapter, so that students can use them in writing chemical equations and in their laboratory work.

Because many students have difficulty in *systematizing* and *using* information, we have done our utmost to assist them. At many points throughout the text we summarize the results of recent discussions or illustrative examples in tabular form to help students see the "big picture." The basic ideas on chemical periodicity are introduced early (**Chapters 4** and **6**) and are used throughout the text. A detailed discussion of inorganic nomenclature is included at the end of **Chapter 7.** The simplified classification of acids and bases introduced in **Chapter 4** is expanded in **Chapter 10,** acids, bases, and salts, after the appropriate background on structure and bonding.

References are made to the classification of acids and bases and to the solubility rules throughout the text to emphasize the importance of systematizing and using previously covered information.

Chapter 11 covers solution stoichiometry for both acid–base and redox reactions. The qualitative aspects of redox reactions were presented in **Chapter 4.**

After our excursion through Gases and the Kinetic–Molecular Theory (**Chapter 12**), Liquids and Solids (**Chapter 13**), and Solutions (**Chapter 14**), we have covered sufficient material that students have appropriate background for a wide variety of laboratory experiments.

Comprehensive chapters are presented on Chemical Thermodynamics (**Chapter 15**) and Chemical Kinetics (**Chapter 16**). The distinction between the roles of standard and nonstandard Gibbs free energy change in predicting reaction spontaneity is clearly discussed. Chapter 16, Chemical Kinetics, has been thoroughly reorganized to provide early and consistent emphasis on the experimental basis of kinetics.

These chapters provide the necessary background for a strong introduction to Chemical Equilibrium in **Chapter 17.** This is followed by three chapters on Equilibria in Aqueous Solutions. A chapter on Electrochemistry (**Chapter 21**) completes the "common core" of the text except for Nuclear Chemistry, (**Chapter 30**), which is self-contained and may be studied at any point in the course.

A group of basically descriptive chapters follow. However, we have been careful to include appropriate applications of the principles that have been evolved in the first part of the text to explain descriptive chemistry. **Chapters 22,** The Metals and Metallurgy, **23,** The Representative Metals, and **28,** The Transition Metals, give broad coverage to the chemistry of the metals. **Chapter 29,** Coordination Compounds, is a sound introduction to that field.

Chapters 24–27 give a comprehensive introduction to the chemistry of the nonmetals. Again, care has been taken to explain descriptive chemistry in terms of the principles that have been developed earlier.

The section on organic chemistry has been reorganized and rewritten. **Chapter 31 (Organic Chemistry I: Compounds)** presents the classes of compounds, their structures, and nomenclature with major emphasis on the principal functional groups. **Chapter 32 (Organic Chemistry II: Molecular Geometry and Reactions)** is a highly structured, concise, well-illustrated discussion of the geometries of organic molecules, the three fundamental classes of organic reactions, and some reactions of key functional groups. This material provides a broad overview for students who will not take a course in organic chemistry. It also provides the introduction to important concepts for those who will study organic chemistry.

Eight additional chapters are included in GENERAL CHEMISTRY WITH QUALITATIVE ANALYSIS. In **Chapter 33,** the important properties of the metals of the five cation groups are tabulated, their properties are discussed, the sources of the elements are listed, their metallurgies are described, and a few uses of each metal are given.

Chapter 34 is a detailed introduction to the laboratory procedures used in semimicro qualitative analysis.

Chapters 35–39 cover the analysis of the five groups of cations. Each chapter includes a discussion of the important oxidation states of the metals, an introduction to the analytical procedures, and comprehensive discussions of the chemistry of each cation group. Detailed laboratory directions, set off in color, follow. Students are alerted to pitfalls in advance, and alternate confirmatory tests and "clean-up" procedures are described for troublesome cations. A set of exercises accompanies each chapter.

Chapter 40 contains a discussion of some of the more sophisticated ionic equilibria of qualitative analysis. The material is presented in a single chapter for the convenience of the instructor.

A Complete Ancillary Package

A number of ancillary materials have been prepared to assist the student in his or her study of GENERAL CHEMISTRY and to aid the instructor in presenting the course. Each supplement can be used with either version of this text.

1. LECTURE OUTLINE FOR GENERAL CHEMISTRY, 4th ed., Kenneth W. Whitten, Kenneth D. Gailey, and Richard M. Hedges (Texas A&M University). A comprehensive lecture outline that allows professors to use valuable classroom time more effectively. It provides great flexibility for the professor and makes available more time for special topics, increased drill, or whatever the professor chooses to do.

2. SOLUTIONS MANUAL FOR GENERAL CHEMISTRY, 4th ed., Yi-Noo Tang and Wendy Keeney-Kennicutt (both of Texas A&M Univer-

University of California, Lawrence Livermore Laboratory

sity). It includes detailed answers and solutions for *all even-numbered* end-of-chapter exercises. In-depth answers for discussion questions and helpful comments that reinforce basic concepts are included, as well as references to specific examples and appropriate sections of chapters in the text.

3. STUDY GUIDE FOR GENERAL CHEMISTRY, 4th ed., Raymond E. Davis. It includes brief summaries of important ideas in each chapter, study goals with references to text sections and exercises, and simple preliminary tests (averaging more than 80 short questions per chapter, all with answers) that reinforce basic skills and vocabulary and encourage students to think about important ideas.

4. INSTRUCTOR'S MANUAL TO ACCOMPANY GENERAL CHEMISTRY, 4th ed., Raymond E. Davis. Also includes solutions to *odd-numbered* end-of-chapter exercises and may be made available to students, if the professor chooses.

5. EXPERIMENTAL GENERAL CHEMISTRY, 2nd Edition, Carl B. Bishop, Muriel B. Bishop (Clemson University), K. D. Gailey, and K. W. Whitten. A modern laboratory manual with excellent variety that includes descriptive, quantitative, and instrumental experiments. Designed for mainstream courses for science majors.

6. PROBLEM-SOLVING IN GENERAL CHEMISTRY, 4th ed., Leslie N. Kinsland (University of Southwestern Louisiana), K. W. Whitten, and K. D. Gailey. Covers the common core of general chemistry courses for science majors.

7. OVERHEAD TRANSPARENCIES. One hundred twenty-five four-color figures from the text.

8. TEST BANK AND COMPUTERIZED TEST BANK, Steven H. Albrecht (Ball State University). Over 1,100 questions, all completely new, and available in computerized versions for both IBM and Macintosh computers.

9. SHAKHASHIRI VIDEO TAPES, Bassam Shakhashiri (University of Wisconsin, Madison). Fifty 3–5 minute classroom experiments.

10. VIDEO DISK AND BARCODE MANUAL, containing all the Shakhashiri demonstrations and over 600 images drawn from the text and other sources. Barcode manual allows easy access.

11. PERIODIC TABLE VIDEODISC: REACTIONS OF THE ELEMENTS, JCE: Software. Alton J. Banks (Southwest Texas State University). A visual compilation of information about the chemical elements.

Acknowledgments

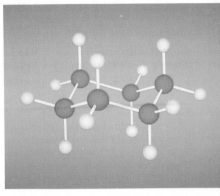

R. E. Davis

The list of individuals who contributed to the evolution of this book is long indeed. First, we would like to express our appreciation to the professors who contributed so much to our scientific education: Professors Arnold Gilbert, M. L. Bryant, the late W. N. Pirkle and Alta Sproull, C. N. Jones, S. F. Clark, R. S. Drago (KWW); C. R. Russ, R. D. Dunlap, H. H. Patterson, R. L. Wells, A. L. Crumbliss, P. Smith, D. B. Chestnut, R. A. Palmer, and B. E. Douglas (KDG); the late Dorothy Vaughn, David Harker, and Calvin Vanderwerf, Professors Ralph N. Adams, F. S. Rowland, A. Tulinsky, and Wm. von E. Doering (RED).

The staff at Saunders College Publishing has contributed immeasurably to the evolution of this book. Our developmental editor, Richard Koreto, has done an outstanding job. With ingenuity, persistence and patience, our photo researcher Dena Digilio-Betz, has gathered many excellent photographs. Carol Bleistine has again given us high-quality design and artwork that contribute to the appearance and the substance of the book. Additionally, we have drawn from the excellent artwork in other Saunders texts. With remarkable good cheer our project editor, Martha Brown, has handled innumerable details with skill, insight, and imagination. Her keen eye for detail has improved consistency and spared us many embarrassments. We express our continued deep appreciation to our editor and friend, John Vondeling, the best editor in the business. John has guided us at every step in the development of each edition of these books, and our respect and admiration for him have grown with each passing day.

Our secretary, Martha Dove, has been patient and skillful through the many revisions. We are indeed grateful for her patience, her skill, and her dedication. Jim Morgenthaler (Athens) and Charles Steele (Austin) did the original photography. We thank them for the tremendous contribution that their enthusiasm and their many hours of patient work and professional expertise have brought to these editions.

Finally, we are deeply indebted to our families, Betty, Andy, and Kathryn Whitten, Kathy, Kristen, and Karen Gailey, and Sharon, Angela, Laura, and Brian Davis, who have supported us during the many years we have worked on this project. Their understanding, encouragement, and moral support have "kept us going."

We are deeply indebted to Professor Gary F. Riley of the St. Louis College of Pharmacy who has again worked all of the end-of-chapter exercises. Only the authors fully appreciate the magnitude of this important contribution.

Carolyn and Steven Albrecht have also made a major contribution to this edition by reading critically both galleys and page proofs.

Reviewers of the Fourth Edition

The following individuals have reviewed the manuscripts for the fourth edition of GENERAL CHEMISTRY and GENERAL CHEMISTRY WITH QUALITATIVE ANALYSIS; several also provided detailed comments on earlier editions. Their suggestions have improved the texts significantly.

Ed Acheson, *Millikin University*
Steven Albrecht, *Ball State University*
Barbara Burke, *California State Polytechnic University*
L. A. Burns, *St. Clair County Community College*
James Carr, *University of Nebraska, Lincoln*
Elaine Carter, *Los Angeles City College*
John DeKorte, *Northern Arizona University*
George Eastland, Jr., *Saginaw Valley State University*
Sandra Etheridge, *Gulf Coast Community College*
Gary Gray, *University of Alabama, Birmingham*
Forrest Hentz, Jr., *North Carolina State University*
Albert Jache, *Marquette University*
William Jensen, *South Dakota State University*

Andrew Jorgensen, *University of Toledo*
Leslie Kinsland, *University of Southwestern Louisiana*
James Krueger, *Oregon State University*
Robert Lamb, *Ohio Northern University*
Alfred Lee, *City College of San Francisco*
Barbara O'Brien, *Texas A&M University*
Christopher Ott, *Assumption College*
William Pietro, *University of Wisconsin, Madison*
Susanne Raynor, *Rutgers University*
Diane Sedney, *George Washington University*
Margaret Tierney, *Prince George's Community College*
James Valentini, *University of California, Irvine*
David Winters, *Tidewater Community College*

Reviewers of the First Three Editions of General Chemistry

Edwin Abbott, *Montana State University*
David R. Adams, *North Shore Community College*
Dale Arrington, *South Dakota School of Mines*
George Atkinson, *Syracuse University*
Jerry Atwood, *University of Alabama*
William G. Bailey, *Broward Community College*
J. M. Bellama, *University of Maryland*
Carl B. Bishop, *Clemson University*
Muriel B. Bishop, *Clemson University*
George Bodner, *Purdue University*
Greg Brewer, *The Citadel*
Clark Bricker, *University of Kansas*
Robert Broman, *University of Missouri*
Robert F. Bryan, *University of Virginia*
Thomas Cassen, *University of North Carolina*
Evelyn A. Clarke, *Community College of Philadelphia*
Lawrence Conroy, *University of Minnesota*
John M. DeKorte, *Northern Arizona University*
Harry A. Eick, *Michigan State University*
Lawrence Epstein, *University of Pittsburgh*
Darrell Eyman, *University of Iowa*
Wade A. Freeman, *University of Illinois, Chicago Circle*
Richard Gaver, *San Jose State University*
Robert Hanrahan, *University of Florida*
Henry Heikkinen, *University of Maryland*
Forrest C. Hentz, *North Carolina State University*
R. K. Hill, *University of Georgia*
Larry W. Houk, *Memphis State University*
Arthur Hufnagel, *Erie Community College (North Campus)*
Wilbert Hutton, *Iowa State University*
M. D. Joesten, *Vanderbilt University*

Philip Kinsey, *University of Evansville*
Leslie N. Kinsland, *The University of Southwestern Louisiana*
Marlene Kolz
Robert Kowerski, *College of San Mateo*
Norman Kulevsky, *University of North Dakota*
Patricia Lee, *Bakersfield College*
William Litchman, *University of New Mexico*
Gilbert J. Mains, *Oklahoma State University*
Ronald Marks, *Indiana University of Pennsylvania*
William Masterton, *University of Connecticut*
Clinton Medbery, *The Citadel*
Joyce Miller, *San Jacinto College*
Joyce Neiburger, *Purdue University*
James L. Pauley, *Pittsburgh State University*
Ronald O. Ragsdale, *University of Utah*
Randal Remmel
Gary F. Riley, *St. Louis College of Pharmacy*
Eugene Rochow, *Harvard University*
John Ruff, *University of Georgia*
Don Roach, *Miami Dade Community College*
George Schenk, *Wayne State University*
William Scroggins, *El Camino College*
Curtis Sears, *Georgia State University*
Mahesh Sharma, *Columbus College*
C. H. Stammer, *University of Georgia*
Yi-Noo Tang, *Texas A&M University*
Janice Turner, *Augusta College*
W. H. Waggoner, *University of Georgia*
Susan Weiner, *West Valley College*
Steve Zumdahl, *University of Illinois*
Douglas Vaughan

Kenneth D. Whitten
Raymond E. Davis
Kenneth D. Gailey

To the Student

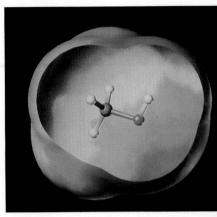

J. Weber

We have written this text to assist you as you study chemistry. Chemistry is a fundamental science—some call it the central science. As you and your classmates pursue diverse career goals you will find that the vocabulary and ideas presented in this text will be useful in more places and in more ways than you may imagine now.

We begin with the most basic vocabulary and ideas. We then carefully evolve increasingly sophisticated ideas that are necessary and useful in all the other physical sciences, the biological sciences, and the applied sciences such as medicine, dentistry, engineering, agriculture, and home economics.

We have made the early chapters as nearly self-contained as possible. The material can be presented in the order considered most appropriate by your professor. Some professors will cover chapters in different orders or will omit some chapters completely—the text was designed to accommodate this.

Early in each section we have attempted to provide the experimental basis for the ideas we evolve. By *experimental basis* we mean the observations and experiments on the phenomena that have been most important in developing concepts. We then present an explanation of the experimental observations.

Chemistry is an experimental science. We know what we know because we (literally thousands of scientists) have observed it to be true. Theories have been evolved to explain experimental observations (facts). Successful theories explain observations fully and accurately. More importantly, they enable us to predict the results of experiments that have not yet been performed. Thus, we should always keep in mind the fact that experiment and theory go hand-in-hand. They are intimately related parts of our attempt to understand and explain natural phenomena.

"*What is the best way to study chemistry?*" is a question we are asked often by our students. While there is no single answer to this question, the following suggestions may be helpful. Your professor may provide additional suggestions. A number of supplementary materials accompany this text. All are designed to assist you as you study chemistry. Your professor may suggest that you use some of them.

Students often underestimate the importance of the act of *writing* as a tool for learning. Whenever you read, do not just highlight passages in the text, but also *take notes*. Whenever you work problems or answer questions *write yourself explanations* of why each step was done or how you reasoned out the answer. Keep a special section of your notebook for working out problems or answering questions. The very act of writing forces you to concentrate more on what you are doing, and you learn more. This is true even if you never go back to review what you wrote earlier. Of course, these notes will also help you to review for an examination.

You should always read over the assigned material before it is covered in class. This helps you to recognize the ideas as your professor discusses them. Take careful class notes. *At the first opportunity,* and certainly the same day, you should recopy your class notes. As you do this, fill in more detail where you can. Try to work the illustrative examples that your professor solved in class, without looking at the solution in your notes. If you must look at the solution, look at only one line (step), and then try to figure out the next step. Read the assigned material again and take notes, integrating these with your class notes. Reading should be much more informative the second time.

Review the "key terms" at the end of the chapter to be sure that you know the exact meaning of each. Work the illustrative examples in the text while covering the solutions with a sheet of paper. If you find it necessary to look at the solutions, look at only one line at a time and try to figure out the next step. Answers to illustrative Examples are displayed on blue backgrounds. At the end of most Examples, we suggest related questions from the end-of-chapter exercises (EOC). You should work these suggested EOC's as you come to them.

This is a good time to work through the appropriate chapter in the STUDY GUIDE TO GENERAL CHEMISTRY. This will help you to see an overview of the chapter, to set specific study goals, and then to check and improve your grasp of basic vocabulary, concepts, and skills. Next, work the assigned exercises at the end of the chapter.

The Appendices contain much useful information. You should become familiar with them and their contents so that you may use them whenever necessary. Answers to all even-numbered numerical exercises are given at the end of the text so that you may check your work.

We heartily recommend the STUDY GUIDE TO GENERAL CHEMISTRY by Raymond E. Davis and the SOLUTIONS MANUAL by Professor Yi-Noo Tang and Dr. Wendy Keeney-Kennicutt, which were written to accompany this text. The STUDY GUIDE provides an overview of each chapter and emphasizes the threads of continuity that run through chemistry. It lists study goals, tells you which ideas are most important and why they are important, and provides many forward and backward references. Additionally, the STUDY GUIDE contains many easy to moderately difficult questions that enable you to gauge your progress. These short questions provide excellent practice in preparing for examinations. Answers are provided for all questions, and many have explanations or references to appropriate sections in the text.

The SOLUTIONS MANUAL contains detailed solutions and answers to all even-numbered end-of-chapter exercises. It also has many helpful references to appropriate sections and illustrative examples in the text.

If you have suggestions for improving this text, please write to us and tell us about them.

Keys for Color Codes

In addition to four-color photography and art, we have used color to help you identify and organize important ideas, techniques, and concepts as you study this book. (Some of the following passages and pieces are displayed for example only, and may not make scientific sense without surrounding text and art.)

1. Important ideas, mathematical relationships, and summaries are displayed on pale blue screens, the width of the text.

> Different pure samples of a compound always contain the same elements in the same proportion by mass; this corresponds to atoms of these elements combined in fixed numerical ratios.

2. Answers to examples are shown on pale blue screens.

Solution

$$? \frac{lb}{ft^3} = 13.59 \frac{g}{cm^3} \times \frac{1\ lb}{453.6\ g} \times \left(\frac{2.54\ cm}{1\ in}\right)^3 \times \left(\frac{12\ in}{1\ ft}\right)^3 = \boxed{848.4\ lb/ft^3}$$

3. Intermediate steps (logic, guidance, and so on) are shown on gold screens.

Example 1-10
Express 1.0 gallon in milliliters.

Plan
We ask ? mL = 1.0 gal and multiply by the appropriate factors.

$$\boxed{\text{gallons}} \rightarrow \boxed{\text{quarts}} \rightarrow \boxed{\text{liters}} \rightarrow \boxed{\text{milliliters}}$$

Solution

$$? \text{ mL} = 1.0 \text{ gal} \times \frac{4\ qt}{1\ gal} \times \frac{1\ L}{1.06\ qt} \times \frac{1000\ mL}{1\ L} = \boxed{3.8 \times 10^3 \text{ mL}}$$

EOC 31, 44

4. Hybridization schemes and hybrid orbitals are emphasized in green.

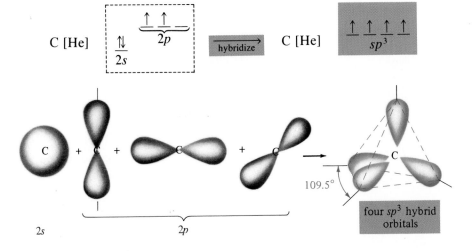

5. Acidic and basic properties are contrasted by using red and blue, respectively. Neutral solutions are indicated in pale purple.

Table 4-9
Bonding, Solubility, Electrolyte Characteristics, and Predominant Forms of Solutes in Contact with Water

	Acids		Bases			Salts	
	Strong acids	Weak acids	Strong soluble bases	Insoluble bases	Weak bases	Soluble salts	Insoluble salts
Examples	HCl HNO$_3$	CH$_3$COOH HF	NaOH Ca(OH)$_2$	Mg(OH)$_2$ Al(OH)$_3$	NH$_3$ CH$_3$NH$_2$	KCl, NaNO$_3$, NH$_4$Br	BaSO$_4$, AgCl, Ca$_3$(PO$_4$)$_2$
Pure compound ionic or covalent?	Covalent	Covalent	Ionic	Ionic	Covalent	Ionic	Ionic
Water soluble or insoluble?	Soluble*	Soluble*	Soluble	Insoluble	Soluble†	Soluble	Insoluble
~100% ionized or dissociated in dilute aqueous solution?	Yes	No	Yes	(footnote ‡)	No	Yes§	(footnote ‡)
Written in ionic equations as	Separate ions	Molecules	Separate ions	Complete formulas	Molecules	Separate ions	Complete formulas

* Most common inorganic acids and the low-molecular-weight organic acids (—COOH) are water soluble.
† The low-molecular-weight amines are water-soluble.
‡ The *very small concentrations* of "insoluble" metal hydroxides and insoluble salts in saturated aqueous solutions are nearly completely dissociated.
§ There are a few exceptions. A few soluble salts are molecular (and not ionic) compounds.

Table 18-1
Common Strong Acids and Strong Soluble Bases

Strong Acids	
HCl	HNO$_3$
HBr	HClO$_4$
HI	HClO$_3$
	H$_2$SO$_4$

Strong Soluble Bases	
LiOH	
NaOH	
KOH	Ca(OH)$_2$
RbOH	Sr(OH)$_2$
CsOH	Ba(OH)$_2$

$$NH_3(aq) + H_2O(\ell) \rightleftharpoons NH_4^+(aq) + OH^-(aq)$$

base$_1$ acid$_2$ acid$_1$ base$_2$

6. Red and blue are also used in oxidation-reduction reactions and electrochemistry (Chapter 21).
   (a) Oxidation numbers are shown in red in red circles to avoid confusion with ionic charges.

Nonmetal Oxide	+ Water	→	Ternary Acid	
carbon dioxide	CO$_2$(g) (+4)	+ H$_2$O(ℓ)	→	H$_2$CO$_3$(aq) (+4)	carbonic acid
sulfur dioxide	SO$_2$(g) (+4)	+ H$_2$O(ℓ)	→	H$_2$SO$_3$(aq) (+4)	sulfurous acid
sulfur trioxide	SO$_3$(g) (+6)	+ H$_2$O(ℓ)	→	H$_2$SO$_4$(aq) (+6)	sulfuric acid

   (b) Oxidation is indicated by blue and reduction is indicated by red.

$$2KClO_3(s) \longrightarrow 2KCl(s) + 3O_2(g)$$

(+1)(+5)(−2) → (+1)(−1) (0)

−6
+2

(c) In electrochemistry (Chapter 21) we learn that oxidation occurs at
the *anode*; we use blue to indicate the anode and its half-reaction.
Similarly, reduction occurs at the *cathode*; we use red to indicate
the cathode and its half-reaction.

$$\text{Cu} \longrightarrow \text{Cu}^{2+} + 2e^- \quad \text{(oxidation, anode)}$$
$$2(\text{Ag}^+ + e^- \longrightarrow \text{Ag}) \quad \text{(reduction, cathode)}$$
$$\overline{\text{Cu} + 2\text{Ag}^+ \longrightarrow \text{Cu}^{2+} + 2\text{Ag}} \quad \text{(overall cell reaction)}$$

7. In discussions of molecular orbitals (Chapter 9), bonding and antibonding
molecular orbitals are shown in blue and red, respectively.

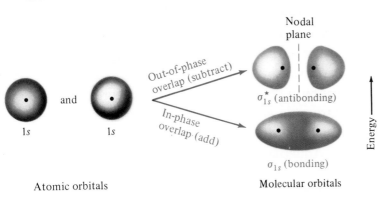

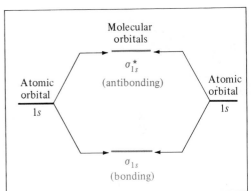

8. Color-coded periodic tables emphasize the classification of the elements as metals (blue), nonmetals (yellow), and metalloids (green).
Please study the periodic table inside the front cover carefully so
that you recognize this color scheme.

Table 4-2
The Periodic Table*

IA												IIIA	IVA	VA	VIA	VIIA	0
1 H																	2 He
3 Li	4 Be											5 B	6 C	7 N	8 O	9 F	10 Ne
11 Na	12 Mg					VIIIB			IB	IIB		13 Al	14 Si	15 P	16 S	17 Cl	18 Ar
19 K	20 Ca	21 Sc	22 Ti	23 V	24 Cr	25 Mn	26 Fe	27 Co	28 Ni	29 Cu	30 Zn	31 Ga	32 Ge	33 As	34 Se	35 Br	36 Kr
37 Rb	38 Sr	39 Y	40 Zr	41 Nb	42 Mo	43 Tc	44 Ru	45 Rh	46 Pd	47 Ag	48 Cd	49 In	50 Sn	51 Sb	52 Te	53 I	54 Xe
55 Cs	56 Ba	57 La *	72 Hf	73 Ta	74 W	75 Re	76 Os	77 Ir	78 Pt	79 Au	80 Hg	81 Tl	82 Pb	83 Bi	84 Po	85 At	86 Rn
87 Fr	88 Ra	89 Ac †															

Contents Overview

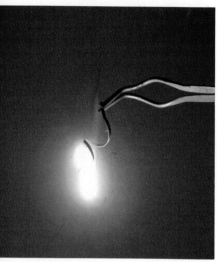

Charles D. Winters

Table of Contents

Charles D. Winters

Charles Steele

Robert W. Metz

Kip Peticolas, Fundamental Photographs, New York

Atlanta Gas Light Company

Charles Steele

Paul Silverman, Fundamental Photographs, New York

Courtesy American Cyanamid Company

Paul Silverman, Fundamental Photographs, New York

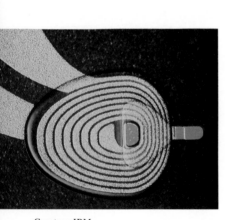

Courtesy IBM

U.S. Dept. Energy/Science Photo Library,
Photo Researchers, Inc.

1 The Foundations of Chemistry

The earth is a huge chemical system, including innumerable reactions taking place constantly with some energy input from sunlight. The earth serves as the source of raw materials for *all* human activities, as well as the depository for the products of these activities. Maintaining life on the planet requires significantly improved understanding and wise use of these resources. Scientists can provide important information about the processes, but each of us must share in the responsibility for our environment.

Outline

Objectives

As you study this chapter, you should learn to

☐ Use the basic vocabulary of matter and energy
☐ Distinguish between chemical and physical properties and between chemical and physical changes
☐ Recognize various forms of matter: homogeneous and heterogeneous mixtures, substances, compounds, and elements
☐ Recognize some common methods of separation used in chemistry
☐ Apply the concept of significant figures

☐ Apply appropriate units to describe the results of measurement
☐ Use the unit factor method to carry out conversions among units
☐ Describe temperature measurements on various common scales, and convert between these scales
☐ Carry out calculations relating temperature change to heat absorbed or liberated

T housands of practical questions are studied by chemists. A few of them are

How can we modify a useful drug so as to improve its effectiveness while minimizing harmful or unpleasant side effects?

How can we develop better materials to be used as synthetic bone in transplants?

Which substances could help to avoid rejection of foreign tissue in organ transplants?

What improvements in fertilizers or pesticides can improve agricultural yields? How can this be done with minimal environmental danger?

How can we get the maximum work from a fuel while producing the least harmful emissions possible?

Which really poses the greater environmental threat—the burning of fossil fuels and its contribution to the greenhouse effect and climatic change, or the use of nuclear power and the related radiation and disposal problems?

How can we develop suitable materials for the semiconductor and microelectronics industry? Can we develop a battery that is cheaper, lighter, and more powerful?

What changes in structural materials could help to make aircraft lighter and more economical, yet at the same time stronger and safer?

What relation is there between the substances we eat, drink, or breathe and the possibility of developing cancer? How can we develop substances that are effective in killing cancer cells preferentially over normal cells?

Can we economically produce fresh water from sea water for irrigation or consumption?

How can we slow down unfavorable reactions, such as corrosion of metals, while speeding up favorable ones, such as the growth of foodstuffs?

Chemistry touches almost every aspect of our lives, our culture, and our environment. Its scope encompasses the air we breathe, the food we eat, the fluids we drink, our clothing, dwellings, transportation and fuel supplies, and our fellow creatures.

Chemistry is the science that describes matter—its chemical and physical properties, the chemical and physical changes it undergoes, and the energy changes that accompany those processes. Matter includes everything that is tangible, from our bodies and the stuff of our everyday lives to the grandest objects in the universe. Some call chemistry the central science. It rests on the foundation of mathematics and physics and in turn underlies the life sciences—biology and medicine. To understand living systems fully, we must first understand the chemical reactions and chemical influences that operate within them. The chemicals of our bodies profoundly affect even the personal world of our thoughts and emotions.

We understand simple chemical systems well; they lie near chemistry's fuzzy boundary with physics. They can often be described exactly by mathematical equations. We fare less well with more complicated systems. Even where our understanding is fairly thorough, we must make approximations, and often our knowledge is far from complete. Each year researchers provide new insights into the nature of matter and its interactions. As chemists find answers to old questions, they learn to ask new ones. Our scientific knowledge has been described as an expanding sphere that, as it grows, encounters an ever-enlarging frontier.

In our search for understanding, we eventually must ask fundamental questions such as the following:

How do substances combine to form other substances? How are energy changes involved in chemical and physical changes?

How is matter constructed, in its intimate detail? How are atoms and the ways that they combine related to the properties of the matter that we can measure, such as color, hardness, chemical reactivity, and electrical conductivity?

What fundamental factors influence the stability of a substance? How can we force a desired (but energetically unfavorable) change to take place? What factors control the rate at which a chemical change takes place?

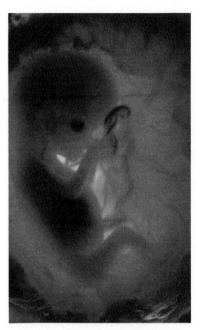

Enormous numbers of chemical reactions are necessary to produce a human embryo (here at 10 weeks, 6 cm long).

In your study of chemistry, you will learn about these and many other basic ideas that chemists have developed to help them describe and understand the behavior of matter. Along the way, we hope that you come to appreciate the development of this science, one of the grandest intellectual achievements of human endeavor. You will also learn how to apply these fundamental principles to solve real problems. One of your major goals in the study of chemistry should be to develop your ability to think critically and to solve problems (not just do numerical calculations!). In other words, you need to learn to manipulate not only numbers, but also quantitative ideas, words, and concepts.

In the first chapter, our main goals are (1) to begin to get an idea of what chemistry is about and the ways in which chemists view and describe the material world and (2) to acquire some skills that are useful and necessary in the understanding of chemistry and of science.

1-1 Matter and Energy

Matter is anything that has mass and occupies space. Mass is a measure of the quantity of matter in a sample of any material. The more massive an object is, the more force is required to put it in motion. All bodies consist of matter. Our senses of sight and touch usually tell us that an object occupies space. In the case of colorless, odorless, tasteless gases (such as air), our senses may fail us.

Energy is defined as the capacity to do work or to transfer heat. We are familiar with many forms of energy, including mechanical energy, electrical energy, heat energy, and light energy. Light energy from the sun is used by plants as they grow; electrical energy allows us to light a room by flicking a switch; and heat energy cooks our food and warms our homes. Energy can be classified into two principal types: kinetic energy and potential energy.

We might say that we can "touch" air when it blows in our faces, but we depend on other evidence to show that a still body of air fits our definition of matter.

Figure 1-1
Magnesium burns in the oxygen of the air to form magnesium oxide, a white solid. This reaction occurs in photographic flashbulbs. There is no gain or loss of mass as the reaction occurs.

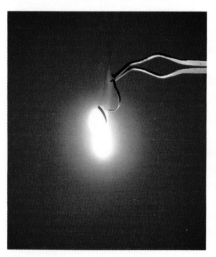
(a)

(b)

(c)

A body in motion, such as a rolling boulder, possesses energy because of its motion. Such energy is called **kinetic energy**. Kinetic energy represents the capacity for doing work directly. It is easily transferred between objects. **Potential energy** is the energy an object possesses because of its position or composition. Coal, for example, possesses chemical energy, a form of potential energy, because of its composition. Many electrical generating plants burn coal, producing heat and subsequently electrical energy. A boulder located atop a mountain possesses potential energy because of its height. It can roll down the mountainside and convert its potential energy into kinetic energy. We discuss energy because all chemical processes are accompanied by energy changes. As some processes occur, energy is released to the surroundings, usually as heat energy. We call such processes **exothermic**. Any combustion (burning) reaction is exothermic. However, some chemical reactions and physical changes are **endothermic**; i.e., they absorb energy from their surroundings. An example of a physical change that is endothermic is the melting of ice.

The term comes from the Greek word *kinein*, meaning "to move." The word "cinema" is derived from the same word.

Nuclear energy is an important kind of potential energy.

The Law of Conservation of Matter

When we burn a sample of metallic magnesium in the air, the magnesium combines with oxygen from the air (Figure 1-1) to form magnesium oxide, a white powder. This chemical reaction is accompanied by the release of large amounts of heat energy and light energy. When we weigh the product of the reaction, magnesium oxide, we find that it is heavier than the original piece of magnesium. The increase in mass of the solid is due to the combination of oxygen with magnesium to form magnesium oxide. Many experiments have shown that the mass of the magnesium oxide is exactly the sum of the masses of magnesium and oxygen that combined to form it. Similar statements can be made for all chemical reactions. These observations are summarized in the **Law of Conservation of Matter**:

> There is no observable change in the quantity of matter during a chemical reaction or during a physical change.

This statement is an example of a **scientific (natural) law**, a general statement based on the observed behavior of matter to which no exceptions are known. A nuclear reaction is *not* a chemical reaction.

The Law of Conservation of Energy

In exothermic chemical reactions, *chemical energy* usually is converted into *heat energy*. Some exothermic processes involve other kinds of energy changes. For example, some liberate light energy without heat, and others produce electrical energy without heat or light. In *endothermic* reactions, heat energy, light energy, or electrical energy is converted into chemical energy. Although chemical changes always involve energy changes, some energy transformations do not involve chemical changes at all. For example, heat energy may be converted into electrical energy or into mechanical energy without any simultaneous chemical changes. Many experiments have demonstrated that all of the energy involved in any chemical or physical change appears

Electricity is produced in hydroelectric plants by the conversion of mechanical energy (from flowing water) into electrical energy.

in some form after the change. These observations are summarized in the **Law of Conservation of Energy**:

> Energy cannot be created or destroyed in a chemical reaction or in a physical change. It can only be converted from one form to another.

The Law of Conservation of Matter and Energy

With the dawn of the nuclear age in the 1940s, scientists, and then the world, became aware that matter can be converted into energy. In nuclear reactions (Chapter 30) matter is transformed into energy. The relationship between matter and energy is given by Albert Einstein's now famous equation

$$E = mc^2$$

This equation tells us that the amount of energy released when matter is transformed into energy is the product of the mass of matter transformed and the speed of light squared. At the present time, man has not (knowingly) observed the transformation of energy into matter on a large scale. It does, however, happen on an extremely small scale in "atom smashers," or particle accelerators, used to induce nuclear reactions. Now that the equivalence of matter and energy is recognized, the **Law of Conservation of Matter and Energy** can be stated in a single sentence:

> The combined amount of matter and energy in the universe is fixed.

Einstein formulated this equation in 1905 as a part of his theory of relativity. Its validity was demonstrated in 1939 with the first controlled nuclear reaction.

1-2 States of Matter

Matter can be classified into three states (Figure 1-2), although everyone can think of examples that do not fit neatly into any of the three categories. In the **solid state**, substances are rigid and have definite shapes. Volumes of

Figure 1-2
(a) Iodine, a solid element. (b) Bromine, a liquid element. (c) Chlorine, a gaseous element.

(a)   (b)   (c)

solids do not vary much with changes in temperature and pressure. In many solids, called crystalline solids, the individual particles that make up the solid occupy definite positions in the crystal structure. The strengths of interaction between the individual particles determine how hard and how strong the crystals are. In the **liquid state**, the individual particles are confined to a given volume. A liquid flows and assumes the shape of its container up to the volume of the liquid. Liquids are very hard to compress. **Gases** are much less dense than liquids and solids. They occupy all parts of any vessel in which they are confined. Gases are capable of infinite expansion and are compressed easily. We conclude that they consist primarily of empty space; i.e., the individual particles are quite far apart.

1-3 Chemical and Physical Properties

To distinguish among samples of different kinds of matter, we determine and compare their **properties**. We recognize different kinds of matter by their properties, which are broadly classified into chemical properties and physical properties.

The properties of a person include height, weight, sex, skin and hair color, and the many subtle features that constitute that person's general appearance.

Chemical properties are properties exhibited by matter as it undergoes changes in composition. These properties of substances are related to the kinds of chemical changes that the substances undergo. For instance, we have already described the combination of metallic magnesium with gaseous oxygen to form magnesium oxide, a white powder. A chemical property of magnesium is that it can combine with oxygen, releasing energy in the process. A chemical property of oxygen is that it can combine with magnesium.

All substances also exhibit **physical properties** that can be observed in the *absence of any change in composition*. Color, density, hardness, melting point, boiling point, and electrical and thermal conductivities are physical properties. Some physical properties of a substance depend on the conditions, such as temperature and pressure, under which they are measured. For instance, water is a solid (ice) at low temperatures but is a liquid at higher temperatures. At still higher temperatures, it is a gas (steam). As water is converted from one state to another, its composition is constant. Its chemical properties change very little. On the other hand, the physical properties of ice, liquid water, and steam are different (Figure 1-3).

Properties of matter can be further classified according to whether or not they depend on the *amount* of substance present. The volume and the mass of a sample depend on, and are directly proportional to, the amount of matter in that sample. Such properties, which depend on the amount of material examined, are called **extensive properties**. By contrast, the color and the melting point of a substance are the same for a small sample and for a large one. Properties such as these, which are independent of the amount of material examined, are called **intensive properties**. All chemical properties are intensive properties.

Because no two substances have identical sets of chemical and physical properties under the same conditions, we are able to identify and distinguish among different substances. For instance, water is the only clear, colorless liquid that freezes at 0°C, boils at 100°C at one atmosphere of pressure, dissolves a wide variety of substances (including copper(II) sulfate), and

One atmosphere of pressure is the average atmospheric pressure at sea level.

Figure 1-3
A comparison of some physical
properties of the three states of matter
(for water).

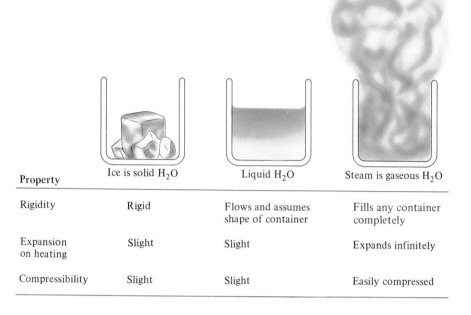

Property	Ice is solid H_2O	Liquid H_2O	Steam is gaseous H_2O
Rigidity	Rigid	Flows and assumes shape of container	Fills any container completely
Expansion on heating	Slight	Slight	Expands infinitely
Compressibility	Slight	Slight	Easily compressed

(a)   (b)   (c)   (d)

Figure 1-4
Some physical and chemical properties of water. *Physical*: (a) It melts at 0°C; (b) it boils at
100°C (at normal atmospheric pressure); (c) it dissolves a wide range of substances,
including copper(II) sulfate, a blue solid. *Chemical*: (d) It reacts with sodium to form
hydrogen gas and a solution of sodium hydroxide. The solution is pink in the presence of the
indicator phenolphthalein.

Table 1-1
Physical Properties of a Few Common Substances
(at one atmosphere pressure)

Substance	Melting Pt. (°C)	Boiling Pt. (°C)	Solubility at 25°C (g/100 g)		Density (g/cm³)
			In water	In ethyl alcohol	
acetic acid	16.6	118.1	infinite	infinite	1.05
benzene	5.5	80.1	0.07	infinite	0.879
bromine	−7.1	58.8	3.51	infinite	3.12
iron	1530	3000	insoluble	insoluble	7.86
methane	−182.5	−161.5	0.0022	0.033	6.67×10^{-4}
oxygen	−218.8	−183.0	0.0040	0.037	1.33×10^{-3}
sodium chloride	801	1473	36.5	0.065	2.16
water	0	100	—	infinite	1.00

reacts violently with sodium (Figure 1-4). Table 1-1 compares several physical properties of a few substances. A sample of any of these substances can be distinguished from the others by measurement of their properties.

1-4 Chemical and Physical Changes

We described the reaction of magnesium as it burns in the oxygen of the air (Figure 1-1). This reaction is a *chemical change*. In any **chemical change**, (1) one or more substances are used up (at least partially), (2) one or more new substances are formed, and (3) energy is absorbed or released. As substances undergo chemical changes they demonstrate their chemical properties. A **physical change**, on the other hand, occurs with *no change in chemical composition*. Physical properties are usually altered significantly as matter undergoes physical changes. In addition, a physical change *may* suggest that a chemical change has also taken place. For instance, a color change, a warming, or the formation of a solid when two solutions are mixed could indicate a chemical change.

Energy is always released or absorbed when chemical or physical changes occur. Energy is required to melt ice, and energy is required to boil water. Conversely, the condensation of steam to form liquid water always liberates

Figure 1-5
Changes in energy that accompany some physical changes for water. The energy unit joules (J) is defined in Section 1-13. The positive signs preceding joules above the arrows tell us that heat is *absorbed*. Negative signs below the arrows tell us that heat is *liberated*.

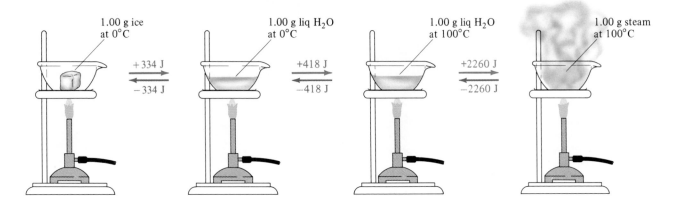

energy, as does the freezing of liquid water to form ice. The changes in energy that accompany these physical changes for water are shown in Figure 1-5. At a pressure of one atmosphere, ice always melts at the same temperature (0°C) and pure water always boils at the same temperature (100°C).

1-5 Mixtures, Substances, Compounds, and Elements

Mixtures are combinations of two or more pure substances in which each substance retains its own composition and properties. Almost every sample of matter that we ordinarily encounter is a mixture. The most easily recognized type of mixture is one in which different portions of the sample have recognizably different properties. Such a mixture, which is not uniform throughout, is called **heterogeneous**. Examples include mixtures of salt and charcoal (in which two components with different colors can be distinguished readily from one another by sight), foggy air (which includes a suspended mist of water droplets), and vegetable soup. Another kind of mixture has uniform properties throughout; such a mixture is described as a **homogeneous mixture** and is also called a **solution**. Examples include saltwater; some **alloys**, which are homogeneous mixtures of metals in the solid state; and air (free of particulate matter or mists). Air is a mixture of gases. It is mainly nitrogen, oxygen, argon, carbon dioxide, and water vapor. There are only trace amounts of other substances in the atmosphere.

An important characteristic of all mixtures is that they can have variable composition. (For instance, we can make an infinite number of different mixtures of salt and sugar by varying the relative amounts of the two components used.) Consequently, performing the same experiment over again on mixtures from different sources may give different results, whereas the same treatment of a pure sample will always give the same results. When the distinction between homogeneous mixtures and pure substances was realized and methods were developed (in the late 1700s) for separating mixtures and studying pure substances, consistent results could be obtained. This resulted in reproducible study of chemical properties, which formed the basis of real progress in the development of chemical theory.

Mixtures can be separated by physical means because each component retains its properties (see Figures 1-6 and 1-7). For example, a mixture of

By "composition of a mixture" we mean both the identities of the substances present and their relative amounts in the mixture.

Figure 1-6
(a) A mixture of iron and sulfur is a *heterogeneous* mixture. (b) Like any mixture, it can be separated by physical means, such as removing the iron with a magnet.

(a)   (b)

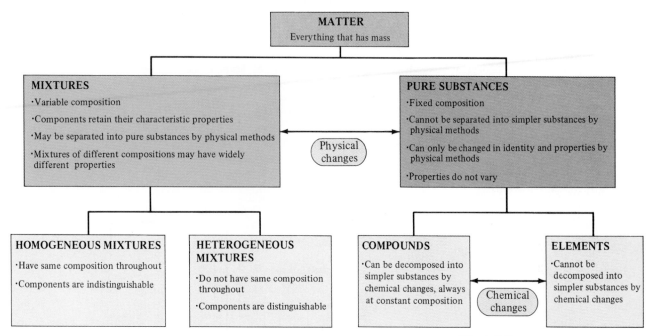

Figure 1-7
One scheme for classification of matter. Arrows indicate the general means by which matter can be separated.

salt and water can be separated by evaporating the water and leaving the solid salt behind. To separate a mixture of sand and salt, we could treat it with water to dissolve the salt, collect the sand by filtration, and then evaporate the water to obtain the solid salt. Very fine iron powder can be mixed with powdered sulfur to give what appears to the naked eye to be a homogeneous mixture of the two. However, separation of the components of this mixture is easy. The iron may be removed by a magnet, or the sulfur may be dissolved in carbon disulfide, which does not dissolve iron (Figure 1-6).

In *any* mixture, (1) the composition can be varied and (2) each component of the mixture retains its own properties.

Imagine that we have a sample of muddy river water (a heterogeneous mixture). We might first separate the suspended dirt from the liquid by filtration (Section 1-6). Then we could remove dissolved air by warming the water. Dissolved solids might be removed by cooling the sample until some of it freezes, pouring off the liquid, and then melting the ice. Other dissolved components might be separated by distillation (Section 1-6) or other methods. Eventually we would obtain a sample of pure water; it could not be further separated by any physical separation methods. No matter what the original source of the impure water—the ocean, the Mississippi River, a can of tomato juice, and so on—water samples obtained by purification all have identical composition and, under identical conditions, they all have identical properties. Any such sample is called a substance, or sometimes a pure substance.

The first ice that forms is quite pure. The dissolved solids tend to stay behind in the remaining liquid.

A **substance** cannot be further broken down by physical means.

If we use the definition given here of a *substance*, the phrase *pure substance* may appear to be redundant.

Figure 1-8
Electrolysis apparatus for small-scale decomposition of water by electrical energy. The volume of hydrogen produced (right) is twice that of oxygen (left). Some dilute sulfuric acid is added to increase the conductivity.

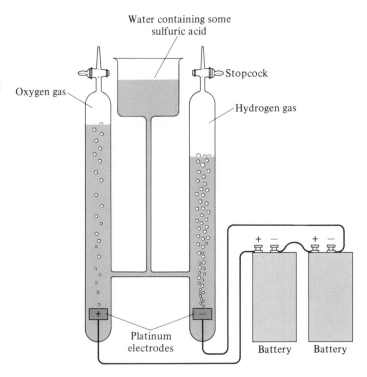

Water containing some sulfuric acid

Oxygen gas

Stopcock

Hydrogen gas

Platinum electrodes

Battery   Battery

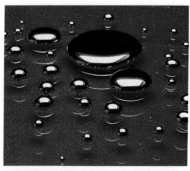

(a)

(b)

(a) Mercury is the only metal that is a liquid at room temperature. (b) The stable form of sulfur at room temperature is a solid.

Now suppose we decompose some water by passing electricity through it (Figure 1-8). (An *electrolysis* process is a chemical reaction.) We find that the water is converted into two simpler substances, hydrogen and oxygen; more significantly, hydrogen and oxygen are *always* present in the same ratio by mass, 11.1% to 88.9%. These observations allow us to identify water as a compound.

> A **compound** is a substance that can be decomposed by chemical means into simpler substances, always in the same ratio by mass.

As we continue this process, starting with any substance, we eventually reach a stage at which the new substances formed cannot be further broken down by chemical means. The substances at the end of this chain are called elements.

> An **element** is a substance that cannot be decomposed into simpler substances by chemical changes.

For instance, neither of the two gases obtained by the electrolysis of water hydrogen and oxygen—can be further decomposed, so they are elements.

As another illustration (see Figure 1-9), pure calcium carbonate (a white solid present in limestone and seashells) can be broken down by heating to give another white solid (call it A) and a gas (call it B) in the mass ratio 56.0:44.0. This observation tells us that calcium carbonate is a compound. The white solid A obtained from calcium carbonate can be further broken

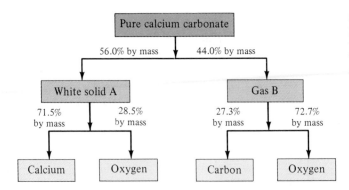

Figure 1-9
Diagram of the decomposition of calcium carbonate to give a white solid A (56.0% by mass) and a gas B (44.0% by mass). This decomposition into simpler substances at fixed ratio proves that calcium carbonate is a compound. The white solid A further decomposes to give the elements calcium (71.5% by mass) and oxygen (28.5% by mass). This proves that the white solid A is a compound; it is known as calcium oxide. The gas B also can be broken down to give the elements carbon (27.3% by mass) and oxygen (72.7% by mass). This establishes that gas B is a compound; it is known as carbon dioxide.

down into a solid and a gas in a definite ratio by mass, 71.5:28.5. But neither of these can be further decomposed, so they must be elements. The gas is identical to the oxygen obtained from the electrolysis of water; the solid is a metallic element called calcium. Similarly, the gas B originally obtained from calcium carbonate can be decomposed into two elements, carbon and oxygen in fixed mass ratio, 27.3:72.7. This sequence illustrates that a compound can be broken apart into simpler substances at fixed mass ratio, but those simpler substances may be either elements or simpler compounds.

How would you know that the white solid we called A is a compound?

Further, we may say that *a compound is a pure substance consisting of two or more different elements in a fixed ratio.* Water is 11.1% hydrogen and 88.9% oxygen by mass. Similarly, carbon dioxide is 27.3% carbon and 72.7% oxygen by mass, and calcium oxide (the white solid A above) is 71.5% calcium and 28.5% oxygen by mass. As we shall see presently, we could also combine the numbers in the previous paragraph to show that calcium carbonate is 40.1% calcium, 12.0% carbon, and 47.9% oxygen by mass. Observations such as these on innumerable pure compounds led to the statement of the **Law of Definite Proportions** (also known as the **Law of Constant Composition**):

> Different samples of any pure compound contain the same elements in the same proportions by mass.

The physical and chemical properties of a compound are different from the properties of its constituent elements. Sodium chloride is a white solid that we ordinarily use as table salt. This compound is produced by the combination of sodium (a soft, silvery white metal that reacts violently with water; Figure 1-4d) and chlorine (a pale green, corrosive, poisonous gas; Figure 1-2c). See Figure 1-10.

Recall that elements are substances that cannot be decomposed into simpler substances by chemical changes. Nitrogen, silver, aluminum, copper, gold, and sulfur are other examples of elements.

We use a set of **symbols** to represent the elements. These symbols can be written more quickly than names, and they occupy less space. The symbols for the first 103 elements consist of either a capital letter *or* a capital letter and a lowercase letter, such as C (carbon) or Ca (calcium). Symbols for elements beyond number 103 consist of three letters. A list of the known elements and their symbols is inside the front cover.

Figure 1-10
The reaction of sodium and chlorine to produce table salt, sodium chloride.

Table 1-2
Some Common Elements and Their Symbols

Symbol	Element	Symbol	Element	Symbol	Element
Ag	silver (*argentum*)	F	fluorine	Ni	nickel
Al	aluminum	Fe	iron (*ferrum*)	O	oxygen
Au	gold (*aurum*)	H	hydrogen	P	phosphorus
B	boron	He	helium	Pb	lead (*plumbum*)
Ba	barium	Hg	mercury	Pt	platinum
Bi	bismuth	I	iodine	Rb	rubidium
Br	bromine	K	potassium (*kalium*)	S	sulfur
C	carbon	Kr	krypton	Sb	antimony (*stibium*)
Ca	calcium	Li	lithium	Si	silicon
Cd	cadmium	Mg	magnesium	Sn	tin (*stannum*)
Cl	chlorine	Mn	manganese	Sr	strontium
Co	cobalt	N	nitrogen	U	uranium
Cr	chromium	Na	sodium (*natrium*)	W	tungsten (*Wolfram*)
Cu	copper (*cuprum*)	Ne	neon	Zn	zinc

A short list of symbols of common elements is given in Table 1-2. Learning this list will be helpful. Many symbols consist of the first one or two letters of the element's English name. Some are derived from the element's Latin name (indicated in parentheses in Table 1-2) and one, W for tungsten, is from the German *Wolfram*. Names and symbols for additional elements should be learned as they are encountered.

Most of the earth's crust is made up of a relatively small number of elements. Only 10 of the 88 naturally occurring elements make up more than 99% by mass of the earth's crust, oceans, and atmosphere (Table 1-3). Oxygen accounts for roughly half and silicon for approximately one fourth of the whole. Relatively few elements, approximately one fourth of the naturally occurring ones, occur in nature as free elements. The rest are always found chemically combined with other elements.

The other known elements have been made artificially in laboratories, as described in Chapter 30.

Figure 1-7 summarizes the classification of matter and the general means by which separations can be achieved.

Table 1-3
Abundance of Elements in the Earth's Crust, Oceans, and Atmosphere

Element	Symbol	% by Mass		Element	Symbol	% by Mass	
oxygen	O	49.5%		chlorine	Cl	0.19%	
silicon	Si	25.7		phosphorus	P	0.12	
aluminum	Al	7.5		manganese	Mn	0.09	
iron	Fe	4.7		carbon	C	0.08	
calcium	Ca	3.4	99.2%	sulfur	S	0.06	0.7%
sodium	Na	2.6		barium	Ba	0.04	
potassium	K	2.4		chromium	Cr	0.033	
magnesium	Mg	1.9		nitrogen	N	0.030	
hydrogen	H	0.87		fluorine	F	0.027	
titanium	Ti	0.58		zirconium	Zr	0.023	
				All others combined		~0.1%	

Chemistry in Use. . .
The Resources of the Ocean

As is apparent to anyone who has swum in the ocean, sea water is not pure water but contains a large amount of dissolved solids. In fact, each cubic kilometer of seawater contains about 3.6×10^{10} kilograms of dissolved solids. Nearly 71% of the earth's surface is covered with water. The oceans cover an area of 361 million square kilometers at an average depth of 3729 meters, and hold approximately 1.35 billion cubic kilometers of water. This means that the oceans contain a total of more than 4.8×10^{21} kilograms of dissolved material (or more than 100,000,000,000,000,000,000,000 pounds). Rivers flowing into the oceans and submarine volcanoes constantly add to this storehouse of minerals. The formation of sediment and the biological demands of organisms constantly remove a similar amount.

Sea water is a very complicated solution of many substances. The main dissolved component of sea water is sodium chloride, common salt. Besides sodium and chlorine, the main elements in sea water are magnesium, sulfur, calcium, potassium, bromine, carbon, nitrogen, and strontium. Together these ten elements make up more than 99% of the dissolved materials in the oceans. In addition to sodium chloride, they combine to form such compounds as magnesium chloride, potassium sulfate, and calcium carbonate (lime). Animals absorb the latter from the sea and build it into bones and shells.

Many other substances exist in smaller amounts in sea water. In fact, most of the 92 naturally occurring elements have been measured or detected in sea water, and the remainder will probably be found as more sensitive analytical techniques become available. From an economic standpoint, there are staggering amounts of valuable metals in sea water, including approximately 1.3×10^{11} kilograms of copper, 4.2×10^{12} kilograms of uranium, 5.3×10^9 kilograms of gold, 2.6×10^9 kilograms of silver, and 6.6×10^8 kilograms of lead. Other elements of economic importance include 2.6×10^{12} kilograms of aluminum, 1.3×10^{10} kilograms of tin, 2.6×10^{11} kilograms of manganese, and 4.0×10^{10} kilograms of mercury.

One would think that with such a large reservoir of dissolved solids, considerable "chemical mining" of the ocean would occur. At present (1991) only four elements are commercially extracted in large quantities. They are sodium and chlorine, which are produced from the sea by solar evaporation; magnesium; and bromine. In fact, most of the U.S. production of magnesium is derived from sea water, and the ocean is one of the principal sources of bromine. Most of the other elements are so thinly scattered through the ocean that the cost of their recovery would be much higher than their economic value. However, it is probable that as resources become more and more depleted from the continents, and as recovery techniques become more efficient, mining of sea water will become a much more desirable and feasible prospect.

One promising method of extracting elements from sea water uses marine organisms. Many marine animals concentrate certain elements in their bodies at levels many times higher than the levels in sea water. Vanadium, for example, is taken up by the mucus of certain tunicates and can be concentrated in these animals to more than 280,000 times its concentration in sea water. Other marine organisms can concentrate copper and zinc by a factor of about 1 million. If these animals could be cultivated in large quantities without endangering the ocean ecosystem, they could become a valuable source of trace metals.

In addition to dissolved materials, sea water holds a great store of suspended particulate matter that floats through the water. Some 15% of the manganese in sea water is present in particulate form, as are appreciable amounts of lead and iron. Similarly, most of the gold in sea water is thought to adhere to the surfaces of clay minerals in suspension. As in the case of dissolved solids, the economics of filtering these very fine particles from sea water is not favorable at present. However, because many of the particles suspended in sea water carry an electric charge, ion exchange techniques and modifications of electrostatic processes may someday provide important methods for the recovery of trace metals.

Beth A. Trust
Graduate student in chemistry
University of Texas Marine Sciences
Institute

A large distillation tower used to separate complex mixtures such as petroleum.

1-6 Separation of Mixtures

Pure, or even nearly pure, specimens of elements and compounds seldom occur in nature. It is usually necessary to separate them from the mixtures in which they occur. When a compound is prepared in the laboratory, several steps are usually necessary to separate the pure compound from the reaction mixture (by-products, unreacted starting materials, and solvent). Thus we see that separation of mixtures is very important. We shall describe a few methods of separating pure substances from mixtures.

Filtration

Filtration is the process of separating solids that are suspended in liquids by pouring the mixture into a filter funnel. As the liquid passes through the filter, the solid particles are held on the filter (Figure 1-11). The amount of silver in a solution can be found by adding hydrochloric acid to form solid silver chloride, which is collected on a filter, dried, and weighed.

Distillation

A liquid that vaporizes readily is called a *volatile liquid*. When a liquid is heated to a sufficiently high temperature it boils; i.e., it is converted to the gaseous, or vapor, state. **Distillation** is one method by which a mixture containing volatile substances can be separated into its components. For example, if a salt solution is heated, the more volatile water boils off, leaving behind the solid salt. A simple laboratory distillation apparatus is shown in Figure 1-12a.

The container in which the mixture is heated is called the distilling flask. The condenser is a double-walled glass tube. Cold water passes through the outer chamber to condense the hot vapor (gas) to a liquid.

Figure 1-11
(a) Silver chloride precipitates when silver ions are added to a solution containing chloride ions. (b) The solution, which is mixed with solid silver chloride, is poured through filter paper in the shape of a cone. Solid silver chloride remains on the paper. (On exposure to light, silver chloride turns dark.)

(a)

(b)

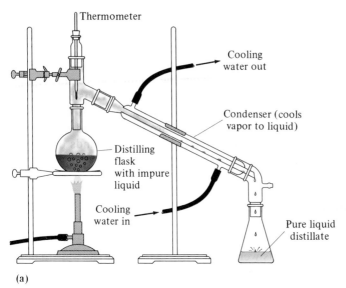

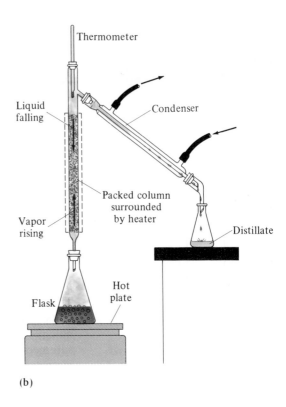

Figure 1-12

(a) A laboratory set-up for distillation. During distillation of an impure liquid, nonvolatile substances remain in the distilling flask. The liquid is vaporized and condensed before being collected in the receiving flask. (b) A fractional distillation apparatus. The vapor phase rising in the column is in equilibrium with the liquid phase that has condensed and is flowing slowly back down the column.

When a liquid mixture consists of two or more volatile liquids, *fractional distillation* can be used (Figure 1-12b). The liquid with the lowest boiling point usually distills out first. One kind of *fractionating column* is packed with glass beads that provide a large surface area on which the less volatile part of the vapor condenses. The temperature is higher at the bottom and lower at the top of the fractionating column. Thus the liquid with the lowest boiling point boils off first, while less volatile liquid condenses on the beads and falls back into the distilling flask. When the most volatile liquid has boiled off, the temperature in the distilling flask rises and the next most volatile liquid boils off.

Chromatography

Chromatography refers to several similar techniques used to separate mixtures. They involve a stationary phase and a moving, or mobile, phase.

Consider *paper chromatography*. Figure 1-13 shows a piece of filter paper on which a line of ink has been placed. The various dyes that make up the ink have been separated. When the end of the dry filter paper was dipped in water, each dye moved away from the line at its own characteristic rate. The wet fibers of the paper are the stationary phase and water–ink solution is the mobile phase. Because different dyes have different attractions for the wet fibers in the paper, they move along the paper at different rates. Such a separation of colored components is responsible for the name "chromatography" (based on Greek *chromos*, meaning "color," and *graphe*, meaning "to write"—literally "color writing"), although many separations involve colorless substances.

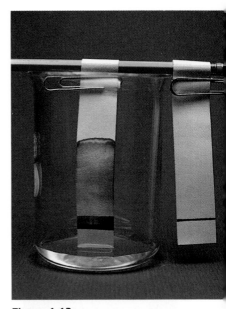

Figure 1-13

The components of an ink have been partially separated on a piece of filter paper.

Figure 1-16
Duplicates of the SI kilogram (2.205 pounds) standard of mass and the meter bar (39.37 inches).

**Table 1-6
Some SI Units of Mass**

*kilo*gram, kg	base unit
gram, g	1000 g = 1 kg
*milli*gram, mg	1000 mg = 1 g
*micro*gram, μg	1,000,000 μg = 1 g

chemical reactions at constant gravity, weight relationships are just as valid as mass relationships. We should keep in mind that the two are not identical.

The basic unit of mass in the SI system is the **kilogram** (Table 1-6). The kilogram is defined as the mass of a platinum–iridium cylinder stored in a vault in Sevres, near Paris, France (Figure 1-16). A one-pound object has a mass of 0.4536 kilogram. The basic mass unit in the *metric system* is the gram.

Length

The **meter** is the standard unit of length (distance) in both SI and metric systems. The meter is *defined* as the distance light travels in a vacuum in 1/299,792,468 second. It is approximately 39.37 inches. In situations where the English system would use inches, the metric centimeter (1/100 meter) is convenient. The relationship between inches and centimeters is shown in Figure 1-17.

Figure 1-17
The relationship between inches and centimeters (1 in = 2.54 cm).

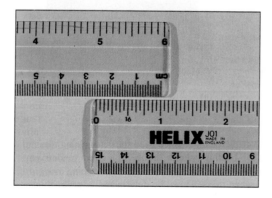

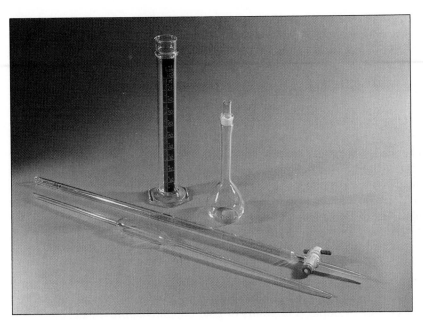

Figure 1-18
Some laboratory apparatus used to measure volumes of liquids: 100-mL graduated cylinder (left rear), 100-mL volumetric flask (right rear), 25-mL buret (center), and 25-mL volumetric pipet (front).

Volume

Volumes are measured in liters or milliliters in the metric system. One liter (1 L) is one cubic decimeter (1 dm^3), or 1000 cubic centimeters (1000 cm^3). One milliliter (1 mL) is 1 cm^3. In the SI the cubic meter is the basic volume unit, and the cubic decimeter replaces the metric unit, liter. Different kinds of glassware are used to measure the volume of liquids. The one we choose depends on the accuracy we desire. For example, the volume of a liquid dispensed can be measured more accurately with a buret than with a small graduated cylinder (Figure 1-18). Equivalences between common English units and metric units are summarized in Table 1-7.

Table 1-7
Conversion Factors Relating Length, Volume, and Mass (weight) Units

	Metric		English		Metric–English Equivalents	
Length	1 km	= 10^3 m	1 ft	= 12 in	2.54 cm	= 1 in
	1 cm	= 10^{-2} m	1 yd	= 3 ft	39.37 in*	= 1 m
	1 mm	= 10^{-3} m	1 mile	= 5280 ft	1.609 km*	= 1 mile
	1 nm	= 10^{-9} m				
	1 Å	= 10^{-10} m				
Volume	1 mL	= 1 cm^3 = 10^{-3} L	1 gal	= 4 qt = 8 pt	1 L	= 1.057 qt*
	1 m^3	= 10^6 cm^3 = 10^3 L	1 qt	= 57.75 in^3*	28.32 L	= 1 ft^3
Mass	1 kg	= 10^3 g	1 lb	= 16 oz	453.6 g*	= 1 lb
	1 mg	= 10^{-3} g			1 g	= 0.03527 oz*
	1 metric tonne	= 10^3 kg	1 short ton	= 2000 lb	1 metric tonne	= 1.102 short ton

* These conversion factors, unlike the others listed, are inexact. They are quoted to four significant figures, which is ordinarily more than sufficient.

Sometimes we must combine two or more units to describe a quantity. For instance, we might express the speed of a car as 60 mi/hr. Recall that the algebraic notation x^{-1} means $1/x$; applying this notation to units, we see that hr^{-1} means 1/hr, or "per hour." So the unit of speed could also be expressed as $mi \cdot hr^{-1}$.

1-9 Use of Numbers

In chemistry, we measure and calculate many things, so we must be sure we understand how to use numbers. In this section we discuss two aspects of the use of numbers: (1) the notation of very large and very small numbers and (2) an indication of how well we actually know the numbers we are using. You will carry out many calculations with calculators. Please refer to Appendix A for some instructions about the use of electronic calculators.

Scientific Notation

We use **scientific notation** when we deal with very large and very small numbers. For example, 197 grams of gold contains approximately

$$602,000,000,000,000,000,000,000 \text{ gold atoms.}$$

The mass of one gold atom is approximately

$$0.000\ 000\ 000\ 000\ 000\ 000\ 000\ 327 \text{ gram.}$$

In exponential form these numbers are
6.02×10^{23} gold atoms

3.27×10^{-22} gram

In using such large and small numbers, it is inconvenient to write down all the zeroes. In scientific (exponential) notation, we place one nonzero digit to the left of the decimal.

$$4,300,000. = 4.3 \times 10^6$$

6 places to the left, $\therefore$ exponent of 10 is 6

$$0.000348 = 3.48 \times 10^{-4}$$

4 places to the right, $\therefore$ exponent of 10 is -4

The reverse process converts numbers from exponential to decimal form. See Appendix A for more detail, if necessary.

Significant Figures

An *exact* number may be thought of as containing an *infinite* number of significant figures.

There are two kinds of numbers. **Exact numbers** may be *counted* or *defined*. They are known to be absolutely accurate. For example, the exact number of people in a closed room can be counted, and there is no doubt about the number of people. A dozen eggs is defined as exactly 12 eggs, no more, no fewer (Figure 1-19).

Numbers obtained from measurements are not exact. Every measurement involves an estimate. For example, suppose you are asked to measure the length of this page to the nearest 0.1 mm. How do you do it? The smallest divisions (calibration lines) on a meter stick are 1 mm apart (Figure 1-17). An attempt to measure to 0.1 mm requires estimation. If three different people measure the length of the page to 0.1 mm, will they get the same answer? Probably not. We deal with this problem by using significant figures.

There is some uncertainty in all measurements.

Significant figures indicate the *uncertainty* in measurements.

Significant figures are digits believed to be correct by the person who makes a measurement. We assume that the person is competent to use the measuring device. Suppose one measures a distance with a meter stick and reports the distance as 343.5 mm. What does this number mean? In this person's judgment, the distance is greater than 343.4 mm but less than 343.6 mm, and the best estimate is 343.5 mm. The number 343.5 mm contains four significant figures. The last digit, 5, is a *best estimate* and is therefore doubtful, but it is considered to be a significant figure. In reporting numbers obtained from measurements, *we report one estimated digit, and no more.* Because the person making the measurement is not certain that the 5 is correct, it would be meaningless to report the distance as 343.53 mm.

To see more clearly the part significant figures play in reporting the results of measurements, consider Figure 1-20a. Graduated cylinders are used to measure volumes of liquids when a high degree of accuracy is not necessary. The calibration lines on a 50-mL graduated cylinder represent 1-mL increments. Estimation of the volume of liquid in a 50-mL cylinder to within 0.2 mL ($\frac{1}{5}$ of one calibration increment) with reasonable certainty is possible. We might measure a volume of liquid in such a cylinder and report the volume as 38.6 mL, i.e., to three significant figures.

Burets are used to measure volumes of liquids when higher accuracy is required. The calibration lines on a 50-mL buret represents 0.1-mL increments, allowing us to make estimates to within 0.02 mL ($\frac{1}{5}$ of one calibration increment) with reasonable certainty (Figure 1-20b). Experienced individuals estimate volumes in 50-mL burets to 0.01 mL with considerable reproducibility. For example, using a 50-mL buret, we can measure out 38.57 mL (four significant figures) of liquid with reasonable accuracy.

Accuracy refers to how closely a measured value agrees with the correct value. **Precision** refers to how closely individual measurements agree with each other. Ideally, all measurements should be both accurate and precise. Measurements may be quite precise, yet quite inaccurate, because of some *systematic error,* which is an error repeated in each measurement. (A faulty balance, for example, might produce a systematic error.) Very accurate measurements are seldom imprecise.

(a)

(b)

Figure 1-19
(a) A dozen eggs is exactly 12 eggs.
(b) A swarm of honey bees contains an *exact* number of live bees. Could you count them easily?

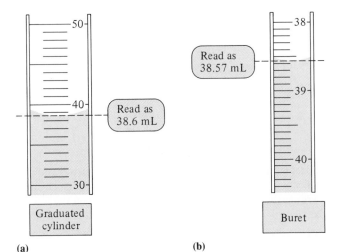

(a)   (b)

Figure 1-20
Measurement of the volume of water using two different pieces of volumetric glassware. For consistency, we always read the bottom of the meniscus (the curved surface of the water). (a) The level in a 50-mL graduated cylinder can be estimated to within 0.2 mL. The level here is 38.6 mL (three significant figures). (b) The level in a 50-mL buret can be read to within 0.02 mL. The level here is 38.57 mL (four significant figures).

Measurements are frequently repeated to improve accuracy and precision. Average values obtained from several measurements are usually more reliable than individual measurements. Significant figures indicate how accurately measurements have been made (assuming the person who made the measurements was competent).

Some simple rules govern the use of significant figures in calculations.

> 1. Zeroes used just to position the decimal point are not significant figures.

For example, the number 0.0234 g contains only three significant figures, because the two zeroes are used to place the decimal point. The number could also be reported as 2.34×10^{-2} g in scientific notation (Appendix A). When zeroes precede the decimal point, but come after other digits, we may have some difficulty in deciding whether the zeroes are significant figures or not. How many significant figures does the number 23,000 contain? We are given insufficient information to answer the question. If all three of the zeroes are used just to place the decimal point, the number should appear as 2.3×10^4 (two significant figures). If only two of the zeroes are being used to place the decimal point, the number is 2.30×10^4 (three significant figures). In the unlikely event that the number is actually known to be 23,000 $\pm$ 1, it should be written as 2.3000×10^4 (five significant figures).

> 2. In multiplication and division, an answer contains no more significant figures than the least number of significant figures used in the operation.

When we wish to specify that all of the zeroes in such a number *are* significant, we may indicate this by placing a decimal point after the number. For instance, 130. grams can represent a mass known to *three* significant figures, that is, 130 $\pm$ 1 gram.

Example 1-1

What is the area of a rectangle 1.23 cm wide and 12.34 cm long?

Plan

The area of a rectangle is its length times its width. We should first check to see that the width and length are expressed in the same units. (They are—but if they were not, one would first have to be converted to the units of the other.) Then we multiply the width by the length. We then follow Rule 2 for significant figures to find the correct number of significant figures. The units for the result are equal to the product of the units for the individual terms in the multiplication.

Solution

$$A = \ell \times w = (12.34 \text{ cm})(1.23 \text{ cm}) = \boxed{15.2 \text{ cm}^2}$$

(calculator result = 15.1782)

Because three is the smallest number of significant figures used, the answer should contain only three significant figures. The number generated by an electronic calculator (15.1782) is wrong; the result cannot be more accurate than the information that led to it. Calculators have no judgment, so you must exercise yours.

The step-by-step calculation in the margin demonstrates why the area is reported as 15.2 cm² rather than 15.1782 cm². The length, 12.34 cm, contains four significant figures, whereas the width, 1.23 cm, contains only three. If we underline each uncertain figure, as well as each figure obtained from an uncertain figure, the step-by-step multiplication gives the result reported in Example 1-1. We see that there are only two certain figures (15) in the result. We report the first doubtful figure (.2), but no more. Division is just the reverse of multiplication, and the same rules apply.

$$
\begin{array}{r}
12.3\underline{4} \text{ cm} \\
\times\quad 1.2\underline{3} \text{ cm} \\
\hline
37\ 0\underline{2} \\
2\ 4\underline{6}\ \underline{8} \\
12\ \underline{34} \\
\hline
15.1\underline{7}\ \underline{82} \text{ cm}^2 = 15.2 \text{ cm}^2
\end{array}
$$

> 3. In addition and subtraction, the last digit retained in the sum or difference is determined by the position of the first doubtful digit.

Example 1-2

(a) Add 37.24 mL and 10.3 mL. (b) Subtract 21.2342 g from 27.87 g.

Plan

Again we first check to see that the quantities to be added or subtracted are expressed in the same units. We carry out the addition or subtraction. Then we follow Rule 3 for significant figures to express the answer to the correct number of significant figures.

Solution

(a)
$$
\begin{array}{r}
37.2\underline{4} \text{ mL} \\
+10.\underline{3}\ \ \text{ mL} \\
\hline
\end{array}
$$
47.5$\underline{4}$ mL is reported as ⎢ 47.5 mL ⎥ (calculator gives 47.54)

Doubtful digits are underlined in this example.

(b)
$$
\begin{array}{r}
27.8\underline{7}\ \ \text{ g} \\
-21.234\underline{2} \text{ g} \\
\hline
\end{array}
$$
6.6$\underline{358}$ g is reported as ⎢ 6.64 g ⎥ (calculator gives 6.6358)

With many examples we suggest selected exercises from the "end of chapter" (EOC). These exercises use the skills or concepts from that example. Now you should work Exercises 25 and 26 from the end of this chapter.

EOC 25, 26

In the three simple arithmetic operations we have performed, the number combination generated by an electronic calculator is not the "answer" in a single case! However, the correct result of each calculation can be obtained by "rounding off." The rules of significant figures tell us where to round off.

In rounding off, certain conventions have been adopted. When the number to be dropped is less than 5, the preceding number is left unchanged (e.g., 7.34 rounds off to 7.3). When it is more than 5, the preceding number is increased by 1 (e.g., 7.37 rounds off to 7.4). When the number to be dropped is 5, the preceding number is not changed when it (the preceding number) is even (e.g., 7.45 rounds off to 7.4). When the preceding number is odd, it is increased by one (e.g., 7.35 rounds off to 7.4).

The even–odd rules are intended to reduce the accumulation of errors in chains of calculations.

1-10 The Unit Factor Method (Dimensional Analysis)

Many chemical and physical processes can be described by numerical relationships. In fact, many of the most useful ideas in science must be treated mathematically. Let us devote a little time to reviewing problem-solving skills.

First, multiplication by unity (by one) does not change the value of an expression. If we represent "one" in the right way, we can do many conversions by just "multiplying by one." This method of performing calculations is known as **dimensional analysis**, the **factor-label method**, or the **unit factor method**. Regardless of the name chosen, it is a very powerful mathematical tool that is almost foolproof.

Unit factors may be constructed from any two terms that describe the same or equivalent "amounts" of whatever we may consider. For example, 1 foot is equal to exactly 12 inches, by definition. We may write an equation to describe this equality:

$$1 \text{ ft} = 12 \text{ in}$$

Dividing both sides of the equation by 1 ft gives

$$\frac{1 \text{ ft}}{1 \text{ ft}} = \frac{12 \text{ in}}{1 \text{ ft}} \qquad \text{or} \qquad 1 = \frac{12 \text{ in}}{1 \text{ ft}}$$

The factor (fraction) 12 in/1 ft is a unit factor because the numerator and denominator describe the same distance. Dividing both sides of the original equation by 12 in gives 1 = 1 ft/12 in, a second unit factor that is the reciprocal of the first. *The reciprocal of any unit factor is also a unit factor.* Stated differently, division of an amount by the same amount always yields one!

In the English system we can write many unit factors, such as

$$\frac{1 \text{ yd}}{3 \text{ ft}} \text{ , } \frac{1 \text{ yd}}{36 \text{ in}} \text{ , } \frac{1 \text{ mile}}{5280 \text{ ft}} \text{ , } \frac{4 \text{ qt}}{1 \text{ gal}} \text{ , } \frac{2000 \text{ lb}}{1 \text{ ton}}$$

Unless otherwise indicated, a "ton" refers to a "short ton," 2000 lb. There are also the "long ton," which is 2240 lb, and the metric tonne, which is 1000 kg.

The reciprocal of each of these is also a unit factor. Items in retail stores are frequently priced with unit factors, such as 39¢/lb and $3.98/gal. When all the quantities in a unit factor come from definitions, the unit is known to an unlimited (infinite) number of significant figures. For instance, if you bought eight 1-gallon jugs of something priced at $3.98/gal, the total cost would be 8 × $3.98, or $31.84; the merchant would not round this to $31.80, let alone to $30.

In science, nearly all numbers have units. What does 12 mean? Usually we must supply appropriate units, such as 12 eggs or 12 people. In the unit factor method, the units guide us through calculations in a step-by-step process, because all units except those in the desired result cancel.

Example 1-3
Express 1.47 miles in inches.

Plan
First we write down the units of what we wish to know preceded by a question mark. Then we set it equal to whatever we are given:

$$\underline{?}\ \text{in} = 1.47\ \text{miles}$$

Then we choose unit factors to convert the given units (miles) to the desired units (inches):

$$\boxed{\text{miles}} \rightarrow \boxed{\text{feet}} \rightarrow \boxed{\text{inches}}$$

We relate (a) miles to feet and then (b) feet to inches.

Solution

$$\underline{?}\ \text{in} = 1.47\ \text{miles} \times \frac{5280\ \text{ft}}{1\ \text{mile}} \times \frac{12\ \text{in}}{1\ \text{ft}} = \boxed{9.31 \times 10^4\ \text{in}} \quad \begin{array}{l}\text{(calculator} \\ \text{gives 93139.2)}\end{array}$$

Note that both miles and feet cancel, leaving only inches, the desired unit. Thus there is no ambiguity as to how the unit factors should be written. The answer contains three significant figures because there are three significant figures in 1.47 miles. The quantities 5280 ft, 1 mile, 12 in, and 1 ft all come from definitions, so each of them is known to an unlimited number of significant figures.

In the interest of clarity, cancellation of units will be omitted in the remainder of this book. You may find it useful to continue the practice.

EOC 30

It is often helpful to consider whether an answer is reasonable. In Example 1-3, the distance involved is more than a mile; we expect this distance to be many inches, so a large answer is not surprising. Suppose we had mistakenly multiplied by the unit factor $\dfrac{1\ \text{mile}}{5280\ \text{ft}}$ (and not noticed that the units did not cancel properly); we would have gotten the answer 3.34×10^{-3} inches (0.00334 inches), which we would have immediately recognized as nonsense!

Conversions within the SI and metric systems are easy, because measurements of a particular kind are related to each other by powers of ten.

Example 1-4

The Ångstrom (Å) is a unit of length, 1×10^{-10} meter, that provides a convenient scale on which to express the radii of atoms. Radii of atoms are often expressed in nanometers. The radius of a phosphorus atom is 1.10 Å. What is the distance expressed in centimeters and nanometers?

Plan

We use the equalities $1\ \text{Å} = 1 \times 10^{-10}\ \text{m}$, $1\ \text{cm} = 1 \times 10^{-2}\ \text{m}$, and $1\ \text{nm} = 1 \times 10^{-9}\ \text{m}$ to construct the unit factors that convert 1.10 Å to the desired units.

Solution

$$\underline{?}\ \text{cm} = 1.10\ \text{Å} \times \frac{1 \times 10^{-10}\ \text{m}}{1\ \text{Å}} \times \frac{1\ \text{cm}}{1 \times 10^{-2}\ \text{m}} = \boxed{1.10 \times 10^{-8}\ \text{cm}}$$

$$\underline{?}\ \text{nm} = 1.10\ \text{Å} \times \frac{1.0 \times 10^{-10}\ \text{m}}{1\ \text{Å}} \times \frac{1\ \text{nm}}{1 \times 10^{-9}\ \text{m}} = \boxed{0.110\ \text{nm}}$$

All the unit factors used in this example contain only exact numbers.

EOC 34

Example 1-5

Assuming a phosphorus atom is spherical, calculate its volume in Å^3, cm^3, and nm^3. The volume of a sphere is $V = (\frac{4}{3})\pi r^3$. Refer to Example 1-4.

Plan

We use the results of Example 1-4 to calculate the volume in each of the desired units.

$1 \text{ Å} = 10^{-10} \text{ m} = 10^{-8} \text{ cm}$

Solution

$$\underline{?} \text{ Å}^3 = (\tfrac{4}{3})\pi(1.10 \text{ Å})^3 = \boxed{5.58 \text{ Å}^3}$$

$$\underline{?} \text{ cm}^3 = (\tfrac{4}{3})\pi(1.10 \times 10^{-8} \text{ cm})^3 = \boxed{5.58 \times 10^{-24} \text{ cm}^3}$$

$$\underline{?} \text{ nm}^3 = (\tfrac{4}{3})\pi(1.10 \times 10^{-1} \text{ nm})^3 = \boxed{5.58 \times 10^{-3} \text{ nm}^3}$$

EOC 42

Example 1-6

A sample of gold has a mass of 0.234 mg. What is its mass in g? in cg?

Plan

We use the relationships 1 g = 1000 mg and 1 cg = 10 mg to write the required unit factors.

Solution

$$\underline{?} \text{ g} = 0.234 \text{ mg} \times \frac{1 \text{ g}}{1000 \text{ mg}} = \boxed{2.34 \times 10^{-4} \text{ g}}$$

$$\underline{?} \text{ cg} = 0.234 \text{ mg} \times \frac{1 \text{ cg}}{10 \text{ mg}} = \boxed{0.0234 \text{ cg}} \quad \text{or} \quad \boxed{2.34 \times 10^{-2} \text{ cg}}$$

Again, we have used unit factors that contain only exact numbers.

EOC 43

Unity raised to *any* power is one. *Any* unit factor raised to a power is still a unit factor, as the next two examples show.

Example 1-7

How many square decimeters are there in 215 square centimeters?

Plan

We would multiply by the unit factor $\frac{1 \text{ dm}}{10 \text{ cm}}$ to convert cm to dm. Here we require the *square* of this unit factor.

Plan

From the percentage information given, we may write the required unit factor

$$\frac{97.6 \text{ g zinc}}{100 \text{ g sample}}$$

Solution

$$? \text{ g Zn} = 1.494 \text{ grams sample} \times \frac{97.6 \text{ g zinc}}{100 \text{ g sample}} = \boxed{1.46 \text{ g zinc}}$$

The number of significant figures in the result is limited by the three significant figures in 97.6%. Because the definition of percentage involves *exactly* 100 parts, the number 100 is known to an infinite number of significant figures.

EOC 35, 36

Examples 1-1 through 1-11 show that multiplication by one or more unit factors changes the units and the number of units, but not the amount of whatever we are concerned with.

1-11 Density and Specific Gravity

The **density** of a sample of matter is defined as the mass per unit volume:

$$\text{density} = \frac{\text{mass}}{\text{volume}} \quad \text{or} \quad D = \frac{m}{V}$$

Densities may be used to distinguish between two substances or to assist in identifying a particular substance. They are usually expressed as g/cm^3 or g/mL for liquids and solids and as g/L for gases. These units can also be expressed as $g \cdot cm^{-3}$, $g \cdot mL^{-1}$, and $g \cdot L^{-1}$, respectively. Densities of several substances are listed in Table 1-8.

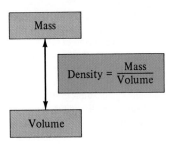

The two extensive properties *mass* and *volume* are related by the intensive property *density*.

Table 1-8
Densities of Common Substances*

Substance	Density g/cm³	Substance	Density g/cm³
hydrogen (gas)	0.000089	sand*	2.32
carbon dioxide (gas)	0.0019	aluminum	2.70
cork*	0.21	iron	7.86
oak wood*	0.71	copper	8.92
ethyl alcohol	0.789	lead	11.34
water	1.00	mercury	13.59
magnesium	1.74	gold	19.3
table salt	2.16		

These densities are given at room temperature and *one atmosphere* pressure, the average atmospheric pressure at sea level. Densities of solids and liquids change only slightly, but densities of gases change greatly, with changes in temperature and pressure.

* Cork, oak wood, and sand are common materials that have been included to provide familiar reference points. They are *not* pure elements or compounds as are the other substances listed here.

Six substances with different densities. The liquid layers are gasoline (top), water (middle), and mercury (bottom). A cork floats on gasoline. A piece of oak wood sinks in gasoline, but floats on water. Brass sinks in water, but floats on mercury.

Example 1-12

A 47.3-mL sample of ethyl alcohol (ethanol) has a mass of 37.32 g. What is its density?

Plan

We use the definition of density.

Solution

$$D = \frac{m}{V} = \frac{37.32 \text{ g}}{47.3 \text{ mL}} = \boxed{0.789 \text{ g/mL}}$$

EOC 50

Example 1-13

If 103 g of ethanol is needed for a chemical reaction, what volume of liquid would you use?

Plan

We determined the density of ethanol in Example 1-12. Here we are given the mass, m, of a sample of ethanol. So we know values for D and m in the relationship

$$D = \frac{m}{V}$$

We rearrange this relationship to solve for V, put in the known values, and carry out the calculation. Alternatively, we can use the unit factor method to solve the problem.

Solution

The density of ethanol is 0.789 g/mL (Table 1-8).

$$D = \frac{m}{V}, \quad \text{so} \quad V = \frac{m}{D} = \frac{103 \text{ g}}{0.789 \text{ g/mL}} = \boxed{130 \text{ mL}}$$

Alternatively,

$$\underline{?} \text{ mL} = 103 \text{ g} \times \frac{1 \text{ mL}}{0.789 \text{ g}} = \boxed{130 \text{ mL}}$$

EOC 53

Example 1-14

Express the density of mercury in lb/ft³.

Plan

The density of mercury is 13.59 g/cm³ (Table 1-8). To convert this value to the desired unit, we can use unit factors constructed from the conversion factors in Table 1-7.

Solution

$$\underline{?} \frac{\text{lb}}{\text{ft}^3} = 13.59 \frac{\text{g}}{\text{cm}^3} \times \frac{1 \text{ lb}}{453.6 \text{ g}} \times \left(\frac{2.54 \text{ cm}}{1 \text{ in}}\right)^3 \times \left(\frac{12 \text{ in}}{1 \text{ ft}}\right)^3 = \boxed{848.4 \text{ lb/ft}^3}$$

It would take a very strong person to lift a cubic foot of mercury!

The **specific gravity** (Sp. Gr.) of a substance is the ratio of its density to the density of water, both at the same temperature. Specific gravities are dimensionless numbers.

$$\text{Sp. Gr.} = \frac{D_{substance}}{D_{water}}$$

The density of water is 1.000 g/mL at 3.98°C, the temperature at which the density of water is greatest. However, variations in the density of water with changes in temperature are small enough that we may use 1.00 g/mL up to 25°C without introducing significant errors into our calculations.

Density and specific gravity are both intensive properties; i.e., they do not depend upon the size of the sample.

Example 1-15
The density of table salt is 2.16 g/mL at 20°C. What is its specific gravity?

Plan

We use the definition of specific gravity given above. The numerator and denominator have the same units, so the result is dimensionless.

Solution

$$\text{Sp. Gr.} = \frac{D_{salt}}{D_{water}} = \frac{2.16 \text{ g/mL}}{1.00 \text{ g/mL}} = \boxed{2.16}$$

EOC 57

This example also demonstrates that the density and specific gravity of a substance are numerically equal near room temperature if density is expressed in g/mL (g/cm³).

Labels on commercial solutions of acids and bases give specific gravities and the percentage by mass of the acid or base present in the solution. From this information, the amount of acid or base present in a given volume of the solution can be calculated.

At this point you need not be concerned if you do not know the chemical characteristics of an acid or a base. They are described in Chapters 4 and 10.

Example 1-16
Battery acid is 40.0% sulfuric acid, H_2SO_4, and 60.0% water by mass. Its specific gravity is 1.31. Calculate the mass of pure H_2SO_4 in 100.0 mL of battery acid.

Plan

The percentages given are on a mass basis, so we must first convert the 100.0 mL of acid solution to mass. To do this, we need a value for the density. We have demonstrated that density and specific gravity are numerically equal at 20°C because the density of water is 1.00 g/mL. We can use the density as a unit factor to convert the given volume of solution to mass of solution. Then we use the percentage by mass to convert the mass of solution to mass of acid.

Solution

From the given value for specific gravity, we may write

$$\text{density} = 1.31 \text{ g/mL}$$

The solution is 40.0% H_2SO_4 and 60.0% H_2O by mass. From this information we may construct the desired unit factor:

$$\frac{40.0 \text{ g } H_2SO_4}{100 \text{ g soln}} \leftarrow \boxed{\text{because 100 g of solution contains 40.0 g of } H_2SO_4}$$

We can now solve the problem:

$$\underline{?} \ H_2SO_4 = 100.0 \text{ mL soln} \times \frac{1.31 \text{ g soln}}{1 \text{ mL soln}} \times \frac{40.0 \text{ g } H_2SO_4}{100 \text{ g soln}} = \boxed{52.4 \text{ g } H_2SO_4}$$

EOC 59

1-12 Heat and Temperature

In Section 1-1 you learned that heat is one form of energy. You also learned that the many different forms of energy can be interconverted and that in chemical processes, chemical energy is converted to heat energy or vice versa. The amount of heat a process uses (*endothermic*) or gives off (*exothermic*) can tell us a great deal about that process (Chapters 13 and 15). For this reason it is important for us to be able to measure intensity of heat.

Temperature measures the intensity of heat, the "hotness" or "coldness" of a body. A piece of metal at 100°C feels hot to the touch, while an ice cube at 0°C feels cold. Why? Because the temperature of the metal is higher, and that of the ice cube lower, than body temperature. *Heat always flows spontaneously from a hotter body to a colder body*—never in the reverse direction.

Temperatures are commonly measured with mercury-in-glass thermometers. A mercury thermometer consists of a reservoir of mercury at the base of a glass tube, open to a very thin (capillary) column extending upward. Mercury expands more than most other liquids as its temperature rises. As it expands, its movement up into the evacuated column can be seen.

Anders Celsius, a Swedish astronomer, developed the Celsius temperature scale, formerly called the centigrade temperature scale. When we place a Celsius thermometer in a beaker of crushed ice and water, the mercury level stands at exactly 0°C, the lower reference point. In a beaker of water boiling at one atmosphere pressure, the mercury level stands at 100°C, the higher reference point. There are 100 equal steps between these two mercury levels. They correspond to an interval of 100 degrees between the melting point of ice and the boiling point of water at one atmosphere. Figure 1-21 shows how temperature marks between the reference points are established.

In the United States, temperatures are frequently measured on the temperature scale devised by Gabriel Fahrenheit, a German instrument maker. On this scale the freezing and boiling points of water are defined as 32°F and 212°F, respectively. In scientific work, temperatures are often expressed on the **Kelvin** (absolute) temperature scale. As we shall see in Section 12-5, the zero point of the Kelvin temperature scale is *derived* from the observed behavior of all matter.

Relationships among the three temperature scales are illustrated in Figure 1-22. Between the freezing point of water and the boiling point of water, there are 100 steps (degrees C or kelvins, respectively) on the Celsius and Kelvin scales. Thus the "degree" is the same size on the Celsius and Kelvin

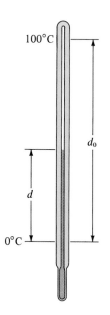

Figure 1-21
At 45°C, as read on a mercury-in-glass thermometer, d equals $0.45d_0$ where d_0 is the distance from the mercury level at 0°C to the level at 100°C.

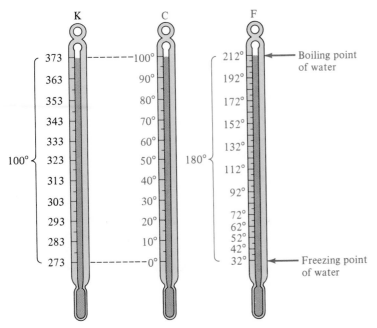

Figure 1-22
The relationships among the Kelvin, Celsius (centigrade), and Fahrenheit temperature scales.

scales. But every Kelvin temperature is 273.15 units above the corresponding Celsius temperature. The relationship between these two scales is as follows:

$$\underline{?}\ K = {}^{\circ}C + 273.15^{\circ} \qquad \text{or} \qquad \underline{?}{}^{\circ}C = K - 273.15^{\circ}$$

We shall usually round 273.15 to 273.

In the SI system, "degrees Kelvin" are abbreviated simply as K rather than °K and are called **kelvins**.

Please recognize that any temperature *change* has the same numerical value whether expressed on the Celsius scale or on the Kelvin scale. For example, a change from 25°C to 59°C represents a *change* of 34 Celsius degrees. Converting these to the Kelvin scale, the same change is expressed as (273 + 25) = 298 K to (59 + 273) = 332 K, or a *change* of 34 kelvins.

Comparing the Fahrenheit and Celsius scales, we find that the intervals between the same reference points are 180 Fahrenheit degrees and 100 Celsius degrees, respectively. Thus a Fahrenheit degree must be smaller than a Celsius degree. It takes 180 Fahrenheit degrees to cover the same temperature *interval* as 100 Celsius degrees. From this information, we can construct the unit factors for temperature *changes*:

$$\frac{180{}^{\circ}F}{100{}^{\circ}C} \quad \text{or} \quad \frac{1.8{}^{\circ}F}{1.0{}^{\circ}C} \qquad \text{and} \qquad \frac{100{}^{\circ}C}{180{}^{\circ}F} \quad \text{or} \quad \frac{1.0{}^{\circ}C}{1.8{}^{\circ}F}$$

But the starting points of the two scales are different, so we *cannot convert* a temperature on one scale to a temperature on the other just by multiplying by the unit factor. In converting from °F to °C, we must add 32 Fahrenheit degrees to reach the zero point on the Celsius scale (Figure 1-22).

These are often remembered in abbreviated form:

$$°F = 1.8°C + 32°$$
$$°C = \frac{(°F - 32°)}{1.8}$$

$$\underline{?}{}^{\circ}F = \left(x{}^{\circ}C \times \frac{1.8{}^{\circ}F}{1.0{}^{\circ}C} \right) + 32{}^{\circ}F \qquad \text{and} \qquad \underline{?}{}^{\circ}C = \frac{1.0{}^{\circ}C}{1.8{}^{\circ}F}\, (x{}^{\circ}F - 32{}^{\circ}F)$$

Figure 1-23
A graphical representation of the relationship between the Fahrenheit and Celsius temperature scales.

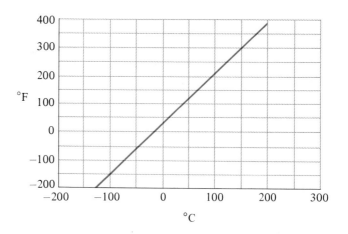

The relationship between temperatures on the Fahrenheit and Celsius scales is shown graphically in Figure 1-23.

Example 1-17
When the temperature reaches "100.°F in the shade," it's hot. What is this temperature on the Celsius scale?

Plan

We use the relationship $\underline{?}°C = \dfrac{1.0°C}{1.8°F} (x°F - 32°F)$ to carry out the desired conversion.

Solution

A temperature of 100°F is 38°C.

$$\underline{?}°C = \frac{1.0°C}{1.8°F} (100.°F - 32°F) = \frac{1.0°C}{1.8°F} (68°F) = \boxed{38°C}$$

Example 1-18
When the absolute temperature is 400 K, what is the Fahrenheit temperature?

Plan
We first use the relationship $\underline{?}°C = K - 273.15°$ to convert from kelvins to degrees Celsius; then we carry out the further conversion from degrees Celsius to degrees Fahrenheit.

Solution

$$\underline{?}°C = (400 \text{ K} - 273 \text{ K}) \frac{1.0°C}{1.0 \text{ K}} = 127°C$$

$$\underline{?}°F = \left(127°C \times \frac{1.8°F}{1.0°C} \right) + 32°F = \boxed{261°F}$$

EOC 62

1-13 Heat Transfer and the Measurement of Heat

Chemical reactions and physical changes occur with either the simultaneous evolution of heat (**exothermic processes**) or the absorption of heat (**endothermic processes**). The amount of heat transferred in a process is usually expressed in calories or in the SI unit joules. The **calorie** was originally defined as the amount of heat necessary to raise the temperature of one gram of water at one atmosphere from 14.5°C to 15.5°C. One calorie is *now defined* as exactly 4.184 joules. The amount of heat necessary to raise the temperature of one gram of liquid water varies slightly with temperature and pressure, so it was necessary to specify a particular temperature increment and constant pressure in describing the calorie. For our purposes the variations are sufficiently small that we are justified in ignoring them. The so-called "large calorie," used to indicate the energy content of foods, is really one kilocalorie, or 1000 calories. We shall do most calculations in joules.

The SI unit of energy and work is the **joule** (J), which is defined as $1 \text{ kg} \cdot \text{m}^2/\text{s}^2$. The kinetic energy (KE) of a body of mass m moving at speed v is given by $\frac{1}{2}mv^2$. A 2-kg object moving at one meter per second has KE $= \frac{1}{2}(2 \text{ kg})(1 \text{ m/s})^2 = 1 \text{ kg} \cdot \text{m}^2/\text{s}^2 = 1$ joule. You may find it more convenient to think in terms of the amount of heat required to raise the temperature of one gram of water from 14.5°C to 15.5°C, which is 4.184 joules.

The **specific heat** of a substance is the amount of heat required to raise the temperature of one gram of the substance one degree C (also one kelvin) with no change in phase. Changes in phase (physical state) absorb or liberate relatively large amounts of energy (Figure 1-5). The specific heat of each substance, a physical property, is different for the solid, liquid, and gaseous phases of the substance. For example, the specific heat of ice is 2.09 J/g·°C near 0°C; for liquid water it is 4.18 J/g·°C; and for steam it is 2.03 J/g·°C near 100°C. The specific heat for water is quite high. A table of specific heats is provided in Appendix E.

$$\text{specific heat} = \frac{\text{(amount of heat in J)}}{\text{(mass of substance in g) (temperature change in °C)}}$$

The **heat capacity** of a body is the amount of heat required to raise its temperature 1°C. The heat capacity of a body is its mass in grams times its specific heat.

> In terms of electrical energy, one joule is equal to one watt/second. Thus one joule is enough energy to operate a 10-watt light bulb for 1/10 second.

> In English units this corresponds to a 4.4-pound object moving at 197 feet per minute, or 2.2 miles per hour.

> The specific heat of a substance varies *slightly* with temperature and pressure. These variations can be ignored for calculations in this text.

Example 1-19

How much heat, in joules, is required to raise the temperature of 205 g of water from 21.2°C to 91.4°C?

Plan

The specific heat of a substance is the amount of heat required to raise the temperature of 1 g of substance 1°C:

$$\text{specific heat} = \frac{\text{(amount of heat in J)}}{\text{(mass of substance in g) (temperature change in °C)}}$$

We can rearrange the equation so that

$$\text{amount of heat} = \text{(mass of substance) (specific heat) (temperature change)}$$

> In this example, we calculate the amount of heat needed to prepare a cup of hot tea.

Alternatively, we can use the unit factor approach.

Solution

$$\text{amount of heat} = (205 \text{ g}) (4.18 \text{ J/g} \cdot {}^\circ\text{C}) (70.2 {}^\circ\text{C}) = \boxed{6.02 \times 10^4 \text{ J}}$$

By the unit factor approach,

$$\underline{?} \text{ J} = (205 \text{ g}) (4.18 \text{ J/g} \cdot {}^\circ\text{C}) (70.2 {}^\circ\text{C}) = \boxed{6.02 \times 10^4 \text{ J}} \quad \text{or} \quad \boxed{60.2 \text{ kJ}}$$

All units except joules cancel. To cool 205 g of water from 91.4°C to 21.2°C, it would be necessary to remove exactly the same amount of heat, 60.2 kJ.

EOC 65

Example 1-20

How much heat, in calories, kilocalories, joules, and kilojoules, is required to raise the temperature of 205 g of iron from 294.2 K to 364.4 K? The specific heat of iron is 0.106 cal/g · °C, or 0.444 J/g · °C.

This is the same temperature change for the same mass of iron that we used for water in Example 1-19, except that here the temperatures are given in kelvins. You may wish to check the result of this example by converting the starting and ending temperatures to °C and reworking the problem on the Celsius scale.

Plan

First we recall (Section 1-12) that a temperature *change* expressed in kelvins has the *same* numerical value expressed in degrees Celsius. Remembering that specific heat is in terms of temperature change, we can write the specific heat of iron as 0.106 cal/g · K or 0.444 J/g · K. Then we can solve this problem with the temperature change expressed in kelvins, and avoid the work of converting temperatures to °C.

Solution

$$\text{temperature change} = 364.4 \text{ K} - 294.2 \text{ K} = 70.2 \text{ K}$$

$$\underline{?} \text{ cal} = (205 \text{ g}) (0.106 \text{ cal/g} \cdot \text{K}) (70.2 \text{ K}) = \boxed{1.52 \times 10^3 \text{ cal}} \quad \text{or} \quad \boxed{1.52 \text{ kcal}}$$

$$\underline{?} \text{ J} = (205 \text{ g}) (0.444 \text{ J/g} \cdot \text{K}) (70.2 \text{ K}) = \boxed{6.39 \times 10^3 \text{ J}} \quad \text{or} \quad \boxed{6.39 \text{ kJ}}$$

EOC 66, 67

The specific heat of iron is much smaller than the specific heat of water.

$$\frac{\text{specific heat of iron}}{\text{specific heat of water}} = \frac{0.444 \text{ J/g} \cdot \text{K}}{4.18 \text{ J/g} \cdot \text{K}} = 0.106$$

As a result, the amount of heat required to raise the temperature of 205 g of iron by 70.2 K (70.2°C) is less than that required to do the same for 205 g of water, by the same ratio.

$$\frac{\text{amount of heat for iron}}{\text{amount of heat for water}} = \frac{6.39 \text{ kJ}}{60.2 \text{ kJ}} = 0.106$$

Key Terms

Accuracy   How closely a measured value agrees with the correct value.

Calorie   The amount of heat required to raise the temperature of one gram of water from 14.5°C to 15.5°C. 1 calorie = 4.184 joules.

Chemical change   A change in which one or more new substances are formed.

Chemical property   See *Properties*.

Compound   A substance composed of two or more elements in fixed proportions. Compounds can be decomposed into their constituent elements.

Density   Mass per unit volume, $D = m/V$.

Element   A substance that cannot be decomposed into simpler substances by chemical means.

Endothermic   Describes processes that absorb heat energy.

Energy   The capacity to do work or transfer heat.

Exothermic   Describes processes that release heat energy.

Extensive property   A property that depends upon the amount of material in a sample.

Heat   A form of energy that flows between two samples of matter because of their difference in temperature.

Heat capacity   The amount of heat required to raise the temperature of a body (of whatever mass) one degree Celsius.

Heterogeneous mixture   A mixture that does not have uniform composition and properties throughout.

Homogeneous mixture   A mixture that has uniform composition and properties throughout.

Intensive property   A property that is independent of the amount of material in a sample.

Joule   A unit of energy in the SI system. One joule is $1 \text{ kg} \cdot \text{m}^2/\text{s}^2$, which is also 0.2390 calorie.

Kinetic energy   Energy that matter possesses by virtue of its motion.

Law of Conservation of Energy   Energy cannot be created or destroyed in a chemical reaction or in a physical change; it may be changed from one form to another.

Law of Conservation of Matter   There is no detectable change in the quantity of matter during a chemical reaction or during a physical change.

Law of Conservation of Matter and Energy   The combined amount of matter and energy available in the universe is fixed.

Law of Constant Composition   See *Law of Definite Proportions*.

Law of Definite Proportions   Different samples of any pure compound contain the same elements in the same proportions by mass; also known as the *Law of Constant Composition*.

Mass   A measure of the amount of matter in an object. Mass is usually measured in grams or kilograms.

Matter   Anything that has mass and occupies space.

Mixture   A sample of matter composed of variable amounts of two or more substances, each of which retains its identity and properties.

Physical change   A change in which a substance changes from one physical state to another, but no substances with different compositions are formed.

Physical property   See *Properties*.

Potential energy   Energy that matter possesses by virtue of its position, condition, or composition.

Precision   How closely repeated measurements of the same quantity agree with each other.

Properties   Characteristics that describe samples of matter. Chemical properties are exhibited as matter undergoes chemical changes. Physical properties are exhibited by matter with no changes in chemical composition.

Significant figures   Digits that indicate the precision of measurements—digits of a measured number that have uncertainty only in the last digit.

Specific gravity   The ratio of the density of a substance to the density of water at the same temperature.

Specific heat   The amount of heat required to raise the temperature of one gram of a substance one degree Celsius.

Substance   Any kind of matter all specimens of which have the same chemical composition and physical properties.

Symbol   A letter or group of letters that represents (identifies) an element.

Temperature   A measure of the intensity of heat, i.e., the hotness or coldness of a sample or object.

Unit factor   A factor in which the numerator and denominator are expressed in different units but represent the same or equivalent amounts. Multiplying by a unit factor is the same as multiplying by one.

Weight   A measure of the gravitational attraction of the earth for a body.

Exercises

Basic Ideas

1. Define the following terms and illustrate each with a specific example: (a) matter; (b) energy; (c) heat energy; (d) exothermic process.

2. Define the following terms and illustrate each with a specific example: (a) mass; (b) potential energy; (c) kinetic energy; (d) endothermic process.

3. State the following laws and illustrate each.
   (a) the Law of Conservation of Matter
   (b) the Law of Conservation of Energy
   (c) the Law of Conservation of Matter and Energy

4. List the three states of matter and some characteristics of each. How are they alike? different?

5. Distinguish between the following pairs of terms and give two specific examples of each.
   (a) chemical properties and physical properties
   (b) intensive properties and extensive properties
   (c) chemical changes and physical changes

6. Are the following chemical properties or physical properties? How can you tell? (a) the color of a liquid; (b) the odor of a gas; (c) the boiling point of water; (d) the ability to react with nitric acid; (e) the softness of paraffin.

7. Are the following chemical properties or physical properties? How can you tell? (a) the volume of a metal sphere; (b) the density of a gas; (c) ease of corrosion; (d) flammability; (e) electrical conductivity.

8. Are the following chemical changes or physical changes? How can you tell? (a) freezing water to make ice; (b) emission of light by an electric light bulb; (c) emission of light by a candle; (d) dissolving a penny in nitric acid.

9. Are the following chemical changes or physical changes? How can you tell? (a) corrosion of steel; (b) burning of gasoline; (c) boiling of water; (d) condensation of moisture on a cold window.

10. Which of the following processes are exothermic? endothermic? How can you tell? (a) combustion; (b) freezing water; (c) melting ice; (d) boiling water; (e) condensing steam.

11. Which of the following properties of a sample of matter are extensive? Which are intensive? (a) density; (b) melting point; (c) volume; (d) mass; (e) ability to conduct electricity; (f) temperature.

12. Define the following terms clearly and concisely. Give two illustrations of each. (a) substance; (b) mixture; (c) element; (d) compound.

13. Does each of the following describe a mixture or a pure substance? Justify your classification.
   (a) A piece of "dry ice," commonly used as a coolant, sublimes (is transformed directly from solid to gas) at −78.5°C under normal atmospheric pressure. As the dry ice continues to sublime, it maintains a steady −78.5°C temperature.
   (b) An alcoholic liquid labeled "100 proof neutral spirit" is divided into two equal portions, each of which is placed in an open dish. When a lighted match is applied to the first dish, the liquid burns with a pale blue flame. The liquid in the second dish is allowed to stand undisturbed until much of it has evaporated. This liquid residue cannot be ignited with a match.
   (c) The label on a bottle of "hydrogen peroxide" includes the statement, "Contains 3% hydrogen peroxide. Inert ingredients 97%."

14. Classify each of the following as an element, a compound, or a mixture. Justify your classification. (a) coffee; (b) silver; (c) calcium carbonate; (d) a piece of concrete from a sidewalk; (e) shoe polish.

15. Classify each of the following as an element, a compound, or a mixture. Justify your classification. (a) a soft drink; (b) water; (c) air; (d) chicken noodle soup; (e) table salt; (f) popcorn.

16. What is a heterogeneous mixture? Which of the following are heterogeneous mixtures? Explain your answers. (a) salt and sulfur; (b) milk; (c) clean air; (d) gasoline; (e) a chocolate chip cookie.

17. What is a homogeneous mixture? Which of the following are homogeneous mixtures? Explain your answers. (a) sugar dissolved in water; (b) coffee; (c) french onion soup; (d) mud; (e) a clear liquid with no internal boundaries, consisting of corn oil and olive oil.

18. How could you separate the components of each of the following mixtures? Explain. (a) vinegar and oil; (b) sugar dissolved in water; (c) a suspension of sand in water.

19. How could you separate the components of each of the following mixtures? Explain. (a) charcoal and sugar; (b) baking soda dissolved in water; (c) ether (boiling point 36°C) and acetone (boiling point 56°C).

20. Describe paper chromatography. Describe column chromatography.

Scientific Notation and Significant Figures

21. Express the following numbers in scientific notation. (a) 6500.; (b) 0.0041; (c) 860 (assume that this number is measured to ± 10); (d) 860 (assume that this number is measured to ± 1); (e) 186,000; (f) 0.0516.

22. Express the following exponentials as ordinary numbers. (a) 5.26×10^4; (b) 4.10×10^{-6}; (c) 1.00×10^2; (d) 8.206×10^{-2}; (e) 9.346×10^3; (f) 9.346×10^{-3}.

23. Which of the following are likely to be exact numbers? Why? (a) 227 inches; (b) 7 computers; (c) $20,335.47; (d) 5 pounds of sugar; (e) 14.7 gallons of diesel fuel; (f) 25,446 ants.

24. To which of the quantities appearing in the following statements would the concept of significant figures apply? Where it would apply, indicate the number of significant figures. (a) The density of platinum at 20°C is 21.45 g/cm³. (b) Wilbur Shaw won the Indianapolis 500-mile race in 1940 with an average speed of 114.277 mi/h. (c) A mile is defined as 5280 feet. (d) The International Committee for Weights and Measures "accepts that the curie be . . . retained as a unit of radioactivity, with the value 3.7×10^{10} s⁻¹." (This resolution was passed in 1964.)

In Exercises 25–26, perform the indicated operations, and round off your answers to the proper number of significant figures. Apply appropriate units to your answers. Assume that all numbers were obtained from measurements.

25. (a) 2.68 ft + 11.4 ft; (b) 142 mi/2.2 h; (c) Three men work for 1.25 hours each. How many man-hours of work did they do? (d) $(1.54 \times 10^2 \text{ cm})(2.336 \times 10^3 \text{ cm})$; (e) What is the area of a square 4.62 m on edge?

26. (a) $0.3135 \text{ ft} \times 0.0669 \text{ ft}$; (b) $423.1 \text{ in} + 0.256 \text{ in} - 116 \text{ in}$; (c) The volume of a sphere is given by $\frac{4}{3}\pi r^3$, where r is the radius of the sphere. Note that the numbers 4 and 3 are exact numbers. Calculate the volume of a sphere with radius 2.27 in. Take the value of π as 3.141593. (d) $6.057 \times 10^3 \text{ m} - 9.35 \text{ m}$; (e) $(8.54 \times 10^5 \text{ mi})/(22 \text{ days})$.

27. How many significant figures are there in each of the following measured quantities? (a) 5.0045 g; (b) 0.087000 L; (c) 5.5×10^{22} atoms; (d) 2.25×10^3 kg; (e) 0.001 mL.

28. Which word prefix indicates the SI multiplier in each of the following numbers? (a) 1×10^3; (b) 1×10^{-3}; (c) 1×10^6; (d) 1×10^{-1}; (e) 0.01; (f) 0.1; (g) 0.001; (h) 1×10^{-6}.

29. Indicate the multiple or fraction of 10 by which a quantity is multiplied when it is preceded by each of the following prefixes. (a) M; (b) m; (c) c; (d) d; (e) k; (f) μ.

Conversions and Dimensional Analysis

30. Carry out each of the following conversions. (a) 10.3 m to km; (b) 10.3 km to m; (c) 247 kg to g; (d) 4.32 L to mL; (e) 85.9 dL to L; (f) 4567 L to cm³.

31. Express 3.72 yards in millimeters, centimeters, meters, and kilometers.

32. (a) Express 65 miles per hour in kilometers per hour. (b) If you are traveling at 65 miles per hour, how many feet are you traveling per second?

33. Express: (a) 2.00 gallons in cm³; (b) 1.00 pint in milliliters; (c) 3.75 cubic yards in cubic inches; (d) 3.75 cubic feet in milliliters.

34. For each of the following pairs, determine which quantity is larger. (a) 24.0 mg or 24.0 cg; (b) 500 cm or 0.5 m; (c) 0.8 nm or 8 Å; (d) 5 L or 3.2 m³.

35. A sample is marked as containing 27.3% calcium carbonate by mass. (a) How many grams of calcium carbonate are contained in 64.33 grams of the sample? (b) How many grams of the sample would contain 11.4 grams of calcium carbonate?

36. An iron ore is found to contain 8.77% hematite (a compound that contains iron). (a) How many tons of this ore would contain 5.50 tons of hematite? (b) How many kilograms of this ore would contain 5.50 kilograms of hematite?

*37. A foundry releases 5.0 tons of gas into the atmosphere each day. The gas contains 2.4% sulfur dioxide by mass. What mass of sulfur dioxide is released in one week?

*38. A certain chemical process requires 75 gallons of pure water each day. The available water contains 11 parts per million by mass of salt (i.e., for every 1,000,000 parts of available water, 11 parts of salt). What mass of salt must be removed each day? A gallon of water weighs 3.67 kg.

39. The radius of a hydrogen atom is about 0.37 Å, and the average radius of the earth's orbit around the sun is about 1.5×10^8 km. Find the ratio of the average radius of the earth's orbit to the radius of the hydrogen atom.

40. If the price of gasoline is $1.059 per gallon, what is its price in cents per liter?

41. Suppose your automobile gas tank holds 18 gallons and the price of gasoline is $0.293 per liter. How much would it cost to fill your gas tank?

42. A particular medicine dropper delivers 25 drops of water to make 1.0 mL. (a) What is the volume of one drop in cubic centimeters? In microliters? (b) If the drop were spherical, what would be its diameter in millimeters? (Volume of a sphere = $\frac{4}{3}\pi r^3$.)

43. Express the following masses or weights in grams and in kilograms. (a) 4.2 ounces; (b) 3.15 short tons; (c) 2.7×10^6 milligrams; (d) 7.33×10^4 centigrams.

44. Express: (a) 215 kilograms in pounds; (b) 215 pounds in kilograms; (c) 215 ounces in centigrams.

*45. At a given point in its orbit, the earth is 92.98 million miles from the sun (center to center). The radius of the sun is 432,000 miles and the radius of the earth is 3960 miles. How long does it take for light from the surface of the sun to reach the earth's surface? The speed of light is 3.00×10^8 m/s.

*46. Cesium atoms are the largest naturally occurring atoms. The radius of a cesium atom is 2.62 Å. How many cesium atoms would have to be laid side by side to give a row of cesium atoms 1.00 inch long? Assume that the atoms are spherical.

*47. The radius of a bromine atom is 1.04 Å. If 1.00 million bromine atoms were laid side by side to make a row, how long would the row be in inches? Assume that the atoms are spherical.

Density and Specific Gravity

48. Mercury poured into a glass of water sinks to the bottom, while gasoline poured into the same glass floats on the surface of the water. A piece of paraffin dropped into the mixture comes to rest between the water and the gasoline, while a piece of iron comes to rest between the water and the mercury. List these five substances in order of increasing density (least dense first).

49. Which is more dense at 0°C, ice or water? Which has the higher specific gravity? How do you know?

50. A 20.4-cubic-centimeter piece of a metal has a mass of 124.3 grams. What is its density?

51. What is the mass of a rectangular piece of copper 21.3 cm × 11.4 cm × 7.9 cm? The density of copper is 8.92 g/cm³.

52. A small crystal of sucrose (table sugar) had a mass of 2.236 mg. The dimensions of the box-like crystal were 1.11 × 1.09 × 1.12 mm. What is the density of sucrose expressed in g/cm³?

53. Vinegar has a density of 1.0056 g/cm³. What is the mass of one liter of vinegar?

*54. The radius of a neutron is approximately 1.5×10^{-15} m and its mass is 1.675×10^{-24} g. Find the density of a neutron, in g/cm³.

*55. A container has a mass of 68.31 g when empty and 93.34 g when filled with water. The density of water is 1.0000 g/cm³. (a) Calculate the volume of the container. (b) When filled with an unknown liquid, the container had a mass of 88.42 g. Calculate the density of the unknown liquid.

*56. The mass of an empty container is 66.734 g. The mass of the container filled with water is 91.786 g. (a) Calculate the volume of the container, using a density of 1.0000 g/cm³ for water. (b) A piece of metal was added to the empty container and the combined mass was 87.807 g. Calculate the mass of the metal. (c) The container with the metal was filled with water and the mass of the entire system was 105.408 g. What mass of water was added? (d) What volume of water was added? (e) What is the volume of the piece of metal? (f) Calculate the density of the metal.

57. What is the specific gravity of a liquid if 225 mL of the liquid has the same mass as 396 mL of water?

58. The specific gravity of silver is 10.5. (a) What is the volume, in cm³, of an ingot of silver with mass 0.555 kg? (b) If this sample of silver is a cube, how long is each edge in cm? (c) How long is the edge of this cube in inches?

*59. The acid in an automobile battery is sulfuric acid. The density of a particular sulfuric acid solution is 1.36 g/mL. This solution is 46.3% H_2SO_4 by mass, the remainder being water.
(a) 245 grams of the solution contains _____ grams of H_2SO_4 and _____ grams of water.
(b) 245 mL of the solution contains _____ grams of H_2SO_4 and _____ grams of water.

Temperature Scales

60. Which represents a larger temperature interval: (a) a Celsius degree or a Fahrenheit degree? (b) a kelvin or a Fahrenheit degree?

61. Express: (a) 373°C in K; (b) 10.15 K in °C; (c) −32.0°C in °F; (d) 100.0°F in K.

62. Express: (a) 0°F in °C; (b) 98.6°F in K; (c) 298 K in °F; (d) 18.5°C in °F.

*63. Use the graph in Figure 1-23 to make each of the following temperature conversions: (a) 50°C to °F, (b) −50°C to °F, and (c) 100°F to °C. Then check each of the conversions by calculation.

*64. On the Réamur scale, which is no longer used, water freezes at 0°R and boils at 80°R. (a) Derive an equation that relates this to the Celsius scale. (b) Derive an equation that relates this to the Fahrenheit scale. (c) Mercury is a liquid metal at room temperature. It boils at 356.6°C (673.9°F). What is the boiling point of mercury on the Réamur scale?

Heat and Heat Transfer

65. Calculate the amount of heat required to raise the temperature of 35.0 grams of water from 10.0°C to 35.0°C.

66. The specific heat of aluminum is 0.895 J/g · °C. Calculate the amount of heat required to raise the temperature of 15.8 grams of aluminum from 27.0°C to 41.0°C.

67. How much heat must be removed from 75.0 grams of water at 90.0°C to cool it to 23.0°C?

*68. In some solar-heated homes, heat from the sun is stored in rocks during the day, and then released during the cooler night. (a) Calculate the amount of heat required to raise the temperature of 85.0 kg of rocks from 25.0°C to 45.0°C. Assume that the rocks are limestone, which is essentially pure calcium carbonate. The specific heat of calcium carbonate is 0.818 J/g · °C. (b) Suppose that when the rocks in part (a) cool to 30.0°C, all the heat released goes to warm the 10,000 cubic feet (2.83×10^5 liters) of air in the house, originally at 10.0°C. To what final temperature would the air be heated? The specific heat of air is 1.004 J/g · °C, and its density is 1.20×10^{-3} g/mL.

*69. A small immersion heater is used to heat water for a cup of coffee. We wish to use it to heat 225 mL of water (about a teacupful) from 25°C to 90°C in 2.00 minutes. What must be the heat rating of the heater, in kJ/min, to accomplish this? Neglect the heat that goes to heat the teacup itself. The density of water is 0.997 g/mL.

Mixed Exercises

70. A 20.0-gram sample of Ca initially at 26.1°C absorbs 905 J. What is the final temperature of the sample? The specific heat of calcium is 0.628 J/g · °C.

71. If you ran a mile in 4.00 minutes, what would be your average speed in (a) km/h, (b) cm/s, and (c) mi/h?

72. A student found the following list of properties of iodine in an encyclopedia: (a) grayish black granules, (b) metallic luster, (c) characteristic odor, (d) forms a purple vapor, (e) density = 4.93 g/cm³, (f) melting point = 113.5°C, (g) soluble in alcohol, (h) insoluble in water, (i) noncombustible, (j) forms ions in aqueous solutions, (k) poisonous. Which of these properties are chemical properties?

73. Which of the properties listed in Exercise 72 are intensive properties?

*74. (a) Draw a graph (similar to that in Figure 1-23) that relates temperatures on the Kelvin and Fahrenheit scales. (Use the Kelvin scale as the horizontal axis.) (b) Use this graph to convert 300 K to the Fahrenheit scale. (c) Use the graph to convert 300°F to the Kelvin scale. (d) Calculate the conversions in parts (b) and (c).

*75. At what temperature will a Fahrenheit thermometer give (a) the same reading as a Celsius thermometer? (b) a reading that is twice that on the Celsius thermometer? (c) a reading that is numerically the same but opposite in sign from that on the Celsius thermometer?

76. Use data from Table 1-8. (a) What would be the mass of a rectangular block of aluminum 1.70 in × 6.25 in × 12.00 in? (b) Calculate the volume in cm³ of 1.00 lb of mercury.

77. At Angel Falls in Venezuela, the water falls 3212 feet. How many meters is this?

*78. The lethal dose of potassium cyanide (KCN), taken orally is 1.6 milligrams per kilogram of body weight. Calculate the lethal dose of potassium cyanide taken orally by a 145-pound person.

79. The mass of a glass object is 395 grams. When the object is immersed in water at 25°C, it displaces 134 mL. What is the density of the glass?

80. The distance light travels through space in one year is called one light-year. Using the speed of light in vacuum listed in Appendix D, and assuming that one year is 365 days, determine the distance of a light-year in kilometers and in miles.

2 Chemical Formulas and Composition Stoichiometry

Objectives

As you study this chapter, you should learn to

☐ Understand an early concept of atoms

☐ Use chemical formulas to solve various kinds of chemical problems

☐ Relate names to formulas and charges of simple ions

☐ Combine simple ions to write formulas and names of ionic compounds

☐ Recognize and use formula weights and mole relationships

☐ Interconvert masses, moles, and formulas in problems

☐ Determine percent compositions in compounds

☐ Determine formulas from composition

☐ Perform calculations about purity of substances

Pure quartz crystals (silicon dioxide, SiO_2) are clear and colorless. Many gemstones consist of quartz that contains traces of certain metal ions. Amethyst, shown here, is quartz that contains small amounts of Fe^{3+} ions. Its color can range from pale lilac to royal purple depending on the amounts of Fe^{3+} ions present. Quartz is widely distributed over the earth's surface. White sand is relatively pure quartz that has been weathered into small pieces.

The language that chemists use to describe the forms of matter and the possible changes in its composition appears throughout the scientific world. Chemical symbols, formulas, and equations are used in such diverse areas as agriculture, home economics, engineering, geology, physics, biology, medicine, and dentistry. In this chapter we shall describe the simplest atomic theory. We shall use it as we represent chemical formulas of elements and compounds. Later this theory will be expanded when we discuss chemical changes.

It is important to learn this fundamental material well so that you can use it correctly and effectively.

The word "stoichiometry" is derived from the Greek *stoicheion,* which means "first principle or element," and *metron,* which means "measure." **Stoichiometry** describes the quantitative relationships among elements in compounds (composition stoichiometry) and among substances as they undergo chemical changes (reaction stoichiometry). In this chapter we shall be concerned with chemical formulas and composition stoichiometry. In Chapter 3 we shall discuss chemical equations and reaction stoichiometry.

2-1 Atoms and Molecules

Around 400 BC, the Greek philosopher Democritus suggested that all matter is composed of tiny, discrete, indivisible particles that he called atoms. His ideas, based entirely on philosophical speculation rather than experimental

The term "atom" comes from the Greek language and means "not divided" or "indivisible."

45

Figure 2-1
Relative sizes of atoms of the noble gases.

evidence, were rejected for 2000 years. By the late 1700s, scientists began to realize that the concept of atoms provided an explanation for many experimental observations about the nature of matter.

By the early 1800s, the Law of Conservation of Matter (Section 1-1) and the Law of Definite Proportions (Section 1-5) were both accepted as general descriptions of how matter behaves. John Dalton, an English schoolteacher, tried to explain why matter behaves in such simple and systematic ways as those expressed above. In 1808, he published the first "modern" ideas about the existence and nature of atoms. He summarized and expanded the nebulous concepts of early philosophers and scientists; more importantly, his ideas were based on *reproducible experimental results* of measurements by many scientists. Taken together, these ideas form the core of **Dalton's Atomic Theory**, one of the highlights of scientific thought. In condensed form, Dalton's ideas may be stated as follows:

The radius of a calcium atom is only 1.97×10^{-8} cm, and its mass is 6.66×10^{-23} g.

Statement 3 is true for *chemical* reactions. However, it is not true for *nuclear* reactions (Chapter 30).

1. An element is composed of extremely small indivisible particles called atoms.
2. All atoms of a given element have identical properties, which differ from those of other elements.
3. Atoms cannot be created, destroyed, or transformed into atoms of another element.
4. Compounds are formed when atoms of different elements combine with each other in small whole-number ratios.
5. The relative numbers and kinds of atoms are constant in a given compound.

Dalton believed that atoms were solid indivisible spheres, an idea we now reject. But he showed remarkable insight into the nature of matter and its interactions. Some of his ideas could not be verified (or refuted) experimentally at the time. They were based on the limited experimental observations of his day. Even with their shortcomings, Dalton's ideas provided a framework that could be modified and expanded by later scientists. Thus John Dalton is the father of modern atomic theory.

For Group 0 elements, the noble gases, a molecule contains only one atom and so an atom and a molecule are the same (Figure 2-1).

The smallest particle of an element that maintains its chemical identity through all chemical and physical changes is called an **atom** (Figure 2-1). In nearly all **molecules**, two or more atoms are bonded together in very small, discrete units (particles) that are electrically neutral. A **molecule** is the small-

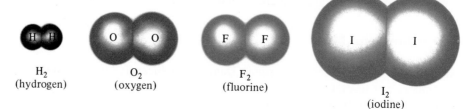

Figure 2-2
Models of diatomic molecules of some elements, approximately to scale.

H_2 (hydrogen) O_2 (oxygen) F_2 (fluorine) I_2 (iodine)

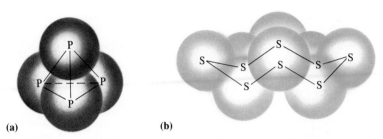

(a)   **(b)**

Figure 2-3
(a) A model of the P_4 molecule of white phosphorus. (b) A model of the S_8 ring found in rhombic sulfur.

est particle of an element or compound that can have a stable independent existence.

Individual oxygen atoms are not stable at room temperature and atmospheric pressure. Hence, single atoms of oxygen mixed under these conditions quickly combine to form pairs. The oxygen with which we are all familiar is made up of two atoms of oxygen; it is a *diatomic* molecule, O_2. Hydrogen, nitrogen, fluorine, chlorine, bromine, and iodine are other examples of diatomic molecules (Figure 2-2).

Some other elements exist as more complex molecules. Phosphorus molecules consist of four atoms, while sulfur exists as eight-atom molecules at ordinary temperatures and pressures. Molecules that contain two or more atoms are called *polyatomic* molecules. See Figure 2-3.

In modern terminology, O_2 is named dioxygen, H_2 is dihydrogen, P_4 is tetraphosphorus, and so on. Even though such terminology is officially preferred, it has not yet gained wide acceptance. Most chemists still refer to O_2 as oxygen, H_2 as hydrogen, P_4 as phosphorus, and so on.

Molecules of compounds are composed of more than one kind of atom. A water molecule consists of two atoms of hydrogen and one atom of oxygen. A molecule of methane consists of one carbon atom and four hydrogen atoms. The shapes of these and a few other molecules are shown in Figure 2-4.

Atoms are the components of molecules, and molecules are the components of elements and most compounds. We are able to see samples of compounds and elements that consist of large numbers of atoms and mol-

> You should remember the common elements that occur as diatomic molecules: H_2, N_2, O_2, F_2, Cl_2, Br_2, I_2.

> The compound water, H_2O, is made up of water molecules, which are made up of hydrogen and oxygen atoms. Methane is the principal component of natural gas.

H_2O
(water)

CO_2
(carbon dioxide)

CH_4
(methane)

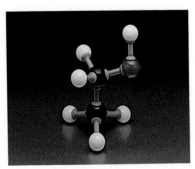

C_2H_5OH
(ethyl alcohol)

Figure 2-4
Formulas and models for some molecules.

Figure 2-5
A computer reconstruction of the surface of a sample of highly ordered graphite, as observed with a scanning tunnelling electron microscope (STM), reveals the regular pattern of individual carbon atoms. Many important reactions occur on the surfaces of solids. Observations of the atomic arrangements on surfaces help chemists understand such reactions. New information available with the STM will give many details about chemical bonding in solids.

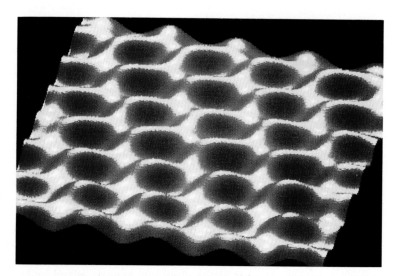

ecules. With the scanning tunnelling microscope it is now possible to "see" atoms (Figure 2-5). It would take 217 million silicon atoms to make a row 1 inch long.

2-2 Chemical Formulas

An O_2 molecule.

The **chemical formula** for a substance shows its chemical composition. This represents the elements present as well as the ratio in which the atoms of the elements occur. The formula for a single atom is the same as the symbol for the element. Thus, Na can represent a single sodium atom. It is unusual to find such isolated atoms in nature, with the exception of the noble gases (He, Ne, Ar, Kr, Xe, and Rn). A subscript following the symbol of an element indicates the number of atoms in a molecule. For instance, F_2 indicates a molecule containing two fluorine atoms, and P_4 a molecule containing four phosphorus atoms.

Some elements exist in more than one form. Familiar examples include (1) oxygen, found as O_2 molecules, and ozone, found as O_3 molecules, and (2) two different crystalline forms of carbon, diamond, and graphite (Figure 13-31). Different forms of the same element in the same physical state are called **allotropic modifications** or **allotropes**.

Compounds contain two or more elements in chemical combination in fixed proportions. Hence, each molecule of hydrogen chloride, HCl, contains one atom of hydrogen and one atom of chlorine; each molecule of carbon tetrachloride, CCl_4, contains one carbon atom and four chlorine atoms. An aspirin molecule, $C_9H_8O_4$, contains nine carbon atoms, eight hydrogen atoms, and four oxygen atoms.

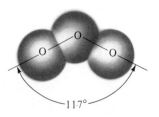

An O_3 molecule.

Some groups of atoms behave chemically as single entities. For instance, one nitrogen atom and two oxygen atoms may combine to form a *nitro* group that is a part of a molecule. In formulas of compounds containing two or more of the same group, the group formula is enclosed in parentheses. Thus, 2,4,6-trinitrotoluene (often abbreviated TNT) contains three *nitro* groups,

Table 2-1
Formulas and Names of Some Common Molecular Compounds

Formula	Name	Formula	Name
HCl	hydrogen chloride (or hydrochloric acid if dissolved in water)	SO_2	sulfur dioxide
		SO_3	sulfur trioxide
		CO	carbon monoxide
H_2SO_4	sulfuric acid	CO_2	carbon dioxide
HNO_3	nitric acid	CH_4	methane
CH_3COOH	acetic acid	C_2H_6	ethane
NH_3	ammonia	C_3H_8	propane

and its formula is $C_7H_5(NO_2)_3$ (see margin). When you count up the number of atoms in this molecule from its formula, you must multiply the numbers of nitrogen and oxygen atoms in the NO_2 group by 3. There are *seven* carbon atoms, *five* hydrogen atoms, *three* nitrogen atoms, and *six* oxygen atoms in a molecule of TNT.

Compounds were first recognized as distinct substances because of their different physical properties and because they could be separated from one another. Once the concept of atoms and molecules was established, the reason for these differences in properties could be understood: Two compounds differ from one another because their molecules are different. Conversely, if two molecules contain the same number of the same kinds of atoms, arranged the same way, then both are molecules of the same compound. Thus the atomic theory explains the **Law of Definite Proportions** (Section 1-5).

This law, also known as the **Law of Constant Composition**, can now be extended to include its interpretation in terms of atoms. It is so important for performing the calculations in this chapter that we restate it here:

> Different pure samples of a compound always contain the same elements in the same proportion by mass; this corresponds to atoms of these elements combined in fixed numerical ratios.

Throughout your study of chemistry you will have many occasions to refer to compounds by name. Table 2-1 includes a few examples for molecular compounds.

So we see that for a substance composed of molecules, the *chemical formula* gives the number of atoms of each type in the molecule. But this formula does not express the order in which the atoms in the molecules are bonded together. The **structural formula** shows how the atoms are connected. The lines connecting atomic symbols represent chemical bonds between atoms. The bonds are actually forces that tend to hold atoms at certain distances and angles from one another. For instance, the structural formula of propane shows that the three C atoms are linked in a chain, with three H atoms bonded to each of the end C atoms and two H atoms bonded to the center C. **Ball-and-stick** molecular models and **space-filling** molecular models help us to see the shapes and relative sizes of molecules. These four representations are shown in Figure 2-6.

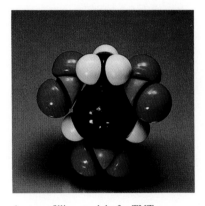

A space-filling model of a TNT molecule, $C_7H_5(NO_2)_3$.

Chemical Formula	Structural Formula	Ball-and-Stick Model	Space-Filling Model
H_2O, water	H—O—H		
H_2O_2, hydrogen peroxide	H—O—O—H		
CCl_4, carbon tetrachloride	Cl | Cl—C—Cl | Cl		
C_3H_8, propane	H H H | | | H—C—C—C—H | | | H H H		
C_2H_5OH, ethanol	H H | | H—C—C—O—H | | H H		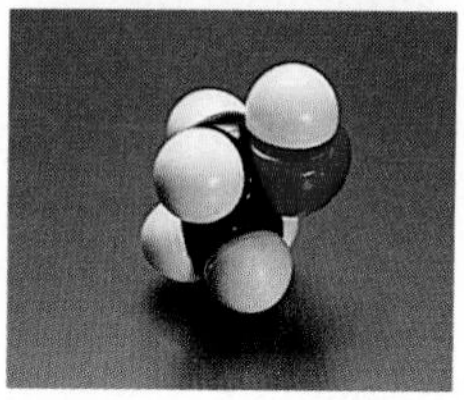

Figure 2-6
Formulas and models for some molecules. Structural formulas show the order in which atoms are connected, but do not represent true molecular shapes. Ball-and-stick models use balls of different colors to represent atoms and sticks to represent bonds; they show the three-dimensional shapes of molecules. Space-filling models show the (approximate) relative sizes of atoms and the shapes of molecules.

2-3 Ions and Ionic Compounds

So far we have discussed only compounds that exist as discrete molecules. Some compounds, such as sodium chloride, NaCl, consist of ions. An **ion** is an atom or group of atoms that carries an electrical charge. Ions that possess a *positive* charge, such as the sodium ion, Na^+, are called **cations**. Those carrying a *negative* charge, such as the chloride ion, Cl^-, are called

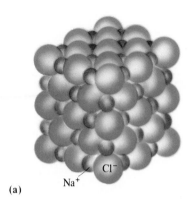

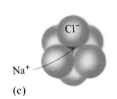

(a)   (b)   (c)

Figure 2-7
The arrangement of ions in NaCl. (a) A crystal of sodium chloride consists of an extended array that contains equal numbers of sodium ions (small spheres) and chloride ions (large spheres). Within the crystal, each chloride ion is surrounded by six sodium ions (b), and each sodium ion is surrounded by six chloride ions (c).

anions. The charge on an ion *must* be included as a superscript on the right side of the chemical symbol(s) when we write the formula for the individual ion.

As we shall see in Chapter 5, an atom consists of a very small, very dense, positively charged *nucleus* surrounded by a diffuse distribution of negatively charged particles called *electrons*. The number of positive charges in the nucleus defines the identity of the element to which the atom corresponds. Electrically neutral atoms contain the same number of electrons outside the nucleus as positive charges (protons) within the nucleus. Ions are formed when neutral atoms lose or gain electrons. An Na^+ ion is formed when a sodium atom loses one electron, and a Cl^- ion is formed when a chlorine atom gains one electron.

The compound NaCl consists of an extended array of Na^+ and Cl^- ions (Figure 2-7). Within the crystal (though not on the surface) each Na^+ ion is surrounded at equal distances by six Cl^- ions, and each Cl^- ion is similarly surrounded by six Na^+ ions. *Any* compound, whether ionic or molecular, is electrically neutral; i.e., it has no net charge. In NaCl this means that the Na^+ and Cl^- ions are present in a 1:1 ratio, and this is indicated by the formula NaCl.

Because there are no "molecules" of ionic substances, we should not refer to "a molecule of NaCl." Instead, we refer to a **formula unit (FU)** of NaCl, consisting of one Na^+ ion and one Cl^- ion. Similarly, we speak of the formula unit of all ionic compounds. It is also acceptable to refer to a molecule of a molecular compound as a formula unit. One formula unit of C_3H_8, which is the same as one molecule, contains three C atoms and eight H atoms.

For the present, we shall tell you which substances are ionic and which are covalent when it is important to know. Later you will learn to make the distinction yourself.

Polyatomic ions are groups of atoms that bear an electrical charge. Examples include the ammonium ion, NH_4^+, the sulfate ion, SO_4^{2-}, and the nitrate ion, NO_3^-. Table 2-2 shows the formulas, ionic charges, and names of some common ions. You should learn the formulas and names of these frequently encountered ions. They can be used to write the formulas and names of many ionic compounds. We write the formula of an ionic compound by adjusting the relative numbers of positive and negative ions so their total charges cancel (i.e., add to zero). The name is formed by giving the names of the ions, with the positive ion named first (by convention).

The general term "formula unit" applies to molecular or ionic compounds, whereas the more specific term "molecule" applies only to elements and compounds that exist as discrete molecules.

Electron pairs are shared by atoms in *covalent* compounds.

As we shall see, some metals can form more than one kind of ion with positive charge. For such metals we specify which ion we mean with a Roman numeral—e.g., iron(II) or iron(III). Because zinc forms no stable ions other than Zn^{2+}, we do not need to use Roman numerals in its name.

Table 2-2
Formulas, Ionic Charges, and Names of Some Common Ions

Common Cations (positive ions)			Common Anions (negative ions)		
Formula	Charge	Name	Formula	Charge	Name
Na^+	1+	sodium ion	F^-	1–	fluoride ion
K^+	1+	potassium ion	Cl^-	1–	chloride ion
NH_4^+	1+	ammonium ion	Br^-	1–	bromide ion
Ag^+	1+	silver ion	OH^-	1–	hydroxide ion
Mg^{2+}	2+	magnesium ion	CH_3COO^-	1–	acetate ion
Ca^{2+}	2+	calcium ion	NO_3^-	1–	nitrate ion
Zn^{2+}	2+	zinc ion			
Cu^+	1+	copper(I) or cuprous ion	O^{2-}	2–	oxide ion
			S^{2-}	2–	sulfide ion
Cu^{2+}	2+	copper(II) or cupric ion	SO_3^{2-}	2–	sulfite ion
			SO_4^{2-}	2–	sulfate ion
Fe^{2+}	2+	iron(II) or ferrous ion	CO_3^{2-}	2–	carbonate ion
Fe^{3+}	3+	iron(III) or ferric ion			
Al^{3+}	3+	aluminum ion	PO_4^{3-}	3–	phosphate ion

Example 2-1

Write the formulas for the following ionic compounds: (a) sodium fluoride, (b) calcium fluoride, (c) iron(II) sulfate, (d) zinc phosphate.

Plan

In each case, we identify the chemical formulas of the ions from Table 2-2. These ions must be present in a ratio that gives the compound *no net charge*. The formulas and names of ionic compounds are written by giving the positively charged ion first.

Solution

(a) The formula for the sodium ion is Na^+ and the formula for fluoride ion is F^- (Table 2-2). Because the charges on these two ions are equal in magnitude, the ions must be present in equal numbers, or in a 1:1 ratio. Thus the formula for sodium fluoride is   NaF.

(b) The formula for the calcium ion is Ca^{2+} and the formula for fluoride ion is F^-. Now each positive ion (Ca^{2+}) provides twice as much charge as each negative ion (F^-). So there must be twice as many F^- ions as Ca^{2+} ions to equalize the charge. This means that the ratio of calcium to fluoride ions is 1:2. So the formula for calcium fluoride is CaF_2.

(c) The iron(II) ion is Fe^{2+} and the sulfate ion is SO_4^{2-}. As in (a), the equal magnitudes of positive and negative charges tell us that the ions must be present in equal numbers, or in a 1:1 ratio. The formula for iron(II) sulfate is $FeSO_4$.

(d) The zinc ion is Zn^{2+} and the phosphate ion is PO_4^{3-}. Now it will take *three* Zn^{2+} ions to account for as much charge (6+ total) as would be present in *two* PO_4^{3-} ions (6– total). So the formula for zinc phosphate is $Zn_3(PO_4)_2$.

EOC 12,14

Example 2-2

Name the following ionic compounds: (a) $(NH_4)_2S$, (b) $Cu(NO_3)_2$, (c) $ZnCl_2$, (d) $Fe_2(CO_3)_3$.

Plan

In naming ionic compounds, it is helpful to inspect the formula for atoms or groups of atoms that we recognize as representing familiar ions.

Solution

(a) The presence of the polyatomic grouping NH_4 in the formula suggests to us the presence of the ammonium ion, NH_4^+. There are two of these, each accounting for 1+ in charge. To balance this, the single S must account for 2− in charge, or S^{2-}, which we recognize as the sulfide ion. Thus the name of the compound is

ammonium sulfide.

(b) The NO_3 grouping in the formula tells us that the nitrate ion, NO_3^-, is present. Two of these nitrate ions account for $2 \times 1- = 2-$ in negative charge. To balance this, copper must account for 2+ charge and be the copper(II) ion. The name of

the compound is   copper(II) nitrate   or, alternatively,   cupric nitrate.

(c) The positive ion present is zinc ion, Zn^{2+}, and the negative ion is chloride,

Cl^-. The name of the compound is   zinc chloride.

(d) Each CO_3 grouping in the formula must represent the carbonate ion, CO_3^{2-}. The presence of *three* such ions accounts for a total of 6− in negative charge, so there must be a total of 6+ present in positive charge to balance this. It takes *two* iron ions to provide this 6+, so each ion must have a charge of 3+ and be Fe^{3+}, the iron(III) ion or ferric ion. The name of the compound is

iron(III) carbonate   or   ferric carbonate.

EOC 11,13

A more extensive discussion on naming compounds appears in Sections 7-11 and 7-12.

2-4 Atomic Weights

As the chemists of the eighteenth and nineteenth centuries painstakingly sought information about the compositions of compounds and tried to systematize their knowledge, it became apparent that each element has a characteristic mass relative to every other element. Although these early scientists did not have the experimental means to measure the mass of each kind of atom, they succeeded in defining a *relative* scale of atomic masses.

An early observation was that carbon and hydrogen have relative atomic masses, also traditionally called **atomic weights**, **AW**, of approximately 12 and 1, respectively. Thousands of experiments on the compositions of compounds have resulted in the establishment of a scale of relative atomic weights based on the **atomic mass unit (amu)**, which is defined as *exactly $\frac{1}{12}$ of the mass of an atom of a particular kind of carbon atom, called carbon-12.*

The term "atomic weight" is widely accepted because of its traditional use, although it is properly a mass rather than a weight. "Atomic mass" is often used.

On this scale, the atomic weight of hydrogen (H) is 1.00794 amu, that of sodium (Na) is 22.989768 amu, and that of magnesium (Mg) is 24.3050 amu. This tells us that Na atoms have nearly 23 times the mass of H atoms, while Mg atoms are about 24 times heavier than H atoms.

2-5 The Mole

Even the smallest bit of matter that can be handled reliably contains an enormous number of atoms. So we must deal with large numbers of atoms in any real situation, and some unit for conveniently describing a large number of atoms is desirable. The idea of using a unit to describe a particular number (amount) of objects has been around for a long time. You are no doubt already familiar with the dozen (12 items) and the gross (144 items).

"Mole" is derived from the Latin word moles, which means "a mass." "Molecule" is the diminutive form of this word and means "a small mass."

The SI unit for amount is the **mole**, abbreviated mol. It is *defined* as the amount of substance that contains as many entities (atoms, molecules, or other particles) as there are atoms in 0.012 kg of pure carbon-12 atoms. Many experiments have refined the number, and the currently accepted value is

$$1 \text{ mole} = 6.022045 \times 10^{23} \text{ particles}$$

This number, often rounded off to 6.022×10^{23}, is called **Avogadro's number** in honor of Amedeo Avogadro (1776–1856), whose contributions to chemistry are discussed in Section 12-8.

According to its definition, the mole unit refers to a fixed number of entities, whose identities must be specified. Just as we speak of a dozen eggs or a dozen automobiles, we refer to a mole of atoms or a mole of molecules (or a mole of ions, electrons, or other particles). We could even think about a mole of eggs, although the size of the required carton staggers the imagination! Helium exists as discrete He atoms, so one mole of helium consists of 6.022×10^{23} He *atoms*. Hydrogen commonly exists as diatomic (two-atom) molecules, so one mole of hydrogen contains 6.022×10^{23} H_2 *molecules* and $2(6.022 \times 10^{23})$ H atoms.

Every kind of atom, molecule, and ion has a definite characteristic mass. It follows that one mole of a given pure substance also has a definite mass, regardless of the source of the sample. This idea is of central importance in many calculations throughout the study of chemistry and the related sciences.

Because the mole is defined as the number of atoms in 0.012 kg (or 12 grams) of carbon-12, and the atomic mass unit is defined as $\frac{1}{12}$ of the mass of a carbon-12 atom, the following convenient relationship is true:

Atomic weights of the elements are listed inside the front cover.

The mass of one mole of atoms of a pure element in grams is numerically equal to the atomic weight of that element in amu. This is also called the **molar mass** of the element; its units are grams/mole.

For instance, if you obtain a pure sample of the metallic element titanium (Ti), whose atomic weight is 47.88 amu, and measure out 47.88 grams of it, you will have one mole, or 6.022×10^{23} atoms, of titanium.

The symbol for an element can (1) identify the element, (2) represent one atom of the element, or (3) represent one mole of atoms of the element. The last interpretation will be extremely useful in calculations in the next chapter.

12 eggs
or
1 dozen eggs
or
24 ounces of eggs

6.022×10^{23} Fe atoms
or
1 mole of Fe atoms
or
55.847 grams of iron

Figure 2-8
Three different ways of representing amounts.

The atomic weight of iron (Fe) is 55.847 amu.

A quantity of a substance may be expressed in a variety of ways. For example, consider a dozen eggs and 55.847 grams of iron filings, or one mole of iron (Figure 2-8). We can express the amount of eggs or iron filings present in any of several different units. We can then construct unit factors to relate an amount of the substance expressed in one kind of unit to the same amount expressed in another unit.

Unit Factors for Eggs

$$\frac{12 \text{ eggs}}{1 \text{ doz eggs}}$$

$$\frac{12 \text{ eggs}}{24 \text{ ounces of eggs}}$$

and so on

Unit Factors for Iron

$$\frac{6.022 \times 10^{23} \text{ Fe atoms}}{1 \text{ mol Fe atoms}}$$

$$\frac{6.022 \times 10^{23} \text{ Fe atoms}}{55.847 \text{ g Fe}}$$

and so on

As Table 2-3 suggests, the concept of a mole as applied to atoms is especially useful. It provides a convenient basis for comparing equal numbers of atoms of different elements.

Table 2-3
Mass of One Mole of Atoms of Some Common Elements

Element	A Sample with a Mass of	Contains
carbon	12.011 g C	6.022×10^{23} C atoms or 1 mole of C atoms
titanium	47.88 g Ti	6.022×10^{23} Ti atoms or 1 mole of Ti atoms
gold	196.96654 g Au	6.022×10^{23} Au atoms or 1 mole of Au atoms
hydrogen	1.00794 g H_2	6.022×10^{23} H atoms or 1 mole of H atoms $(3.011 \times 10^{23}$ H_2 molecules or 0.5 mole of H_2 molecules)
sulfur	32.066 g S_8	6.022×10^{23} S atoms or 1 mole of S atoms $(0.7528 \times 10^{23}$ S_8 molecules or 0.1250 mole of S_8 molecules)

Chemistry in Use...
Names of the Elements

If you were to discover a new element, how would you name it? Throughout history, scientists have answered this question in different ways. Most have chosen to honor a person or place or to describe the new substance. Even elements known long ago, whose discoverers are unknown, have names with etymological significance. Looked at from a historical point of view, these names tell us much about the nature of chemistry and scientific discovery. They also tell us about the nature of scientists—their values, heroes, and practices. Many elements were unearthed by teams rather than individuals. Chemists of different nationalities or schools of thought who had worked cooperatively were sometimes reduced to bickering enemies when the time came to choose a name for their discovery!

Until the Middle Ages only nine elements were known: gold, silver, tin, mercury, copper, lead, iron, sulfur, and carbon. The metals' chemical symbols are taken from descriptive Latin names: *aurum* ("yellow"), *argentum* ("shining"), *stannum* ("dripping" or "easily melted"), *hydrargyrum* ("silvery water"), *cuprum* (Cyprus, where many copper mines were located), *plumbum* (exact meaning unknown—possibly "heavy"), and *ferrum* (also unknown). Some of these were derived from even earlier Sanskrit words. The English names are derived from old Anglo-Saxon terms. Mercury is named after the planet, one reminder that the ancients associated metals with gods and celestial bodies. In turn, both the planet, which moves rapidly across the sky, and the element, which is the only metal that is liquid at room temperature and thus flows rapidly, are named for the fleet god of messengers in Roman mythology. In English, mercury is nicknamed "quicksilver."

Prior to the reforms of Antoine Lavoisier, chemistry was a largely nonquantitative, unsystematic science in which experimenters had little contact with each other. There were few rules for documenting and sharing information. Thus, elements discovered prior to Lavoisier's contributions in the late 18th century have names whose sources are hard to identify. They include the following. "Zinc" may have originated from the Persian *seng* ("stone") or the German *Zinke* ("spike"). "Antimony" is thought to have come from the Arabic *al ithmid*, the name for the compound Sb_2S_3, which was used to darken women's eyebrows (its Latin name, *stibium*, means mark). Arsenic is another element with an ambiguous etymology. The Greek word *arsenikos*, meaning male, is one possible source, as alchemists believed that metals were either male or female. The Persian *zarnik* ("golden") is another.

In 1787 Lavoisier published his *Methode de Nomenclature Chimique*, which proposed, among other changes, that all new elements be named descriptively. For the next 125 years, most elements were given names that corresponded to their properties. Greek roots were one popular source, as evidenced by hydrogen (*hydros-gen*, "water-producing"), oxygen (*oksys-gen*, "acid-producing"), nitrogen (*nitron-gen*, "soda-producing"), bromine (*bromos*, "stink"), and argon (*a-er-gon*, "no reaction"). The discoverers of argon, Ramsay and Rayleigh, originally proposed the name *aeron* (from *aer* or air) but critics thought is was too close to the biblical name Aaron! Latin roots such as *radius* ("ray") were also used (radium and radon are both naturally radioactive elements that emit "rays"). Color was often the determining property, especially after the invention of the spectroscope in 1859, because different elements (or the light that they emit) have prominent characteristic colors. Cesium, indium, iodine, rubidium, and thallium were all named in this manner. Their respective Greek and Latin roots denote blue-gray, indigo, violet, red, and green (*thallus* means "tree sprout"). Because of the great variety of colors of its compounds. iridium takes its name from the Latin *iris*, meaning rainbow. Alternatively, an element name might suggest a mineral or the ore that contained it. One example is wolfram or tungsten (W), which was isolated from wolframite. Two other "inconsistent" elemental symbols, K and Na, arose from occurrence as well. *Kalium* was first obtained from the saltwort plant, *Salsola kali*, and *natrium* from niter. Their English names, potassium and sodium, are derived from the ores potash and soda.

Other elements, contrary to Lavoisier's suggestion, were named after planets, mythological figures, places, or superstitions, "Celestial elements" include helium (sun), tellurium (earth), selenium (moon—the element was discovered in close proximity to tellurium), cerium (the asteroid Ceres, which was discovered only two years before the element), and uranium (the planet Uranus, discovered a few years earlier). The first two transuranium elements (those *beyond* uranium) to be produced were named neptunium and plutonium for the next two planets, Neptune and Pluto. The names promethium (Prometheus, who stole fire from heaven), vanadium (Scandinavian goddess, Vanadis), titanium (Titans, the first sons of the earth), tantalum (Tantalos, father of Niobe), and thorium (Thor, Scandinavian god of war) all arise from Greek or Norse mythology. Cobalt was named for Kobold, German evil spirit, when its presence interfered with the mining of copper (as did nickel, from *Kupfernickel*, or false copper).

"Geographical elements," shown on the map, sometimes honored the discoverer's native country or workplace. The Latin names for Russia (*ruthenium*), France (*gallium*), Paris (*lutetium*), and Germany (*germanium*) were among those used. Marie Sklodowska Curie named one of the elements that she discovered polonium, after her native Poland. Often the locale of discovery lends its name to the element; the record holder is certainly the Swedish village Ytterby, the site of ores from which the four elements terbium, erbium, ytterbium, and yttrium were isolated. Elements honoring important scientists include curium, einsteinium, nobelium, fermium, and lawrencium.

Sometimes the name of an element contains a history of its discovery. In 1839 Mosander gave a new element he had extracted as a minor component in a cerium compound the name lanthanum (Greek,

"to lie hidden"). Two years later he thought that he had found another new element from the same source, and named it didymium (Greek, "twin") because it was "an inseparable twin brother of lanthanum." But in 1885, von Welsbach separated didymium into two new elements, which he named neodymium (Greek, "new twin") and praseodymium (Greek, "green twin").

Most of the 109 elements now known were given titles peacefully, but a few were not. Niobium, isolated in 1803 by Ekeberg from an ore that also contained tantalum, and named after the Greek goddess Niobe (daughter of Tantalus), was later found to be identical to an 1802 discovery of Hatchett, columbium. (Interestingly, Hatchett first found the element in an ore sample that had been sent to England more than a century earlier by John Winthrop, the first governor of Connecticut.) While "niobium" became the accepted designation in Europe,

the Americans, not surprisingly, chose "columbium." It was not until 1949—when the International Union of Pure and Applied Chemistry (IUPAC) ended more than a century of controversy by ruling in favor of mythology—that element 41 received a unique name. Current arguments over the proper names of elements 104 and 105 have prompted the IUPAC to begin hearing claims of priority to numbers 104 through 110. Some Russian and Scandinavian texts refer to 104 as kurchatovium; some American and English texts, as rutherfordium. Similarly, 105 is known as both hahnium and nielsbohrium. In 1978, the IUPAC recommended that, at least for now, the elements beyond element 103 be known by systematic names based on numerical roots; element 104 is unnilquadium (*un* for 1, *nil* for 0, *quad* for 4, plus the *-ium* ending), followed by unnilpentium, unnilhexium, and so on.

The number and variety of element names indicate something of the long and colorful history of chemistry. Examining them shows us how scientists' values and beliefs have changed over the years. As there will probably not be many more new elements synthesized, we might hope that the IUPAC will allow nonsystematic titles to be given to the new, artificially created elements. It would be a shame if discoverers were not allowed to participate in naming their discoveries. Even when the last possible transuranium element has been made, synthetic chemists will still be in need of monikers for their products. Judging from the abundance of names such as "buckminsterfullerene" (C_{60}) for novel substances, it is doubtful that scientists' creativity and sense of humor in such matters will be easily stifled.

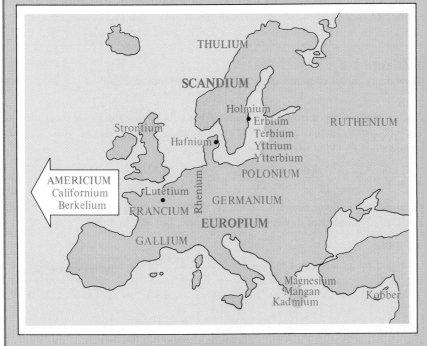

Many chemical elements were named after places.

Lisa L. Saunders
Chemistry major
University of Texas at Austin

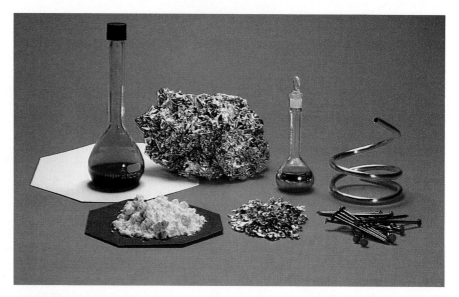

Figure 2-9
One mole of atoms of some common elements. Back row (left to right): bromine, aluminum, mercury, copper. Front row (left to right): sulfur, zinc, iron.

Figure 2-9 shows what one mole of atoms of each of some common elements looks like. Each of the examples in Figure 2-9 represents 6.022×10^{23} *atoms* of the element.

The relationship between the mass of a sample of an element and the number of moles of atoms in the sample is illustrated in Example 2-3.

Example 2-3
How many moles of atoms does 245.2 g of iron metal contain?

Plan
The atomic weight of iron is 55.85 amu. This tells us that the molar mass of iron is 55.85 g/mol, or that one mole of iron atoms is 55.85 grams of iron. We can express this as either of two unit factors:

$$\frac{1 \text{ mol Fe atom}}{55.85 \text{ g Fe}} \quad \text{or} \quad \frac{55.85 \text{ g Fe}}{1 \text{ mol Fe atom}}$$

Because one mole of iron has a mass of 55.85 g, we expect that 245.2 g will be a fairly small number of moles (greater than one, but less than ten).

Solution

$$\underline{?} \text{ mol Fe atoms} = 245.2 \text{ g Fe} \times \frac{1 \text{ mol Fe atom}}{55.85 \text{ g Fe}} = \boxed{4.390 \text{ mol Fe atoms}}$$

Once the number of moles of atoms of an element is known, the number of atoms in the sample can be calculated, as Example 2-4 illustrates.

Example 2-4
How many atoms are contained in 4.390 moles of iron atoms?

Plan

One mole of atoms of an element contains Avogadro's number of atoms, or 6.022 $\times 10^{23}$ atoms. This lets us generate the two unit factors

$$\frac{6.022 \times 10^{23} \text{ atoms}}{1 \text{ mol atoms}} \quad \text{or} \quad \frac{1 \text{ mol atoms}}{6.022 \times 10^{23} \text{ atoms}}$$

We expect that the number of atoms in more than four moles of atoms is a very large number.

Solution

$$\underline{?} \text{ Fe atoms} = 4.390 \text{ mol Fe atoms} \times \frac{6.022 \times 10^{23} \text{ Fe atoms}}{1 \text{ mol Fe atoms}}$$

$$= 2.644 \times 10^{24} \text{ Fe atoms}$$

EOC 26

If we know the atomic weight of an element on the carbon-12 scale, we can use the mole concept and Avogadro's number to calculate the *average* mass of one atom of that element in grams (or any other mass unit we choose).

Example 2-5

Calculate the mass of one iron atom in grams.

Plan

We expect that the mass of a single atom in grams would be a *very* small number. We know that one mole of Fe atoms has a mass of 55.85 g and contains 6.022 $\times$ 10^{23} Fe atoms. We use this information to generate unit factors to carry out the desired conversion.

Solution

$$\frac{\underline{?} \text{ g Fe}}{\text{Fe atom}} = \frac{55.85 \text{ g Fe}}{1 \text{ mol Fe atoms}} \times \frac{1 \text{ mol Fe atoms}}{6.022 \times 10^{23} \text{ Fe atoms}}$$

$$= 9.274 \times 10^{-23} \text{ g Fe/Fe atom}$$

Thus, we see that the mass of one Fe atom is only 9.274×10^{-23} g.

To gain some appreciation of how little this is, write 9.274×10^{-23} gram as a decimal fraction and try to name the fraction.

This example demonstrates how small atoms are and why it is necessary to use large numbers of atoms in practical work. Let's calculate the number of atoms in an iron micrometeorite that weighs only one billionth of a gram. The radius of a spherical micrometeorite of this mass is 3.0×10^{-4} cm.

Micrometeorites (mostly Ni and Fe) of this size constantly shower down on the earth from space.

Example 2-6

Calculate the number of atoms in 1.0 billionth of a gram of iron metal to two significant figures.

Plan

We use unit factors that we can construct from the atomic weight of iron and Avogadro's number.

$$\boxed{\text{g of Fe}} \longrightarrow \boxed{\text{mol of Fe atoms}} \longrightarrow \boxed{\text{number of Fe atoms}}$$

Solution

Even though the number 1.0×10^{-9} has two significant figures, we carry the other numbers to more significant figures. Then we round at the end to the appropriate number of significant figures.

$$\underline{?} \text{ Fe atoms} = 1.0 \times 10^{-9} \text{ g} \times \frac{1 \text{ mol Fe atoms}}{55.85 \text{ g Fe}} \times \frac{6.022 \times 10^{23} \text{ Fe atoms}}{1 \text{ mol Fe atoms}}$$

$$= \boxed{1.1 \times 10^{13} \text{ Fe atoms}}$$

One billionth of a gram of iron contains about 11,000,000,000,000 Fe atoms.

2-6 Formula Weights, Molecular Weights, and Moles

The **formula weight (FW)** of a substance is the sum of the atomic weights (AW) of the elements in the formula, each taken the number of times the element occurs. Hence a formula weight gives the mass of one formula unit in amu.

Formula weights, like the atomic weights on which they are based, are relative masses. The formula weight for sodium hydroxide, NaOH, (rounded to the nearest 0.01 amu) is found as follows.

No. of Atoms of Stated Kind		× Mass of One Atom	= Mass Due to Elements
$1 \times$ Na =	1	× 23.00 amu	= 23.00 amu of Na
$1 \times$ H =	1	× 1.01 amu	= 1.01 amu of H
$1 \times$ O =	1	× 16.00 amu	= 16.00 amu of O

Formula weight of NaOH = 40.01 amu

The term **molecular weight (MW)** is used interchangeably with "formula weight" when reference is made to molecular (nonionic) substances, i.e., substances that exist as discrete molecules.

Example 2-7

Calculate the formula weight (molecular weight) of 2,4,6-trinitrotoluene (TNT), $C_7H_5(NO_2)_3$, using the precisely known values for atomic weights given in the International Table of Atomic Weights (1987) inside the front cover of the text.

Plan

We add the atomic weights of the elements in the formula, each multiplied by the number of times the element occurs. Because the least precisely known atomic weight (12.011 amu for C) is known to three significant figures past the decimal point, the result is known to only that number of significant figures.

Solution

No. of Atoms of Stated Kind		× Mass of One Atom	= Mass Due to Element
$7 \times C =$	7	× 12.011 amu	= 84.077 amu of C
$5 \times H =$	5	× 1.00794 amu	= 5.09370 amu of H
$3 \times N =$	3	× 14.00674 amu	= 42.02022 amu of N
$6 \times O =$	6	× 15.9994 amu	= 95.9964 amu of O

Formula weight of 2,4,6-trinitrotoluene (TNT) = 227.187 amu

EOC 22

The term "formula weight" is correctly used for either ionic or molecular substances. When we refer specifically to molecular (nonionic) substances, i.e., substances that exist as discrete molecules, we often substitute the term **molecular weight (MW)**. We could say that the *molecular weight* of 2,4,6-trinitrotoluene is 227.187 amu.

The amount of substance that contains the mass in grams numerically equal to its formula weight in amu contains 6.022×10^{23} formula units, or *one mole* of the substance. This is sometimes called the **molar mass** of the substance. Molar mass is *numerically equal* to the formula weight of the substance (the atomic weight for atoms of elements), and has the units grams/mole.

One mole of sodium hydroxide is 40.01 g of NaOH, and one mole of TNT is 227.187 g of $C_7H_5(NO_2)_3$. One mole of any molecular substance contains 6.022×10^{23} molecules of the substance, as Table 2-4 illustrates.

Table 2-4
One Mole of Some Common Molecular Substances

Substance	A Sample with a Mass of	Contains
hydrogen	2.016 g H_2	6.022×10^{23} H_2 molecules or 1 mol of H_2 molecules (contains $2 \times 6.022 \times 10^{23}$ H atoms or 2 mol of H atoms)
oxygen	32.00 g O_2	6.022×10^{23} O_2 molecules or 1 mol of O_2 molecules (contains $2 \times 6.022 \times 10^{23}$ O atoms or 2 mol of O atoms)
methane	16.04 g CH_4	6.022×10^{23} CH_4 molecules or 1 mol of CH_4 molecules (contains 6.022×10^{23} C atoms and $4 \times 6.022 \times 10^{23}$ H atoms)
2,4,6-trinitro-toluene (TNT)	227.19 g $C_7H_5(NO_2)_3$	6.022×10^{23} $C_7H_5(NO_2)_3$ molecules or 1 mol of $C_7H_5(NO_2)_3$ molecules

Figure 2-10
One mole of some compounds. The clear liquid is water, H_2O (1 mol = 18.0 g = 18.0 mL). The white solid (left) is *anhydrous* oxalic acid, $(COOH)_2$ (1 mol = 90.0 g). The second white solid is *hydrated* oxalic acid, $(COOH)_2 \cdot 2H_2O$ (1 mol = 126.0 g). The blue solid is hydrated copper(II) sulfate, $CuSO_4 \cdot 5H_2O$ (1 mol = 249.68 g). The red solid is mercury(II) oxide (1 mol = 216.59 g).

Heating blue $CuSO_4 \cdot 5H_2O$ forms anhydrous $CuSO_4$, which is white. Some blue $CuSO_4 \cdot 5H_2O$ is visible in the cooler center portion of the crucible.

The physical appearance of one mole of each of some compounds is illustrated in Figure 2-10. Two different forms of oxalic acid are shown. The formula unit (molecule) of oxalic acid is $(COOH)_2$ (FW = 90.04 amu; molar mass = 90.04 g/mol). However, when oxalic acid is obtained by crystallization from a water solution, two molecules of water are present for each molecule of oxalic acid, even though it appears dry. The formula of this **hydrate** is $(COOH)_2 \cdot 2H_2O$ (FW = 126.06 amu; molar mass = 126.06 g/mol). The dot shows that the crystals contain two H_2O molecules per $(COOH)_2$ molecule. The water can be driven out of the crystals by heating to leave **anhydrous** oxalic acid, $(COOH)_2$. Anhydrous means "without water." Copper(II) sulfate, an *ionic* compound, shows similar behavior. Anhydrous copper(II) sulfate ($CuSO_4$; FW = 159.60 amu; molar mass = 159.60 g/mol) is almost white. Hydrated copper(II) sulfate ($CuSO_4 \cdot 5H_2O$; FW = 249.68 amu; molar mass = 249.68 g/mol) is deep blue.

Because there are no simple NaCl molecules at ordinary temperatures, it is inappropriate to refer to the "molecular weight" of NaCl or any ionic compound. One mole of an ionic compound contains 6.022×10^{23} *formula units* (FU) of the substance. Recall that one formula unit of sodium chloride consists of one sodium ion, Na^+, and one chloride ion, Cl^-. One mole, or 58.44 grams, of NaCl contains 6.022×10^{23} Na^+ ions and 6.022×10^{23} Cl^- ions. See Table 2-5.

Table 2-5
One Mole of Some Ionic Compounds

Compound	A Sample with a Mass of 1 Mole	Contains
sodium chloride	58.44 g NaCl	6.022×10^{23} Na^+ ions or 1 mole of Na^+ ions 6.022×10^{23} Cl^- ions or 1 mole of Cl^- ions
calcium chloride	111.0 g $CaCl_2$	6.022×10^{23} Ca^{2+} ions or 1 mole of Ca^{2+} ions $2(6.022 \times 10^{23})$ Cl^- ions or 2 moles of Cl^- ions
aluminum sulfate	342.1 g $Al_2(SO_4)_3$	$2(6.022 \times 10^{23})$ Al^{3+} ions or 2 moles of Al^{3+} ions $3(6.022 \times 10^{23})$ SO_4^{2-} ions or 3 moles of SO_4^{2-} ions

The mole concept, together with Avogadro's number, provides important connections among the extensive properties mass of substance, number of moles of substance, and number of molecules or ions. These are summarized as follows.

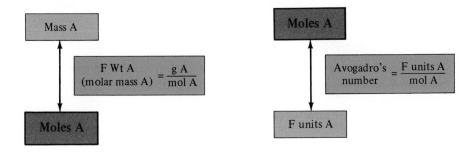

The following examples show the relations between numbers of molecules, atoms, or formula units and their masses.

Example 2-8

What is the mass in grams of 10.0 billion SO_2 molecules?

Plan

One mole of SO_2 contains 6.02×10^{23} SO_2 molecules and has a mass of 64.1 grams.

Solution

$$? \text{ g } SO_2 = 10.0 \times 10^9 \text{ } SO_2 \text{ molecules} \times \frac{64.1 \text{ g } SO_2}{6.02 \times 10^{23} \text{ } SO_2 \text{ molecules}}$$

$$= 1.06 \times 10^{-12} \text{ g } SO_2$$

Ten billion SO_2 molecules have a mass of only 0.00000000000106 gram. The most commonly used analytical balances are capable of weighing to ± 0.0001 gram.

EOC 30

When fewer than four significant figures are used in calculations, Avogadro's number is rounded off to 6.02×10^{23}.

Example 2-9

How many (a) moles of O_2, (b) O_2 molecules, and (c) O atoms are contained in 40.0 grams of oxygen gas (dioxygen) at 25°C?

Plan

We construct the needed unit factors from the following equalities: (a) the mass of one mole of O_2 is 32.0 g (molar mass O_2 = 32.0 g/mol); (b) one mole of O_2 contains 6.02×10^{23} O_2 molecules; (c) one O_2 molecule contains two O atoms.

Solution

One mole of O_2 contains 6.02×10^{23} O_2 molecules, and its mass is 32.0 g.

(a) $$? \text{ mol } O_2 = 40.0 \text{ g } O_2 \times \frac{1 \text{ mol } O_2}{32.0 \text{ g } O_2} = 1.25 \text{ mol } O_2$$

(b) $\underline{?}$ O_2 molecules = 40.0 g O_2 $\times$ $\dfrac{6.02 \times 10^{23} \text{ } O_2 \text{ molecules}}{32.0 \text{ g } O_2}$

$$= 7.52 \times 10^{23} \text{ molecules}$$

Or, we can use the number of moles of O_2 calculated in (a) to find the number of O_2 molecules.

$$\underline{?} \text{ } O_2 \text{ molecules} = 1.25 \text{ mol } O_2 \times \dfrac{6.02 \times 10^{23} \text{ } O_2 \text{ molecules}}{1 \text{ mol } O_2}$$

$$= 7.52 \times 10^{23} \text{ } O_2 \text{ molecules}$$

(c) $\underline{?}$ O atoms = 40.0 g O_2 $\times$ $\dfrac{6.02 \times 10^{23} \text{ } O_2 \text{ molecules}}{32.0 \text{ g } O_2}$ $\times$ $\dfrac{2 \text{ O atoms}}{1 \text{ } O_2 \text{ molecule}}$

$$= 1.50 \times 10^{24} \text{ O atoms}$$

EOC 32

Example 2-10

Calculate the number of hydrogen atoms in 39.6 grams of ammonium sulfate, $(NH_4)_2SO_4$.

Plan

One mole of $(NH_4)_2SO_4$ is 6.02×10^{23} formula units (FU) and has a mass of 132 g.

| g of $(NH_4)_2SO_4$ | $\rightarrow$ | mol of $(NH_4)_2SO_4$ | $\rightarrow$ | FU of $(NH_4)_2SO_4$ | $\rightarrow$ | H atoms |

In Example 2-10, we relate (a) g to mol, (b) mol to FU, and (c) FU to H atoms.

Solution

$$\underline{?} \text{ H atoms} = 39.6 \text{ g } (NH_4)_2SO_4 \times \dfrac{1 \text{ mol } (NH_4)_2SO_4}{132 \text{ g } (NH_4)_2SO_4}$$

$$\times \dfrac{6.02 \times 10^{23} \text{ FU } (NH_4)_2SO_4}{1 \text{ mol } (NH_4)_2SO_4} \times \dfrac{8 \text{ H atoms}}{1 \text{ FU } (NH_4)_2SO_4}$$

$$= 1.44 \times 10^{24} \text{ H atoms}$$

EOC 31

Table 2-6
Comparison of Moles and Millimoles

Compound	1 Mole	1 Millimole
NaOH	40.0 g	40.0 mg or 0.0400 g
H_3PO_4	98.1 g	98.1 mg or 0.0981 g
SO_2	64.1 g	64.1 mg or 0.0641 g
C_3H_8	44.1 g	44.1 mg or 0.0441 g

Grams can be converted to milligrams by shifting the decimal three places to the right.

The term ''millimole'' (mmol) is useful in laboratory work. As the prefix indicates, one **mmol** is 1/1000 of a mole. Small masses are frequently expressed in milligrams (mg) rather than grams. The relation between millimoles and milligrams is the same as that between moles and grams (Table 2-6).

Example 2-11

Calculate the number of millimoles of sulfuric acid in 0.147 gram of H_2SO_4.

Plan

1 mol H_2SO_4 = 98.1 g H_2SO_4; 1 mmol H_2SO_4 = 98.1 mg H_2SO_4, or 0.0981 g H_2SO_4. We can use these equalities to solve this problem by either of two methods. Method 1: Express formula weight in g/mmol, then convert g H_2SO_4 to mmol H_2SO_4. Method 2: Convert g H_2SO_4 to mg H_2SO_4, then use unit factor mg/mmol to convert to mmol H_2SO_4.

Solution

Method 1:

$$\underline{?}\text{ mmol }H_2SO_4 = 0.147\text{ g }H_2SO_4 \times \frac{1\text{ mmol }H_2SO_4}{0.0981\text{ g }H_2SO_4} = \boxed{1.50\text{ mmol }H_2SO_4}$$

Method 2: Using 0.147 g H_2SO_4 = 147 mg H_2SO_4, we have

$$\underline{?}\text{ mmol }H_2SO_4 = 147\text{ mg }H_2SO_4 \times \frac{1\text{ mmol }H_2SO_4}{98.1\text{ mg }H_2SO_4} = \boxed{1.50\text{ mmol }H_2SO_4}$$

EOC 33

2-7 Percent Composition and Formulas of Compounds

If the formula of a compound is known, its chemical composition can be expressed as the mass percent of each element in the compound. For example, one carbon dioxide molecule, CO_2, contains one C atom and two O atoms. Percentage is the part divided by the whole times 100 percent (or simply parts per 100), so we can represent the percent composition of carbon dioxide as follows:

$$\% \text{ C} = \frac{\text{mass of C}}{\text{mass of }CO_2} \times 100\% = \frac{\text{AW of C}}{\text{MW of }CO_2} \times 100\%$$

$$= \frac{12.0\text{ amu}}{44.0\text{ amu}} \times 100\% = \boxed{27.3\%}$$

$$\% \text{ O} = \frac{\text{mass of O}}{\text{mass of }CO_2} \times 100\% = \frac{2 \times \text{AW of O}}{\text{MW of }CO_2} \times 100\%$$

$$= \frac{2(16.0\text{ amu})}{44.0\text{ amu}} \times 100\% = \boxed{72.7\%\text{ O}}$$

One *mole* of CO_2 (44.0 g) contains one *mole* of C atoms (12.0 g) and two *moles* of O atoms (32.0 g). Therefore, we could have used these masses in the preceding calculation. These numbers are the same as the ones used—

only the units are different. In Example 2-12 we shall base our calculation on one *mole* rather than one *molecule*.

Example 2-12

Calculate the percent composition of HNO_3 by mass.

Plan

We first calculate the mass of one mole as in Example 2-7. Then we express the mass of each element as a percent of the total.

Solution

The molar mass of HNO_3 is calculated first.

No. of Mol of Atoms		× Mass of One Mol of Atoms	= Mass Due to Element
$1 \times H =$	1	× 1.0 g	= 1.0 g of H
$1 \times N =$	1	× 14.0 g	= 14.0 g of N
$3 \times O =$	3	× 16.0 g	= 48.0 g of O

$$\text{Mass of 1 mol of } HNO_3 = 63.0 \text{ g}$$

Now, its percent composition is

$$\% \text{ H} = \frac{\text{mass of H}}{\text{mass of } HNO_3} \times 100\% = \frac{1.0 \text{ g}}{63.0 \text{ g}} \times 100\% = \boxed{1.6\% \text{ H}}$$

$$\% \text{ N} = \frac{\text{mass of N}}{\text{mass of } HNO_3} \times 100\% = \frac{14.0 \text{ g}}{63.0 \text{ g}} \times 100\% = \boxed{22.2\% \text{ N}}$$

$$\% \text{ O} = \frac{\text{mass of O}}{\text{mass of } HNO_3} \times 100\% = \frac{48.0 \text{ g}}{63.0 \text{ g}} \times 100\% = \boxed{76.2\% \text{ O}}$$

$$\text{Total} = 100.0\%$$

EOC 39

When chemists use the % notation, they mean percent by mass unless they specify otherwise.

Percentages must add to 100%. However, round-off errors may not cancel, and totals such as 99.9% or 100.1% may be obtained in calculations.

Nitric acid is 1.6% H, 22.2% N, and 76.2% O by mass. All samples of pure HNO_3 have this composition, according to the Law of Definite Proportions.

2-8 Derivation of Formulas from Elemental Composition

Each year thousands of new compounds are made in laboratories or discovered in nature. One of the first steps in characterizing a new compound is the determination of its percent composition. A *qualitative* analysis is performed to determine *which* elements are present in the compound. Then a *quantitative* analysis is performed to determine the *amount* of each element.

Once the percent composition of a compound (or its elemental composition by mass) is known, the simplest formula can be determined. The **simplest** or **empirical formula** for a compound is the smallest whole-number ratio of atoms present. For molecular compounds the **molecular formula** indicates the *actual* numbers of atoms present in a molecule of the compound. It may be the same as the simplest formula or else some whole-number multiple of it. For example, the simplest and molecular formulas for water are both H_2O. However, for hydrogen peroxide, they are HO and H_2O_2, respectively.

Example 2-13

Compounds containing sulfur and oxygen are serious air pollutants; they represent the major cause of acid rain. Analysis of a sample of a pure compound reveals that it contains 50.1% sulfur and 49.9% oxygen by mass. What is the simplest formula of the compound?

Plan

The ratio of moles of atoms in any sample of a compound is the same as the ratio of atoms in that compound, because one mole of atoms of any element is 6.022×10^{23} atoms. This calculation is carried out in two steps. Step 1: Let's consider 100.0 grams of compound, which contains 50.1 grams of S and 49.9 grams of O. We calculate the number of moles of atoms of each. Step 2: We then obtain a whole-number ratio between these numbers that gives the ratio of atoms in the sample, and hence in the simplest formula for the compound.

Solution

Step 1: $? \text{ mol S atoms} = 50.1 \text{ g S} \times \dfrac{1 \text{ mol S atoms}}{32.1 \text{ g S}} = 1.56 \text{ mol S atoms}$

$? \text{ mol O atoms} = 49.9 \text{ g O} \times \dfrac{1 \text{ mol O atoms}}{16.0 \text{ g O}} = 3.12 \text{ mol O atoms}$

Step 2: Now we know that 100.0 grams of compound contains 1.56 moles of S atoms and 3.12 moles of O atoms. We obtain a whole-number ratio between these numbers that gives the ratio of atoms in the simplest formula.

$$\dfrac{1.56}{1.56} = 1 \text{ S}$$
$$\dfrac{3.12}{1.56} = 2 \text{ O}$$
$$SO_2$$

A simple and useful way to obtain whole-number ratios among several numbers follows. (1) Divide each number by the smallest, and then, (2) if necessary, multiply all of the resulting numbers by the smallest whole number that will eliminate fractions.

EOC 41, 42

The solution for Example 2-13 can be set up in tabular form:

Element	Relative Mass of Element	Relative Number of Atoms (divide mass by AW)	Divide by Smaller Number	Smallest Whole-Number Ratio of Atoms
S	50.1	$\dfrac{50.1}{32.1} = 1.56$	$\dfrac{1.56}{1.56} = 1.00$	SO_2
O	49.9	$\dfrac{49.9}{16.0} = 3.12$	$\dfrac{3.12}{1.56} = 2.00$	

The "Relative Mass" column is proportional to the mass of each element in grams. With this interpretation, the next column could be headed "Relative Number of *Moles* of Atoms." Then the last column would represent the smallest whole-number ratios of *moles* of atoms. But because a mole is always the same number of items (atoms), that ratio is the same as the smallest whole-number ratio of atoms.

This tabular format provides a convenient way to solve simplest-formula problems, as the next example illustrates.

Example 2-14

A 20.882-gram sample of an ionic compound is found to contain 6.072 grams of Na, 8.474 grams of S, and 6.336 grams of O. What is its simplest formula?

Plan

We reason as in Example 2-13, calculating the number of moles of each element and the ratio among them. Here we use the tabular format that was introduced above.

Solution

Element	Relative Mass of Element	Relative Number of Atoms (divide mass by AW)	Divide by Smallest Number	Convert Fractions to Whole Numbers (multiply by integer)	Smallest Whole-Number Ratio of Atoms
Na	6.072	$\dfrac{6.072}{23.0} = 0.264$	$\dfrac{0.264}{0.264} = 1.00$	$1.00 \times 2 = 2$	
S	8.474	$\dfrac{8.474}{32.1} = 0.264$	$\dfrac{0.264}{0.264} = 1.00$	$1.00 \times 2 = 2$	$Na_2S_2O_3$
O	6.336	$\dfrac{6.336}{16.0} = 0.396$	$\dfrac{0.396}{0.264} = 1.50$	$1.50 \times 2 = 3$	

In this procedure we often obtain numbers such as 0.99 and 1.52. Because there is always some error in results obtained by analysis of samples (as well as round-off errors), we would interpret 0.99 as 1.0 and 1.52 as 1.5.

The ratio of atoms in the simplest formula *must be a whole-number ratio* (by definition). To convert the ratio 1:1:1.5 to a whole-number ratio, each number in the ratio was multiplied by 2, which gave the simplest formula $Na_2S_2O_3$.

EOC 43

2-9 Determination of Molecular Formulas

Percent composition data yield only simplest formulas. To determine the molecular formula for a molecular compound, *both* its simplest formula and its molecular weight must be known. Methods for experimental determination of molecular weights are introduced in Chapter 12.

Millions of compounds are composed of carbon, hydrogen, and oxygen. Analyses for C and H can be performed in a C-H combustion system (Figure 2-11). An accurately known mass of a compound is burned in a furnace in a stream of oxygen. The carbon and hydrogen in the sample are converted to carbon dioxide and water vapor, respectively. The resulting increases in

Figure 2-11
A combustion train used for carbon-hydrogen analysis. The absorbent for water is magnesium perchlorate, $Mg(ClO_4)_2$. Carbon dioxide is absorbed by finely divided sodium hydroxide supported on glass wool. Only a few milligrams of sample are needed for an analysis.

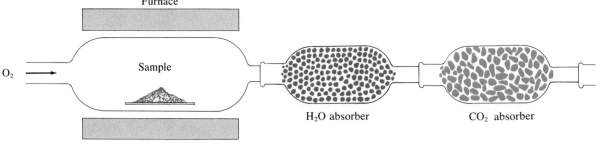

O$_2$ →   Furnace   Sample   H₂O absorber   CO₂ absorber

masses of the CO_2 and H_2O absorbers can then be related to the masses and percentages of carbon and hydrogen in the original sample.

Example 2-15

Hydrocarbons are organic compounds composed entirely of hydrogen and carbon. A 0.1647-gram sample of a pure hydrocarbon was burned in a C-H combustion train to produce 0.5694 gram of CO_2 and 0.0826 gram of H_2O. Determine the masses of C and H in the sample and the percentages of these elements in this hydrocarbon.

Plan

Step 1: We use the observed mass of CO_2, 0.5694 grams, to determine the mass of carbon in the original sample. There is one mole of carbon atoms, 12.01 grams, in each mole of CO_2, 44.01 grams; we use this information to construct the unit factor

$$\frac{12.01 \text{ g C}}{44.01 \text{ g } CO_2}$$

Step 2: Likewise, we can use the observed mass of H_2O, 0.0826 grams, to calculate the amount of hydrogen in the original sample. We use the fact that there are two moles of hydrogen atoms, 2.016 grams, in each mole of H_2O, 18.02 grams, to construct the unit factor

$$\frac{2.016 \text{ g H}}{18.02 \text{ g } H_2O}$$

Step 3: Then we calculate the percentages by mass of each element in turn, using the relationship

$$\% \text{ element} = \frac{\text{g element}}{\text{g sample}} \times 100\%$$

We could calculate the mass of H by subtracting mass of C from mass of sample. However, it is good experimental practice, if possible, to base both on experimental measurements as we have done here. This would help to check for errors in the analysis or calculation.

Solution

Step 1: $? \text{ g C} = 0.5694 \text{ g } CO_2 \times \dfrac{12.01 \text{ g C}}{44.01 \text{ g } CO_2} =$ 0.1554 g C

Step 2: $? \text{ g H} = 0.0826 \text{ g } H_2O \times \dfrac{2.016 \text{ g H}}{18.02 \text{ g } H_2O} =$ 0.00924 g H

Step 3: $\% \text{ C} = \dfrac{0.1554 \text{ g C}}{0.1647 \text{ g sample}} \times 100\% =$ $94.4\% \text{ C}$

$\% \text{ H} = \dfrac{0.00924 \text{ g H}}{0.1647 \text{ g sample}} \times 100\% =$ $5.61\% \text{ H}$

Total $=$ 100.0%

EOC 43

When the compound to be analyzed contains oxygen, the calculation of the amount or percentage of oxygen in the sample is somewhat different. Part of the oxygen that goes to form CO_2 and H_2O comes from the sample and part comes from the oxygen stream supplied. Therefore we cannot directly determine the amount of oxygen already in the sample. The approach

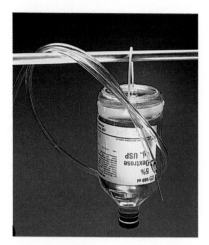

Glucose, a simple sugar, is the main component of intravenous feeding liquids. Its common name is dextrose. It is also one of the products of carbohydrate metabolism.

is to analyze as we did in Example 2-15 for all elements *except* oxygen. Then we subtract the sum of their masses from the mass of the original sample. The next example illustrates such a calculation.

Example 2-16

A 0.1014-gram sample of purified glucose was burned in a C-H combustion train to produce 0.1486 gram of CO_2 and 0.0609 gram of H_2O. An elemental analysis showed that glucose contains only carbon, hydrogen, and oxygen. Determine the masses of C, H, and O in the sample and the percentages of these elements in glucose.

Plan

Steps 1 and 2: We first calculate the masses of carbon and hydrogen as we did in Example 2-15. Step 3: The rest of the sample must be oxygen because glucose has been shown to contain only C, H, and O. So we subtract the masses of C and H from the total mass of sample. Step 4: Then we calculate the percentage by mass for each element.

Solution

Step 1:
$$? \text{ g C} = 0.1486 \text{ g CO}_2 \times \frac{12.01 \text{ g C}}{44.01 \text{ g CO}_2} = \boxed{0.0406 \text{ g C}}$$

Step 2:
$$? \text{ g H} = 0.0609 \text{ g H}_2\text{O} \times \frac{2.016 \text{ g H}}{18.02 \text{ g H}_2\text{O}} = \boxed{0.00681 \text{ g H}}$$

Step 3:
$$? \text{ g O} = 0.1014 \text{ g sample} - [0.0406 \text{ g C} + 0.00681 \text{ g H}]$$

$$= \boxed{0.0540 \text{ g O}}$$

We say that the mass of O in the sample is calculated by *difference*.

Step 4: Now we can calculate the percentages by mass for each element:

$$\% \text{ C} = \frac{0.0406 \text{ g C}}{0.1014 \text{ g}} \times 100\% = \boxed{40.0\% \text{ C}}$$

$$\% \text{ H} = \frac{0.00681 \text{ g H}}{0.1014 \text{ g}} \times 100\% = \boxed{6.72\% \text{ H}}$$

$$\% \text{ O} = \frac{0.0540 \text{ g O}}{0.1014 \text{ g}} \times 100\% = \boxed{53.3\% \text{ O}}$$

Total = 100.0%

EOC 44

For many compounds the molecular formula is a multiple of the simplest formula. Consider butane, C_4H_{10}. The simplest formula for butane is C_2H_5, but the molecular formula contains twice as many atoms; i.e., $(C_2H_5)_2 = C_4H_{10}$. Benzene, C_6H_6, is another example. The simplest formula for benzene is CH, but the molecular formula contains six times as many atoms; i.e., $(CH)_6 = C_6H_6$.

The molecular formula for a compound is *either* the same as, *or* a whole-number multiple of, the simplest formula. So we can write

$$\text{molecular weight} = n \times \text{simplest formula weight}$$

$$n = \frac{\text{molecular weight}}{\text{simplest formula weight}}$$

where n is the number of simplest formula units in a molecule of the compound. The subscripts in the molecular formula are obtained by multiplying the subscripts in the simplest formula by n.

Example 2-17

In Example 2-16 we found the elemental composition of glucose. Other experiments show that its molecular weight is approximately 180 amu. Determine the simplest formula and the molecular formula of glucose.

Plan
Step 1: We first use the masses of C, H, and O found in Example 2-16 to determine the simplest formula. Step 2: We can use the simplest formula to calculate the simplest formula weight. Because the molecular weight of glucose is known (approximately 180 amu), we can determine the molecular formula by dividing the molecular weight by the simplest formula weight.

As an alternative, we could have used the percentages by mass from Example 2-16. Using the earliest available numbers helps to minimize the effects of rounding-off errors.

$$n = \frac{\text{molecular weight}}{\text{simplest formula weight}}$$

The molecular weight is n times the simplest formula weight, so the molecular formula of glucose is n times the simplest formula.

Solution
Step 1:

Element	Mass of Element	Moles of Element (divide mass by AW)	Divide by Smallest	Smallest Whole-Number Ratio of Atoms
C	0.0406 g	$\frac{0.0406}{12.01} = 0.00338$ mol	$\frac{0.00338}{0.00338} = 1.00$	
H	0.00681 g	$\frac{0.00681}{1.008} = 0.00676$ mol	$\frac{0.00676}{0.00338} = 2.00$	CH_2O
O	0.0540 g	$\frac{0.0540}{16.00} = 0.00338$ mol	$\frac{0.00338}{0.00338} = 1.00$	

Step 2: The simplest formula is CH_2O, which has a formula weight of 30.02 amu. Because the molecular weight of glucose is approximately 180 amu, we can determine the molecular formula by dividing the molecular weight by the simplest formula weight.

$$n = \frac{180 \text{ amu}}{30.02 \text{ amu}} = 6.00$$

The molecular weight is six times the simplest formula weight, $(CH_2O)_6 = C_6H_{12}O_6$,

so the molecular formula of glucose is $C_6H_{12}O_6$.

Traces of impurities and round-off errors may give numbers such as 5.96, 6.02, and so on, which are very close to integers.

EOC 50

As we shall see when we discuss the composition of compounds in some detail, two (and sometimes more) elements may form more than one compound. The **Law of Multiple Proportions** summarizes many experiments on such compounds. It is usually stated: When two elements, A and B, form more than one compound, the ratio of the masses of element B that combine with a given mass of element A in each of the compounds can be expressed by small whole numbers. Water, H_2O, and hydrogen peroxide, H_2O_2, provide an example. The ratio of masses of oxygen that combine with a given mass of hydrogen is 1:2 in H_2O and H_2O_2. Many similar examples, such as CO and CO_2 (1:2 ratio) and SO_2 and SO_3 (2:3 ratio), are known. The Law of Multiple Proportions had been recognized from studies of elemental composition before the time of Dalton. It provided additional support for his atomic theory.

Example 2-18

What is the ratio of the masses of oxygen that are combined with 1.00 gram of nitrogen in the compounds NO and N_2O_3?

Plan

First we calculate the mass of O that combines with one gram of N in each compound. Then we determine the ratio of the values of $\dfrac{\text{g O}}{\text{g N}}$ for the two compounds.

Solution

$$\text{In NO:} \qquad \frac{\underline{?}\ \text{g O}}{\text{g N}} = \frac{16.0\ \text{g O}}{14.0\ \text{g N}} = 1.14\ \text{g O/g N}$$

$$\text{In } N_2O_3\text{:} \qquad \frac{\underline{?}\ \text{g O}}{\text{g N}} = \frac{48.0\ \text{g O}}{28.0\ \text{g N}} = 1.71\ \text{g O/g N}$$

$$\text{The ratio is} \begin{cases} \dfrac{\text{g O}}{\text{g N}}\ (\text{in } N_2O_3) \\[2em] \dfrac{\text{g O}}{\text{g N}}\ (\text{in NO}) \end{cases} \frac{1.71\ \text{g O/g N}}{1.14\ \text{g O/g N}} = \frac{1.5}{1.0} = \frac{3}{2}$$

We see that the ratio is 3 mass units of O (in N_2O_3) to 2 mass units of O (in NO).

EOC 55, 56

2-10 Some Other Interpretations of Chemical Formulas

Once we master the mole concept and the meaning of chemical formulas, we can use them in many other ways. The examples in this section illustrate a few additional kinds of information we can get from a chemical formula and the mole concept.

Example 2-19

What mass of chromium is contained in 35.8 grams of $(NH_4)_2Cr_2O_7$?

Plan

Let us first solve the problem in several steps. Step 1: The formula tells us that each mole of $(NH_4)_2Cr_2O_7$ contains two moles of Cr atoms, so we first find the number of moles of $(NH_4)_2Cr_2O_7$, using the unit factor

$$\frac{1 \text{ mol } (NH_4)_2Cr_2O_7}{252.0 \text{ g } (NH_4)_2Cr_2O_7}$$

Step 2: Then we convert the number of moles of $(NH_4)_2Cr_2O_7$ into the number of moles of Cr atoms it contains, using the unit factor

$$\frac{2 \text{ mol Cr atoms}}{1 \text{ mol } (NH_4)_2Cr_2O_7}$$

Step 3: We then use the atomic weight of Cr to convert the number of moles of chromium atoms to mass of chromium.

$$\boxed{\text{Mass } (NH_4)_2Cr_2O_7} \longrightarrow \boxed{\text{mol } (NH_4)_2Cr_2O_7} \longrightarrow \boxed{\text{mol Cr}} \longrightarrow \boxed{\text{Mass Cr}}$$

Solution

Step 1: $\underline{?}$ mol $(NH_4)_2Cr_2O_7 = 35.8$ g $(NH_4)_2Cr_2O_7 \times \dfrac{1 \text{ mol } (NH_4)_2Cr_2O_7}{252.0 \text{ g } (NH_4)_2Cr_2O_7}$

$= \boxed{0.142 \text{ mol } (NH_4)_2Cr_2O_7}$

Step 2: $\underline{?}$ mol Cr atoms $= 0.142$ mol $(NH_4)_2Cr_2O_7 \times \dfrac{2 \text{ mol Cr atoms}}{1 \text{ mol } (NH_4)_2Cr_2O_7}$

$= \boxed{0.284 \text{ mol Cr atoms}}$

Step 3: $\underline{?}$ g Cr $= 0.284$ mol Cr atoms $\times \dfrac{52.0 \text{ g Cr}}{1 \text{ mol Cr atoms}} = \boxed{14.8 \text{ g Cr}}$

If you understand the reasoning in these conversions, you should be able to solve this problem in a single setup:

$\underline{?}$ g Cr $= 35.8$ g $(NH_4)_2Cr_2O_7 \times \dfrac{1 \text{ mol } (NH_4)_2Cr_2O_7}{252.0 \text{ g } (NH_4)_2Cr_2O_7}$

$\times \dfrac{2 \text{ mol Cr atoms}}{1 \text{ mol } (NH_4)_2Cr_2O_7} \times \dfrac{52.0 \text{ g Cr}}{1 \text{ mol Cr}} = \boxed{14.8 \text{ g Cr}}$

EOC 60

Example 2-20

What mass of potassium chlorate, $KClO_3$, would contain 40.0 grams of oxygen?

Plan

The formula $KClO_3$ tells us that each mole of $KClO_3$ contains three moles of oxygen atoms. Each mole of oxygen atoms weighs 16.0 grams. So we can set up the solution to convert:

$$\boxed{\text{Mass O}_2} \longrightarrow \boxed{\text{mol O}_2} \longrightarrow \boxed{\text{mol KClO}_3} \longrightarrow \boxed{\text{Mass KClO}_3}$$

Solution

$$\underline{?} \text{ g KClO}_3 = 40.0 \text{ g O} \times \frac{1 \text{ mol O atoms}}{16.0 \text{ g O atoms}} \times \frac{1 \text{ mol KClO}_3}{3 \text{ mol O}} \times \frac{122.6 \text{ g KClO}_3}{1 \text{ mol KClO}_3}$$

$$= \boxed{102 \text{ g KClO}_3}$$

EOC 62

Example 2-21

(a) What mass of sulfur dioxide, SO_2, would contain the same mass of oxygen as is contained in 33.7 g of arsenic pentoxide, As_2O_5? (b) What mass of calcium chloride, $CaCl_2$, would contain the same number of chloride ions as are contained in 48.6 g of sodium chloride, $NaCl$?

Plan

(a) We could find explicitly the number of grams of O in 33.7 g of As_2O_5, and then find the mass of SO_2 that contains that same number of grams of O. But this method includes some unnecessary calculation. We need only convert to *moles* of O (because this is the same mass of O regardless of its environment) and then to SO_2.

$$\boxed{\text{Mass As}_2\text{O}_5} \longrightarrow \boxed{\text{mol As}_2\text{O}_5} \longrightarrow \boxed{\text{mol O atoms}} \longrightarrow \boxed{\text{mol SO}_2} \longrightarrow \boxed{\text{Mass SO}_2}$$

(b) Because one mole always consists of the same number (Avogadro's number) of items, we can reason in terms of *moles* of Cl^- ions:

$$\boxed{\text{Mass NaCl}} \longrightarrow \boxed{\text{mol NaCl}} \longrightarrow \boxed{\text{mol Cl}^- \text{ ions}} \longrightarrow \boxed{\text{mol CaCl}_2} \longrightarrow \boxed{\text{Mass CaCl}_2}$$

Solution

(a) $$\underline{?} \text{ g SO}_2 = 33.7 \text{ g As}_2\text{O}_5 \times \frac{1 \text{ mol As}_2\text{O}_5}{229.8 \text{ g As}_2\text{O}_5} \times \frac{5 \text{ mol O atoms}}{1 \text{ mol As}_2\text{O}_5}$$

$$\times \frac{1 \text{ mol SO}_2}{2 \text{ mol O atoms}} \times \frac{64.1 \text{ g SO}_2}{1 \text{ mol SO}_2} = \boxed{23.5 \text{ g SO}_2}$$

(b) $$\underline{?} \text{ g CaCl}_2 = 48.6 \text{ g NaCl} \times \frac{1 \text{ mol NaCl}}{58.4 \text{ g NaCl}} \times \frac{1 \text{ mol Cl}^-}{1 \text{ mol NaCl}}$$

$$\times \frac{1 \text{ mol CaCl}_2}{2 \text{ mol Cl}^-} \times \frac{111.0 \text{ g CaCl}_2}{1 \text{ mol CaCl}_2} = \boxed{46.2 \text{ g CaCl}_2}$$

EOC 64

We have already mentioned the existence of hydrates (for example, $(COOH)_2 \cdot 2H_2O$ and $CuSO_4 \cdot 5H_2O$ in Section 2-6). In such hydrates, two components, water and another compound are present in a definite integer ratio by moles. The following example illustrates how we might find and use the formula of a hydrate.

Example 2-22

A reaction requires pure anhydrous calcium sulfate, $CaSO_4$. Only an unidentified hydrate of calcium sulfate, $CaSO_4 \cdot xH_2O$, is available. (a) We heat 67.5 g of the unknown hydrate until all the water has been driven off. The resulting mass of pure $CaSO_4$ is 53.4 g. What is the formula of the hydrate, and what is its formula weight? (b) Suppose we wish to obtain enough of this hydrate to supply 95.5 grams of $CaSO_4$. How many grams should we weigh out?

Plan

(a) To find the formula of the hydrate, we must figure out the value of x in the formula $CaSO_4 \cdot xH_2O$. The mass of water removed from the sample is equal to the difference in the two masses given. The value of x is the number of moles of H_2O per mole of $CaSO_4$ in the hydrate.
(b) We use the formula weights of $CaSO_4$, 136.2 g/mol, and of $CaSO \cdot xH_2O$, $(136.2 + x18.0)$ g/mol, to write the conversion factor required for the calculation.

Solution

(a) $\underline{?}$ g water driven off = 67.5 g $CaSO_4 \cdot xH_2O$ − 53.4 g $CaSO_4$ = 14.1 g H_2O

$$x = \underline{?} \ \frac{\text{mol } H_2O}{\text{mol } CaSO_4} = \frac{14.1 \text{ g } H_2O}{53.4 \text{ g } CaSO_4} \times \frac{1.0 \text{ mol } H_2O}{18.0 \text{ g } H_2O} \times \frac{136.2 \text{ g } CaSO_4}{1 \text{ mol } CaSO_4}$$

$$= \frac{2.00 \text{ mol } H_2O}{\text{mol } CaSO_4}$$

Thus the formula of the hydrate is $\boxed{CaSO_4 \cdot 2H_2O.}$ Its formula weight is

$$FW = 1 \times (\text{formula weight } CaSO_4) + 2 \times (\text{formula weight } H_2O)$$

$$= 136.2 \text{ g/mol} + 2(18.0 \text{ g/mol}) = \boxed{172.2 \text{ g/mol.}}$$

(b) The formula weights of $CaSO_4$ (136.2 g/mol) and of $CaSO_4 \cdot 2H_2O$ (172.2 g/mol) allow us to write the unit factor

$$\frac{172.2 \text{ g } CaSO_4 \cdot 2H_2O}{136.2 \text{ g } CaSO_4}$$

We use this factor to perform the required conversion:

$$\underline{?} \text{ g } CaSO_4 \cdot 2H_2O = 95.5 \text{ g } CaSO_4 \text{ desired} \times \frac{172.2 \text{ g } CaSO_4 \cdot 2H_2O}{136.2 \text{ g } CaSO_4}$$

$$= \boxed{121 \text{ g } CaSO_4 \cdot 2H_2O}$$

EOC 68

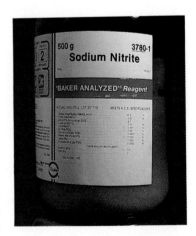

A label from a bottle of sodium nitrite.

2-11 Purity of Samples

Most substances obtained from laboratory reagent shelves are not 100% pure. When impure samples are used for precise work, account must be taken of impurities. The figure shows the label of a jar of reagent-grade sodium nitrite, $NaNO_2$, which is 99.4% pure by mass. From this information we know that total impurities represent 0.6% of the total mass of any sample. We can write several unit factors:

$$\frac{99.4 \text{ g NaNO}_2}{100 \text{ g sample}} \text{ , } \quad \frac{0.6 \text{ g impurities}}{100 \text{ g sample}} \text{ , } \quad \text{and} \quad \frac{0.6 \text{ g impurities}}{99.4 \text{ g NaNO}_2}$$

The inverse of each of these gives us a total of six unit factors.

Example 2-23

Calculate the masses of $NaNO_2$ and impurities in 45.2 g of 99.4% pure $NaNO_2$.

Plan

The percentage of $NaNO_2$ in the sample gives the unit factor $\dfrac{99.4 \text{ g NaNO}_2}{100 \text{ g sample}}$. The remainder of the sample is 100% − 99.4% = 0.6% impurities; this gives the unit factor $\dfrac{0.6 \text{ g impurities}}{100 \text{ g sample}}$.

Solution

$$\underline{?} \text{ g NaNO}_2 = 45.2 \text{ g sample} \times \frac{99.4 \text{ g NaNO}_2}{100 \text{ g sample}} = \boxed{44.9 \text{ g NaNO}_2}$$

$$\underline{?} \text{ g impurities} = 45.2 \text{ sample} \times \frac{0.6 \text{ g impurities}}{100 \text{ g sample}} = \boxed{0.3 \text{ g impurities}}$$

EOC 69, 72

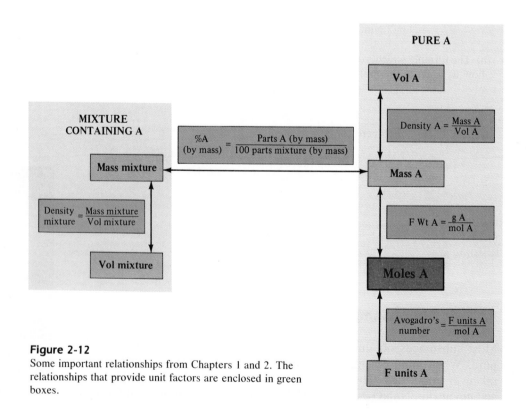

Figure 2-12
Some important relationships from Chapters 1 and 2. The relationships that provide unit factors are enclosed in green boxes.

Observe the beauty of the unit-factor approach to problem solving! Such questions as "Do we multiply by 0.994 or divide by 0.994?" never arise. The units always point toward the correct answer because we use unit factors constructed so that units *always* cancel until we arrive at the desired unit.

Many important relationships have been introduced in this chapter. Some of the most important transformations you have seen in Chapters 1 and 2 are summarized in Figure 2-12.

Key Terms

Allotropic modifications (allotropes)   Different forms of the same element in the same physical state.

Atom   The smallest particle of an element that maintains its chemical identity through all chemical and physical changes.

Atomic mass unit (amu)   One twelfth of the mass of an atom of the carbon-12 isotope; a unit used for stating atomic and formula weights; also called dalton.

Atomic weight   Weighted average of the masses of the constituent isotopes of an element; the relative masses of atoms of different elements.

Avogadro's number 6.022×10^{23} of the specified items. See *Mole*.

Composition stoichiometry   Describes the quantitative (mass) relationships among elements in compounds.

Empirical formula   See *Simplest formula*.

Formula   Combination of symbols that indicates the chemical composition of a substance.

Formula unit   The smallest repeating unit of a substance— for nonionic substances, the molecule.

Formula weight   The mass, in atomic weight units, of one formula unit of a substance. Numerically equal to the mass, in grams, of one mole of the substance (see *Molar mass*). This number is obtained by adding the atomic weights of the atoms specified in the formula.

Hydrate   A crystalline sample that contains water, H_2O, and another compound in a fixed mole ratio. Examples include $CuSO_4 \cdot 5H_2O$ and $(COOH)_2 \cdot 2H_2O$.

Ion   An atom or group of atoms that carries an electrical charge. A positive ion is a *cation;* a negative ion is an *anion*.

Ionic compound   A compound that is composed of ions. An example is sodium chloride, NaCl.

Law of Constant Composition   See *Law of Definite Proportions*.

Law of Definite Proportions   Different samples of a pure compound always contain the same elements in the same proportions by mass; this corresponds to atoms of these elements in fixed numerical ratios. Also called Law of Constant Composition.

Law of Multiple Proportions   When two elements, A and B, form more than one compound, the ratio of the masses of element B that combine with a given mass of element A in each of the compounds can be expressed by small whole numbers.

Molar mass   The mass of substance in one mole of the substance; numerically equal to the formula weight of the substance. See *Formula weight*; see *Molecular weight*.

Mole 6.022×10^{23} (Avogadro's number of) formula units (or molecules, for a nonionic substance) of the substance under discussion. The mass of one mole, in grams, is numerically equal to the formula (molecular) weight of the substance.

Molecular formula   A formula that indicates the actual number of atoms present in a molecule of a molecular substance. Compare with *Simplest formula*.

Molecule   The smallest particle of an element or compound that can have a stable independent existence.

Molecular weight   The mass, in atomic mass units, of one molecule of a nonionic (molecular) substance. Numerically equal to the mass, in grams, of one mole of such a substance. This number is obtained by adding the atomic weights of the atoms specified in the formula.

Percent composition   The mass percentage of each element in a compound.

Percent purity   The percentage of a specified compound or element in an impure sample.

Simplest formula   The smallest whole-number ratio of atoms present in a compound; also called empirical formula. Compare with *Molecular formula*.

Stoichiometry   Description of the quantitative relationships among elements in compounds (composition stoichiometry) and among substances as they undergo chemical changes (reaction stoichiometry).

Exercises

Basic Ideas

1. (a) What is the origin of the word "stoichiometry"? (b) Distinguish between composition stoichiometry and reaction stoichiometry.
2. List the basic ideas of Dalton's atomic theory.
3. Give examples of molecules that contain (a) two atoms, (b) three atoms, (c) four atoms, and (d) eight atoms.
4. Give two examples of diatomic molecules and two examples of polyatomic molecules.
5. Write formulas for the following compounds: (a) sulfuric acid; (b) methane; (c) sulfur dioxide; (d) acetic acid.
6. Name the following compounds: (a) HNO_3; (b) C_2H_6; (c) NH_3; (d) CO_2.

Ions and Ionic Compounds

7. Define and illustrate the following: (a) ion; (b) cation; (c) anion; (d) polyatomic ion.
8. (a) There are no *molecules* in ionic compounds. Why not? (b) What is the difference between a formula unit of an ionic compound and a polyatomic molecule?
9. Name each of the following ions. Classify each as a monatomic or polyatomic ion. Classify each as a cation or an anion. (a) Na^+; (b) OH^-; (c) NO_3^-; (d) S^{2-}; (e) Fe^{2+}.
10. Write the chemical symbol for each of the following ions. Classify each as a monatomic or polyatomic ion. Classify each as a cation or an anion. (a) potassium ion; (b) sulfate ion; (c) copper(II) ion; (d) ammonium ion; (e) carbonate ion.
11. Name each of the following compounds: (a) $AgCl$; (b) $FeCl_2$; (c) K_2SO_4; (d) $Ca(OH)_2$; (e) $Fe_2(SO_4)_3$.
12. Write the chemical formula for each of the following ionic compounds: (a) calcium acetate; (b) ammonium carbonate; (c) zinc phosphate; (d) sodium hydroxide; (e) aluminum nitrate.
13. Write the chemical formula for the ionic compound formed between each of the following pairs of ions. Name each compound. (a) Ca^{2+} and Cl^-; (b) Al^{3+} and SO_4^{2-}; (c) K^+ and PO_4^{3-}; (d) Mg^{2+} and NO_3^-; (e) Fe^{3+} and CO_3^{2-}.
14. Write the chemical formula for the ionic compound formed between each of the following pairs of ions. Name each compound. (a) Cu^{2+} and SO_4^{2-}; (b) Mg^{2+} and OH^-; (c) NH_4^+ and CO_3^{2-}; (d) Zn^{2+} and Cl^-; (e) Fe^{3+} and CH_3COO^-.
15. Convert each of the following into a correct formula represented with correct notation. (a) $CaOH_2$; (b) $Mg(CO_3)$; (c) $Zn(CO_3)_2$; (d) $(NH_4)^2SO_4$; (e) $Mg_2(SO_4)_2$.

Atomic Weights

16. What is the mass ratio (4 significant figures) of one atom of P to 1 atom of F?

17. 4.05 g of magnesium combines exactly with 6.33 g of fluorine, forming magnesium fluoride, MgF_2. Find the relative masses of the atoms of magnesium and fluorine. Check your answer using a table of atomic weights. If the formula were not known, could you still do this calculation?

The Mole Concept

18. The mass of one round-head wood screw is 3.71 g. Find the mass of one dozen, one gross, and one mole of these wood screws. Compare the latter value with the mass of the earth, 5.98×10^{24} kg.
19. Sulfur molecules exist under various conditions as S_8, S_6, S_4, S_2, and S. (a) Is the mass of one mole of each of these molecules the same? (b) Is the number of molecules in one mole of each of these molecules the same? (c) Is the mass of sulfur in one mole of each of these molecules the same? (d) Is the number of atoms of sulfur in one mole of each of these molecules the same?
20. Complete the following table. You may refer to a table of atomic weights.

	Element	Atomic Weight	Mass of One Mole of Atoms
(a)	B		
(b)		74.922 amu	
(c)	Al		
(d)			51.9961 g

21. Complete the following table. You may refer to a table of atomic weights.

	Element	Formula	Mass of One Mole of Molecules
(a)	Cl	Cl_2	
(b)		H_2	
(c)		P_4	
(d)			4.0026 g
(e)	S		256.528 g
(f)	O		

22. Determine the formula weight of each of the following substances: (a) chlorine, Cl_2; (b) water, H_2O; (c) saccharin, $C_7H_5NSO_3$; (d) sodium dichromate, $Na_2Cr_2O_7$.
23. Determine the formula weight of each of the following substances: (a) calcium sulfate, $CaSO_4$; (b) benzene, C_6H_6; (c) the sulfa drug sulfanilamide, $C_6H_4SO_2(NH_2)_2$; (d) uranyl phosphate, $(UO_2)_3(PO_4)_2$.
24. How many moles of substance are contained in each of the following samples? (a) 16.8 g of NH_3; (b) 3.25 kg of ammonium bromide; (c) 5.6 g of PCl_5; (d) 126.5 g of Fe.
25. How many moles of substance are contained in each of the following samples? (a) 24.5 g of formaldehyde, H_2CO; (b) 10.03 g of calcium carbonate; (c) 33.5 g of acetic acid; (d) 19.4 g of ethyl alcohol, C_2H_5OH.

26. The atomic weight of chlorine is 35.453 amu. Calculate the number of moles of chlorine atoms in (a) 1.00 g of chlorine, (b) 35.453 atomic mass units of chlorine, and (c) 5.66×10^{20} chlorine atoms.

27. Calculate the number of moles equivalent to each of the following: (a) 9.5×10^{21} atoms of Cs; (b) 4.7×10^{27} molecules of carbon dioxide; (c) 1.63×10^{23} formula units of $BaCl_2$; (d) 1.2×10^{22} atoms of Cu.

28. Calculate the number of moles equivalent to each of the following: (a) 5.5×10^{16} atoms of Fe; (b) 3.92×10^{18} molecules of CH_4; (c) 4.61×10^{25} molecules of O_2; (d) 4.61×10^{25} formula units of iron(III) nitrate.

29. What is the mass, in grams, of each of the samples of Exercise 27?

30. What is the mass, in grams, of each of the samples of Exercise 28?

31. How many atoms of C, H, and O are in each of the following? (a) 0.744 mol of glucose, $C_6H_{12}O_6$; (b) 2.50×10^{19} glucose ($C_6H_{12}O_6$) molecules; (c) 0.300 g of glucose.

32. How many molecules are in 28.0 g of each of the following substances? (a) CO; (b) N_2; (c) P_4; (d) P_2. (e) Do parts (c) and (d) contain the same number of atoms of phosphorus?

33. What mass, in grams, should be weighed for an experiment that requires 135 mmol $(NH_4)_2HPO_4$?

34. What mass, in grams, corresponds to each of the following? (a) 0.503 mol phenol, C_6H_5OH; (b) 1.01 mol of quartz, SiO_2; (c) 422 mmol of saccharin, $C_7H_5NSO_3$; (d) 125 μmol of saltpeter, KNO_3.

*35. Which of the following samples contains the smallest mass of silver? Which contains the largest? 0.0100 mol Ag; 0.0100 mol Ag_2O; 200 mg Ag; 3.01×10^{21} Ag atoms; 3.01×10^{21} Ag_2 molecules.

36. Complete the following table.

Moles of Compound	Moles of Cations	Moles of Anions
1 mol NaCl	_____	_____
2 mol Na_2SO_4	_____	_____
0.1 mol calcium nitrate	_____	_____
_____	0.75 mol NH_4^+	0.25 mol PO_4^{3-}

*37. What volume of glycerine, $C_3H_8O_3$, density 1.26 g/mL, should be taken to obtain 2.50 moles of glycerine?

Percent Composition and Simplest Formulas

38. A 3.56-gram sample of iron powder was heated in gaseous chlorine, and 10.39 g of an iron chloride was formed. What is the percent composition of this compound?

39. Calculate the percent composition of each of the following compounds: (a) acetone, CH_3COCH_3; (b) carborundum, SiC; (c) aspirin, $CH_3COOC_6H_4COOH$.

*40. Copper is obtained from ores containing the following minerals: azurite, $Cu_3(CO_3)_2(OH)_2$; chalcocite, Cu_2S; chalcopyrite, $CuFeS_2$; covelite, CuS; cuprite, Cu_2O; and

malachite, $Cu_2CO_3(OH)_2$. Which mineral has the highest copper content on a percent-by-mass basis?

41. Determine the simplest formula for each of the following compounds.
    (a) copper(II) tartrate: 30.03% Cu; 22.70% C; 1.91% H; 45.37% O.
    (b) nitrosyl fluoroborate: 11.99% N; 13.70% O; 9.25% B; 65.06% F.

42. The hormone epinephrine is released in the human body during stress and increases the body's metabolic rate. Like many biochemical compounds, epinephrine is composed of carbon, hydrogen, oxygen, and nitrogen. The percent composition of this hormone is 56.8% C, 6.56% H, 28.4% O, and 8.28% N. What is the simplest formula of epinephrine?

43. (a) A compound is found to contain 5.60 g N, 14.2 g Cl, and 0.800 g H. What is the simplest formula of this compound? (b) Another compound containing the same elements is found to be 26.2% N, 66.4% Cl, and 7.5% H. What is the simplest formula of this compound?

44. Combustion of 0.5707 mg of a hydrocarbon produces 1.759 mg of CO_2. What is the simplest formula of the hydrocarbon.

45. A 2.00-gram sample of a compound gave 4.86 g of CO_2 and 2.03 g of H_2O upon combustion in oxygen. The compound is known to contain only C, H, and O. What is its simplest formula?

46. A 1.000-gram sample of an alcohol was burned in oxygen to produce 1.913 g of CO_2 and 1.174 g of H_2O. The alcohol contained only C, H, and O. What is the simplest formula of the alcohol?

*47. Complicated chemical reactions occur at hot springs on the ocean floor. One compound obtained from such a hot spring consists of Mg, Si, H, and O. From a 0.334-gram sample, the Mg is recovered as 0.115 g of MgO; H is recovered as 25.7 mg of H_2O; and Si is recovered as 0.172 g of SiO_2. What is the simplest formula of this compound?

*48. A 2.31-gram sample of an oxide of iron, heated in a stream of H_2, produces 0.720 g of H_2O. What is the simplest formula of the oxide.

49. A 0.2360-gram sample of a white compound was analyzed and found to contain 0.0944 g of Ca, 0.0283 g of C, and 0.1133 g of O. (a) What is the percent composition by mass of this compound? (b) What is its simplest formula?

Determination of Molecular Formulas

50. Skatole is found in coal tar and in human feces. It contains three elements: C, H, and N. It is 82.40% C and 6.92% H by mass. Its simplest formula is its molecular formula. What are (a) the formula and (b) the molecular weight of skatole?

51. Testosterone, the male sex hormone, contains only C, H, and O. It is 9.79% H and 11.09% O by mass. Each molecule contains two O atoms. What are (a) the molecular weight and (b) the molecular formula for testosterone?

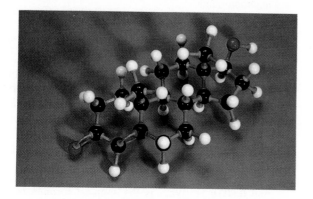

*52. More than 1 billion pounds of adipic acid (MW 146.1 g/mol) is manufactured in the United States each year. Most of it is used to make synthetic fabrics. Adipic acid contains only C, H, and O. Combustion of a 1.6380-gram sample of adipic acid gives 2.960 g of CO_2 and 1.010 g of H_2O. (a) What is the simplest formula for adipic acid? (b) What is its molecular formula?

53. Three allotropes of phosphorus are observed, with molecular weights of 62.0, 31.0, and 124.0. Write the molecular formula for each allotrope.

*54. The β-blocker drug, timolol, is expected to reduce the need for heart bypass surgery. Its composition by mass is 47.2% C, 6.55% H, 13.0% N, 25.9% O, and 7.43% S. The mass of 0.0100 mol of timolol is 4.32 g. (a) What is the simplest formula of timolol? (b) What is the molecular formula of timolol?

The Law of Multiple Proportions

55. Show that the compounds water, H_2O, and hydrogen peroxide, H_2O_2, obey the Law of Multiple Proportions.

56. Nitric oxide, NO, is produced in internal combustion engines. When NO comes in contact with air, it is quickly converted into nitrogen dioxide, NO_2, a very poisonous, corrosive gas. What mass of O is combined with 1.00 g of N in (a) NO and (b) NO_2? Show that NO and NO_2 obey the Law of Multiple Proportions.

57. Phosphorus forms two chlorides. A 30.00-gram sample of one chloride decomposes to give 4.35 g of P and 25.65 g of Cl. A 30.00-gram sample of the other chloride decomposes to give 6.61 g of P and 23.39 g of Cl. Show that these compounds obey the Law of Multiple Proportions.

58. What mass of oxygen is combined with 1.00 g of sulfur in (a) sulfur dioxide, SO_2, and in (b) sulfur trioxide, SO_3?

Interpretation of Chemical Formulas

59. One prominent ore of copper is chalcopyrite, $CuFeS_2$. How many tons of copper are contained in 255 tons of pure $CuFeS_2$?

60. Mercury occurs as a sulfide ore called *cinnabar*, HgS. How many grams of mercury are contained in 175.0 g of pure HgS?

61. (a) How many grams of copper are contained in 155 g of $CuSO_4$? (b) How many grams of copper are contained in 155 g of $CuSO_4 \cdot 5H_2O$?

62. What mass of $KMnO_4$ would contain 27.5 g of manganese?

63. What mass of azurite, $Cu_3(CO_3)_2(OH)_2$, would contain 435 g of copper?

64. Two ores that contain copper are chalcopyrite, $CuFeS_2$, and chalcocite, Cu_2S. What mass of chalcocite would contain the same mass of copper as is contained in 375 tons of chalcopyrite?

65. Tungsten is a very dense metal (19.3 g/cm³) with extremely high melting and boiling points (3370°C and 5900°C). When a small amount of it is included in steel, the resulting alloy is far harder and stronger than ordinary steel. Two important ores of tungsten are $FeWO_4$ and $CaWO_4$. How many grams of $CaWO_4$ would contain the same mass of tungsten that is contained in 874 g of $FeWO_4$?

66. What mass of NaCl would contain the same total number of ions as 245 g of $CaCl_2$?

67. Suppose we have equal masses of ammonium sulfate, $(NH_4)_2SO_4$, and aluminum sulfate, $Al_2(SO_4)_3$. (a) What is the ratio of the numbers of sulfate ions contained in these two samples? (b) What is the ratio of the masses of sulfur contained in these two samples?

*68. When $CuSO_4 \cdot 5H_2O$ is heated to 110°C, it loses only four moles of H_2O per mole of $CuSO_4$, to form $CuSO_4 \cdot H_2O$. When it is heated to temperatures above 150°C, the other mole of H_2O is lost. (a) How many grams of $CuSO_4 \cdot H_2O$ could be obtained by heating 665 g of $CuSO_4 \cdot 5H_2O$ to 110°C? (b) How many grams of anhydrous $CuSO_4$ could be obtained by heating 665 g of $CuSO_4 \cdot 5H_2O$ to 180°C?

Percent Purity

69. (a) How many grams of sodium chloride, NaCl, are contained in 33.0 g of saline solution that is 5.0% NaCl by mass? (b) Vinegar is 5.0% acetic acid, CH_3COOH, by mass. How many grams of acetic acid are contained in 33.0 g of vinegar? (c) How many pounds of acetic acid are contained in 33.0 pounds of vinegar?

70. (a) What is the percent by mass of oxalic acid, $(COOH)_2$, in a sample of pure oxalic acid dihydrate, $(COOH)_2 \cdot 2H_2O$? (b) What is the percent by mass of $(COOH)_2$ in a sample that is 72.4% $(COOH)_2 \cdot 2H_2O$ by mass?

71. What mass of each of the following elements is contained in 64.4 g of 88.2% pure $Ca(NO_3)_2$? Assume that the impurities do not contain the element mentioned. (a) Ca; (b) N

72. What weight of magnesium carbonate is contained in 775 pounds of an ore that is 24.3% magnesium carbonate by weight? (b) What weight of impurities is contained in the sample? (c) What weight of magnesium is contained in the sample? (Assume that no magnesium is present in the impurities.)

Mixed Examples

73. How many moles of bromine atoms are contained in each of the following? (a) 79.9×10^{23} Br atoms; (b) 79.9×10^{23} Br_2 molecules; (c) 79.9 g of bromine; (d) 79.9 mol of Br_2.

74. What is the *maximum* number of moles of CO_2 that could be obtained from the carbon in each of the following? (a) 1.00 mol of $Fe_2(CO_3)_3$; (b) 2.00 mol of $CaCO_3$; (c) 1.00 mol of $Ni(CN)_4$.

75. (a) How many formula units are contained in 238.1 g of

K$_2$MoO$_4$? (b) How many potassium ions? (c) How many MoO_4^{2-} ions? (d) How many atoms of all kinds?

76. (a) How many moles of ozone molecules are contained in 64.0 g of ozone, O_3? (b) How many moles of oxygen atoms are contained in 64.0 g of ozone? (c) What mass of O_2 would contain the same number of oxygen atoms as 64.0 g of ozone? (d) What mass of oxygen gas, O_2, would contain the same number of molecules as 64.0 g of ozone?

77. Cocaine has the following percent composition by mass: 67.30% C, 6.930% H, 21.15% O, and 4.62% N. What is the simplest formula of cocaine?

78. What mass corresponds to each of the following? (a) 5.3 mol of C; (b) 0.12173 mol of N_2O_5; (c) 1.3 mmol of $Al_2(SO_4)_3$; (d) 1.0×10^{-10} mol of HCl.

79. How many moles of chloroform, $CHCl_3$, are contained in 338 mL of chloroform? The density of chloroform is 1.84 g/mL.

80. Find the number of moles of Ag needed to form each of the following: (a) 0.263 mol Ag_2S; (b) 0.263 mol Ag_2O; (c) 0.263 g Ag_2S; (d) 2.63×10^{20} formula units of Ag_2S.

*81. (a) A sample contains 50.0% NaCl and 50.0% KCl by mass. What is the percent Cl, by mass, in this sample? (b) A second sample of NaCl and KCl contains 50.0% Cl by mass. What is the mass percent of NaCl in this sample?

*82. Analysis of a 20.0-mg sample of an organic compound for H, N, and C yields 1.99 mg H_2O, 1.25 mg NH_3, and 6.47 mg CO_2. The Cl and Br are recovered as a mixture of AgCl and AgBr with a mass of 48.8 mg. Finally, when all the AgBr is converted to AgCl (by adding chloride from an outside source), the mass of AgCl formed from the AgBr plus the mass of AgCl originally present in the mixture are 42.3 mg. Calculate the empirical formula of the compound.

83. A metal, M, forms an oxide having the empirical formula M_2O_3. This oxide contains 68.4% of the metal by mass. (a) Calculate the atomic weight of the metal. (b) Identify the metal.

84. We have a 85.0-gram sample of "red lead," Pb_3O_4. (a) How many moles of lead are in the sample? (b) How many grams of lead are in the sample? (c) How many lead atoms are in the sample?

*85. A 23.4-gram sample of a nickel sulfide is converted to 17.20 g of NiO. (a) What is the mass composition of the sulfide? (b) What is the empirical formula of the sulfide?

86. Three samples of magnesium oxide were analyzed to determine the mass ratios O/Mg, giving the following results: $\dfrac{1.60 \text{ g O}}{2.43 \text{ g Mg}}$, $\dfrac{0.658 \text{ g O}}{1.00 \text{ g Mg}}$, $\dfrac{2.29 \text{ g O}}{3.48 \text{ g Mg}}$. Which law of chemical combination is illustrated by these data?

*87. The molecular weight of hemoglobin is about 65,000 g/mol. Hemoglobin contains 0.35% Fe by mass. How many iron atoms are in a hemoglobin molecule?

*88. (a) We can drive off the water from copper sulfate pentahydrate, $CuSO_4 \cdot 5H_2O$, by heating. How many grams of anhydrous (meaning "without water") $CuSO_4$ could be obtained by heating 173 g of $CuSO_4 \cdot 5H_2O$? (b) An experiment calls for 2.50 mol of $CuSO_4$. How many grams of $CuSO_4 \cdot 5H_2O$ should we use to supply the required amount of $CuSO_4$?

*89. During volcanic action, S_8 is converted to S, which is then converted to H_2S. In turn, the H_2S reacts with Fe,

forming FeS_2. In water containing O_2, the FeS_2 reacts to form "mine acid," H_2SO_4. Find the maximum mass, in grams, of H_2SO_4 that can be formed from 0.366 mol of S_8.

90. One method of analyzing for the amount of Cr_2O_3 in a sample involves converting the chromium to $BaCrO_4$, and then weighing the amount of $BaCrO_4$ formed. Suppose that this process could be carried out with no loss of chromium. How many grams of Cr_2O_3 are present in the original sample for every gram of $BaCrO_4$ that could be isolated and weighed?

91. A 475-mg sample was found to contain 32.6% $PbSO_4$ (and no other lead). (a) How many milligrams of lead were contained in the sample? (b) What is the percent lead in the sample?

3 Chemical Equations and Reaction Stoichiometry

Sulfur burns in oxygen with a bright blue flame.

$$S_8 + 8O_2 \longrightarrow 8SO_2$$

This reaction is the first step in the commercial production of sulfuric acid, H_2SO_4, the most widely used industrial chemical.

Objectives

As you study this chapter, you should learn to

☐ Write a balanced chemical equation to describe a chemical reaction

☐ Interpret a balanced chemical equation to calculate the *moles* of reactants and products involved in the reaction

☐ Interpret a balanced chemical equation to calculate the *masses* of reactants and products involved in the reaction

☐ Determine which is the limiting reactant

☐ Use the limiting reactant concept in calculations with chemical equations

☐ Compare the amount of substance actually formed in a reaction (actual yield) with the predicted amount (theoretical yield), and determine the percent yield

☐ Understand sequential reactions

☐ Use the terminology of solutions—solute, solvent, concentration

☐ Calculate concentrations of solutions when they are diluted

☐ Understand the concept of, and calculations for, titrations

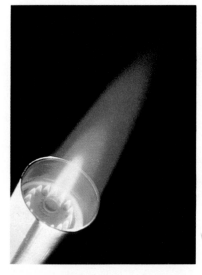

Methane, CH_4, is the main component of natural gas.

I n the last chapter we studied composition stoichiometry, the quantitative relationships among elements in compounds. In this chapter we shall study reaction stoichiometry, the quantitative relationships among substances as they participate in chemical reactions. We ask several important questions. *How* can we describe the reaction of one substance with another? *How much* of one substance reacts with a given amount of another substance? *Which reactant* determines the amounts of products formed in a chemical reaction? *How* can we describe reactions in aqueous solutions?

Whether we are concerned with describing a reaction used in a chemical analysis, one used industrially in the production of a plastic, or one that occurs during metabolism in the body, we must describe it accurately. Chemical equations represent a very precise, yet a very versatile, language that describes chemical changes. We shall begin by studying chemical equations.

3-1 Chemical Equations

Chemical reactions always involve changing one or more substances into one or more different substances. That is, they involve regrouping atoms or ions to form other substances.

Chemical equations are used to describe chemical reactions, and they show (1) *the substances that react,* called **reactants**, (2) *the substances formed,* called **products**, and (3) *the relative amounts of the substances involved.* As a typical example, let's consider the combustion (burning) of natural gas, a reaction used to heat buildings and cook foods. Natural gas is a mixture of several substances, but the principal component is methane, CH_4. The equation that describes the reaction of methane with excess oxygen is

$$CH_4 + 2O_2 \longrightarrow CO_2 + 2H_2O$$

$$\underbrace{\qquad\qquad}_{\text{reactants}} \qquad \underbrace{\qquad\qquad}_{\text{products}}$$

What does this equation tell us? In the simplest terms, it tells us that methane reacts with oxygen to produce carbon dioxide, CO_2, and water. More specifically, it says that for every CH_4 molecule that reacts, two molecules of O_2 also react, and that one CO_2 molecule and two H_2O molecules are formed. That is,

$$CH_4 \quad + \quad 2O_2 \quad \xrightarrow{\Delta} \quad CO_2 \quad + \quad 2H_2O$$
$$\text{1 molecule} \quad \text{2 molecules} \qquad \text{1 molecule} \quad \text{2 molecules}$$

Special conditions required for some reactions are indicated by notation over the arrow. The capital Greek letter delta (Δ) means that heat is necessary to start this reaction. Figure 3-1 shows the rearrangement of atoms described by this equation.

As we pointed out in Section 1-1, *there is no detectable change in the quantity of matter during an ordinary chemical reaction.* This guiding principle, the **Law of Conservation of Matter**, provides the basis for "balancing" chemical equations and for calculations based on those equations. Because matter is neither created nor destroyed during a chemical reaction, a balanced chemical equation must always include the same number of each kind of atom on both sides of the equation. Chemists usually write equations with the smallest possible whole-number coefficients.

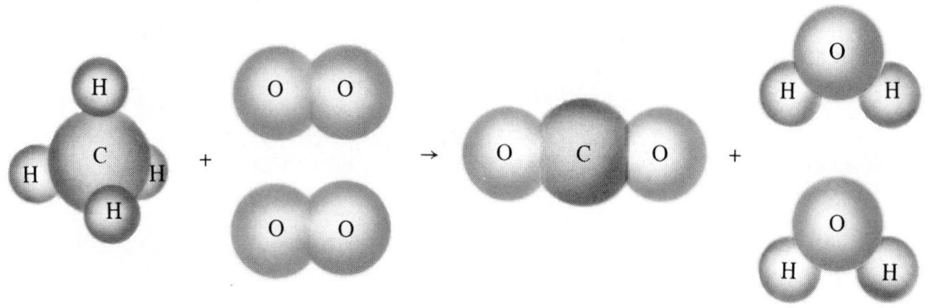

Sometimes it is not possible to represent a chemical change with a single chemical equation. For example, under certain conditions, both CO_2 and CO are found as products, and a second chemical equation must be used. In the present case (excess oxygen), only one equation is required.

The arrow may be read "yields."

Figure 3-1
Representation of the reaction of methane with oxygen to form carbon dioxide and water. Some chemical bonds are broken and some new ones are formed.

$$CH_4 \quad + \quad 2O_2 \quad \rightarrow \quad CO_2 \quad + \quad 2H_2O$$

Chemistry in Use. . .
Alchemy

During the Dark Ages, a pseudo-science known as *alchemy* flourished in Europe and the Middle East. Its practitioners, the predecessors of modern chemists, sought to turn base metals into gold and silver. They believed they could accomplish this by means of the Philosopher's Stone, which would transmute one element into another. Although their efforts proved fruitless, in the process of trying to make this magical substance the alchemists became very skillful at refining and alloying metals. They are credited with the discovery of arsenic, antimony, bismuth, zinc, and phosphorus, and they developed useful purification techniques such as distillation, sublimation, and crystallization. We also owe to alchemy the concept of scientific laboratories and the experimental approach to solving problems.

Practiced during a period when religion and mysticism affected all areas of life, the philosophy of alchemy dealt dually with matter and spirit. The desire to change ordinary metals into gold also represented man's wish to become better. Thus, alchemical processes and materials were frequently described in terms of astrology, mythology, the Church, or nature. Metals were identified with planets and gods; other substances were depicted as animals or people. Treatises were thick with allegory and had titles such as *The Crowne of Nature* and *The Triumphal Chariot of Antimony*. Further obscurity resulted from alchemists' desire to keep their "secrets" to themselves. Only *adepts,* or initiated experimenters, were able to decipher the complex codes and references. Because they often made up their own personal symbols and wrote in several languages, including Greek, Latin, Arabic, and Hebrew, it is unlikely that even an adept could have understood a recipe for the Philosopher's Stone written by a colleague!

One way of transmitting alchemical information was through pictures. In the figure, a method for refining gold is described. Antimony (wolf) is added to the impure gold (dead king) to "devour" any impurities present (foreground). When the two are thrown into a fire (background), the gold becomes pure (resurrected king).

One of the earliest alchemists was the Arab Jābir ibn Hayyān (also known as Geber), who originated a theory that all metals are composed of mercury and sulfur. This theory lasted for many centuries in Europe, where it was taken up by Christian scholars after being translated into Latin. (One of those scholars, a monk named Basil Valentine, authored the pompously named *The Triumphal Chariot of Antimony*.)

Other alchemists made great advances in science. Agricola, a German physician, fathered mineralogy with *De Re Metallica,* a clear and scholarly work classifying metals and describing their uses. His contemporary, Paracelsus, began the application of chemistry to medicine, called iatrochemistry, by noting the curative properties of several elements. As alchemists applied their chemical expertise in new directions, desires to synthesize gold slowly faded away. New findings were recorded clearly, and references to religion and the occult gradually vanished. A "scientific approach" prevailed, and in the 17th century chemistry came into being as an experimental science.

It can be said that in recent times the Philosopher's Stone *has* been discovered—through atomic reactions that split and fuse nuclei. As they explore ways to make new elements and transmute others, today's scientists could be considered "modern alchemists!"

Lisa L. Saunders
Chemistry major
University of Texas at Austin

The Death and Resurrection of the King (Atalanta Fugiens, 1617). Alchemists often represented chemicals with animals and mythological symbols.

All substances must be represented by formulas that describe them *as they exist* before we attempt to balance an equation. For instance, we must write H_2 to represent diatomic hydrogen molecules—not H, which represents hydrogen atoms. Once formulas are correct, the subscripts in the formulas may not be changed. Different subscripts in formulas specify different compounds, so the equation would no longer describe the same reaction if formulas were changed.

Let's generate the balanced equation for the reaction of aluminum metal with hydrochloric acid (hydrogen chloride dissolved in water) to produce aluminum chloride and hydrogen. The unbalanced "equation" is

$$Al + HCl \longrightarrow AlCl_3 + H_2$$

Atoms of	In reactants	In products
Al	1	1
H	1	2
Cl	1	3

Equation is not balanced.

As it now stands, the "equation" does not satisfy the Law of Conservation of Matter because there are two H atoms in the H_2 molecule and three Cl atoms in one formula unit of $AlCl_3$ (right side), but only one H atom and one Cl atom in the HCl molecule (left side).

Let us first balance chlorine by putting a coefficient of 3 in front of HCl.

$$Al + \boxed{3HCl} \longrightarrow AlCl_3 + H_2$$

Atoms of	In reactants	In products
Al	1	1
H	3	2
Cl	3	3

Equation is not balanced.

Now there are 3H on the left and 2H on the right. The least common multiple of 3 and 2 is 6; to balance H, we multiply the 3HCl by 2 and the H_2 by 3.

$$Al + \boxed{6HCl} \longrightarrow AlCl_3 + \boxed{3H_2}$$

Atoms of	In reactants	In products
Al	1	1
H	6	6
Cl	6	3

Equation is still not balanced.

Now Cl is again unbalanced (6Cl on the left, 3 on the right), but we can fix this by putting a coefficient of 2 in front of $AlCl_3$ on the right.

$$Al + 6HCl \longrightarrow \boxed{2AlCl_3} + 3H_2$$

Atoms of	In reactants	In products
Al	1	2
H	6	6
Cl	6	6

Equation is still not balanced.

Now all elements except Al are balanced (1 on the left, 2 on the right); we complete the balancing by putting a coefficient of 2 in front of Al on the left.

$$\boxed{2Al} + 6HCl \longrightarrow 2AlCl_3 + 3H_2$$

aluminum   hydrochloric acid   aluminum chloride   hydrogen

Atoms of	In reactants	In products
Al	2	2
H	6	6
Cl	6	6

Now the equation is balanced.

When we think that we have finished the balancing, we should *always* do a complete check for each element, as shown in red in the margin.

Dimethyl ether, C_2H_6O, burns in an excess of oxygen to give carbon dioxide and water. Let's balance the equation for this reaction. In unbalanced form,

$$C_2H_6O + O_2 \longrightarrow CO_2 + H_2O$$

Atoms of	In reactants	In products
C	2	1
H	6	2
O	3	3

Equation is not balanced.

Carbon appears in only one compound on each side, and the same is true for hydrogen. We begin by balancing these elements:

$$C_2H_6O + O_2 \longrightarrow \boxed{2CO_2} + \boxed{3H_2O}$$

Atoms of	In reactants	In products
C	2	2
H	6	6
O	3	7

Equation is still not balanced.

Now we have an odd number of atoms of O on each side. The single O in C_2H_6O balances one of the atoms of O on the right. We balance the other six by placing a coefficient of 3 before O_2 on the left.

Atoms of	In reactants	In products
C	2	2
H	6	6
O	7	7

Now the equation is balanced.

$$C_2H_6O + 3O_2 \longrightarrow 2CO_2 + 3H_2O$$

Balancing chemical equations ''by inspection'' is a *trial-and-error* approach. It requires a great deal of practice, but it is *very important!* Remember that we use the smallest whole-number coefficients.

3-2 Calculations Based on Chemical Equations

As we indicated earlier, chemical equations represent a very precise and versatile language. We are now ready to use them to calculate the relative *amounts* of substances involved in chemical reactions. Let us again consider the combustion of methane in excess oxygen. The balanced chemical equation for that reaction is

$$CH_4 + 2O_2 \xrightarrow{\Delta} CO_2 + 2H_2O$$

On a quantitative basis, at the molecular level, the equation says

A balanced chemical equation may be interpreted on a *molecular* basis.

$$\begin{array}{ccccccc} CH_4 & + & 2O_2 & \longrightarrow & CO_2 & + & 2H_2O \\ \text{1 molecule} & & \text{2 molecules} & & \text{1 molecule} & & \text{2 molecules} \\ \text{of methane} & & \text{of oxygen} & & \text{of carbon dioxide} & & \text{of water} \end{array}$$

Example 3-1

How many O_2 molecules are required to react with 47 CH_4 molecules according to the above equation?

Plan

The *balanced* equation tells us that *one* CH_4 molecule reacts with *two* O_2 molecules. We can construct two unit factors from this fact:

$$\frac{1\ CH_4\ \text{molecule}}{2\ O_2\ \text{molecules}} \quad \text{and} \quad \frac{2\ O_2\ \text{molecules}}{1\ CH_4\ \text{molecule}}$$

These are unit factors for *this* reaction because the numerator and denominator are *chemically equivalent.* In other words, the numerator and the denominator represent the same amount of reaction. We convert CH_4 molecules to O_2 molecules.

Solution

$$\underline{?}\ O_2\ \text{molecules} = 47\ CH_4\ \text{molecules} \times \frac{2\ O_2\ \text{molecules}}{1\ CH_4\ \text{molecule}} = \boxed{94\ O_2\ \text{molecules}}$$

EOC 8, 10

A chemical equation also indicates the relative amounts of each reactant and product in a given chemical reaction. We showed earlier that formulas can represent moles of substances. Suppose Avogadro's number of CH_4 molecules, rather than just one CH_4 molecule, undergo this reaction. Then the equation can be written as follows:

$$\begin{array}{ccccccc} CH_4 & + & 2O_2 & \xrightarrow{\Delta} & CO_2 & + & 2H_2O \\ 6.02 \times 10^{23}\ \text{molecules} & & 2(6.02 \times 10^{23}\ \text{molecules}) & & 6.02 \times 10^{23}\ \text{molecules} & & 2(6.02 \times 10^{23}\ \text{molecules}) \\ \text{1 mol} & & \text{2 mol} & & \text{1 mol} & & \text{2 mol} \end{array}$$

This tells us that *one* mole of methane reacts with *two* moles of oxygen to produce *one* mole of carbon dioxide and *two* moles of water.

A balanced chemical equation may be interpreted in terms of *moles* of reactants and products.

Example 3-2

How many moles of water could be produced by the reaction of 3.5 moles of methane with excess oxygen?

Plan

The equation for the combustion of methane

$$CH_4 + 2O_2 \longrightarrow CO_2 + 2H_2O$$
$$\text{1 mol} \quad \text{2 mol} \quad \quad \text{1 mol} \quad \text{2 mol}$$

shows that one mole of methane reacts with two moles of oxygen to produce two moles of water. From this information we construct two *unit factors*:

$$\frac{\text{1 mol } CH_4}{\text{2 mol } H_2O} \quad \text{and} \quad \frac{\text{2 mol } H_2O}{\text{1 mol } CH_4}$$

We use the second factor in this calculation.

Solution

$$\underline{?} \text{ mol } H_2O = 3.5 \text{ mol } CH_4 \times \frac{\text{2 mol } H_2O}{\text{1 mol } CH_4} = \boxed{7.0 \text{ mol } H_2O}$$

EOC 18, 20

We know the mass of one mole of each of these substances, so we can also write

$$CH_4 + 2O_2 \longrightarrow CO_2 + 2H_2O$$

CH₄	2O₂	CO₂	2H₂O
1 mol	2 mol	1 mol	2 mol
16 g	2(32 g)	44 g	2(18 g)
16 g	64 g	44 g	36 g

80 g reactants   80 g products

The equation now tells us that 16 grams of CH_4 reacts with 64 grams of O_2 to form 44 grams of CO_2 and 36 grams of H_2O. The Law of Conservation of Matter is satisfied. Chemical equations describe **reaction ratios**, i.e., the *mole ratios* of reactants and products as well as the *relative masses* of reactants and products.

A balanced equation may be interpreted on a *mass* basis. We have rounded molecular weights to the nearest whole number of grams here.

Example 3-3

What mass of oxygen is required to react completely with 1.2 moles of CH_4?

Plan

The balanced equation

$$CH_4 + 2O_2 \longrightarrow CO_2 + 2H_2O$$
$$\text{1 mol} \quad \text{2 mol} \quad \quad \text{1 mol} \quad \text{2 mol}$$
$$\text{16 g} \quad \text{2(32 g)} \quad \quad \text{44 g} \quad \text{2(18 g)}$$

gives the relationships among moles and grams of reactants and products.

$$\boxed{\text{mol } CH_4} \longrightarrow \boxed{\text{mol } O_2} \longrightarrow \boxed{\text{g } O_2}$$

Solution

$$? \text{ g } O_2 = 1.2 \text{ mol } CH_4 \times \frac{2 \text{ mol } O_2}{1 \text{ mol } CH_4} \times \frac{32 \text{ g } O_2}{1 \text{ mol } O_2} = \boxed{77 \text{ g } O_2}$$

EOC 22, 24

Example 3-4

What mass of oxygen is required to react completely with 24 grams of CH_4?

Plan

Recall the balanced equation

$$CH_4 + 2O_2 \longrightarrow CO_2 + 2H_2O$$

| 1 mol | 2 mol | 1 mol | 2 mol |
| 16 g | 64 g | 44 g | 36 g |

This shows that 16 grams of CH_4 reacts with 64 grams of O_2. These two quantities are chemically equivalent, so we can construct two *unit factors:*

$$\frac{16 \text{ g } CH_4}{64 \text{ g } O_2} \quad \text{and} \quad \frac{64 \text{ g } O_2}{16 \text{ g } CH_4}$$

Solution

$$? \text{ g } O_2 = 24 \text{ g } CH_4 \times \frac{64 \text{ g } O_2}{16 \text{ g } CH_4} = \boxed{96 \text{ g } O_2}$$

EOC 26, 27

Another approach to the problem we have just solved is known as the **mole method**. In this method, the number of moles of reactant or product is calculated and then converted to grams (or other desired unit). Example 3-4 asked, "What mass of oxygen is required to react with 24 grams of CH_4?" The balanced equation and the calculation by the mole method follow:

$$CH_4 + 2O_2 \longrightarrow CO_2 + 2H_2O$$

| 1 mol | 2 mol | 1 mol | 2 mol |

We convert

1. g CH_4 → mol CH_4

$$? \text{ mol } CH_4 = 24 \text{ g } CH_4 \times \frac{1 \text{ mol } CH_4}{16 \text{ g } CH_4} = 1.5 \text{ mol } CH_4$$

2. mol CH_4 → mol O_2

$$? \text{ mol } O_2 = 1.5 \text{ mol } CH_4 \times \frac{2 \text{ mol } O_2}{1 \text{ mol } CH_4} = 3.0 \text{ mol } O_2$$

3. mol O_2 → g O_2

$$? \text{ g } O_2 = 3.0 \text{ mol } O_2 \times \frac{32 \text{ g } O_2}{1 \text{ mol } O_2} = \boxed{96 \text{ g } O_2}$$

All these steps could be combined into one setup in which we convert:

$$\text{g of } CH_4 \longrightarrow \text{mol of } CH_4 \longrightarrow \text{mol of } O_2 \longrightarrow \text{g of } O_2$$

$$? \text{ g } O_2 = 24 \text{ g } CH_4 \times \frac{1 \text{ mol } CH_4}{16 \text{ g } CH_4} \times \frac{2 \text{ mol } O_2}{1 \text{ mol } CH_4} \times \frac{32 \text{ g } O_2}{1 \text{ mol } O_2} = \boxed{96 \text{ g } O_2}$$

All valid methods of calculations are based on balanced chemical equations.

The same answer, 96 grams of O_2, is obtained by both of these methods. The question may be reversed, as in Example 3-5.

Example 3-5

What mass of CH_4, in grams, is required to react with 96 grams of O_2?

Plan

We recall that one mole of CH_4 reacts with two moles of O_2.

Solution

$$\underline{?}\text{ g CH}_4 = 96\text{ g O}_2 \times \frac{1\text{ mol O}_2}{32\text{ g O}_2} \times \frac{1\text{ mol CH}_4}{2\text{ mol O}_2} \times \frac{16\text{ g CH}_4}{1\text{ mol CH}_4} = \boxed{24\text{ g CH}_4}$$

These unit factors are the reciprocals of those used in Example 3-4.

or, more simply,

$$\underline{?}\text{ g CH}_4 = 96\text{ g O}_2 \times \frac{16\text{ g CH}_4}{64\text{ g O}_2} = \boxed{24\text{ g CH}_4}$$

EOC 28

This is the amount of CH_4 in Example 3-4 that reacted with 96 grams of O_2.

Example 3-6

Most combustion reactions occur in excess O_2, i.e., more than enough O_2 to burn the substance completely. Calculate the mass of CO_2, in grams, that can be produced by burning 6.0 moles of CH_4 in excess O_2.

Plan

Recall the balanced equation

$$CH_4 + 2O_2 \longrightarrow CO_2 + 2H_2O$$

| 1 mol | 2 mol | 1 mol | 2 mol |
| 16 g | 2(32 g) | 44 g | 2(18 g) |

It is important to recognize that the reaction must stop when the 6.0 mol of CH_4 has been used up. Some O_2 will remain unreacted.

which tells us that one mole of CH_4 produces one mole (44 g) of CO_2.

Solution

$$\underline{?}\text{ g CO}_2 = 6.0\text{ mol CH}_4 \times \frac{1\text{ mol CO}_2}{1\text{ mol CH}_4} \times \frac{44\text{ g CO}_2}{1\text{ mol CO}_2} = \boxed{2.6 \times 10^2\text{ g CO}_2}$$

From the mole interpretation of the chemical equation for the combustion of methane, we can see many chemically equivalent pairs of terms. Each pair gives a unit factor relating substances. Some of these factors are

$$\frac{1\text{ mol CH}_4}{2\text{ mol O}_2} \qquad \frac{1\text{ mol CH}_4}{64\text{ g O}_2} \qquad \frac{1\text{ mol CH}_4}{2(6.02 \times 10^{23})\text{ O}_2\text{ molecules}}$$

$$\frac{16\text{ g CH}_4}{2\text{ mol O}_2} \qquad \frac{16\text{ g CH}_4}{64\text{ g O}_2} \qquad \frac{16\text{ g CH}_4}{2(6.02 \times 10^{23})\text{ O}_2\text{ molecules}}$$

$$\frac{6.02 \times 10^{23}\text{ CH}_4\text{ molecules}}{2\text{ mol O}_2} \qquad \frac{6.02 \times 10^{23}\text{ CH}_4\text{ molecules}}{64\text{ g O}_2} \qquad \frac{6.02 \times 10^{23}\text{ CH}_4\text{ molecules}}{2(6.02 \times 10^{23})\text{ O}_2\text{ molecules}}$$

We have written down nine unit factors relating CH_4 and O_2 for this particular reaction. The nine factors obtained by inverting each of these give

Please don't try to memorize unit factors for chemical reactions; rather, learn the general method *for constructing them from balanced chemical equations.*

a total of 18 unit factors relating CH_4 and O_2 for *this* reaction in three kinds of units. In fact, we can write down many factors relating *any* two substances involved in any chemical reaction! Try writing down some that involve reactants and products.

Reaction stoichiometry usually involves interpreting a balanced chemical equation to relate a *given* bit of information to the *desired* bit of information.

Example 3-7

What mass of CH_4 produces 3.01×10^{23} H_2O molecules when burned in excess O_2?

Plan

The balanced equation tells us that one mole of CH_4 produces two moles of H_2O.

$$CH_4 + 2O_2 \longrightarrow CO_2 + 2H_2O$$
$$\text{1 mol} \quad \text{2 mol} \quad\quad \text{1 mol} \quad \text{2 mol}$$

$$\boxed{H_2O \text{ molecules}} \longrightarrow \boxed{\text{mol of } H_2O} \longrightarrow \boxed{\text{mol of } CH_4} \longrightarrow \boxed{\text{g of } CH_4}$$

Solution

$$\underline{?} \text{ g } CH_4 = 3.01 \times 10^{23} \text{ } H_2O \text{ molecules} \times \frac{1 \text{ mol } H_2O}{6.02 \times 10^{23} \text{ } H_2O \text{ molecules}}$$

$$\times \frac{1 \text{ mol } CH_4}{2 \text{ mol } H_2O} \times \frac{16 \text{ g } CH_4}{1 \text{ mol } CH_4} = \boxed{4.0 \text{ g } CH_4}$$

The possibilities for this kind of problem-solving go on and on. Before you continue, you should work Exercises 16–31 at the end of the chapter.

3-3 The Limiting Reactant Concept

In the problems we have worked thus far, the presence of an excess of one reactant was stated or implied. The calculations were based on the substance that was used up first, called the **limiting reactant**. Before we study the concept of the limiting reactant in stoichiometry, let's develop the basic idea by considering a simple but analogous nonchemical example.

Suppose you have four slices of ham and six slices of bread and you wish to make as many ham sandwiches as possible using only one slice of ham and two slices of bread per sandwich. Obviously, you can make only three sandwiches, at which point you run out of bread. (In a chemical reaction this would correspond to one of the reactants being used up—so the reaction would stop.) Therefore the bread is the "limiting reactant" and the extra slice of ham is the "excess reactant." The amount of product, ham sandwiches, is determined by the amount of the limiting reactant, bread in this case.

Example 3-8

What mass of CO_2 could be formed by the reaction of 8.0 g of CH_4 with 48 g of O_2?

Plan

Recall the balanced equation:

$$CH_4 + 2O_2 \longrightarrow CO_2 + 2H_2O$$

| 1 mol | 2 mol | 1 mol | 2 mol |
| 16 g | 2(32 g) | 44 g | 2(18 g) |

This tells us that *one* mole of CH_4 reacts with *two* moles of O_2. We are given masses of both CH_4 and O_2, so we calculate the number of moles of each reactant, and then determine the number of moles of each reactant required to react with the other. From these calculations we can identify the limiting reactant. We base the calculation on it.

Solution

$$\underline{?} \text{ mol } CH_4 = 8.0 \text{ g } CH_4 \times \frac{1 \text{ mol } CH_4}{16 \text{ g } CH_4} = \underline{0.50 \text{ mol } CH_4}$$

$$\underline{?} \text{ mol } O_2 = 48 \text{ g } O_2 \times \frac{1 \text{ mol } O_2}{32 \text{ g } O_2} = \underline{1.5 \text{ mol } O_2}$$

Now we return to the balanced equation. First we calculate the number of moles of O_2 required to react with 0.50 mole of CH_4.

$$\underline{?} \text{ mol } O_2 = 0.50 \text{ mol } CH_4 \times \frac{2 \text{ mol } O_2}{1 \text{ mol } CH_4} = 1 \text{ mol } O_2$$

We have 1.5 moles of O_2, but only 1 mole of O_2 is required, so CH_4 is the limiting reactant. Or we can calculate the number of moles of CH_4 required to react with 1.5 moles of O_2.

$$\underline{?} \text{ mol } CH_4 = 1.5 \text{ mol } O_2 \times \frac{1 \text{ mol } CH_4}{2 \text{ mol } O_2} = 0.75 \text{ mol } CH_4$$

This tells us that 0.75 mole of CH_4 would be required to react with 1.5 moles of O_2. But we have only 0.50 mole of CH_4, so we see again that CH_4 is the limiting reactant. The reaction must stop when the limiting reactant, CH_4, is used up; we base the calculation on CH_4.

$$\boxed{\text{g of } CH_4} \longrightarrow \boxed{\text{mol of } CH_4} \longrightarrow \boxed{\text{mol of } CO_2} \longrightarrow \boxed{\text{g of } CO_2}$$

$$\underline{?} \text{ g } CO_2 = 8.0 \text{ g } CH_4 \times \frac{1 \text{ mol } CH_4}{16 \text{ g } CH_4} \times \frac{1 \text{ mol } CO_2}{1 \text{ mol } CH_4} \times \frac{44 \text{ g } CO_2}{1 \text{ mol } CO_2} = \boxed{22 \text{ g } CO_2}$$

Thus, 22 grams of CO_2 is the most CO_2 that can be produced from 8.0 grams of CH_4 and 48 grams of O_2. If the calculation had been based on O_2 rather than CH_4, the answer would be too big and *wrong*. This would require more CH_4 than we had.

Another approach to problems like Example 3-8 is to calculate the number of moles of each reactant:

$$\underline{?} \text{ mol } CH_4 = 8.0 \text{ g } CH_4 \times \frac{1 \text{ mol } CH_4}{16 \text{ g } CH_4} = \underline{0.50 \text{ mol } CH_4}$$

$$\underline{?} \text{ mol } O_2 = 48 \text{ g } O_2 \times \frac{1 \text{ mol } O_2}{32 \text{ g } O_2} = \underline{1.5 \text{ mol } O_2}$$

A precipitate of solid $Ni(OH)_2$ forms when colorless NaOH solution is added to green $NiCl_2$ solution.

Note that even though the reaction occurs in aqueous solution, this calculation is similar to earlier examples because we are given the amounts of both reactants.

Then we return to the balanced equation. We first calculate the *required ratio* of reactants as indicated by the balanced chemical equation. We then calculate the *available ratio* of reactants and compare the two:

Required Ratio	Available Ratio
$\dfrac{1 \text{ mol } CH_4}{2 \text{ mol } O_2} = \dfrac{0.50 \text{ mol } CH_4}{1.00 \text{ mol } O_2}$	$\dfrac{0.50 \text{ mol } CH_4}{1.50 \text{ mol } O_2} = \dfrac{0.33 \text{ mol } CH_4}{1.00 \text{ mol } O_2}$

We see that each mole of O_2 would require exactly 0.50 mole of CH_4 to be completely used up. But we have only 0.33 mole of CH_4 for each mole of O_2, so there is *insufficient* CH_4 to react with all of the available O_2. The reaction must stop when the CH_4 is gone; CH_4 is the "limiting reactant" and we must base the calculation on it.

Example 3-9

What is the maximum mass of $Ni(OH)_2$ that could be prepared by mixing two solutions that contain 26.0 grams of $NiCl_2$ and 10.0 grams of NaOH, respectively?

$$NiCl_2 + 2NaOH \longrightarrow Ni(OH)_2 + 2NaCl$$

Plan

Interpreting the balanced equation as usual, we have

$$NiCl_2 + 2NaOH \longrightarrow Ni(OH)_2 + 2NaCl$$

1 mol	2 mol	1 mol	2 mol
129.7 g	2(40.0 g)	92.7 g	2(58.4 g)

We determine the number of moles of $NiCl_2$ and NaOH present. Then we find the number of moles of each reactant required to react with the other reactant. These calculations identify the limiting reactant. We base the calculation on it.

Solution

$$\underline{?} \text{ mol } NiCl_2 = 26.0 \text{ g } NiCl_2 \times \frac{1 \text{ mol } NiCl_2}{129.7 \text{ g } NiCl_2} = 0.200 \text{ mol } NiCl_2$$

$$\underline{?} \text{ mol } NaOH = 10.0 \text{ g } NaOH \times \frac{1 \text{ mol } NaOH}{40.0 \text{ g } NaOH} = 0.250 \text{ mol } NaOH$$

We return to the balanced equation and calculate the number of moles of NaOH required to react with 0.200 mole of $NiCl_2$.

$$\underline{?} \text{ mol } NaOH = 0.200 \text{ mol } NiCl_2 \times \frac{2 \text{ mol } NaOH}{1 \text{ mol } NiCl_2} = 0.400 \text{ mol } NaOH$$

But we have only 0.250 mole of NaOH, so NaOH is the limiting reactant.

If we calculate the number of moles of $NiCl_2$ required to react with 0.250 mole of NaOH, we get

$$\underline{?} \text{ mol } NiCl_2 = 0.250 \text{ mol } NaOH \times \frac{1 \text{ mol } NiCl_2}{2 \text{ mol } NaOH} = 0.125 \text{ mol } NiCl_2$$

We see that 0.250 mole of NaOH can react with only 0.125 mole of $NiCl_2$. But we have 0.200 mole of $NiCl_2$. This also tells us that NaOH is the limiting reactant, and so the calculation must be based on NaOH. The reaction must stop when all of the NaOH has been used up.

$$\boxed{\text{g of NaOH}} \longrightarrow \boxed{\text{mol of NaOH}} \longrightarrow \boxed{\text{mol Ni(OH)}_2} \longrightarrow \boxed{\text{g of Ni(OH)}_2}$$

$$\underline{?} \text{ g Ni(OH)}_2 = 10.0 \text{ g NaOH} \times \frac{1 \text{ mol NaOH}}{40.0 \text{ g NaOH}} \times \frac{1 \text{ mol Ni(OH)}_2}{2 \text{ mol NaOH}} \times \frac{92.7 \text{ g Ni(OH)}_2}{1 \text{ mol Ni(OH)}_2}$$

$$= \boxed{11.6 \text{ g Ni(OH)}_2}$$

EOC 32, 36

3-4 Percent Yields from Chemical Reactions

The **theoretical yield** from a chemical reaction is the yield calculated by assuming that the chemical reaction goes to completion. In practice we often do not isolate as much product from a reaction mixture as is theoretically possible. There are several reasons. (1) Many reactions do not go to completion; i.e., the reactants are not completely converted to products. (2) In some cases, a particular set of reactants undergoes two or more reactions simultaneously, forming undesired products as well as desired products. Reactions other than the desired one are called "side reactions." (3) In some cases, separation of the desired product from the reaction mixture is so difficult that not all of the product formed is successfully isolated.

> *In the examples we have worked to this point, the amounts of products that we calculated were theoretical yields.*

The term **percent yield** is used to indicate how much of a desired product is obtained from a reaction.

$$\text{percent yield} = \frac{\text{actual yield of product}}{\text{theoretical yield of product}} \times 100\%$$

Consider the preparation of nitrobenzene, $C_5H_6NO_2$, by the reaction of a limited amount of benzene, C_6H_6, with excess nitric acid, HNO_3. The balanced equation for the reaction may be written

$$\underset{\substack{1 \text{ mol} \\ 78.1 \text{ g}}}{C_6H_6} + \underset{\substack{1 \text{ mol} \\ 63.0 \text{ g}}}{HNO_3} \longrightarrow \underset{\substack{1 \text{ mol} \\ 123.1 \text{ g}}}{C_6H_5NO_2} + \underset{\substack{1 \text{ mol} \\ 18.0 \text{ g}}}{H_2O}$$

Example 3-10
A 15.6-gram sample of C_6H_6 is mixed with excess HNO_3. We isolate 18.0 grams of $C_6H_5NO_2$. What is the percent yield of $C_6H_5NO_2$ in this reaction?

Plan

First we interpret the balanced chemical equation to calculate the theoretical yield of $C_6H_5NO_2$. Then we use the actual (isolated) yield with the definition given above to calculate the percent yield.

Solution

We calculate the theoretical yield of $C_6H_5NO_2$.

It is not necessary to know the mass of one mole of HNO_3 to solve this problem.

$$? \text{ g } C_6H_5NO_2 = 15.6 \text{ g } C_6H_6 \times \frac{1 \text{ mol } C_6H_6}{78.1 \text{ g } C_6H_6} \times \frac{1 \text{ mol } C_6H_5NO_2}{1 \text{ mol } C_6H_6} \times \frac{123.1 \text{ g } C_6H_5NO_2}{1 \text{ mol } C_6H_5NO_2}$$

$$= 24.6 \text{ g } C_6H_5NO_2 \leftarrow \text{ theoretical yield}$$

This tells us that if *all* the C_6H_6 were converted to $C_6H_5NO_2$ and isolated, we should obtain 24.6 grams of $C_6H_5NO_2$ (100% yield). However, we isolate only 18.0 grams of $C_6H_5NO_2$.

$$\text{percent yield} = \frac{\text{actual yield of product}}{\text{theoretical yield of product}} \times 100\% = \frac{18.0 \text{ g}}{24.6 \text{ g}} \times 100\%$$

$$= \boxed{73.2 \text{ percent yield}}$$

EOC 44

The amount of nitrobenzene obtained *in this experiment* is 73.2% of the amount that would be expected *if* the reaction had gone to completion, *if* there were no side reactions, and *if* we could have recovered all of the product.

In many important chemical processes, especially those encountered in the chemical industry, several equations are required to describe the chemical change. An analysis of the products often lets us describe the fraction of the change that occurs by each reaction.

3-5 Sequential Reactions

Many chemical reactions occur in a series of steps. They are called **sequential reactions**. Often more than one step (reaction) is required to change starting materials into the desired product. This is true for many reactions that we carry out in the laboratory and for many industrial processes. The amount of desired product from the first reaction is taken as the starting material for the second reaction.

Example 3-11
At high temperatures carbon reacts with water to produce a mixture of carbon monoxide, CO, and hydrogen, H_2.

$$C + H_2O \xrightarrow{\text{red heat}} CO + H_2$$

Carbon monoxide is separated from H_2 and then used to separate nickel from cobalt by forming a volatile compound, nickel tetracarbonyl, $Ni(CO)_4$.

$$Ni + 4CO \longrightarrow Ni(CO)_4$$

What mass of $Ni(CO)_4$ could be obtained from the CO produced by the reaction of 75.0 grams of carbon? Assume 100% reaction and 100% recovery in both steps.

Plan
We interpret both chemical equations in the usual way, and solve the problem in two steps. They tell us that one mole of C produces one mole of CO and that four moles of CO are required to produce one mole of $Ni(CO)_4$.

1. We determine the amount of CO formed in the first reaction. It is most conveniently expressed in moles of CO.
2. We determine the number of grams of $Ni(CO)_4$ that would be formed, in the second reaction, from the number of moles of CO produced in Step 1.

Solution

1.
$$C + H_2O \longrightarrow CO + H_2$$

1 mol   1 mol   1 mol   1 mol
12.0 g

$$? \text{ mol CO} = 75.0 \text{ g} \times \frac{1 \text{ mol C}}{12.0 \text{ g C}} \times \frac{1 \text{ mol CO}}{1 \text{ mol C}} = 6.25 \text{ mol CO}$$

2.
$$Ni + 4CO \longrightarrow Ni(CO)_4$$

1 mol   4 mol   1 mol
96.0 g

$$? \text{ g } Ni(CO)_4 = 6.25 \text{ mol CO} \times \frac{1 \text{ mol } Ni(CO)_4}{4 \text{ mol CO}} \times \frac{96.0 \text{ g } Ni(CO)_4}{1 \text{ mol } Ni(CO)_4}$$

$$= \boxed{150 \text{ g } Ni(CO)_4}$$

Alternatively, we can set up a series of unit factors based on the conversions in the reaction sequence and solve the problem in one step.

$$\boxed{\text{g C}} \longrightarrow \boxed{\text{mol C}} \longrightarrow \boxed{\text{mol CO}} \longrightarrow \boxed{\text{mol } Ni(CO)_4} \longrightarrow \boxed{\text{g } Ni(CO)_4}$$

$$? \text{ g } Ni(CO)_4 = 75.0 \text{ g C} \times \frac{1 \text{ mol C}}{12.0 \text{ g C}} \times \frac{1 \text{ mol CO}}{1 \text{ mol C}} \times \frac{1 \text{ mol } Ni(CO)_4}{4 \text{ mol CO}} \times \frac{96.0 \text{ g } Ni(CO)_4}{1 \text{ mol } Ni(CO)_4}$$

$$= \boxed{150 \text{ g } Ni(CO)_4}$$

EOC 50, 52

Example 3-12

The Grignard reaction is a two-step reaction that is used to prepare pure hydrocarbons. Consider the preparation of pure ethane, C_2H_6, from ethyl chloride, CH_3CH_2Cl.

Step 1: $CH_3CH_2Cl + Mg \longrightarrow CH_3CH_2MgCl$

Step 2: $CH_3CH_2MgCl + HOH \longrightarrow CH_3CH_3 + Mg(OH)Cl$

We allow 27.2 grams of ethyl chloride to react with excess magnesium. From the first step reaction, CH_3CH_2MgCl is obtained in 79.5% yield. In the second step reaction, a 78.8% yield of CH_3CH_3 is obtained. What mass of CH_3CH_3 is obtained?

Plan

1. We interpret the first step equation as usual and calculate the amount of CH_3CH_2MgCl *obtained*.

$$CH_3CH_2Cl + Mg \longrightarrow CH_3CH_2MgCl$$

1 mol   1 mol   1 mol
64.4 g   24.3 g   88.7 g

$$\boxed{\text{g } CH_3CH_2Cl} \rightarrow \boxed{\text{mol } CH_3CH_2Cl} \rightarrow \boxed{\text{mol } CH_3CH_2MgCl} \rightarrow \boxed{\text{g } CH_3CH_2MgCl}$$

2. Then we interpret the second step equation and calculate the amount of CH_3CH_3 obtained.

$$CH_3CH_2MgCl + HOH \longrightarrow CH_3CH_3 + Mg(OH)Cl$$

1 mol	1 mol	1mol	1 mol
88.7 g		30.0 g	

$$\boxed{\text{g } CH_3CH_2MgCl} \rightarrow \boxed{\text{mol } CH_3CH_2MgCl} \rightarrow \boxed{\text{mol } CH_3CH_3} \rightarrow \boxed{\text{g } CH_3CH_3}$$

Solution

The first step reaction gives a 79.5% yield. This gives the unit factor

$$\frac{79.5 \text{ g } CH_3CH_2MgCl \quad actual}{100 \text{ g } CH_3CH_2MgCl \quad theor.}$$

1. $\underline{?}$ g CH_3CH_2MgCl = 27.2 g $CH_3CH_2Cl \times \dfrac{1 \text{ mol } CH_3CH_2Cl}{64.4 \text{ g } CH_3CH_2Cl}$

$\times \dfrac{1 \text{ mol } CH_3CH_2MgCl \text{ theor.}}{1 \text{ mol } CH_3CH_2Cl} \times \dfrac{88.7 \text{ g } CH_3CH_2MgCl \text{ theor.}}{1 \text{ mol } CH_3CH_2MgCl \text{ theor.}}$

$\times \dfrac{79.5 \text{ g } CH_3CH_2MgCl \text{ actual}}{100 \text{ g } CH_3CH_2MgCl \text{ theor.}} = \boxed{29.8 \text{ g } CH_3CH_2MgCl}$

2. $\underline{?}$ g CH_3CH_3 = 29.8 g $CH_3CH_2MgCl \times \dfrac{1 \text{ mol } CH_3CH_2MgCl}{88.7 \text{ g } CH_3CH_2Cl}$

$\times \dfrac{1 \text{ mol } CH_3CH_3 \text{ theor.}}{1 \text{ mol } CH_3CH_2MgCl} \times \dfrac{30.0 \text{ g } CH_3CH_3 \text{ theor.}}{1 \text{ mol } CH_3CH_3 \text{ theor.}}$

$\times \dfrac{78.8 \text{ g } CH_3CH_3 \text{ actual}}{100 \text{ g } CH_3CH_3 \text{ theor.}} = \boxed{7.94 \text{ g } CH_3CH_3}$

Alternatively, we could set the calculation up in a single step.

2. $\underline{?}$ g CH_3CH_3 = 27.2 g $CH_3CH_2Cl \times \dfrac{1 \text{ mol } CH_3CH_2Cl}{64.4 \text{ g } CH_3CH_2Cl} \times \dfrac{1 \text{ mol } CH_3CH_2MgCl}{1 \text{ mol } CH_3CH_2Cl}$

$\times \dfrac{79.5 \text{ g } CH_3CH_2MgCl \text{ actual}}{100 \text{ g } CH_3CH_2MgCl \text{ theor.}} \times \dfrac{1 \text{ mol } CH_3CH_3}{1 \text{ mol } CH_3CH_2MgCl}$

$\times \dfrac{30.0 \text{ g } CH_3CH_3 \text{ theor.}}{1 \text{ mol } CH_3CH_3 \text{ theor.}} \times \dfrac{78.8 \text{ g } CH_3CH_3 \text{ actual}}{100 \text{ g } CH_3CH_3 \text{ theor.}}$

$= \boxed{7.94 \text{ g } CH_3CH_3}$

EOC 54, 55

The sodium hydroxide and aluminum in some drain cleaners do not react while they are stored in solid form. When water is added, the NaOH dissolves and begins to act on trapped grease. At the same time, NaOH and Al react to produce H_2 gas; the resulting turbulence helps to dislodge the blockage. Do you see why the container should be kept tightly closed?

Chemists have determined the structures of many naturally occurring compounds. One way of proving the structure involves the synthesis of the natural product (compound) from available starting materials. Professor Grieco, now at Indiana University, was assisted by Majetich and Ohfune in the synthesis of helenalin, a powerful cancer drug, in a forty-step process. This forty-step synthesis gave a remarkable average yield of about 90% for each step, which resulted in an overall yield of about 1.5%.

3-6 Concentrations of Solutions

Many chemical reactions are more conveniently carried out with the reactants in solution rather than as pure solids, liquids, or gases. A **solution** is a homogeneous mixture, at the molecular level, of two or more substances. Simple solutions usually consist of one substance, the **solute**, dissolved in another substance, the **solvent**. The solutions used in the laboratory are usually liquids, and the solvent is often water. These are called **aqueous solutions**. For example, solutions of hydrochloric acid are prepared by dissolving hydrogen chloride (HCl, a gas at room temperature and atmospheric pressure) in water. Solutions of sodium hydroxide are prepared by dissolving solid NaOH in water.

In some solutions, such as a nearly equal mixture of ethyl alcohol and water, the distinction between *solute* and *solvent* is arbitrary.

We often use solutions to supply the reactants for chemical reactions. Solutions allow the most intimate mixing of the reacting substances at the molecular level, much more than would be possible in solid form. (A practical example is drain cleaner, shown in the photo.) Furthermore, the rate of the reaction can often be controlled by adjusting the concentrations of the solutions. In this section we shall study methods for expressing the quantities of the various components present in a given amount of solution.

Concentrations of solutions are expressed in terms of *either* the amount of solute present in a given mass or volume of *solution*, or the amount of solute dissolved in a given mass or volume of *solvent*.

Percent by Mass

Concentrations of solutions may be expressed in terms of percent by mass of solute, which gives the mass of solute per 100 mass units of solution. The gram is the usual mass unit.

$$\% \text{ solute} = \frac{\text{mass of solute}}{\text{mass of solution}} \times 100\%$$

Thus, a solution that is 10.0% calcium gluconate, $Ca(C_6H_{11}O_7)_2$, by mass contains 10.0 grams of calcium gluconate in 100.0 grams of *solution*. This could be described as 10.0 grams of calcium gluconate in 90.0 grams of water. The density of a 10.0% solution of calcium gluconate is 1.07 g/mL, so 100 mL of a 10.0% solution of calcium gluconate has a mass of 107 grams. Observe that 100 grams of a solution usually does *not* occupy 100 mL. Unless otherwise specified, percent means percent *by mass*, and water is the solvent.

A 10.0% solution of $Ca(C_6H_{11}O_7)_2$ is sometimes administered intravenously in emergency treatment for black widow spider bites.

Example 3-13

Calculate the mass of nickel(II) sulfate, $NiSO_4$, contained in 200 grams of a 6.00% solution of $NiSO_4$.

Plan

The percentage information tells us that the solution contains 6.00 grams of $NiSO_4$ per 100 grams of solution. The desired information is the mass of $NiSO_4$ in 200 grams of solution. A unit factor is constructed by placing 6.00 grams of $NiSO_4$ over 100 grams of solution. Multiplication of the mass of the solution, 200 grams, by the unit factor gives the mass of $NiSO_4$ in the solution.

Solution

$$\underline{?} \text{ g NiSO}_4 = 200 \text{ g soln} \times \frac{6.00 \text{ g NiSO}_4}{100 \text{ g soln}} = \boxed{12.0 \text{ g NiSO}_4}$$

Example 3-14

Calculate the mass of 6.00% $NiSO_4$ solution that contains 40.0 grams of $NiSO_4$.

Plan

Placing 100 grams of solution over 6.00 grams of $NiSO_4$ gives another unit factor.

Solution

$$\underline{?}\text{ g soln} = 40.0 \text{ g NiSO}_4 \times \frac{100 \text{ g soln}}{6.00 \text{ g NiSO}_4} = \boxed{667 \text{ g soln}}$$

EOC 57, 58

Example 3-15

Calculate the mass of $NiSO_4$ contained in 200 mL of a 6.00% solution of $NiSO_4$. The density of the solution is 1.06 g/mL at 25°C.

Plan

The volume of a solution multiplied by its density gives the mass of solution (Section 1-11). The mass of solution is then multiplied by the fraction of that mass due to $NiSO_4$ (6.00 g $NiSO_4$/100 g soln) to give the mass of $NiSO_4$ in 200 mL of solution.

Solution

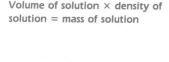
Volume of solution × density of solution = mass of solution

$$\underline{?}\text{ g NiSO}_4 = \underbrace{200 \text{ mL soln} \times \frac{1.06 \text{ g soln}}{1.00 \text{ mL soln}}}_{212 \text{ g soln}} \times \frac{6.00 \text{ g NiSO}_4}{100 \text{ g soln}} = \boxed{12.7 \text{ g NiSO}_4}$$

EOC 59, 60

Example 3-16

What volume of a solution that is 15.0% iron(III) nitrate contains 30.0 grams of $Fe(NO_3)_3$? The density of the solution is 1.16 g/mL at 25°C.

Plan

Two unit factors relate mass of $Fe(NO_3)_3$ and mass of solution, 15.0 g $Fe(NO_3)_3$/100 g soln and 100 g soln/15.0 g $Fe(NO_3)_3$. The second factor converts grams of $Fe(NO_3)_3$ to grams of solution.

Solution

$$\underline{?}\text{ mL soln} = \underbrace{30.0 \text{ g Fe(NO}_3)_3 \times \frac{100 \text{ g soln}}{15.0 \text{ g Fe(NO}_3)_3}}_{200 \text{ g soln}} \times \frac{1.00 \text{ mL soln}}{1.16 \text{ g soln}} = \boxed{172 \text{ mL}}$$

Note that the answer is not 200 mL but considerably less because 1.00 mL of solution has a mass of 1.16 grams. However, 172 mL of the solution has a mass of 200 grams.

EOC 62, 63

Molarity (molar concentration)

Molarity (*M*), or molar concentration, is a common unit for expressing the concentrations of solutions. **Molarity** is defined as the number of moles of solute per liter of solution:

$$\text{molarity} = \frac{\text{number of moles of solute}}{\text{number of liters of solution}}$$

To prepare one liter of a one molar solution, one mole of solute is placed in a one-liter volumetric flask, enough solvent is added to dissolve the solute, and solvent is then added until the volume of the solution is exactly one liter. Students sometimes make the mistake of assuming that a one molar solution contains one mole of solute in a liter of solvent. This is *not* the case; one liter of solvent *plus* one mole of solute usually has a total volume of more than one liter. A $0.100\,M$ solution contains 0.100 mole of solute per liter, and a $0.0100\,M$ solution contains 0.0100 mole of solute per liter (Figure 3-2).

We often express the volume of a solution in milliliters rather than in liters. Likewise, we may express the amount of solute in millimoles (mmol) rather than in moles. Because one milliliter is 1/1000 of a liter and one millimole is 1/1000 of a mole, molarity also may be expressed as the number of millimoles of solute per milliliter of solution:

$$\text{molarity} = \frac{\text{number of millimoles of solute}}{\text{number of milliliters of solution}}$$

Water is the solvent in *most* of the solutions that we encounter. Unless otherwise indicated, we assume that water is the solvent. When the solvent is other than water, we state this explicitly.

The definition of molarity specifies the amount of solute *per unit volume of solution,* whereas percent specifies the amount of solute *per unit mass of solution.* Therefore, molarity depends on temperature and pressure, whereas percent by mass does not.

Figure 3-2

Preparation of $0.0100\,M$ solution of $KMnO_4$, potassium permanganate. 250 mL of $0.0100\,M\ KMnO_4$ solution contains 0.395 g of $KMnO_4$ (1 mol = 158 g). (a) 0.395 g of $KMnO_4$ (0.00250 mole) is weighed out carefully and transferred into a 250-mL volumetric flask. (b) The $KMnO_4$ is dissolved in water. (c) Distilled H_2O is added to the volumetric flask until the volume of solution is 250 mL. The flask is then stoppered, and its contents are mixed thoroughly to give a homogeneous solution.

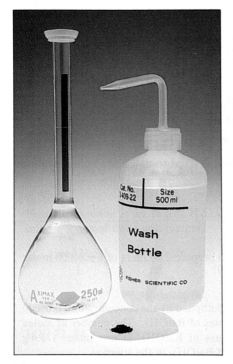

(a)

(b)

(c)

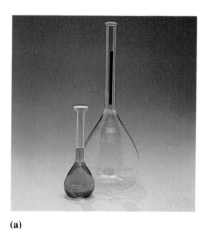

(a)

(b)

(c)

Figure 3-3
Dilution of solution. (a) A 100-mL volumetric flask is filled to the calibration line with 0.100 M potassium dichromate, $K_2Cr_2O_7$, solution. (b) The 0.100 M $K_2Cr_2O_7$ solution is transferred into a 1.00-L volumetric flask. The small flask is rinsed with distilled H_2O several times, and the rinse solutions are added to the larger flask. (c) Distilled water is added until the 1.00-L flask contains 1.00 L of solution. The flask is stoppered and its contents are mixed thoroughly. The new solution is 0.0100 M $K_2Cr_2O_7$. (100 mL of 0.100 M $K_2Cr_2O_7$ solution has been diluted to 1000 mL.)

3-7 Dilution of Solutions

Recall that the definition of molarity is the number of moles of solute divided by the volume of the solution in liters:

$$\text{molarity} = \frac{\text{number of moles of solute}}{\text{number of liters of solution}}$$

Multiplying both sides of the equation by the volume, we obtain

$$\text{volume (in L)} \quad \times \text{molarity} = \text{number of moles of solute}$$

or $\text{volume (in mL)} \times \text{molarity} = \text{number of mmol of solute}$

> Multiplication of the volume of a solution by its molar concentration gives the amount of solute in the solution.

When we dilute a solution by mixing it with more solvent, the number of moles of solute present does not change. But the volume and the concentration of the solution *do* change. Because the same number of moles of solute is divided by a larger number of liters of solution, the molarity decreases. Using a subscript 1 to represent the original concentrated solution and a subscript 2 to represent the dilute solution, we obtain

$$\text{volume}_1 \times \text{molarity}_1 = \text{number of moles of solute} = \text{volume}_2 \times \text{molarity}_2$$

or

This relationship also applies when the concentration is changed by evaporating some solvent.

$$V_1 \times M_1 = V_2 \times M_2 \quad \text{(for dilution only)}$$

This expression can be used to calculate any one of four quantities when the other three are known (Figure 3-3). Suppose a certain volume of dilute solution of a given molarity is required for use in the laboratory, and we know the concentration of the stock solution available. Then we can calculate the amount of stock solution that must be used to make the dilute solution.

Caution

Dilution of a concentrated solution, especially of a strong acid or base, frequently liberates a great deal of heat. This can vaporize drops of water as they hit the concentrated solution and can cause dangerous spattering. As a safety precaution, *concentrated solutions of acids or bases are always poured slowly into water*, allowing the heat to be absorbed by the larger quantity of water. Calculations are usually simpler to visualize by assuming that water is added to the concentrated solution.

Example 3-20

Calculate the volume of $18.0\,M$ H_2SO_4 required to prepare 1.00 liter of a $0.900\,M$ solution of H_2SO_4.

Plan

The volume (1.00 L) and molarity ($0.900\,M$) of the final solution, as well as the molarity ($18.0\,M$) of the original solution, are given. Therefore, the relation $V_1 \times M_1 = V_2 \times M_2$ can be used, with subscript 1 for the commercial acid solution and subscript 2 for the dilute solution. We solve

$$V_1 \times M_1 = V_2 \times M_2 \qquad \text{for } V_1$$

Solution

$$V_1 = \frac{V_2 \times M_2}{M_1} = \frac{1.00\ \text{L} \times 0.900\,M}{18.0\,M} = 0.0500\ \text{L} = \boxed{50.0\ \text{mL}}$$

The dilute solution contains $1.00\ \text{L} \times 0.900\,M = 0.900$ mol of H_2SO_4, so 0.900 mole of H_2SO_4 must be present in the original concentrated solution. Indeed, $0.0500\ \text{L} \times 18.0\,M = 0.900$ mol of H_2SO_4.

EOC 79, 80

3-8 Using Solutions in Chemical Reactions

If we plan to carry out a reaction in a solution, we must calculate the amounts of solutions required. If we know the molarity of a solution, we can calculate the amount of solute contained in a specified volume of that solution. This is illustrated in Example 3-21.

Example 3-21

Calculate (a) the number of moles of H_2SO_4, (b) the number of millimoles of H_2SO_4, and (c) the mass of H_2SO_4 in 500 mL of $0.324\,M$ H_2SO_4 solution.

Plan

Because we have two parallel calculations in this example, we shall state the plan for each step just before the calculation is done.

Solution

(a) The volume of a solution in liters multiplied by its molarity gives the number of moles of solute, H_2SO_4 in this case.

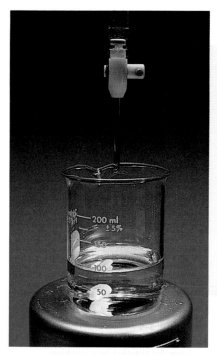 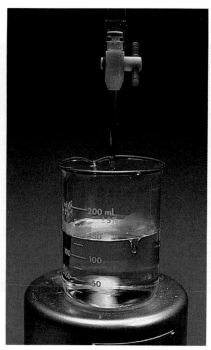 

The indicator phenolphthalein changes from colorless, its color in acidic solutions, to pink, its color in basic solutions, when the reaction in Example 3-23 reaches completion. Note the first appearance of a faint pink coloration in the middle beaker; this signals that the end point is near.

Plan

The balanced equation tells us that the reaction ratio is one mole of HCl to one mole of NaOH, which gives the unit factor, 1 mol HCl/1 mol NaOH.

$$HCl + NaOH \longrightarrow NaCl + H_2O$$
$$1 \text{ mol} \quad 1 \text{ mol} \quad\quad 1 \text{ mol} \quad 1 \text{ mol}$$

First we find the number of moles of NaOH. The reaction ratio is one mole of HCl to one mole of NaOH, so the HCl solution must contain the same number of moles of HCl. Then we can calculate the molarity of the HCl solution because we know its volume.

Solution

The volume of a solution (in liters) multiplied by its molarity gives the number of moles of solute.

$$\underline{?} \text{ mol NaOH} = 0.0432 \text{ L NaOH soln} \times \frac{0.236 \text{ mol NaOH}}{1 \text{ L NaOH soln}} = 0.0102 \text{ mol NaOH}$$

Because the reaction ratio is one mole of NaOH to one mole of HCl, the HCl solution must contain 0.0102 mole of HCl.

$$\underline{?} \text{ mol HCl} = 0.0102 \text{ mol NaOH} \times \frac{1 \text{ mol HCl}}{1 \text{ mol NaOH}} = 0.0102 \text{ mol HCl}$$

We know the volume of the HCl solution, so we can calculate its molarity.

$$\frac{\underline{?} \text{ mol HCl}}{\text{L HCl soln}} = \frac{0.0102 \text{ mol HCl}}{0.0367 \text{ L HCl soln}} = \boxed{0.278 \ M \text{ HCl}}$$

EOC 95

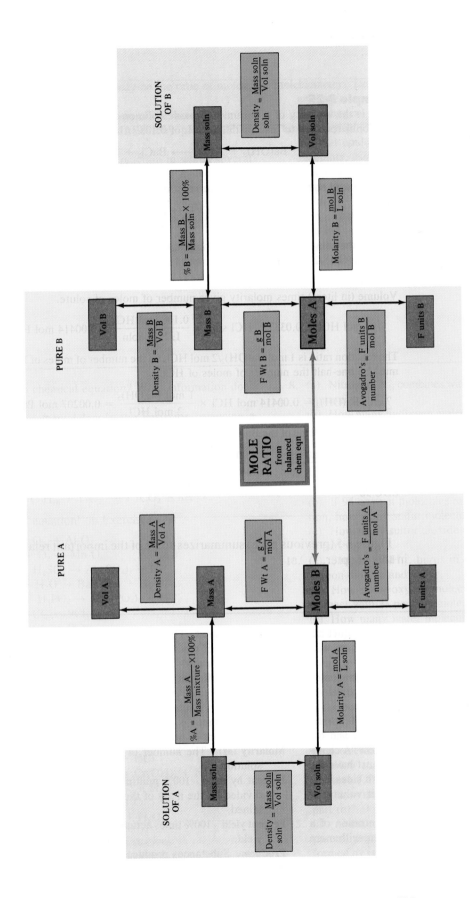

Figure 3-5
Some important relationships in reaction stoichiometry, Chapter 3. The left-hand portion of this diagram relates various ways of describing a single substance, A. It is similar to Figure 2-12, with the addition of the molarity calculation. Similarly, the right-hand portion applies to substance B. The mole concept (red boxes and arrows) relates the amounts of A and B involved in a chemical reaction.

(b) How many moles of H_2SO_4, are required to react with 3.0 mol of Al_2O_3?

(c) How many moles of $Al_2(SO_4)_3$ are formed in part (b)?

14. (a) Nitromethane, CH_3NO_2, often called "nitro," is used as a fuel additive in some racing vehicles. When CH_3NO_2 burns in excess oxygen, it forms carbon dioxide, nitrogen dioxide, and water.

(b) How many moles of oxygen are required to burn 5.0 mol of CH_3NO_2?

(c) How many moles of water are formed in part (b)?

15. (a) Butane, C_4H_{10}, burns in excess oxygen to form carbon dioxide and water.

(b) If 3.9 mol of oxygen are used up in this reaction, how many moles of butane were burned?

(c) How many moles of water were formed in part (b)?

16. The equation that describes the commercial "roasting" of zinc sulfide is

$$2ZnS + 3O_2 \xrightarrow{\Delta} 2ZnO + 2SO_2$$

What is the mole ratio of (a) O_2 to ZnS, (b) ZnO to ZnS, and (c) SO_2 to ZnS?

17. The reaction between dilute nitric acid and copper is given by the equation

$$3Cu + 8HNO_3 \longrightarrow 3Cu(NO_3)_2 + 2NO + 4H_2O$$

What is the mole ratio of (a) HNO_3 to Cu, (b) NO to Cu, and (c) $Cu(NO_3)_2$ to Cu?

18. How many moles of oxygen can be obtained by the decomposition of 1.00 mol of reactant in each of the following reactions?

(a) $2KClO_3 \rightarrow 2KCl + 3O_2$

(b) $2H_2O_2 \rightarrow 2H_2O + O_2$

(c) $2HgO \rightarrow 2Hg + O_2$

(d) $2NaNO_3 \rightarrow 2NaNO_2 + O_2$

(e) $KClO_4 \rightarrow KCl + 2O_2$

19. For the formation of 1.00 mol of water, which reaction uses the most nitric acid?

(a) $3Cu + 8HNO_3 \rightarrow 3Cu(NO_3)_2 + 2NO + 4H_2O$

(b) $Al_2O_3 + 6HNO_3 \rightarrow 2Al(NO_3)_3 + 3H_2O$

(c) $4Zn + 10HNO_3 \rightarrow 4Zn(NO_3)_2 + NH_4NO_3 + 3H_2O$

20. Consider the reaction

$$NH_3 + O_2 \xrightarrow{\text{not balanced}} NO + H_2O$$

For every 1.50 mol of NH_3, (a) how many moles of O_2 are required, (b) how many moles of NO are produced, and (c) how many moles of H_2O are produced?

21. Consider the reaction

$$2NO + Br_2 \longrightarrow 2NOBr$$

For every 3.00 mol of bromine that reacts, how many moles of (a) NO react and (b) NOBr are produced?

22. Find the mass of chlorine that will combine with 3.18 g of hydrogen to form hydrogen chloride:

$$H_2 + Cl_2 \longrightarrow 2HCl$$

23. What mass of solid AgCl will precipitate from a solution containing 1.50 g of $CaCl_2$ if an excess amount of $AgNO_3$ is added?

$$CaCl_2 + 2AgNO_3 \longrightarrow 2AgCl + Ca(NO_3)_2$$

24. A sample of magnetic iron oxide, Fe_3O_4, reacted completely with hydrogen at red heat. The water vapor formed by the reaction

$$Fe_3O_4 + 4H_2 \xrightarrow{\Delta} 3Fe + 4H_2O$$

was condensed and found to weigh 7.5 g. Calculate the mass of Fe_3O_4 that reacted.

25. What masses of cobalt(II) chloride and of hydrogen fluoride are needed to prepare 10.0 moles of cobalt(II) fluoride by the following reaction?

$$CoCl_2 + 2HF \longrightarrow CoF_2 + 2HCl$$

26. Gaseous chlorine and gaseous fluorine undergo a combination reaction to form the interhalogen compound ClF. Write the chemical equation for this reaction and calculate the mass of fluorine needed to react with 3.27 g of Cl_2.

27. Dinitrogen pentoxide, N_2O_5, undergoes a decomposition reaction to form nitrogen dioxide, NO_2, and oxygen. Write the chemical equation for this reaction. A 0.165-g sample of O_2 was produced by the reaction. What mass of NO_2 was produced?

28. Gaseous chlorine will displace bromide ion from an aqueous solution of potassium bromide to form aqueous potassium chloride and aqueous bromine. Write the chemical equation for this reaction. What mass of bromine will be produced if 0.289 g of chlorine undergoes reaction?

29. Solid zinc sulfide reacts with hydrochloric acid to form a mixture of aqueous zinc chloride and hydrogen sulfide, H_2S. Write the chemical equation for this reaction. What mass of zinc sulfide is needed to react with 10.65 g of HCl?

*30. An impure sample of $CuSO_4$ weighing 5.52 g was dissolved in water and allowed to react with excess zinc.

$$CuSO_4 + Zn \longrightarrow ZnSO_4 + Cu$$

What was the percent $CuSO_4$ in the sample if 1.49 g of Cu was produced?

31. You are designing an experiment for the preparation of hydrogen. For the production of equal amounts of hydrogen, which metal, Zn or Al, is less expensive if Zn costs about half as much as Al on a mass basis?

$$Zn + 2HCl \longrightarrow ZnCl_2 + H_2$$
$$2Al + 6HCl \longrightarrow 2AlCl_3 + 3H_2$$

Limiting Reactant

32. How many grams of NH_3 can be prepared from 77.3 grams of N_2 and 14.2 grams of H_2?

$$N_2 + 3H_2 \longrightarrow 2NH_3$$

33. Silver nitrate solution reacts with barium chloride solution according to the equation

$$2AgNO_3 + BaCl_2 \longrightarrow Ba(NO_3)_2 + 2AgCl$$

All of the substances involved in this reaction are soluble in water except silver chloride, $AgCl$, which forms a solid (precipitate) at the bottom of the flask. Suppose we mix together a solution containing 12.6 g of $AgNO_3$ and 8.4 g of $BaCl_2$. What mass of $AgCl$ would be formed?

*34. "Superphosphate," a water-soluble fertilizer, is a mixture of $Ca(H_2PO_4)_2$ and $CaSO_4$ on a 1:2 *mole* basis. It is formed by the reaction

$$Ca_3(PO_4)_2 + 2H_2SO_4 \longrightarrow Ca(H_2PO_4)_2 + 2CaSO_4$$

We treat 250 g of $Ca_3(PO_4)_2$ with 150 g of H_2SO_4. How many grams of superphosphate could be formed?

35. Silicon carbide, an abrasive, is made by the reaction of silicon dioxide with graphite:

$$SiO_2 + C \xrightarrow{\Delta} SiC + CO \quad \text{(balanced?)}$$

We mix 377 g of SiO_2 and 44.6 g of C. If the reaction proceeds as far as possible, which reactant will be left over? How much of this reactant will remain?

36. What mass of potassium can be produced by the reaction of 100.0 g of Na with 100.0 g of KCl?

$$Na + KCl \xrightarrow{\Delta} NaCl + K$$

37. A reaction mixture contains 25.0 g of PCl_3 and 45.0 g of PbF_2. What mass of $PbCl_2$ can be obtained from the following reaction?

$$3PbF_2 + 2PCl_3 \longrightarrow 2PF_3 + 3PbCl_2$$

How much of which reactant will be left unchanged?

38. What mass of $BaSO_4$ will be produced by the reaction of 33.2 g of Na_2SO_4 with 43.5 g of $Ba(NO_3)_2$?

$$Ba(NO_3)_2 + Na_2SO_4 \longrightarrow BaSO_4 + 2NaNO_3$$

39. Consider the reaction

$$3HCl + 3HNF_2 \longrightarrow 2ClNF_2 + NH_4Cl + 2HF$$

A mixture of 8.00 g of HCl and 10.00 g of HNF_2 is allowed to react. If only 15% of the limiting reactant does react, what is the composition of the final mixture?

Percent Yield from Chemical Reactions

40. What mass of chromium is present in 150 grams of an ore of chromium that is 67.0% chromite, $FeCr_2O_4$, and 33.0% impurities by mass? If 87.5% of the chromium can be recovered from 125 grams of the ore, what mass of pure chromium is obtained?

41. A particular ore of lead, galena, is 10% lead sulfide, PbS, and 90% impurities by weight. What mass of lead is contained in 75 grams of this ore?

42. The percent yield for the reaction

$$PCl_3 + Cl_2 \longrightarrow PCl_5$$

is 85.0%. What mass of PCl_5 would be expected from the reaction of 38.5 g of PCl_3 with excess chlorine?

43. The percent yield for the following reaction carried out in carbon tetrachloride solution

$$Br_2 + Cl_2 \longrightarrow 2BrCl$$

is 57.0%. (a) What amount of BrCl would be formed from the reaction of 0.0100 mol Br_2 with 0.0100 mol Cl_2? (b) What amount of Br_2 is left unchanged?

44. Solid silver nitrate undergoes thermal decomposition to form silver metal, nitrogen dioxide, and oxygen. Write the chemical equation for this reaction. A 0.362-g sample of silver metal was obtained from the decomposition of a 0.575-g sample of $AgNO_3$. What is the percent yield of the reaction?

45. Gaseous nitrogen and hydrogen undergo a reaction to form gaseous ammonia (the Haber process). Write the chemical equation for this reaction. At a temperature of 400°C and a total pressure of 250 atm, 0.720 g of NH_3 was produced by the reaction of 2.80 g of N_2 with excess H_2. What is the percent yield of the reaction?

46. Ethylene oxide, C_2H_4O, a fumigant sometimes used by exterminators, is synthesized in 89% yield by reaction of ethylene bromohydrin, C_2H_5OBr, with sodium hydroxide:

$$C_2H_5OBr + NaOH \longrightarrow C_2H_4O + NaBr + H_2O$$

How many grams of ethylene bromohydrin would be consumed in the production of 255 g of ethylene oxide, at 89% yield?

*47. How much 68% Na_2SO_4 could be produced from 375 g of 88% pure NaCl?

$$2NaCl + H_2SO_4 \longrightarrow Na_2SO_4 + 2HCl$$

*48. Calcium carbide is made in an electric furnace by the reaction

$$CaO + 3C \longrightarrow CaC_2 + CO$$

The crude product is usually 85% CaC_2 and 15% unreacted CaO. (a) How much CaO should we start with to produce 250 kg of crude product? (b) How much CaC_2 would this crude product contain?

49. Ethylene glycol, $C_2H_6O_2$, is used as antifreeze in automobile radiators. A method of producing small amounts of ethylene glycol in the laboratory is by reaction of 1,2-dichloroethane with sodium carbonate in a water solution, followed by distillation of the reaction mixture to purify the ethylene glycol.

$$C_2H_4Cl_2 + Na_2CO_3 + H_2O \longrightarrow$$
$$C_2H_6O_2 + 2NaCl + CO_2$$

When 27.4 g of 1,2-dichloroethane is used in this reaction, 10.3 g of ethylene glycol is obtained. (a) Calculate the theoretical yield of ethylene glycol. (b) What is the percent yield of ethylene glycol in this process? (c) What mass of Na_2CO_3 is consumed?

Sequential Reactions

50. Consider the two-step process for the formation of tellurous acid described by the following equations:

$$TeO_2 + 2OH^- \longrightarrow TeO_3^{2-} + H_2O$$
$$TeO_3^{2-} + 2H^+ \longrightarrow H_2TeO_3$$

What mass of H_2TeO_3 would be formed from 62.1 g of TeO_2, assuming 100% yield?

51. Consider the formation of cyanogen, C_2N_2, and its subsequent decomposition in water given by the equations

$$2Cu^{2+} + 6CN^- \longrightarrow 2[Cu(CN)_2]^- + C_2N_2$$
$$C_2N_2 + H_2O \longrightarrow HCN + HOCN$$

How much hydrocyanic acid, HCN, can be produced from 10.00 g of KCN, assuming 100% yield?

52. What mass of potassium chlorate would be required to supply the proper amount of oxygen needed to burn 35.0 g of methane, CH_4?

$$2KClO_3 \longrightarrow 2KCl + 3O_2$$
$$CH_4 + 2O_2 \longrightarrow CO_2 + 2H_2O$$

53. Hydrogen, obtained by the electrical decomposition of water, was combined with chlorine to produce 51.0 g of hydrogen chloride. Calculate the mass of water decomposed.

$$2H_2O \longrightarrow 2H_2 + O_2$$
$$H_2 + Cl_2 \longrightarrow 2HCl$$

*54. About half of the world's production of pigments for paints involves the formation of white TiO_2. In the United States, it is made on a large scale by the *chloride process*, starting with ores containing only small amounts of rutile, TiO_2. The ore is treated with chlo-

rine and carbon (coke). This produces $TiCl_4$ and gaseous products:

$$2TiO_2 + 3C + 4Cl_2 \longrightarrow 2TiCl_4 + CO_2 + 2CO$$

The $TiCl_4$ is then converted into TiO_2 of high purity:

$$TiCl_4 + O_2 \longrightarrow TiO_2 + 2Cl_2$$

Suppose the first process can be carried out with 65.0% yield and the second with 92.0% yield. How many kg of TiO_2 could be produced starting with 1.00 metric ton (1.00×10^6 g) of an ore that is 0.25% rutile?

55. When sulfuric acid dissolves in water, the following reactions take place:

$$H_2SO_4 \longrightarrow H^+ + HSO_4^-$$
$$HSO_4^- \longrightarrow H^+ + SO_4^{2-}$$

The first reaction is 100.0% complete and the second reaction is 10.0% complete. Calculate the concentrations of the various ions in a 0.100 M aqueous solution of H_2SO_4.

*56. The chief ore of zinc is the sulfide, ZnS. The ore is concentrated by flotation and then heated in air, which converts the ZnS to ZnO.

$$2ZnS + 3O_2 \longrightarrow 2ZnO + 2SO_2$$

The ZnO is then treated with dilute H_2SO_4

$$ZnO + H_2SO_4 \longrightarrow ZnSO_4 + H_2O$$

to produce an aqueous solution containing the zinc as $ZnSO_4$. An electrical current is passed through the solution to produce the metal.

$$2ZnSO_4 + 2H_2O \longrightarrow 2Zn + 2H_2SO_4 + O_2$$

What mass of Zn will be obtained from an ore containing 100. kg of ZnS? Assume the flotation process to be 91% efficient, the electrolysis step to be 98% efficient, and the other steps to be 100% efficient.

Concentrations of Solutions—Percent by Mass

57. What mass of an 8.65% solution of potassium dichromate contains 60.0 g of $K_2Cr_2O_7$? What mass of water does this amount of solution contain?

58. Calculate the mass of an 8.30% solution of ammonium chloride, NH_4Cl, that contains 100 g of water. What mass of NH_4Cl does this amount of solution contain?

59. The density of an 18.0% solution of ammonium chloride, NH_4Cl, solution is 1.05 g/mL. What mass of NH_4Cl does 350 mL of this solution contain?

60. The density of an 18.0% solution of ammonium sulfate, $(NH_4)_2SO_4$, is 1.10 g/mL. What mass of $(NH_4)_2SO_4$ would be required to prepare 350 mL of this solution?

61. What volume of the solution of NH_4Cl described in Exercise 59 contains 80.0 g of NH_4Cl?

62. What volume of the solution of $(NH_4)_2SO_4$ described in Exercise 60 contains 80.0 g of $(NH_4)_2SO_4$?

*63. A reaction requires 37.8 g of NH_4Cl. What volume of the solution described in Exercise 59 would you use if you wished to use a 20.0% excess of NH_4Cl?

Concentrations of Solutions—Molarity

64. What is the molarity of a solution that contains 490 g of phosphoric acid, H_3PO_4, in 2.00 L of solution?

65. What is the molarity of a solution that contains 1.37 g of sodium chloride in 25.0 mL of solution?

66. What is the molarity of a solution containing 0.155 mol H_3PO_4 in 200 mL of solution?

67. A solution contains 0.100 mole per liter of each of the following acids: HCl, H_2SO_4, H_3PO_4.
    (a) Is the molarity the same for each acid?
    (b) Is the number of molecules per liter the same for each acid?
    (c) Is the mass per liter the same for each acid?

68. A solution contains 1.05 g of a rubbing alcohol, $(CH_3)_2CHOH$, in 100 mL of solution. Find (a) the molarity of the solution and (b) the number of moles in 1.00 mL of solution.

69. (a) Calculate the molarity of caffeine in a 12-oz cola drink containing 50 mg caffeine, $C_8H_{10}N_4O_2$.
    (b) Cola drinks are usually $5.06 \times 10^{-3} M$ with respect to H_3PO_4. How much of this acid is in a 250-mL drink? (1 oz = 29.6 mL)

70. How many grams of the cleansing agent Na_3PO_4 (a) are needed to prepare 200 mL of 0.25 M solution, and (b) are in 200 mL of 0.25 M solution?

71. How many kg of ethylene glycol, $C_2H_6O_2$, are needed to prepare a 9.00 M solution to protect a 15.0-L car radiator against freezing? What is the mass of $C_2H_6O_2$ in 15.0 L of 9.00 M solution?

72. A solution made by dissolving 18.0 g of $CaCl_2$ in 72.0 g of water has a density of 1.180 g/mL at 20°C.
    (a) What is the percent by mass of $CaCl_2$ in the solution?
    (b) What is the molarity of $CaCl_2$ in the solution?

73. Stock phosphoric acid solution is 85.0% H_3PO_4 and has a specific gravity of 1.70. What is the molarity of the solution?

74. Stock hydrofluoric acid solution is 49.0% HF and has a specific gravity of 1.17. What is the molarity of the solution?

75. What mass of sodium sulfate, Na_2SO_4, is contained in 750 mL of a 2.00 molar solution?

76. What is the molarity of a barium chloride solution prepared by dissolving 3.50 g of $BaCl_2 \cdot 2H_2O$ in enough water to make 500 mL of solution?

77. What mass of potassium benzoate trihydrate, $KC_7H_5O_2 \cdot 3H_2O$, is needed to prepare one liter of a 0.125 molar solution of potassium benzoate?

78. What volume of 0.850 M $CuSO_4$ solution can be prepared from 75.0 g of copper(II) sulfate pentahydrate, $CuSO_4 \cdot 5H_2O$?

Dilution of Solutions

79. Commercially available concentrated sulfuric acid is 18.0 M H_2SO_4. Calculate the volume of concentrated sulfuric acid required to prepare 2.50 L of 0.150 M H_2SO_4 solution.

80. Commercial concentrated hydrochloric acid is 12.0 M HCl. What volume of concentrated hydrochloric acid is required to prepare 3.50 L of 2.40 M HCl solution?

81. Calculate the volume of 2.00 M NaOH solution required to prepare 100 mL of a 0.500 M solution of NaOH.

82. Calculate the volume of 0.0500 M $Ba(OH)_2$ solution that contains the same number of moles of $Ba(OH)_2$ as 120 mL of 0.0800 M $Ba(OH)_2$ solution.

*83. Calculate the resulting molarity when 50.0 mL of 2.30 M NaCl solution is mixed with 80.0 mL of 1.40 M NaCl.

*84. Calculate the resulting molarity when 125 mL of 6.00 M H_2SO_4 solution is mixed with 225 mL of 3.00 M H_2SO_4.

Using Solutions in Chemical Reactions

85. What volume of 0.50 M HBr is required to react completely with 0.75 mol of $Ca(OH)_2$?

$$2HBr + Ca(OH)_2 \longrightarrow CaBr_2 + 2H_2O$$

86. What volume of 0.324 M HNO_3 solution is required to react completely with 22.0 mL of 0.0612 M $Ba(OH)_2$?

$$Ba(OH)_2 + 2HNO_3 \longrightarrow Ba(NO_3)_2 + 2H_2O$$

87. What is the concentration, in mol/L, of an HCl solution if 23.65 mL reacts completely with (neutralizes) 25.00 mL of a 0.1037 M solution of NaOH?

$$HCl + NaOH \longrightarrow NaCl + H_2O$$

88. An excess of $AgNO_3$ reacts with 100.0 mL of an $AlCl_3$ solution to give 0.275 g of AgCl. What is the concentration, in mol/L, of the $AlCl_3$ solution?

$$AlCl_3 + 3AgNO_3 \longrightarrow 3AgCl + Al(NO_3)_3$$

89. An impure sample of solid Na_2CO_3 was allowed to react with 0.1026 M HCl.

$$Na_2CO_3 + 2HCl \longrightarrow 2NaCl + CO_2 + H_2O$$

A 0.1247-g sample of sodium carbonate required 14.78 mL of HCl. What is the purity of the sodium carbonate?

90. Calculate the theoretical yield of AgCl formed from the reaction of an aqueous solution containing excess $ZnCl_2$ with 35.0 mL of 0.325 M $AgNO_3$.

$$ZnCl_2 + 2AgNO_3 \longrightarrow Zn(NO_3)_2 + 2AgCl$$

Titrations

91. Define and illustrate the following terms clearly and concisely: (a) standard solution; (b) titration.

92. Distinguish between the *equivalence point* and *end point* of a titration.

93. What volume of 0.275 molar hydrochloric acid solution reacts with 36.4 mL of 0.150 molar sodium hydroxide solution? (See Example 3-24.)

94. What volume of 0.112 molar sodium hydroxide solution would be required to react with 25.3 mL of 0.400 molar sulfuric acid solution? (See Example 3-23.)

95. A 0.08964 M solution of NaOH was used to titrate a solution of unknown concentration of HCl. A 30.00-mL sample of the HCl solution required 24.21 mL of the NaOH solution for complete reaction. What is the molarity of the HCl solution? (See Example 3-24.)

96. A 34.53-mL sample of a solution of sulfuric acid, H_2SO_4, reacts with 27.86 mL of 0.08964 M NaOH solution. Calculate the molarity of the sulfuric acid solution. (See Example 3-23.)

97. Benzoic acid, C_6H_5COOH, is sometimes used for the standardization of solutions of bases. A 1.862-g sample of the acid reacts with 31.62 mL of an NaOH solution. What is the molarity of the base solution?

$$C_6H_5COOH + NaOH \longrightarrow C_6H_5COONa + H_2O$$

98. An antacid tablet containing calcium carbonate as an active ingredient requires 22.6 mL of 0.0932 M HCl for complete reaction. What mass of $CaCO_3$ did the tablet contain?

$$2HCl + CaCO_3 \longrightarrow CaCl_2 + H_2O + CO_2$$

Mixed Exercises

*99. What mass of sulfuric acid can be obtained from 1.00 kg of sulfur by the following series of reactions?

$$S + O_2 \xrightarrow{98\% \text{ yield}} SO_2$$

$$2SO_2 + O_2 \xrightarrow{96\% \text{ yield}} 2SO_3$$

$$SO_3 + H_2SO_4 \xrightarrow{100\% \text{ yield}} H_2S_2O_7$$

$$H_2S_2O_7 + H_2O \xrightarrow{97\% \text{ yield}} 2H_2SO_4$$

*100. What is the total mass of products formed when 33.8 g of carbon disulfide is burned in air? What mass of carbon disulfide would have to be burned to produce a mixture of carbon dioxide and sulfur dioxide that has a mass of 54.2 g?

$$CS_2 + 3O_2 \xrightarrow{\Delta} CO_2 + 2SO_2$$

*101. A mixture of calcium oxide, CaO, and calcium carbonate, $CaCO_3$, that had a mass of 1.844 g was heated until all the calcium carbonate was decomposed according to the following equation. After heating, the sample weighed 1.462 g. Calculate the masses of CaO and $CaCO_3$ present in the original sample.

$$CaCO_3 \xrightarrow{\Delta} CaO + CO_2$$

Pouring ammonium sulfide solution into a solution of cadmium nitrate gives a precipitate of cadmium sulfide.

$$(NH_4)_2S + Cd(NO_3)_2 \longrightarrow$$

$$CdS(s) + 2NH_4NO_3$$

Cadmium sulfide is used as a pigment in artists' oil-based paints.

Objectives

As you study this chapter, you should learn

☐ About the periodic table and the classification of elements

☐ About reactions of solutes in aqueous solutions

☐ To recognize nonelectrolytes, strong electrolytes, and weak electrolytes

☐ The classification of acids, bases, and salts

☐ Which kinds of compounds are soluble and which kinds are insoluble in water

☐ How to describe reactions in aqueous solutions by writing formula unit equations as well as total ionic and net ionic equations

☐ About displacement reactions and the activity series

☐ About precipitation reactions

☐ About oxidation numbers

☐ To balance equations for oxidation–reduction reactions

W e observe that some elements form compounds with only a few (or no) other elements, while others combine with nearly every other element (oxygen) or form millions of compounds (carbon). Some elements are metals, and others obviously lack metallic properties. Some substances are gases, others are liquids, and still others are solids. Some solids are soft (paraffin), others are quite hard (diamond), and others are quite strong (steel). Clearly, there are significant differences among the chemical bonds and other attractive forces that hold atoms together in different substances.

Compounds are formed when atoms of different elements are joined together by chemical bonds. Chemical bonds are broken when atoms are separated and the original compounds cease to exist. The attractive forces between atoms are electrical in nature, and chemical reactions between atoms involve *changes* in their electronic structures.

As we shall see in the next few chapters,

> the positions of elements in the periodic table are related to the arrangements of their electrons; these determine the chemical and physical properties of the elements.

Let us now turn our attention to the periodic table and to **chemical periodicity**, the variation in properties of elements with their positions in the periodic table.

4-1 The Periodic Table: Metals, Nonmetals, and Metalloids

In 1869 the Russian chemist Dimitri Mendeleev and the German chemist Lothar Meyer independently published arrangements of known elements that are much like the periodic table in use today. Mendeleev's classification was based primarily on chemical properties of the elements, whereas Meyer's classification was based largely on physical properties. The tabulations were surprisingly similar. Both emphasized the *periodicity*, or regular periodic repetition, of properties with increasing atomic weight.

Mendeleev arranged the known elements in order of increasing atomic weight in successive sequences so that elements with similar chemical properties fell in the same column. He noted that both physical and chemical properties of the elements vary in a periodic fashion with atomic weight. His periodic table of 1872 contained the 62 elements that were known then (Figure 4-1).

Consider H, Li, Na, and K, all of which appear in "Gruppe I" of Mendeleev's table. All were known to combine with F, Cl, Br, and I of "Gruppe VII" to produce compounds that have similar formulas such as HF, LiCl, NaCl, and KI. All these compounds dissolve in water to produce solutions that conduct electricity. The "Gruppe II" elements form compounds such as $BeCl_2$, $MgBr_2$, and $CaCl_2$, as well as compounds with O and S from "Gruppe VI" such as MgO, CaO, MgS, and CaS.

In most areas of human endeavor progress is slow and faltering. However, there is an occasional individual who develops concepts and techniques that clarify confused situations. Mendeleev was such an individual. One of the brilliant successes of his periodic table was that it provided for elements that were unknown at the time. When he encountered "missing" elements, Mendeleev left blank spaces. Some appreciation of his genius in constructing the table as he did can be gained by comparing the predicted (1871) and observed properties of germanium, which was not discovered until 1886. Mendeleev called the undiscovered element eka-silicon because it fell below

Pronounced "men-del-*lay*-ev."

REIHEN	GRUPPE I – R^2O	GRUPPE II – RO	GRUPPE III – R^2O^3	GRUPPE IV RH^4 RO^2	GRUPPE V RH^3 R^2O^5	GRUPPE VI RH^2 RO^3	GRUPPE VII RH R^2O^7	GRUPPE VIII – RO^4
1	H = 1							
2	Li = 7	Be = 9,4	B = 11	C = 12	N = 14	O = 16	F = 19	
3	Na = 23	Mg = 24	Al = 27,3	Si = 28	P = 31	S = 32	Cl = 35,5	
4	K = 39	Ca = 40	– = 44	Ti = 48	V = 51	Cr = 52	Mn = 55	Fe = 56, Co = 59, Ni = 59, Cu = 63.
5	(Cu = 63)	Zn = 65	– = 68	– = 72	As = 75	Se = 78	Br = 80	
6	Rb = 85	Sr = 87	?Yt = 88	Zr = 90	Nb = 94	Mo = 96	– = 100	Ru = 104, Rh = 104, Pd = 106, Ag = 108.
7	(Ag = 108)	Cd = 112	In = 113	Sn = 118	Sb = 122	Te = 125	J = 127	
8	Cs = 133	Ba = 137	?Di = 138	?Ce = 140	–	–	–	– – – –
9	(–)	–	–	–	–	–	–	
10	–	–	?Er = 178	?La = 180	Ta = 182	W = 184	–	Os = 195, Ir = 197, Pt = 198, Au = 199.
11	(Au = 199)	Hg = 200	Tl = 204	Pb = 207	Bi = 208	–	–	
12	–	–	–	Th = 231	–	U = 240	–	– – – –

Figure 4-1
Mendeleev's early periodic table (1872). "J" is the German symbol for iodine.

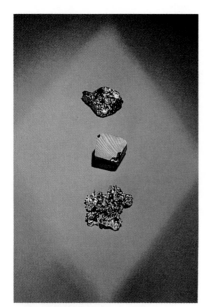

Silicon (top), germanium, and tin (bottom).

silicon in his table. He was familiar with the properties of germanium's neighboring elements. They served as the basis for his predictions of properties of germanium (Table 4-1). Some modern values for properties of germanium differ significantly from those reported in 1886. But many of the values upon which Mendeleev based his predictions were inaccurate, as were most of the 1886 values for Ge.

Because Mendeleev's arrangement of the elements was based on increasing *atomic weights*, several elements appeared to be out of place in his table. Mendeleev put the controversial elements (Te and I, Co and Ni) in locations consistent with their properties. He thought the apparent reversal of atomic weights was due to inaccurate values for those weights. Careful redetermination showed that the values were correct. Resolution of the problem of these "out-of-place" elements had to await the development of the concept of *atomic number*, approximately 50 years after Mendeleev's work. The **atomic number** of an element is the number of protons in the nucleus of its atoms. (It is also the number of electrons in an atom of an element.) This quantity is fundamental to the identity of each element because it is related to the electrical make-up of atoms. Elements are arranged in the periodic table in order of increasing atomic number. With the development of this concept, the **periodic law** attained essentially its present form:

> The properties of the elements are periodic functions of their atomic numbers.

The periodic law tells us that if we arrange the elements in order of increasing atomic number, we periodically encounter elements that have similar chemical and physical properties. The presently used "long form" of the period table (Table 4-2) is such an arrangement. The vertical columns are referred to as **groups** or **families**, and the horizontal rows are called

Table 4-1
Predicted and Observed Properties of Germanium

Property	Eka-Silicon Predicted, 1871	Germanium Reported, 1886	Modern Values
Atomic weight	72	72.32	72.61
Atomic volume	13 cm³	13.22 cm³	13.5 cm³
Specific gravity	5.5	5.47	5.35
Specific heat	0.073 cal/g°C	0.076 cal/g°C	0.074 cal/g°C
Maximum valence*	4	4	4
Color	Dark gray	Grayish white	Grayish white
Reaction with water	Will decompose steam with difficulty	Does not decompose water	Does not decompose water
Reactions with acids and alkalis	Slight with acids; more pronounced with alkalis	Not attacked by HCl or dilute aqueous NaOH; reacts vigorously with molten NaOH	Not dissolved by HCl or H_2SO_4 or dilute NaOH; dissolved by concentrated NaOH
Formula of oxide	EsO_2	GeO_2	GeO_2
Specific gravity of oxide	4.7	4.703	4.228
Specific gravity of tetrachloride	1.9 at 0°C	1.887 at 18°C	1.8443 at 30°C
Boiling point of tetrachloride	100°C	86°C	84°C
Boiling point of tetraethyl derivative	160°C	160°C	186°C

* "Valence" refers to the combining power of a specific element.

periods. Elements in a *group* have similar chemical and physical properties, while those within a *period* have properties that change progressively across the table. Several groups of elements have common names that are used so frequently they should be learned. The Group IA elements, except H, are referred to as **alkali metals**, and the Group IIA elements are called the **alkaline earth metals**. The Group VIIA elements are called **halogens**, which means "salt formers," and the Group 0 elements are called **noble** (or **rare**) **gases**.

Alkaline means basic. The character of basic compounds is described in Section 10-4.

Three of the halogens: (left to right) chlorine, bromine, iodine.

Table 4-2
The Periodic Table*

IA																		0
1 H	IIA											IIIA	IVA	VA	VIA	VIIA		2 He
3 Li	4 Be											5 B	6 C	7 N	8 O	9 F		10 Ne
11 Na	12 Mg	IIIB	IVB	VB	VIB	VIIB	VIIIB			IB	IIB	13 Al	14 Si	15 P	16 S	17 Cl		18 Ar
19 K	20 Ca	21 Sc	22 Ti	23 V	24 Cr	25 Mn	26 Fe	27 Co	28 Ni	29 Cu	30 Zn	31 Ga	32 Ge	33 As	34 Se	35 Br		36 Kr
37 Rb	38 Sr	39 Y	40 Zr	41 Nb	42 Mo	43 Tc	44 Ru	45 Rh	46 Pd	47 Ag	48 Cd	49 In	50 Sn	51 Sb	52 Te	53 I		54 Xe
55 Cs	56 Ba	57 La	72 Hf	73 Ta	74 W	75 Re	76 Os	77 Ir	78 Pt	79 Au	80 Hg	81 Tl	82 Pb	83 Bi	84 Po	85 At		86 Rn
87 Fr	88 Ra	89 Ac																

Metals ▭
Nonmetals ▭
Metalloids ▭

*	58 Ce	59 Pr	60 Nd	61 Pm	62 Sm	63 Eu	64 Gd	65 Tb	66 Dy	67 Ho	68 Er	69 Tm	70 Yb	71 Lu

†	90 Th	91 Pa	92 U	93 Np	94 Pu	95 Am	96 Cm	97 Bk	98 Cf	99 Es	100 Fm	101 Md	102 No	103 Lr

* There are other systems for numbering the groups in the periodic table. We use the standard American system.

About 80% of the elements are metals.

The physical and chemical properties that distinguish metals from nonmetals are summarized in Tables 4-3 and 4-4. The general properties of metals and nonmetals are opposite. Not all metals and nonmetals possess all these properties, but they share most of them to varying degrees. The physical properties of metals can be explained on the basis of metallic bonding in solids (Section 13-17).

Table 4-2, the periodic table, shows how we divide the known elements into *metals* (shown in blue), *nonmetals* (yellow), and *metalloids* (green). The elements to the left of those touching the heavy stairstep line are *metals* (except hydrogen), while those to the right are *nonmetals*. Such a classification is somewhat arbitrary, and several elements do not fit neatly into either class. The elements adjacent to the heavy line are often called *metalloids* (or semimetals), because they are metallic (or nonmetallic) only to a limited degree.

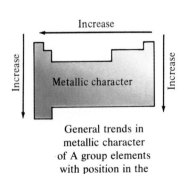

Increase

Increase   Metallic character   Increase

General trends in metallic character of A group elements with position in the periodic table

Metallic character increases from top to bottom and from right to left with respect to position in the periodic table.

Cesium, atomic number 55, is the most active naturally occurring metal.

Table 4-3
Some Physical Properties of Metals and Nonmetals

Metals	Nonmetals
1. High electrical conductivity that decreases with increasing temperature	1. Poor electrical conductivity (except carbon in the form of graphite)
2. High thermal conductivity	2. Good heat insulators (except carbon in the form of diamond)
3. Metallic gray or silver luster*	3. No metallic luster
4. Almost all are solids†	4. Solids, liquids, or gases
5. Malleable (can be hammered into sheets)	5. Brittle in solid state
6. Ductile (can be drawn into wires)	6. Nonductile
7. Solid state characterized by metallic bonding	7. Covalently bonded molecules; noble gases are monatomic

* Except copper and gold.
† Except mercury; cesium and gallium melt in protected hand.

Copper is rolled into sheets.

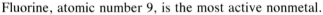

Nonmetallic character increases from bottom to top and from left to right in the periodic table.

Fluorine, atomic number 9, is the most active nonmetal.

Metalloids show some properties that are characteristic of both metals and nonmetals. Many of the metalloids, such as silicon, germanium, and antimony, act as semiconductors, which are important in solid-state electronic circuits. **Semiconductors** are insulators at lower temperatures, but become conductors at higher temperatures (Section 13-17).

Aluminum is the most metallic of the metalloids and is sometimes classified as a metal. It is metallic in appearance, and an excellent conductor of electricity, but its electrical conductivity *increases with increasing temperature*. The conductivities of metals, by contrast, decrease with increasing temperature.

Two metals (left), copper and magnesium; two metalloids (center), aluminum and silicon; two nonmetals (right), bromine and carbon.

Table 4-4
Some Chemical Properties of Metals and Nonmetals

Metals	Nonmetals
1. Outer shells contain few electrons— usually three or fewer	1. Outer shells contain four or more electrons*
2. Form cations by losing electrons	2. Form anions by gaining electrons†
3. Form ionic compounds with nonmetals	3. Form ionic compounds with metals† and molecular (covalent) compounds with other nonmetals

* Except hydrogen and helium.
† Except the noble gases.

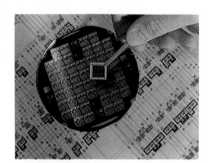

Silicon, a metalloid, is widely used in the manufacture of electronic chips.

Chemistry in Use. . .
The Periodic Table

The periodic table is almost the first thing a student of chemistry encounters. It appears invariably in textbooks, in lecture halls, and in laboratories. The most knowledgeable scientists consider it an indispensable reference. And yet, less than 150 years ago, the idea of arranging the elements by atomic weight or number was considered absurd. At an 1866 meeting of the Chemical Society at Burlington House, England, J. A. R. Newlands presented a theory he called the Law of Octaves. It stated that when the known elements were listed by increasing atomic weights, those that were eight places apart would be similar, much like notes on a piano keyboard. His colleagues' reactions are probably summed up best by the remark of a Professor Foster: "Have you thought of arranging the elements according to their initial letters? Maybe some better connections would come to light that way."

It is not surprising that poor Newlands was not taken seriously. In the 1860s, little information was available to illustrate relationships among the elements. Only 62 of them had been distinguished from more complex substances when Mendeleev first announced his discovery of the Periodic Law in 1869. However, as advances in atomic theory were made and as new experiments contributed to the understanding of chemical behavior, some scientists had begun to see similarities and patterns among the elements. In 1829 Johann Dobereiner first noticed *triads,* series of three elements in which the characteristics of the middle element are an average of the characteristics of the others. One such triad consists of Li, Na, and K (atomic weights 6.9, 23.0, and 39.1 g/mol). Beguyer de Chantcourtois, in 1862, assigned them to a repeating pattern known as a Telluric Screw, based on the observation that elements whose atomic weights differ by 16 g/mol are often similar. And in 1869 Lothar Meyer and Dmitri Mendeleev independently published similar versions of the now-famous periodic table.

Mendeleev's discovery was the result of many years of hard work. He gathered information on the elements from all corners of the earth—by corresponding with colleagues, studying books and papers, and redoing experiments to confirm data. He put the statistics of each element on a small card and pinned the cards to his laboratory wall, where he arranged and rearranged them many times until he was sure that they were in the right order. One especially farsighted feature of Mendeleev's accomplishment was his realization that some elements were missing from the table. His predictions for the properties of these substances (gallium, scandium, and germanium) were found to be remarkably accurate considering the data available to him. (It is important to remember that Mendeleev's periodic table organization was devised more than 50 years before the discovery and characterization of subatomic particles.)

Since its birth in 1869, the periodic table has been discussed and

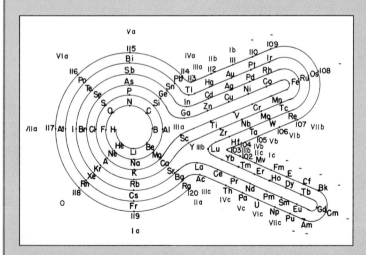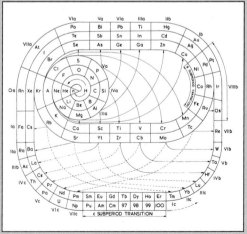

Alternative representations of the periodic table, as proposed by Charles Janet, 1928 (left), and John D. Clark, 1950 (right).

revised many times. Spectroscopic and other discoveries have filled in the blanks left by Mendeleev and added a new column consisting of the noble gases. As scientists learned more about atomic structure, the basis for ordering was changed from atomic weight to atomic number. The perplexing rare earths were sorted out and given a special place, along with many of the elements created by atomic bombardment. Even the form of the table has been experimented with, resulting in everything from spiral and circular tables to exotic shapes such as those suggested by Janet and by Clark. Recently a three-dimensional periodic table that takes into account valence-shell energies has been proposed by Professor Leland C. Allen of Princeton University.

It is certain that the future will disclose new and exciting information on the elements that make up our universe. During the past century, chemistry has become a fast-moving science in which methods and instruments are often outdated within a few years. But it is doubtful that our old friend, the periodic table, will ever become obsolete. It may be modified, but it will always stand as a statement of basic relationships in chemistry and as a monument to the wisdom and insight of its creator, Dmitri Mendeleev.

Lisa L. Saunders
Chemistry major
University of Texas at Austin

4-2 Aqueous Solutions—An Introduction

Approximately three fourths of the earth's surface is covered with water. The body fluids of all plants and animals are mainly water. Thus we can see that many important chemical reactions occur in aqueous (water) solutions, or in contact with water. In Chapter 3, we introduced solutions and methods of expressing concentrations of solutions. It is useful to know the kinds of substances that are soluble in water, and the forms in which they exist, before we begin our systematic study of reactions.

1 Electrolytes and Extent of Ionization

Solutes that are water-soluble can be classified as either electrolytes or nonelectrolytes. **Electrolytes** are substances whose aqueous solutions conduct electrical current. **Strong electrolytes** are substances that conduct electricity well in dilute aqueous solution. **Weak electrolytes** conduct electricity poorly in dilute aqueous solution. Aqueous solutions of **nonelectrolytes** do not conduct electricity. Electrical current is carried through aqueous solution by the movement of ions. The strength of an electrolyte depends upon the number of ions in solution and also on the charges on these ions (see Figure 4-2).

Recall that *ions* are charged particles. The movement of charged particles conducts electricity.

Dissociation refers to the process in which a solid *ionic compound*, such as NaCl, separates into its ions in solution. **Ionization** refers to the process in which a *molecular compound* separates to form ions in solution. Molecular compounds, for example *pure* HCl, exist as discrete molecules and do not contain ions. However, many such compounds ionize in solution to form ions.

Three major classes of solutes are strong electrolytes: (1) strong acids, (2) strong soluble bases, and (3) most soluble salts. *These compounds are completely or nearly completely ionized (or dissociated) in dilute aqueous solutions*, and therefore are strong electrolytes.

Acids and bases are further identified in Subsections 2, 3, and 4.

(a)

(b)

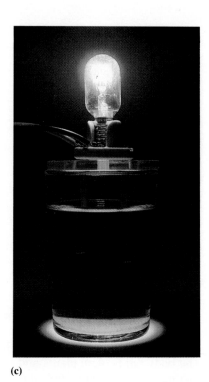
(c)

Figure 4-2

An experiment to demonstrate the presence of ions in solution. Two copper electrodes dip into a liquid in a beaker. When the liquid contains significant concentrations of ions, the ions move between the electrodes to complete the circuit (which includes a light bulb). (a) Pure water is a nonelectrolyte. (b) A solution of a weak electrolyte, acetic acid (CH_3COOH); it contains low concentrations of ions, and so the bulb glows dimly. (c) A solution of a strong electrolyte, potassium chromate (K_2CrO_4); it contains a high concentration of ions, and so the bulb glows brightly.

Recall that positively charged ions are called *cations* while negatively charged ions are called *anions.* Even though the formula for a salt may include H or OH, it must contain another cation *and* another anion. For example, $NaHSO_4$ and $Al(OH)_2Cl$ are salts.

Many properties of aqueous solutions of acids are due to $H^+(aq)$ ions. These are described in Section 10-4.

An **acid** can be defined as a substance that produces hydrogen ions, H^+, in aqueous solutions. A **base** is a substance that produces hydroxide ions, OH^-, in aqueous solutions. A **salt** is a compound that contains a cation other than H^+ and an anion other than hydroxide ion, OH^-, or oxide ion, O^{2-}. As we shall see later in this chapter, salts are formed when acids react with bases.

2 Strong and Weak Acids

As a matter of convenience we place acids into two classes: strong acids and weak acids. **Strong acids** ionize (separate into ions) completely, or very nearly completely, in dilute aqueous solution. The seven common strong acids and their anions are listed in Table 4-5. Please learn this list because it is short. (The list of common weak acids is long.)

Because strong acids ionize completely or very nearly completely in dilute solutions, their solutions contain (predominantly) the ions of the acid rather than acid molecules. Consider the ionization of hydrochloric acid. Pure hydrogen chloride, HCl, is a molecular compound that is a gas at room

Table 4-5
Common Strong Acids and Their Anions

Common Strong Acids		Anions of These Strong Acids	
Formula	Name	Formula	Name
HCl	hydrochloric acid	Cl^-	chloride ion
HBr	hydrobromic acid	Br^-	bromide ion
HI	hydroiodic acid	I^-	iodide ion
HNO_3	nitric acid	NO_3^-	nitrate ion
$HClO_4$	perchloric acid	ClO_4^-	perchlorate ion
$HClO_3$	chloric acid	ClO_3^-	chlorate ion
H_2SO_4	sulfuric acid	$\begin{cases} HSO_4^- \\ SO_4^{2-} \end{cases}$	hydrogen sulfate ion sulfate ion

temperature and atmospheric pressure. When it dissolves in water, it reacts nearly 100% to produce a solution that contains hydrogen ions and chloride ions:

$$HCl(g) \xrightarrow{H_2O} H^+(aq) + Cl^-(aq) \quad \text{(to completion)}$$

Similar equations can be written for all strong acids.

The species we have shown as $H^+(aq)$ is sometimes shown as H_3O^+ or $[H(H_2O)^+]$ to emphasize hydration. However, it really exists in varying degrees of hydration, such as $H(H_2O)^+$, $H(H_2O)_2^+$, and $H(H_2O)_3^+$.

Weak acids ionize only slightly (usually less than 5%) in dilute aqueous solution; the list of weak acids is very long. Many common ones are listed in Appendix F, and a few of them and their anions are given in Table 4-6.

To give a more complete description of reactions, we indicate the physical states of reactants and products: (g) for gases, (ℓ) for liquids, and (s) for solids. The notation (aq) following ions indicates that they are hydrated in aqueous solution; that is, they interact with water molecules in solution. The complete ionization of a strong electrolyte is indicated by a single arrow (→).

Because there are so many weak acids, please learn the list of common strong acids (Table 4-5). Then assume that the other acids you encounter are weak.

Table 4-6
Some Common Weak Acids and Their Anions

Common Weak Acids		Anions of These Weak Acids	
Formula	Name	Formula	Name
HF*	hydrofluoric acid	F^-	fluoride ion
CH_3COOH	acetic acid	CH_3COO^-	acetate ion
HCN	hydrocyanic acid	CN^-	cyanide ion
HNO_2†	nitrous acid	NO_2^-	nitrite ion
H_2CO_3†	carbonic acid	$\begin{cases} HCO_3^- \\ CO_3^{2-} \end{cases}$	hydrogen carbonate ion carbonate ion
H_2SO_3†	sulfurous acid	$\begin{cases} HSO_3^- \\ SO_3^{2-} \end{cases}$	hydrogen sulfite ion sulfite ion
H_3PO_4	phosphoric acid	$\begin{cases} H_2PO_4^- \\ HPO_4^{2-} \\ PO_4^{3-} \end{cases}$	dihydrogen phosphate ion hydrogen phosphate ion phosphate ion
$(COOH)_2$	oxalic acid	$\begin{cases} H(COO)_2^- \\ (COO)_2^{2-} \end{cases}$	hydrogen oxalate ion oxalate ion

* HF is a weak acid, whereas HCl, HBr, and HI are strong acids.
† Free acid molecules exist only in dilute aqueous solution or not at all. However, many salts of these acids are common, stable compounds.

Citrus fruits contain citric acid, and so their juices are acidic. This is shown here by the color changes on the indicator paper.

We usually write the formulas of inorganic acids with hydrogen written first. Organic acids can often be recognized by the presence of the COOH group in the formula.

The equation for the ionization of acetic acid, CH_3COOH, in water is typical of weak acids:

$$CH_3COOH(aq) \rightleftharpoons H^+(aq) + CH_3COO^-(aq) \quad \text{(reversible)}$$

The double arrow ($\rightleftharpoons$) generally signifies that the reaction occurs in *both* directions and that the forward reaction does not go to completion. All of us are familiar with solutions of acetic acid. Vinegar is 5% acetic acid by mass. Our use of oil and vinegar as a salad dressing tells us that acetic acid is a weak acid. To be specific, acetic acid is 0.5% ionized (and 99.5% non-ionized) in 5% solution.

Acetic acid is one of the organic acids, most of which are weak. A multitude of organic acids occur in living systems. Organic acids contain the carboxylate grouping of atoms, —COOH. They can ionize slightly by breakage of the O—H bond, as shown here for acetic acid:

Organic acids are discussed in Chapter 31. Carbonic acid, H_2CO_3, and hydrocyanic acid, HCN(aq), are two common acids that contain carbon but that are considered to be *inorganic* acids. Inorganic acids are often called **mineral acids** because they are obtained primarily from mineral (nonliving) sources.

> Our stomachs have linings that are much more resistant to attack by acids than are our other tissues.

> The carboxylate group —COOH is

> Other organic acids have other groups in the position of the H_3C— group in acetic acid.

> Inorganic acids may be strong or weak; most organic acids are weak.

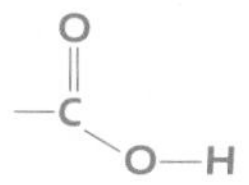

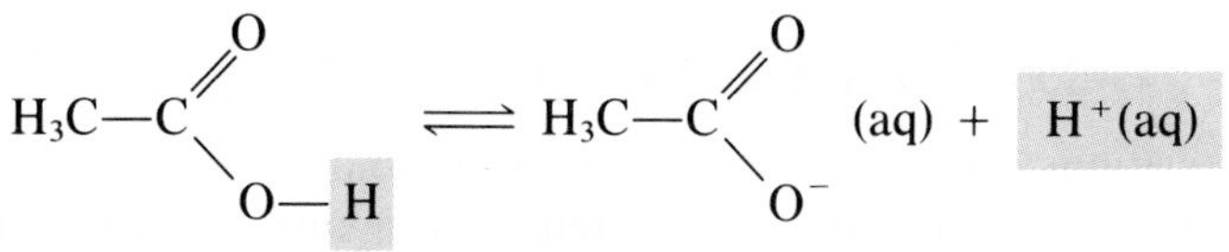

Example 4-1

In the following lists of common acids, which are strong and which are weak?

(a) H_3PO_4, HCl, H_2CO_3, HNO_3; (b) $HClO_4$, H_2SO_4, HClO, HF

Plan
We recall that Table 4-5 lists the common strong acids. Other *common* acids are assumed to be weak.

Solution

(a) HCl and HNO_3 are strong acids; H_3PO_4 and H_2CO_3 are weak acids.
(b) $HClO_4$ and H_2SO_4 are strong acids; HClO and HF are weak acids.

---

3 Reversible Reactions

Reactions that can occur in both directions are **reversible reactions**. We use a double arrow ($\rightleftharpoons$) to indicate that a reaction is *reversible*. What is the fundamental difference between reactions that go to completion and those that are reversible? We have seen that the ionization of HCl in water is nearly complete. Suppose we dissolve some table salt, NaCl, in water and then add some dilute nitric acid to it. The resulting solution contains hydrogen ions and chloride ions, the products of the ionization of HCl, as well as sodium ions and nitrate ions. The H^+ and Cl^- ions do *not* react significantly to form nonionized molecules of HCl, the reverse of the ionization of HCl.

$$H^+(aq) + Cl^-(aq) \longrightarrow \text{no reaction}$$

In contrast, when a sample of sodium acetate, $NaCH_3COO$, is dissolved in H_2O and mixed with nitric acid, the resulting solution initially contains Na^+, CH_3COO^-, H^+, and NO_3^- ions. But most of the H^+ and CH_3COO^- ions combine to produce nonionized molecules of acetic acid, the reverse of the ionization of the acid. Thus, the ionization of acetic acid, like that of any other weak electrolyte, is reversible.

Na^+ and NO_3^- ions do not combine because $NaNO_3$ is soluble in water.

$$H^+(aq) + CH_3COO^-(aq) \rightleftharpoons CH_3COOH(aq) \qquad \text{(reversible)}$$

4 Strong Soluble Bases, Insoluble Bases, and Weak Bases

Most common bases are ionic metal hydroxides. Most are insoluble in water. **Strong soluble bases** are soluble in water and are dissociated completely in dilute aqueous solution. The common strong soluble bases are listed in Table 4-7. They are the hydroxides of the Group IA metals and the heavier members of Group IIA. The equation for the dissociation of sodium hydroxide in water is typical. Similar equations can be written for other strong soluble bases.

Solutions of bases have a set of common properties due to the OH^- ion. These are described in Section 10-4.

$$NaOH(s) \xrightarrow{H_2O} Na^+(aq) + OH^-(aq) \qquad \text{(to completion)}$$

Other metals form ionic hydroxides, but these are so sparingly soluble in water that they cannot produce strongly basic solutions. They are called **insoluble bases**. Typical examples include $Cu(OH)_2$, $Zn(OH)_2$, $Fe(OH)_2$, and $Fe(OH)_3$.

Strong soluble bases are ionic compounds in the solid state.

Many **weak bases**, such as ammonia, NH_3, are very soluble in water but ionize only slightly in solution.

The weak bases are *molecular* substances that ionize only slightly in water; they are sometimes called molecular bases.

$$NH_3(aq) + H_2O(\ell) \rightleftharpoons NH_4^+(aq) + OH^-(aq) \qquad \text{(reversible)}$$

Table 4-7
Common Strong Soluble Bases

Group IA		Group IIA	
LiOH	lithium hydroxide		
NaOH	sodium hydroxide		
KOH	potassium hydroxide	$Ca(OH)_2$	calcium hydroxide
RbOH	rubidium hydroxide	$Sr(OH)_2$	strontium hydroxide
CsOH	cesium hydroxide	$Ba(OH)_2$	barium hydroxide

Example 4-2

From the following lists, choose (i) the strong soluble bases, (ii) the insoluble bases, and (iii) the weak bases. (a) NaOH, Cu(OH)$_2$, Pb(OH)$_2$, Ba(OH)$_2$; (b) Fe(OH)$_3$, KOH, Mg(OH)$_2$, Sr(OH)$_2$, NH$_3$

Plan

(i) We recall that Table 4-7 lists the *common strong soluble bases*. (ii) Other common metal hydroxides are assumed to be *insoluble* bases. (iii) Ammonia and its derivatives, the amines, are the common *weak bases*.

Solution

(a) (i) The strong soluble bases are NaOH and Ba(OH)$_2$, so
     (ii) the insoluble bases are Cu(OH)$_2$ and Pb(OH)$_2$.
(b) (i)The strong soluble bases are KOH and Sr(OH)$_2$, so
     (ii) the insoluble bases are Fe(OH)$_3$ and Mg(OH)$_2$, and
     (iii) the weak base is NH$_3$.

EOC 30–32

5 Solubility Rules for Compounds in Aqueous Solution

Solubility is a complex phenomenon, and it is not possible to state simple rules that cover all cases. The following rules for solutes in aqueous solutions will be very useful for nearly all acids, bases, and salts encountered in general chemistry. Compounds whose solubility in water is less than about 0.02 mole per liter are usually classified as insoluble compounds, whereas those that are more soluble are classified as soluble compounds. No gaseous or solid substances are infinitely soluble in water. You may wish to review Table 2-2 on page 52, a list of some common ions. Table 7-4 on page 285 contains a more comprehensive list.

There is no sharp dividing line between "soluble" and "insoluble" compounds. Compounds whose solubilities fall near the arbitrary dividing line are called "moderately soluble" compounds.

1. The common inorganic acids are soluble in water. Low molecular weight organic acids are soluble.
2. The common compounds of the Group IA metals (Li, Na, K, Rb, Cs) and the ammonium ion, NH$_4^+$, are soluble in water.
3. The common nitrates, NO$_3^-$; acetates, CH$_3$COO$^-$; chlorates, ClO$_3^-$; and perchlorates, ClO$_4^-$, are soluble in water.
4. (a) The common chlorides, Cl$^-$, are soluble in water except AgCl, Hg$_2$Cl$_2$, and PbCl$_2$.
    (b) The common bromides, Br$^-$, and iodides, I$^-$, show approximately the same solubility behavior as chlorides, but there are some exceptions. As these halide ions (Cl$^-$, Br$^-$, I$^-$) increase in size, the solubilities of their slightly soluble compounds decrease.
5. The common sulfates, SO$_4^{2-}$, are soluble in water except PbSO$_4$, BaSO$_4$, and HgSO$_4$; CaSO$_4$ and Ag$_2$SO$_4$ are moderately soluble.
6. The common metal hydroxides, OH$^-$, are *insoluble* in water except those of the Group IA metals and the heavier members of the Group IIA metals, beginning with Ca(OH)$_2$.

7. The common carbonates, CO_3^{2-}, phosphates, PO_4^{3-}, and arsenates, AsO_4^{3-}, are *insoluble* in water except those of the Group IA metals and NH_4^+. $MgCO_3$ is moderately soluble.
8. The common sulfides, S^{2-}, are *insoluble* in water except those of the Group IA and Group IIA metals and the ammonium ion.

Table 4-8 summarizes much of the information about the solubility rules.

We shall now discuss chemical reactions in some detail. Because millions of reactions are known, it is useful to group them into classes, or types, so that we can deal with such massive amounts of information systematically. We shall classify them as (1) precipitation reactions, (2) acid–base reactions, (3) displacement reactions, and (4) oxidation–reduction reactions. We shall also distinguish among (1) formula unit, (2) total ionic, and (3) net ionic equations for chemical reactions and indicate the advantages and disadvantages of these methods for representing chemical reactions. As we study different kinds of chemical reactions, we shall learn to predict the products of other similar reactions.

In Chapter 6 we shall describe typical reactions of hydrogen, oxygen, and their compounds. These reactions will illustrate periodic relationships with

The fact that a substance is insoluble in water does not mean that it cannot take part in a reaction in contact with water.

Table 4-8
Solubility of Common Ionic Compounds in Water

Generally Soluble	Exceptions
Na^+, K^+, NH_4^+ compounds	No common exceptions
chlorides (Cl^-)	Insoluble: $AgCl$, Hg_2Cl_2 Soluble in hot water: $PbCl_2$
bromides (Br^-)	Insoluble: $AgBr$, Hg_2Br_2, $PbBr_2$ Moderately soluble: $HgBr_2$
iodides (I^-)	Insoluble: many heavy metal iodides
sulfates (SO_4^{2-})	Insoluble: $BaSO_4$, $PbSO_4$, $HgSO_4$ Moderately soluble: $CaSO_4$, $SrSO_4$, Ag_2SO_4
nitrates (NO_3^-), nitrites (NO_2^-)	Moderately soluble: $AgNO_2$
chlorates (ClO_3^-), perchlorates (ClO_4^-), permanganates (MnO_4^-)	Moderately soluble: $KClO_4$
acetates (CH_3COO^-)	Moderately soluble: $AgCH_3COO$

Generally Insoluble	Exceptions
sulfides (S^{2-})	Soluble: those of NH_4^+, Na^+, K^+, Mg^{2+}, Ca^{2+}
oxides (O^{2-}), hydroxides (OH^-)	Soluble: Li_2O^*, $LiOH$, Na_2O^*, $NaOH$, K_2O^*, KOH, BaO^*, $Ba(OH)_2$ Moderately soluble: CaO^*, $Ca(OH)_2$, SrO^*, $Sr(OH)_2$
carbonates (CO_3^{2-}), phosphates (PO_4^{3-}), arsenates (AsO_4^{3-})	Soluble: those of NH_4^+, Na^+, K^+

* Dissolves with evolution of heat and formation of hydroxides.

water are also strong electrolytes. Exceptions such as lead acetate, $Pb(CH_3COO)_2$, which is soluble but predominantly nonionized, will be noted as they are encountered.

> The only common substances that should be written in ionized or dissociated form in ionic equations are (1) strong acids, (2) strong soluble bases, and (3) soluble ionic salts.

4-4 Precipitation Reactions

To understand the discussion of precipitation reactions, you must know the solubility rules (page 132) and Table 4-8.

In **precipitation reactions** an insoluble solid, a **precipitate**, forms and then settles out of solution. Our teeth and bones were formed by very slow precipitation reactions in which mostly calcium phosphate $Ca_3(PO_4)_2$ was deposited in the correct geometric arrangements.

An example of a precipitation reaction is the formation of bright yellow insoluble lead(II) chromate as a result of mixing solutions of the soluble ionic compounds lead(II) nitrate and potassium chromate (Figure 4-4). The other product of the reaction is KNO_3, a soluble ionic salt.

The balanced formula unit, total ionic, and net ionic equations for this reaction follow.

$$Pb(NO_3)_2(aq) + K_2CrO_4(aq) \longrightarrow PbCrO_4(s) + 2KNO_3(aq)$$

$$[Pb^{2+}(aq) + 2\,NO_3^-(aq)] + [2K^+(aq) + CrO_4^{2-}(aq)] \longrightarrow PbCrO_4(s) + 2[K^+(aq) + NO_3^-(aq)]$$

$$Pb^{2+}(aq) + CrO_4^{2-}(aq) \longrightarrow PbCrO_4(s)$$

Another important precipitation reaction involves the formation of insoluble carbonates (solubility rule 7). Limestone deposits are mostly calcium carbonate, $CaCO_3$, although many also contain significant amounts of magnesium carbonate, $MgCO_3$.

Suppose we mix together aqueous solutions of sodium carbonate, Na_2CO_3, and calcium chloride, $CaCl_2$. We recognize that *both* Na_2CO_3 and $CaCl_2$ (solubility rules 2, 4a, and 7) are soluble ionic compounds. At the instant of mixing, the resulting solution contains four ions:

$$Na^+(aq), \quad CO_3^{2-}(aq), \quad Ca^{2+}(aq), \quad Cl^-(aq)$$

Figure 4-4
A precipitation reaction. When K_2CrO_4 solution is added to aqueous $Pb(NO_3)_2$ solution, the yellow compound $PbCrO_4$ precipitates. The resulting solution contains K^+ and NO_3^- ions, the ions of KNO_3.

Seashells, which are formed in very slow precipitation reactions, are mostly calcium carbonate ($CaCO_3$), a white compound. Traces of transition metal ions give them color.

Because \
trends in \
metals, it \
predict t \
The fact \
(see Figu

Table \
Bondl

Exam

Pure \
   co \
Wat \
   in \
~ 10 \
   di \
   ac \
Writ \
   ec

* Most \
† The l \
‡ The *v* \
  dissoc \
§ There

Na$^+$, and t
that NaCl i

[H$^+$(aq) +

The net i
soluble ba

One pair of ions, Na$^+$ and Cl$^-$, *cannot* form an insoluble compound (solubility rules 2 and 4). We look for a pair of ions that would form an insoluble compound. Ca^{2+} ions and CO$_3^{2-}$ ions are such a combination; they form insoluble CaCO$_3$ (solubility rule 7). The equations for the reaction follow.

$$CaCl_2(aq) + Na_2CO_3(aq) \longrightarrow CaCO_3(s) + 2\ NaCl(aq)$$

$$[Ca^{2+}(aq) + 2\ Cl^-(aq)] + [2Na^+(aq) + CO_3^{2-}(aq)] \longrightarrow CaCO_3(s) + 2[Na^+(aq) + Cl^-(aq)]$$

$$Ca^{2+} + CO_3^{2-}(aq) \longrightarrow CaCO_3(s)$$

Example 4-3

Will a precipitate form when aqueous solutions of Ca(NO$_3$)$_2$ and NaCl are mixed in reasonable concentrations?

Plan

We recognize that both Ca(NO$_3$)$_2$ (solubility rule 3) and NaCl (solubility rules 2 and 4) are soluble compounds. At the instant of mixing, the resulting solution contains four ions:

$$Ca^{2+}(aq), \quad NO_3^-(aq), \quad Na^+(aq), \quad Cl^-(aq)$$

New combinations of ions *could* be CaCl$_2$ and NaNO$_3$. Solubility rule 4 tells us that CaCl$_2$ is a soluble compound, while solubility rules 2 and 3 tell us that NaNO$_3$ is a soluble compound.

Solution

Therefore no precipitate forms in this solution.

Example 4-4

Will a precipitate form when aqueous solutions of CaCl$_2$ and K$_3$PO$_4$ are mixed in reasonable concentrations? Write the appropriate equations for any reaction.

Plan

Both CaCl$_2$ (solubility rule 4) and K$_3$PO$_4$ (solubility rule 2) are soluble compounds. At the instant of mixing, four ions are present in the solution:

$$Ca^{2+}(aq), \quad Cl^-(aq), \quad K^+(aq), \quad PO_4^{3-}(aq)$$

New combinations of these ions *could* be KCl and Ca$_3$(PO$_4$)$_2$. Solubility rules 2 and 4 tell us that potassium chloride, KCl, is a soluble compound.

Solution

Solubility rule 7 tells us that calcium phosphate, Ca$_3$(PO$_4$)$_2$, is an insoluble compound and so it forms a precipitate.

The equations for the formation of calcium phosphate follow.

$$3CaCl_2(aq) + 2K_3PO_4(aq) \longrightarrow Ca_3(PO_4)_2(s) + 6KCl(aq)$$

$$3[Ca^{2+}(aq) + 2\ Cl^-(aq)] + 2[3K^+(aq) + PO_4^{3-}(aq)] \longrightarrow Ca_3(PO_4)_2(s) + 6[K^+(aq) + Cl^-(aq)]$$

$$3Ca^{2+}(aq) + 2PO_4^{3-}(aq) \longrightarrow Ca_3(PO_4)_2(s)$$

EOC 40, 42

Example 4
Predict the
balanced for

Plan
This is an ac
that contains
a soluble salt
strong soluble
in ionic form.

Solution

2[H$^+$(aq) +

We cancel the

Dividing by 2

EOC 45

Reactions of w
water, but ther
because weak a

Example 4-6
Write balanced
of acetic acid w

Plan
Neutralization
(Table 4-6) and

(Table 4-7) and KCH_3COO is a soluble salt (solubility rules 2 and 3), and so both are written in ionic form.

Solution

$$CH_3COOH(aq) + KOH(aq) \longrightarrow KCH_3COO(aq) + H_2O(\ell)$$

$$CH_3COOH(aq) + [K^+(aq) + OH^-(aq)] \longrightarrow [K^+(aq) + CH_3COO^-(aq)] + H_2O(\ell)$$

The spectator ion is K^+, the cation of the strong soluble base, KOH.

$$CH_3COOH(aq) + OH^-(aq) \longrightarrow CH_3COO^-(aq) + H_2O(\ell)$$

Thus, we see that *this* net ionic equation includes *molecules* of the weak acid and *anions* of the weak acid.

EOC 46, 47

A *monoprotic acid* contains one acidic H per formula unit.

The reactions of *weak monoprotic acids* with *strong soluble bases* that form *soluble salts* can be represented in general terms as

$$HA(aq) + OH^-(aq) \longrightarrow A^-(aq) + H_2O(\ell)$$

where HA represents the weak acid and A^- represents its anion.

The manufacture [...]
sumes more H_2SO_4 [...]
than any other s[...]

Table 4-10
1990 Production

Formula
H_2SO_4
CaO, $Ca(OH)_2$
NH_3
H_3PO_4
NaOH
Na_2CO_3
HNO_3
NH_4NO_3
$C_6H_4(COOH)_2$*
$(NH_4)_2SO_4$
HCl
CH_3COOH*
KOH, K_2CO_3
$Al_2(SO_4)_2$
Na_2SiO_3
$C_4H_8(COOH)_2$*
Na_2SO_4
$CaCl_2$

* Organic compound.

Blackboard chalk is mostly calcium carbonate, $CaCO_3$. Bubbles of carbon dioxide, CO_2, are clearly visible in this photograph of $CaCO_3$ dissolving in HCl.

Removal of Ions from Aqueous Solutions
Many reactions that occur in aqueous solution result in the removal of ions from the solution. This happens in one of three ways: (1) formation of predominantly nonionized molecules (weak or nonelectrolyte) in solution, (2) formation of a precipitate, or (3) formation of a gas that escapes. We have discussed examples of the first two in acid–base neutralization and precipitation reactions.

Let us illustrate the third case. When an acid—for example, hydrochloric acid—is added to solid calcium carbonate, a reaction occurs in which carbonic acid, a weak acid, is produced.

$$2HCl(aq) + CaCO_3(s) \longrightarrow H_2CO_3(aq) + CaCl_2(aq)$$

$$2[H^+(aq) + Cl^-(aq)] + CaCO_3(s) \longrightarrow H_2CO_3(aq) + [Ca^{2+}(aq) + 2Cl^-(aq)]$$

$$2H^+(aq) + CaCO_3(s) \longrightarrow H_2CO_3(aq) + Ca^{2+}(aq)$$

The heat generated in the reaction causes thermal decomposition of carbonic acid to gaseous carbon dioxide and water:

$$H_2CO_3(aq) \longrightarrow CO_2(g) + H_2O(\ell)$$

Most of the CO_2 bubbles off and the reaction goes to completion (with respect to the limiting reactant). The net effect is the conversion of ionic species into nonionized molecules of a gas (CO_2) and water.

4-6 Displacement Reactions

Reactions in which one element displaces another from a compound are called **displacement reactions**. Active metals displace less active metals or hydrogen from their compounds in aqueous solution. Active metals are those that readily lose electrons to form cations (see Table 4-11).

1 $\left[\begin{array}{l}\text{More Active Metal +}\\ \text{Salt of Less Active Metal}\end{array}\right] \rightarrow \left[\begin{array}{l}\text{Less Active Metal +}\\ \text{Salt of More Active Metal}\end{array}\right]$

The reaction of copper with silver nitrate that was described in detail in Section 4-3 is typical. Please refer to it.

Example 4-7
A large piece of zinc metal is placed in a copper(II) sulfate, $CuSO_4$, solution. The blue solution becomes colorless as copper metal falls to the bottom of the container. The resulting solution contains zinc sulfate, $ZnSO_4$. Write balanced formula unit, total ionic, and net ionic equations for the reaction.

Plan
The metals zinc and copper are *not* ionized or dissociated in contact with H_2O. Both $CuSO_4$ and $ZnSO_4$ are soluble salts (solubility rule 5), and so they are written in ionic form.

Solution

$$CuSO_4(aq) + Zn(s) \longrightarrow Cu(s) + ZnSO_4(aq)$$

$$[Cu^{2+}(aq) + SO_4^{2-}(aq)] + Zn(s) \longrightarrow Cu(s) + [Zn^{2+}(aq) + SO_4^{2-}(aq)]$$

$$Cu^{2+}(aq) + Zn(s) \longrightarrow Cu(s) + Zn^{2+}(aq)$$

In this *displacement reaction*, the more active metal, zinc, displaces the ions of the less active metal, copper, from aqueous solution.

2 [Active Metal + Nonoxidizing Acid] → [Hydrogen + Salt of Acid]

A common method for the preparation of small amounts of hydrogen involves the reaction of active metals with nonoxidizing acids, such as HCl and H_2SO_4. For example, when zinc is dissolved in H_2SO_4, the reaction produces zinc sulfate; hydrogen is displaced from the acid, and it bubbles off as gaseous H_2. The formula unit equation for this reaction is

$$\underset{\text{strong acid}}{Zn(s) + H_2SO_4(aq)} \longrightarrow \underset{\text{soluble salt}}{ZnSO_4(aq)} + H_2(g)$$

Both sulfuric acid (in dilute solution) and zinc sulfate exist primarily as ions, so the total ionic equation is

$$Zn(s) + [2H^+(aq) + SO_4^{2-}(aq)] \longrightarrow [Zn^{2+}(aq) + SO_4^{2-}(aq)] + H_2(g)$$

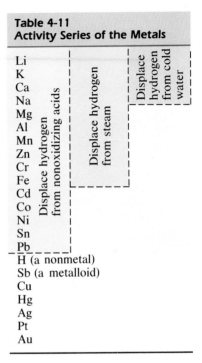

Table 4-11
Activity Series of the Metals

	Displace hydrogen from nonoxidizing acids	Displace hydrogen from steam	Displace hydrogen from cold water
Li			
K			
Ca			
Na			
Mg			
Al			
Mn			
Zn			
Cr			
Fe			
Cd			
Co			
Ni			
Sn			
Pb			
H (a nonmetal)			
Sb (a metalloid)			
Cu			
Hg			
Ag			
Pt			
Au			

A strip of zinc was placed in a blue solution of copper(II) sulfate, $CuSO_4$. The copper has been displaced from solution and has fallen to the bottom of the beaker. The resulting zinc sulfate solution is colorless.

Zinc dissolves in dilute H_2SO_4 to produce H_2 and a solution that contains $ZnSO_4$.

Elimination of unreacting species common to both sides of the total ionic equation gives the net ionic equation:

$$Zn(s) + 2H^+(aq) \longrightarrow Zn^{2+}(aq) + H_2(g)$$

Table 4-11 lists the **activity series**. When any metal listed above hydrogen in this series is added to solutions of *nonoxidizing* acids such as hydrochloric acid, HCl, and sulfuric acid, H_2SO_4, the metal dissolves to produce hydrogen, and a salt is formed. HNO_3 is the common *oxidizing acid*. It reacts with active metals to produce oxides of nitrogen, but *not* hydrogen, H_2.

Example 4-8

Which of the following metals can displace hydrogen from hydrochloric acid solution? Write appropriate equations for any reactions that can occur.

$$Al, \quad Cu, \quad Ag$$

Plan

The activity series of the metals, Table 4-11, tells us that copper and silver *do not* displace hydrogen from solutions of nonoxidizing acids. Aluminum is an active metal that can displace H_2 from HCl and form aluminum chloride (photo on page 150).

Solution

$$6HCl(aq) + 2Al(s) \longrightarrow 3H_2(g) + 2AlCl_3(aq)$$

$$6[H^+(aq) + Cl^-(aq)] + 2Al(s) \longrightarrow 3H_2(g) + 2[Al^{3+}(aq) + 3Cl^-(aq)]$$

$$6H^+(aq) + 2Al(s) \longrightarrow 3H_2(g) + 2Al^{3+}(aq)$$

EOC 55, 56

Very active metals can even displace hydrogen from water. However, such reactions of very active metals of Group IA are dangerous because they generate enough heat to cause explosive ignition of the hydrogen (Figure 4-5). The reaction of potassium, or another metal of Group IA, with water is also a *displacement reaction*:

$$2K(s) + 2H_2O(\ell) \longrightarrow 2[K^+(aq) + OH^-(aq)] + H_2(g)$$

Example 4-9

Which of the following metals can displace hydrogen from water at room temperature? Write appropriate equations for any reactions that can occur.

$$Sn, \quad Ca, \quad Hg$$

Plan

The activity series, Table 4-11, tells us that tin and mercury *cannot* displace hydrogen from water. Calcium is a very active metal (Table 4-11) that displaces hydrogen from cold water and forms calcium hydroxide, a strong soluble base.

Figure 4-5

Potassium, like other Group IA metals, reacts vigorously with water. The room was completely dark, and all the light for this photograph was produced by dropping a small piece of potassium into a beaker of water.

Solution

$$Ca(s) + 2H_2O(\ell) \longrightarrow H_2(g) + Ca(OH)_2(aq)$$

$$Ca(s) + 2H_2O(\ell) \longrightarrow H_2(g) + [Ca^{2+}(aq) + 2OH^-(aq)]$$

$$Ca(s) + 2H_2O(\ell) \longrightarrow H_2(g) + Ca^{2+}(aq) + 2OH^-(aq)$$

EOC 59, 60

The reaction of calcium with water at room temperature produces a lazy stream of bubbles of hydrogen.

3 [Active Nonmetal + Salt of Less Active Nonmetal] → [Less Active Nonmetal + Salt of Active Nonmetal]

Many *nonmetals* displace less active nonmetals from combination with a metal or other cation. For example, when chlorine is bubbled through a solution containing bromide ions (derived from a soluble ionic salt such as sodium bromide, NaBr), chlorine displaces bromide ions to form elemental bromine and chloride ions (as aqueous sodium chloride):

$$Cl_2(g) + 2[Na^+(aq) + Br^-(aq)] \longrightarrow 2[Na^+(aq) + Cl^-(aq)] + Br_2(\ell)$$
chlorine   sodium bromide   sodium chloride   bromine

Similarly, when bromine is added to a solution containing iodide ions, the iodide ions are displaced by bromine to form iodine and bromide ions:

$$Br_2(\ell) + 2[Na^+(aq) + I^-(aq)] \longrightarrow 2[Na^+(aq) + Br^-(aq)] + I_2(s)$$
bromine   sodium iodide   sodium bromide   iodine

Each halogen will displace less active (heavier) halogens from their binary salts; i.e., the order of increasing activities is

$$I_2 < Br_2 < Cl_2 < F_2$$

Activity of the halogens decreases as the group is descended.

Conversely, a halogen will *not* displace more active (lighter) members from their salts:

$$I_2(s) + 2F^- \longrightarrow \text{no reaction}$$

Example 4-10

Which of the following combinations would result in a displacement reaction? Write appropriate equations for any reactions that occur.
(a) $I_2(s) + NaBr(aq) \longrightarrow$
(b) $Cl_2(g) + NaI(aq) \longrightarrow$
(c) $Br_2(\ell) + NaCl(aq) \longrightarrow$

Plan

The activity of the halogens decreases from top to bottom in the periodic table. We see (a) that Br is above I and (c) that Cl is above Br in the periodic table. Therefore neither combination (a) nor combination (c) could result in reaction. Cl is above I in the periodic table, and so combination (b) results in a displacement reaction.

Solution

The more active halogen, Cl_2, displaces the less active halogen, I_2, from its compounds.

$$2NaI(aq) + Cl_2(g) \longrightarrow I_2(s) + 2NaCl(aq)$$

$$2[Na^+(aq) + I^-(aq)] + Cl_2(g) \longrightarrow I_2(s) + 2[Na^+(aq) + Cl^-(aq)]$$

$$2I^-(aq) + Cl_2(g) \longrightarrow I_2(s) + 2Cl^-(aq)$$

EOC 61, 62

4-7 Oxidation Numbers

Many reactions involve the transfer of electrons from one species to another. They are called **oxidation–reduction reactions** or simply **redox reactions**. Some of the reactions discussed earlier in this chapter are also redox reactions, although they were not identified as such. We use oxidation numbers to keep track of electron transfers.

Table 4-12
Common Oxidation Numbers for Group A Elements in Compounds and Ions

Element(s)	Common Ox. Nos.	Examples	Other Ox. Nos.
H	+1	H_2O, CH_4, NH_4Cl	−1 in metal hydrides, e.g., NaH, CaH_2
Group IA	+1	KCl, NaH, $RbNO_3$, K_2SO_4	None
Group IIA	+2	$CaCl_2$, MgH_2, $Ba(NO_3)_2$, $SrSO_4$	None
Group IIIA	+3	$AlCl_3$, BF_3, $Al(NO_3)_3$, GaI_3	None in common compounds
Group IVA	+2 +4	CO, PbO, $SnCl_2$, $Pb(NO_3)_2$ CCl_4, SiO_2, SiO_3^{2-}, $SnCl_4$	Many others are also seen, including −4, −3, −2, −1, +1, +3
Group VA	−3 in binary compounds with metals −3 in NH_4^+, binary compounds with H	Mg_3N_2, Na_3P, Cs_3As NH_3, PH_3, AsH_3, NH_4^+	+3, e.g., NO_2^-, PCl_3 +5, e.g., NO_3^-, PO_4^{3-}, AsF_5, P_4O_{10}
O	−2	H_2O, P_4O_{10}, Fe_2O_3, CaO, ClO_3^-	+2 in OF_2 −1 in peroxides, e.g., H_2O_2, Na_2O_2 $-\frac{1}{2}$ in superoxides, e.g., KO_2, RbO_2
Group VIA (other than O)	−2 in binary compounds with metals and H −2 in binary compounds with NH_4^+	H_2S, CaS, Fe_2S_3, Na_2Se $(NH_4)_2S$, $(NH_4)_2Se$	+4 with O and the lighter halogens, e.g., SO_2, SeO_2, Na_2SO_3, SO_3^{2-}, SF_4 +6 with O and the lighter halogens, e.g., SO_3, TeO_3, H_2SO_4, SO_4^{2-}, SF_6
Group VIIA	−1 in binary compounds with metals and H −1 in binary compounds with NH_4^+	MgF_2, KI, $ZnCl_2$, $FeBr_3$ NH_4Cl, NH_4Br	Except F, with O and the lighter halogens +1, e.g., BrF, ClO^-, BrO^- +3, e.g., ICl_3, ClO_2^-, BrO_2^- +5, e.g., BrF_5, ClO_3^-, BrO_3^- +7, e.g., IF_7, ClO_4^-, BrO_4^-

The **oxidation number**, or **oxidation state**, of an element in a simple *binary* ionic compound is the number of electrons gained or lost by an atom of that element when it forms the compound. In the case of a single-atom ion, it corresponds to the actual charge on the ion. In molecular compounds, oxidation numbers do not have the same physical significance they have in ionic compounds. However, they are very useful mechanical aids in writing formulas and in balancing equations. In molecular species, the oxidation numbers are assigned according to an arbitrary, but useful, set of rules. The element farther to the right and higher up in the periodic table is assigned a negative oxidation number, and the element farther to the left and lower down in the periodic table is assigned a positive oxidation number.

The general rules for assigning oxidation numbers follow. These rules are not comprehensive, but they cover most cases.

1. The oxidation number of any free, uncombined element is zero. This includes multiatomic elements such as H_2, O_2, O_3, and S_8.

> Binary means two. Binary compounds contain two elements.

> The terms "oxidation number" and "oxidation state" are used interchangeably.

Table 4-13
The Most Common Nonzero Oxidation States (numbers) of the Elements

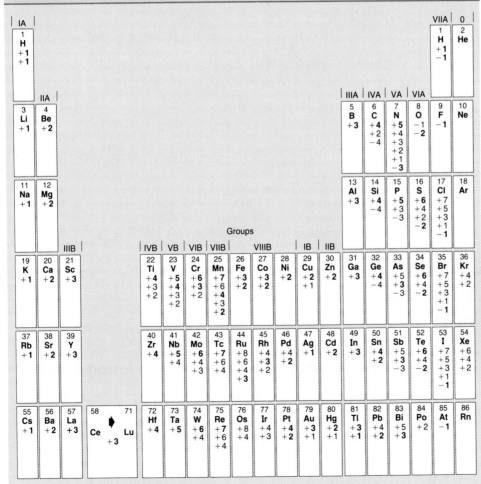

2. The oxidation number of an element in a simple (monatomic) ion is the charge on the ion. In a polyatomic ion, the sum of the oxidation numbers of the constituent atoms is equal to the charge on the ion.
3. The sum of the oxidation numbers of all atoms in the compound is zero.

Tables 4-12 and 4-13 (see pp. 144–145) summarize the common nonzero oxidation numbers for the Group A elements.

Example 4-11

Determine the oxidation numbers of nitrogen in the following species: (a) N_2O_4, (b) NH_3, (c) NO_3^-, (d) N_2.

Plan

We recall that oxidation numbers are represented *per atom* and that the sum of the oxidation numbers in a molecule is zero, while the sum of the oxidation numbers in an ion equals the charge on the ion.

Solution

By convention, *oxidation numbers* are represented as $+n$ and $-n$, while ionic charges are represented as $n+$ and $n-$. We shall circle oxidation numbers associated with formulas and show them in red. Both oxidation numbers and ionic charges can be combined algebraically.

(a) The oxidation number of O is -2. The sum of the oxidation numbers for all atoms in a molecule must be zero:

ox. no./atom: $\textcircled{x}\ \textcircled{-2}$
$\qquad\qquad\qquad N_2O_4$

total ox. no.: $2x + 4(-2) = 0$ or $x = \boxed{+4}$

(b) The oxidation number of H is $+1$:

The formula for ammonia, NH_3, is usually written with the more electronegative element first—the only reason is that it has been written this way for many years.

ox. no./atom: $\textcircled{x}\ \textcircled{+1}$
$\qquad\qquad\qquad NH_3$

total ox. no.: $x + 3(1) = 0$ or $x = \boxed{-3}$

(c) The sum of the oxidation numbers for all atoms in an ion equals the charge on the ion:

ox. no./atom: $\textcircled{x}\ \textcircled{-2}$
$\qquad\qquad\qquad NO_3^-$

total ox. no.: $x + 3(-2) = -1$ or $x = \boxed{+5}$

(d) The oxidation number of any free element is $\boxed{\text{zero}}$.

EOC 67, 68

Aqueous solutions of some compounds that contain chromium. Left to right: chromium(II) chloride ($CrCl_2$) is blue; chromium(III) chloride ($CrCl_3$) is green; potassium chromate (K_2CrO_4) is yellow; potassium dichromate ($K_2Cr_2O_7$) is orange.

4-8 Oxidation–Reduction Reactions—An Introduction

Several reactions that we discussed earlier are redox reactions.

> Displacement reactions are *always* redox reactions. Acid–base reactions are *never* redox reactions.

We now describe the important terms and illustrate their meanings.

Oxidation–reduction reactions occur in every area of chemistry and biochemistry. We learn to identify oxidizing agents and reducing agents and to balance oxidation–reduction equations. These skills are necessary for the study of electrochemistry in Chapter 21. Electrochemistry involves electron transfer between physically separated oxidizing and reducing agents and interconversions between chemical energy and electrical energy. These skills are also fundamental to biology and biochemistry, because many reactions associated with metabolism are redox reactions.

The term "oxidation" originally referred to the combination of a substance with oxygen. This results in an increase in the oxidation number of an element in that substance. According to the original definition, the following reactions involve oxidation of the substance shown on the far left of each equation. Oxidation numbers are shown for *one* atom of the indicated kind.

1. The formation of rust, Fe_2O_3, iron(III) oxide: oxidation state of Fe

$$4Fe(s) + 3O_2(g) \longrightarrow 2Fe_2O_3(s) \qquad 0 \longrightarrow +3$$

2. Combustion reactions: oxidation state of C

$$C(s) + O_2(g) \longrightarrow CO_2(g) \qquad 0 \longrightarrow +4$$

$$2CO(g) + O_2(g) \longrightarrow 2CO_2(g) \qquad +2 \longrightarrow +4$$

$$C_3H_8(g) + 5O_2(g) \longrightarrow 3CO_2(g) + 4H_2O(g) \qquad -8/3 \longrightarrow +4$$

> Oxidation number is a formal concept adopted for our convenience. The numbers are determined solely by reliance upon rules. These rules can result in a fractional oxidation number, as shown here. This does not mean that electronic charges are split.

Originally *reduction* described the removal of oxygen from a compound. Oxide ores are reduced to metals (a very real reduction in mass). For example, tungsten for use in light bulb filaments can be prepared by reduction of tungsten(VI) oxide with hydrogen at 1200°C:

oxidation number of W

$$WO_3(s) + 3H_2(g) \longrightarrow W(s) + 3H_2O(g) \qquad +6 \longrightarrow 0$$

Tungsten is reduced and its oxidation state decreases from +6 to zero. Hydrogen is oxidized from zero to the +1 oxidation state. The terms "oxidation" and "reduction" are now applied much more broadly.

> In biological systems *oxidation* usually corresponds to the removal of hydrogen.

> **Oxidation** is an algebraic increase in oxidation number and corresponds to the loss, or apparent loss, of electrons. **Reduction** is an algebraic decrease in oxidation number and corresponds to a gain, or apparent gain, of electrons.

> In biological systems *reduction* usually corresponds to the addition of hydrogen to molecules or polyatomic ions.

Electrons are neither created nor destroyed in chemical reactions. So oxidation and reduction always occur simultaneously, and to the same extent, in ordinary chemical reactions. In the four equations cited previously as *examples of oxidation*, the oxidation numbers of iron and carbon atoms increase as they are oxidized. In each case oxygen is reduced as its oxidation number decreases from zero to −2.

> Oxidizing agents are species that (1) oxidize other substances, (2) are reduced, and (3) gain (or appear to gain) electrons. Reducing agents are species that (1) reduce other substances, (2) are oxidized, and (3) lose (or appear to lose) electrons.

The equations below represent examples of redox reactions. Oxidation numbers are shown above the formulas, and oxidizing and reducing agents are indicated:

$$\overset{(0)}{2Fe(s)} + \overset{(0)}{3Cl_2(g)} \longrightarrow \overset{(+3)(-1)}{2FeCl_3(s)}$$
red. agt.   ox. agt.

$$\overset{(+3)(-1)}{2FeBr_3(aq)} + \overset{(0)}{3Cl_2(g)} \longrightarrow \overset{(+3)(-1)}{2FeCl_3(aq)} + \overset{(0)}{3Br_2(\ell)}$$
red. agt.   ox. agt.

Equations for redox reactions can also be written as total ionic and net ionic equations. For example, the previous equation may also be written as shown below. We distinguish between oxidation numbers and actual charges on ions by denoting oxidation numbers as $+n$ or $-n$ *in red circles just above the symbols of the elements*, and actual charges as $n+$ or $n-$ above and to the right of formulas of ions.

$$2[Fe^{3+}(aq) + 3Br^-(aq)] + 3Cl_2(g) \longrightarrow 2[Fe^{3+}(aq) + 3Cl^-(aq)] + 3Br_2(\ell)$$

The spectator ions, Fe^{3+}, do not participate in electron transfer. Their cancellation allows us to focus on the oxidizing agent, $Cl_2(g)$, and the reducing agent, $Br^-(aq)$.

$$2Br^-(aq) + Cl_2(g) \longrightarrow 2Cl^-(aq) + Br_2(\ell)$$

A **disproportionation reaction** is a redox reaction in which the same species is oxidized and reduced.

Iron reacting with chlorine to form iron(III) chloride.

Example 4-12

Write each of the following formula unit equations as a net ionic equation if the two differ. Which ones are redox reactions? For the redox reactions, identify the oxidizing agent, the reducing agent, the species oxidized, and the species reduced.

(a) $2AgNO_3(aq) + Cu(s) \longrightarrow Cu(NO_3)_2(aq) + 2Ag(s)$

(b) $2KClO_3(s) \xrightarrow{\Delta} 2KCl(s) + 3O_2(g)$

(c) $3AgNO_3(aq) + K_3PO_4(aq) \longrightarrow Ag_3PO_4(s) + 3KNO_3(aq)$

Plan

To write ionic equations, we must recognize compounds that are (1) soluble in water and (2) ionized or dissociated in aqueous solutions. To determine which are oxidation–reduction reactions, we must assign an oxidation number to each element.

Solution

(a) According to the solubility rules (page 132), both silver nitrate, $AgNO_3$, and copper(II) nitrate, $Cu(NO_3)_2$, are water-soluble ionic compounds. The total ionic equation and oxidation numbers are

$$\overset{(+1)}{2[Ag^+(aq)} + \overset{(+5)(-2)}{NO_3^-(aq)]} + \overset{(0)}{Cu(s)} \longrightarrow [\overset{(+2)}{Cu^{2+}(aq)} + \overset{(+5)(-2)}{2NO_3^-(aq)}] + \overset{(0)}{2Ag(s)}$$

Metallic silver formed by immersing a spiral of copper wire in a silver nitrate solution (Example 4-12a).

The nitrate ions, NO_3^-, are spectator ions. Canceling them from both sides gives the net ionic equation:

$$\overset{+1}{2Ag^+}(aq) + \overset{0}{Cu}(s) \longrightarrow \overset{+2}{Cu^{2+}}(aq) + \overset{0}{2Ag}(s)$$

This is a redox equation. The oxidation number of silver decreases from + 1 to zero; silver ion is reduced and is the oxidizing agent. The oxidation number of copper increases from zero to +2; copper is oxidized and is the reducing agent.

(b) This reaction involves two solids and a gas, so the formula unit and net ionic equations are identical. It is a redox reaction:

$$\overset{+1\ +5\ -2}{2KClO_3}(s) \longrightarrow \overset{+1\ -1}{2KCl}(s) + \overset{0}{3O_2}(g)$$

The oxidizing agent is $KClO_3$. The chlorine in $KClO_3$ is reduced from the +5 to the −1 oxidation state. The oxidation number of oxygen increases from −2 to zero; oxygen is oxidized and is the reducing agent. We might also say that $KClO_3$ is both oxidizing agent and reducing agent; this is a disproportionation reaction.

(c) The solubility rules indicate that all these salts are soluble and ionic except for silver phosphate, Ag_3PO_4. The total ionic equation is

$$3[Ag^+(aq) + NO_3^-(aq)] + [3K^+(aq) + PO_4^{3-}(aq)] \longrightarrow$$
$$Ag_3PO_4(s) + 3[K^+(aq) + NO_3^-(aq)]$$

Eliminating the spectator ions gives the net ionic equation:

$$\overset{+1}{3Ag^+}(aq) + \overset{+5\ -2}{PO_4^{3-}}(aq) \longrightarrow \overset{+1\ +5\ -2}{Ag_3PO_4}(s)$$

There are no changes in oxidation numbers; this is not a redox reaction.

EOC 73, 74

The reaction of $AgNO_3$(aq) and K_3PO_4(aq) is a precipitation reaction (Example 4-12c).

Balancing Oxidation–Reduction Equations

Our rules for assigning oxidation numbers are constructed so that

the total increase in oxidation numbers must equal the total decrease in oxidation numbers in all redox reactions.

This equivalence provides the basis for balancing redox equations. Although there is no single "best method" for balancing all redox equations, two methods are particularly useful: (1) the change-in-oxidation-number method and (2) the ion–electron method, which is used extensively in electrochemistry (Chapter 21).

Most redox equations can be balanced by both methods, but in some instances one may be easier to use than the other.

4-9 Change-in-Oxidation-Number Method

The next few examples illustrate this method, which is based on *equal total increases and decreases in oxidation numbers*. While many redox equations can be balanced by simple inspection, you should learn the method because it can be used to balance difficult equations. The general procedure follows.

1. Write as much of the overall *unbalanced* equation as possible.
2. Assign oxidation numbers to find the elements that undergo changes in oxidation numbers.
3. a. Draw a bracket to connect atoms of the element that are oxidized. Show the increase in oxidation number *per atom*. Draw a bracket to connect atoms of the element that are reduced. Show the decrease in oxidation number *per atom*.
   b. Determine the factors that will make the *total* increase and decrease in oxidation numbers equal.
4. Insert coefficients into the equation to make the total increase and decrease in oxidation numbers equal.
5. Balance the other atoms by inspection.

Aluminum wire reacting with hydrochloric acid.

Example 4-13

Aluminum reacts with hydrochloric acid to form aqueous aluminum chloride and gaseous hydrogen. Balance the formula unit equation and identify the oxidizing and reducing agents.

Plan

We follow the five-step procedure, one step at a time.

Solution

The unbalanced formula unit equation and oxidation numbers (Steps 1 and 2) are

$$\overset{(+1)(-1)}{HCl(aq)} + \overset{(0)}{Al(s)} \longrightarrow \overset{(+3)(-1)}{AlCl_3(aq)} + \overset{(0)}{H_2(g)}$$

The oxidation number of Al increases from 0 to +3. Al is the reducing agent; it is oxidized. The oxidation number of H decreases from +1 to 0. HCl is the oxidizing agent; it is reduced.

$$\overset{(+1)}{HCl} + \overset{(0)}{Al} \longrightarrow \overset{(+3)}{AlCl_3} + \overset{(0)}{H_2} \quad \text{(Step 3a)}$$

We make the *total* increase and decrease in oxidation numbers equal (Step 3b):

Oxidation Numbers	Change/Atom	Equalizing Changes Gives
Al = 0 ⟶ Al = +3	+3	1(+3) = +3
H = +1 ⟶ H = 0	−1	3(−1) = −3

Each change must be multiplied by two because there are two H's in each H_2.

$$2(+3) = +6 \text{ (total increase)} \qquad 2(-3) = -6 \text{ (total decrease)}$$

We need 2 Al's and 6 H's on each side of the equation (Step 4):

$$6HCl(aq) + 2Al(s) \longrightarrow 2AlCl_3(aq) + 3H_2(g)$$

EOC 77, 78

All balanced equations must satisfy two criteria:

1. **There must be mass balance.** That is, the same number of atoms of each kind must be shown as reactants and products.
2. **There must be charge balance.** The sums of actual charges on the left and right sides of the equation must be equal.

Example 4-14

Copper is a widely used metal. Before it is welded (brazed), copper is cleaned by dipping it into nitric acid. HNO_3 oxidizes Cu to Cu^{2+} ions and is reduced to NO. The other product is H_2O. Write the balanced net ionic and formula unit equations for the reaction. Excess HNO_3 is present.

Plan

In writing ionic equations, we recall that strong acids, strong soluble bases, and most soluble salts are strong electrolytes. Then we apply our five-step procedure for redox equations.

Solution

We write the unbalanced net ionic equation and assign oxidation numbers. HNO_3 is a strong acid.

$$\overset{(+1)}{H^+}(aq) + \overset{(+5)}{NO_3^-}(aq) + \overset{(0)}{Cu}(s) \longrightarrow \overset{(+2)}{Cu^{2+}}(aq) + \overset{(+2)}{NO}(g) + H_2O(\ell)$$

We see that copper is oxidized; it is the reducing agent. Nitrate ions are reduced; they are the oxidizing agent.

$$\overset{(+1)}{H^+}(aq) + \overset{(+5)}{NO_3^-}(aq) + \overset{(0)}{Cu}(s) \longrightarrow \overset{(+2)}{Cu^{2+}}(aq) + \overset{(+2)}{NO}(g) + H_2O(\ell)$$

$$+2$$
$$-3$$

We make the *total* increase and decrease in oxidation numbers equal:

Oxidation Numbers	Change/Atom	Equalizing Changes Gives
Cu = 0 ⟶ Cu = +2	+2	3(+2) = +6
N = +5 ⟶ N = +2	−3	2(−3) = −6

Now we balance the *redox part* of the reaction:

$$H^+ + 2\,NO_3^- + 3\,Cu \longrightarrow 3\,Cu^{2+} + 2\,NO + H_2O$$

There are six O's on the left in NO_3^- ions. A coefficient of 4 before H_2O balances O and gives eight H's on the right. So we need eight H^+ ions on the left to balance the net ionic equation.

$$8H^+(aq) + 2NO_3^-(aq) + 3Cu(s) \longrightarrow 3Cu^{2+}(aq) + 2NO(g) + 4H_2O(\ell)$$

This solution contains excess HNO_3, so NO_3^- is the only anion present in significant concentration. Therefore, we add six more NO_3^- ions on each side to give the balanced formula unit equation:

$$8HNO_3(aq) + 3Cu(s) \longrightarrow 3Cu(NO_3)_2(aq) + 2NO(g) + 4H_2O(\ell)$$

Copper is cleaned by dipping it into nitric acid.

4-10 Adding H^+, OH^-, or H_2O to Balance Oxygen or Hydrogen

Frequently we need more oxygen or hydrogen to complete the mass balance for a reaction in aqueous solution. However, we must be careful not to introduce other changes in oxidation number or to use species that could not actually be present in the solution. (We cannot add H_2 or O_2 to equations because these species are not present in aqueous solutions. Acidic solutions do not contain significant concentrations of OH^- ions. Basic solutions do not contain significant concentrations of H^+ ions.) We accomplish this balance as follows:

> In acidic solution: We add only H^+ or H_2O (*not* OH^- in acidic solution)
>
> In basic solution: We add only OH^- or H_2O (*not* H^+ in basic solution)

The following chart shows how to balance hydrogen and oxygen.

Type of solution		To balance O:	and then	To balance H:
	Acidic	Add H_2O	→	Add H^+
	Basic	To balance O: For *each* O needed, (1) add *two* OH^- to side needing O and (2) add *one* H_2O to other side	and then →	To balance H: For *each* H needed, (1) add *one* H_2O to side needing H and (2) add *one* OH^- to other side

---

Example 4-15

"Drāno" drain cleaner is solid sodium hydroxide that contains some aluminum turnings. When Drāno is added to water, the NaOH dissolves rapidly with the evolution of a lot of heat. The Al reduces H_2O in the basic solution to produce $[Al(OH)_4]^-$ ions and H_2 gas, which gives the bubbling action. Write the balanced net ionic and formula unit equations for this reaction.

Plan

We are given formulas for reactants and products. Recall that NaOH is a strong soluble base (OH^- and H_2O can be added to either side as needed). We apply our five-step procedure.

Solution

We write the unbalanced net ionic equation and assign oxidation numbers:

$$\overset{(0)}{OH^-(aq)} + \overset{(+1)}{Al(s)} + \overset{}{H_2O(\ell)} \longrightarrow \overset{(+3)}{[Al(OH)_4]^-(aq)} + \overset{(0)}{H_2(g)}$$

Aluminum is oxidized; it is the reducing agent. H_2O is reduced; it is the oxidizing agent.

$$OH^-(aq) + \overset{(+1)}{H_2O}(\ell) + \overset{(0)}{Al}(s) \longrightarrow \overset{(+3)}{[Al(OH)_4]^-}(aq) + \overset{(0)}{H_2}(g)$$

We make the *total* increase and decrease in oxidation numbers equal:

Oxidation Numbers	Change/Atom	Equalizing Changes Gives
Al = 0 $\longrightarrow$ Al = +3	+3	1(+3) = +3
H = +1 $\longrightarrow$ H = 0	−1	3(−1) = −3

Each change must be multiplied by two because there are 2 H's in each H_2.

$$2(+3) = +6 \text{ (total increase)} \qquad 2(-3) = -6 \text{ (total decrease)}$$

Now we balance the redox part of the equation. We need 2 Al's on each side. Because only one H in each H_2O molecule is reduced, we show six H_2O on the left and three H_2 on the right:

$$OH^- + 6H_2O + 2Al \longrightarrow 2[Al(OH)_4]^- + 3H_2$$

The net charge on the right is 2−, and so we need two OH^- on the left to balance the net ionic equation:

$$2OH^-(aq) + 6H_2O(\ell) + 2Al(s) \longrightarrow 2[Al(OH)_4]^-(aq) + 3H_2(g)$$

This reaction occurs in excess NaOH solution. We need two $Na^+(aq)$ on each side to balance the negative charges:

$$2NaOH(aq) + 6H_2O(\ell) + 2Al(s) \longrightarrow 2Na[Al(OH)_4](aq) + 3H_2(g)$$

EOC 79

The Drāno reaction.

Example 4-16

The breathalyzer detects the presence of ethanol (ethyl alcohol) in the breath of persons suspected of drunken driving. It utilizes the oxidation of ethanol to acetaldehyde by dichromate ions in acidic solution. The $Cr_2O_7^{2-}(aq)$ ion is orange (see page 391). The $Cr^{3+}(aq)$ ion is green. The appearance of a green color signals alcohol in the breath that exceeds the legal limit. Balance the net ionic equation for this reaction.

$$H^+(aq) + Cr_2O_7^{2-}(aq) + C_2H_5OH(\ell) \longrightarrow Cr^{3+}(aq) + C_2H_4O(\ell) + H_2O(\ell)$$

Plan

We are given the unbalanced equation, which includes H^+. This tells us that the reaction occurs in acidic solution. We apply our five-step procedure.

Solution

We first assign oxidation numbers to the elements that change:

$$H^+ + \overset{(-2)}{C_2H_5OH} + \overset{(+6)}{Cr_2O_7^{2-}} \longrightarrow \overset{(+3)}{Cr^{3+}} + \overset{(-1)}{C_2H_4O} + H_2O$$

We see that ethanol is oxidized; it is the reducing agent. $Cr_2O_7^{2-}$ ions are reduced; they are the oxidizing agent.

$$\text{H}^+ + \overset{\overset{\scriptstyle(-2)}{}}{\text{C}_2\text{H}_5\text{OH}} + \overset{\overset{\scriptstyle(+6)}{}}{\text{Cr}_2\text{O}_7^{2-}} \longrightarrow \overset{\overset{\scriptstyle(+3)}{}}{\text{Cr}^{3+}} + \overset{\overset{\scriptstyle(-1)}{}}{\text{C}_2\text{H}_4\text{O}} + \text{H}_2\text{O}$$

Oxidation Numbers	Change/Atom	Equalizing Changes Gives
Cr = +6 ⟶ Cr = +3	−3	1(−3) = −3
C = −2 ⟶ C = −1	+1	3(+1) = +3

Each change must be multiplied by two because there are two Cr's in each $\text{Cr}_2\text{O}_7^{2-}$ and two C's in $\text{C}_2\text{H}_5\text{OH}$.

$$2(-3) = -6 \text{ (total decrease)} \qquad 2(+3) = +6 \text{ (total increase)}$$

We need 2 Cr's and 6 C's on each side of the equation to balance the redox part:

$$\text{H}^+ + 3\text{C}_2\text{H}_5\text{OH} + \text{Cr}_2\text{O}_7^{2-} \longrightarrow 2\text{Cr}^{3+} + 3\text{C}_2\text{H}_4\text{O} + \text{H}_2\text{O}$$

Now we balance H and O using our chart. There are 10 O's on the left and only 4 O's on the right. So we add 6 *more* H_2O molecules on the right.

$$\text{H}^+ + 3\text{C}_2\text{H}_5\text{OH} + \text{Cr}_2\text{O}_7^{2-} \longrightarrow 2\text{Cr}^{3+} + 3\text{C}_2\text{H}_4\text{O} + 7\text{H}_2\text{O}$$

Now there are 26 H's on the right and only 19 on the left. So we add 7 *more* H^+ ions on the left to give the balanced net ionic equation.

$$8\text{H}^+(aq) + 3\text{C}_2\text{H}_5\text{OH}(\ell) + \text{Cr}_2\text{O}_7^{2-}(aq) \longrightarrow 2\text{Cr}^{3+}(aq) + 3\text{C}_2\text{H}_4\text{O}(\ell) + 7\text{H}_2\text{O}(\ell)$$

EOC 80

Every balanced equation must have both mass balance and charge balance. Once the redox part of an equation has been balanced, we may count *either* atoms or charges. After we balanced the redox part in Example 4-16, we had

$$\text{H}^+ + 3\text{C}_2\text{H}_5\text{OH} + \text{Cr}_2\text{O}_7^{2-} \longrightarrow 2\text{Cr}^{3+} + 3\text{C}_2\text{H}_4\text{O} + \text{H}_2\text{O}$$

The net charge on the left side is (1 + 2−) = 1−. On the right, it is 2(3+) = 6+. Because H^+ is the *only charged species whose coefficient isn't known,* we add 7 *more* H^+ to give a net charge of 6+ on both sides.

$$8\text{H}^+ + 3\text{C}_2\text{H}_5\text{OH} + \text{Cr}_2\text{O}_7^{2-} \longrightarrow 2\text{Cr}^{3+} + 3\text{C}_2\text{H}_4\text{O} + \text{H}_2\text{O}$$

Now we have 10 O's on the left and only 4 O's on the right. We add six *more* H_2O molecules to give the balanced net ionic equation.

$$8\text{H}^+(aq) + 3\text{C}_2\text{H}_5\text{OH}(\ell) + \text{Cr}_2\text{O}_7^{2-}(aq) \longrightarrow$$
$$2\text{Cr}^{3+}(aq) + 3\text{C}_2\text{H}_4\text{O}(\ell) + 7\text{H}_2\text{O}(\ell)$$

How can you tell whether to balance atoms or charges first? Look at the equation *after you have balanced the redox part.* Decide which is simpler, and do that. In the preceding equation, it is easier to balance charges than to balance atoms.

4-11 The Ion–Electron Method (Half-Reaction Method)

In the ion–electron method we separate and completely balance equations describing oxidation and reduction **half-reactions**. This is followed by equalizing the numbers of electrons gained and lost in each. Finally, the resulting half-reactions are added to give the overall balanced equation. The general procedure follows.

1. Write as much of the overall unbalanced equation as possible.
2. Construct unbalanced oxidation and reduction half-reactions (these are usually incomplete as well as unbalanced). Show complete formulas for polyatomic ions and molecules.
3. Balance by inspection all elements in each half-reaction, except H and O. Then use the chart on page 152 to balance H and O in each half-reaction.
4. Balance the charge in each half-reaction by adding electrons as "products" or "reactants."
5. Balance the electron transfer by multiplying the balanced half-reactions by appropriate integers.
6. Add the resulting half-reactions and eliminate any common terms to obtain the balanced equation.

Example 4-17

A useful analytical procedure involves the oxidation of iodide ions to free iodine. The free iodine is then titrated with a standard solution of sodium thiosulfate, $Na_2S_2O_3$. Iodine oxidizes $S_2O_3^{2-}$ ions to tetrathionate ions, $S_4O_6^{2-}$, and is reduced to I^- ions. Write the balanced net ionic equation for this reaction.

Plan

We are given the formulas for two reactants and two products. We use these to write as much of the equations as possible (the skeletal equation). We construct and balance the appropriate half-reactions using the rules just described. Then we add the half-reactions.

Solution

$$I_2 + S_2O_3^{2-} \longrightarrow I^- + S_4O_6^{2-} \quad \text{(skeletal equation)}$$

$$I_2 \longrightarrow I^- \quad \text{(red. half-reaction)}$$

$$I_2 \longrightarrow 2I^-$$

$$I_2 + 2e^- \longrightarrow 2I^- \quad \text{(balanced red. half-reaction)}$$

Each I_2 gains $2e^-$. I_2 is reduced; it is the oxidizing agent.

$$S_2O_3^{2-} \longrightarrow S_4O_6^{2-} \quad \text{(ox. half-reaction)}$$

$$2S_2O_3^{2-} \longrightarrow S_4O_6^{2-}$$

$$2S_2O_3^{2-} \longrightarrow S_4O_6^{2-} + 2e^- \quad \text{(balanced ox. half-reaction)}$$

Each $S_2O_3^{2-}$ ion loses e^-. $S_2O_3^{2-}$ is oxidized; it is the reducing agent.

Each half-reaction has two electrons. We can add these half-reactions term by term and cancel the electrons:

$$I_2 + 2e^- \longrightarrow 2I^-$$

$$2S_2O_3^{2-} \longrightarrow S_4O_6^{2-} + 2e^-$$

$$I_2(s) + 2S_2O_3^{2-}(aq) \longrightarrow 2I^-(aq) + S_4O_6^{2-}(aq)$$

These common household chemicals should never be mixed because they react to form chloramine (NH_2Cl), a very poisonous volatile compound.

$$NH_3(aq) + ClO^-(aq) \longrightarrow$$
$$NH_2Cl(aq) + OH^-(aq)$$

These common household chemicals should never be mixed because they react to form chlorine, a very poisonous gas.

$$2H^+(aq) + ClO^-(aq) + Cl^-(aq) \longrightarrow$$
$$Cl_2(g) + H_2O(\ell)$$

Bleaches sold under trade names such as Clorox and Purex are 5% solutions of sodium hypochlorite. The hypochlorite ion is a very strong oxidizing agent in basic solution. It oxidizes many stains on fabrics to colorless substances.

Example 4-18

In basic solution, hypochlorite ions, ClO^-, oxidize chromite ions, CrO_2^-, to chromate ions, CrO_4^{2-}, and are reduced to chloride ions. Write the balanced net ionic equation for this reaction.

Plan

We are given the formulas for two reactants and two products; we can write the skeletal equation. The reaction occurs in basic solution; we can add OH^- and H_2O as needed. We construct and balance the appropriate half-reactions, equalize the electron transfer, add the half-reactions, and eliminate common terms.

Solution

$$CrO_2^- + ClO^- \longrightarrow CrO_4^{2-} + Cl^- \qquad \text{(skeletal equation)}$$

$$CrO_2^- \longrightarrow CrO_4^{2-} \qquad \text{(ox. half-rxn)}$$

$$CrO_2^- + 4OH^- \longrightarrow CrO_4^{2-} + 2H_2O$$

$$CrO_2^- + 4OH^- \longrightarrow CrO_4^{2-} + 2H_2O + 3e^- \qquad \text{(balanced ox. half-rxn)}$$

$$ClO^- \longrightarrow Cl^- \qquad \text{(red. half-rxn)}$$

$$ClO^- + H_2O \longrightarrow Cl^- + 2OH^-$$

$$ClO^- + H_2O + 2e^- \longrightarrow Cl^- + 2OH^- \qquad \text{(balanced red. half-rxn)}$$

One half-reaction involves three electrons and the other involves two electrons. We balance the electron transfer and add the half-reactions term by term.

$$2(CrO_2^- + 4OH^- \longrightarrow CrO_4^{2-} + 2H_2O + 3e^-)$$

$$3(ClO^- + H_2O + 2e^- \longrightarrow Cl^- + 2OH^-)$$

$$2CrO_2^- + 8OH^- + 3ClO^- + 3H_2O \longrightarrow 2CrO_4^{2-} + 4H_2O + 3Cl^- + 6OH^-$$

We see six OH^- and three H_2O that can be eliminated from both sides to give the balanced net ionic equation.

$$2CrO_2^-(aq) + 2OH^-(aq) + 3ClO^-(aq) \longrightarrow 2CrO_4^{2-}(aq) + H_2O(\ell) + 3Cl^-(aq)$$

Example 4-19

Potassium permanganate oxidizes iron(II) sulfate to iron(III) sulfate in sulfuric acid solution. Permanganate ions are reduced to manganese(II) ions. Write the balanced net ionic and formula unit equations for this reaction.

Plan

We use the given information to write the skeletal equation. The reaction occurs in H_2SO_4 solution; we can add H^+ and H_2O as needed to construct and balance the appropriate half-reactions. Then we proceed as in earlier examples.

Solution

$$Fe^{2+} + MnO_4^- \longrightarrow Fe^{3+} + Mn^{2+} \qquad \text{(skeletal equation)}$$

$$Fe^{2+} \longrightarrow Fe^{3+} \qquad \text{(ox. half-reaction)}$$

$$Fe^{2+} \longrightarrow Fe^{3+} + \boxed{1e^-} \qquad \text{(balanced ox. half-reaction)}$$

$$MnO_4^- \longrightarrow Mn^{2+} \qquad \text{(red. half-reaction)}$$

$$MnO_4^- + \boxed{8H^+} \longrightarrow Mn^{2+} + \boxed{4H_2O}$$

$$MnO_4^- + 8H^+ + \boxed{5e^-} \longrightarrow Mn^{2+} + 4H_2O \qquad \text{(balanced red. half-reaction)}$$

One half-reaction involves one electron and the other involves five electrons. Now we balance the electron transfer and then add the two equations term by term. This gives the balanced net ionic equation.

$$5(Fe^{2+} \longrightarrow Fe^{3+} + 1e^-)$$

$$\underline{1(MnO_4^- + 8H^+ + 5e^- \longrightarrow Mn^{2+} + 4H_2O)}$$

$$5Fe^{2+}(aq) + MnO_4^-(aq) + 8H^+(aq) \longrightarrow 5Fe^{3+}(aq) + Mn^{2+}(aq) + 4H_2O(\ell)$$

The reaction occurs in H_2SO_4 solution. The SO_4^{2-} ion is the counter anion in the formula unit equation. Because the Fe^{3+} ion occurs twice in $Fe_2(SO_4)_3$, there must be an even number of Fe. So the net ionic equation is multiplied by two. Then we add $18SO_4^{2-}$ to each side to give complete formulas in the balanced formula unit equation.

$$10FeSO_4(aq) + 2KMnO_4(aq) + 8H_2SO_4(aq) \longrightarrow$$
$$5Fe_2(SO_4)_3(aq) + 2MnSO_4(aq) + K_2SO_4(aq) + 8H_2O(\ell)$$

EOC 81, 82

Key Terms

Acid   A substance that produces $H^+(aq)$ ions in aqueous solution. Strong acids ionize completely or almost completely in dilute aqueous solution. Weak acids ionize only slightly.

Active metal   A metal that loses electrons readily to form cations.

Activity series   A listing of metals (and hydrogen) in order of decreasing activity.

Amphoterism   The ability to react with both acids and bases.

Atomic number   The number of protons in the nucleus of an atom of an element.

Base   A substance that produces OH^-(aq) ions in aqueous solution. Strong soluble bases are soluble in water and are completely *dissociated*. Weak bases ionize only slightly.

Chemical periodicity   The variation in properties of elements with their positions in the periodic table.

Combustion reaction   A highly exothermic reaction of a substance with oxygen, usually with a visible flame.

Displacement reaction   A reaction in which one element displaces another from a compound.

Disproportionation reaction   A redox reaction in which the oxidizing agent and the reducing agent are the same species.

Dissociation   In aqueous solution, the process in which a *solid ionic compound* separates into its ions.

Electrolyte   A substance whose aqueous solutions conduct electricity.

Formula unit equation   An equation for a chemical reaction in which all formulas are written as complete formulas.

Group (family)   The elements in a vertical column of the periodic table.

Ionization   In aqueous solution, the process in which a *molecular* compound reacts with water and forms ions.

Metal   An element below and to the left of the stepwise division (metalloids) in the upper right corner of the periodic table; about 80% of the known elements are metals.

Metalloids   Elements with properties intermediate between metals and nonmetals: B, Al, Si, Ge, As, Sb, Te, Po, and At.

Net ionic equation   An equation that results from canceling spectator ions and eliminating brackets from a total ionic equation.

Neutralization   The reaction of an acid with a base to form a salt and water. Usually, the reaction of hydrogen ions with hydroxide ions to form water molecules.

Nonelectrolyte   A substance whose aqueous solutions do not conduct electricity.

Nonmetals   Elements above and to the right of the metalloids in the periodic table.

Oxidation   An algebraic increase in oxidation number; may correspond to a loss of electrons.

Oxidation numbers   Arbitrary numbers that can be used as mechanical aids in writing formulas and balancing equations; for single-atom ions they correspond to the charge on the ion; more electronegative atoms are assigned negative oxidation numbers.

Oxidation states   See *Oxidation numbers*.

Oxidation–reduction reaction   A reaction in which oxidation and reduction occur; also called redox reactions.

Oxidizing agent   The substance that oxidizes another substance and is reduced.

Period   The elements in a horizontal row of the periodic table.

Periodicity   Regular periodic variations of properties of elements with atomic number (and position in the periodic table).

Periodic law   The properties of the elements are periodic functions of their atomic numbers.

Periodic table   An arrangement of elements in order of increasing atomic number that also emphasizes periodicity.

Precipitate   An insoluble solid that forms and separates from a solution.

Precipitation reaction   A reaction in which a precipitate forms.

Reducing agent   The substance that reduces another substance and is oxidized.

Reduction   An algebraic decrease in oxidation number; may correspond to a gain of electrons.

Reversible reaction   A reaction that occurs in both directions; indicated by double arrows ($\rightleftharpoons$).

Salt   A compound that contains a cation other than H^+ and an anion other than OH^- or O^{2-}.

Spectator ions   Ions in solution that do not participate in a chemical reaction.

Strong electrolyte   A substance that conducts electricity well in dilute aqueous solution.

Semiconductor   A substance that does not conduct electricity at low temperatures but does so at higher temperatures.

Ternary acid   An acid containing three elements: H, O, and (usually) another nonmetal.

Total ionic equation   An equation for a chemical reaction written to show the predominant form of all species in aqueous solution or in contact with water.

Weak electrolyte   A substance that conducts electricity poorly in dilute aqueous solution.

Exercises

The Periodic Table

1. State the periodic law. What does it mean?
2. What was Mendeleev's contribution to the construction of the modern periodic table?
3. Consult a handbook of chemistry and look up melting points of the elements of periods 2 and 3. Show that melting point is a property that varies periodically for these elements.
*4. Mendeleev's periodic table was based on increasing atomic weight, whereas the modern periodic table is

based on increasing atomic number. In the modern table argon comes before potassium, yet it has a higher atomic weight. Explain how this can be.

5. Estimate the density of antimony from the following densities (g/cm³): As, 5.72; Bi, 9.8; Sn, 7.30; Te, 6.24. Show how you arrived at your answer.

6. Estimate the specific heat of antimony from the following specific heats (J/g · °C): As, 0.34; Bi, 0.14; Sn, 0.23; Te, 0.20. Show how you arrived at your answer.

7. Estimate the density of selenium from the following densities (g/cm³): S, 2.07; Te, 6.24; As, 5.72; Br, 3.12. Show how you arrived at your answer.

8. Given the following melting points in °C, estimate the value for CBr_4: CF_4, −184; CCl_4, −23; CI_4, 171 (decomposes).

9. Calcium and magnesium form the following compounds: $CaCl_2$, $MgCl_2$, CaO, MgO, Ca_3N_2, and Mg_3N_2. Predict the formula for a compound of (a) barium and sulfur, (b) strontium and iodine.

10. The formulas of some hydrides of second-period representative elements are as follows: BeH_2, BH_3, CH_4, NH_3, H_2O, HF. A famous test in criminology laboratories for the presence of arsenic (As) involves the formation of arsine, the hydride of arsenic. Predict the formula of arsine.

11. Distinguish between the following terms clearly and concisely, and provide specific examples of each: groups (families) of elements, and periods of elements.

12. Write names and symbols for (a) the alkaline earth metals, (b) the Group IIIA elements, (c) the Group VIB elements.

13. Write names and symbols for (a) the alkali metals, (b) the noble gases, (c) the Group IB elements.

14. Define and illustrate the following terms clearly and concisely: (a) metals, (b) nonmetals, (c) noble gases.

Aqueous Solutions

15. Define and distinguish among (a) strong electrolytes, (b) weak electrolytes, and (c) nonelectrolytes.

16. Three common classes of compounds are electrolytes. Name them and give an example of each.

17. Define (a) acids, (b) bases, and (c) salts.

18. How can a salt be related to a particular acid and a particular base?

19. List the names and formulas of the common strong acids.

20. Write equations for the ionization of the following acids: (a) hydrochloric acid, (b) nitric acid, (c) perchloric acid.

21. List names and formulas of five weak acids.

22. What are reversible reactions? Give some examples.

23. Write equations for the ionization of the following acids. Which ones ionize only slightly? (a) HF, (b) HNO_2, (c) CH_3COOH.

24. List names and formulas of the common strong soluble bases.

25. What is the difference between ionization and dissociation in aqueous solution?

26. The most common weak base is present in a common household chemical. Write the equation for the ionization of this weak base.

27. Summarize the electrical properties of strong electrolytes, weak electrolytes, and nonelectrolytes.

28. Write the formulas of two soluble and two insoluble chlorides, sulfates, and hydroxides.

29. Describe an experiment for classifying each of these compounds as a strong electrolyte, a weak electrolyte, or a nonelectrolyte: K_2CO_3, HCN, CH_3OH, H_2S, H_2SO_4, NH_3.

30. (a) Which of these are acids? HBr, NH_3, H_2SeO_4, BF_3, H_3SbO_4, $Al(OH)_3$, H_2S, C_6H_6, $CsOH$, H_3BO_3, HCN. (b) Which of these are bases? $NaOH$, H_2Se, BCl_3, NH_3.

*31. Classify each substance as either an electrolyte or a nonelectrolyte: NH_4Cl, HI, C_6H_6, $Zn(CH_3COO)_2$, $Cu(NO_3)_2$, CH_3COOH, $C_{12}H_{22}O_{11}$, $LiOH$, $KHCO_3$, CCl_4, $La_2(SO_4)_3$, I_2.

*32. Classify each substance as either a strong or weak electrolyte, and then list (a) the strong acids, (b) the strong bases, (c) the weak acids, and (d) the weak bases. $NaCl$, $MgSO_4$, HCl, $H_2C_2O_4$, $Ba(NO_3)_2$, H_3PO_4, $Sr(OH)_2$, HNO_3, HI, $Ba(OH)_2$, $LiOH$, C_2H_5COOH, NH_3, KOH, $MgMoO_4$, HCN, $HClO_4$.

33. Based on the solubility rules given in Table 4-8, how would you write the formulas for the following substances in a net ionic equation? (a) $PbSO_4$, (b) $Na(CH_3COO)$, (c) $(NH_4)_2CO_3$, (d) MnS, (e) $BaCl_2$.

34. Repeat Exercise 33 for the following: (a) $(NH_4)_2SO_4$, (b) $NaBr$, (c) $Ba(OH)_2$, (d) $Mg(OH)_2$, (e) K_2CO_3.

Precipitation Reactions

Refer to the solubility rules on page 132. Classify the compounds in Exercises 35 through 38 as soluble, moderately soluble, or insoluble in water.

35. (a) $NaClO_4$, (b) $AgCl$, (c) $Pb(NO_3)_2$, (d) KOH, (e) $MgSO_4$

36. (a) $BaSO_4$, (b) $Al(NO_3)_3$, (c) CuS, (d) Na_2S, (e) $Ca(CH_3COO)_2$

37. (a) $Fe(NO_3)_3$, (b) $Hg(CH_3COO)_2$, (c) $BeCl_2$, (d) $NiSO_4$, (e) $CaCO_3$

38. (a) $KClO_3$, (b) NH_4Br, (c) NH_3, (d) HNO_2, (e) PbS

Exercises 39 and 40 describe precipitation reactions *in aqueous solutions*. For each, write balanced (i) formula unit, (ii) total ionic, and (iii) net ionic equations. Refer to the solubility rules as necessary.

39. (a) Black-and-white photographic film contains some silver bromide, which can be formed by the reaction of sodium bromide with silver nitrate.
(b) Barium sulfate is used when X-rays of the gastrointestinal tract are made. Barium sulfate can be prepared by reacting barium chloride with dilute sulfuric acid.
(c) In water purification small solid particles are often

"trapped" as aluminum hydroxide precipitates and falls to the bottom of the sedimentation pool. Aluminum sulfate reacts with calcium hydroxide (from lime) to form aluminum hydroxide and calcium sulfate.

*40. (a) Our bones are mostly calcium phosphate. Calcium chloride reacts with potassium phosphate to form calcium phosphate and potassium chloride.

(b) Mercury compounds are very poisonous. Mercury(II) nitrate reacts with sodium sulfide to form mercury(II) sulfide, which is very insoluble, and sodium nitrate.

(c) Chromium(III) ions are very poisonous. They can be removed from solution by precipitating very insoluble chromium(III) hydroxide. Chromium(III) chloride reacts with calcium hydroxide to form chromium(III) hydroxide and calcium chloride.

In Exercises 41 and 42, write balanced (i) formula unit, (ii) total ionic, and (iii) net ionic equations for the reactions that occur when *aqueous solutions* of the compounds are mixed.

41. (a) $Ba(NO_3)_2 + K_2CO_3 \rightarrow$
    (b) $NaOH + CoCl_2 \rightarrow$
    (c) $Al_2(SO_4)_3 + NaOH \rightarrow$
42. (a) $Cu(NO_3)_2 + Na_2S \rightarrow$
    (b) $CdSO_4 + H_2S \rightarrow$
    (c) $Bi_2(SO_4)_3 + (NH_4)_2S \rightarrow$
43. Use the solubility rules to determine whether or not reactions will occur when aqueous solutions of the following compounds are mixed.
    (a) $Hg(NO_3)_2(aq) + Na_2S(aq) \rightarrow$
    (b) $Al(NO_3)_3(aq) + LiOH(aq) \rightarrow$
    (c) $Li_2SO_3(aq) + NaCl(aq) \rightarrow$
    (d) $Fe(OH)_3(s) + KNO_3(aq) \rightarrow$
    Write net ionic equations for those reactions that occur.
44. Repeat Exercise 43 for
    (a) $Al(OH)_3(s) + NaNO_3(aq) \rightarrow$
    (b) $NaBr(aq) + NH_4I(aq) \rightarrow$
    (c) $AgNO_3(aq) + HCl(aq) \rightarrow$
    (d) $CaCl_2(aq) + Na_2CO_3(aq) \rightarrow$

Acid–Base Reactions

In Exercises 45 through 48, write balanced (i) formula unit, (ii) total ionic, and (iii) net ionic equations for the reactions that occur between the acid and the base. Assume that all reactions occur in water or in contact with water.

45. (a) hydrochloric acid + barium hydroxide
    (b) dilute sulfuric acid + potassium hydroxide
    (c) perchloric acid + aqueous ammonia
46. (a) acetic acid + calcium hydroxide
    (b) sulfurous acid + sodium hydroxide
    (c) hydrofluoric acid + lithium hydroxide
*47. (a) sodium hydroxide + hydrosulfuric acid
    (b) barium hydroxide + hydrosulfuric acid
    (c) lead(II) hydroxide + hydrosulfuric acid

48. (a) sodium hydroxide + sulfuric acid
    (b) calcium hydroxide + phosphoric acid
    (c) copper(II) hydroxide + nitric acid

In Exercises 49 through 52, write balanced (i) formula unit, (ii) total ionic, and (iii) net ionic equations for the reaction of an acid and a base that will produce the indicated salts.

49. (a) potassium chloride, (b) sodium phosphate, (c) barium acetate
50. (a) calcium perchlorate, (b) ammonium sulfate, (c) copper(II) sulfide
*51. (a) sodium carbonate, (b) barium carbonate, (c) nickel(II) nitrate
*52. (a) sodium sulfide, (b) barium phosphate, (c) lead(II) arsenate
53. Write a balanced equation for the preparation of each of the following salts by a neutralization reaction. SrC_2O_4 is insoluble in water. $Ca(NO_3)_2$, SrC_2O_4, $ZnSO_3$, $(NH_4)_2CO_3$.
54. Write the formulas for the acid and the base that could react to form each of the following *insoluble* salts. (a) $CuCO_3$, (b) Ag_2CrO_4, (c) $Hg_3(PO_4)_2$.

Displacement Reactions

55. Which of the following would displace hydrogen when a piece of the metal is dropped into dilute H_2SO_4 solution? Write balanced net ionic equations for the reactions: Zn, Cu, Fe, Ag.
56. Which of the following metals would displace copper from an aqueous solution of copper(II) sulfate? Write balanced net ionic equations for the reactions: Hg, Zn, Fe, Ag.
57. Arrange the metals listed in Exercise 55 in order of increasing activity.
58. Arrange the metals listed in Exercise 56 in order of increasing activity.
59. Which of the following metals would displace hydrogen from cold water? Write balanced net ionic equations for the reactions: Zn, Na, Ca, Fe.
60. Arrange the metals listed in Exercise 59 in order of increasing activity.
61. What is the order of increasing activity of the halogens?
62. Of the possible displacement reactions shown, which one(s) could occur?
    (a) $2Cl^-(aq) + Br_2(\ell) \rightarrow 2Br^-(aq) + Cl_2(g)$
    (b) $2Br^-(aq) + F_2(g) \rightarrow 2F^-(aq) + Br_2(\ell)$
    (c) $2I^-(aq) + Cl_2(g) \rightarrow 2Cl^-(aq) + I_2(s)$
    (d) $2Br^-(aq) + Cl_2(g) \rightarrow 2Cl^-(aq) + Br_2(\ell)$
63. (a) Name two common metals—one that *does not* displace hydrogen from water, and one that *does not* displace hydrogen from water or acid solutions.
    (b) Name two common metals—one that *does* displace hydrogen from water, and one that displaces hydrogen from acid solutions but not from water. Write net ionic equations for the reactions that occur.

64. Predict the products of each mixture. If a reaction occurs, write the net ionic equation. If no reaction occurs, write "no reaction."
    (a) $Cd^{2+}(aq) + Al \rightarrow$
    (b) $Ca + H_2O \rightarrow$
    (c) $Ni + H_2O \rightarrow$
    (d) $Hg + HCl(aq) \rightarrow$
    (e) $Ni + H_2SO_4(aq) \rightarrow$
    (f) $Fe + H_2SO_4(aq) \rightarrow$

65. Use the activity series to predict whether or not the following reactions will occur:
    (a) $Fe(s) + Mg^{2+} \rightarrow Mg(s) + Fe^{2+}$
    (b) $Ni(s) + Cu^{2+} \rightarrow Ni^{2+} + Cu(s)$
    (c) $Cu(s) + 2H^+ \rightarrow Cu^{2+} + H_2(g)$
    (d) $Mg(s) + H_2O(g) \rightarrow MgO(s) + H_2(g)$

66. Repeat Exercise 65 for
    (a) $Sn(s) + Ca^{2+} \rightarrow Sn^{2+} + Ca(s)$
    (b) $Al_2O_3(s) + 3H_2(g) \xrightarrow{\Delta} 2Al(s) + 3H_2O(g)$
    (c) $Ca(s) + 2H^+ \rightarrow Ca^{2+} + H_2(g)$
    (d) $Cu(s) + Pb^{2+} \rightarrow Cu^{2+} + Pb(s)$

Oxidation Numbers

67. Assign oxidation numbers to the element specified in each group of compounds.
    (a) N in NO, N_2O_3, N_2O_4, NH_3, N_2H_4, NH_2OH, HNO_3
    (b) C in CO, CO_2, CH_2O, CH_4O, C_2H_6O, $(COOH)_2$, Na_2CO_3
    (c) S in S_8, H_2S, SO_2, SO_3, Na_2SO_3, H_2SO_4, K_2SO_4

68. Assign oxidation numbers to the element specified in each group of compounds.
    (a) P in PCl_3, P_4O_6, P_4O_{10}, HPO_3, H_3PO_4, $POCl_3$, $H_4P_2O_7$, $Mg_3(PO_4)_2$
    (b) Cl in Cl_2, HCl, $HClO$, $HClO_2$, $KClO_3$, Cl_2O_7, $Ca(ClO_4)_2$
    (c) Mn in MnO, MnO_2, $Mn(OH)_2$, K_2MnO_4, $KMnO_4$, Mn_2O_7
    (d) O in OF_2, Na_2O, Na_2O_2, KO_2

69. Assign oxidation numbers to the element specified in each group of ions.
    (a) S in S^{2-}, SO_3^{2-}, SO_4^{2-}, $S_2O_3^{2-}$, $S_4O_6^{2-}$
    (b) Cr in CrO_2^-, $Cr(OH)_4^-$, CrO_4^{2-}, $Cr_2O_7^{2-}$
    (c) B in BO_2^-, BO_3^{3-}, $B_4O_7^{2-}$

70. Assign oxidation numbers to the element specified in each group of ions.
    (a) N in N^{3-}, NO_2^-, NO_3^-, N_3^-, NH_4^+
    (b) Br in Br^-, BrO^-, BrO_3^-, BrO_4^-

Oxidation–Reduction Reactions

71. Define and illustrate the following terms: (a) oxidation, (b) reduction, (c) oxidizing agent, (d) reducing agent.

72. Why must oxidation and reduction always occur simultaneously in chemical reactions?

73. Determine which of the following are oxidation–reduction reactions. For those that are, identify the oxidizing and reducing agents.
    (a) $3Zn(s) + 2CoCl_3(aq) \rightarrow 3ZnCl_2(aq) + 2Co(s)$
    (b) $ICl(s) + H_2O(\ell) \rightarrow HCl(aq) + HOI(aq)$
    (c) $3HCl(aq) + HNO_3(aq) \rightarrow$
    $$Cl_2(g) + NOCl(g) + 2H_2O(\ell)$$
    (d) $Fe_2O_3(s) + 3CO(g) \xrightarrow{\Delta} 2Fe(s) + 3CO_2(g)$

74. Determine which of the following are oxidation–reduction reactions. For those that are, identify the oxidizing and reducing agents.
    (a) $HgCl_2(aq) + 2KI(aq) \rightarrow HgI_2(s) + 2KCl(aq)$
    (b) $4NH_3(g) + 3O_2(g) \rightarrow 2N_2(g) + 6H_2O(g)$
    (c) $CaCO_3(s) + 2HNO_3(aq) \rightarrow$
    $$Ca(NO_3)_2(aq) + CO_2(g) + H_2O(\ell)$$
    (d) $PCl_3(\ell) + 3H_2O(\ell) \rightarrow 3HCl(aq) + H_3PO_3(aq)$

75. What is oxidized, what is reduced, what is the oxidizing agent, and what is the reducing agent in each reaction?
    (a) $Mg(s) + Sn^{2+}(aq) \rightarrow Sn(s) + Mg^{2+}(aq)$
    (b) $2H_2O_2(\ell) \rightarrow 2H_2O(\ell) + O_2(g)$
    (c) $3H_2SO_3(aq) + HIO_3(aq) \rightarrow 3H_2SO_4(aq) + HI(aq)$
    (d) $CH_4(g) + 4Cl_2(g) \rightarrow CCl_4(\ell) + 4HCl(g)$

76. What mass of Zn is needed to displace 12.5 g of Cu from $CuSO_4 \cdot 5H_2O$?

Change-in-Oxidation-Number Method

In Exercises 77 and 78, write balanced formula unit equations for the reactions described by words.

77. (a) Carbon reacts with hot concentrated nitric acid to form carbon dioxide, nitrogen dioxide, and water.
    (b) Sodium reacts with water to form aqueous sodium hydroxide and gaseous hydrogen.
    (c) Zinc reacts with sodium hydroxide solution to form aqueous sodium tetrahydroxozincate and gaseous hydrogen. (The tetrahydroxozincate ion is $[Zn(OH)_4]^{2-}$.)

*78. (a) Iron reacts with hydrochloric acid to form aqueous iron(II) chloride and gaseous hydrogen.
    (b) Chromium reacts with sulfuric acid to form aqueous chromium(III) sulfate and gaseous hydrogen.
    (c) Tin reacts with concentrated nitric acid to form tin(IV) oxide, nitrogen dioxide, and water.

79. Balance the following ionic equations.
    (a) $Cr(OH)_4^-(aq) + OH^-(aq) + H_2O_2(aq) \rightarrow$
    $$CrO_4^{2-}(aq) + H_2O(\ell)$$
    (b) $MnO_2(s) + H^+(aq) + NO_2^-(aq) \rightarrow$
    $$NO_3^-(aq) + Mn^{2+}(aq) + H_2O(\ell)$$
    (c) $Sn(OH)_3^-(aq) + Bi(OH)_3(s) + OH^-(aq) \rightarrow$
    $$Sn(OH)_6^{2-}(aq) + Bi(s)$$

80. Balance the following ionic equations.
    (a) $MnO_4^-(aq) + H^+(aq) + Br^-(aq) \rightarrow$
    $$Mn^{2+}(aq) + Br_2(\ell) + H_2O(\ell)$$
    (b) $Cr_2O_7^{2-}(aq) + H^+(aq) + I^-(aq) \rightarrow$
    $$Cr^{3+}(aq) + I_2(s) + H_2O(\ell)$$
    (c) $MnO_4^-(aq) + SO_3^{2-}(aq) + H^+(aq) \rightarrow$
    $$Mn^{2+}(aq) + SO_4^{2-}(aq) + H_2O(\ell)$$
    (d) $Cr_2O_7^{2-}(aq) + Fe^{2+}(aq) + H^+(aq) \rightarrow$
    $$Cr^{3+}(aq) + Fe^{3+}(aq) + H_2O(\ell)$$

Ion–Electron Method

81. Balance the following ionic equations.

(a) $CrO_4^{2-}(aq) + H_2O(\ell) + HSnO_2^-(aq) \rightarrow$
$$CrO_2^-(aq) + OH^-(aq) + HSnO_3^-(aq)$$

(b) $C_2H_4(g) + MnO_4^-(aq) + H^+(aq) \rightarrow$
$$CO_2(g) + Mn^{2+}(aq) + H_2O(\ell)$$

(c) $H_2S(aq) + H^+(aq) + Cr_2O_7^{2-}(aq) \rightarrow$
$$Cr^{3+}(aq) + S(s) + H_2O(\ell)$$

(d) $ClO_3^-(aq) + H_2O(\ell) + I_2(s) \rightarrow$
$$IO_3^-(aq) + Cl^-(aq) + H^+(aq)$$

(e) $Cu(s) + H^+(aq) + SO_4^{2-}(aq) \rightarrow$
$$Cu^{2+}(aq) + H_2O(\ell) + SO_2(g)$$

82. Balance the following ionic equations.

(a) $Al(s) + NO_3^-(aq) + OH^-(aq) + H_2O \rightarrow$
$$Al(OH)_4^-(aq) + NH_3(g)$$

(b) $NO_2(g) + OH^-(aq) \rightarrow$
$$NO_3^-(aq) + NO_2^-(aq) + H_2O(\ell)$$

(c) $MnO_4^-(aq) + H_2O(\ell) + NO_2^-(aq) \rightarrow$
$$MnO_2(s) + NO_3^-(aq) + OH^-(aq)$$

(d) $I^-(aq) + H^+(aq) + NO_2^-(aq) \rightarrow$
$$NO(g) + H_2O(\ell) + I_2(s)$$

(e) $Hg_2Cl_2(s) + NH_3(aq) \rightarrow$
$$Hg(\ell) + HgNH_2Cl(s) + NH_4^+(aq) + Cl^-(aq)$$

83. Balance the following ionic equations for reactions in acidic solution. H^+ or H_2O (but not OH^-) may be added as necessary.

(a) $P_4(s) + NO_3^-(aq) \rightarrow H_3PO_4(aq) + NO(g)$

(b) $H_2O_2(aq) + MnO_4^-(aq) \rightarrow Mn^{2+}(aq) + O_2(g)$

(c) $HgS(s) + Cl^-(aq) + NO_3^-(aq) \rightarrow$
$$HgCl_4^{2-}(aq) + NO_2(g) + S(s)$$

(d) $HBrO(aq) \rightarrow Br^-(aq) + O_2(g)$

(e) $Cl_2(g) \rightarrow ClO_3^-(aq) + Cl^-(aq)$

84. Balance the following ionic equations for reactions in acidic solution. H^+ or H_2O (but not OH^-) may be added as necessary.

(a) $Fe^{2+}(aq) + MnO_4^-(aq) \rightarrow Fe^{3+}(aq) + Mn^{2+}(aq)$

(b) $Br_2(\ell) + SO_2(g) \rightarrow Br^-(aq) + SO_4^{2-}(aq)$

(c) $Cu(s) + NO_3^-(aq) \rightarrow Cu^{2+}(aq) + NO_2(g)$

(d) $PbO_2(s) + Cl^-(aq) \rightarrow PbCl_2(s) + Cl_2(g)$

(e) $Zn(s) + NO_3^-(aq) \rightarrow Zn^{2+}(aq) + N_2(g)$

85. Balance the following ionic equations in basic solution. OH^- or H_2O (but not H^+) may be added as necessary.

(a) $Mn(OH)_2(s) + H_2O_2(aq) \rightarrow MnO_2(s)$

(b) $CN^-(aq) + MnO_4^-(aq) \rightarrow CNO^-(aq) + MnO_2(s)$

(c) $As_2S_3(s) + H_2O_2(aq) \rightarrow AsO_4^{3-}(aq) + SO_4^{2-}(aq)$

(d) $CrI_3(aq) + H_2O_2(aq) \rightarrow CrO_4^{2-}(aq) + IO_4^-(aq)$

86. Balance the following ionic equations in basic solution. OH^- or H_2O (but not H^+) may be added as necessary.

(a) $MnO_4^-(aq) + NO_2^-(aq) \rightarrow MnO_2(s) + NO_3^-(aq)$

(b) $Zn(s) + NO_3^-(aq) \rightarrow NH_3(aq) + Zn(OH)_4^{2-}(aq)$

(c) $N_2H_4(aq) + Cu(OH)_2(s) \rightarrow N_2(g) + Cu(s)$

(d) $Mn^{2+}(aq) + MnO_4^-(aq) \rightarrow MnO_2(s)$

(e) $Cl_2(g) \rightarrow ClO_3^-(aq) + Cl^-(aq)$

Mixed Exercises

The following reactions apply to Exercises 87 through 93.

a. $H_2SO_4(aq) + 2KOH(aq) \rightarrow K_2SO_4(aq) + 2H_2O(\ell)$

b. $2Rb(s) + Br_2(\ell) \xrightarrow{\Delta} 2RbBr(s)$

c. $2KI(aq) + F_2(g) \rightarrow 2KF(aq) + I_2(s)$

d. $CaO(s) + SiO_2(s) \xrightarrow{\Delta} CaSiO_3(s)$

e. $S(s) + O_2(g) \xrightarrow{\Delta} SO_2(g)$

f. $BaCO_3(s) \xrightarrow{\Delta} BaO(s) + CO_2(g)$

g. $HgS(s) + O_2(g) \xrightarrow{\Delta} Hg(\ell) + SO_2(g)$

h. $AgNO_3(aq) + HCl(aq) \rightarrow AgCl(s) + HNO_3(aq)$

i. $Pb(s) + 2HBr(aq) \rightarrow PbBr_2(s) + H_2(g)$

j. $2HI(aq) + H_2O_2(aq) \rightarrow I_2(s) + 2H_2O(\ell)$

k. $RbOH(aq) + HNO_3(aq) \rightarrow RbNO_3(aq) + H_2O(\ell)$

l. $N_2O_5(s) + H_2O(\ell) \rightarrow 2HNO_3(aq)$

m. $H_2O(g) + CO(g) \xrightarrow{\Delta} H_2(g) + CO_2(g)$

n. $MgO(s) + H_2O(\ell) \rightarrow Mg(OH)_2(s)$

o. $PbSO_4(s) + PbS(s) \xrightarrow{\Delta} 2Pb(s) + 2SO_2(g)$

87. Identify the precipitation reactions.

88. Identify the acid–base reactions.

89. Identify the oxidation–reduction reactions.

90. Identify the oxidizing agent and reducing agent for each oxidation–reduction reaction.

91. Identify the oxidation–reduction reactions that are also displacement reactions.

92. Why can some reactions fit into more than one class?

93. Which of these reactions do not fit into any of our classes of reactions?

94. How many moles of oxygen can be obtained by the decomposition of 10.0 grams of reactant in each of the following reactions?

(a) $2KClO_3(s) \rightarrow 2KCl(s) + 3O_2(g)$

(b) $2H_2O_2(aq) \rightarrow 2H_2O(\ell) + O_2(g)$

(c) $2HgO(s) \rightarrow 2Hg(\ell) + O_2(g)$

95. For the formation of 1.00 mol of water, which reaction uses the most nitric acid?

(a) $3Cu(s) + 8HNO_3(aq) \rightarrow$
$$3Cu(NO_3)_2(aq) + 2NO(g) + 4H_2O(\ell)$$

(b) $Al_2O_3(s) + 6HNO_3(aq) \rightarrow 2Al(NO_3)_3(aq) + 3H_2O(\ell)$

(c) $4Zn(s) + 10HNO_3(aq) \rightarrow$
$$4Zn(NO_3)_2(aq) + NH_4NO_3(aq) + 3H_2O(\ell)$$

96. Balance these equations for reactions in acidic solutions by the ion–electron method.

(a) $MnO_4^- + H_2C_2O_4 \rightarrow Mn^{2+} + CO_2$

(b) $IO_3^- + Cl^- + N_2H_4 \rightarrow ICl_2^- + N_2$

(c) $Zn + NO_3^- \rightarrow Zn^{2+} + NH_4^+$

(d) $I_2 + S_2O_3^{2-} \rightarrow S_4O_6^{2-} + I^-$

(e) $NO_2^- + I^- \rightarrow I_2 + NO$

(f) $Ag^+ + AsH_3 \rightarrow H_3AsO_4 + Ag$

97. Balance these equations for reactions in basic solutions by the ion–electron method.

(a) $MnO_4^- + IO_3^- \rightarrow IO_4^- + MnO_2$

(b) $SO_3^{2-} + MnO_4^- \rightarrow SO_4^{2-} + MnO_4^{2-}$

(c) $Cl_2 \rightarrow Cl^- + ClO_3^-$

5 The Structure of Atoms

The Rosette Nebula. The red light is given off by ionized atoms, mostly interstellar hydrogen. A study of the wavelengths of light emitted and absorbed gives information about the composition of stars, interstellar gas, and other astronomical objects.

Outline

Objectives

As you study this chapter, you should learn about

- [ ] The evidence for the existence and properties of electrons, protons, and neutrons
- [ ] The arrangements of these particles in atoms
- [ ] Isotopes and their composition
- [ ] The relation between isotopic abundance and observed atomic weights
- [ ] The wave view of light and how wavelength, frequency, and speed are related
- [ ] The particle description of light, and how it is related to the wave description
- [ ] Atomic emission and absorption spectra, and how these were the basis for an important advance in atomic theory
- [ ] The quantum mechanical picture of the atom
- [ ] The four quantum numbers and possible combinations of their values
- [ ] The shapes of orbitals and the usual order of their relative energies
- [ ] Ways to determine electronic configurations of atoms
- [ ] The relation between electronic configurations and the positions of elements in the periodic table

The Dalton theory of the atom and related ideas were the basis for our study of *composition stoichiometry* (Chapter 2) and *reaction stoichiometry* (Chapter 3). But that level of atomic theory leaves many questions unanswered. *Why* do atoms combine to form compounds? *Why* do they combine only in simple numerical ratios? *Why* are particular numerical ratios of atoms observed in compounds? *Why* do different elements have such different properties—gases, liquids, solids, metals, nonmetals, and so on? *Why* do some groups of elements have such similar properties, and form compounds with similar formulas? The answers to these and many other fascinating questions in chemistry are supplied by our modern understanding of the nature of atoms. But how can we study something as small as an atom?

Much of the development of modern atomic theory was based on two broad types of research carried out by dozens of scientists just before and after 1900. The first type dealt with the electrical nature of matter, studies of which led scientists to recognize that atoms are composed of still more fundamental particles, and helped to describe the approximate arrangements of these particles in atoms. The second broad area of research dealt with the interaction of matter with energy in the form of light. Such research included studies of the colors of light that substances give off or absorb. These studies led to a much more detailed understanding of the arrangements of particles in atoms. It became clear that the arrangement of the particles determines the chemical and physical properties of each element. As we learn more about the structures of atoms, we are able to collect chemical facts in ways that help us to understand the behavior of matter.

We shall first study the particles that make up atoms and the basic structure of atoms. Then we shall trace the development of the quantum mechanical theory of atoms and see how this theory describes the arrangement of the electrons in atoms. This will give us the background necessary to describe, in the next few chapters, the forces responsible for chemical bonding. Current atomic theory is considerably less than complete. Even so, it is a powerful tool that helps us to understand the forces holding atoms in chemical combination with each other.

Subatomic Particles

5-1 Fundamental Particles

In our study of atomic theory, we look first at the **fundamental particles**. These are the basic building blocks of all atoms. Atoms, and hence *all* matter, consist principally of three fundamental particles: *electrons, protons,* and *neutrons*. Knowledge of the nature and functions of these particles is essential to understanding chemical interactions. The masses and charges of the three fundamental particles are shown in Table 5-1. The mass of an electron is very small compared with the mass of either a proton or a neutron. The charge on a proton is equal in magnitude, but opposite in sign, to the charge on an electron. Let's examine these particles in more detail.

Many other particles, such as quarks, positrons, neutrinos, pions, and muons, have also been discovered. It is not necessary to study their characteristics to learn the fundamentals of atomic structure that are important in chemical reactions.

Table 5-1
Fundamental Particles of Matter

Particle	Isolated Rest Mass	Charge (relative scale)
electron (e^-)	0.00054858 amu	1−
proton (p or p^+)	1.0073 amu	1+
neutron (n or n^0)	1.0087 amu	none

5-2 The Discovery of Electrons

The process is called chemical electrolysis. Lysis means "splitting apart."

Some of the earliest evidence about atomic structure was supplied in the early 1800s by the English chemist Humphrey Davy. He found that when he passed electrical current through some substances, the substances decomposed. This led him to propose that the elements of a chemical compound are held together by electrical forces. In 1832–33, Michael Faraday, Davy's protégé, determined the quantitative relationship between the amount of electricity used in electrolysis and the amount of chemical reaction that occurs. Studies of Faraday's work by George Stoney led him to suggest in 1874 that units of electrical charge are associated with atoms. In 1891 he suggested that they be named *electrons*.

Study Figures 5-1 and 5-2 carefully as you read this section.

The most convincing evidence for the existence of electrons came from experiments using *cathode ray tubes* (Figure 5-1). Two electrodes are sealed in a glass tube containing gas at a very low pressure. When a high voltage is applied, current flows and rays are given off by the cathode (negative electrode). These rays travel in straight lines to the anode (positive electrode)

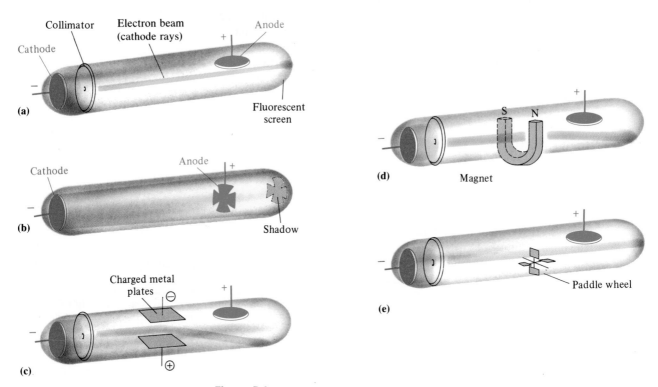

Figure 5-1
Some experiments with cathode ray tubes that show the nature of cathode rays. (a) A cathode ray (discharge) tube, showing the production of a beam of electrons (cathode rays). The beam is detected by observing the glow of a fluorescent screen. (b) A small object placed in a beam of cathode rays casts a shadow. This shows that cathode rays travel in straight lines. (c) Cathode rays have negative electrical charge, as demonstrated by their deflection in an electric field. (The electrically charged plates produce an electric field.) (d) Interaction of cathode rays with a magnetic field is also consistent with negative charge. The magnetic field goes from one pole to the other. (e) Cathode rays have mass, as shown by their ability to turn a small paddle wheel in their path.

and cause the walls opposite the cathode to glow. An object placed in the path of the cathode rays casts a shadow on a zinc sulfide screen placed near the anode. The shadow shows that the rays travel from the cathode toward the anode. Therefore the rays must be negatively charged. Additionally, they are deflected by both magnetic and electrical fields in the directions expected for negatively charged particles.

In 1897, J. J. Thomson studied these negatively charged particles more carefully. He called them **electrons**, the name Stoney had suggested in 1891. By studying the degree of deflections of cathode rays in different magnetic and electric fields, Thomson determined the charge (e) to mass (m) ratio for electrons. The modern value for this ratio is

$$e/m = 1.75881 \times 10^8 \text{ coulomb (C) per gram}$$

This ratio is the same regardless of the type of gas in the tube, the composition of the electrodes, or the nature of the electrical power source. The clear implication of Thomson's work was that electrons are fundamental particles present in all atoms. We now know that this is true and that all atoms contain integral numbers of electrons.

Once the charge-to-mass ratio for the electron had been determined, additional experiments were necessary to determine the value of either its mass or its charge, so that the other could be calculated. In 1909 Robert Millikan very nicely solved this dilemma with his famous "oil-drop experiment," in which he determined the charge on the electron. This experiment is described in Figure 5-2. All of the charges measured by Millikan turned out to be

The coulomb (C) is the standard unit of *quantity* of electrical charge. It is defined as the quantity of electricity transported in one second by a current of one ampere. It corresponds to the amount of electricity that will deposit 0.001118 g of silver in an apparatus set up for plating silver.

X-rays are radiations of much shorter wavelength than visible light (Section 5-9). They are sufficiently energetic to knock electrons out of the atoms in the air. In Millikan's experiment these free electrons became attached to some of the oil droplets.

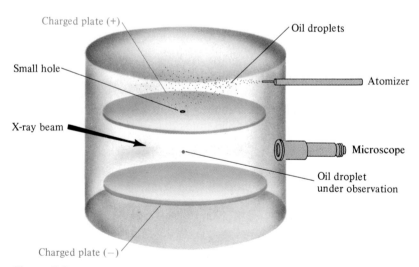

Figure 5-2
The Millikan oil-drop experiment. Tiny oil droplets are produced by an atomizer. A few of them fall through the hole in the upper plate. Irradiation with X-rays gives some of these oil droplets a negative charge. When the voltage between the plates is increased, a negatively charged drop falls more slowly because it is attracted by the positively charged upper plate and repelled by the negatively charged lower plate. At one particular voltage, the electrical force (up) and the gravitational force (down) on the drop are exactly balanced, and the drop remains stationary. If we know this voltage and the mass of the drop, we can calculate the charge on the drop. The mass of the spherical drop can be calculated from its volume (obtained from a measurement of the radius of the drop with a microscope) and the known density of the oil.

Robert A. Millikan (1868–1953) was an American physicist who was a physics professor at the University of Chicago and later director of the physics laboratory at the California Institute of Technology. For his investigations into photoelectric phenomena and the determination of the charge on the electron, he won the 1923 Nobel Prize in physics.

The charge on one mole (Avogadro's number) of electrons is 96,487 coulombs.

integral multiples of the same number. He assumed that this smallest charge was the charge on one electron. This value is 1.60219×10^{-19} coulomb (modern value).

The charge-to-mass ratio, $e/m = 1.75881 \times 10^8$ C/g, can be used in inverse form to calculate the mass of the electron:

$$m = \frac{1 \text{ g}}{1.75881 \times 10^8 \text{ C}} \times 1.60219 \times 10^{-19} \text{ C}$$

$$= 9.10952 \times 10^{-28} \text{ g per electron}$$

The value of e/m obtained by Thomson and the values of e and m obtained by Millikan differ slightly from the modern values given in this text because early measurements were not as accurate as modern ones.

This is only about 1/1836 the mass of a hydrogen atom, the lightest of all atoms. Millikan's simple oil-drop experiment stands as one of the most clever, yet most fundamental, of all classic scientific experiments. It was the first experiment to suggest that atoms contain integral numbers of electrons, a fact we now know to be true.

5-3 Canal Rays and Protons

In 1886 Eugen Goldstein first observed that a cathode ray tube also generates a stream of positively charged particles that moves toward the cathode. These were called **canal rays** because they were observed occasionally to pass through a channel, or "canal," drilled in the negative electrode (Figure 5-3). These *positive rays,* or *positive ions,* are created when cathode rays knock electrons from the gaseous atoms in the tube, forming positive ions by processes such as

$$\text{atom} \longrightarrow \text{cation}^+ + e^- \qquad \text{or} \qquad X \longrightarrow X^+ + e^-$$

The proton was observed by Rutherford and Chadwick in 1919 as a particle that is emitted by bombardment of certain atoms with alpha particles.

Different elements give positive ions with different e/m ratios. The regularity of the e/m values for different ions led to the idea that there is a unit of positive charge and that it resides in the **proton**. The proton is a fundamental particle with a charge equal in magnitude but opposite in sign to the charge on the electron. Its mass is almost 1836 times that of the electron.

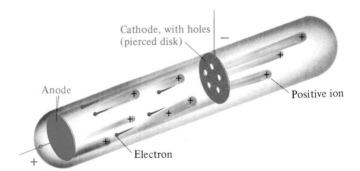

Figure 5-3
A cathode ray tube with a different design and with a perforated cathode. Such a tube was used to produce canal rays and to demonstrate that they travel toward the cathode. Like cathode rays, these *positive* rays are deflected by magnetic or electric fields, but in the opposite direction from cathode rays. Canal ray particles have e/m ratios many times smaller than those of electrons due to their much greater masses. When different elements are in the tube, positive ions with different e/m ratios are observed.

5-4 Rutherford and the Nuclear Atom: Atomic Number

By the first decade of this century, it was clear that each atom contained regions of both positive and negative charge. The question was, how are these charges distributed? The dominant view of that time was summarized in J. J. Thomson's model of the atom, in which the positive charge was assumed to be distributed evenly throughout the atom. The negative charges were pictured as being imbedded in the atom like plums in a pudding (hence the name "plum pudding model").

Soon after Thomson developed his model, tremendous insight into atomic structure was provided by one of Thomson's former students, Ernest Rutherford, certainly the outstanding experimental physicist of his time.

By 1909 Ernest Rutherford had established that alpha (α) particles are positively charged particles. They can be emitted by some radioactive atoms, i.e., atoms that undergo spontaneous disintegration. In 1910 Rutherford's research group carried out a series of experiments that had enormous impact on the scientific world. They bombarded a very thin gold foil with α-particles from a radioactive source. A fluorescent zinc sulfide screen was placed behind the foil to observe the scattering of the α-particles by the gold foil (Figure 5-4). Scintillations (flashes) on the screen, caused by the individual α-particles, were counted to determine the relative numbers of α-particles deflected at various angles. Alpha particles were known to be extremely dense, much denser than gold. Furthermore, they were known to be emitted at high kinetic energies.

If the Thomson model of the atom were correct, any α-particles passing through the foil would be expected to be deflected by very small angles. Quite unexpectedly, nearly all of the α-particles passed through the foil with little or no deflection. However, a few were deflected through large angles. A very few α-particles even returned from the gold foil in the direction from which they had come! Rutherford was astounded. In his own words.

> It was quite the most incredible event that has ever happened to me in my life. It was almost as if you fired a 15-inch shell into a piece of tissue paper and it came back and hit you.

Alpha particles are now known to be helium atoms minus their two electrons, or helium nuclei, which have 2+ charges (see Chapter 30).

Radioactivity is contrary to the Daltonian idea of the indivisibility of atoms.

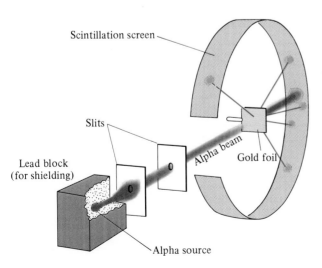

Scintillation screen

Slits

Lead block (for shielding)

Alpha beam

Gold foil

Alpha source

Figure 5-4
The Rutherford scattering experiment. A narrow beam of alpha particles (helium atoms stripped of their electrons) from a radioactive source was directed at a very thin gold foil. Most of the particles passed right through the foil (gold). Many were deflected through moderate angles (shown in red). These deflections were surprises, but the 0.001% of the total that were reflected at acute angles (shown in blue) were totally unexpected. Similar results were observed using foils of other metals.

Ernest Rutherford (1871–1937) was one of the giants in the development of our understanding of atomic structure. A native of New Zealand, Rutherford traveled to England in 1895 where he worked for much of his life. While working with J. J. Thomson at Cambridge University, he discovered α and β radiation. The years 1899–1907 were spent at McGill University in Canada where he proved the nature of these two radiations, for which he received the Nobel Prize in Chemistry in 1908. He returned to England in 1908, and it was there, at Manchester University, that he and his coworkers Geiger and Marsden performed the famous gold foil experiments that revolutionized our view of the atom. Not only did he perform much important research in physics and chemistry, but he also guided the work of ten future recipients of the Nobel Prize.

Rutherford's mathematical analysis of his results showed that the scattering of positively charged α-particles was caused by repulsion from very dense regions of positive charge in the gold foil. He concluded that the mass of one of these regions is nearly equal to that of a gold atom, but that the diameter is no more than 1/10,000 that of an atom. Many experiments with foils of different metals yielded similar results. Realizing that these observations were inconsistent with previous theories about atomic structure, Rutherford discarded the old theory and proposed a better one. He suggested that each atom contains a *tiny, positively charged, massive center* that he called an **atomic nucleus**. Most α-particles pass through metal foils undeflected because atoms are *primarily* empty space populated only by the very light electrons. The few particles that are deflected are the ones that come close to the heavy, highly charged metal nuclei (Figure 5-5).

As a somewhat similar experiment, imagine that we shoot pellets from a BB gun at a chain-link fence. Because the "particles" are much smaller than the "empty spaces" in such a fence, most of the pellets would go through the fence undeflected. However, some would glance off the wires of the fence and be deflected through moderate angles; a few would hit a wire directly enough to bounce back to our side of the fence. If we fired a very large number, say a million such "particles," at the fence, and counted how

This representation is *not* to scale. If nuclei were as large as the black dots that represent them, each white region, which represents the size of an atom, would have a diameter of more than 30 feet!

Figure 5-5

An interpretation of the Rutherford scattering experiment. The atom is pictured as consisting mostly of "open" space. At the center is a tiny and extremely dense nucleus that contains all of the atom's positive charge and nearly all of the mass. The electrons are thinly distributed throughout the "open" space. Most of the positively charged alpha particles (shown in black) pass through the open space undeflected, not coming near any gold nuclei. The few that pass fairly close to a nucleus (shown in red) are repelled by electrostatic force and thereby deflected. The very few particles that are on a "collision course" with gold nuclei are repelled backward at acute angles (shown in blue). Calculations based on the results of the experiment indicated that the diameter of the open-space portion of the atom is from 10,000 to 100,000 times greater than the diameter of the nucleus.

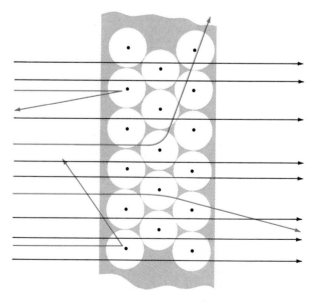

many went through undeflected and how many were deflected by various angles, we could perhaps calculate the fraction of the area of the fence that is open space, and even discover something about the size and distribution of the wires. (Gold foil is many atomic layers thick, so we could suppose that there were several such fences, one behind the other.)

Rutherford was able to determine the magnitudes of the positive charges on the atomic nuclei. The picture of atomic structure that he developed is called the Rutherford model of the atom.

> Atoms consist of very small, very dense positively charged nuclei surrounded by clouds of electrons at relatively great distances from the nuclei.

We now know that every nucleus contains an integral number of protons exactly equal to the number of electrons in a neutral atom of element. Every hydrogen atom contains one proton, every helium atom contains two protons, and every lithium atom contains three protons. The number of protons in the nucleus of an atom determines its identity; this number is known as the **atomic number** of that element.

5-5 Neutrons

The third fundamental particle, the neutron, eluded discovery until 1932. James Chadwick correctly interpreted experiments on the bombardment of beryllium with high-energy alpha particles. Later experiments showed that nearly all elements up to potassium, element 19, produce neutrons when they are bombarded with high-energy alpha particles. The **neutron** is an uncharged particle with a mass slightly greater than that of the proton. With its discovery, the picture of the nuclear atom was complete:

This does not mean that elements above number 19 do not have neutrons, only that neutrons are not generally knocked out of atoms of higher atomic number by alpha particle bombardment.

> Atoms consist of very small, very dense nuclei surrounded by clouds of electrons at relatively great distances from the nuclei. All nuclei contain protons; nuclei of all atoms except the common form of hydrogen also contain neutrons.

Nuclear diameters are about 10^{-4} Ångstroms (10^{-5} nanometers); atomic diameters are about 1 Ångstrom (10^{-1} nanometers). To put this difference in perspective, suppose that you wish to build a model of an atom using a basketball (diameter about 9.5 inches) as the nucleus; on this scale, the atomic model would be nearly 6 miles across!

5-6 Mass Number and Isotopes

Only a few years after Rutherford's scattering experiments, H. G. J. Moseley studied X-rays given off by various elements. Max von Laue had shown that X-rays could be diffracted by crystals into a spectrum in much the same way that visible light can be separated into its component colors. Moseley generated X-rays by aiming a beam of high-energy electrons at a solid target made of a single pure element (Figure 5-6).

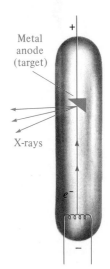

Figure 5-6
A simplified representation of the production of X-rays by bombardment of a solid target with a high-energy beam of electrons.

Chemistry in Use. . .
Stable Isotope Ratio Analysis

Many elements exist as two or more stable isotopes, although one isotope is usually present in far greater abundance. For example, there are two stable isotopes of carbon, ^{13}C and ^{12}C, of which ^{12}C is the more abundant, constituting 98.89% of all carbon. Similarly, there are two stable isotopes of nitrogen, ^{14}N and ^{15}N, of which ^{14}N makes up 99.63% of all nitrogen.

Differences in chemical and physical properties that arise from differences in atomic mass of an element are known as isotope effects. We know that the extranuclear structure of an element (the number of electrons and their arrangement) essentially determines its chemical behavior, whereas the nucleus has more influence on many of the physical properties of the element. Because all isotopes of a given element contain the same number and arrangement of electrons, it was assumed for a long time that isotopes would behave identically in chemical reactions. In reality, although isotopes behave very similarly in chemical reactions, the correspondence is not perfect. The mass differences between different isotopes of the same element cause them to have slightly different physical and chemical properties. For example, the presence of only one additional neutron in the nucleus of the heavier isotope can cause it to react a little more slowly than its lighter counterpart. Such an effect often results in a ratio of heavy isotope to light isotope in the product of a reaction that is different from the ratio found in the reactant.

Stable isotope ratio analysis (SIRA) is an analytical technique that takes advantage of the different chemical and physical properties of isotopes. In SIRA the isotopic composition of a sample is measured using a mass spectrometer. This composition is then expressed as the relative ratios of two or more of the stable isotopes of a specific element. For instance, the ratio of ^{13}C to ^{12}C in a sample can be determined. This ratio is then compared to the isotope ratio of a defined standard. Because mass differences are most pronounced among the lightest elements, those elements experience the greatest isotope effects. Thus, the isotopes of the elements H, C, N, O, and S are used most frequently for SIRA. These elements have further significance because they are among the most abundant elements in biological systems.

The isotopic composition of a sample is usually expressed as a "del" value (∂), defined as

$$\partial X_{\text{sample}} \ (\permil) = \frac{(R_{\text{sample}} - R_{\text{standard}})}{R_{\text{standard}}} \times 1000$$

where, $\partial X_{\text{sample}}$ is the isotope ratio relative to a standard, and R_{sample} and R_{standard} are the absolute isotope ratios of the sample and standard, respectively. Multiplying by 1000 allows the values to be expressed in parts per thousand ($\permil$). If the del value is a positive number, the sample has a greater amount of the heavier isotope than does the standard. In such cases the sample is said to be "heavier" than the standard, or to have been "enriched" in the heavy isotope. Similarly, if the del value is negative, the sample has a higher proportion of the lighter isotope and thus is described as "lighter" than the standard.

The most frequently used element for SIRA is carbon. The first limited data on $^{13}C/^{12}C$ isotope ratios in natural materials were published in 1939. At that time it was established that limestones, atmospheric CO_2, marine plants, and terrestrial plants each possessed characteristic carbon isotope ratios. In the succeeding years, $^{13}C/^{12}C$ ratios were determined for a wide variety of things, including petroleum, coal, diamonds, marine organisms, and terrestrial organisms. Such data led to the important conclusion that a biological organism has an isotope

ratio that depends on the main source of carbon to that organism—that is, its food source. For example, if an herbivore (an animal that feeds on plants) feeds exclusively on one type of plant, that animal's carbon isotope ratio will be almost identical to that of the plant. If another animal were to feed exclusively on that herbivore, it would also have a similar carbon isotope ratio. Suppose now that an animal, say a rabbit, has a diet comprising two different plants, A and B. Plant A has a $\partial^{13}C$ value of $-24‰$, and plant B has a del value of $-10‰$. If the rabbit eats equal amounts of the two plants, then the $\partial^{13}C$ value of the rabbit will be the average of the two values, or $-17‰$. Values more positive than $-17‰$ would indicate a higher consumption of plant B than of plant A, whereas more negative values would reflect a preference for plant A.

Similar studies have been conducted with the stable isotopes of nitrogen. A major way in which nitrogen differs from carbon in isotopic studies relates to how $\partial^{13}C$ and $\partial^{15}N$ values change as organic matter moves along the food chain—from inorganic nutrient to plant, then to herbivore, to carnivore, and on to higher carnivores. It has been pointed out that $\partial^{13}C$ remains nearly constant throughout successive levels of the food chain. In contrast, on average there is a $+3$ to $+5‰$ shift in the value of $\partial^{15}N$ at each successive level of the food chain. For instance, suppose a plant has a $\partial^{15}N$ value of $1‰$. If an herbivore, such as a rabbit, feeds exclusively on that one type of plant, it will have a $\partial^{15}N$ value of $4‰$. If another animal, such as a fox, feeds exclusively on that particular type of rabbit, it in turn will have a $\partial^{15}N$ value of $7‰$. An important implication of this phenomenon is that an organism's nitrogen isotope ratio can be used as an indi-

cator of the level in the food chain at which that species of animal feeds.

An interesting application of SIRA is the determination of the adulteration of food. As already mentioned, the isotope ratios of different plants and animals have been determined. For instance, corn has a $\partial^{13}C$ value of about $-12‰$ and most flowering plants have $\partial^{13}C$ values of about $-26‰$. The difference in these $\partial^{13}C$ values arises because these plants carry out photosynthesis by slightly different chemical reactions. In the first reaction of photosynthesis, corn produces a molecule that contains four carbons, whereas flowering plants produce a molecule that has only three carbons. High-fructose corn syrup (HFCS) is thus derived from a "C_4" plant, whereas the nectar that bees gather comes from "C_3" plants. The slight differences in the photosynthetic pathways of C_3 and C_4 plants create the large differences in their $\partial^{13}C$ values. Brokers who buy and sell huge quantities of "sweet" products are able to monitor HFCS adulteration of honey, maple syrup, apple juice, and so on by taking advantage of the SIRA technique. If

the $\partial^{13}C$ value of one of these products is not appropriate, then the product obviously has had other substances added to it, i.e., has been adulterated. The U.S. Department of Agriculture conducts routine isotope analyses to ensure the purity of those products submitted for subsidy programs. Similarly, the honey industry monitors itself with the SIRA technique.

Another interesting use of SIRA is in the determination of the diets of prehistoric human populations. It is known that marine plants have higher $\partial^{15}N$ values than terrestrial plants. This difference in $\partial^{15}N$ is carried up food chains, causing marine animals to have higher $\partial^{15}N$ values than terrestrial animals. The $\partial^{15}N$ values of humans feeding on marine food sources are therefore higher than those of people feeding on terrestrial food. This phenomenon has been used to estimate the marine and terrestrial components of the diets of historic and prehistoric human groups through the simple determination of the $\partial^{15}N$ value of bone collagen collected from excavated skeletons.

Stable isotope ratio analysis is a powerful tool; many of its potential uses are only slowly being recognized by researchers. In the meantime, the use of stable isotope methods in research is becoming increasingly common, and through these methods scientists are attaining new levels of understanding of chemical, biological, and geological processes.

Beth A. Trust
Graduate student in chemistry
University of Texas Marine Sciences Institute

H. G. J. Moseley was one of the many remarkable scientists who worked with Ernest Rutherford. In 1913 Moseley found that the wavelengths of X-rays emitted by an element are related in a precise way to the atomic number of the element. This discovery led to the realization that atomic number, related to electrical properties of the atom, was more fundamental to determining the properties of the elements than atomic weight. This put the ideas of the periodic table on a more fundamental footing. Moseley's scientific career was very short. He was enlisted in the British army during World War I, and died in battle in the Gallipoli campaign in 1915. In subsequent wars, most countries have not allowed their promising scientists to take part in front-line service.

The spectra of X-rays produced by targets of different elements were recorded photographically. Each photograph consisted of a series of lines representing X-rays at various wavelengths. Comparison of results from different elements revealed that corresponding lines were displaced toward shorter wavelengths as atomic weights of the target materials increased, with three exceptions. Moseley showed that the X-ray wavelengths could be better correlated with the atomic number (Section 5-4). On the basis of his mathematical analysis of these X-ray data, he concluded that

> each element differs from the preceding element by having one more positive charge in its nucleus.

For the first time it was possible to arrange all known elements in order of increasing nuclear charge. A plot summarizing this interpretation of Moseley's data appears in Figure 5-7.

Most elements consist of atoms of different masses, called **isotopes**. The isotopes of a given element contain the same number of protons (and also the same number of electrons) because they are atoms of the same element. They differ in mass because they contain different numbers of neutrons in their nuclei.

Figure 5-7

A plot of some of Moseley's X-ray data. The atomic number of an element is found to be directly proportional to the square root of the reciprocal of the wavelength of a particular X-ray spectral line. Wavelength (Section 5-9) is represented by λ.

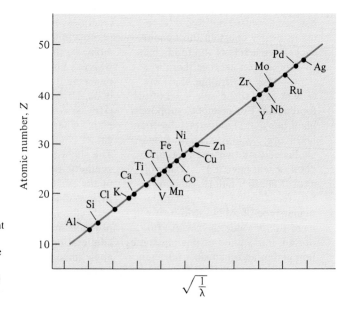

Table 5-2
Make-Up of the Three Isotopes of Hydrogen (neutral atoms)

Name	Symbol	Nuclide Symbol	Atomic Abundance in Nature	No. of Protons	No. of Neutrons	No. of Electrons (in neutral atoms)
hydrogen	H	^1_1H	99.985%	1	0	1
deuterium	D	^2_1H	0.015%	1	1	1
tritium*	T	^3_1H	0.000%	1	2	1

*No known natural sources; produced by decomposition of artificial isotopes.

For example, there are three distinct kinds of hydrogen atoms, commonly called hydrogen, deuterium, and tritium. (This is the only element for which we give each isotope a different name.) Each of these three contains one proton in the atomic nucleus. The predominant form of hydrogen contains no neutrons, but each deuterium atom contains one neutron and each tritium atom contains two neutrons in its nucleus (Table 5-2). All three forms of hydrogen display very similar chemical properties.

The **mass number** of an atom is the sum of the number of protons and the number of neutrons in its nucleus; i.e.,

$$\text{mass number} = \text{number of protons} + \text{number of neutrons}$$
$$= \text{atomic number} \quad + \text{neutron number}$$

The mass number for normal hydrogen atoms is 1, for deuterium 2, and for tritium 3. The composition of a nucleus is indicated by its **nuclide symbol**. This consists of the symbol for the element (E), with the atomic number (Z) written as a subscript at the lower left and the mass number (A) as a superscript at the upper left, ^A_ZE. By this system, the three isotopes of hydrogen are designated as ^1_1H, ^2_1H, and ^3_1H.

A mass number is a count of the number of things present, so it must be a whole number. Because the masses of the proton and the neutron are both about 1 amu, the mass number is *approximately* equal to the actual mass of the isotope (which is not a whole number).

5-7 Mass Spectrometry and Isotopic Abundance

Mass spectrometers are instruments that measure the charge-to-mass ratio of charged particles (Figure 5-8). A gas sample at very low pressure is bombarded with high-energy electrons. This causes electrons to be ejected from some of the gas molecules, creating positive ions. The positive ions are then focused into a very narrow beam and accelerated by an electric field toward a magnetic field. The magnetic field deflects the ions from their straight-line path. The extent to which the beam of ions is deflected depends upon four factors:

1. *Magnitude of the accelerating voltage (electric field strength).* The range varies from about 500 to about 2000 volts. Higher voltages result in beams of more rapidly moving particles that are deflected less than the beams of the more slowly moving particles produced by lower voltages.
2. *Magnetic field strength.* Stronger fields deflect a given beam more than weaker fields.
3. *Masses of the particles.* Because of their inertia, heavier particles are deflected less than lighter particles that carry the same charge.

Figure 5-8
The mass spectrometer. In the mass spectrometer, gas molecules at low pressure are ionized and accelerated by an electric field. The ion beam is then passed through a magnetic field. In that field the beam is resolved into components, each containing particles of equal charge-to-mass ratio. Lighter particles are deflected more strongly than heavy ones with the same charge. In a beam containing $^{12}_{6}C^+$ and $^{4}_{2}He^+$ ions, the lighter $^{4}_{2}He^+$ ions would be deflected more than the heavier $^{12}_{6}C^+$ ions. The spectrometer shown is adjusted to detect the $^{12}_{6}C^+$ ions. By changing the magnitude of the magnetic or electric field, we can move the beam of $^{4}_{2}He^+$ ions striking the collector from B to A, where it would be detected.

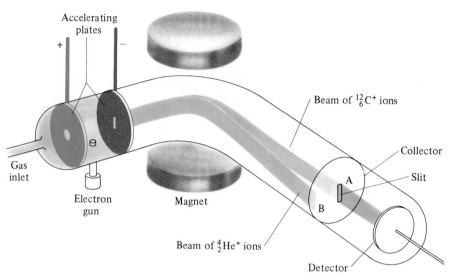

4. *Charges on the particles.* Highly charged particles interact more strongly with magnetic and electric fields and are thus deflected more than particles of equal mass with smaller charges.

The mass spectrometer is used to measure masses of isotopes as well as isotopic abundances. Helium occurs in nature almost exclusively as $^{4}_{2}He$. Let's see how its atomic mass is measured. To simplify the picture, we'll assume that only ions with 1+ charge are formed in the experiment illustrated in Figure 5-8.

As we saw in Section 2-4, $^{12}_{6}C$ is the (arbitrary) reference point on the atomic weight scale. We first use carbon-12 to calibrate the instrument. A carefully measured, fixed accelerating voltage is applied and a sample of vaporized $^{12}_{6}C$ is fed into the mass spectrometer. The magnetic field strength that causes $^{12}_{6}C^+$ ions to arrive at point A is measured. A sample of helium is then fed into the mass spectrometer with exactly the same accelerating voltage. The lighter $^{4}_{2}He^+$ ions are deflected more than the heavier $^{12}_{6}C^+$ ions to some point B. The magnetic field strength is then decreased slowly until the beam of lighter ions falls at point A. Now that the field strengths required to focus the two kinds of ions at the same point are known, the mass of the lighter ion can be calculated.

The relationship between masses of particles and magnetic field strength, $\mathscr{H}$, is

The $\mathscr{H}$'s are the magnetic field strengths required to focus the beams of ions at point A.

$$\frac{\text{mass } ^{4}_{2}He^+}{\text{mass } ^{12}_{6}C^+} = \left(\frac{\mathscr{H} \text{ for } ^{4}_{2}He^+}{\mathscr{H} \text{ for } ^{12}_{6}C^+}\right)^2$$

$(0.5776)^2 = 0.3336$

The ratio of the two magnetic field strengths in the experiment is measured as 0.5776, so the mass of $^{4}_{2}He^+$ is found to be 0.3336 times the mass of the $^{12}_{6}C^+$ ion. The mass of the $^{12}_{6}C^+$ ion is exactly 12 amu, so the mass of the $^{4}_{2}He^+$ ion is 0.3336 × 12 amu = 4.003 amu.

A beam of Ne^+ ions in the mass spectrometer is split into three segments. The mass spectrum of these ions (a graph of the relative numbers of ions of each mass) is shown in Figure 5-9. This indicates that neon occurs in nature

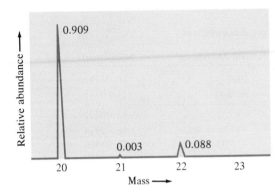

Figure 5-9
Mass spectrum of neon(1+ ions only). Neon consists of three isotopes, of which neon-20 is by far the most abundant (90.9%). The mass of that isotope, to five decimal places, is 19.99244 amu on the carbon-12 scale. The number by each peak represents the number of Ne$^+$ ions corresponding to that isotope, expressed as a fraction of all Ne$^+$ ions.

as three isotopes: $^{20}_{10}$Ne, $^{21}_{10}$Ne, and $^{22}_{10}$Ne. In Figure 5-9 we see that the isotope $^{20}_{10}$Ne, mass 19.99244 amu, is the most abundant isotope (has the tallest peak). It accounts for 90.5% of the atoms. $^{22}_{10}$Ne accounts for 9.2% and $^{21}_{10}$Ne for only 0.3% of the atoms.

Figure 5-10 shows a modern mass spectrometer and a typical spectrum of an element. In nature, some elements, such as fluorine and phosphorus, exist in only one form, but most elements occur as isotopic mixtures. Some examples of natural isotopic abundances are given in Table 5-3. The percentages are based on the numbers of naturally occurring atoms of each isotope, *not* on their masses.

The distribution of isotopic masses, while nearly constant, does vary somewhat depending on the source of the element. For example, the abundance of $^{13}_{6}$C in atmospheric CO_2 is slightly different from that in seashells. The chemical history of a compound can be inferred from the small differences in isotope ratios.

Figure 5-10
(a) A modern mass spectrometer.
(b) The mass spectrum of Xe^{1+} ions, measured on the instrument shown in (a). The isotope ^{126}Xe is at too low an abundance (0.090%) to appear in this experiment.

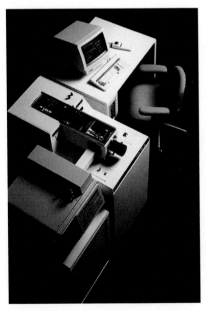

(a)

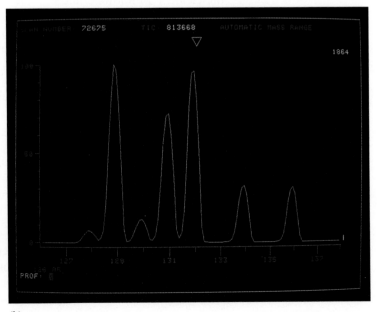

(b)

Table 5-3
Some Naturally Occurring Isotopic Abundances

Element	Isotope	% Natural Abundance	Mass (amu)
boron	$^{10}_{5}B$	20.0	10.01294
	$^{11}_{5}B$	80.0	11.00931
oxygen	$^{16}_{8}O$	99.762	15.99491
	$^{17}_{8}O$	0.038	16.99914
	$^{18}_{8}O$	0.200	17.99916
chlorine	$^{35}_{17}Cl$	75.77	34.96885
	$^{37}_{17}Cl$	24.23	36.9658
uranium	$^{234}_{92}U$	0.0057	234.0409
	$^{235}_{92}U$	0.72	235.0439
	$^{238}_{92}U$	99.27	238.0508

The 20 elements that have only one naturally occurring isotope are $^{9}_{4}Be$, $^{19}_{9}F$, $^{23}_{11}Na$, $^{27}_{13}Al$, $^{31}_{15}P$, $^{45}_{21}Sc$, $^{55}_{25}Mn$, $^{59}_{27}Co$, $^{75}_{33}As$, $^{89}_{39}Y$, $^{93}_{41}Nb$, $^{103}_{45}Rh$, $^{127}_{53}I$, $^{133}_{55}Cs$, $^{141}_{59}Pr$, $^{159}_{65}Tb$, $^{165}_{67}Ho$, $^{169}_{69}Tm$, $^{197}_{79}Au$, and $^{209}_{83}Bi$. However, there are other, artificially produced isotopes of these elements.

5-8 The Atomic Weight Scale and Atomic Weights

We saw in Section 2-4 that the **atomic weight scale** is based on the mass of the carbon-12 isotope. As a result of action taken by the International Union of Pure and Applied Chemistry in 1962,

Described another way, the mass of one atom of $^{12}_{6}C$ is *exactly* 12 amu.

one **amu** is exactly 1/12 of the mass of a carbon-12 atom.

This is approximately the mass of one atom of ^{1}H, the lightest isotope of the element with lowest mass.

In Section 2-5 we said that one mole of atoms contains 6.022×10^{23} atoms. The mass of one mole of atoms of any element, in grams, is numerically equal to the atomic weight of the element. Because the mass of one carbon-12 atom is exactly 12 amu, the mass of one mole of carbon-12 atoms is exactly 12 grams.

Let us now show the relationship between atomic mass units and grams.

$$\underline{?}\ g = 1\ \text{amu} \times \frac{1\ ^{12}_{6}C\ \text{atom}}{12\ \text{amu}} \times \frac{1\ \text{mol}\ ^{12}_{6}C\ \text{atoms}}{6.022 \times 10^{23}\ ^{12}_{6}C\ \text{atoms}} \times \frac{12\ g\ ^{12}_{6}C}{1\ \text{mol}\ ^{12}_{6}C\ \text{atoms}} = 1.660 \times 10^{-24}\ g$$

Thus we see that *1 amu = 1.660×10^{-24} g*. Multiplying both sides by Avogadro's number, we see that *1 g = 6.022×10^{23} amu*.

At this point, we should clearly emphasize the differences among the following quantities:

1. The *atomic number, Z,* is an integer equal to the number of protons in the nucleus of an atom of the element. It is also the number of electrons in a neutral atom. It is the same for all atoms of an element.
2. The *mass number, A,* is an integer equal to the *sum* of the number of protons and the number of neutrons in the nucleus of an atom of a

You may wish to verify that the same result is obtained regardless of the element or isotope chosen.

particular isotope of an element. It is different for different isotopes of the same element.

3. The *atomic weight* of an element is the weighted average of the masses of its constituent isotopes. Atomic weights are fractional numbers, not integers.

The atomic weight that we determine experimentally (for an element that consists of more than one isotope) is such a weighted average. The following example shows how such an atomic weight can be calculated from measured isotopic abundances.

Example 5-1

Three isotopes of magnesium occur in nature. Their abundances and masses, determined by mass spectrometry, are listed below. Use this information to calculate the atomic weight of magnesium.

Isotope	% Abundance	Mass (amu)
$^{24}_{12}Mg$	78.99	23.98504
$^{25}_{12}Mg$	10.00	24.98584
$^{26}_{12}Mg$	11.01	25.98259

Plan

We multiply the fraction of each isotope by its mass and add these numbers to obtain the atomic weight of magnesium.

Solution

atomic weight = 0.7899(23.98504 amu) + 0.1000(24.98584 amu) + 0.1101(25.98259 amu)

= 18.94 amu   + 2.498 amu   + 2.861 amu

= 24.30 amu   (to four significant figures)

The two heavier isotopes make small contributions to the atomic weight of magnesium, because most magnesium atoms are the lightest isotope.

EOC 33, 34

Example 5-2 shows how the process can be reversed. Percent abundances can be calculated from isotopic masses and from the atomic weight of an element that occurs in nature as a mixture of only two isotopes.

Example 5-2

The atomic weight of gallium is 69.72 amu. The masses of the naturally occurring isotopes are 68.9257 amu for $^{69}_{31}Ga$ and 70.9249 amu for $^{71}_{31}Ga$. Calculate the percent abundance of each isotope.

Plan

We can represent the fraction of each isotope algebraically. Atomic weight is the weighted average of the masses of the constituent isotopes. Therefore, the fraction of each isotope is multiplied by its mass and the sum of the results is equal to the atomic weight.

When a quantity is represented by fractions, the sum of the fractions must always be unity. In this case, $x + (1 - x) = 1$.

Solution

Let x = fraction of $^{69}_{31}Ga$. Then $(1 - x)$ = fraction of $^{71}_{31}Ga$.

$$x(68.9257 \text{ amu}) + (1 - x)(70.9249 \text{ amu}) = 69.72 \text{ amu}$$

$$68.9257x + 70.9249 - 70.9249x = 69.72$$

$$-1.9992x = -1.20$$

$$x = 0.600$$

$$x = 0.600 = \text{fraction of } ^{69}_{31}Ga \quad \therefore \quad \boxed{60.0\% \; ^{69}_{31}Ga}$$

$$(1 - x) = 0.400 = \text{fraction of } ^{71}_{31}Ga \quad \therefore \quad \boxed{40.0\% \; ^{71}_{31}Ga}$$

EOC 28, 29

The Electronic Structures of Atoms

The Rutherford model of the atom is consistent with the evidence presented so far, but it has some serious limitations. It does not answer important questions such as the following. *Why* do different elements have such different chemical and physical properties? *Why* does chemical bonding occur at all? *Why* does each element form compounds with characteristic formulas? *How* can atoms of different elements give off or absorb light only of characteristic colors (as was known long before 1900)?

To go further in our understanding, we must first learn more about the arrangements of electrons in atoms. The theory of these arrangements is based largely on the study of the light given off and absorbed by atoms. Then we shall develop a detailed picture of the *electron configurations* of different elements.

5-9 Electromagnetic Radiation

Our ideas about the arrangements of electrons in atoms have evolved slowly. Much of the information has been derived from **atomic emission spectra**. These are the lines, or bands, produced on photographic film by radiation that has passed through a refracting glass prism after being emitted from electrically or thermally excited atoms. To help us understand the nature of atomic spectra, let us first describe electromagnetic radiation in general.

All types of electromagnetic radiation, or radiant energy, can be described in the terminology of waves. To help characterize any wave, we specify its *wavelength* (or its *frequency*). Let us use a familiar kind of wave, that on the surface of water (Figure 5-11), to illustrate these terms. The significant feature of wave motion is its repetitive nature. The **wavelength**, λ, is the distance between any two adjacent identical points of the wave, for instance, two adjacent crests. The **frequency** is the number of wave crests passing a given point per unit time; it is represented by the symbol ν (Greek letter

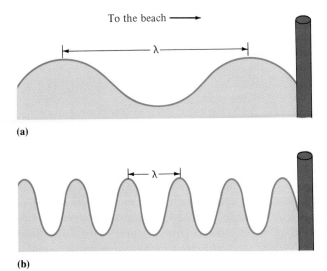

To the beach ⟶

(a)

(b)

Figure 5-11
Illustrations of the wavelength and frequency of water waves. The distance between any two identical points, e.g., crests, is the wavelength, λ. We could measure the frequency, ν, of the wave by observing the frequency at which the level rises and falls at a fixed point in its path—for instance, at the post. (a) and (b) represent two waves that are traveling at the same speed. In (a) the wave has long wavelength and low frequency; in (b) the wave has shorter wavelength and higher frequency.

''nu'') and is usually expressed in cycles/second or, more commonly, simply as 1/s or s^{-1}. For a wave that is ''traveling'' at some speed, the wavelength and the frequency are related to each other by

$$\lambda\nu = \text{speed of propagation of the wave} \quad \text{or} \quad \lambda\nu = c$$

One cycle per second is also called one *hertz* (Hz), after Heinrich Hertz. In 1887 Hertz discovered electromagnetic radiation outside the visible range and measured its speed and wavelengths.

Thus, wavelength and frequency are inversely proportional to each other; for the same wave speed, the shorter the wavelength, the higher the frequency.

For water waves, it is the surface of the water that changes repetitively; for a vibrating violin string, it is the displacement of any point on the string. Electromagnetic radiation consists of a regular, repetitive variation in electrical and magnetic fields. The electromagnetic radiation most obvious to us is visible light. It has wavelengths ranging from about 4×10^{-7} m (violet) to about 7×10^{-7} m (red). Expressed in frequencies, this range is about 7.5×10^{14} Hz (violet) to about 4.3×10^{14} Hz (red).

In a vacuum, the speed of electromagnetic radiation, c, is the same for all wavelengths, 2.9979249×10^8 m/s. The relationship between the wavelength and frequency of electromagnetic radiation, with c rounded to three significant figures, is

$$\lambda\nu = c = 3.00 \times 10^8 \text{ m/s}$$

Example 5-3
The frequency of violet light is 7.31×10^{14} s^{-1}, and that of red light is 4.57×10^{14} s^{-1}. Calculate the wavelength of each color.

Plan
Frequency and wavelength are inversely proportional to each other, $\lambda = c/\nu$. We can substitute the frequencies into this relationship and calculate wavelengths.

Sir Isaac Newton (1642–1727), one of the giants of science. You probably know of him from his theory of gravitation. In addition, he made enormous contributions to the understanding of many other aspects of physics, including the nature and behavior of light, optics, and the laws of motion. He is credited with the discoveries of differential calculus and of expansions into infinite series.

Solution

$$\text{(violet light) } \lambda = \frac{c}{\nu} = \frac{3.00 \times 10^8 \text{ m/s}}{7.31 \times 10^{14} \text{ s}^{-1}} = \boxed{4.10 \times 10^{-7} \text{ m } (4.10 \times 10^3 \text{ Å})}$$

$$\text{(red light) } \lambda = \frac{c}{\nu} = \frac{3.00 \times 10^8 \text{ m/s}}{4.57 \times 10^{14} \text{ s}^{-1}} = \boxed{6.56 \times 10^{-7} \text{ m } (6.56 \times 10^3 \text{ Å})}$$

EOC 39, 40

Isaac Newton first recorded the separation of sunlight into its component colors by allowing it to pass through a prism. Because sunlight (white light) contains all wavelengths of visible light, it gives the *continuous spectrum* observed in a rainbow (Figure 5-12a). Visible light represents only a tiny segment of the electromagnetic radiation spectrum (Figure 5-12b). In addition to all wavelengths of visible light, sunlight also contains shorter wavelength (ultraviolet) radiation as well as longer wavelength (infrared) radiation. Neither of these can be detected by the human eye. Both may be detected and recorded photographically or by detectors designed for that purpose. Many other familiar kinds of radiation are simply electromagnetic radiation of longer or shorter wavelengths.

Thus, we see that light is usually described in terms of wave behavior. Under certain conditions, it is also possible to describe light as composed of *particles,* or **photons**. According to the ideas presented by Max Planck in 1900, each photon of light has a particular amount (a **quantum**) of energy. Furthermore, the amount of energy possessed by a photon depends on the color of the light. The energy of a photon of light is given by Planck's equation

$$E = h\nu \qquad \text{or} \qquad E = \frac{hc}{\lambda}$$

Violet light has a shorter wavelength and a higher frequency than red light.

where h is Planck's constant, 6.6262×10^{-34} J · s, and ν is the frequency of the light. Thus, energy is directly proportional to frequency. Planck's equation is used in Example 5-4 to show that a photon of violet light has more energy than a photon of red light.

Example 5-4

In Example 5-3 we calculated the wavelengths of violet light of frequency 7.31×10^{14} s^{-1} and of red light of frequency 4.57×10^{14} s^{-1}. Calculate the energy, in joules, of an individual photon in each of these two colors of light.

Plan

We use the frequencies to calculate the energy of a photon from the relationship $E = h\nu$.

Solution

$$\text{(violet light) } E = h\nu = (6.63 \times 10^{-34} \text{ J · s})(7.31 \times 10^{14} \text{ s}^{-1}) = \boxed{4.85 \times 10^{-19} \text{ J}}$$

$$\text{(red light) } E = h\nu = (6.63 \times 10^{-34} \text{ J · s})(4.57 \times 10^{14} \text{ s}^{-1}) = \boxed{3.03 \times 10^{-19} \text{ J}}$$

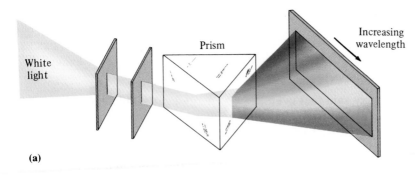

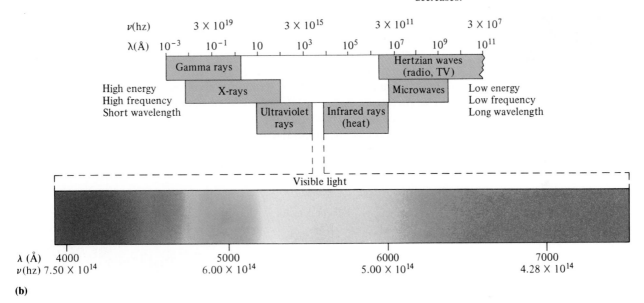

Figure 5-12

(a) Dispersion of visible light by a prism. Light from a source of white light is passed through a slit and then through a prism. It is separated into a continuous spectrum of all wavelengths of visible light. (b) Visible light is only a very small portion of the electromagnetic spectrum. Some radiant energy has longer or shorter wavelengths than our eyes can detect. The upper part shows the approximate ranges of the electromagnetic spectrum on a logarithmic scale. The lower part shows the visible region on an expanded scale. Note that wavelength increases as frequency decreases.

You can check these answers by calculating the energies directly from the wavelengths.

EOC 41

5-10 The Photoelectric Effect

One experiment that had not been satisfactorily explained with the wave model of light was the **photoelectric effect**. The apparatus for the photoelectric effect is shown in Figure 5-13. The negative electrode in the evacuated tube is made of a pure metal such as cesium. When light of a sufficiently high energy strikes the metal, electrons are knocked off its surface. They then travel to the positive electrode and form a current flowing through the circuit. The important observations follow.

1. Electrons can be ejected only if the light is of sufficiently short wavelength (has sufficiently high energy), no matter how long or how brightly the light shines. This wavelength limit is different for different metals.
2. The current (the number of electrons emitted per second) increases with increasing *brightness* (intensity) of the light. However, it does not depend

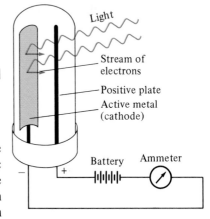

Figure 5-13
The photoelectric effect. When electromagnetic radiation of sufficient minimum energy strikes the surface of a metal (negative electrode) inside an evacuated tube, electrons are stripped off the metal to create an electric current. The current increases with increasing radiation intensity.

The intensity of light is the brightness of the light. In wave terms, it is related to the amplitude of the light waves.

The photoelectric effect is used in the photoelectric sensors that open some supermarket and elevator doors when the shadow of a person interrupts the light beam. Automatic cameras also use photocells.

on the color of the light as long as the wavelength is short enough (has high enough energy).

Classical theory said that even "low" energy light should cause current to flow if the metal is irradiated long enough. Electrons should accumulate energy and be released when they have enough energy to escape from the metal atoms. According to the old theory, if the light is made more energetic, then the current should increase even though the light intensity remains the same. Such is *not* the case.

The answer to the puzzle was provided by Albert Einstein. In 1905 he extended Planck's idea that light behaves as though it were composed of *photons,* each with a particular amount (a quantum) of energy. According to Einstein, each photon can transfer its energy to a single electron during a collision. When we say that the intensity of light is increased, we mean that the number of photons striking a given area per second is increased. The picture is now one of a particle of light striking an electron near the surface of the metal and giving up its energy to the electron. If that energy is equal to or greater than the amount needed to liberate the electron, it can escape to join the photoelectric current. For this explanation, Einstein received the 1921 Nobel prize in physics.

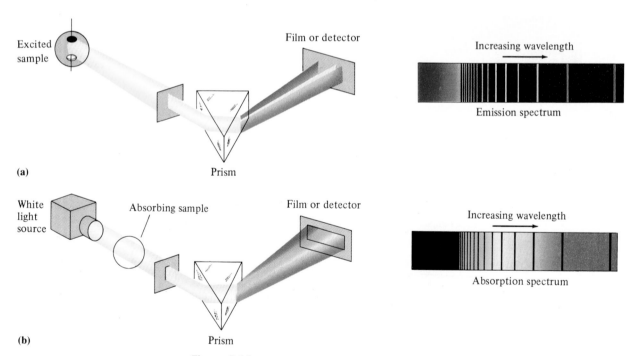

Figure 5-14

(a) *Atomic emission.* The light emitted by a sample of excited hydrogen atoms (or any other element) can be passed through a prism and separated into certain discrete wavelengths. Thus an emission spectrum, which is a photographic recording of the separated wavelengths, is called a line spectrum. Any sample of reasonable size contains an enormous number of atoms. Although a single atom can be in only one excited state at a time, the collection of atoms contains all possible excited states. The light emitted as these atoms fall to lower energy states is responsible for the spectrum. (b) *Atomic absorption.* When white light is passed through unexcited hydrogen and then through a slit and a prism, the transmitted light is lacking in intensity at the same wavelengths as are emitted in (a). The recorded absorption spectrum is also a line spectrum and the photographic negative of the emission spectrum.

5-11 Atomic Spectra and the Bohr Atom

Incandescent ("red hot" or "white hot") solids, liquids, and high-pressure gases give continuous spectra. However, when an electric current is passed through a gas in a vacuum tube at very low pressures, the light that the gas emits is dispersed by a prism into distinct lines (Figure 5-14a). Such **emission spectra** are described as *bright line spectra*. The lines can be recorded photographically, and the wavelength of light that produced each line can be calculated from the position of that line on the photograph.

Similarly, we can shine a beam of white light (containing a continuous distribution of wavelengths) through a gas and analyze the beam that emerges. We find that only certain wavelengths have been absorbed (Figure 5-14b). The wavelengths that are absorbed in this **absorption spectrum** are the same as those given off in the emission experiment. Each element displays its own characteristic set of lines in its emission or absorption spectrum (Figure 5-15). These spectra can serve as "fingerprints" to allow us to identify different elements present in a sample, even in trace amounts.

Example 5-5
A green line of wavelength 4.86×10^{-7} m is observed in the emission spectrum of hydrogen. Calculate the energy of one photon of this green light.

Plan
We know the wavelength of the light, and we calculate its frequency so that we can then calculate the energy of each photon.

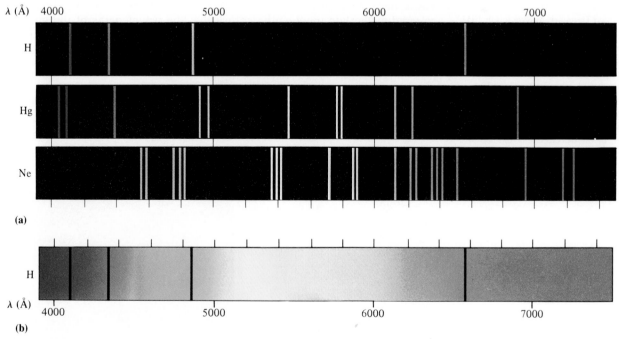

Figure 5-15
Atomic spectra in the visible region for some elements. Figure 5-14(a) shows how such spectra are produced. (a) Emission spectra for some elements. (b) Absorption spectrum for hydrogen. Compare the positions of these lines with those in the emission spectrum for H in (a).

The Danish physicist Niels Bohr (1885–1962) was one of the most influential scientists of the twentieth century. Like many other now-famous physicists of his time, he worked for a time in England with J. J. Thomson and later with Ernest Rutherford. During this period, he began to develop the ideas that led to the publication of his explanation of atomic spectra and his theory of atomic structure, for which he received the Nobel Prize in 1922. After escaping from German-occupied Denmark to Sweden in 1943, he helped to arrange the escape of hundreds of Danish Jews from the Hitler regime. He later went to the United States, where, until 1945, he worked with other scientists at Los Alamos, New Mexico on the development of the atomic bomb. From then until his death in 1962, he worked hard for the development and use of atomic energy for peaceful purposes.

Solution

$$\lambda v = c$$

$$v = \frac{c}{\lambda} = \frac{3.00 \times 10^8 \text{ m/s}}{4.86 \times 10^{-7} \text{ m}} = 6.17 \times 10^{14} \text{ s}^{-1}$$

$$E = hv = (6.63 \times 10^{-34} \text{ J} \cdot \text{s})(6.17 \times 10^{14} \text{ s}^{-1}) = \boxed{4.09 \times 10^{-19} \text{ J/photon}}$$

To gain a better appreciation of the amount of energy involved, let's calculate the total energy, in kilojoules, emitted by one mole of atoms. (Each atom emits one photon.)

$$\frac{? \text{ kJ}}{\text{mol}} = 4.09 \times 10^{-19} \frac{\text{J}}{\text{atom}} \times \frac{1 \text{ kJ}}{1 \times 10^3 \text{ J}} \times \frac{6.02 \times 10^{23} \text{ atoms}}{\text{mol}}$$

$$= \boxed{2.46 \times 10^2 \text{ kJ/mol}}$$

This calculation shows that when each atom in one mole of hydrogen atoms emits light of wavelength 4.86×10^{-7} m, the mole of atoms loses $\boxed{246 \text{ kJ}}$ of energy as green light. (This would be enough energy to operate a 100-watt light bulb for more than 40 minutes.)

EOC 44, 45, 46

When an electric current is passed through hydrogen gas at very low pressures, several series of lines in the spectrum of hydrogen are produced. These lines were studied intensely by many scientists. J. R. Rydberg discovered in the late nineteenth century that the wavelengths of the various lines in the hydrogen spectrum can be related by a mathematical equation:

$$\frac{1}{\lambda} = R \left(\frac{1}{n_1^2} - \frac{1}{n_2^2} \right)$$

Here R is 1.097×10^7 m^{-1} and is known as the Rydberg constant. The n's are positive integers, and n_1 is smaller than n_2. The Rydberg equation was derived from numerous observations, not theory. It is thus an empirical equation.

The lightning flashes produced in electrical storms and the light produced by neon gas in neon signs are two familiar examples of visible light produced by electronic transitions.

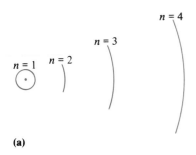

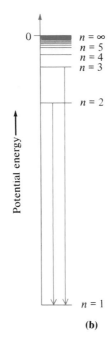

(a)

(b)

Figure 5-16
(a) The radii of the first four Bohr orbits for a hydrogen atom. The dot at the center represents the nuclear position. The radius of each orbit is proportional to n^2, so these four are in the ratio $1:4:9:16$. (b) Relative values for the energies associated with the various energy levels in a hydrogen atom. The energies become closer together as n increases. They are so close together for large values of n that they form a continuum. By convention, potential energy is defined as zero when the electron is at an infinite distance from the atom. Any more stable arrangement would have a lower energy. Therefore, potential energies of electrons in atoms are always negative. Some possible electronic transitions corresponding to lines in the hydrogen emission spectrum are indicated by arrows. Transitions in the opposite directions account for lines in the absorption spectrum.

In 1913 Niels Bohr, a Danish physicist, provided an explanation for Rydberg's observations. He wrote equations that described the electron of a hydrogen atom as revolving around the nucleus of an atom in circular orbits. He included the assumption that the electronic energy is *quantized;* that is, only certain values of electron energy are possible. This led him to the suggestion that electrons can only be in certain discrete orbits, and that they absorb or emit energy in discrete amounts as they move from one orbit to another. Each orbit thus corresponds to a definite *energy level* for the electron. When an electron is promoted from a lower energy level to a higher one, it absorbs a definite (or quantized) amount of energy. When the electron falls back to the original energy level, it emits exactly the same amount of energy it absorbed in moving from the lower to the higher energy level. Figure 5-16 illustrates these transitions schematically. The values of n_1 and n_2 in the Rydberg equation identify the lower and higher levels, respectively, of these electronic transitions.

The Bohr Theory and the Rydberg Equation

Enrichment

From mathematical equations describing the orbits for the hydrogen atoms, together with the assumption of quantization of energy, Bohr was able to determine two significant aspects of each allowed orbit:

1. *Where* (with respect to the nucleus) the electron can be—that is, the radius, r, of the circular orbit. This is given by

$$r = \frac{n^2 h^2}{4\pi^2 m e^2}$$

Note: r is proportional to n^2.

where h = Planck's constant, e = the charge of the electron, m = the mass of the electron, and n is a positive integer (1, 2, 3, ...) that tells us which orbit is being described.

2. *How stable* the electron would be in that orbit—that is, its potential energy, E. This is given by

Note: E is proportional to $-\dfrac{1}{n^2}$.

$$E = -\frac{2\pi m e^4}{n^2 h^2}$$

where the symbols have the same meaning as before. Note that E is always negative.

Results of evaluating these equations for some of the possible values of n (1, 2, 3, ...) are shown in Figure 5-17. The larger the value of n, the farther from the nucleus is the orbit being described, and the radius of this orbit increases as the *square of n* increases. As n increases, n^2 increases, $1/n^2$ decreases, and thus the electronic energy increases (becomes less negative and smaller in magnitude). For orbits farther from the nucleus, the electronic potential energy is higher (less negative—the electron is in a *higher* energy level or in a less stable state). Going away from the nucleus, the allowable orbits are farther apart in distance, but closer together in energy. Consider the two possible limits of these equations. One limit is when $n = 1$; this describes the electron at the smallest possible distance from the nucleus and at its lowest (most negative) energy. The other limit is for very large values of n, i.e., as n approaches infinity. As this limit is approached, the electron is very far from the nucleus, or effectively removed from the atom; the energy is as high as possible, approaching zero.

With these equations and the relationship that the Planck equation provides between energy and the frequency or wavelength, Bohr was able to predict the wavelengths observed in the hydrogen emission spectrum. Figure 5-17 illustrates the relationship between lines in the emission spectrum and the electronic transitions (changes of energy level) that occur in hydrogen atoms.

Each line in the emission spectrum represents the *difference in energies* between two allowed energy levels for the electron. When the electron goes from energy level n_2 to energy level n_1, the difference in energy is given off as a single photon. The energy of this photon can be calculated from Bohr's equation for the energy, as follows.

$$E \text{ of photon} = E_2 - E_1 = -\frac{2\pi m e^4}{n_2^2 h^2} - \left(-\frac{2\pi m e^4}{n_1^2 h^2}\right)$$

Factoring out the quantity $2\pi m e^4 / h^2$ and rearranging, we get

$$E \text{ of photon} = \frac{2\pi m e^4}{h^2}\left(\frac{1}{n_1^2} - \frac{1}{n_2^2}\right)$$

The Planck equation, $E = hc/\lambda$, relates the energy of the photon to the wavelength of the light, so

$$\frac{hc}{\lambda} = \frac{2\pi m e^4}{h^2}\left(\frac{1}{n_1^2} - \frac{1}{n_2^2}\right)$$

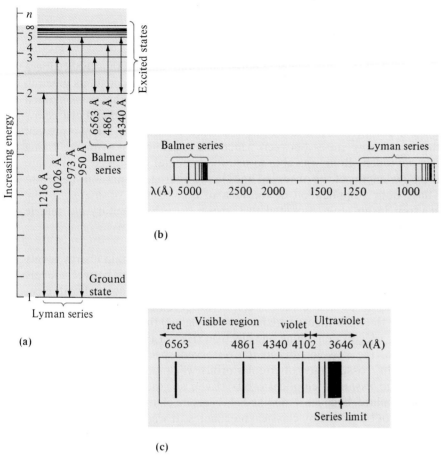

Figure 5-17

(a) The energy levels that the electron can occupy in a hydrogen atom and a few of the transitions that cause the emission spectrum of hydrogen. The numbers on the vertical lines show the wavelengths of light emitted when the electron falls to a lower energy level. (Light of the same wavelength is absorbed when the electron is promoted to the higher energy level.) The difference in energy between two given levels is exactly the same for all hydrogen atoms, so it corresponds to a specific wavelength and to a specific line in the emission spectrum of hydrogen. In a given sample, some hydrogen atoms could have their electrons excited to the $n = 2$ level. Some of these electrons could then fall to the $n = 1$ energy level, giving off the *difference* in energy in the form of light (the 1216-Å transition). Other hydrogen atoms might have their electrons excited to the $n = 3$ level; subsequently some could fall to the $n = 1$ level (the 1026-Å transition). Because higher energy levels become closer and closer in energy, *differences* in energy between successive transitions become smaller and smaller. The corresponding lines in the emission spectrum become closer together and eventually result in a continuum, a series of lines so close together that they are indistinguishable. (b) The emission spectrum of hydrogen. The series of lines produced by the electron falling to the $n = 1$ level is known as the *Lyman series;* it is in the ultraviolet region. A transition in which the electron falls to the $n = 2$ level gives rise to a similar set of lines in the visible region of the spectrum, known as the *Balmer series.* Not shown are series involving transitions to energy levels with higher values of n. (c) The Balmer series shown on an expanded scale. The line of 6563 Å (the $n = 3 \rightarrow n = 2$ transition) is much more intense than the line at 4861 Å (the $n = 4 \rightarrow n = 2$ transition) because the first transition occurs much more frequently than the second. Successive lines in the spectrum become less intense as the series limit is approached because the transitions that correspond to these lines are less probable.

Rearranging for $1/\lambda$, we obtain

$$\frac{1}{\lambda} = \frac{2\pi m e^4}{h^3 c} \left(\frac{1}{n_1^2} - \frac{1}{n_2^2} \right)$$

Comparing this to the Rydberg equation, Bohr showed that the Rydberg constant is equivalent to $2\pi m e^4/h^3 c$. We could use the constants given in Appendix D to obtain the same value, 1.097×10^7 m^{-1}, that was obtained by Rydberg on a solely empirical basis. Further, Bohr could also show the physical meaning of the two whole numbers n_1 and n_2; they represent the two energy states between which the transition takes place. Using this approach, Bohr was able to use fundamental constants to calculate the wavelengths of the observed lines in the hydrogen emission spectrum. Thus, Bohr's application of the idea of quantization of energy to the electron in an atom provided the answer to a half-century-old puzzle concerning the discrete colors given off in the spectrum.

We now accept the fact that electrons occupy only certain energy levels in atoms. In most atoms, some of the energy differences between levels correspond to the energy of visible light. Thus, colors associated with electronic transitions in such elements can be observed by the human eye.

Although the Bohr theory satisfactorily explained the spectra of hydrogen and of other species containing one electron (He$^+$, Li^{2+}, and so on) it could not calculate the wavelengths in the observed spectra of more complex species. Bohr's assumption of circular orbits was modified in 1916 by Sommerfeld, who assumed elliptical orbits. Even so, the Bohr approach was doomed to failure, because it modified classical mechanics to solve a problem that could not be solved by classical mechanics. It was a contrived solution. There was a need literally "to invent" a new physics, quantum mechanics, to deal with small particles, and Bohr's failure set the stage for the development of this new physics. However, the Bohr theory did introduce the ideas that only certain energy levels are possible, that these energy levels are described by quantum numbers that can have only certain allowed values, and that the quantum numbers indicate something about where and how stable the electrons are in these energy levels. The ideas of modern atomic theory have superseded Bohr's original theory. But his achievement in showing a link between electronic arrangements and Rydberg's empirical description of light absorption, and in establishing the quantization of electronic energy, was a very important step toward an understanding of atomic structure.

Two big questions remained about electrons in atoms: (1) How are electrons arranged in atoms? (2) How do these electrons behave? We now have the background to consider how modern atomic theory answers these questions.

5-12 The Wave Nature of the Electron

The idea that light can exhibit both wave properties and particle properties suggested to Louis de Broglie that very small particles, such as electrons, might also display wave properties under the proper circumstances. In his doctoral thesis in 1925, de Broglie predicted that a particle with a mass m and velocity v should have a wavelength associated with it. The numerical value of this de Broglie wavelength is given by

$$\lambda = h/mv \quad \text{(where } h = \text{Planck's constant)}$$

Two years after de Broglie's prediction, C. Davisson and L. H. Germer at the Bell Telephone Laboratory demonstrated diffraction of electrons by a crystal of nickel. This behavior is an important characteristic of waves. It shows conclusively that electrons do have wave properties. Davisson and Germer found that the wavelength associated with electrons of known energy is exactly that predicted by de Broglie. Similar diffraction experiments have been successfully performed with other particles, such as neutrons.

Be sure to distinguish between velocity, represented by the letter v, and frequency (Section 5-9), represented by v (Greek letter *nu*).

Example 5-6

(a) Calculate the wavelength of an electron traveling at 1.24×10^7 m/s. The mass of an electron is 9.11×10^{-28} g. (b) Calculate the wavelength of a baseball of mass 5.25 oz traveling at 92.5 mi/h. Recall that $1 \text{ J} = 1 \text{ kg} \cdot \text{m}^2/\text{s}^2$.

Plan

For each calculation, we use the de Broglie equation

$$\lambda = \frac{h}{mv}$$

where

$$h \text{ (Planck's constant)} = 6.63 \times 10^{-34} \text{ J} \cdot \text{s} \times \frac{1 \dfrac{\text{kg} \cdot \text{m}^2}{\text{s}^2}}{1 \text{ J}}$$

$$= 6.63 \times 10^{-34} \frac{\text{kg} \cdot \text{m}^2}{\text{s}}$$

For consistency of units, mass must be expressed in kilograms. In part (b), we must also convert the speed to meters per second.

Solution

(a)
$$m = 9.11 \times 10^{-28} \text{ g} \times \frac{1 \text{ kg}}{1000 \text{ g}} = 9.11 \times 10^{-31} \text{ kg}$$

Substituting into the de Broglie equation,

$$\lambda = \frac{h}{mv} = \frac{6.63 \times 10^{-34} \dfrac{\text{kg} \cdot \text{m}^2}{\text{s}}}{(9.11 \times 10^{-31} \text{ kg})\left(1.24 \times 10^7 \dfrac{\text{m}}{\text{s}}\right)} = \boxed{5.87 \times 10^{-11} \text{ m}}$$

While this seems like a very short wavelength, it is of the same order as the spacing between atoms in many crystals. A stream of such electrons hitting a crystal gives easily measurable diffraction effects.

(b)
$$m = 5.25 \text{ oz} \times \frac{1 \text{ lb}}{16 \text{ oz}} \times \frac{1 \text{ kg}}{2.205 \text{ lb}} = 0.149 \text{ kg}$$

$$v = \frac{92.5 \text{ mi}}{\text{h}} \times \frac{1 \text{ h}}{3600 \text{ s}} \times \frac{1.609 \text{ km}}{1 \text{ mi}} \times \frac{1000 \text{ m}}{1 \text{ km}} = 41.3 \frac{\text{m}}{\text{s}}$$

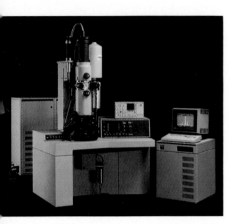

A modern electron microscope.

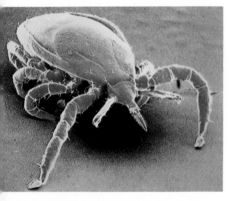

A color-enhanced scanning electron micrograph of a deer tick, the carrier of Lyme disease.

Now we substitute into the de Broglie equation.

$$\lambda = \frac{h}{mv} = \frac{6.63 \times 10^{-34} \, \frac{\text{kg} \cdot \text{m}^2}{\text{s}}}{(0.149 \, \text{kg})\left(41.3 \, \frac{\text{m}}{\text{s}}\right)} = \boxed{1.08 \times 10^{-34} \, \text{m}}$$

This wavelength is far too short to give any measurable effects. Recall that atomic diameters are of the order of 10^{-10} m, which is 24 powers of 10 greater than the baseball "wavelength."

EOC 68

As you can see from the results of Example 5-6, the particles of the subatomic world behave very differently from the macroscopic objects with which we are familiar. To talk about behavior of atoms and their particles, we must give up many of our long-held prejudices about the behavior of matter. We must be willing to visualize a world of new and unfamiliar properties, such as the ability to act in some ways like a particle and in other ways like a wave.

The wave behavior of electrons is exploited in the electron microscope. This instrument allows magnification of objects far too small to be seen with an ordinary light microscope.

5-13 The Quantum Mechanical Picture of the Atom

Through the work of de Broglie, Davisson and Germer, and others, we now know that electrons in atoms can be treated as waves more effectively than as small compact particles traveling in circular or elliptical orbits. Large objects such as golf balls and moving automobiles obey the laws of classical mechanics (Isaac Newton's laws), but very small particles such as electrons, atoms, and molecules do not. A different kind of mechanics, called **quantum mechanics**, describes the behavior of very small particles much better because it is based on the *wave* properties of matter. Quantization of energy is a consequence of these properties.

One of the underlying principles of quantum mechanics is that we cannot determine precisely the paths that electrons follow as they move about atomic nuclei. The **Heisenberg Uncertainty Principle**, stated in 1927 by Werner Heisenberg, is a theoretical assertion that is consistent with all experimental observations.

This is like trying to locate the position of a moving automobile by driving another automobile into it.

> It is impossible to determine accurately both the momentum and the position of an electron (or any other very small particle) simultaneously.

Momentum is mass times velocity, mv. Because electrons are so small and move so rapidly, their motion is usually detected by electromagnetic radiation. Photons that interact with electrons have about the same energies as the electrons. Consequently, the interaction of a photon with an electron severely disturbs the motion of the electron. It is not possible to determine simultaneously both the position and the velocity of an electron, so we resort to a statistical approach and speak of the probability of finding an electron within specified regions in space.

With these ideas in mind, we can now list some basic ideas of quantum mechanics.

1. Atoms and molecules can exist only in certain energy states. In each energy state, the atom or molecule has a definite energy. When an atom or molecule changes its energy state, it must emit or absorb just enough energy to bring it to the new energy state (the quantum condition).

Atoms and molecules possess various forms of energy. Let us focus our attention on their *electronic energies.*

2. Atoms or molecules emit or absorb radiation (light) as they change their energies. The frequency of the light emitted or absorbed is related to the energy change by a single equation:

$$\Delta E = h\nu \qquad \text{or} \qquad \Delta E = hc/\lambda$$

Recall that $\lambda\nu = c$, so $\nu = c/\lambda$.

This gives a relationship between the energy change, ΔE, and the wavelength, λ, of the radiation emitted or absorbed. *The energy lost (or gained) by an atom as it goes from higher to lower (or lower to higher) energy states is equal to the energy of the photon emitted (or absorbed) during the transition.*

3. The allowed energy states of atoms and molecules can be described by sets of numbers called *quantum numbers.*

The mathematical approach of quantum mechanics involves treating the electron in an atom as a *standing wave*. A standing wave is a wave that does not travel and therefore has at least one point at which it has zero amplitude, called a node. As an example, consider the various ways that a guitar string can vibrate when it is plucked (Figure 5-18). Because both ends are fixed (nodes), the string can vibrate only in ways in which there is a whole number of *half-wavelengths* in the length of the string (Figure 5-18a). Any actual motion of the string can be described as some combination of these allowed vibrations. In a similar way, we can imagine that the electron in the hydrogen atom behaves as a wave (recall the de Broglie relationship in the last section). The electron can be described by the same kind of standing-wave mathematics that is applied to the vibrating guitar string. In a given space around the nucleus, only certain "waves" can exist. Each "allowed wave" corresponds to a stable energy state for the electron, and is described by a particular set of quantum numbers.

The quantum mechanical treatment of atoms and molecules is highly mathematical. The important point is that each solution of the Schrödinger wave equation (see Enrichment section) describes a possible energy state for the electrons in the atom. Each solution is described by a set of three **quantum numbers**. These numbers are in accord with those deduced from experiment and from empirical equations such as the Rydberg equation. Solutions of the Schrödinger equation also tell us about the shapes and orientations of the statistical probability distributions of the electrons. (The Heisenberg Principle implies that this is how we must describe the positions of the electrons.) These *atomic orbitals* (which are described in Section 5-15) are deduced from the solutions of the Schrödinger equation. The orbitals

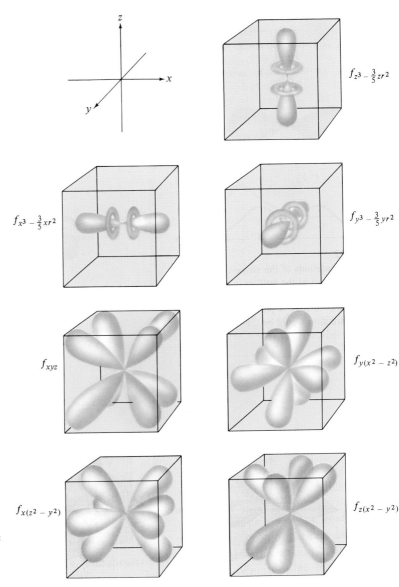

Figure 5-25
Relative directional character of atomic
f orbitals. The seven orbitals are
shown within cubes as an aid to
visualization.

axis along which each of the three two-lobed orbitals is directed. A set of three p atomic orbitals may be represented as in Figure 5-23b.

Beginning at the third energy level, each level contains a third sublevel ($\ell = 2$) composed of a set of *five d* atomic orbitals ($m_\ell = -2, -1, 0, +1, +2$). They are designated $3d, 4d, 5d, \ldots$ to indicate the energy level in which they are found. The shapes of the members of a set are indicated in Figure 5-24.

In each of the fourth and higher energy levels, there is also a fourth sublevel, containing a set of *seven f* atomic orbitals ($\ell = 3, m_\ell = -3, -2, -1, 0, +1, +2, +3$). These are shown in Figure 5-25.

Thus we see the first energy level contains only the $1s$ orbital; the second energy level contains the $2s$ and three $2p$ orbitals; the third energy level contains the 3s, three 3p, and five 3d orbitals; and the fourth energy level consists of a 4s, three 4p, five 4d, and seven 4f orbitals. All subsequent energy levels contain s, p, d, and f sublevels as well as others that are not occupied in any presently known elements in their lowest energy states.

The sizes of orbitals increase with increasing n, as shown in Figure 5-26. Slender representations of orbitals, such as those in Figures 5-23 through 5-25, are usually used for convenience. The true shapes are actually more like those in Figure 5-26.

Let us summarize in tabular form some of the information we have developed to this point. The principal quantum number n indicates the energy level. The number of sublevels per energy level is equal to n, the number of atomic orbitals per energy level is n^2, and the maximum number of electrons per energy level is $2n^2$, because each atomic orbital can hold two electrons.

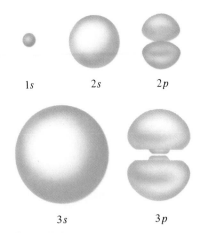

Energy Level n	Number of Sublevels per Energy Level n	Number of Atomic Orbitals n^2	Maximum Number of Electrons $2n^2$
1	1	1 (1s)	2
2	2	4 (2s, 2p_x, 2p_y, 2p_z)	8
3	3	9 (3s, three 3p's, five 3d's)	18
4	4	16	32
5	5	25	50

Figure 5-26
Shapes and approximate relative sizes of several orbitals in an atom.

In this section, we haven't yet discussed the fourth quantum number, the spin quantum number, m_s. Because m_s has two possible values, $+\frac{1}{2}$ and $-\frac{1}{2}$, each atomic orbital, defined by the values of n, ℓ, and m_ℓ, has a capacity of two electrons. Electrons are negatively charged, and they behave as though they were spinning about axes through their centers, so they act like tiny magnets. The motions of electrons produce magnetic fields, and these can interact with one another. Two electrons in the same orbital having opposite m_s values are said to be **spin-paired**, or simply **paired** (Figure 5-27).

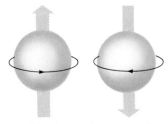

One electron has $m_s = +\frac{1}{2}$; the other has $m_s = -\frac{1}{2}$.

Figure 5-27
Electron spin. Electrons act as though they spin about an axis through their centers. Because there are two directions in which an electron may spin, the spin quantum number has two possible values, $+\frac{1}{2}$ and $-\frac{1}{2}$. Each electron spin produces a magnetic field. When two electrons have opposite spins, the attraction due to their opposite magnetic fields helps to overcome the repulsion of their like charges. This permits two electrons to occupy the same region (orbital).

5-16 Electron Configurations

The wave function for an atom simultaneously depends on (describes) all of the electrons in the atom. The Schrödinger equation is much more complicated for atoms with more than one electron than for a one-electron species such as hydrogen, and an explicit solution to this equation is not possible even for helium, let alone for more complicated atoms. Therefore we must rely on approximations to solutions of the many-electron Schrödinger equation. We shall use one of the most common and useful, called the **orbital approximation**. In this approximation, the electron cloud of an atom is assumed to be the superposition of charge clouds, or orbitals, arising from the individual electrons; these orbitals resemble the atomic orbitals of hydrogen (for which exact solutions are known), which we described in some detail in the last section. Each electron is described by the same allowed combinations of quantum numbers (n, ℓ, m_ℓ, and m_s) that we used for the hydrogen atom; however, the order of energies of the orbitals is often different from that in hydrogen.

The great power of modern computers has allowed scientists to make numerical approximations to this solution to very high accuracy for simple atoms such as helium. However, as the number of electrons increases, even such numerical approaches become quite difficult to apply and interpret. For many purposes, more qualitative approximations are suitable.

Let us now examine the electronic structures of atoms of different elements. The electronic arrangement that we shall describe for each atom is called the **ground state electron configuration**. This corresponds to the isolated atom in its lowest-energy, or unexcited, state. We shall consider the elements in order of increasing atomic number, using as our guide the periodic table inside the front cover of this text. For simplicity, we shall indicate atomic orbitals as ⎯ and show an unpaired electron as ↿ and spin-paired electrons as ↿⇂. By "unpaired electron" we mean an electron that occupies an orbital singly.

In building up ground state electron configurations, the guiding idea is that the *total energy* of the atom is as low as possible. To determine these configurations, we use the **Aufbau Principle** as a guide:

The German verb *aufbauen* means "to build up."

> Each atom is "built up" by (1) adding the appropriate numbers of protons and neutrons as specified by the atomic number and the mass number, and (2) adding the necessary number of electrons into orbitals in the way that gives the lowest *total* energy for the atom.

As we apply this principle, we shall focus on the difference in electronic arrangement between a given element and the element with an atomic number

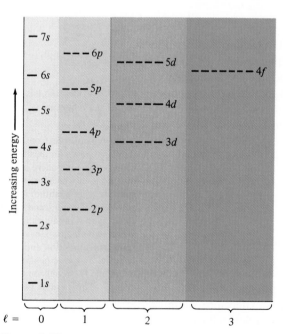

Figure 5-28
The usual order of filling (Aufbau order) of the orbitals of an atom. The energy scale varies for different elements, but the following main features should be noted: (1) The largest energy gap is between the 1s and 2s orbitals. (2) The energies of orbitals are generally closer together at higher energies. (3) The gap between *np* and (*n* + 1)*s* (e.g., between 2*p* and 3*s* or between 3*p* and 4*s*) is usually fairly large. (4) The gap between (*n* − 1)*d* and *ns* (e.g., between 3*d* and 4*s*) is quite small. (5) The gap between (*n* − 2)*f* and *ns* (e.g., between 4*f* and 6*s*) is even smaller.

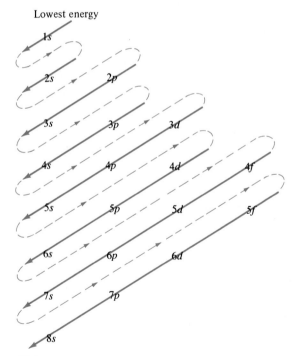

Figure 5-29
An aid to remembering the Aufbau order of atomic orbitals. Write all sublevels in the same major energy level on the same horizontal line. Write all like sublevels in the same vertical column. Draw parallel arrows diagonally from upper right to lower left. The arrows are read from top to bottom, and tail to head. The order is 1*s*, 2*s*, 2*p*, 3*s*, 3*p*, 4*s*, 3*d*, 4*p*, 5*s*, 4*d*, . . . and so on.

that is one lower. Though we do not always point it out, we *must* keep in mind that the atomic number (the charge on the nucleus) is different. We also emphasize the particular electron that distinguishes each element from the previous one; however, we should remember that this distinction is artificial, because electrons are not really distinguishable.

The orbitals increase in energy with increasing value of the quantum number n. For a given value of n, energy increases with increasing value of ℓ. In other words, within a particular major energy level, the s sublevel is lowest in energy, the p sublevel is the next lowest, then the d, then the f, and so on. As a result of changes in the nuclear charge and interactions among the electrons in the atom, the order of energies of the orbitals can vary somewhat from atom to atom. The *usual* order of energies of the orbitals of an atom and a helpful device for remembering this order are shown in Figures 5-28 and 5-29.

The electronic structures of atoms are governed by the **Pauli Exclusion Principle**:

> No two electrons in an atom may have identical sets of four quantum numbers.

An orbital is described by a particular allowed set of values for n, ℓ, and m_ℓ. Thus, two electrons can occupy the same orbital only if they have opposite spins, m_s. Two such electrons in the same orbital are *paired*.

Row 1 The first energy level consists of only one atomic orbital, $1s$. This can hold a maximum of two electrons. Hydrogen, as we have already noted, contains just one electron. Helium, a noble gas, has a filled first energy level (two electrons). The atom is so stable that no chemical reactions of helium are known.

Helium's electrons can be displaced only by electrical forces, as in excitation by high-voltage discharge.

	Orbital Notation	**Simplified Notation**
	$1s$	
$_1$H	↑	$1s^1$
$_2$He	↑↓	$1s^2$

In the simplified notation, we indicate with superscripts the number of electrons in each sublevel.

Row 2 Elements of atomic numbers 3 through 10 occupy the second period, or horizontal row, in the periodic table. In neon atoms the second energy level is filled completely. Neon, a noble gas, is extremely stable. No reactions of it are known.

	Orbital Notation			**Simplified Notation**		
	$1s$	$2s$	$2p$			
$_3$Li	↑↓	↑		$1s^2 2s^1$	or	[He] $2s^1$
$_4$Be	↑↓	↑↓		$1s^2 2s^2$		[He] $2s^2$
$_5$B	↑↓	↑↓	↑ _ _	$1s^2 2s^2 2p^1$		[He] $2s^2 2p^1$
$_6$C	↑↓	↑↓	↑ ↑ _	$1s^2 2s^2 2p^2$		[He] $2s^2 2p^2$
$_7$N	↑↓	↑↓	↑ ↑ ↑	$1s^2 2s^2 2p^3$		[He] $2s^2 2p^3$
$_8$O	↑↓	↑↓	↑↓ ↑ ↑	$1s^2 2s^2 2p^4$		[He] $2s^2 2p^4$
$_9$F	↑↓	↑↓	↑↓ ↑↓ ↑	$1s^2 2s^2 2p^5$		[He] $2s^2 2p^5$
$_{10}$Ne	↑↓	↑↓	↑↓ ↑↓ ↑↓	$1s^2 2s^2 2p^6$		[He] $2s^2 2p^6$

In writing electronic structures of atoms, we frequently simplify notations. The abbreviation [He] indicates that the $1s$ orbital is completely filled, $1s^2$, as in helium.

As with helium, neon's electrons can be displaced by high-voltage electrical discharge, as is observed in neon signs.

We see that some atoms have unpaired electrons in the same set of energetically equivalent, or **degenerate**, orbitals. We have already seen that two electrons can occupy a given atomic orbital (with the same values of n, ℓ, and m_ℓ) *only* if their spins are paired (have opposite values of m_s). Even with pairing of spins, however, two electrons that are in the same orbital repel each other more strongly than do two electrons in different (but equal-energy) orbitals. Thus, both theory and experimental observations (see Enrichment section) lead to **Hund's Rule**:

> Electrons must occupy all the orbitals of a given sublevel singly before pairing begins. These unpaired electrons have parallel spins.

Thus, carbon has two unpaired electrons in its $2p$ orbitals, and nitrogen has three.

Enrichment

Both paramagnetism and diamagnetism are hundreds to thousands of times weaker than *ferromagnetism*, the effect seen in iron bar magnets.

Paramagnetism and Diamagnetism

Substances that contain unpaired electrons are weakly *attracted* into magnetic fields and are said to be **paramagnetic**. By contrast, those in which all electrons are paired are very weakly repelled by magnetic fields and are called **diamagnetic**. The magnetic effect can be measured by hanging a test tube full of a substance on a balance by a long thread and suspending it above the gap of an electromagnet (Figure 5-30). When the current is switched on, a paramagnetic substance such as copper(II) sulfate is pulled into the strong field. The paramagnetic attraction per mole of substance can be measured by weighing the sample before and after energizing the magnet. The paramagnetism per mole increases with increasing number of unpaired electrons per formula unit.

Row 3 The next element beyond neon is sodium. Here we begin to add electrons to the third energy level. Elements 11 through 18 occupy the third period in the periodic table.

	Orbital Notation		
	3s	3p	Simplified Notation
$_{11}$Na	[Ne] ↿		[Ne] $3s^1$
$_{12}$Mg	[Ne] ⇅		[Ne] $3s^2$
$_{13}$Al	[Ne] ⇅	↿ _ _	[Ne] $3s^2 3p^1$
$_{14}$Si	[Ne] ⇅	↿ ↿ _	[Ne] $3s^2 3p^2$
$_{15}$P	[Ne] ⇅	↿ ↿ ↿	[Ne] $3s^2 3p^3$
$_{16}$S	[Ne] ⇅	⇅ ↿ ↿	[Ne] $3s^2 3p^4$
$_{17}$Cl	[Ne] ⇅	⇅ ⇅ ↿	[Ne] $3s^2 3p^5$
$_{18}$Ar	[Ne] ⇅	⇅ ⇅ ⇅	[Ne] $3s^2 3p^6$

Although the third energy level is not yet filled (the d orbitals are still empty), argon is a noble gas. All noble gases except helium have $ns^2 np^6$ electronic configurations (where n indicates the highest occupied energy level). The noble gases are quite unreactive elements.

Rows 4 and 5 It is an experimentally observed fact that *an electron occupies the available orbital that gives the atom the lowest total energy*. It is observed that filling the $4s$ orbitals before electrons enter the $3d$ orbitals *usually* leads to a lower total energy for the atom than some other arrangement.

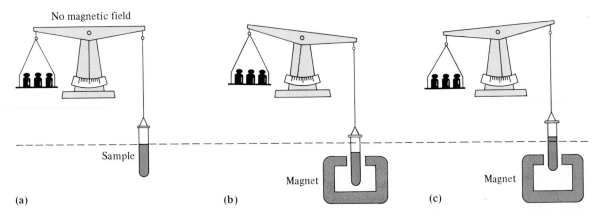

No magnetic field

Sample

(a)

Magnet

(b)

Magnet

(c)

Therefore we fill the orbitals in this order (see Figure 5-28). According to the normal Aufbau order (recall Figures 5-28 and 5-29), 4s fills before 3d. In general, *the (n+1)s orbital fills before the nd orbital.* This is sometimes referred to as the *n + 1 rule.*

After the 3d sublevel is filled to its capacity of 10 electrons, the 4p orbitals fill next, taking us to the noble gas krypton. Then the 5s orbital, the five 4d orbitals, and the three 5p orbitals fill to take us to xenon, a noble gas.

Let us now examine the electronic structures of the 18 elements in the fourth period in some detail. Some of these have electrons in d orbitals.

Figure 5-30

Diagram of an apparatus for measuring the paramagnetism of a substance. The tube contains a measured amount of the substance, often in solution. (a) Before the magnetic field is turned on, the position and mass of the sample are determined. (b) When the field is on, a paramagnetic substance is attracted *into* the field. (c) A diamagnetic substance would be repelled *very weakly* by the field.

		Orbital Notation			Simplified Notation
		3d	4s	4p	
$_{19}$K	[Ar]		↑		[Ar] $4s^1$
$_{20}$Ca	[Ar]		↑↓		[Ar] $4s^2$
$_{21}$Sc	[Ar]	↑ _ _ _ _	↑↓		[Ar] $3d^14s^2$
$_{22}$Ti	[Ar]	↑ ↑ _ _ _	↑↓		[Ar] $3d^24s^2$
$_{23}$V	[Ar]	↑ ↑ ↑ _ _	↑↓		[Ar] $3d^34s^2$
$_{24}$Cr	[Ar]	↑ ↑ ↑ ↑ ↑	↑		[Ar] $3d^54s^1$
$_{25}$Mn	[Ar]	↑ ↑ ↑ ↑ ↑	↑↓		[Ar] $3d^54s^2$
$_{26}$Fe	[Ar]	↑↓ ↑ ↑ ↑ ↑	↑↓		[Ar] $3d^64s^2$
$_{27}$Co	[Ar]	↑↓ ↑↓ ↑ ↑ ↑	↑↓		[Ar] $3d^74s^2$
$_{28}$Ni	[Ar]	↑↓ ↑↓ ↑↓ ↑ ↑	↑↓		[Ar] $3d^84s^2$
$_{29}$Cu	[Ar]	↑↓ ↑↓ ↑↓ ↑↓ ↑↓	↑		[Ar] $3d^{10}4s^1$
$_{30}$Zn	[Ar]	↑↓ ↑↓ ↑↓ ↑↓ ↑↓	↑↓		[Ar] $3d^{10}4s^2$
$_{31}$Ga	[Ar]	↑↓ ↑↓ ↑↓ ↑↓ ↑↓	↑↓	↑ _ _	[Ar] $3d^{10}4s^24p^1$
$_{32}$Ge	[Ar]	↑↓ ↑↓ ↑↓ ↑↓ ↑↓	↑↓	↑ ↑ _	[Ar] $3d^{10}4s^24p^2$
$_{33}$As	[Ar]	↑↓ ↑↓ ↑↓ ↑↓ ↑↓	↑↓	↑ ↑ ↑	[Ar] $3d^{10}4s^24p^3$
$_{34}$Se	[Ar]	↑↓ ↑↓ ↑↓ ↑↓ ↑↓	↑↓	↑↓ ↑ ↑	[Ar] $3d^{10}4s^24p^4$
$_{35}$Br	[Ar]	↑↓ ↑↓ ↑↓ ↑↓ ↑↓	↑↓	↑↓ ↑↓ ↑	[Ar] $3d^{10}4s^24p^5$
$_{36}$Kr	[Ar]	↑↓ ↑↓ ↑↓ ↑↓ ↑↓	↑↓	↑↓ ↑↓ ↑↓	[Ar] $3d^{10}4s^24p^6$

As you study these electronic configurations, you should be able to see how most of them are predicted from the Aufbau order. However, as we fill the 3d set of orbitals, from $_{21}$Sc to $_{30}$Zn, we see that these orbitals are not filled quite regularly. Some sets of orbitals are so close in energy (e.g., 4s and 3d) that minor changes in their relative energies may occasionally change the order of filling.

End-of-chapter Exercises 84–107 provide much valuable practice in writing electron configurations.

These two elements illustrate an exception to the $n + 1$ rule. You should realize that such statements as the Aufbau Principle and the $n + 1$ rule merely represent general guidelines and should not be viewed as hard-and-fast rules. It is the *total energy* of the atom that is as low as possible. The Aufbau order of orbital energies is based on calculations for the hydrogen atom. The orbital energies also depend on additional factors such as the nuclear charge and interactions of different occupied orbitals.

Chemical and spectroscopic evidence indicates that the configurations of Cr and Cu have only one electron in the $4s$ orbital. Their $3d$ sets are half-filled and filled, respectively, in the ground state. Calculations from the quantum mechanical equations also indicate that *half-filled and filled sets of equivalent orbitals have a special stability*. In $_{24}$Cr, for example, this increased stability is apparently sufficient to make the *total* energy of [Ar] $3d$ ↿ ↿ ↿ ↿ ↿ $4s$ ↿ lower than that of [Ar] $3d$ ↿ ↿ ↿ ↿ _ $4s$ ↿⇂. Similar reasoning helps us understand the apparent exception of the configuration of $_{29}$Cu from that predicted by the Aufbau Principle.

You may wonder why such an exception does not occur in, for example, $_{32}$Ge or $_{14}$Si, where we could have an s^1p^3 configuration that would have half-filled sets of s and p orbitals. It does not occur because of the very large energy gap between ns and np orbitals. We shall see evidence in Chapter 6 that does, however, illustrate the enhanced stability of half-filled sets of p orbitals.

Let us now write the quantum numbers to describe each electron in an atom of nitrogen. Keep in mind the fact that Hund's Rule must be obeyed. Thus, there is only one (unpaired) electron in each $2p$ orbital in a nitrogen atom.

Example 5-7

Write an acceptable set of four quantum numbers for each of the electrons of a nitrogen atom.

Plan

Nitrogen has seven electrons, which occupy the lowest-energy orbitals available. Two electrons can occupy the first energy level, $n = 1$, in which there is only one s orbital; when $n = 1$, then ℓ must be zero, and therefore $m_\ell = 0$. The two electrons differ only in spin quantum number, m_s. The next five electrons can all fit into the second energy level, for which $n = 2$ and ℓ may be either 0 or 1. The $\ell = 0$ (s) sublevel fills first, and the $\ell = 1$ (p) sublevel is occupied next.

Solution

Electrons are indistinguishable. We have numbered them 1, 2, 3, and so on as an aid to counting them.

In the lowest-energy configurations, the three $2p$ electrons either all have $m_s = +\frac{1}{2}$ or all have $m_s = -\frac{1}{2}$.

Electron	n	ℓ	m_ℓ	m_s	e^- Configuration
1, 2	$\begin{cases}1\\1\end{cases}$	0 0	0 0	$+\frac{1}{2}$ $-\frac{1}{2}$	$1s^2$
3, 4	$\begin{cases}2\\2\end{cases}$	0 0	0 0	$+\frac{1}{2}$ $-\frac{1}{2}$	$2s^2$
5, 6, 7	$\begin{cases}2\\2\\2\end{cases}$	1 1 1	-1 0 $+1$	$+\frac{1}{2}$ or $-\frac{1}{2}$ $+\frac{1}{2}$ or $-\frac{1}{2}$ $+\frac{1}{2}$ or $-\frac{1}{2}$	$2p_x^{\ 1}$ $2p_y^{\ 1}$ or $2p^3$ $2p_z^{\ 1}$

Example 5-8

Write an acceptable set of four quantum numbers that decribe each electron in a chlorine atom.

Plan

Chlorine is element number 17. Its first seven electrons have the same quantum numbers as those of nitrogen in Example 5-7. Electrons 8, 9, and 10 complete the filling of the 2p sublevel ($n = 2$, $\ell = 1$) and therefore also the second energy level. Electrons 11 through 17 fill the 3s sublevel ($n = 3$, $\ell = 0$) and partially fill the 3p sublevel ($n = 3$, $\ell = 1$).

Solution

Electron	n	ℓ	m_ℓ	m_s	e^- Configuration
1, 2	1	0	0	$\pm\frac{1}{2}$	$1s^2$
3, 4	2	0	0	$\pm\frac{1}{2}$	$2s^2$
5–10	$\begin{cases}2\\2\\2\end{cases}$	1 1 1	-1 0 $+1$	$\left.\begin{matrix}\pm\frac{1}{2}\\\pm\frac{1}{2}\\\pm\frac{1}{2}\end{matrix}\right\}$	$2p^6$
11, 12	3	0	0	$\pm\frac{1}{2}$	$3s^2$
13–17	$\begin{cases}3\\3\\3\end{cases}$	1 1 1	-1 0 $+1$	$\left.\begin{matrix}\pm\frac{1}{2}\\\pm\frac{1}{2}\\+\frac{1}{2}\text{ or }-\frac{1}{2}*\end{matrix}\right\}$	$3p^5$

*The 3p orbital with only a single electron can be any one of the set, not *necessarily* the one with $m_\ell = +1$.

EOC 101, 105

5-17 The Periodic Table and Electron Configurations

In this section, we view the *periodic table* (see inside front cover and Section 4-1) from a modern, much more useful perspective—as a systematic representation of the electronic configurations of the elements. In the long form of the periodic table, used throughout this text, elements are arranged in blocks based on the kinds of atomic orbitals that are being filled (Figure 5-31). The periodic tables in this text are divided into "A" and "B" groups. The A groups contain elements in which s and p orbitals are being filled. Elements within any particular A group have similar electronic configurations and chemical properties, as we shall see in the next chapter. The B groups are those in which there are one or two electrons in the s orbital of the highest occupied energy level, and the d orbitals one energy level lower are being filled.

Lithium, sodium, and potassium, elements of the leftmost column of the periodic table (Group IA), have a single electron in their outermost s orbital (ns^1). Beryllium and magnesium, of Group IIA, have two electrons in their highest energy level, ns^2, while boron and aluminum (Group IIIA) have three electrons in their highest energy level, ns^2np^1. Similar observations can be made for each A group.

The electronic configurations of the A group elements and the noble gases can be predicted reliably from Figures 5-28 and 5-29. However, there are some irregularities in the B groups below the fourth period that require special attention. In the heavier B group elements, the higher-energy sublevels in different principal energy levels have energies that are very nearly equal (Figure 5-29). It is easy for an electron to jump from one orbital to

H is shown in the 1s block in Figure 5-31. It is usually shown in Group IA.

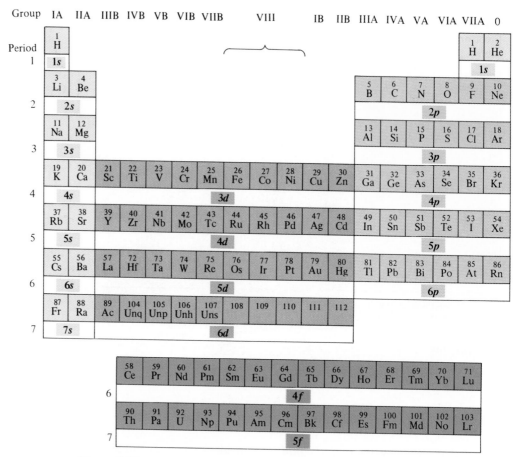

Figure 5-31

A periodic table colored to show the kinds of atomic orbitals (sublevels) being filled, below the symbols of blocks of elements. The electronic structures of the A group and 0 group elements are perfectly regular and can be predicted from their positions in the periodic table, but there are many exceptions in the *d* and *f* blocks. The colors in this figure are the same as those in Figure 5-28.

Hydrogen and helium are shown here in their usual positions in the periodic table. These may seem somewhat unusual based just on their electronic configurations. However, we should remember that the first main energy level ($n = 1$) can hold a maximum of only two electrons. This shell is entirely filled in helium, so He behaves as a noble gas, and we put it in the column with the other noble gases (Group 0). Hydrogen has one electron that is easily lost, like the metals in Group IA, so we put it in Group IA even though it is not a metal. Furthermore, hydrogen is one electron short of a noble gas configuration (He), so we could also place it with the other such elements in Group VIIA.

another of nearly the same energy in a different set. This is because the orbital energies are *perturbed* (they change slightly) as the nuclear charge changes, and an extra electron is added in going from one element to the next. This phenomenon gives rise to other irregularities that are analogous to those of Cr and Cu, described earlier.

We can extend the information in Figure 5-31 to indicate the electron configurations that are represented by each *group* (column) of the periodic table. Table 5-5 shows this interpretation of the periodic table, along with the most important exceptions. (A more complete listing of electron config-

Table 5-5
The *s*, *p*, *d*, and *f* Blocks of the Periodic Table*

**n is the principal quantum number. The d^1s^2, d^2s^2, . . . designations represent *known* configurations. They refer to $(n-1)d$ and ns orbitals. Several exceptions to the configurations indicated above each group are shown in gray.

urations is given in Appendix B.) We can use this interpretation of the periodic table to write, quickly and reliably, the electron configurations for elements.

Example 5-9

Use Table 5-5 to determine the electronic configurations of (a) germanium, Ge; (b) magnesium, Mg; and (c) molybdenum, Mo.

Plan

We use the electron configurations indicated in Table 5-5 for each group. Each *period* (row) begins filling a new shell (new value of *n*). Elements to the right of the *d* orbital block have the *d* orbitals in the $(n-1)$ shell already filled. We often find it convenient to collect all sets of orbitals with the same value of *n* together, to emphasize the number of electrons in the *outermost* shell, i.e., the shell with the highest value of *n*.

Solution

(a) Germanium, Ge, is in Group IVA, for which Table 5-5 shows the general configuration s^2p^2. It is in Period 4 (the 4th row), so we interpret this as $4s^24p^2$. The last filled noble gas configuration is that of argon, Ar, accounting for 18 electrons. In addition, Ge lies beyond the *d* orbital block, so we know that the 3*d* orbitals are completely filled. The electron configuration of Ge is [Ar] $4s^23d^{10}4p^2$ or   [Ar] $3d^{10}4s^24p^2$.

(b) Magnesium, Mg, is in Group IIA, which has the general configuration s^2; it is in Period 3 (3rd row). The last filled noble gas configuration is that of neon, or [Ne]. The electron configuration of Mg is [Ne] $3s^2$.

(c) Molybdenum, Mo, is in Group VIB, with general configuration d^5s^1; it is in Period 5, which begins with 5*s* and is beyond the noble gas krypton. The electron configuration of Mo is [Kr] $5s^14d^5$ or   [Kr] $4d^55s^1$. The electron configuration of molybdenum is analogous to that of chromium, Cr, the element just above it. The configuration of Cr was discussed in Section 5-16 as one of the exceptions to the Aufbau order of filling.

EOC 102

Example 5-10

Determine the number of unpaired electrons in an atom of tellurium, Te.

Plan

Te is in Group IVA in the periodic table, which tells us that its configuration is s^2p^4. All other shells are completely filled, so they contain only paired electrons. We need only to find out how many unpaired electrons are represented by s^2p^4.

Solution

The notation s^2p^4 is a short representation for $s^{\underline{\Updownarrow}} p^{\underline{\Updownarrow}} \uparrow \uparrow$. This shows that an atom of Te contains two unpaired electrons.

The periodic table has been described as "the chemist's best friend." Notice how easy it is to use the periodic table to determine many important aspects of the electron configurations of atoms. You should practice until you can use the periodic table with confidence to answer many questions about electron configurations. As we continue our study of chemistry, we shall learn many other useful ways to interpret the periodic table.

Key Terms

Absorption spectrum   The spectrum associated with absorption of electromagnetic radiation by atoms (or other species) resulting from transitions from lower to higher energy states.

Alpha (α) particle   A helium ion with 2+ charge; an assembly of two protons and two neutrons.

Anode   In a cathode ray tube, the positive electrode.

Atomic mass unit   An arbitrary mass unit defined to be exactly one-twelfth the mass of the carbon-12 isotope.

Atomic number   The integral number of protons in the nucleus; defines the identity of an element.

Atomic orbital   The region or volume in space in which the probability of finding electrons is highest.

Aufbau ("building up") Principle   Describes the order in which electrons fill orbitals in atoms.

Canal ray   A stream of positively charged particles (cations) that moves toward the negative electrode in a cathode ray tube; observed to pass through canals in the negative electrode.

Cathode   In a cathode ray tube, the negative electrode.

Cathode ray   The beam of electrons going from the negative electrode toward the positive electrode in a cathode ray tube.

Cathode ray tube   A closed glass tube containing a gas under low pressure, with electrodes near the ends and a luminescent screen at the end near the positive electrode; produces cathode rays when high voltage is applied.

Continuous spectrum   The spectrum that contains all wavelengths in a specified region of the electromagnetic spectrum.

Degenerate   Of the same energy.

Diamagnetism   *Weak* repulsion by a magnetic field.

d orbitals   Beginning in the third energy level, a set of five degenerate orbitals per energy level, higher in energy than s and p orbitals of the same energy level.

Electromagnetic radiation   Energy that is propagated by means of electric and magnetic fields that oscillate in directions perpendicular to the direction of travel of the energy.

Electron   A subatomic particle having a mass of 0.00054858 amu and a charge of $1-$.

Electron configuration   The specific distribution of electrons in the atomic orbitals of atoms or ions.

Electronic transition   The transfer of an electron from one energy level to another.

Emission spectrum   The spectrum associated with emission of electromagnetic radiation by atoms (or other species) resulting from electronic transitions from higher to lower energy states.

Excited state   Any state other than the ground state of an atom or molecule.

f orbitals   Beginning in the fourth energy level, a set of seven degenerate orbitals per energy level, higher in energy than s, p, and d orbitals of the same energy level.

Frequency   The number of repeating corresponding points on a wave that pass a given observation point per unit time.

Ground state   The lowest energy state or most stable state of an atom, molecule, or ion.

Group   A vertical column in the periodic table; also called a family.

Heisenberg Uncertainty Principle   It is impossible to determine accurately both the momentum and position of an electron simultaneously.

Hund's Rule   All orbitals of a given sublevel must be occupied by single electrons before pairing begins. See *Aufbau Principle*.

Isotopes   Two or more forms of atoms of the same element with different masses; atoms containing the same number of protons but different numbers of neutrons.

Line spectrum   An atomic emission or absorption spectrum.

Magnetic quantum number (m_ℓ)   Quantum mechanical solution to a wave equation that designates the particular orbital within a given set (s, p, d, f) in which an electron resides.

Mass number   The integral sum of the numbers of protons and neutrons in an atom.

Mass spectrometer   An instrument that measures the charge-to-mass ratios of charged particles.

Natural radioactivity   Spontaneous decomposition of an atom.

Neutron   A neutral subatomic nuclear particle having a mass of 1.0087 amu.

Nucleus   The very small, very dense, positively charged center of an atom containing protons and neutrons, as well as other subatomic particles.

Nuclide symbol   The symbol for an atom, ${}_{Z}^{A}E$, in which E is the symbol for an element, Z is its atomic number, and A is its mass number.

Pairing of electrons   Favorable interaction of two electrons with opposite m_s values in the same orbital.

Paramagnetism   Attraction toward a magnetic field, stronger than diamagnetism, but still weak compared with ferromagnetism.

Pauli Exclusion Principle   No two electrons in the same atom may have identical sets of four quantum numbers.

Period   A horizontal row in the periodic table.

Photoelectric effect   Emission of an electron from the surface of a metal, caused by impinging electromagnetic radiation of certain minimum energy; current increases with increasing intensity of radiation.

Photon   A "packet" of light or electromagnetic radiation; also called a quantum of light.

p orbitals   Beginning with the second energy level, a set of three mutually perpendicular, equal-arm, dumbbell-shaped atomic orbitals per energy level.

Principal quantum number (n)   The quantum mechanical solution to a wave equation that designates the major energy level, or shell, in which an electron resides.

Proton   A subatomic particle having a mass of 1.0073 amu and a charge of $+1$, found in the nuclei of atoms.

Quantum   A "packet" of energy.

Quantum mechanics   A mathematical method of treating particles on the basis of quantum theory, which assumes that energy (of small particles) is not infinitely divisible.

Quantum numbers   Numbers that describe the energies of electrons in atoms; derived from quantum mechanical treatment.

Radiant energy   See *Electromagnetic radiation*.

Rydberg equation   An empirical equation that relates wavelengths in the hydrogen emission spectrum to integers.

s orbital   A spherically symmetrical atomic orbital; one per energy level.

Spectral line   Any of a number of lines corresponding to definite wavelengths in an atomic emission or absorption spectrum; represents the energy difference between two energy levels.

Spectrum   Display of component wavelengths (colors) of electromagnetic radiation.

Spin quantum number (m_s)   The quantum mechanical solution to a wave equation that indicates the relative spins of electrons.

Subsidiary quantum number (ℓ)   The quantum mechanical solution to a wave equation that designates the sublevel, or set of orbitals (s, p, d, f), within a given major energy level in which an electron resides.

Wavelength   The distance between two corresponding points of a wave.

Exercises

Particles and the Nuclear Atom

1. List the three fundamental particles of matter and indicate the mass and charge associated with each.
2. (a) Describe the cathode ray experiment. How can we detect where the rays strike?
   (b) Describe an experiment in which it can be established that the streams of electrons in an operating cathode ray tube travel toward the anode rather than the cathode.
3. Describe how Thomson determined the charge-to-mass ratio for the electron.
4. In the oil-drop experiment, how did Millikan know that none of the oil droplets he observed were ones that had a deficiency of electrons rather than an excess?
5. How many electrons carry a total charge of 1.00 coulomb?
6. (a) How do we know that canal rays have charges opposite in sign to cathode rays? What are canal rays?
   (b) Why are cathode rays from all samples of gases identical, whereas canal rays are not?
*7. The following data are measurements of the charges on oil droplets using an apparatus similar to that used by Millikan:

$$11.215 \times 10^{-19} \text{ C} \qquad 14.423 \times 10^{-19} \text{ C}$$
$$12.811 \times 10^{-19} \text{ C} \qquad 24.037 \times 10^{-19} \text{ C}$$
$$14.419 \times 10^{-19} \text{ C} \qquad 9.621 \times 10^{-19} \text{ C}$$
$$12.815 \times 10^{-19} \text{ C} \qquad 16.012 \times 10^{-19} \text{ C}$$

   Each should be a whole-number ratio of some fundamental charge. Using these data, determine the value of the fundamental charge.
*8. Suppose we discover a new positively charged particle, which we call the "whizatron." We want to determine its charge.
   (a) What modifications would we have to make to the Millikan oil-drop apparatus to carry out the corresponding experiment on whizatrons?
   (b) In such an experiment, we observe the following charges on five different droplets:

$$3.26 \times 10^{-19} \text{ C}$$
$$4.08 \times 10^{-19} \text{ C}$$
$$1.63 \times 10^{-19} \text{ C}$$
$$5.70 \times 10^{-19} \text{ C}$$
$$4.89 \times 10^{-19} \text{ C}$$

   What is the charge on the whizatron?
9. What are alpha particles? Characterize them as to mass and charge.
10. (a) What do we mean when we refer to the nuclear atom?
    (b) Outline Rutherford's contribution to understanding the nature of atoms.
11. If the mass and electrical charge were uniformly distributed throughout an atom, what would be the expected results of an α-particle scattering experiment? What was the major conclusion drawn from the results of the α-particle scattering experiments?
12. The approximate radius of a hydrogen atom is 0.0529 nm, and that of a proton is 1.5×10^{-15} m. Assuming both the hydrogen atom and the proton to be spherical, calculate the fraction of the space in an atom of hydrogen that is occupied by the nucleus. $V = (4/3)\pi r^3$ for a sphere.
13. The approximate radius of a neutron is 1.5×10^{-15} m, and the mass is 1.675×10^{-27} kg. Calculate the density of a neutron. $V = (4/3)\pi r^3$ for a sphere.
14. Arrange the following in order of increasing ratio of charge to mass: $^{12}C^+$, $^{12}C^{2+}$, $^{13}C^+$, $^{13}C^{2+}$.
15. Refer to Exercise 14. Suppose all of these high-energy ions are present in a mass spectrometer. For which one will its path be changed (a) the most and (b) the least by increasing the external magnetic field?

Atom Composition, Isotopes, and Atomic Weights

16. Estimate the percentage of the total mass of a $^{195}_{78}Pt$ atom that is due to (a) electrons, (b) protons, and (c) neutrons by *assuming* that the mass of the atom is simply the sum of the masses of the appropriate numbers of subatomic particles.
17. (a) How are isotopic abundances determined experimentally?
    (b) How do the isotopes of a given element differ?
18. Define and illustrate the following terms clearly and concisely: (a) atomic number, (b) isotope, (c) mass number, (d) nuclear charge.
19. Write the composition of one atom of each of the three isotopes of silicon: ^{28}Si, ^{29}Si, and ^{30}Si.
20. Write the composition of one atom of each of the four isotopes of iron: ^{54}Fe, ^{56}Fe, ^{57}Fe, and ^{58}Fe.
21. Complete Chart A for neutral atoms.
22. Complete Chart B for neutral atoms.
23. The element iodine (I) occurs naturally as a single isotope of mass number 127; its atomic number is 53. How many protons and how many neutrons does it have in its nucleus?
24. On the average, a silver atom weighs 107.02 times as much as an atom of hydrogen. Using 1.0079 amu as the atomic weight of hydrogen, calculate the atomic weight of silver in atomic mass units.
25. Prior to 1962 the atomic weight scale was based on the assignment of an atomic weight of exactly 16 amu to the *naturally occurring* mixture of oxygen. The atomic weight of cadmium is 112.411 amu on the carbon-12 scale. What was it on the older scale?
26. What is the nuclear composition of each of the following isotopes of platinum, $_{78}Pt$—mass numbers 192, 194,

Chart A

Kind of Atom	Atomic Number	Mass Number	Isotope	Number of Protons	Number of Electrons	Number of Neutrons
			$^{44}_{20}Ca$			
potassium		39				
	16	32				
		174		70		

Chart B

Kind of Atom	Atomic Number	Mass Number	Isotope	Number of Protons	Number of Electrons	Number of Neutrons
nickel						32
			$^{191}_{77}Ir$			
					26	28
		195			78	

195, 196, 198? How many electrons are there in each of these atoms?

27. How many neutrons, protons, and electrons are in each of the following atoms or ions? $^{209}_{83}Bi^{3+}$, $^{193}_{77}Ir$, $^{51}_{23}V^{5+}$, $^{81}_{35}Br^-$, $^{98}_{42}Mo^{4+}$, $^{32}_{16}S^{2-}$.

28. The atomic weight of copper is 63.546 amu. The two naturally occurring isotopes of copper have the following masses: ^{63}Cu, 62.9298 amu; ^{65}Cu, 64.9278 amu. Calculate the percent of ^{63}Cu in naturally occurring copper.

29. The atomic weight of chlorine is 35.453 amu. There are only two isotopes in naturally occurring chlorine: ^{35}Cl (34.96885 amu) and ^{37}Cl (36.9658 amu). Calculate the percent composition of naturally occurring chlorine.

30. Determine the charge on each of the following species: (a) Ca with 18 electrons, (b) Cu with 28 electrons, (c) Cu with 26 electrons, (d) Pt with 74 electrons, (e) F with 10 electrons, (f) C with 6 electrons, (g) C with 7 electrons, (h) C with 5 electrons.

31. The following is a mass spectrum of the 1+ charged ions of an element. Calculate the atomic weight of the element. What is the element?

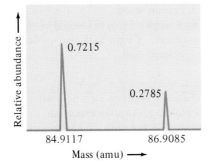

32. Suppose you measure the mass spectrum of the 1+ charged ions of germanium, atomic weight 72.61 amu.

Unfortunately, the recorder on the mass spectrometer jams at the beginning and again at the end of your experiment. You obtain only the partial spectrum shown below, which *may or may not be complete*. From the information given here, can you tell whether one of the germanium isotopes is missing? If one is missing, at which end of the plot should it appear?

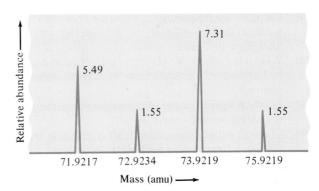

33. Calculate the atomic weight of zinc using the following percent of natural abundance and mass of each isotope: 48.6% $^{64}_{30}Zn$ (63.929 amu), 27.9% $^{66}_{30}Zn$ (65.9260 amu), 4.1% $^{67}_{30}Zn$ (66.9271 amu), and 18.8% $^{68}_{30}Zn$ (67.9298 amu).

34. Calculate the atomic weight of strontium using the following data for the percent of natural abundance and mass of each isotope: 0.5% of ^{84}Sr (83.9134 amu), 9.9% of ^{86}Sr (85.9094 amu), 7.0% of ^{87}Sr (86.9089 amu), and 82.6% of ^{88}Sr (87.9056 amu).

*35. There are two naturally occurring isotopes of hydrogen (^1H, >99%, and ^2H, <1%) and two of chlorine (^{35}Cl, 76%, and ^{37}Cl, 24%).
(a) How many different HCl molecules can be formed from these isotopes?
(b) What is the approximate mass of each of the molecules, expressed in atomic mass units?

(c) List the molecules in order of decreasing relative abundance—as would be observed on a mass spectrometer—assuming only the formation of $(HCl)^+$ ions.

*36. In a suitable reference such as the Table of Isotopes in the *Handbook of Chemistry and Physics* (The Chemical Rubber Co.), look up the following information for selenium: (a) the total number of known isotopes, (b) the atomic mass, and (c) the percentage of natural abundance and mass of each of the stable isotopes. (d) Calculate the atomic weight of selenium.

37. Consider the ions $^{16}_{8}O^+$, $^{17}_{8}O^+$, $^{16}_{8}O^{2+}$, $^{17}_{8}O^{2+}$ produced in a mass spectrometer. Which ion's path would be deflected (a) most and (b) least by a magnetic field?

38. Consider the ions in Exercise 36. Which one would travel (a) most rapidly and (b) least rapidly under the influence of a particular accelerating voltage? Justify your answers.

Electromagnetic Radiation

39. Calculate the wavelengths, in meters, of radiation of the following frequencies:
(a) 4.80×10^{15} s^{-1}
(b) 1.18×10^{14} s^{-1}
(c) 5.44×10^{12} s^{-1}

40. Calculate the frequencies of radiation of the following wavelengths: (a) 9774 Å, (b) 492 nm, (c) 4.92 cm, (d) 4.92×10^{-9} cm.

41. What is the energy of a photon of each of the radiations in Exercise 39? Express your answer in joules per photon.

42. In which regions of the electromagnetic spectrum do the radiations in Exercise 39 fall?

43. Classical music radio station KMFA in Austin broadcasts at a frequency of 89.5 MHz. What is the wavelength of its signal in meters?

44. Excited lithium ions emit radiation at a wavelength of 670.8 nm in the visible range of the spectrum. (This characteristic color is often used as a qualitative analysis test for the presence of Li^+.) Calculate (a) the frequency and (b) the energy of a photon of this radiation. (c) What color is this light?

45. Find the energy of the photons corresponding to the red line, 6573 Å, in the spectrum of the Ca atom.

46. Ozone in the upper atmosphere absorbs ultraviolet radiation, which induces the following chemical reaction:

$$O_3(g) \longrightarrow O_2(g) + O(g)$$

What is the energy of a 3400-Å photon that is absorbed? What is the energy of a mole of these photons?

*47. During photosynthesis, chlorophyll-a absorbs light of wavelength 440 nm and emits light of wavelength 670 nm. What is the energy available for photosynthesis from the absorption–emission of a mole of photons?

*48. Assume that 10^{-17} J of light energy is needed by the interior of the human eye to "see" an object. How many photons of green light (wavelength = 550 nm) are needed to generate this minimum energy?

*49. The human eye receives a 2.500×10^{-14} J-signal consisting of photons of blue light, $\lambda = 4700$ Å. How many photons reach the eye?

*50. Water absorbs microwave radiation of wavelength 3 mm. How many photons are needed to raise the temperature of a cup of water (250 g) from 25°C to 85°C in a microwave oven, using this radiation? The specific heat of water is 4.184 J/g · °C.

The Photoelectric Effect

51. What evidence supports the idea that electromagnetic radiation is (a) wave-like; (b) particle-like?

52. Describe the influence of frequency and intensity of electromagnetic radiation on the current in the photoelectric effect.

*53. Cesium is often used in "electric eyes" for self-opening doors in an application of the photoelectric effect. The amount of energy required to ionize (remove an electron from) a cesium atom is 3.89 electron volts (1 eV = 1.60×10^{-19} J). Show by calculation whether a beam of yellow light with wavelength 5230 Å would ionize a cesium atom.

*54. Refer to Exercise 53. What would be the wavelength, in nanometers, of light with just sufficient energy to ionize a cesium atom?

Atomic Spectra and the Bohr Theory

55. (a) Distinguish between an atomic emission spectrum and an atomic absorption spectrum.
(b) Distinguish between a continuous spectrum and a line spectrum.

56. Prepare a sketch similar to Figure 5-16b that shows a ground energy state and three excited energy states. Using vertical arrows, indicate the transitions that would correspond to the absorption spectrum for this system.

57. What is the Rydberg equation? Why is it called an empirical equation?

58. Hydrogen atoms absorb energy so that the electrons are excited to the energy level $n = 7$. Electrons then undergo these transitions: (1) $n = 7 \rightarrow n = 1$, (2) $n = 7 \rightarrow n = 6$, (3) $n = 2 \rightarrow n = 1$. Which transition will produce the photon with (a) the smallest energy; (b) the highest frequency; (c) the shortest wavelength? (d) What is the frequency of a photon resulting from the transition $n = 6 \rightarrow n = 1$?

*59. Five energy levels of the He atom are given in J/atom above an *arbitrary* reference energy: (1) 6.000×10^{-19}, (2) 8.812×10^{-19}, (3) 9.381×10^{-19}, (4) 10.443×10^{-19}, (5) 10.934×10^{-19}. Construct an energy-level diagram for He and find the energy of the photon (a) absorbed for the electron transition from level 1 to level 4 and (b) emitted for the electron transition from level 5 to level 2.

*60. The *Lyman series* is the name given to the series of lines in the ultraviolet portion of the emission spectrum of hydrogen. These lines correspond to transitions of electrons from higher energy states to the lowest ($n_1 = 1$) energy state. Use the Rydberg equation to calculate the wavelengths, in nm, of the three lowest energy lines in the Lyman series.

*61. The Balmer series of lines in the hydrogen emission spectrum is in the visible to ultraviolet range. These lines correspond to transitions of electrons from a higher energy state to the second-lowest ($n_1 = 2$) energy state. One line of the Balmer series has a wavelength of 410.2 nm. What is the quantum number of the upper energy state?

62. The following are prominent lines in the visible region of the emission spectra of the elements listed. The lines can be used to identify the elements. What color is the light responsible for each line? (a) lithium, 6708 Å; (b) neon, 616.0 nm; (c) mercury, 4540 Å; (d) cesium, $\nu = 3.45 \times 10^{14}$ Hz; (e) potassium, $\nu = 3.90 \times 10^{14}$ Hz.

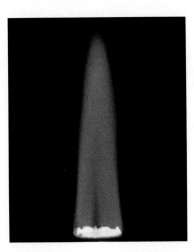

63. Hydrogen atoms have an absorption line at 973 Å. What is the frequency of the photons absorbed, and what is the energy difference, in joules, between the ground state and this excited state of the atom?

*64. If each atom in one mole of atoms emits a photon of wavelength 6.24×10^3 Å, how much energy is lost? Express the answer in kJ/mol. As a reference point, burning one mole (16 g) of CH_4 produces 819 kJ of heat.

*65. Suppose we could excite all of the electrons in a sample of hydrogen atoms to the $n = 5$ level. They would then emit light as they relaxed to lower energy states. Some atoms might undergo the transition $n = 5$ to $n = 1$, while others might go from $n = 5$ to $n = 4$, then from $n = 4$ to $n = 3$, and so on. How many lines would we expect to observe in the resulting emission spectrum?

*66. An argon laser emits blue light with a wavelength of 488.0 nm. How many photons are emitted by this laser in 2.00 seconds, operating at a power of 515 milliwatts? One watt (a unit of power) is equal to 1 joule/second.

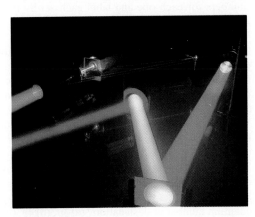

The Wave–Particle View of Matter

67. (a) What evidence supports the idea that electrons are particle-like?
(b) What evidence supports the idea that electrons are wave-like?

68. (a) What is the de Broglie wavelength of a proton moving at a speed of 3.00×10^7 m/s? The proton mass is 1.67×10^{-24} g.
(b) What is the de Broglie wavelength of a stone with a mass of 30.0 g moving at 2.00×10^3 m/h (≈ 100 mi/h)?
(c) How do the wavelengths in (a) and (b) compare with the typical radii of atoms? (See the atomic radii in Figure 6-2.)

69. What is the wavelength corresponding to a neutron of mass 1.67×10^{-27} kg moving at 2200 m/s?

*70. The energy of a photon in the X-ray region of the spectrum is 7×10^{-16} J. According to de Broglie's equation, what is the mass of this photon?

Quantum Numbers and Atomic Orbitals

71. (a) What is a quantum number? What is an atomic orbital?

    (b) How many quantum numbers are required to specify a single atomic orbital? What are they?

72. How are the possible values for the subsidiary quantum number for a given electron restricted by the value of n?

73. Without giving the ranges of possible values of the four quantum numbers, n, ℓ, m_ℓ and m_s, describe briefly what information each one gives.

74. (a) How are the values of m_ℓ for a particular electron restricted by the value of ℓ?

    (b) What are the letter designations for the values $n = 1, 2, 3, 4$?

    (c) What are the letter designations for the values $\ell = 0, 1, 2, 3$?

75. What are the values of n and ℓ for the following sublevels? (a) $2s$, (b) $3d$, (c) $4p$, (d) $5s$, (e) $4f$.

76. How many individual orbitals are there in the third major energy level? Write out n, ℓ and m_ℓ quantum numbers for each one and label each set by the s, p, d, f designations.

77. (a) Write the possible values of ℓ when $n = 5$.

    (b) Write the allowed number of orbitals (1) with the quantum numbers $n = 4$, $\ell = 3$; (2) with the quantum number $n = 4$; (3) with the quantum numbers $n = 7$, $\ell = 6$, $m_\ell = 6$; (4) with quantum numbers $n = 6$, $\ell = 5$.

78. Write the subshell notations that correspond to (a) $n = 2$, $\ell = 0$; (b) $n = 4$, $\ell = 2$; (c) $n = 7$, $\ell = 0$; (d) $n = 5$, $\ell = 3$.

79. What values can m_ℓ take for (a) a $4d$ orbital, (b) a $1s$ orbital, and (c) a $3p$ orbital?

80. How many orbitals in any atom can have the given quantum number or designation? (a) $3p$, (b) $4p$, (c) $4p_x$, (d) $n = 5$, (e) $6d$, (f) $5d$, (g) $5f$, (h) $7s$.

81. The following incorrect sets of quantum numbers in the order n, ℓ, m_ℓ, m_s are written for paired electrons or for one electron in an orbital. Correct them, assuming n values are correct. (a) $1, 0, 0, +\frac{1}{2}, +\frac{1}{2}$; (b) $2, 2, 1, \pm\frac{1}{2}$; (c) $3, 2, 3, \pm\frac{1}{2}$; (d) $3, 1, 2, +\frac{1}{2}$; (e) $2, 1, -1, 0$; (f) $3, 0, -1, -\frac{1}{2}$.

82. (a) What do we mean when we refer to the "spin" of an electron?

    (b) What are spin-paired electrons?

83. In an atom, how many electrons could have principal quantum number $n = 5$?

84. (a) How are a $1s$ orbital and a $2s$ orbital in an atom similar? How do they differ?

    (b) How are a $2p_x$ orbital and a $2p_y$ orbital in an atom similar? How do they differ?

Electron Configurations and the Periodic Table

You should be able to use the positions of elements in the periodic table to answer the remaining exercises.

85. Draw representations of ground state electron configurations using the orbital notation ($\underline{\uparrow\downarrow}$) for the following elements. (a) C, (b) Fe, (c) P, (d) Rh.

86. Draw representations of ground state electron configurations using the orbital notation ($\underline{\uparrow\downarrow}$) for the following elements. (a) S, (b) Ni, (c) Mg, (d) Zr.

87. Give the ground state electron configurations for the elements of Exercise 85 using shorthand notation— that is, $1s^2 2s^2 2p^6$, and so on.

88. Give the ground state electron configurations for the elements of Exercise 86 using shorthand notation— that is, $1s^2 2s^2 2p^6$, and so on.

89. State the Pauli Exclusion Principle. Would any of the following electron configurations violate this rule: (a) $1s^2$, (b) $1s^2 2p^1$, (c) $1s^3$? Explain.

90. State Hund's Rule. Would any of the following electron configurations violate this rule: (a) $1s^2$, (b) $1s^2 2s^2 2p_x^2$, (c) $1s^2 2s^2 2p_x^1 2p_y^1$, (d) $1s^2 2s^2 2p_x^1 2p_z^1$, (e) $1s^2 2s^2 2p_x^2 2p_y^1 2p_z^1$? Explain.

*91. Classify each of the following atomic electron configurations as (i) a ground state, (ii) an excited state, or (iii) a forbidden state: (a) $1s^2 2s^2 2p^5 3s^1$, (b) $[Kr] 4d^{10} 5s^3$, (c) $1s^2 2s^2 2p^6 3s^2 3p^6 3d^8 4s^2$, (d) $1s^2 2s^2 2p^6 3s^2 3p^6 3d^1$, (e) $1s^2 2s^2 2p^{10} 3s^2 3p^5$.

92. Which elements are represented by the following electron configurations?

    (a) $1s^2 2s^2 2p^6 3s^2 3p^6 3d^{10} 4s^2 4p^3$

    (b) $[Kr] 4d^{10} 4f^{14} 5s^2 5p^6 5d^{10} 5f^{14} 6s^2 6p^6 6d^2 7s^2$

    (c) $[Kr] 4d^{10} 4f^{14} 5s^2 5p^6 5d^{10} 6s^2 6p^4$

    (d) $[Kr] 4d^5 5s^2$

    (e) $1s^2 2s^2 2p^6 3s^2 3p^6 3d^3 4s^2$

93. Repeat Exercise 92 for

    (a) $1s^2 2s^2 2p^6 3s^2 3p^6 3d^5 4s^1$

    (b) $[Kr] 4d^{10} 4f^{14} 5s^2 5p^6 5d^{10} 6s^2 6p^1$

    (c) $1s^2 2s^2 2p^6 3s^2 3p^6$

    (d) $[Kr] 4d^{10} 4f^{14} 5s^2 5p^6 5d^{10} 6s^2 6p^6 7s^2$

94. Find the total number of s, p, and d electrons in each of the following: (a) Si, (b) Ar, (c) Ni, (d) Zn, (e) Rb.

95. (a) Distinguish between the terms "diamagnetic" and "paramagnetic," and provide an example that illustrates the meaning of each.

    (b) How is paramagnetism measured experimentally?

96. How many unpaired electrons are in atoms of Na, Ne, B, Be, Se, and Ti?

97. Which of the following ions or atoms possess paramagnetic properties? (a) F^-, (b) Na^+, (c) Co, (d) Ar^-, (e) S.

98. Which of the following ions or atoms possess paramagnetic properties? (a) F, (b) Ar, (c) Ar^+, (d) Zn, (e) S^{2-}.

99. Write the electron configurations of the Group IA elements Li, Na, and K (see inside front cover). What similarity do you observe?

100. Construct a table in which you list a possible set of values for the four quantum numbers for each electron

in the following atoms in their ground states. (a) N, (b) P, (c) Cu.

101. Construct a table in which you list a possible set of values for the four quantum numbers for each electron in the following atoms in their ground states. (a) B, (b) Cl, (c) Ni.

102. Draw general electronic structures for the A group elements using the $\underline{\uparrow\downarrow}$ notation, where n is the principal quantum number for the highest occupied energy level.

	ns	*np*
IA	__	__ __ __
IIA	__	__ __ __
and so on		

103. Repeat Exercise 102 using $ns^x np^y$ notation.

104. List n, ℓ, and m_ℓ quantum numbers for the highest-energy electron (or one of the highest-energy electrons if there are more than one) in the following atoms in their ground states (a) Si, (b) Nb, (c) Br, (d) Pr.

105. List n, ℓ, and m_ℓ quantum numbers for the highest-energy electron (or one of the highest-energy electrons if there are more than one) in the following atoms in their ground states. (a) As, (b) Ag, (c) Mg, (d) Pu.

106. Write the ground state electron configurations for elements A–E.

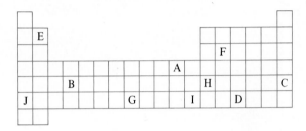

107. Repeat Exercise 105 for elements F–J.

6 Chemical Periodicity

Objectives

As you study this chapter, you should learn

☐ More about the periodic table (it is so useful)
☐ About chemical periodicity in physical properties:
    Atomic radii
    Ionization energy
    Electron affinity
    Ionic radii
    Electronegativity

☐ About chemical periodicity in the reactions of
    Hydrogen
    Oxygen
☐ About chemical periodicity in the compounds of
    Hydrogen
    Oxygen

The properties of elements are correlated with their positions in the periodic table. Chemists use the periodic table as an invaluable guide in their search for new, useful materials. A barium sodium niobate crystal can convert infrared laser light into visible light. This harmonic generation or "frequency doubling" is very important in chemical research using lasers and in the telecommunications industry.

6-1 More about the Periodic Table

In Chapter 4 we described the development of the periodic table, some terminology for it, and its guiding principle, the *periodic law*.

The properties of the elements are periodic functions of their atomic numbers.

In Chapter 5 we described electronic configurations of the elements. In the long form of the periodic table, elements are arranged in blocks based on the kinds of atomic orbitals being filled. (You may wish to review Table 5-5 and Figure 5-31 carefully.) We saw that electronic configurations of elements in the A groups and in Group 0 are entirely predictable from their positions in the periodic table. There are some irregularities among the transition elements.

Now we classify the elements according to their electronic configurations, which is a very useful system.

Noble Gases   For many years the Group 0 elements—the noble gases—were called inert gases because no chemical reactions were known for them. We

[He] = $1s^2$

now know that the heavier members do form compounds, mostly with fluorine and oxygen. Except for helium, each of these elements has eight electrons in its highest occupied energy level. Their structures may be represented as ... ns^2np^6.

Representative Elements   The A group elements in the periodic table are called representative elements. They have partially occupied highest energy levels. Their "last" electron was added to an s or p orbital. These elements show distinct and fairly regular variations in their properties with changes in atomic number.

d-Transition Elements   Elements in the B groups (except IIB) in the periodic table are known as the d-transition elements or, more simply, as transition elements or transition metals. They were considered to be transitions between the alkaline elements (base-formers) on the left and the acid-formers on the right. All are metals and are characterized by electrons being added to d orbitals. Stated differently, the d-transition elements are building an inner (next to highest occupied) energy level from 8 to 18 electrons. They are referred to as

First Transition Series: $_{21}$Sc through $_{29}$Cu
Second Transition Series: $_{39}$Y through $_{47}$Ag
Third Transition Series: $_{57}$La and $_{72}$Hf through $_{79}$Au
Fourth Transition Series: (not complete) $_{89}$Ac and elements 104 through 111

Strictly speaking, the Group IIB elements—zinc, cadmium, and mercury—are not d-transition metals because their "last" electrons go into s orbitals. They are usually discussed with the d-transition metals because their chemical properties are similar.

Inner Transition Elements   Sometimes known as *f-transition elements*, these are elements in which electrons are being added to f orbitals. In these

Some transition metals (left to right):
Ti, V, Cr, Mn, Fe, Co, Ni, Cu.

The elements of Period 3. Properties progress (left to right) from solids (Na, Mg, Al, Si, P, S) to gases (Cl, Ar) and from the most metallic (Na) to the most nonmetallic (Ar).

elements, the second from the highest occupied energy level is building from 18 to 32 electrons. All are metals. The inner transition elements are located between Groups IIIB and IVB in the periodic table. They are

First Inner Transition Series (lanthanides): $_{58}$Ce through $_{71}$Lu
Second Inner Transition Series (actinides): $_{90}$Th through $_{103}$Lr

The A and B designations for groups of elements in the periodic table are arbitrary, and they are reversed in some periodic tables. In another designation, the groups are numbered 1 through 18. The system used in this text is the one commonly used in the United States. Elements with the same group numbers, but with different letters, have relatively few similar properties. The origin of the A and B designations is the fact that some compounds of elements with the same group numbers have similar formulas but quite different properties, e.g., NaCl (IA) and AgCl (IB), $MgCl_2$ (IIA) and $ZnCl_2$ (IIB). As we shall see, variations in the properties of the B groups across a row are not nearly as dramatic as the variations observed across a row of A group elements. In the B groups, electrons are being added to $(n - 1)d$ orbitals, where n represents the highest energy level that contains electrons.

The *outermost* electrons have the greatest influence on the properties of elements. Adding an electron to an *inner d* orbital results in less striking changes in properties than adding an electron to an *outer s* or *p* orbital.

Periodic Properties of the Elements

We shall now investigate the nature of periodicity in some detail. Some knowledge of periodicity is valuable in understanding bonding in simple compounds. Many physical properties, such as melting points, boiling points, and atomic volumes, show periodic variations. For now, we describe the variations that are most useful in predicting chemical properties. The variations in these properties depend on electronic configurations, especially the configurations in the outermost occupied shell, and on how far away that shell is from the nucleus.

6-2 Atomic Radii

In Section 5-15 we described atomic orbitals in terms of probabilities of distributions of electrons over certain regions in space. We can visualize the electron cloud that surrounds an atomic nucleus in a similar way, i.e., as somewhat indefinite. Further, we cannot isolate a single atom and measure its diameter the way we can measure the diameter of a golf ball. For all practical purposes, the size of an individual atom cannot be uniquely defined. An indirect approach is required. The size of an atom is determined by its immediate environment, especially its interaction with surrounding atoms. By analogy, suppose we arrange some golf balls in an orderly array in a box. If we know how the balls are positioned, the number of balls, and the dimensions of the box, we can calculate the diameter of an individual ball. Application of this reasoning to solids and their densities leads us to values for the atomic sizes of many elements. In other cases, we derive atomic radii from the observed distances between atoms that are combined with each other. For example, the distance between atomic centers (nuclei) in the Cl_2 molecule is measured to be 1.98 Å. We take the radius of *each* Cl atom to be half the interatomic distance, or 0.99 Å. We collect the data obtained from many such measurements to indicate the *relative* sizes of individual atoms.

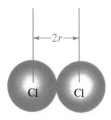

The radius of an atom, *r*, is taken as half of the distance between nuclei in *homonuclear* molecules such as Cl_2.

The top of Figure 6-1 displays the relative sizes of atoms of the representative elements and the noble gases. It shows the periodicity in atomic radii. (The ionic radii at the bottom of Figure 6-1 are discussed in Section 6-5.)

The **effective nuclear charge**, Z_{eff}, experienced by an electron in an outer energy level is less than the actual nuclear charge, Z. This is because the *attraction* of outer shell electrons by the nucleus is partly counteracted by the *repulsion* of these outer shell electrons by electrons in filled inner shells. We might say that the electrons in filled sets of energy levels *screen* or *shield* electrons in outer energy levels from the full effect of the nuclear charge. This **screening**, or **shielding**, effect helps us to understand many periodic trends in atomic properties.

Consider the Group IA metals, Li–Cs. Lithium, element number 3, has two electrons in a filled energy level, $1s^2$, and one electron in the $2s$ orbital, $2s^1$. The electron in the $2s$ orbital is fairly effectively screened from the nucleus by the two electrons in the filled $1s$ orbital, the He configuration. The electron in the $2s$ orbital "feels" an effective nuclear charge of about 1+, rather than the full nuclear charge of 3+. Sodium, element number 11, has ten electrons in filled sets of orbitals, $1s^2 2s^2 2p^6$, the Ne configuration. These ten electrons in a noble gas configuration fairly effectively screen (shield) its outer electron ($3s^1$) from the nucleus. Thus, the $3s$ electron in sodium also "feels" an effective nuclear charge of about 1+ rather than 11+. A sodium atom is larger than a lithium atom because (1) the effective nuclear charge is *approximately* the same for both atoms, and (2) the "outer" electron in a sodium atom is in the third energy level. A similar argument explains why potassium atoms are larger than sodium atoms.

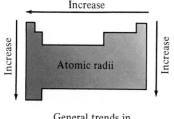

General trends in atomic radii of A group elements with position in the periodic table

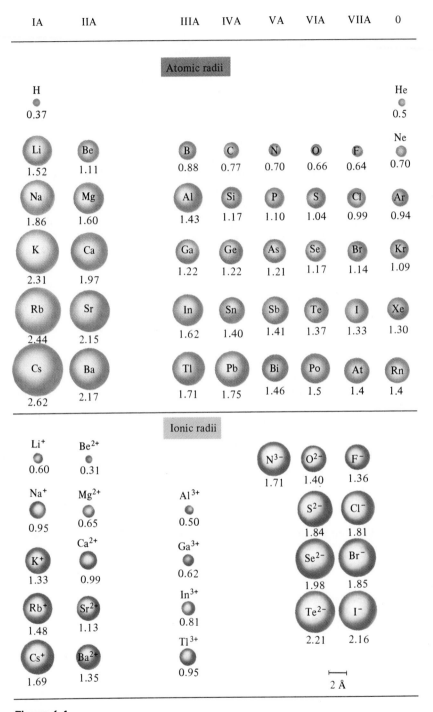

Atomic radii are often stated in **angstroms** (1 Å = 10^{-10} m) or in the SI units **nanometers** (1 nm = 10^{-9} m) or **picometers** (1 pm = 10^{-12} m). To convert from Å to nm, move the decimal point to the left one place (1 Å = 0.1 nm). For example, the atomic radius of Li is 1.52 Å, or 0.152 nm.

Figure 6-1

(Top) Atomic radii of the A group (representative) elements and the noble gases, in angstroms. Atomic radii *increase as a group is descended* because electrons are being added to shells farther from the nucleus. Atomic radii *decrease from left to right within a given row* owing to increasing effective nuclear charge. Hydrogen atoms are the smallest and cesium atoms are the largest naturally occurring atoms. 1 Å = 1 × 10^{-10} m.

(Bottom) Sizes of ions of the A group elements, in angstroms. Positive ions (cations) are always *smaller* than the neutral atoms from which they are formed. Negative ions (anions) are always *larger* than the neutral atoms from which they are formed.

Within a group of representative elements, atomic radii *increase* from top to bottom as electrons are added to higher energy levels.

As we move *across* the periodic table, atoms become smaller due to increasing effective nuclear charges. Consider the elements B ($Z = 5$, $1s^22s^22p^1$) to F ($Z = 9$, $1s^22s^22p^5$). In B there are two electrons in a noble gas configuration, $1s^2$, and three electrons in the second energy level, $2s^22p^1$. The two electrons in the noble gas configuration fairly effectively screen out the effect of two protons in the nucleus. So the three electrons in the second energy level of B ''feel'' an effective nuclear charge of approximately $3+$. By similar arguments, we see that in carbon ($Z = 6$, $1s^22s^22p^2$) the four electrons in the second energy level ''feel'' an effective nuclear charge of approximately $4+$. So we expect C atoms to be smaller than B atoms, and they are. In nitrogen ($Z = 7$, $1s^22s^22p^3$) the five electrons in the second energy level ''feel'' an effective nuclear charge of approximately $5+$, and so N atoms are smaller than C atoms.

As we move from left to right *across a period* in the periodic table, atomic radii of representative elements *decrease* as a proton is added to the nucleus and an electron is added to a particular energy level.

For the transition elements, the variations are not so regular because electrons are being added to an inner shell. All transition elements have smaller radii than the preceding Group IA and IIA elements in the same period.

Example 6-1
Arrange the following elements in order of increasing atomic radii. Justify your order.

$$Cs, \quad F, \quad Li, \quad Cl$$

Plan
Both Li and Cs are Group IA metals, while F and Cl are halogens (VIIA nonmetals). Figure 6-1 shows that atomic radii increase as a group is descended, so Li < Cs and F < Cl. Atomic radii decrease from left to right.

Solution
The order of increasing atomic radii is

$$F < Cl < Li < Cs$$

EOC 19, 20

6-3 Ionization Energy

The **first ionization energy (IE₁)**, also called **first ionization potential**, is

the minimum amount of energy required to remove the most loosely bound electron from an isolated gaseous atom to form an ion with a $1+$ charge.

Table 6-1
First Ionization Energies (kJ/mol of atoms) of Some Elements

H 1312																	He 2372
Li 520	Be 899											B 801	C 1086	N 1402	O 1314	F 1681	Ne 2081
Na 497	Mg 738											Al 578	Si 786	P 1012	S 1000	Cl 1251	Ar 1521
K 419	Ca 590	Sc 631	Ti 658	V 650	Cr 653	Mn 717	Fe 759	Co 758	Ni 737	Cu 745	Zn 906	Ga 579	Ge 762	As 947	Se 941	Br 1140	Kr 1351
Rb 403	Sr 549	Y 616	Zr 660	Nb 664	Mo 685	Tc 702	Ru 711	Rh 720	Pd 805	Ag 731	Cd 868	In 558	Sn 709	Sb 834	Te 869	I 1008	Xe 1170
Cs 376	Ba 503	La 538	Hf 675	Ta 761	W 770	Re 760	Os 840	Ir 878	Pt 870	Au 890	Hg 1007	Tl 589	Pb 716	Bi 703	Po 812	At 920	Rn 1037

For calcium, for example, the first ionization energy, IE_1, is 590 kJ/mol:

$$Ca(g) + 590 \text{ kJ} \longrightarrow Ca^+(g) + e^-$$

The **second ionization energy (IE_2)** is the amount of energy required to remove the second electron. For calcium, it may be represented as

$$Ca^+(g) + 1145 \text{ kJ} \longrightarrow Ca^{2+}(g) + e^-$$

For a given element, *IE_2 is always greater than IE_1* because it is always more difficult to remove an electron from a positively charged ion than from the corresponding neutral atom. Table 6-1 gives first ionization energies.

Ionization energies measure how tightly electrons are bound to atoms. Ionization always requires energy to remove an electron from the attractive force of the nucleus. Low ionization energies indicate ease of removal of electrons, and hence ease of positive ion (cation) formation. Figure 6-2 shows a plot of first ionization energy versus atomic number for several elements.

We see that in each period of Figure 6-2, the noble gases have the highest first ionization energies. This should not be surprising, because the noble

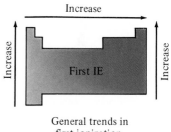

General trends in first ionization energies of A group elements with position in the periodic table. Exceptions occur at Groups IIIA and VIA.

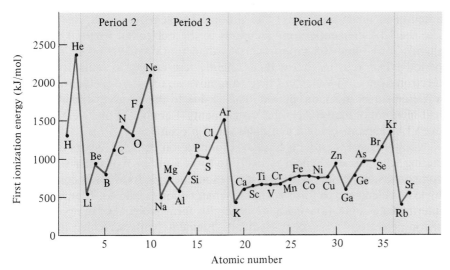

Figure 6-2
A plot of first ionization energies for the first 38 elements versus atomic number. The noble gases have very high first ionization energies, and the IA metals have low first ionization energies. Note the similarities in the variations for the Period 2 elements, 3 through 10, to those for the Period 3 elements, 11 through 18, as well as for the later A group elements. Variations for B group elements are not nearly so pronounced as those for A group elements.

gases are known to be very unreactive elements. It requires more energy to remove an electron from a helium atom (slightly less than 4.0×10^{-18} J/atom or 2372 kJ/mol) than to remove one from a neutral atom of any other element:

$$He(g) + 2372 \text{ kJ} \longrightarrow He^+(g) + e^-$$

The Group IA metals (Li, Na, K, Rb, Cs) have very low first ionization energies. Each of these elements has only one electron in its highest energy level ($\ldots ns^1$), and they are the largest atoms in their periods. The first electron added to a principal energy level is easily removed to form a noble gas configuration. As we move down the group, the first ionization energies become smaller. The force of attraction of the positively charged nucleus for electrons decreases as the square of the separation between them increases. So as atomic radii increase in a given group, first ionization energies decrease because the valence electrons are farther from the nucleus. The shielding effect of the electrons in filled inner shells, and the decreased Z_{eff} as the group is descended, further weaken the attraction for the outer shell electrons.

The first ionization energies of the Group IIA elements (Be, Mg, Ca, Sr, Ba) are significantly higher than those of the Group IA elements in the same periods. This is because the Group IIA elements have smaller atomic radii and higher Z_{eff} values. Thus, their valence electrons are held more tightly than those of the neighboring IA metals. It is harder to remove an electron from a pair in the filled outermost s orbitals of the Group IIA elements than to remove the single electron from the half-filled outermost s orbitals of the Group IA elements.

The first ionization energies for the Group IIIA elements (B, Al, Ga, In, Tl) are exceptions to the general horizontal trends. They are *lower* than those of the IIA elements in the same periods because the IIIA elements have only a single electron in their outermost p orbitals. It requires less energy to remove the first p than to remove the second s electron from the same principal energy level because an ns orbital is lower in energy (more stable) than an np orbital.

The second peak for each period in the ionization energy curve occurs at the Group VA elements (N, P, As, Sb, Bi). These elements have three unpaired electrons in the three outermost p orbitals, that is, $ns \; \underline{\uparrow\downarrow} \; np \; \underline{\uparrow} \; \underline{\uparrow} \; \underline{\uparrow}$, a half-filled set of p orbitals. The Group VIA elements (O, S, Se, Te, Po), like the IIIA elements, are exceptions to the horizontal trend. They have slightly *lower* first ionization energies than the VA elements in the same periods. This tells us that it takes slightly less energy to remove a paired electron from a VIA element than to remove an unpaired p electron from a VA element in the same period. This is due to the relative stability of the half-filled set of p orbitals in the VA elements. Removal of one electron from the VIA elements gives a half-filled set of p orbitals.

Knowledge of the relative values of ionization energies assists us in predicting whether an element is likely to form ionic or molecular (covalent) compounds. Elements with low ionization energies form ionic compounds by losing electrons to form positively charged ions (**cations**). Elements with intermediate ionization energies generally form molecular compounds by sharing electrons with other elements. Elements with very high ionization energies, e.g., Groups VIA and VIIA, often gain electrons to form negatively charged ions (**anions**).

By Coulomb's Law, $F \propto \dfrac{(q^+)(q^-)}{d^2}$, the attraction for the outer shell electrons is directly proportional to the *effective* charges and inversely proportional to the square of the distance between the charges.

We have seen other consequences of the special stability of half-filled sets of equivalent orbitals, i.e., exceptions to the Aufbau Principle (Section 5-16).

Here is one reason why trends in ionization energies are important.

One factor that favors an atom of a *representative* element forming a monatomic ion in a compound is the formation of a stable noble gas configuration. Energy considerations are consistent with this observation. For example, as one mole of Li from Group IA forms one mole of Li^+ ions, it absorbs 520 kJ per mole of Li atoms. The IE_2 value is 14 times greater, 7298 kJ/mol, and is prohibitively large for the formation of Li^{2+} ions under ordinary conditions. For Li^{2+} ions to form, an electron would have to be removed from the filled first energy level. We recognize that this is unlikely. The other alkali metals behave in the same way, for the same reason.

Likewise, the first two ionization energies of Be are 899 and 1757 kJ/mol, but IE_3 is more than eight times larger, 14,849 kJ/mol. So Be forms Be^{2+} ions, but not Be^{3+} ions. The other alkaline earth metals—Mg, Ca, Sr, Ba, and Ra—behave in a similar way. Owing to the high energy required, *simple monatomic cations with charges greater than 3+ do not form under ordinary circumstances.* Only the lower members of Group IIIA, beginning with Al, form 3+ ions. Bi and some *d*- and *f*-transition metals do so, too. We see that the magnitudes of successive ionization energies support the ideas of electronic configurations discussed in Chapter 5.

> Noble gas configurations are stable only for ions in *compounds*. In fact, $Li^+(g)$ is less stable than $Li(g)$ by 520 kJ/mol.

Example 6-2

Arrange the following in order of increasing first ionization energy. Justify your order.

Na,   Mg,   Al,   Si

Plan

Table 6-1 shows that first ionization energies generally increase from left to right in the periodic table, but there are exceptions at Groups IIIA and VIA. Al is a IIIA element with only one electron in its outer *p* orbitals, $1s^2 2s^2 2p^1$.

Solution

There is a slight dip at Group IIIA in the plot of first IE versus atomic number. The order of increasing first ionization energy is

Na < Al < Mg < Si

EOC 28–30

6-4 Electron Affinity

The **electron affinity (EA)** of an element is defined as

> the amount of energy *absorbed* when an electron is added to an isolated gaseous atom to form an ion with a 1− charge.

The convention is to assign a positive value when energy is absorbed and a negative value when energy is released. For most elements, energy is absorbed. We can represent the electron affinities of beryllium and chlorine as follows:

> This is consistent with thermodynamic convention.

The value of *EA* for Cl can also be represented as -5.78×10^{-19} J/atom or -3.61 eV/atom. The electron volt (eV) is a unit of energy (1 eV = 1.60222×10^{-19} J).

$$Be(g) + e^- + 241 \text{ kJ} \longrightarrow Be^-(g) \qquad EA = \quad 241 \text{ kJ/mol}$$

$$Cl(g) + e^- \longrightarrow Cl^-(g) + 348 \text{ kJ} \qquad EA = -348 \text{ kJ/mol}$$

The first equation tells us that when one mole of gaseous beryllium atoms gain one electron each to form gaseous Be^- ions, 241 kJ/mol of ions is *absorbed* (*endothermic*). The second equation tells us that when one mole of gaseous chlorine atoms gain one electron each to form gaseous chloride ions, 348 kJ of energy is *released* (*exothermic*). Figure 6-3 shows a plot of electron affinity versus atomic number for several elements.

Electron affinity involves the *addition* of an electron to a neutral gaseous atom. The process by which a neutral atom X gains an electron (EA),

$$X(g) + e^- \longrightarrow X^-(g) \qquad (EA)$$

is *not* the reverse of the ionization process,

$$X^+(g) + e^- \longrightarrow X(g) \qquad (\text{reverse of } IE_1)$$

The first process begins with a neutral atom, whereas the second begins with a positive ion. Thus, IE_1 and EA are *not* simply equal in value with the signs reversed.

Elements with very negative electron affinities gain electrons easily to form negative ions (anions). We see from Figure 6-3 that electron affinities generally become more negative from left to right across a row in the periodic table (excluding the noble gases). This means that most representative elements in Groups IA to VIIA show a greater attraction for an extra electron from left to right. The halogens, which have the outer electronic configuration ns^2np^5, have the most negative electron affinities. They form stable anions with noble gas configurations, ... ns^2np^6, by gaining one electron.

"Electron affinity" is a precise and quantitative term, like "ionization energy," but it is difficult to measure. Table 6-2 shows electron affinities for the representative elements.

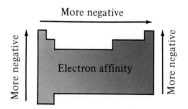

General trends in electron affinities of A group elements with position in the periodic table. There are many exceptions.

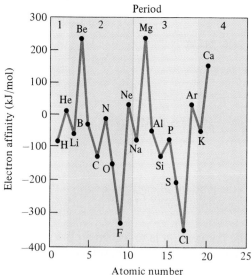

Figure 6-3

A plot of electron affinity versus atomic number for the first 20 elements. The general horizontal trend is that electron affinities become more negative (more energy is released as an extra electron is added) from Group IA through Group VIIA for a given period. Exceptions occur at the IIA and VA elements.

Table 6-2
Electron Affinity Values (kJ/mol) of Some Elements*

	IA	IIA			IIIA	IVA	VA	VIA	VIIA	
1	H −72									He (21)
2	Li −60	Be (241)			B −23	C −122	N 0	O −142	F −322	Ne (29)
3	Na −53	Mg (231)		Cu −123	Al −44	Si −119	P −74	S −200	Cl −348	Ar (35)
4	K −48	Ca (156)		Ag −125	Ga (−36)	Ge −116	As −77	Se −194	Br −323	Kr (39)
5	Rb −47	Sr (119)		Au −222	In (−34)	Sn −120	Sb −101	Te −190	I −295	Xe (40)
6	Cs −45	Ba (52)			Tl (−48)	Pb −101	Bi −101	Po (−173)	At (−270)	Rn (40)
7	Fr (−44)									

* Estimated values are in parentheses.

For many reasons, the variations in electron affinities are not regular across a period. The general trend is for the electron affinities of the elements to become more negative from left to right in each period. Noteworthy exceptions are the elements of Groups IIA and VA, which have less negative (more positive) values than the trends suggest (Figure 6-3). It is very difficult to add an electron to a IIA metal atom because its outer s subshell is filled. The values for the VA elements are slightly less negative than expected because they apply to the addition of an electron to a relatively stable half-filled set of np orbitals ($ns^2np^3 \rightarrow ns^2np^4$).

The addition of a second electron to form an ion with a 2− charge is always endothermic. So electron affinities of anions are always positive.

Example 6-3

Arrange the following elements in order of increasing values of electron affinity, i.e., from most negative to most positive.

$$\text{Be,} \quad \text{N,} \quad \text{Na,} \quad \text{Cl}$$

Plan

Table 6-2 shows that electron affinity values generally become more negative from left to right across a period with major exceptions at Groups IIA (Be) and VA (N). They generally become more negative from bottom to top.

Solution

The order of increasing values of electron affinity is

(most negative EA)   Cl < Na < N < Be   (most positive EA)

EOC 39

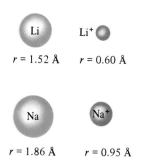

Li   Li⁺
$r = 1.52$ Å $r = 0.60$ Å

Na   Na⁺
$r = 1.86$ Å $r = 0.95$ Å

The nuclear charge remains constant when the ion is formed.

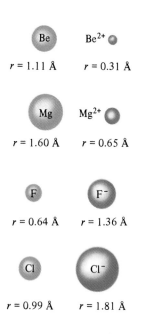

Be   Be²⁺
$r = 1.11$ Å $r = 0.31$ Å

Mg   Mg²⁺
$r = 1.60$ Å $r = 0.65$ Å

F   F⁻
$r = 0.64$ Å $r = 1.36$ Å

Cl   Cl⁻
$r = 0.99$ Å $r = 1.81$ Å

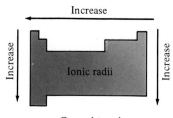

General trends in ionic radii of A group elements with position in the periodic table

6-5 Ionic Radii

Many elements on the left side of the periodic table react with other elements by *losing* electrons to form positively charged ions. Each of the Group IA elements (Li, Na, K, Rb, Cs) has only one electron in its highest energy level (electronic configuration ... ns^1). These elements react with other elements by losing one electron to attain noble gas configurations. They form the ions Li⁺, Na⁺, K⁺, Rb⁺, and Cs⁺. A neutral lithium atom, Li, contains three protons in its nucleus and three electrons, with its outermost electron in the 2s orbital. However, a lithium ion, Li⁺, contains three protons in its nucleus but only two electrons, both in the 1s orbital. So a Li⁺ ion is much smaller than a neutral Li atom (see margin). Likewise, a sodium ion, Na⁺, is smaller than a sodium atom, Na. The relative sizes of atoms and common ions of some representative elements are shown in Figure 6-1.

Isoelectronic species have the same number of electrons. We see that the ions formed by the Group IIA elements (Be²⁺, Mg²⁺, Ca²⁺, Sr²⁺, Ba²⁺) are significantly smaller than the *isoelectronic* ions formed by the Group IA elements in the same period. The radius of the Li⁺ ion is 0.60 Å, while the radius of the Be²⁺ ion is only 0.31 Å. This is just what we might expect. A beryllium ion, Be²⁺, is formed when a beryllium atom, Be, loses both of its 2s electrons while the 4+ nuclear charge remains constant. We expect the 4+ nuclear charge in Be²⁺ to attract the remaining two electrons quite strongly. Comparison of the ionic radii of the IIA elements with their atomic radii indicates the validity of our reasoning. Similar reasoning indicates that the ions of the Group IIIA metals (Al³⁺, Ga³⁺, In³⁺, Tl³⁺) should be even smaller than the ions of Group IA and Group IIA elements in the same periods.

Now consider the Group VIIA elements (F, Cl, Br, I). These have the outermost electronic configuration ... ns^2np^5. These elements can completely fill their outermost p orbitals by *gaining* one electron each to attain noble gas configurations. Thus, when a fluorine atom (with seven electrons in its highest energy level) gains one electron, it becomes a fluoride ion, F⁻, with eight electrons in its highest energy level. These eight electrons repel one another more strongly than the original seven, so the electron cloud expands. The F⁻ ion is much larger than the neutral F atom (see margin). Similar reasoning indicates that a chloride ion, Cl⁻, should be larger than a neutral chlorine atom, Cl. Observed ionic radii (see Figure 6-1) verify this prediction.

Comparing the sizes of an oxygen atom (Group VIA) and an oxide ion, O²⁻, we find that the negatively charged ion is larger than the neutral atom. The oxide ion is also larger than the fluoride ion because the oxide ion contains ten electrons held by a nuclear charge of only 8+, whereas the fluoride ion has ten electrons held by a nuclear charge of 9+.

1. Simple positively charged ions (cations) are always smaller than the neutral atoms from which they are formed.
2. Simple negatively charged ions (anions) are always larger than the neutral atoms from which they are formed.
3. Within an isoelectronic series of ions, ionic radii decrease with increasing atomic number.

An isoelectronic series of ions					
N^{3-}	O^{2-}	F^-	Na^+	Mg^{2+}	Al^{3+}
Ionic radius (Å)					
1.71	1.40	1.36	0.95	0.65	0.50
No. of electrons					
10	10	10	10	10	10
Nuclear charge					
+7	+8	+9	+11	+12	+13

Example 6-4

Arrange the following ions in order of increasing ionic radii: (a) Ca^{2+}, K^+, Al^{3+}; (b) S^{2-}, Cl^-, Te^{2-}.

Plan

Figure 6-1 shows that ionic radii increase from right to left across a period and from top to bottom within a group.

Solution

(a) Cations are always smaller than the neutral atoms from which they are formed.

$$Al^{3+} < Ca^{2+} < K^+$$

(b) Anions are always larger than the neutral atoms from which they are formed.

$$Cl^- < S^{2-} < Te^{2-}$$

EOC 43–45

6-6 Electronegativity

The **electronegativity** of an element is a measure of the relative tendency of an atom to attract electrons to itself when it is chemically combined with another atom. Electronegativities of the elements are expressed on a somewhat arbitrary scale, called the Pauling scale (Table 6-3). The electronegativity of fluorine (4.0) is higher than that of any other element. This tells us that when fluorine is chemically bonded to other elements, it has a greater tendency to attract electron density to itself than does any other element. Oxygen is the second most electronegative element.

Because the noble gases form few compounds, they are not included in this discussion.

> For the representative elements, electronegativities usually increase from left to right across periods and from bottom to top within groups.

Variations among the transition elements are not as regular. In general, both ionization energies and electronegativities are low for elements at the lower left of the periodic table and high for those at the upper right.

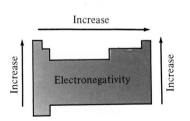

General trends in electronegativities of A group elements with position in the periodic table

Table 6-3
Electronegativity Values of the Elements*

Metals
Nonmetals
Metalloids

IA							VIIIB					IIIA	IVA	VA	VIA	VIIA	0
1 H 2.1	IIA																2 He
3 Li 1.0	4 Be 1.5	IIIB	IVB VB VIB VIIB						IB	IIB		5 B 2.0	6 C 2.5	7 N 3.0	8 O 3.5	9 F 4.0	10 Ne
11 Na 1.0	12 Mg 1.2											13 Al 1.5	14 Si 1.8	15 P 2.1	16 S 2.5	17 Cl 3.0	18 Ar
19 K 0.9	20 Ca 1.0	21 Sc 1.3	22 Ti 1.4	23 V 1.5	24 Cr 1.6	25 Mn 1.6	26 Fe 1.7	27 Co 1.7	28 Ni 1.8	29 Cu 1.8	30 Zn 1.6	31 Ga 1.7	32 Ge 1.9	33 As 2.1	34 Se 2.4	35 Br 2.8	36 Kr
37 Rb 0.9	38 Sr 1.0	39 Y 1.2	40 Zr 1.3	41 Nb 1.5	42 Mo 1.6	43 Tc 1.7	44 Ru 1.8	45 Rh 1.8	46 Pd 1.8	47 Ag 1.6	48 Cd 1.6	49 In 1.6	50 Sn 1.8	51 Sb 1.9	52 Te 2.1	53 I 2.5	54 Xe
55 Cs 0.8	56 Ba 1.0	57 La* 1.1	72 Hf 1.3	73 Ta 1.4	74 W 1.5	75 Re 1.7	76 Os 1.9	77 Ir 1.9	78 Pt 1.8	79 Au 1.9	80 Hg 1.7	81 Tl 1.6	82 Pb 1.7	83 Bi 1.8	84 Po 1.9	85 At 2.1	86 Rn
87 Fr 0.8	88 Ra 1.0	89 Ac† 1.1															

*

58 Ce 1.1	59 Pr 1.1	60 Nd 1.1	61 Pm 1.1	62 Sm 1.1	63 Eu 1.1	64 Gd 1.1	65 Tb 1.1	66 Dy 1.1	67 Ho 1.1	68 Er 1.1	69 Tm 1.1	70 Yb 1.0	71 Lu 1.2

†

90 Th 1.2	91 Pa 1.3	92 U 1.5	93 Np 1.3	94 Pu 1.3	95 Am 1.3	96 Cm 1.3	97 Bk 1.3	98 Cf 1.3	99 Es 1.3	100 Fm 1.3	101 Md 1.3	102 No 1.3	103 Lr 1.5

* Electronegativity values are given at the bottoms of the boxes. The noble gases are not included in this discussion. The heavy "stair step" line approximately separates the metallic elements, to the left, from the nonmetallic elements, to the right.

Example 6-5
Arrange the following elements in order of increasing electronegativity.

B, Na, F, S

Plan
Table 6-3 shows that electronegativities increase from left to right across a period and from bottom to top within a group.

Solution
The order of increasing electronegativity is

$$Na < B < S < F$$

EOC 50

Although the electronegativity scale is somewhat arbitrary, we can use it with reasonable confidence to make predictions about bonding. Two elements with quite different electronegativities tend to react with each other to form ionic compounds. The less electronegative element gives up its electron(s) to the more electronegative element. Two elements with similar electronegativities tend to form covalent bonds with each other. That is, they share their electrons. In this sharing, the more electronegative element attains a greater share. This is discussed in detail in Chapters 7 and 8.

Chemical Reactions and Periodicity

Now we shall illustrate the periodicity of chemical properties by considering some reactions of hydrogen, oxygen, and their compounds. We choose to discuss hydrogen and oxygen because, of all the elements, they form the most kinds of compounds with other elements. Additionally, compounds of hydrogen and oxygen are very important in such diverse phenomena as all life processes and most corrosion processes.

6-7 Hydrogen and the Hydrides

1 Hydrogen

Elemental hydrogen is a colorless, odorless, tasteless diatomic gas with the lowest atomic weight and density of any known substance. Discovery of the element is attributed to the Englishman Henry Cavendish, who prepared it in 1766 by passing steam through a red-hot gun barrel (mostly iron) and by the reaction of acids with active metals. The latter is still the method commonly used for the preparation of small amounts of H_2 in the laboratory. In each case, H_2 is liberated by a displacement (and redox) reaction, of the kind described in Section 4-6. (See also the activity series, Table 4-11.)

The name "hydrogen" means "water former."

$$3Fe(s) + 4H_2O(g) \xrightarrow{\Delta} Fe_3O_4(s) + 4H_2(g)$$

$$Zn(s) + 2HCl(aq) \longrightarrow ZnCl_2(aq) + H_2(g)$$

Can you write the net ionic equation for the reaction of Zn with HCl(aq)?

Hydrogen also can be prepared by electrolysis of water.

$$2H_2O(\ell) \xrightarrow{electricity} 2H_2(g) + O_2(g)$$

In the future, if it becomes economical to convert solar energy into electrical energy that can be used to electrolyze water, H_2 could become an important fuel (although the dangers of storage and transportation would have to be overcome). The *combustion* of H_2 liberates a great deal of heat. **Combustion** is the highly exothermic combination of a substance with oxygen, usually with a flame. (See Section 6-8, part 3.)

$$2H_2(g) + O_2(g) \xrightarrow[\text{or } \Delta]{\text{spark}} 2H_2O(\ell) + \text{energy}$$

This is the reverse of the decomposition of H_2O.

Hydrogen is very flammable; it was responsible for the Hindenburg airship disaster in 1937. A spark is all it takes to initiate the **combustion reaction**, which is exothermic enough to provide the heat necessary to sustain the reaction.

Hydrogen is no longer used in blimps and dirigibles. It has been replaced by helium, which is slightly denser, nonflammable, and much safer.

Figure 6-6
Ammonia may be applied directly to the soil as a fertilizer.

The primary industrial use of H_2 is in the synthesis of ammonia, a molecular hydride, by the Haber process (Section 17-6). Most of the NH_3 is used as liquid ammonia as a fertilizer (Figure 6-6) or to make other fertilizers, such as ammonium nitrate, NH_4NO_3, and ammonium sulfate, $(NH_4)_2SO_4$:

$$N_2(g) + 3H_2(g) \xrightarrow[\Delta, \text{ high pressure}]{\text{catalysts}} 2NH_3(g)$$

As we shall see in Chapter 10, even H_2O is weakly acidic.

Many of the molecular (nonmetal) hydrides are acidic; their aqueous solutions produce hydrogen ions. These include HF, HCl, HBr, HI, H_2S, H_2Se, and H_2Te.

Example 6-6

Predict the products of the reactions involving the reactants shown. Write a balanced formula unit equation for each one.

(a) $H_2(g) + I_2(g) \xrightarrow{\Delta}$
(b) $K(\ell) + H_2(g) \xrightarrow{\Delta}$
(c) $NaH(s) + H_2O(\ell)$ (excess) $\longrightarrow$

Plan

(a) Hydrogen reacts with the halogens (Group VIIA) to form hydrogen halides—in this example, HI.
(b) Hydrogen reacts with active metals to produce hydrides—in this case, KH.
(c) Active metal hydrides reduce water to produce H_2 and a metal hydroxide.

Remember that hydride ions, H^-, react with (reduce) water to produce OH^- ions and $H_2(g)$.

Solution

(a) $H_2(g) + I_2(g) \xrightarrow{\Delta} 2HI(g)$
(b) $2K(\ell) + H_2(g) \xrightarrow{\Delta} 2KH(s)$
(c) $NaH(s) + H_2O(\ell) \longrightarrow NaOH(aq) + H_2(g)$

Example 6-7

Predict the ionic or molecular character of the products in Example 6-6.

Plan

We refer to Figure 6-4, which displays the nature of hydrides.

Solution

Reaction (a) is a reaction between hydrogen and another nonmetal. The product, HI, must be molecular. Reaction (b) is the reaction of hydrogen with an active Group IA metal. Thus, KH must be ionic. The products of reaction (c) are molecular $H_2(g)$ and the strong soluble base, NaOH, which is ionic.

EOC 65–67

6-8 Oxygen and the Oxides

1 Oxygen and Ozone

Oxygen was discovered in 1774 by an English minister and scientist, Joseph Priestley. He observed the thermal decomposition of mercury(II) oxide, a red powder:

$$2HgO(s) \xrightarrow{\Delta} 2Hg(\ell) + O_2(g)$$

That part of the earth we see—land, water, and air—is approximately 50% oxygen by mass. About two thirds of the mass of the human body is due to oxygen in H_2O. Elemental oxygen, O_2, is an odorless and nearly colorless gas that makes up about 21% by volume of dry air. In the liquid and solid states it is pale blue. Oxygen is only very slightly soluble in water; only about 0.04 gram dissolves in 1 liter of water at 25°C. This is sufficient to sustain fish and other marine organisms. The greatest single industrial use of O_2 is for oxygen-enrichment in blast furnaces for the conversion of pig iron to steel. Oxygen is obtained commercially by the fractional distillation of liquid air.

Oxygen also exists in a second allotropic form, ozone, O_3. Ozone is an unstable, pale blue gas at room temperature. It is formed by passing an electrical discharge through gaseous oxygen. Its unique, pungent odor is often noticed during electrical storms and in the vicinity of electrical equipment. Not surprisingly, its density is about $1\frac{1}{2}$ times that of O_2. At −112°C it condenses to a deep blue liquid. It is a very strong oxidizing agent. As a concentrated gas or a liquid, ozone can easily decompose explosively:

$$2O_3(g) \longrightarrow 3O_2(g)$$

Oxygen atoms, or **radicals**, are intermediates in this exothermic decomposition of O_3 to O_2. They act as strong oxidizing agents in such applications as destroying bacteria in water purification.

The ozone molecule is angular and diamagnetic. Both oxygen—oxygen bond lengths (1.28 Å) are identical and are intermediate between typical single and double bond lengths.

Ozone is formed in the upper atmosphere as O_2 molecules absorb high-energy electromagnetic radiation from the sun. Its concentration in the stratosphere is about 10 ppm, whereas it is only about 0.04 ppm near the earth's surface. The ozone layer is responsible for absorbing some of the ultraviolet light from the sun that, if it reached the surface of the earth in higher intensity, could cause damage to plants and animals (including humans). It has been

The name "oxygen" means "acid former."

Liquid O_2 is used as an oxidizer for rocket fuels. O_2 also is used in the health fields for oxygen-enriched air.

Allotropes are different forms of the same element in the same physical state (Section 2-2).

A *radical* is a species containing one or more unpaired electrons; many radicals are very reactive.

The concentration unit "ppm" stands for *parts per million*.

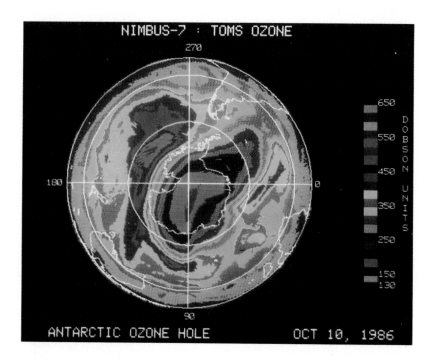

Figure 6-7
A computer map of ozone concentrations over the southern hemisphere. Dobson units are used to show the concentration of ozone. Measurements were made by the Total Ozone Mapping Spectrometer (TOMS) aboard the satellite Nimbus-7 on October 10, 1986. The depletion of the ozone layer by man-made pollutants is a major environmental concern. This map shows the hole (purple) in the ozone layer over Antarctica.

Small amounts of O_3 at the surface of the earth decompose rubber and plastic products by oxidation.

estimated that the incidence of skin cancer would increase by 2% for every 1% decrease in the concentration of ozone in the stratosphere (Figure 6-7). Although it decomposes rapidly in the upper atmosphere, the ozone supply is constantly replenished.

2 Reactions of Oxygen and the Oxides

Oxygen combines directly with all other elements except the noble gases and noble (unreactive) metals (Au, Pd, Pt) to form **oxides**, binary compounds that contain oxygen. Although such reactions are generally very exothermic, many proceed quite slowly and require heating to supply the energy necessary to break the strong bonds in O_2 molecules. Once these reactions are initiated, most release more than enough energy to be self-sustaining and sometimes become "red hot."

Reactions of O_2 with Metals
In general, metallic oxides (and peroxides and superoxides) are ionic solids. The Group IA metals combine with oxygen to form three kinds of solid ionic products called oxides, peroxides, and superoxides. Lithium combines with oxygen to form lithium oxide:

$$4Li(s) + O_2(g) \longrightarrow 2Li_2O(s) \qquad \text{lithium oxide (mp} > 1700°C)$$

By contrast, sodium reacts with an excess of oxygen to form sodium peroxide, Na_2O_2, rather than sodium oxide, Na_2O, as the *major* product:

$$2Na(s) + O_2(g) \longrightarrow Na_2O_2(g) \qquad \text{sodium peroxide (decomposes at 460°C)}$$

Peroxides contain the $O-O^{2-}$, O_2^{2-} group, in which the oxidation number of oxygen is -1, whereas *normal oxides* such as lithium oxide, Li_2O, contain oxide ions, O^{2-}. The heavier members of the family (K, Rb, Cs) react with

Figure 6-8
Iron powder burns brilliantly to form iron(III) oxide, Fe_2O_3.

Table 6-4
Oxygen Compounds of the IA and IIA Metals*

	IA					IIA				
	Li	Na	K	Rb	Cs	Be	Mg	Ca	Sr	Ba
oxide	Li_2O	Na_2O	K_2O	Rb_2O	Cs_2O	BeO	MgO	CaO	SrO	BaO
peroxide	Li_2O_2	Na_2O_2	K_2O_2	Rb_2O_2	Cs_2O_2			CaO_2	SrO_2	BaO_2
superoxide		NaO_2	KO_2	RbO_2	CsO_2					

* The shaded compounds represent the principal products of the direct reaction of the metal with oxygen.

excess oxygen to form **superoxides**. These contain the superoxide ion, O_2^-, in which the oxidation number of oxygen is $-\frac{1}{2}$. The reaction with K is

$$K(s) + O_2(g) \longrightarrow KO_2(s) \qquad \text{potassium superoxide (mp 430°C)}$$

The tendency of the Group IA metals to form oxygen-rich compounds increases as the group is descended. This is because cation radii increase going down the group. A similar trend is observed in the reactions of the Group IIA metals with oxygen. You can recognize these classes of compounds as follows:

Class	Contains Ions	Oxidation No. of Oxygen
oxide	O^{2-}	-2
peroxide	O_2^{2-}	-1
superoxide	O_2^-	$-\frac{1}{2}$

With the exception of Be, the Group IIA metals react with oxygen at moderate temperatures to form normal ionic oxides, MO, and at high pressures of oxygen the heavier ones form ionic peroxides, MO_2 (Table 6-4).

Beryllium reacts with oxygen only at elevated temperatures and forms only the normal oxide, BeO.

$$2M(s) + O_2(g) \longrightarrow 2(M^{2+}, O^{2-})(s) \qquad M = Be, Mg, Ca, Sr, Ba$$

$$M(s) + O_2(g) \longrightarrow (M^{2+}, O_2^{2-})(s) \qquad M = Ca, Sr, Ba$$

For example, the equations for the reactions of calcium and oxygen are

$$2Ca(s) + O_2(g) \longrightarrow 2CaO(s) \qquad \text{calcium oxide (mp 2580°C)}$$

$$Ca(s) + O_2(g) \longrightarrow CaO_2(s) \qquad \text{calcium peroxide (decomposes at 275°C)}$$

The other metals, with the exceptions noted previously (Au, Pd, and Pt), react with oxygen to form solid metal oxides. Many metals to the right of Group IIA show variable oxidation states, so they may form several oxides. For example, iron combines with oxygen in the following series of reactions to form three different oxides (Figure 6-8).

$$2Fe(s) + O_2(g) \xrightarrow{\Delta} 2FeO(s) \qquad \text{iron(II) oxide } or \text{ ferrous oxide}$$

$$6FeO(s) + O_2(g) \xrightarrow{\Delta} 2Fe_3O_4(s) \qquad \text{magnetic iron oxide (a mixed oxide)}$$

$$4Fe_3O_4(s) + O_2(g) \xrightarrow{\Delta} 6Fe_2O_3(s) \qquad \text{iron(III) oxide } or \text{ ferric oxide}$$

Figure 6-9
The normal oxides of the representative elements in their maximum oxidation states. Acidic oxides (acid anhydrides) are shaded red, amphoteric oxides are shaded purple, and basic oxides (basic anhydrides) are shaded blue. An amphoteric oxide is one that shows some acidic and some basic properties.

Increasing acidic character ⟶

Increasing base character ↓

IV	IIA	IIIA	IVA	VA	VIA	VIIA
Li_2O	BeO	B_2O_3	CO_2	N_2O_5		F_2O
Na_2O	MgO	Al_2O_3	SiO_2	P_4O_{10}	SO_3	Cl_2O_7
K_2O	CaO	Ga_2O_3	GeO_2	As_2O_5	SeO_3	Br_2O_7
Rb_2O	SrO	In_2O_3	SnO_2	Sb_2O_5	TeO_3	I_2O_7
Cs_2O	BaO	Tl_2O_3	PbO_2	Bi_2O_5	PoO_3	At_2O_7

Copper reacts with a limited amount of oxygen to form red Cu_2O, whereas with excess oxygen it forms black CuO.

$$4Cu(s) + O_2(g) \xrightarrow{\Delta} 2Cu_2O(s) \qquad \text{copper(I) oxide } or \text{ cuprous oxide}$$

$$2Cu(s) + O_2(g) \xrightarrow{\Delta} 2CuO(s) \qquad \text{copper(II) oxide } or \text{ cupric oxide}$$

> Metals that exhibit variable oxidation states react with a limited amount of oxygen to give lower oxidation state oxides (such as FeO and Cu_2O). They react with an excess of oxygen to give higher oxidation state oxides (such as Fe_2O_3 and CuO).

Reactions of Metal Oxides with Water

Oxides of metals are called **basic anhydrides** because many of them combine with water to form bases with no change in oxidation state of the metal (Figure 6-9). "Anhydride" means "without water"; in a sense, the metal oxide is a hydroxide base with the water "removed." Metal oxides that are soluble in water react to produce the corresponding hydroxides.

	Metal Oxide	**+ Water**	⟶	**Metal Hydroxide (base)**	
sodium oxide	$Na_2O(s)$	$+ H_2O(\ell)$	⟶	$2\ NaOH(aq)$	sodium hydroxide
calcium oxide	$CaO(s)$	$+ H_2O(\ell)$	⟶	$Ca(OH)_2(aq)$	calcium hydroxide
barium oxide	$BaO(s)$	$+ H_2O(\ell)$	⟶	$Ba(OH)_2(aq)$	barium hydroxide

The oxides of the Group IA metals and the heavier Group IIA metals dissolve in water to give solutions of strong soluble bases. Most other metal oxides are insoluble in water.

Reactions of O_2 with Nonmetals

Oxygen combines with many nonmetals to form molecular oxides. For example, carbon burns in oxygen to form carbon monoxide or carbon dioxide, depending on the relative amounts of carbon and oxygen, as the following equations show:

$$2C(s) + O_2(g) \longrightarrow 2\overset{+2}{C}O(s) \qquad \text{(excess C and limited } O_2\text{)}$$

$$C(s) + O(g) \longrightarrow \overset{+4}{C}O_2(g) \qquad \text{(limited C and excess } O_2\text{)}$$

Carbon burns brilliantly in pure O_2 to form CO_2.

Carbon monoxide is also produced by the incomplete combustion of carbon-containing compounds such as gasoline and diesel fuel. It is a very poisonous gas because it forms a stronger bond than oxygen molecules form with the iron atom in hemoglobin. Attachment of the CO molecule to the iron atom destroys the ability of hemoglobin to pick up oxygen in the lungs and carry it to the brain and muscle tissues. Carbon monoxide poisoning is particularly insidious because the gas has no odor and because the victim first becomes drowsy.

Unlike carbon monoxide, carbon dioxide is not toxic. It is one of the products of the respiratory process. It is used to make carbonated beverages, which are mostly saturated solutions of carbon dioxide in water; a small amount of the carbon dioxide combines with the water to form carbonic acid (H_2CO_3), a very weak acid.

Phosphorus reacts with a limited amount of oxygen to form tetraphosphorus hexoxide, P_4O_6,

$$P_4(s) + 3O_2(g) \longrightarrow \overset{(+3)}{P_4O_6}(s) \quad \text{tetraphosphorus hexoxide}$$

while an excess of oxygen reacts with phosphorus to form tetraphosphorus decoxide, P_4O_{10}:

$$P_4(s) + 5O_2(g) \longrightarrow \overset{(+5)}{P_4O_{10}} \quad \text{tetraphosphorus decoxide}$$

Sulfur burns in oxygen to form primarily sulfur dioxide (Figure 6-10) and only very small amounts of sulfur trioxide.

$$S_8(s) + 8O_2(g) \longrightarrow 8\overset{(+4)}{S}O_2(g) \quad \text{sulfur dioxide (mp } -73°C)$$

$$S_8(s) + 12O_2(g) \longrightarrow 8\overset{(+6)}{S}O_3(g) \quad \text{sulfur trioxide (mp } 32.5°C)$$

Figure 6-10
Sulfur burns in oxygen to form sulfur dioxide.

The production of SO_3 at a reasonable rate requires the presence of a catalyst.

Oxidation States of Nonmetals

Nonmetals exhibit more than one oxidation state in their compounds. In general, the *most* common oxidation states of a nonmetal are (1) its periodic group number, (2) its periodic group number minus two, and (3) its periodic group number minus eight. The reactions of nonmetals with a limited amount of oxygen usually give products that contain the nonmetals (other than oxygen) in lower oxidation states, usually case (2). Reactions with excess oxygen give products in which the nonmetals exhibit higher oxidation states, case (1). The examples we have cited are CO and CO_2, P_4O_6 and P_4O_{10}, and SO_2 and SO_3. The molecular formulas of the oxides are sometimes not easily predictable, but the *simplest* formulas are. For example, the two most common oxidation states of phosphorus in molecular compounds are +3 and +5. The simplest formulas for the corresponding phosphorus oxides therefore are P_2O_3 and P_2O_5, respectively. The molecular formulas are twice these, P_4O_6 and P_4O_{10}.

Reactions of Nonmetal Oxides with Water

Nonmetal oxides are called **acid anhydrides** because many of them dissolve in water to form acids *with no change in oxidation state of the nonmetal* (Figure 6-9). Several **ternary acids** can be prepared by reaction of the appropriate nonmetal oxides with water. Ternary acids contain three elements, usually H, O, and another nonmetal.

	Nonmetal Oxide	+ Water	$\longrightarrow$	Ternary Acid	
carbon dioxide	$\overset{+4}{C}O_2(g)$	$+ H_2O(\ell)$	$\longrightarrow$	$H_2\overset{+4}{C}O_3(aq)$	carbonic acid
sulfur dioxide	$\overset{+4}{S}O_2(g)$	$+ H_2O(\ell)$	$\longrightarrow$	$H_2\overset{+4}{S}O_3(aq)$	sulfurous acid
sulfur trioxide	$\overset{+6}{S}O_3(g)$	$+ H_2O(\ell)$	$\longrightarrow$	$H_2\overset{+6}{S}O_4(aq)$	sulfuric acid
dinitrogen pentoxide	$\overset{+5}{N}_2O_5(s)$	$+ H_2O(\ell)$	$\longrightarrow$	$2H\overset{+5}{N}O_3(aq)$	nitric acid
tetraphosphorus decoxide	$\overset{+5}{P}_4O_{10}(s)$	$+ 6H_2O(\ell)$	$\longrightarrow$	$4H_3\overset{+5}{P}O_4(aq)$	phosphoric acid

Nearly all oxides of nonmetals dissolve in water to give solutions of ternary acids. The oxides of boron and silicon, which are insoluble, are two exceptions.

Reactions of Metal Oxides with Nonmetal Oxides

Another common kind of reaction of oxides is the *combination of metal oxides (basic anhydrides) with nonmetal oxides (acid anhydrides), with no change in oxidation states, to form salts.*

Metal Oxide	+	Nonmetal Oxide	$\longrightarrow$	Salt	

calcium oxide + sulfur trioxide

$\overset{+2}{\text{CaO}}(s) + \overset{+6}{\text{SO}_3}(g) \longrightarrow \overset{+2}{\text{Ca}}\overset{+6}{\text{SO}_4}(s)$ calcium sulfate

magnesium oxide + carbon dioxide

$\overset{+2}{\text{MgO}}(s) + \overset{+4}{\text{CO}_2}(g) \longrightarrow \overset{+2}{\text{Mg}}\overset{+4}{\text{CO}_3}(s)$ magnesium carbonate

sodium oxide + tetraphosphorus decoxide

$6\overset{+1}{\text{Na}_2\text{O}}(s) + \overset{+5}{\text{P}_4\text{O}_{10}}(s) \longrightarrow 4\overset{+1}{\text{Na}_3}\overset{+5}{\text{PO}_4}(s)$ sodium phosphate

Example 6-8

Arrange the following oxides in order of increasing molecular (acidic) character: SO_3, Cl_2O_7, CaO, and PbO_2.

Plan

Molecular (acidic) character of oxides increases in the same direction as nonmetallic character of the element that is combined with oxygen (Figure 6-9).

Increasing nonmetallic character

$$\text{Ca} < \text{Pb} < \text{S} < \text{Cl}$$

Periodic group: IIA   IVA   VIA   VIIA

Solution

Thus, the order is

Increasing molecular character

$$\text{CaO} < \text{PbO}_2 < \text{SO}_3 < \text{Cl}_2\text{O}_7$$

Example 6-9

Arrange the oxides in Example 6-8 in order of increasing basicity.

Plan

The greater the molecular character of an oxide, the more acidic it is. Thus, the most basic oxides have the least molecular (most ionic) character (Figure 6-9).

Solution

Increasing basic character

molecular $Cl_2O_7 < SO_3 < PBO_2 < CaO$ ionic

Example 6-10

Predict the products of the reactions involving the following reactants. Write a balanced formula unit equation for each one.

(a) $Cl_2O_7(\ell) + H_2O(\ell) \longrightarrow$

(b) $As_4(s) + O_2(g)$ (excess) $\xrightarrow{\Delta}$

(c) $Mg(s) + O_2(g)$ (low pressure) $\xrightarrow{\Delta}$

Plan

(a) The reaction of a nonmetal oxide (acid anhydride) with water forms a ternary acid in which the nonmetal (Cl) has the same oxidation state (+7) as in the oxide. Thus, the acid is perchloric acid, $HClO_4$.

(b) Arsenic, a nonmetal of Group VA, exhibits common oxidation states of +5 and $+5 - 2 = +3$. Reaction of arsenic with *excess* oxygen produces the higher-oxidation-state oxide, As_2O_5. By analogy with the oxide of phosphorus in the +5 oxidation state, P_4O_{10}, we might write the formula as As_4O_{10}, but this oxide is usually represented as As_2O_5.

(c) The reaction of a Group IIA metal with oxygen (at low pressure) produces the normal metal oxide—MgO in this case.

Solution

(a) $$Cl_2O_7(\ell) + H_2O(\ell) \longrightarrow 2HClO_4(aq)$$

(b) $$As_4(s) + 5O_2(g) \xrightarrow{\Delta} 2As_2O_5(s)$$

(c) $$2Mg(s) + O_2(g) \xrightarrow{\Delta} 2MgO(s)$$

Example 6-11

Predict the products of the following pairs of reactants. Write a balanced formula unit equation for each reaction.

(a) $CaO(s) + H_2O(\ell) \longrightarrow$

(b) $Li_2O(s) + SO_3(g) \longrightarrow$

Plan

(a) The reaction of a metal oxide with water produces the metal hydroxide.

(b) The reaction of a metal oxide with a nonmetal oxide produces a salt containing the cation of the metal oxide and the anion of the acid for which the nonmetal oxide is the anhydride. SO_3 is the acid anhydride of sulfuric acid, H_2SO_4.

Solution

(a) Calcium oxide reacts with water to form calcium hydroxide.

$$CaO(s) + H_2O(\ell) \longrightarrow Ca(OH)_2(aq)$$

CaO is called quicklime. $Ca(OH)_2$ is called slaked lime.

(b) Lithium oxide reacts with sulfur trioxide to form lithium sulfate.

$$Li_2O(s) + SO_3(g) \longrightarrow Li_2SO_4(s)$$

EOC 81–83

---

3 Combustion Reactions

Combustion, or burning, is an oxidation–reduction reaction in which oxygen combines rapidly with oxidizable materials in highly exothermic reactions, usually with a visible flame. The complete combustion of **hydrocarbons**, in fossil fuels for example, produces carbon dioxide and water (steam) as the major products:

Hydrocarbons are compounds that contain only hydrogen and carbon.

$$\overset{(-4)(+1)}{CH_4(g)} + \underset{\text{excess}}{\overset{(0)}{2O_2(g)}} \xrightarrow{\Delta} \overset{(+4)(-2)}{CO_2(g)} + \overset{(+1)(-2)}{2H_2O(g)} + \text{heat}$$

$$\underset{\text{cyclohexane}}{\overset{(-2)(+1)}{C_6H_{12}(g)}} + \underset{\text{excess}}{\overset{(0)}{9O_2(g)}} \xrightarrow{\Delta} \overset{(+4)(-2)}{6CO_2(g)} + \overset{(+1)(-2)}{6H_2O(g)} + \text{heat}$$

As we have seen, the origin of the term "oxidation" lies in just such reactions, in which oxygen "oxidizes" another species.

4 Combustion of Fossil Fuels and Air Pollution

Fossil fuels are mixtures of variable composition that consist primarily of hydrocarbons. We burn them because they release energy, rather than to obtain chemical products (Figure 6-11). The incomplete combustion of hydrocarbons yields undesirable products, carbon monoxide and elemental carbon (soot), which pollute the air. Unfortunately, all fossil fuels—natural gas, coal, gasoline, kerosene, oil, and so on—also have undesirable non-hydrocarbon impurities that undergo combustion to produce oxides that act as additional air pollutants. At this time it is not economically feasible to remove all of these impurities.

Fossil fuels result from the decay of animal and vegetable matter (Figure 6-12). All living matter contains some sulfur and nitrogen, so fossil fuels also contain sulfur and nitrogen impurities to varying degrees. Table 6-5 gives composition data for some common kinds of coal.

Combustion of sulfur produces sulfur dioxide, SO_2, probably the single most harmful pollutant:

$$\overset{(0)}{S_8}(s) + 8O_2(g) \overset{\Delta}{\longrightarrow} 8\overset{(+4)}{SO_2}(g)$$

Large amounts of SO_2 are produced by the burning of sulfur-containing coal.

Many metals occur in nature as sulfides. The process of extracting the free (elemental) metals involves **roasting**—heating an ore in the presence of air. For many metal sulfides this produces a metal oxide and SO_2. The metal oxides are then reduced to the free metals. Consider lead sulfide, PbS, as an example:

$$2PbS(s) + 3O_2(g) \longrightarrow 2PbO(s) + 2SO_2(g)$$

Sulfur dioxide is corrosive; it damages plants, structural materials, and humans. It is a nasal, throat, and lung irritant. Sulfur dioxide is slowly oxidized to sulfur trioxide, SO_3, by oxygen in air:

$$2SO_2(g) + O_2(g) \longrightarrow 2SO_3(\ell)$$

Sulfur trioxide combines with moisture in the air to form the strong, corrosive acid, sulfuric acid:

$$SO_3 + H_2O(\ell) \longrightarrow H_2SO_4(\ell)$$

Oxides of sulfur are the main cause of acid rain.

Table 6-5
Some Typical Coal Compositions in Percent (dry, ash-free)

	C	H	O	N	S
lignite	70.59	4.47	23.13	1.04	0.74
subbituminous	77.2	5.01	15.92	1.30	0.51
bituminous	80.2	5.80	7.53	1.39	5.11
anthracite	92.7	2.80	2.70	1.00	0.90

Carbon in the form of soot is one of many kinds of *particulate matter* in polluted air.

Figure 6-11
Georgia Power Company's Plant Bowen at Taylorsville, Georgia. In 1986 it burned 8,376,726 tons of coal and produced 21,170,999 megawatt-hours of electricity, a national record.

Figure 6-12
The luxuriant growth of vegetation that occurred during the carboniferous age is the source of our coal deposits.

Chemistry in Use. . .
Acid Rain

During the 1980s, the phenomenon of acid deposition, commonly known as acid rain, gained considerable attention from the public and the scientific community. It is the subject of intense research for thousands of environmental scientists. It has also been a source of political conflict and debate within and between nations in North America and Europe.

The term "acid rain" is used to describe all naturally occurring precipitation, including rain, snow, sleet, and hail, that has become acidified. The *acidity* or *alkalinity* of a substance is determined by the relative concentrations of hydrogen ions and hydroxide ions that it contains. Acidity is expressed on a logarithmic scale called the pH scale. Pure water, which has equal concentrations of hydrogen ions and hydroxide ions, has a pH of 7.0 and is said to be *neutral*. A substance with a higher concentration of hydrogen ions than hydroxide ions is acidic and has a value less than 7.0 on the pH scale. Conversely, a substance with a higher concentration of hydroxide ions than hydrogen ions is alkaline, or *basic,* and has a pH value greater than 7.0. The farther a reading is from 7.0, below or above, the more acidic or basic (respectively) the substance is. Each full pH unit decrease represents a tenfold increase in acidity. For example, a solution whose pH value is 6.0 contains ten times more hydrogen ions than pure water and is ten times more acidic. A substance with a pH of 5.0 is 100 times more acidic than pure water.

Normal, uncontaminated precipitation is naturally slightly acidic, having a pH value of about 5.6. This natural acidity is the result of the combination of carbon dioxide in the atmosphere with water vapor to form carbonic acid, H_2CO_3, a very weak acid. Any rainfall with pH lower than 5.6 is considered excessively acidic. Unfortunately, as a result of human activities such as burning fossil fuels and smelting ores, reports of rain with pH values of 3.8 to 4.5 are common. The most acidic rainfall recorded in the United States to date (April 1991) had a pH value of 1.5; this is 12,600 times more acidic than normal uncontaminated rain, and almost one-third as acidic as battery acid, a very strong acid. Widespread acid rain has been known in northern Europe and eastern North America for some time. More recent work has led to the discovery of acid rain in western North America, Japan, China, the Soviet Union, and South America.

The main compounds that produce acid rain are the oxides of sulfur and nitrogen, which quickly react with water to form acids such as sulfurous (H_2SO_3), sulfuric

Most of the oxides of sulfur that contribute to acid rain in North America come from midwestern states. Prevailing winds carry the resulting acid droplets to the north and east, as far as Canada. Oxides of nitrogen also contribute to acid rain formation.

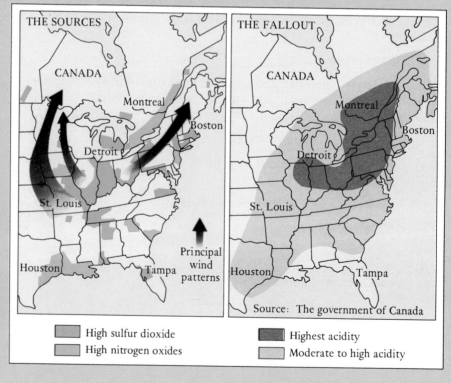

THE SOURCES

THE FALLOUT

Source: The government of Canada

High sulfur dioxide
High nitrogen oxides

Highest acidity
Moderate to high acidity

(H_2SO_4), nitrous (HNO_2), and nitric (HNO_3) acids. Sulfur oxides are produced mainly by the combustion of high-sulfur coal and oil and by the smelting of metal ores. Coal-burning utilities in the midwestern United States appear to be largely responsible for the presence of sulfur oxides in North American acid rain. Nitrogen oxides are by-products of gasoline combustion in automobile and airplane engines and of some processes that generate electricity. After these compounds are emitted into the atmosphere, they may be transported thousands of miles before returning to earth in solution with rain, sleet, or snow.

As a result of the great distances traveled by sulfur and nitrogen oxides before they result in acid rain, what was once assumed to be a harmless dispersal of contaminants has become an international pollution problem. For instance, about 50% of the sulfates falling in eastern Canada are believed to have originated in the United States, a fact which has been a source of controversy between the two countries.

Similarly, much of the acidic precipitation in Scandinavia originates in industrialized areas of central Europe and the United Kingdom.

The consequences of acid rain depend in part on the characteristics of the soil and underlying rock upon which it falls. In areas where the principal rock is limestone (calcium carbonate), there is a natural buffer system that can prevent acidification of soil, lakes, and streams to some extent. In other areas where the soil and bodies of water do not contain such a natural buffer, the pH drops gradually as a result of acid precipitation. At times, it drops quite suddenly—for instance, when the spring melt causes all of the acids that have accumulated in the winter snows and ice to be released into soils and waters over a short period of time. Although the low pH resulting from the spring melt is usually temporary, it can be devastating to aquatic plant life and to many animals, such as small fish and frogs.

Although there is considerable debate about the extent to which

acid deposition constitutes an environmental risk, the effects of acid rain on aquatic and terrestrial ecosystems have been amply documented. Many lakes are virtually sterile, devoid of fish as well as most plants and invertebrates. For example, in the Adirondacks region of the northeastern United States, more than 200 lakes have been rendered totally sterile by the effects of acid rain. Of the 100,000 lakes in Sweden, 4000 have become fishless. In acidified lakes in which fish still exist, the diversity of species has decreased and the life spans of the fish are greatly reduced. Some food chains have also been shortened due to the elimination of sensitive species at intermediate points in the chain. The disappearance of lower organisms may cause starvation of large predatory animals well before direct toxic action of hydrogen ions is evident. In terrestrial systems, trees and plants are suffering from nutrient deficiency, reduced efficiency of photosynthesis, and lowered resistance to disease. This is thought to be due mainly to the

Effects of acid rain on evergreen forests. The two photographs were taken at Camel's Hump, Vermont, 15 years apart.

Effects of acid rain on statues. The photo at the left was taken at the Lincoln Cathedral in England in 1910; the one at the right was taken in 1984.

The effects of acid rain on pH can be offset by adding calcium carbonate, $CaCO_3$, to react with excess hydrogen ions. Here a helicopter sprays finely ground limestone (mostly $CaCO_3$) over forests affected by acid rain.

leaching of important nutrients, such as calcium and magnesium, from the soils. Other elements, such as aluminum and trace metals, are leached from the soil and are subsequently washed into lakes and streams, where they are quite toxic to fishes and other organisms. The acidification of ground water also causes public water supplies to become acidified. Acidic water supplies dissolve metals from plumbing, creating a drinking-water hazard in some areas.

Acid rain has also done considerable damage to structures. Rates of corrosion of metal structures are greatly increased by acid rain. Exfoliation (flaking) of marble and limestone monuments and buildings is common, as is the pitting of granitic stonework. Ancient ruins such as the Acropolis in Greece are also showing signs of erosion by acid rain.

Acid rain is a serious worldwide pollution problem. Beyond its detrimental effect on other species, its potential consequences for humans are enormous: lowered crop yields, decreased timber production, and loss of important fishing and recreational areas as well as public water supplies. It is interesting to note that 21 European countries agreed in 1985 to lower their sulfur dioxide emissions by 30% or more over a 10-year period. By 1989, more than half of those countries had already reached that goal. In 1988, the Canadian government announced a goal of lowering its sulfur dioxide emissions by half by 1994. In the United States, progress has not been so rapid.

Measures to counteract the effects of acid rain have only limited effectiveness. The available processes that remove sulfur and nitrogen oxides *at the source,* before they enter the atmosphere to form acid rain, are expensive and so meet with resistance. But the monetary and ecological costs of allowing the conditions that create acid rain to continue are also very great. Although scientists can provide the information upon which to base important decisions regarding the acid rain problem, the ultimate decisions about alleviating the problem will have to be made in the political arena.

Beth A. Trust
Graduate student in chemistry
University of Texas Marine Sciences
Institute

Figure 6-13
Photochemical pollution (a brown haze)
enveloping a city.

Compounds of nitrogen are also impurities in fossil fuels, and they undergo combustion to form nitric oxide, NO. However, most of the nitrogen in the NO in exhaust gases from furnaces, automobiles, airplanes, and so on comes from the air that is mixed with the fuel:

$$\overset{(0)}{N_2}(g) + O_2(g) \longrightarrow 2\overset{(+2)}{N}O(g)$$

Remember that "clean air" is *about* 80% N_2 and 20% O_2 by mass. This reaction does *not* occur at room temperature but does occur at the high temperatures of furnaces, internal combustion engines, and jet engines.

NO can be further oxidized by oxygen to nitrogen dioxide, NO_2; this reaction is enhanced in the presence of ultraviolet light from the sun:

$$2\overset{(+2)}{N}O(g) + O_2(g) \xrightarrow[\text{light}]{\text{uv}} 2\overset{(+4)}{N}O_2(g) \qquad \text{(a reddish-brown gas)}$$

NO_2 is responsible for the reddish-brown haze that hangs over many cities on sunny afternoons (Figure 6-13) and probably for most of the respiratory problems associated with this kind of air pollution. It can react to produce other oxides of nitrogen and other secondary pollutants.

In addition to being a pollutant itself, nitrogen dioxide reacts with water in the air to form nitric acid, another major contributor to acid rain:

$$3NO_2(g) + H_2O(\ell) \longrightarrow 2HNO_3(\ell) + NO(g)$$

Key Terms

Acid anhydride The oxide of a nonmetal that reacts with water to form an acid.

Actinides Elements 90 through 103 (after *actinium*).

Amphoterism The ability to react with both acids and bases.

Angstrom (Å) 10^{-10} meter.

Atomic radius The radius of an atom.

Basic anhydride The oxide of a metal that reacts with water to form a base.

Catalyst A substance that speeds up a chemical reaction without itself being consumed in the reaction.

Combustion reaction The reaction of a substance with oxygen in a highly exothermic reaction, usually with a visible flame.

***d*-Transition elements (metals)** The B group elements except IIB in the periodic table; sometimes called simply transition elements.

Effective nuclear charge (Z_{eff})   The nuclear charge experienced by the outermost electrons of an atom; the actual nuclear charge minus the effects of shielding due to inner shell electrons.

Electron affinity   The amount of energy absorbed in the process in which an electron is added to a neutral isolated gaseous atom to form a gaseous ion with a 1− charge; has a negative value if energy is released.

Electronegativity   A measure of the relative tendency of an atom to attract electrons to itself when chemically combined with another atom.

f-Transition elements   See *Inner transition elements*.

Hydride   A binary compound of hydrogen.

Inner transition elements   Elements 58 through 71 and 90 through 103; also called *f*-transition elements.

Ionic radius   The radius of an ion.

Ionization energy   The minimum amount of energy required to remove the most loosely held electron of an isolated gaseous atom or ion.

Isoelectronic   Having the same electron configurations.

Lanthanides   Elements 58 through 71 (after *lanthanum*).

Nanometer (nm) 10^{-9} meter.

Noble gases   Elements of periodic Group 0; also called rare gases; formerly called inert gases.

Noble gas configuration   The stable electronic configuration of a noble gas.

Normal oxide   A metal oxide containing the oxide ion, O^{2-} (oxygen in the −2 oxidation state).

Nuclear shielding   See *Shielding effect*.

Oxide   A binary compound of oxygen.

Periodicity   Regular periodic variations of properties of elements with atomic number (and position in the periodic table).

Periodic law   The properties of the elements are periodic functions of their atomic numbers.

Peroxide   A compound containing oxygen in the −1 oxidation state. Metal peroxides contain the peroxide ion, O_2^{2-}.

Radical   A species containing one or more unpaired electrons; many radicals are very reactive.

Rare earths   Inner transition elements.

Rare gases   See *Noble gases*.

Representative elements   A group elements in the periodic table.

Roasting   Heating an ore of an element in the presence of air.

Shielding effect   Electrons in filled sets of s and p orbitals between the nucleus and outer shell electrons shield the outer shell electrons somewhat from the effect of protons in the nucleus; also called screening effect.

Superoxide   A compound containing the superoxide ion, O_2^- (oxygen in the $-\frac{1}{2}$ oxidation state).

Ternary acid   An acid containing three elements, H, O, and (usually) another nonmetal.

Exercises

Classification of the Elements

1. Define and illustrate the following terms clearly and concisely: (a) representative elements, (b) d-transition elements, (c) inner transition elements.

2. Explain why Period 1 contains two elements and Period 2 contains eight elements.

3. Explain why Period 4 contains 18 elements.

4. The third major energy level ($n = 3$) has s, p, and d sublevels. Why does Period 3 contain only eight elements?

5. Account for the number of elements in Period 6.

*6. What would be the atomic number of the as-yet-undiscovered alkaline earth element of Period 8?

*7. How many elements are there in Period 7? Suppose that many new elements are discovered and it is found that Period 8 consists of *more* than 32 elements. How many would you predict? What would account for this number?

*8. In what periodic group would the as-yet-undiscovered element 116 be found? Would you classify it as a metal or a nonmetal? Suggest a reasonable electron configuration for the element.

9. Identify the group, family, and/or other periodic table location of each element with the outer electron configuration
   (a) ns^2np^3,
   (b) ns^1,
   (c) $ns^2(n-1)d^{0-2}(n-2)f^{1-14}$.

10. Repeat Exercise 9 for
    (a) ns^2np^5 (c) $ns^2(n-1)d^{1-10}$
    (b) ns^2 (d) ns^2np^1.

11. Write the outer electron configurations for the (a) alkaline earth metals, (b) d-transition metals, and (c) halogens.

12. Repeat Exercise 11 for the (a) noble gases, (b) alkali metals, (c) f-transition metals, and (d) vanadium family.

13. Identify the elements and the part of the periodic table in which the elements with the following configurations are found.
    (a) $1s^22s^22p^63s^23p^64s^2$
    (b) $[Kr]4d^85s^1$
    (c) $[Xe]4f^{14}5d^66s^2$
    (d) $[Xe]4f^{12}6s^2$
    (e) $[Kr]4d^{10}5s^25p^6$
    (f) $[Kr]4d^{10}4f^{14}5s^25p^65d^{10}6s^26p^2$

14. Which of the elements in the following periodic table

is (are) (a) alkali metals, (b) an element with the outer configuration of d^8s^2, (c) lanthanides, (d) p-block representative elements, (e) elements with incompletely filled f-subshells, (f) halogens, (g) s-block representative elements, (h) actinides, (i) d-transition elements, (j) noble gases?

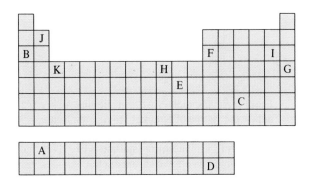

Atomic Radii

15. What is the meaning of the following statement? "It is impossible to describe an atom as having an invariant radius."

16. What is meant by nuclear shielding? What effect does it have on trends in atomic radii?

17. Why do atomic radii decrease from left to right within a period in the periodic table?

18. Why do atomic radii increase from top to bottom within a group in the periodic table?

19. Arrange each of the following sets of atoms in order of increasing atomic volume: (a) N, Mg, Al, Si; (b) O, S, Se, Te; (c) Ca, Sr, Ga, In.

20. Arrange each of the following sets of atoms in order of increasing atomic radii: (a) the alkaline earth elements; (b) the noble gases; (c) the elements in the second period; (d) N, Te, B, Sr, and Sb.

21. Variations in the atomic radii of the transition elements are not so pronounced as those of the representative elements. Why?

Ionization Energy

22. Define (a) first ionization energy and (b) second ionization energy.

23. Why is the second ionization energy for a given element always greater than the first ionization energy?

24. What is the usual relationship between atomic radius and first ionization energy, other factors being equal?

25. What is the usual relationship between nuclear charge and first ionization energy, other factors being equal?

*26. The first ionization energy of potassium, K, is 419 kJ/mol. What is the minimum frequency of light required to ionize gaseous potassium atoms?

27. Write the equation for each of the following, and write the electron configuration for each atom or ion shown:

(a) the first ionization of potassium, (b) the third ionization of aluminum, (c) the second ionization of chlorine.

28. Arrange the members of each of the following sets of elements in order of increasing first ionization energies: (a) the alkali metals; (b) the halogens; (c) the elements in the second period; (d) Br, F, B, Ga, Cs, and H.

29. Explain why there is a general increase in first ionization energy across each period.

30. Explain the trend in first ionization energy upon descending a periodic group.

31. In a plot of first ionization energy versus atomic number for Periods 2 and 3, "dips" occur at the IIIA and VIA elements. Account for these dips.

32. What is the general relationship between the sizes of the atoms of Period 2 and their first ionization energies? Rationalize the relationship.

33. Why must a zinc atom absorb more energy than a calcium atom to ionize a $4s$ electron?

34. On the basis of electronic configurations, would you expect a Mg^{3+} ion to exist in compounds? Why or why not? How about Al^{3+}?

35. How much energy, in kJ, must be absorbed by 1.00 mol of gaseous potassium atoms to convert all of them to gaseous K^+ ions?

36. The second ionization energy for magnesium is 1451 kJ/mol. How much energy, in kJ, must be absorbed by 1.00 g of gaseous magnesium atoms to convert all of them to gaseous Mg^{2+} ions?

37. Would you expect the Ca^+ ion to exist in any compounds? Why or why not?

Electron Affinity

38. What is electron affinity?

39. Arrange the members of each of the following sets of elements in order of increasingly negative electron affinities: (a) the Group IA metals; (b) the Group VIIA elements; (c) the elements in the second period; (d) Li, K, C, F, and I.

40. The electron affinities of the halogens are much more negative than those of the Group VIA elements. Why is this so?

41. The addition of a second electron to form an ion with a 2− charge is always endothermic. Why is this so?

42. Write the equation for the change described by each of the following, and write the electron configuration for each atom or ion shown: (a) the electron affinity of oxygen, (b) the electron affinity of chlorine, (c) the electron affinity of calcium.

Ionic Radii

43. Compare the sizes of cations and the neutral atoms from which they are formed by citing three specific examples.

44. Arrange the members of each of the following sets of cations in order of increasing ionic radii: (a) K^+, Ca^{2+},

Ga^{3+}; (b) Ca^{2+}, Be^{2+}, Ba^{2+}, Mg^{2+}; (c) Al^{3+}, Sr^{2+}, Rb$^+$, K$^+$.

45. Compare the sizes of anions and the neutral atoms from which they are formed by citing three specific examples.

46. Arrange the following sets of anions in order of increasing ionic radii: (a) S^{2-}, Cl$^-$, P^{3-}; (b) S^{2-}, O^{2-}, Se^{2-}; (c) N^{3-}, S^{2-}, Br$^-$, P^{3-}.

47. Compare and explain the relative sizes of H$^+$, H, and H$^-$.

48. Most transition metals can form more than one simple positive ion. For example, iron forms both Fe^{2+} and Fe^{3+} ions, and copper forms both Cu$^+$ and Cu^{2+} ions. Which is the smaller ion of each pair, and why?

Electronegativity

49. What is electronegativity?

50. Arrange the members of each of the following sets of elements in order of increasing electronegativities: (a) B, Ga, Al, In; (b) S, Na, Mg, Cl; (c) P, N, Sb, Bi; (d) Se, Ba, F, Si, Sc.

51. Which of the following statements is better? Why?
(a) Magnesium has a weak attraction for electrons in a chemical bond because it has a low electronegativity.
(b) The electronegativity of magnesium is low because magnesium has a weak attraction for electrons in a chemical bond.

52. Comment on the validity of the following statement: "Chlorine has a high electronegativity because it forms chloride ions, Cl$^-$, readily."

*53. Do you think an element might be discovered (or made) with a lower electronegativity than francium? What would its electronic configuration be?

*54. Some of the second-period elements show a similarity to the element one column to the right and one row down. For instance, Li is similar in many respects to Mg, and Be is similar to Al. This has been attributed to the charge density on the stable ions (Li$^+$ vs. Mg^{2+}; Be^{2+} vs. Al^{3+}). From the values of electronic charge (Chapter 5) and ionic radii (Chapter 6), calculate the charge density for these four ions, in coulombs/Å^3.

Additional Exercises on the Periodic Table

55. The P—Cl bond length in PCl$_3$ is 2.04 Å. The bond length in Cl$_2$ is 1.98 Å. Calculate the atomic radii for these elements. Using the atomic radius for F given in Figure 6-1, predict the P—F bond length in PF$_3$.

56. The bond lengths in F$_2$ and Cl$_2$ molecules are 1.42 Å and 1.98 Å, respectively. Calculate the atomic radii for these elements. Predict the Cl—F bond length. (The actual Cl—F bond length is 1.64 Å.)

*57. The atoms in crystalline nickel are arranged so that they are touching each other in a plane as shown in the sketch:

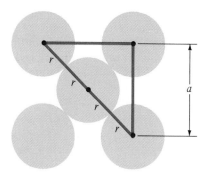

From plane geometry, we can see that $4r = a\sqrt{2}$. Calculate the radius of a nickel atom given that $a = 3.5238$ Å.

*58. The ions in a plane in crystalline KBr, KCl, and LiCl are arranged as shown in the sketch:

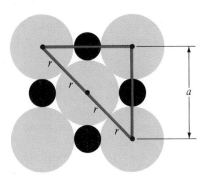

What is the relationship between a and the ionic radii? Using $a = 6.578$ Å for KBr, 6.2931 Å for KCl, and 5.14 Å for LiCl, and 1.96 Å for the ionic radius for Br$^-$, calculate the ionic radii for K$^+$, Cl$^-$, and Li$^+$.

59. How do the values of the ionization energies of the metals compare to those of the nonmetals? How do these values determine the chemical behaviors of the metals and nonmetals?

60. Compare the respective values of the first ionization energy (Table 6-1) and electron affinity (Table 6-2) for several elements. Which energy is greater? Why?

61. Compare the respective values of the first ionization energy (Table 6-1) and electron affinity (Table 6-2) for nitrogen to those for carbon and oxygen. Explain why the nitrogen values are considerably different.

*62. Based on general trends, the electron affinity of fluorine would be expected to be greater than that of chlorine; however, the value is less and is similar to the value for bromine. Explain.

*63. The first ionization energy of oxygen is 1313.9 kJ/mol at 0 K. The second ionization energy is 3388.1 kJ/mol. Why is this second value much larger than the first?

Hydrogen and the Hydrides

64. Summarize the physical properties of hydrogen.

65. Write balanced formula unit equations for (a) the reaction of iron with steam, (b) the reaction of calcium with hydrochloric acid, (c) the electrolysis of water, and (d) the "water gas" reaction.

66. Write a balanced formula unit equation for the preparation of (a) an ionic hydride and (b) a molecular hydride.

67. Classify the following hydrides as molecular or ionic: (a) NaH, (b) H_2S, (c) BaH_2, (d) KH, (e) NH_3.

68. Explain why NaH and H_2S are different kinds of hydrides.

69. Write formula unit equations for the reactions of (a) NaH and (b) BaH_2 with water.

70. Name the following (pure) compounds: (a) H_2S, (b) HF, (c) KH, (d) NH_3, (e) H_2Se, (f) MgH_2.

Oxygen and the Oxides

71. How are O_2 and O_3 similar? Different?

72. Briefly compare and contrast the properties of oxygen with those of hydrogen.

73. Write molecular equations to show how oxygen can be prepared from (a) mercury(II) oxide, HgO, (b) hydrogen peroxide, H_2O_2, and (c) potassium chlorate, $KClO_3$.

74. Which of the following elements form normal oxides as the *major* products of reactions with oxygen? (a) Li, (b) Na, (c) Rb, (d) Mg, (e) Zn (exhibits only one common oxidation state), (f) Al.

75. Write formula unit equations for the primary reactions of oxygen with the following elements: (a) Li, (b) Na, (c) K, (d) Ca.

76. Write formula unit equations for the reactions of the following elements with a *limited* amount of oxygen: (a) Sr, (b) Fe, (c) Mn, (d) Cu.

77. Write formula unit equations for the reactions of the following elements with an *excess* of oxygen: (a) Sr, (b) Fe, (c) Mn, (d) Cu.

78. Write formula unit equations for the reactions of the following elements with a *limited* amount of oxygen: (a) C, (b) As_4, (c) Ge.

79. Write formula unit equations for the reactions of the following elements with an *excess* of oxygen: (a) C, (b) As_4, (c) Ge.

80. Distinguish among normal oxides, peroxides, and superoxides. What is the oxidation state of oxygen in each case?

81. Which of the following can be classified as basic anhydrides? (a) SO_2, (b) Li_2O, (c) SeO_3, (d) CaO, (e) N_2O_5.

82. Write balanced formula unit equations for the following reactions and name the products:
(a) sulfur dioxide, SO_2, with water
(b) sulfur trioxide, SO_3, with water
(c) selenium trioxide, SeO_3, with water
(d) dinitrogen pentoxide, N_2O_5, with water
(e) dichlorine heptoxide, Cl_2O_7, with water

83. Write balanced formula unit equations for the following reactions and name the products:
(a) sodium oxide, Na_2O, with water
(b) calcium oxide, CaO, with water
(c) lithium oxide, Li_2O, with water
(d) magnesium oxide, MgO, with sulfur dioxide, SO_2
(e) barium oxide, BaO, with carbon dioxide, CO_2

84. Identify the acid anhydrides of the following ternary acids: (a) H_2SO_4, (b) H_2CO_3, (c) H_2SO_3, (d) H_3AsO_4, (e) HNO_2.

85. Identify the basic anhydrides of the following metal hydroxides: (a) NaOH, (b) $Mg(OH)_2$, (c) $Fe(OH)_2$ (d) $Al(OH)_3$.

Combustion Reactions

86. Define combustion. Why are all combustion reactions also redox reactions?

87. Write equations for the complete combustion of the following compounds: (a) ethane, $C_2H_6(g)$; (b) propane, $C_3H_8(g)$; (c) ethanol, $C_2H_5OH(\ell)$.

88. Write equations for the *incomplete* combustion of the following compounds to produce carbon monoxide: (a) ethane, $C_2H_6(g)$; (b) propane, $C_3H_8(g)$.

As we have seen, two substances may react to form different products when they are mixed in different proportions under different conditions. In Exercises 89 and 90, write balanced equations for the reactions described. Assign oxidation numbers.

89. (a) Methane burns in excess air to form carbon dioxide and water.
(b) Methane burns in a limited amount of air to form carbon monoxide and water.
(c) Methane burns (poorly) in a very limited amount of air to form elemental carbon and water.

90. (a) Butane (C_4H_{10}) burns in excess air to form carbon dioxide and water.
(b) Butane burns in a limited amount of air to form carbon monoxide and water.
(c) When heated in the presence of *very little* air, butane "cracks" to form acetylene, C_2H_2; carbon monoxide; and hydrogen.

91. (a) How much SO_2 would be formed by burning 1.00 ton of bituminous coal that is 5.11% sulfur by mass? Assume that all of the sulfur is converted to SO_2.
(b) If 27.0% of the SO_2 escaped into the atmosphere and 84.2% of it were converted to H_2SO_4, how many grams of H_2SO_4 would be produced in the atmosphere?

92. Write equations for the complete combustion of the following compounds. Assume that sulfur is converted to SO_2 and nitrogen is converted to NO. (a) $C_6H_5N(\ell)$, (b) $C_2H_5SH(\ell)$, (c) $C_7H_{10}NO_2S(\ell)$.

93. Describe the formation of the reddish-brown haze of some cities experiencing this kind of air pollution.

94. Account for the occurrence of acid rain.

7 Chemical Bonding and Inorganic Nomenclature

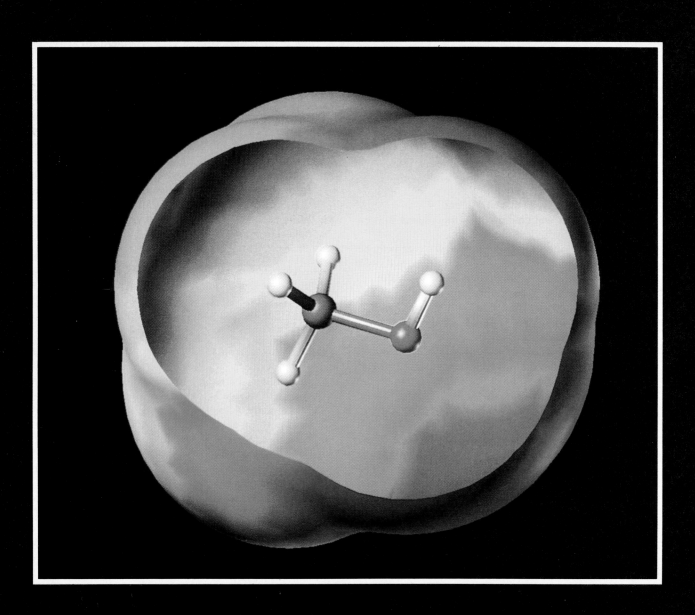

Objectives

As you study this chapter, you should learn

☐ To write Lewis dot representations of atoms
☐ To predict whether bonding will be primarily ionic, covalent, or polar covalent
☐ How the properties of compounds depend on their bonding
☐ How elements bond by electron transfer (ionic bonding)
☐ To predict the formulas of ionic compounds
☐ How elements bond by sharing electrons (covalent bonding)

☐ To write Lewis dot formulas for molecules and polyatomic ions
☐ To recognize exceptions to the octet rule
☐ To relate the nature of the bonding to electronegativity differences
☐ About resonance, when to write resonance structures, and how to do so
☐ How to write formal charges for atoms in covalent structures

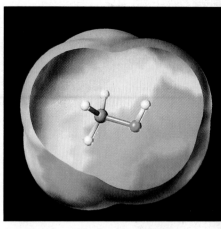

A computer model of wood alcohol, or methanol, CH_3OH (C gray, H white, O red). The ball-and-stick model is shown inside a computer-generated molecular surface. The surface is color-coded to show relative charge densities.

The attractive forces that hold atoms together in compounds are called **chemical bonds**. There are two major classes of bonding. (1) **Ionic bonding** results from electrostatic interactions among ions, which can take the form of the *transfer* of one or more electrons from one atom or group of atoms to another. (2) **Covalent bonding** results from *sharing* one or more electron pairs between two atoms. These two classes represent two extremes; all bonds have at least some degree of both ionic and covalent character. Compounds containing predominantly ionic bonding are called **ionic compounds**. Those containing predominantly covalent bonds are called **covalent compounds**. Some of the properties associated with many simple ionic and covalent compounds in the extreme cases are summarized below. The differences in properties can be accounted for by the differences in bonding between the atoms or ions.

Ionic Compounds

1. They are solids with high melting points (typically >400°C).
2. Many are soluble in polar solvents such as water.
3. Most are insoluble in nonpolar solvents, such as hexane, C_6H_{14}.
4. Molten compounds conduct electricity well because they contain mobile charged particles (ions).
5. Aqueous solutions conduct electricity well because they contain mobile charged particles (ions).

Covalent Compounds

1. They are gases, liquids, or solids with low melting points (typically <300°C).
2. Many are insoluble in polar solvents.
3. Most are soluble in nonpolar solvents, such as hexane, C_6H_{14}.
4. Liquid and molten compounds do not conduct electricity.
5. Aqueous solutions are *usually* poor conductors of electricity because most do not contain charged particles.

The distinction between polar and nonpolar molecules is made in Section 7-4.

As we saw in Section 4-2, aqueous solutions of some covalent compounds do conduct electricity.

7-1 Lewis Dot Representations of Atoms

The number and arrangements of electrons in the outermost shells of atoms determine the chemical and physical properties of the elements as well as the kinds of chemical bonds they form. We write **Lewis dot representations** (or **Lewis dot formulas**) as a convenient bookkeeping method for keeping track of these "chemically important electrons." We now introduce this

Table 7-1
Lewis Electron Dot Formulas for Representative Elements

Group	IA	IIA	IIIA	IVA	VA	VIA	VIIA	0
Number of electrons in outer shell	1	2	3	4	5	6	7	8 (except He)
Row 1	H·							He:
Row 2	Li·	Be:	B·	C·	·N·	·O:	·F:	:Ne:
Row 3	Na·	Mg:	Al·	Si·	·P·	·S:	·Cl:	:Ar:
Row 4	K·	Ca:	Ga·	Ge·	·As·	·Se:	·Br:	:Kr:
Row 5	Rb·	Sr:	In·	Sn·	·Sb·	·Te:	·I:	:Xe:
Row 6	Cs·	Ba:	Tl·	Pb·	·Bi·	·Po:	·At:	:Rn:
Row 7	Fr·	Ra:						

method for atoms of elements; in our discussion of chemical bonding in subsequent sections, we shall frequently use such formulas for atoms, molecules, and ions.

Chemical bonding usually involves only the outermost electrons of atoms, also called **valence electrons**. In Lewis dot representations, only the electrons in the outermost occupied s and p orbitals are shown as dots. Paired and unpaired electrons are also indicated. Table 7-1 shows Lewis dot formulas for the representative elements. All elements in a given group have the same outer shell configuration.

Because of the large numbers of dots, such formulas are not as useful for the transition and inner transition elements.

Ionic Bonding

7-2 Formation of Ionic Compounds

The first kind of chemical bonding we shall describe is the **ionic** or **electrovalent bonding**. Ionic bonding results from the *transfer of one or more electrons from one atom or group of atoms to another*. As our previous discussions of ionization energy, electronegativity, and electron affinity would indicate, ionic bonding occurs most easily when elements that have low ionization energies (metals) react with elements having high electronegativities and high electron affinities (nonmetals). Many metals are easily *oxidized*—that is, they lose electrons; and many nonmetals are readily *reduced*—that is, they gain electrons.

> When the electronegativity difference, ΔEN, between two elements is large, the elements are likely to form a compound by ionic bonding (transfer of electrons).

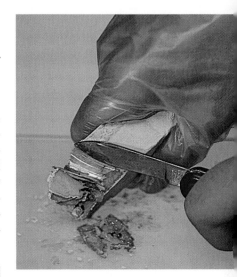

Freshly cut sodium has a metallic luster. A little while after being cut, the sodium metal surface turns white as it reacts with the air.

Group IA Metals and Group VIIA Nonmetals

Consider the reaction of sodium (a Group IA metal) with chlorine (a Group VIIA nonmetal). Sodium is a soft silvery metal (mp 98°C), and chlorine is a yellowish-green corrosive gas at room temperature. Both sodium and chlorine react with water, sodium vigorously. By contrast, sodium chloride is a white solid (mp 801°C) that dissolves in water with no reaction and with the absorption of just a little heat. We can represent the reaction for its formation as

$$2Na(s) + Cl_2(g) \longrightarrow 2NaCl(s)$$
sodium   chlorine   sodium chloride

We can understand this reaction in more detail by showing electron configurations for all species. We represent chlorine as individual atoms rather than molecules, for simplicity.

		3s	**3p**					**3s**	**3p**			
$_{11}$Na	[Ne]	↑				$\Big\}$	Na$^+$ [Ne]	__				1e^- lost
$_{17}$Cl	[Ne]	⇅	⇅	⇅	↑		Cl$^-$ [Ne]	⇅	⇅	⇅	⇅	1e^- gained

Halite crystals (naturally occurring NaCl).

The loss of electrons is *oxidation* (Section 4-8). Na atoms are *oxidized* to form Na^+ ions.

The gain of electrons is *reduction* (Section 4-8). Cl atoms are *reduced* to form Cl^- ions.

In this reaction, Na atoms lose one electron each to form Na^+ ions, which contain only ten electrons, the same number as the *preceding* noble gas, neon. We say that sodium ions have the neon electronic structure; Na^+ is *isoelectronic* with Ne. In contrast, chlorine atoms gain one electron each to form chloride ions, Cl^-, which contain 18 electrons. This is the same number as the *next* noble gas, argon; Cl^- is *isoelectronic* with Ar. Similar observations apply to most ionic compounds formed by reactions between *representative metals and representative nonmetals*.

We can use Lewis dot formulas (Section 7-1) to represent the reaction.

$$Na\cdot \; + \; :\overset{\cdot\cdot}{Cl}\cdot \; \longrightarrow \; Na^+[:\overset{\cdot\cdot}{Cl}:]^-$$

The formula for sodium chloride, NaCl, indicates that the compound contains Na and Cl atoms in a 1:1 ratio. This is the formula we predict based on the fact that each Na atom contains only one electron in its highest energy level and each Cl atom needs only one electron to fill completely its outermost p orbitals.

The chemical formula NaCl does not explicitly indicate the ionic nature of the compound, only the ratio of atoms. So we must learn to recognize, from positions of elements in the periodic table and known trends in electronegativity, when the difference in electronegativity is large enough to favor ionic bonding.

The noble gases are excluded from this generalization.

> The farther apart across the periodic table two elements are, the more likely they are to form an ionic compound.

The greatest difference in electronegativity occurs from lower left to upper right. Thus, CsF is more ionic than LiI.

All the Group IA metals (Li, Na, K, Rb, Cs) will react with the Group VIIA elements (F, Cl, Br, I) to form ionic compounds of the same general formula, MX. The resulting ions, M^+ and X^-, always have noble gas configurations. We can represent the general reaction of the IA metals with the VIIA elements as follows:

$$2M(s) + X_2 \longrightarrow 2MX(s) \qquad M = Li, Na, K, Rb, Cs; X = F, Cl, Br, I$$

The Lewis dot representation for the generalized reaction is

$$2M\cdot \; + \; :\overset{\cdot\cdot}{X}:\overset{\cdot\cdot}{X}: \; \longrightarrow \; 2\,(M^+[:\overset{\cdot\cdot}{X}:]^-)$$

Coulomb's Law is $F \propto \dfrac{q^+q^-}{d^2}$. The symbol $\propto$ means "is proportional to."

Because of the opposite charges on Na^+ and Cl^-, an attractive force is developed. According to Coulomb's Law, the force of attraction, F, between two oppositely charged particles of charge magnitudes q^+ and q^- is directly proportional to the product of the charges and inversely proportional to the square of the distance separating their centers, d. Thus, the greater the charges on the ions and the smaller the ions are, the stronger the resulting ionic bonding. Of course, like-charged ions repel each other, so the distances separating ions in solids are those at which the attractions exceed the repulsions by the greatest amount. The structure of common table salt, sodium chloride (NaCl), is shown in Figure 7-1. Like other simple ionic compounds,

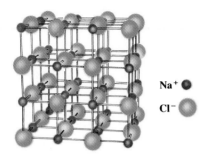

Figure 7-1
The crystal structure of NaCl, expanded for clarity. Each Cl⁻ (green) is surrounded by six sodium ions, and each Na⁺ (gray) is surrounded by six chlorides. The crystal includes billions of ions in the pattern shown. Compare with Figure 2-7, a space-filling drawing of the NaCl structure.

Na⁺ ●

Cl⁻ ●

Figure 7-1 has been expanded to show the spatial arrangement of ions. Adjacent ions actually are in contact with each other. The lines *do not* represent formal chemical bonds. They have been drawn to emphasize the spatial arrangement of ions.

NaCl(s) exists in a regular, extended array of positive and negative ions, Na^+ and Cl^-.

Distinct molecules of solid ionic substances do not exist, so we must refer to *formula units* (Section 2-3) instead of molecules. The sum of all the forces that hold all the particles in an ionic solid is quite large. This explains why such substances have quite high melting and boiling points (a topic that we will discuss more fully in Chapter 13). When an ionic compound is melted or dissolved in water, its charged particles are free to move in an electric field, so such a liquid shows high electrical conductivity (Section 4-2, part 1).

Group IA Metals and Group VIA Nonmetals

Next, consider the reaction of lithium (Group IA) with oxygen (Group VIA) to form lithium oxide, a solid ionic compound (mp >1700°C). We may represent the reaction as

$$4Li(s) + O_2(g) \longrightarrow 2Li_2O(s)$$

lithium   oxygen   lithium oxide

The formula for lithium oxide, Li_2O, indicates that two atoms of lithium combine with one atom of oxygen. If we examine the structures of the atoms

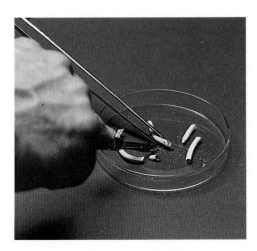

Lithium is a metal, as the shiny surface of freshly cut Li shows. Where it has been exposed to air, the surface is covered with lithium oxide.

Each Li atom has 1 e^- in its valence shell. Each O atom has 6 e^- in its valence shell and needs 2 e^- more to give it a noble gas configuration. The Li^+ ions are formed by oxidation of Li atoms, and the O^{2-} ions are formed by reduction of O atoms.

before reaction, we can see why two lithium atoms react with one oxygen atom.

$$2Li\cdot + \;\;:\!\overset{\cdot\cdot}{\underset{\cdot}{O}}\!\cdot \; \longrightarrow \; 2Li^+[:\overset{\cdot\cdot}{\underset{\cdot\cdot}{O}}:]^{2-}$$

The Lewis dot formulas for the atoms and ions are

Lithium ions, Li^+, are isoelectronic with helium atoms. Oxide ions, O^{2-}, are isoelectronic with neon atoms (10 e^-).

The very small size of the Li^+ ion gives it a much higher *charge density* than that of the larger Na^+ (Figure 6-1). Similarly, O^{2-} is smaller than Cl^-, so its doubly negative charge gives it a much higher charge density. These more concentrated charges and smaller sizes bring the Li^+ and O^{2-} ions closer together than the Na^+ and Cl^- ions are in NaCl. Consequently, the q^+q^- product in the numerator of Coulomb's Law is greater in Li_2O, and the d^2 term in the denominator is smaller. The net result is that the ionic bonding is much stronger in Li_2O than in NaCl. This accounts for the higher melting temperature of Li_2O (>1700°C) compared to NaCl (801°C).

Group IIA Metals and Group VIA Nonmetals

Calcium (Group IIA) reacts with oxygen (Group VIA) to form calcium oxide, a white solid ionic compound with a very high melting point, 2580°C.

$$2Ca(s) + O_2(g) \longrightarrow 2CaO(s)$$

calcium   oxygen   calcium oxide

Again, we write out the electronic structures of the atoms and ions, representing the inner electrons by the symbol of the preceding noble gas (in brackets).

In writing equations in which electrons in different energy levels in different atoms are involved, atomic orbitals are more conveniently labeled *under* the lines that represent them.

The Lewis dot notation for the atoms and ions is

$$Ca\!:\; + \;\;:\!\overset{\cdot\cdot}{\underset{\cdot}{O}}\!\cdot \; \longrightarrow \; Ca^{2+}[:\overset{\cdot\cdot}{\underset{\cdot\cdot}{O}}:]^{2-}$$

Calcium ions, Ca^{2+}, are isoelectronic with argon (18 e^-), the preceding noble gas. Oxide ions, O^{2-}, are isoelectronic with neon (10 e^-), the following noble gas.

Ca^{2+} is about the same size as Na^+ (Figure 6-1) but carries twice the charge, so its charge density is higher. Because the attraction between the two small, highly charged ions Ca^{2+} and O^{2-} is quite high, the ionic bonding is very strong, accounting for the very high melting point of CaO, 2580°C.

Group IIA Metals and Group VA Nonmetals

As our final example of ionic bonding, consider the reaction of magnesium (Group IIA) with nitrogen (Group VA). At elevated temperatures, they form magnesium nitride, Mg_3N_2, a white, solid ionic compound that decomposes at 800°C. We can represent the reaction as

$$3Mg(s) + N_2(g) \xrightarrow{\Delta} Mg_3N_2(s)$$
$$\text{magnesium} \quad \text{nitrogen} \qquad \text{magnesium nitride}$$

The formula Mg_3N_2 indicates that magnesium and nitrogen atoms combine in a 3:2 ratio. We could predict this from the fact that Mg atoms contain two electrons in their highest energy level, whereas N atoms need three electrons to fill theirs completely. Thus, three Mg atoms lose two electrons each for a total of 6 e^-, and two N atoms gain three electrons each, also for a total of 6 e^-. As before, examination of the electronic structures makes the picture clearer.

$$\left. \begin{array}{l} 3 \times \left({}_{12}Mg \quad [Ne] \; \underset{3s}{\uparrow\downarrow} \right) \\[2em] 2 \times \left({}_7N \quad [He] \; \underset{2s}{\uparrow\downarrow} \; \underset{2p}{\uparrow \; \uparrow \; \uparrow} \right) \end{array} \right\} \xrightarrow{\Delta} \left\{ \begin{array}{l} 3 \times \left(Mg^{2+} \quad [Ne] \; \underset{3s}{\underline{}} \right) \quad (2\,e^- \text{ lost}) \times 3 \\[2em] 2 \times \left(N^{3-} \quad [He] \; \underset{2s}{\uparrow\downarrow} \; \underset{2p}{\uparrow\downarrow \; \uparrow\downarrow \; \uparrow\downarrow} \right) \; (3\,e^- \text{ gained}) \times 2 \end{array} \right.$$

Using Lewis dot formulas for the atoms and ions, we have

$$3Mg\!:\, + 2\cdot\overset{\cdot\cdot}{N}\cdot \;\longrightarrow\; 3Mg^{2+}, 2[:\overset{\cdot\cdot}{N}:]^{3-}$$

Both Mg^{2+} and N^{3-} ions are isoelectronic with neon (10 e^-).

Table 7-2 summarizes the general formulas of binary ionic compounds formed by the representative elements. "M" represents metals and "X"

Binary compounds contain two elements.

Table 7-2
Simple Binary Ionic Compounds

Metal		Nonmetal		General Formula	Ions Present	Example	mp (°C)
IA*	+	VIIA	⟶	MX	(M^+, X^-)	LiBr	547
IIA	+	VIIA	⟶	MX_2	$(M^{2+}, 2X^-)$	$MgCl_2$	708
IIIA	+	VIIA	⟶	MX_3	$(M^{3+}, 3X^-)$	GaF_3	800 (subl)
IA*†	+	VIA	⟶	M_2X	$(2M^+, X^{2-})$	Li_2O	>1700
IIA	+	VIA	⟶	MX	(M^{2+}, X^{2-})	CaO	2580
IIIA	+	VIA	⟶	M_2X_3	$(2M^{3+}, 3X^{2-})$	Al_2O_3	2045
IA*	+	VA	⟶	M_3X	$(3M^+, X^{3-})$	Li_3N	840
IIA	+	VA	⟶	M_3X_2	$(3M^{2+}, 2X^{3-})$	Ca_3P_2	~1600
IIIA	+	VA	⟶	MX	(M^{3+}, X^{3-})	AlP	

* Hydrogen is considered a nonmetal. All binary compounds of hydrogen are covalent except certain metal hydrides such as NaH and CaH_2, which contain hydride, H^-, ions.
† As we saw in Section 6-8, part 2, the metals in Groups IA and IIA also commonly form peroxides (containing the O_2^{2-} ion) or superoxides (containing the O_2^- ion). See Table 6-4. The peroxide and superoxide ions contain atoms that are covalently bonded to one another.

represents nonmetals from the indicated groups. In these examples of ionic bonding, each of the metal atoms has lost one, two, or three electrons and each of the nonmetal atoms has gained one, two, or three electrons. *Simple (monatomic) ions rarely have charges greater than 3+ or 3−.* Ions with greater charges interact so strongly with the electron clouds of other ions in compounds that electron clouds are distorted severely, and considerable covalent character in the bonds results. (This point will be developed later.)

The *d*- and *f*-transition elements form many compounds that are essentially ionic in character. Most simple ions of the transition metals do not have noble gas configurations.

The distortion of the electron cloud of an anion by a small, highly charged cation is called *polarization*.

Enrichment

Introduction to Energy Relationships in Ionic Bonding

The following discussion may help you to understand why ionic bonding occurs between elements with low ionization energies and those with high electronegativities. There is a general tendency in nature to achieve stability. One way to do this is by lowering potential energy; *lower* energies generally represent *more stable* arrangements.

A more thorough discussion of these energy changes will be provided in Section 15-10.

Let us use energy relationships to describe why the ionic solid NaCl is more stable than a mixture of individual Na and Cl atoms. Consider a gaseous mixture of one mole of sodium atoms and one mole of chlorine atoms, $Na(g) + Cl(g)$. The energy change associated with the loss of one mole of electrons by one mole of Na atoms to form one mole of Na^+ ions (Step 1 in Figure 7-2) is given by the *first ionization energy* of Na (Section 6-3).

$$Na(g) \longrightarrow Na^+(g) + e^- \qquad \text{first ionization energy} = 496 \, kJ/mol$$

This is a positive value, so the mixture $Na^+(g) + e^- + Cl(g)$ is 496 kJ/mol higher in energy than the original mixture of atoms (the mixture $Na^+ + e^- + Cl$ is *less stable* than the mixture of atoms). The energy change for the gain of one mole of electrons by one mole of Cl atoms to form one mole of Cl^- (Step 2) is given by the *electron affinity* of Cl (Section 6-4).

$$Cl(g) + e^- \longrightarrow Cl^-(g) \qquad \text{electron affinity} = -348 \, kJ/mol$$

This negative value, −348 kJ/mol, lowers the energy of the mixture, but the mixture of separated ions, $Na^+ + Cl^-$, is still *higher* in energy by $(496 − 348) \, kJ/mol = 148 \, kJ/mol$ than the original mixture of atoms (the red arrow in Figure 7-2). Thus, just the formation of ions does not explain why the process occurs. The strong attractive force between ions of opposite charge draws the ions together into the regular array shown in Figure 7-1. The energy associated with this attraction (Step 3) is the *crystal lattice energy* of NaCl, −790 kJ/mol.

$$Na^+(g) + Cl^-(g) \longrightarrow NaCl(s) \qquad \text{crystal lattice energy} = -790 \, kJ/mol$$

The crystal (solid) formation thus further *lowers* the energy to $(148 − 790) \, kJ/mol = -642 \, kJ/mol$. The overall result is that one mole of NaCl(s) is 642 kJ/mol lower in energy (more stable) than the original mixture of atoms (the blue arrow in Figure 7-2). Thus we see that a major driving force for the formation of ionic compounds is the large electrostatic stabilization due to the attraction of the ionic charges (Step 3).

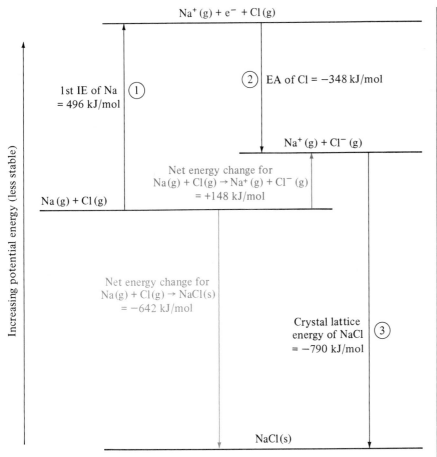

Figure 7-2

A schematic representation of the energy changes that accompany the process $Na^+(g) + Cl^-(g) \rightarrow NaCl(s)$. The red arrow represents the *positive* energy change (unfavorable) for the process of ion formation, $Na(g) + Cl(g) \rightarrow Na^+(g) + Cl^-(g)$. The blue arrow represents the *negative* energy change (favorable) for the overall process, including the formation of the ionic solid.

In this discussion we have not taken into account the fact that sodium is a solid metal or that chlorine actually exists as diatomic molecules. The additional energy changes involved when these are changed to gaseous Na and Cl atoms, respectively, are sufficiently small that the overall energy change starting from Na(s) and $Cl_2(g)$ is still negative. (This will be taken into account in the more thorough discussion in Section 15-10.)

Covalent Bonding

Ionic bonding cannot result from a reaction between two nonmetals, because their electronegativity difference is not great enough for electron transfer to take place. Instead, reactions between two nonmetals result in *covalent bonding*.

> A **covalent bond** is formed when two atoms share one or more pairs of electrons. Covalent bonding occurs when the electronegativity difference, ΔEN, between elements (atoms) is zero or relatively small.

In predominantly covalent compounds the bonds between atoms *within* a molecule (*intra*molecular bonds) are relatively strong, but the forces of attraction *between* molecules (*inter*molecular forces) are relatively weak. As a result, covalent compounds have lower melting and boiling points than ionic compounds. By contrast, distinct molecules of solid ionic substances do not exist. The sum of the attractive forces of all the interactions in an ionic solid is substantial, and such a compound has high melting and boiling points. The relation of bonding types to physical properties of liquids and solids will be developed more fully in Chapter 13.

7-3 Formation of Covalent Bonds

Of course this one electron is un-paired.

Let us look at a simple case of covalent bonding, the reaction of two hydrogen atoms to form the diatomic molecule H_2. As you recall, an isolated hydrogen atom has the ground state electron configuration $1s^1$, with the probability density for this one electron spherically distributed about the hydrogen nucleus (Figure 7-3a). As two hydrogen atoms approach one another, the electron of each hydrogen atom is attracted by the nucleus of the *other* hydrogen atom as well as by its own nucleus (Figure 7-3b). If these two electrons have opposite spins so that they can occupy the same region (orbital), both electrons can now preferentially occupy the region *between* the two nuclei (Figure 7-3c), because they are attracted by both nuclei. The electrons are *shared* between the two hydrogen atoms, and a single covalent bond is formed. In other words, the $1s$ orbitals *overlap* so that both electrons are now in the orbitals of both hydrogen atoms. The closer together the atoms come, the more nearly this is true. In that sense, each hydrogen atom now has the helium configuration, $1s^2$.

The bonded atoms are at lower energy than the separated atoms. This is shown in the plot of energy versus distance in Figure 7-4. However, as the two atoms get closer together, the two nuclei, being positively charged, exert an increasing repulsion on one another. At some distance, a minimum energy, -435 kJ/mol, is reached; it corresponds to the most stable arrangement and occurs at 0.74 Å, the actual distance between two hydrogen nuclei in an H_2 molecule. At greater internuclear separation, the repulsive forces diminish, but the attractive forces decrease even faster. At smaller separations, repulsive forces grow more rapidly than attractive forces.

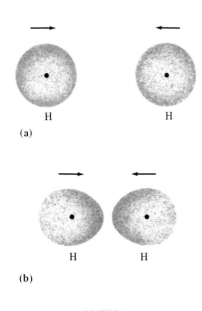

(a)

(b)

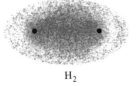

H_2

(c)

Figure 7-3
A representation of the formation of a covalent bond between two hydrogen atoms. The position of each positively charged nucleus is represented by a black dot. Electron density is indicated by the depth of shading. (a) Two hydrogen atoms separated by a large distance (essentially isolated). (b) As the atoms approach one another, the electron of each atom is attracted by the positively charged nucleus of the other atom, so the electron density begins to shift. (c) The two electrons can both occupy the region where the two $1s$ orbitals overlap; the electron density is highest in the region between the two atoms.

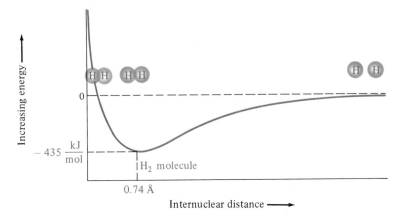

Figure 7-4
The potential energy of the H_2 molecule as a function of the distance between the two nuclei. The lowest point in the curve, -435 kJ/mol, corresponds to the internuclear distance actually observed in the H_2 molecule, 0.74 Å. (The minimum potential energy, -435 kJ/mol, corresponds to a value of -7.23×10^{-19} joule per H_2 molecule.) Energy is compared with that of two separated hydrogen atoms.

Other pairs of nonmetal atoms, and some metal atoms, share electron pairs to form covalent bonds. The result of this sharing is that each atom attains a more stable electron configuration—frequently the same as that of the nearest noble gas. (This is discussed in Section 7-8). Most covalent bonds involve two, four, or six electrons—that is, one, two, or three *pairs* of electrons. Two atoms form a **single covalent bond** when they share one pair of electrons, a **double covalent bond** when they share two electron pairs, and a **triple covalent bond** when they share three electron pairs. These are usually called simply *single, double,* and *triple* bonds. Covalent bonds involving one and three electrons are known, but are relatively rare.

We can represent the sharing of an electron pair by writing the dots in the Lewis formula between the two atom symbols. Thus, the formation of H_2 from two H atoms could be represented as

$$\text{H}\cdot + \cdot\text{H} \longrightarrow \text{H}:\text{H} \qquad \text{or} \qquad \text{H}\!-\!\text{H}$$

where the dash represents a single bond. Similarly, the combination of a hydrogen atom and a fluorine atom to form a hydrogen fluoride (HF) molecule can be shown as

$$\text{H}\cdot + \cdot\ddot{\underset{\cdot\cdot}{\text{F}}}: \longrightarrow \text{H}:\ddot{\underset{\cdot\cdot}{\text{F}}}: \qquad (\text{or } \text{H}\!-\!\ddot{\underset{\cdot\cdot}{\text{F}}}:)$$

We shall see many more examples of this representation.

In our discussion, we have postulated that bonds form by the **overlap** of two atomic orbitals. This is the essence of the **valence bond theory**, which we will describe in more detail in the next chapter. Another theory, **molecular orbital theory**, is discussed in Chapter 9. For now, let us concentrate on the *number* of electron pairs shared and defer the discussion of *which* orbitals are involved in the sharing until the next chapter.

7-4 Polar and Nonpolar Covalent Bonds

Covalent bonds may be either *polar* or *nonpolar.* In a **nonpolar bond** such as that in the hydrogen molecule, H_2, the electron pair is *shared equally* between the two hydrogen nuclei. Recall (Section 6-6) that we defined electronegativity as the tendency of an atom to attract electrons to itself in a chemical bond. Both H atoms have the same electronegativity. This means

$\text{H}:\text{H}$ or $\text{H}\!-\!\text{H}$

that the shared electrons are equally attracted to both hydrogen nuclei and therefore spend equal amounts of time near each nucleus. In this nonpolar covalent bond, the **electron density** is symmetrical about a plane that is perpendicular to a line between the two nuclei. This is true for all homonuclear **diatomic molecules**, such as H_2, O_2, N_2, F_2, and Cl_2, because the two identical atoms have identical electronegativities. We can generalize:

> The covalent bonds in all homonuclear diatomic molecules must be nonpolar.

Let us now consider **heteronuclear diatomic molecules**. Start with the fact that hydrogen fluoride, HF, is a gas at room temperature. This tells us that it is a covalent compound. We also know that the H—F bond has some degree of polarity because H and F are not identical atoms and therefore do not attract the electrons equally. But how polar will this bond be?

The electronegativity of hydrogen is 2.1, and that of fluorine is 4.0 (Table 6-3). Clearly, the F atom, with its higher electronegativity, attracts the shared electron pair much more strongly than does H. We represent the structure of HF as shown in the margin. Notice the unsymmetrical distribution of electron density; the electron density is distorted in the direction of the more electronegative F atom. This small shift of electron density leaves H somewhat positive.

Covalent bonds, such as the one in HF, in which the *electron pairs are shared unequally* are called **polar covalent bonds**. Two kinds of notation used to indicate polar bonds are shown in the margin.

The $\delta-$ over the F atom indicates a "partial negative charge." This means that the F end of the molecule is negative *with respect* to the H end. The $\delta+$ over the H atom indicates a "partial positive charge," or that the H end of the molecule is positive *with respect to* the F end. Please note that we are *not* saying that H has a charge of $1+$ or that F has a charge of $1-$! A second way to indicate the polarity is to draw an arrow so that the head points toward the negative end (F) of the bond and the crossed tail indicates the positive end (H).

The separation of charge in a polar covalent bond creates an electric **dipole**. We expect the dipoles in the covalent molecules HF, HCl, HBr, and HI to be different because F, Cl, Br, and I have different electronegativities. Therefore, atoms of these elements have different tendencies to attract electron pairs shared with hydrogen. We indicate this difference as shown below, where $\Delta(EN)$ is the difference in electronegativity between two atoms that are bonded together.

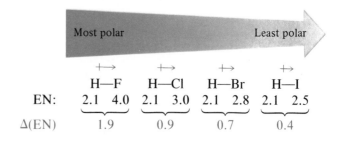

	H—F	H—Cl	H—Br	H—I
EN:	2.1 4.0	2.1 3.0	2.1 2.8	2.1 2.5
$\Delta(EN)$	1.9	0.9	0.7	0.4

Most polar → *Least polar*

A homonuclear molecule contains only one kind of atom. A molecule that contains two or more kinds of atoms is described as heteronuclear.

Remember that ionic compounds are solids at room temperature.

H:F:

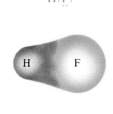

δ+ δ−
H — F   *or*   H—F

The word "dipole" means "two poles." Here it refers to the positive and negative poles that result from the separation of charge within a molecule.

The values of electronegativity are obtained from Table 6-3.

Table 7-3
Dipole Moments and Δ(EN) Values for Some Pure (Gaseous) Substances

Substance	Dipole Moment (μ)*	Δ(EN)
HF	1.91 D	1.9
HCl	1.03 D	0.9
HBr	0.79 D	0.7
HI	0.38 D	0.4
H—H	0 D	0

* The magnitude of a dipole moment is given by the product of charge × distance of separation. Molecular dipole moments are usually expressed in debyes (D).

For comparison, the $\Delta(EN)$ values for some typical $1:1$ ionic compounds are NaBr, 1.8; RbF, 3.1; and KCl, 2.1.

The longest arrow indicates the largest dipole, or greatest separation of electron density in the molecule (see Table 7-3).

7-5 The Continuous Range of Bonding Types

Let us now clarify our classification of bonding types. The degree of electron sharing or transfer depends on the electronegativity difference between the bonding atoms. Nonpolar covalent bonding (involving *equal sharing* of electron pairs) is one extreme, occurring when the atoms are identical (ΔEN is zero). Ionic bonding (involving *complete transfer* of electrons) represents the other extreme, and occurs when two elements with very different electronegativities interact (ΔEN is large).

Polar covalent bonds may be thought of as intermediate between pure (nonpolar) covalent bonds and pure ionic bonds. In fact, bond polarity is sometimes described in terms of **partial ionic character**. This usually increases with increasing difference in electronegativity between bonded atoms. Calculations based on the measured dipole moment (see the next section) of gaseous HCl indicate about 17% "ionic character."

When cations and anions interact strongly, some amount of electron sharing takes place; in such cases we can consider the ionic compound as having some **partial covalent character**. For instance, the high charge density of the very small Li^+ ion causes it to distort large anions that it approaches. The distortion attracts electron density from the anion to the region between it and the Li^+ ion, giving lithium compounds a higher degree of covalent character than in other alkali metal compounds.

Almost all bonds have both ionic and covalent character. By experimental means, a given type of bond can usually be identified as being "closer" to one or the other extreme type. We find it useful and convenient to use the labels for the major classes of bonds to describe simple substances, keeping in mind that they represent ranges of behavior.

In summary, we can describe chemical bonding as a continuum that may be represented as

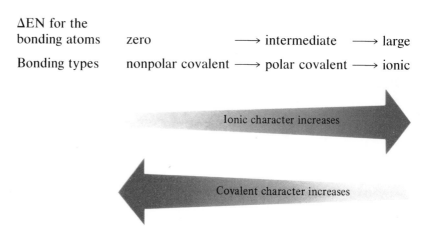

ΔEN for the bonding atoms	zero	$\longrightarrow$	intermediate	$\longrightarrow$	large
Bonding types	nonpolar covalent	$\longrightarrow$	polar covalent	$\longrightarrow$	ionic

Ionic character increases

Covalent character increases

Above all, we must recognize that any classification of a compound that we might suggest based on electronic properties *must* be consistent with the physical properties of ionic and covalent substances described at the beginning of the chapter. For instance, HCl has a rather large electronegativity difference (0.9), and its aqueous solutions conduct electricity. But we know that we cannot view it as an ionic compound because it is a gas, and not a solid, at room temperature. Liquid HCl is a nonconductor.

Let us point out another aspect of the classification of compounds as ionic or covalent. Not all ions consist of single charged atoms. Many are small groups of atoms that are covalently bonded together, yet they still have excess positive or negative charge. Examples of such *polyatomic ions* are ammonium ion, NH_4^+, sulfate ion, SO_4^{2-}, and nitrate ion, NO_3^-. A compound such as potassium sulfate, K_2SO_4, contains potassium ions, K^+, and sulfate ions, SO_4^{2-}, in a 2:1 ratio. We should recognize that this compound contains both covalent bonding (electron sharing *within* the sulfate ions) and ionic bonding (electrostatic attractions *between* potassium and sulfate ions). However, we classify this compound as *ionic,* because it is a high-melting solid (mp 1069°C), it conducts electricity both in molten form and in aqueous solution, and it displays the properties that we generally associate with ionic compounds. Put another way, while covalent bonding holds a part of this substance together (the sulfate ions), the forces that hold the *entire* substance together are ionic.

HCl ionizes in aqueous solution.

7-6 Dipole Moments

It is convenient to express bond polarities on a numerical scale. We indicate the polarity of a molecule by its dipole moment, which measures the separation of charge within the molecule. The **dipole moment**, μ, is defined as the product of the distance, d, separating charges of equal magnitude and opposite sign, and the magnitude of the charge, q. A dipole moment is measured by placing a sample of the substance between two plates and applying a voltage. This causes a small shift in electron density of any molecule, so the applied voltage is diminished very slightly. However, diatomic molecules that contain polar bonds, such as HF, HCl, and CO, tend to orient themselves in the electric field (Figure 7-5). This causes the mea-

$\mu = d \times q$

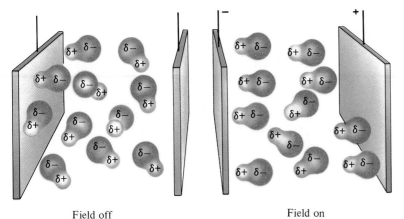

Field off   Field on

Figure 7-5
If polar molecules, such as HF, are subjected to an electric field, they tend to line up very slightly in a direction opposite to that of the field. This minimizes the electrostatic energy of the molecules. Nonpolar molecules are not oriented by an electric field. The effect is greatly exaggerated in this drawing.

sured voltage between the plates to decrease more markedly for these substances, and so we say that these molecules are *polar*. Molecules such as F_2 or N_2 do not reorient, so the change in voltage between the plates remains slight; we say that these molecules are *nonpolar*.

Generally, as electronegativity differences increase in diatomic molecules, the measured dipole moments increase. This can be seen clearly from the data for the hydrogen halides (Table 7-3).

Unfortunately, the dipole moments associated with *individual bonds* can be measured only in simple diatomic molecules. *Entire molecules* rather than selected pairs of atoms must be subjected to measurement. Measured values of dipole moments reflect the *overall* polarities of molecules. For polyatomic molecules they are the result of all the bond dipoles in the molecules. Later we shall see that structural features, such as molecular geometry and the presence of lone (unshared) pairs of electrons, also affect the polarity of a molecule. After we have discussed these additional aspects of covalent bonding, we shall return to the important topic of dipole moments (Section 8-3).

7-7 Lewis Dot Formulas for Molecules and Polyatomic Ions

In Sections 7-1 and 7-2 we drew *Lewis dot formulas* for atoms and monatomic ions. We can also use Lewis dot formulas to show the *valence electrons* in atoms that are covalently bonded in a molecule or a polyatomic ion. A water molecule can be represented by either of the following diagrams.

A polyatomic ion is an ion that contains more than one atom.

$$H:\overset{..}{\underset{..}{O}}: \qquad H—\overset{..}{\underset{|}{O}}:$$
$$\quad H \qquad\qquad\quad H$$
Lewis dot formula   dash formula

An H_2O molecule has two shared electron pairs, i.e., two single covalent bonds. The O atom also has two lone pairs.

In H_2O, the O atom contributes six valence electrons, and each H atom contributes one.

In *dash formulas*, a shared pair of electrons is indicated by a dash. There are two *double* bonds in carbon dioxide, and its Lewis dot formula is

$$\overset{..}{\underset{..}{O}}::C::\overset{..}{\underset{..}{O}} \qquad \overset{..}{\underset{..}{O}}=C=\overset{..}{\underset{..}{O}}$$

A CO_2 molecule has four shared electron pairs, i.e., two double bonds. The central atom (C) has no lone pairs.

In CO_2, the C atom contributes four valence electrons, and each O atom contributes six.

The covalent bonds in a polyatomic ion can be represented in the same way. The Lewis dot formula for the ammonium ion, NH_4^+, shows only eight electrons, even though the N atom has five electrons in its valence shell and each H atom has one, for a total of $5 + 4(1) = 9$ electrons. The NH_4^+ ion, with a charge of $1+$, has one less electron than the original atoms.

The NH₃ molecule, like the NH₄⁺ ion, has eight valence electrons about the N atom.

$$
\begin{array}{ccc}
\overset{\displaystyle\cdot\cdot}{H:N:H} & \text{or} & H-\overset{\displaystyle\cdot\cdot}{\underset{|}{N}}-H \\
\;\;H & & \;\;\;\;H
\end{array}
$$

$$
\left[\begin{array}{c} H \\ \cdot\cdot \\ H:N:H \\ \cdot\cdot \\ H \end{array} \right]^{+} \qquad\qquad
\left[\begin{array}{c} H \\ | \\ H-N-H \\ | \\ H \end{array} \right]^{+}
$$

Lewis dot formula   dash formula

The writing of Lewis formulas is an electron bookkeeping method that is useful as a first approximation to suggest bonding schemes. It is important to remember that Lewis dot formulas only show the number of valence electrons, the number and kinds of bonds, and the order in which the atoms are connected. *They are not intended to depict the three-dimensional shapes of molecules and polyatomic ions.* We will show in Chapter 8, however, that the three-dimensional geometry of a molecule can be predicted from its Lewis structure.

7-8 The Octet Rule

Representative elements usually attain stable noble gas electron configurations when they share electrons. In the water molecule the O has a share in eight outer shell electrons, giving the neon configuration, while H shares two electrons in the helium configuration. Likewise, the C and O of CO_2 and the N of NH_3 and the NH_4^+ ion each have a share in eight electrons in their outer shells. The H atoms in NH_3 and NH_4^+ each share two electrons. Many Lewis formulas are based on the idea that

> the representative elements achieve noble gas configurations in *most* of their compounds.

In some compounds, the central atom does not achieve a noble gas configuration. Such exceptions to the octet rule are discussed later in this chapter.

This statement is usually called the **octet rule**, because the noble gas configurations have $8\ e^-$ in their outermost shells (except for He, which has $2\ e^-$).

For now, we shall restrict our discussion to compounds of the *representative elements*. The octet rule alone does not let us write Lewis formulas. We still need to know how to place the electrons around the bonded atoms—that is, how many of the available valence electrons are **bonding electrons** (shared) and how many are **unshared electrons** (associated with only one atom). A pair of unshared electrons in the same orbital is called a **lone pair**. A simple mathematical relationship is helpful here:

$$S = N - A$$

S is the total number of electrons *shared* in the molecule or polyatomic ion.

N is the number of valence shell electrons *needed* by all the atoms in the molecule or ion to achieve noble gas configurations ($N = 8 \times$ number of atoms not including H, plus $2 \times$ number of H atoms).

> A is the number of electrons *available* in the valence shells of all of the (representative) atoms. This is equal to the sum of their periodic group numbers.

For example, in CO_2, A for each O atom is 6 and A for the carbon atom is 4, so A for CO_2 is $4 + 2(6) = 16$. The following general steps describe the use of this relationship in constructing dot formulas for molecules and polyatomic ions.

Writing Lewis Formulas

1. Select a reasonable (symmetrical) "skeleton" for the molecule or polyatomic ion.
   a. The *least electronegative element* is usually the central element, except that H is never the central element. The least electronegative element is usually the one that needs the most electrons to fill its octet. Example: CS_2 has the skeleton S C S.
   b. Oxygen atoms do not bond to each other except in (1) O_2 and O_3 molecules; (2) the peroxides, which contain the O_2^{2-} group; and (3) the rare superoxides, which contain the O_2^- group. Example: The sulfate ion, SO_4^{2-}, has the skeleton

$$\begin{bmatrix} & O & \\ O & S & O \\ & O & \end{bmatrix}^{2-}$$

   c. In *ternary acids* (oxyacids), hydrogen usually bonds to an O atom, *not* to the central atom. Example: Nitrous acid, HNO_2, has the skeleton H O N O. However, there are a few exceptions to this rule, such as H_3PO_3 and H_3PO_2.

   A ternary acid contains *three* elements—H, O, and another element, often a nonmetal.

   d. For ions or molecules that have more than one central atom, the most symmetrical skeletons possible are used. Examples: C_2H_4 and $P_2O_7^{4-}$ have the following skeletons:

$$\begin{array}{cc} H & H \\ C & C \\ H & H \end{array} \quad \text{and} \quad \begin{bmatrix} & O & & O & \\ O & P & O & P & O \\ & O & & O & \end{bmatrix}^{4-}$$

2. Calculate N, *the number of outer (valence) shell electrons* needed by all atoms in the molecule or ion to achieve noble gas configurations. Examples:

   For compounds containing only representative elements, N is equal to $8 \times$ number of atoms *not* including H, plus $2 \times$ number of H atoms.

   For H_2SO_4,

   $$N = 1 \times 8 \text{ (S atom)} + 4 \times 8 \text{ (O atoms)} + 2 \times 2 \text{ (H atoms)}$$

   $$= 8 + 32 + 4 = 44 \ e^- \text{ needed}$$

   For SO_4^{2-},

   $$N = 8 + 32 = 40 \ e^- \text{ needed}$$

For the representative elements, the number of valence shell electrons in an atom is equal to its periodic group number. Exceptions: 1 for an H atom and 8 for a noble gas (except 2 for He).

3. Calculate A, *the number of electrons available* in the outer (valence) shells of all the atoms. For negatively charged ions, add to the total the number of electrons equal to the charge on the anion; for positively charged ions, subtract the number of electrons equal to the charge on the cation. Examples:

For H_2SO_4,

$$A = 2 \times 1 \text{ (H atoms)} + 1 \times 6 \text{ (S atom)} + 4 \times 6 \text{ (O atoms)}$$

$$= 2 + 6 + 24 = 32 \ e^- \text{ available}$$

For $SO_4{}^{2-}$,

$$A = 1 \times 6 \text{ (S atom)} + 4 \times 6 \text{ (O atoms)} + 2 \text{ (for 2- charge)}$$

$$= 6 + 24 + 2 = 32 \ e^- \text{ available}$$

4. Calculate S, *total number of electrons shared* in the molecule or ion, using the relationship $S = N - A$. Examples:

For H_2SO_4,

$$S = N - A = 44 - 32$$

$$= 12 \text{ electrons shared (6 pairs of } e^- \text{ shared)}$$

For $SO_4{}^{2-}$,

$$S = N - A = 40 - 32$$

$$= 8 \text{ electrons shared (4 pairs of } e^- \text{ shared)}$$

5. Place the S electrons into the skeleton as *shared pairs*. Use double and triple bonds only when necessary. Structures may be shown either by Lewis dot formulas or by dash formulas, in which a dash represents a shared pair of electrons.

C, N, and O often form double and triple bonds. S and Se can form double bonds with C, N, and O.

Please note that a Lewis dot (dash) formula does not indicate the geometry of a molecule or an ion. We will discuss the actual geometries of molecules and ions in Chapter 8.

Formula	Skeleton	Dot Formula ("bonds" in place, but incomplete)	Dash Formula ("bonds" in place, but incomplete)
H_2SO_4	H O S O H (with O above and O below S)	H:O:S:O:H (with O above and O below S)	H—O—S—O—H (with O above and O below S)
$SO_4{}^{2-}$	$\left[\text{O S O (with O above and below)} \right]^{2-}$	$\left[\text{O:S:O (with O above and below)} \right]^{2-}$	$\left[\text{O—S—O (with O above and below)} \right]^{2-}$

6. Place the additional electrons into the skeleton as *unshared (lone) pairs* to fill the octet of every A group element (except H, which can share only 2 e^-). Check that the total number of electrons is equal to A, from Step 3. Examples:

For H_2SO_4,

Check: 16 pairs of e^- have been used. $2 \times 16 = 32\ e^-$ available.

For SO_4^{2-},

$$\begin{bmatrix} \ddot{:}\ddot{O}\ddot{:} \\ :\ddot{O}:S:\ddot{O}: \\ :\ddot{O}: \end{bmatrix}^{2-} \qquad \begin{bmatrix} :\ddot{O}: \\ | \\ :\ddot{O}-S-\ddot{O}: \\ | \\ :\ddot{O}: \end{bmatrix}^{2-}$$

Check: 16 pairs of electrons have been used. $2 \times 16 = 32\ e^-$ available.

Example 7-1

Write the Lewis dot formula and the dash formula for the nitrogen molecule, N_2.

Plan

We follow the stepwise procedure for writing Lewis formulas that was just presented.

Solution

Step 1:   The skeleton is N   N.
Step 2: $N = 2 \times 8 = 16\ e^-$ needed (total) by both atoms
Step 3: $A = 2 \times 5 = 10\ e^-$ available (total) for both atoms
Step 4: $S = N - A = 16\ e^- - 10\ e^- = 6\ e^-$ shared
Step 5: $N:::N$ $6\ e^-$ (3 pairs) are shared; a *triple* bond.
Step 6:   The additional $4\ e^-$ are accounted for by a lone pair on each N. The complete Lewis diagram is

$$:N:::N: \qquad \text{or} \qquad :N\equiv N:$$

Check: $10\ e^-$ (5 pairs) have been used.

Example 7-2

Write the Lewis dot and dash formula for carbon disulfide, CS_2, an ill-smelling liquid.

Plan

Again, we follow the stepwise procedure to apply the relationship $S = N - A$.

Solution

Step 1:   The skeleton is S   C   S.
Step 2: $N = 1 \times 8$ (for C) $+ 2 \times 8$ (for S) $= 24\ e^-$ needed by all atoms
Step 3: $A = 1 \times 4$ (for C) $+ 2 \times 6$ (for S) $= 16\ e^-$ available
Step 4: $S = N - A = 24\ e^- - 16\ e^- = 8\ e^-$ shared
Step 5: $S::C::S$ $8\ e^-$ (4 pairs) are shared; two *double* bonds.
Step 6:   C already has an octet, so the remaining $8e^-$ are distributed as lone pairs on the S atoms to give each S an octet. The complete Lewis formula is

C is the central atom, or the element in the middle of the molecule. It needs four more electrons to acquire an octet, while each S atom needs only two more electrons.

$$\ddot{S}::C::\ddot{S} \qquad \text{or} \qquad \ddot{S}=C=\ddot{S}$$

Check: $16\ e^-$ (8 pairs) have been used. The bonding picture is similar to that of CO_2; this is not surprising, S is below O in Group VIA.

EOC 42, 44

A number of minerals contain the carbonate ion. A very common one is calcium carbonate, $CaCO_3$, the main constituent of limestone and of stalactites and stalagmites.

Example 7-3

Write the Lewis dot formula for the carbonate ion, CO_3^{2-}.

Plan

The same stepwise procedure can be applied to ions. We must remember to adjust A, the total number of electrons, to account for the charge shown on the ion.

Solution

$$O^{2-}$$

Step 1: The skeleton is O C O

Step 2: $N = 1 \times 8$ (for C) $+ 3 \times 8$ (for O) $= 8 + 24 = 32\ e^-$ needed by all atoms

Step 3: $A = 1 \times 4$ (for C) $+ 3 \times 6$ (for O) $+ 2$ (for the $2-$ charge)
$= 4 + 18 + 2 = 24\ e^-$ available

Step 4: $S = N - A = 32\ e^- - 24\ e^- = 8\ e^-$ (4 pairs) shared

Step 5: O^{2-}
$O:\ddot{C}::O$ (Four pairs are shared. At this point it doesn't matter which O is doubly bonded.)

Step 6: The Lewis formula is

$$\left[\begin{array}{c} :\ddot{O}: \\ C \\ :\ddot{O}: \quad :\ddot{O}: \end{array} \right]^{2-} \qquad \text{or} \qquad \left[\begin{array}{c} :\ddot{O}: \\ C \\ :\ddot{O}: \quad \ddot{O}: \end{array} \right]^{2-}$$

Check: $24\ e^-$ (12 pairs) have been used.

EOC 46

7-9 Resonance

In addition to the one shown in Example 7-3, two other Lewis formulas with the same skeleton for the CO_3^{2-} ion are equally acceptable. In these formulas, $4\ e^-$ could be shared between the carbon atom and either of the other two oxygen atoms.

$$\left[\begin{array}{c} \ddot{O} \\ \| \\ C \\ :\ddot{O} \quad \ddot{O}: \end{array} \right]^{2-} \longleftrightarrow \left[\begin{array}{c} :\ddot{O}: \\ C \\ \ddot{O} \quad \ddot{O}: \end{array} \right]^{2-} \longleftrightarrow \left[\begin{array}{c} :\ddot{O}: \\ C \\ :\ddot{O} \quad \ddot{O} \end{array} \right]^{2-}$$

A molecule or polyatomic ion for which two or more dot formulas with the same arrangements of atoms can be drawn to describe the bonding is said to exhibit **resonance**. The three structures above are **resonance structures** of the carbonate ion. The relationship among them is indicated by the double-headed arrows, $\leftrightarrow$. This symbol *does not mean* that the ion flips back and forth among these three structures. The true structure is like an average of the three.

The C—O bonds in CO_3^{2-} are really *neither* double nor single bonds, but are intermediate in bond length (and strength). This has been verified experimentally. Based on measurements in many compounds, the typical C—O single bond length is 1.43 Å, and the typical C=O double bond length is 1.22 Å. The C—O bond length for each bond in the CO_3^{2-} ion is intermediate at 1.29 Å. Another way to represent this situation is by **delocalization** of bonding electrons:

$$\left[\begin{array}{c} O \\ \| \\ O \diagdown C \diagup O \end{array} \right]^{2-}$$

(lone pair on O atoms not shown)

The dashed lines indicate that some of the electrons shared between C and O atoms are *delocalized* among all four atoms; that is, the four pairs of shared electrons are equally distributed among three C—O bonds.

Example 7-4
Draw two resonance structures for the sulfur dioxide molecule, SO_2.

Plan
The stepwise procedure presented in Section 7-8 can be used to write each resonance structure.

Solution

$$\overset{S}{\underset{}{}} \quad \overset{O}{\underset{}{}}$$
$$N = 1(8) + 2(8) = 24 \ e^-$$
$$\underline{A = 1(6) + 2(6) = 18 \ e^-}$$
$$S = \quad N - A \quad = 6 \ e^- \text{ shared}$$

The resonance structures are

$$\ddot{\underset{\cdot\cdot}{O}} :: \ddot{S} : \ddot{\underset{\cdot\cdot}{O}} : \longleftrightarrow : \ddot{\underset{\cdot\cdot}{O}} : \ddot{S} :: \ddot{\underset{\cdot\cdot}{O}} \quad \text{ or } \quad : \ddot{\underset{\cdot\cdot}{O}} = \ddot{S} - \ddot{\underset{\cdot\cdot}{O}} : \longleftrightarrow : \ddot{\underset{\cdot\cdot}{O}} - \ddot{S} = \ddot{\underset{\cdot\cdot}{O}} :$$

We could show delocalization of electrons as follows:

$$O \overset{\cdot\cdot}{=\!\!=\!\!=} \ddot{S} \overset{}{=\!\!=\!\!=} O \qquad \text{(lone pairs on O not shown)}$$

Remember that dot and dash formulas *do not necessarily show shapes*. SO_2 molecules are angular, not linear.

EOC 52, 53, 54

When electrons are shared among more than two atoms, the electrons are said to be *delocalized*. The concept of delocalization is important in molecular orbital theory (Chapter 9).

Trees killed by acid rain in a North Carolina forest. Combustion of sulfur-containing fossil fuels and smelting operations produce sulfur dioxide, SO_2. When this SO_2 is released into the atmosphere, it is one of the major contributors to acid rain. (Carolina Biological Supply Company)

Formal Charges

Enrichment

An experimental determination of the structure of a molecule or polyatomic ion is necessary to establish unequivocally its correct structure. However, we often do not have these results available. The concept of *formal charges* helps us to write Lewis formulas correctly in most cases. This bookkeeping system counts bonding electrons as though they were equally shared between the two bonded atoms. Generally, the most energetically favorable formula for a molecule is usually one in which the formal charge on each atom is zero or as near zero as possible.

Consider the reaction of NH_3 with hydrogen ion, H^+, to form the ammonium ion, NH_4^+.

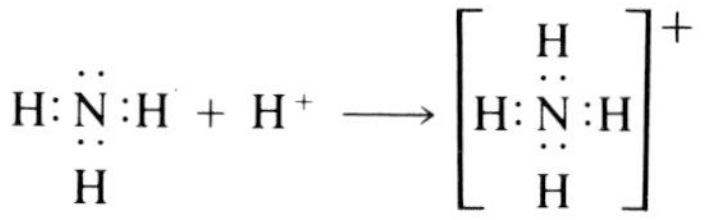

$$H:\overset{..}{\underset{H}{N}}:H + H^+ \longrightarrow \left[H:\overset{..}{\underset{H}{N}}:H\right]^+$$

The unshared pair of electrons on the N atom in the NH_3 molecule is shared with the H^+ ion to form the NH_4^+ ion, in which the N atom has four covalent bonds. Because N is a Group VA element, we expect it to form three covalent bonds to complete its octet. How can we describe the fact that N has four covalent bonds in species like NH_4^+? The answer is obtained by calculating the *formal charge* on each atom in NH_4^+ by the following rules:

Rule for Assigning Formal Charges to Atoms of A Group Elements

1. a. In a molecule, the sum of the formal charges is zero.
   b. In a polyatomic ion, the sum of the formal charges is equal to the charge.

2. The formal charge, abbreviated FC, on an atom in a Lewis formula is given by the relationship

   FC = (group number) − [(number of bonds)
   + (number of unshared e^-)]

   The group number of the noble gases is taken as VIIIA, rather than zero, in calculating formal charges. Formal charges are represented by ⊕ and ⊖ to distinguish between formal charges and real charges on ions.

3. In a Lewis formula, an atom that has the same number of bonds as its periodic group number has no formal charge.

4. When difficulty arises in deciding which atom should be assigned a negative formal charge, the negative formal charge is assigned to the more electronegative element.

5. Atoms that are bonded to each other should not be assigned formal charges with the same sign (the *adjacent charge rule*) if this can be avoided. Lewis formulas in which adjacent atoms have formal charges of the same sign are usually *not* accurate representations.

Let us apply these rules to the ammonia molecule, NH_3, and to the ammonium ion, NH_4^+. Because N is a Group VA element, its group number is 5.

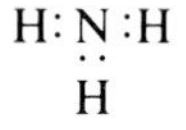

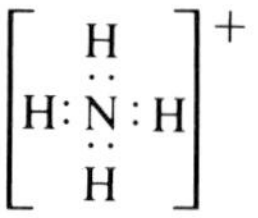

$$H:\overset{..}{\underset{H}{N}}:H \qquad \left[H:\overset{..}{\underset{H}{N}}:H\right]^+$$

In NH_3 the N atom has 3 bonds and 2 unshared e^-, and so for N,

FC = (group number) − [(number of bonds) + (number of unshared e^-)]

$= 5 - (3 + 2) = 0$ (for N)

For H,

FC = (group number) − [(number of bonds) + (number of unshared e^-)]

$= 1 − (1 + 0) = 0$ (for H)

The formal charges of N and H are both zero in NH_3, so the sum of the formal charges is $0 + 3(0) = 0$, consistent with Rule 1a.

In NH_4^+ the N atom has 4 bonds and no unshared e^-, and so for N,

FC = (group number) − [(number of bonds) + (number of unshared e^-)]

$= 5 − (4 + 0) = 1+$ (for N)

Calculation of the FC for H atoms gives zero, as above. The sum of the formal charges in NH_4^+ is $(1+) + 4(0) = 1+$. This is consistent with Rule 1b.

$$\begin{bmatrix} & \overset{\displaystyle H}{\underset{\displaystyle H}{\overset{|}{\underset{|}{H-\overset{\oplus}{N}-H}}}} & \end{bmatrix}^+$$

Thus we see that the octet rule is obeyed in both NH_3 and NH_4^+. The sum of the formal charges in each case is that predicted by Rule 1, even though nitrogen has four covalent bonds in the NH_4^+ ion.

Let us now write a Lewis formula for, and assign formal charges to, the atoms in thionyl chloride, $SOCl_2$, a compound often used in organic synthesis. Both Cl atoms and the O atom are bonded to the S atom. The Lewis formula is

$$\overset{..}{\underset{..}{:Cl}} : \overset{..}{\underset{..}{S}} : \overset{..}{\underset{..}{Cl}} :$$
$$\overset{..}{\underset{..}{:O:}}$$

Formal charges on the atoms are calculated by the usual relationship.

FC = (group number) − [(number of bonds) + (number of unshared e^-)]

For Cl: FC $= 7 − (1 + 6) = 0$

For S: FC $= 6 − (3 + 2) = 1+$

For O: FC $= 6 − (1 + 6) = 1-$

$$\overset{..}{\underset{..}{:Cl}} : \overset{\oplus}{\underset{..}{S}} : \overset{..}{\underset{..}{Cl}} :$$
$$\overset{..}{\underset{\ominus}{:O:}}$$

Dinitrogen oxide (N_2O, also called nitrous oxide) is used as a mild anesthetic. Let us write some possible Lewis formulas for N_2O and assign formal charges to each atom. It is known from experimental molecular structure studies that the oxygen is at the end of the three-atom molecule. We can write three Lewis formulas with O at the end.

(i) $:N \equiv N - \overset{..}{\underset{..}{O}}:$ Left-hand N,   FC $= 5 − (3 + 2) = 0$

N in middle,   FC $= 5 − (4 + 0) = 1+$

O,   FC $= 6 − (1 + 6) = 1-$

The absence of a sign on a number means that the number is *positive*. Therefore, we attached a "+" sign to the 1.

FCs are indicated by $\oplus$ and $\ominus$. The sum of the formal charges in a polyatomic ion is equal to the charge on the ion—1+ in NH_4^+.

(ii) $\ddot{:}N{=}N{=}\ddot{O}\,{\cdot}$ Left-hand N,   FC = 5 − (2 + 4) = 1−

N in middle,   FC = 5 − (4 + 0) = 1+

O,   FC = 6 − (2 + 4) = 0

(iii) $:N{-}\ddot{N}{\equiv}O:$ Left-hand N,   FC = 5 − (1 + 6) = 2−

N in middle,   FC = 5 − (4 + 0) = 1+

O,   FC = 6 − (3 + 2) = 1+

Formula (iii) is less likely than the other two, because it contains one atom that has an unfavorably large formal charge (2− on left-hand N). Thus we write two resonance structures for N_2O:

$$:N{\equiv}N{-}\ddot{O}: \longleftrightarrow \,{\cdot}\,\ddot{N}{=}N{=}\ddot{O}\,{\cdot}$$

Suppose we did not know that the oxygen should be written at the end of the Lewis formula. We might proceed as follows:

(iv) $:N{\equiv}O{-}\ddot{N}:$ Left-hand N,   FC = 5 − (3 + 2) = 0

O,   FC = 6 − (4 + 0) = 2+

Right-hand N,   FC = 5 − (1 + 6) = 2−

(v) $\cdot\,\ddot{N}{=}O{=}\ddot{N}\,{\cdot}$ Left-hand N,   FC = 5 − (2 + 4) = 1−

O,   FC = 6 − (4 + 0) = 2+

Right-hand N,   FC = 5 − (2 + 4) = 1−

Both of these formulas have formal charges that are unnecessarily large. In addition, we should expect that the more electronegative O would have a lower formal charge than any N atom to which it is bonded. On these bases, we might have expected they would be very unlikely formulas. This is consistent with experimental results.

7-10 Limitations of the Octet Rule for Lewis Formulas

Recall that representative elements achieve noble gas electronic configurations in *most* of their compounds. But when the octet rule is not applicable, the relationship $S = N − A$ is not valid without modification. The following are general cases for which the procedure in Section 7-8 *must be modified*— i.e., cases in which there are limitations of the octet rule.

1. Most covalent compounds of beryllium, Be. Because Be contains only two valence shell electrons, it usually forms only two covalent bonds when it bonds to two other atoms. Therefore, we use *four electrons* as the number *needed* by Be in Step 2, Section 7-8. In Steps 5 and 6 we use only two pairs of electrons for Be.

2. Most covalent compounds of the Group IIIA elements, especially boron, B. The IIIA elements contain only three valence shell electrons, so they

often form three covalent bonds when they bond to three other atoms. Therefore, we use *six electrons* as the number *needed* by the IIIA elements in Step 2; and in Steps 5 and 6 we use only three pairs of electrons for the IIIA elements.

3. Compounds or ions containing an odd number of electrons. Examples are NO, with 11 valence shell electrons, and NO_2, with 17 valence shell electrons.

4. Compounds or ions in which the central element needs a share in more than eight valence shell electrons to hold all the available electrons, A. Extra rules are added to Steps 4 and 6 when this is encountered.

   Step 4a: If S, the number of electrons shared, is less than the number needed to bond all atoms to the central atom, then S is increased to the number of electrons needed.

   Step 6a: If S must be increased in Step 4a, then the octets of all the atoms might be satisfied before all A of the electrons have been added. Place the extra electrons on the central element.

Many species that violate the octet rule are quite reactive. For instance, compounds containing atoms with only four valence electrons (limitation 1 above) or six valence electrons (limitation 2 above) frequently react with other species that supply electron pairs. Compounds such as these that accept a share in a pair of electrons are called *Lewis acids;* a *Lewis base* is a species that makes available a share in a pair of electrons. (This kind of behavior will be discussed in detail in Section 10-10.) Molecules with an odd number of electrons often *dimerize* (combine in pairs) to give products that do satisfy the octet rule. Examples are the dimerization of NO to form N_2O_2 (Section 26-8) and of NO_2 to form N_2O_4 (Section 26-10). Examples 7-5 through 7-8 illustrate some limitations and show how such Lewis formulas are constructed.

Lewis formulas are not normally written for compounds containing *d*- and *f*-transition metals. The *d*- and *f*-transition metals utilize *d* and/or *f* orbitals in bonding as well as *s* and *p* orbitals. Thus, they can accommodate more than eight valence electrons.

Example 7-5

Draw the Lewis dot formula and the dash formula for gaseous beryllium chloride, $BeCl_2$, a covalent compound.

Plan

This is an example of limitation 1. So, as we follow the steps in writing the Lewis formula, we must remember to use *four electrons* as the number *needed* by Be in Step 2. Steps 5 and 6 should show only two pairs of electrons for Be.

Solution

Step 1: The skeleton is Cl Be Cl.

Step 2: $N = 2 \times 8$ (for Cl) $+ 1 \times \downarrow$ see limitation 1
4 (for Be) $= 20\ e^-$ needed

Step 3: $A = 2 \times 7$ (for Cl) $+ 1 \times 2$ (for Be) $= 16\ e^-$ available

Step 4: $S = N - A = 20\ e^- - 16\ e^- = 4\ e^-$ shared

Step 5: Cl : Be : Cl

Step 6: : C̈l : Be : C̈l :   or   : C̈l—Be—C̈l :

Calculation of formal charges shows that

$$\text{for Be, FC} = 2 - (2 + 0) = 0 \quad \text{and} \quad \text{for Cl, FC} = 7 - (1 + 6) = 0$$

The chlorine atoms achieve the argon configuration, [Ar], while the beryllium atom has a share of only four electrons. Compounds such as $BeCl_2$, in which the central atom shares fewer than 8 e^-, are sometimes referred to as **electron deficient** compounds. This "deficiency" refers only to satisfying the octet rule for the central atom. The term does not imply that there are fewer electrons than there are protons in the nuclei, as in the case of a cation, because the molecule is neutral.

A Lewis formula can be written for $BeCl_2$ that *does* satisfy the octet rule (see margin). Let us evaluate the formal charges for that formula:

$$\text{for Be, FC} = 2 - (4 + 0) = 2- \quad \text{and} \quad \text{for Cl, FC} = 7 - (2 + 4) = 1+$$

As mentioned earlier, the most favorable structure for a molecule is one in which the formal charge on each atom is zero, if possible. In case some atoms did have nonzero formal charges, we would expect that the more electronegative atoms (Cl) would be the ones with lowest formal charge. Thus, we prefer the Lewis structure shown in Example 7-5 over the one in the margin.

One might expect a similar situation for compounds of the other IIA metals, Mg, Ca, Sr, Ba, and Ra. However, these elements have *lower ionization energies* and *larger radii* than Be, so they usually form ions by losing two electrons.

:Cl̈=Be=C̈l:

Example 7-6
Draw the Lewis dot formula and the dash formula for boron trichloride, BCl_3, a covalent compound.

Plan
This covalent compound of boron is an example of limitation 2. As we follow the steps in writing the Lewis formula, we use *six electrons* as the number *needed* by B in Step 2. Steps 5 and 6 should show only three pairs of electrons for B.

Solution

Step 1:   The skeleton is Cl   B   Cl.
                                  Cl

                                        see limitation 2
                                          ↓
Step 2: $N = 3 \times 8$ (for Cl) $+ 1 \times 6$ (for B) $= 30\ e^-$ needed
Step 3: $A = 3 \times 7$ (for Cl) $+ 1 \times 3$ (for B) $= 24\ e^-$ available
Step 4: $S = N - A = 30\ e^- - 24\ e^- = 6\ e^-$ shared
Step 5:
$$\begin{array}{c} \text{Cl} \\ \text{Cl} : \text{B} : \text{Cl} \end{array}$$

Step 6:   :C̈l:⋅⋅:B̈:C̈l: or   :C̈l—B—C̈l:
                                          :C̈l:

BF_3 and BCl_3 are gases at room temperature. Liquid BBr_3 and solid BI_3 are shown here.

Each chlorine atom achieves the Ne configuration. The boron (central) atom acquires a share of only six valence shell electrons. Calculation of formal charges shows that

$$\text{for B, FC} = 3 - (3 + 0) = 0 \quad \text{and} \quad \text{for Cl, FC} = 7 - (1 + 6) = 0$$

Example 7-7

Write the Lewis dot formula and the dash formula for the covalent compound phosphorus pentafluoride, PF_5.

Plan

We apply the usual stepwise procedure to write the Lewis formula. In PF_5, all five F atoms are bonded to P. This requires the sharing of a minimum of 10 e^-, so this is an example of limitation 4. Therefore we add the extra Step 4a, and increase S from the calculated value of 8 e^- to 10 e^-.

Solution

Step 1: Skeleton is

$$\begin{array}{ccc} & F & F \\ F & P & F \\ & F & \end{array}$$

Step 2: $N = 5 \times 8 \text{ (for F)} + 1 \times 8 \text{ (for P)} = 48 \ e^-$ needed

Step 3: $A = 5 \times 7 \text{ (for F)} + 1 \times 5 \text{ (for P)} = 40 \ e^-$ available

Step 4: $S = N - A = 8 \ e^-$ shared

Five F atoms are bonded to P. This requires the sharing of a minimum of 10 e^-. But only 8 e^- have been calculated in Step 4. Therefore, this is an example of limitation 4.

Step 4a: Increase S from 8 e^- to 10 e^-. The number of electrons available, 40, does not change.

Step 5:

$$\begin{array}{ccc} & F & F \\ F & P & F \\ & F & \end{array}$$

Step 6:

or

When the octets of the five F atoms have been satisfied, all 40 of the available electrons have been added. The phosphorus (central) atom has a share of ten electrons.

Calculation of formal charges shows that

$$\text{for P, FC} = 5 - (5 + 0) = 0 \quad \text{and} \quad \text{for F, FC} = 7 - (1 + 6) = 0$$

EOC 55

When an atom has a share of more than eight electrons, as does P in PF_5, we say that it exhibits an *expanded valence shell*. The electronic basis of the octet rule is that one *s* and three *p* orbitals in the valence shell of an atom can accommodate a maximum of eight electrons. The valence shell of phosphorus has $n = 3$, so it also has available *d* orbitals that can be involved in bonding. It is for this reason that phosphorus (and many other represen-

This is sometimes referred to as *hypervalence*.

tative elements of period 3 and beyond) can exhibit expansion of valence. By contrast, elements in the *second row* of the periodic table can *never* exceed eight electrons in their valence shells, because each atom has only one *s* and three *p* orbitals in that shell. Thus, we understand why NF_3 can exist but NF_5 cannot.

Example 7-8

Write the Lewis dot formula and the dash formula for the triiodide ion, I_3^-.

Plan

We apply the usual stepwise procedure. The calculation of $S = N - A$ in Step 4 shows only $2\ e^-$ shared, but a minimum of $4\ e^-$ are required to bond two I atoms to the central I. Limitation 4 applies, and we proceed accordingly.

Solution

Step 1: The skeleton is $[I\quad I\quad I]^-$.
Step 2: $N = 3 \times 8$ (for I) $= 24\ e^-$ needed
Step 3: $A = 3 \times 7$ (for I) $+ 1$ (for the 1− charge) $= 22\ e^-$ available
Step 4: $S = N - A = 2\ e^-$ shared. Two I atoms are bonded to the central I. This requires a minimum of $4\ e^-$, but only $2\ e^-$ have been calculated in Step 4. Therefore, this is an example of limitation 4.
Step 4a: Increase S from $2\ e^-$ to $4\ e^-$.
Step 5: $[I:I:I]^-$

Step 6: $[:\ddot{I}:\ddot{I}:\ddot{I}:]^-$

Step 6a: Now we have satisfied the octets of all atoms using only 20 of the 22 e^- available. We place the other two electrons on the central I atom.

$$\left[:\ddot{I}:\ddot{\ddot{I}}:\ddot{I}:\right]^-\quad \text{or}\quad \left[:\ddot{I}\!-\!\ddot{\ddot{I}}\!-\!\ddot{I}:\right]^-$$

The central iodine atom in I_3^- has an expanded valence shell.

Calculation of formal charge shows that

for I on ends, FC $= 7 - (1 + 6) = 0$
for I in middle, FC $= 7 - (2 + 6) = 1-$

We have seen that *atoms attached to the central atom nearly always attain noble gas configurations,* even when the central atom does not.

Naming Inorganic Compounds

Because millions of compounds are known, it is important to be able to associate names and formulas in a systematic way.

The rules for naming inorganic compounds were set down in 1957 by the Committee on Inorganic Nomenclature of the International Union of Pure and Applied Chemistry (IUPAC). The concept of oxidation numbers (review Section 4-7) is essential in naming compounds.

7-11 Naming Binary Compounds

Binary compounds consist of two elements; they may be either ionic or covalent. The rule is to name the less electronegative element first and the more electronegative element second. The *more* electronegative element is named by adding an "-ide" suffix to the element's *unambiguous* stem. Stems for the nonmetals follow.

The stem for each element is derived from the name of the element.

IIIA		IVA		VA		VIA		VIIA	
								H	hydr
B	bor	C	carb	N	nitr	O	ox	F	fluor
		Si	silic	P	phosph	S	sulf	Cl	chlor
				As	arsen	Se	selen	Br	brom
				Sb	antimon	Te	tellur	I	iod

Binary ionic compounds contain metal cations and nonmetal anions. The cation is named first and the anion second according to the rule described.

Formula	Name	Formula	Name
KBr	potassium bromide	Rb_2S	rubidium sulfide
$CaCl_2$	calcium chloride	Al_2Se_3	aluminum selenide
NaH	sodium hydride	SrO	strontium oxide

The preceding method is sufficient for naming binary ionic compounds containing metals that exhibit *only one oxidation number* other than zero (Section 4-7). Most transition elements, and a few of the more electronegative representative metals, exhibit more than one oxidation number. These metals may form two or more binary compounds with the same nonmetal. To distinguish among all the possibilities, the oxidation number of the metal is indicated by a Roman numeral in parentheses following its name. This method can be applied to any binary compound of a metal and a nonmetal, whether the compound is ionic or covalent.

Roman numerals are *not* necessary for metals that commonly exhibit only one oxidation number in their compounds.

Formula	Ox. No. of Metal	Name	Formula	Ox. No. of Metal	Name
Cu_2O	+1	copper(I) oxide	$SnCl_2$	+2	tin(II) chloride
CuF_2	+2	copper(II) fluoride	$SnCl_4$	+4	tin(IV) chloride
FeS	+2	iron(II) sulfide	PbO	+2	lead(II) oxide
Fe_2O_3	+3	iron(III) oxide	PbO_2	+4	lead(IV) oxide

The advantage of the IUPAC system is that if you know the formula you can write the exact and unambiguous name; if you are given the name you can write the formula at once.

An older method, still in use but not recommended by the IUPAC, uses "-ous" and "-ic" suffixes to indicate lower and higher oxidation numbers, respectively. This system can distinguish between only two different oxidation numbers for a metal. Therefore, it is not as useful as the Roman numeral system.

Familiarity with the older system is still necessary. It is still widely used in many scientific, engineering, and medical fields.

Formula	Ox. No. of Metal	Name	Formula	Ox. No. of Metal	Name
CuCl	+1	cuprous chloride	SnF_2	+2	stannous fluoride
$CuCl_2$	+2	cupric chloride	SnF_4	+4	stannic fluoride
FeO	+2	ferrous oxide	Hg_2Cl_2	+1	mercurous chloride
$FeBr_3$	+3	ferric bromide	$HgCl_2$	+2	mercuric chloride

Pseudobinary ionic compounds contain more than two elements. In these compounds one or more of the ions consist of more than one element but behave as simple ions. Some common examples of such anions are the hydroxide ion, OH^-; the cyanide ion, CN^-; and the thiocyanate ion, SCN^-. As before, the name of the anion ends in "-ide." The ammonium ion, NH_4^+, is the common cation that behaves like a simple metal cation.

The Common Pseudobinary Anions
hydroxide OH^-
cyanide CN^-
thiocyanate SCN^-
The Common Pseudobinary Cation
ammonium NH_4^+

Formula	Name	Formula	Name
NH_4I	ammonium iodide	NH_4CN	ammonium cyanide
$Ca(CN)_2$	calcium cyanide	$Cu(OH)_2$	copper(II) hydroxide or cupric hydroxide
$NaOH$	sodium hydoxide	$Fe(OH)_3$	iron(III) hydroxide or ferric hydroxide

Nearly all **binary covalent compounds** involve two *nonmetals* bonded together. Although many nonmetals can exhibit different oxidation numbers, their oxidation numbers are *not* properly indicated by Roman numerals or suffixes. Instead, elemental proportions in binary covalent compounds are indicated by using a *prefix* system for both elements. The Greek and Latin prefixes used are mono, di, tri, tetra, penta, hexa, hepta, octa, nona, and deca. The prefix "mono-" is omitted for both elements except in the common name for CO, carbon monoxide. We use the minimum number of prefixes needed to name a compound unambiguously.

If you don't already know them, you should learn these prefixes.

Number	Prefix
2	di
·3	tri
4	tetra
5	penta
6	hexa
7	hepta
8	octa
9	nona
10	deca

Formula	Name	Formula	Name
SO_2	sulfur dioxide	Cl_2O_7	dichlorine heptoxide
SO_3	sulfur trioxide	CS_2	carbon disulfide
N_2O_4	dinitrogen tetroxide	As_4O_6	tetraarsenic hexoxide

Chemists sometimes name binary covalent compounds that contain two nonmetals by the system used to name compounds of metals that show variable oxidation numbers; i.e., the oxidation number of the less electronegative element is indicated by a Roman numeral in parentheses. We do not recommend this procedure because it is incapable of naming compounds *unambiguously,* which is the principal requirement for a system of naming compounds. For example, both NO_2 and N_2O_4 are called nitrogen(IV) oxide by this system, but the name does not distinguish between the two compounds. The compound P_4O_{10} is tetraphosphorus decoxide, which indicates clearly its composition. Using the Roman numeral system, it would be called phosphorus(V) oxide, which could suggest the incorrect formula, P_2O_5. The simplest formula for P_4O_{10} is P_2O_5, but the name for a covalent compound must indicate clearly the composition of its molecules, not just its simplest formula.

Binary acids are compounds in which H is bonded to the more electronegative nonmetals; they act as acids when dissolved in water. The pure compounds are named as typical binary compounds. Their aqueous solutions are named by modifying the characteristic stem of the nonmetal with the prefix "hydro-" and the suffix "-ic" followed by the word "acid." The stem for sulfur in this instance is "sulfur" rather than "sulf."

Formula	Name of Compound	Name of Aqueous Solution
HCl	hydrogen chloride	hydrochloric acid, $HCl(aq)$
HF	hydrogen fluoride	hydrofluoric acid, $HF(aq)$
H_2S	hydrogen sulfide	hydrosulfuric acid, $H_2S(aq)$
HCN	hydrogen cyanide	hydrocyanic acid, $HCN(aq)$

A list of common cations and anions appears in Table 7-4. It will enable you to name many of the ionic compounds you encounter. In later chapters we shall learn the systematic rules for naming some other types of compounds.

Table 7-4
Formulas, Ionic Charges, and Names for Some Common Ions

Common Cations			Common Anions		
Formula	Charge	Name	Formula	Charge	Name
Li^+	1+	lithium ion	F^-	1−	fluoride ion
Na^+	1+	sodium ion	Cl^-	1−	chloride ion
K^+	1+	potassium ion	Br^-	1−	bromide ion
NH_4^+	1+	ammonium ion	I^-	1−	iodide ion
Ag^+	1+	silver ion	OH^-	1−	hydroxide ion
			CN^-	1−	cyanide ion
Mg^{2+}	2+	magnesium ion	ClO^-	1−	hypochlorite ion
Ca^{2+}	2+	calcium ion	ClO_2^-	1−	chlorite ion
Ba^{2+}	2+	barium ion	ClO_3^-	1−	chlorate ion
Cd^{2+}	2+	cadmium ion	ClO_4^-	1−	perchlorate ion
Zn^{2+}	2+	zinc ion	CH_3COO^-	1−	acetate ion
Cu^{2+}	2+	copper(II) ion or	MnO_4^-	1−	permanganate ion
		cupric ion	NO_2^-	1−	nitrite ion
Hg_2^{2+}	2+	mercury(I) ion or	NO_3^-	1−	nitrate ion
		mercurous ion	SCN^-	1−	thiocyanate ion
Hg^{2+}	2+	mercury(II) ion or			
		mercuric ion	O^{2-}	2−	oxide ion
Mn^{2+}	2+	manganese(II) ion or	S^{2-}	2−	sulfide ion
		manganous ion	HSO_3^-	1−	hydrogen sulfite ion
Co^{2+}	2+	cobalt(II) ion or			or bisulfite ion
		cobaltous ion	SO_3^{2-}	2−	sulfite ion
Ni^{2+}	2+	nickel(II) ion or	HSO_4^-	1−	hydrogen sulfate ion or
		nickelous ion			bisulfate ion
Pb^{2+}	2+	lead(II) ion or	SO_4^{2-}	2−	sulfate ion
		plumbous ion	HCO_3^-	1−	hydrogen carbonate ion
Sn^{2+}	2+	tin(II) ion or			or bicarbonate ion
		stannous ion	CO_3^{2-}	2−	carbonate ion
Fe^{2+}	2+	iron(II) ion or	CrO_4^{2-}	2−	chromate ion
		ferrous ion	$Cr_2O_7^{2-}$	2−	dichromate ion
Fe^{3+}	3+	iron(III) ion or	PO_4^{3-}	3−	phosphate ion
		ferric ion	AsO_4^{3-}	3−	arsenate ion
Al^{3+}	3+	aluminum ion			
Cr^{3+}	3+	chromium(III) ion or			
		chromic ion			

7-12 Naming Ternary Acids and Their Salts

A ternary compound consists of three elements. Ternary acids (**oxyacids**) are compounds of hydrogen, oxygen, (usually) and a nonmetal. Nonmetals that exhibit more than one oxidation state form more than one ternary acid. These ternary acids differ in the number of oxygen atoms they contain. The suffixes "-ous" and "-ic" following the stem name of the central element indicate lower and higher oxidation states, respectively. One common ternary acid of each nonmetal is (somewhat arbitrarily) designated as the "-ic" acid." That is, it is named "stem*ic* acid." The common ternary "-ic acids" are shown below. There are no common "-ic" ternary acids for the omitted nonmetals.

The oxyacid with the central element in the highest oxidation state usually contains more O atoms. Oxyacids with their central elements in lower oxidation states usually have fewer O atoms.

It is important to learn the names and formulas of these acids, because the names of all other ternary acids and salts are derived from them.

Periodic Group of Central Elements

IIIA	IVA	VA	VIA	VIIA
(+3) H_3BO_3 boric acid	(+4) H_2CO_3 carbonic acid	(+5) HNO_3 nitric acid		
	(+4) H_4SiO_4 silicic acid	(+5) H_3PO_4 phosphoric acid	(+6) H_2SO_4 sulfuric acid	(+5) $HClO_3$ chloric acid
		(+5) H_3AsO_4 arsenic acid	(+6) H_2SeO_4 selenic acid	(+5) $HBrO_3$ bromic acid
			(+6) H_6TeO_6 telluric acid	(+5) HIO_3 iodic acid

Note that the oxidation state of the central atom is equal to its periodic group number, except in the halogens.

Acids containing *one fewer oxygen atom* per central atom are named in the same way except that the "-ic" suffix is changed to "-ous." The oxidation number of the central element is *lower by 2* in the "-ous" acid than in the "-ic" acid.

Formula	Ox. No.	Name	Formula	Ox. No.	Name
H_2SO_3	+4	sulfur*ous* acid	H_2SO_4	+6	sulfur*ic* acid
HNO_2	+3	nitr*ous* acid	HNO_3	+5	nitr*ic* acid
H_2SeO_3	+4	selen*ous* acid	H_2SeO_4	+6	selen*ic* acid
$HBrO_2$	+3	brom*ous* acid	$HBrO_3$	+5	brom*ic* acid

Ternary acids that have one fewer O atom than the "-ous" acids (two fewer O atoms than the "-ic" acids) are named using the prefix "hypo-" and the suffix "-ous." These are acids in which the oxidation state of the central nonmetal is lower *by 2* than that of the central nonmetal in the "-ous acids."

Formula	Ox. No.	Name
$HClO$	+1	*hypo*chlor*ous* acid
H_3PO_2	+1	*hypo*phosphor*ous* acid
HIO	+1	*hypo*iod*ous* acid
$H_2N_2O_2$	+1	*hypo*nitr*ous* acid

Notice that $H_2N_2O_2$ has a 1:1 ratio of nitrogen to oxygen, as would the hypothetical HNO.

Acids containing *one more oxygen atom* per central nonmetal atom than the normal "-ic acid" are named "*per*stem*ic*" acids.

Formula	Ox. No.	Name
$HClO_4$	+7	*per*chlor*ic* acid
$HBrO_4$	+7	*per*brom*ic* acid
HIO_4	+7	*per*iod*ic* acid

The oxyacids of chlorine follow.

Formula	Ox. No.	Name
HClO	+1	*hypo*chlor*ous* acid
HClO$_2$	+3	chlor*ous* acid
HClO$_3$	+5	chlor*ic* acid
HClO$_4$	+7	*per*chlor*ic* acid

Ternary salts are compounds that result from replacing the hydrogen in a ternary acid with another ion. They usually contain metal cations or the ammonium ion. As with binary compounds, the cation is named first. The name of the anion is based on the name of the ternary acid from which it is derived.

An anion derived from a ternary acid with an "-ic" ending is named by dropping the "-ic acid" and replacing it with "-ate." An anion derived from an "-ous acid" is named by replacing the suffix "-ous acid" with "-ite." The "per-" and "hypo-" prefixes are retained.

Formula	Name
(NH$_4$)$_2$SO$_4$	ammonium sulfate (SO$_4{}^{2-}$, from H$_2$SO$_4$)
KNO$_3$	potassium nitrate (NO$_3{}^-$, from HNO$_3$)
Ca(NO$_2$)$_2$	calcium nitrite (NO$_2{}^-$, from HNO$_2$)
LiClO$_4$	lithium perchlorate (ClO$_4{}^-$, from HClO$_4$)
FePO$_4$	iron(III) phosphate (PO$_4{}^{3-}$, from H$_3$PO$_4$)
NaClO	sodium hypochlorite (ClO$^-$, from HClO)

Acidic salts contain anions derived from ternary acids in which one or more acidic hydrogen atoms remain. These salts are named as if they were the usual type of ternary salt, with the word "hydrogen" or "dihydrogen" inserted after the name of the cation to show the number of acidic hydrogen atoms.

Formula	Name	Formula	Name
NaHSO$_4$	sodium hydrogen sulfate	KH$_2$PO$_4$	potassium dihydrogen phosphate
NaHSO$_3$	sodium hydrogen sulfite	K$_2$HPO$_4$	potassium hydrogen phosphate
		NaHCO$_3$	sodium hydrogen carbonate

An older, commonly used method (which is not recommended by the IUPAC) involves the use of the prefix "bi-" attached to the name of the anion to indicate the presence of an acidic hydrogen. According to this system, NaHSO$_4$ is called sodium bisulfate and NaHCO$_3$ is called sodium bicarbonate.

Key Terms

Binary acid   A binary compound in which H is bonded to one of the more electronegative nonmetals.

Binary compound   A compound consisting of two elements; may be ionic or covalent.

Bonding pair   A pair of electrons involved in a covalent bond.

Chemical bonds   Attractive forces that hold atoms together in elements and compounds.

Covalent bond   A chemical bond formed by the sharing of one or more electron pairs between two atoms.

Covalent compound   A compound containing predominantly covalent bonds.

Debye   The unit used to express dipole moments.

Delocalization of electrons   Refers to bonding electrons distributed among more than two atoms that are bonded together; occurs in species that exhibit resonance.

Dipole   Refers to the separation of charge between two covalently bonded atoms.

Dipole moment (μ)   The product of the distance separating opposite charges of equal magnitude and the magnitude of the charge; a measure of the polarity of a bond or molecule. A measured dipole moment refers to the dipole moment of an entire molecule.

Double bond   A covalent bond resulting from the sharing of four electrons (two pairs) between two atoms.

Electron deficient compound   A compound containing at least one atom (other than H) that shares fewer than eight electrons.

Formal charge   A method of counting electrons in a covalently bonded molecule or ion; bonding electrons are counted as though they were shared equally between the two atoms.

Heteronuclear   Consisting of different elements.

Homonuclear   Consisting of only one element.

Ionic bonding   Chemical bonding resulting from the transfer of one or more electrons from one atom or group of atoms to another.

Ionic compound   A compound containing predominantly ionic bonding.

Isoelectronic   Having the same electronic configurations.

Lewis acid   A substance that accepts a share in a pair of electrons from another species.

Lewis base   A substance that makes available a share in an electron pair.

Lewis dot formula   The representation of a molecule, ion, or formula unit by showing atomic symbols and only outer shell electrons; does not show shape.

Lone pair   A pair of electrons residing on one atom and not shared by other atoms; unshared pair.

Nonpolar bond   A covalent bond in which electron density is symmetrically distributed.

Octet rule   Many representative elements attain at least a share of eight electrons in their valence shells when they form molecular or ionic compounds; there are some limitations.

Oxidation numbers   Arbitrary numbers that can be used as mechanical aids in writing formulas and balancing equations. For single-atom ions they correspond to the charge on the ion; more electronegative atoms are assigned negative oxidation numbers.

Oxidation states   See *Oxidation numbers.*

Polar bond   A covalent bond in which there is an unsymmetrical distribution of electron density.

Pseudobinary ionic compound   A compound that contains more than two elements but is named like a binary compound.

Resonance   A concept in which two or more equivalent dot formulas for the same arrangement of atoms (resonance structures) are necessary to describe the bonding in a molecule or ion.

Single bond   A covalent bond resulting from the sharing of two electrons (one pair) between two atoms.

Ternary acid   A ternary compound containing H, O, and another element, often a nonmetal.

Ternary compound   A compound consisting of three elements; may be ionic or covalent.

Triple bond   A covalent bond resulting from the sharing of six electrons (three pairs) between two atoms.

Unshared pair   See *Lone pair.*

Exercises

Chemical Bonding—Basic Ideas

1. Give a suitable definition of chemical bonding. What type of force is responsible for chemical bonding?

2. List the two basic types of chemical bonding. Give an example of a substance with each type of bonding. What are the differences between these types? What are some of the general properties associated with the two main types of bonding?

3. Why are covalent bonds called directional bonds, whereas ionic bonding is termed nondirectional?

4. The outermost electron configuration of any alkali metal atom, Group IA, is ns^1. How can an alkali metal atom attain a noble gas configuration?

5. The outermost electron configuration of any halogen atom, Group VIIA, is ns^2np^5. How can a halogen atom attain a noble gas configuration?

6. (a) What do Lewis dot representations for atoms show? (b) Draw Lewis dot representations for the following atoms: He, C, S, Ar, Br, Sr.

7. Draw Lewis dot representations for the following atoms: Li, B, P, K, Kr, As.

8. Describe the types of bonding in calcium hydroxide, $Ca(OH)_2$.

$$Ca^{2+} \qquad 2[:\overset{..}{\underset{..}{O}}—H]^-$$

9. Describe the types of bonding in sodium chlorate, $NaClO_3$.

$$Na^+ \qquad \left[\begin{array}{c} :\overset{..}{O}: \\ | \\ :\overset{..}{\underset{..}{O}}—\overset{..}{\underset{..}{Cl}}—\overset{..}{\underset{..}{O}}: \end{array} \right]^-$$

10. Based on the positions in the periodic table of the following pairs of elements, predict whether bonding between the two would be primarily ionic or covalent. Justify your answers. (a) Ba and S, (b) P and O, (c) Cl and Br, (d) Li and I, (e) Si and Cl, (f) Ca and F.

11. Predict whether the bonding between the following pairs of elements would be ionic or covalent. Justify your answers. (a) K and S, (b) N and O, (c) Ca and Br, (d) P and S, (e) C and Br, (f) K and N.

12. Classify the following compounds as ionic or covalent: (a) $MgSO_4$, (b) SO_2, (c) KNO_3, (d) $NiCl_2$, (e) H_2CO_3, (f) NCl_3, (g) Li_2O, (h) H_3PO_4, (i) $SOCl_2$.

Ionic Bonding

13. Describe what happens to the valence electron(s) as a metal atom and a nonmetal atom combine by ionic bonding.

*14. Describe an ionic crystal. What factors might determine the geometrical arrangement of the ions?

15. Why are most solid ionic compounds rather poor conductors of electricity? Why does conductivity increase when an ionic compound is melted or dissolved in water?

16. Write chemical equations for reactions between the following pairs of elements. Draw electronic structures of the atoms before reaction as well as electronic structures of the ions formed in the reactions, using both the orbital and $ns^x np^y$ notations, where n refers to the outermost occupied shell.
   (a) calcium and fluorine
   (b) barium and oxygen
   (c) sodium and sulfur

17. Write chemical equations for reactions between the following pairs of elements. Draw electronic structures of the atoms before reaction as well as electronic structures of the ions formed in the reactions, using both the orbital and $ns^x np^y$ notations, where n refers to the outermost occupied shell.
   (a) lithium and fluorine
   (b) magnesium and sulfur
   (c) calcium and chlorine

18. Write the formula for the ionic compound that forms between each of the following pairs of elements: (a) Cs and Cl_2, (b) Ba and F_2, (c) K and S.

19. Write the formula for the ionic compound that forms between each of the following pairs of elements: (a) Mg and Cl_2, (b) Sr and Cl_2, (c) Na and Se.

20. When a d-transition metal undergoes ionization, it loses its outer s electrons before it loses any d electrons. Using [noble gas]$(n - 1)d^x$ representations, write the outer electron configurations for the following ions: (a) Co^{2+}, (b) Mn^{2+}, (c) Zn^{2+}, (d) Fe^{2+}, (e) Cu^{2+}, (f) Sc^{3+}, (g) Cu^+.

21. Which of the following do not accurately represent stable binary ionic compounds? Why? $BaCl_2$, KF_2, AlF_4, SrS_2, Ca_2O_3, $NaBr_2$, Li_2S.

22. Which of the following do not accurately represent stable binary ionic compounds? Why? MgI, BaO, InF_2, Al_2O_3, $RbCl_2$, $CsSe$, Be_2O.

23. (a) Write Lewis formulas for the positive and negative ions in these salts: $SrBr_2$, Li_2S, Ca_3P_2, PbF_2, Bi_2O_3. (b) Which ions do not have a noble gas configuration?

24. (a) What are isoelectronic species? List three pairs of isoelectronic species. (b) All but one of the following species are isoelectronic. Which one is not isoelectronic with the others? Ne, Al^{3+}, O^{2-}, Na^+, Mg^+, F^-.

25. All but one of the following species are isoelectronic. Which one is not isoelectronic with the others? S^{2-}, Ga^{2+}, Ar, K^+, Ca^{2+}, Sc^{3+}.

26. Write formulas for two cations and two anions that are isoelectronic with neon.

27. Write formulas for two cations and two anions that are isoelectronic with argon.

28. Write formulas for two cations that have the following electron configurations in their *highest* occupied energy level: (a) $4s^2 4p^6$, (b) $5s^2 5p^6$.

29. Write formulas for two anions that have the electron configurations listed in Exercise 28.

Covalent Bonding—General Concepts

30. What does Figure 7-4 tell us about the attractive and repulsive forces in a hydrogen molecule?

31. Distinguish between heteronuclear and homonuclear diatomic molecules.

32. How many electrons are shared between two atoms in (a) a single covalent bond, (b) a double covalent bond, and (c) a triple covalent bond?

33. What is the maximum number of covalent bonds that a second-period element could form? How can the representative elements beyond the second period form more than this number of covalent bonds?

Lewis Dot Formulas for Molecules and Polyatomic Ions

34. What information about chemical bonding can a Lewis formula for a compound or ion give? What information about bonding is not directly represented by a Lewis formula?

35. What is the octet rule? Is it generally applicable to compounds of the transition metals? Why?

36. (a) What is the simple mathematical relationship that is useful in writing Lewis dot formulas? (b) What does each term in the relationship represent?

37. Write Lewis dot formulas for the following: H_2, N_2, Cl_2, HCl, HBr.

38. Write Lewis dot formulas for the following: H_2O, NH_3, OH^-, F^-.

39. Use Lewis formulas to represent the covalent molecules formed by these pairs of elements. Write only structures that satisfy the octet rule. (a) P and H, (b) Se and Br, (c) C and F, (d) Si and F.

40. Use Lewis formulas to represent the covalent molecules formed by these pairs of elements. Write only structures that satisfy the octet rule. (a) O and F, (b) As and Cl, (c) I and Cl, (d) N and F.

41. Find the total number of valence electrons for (a) $SnCl_4$, (b) NH_2^-, (c) CH_5O^+ (d) XeO_3, (e) CN_2H_2.

42. Write Lewis formulas for the following covalent molecules: (a) H_2S, (b) PCl_3, (c) BCl_3, (d) SiH_4, (e) SF_4.

43. Write Lewis formulas for the following molecules: (a) CH_4S, (b) CH_2ClBr, (c) S_2Cl_2, (d) NH_2Cl, (e) C_2H_6S (two possibilities).

44. Write Lewis formulas without formal charges for these multiply bonded compounds: (a) CS_2, (b) $ClNO$, (c) C_2H_2O (two possibilities), (d) C_3H_4 (two possibilities).

45. Write Lewis formulas for (a) ClO_4^-, (b) NOF, (c) SeF_4, (d) $COCl_2$, (e) ClF_3.

46. Write Lewis structures for (a) H_2O_2, (b) IO_4^-, (c) BeH_2, (d) NCl_3, (e) HClO, (f) XeF_4.

47. Write Lewis formulas for (a) H_2NOH (i.e., one H bonded to O), (b) S_8 (a ring of eight atoms), (c) SiH_4, (d) F_2O_2 (O atoms in center, F atoms on outside), (e) CO, (f) $SeCl_6$.

48. Write Lewis formulas for (a) BBr_3, (b) SF_6, (c) CN^-, (d) AlH_4^-, (e) N_2H_4, (f) H_2SO_3.

*49. (a) Write the Lewis formula for $AlCl_3$, a molecular compound. Note that in $AlCl_3$, the aluminum atom is an exception to the octet rule. (b) In the gaseous phase, two molecules of $AlCl_3$ join together (dimerize) to form Al_2Cl_6. (The two molecules are joined by two "bridging" Al—Cl—Al bonds.) Write the Lewis formula for this molecule.

*50. (a) Write the Lewis formula for molecular ClO_2. There is a single unshared electron on the chlorine atom in this molecule. (b) Two of these molecules dimerize to form Cl_2O_4. (The two molecules are joined by a Cl—Cl bond.) Write the Lewis formula for this molecule.

51. What do we mean by the term "resonance"? Do the resonance structures that we draw actually represent the bonding in the substance? Explain your answer.

*52. We can write two resonance structures for toluene, $C_6H_5CH_3$:

How would you expect the carbon–carbon bond length in the six-membered ring to compare with the carbon–carbon bond length between the CH_3 group and the carbon atom on the ring?

53. Write resonance structures for the formate ion, $HCOO^-$.

54. Write resonance structures for the nitrate ion, NO_3^-.

55. Which of the following species contain at least one atom that violates the octet rule?

(a) $: \ddot{F} - \ddot{F} :$ (b) $: \ddot{O} - \ddot{C}l - \ddot{O} :$ (c) $: \ddot{F} - \cdot Xe - \ddot{F} :$

(d) $\left[\begin{array}{c} :\ddot{O}: \\ | \\ :\ddot{O} - S - \ddot{O}: \\ | \\ :\ddot{O}: \end{array} \right]^{2-}$

56. Which of the following species contain at least one atom that violates the octet rule?

(a)

(b) $\ddot{O} = C = \ddot{O}$

(c) $\left[\begin{array}{c} :\ddot{O}: \\ | \\ :\ddot{O} - Cl - \ddot{O}: \end{array} \right]^-$

(d) $: \ddot{C}l - B - \ddot{C}l : \\ \quad\quad | \\ \quad\quad :\ddot{C}l:$

*57. None of the following is known to exist. What is wrong with each one?

(a) $: \ddot{C}l - \ddot{S} = \ddot{O}.$

(b) $H - H - \ddot{O} - P - \ddot{C}l : \\ \quad\quad\quad\quad\quad | \\ \quad\quad\quad\quad\quad :\ddot{C}l:$

(c) $: O \equiv N - \ddot{O}:^-$

(d) $Na - \ddot{S}: \\ \quad\quad | \\ \quad\quad Na$

*58. None of the following is known to exist. What is wrong with each one?

(a) $: \ddot{F} - B - \ddot{F} : \\ \quad\quad | \\ \quad\quad :\ddot{F}:$ with H on top

(b) $: \ddot{O} - \ddot{C}l :^{2-} \\ \quad\quad | \\ \quad\quad :\ddot{O}:$

(c) $H - \ddot{O} - H - C \equiv N :$

(d) $: \ddot{B}r - \ddot{F} : \\ \quad\quad | \\ \quad\quad :\ddot{F}:$

*59. "El" is the general symbol for a representative element. In each case, in which periodic group is El located? Justify your answers and cite a specific example for each one.

(a) $\left[\begin{array}{c} :\ddot{O} - \ddot{E}l - \ddot{O}: \\ | \\ :\ddot{O}: \end{array} \right]^-$

(b) $H - \ddot{O} - El - \ddot{O}: \\ \quad\quad\quad | \\ \quad\quad\quad :\ddot{O}:$ (with $:\ddot{O}:$ on top)

(c) $H - \ddot{O} - \ddot{E}l = \ddot{O}.$

*60. "El" is the general symbol for a representative element. In each case, in which periodic group is El located? Justify your answers and cite a specific example for each one.

(a) $\left[\begin{array}{c} :\!\ddot{O}\!: \\ | \\ :\!\ddot{O}\!-\!El\!-\!\ddot{O}\!: \\ | \\ :\!\ddot{O}\!: \end{array}\right]^{3-}$ (c) $\left[\begin{array}{c} :\!\ddot{O}\!-\!El\!-\!\ddot{O}\!: \\ | \\ :\!\ddot{O}\!: \end{array}\right]^{2-}$

(b) $\left[\begin{array}{c} :\!\ddot{O}\!: \\ | \\ :\!\ddot{O}\!-\!El\!-\!\ddot{O}\!: \\ | \\ :\!\ddot{O}\!: \end{array}\right]^{2-}$

61. Many common stains, such as those of chocolate and other fatty foods, can be removed by dry-cleaning solvents such as tetrachloroethylene, C_2Cl_4. Is C_2Cl_4 ionic or covalent? Draw its Lewis dot formula.

*62. Draw acceptable dot formulas for the following common air pollutants: (a) SO_2, (b) NO_2, (c) CO, (d) O_3 (ozone), (e) SO_3, (f) $(NH_4)_2SO_4$. Which one is a solid? Which ones exhibit resonance?

Formal Charges

63. Assign a formal charge to each atom in the following:

(a) $:\!\ddot{Cl}\!-\!\ddot{O}\!: \\ | \\ :\!\ddot{Cl}\!:$ (d) $\left[\begin{array}{c} :\!\ddot{O}\!=\!C\!-\!\ddot{O}\!: \\ | \\ :\!\ddot{O}\!: \end{array}\right]^{2-}$

(b) $:\!\ddot{O}\!-\!\ddot{S}\!=\!\ddot{O}$

(c) $:\!\ddot{O}\!-\!\ddot{Cl}\!-\!\ddot{O}\!-\!\ddot{Cl}\!-\!\ddot{O}\!: \\ \quad | \qquad\quad | \\ \quad :\!\ddot{O}\!: \quad\;\; :\!\ddot{O}\!:$ (e) $\left[\begin{array}{c} :\!\ddot{O}\!: \\ | \\ :\!\ddot{O}\!-\!\ddot{Cl}\!-\!\ddot{O}\!: \\ | \\ :\!\ddot{O}\!: \end{array}\right]^{-}$

64. Assign a formal charge to each atom in the following:

(a) $:\!\ddot{F}\!-\!Sb\!-\!\ddot{F}\!: \\ \quad\;\; | \\ \quad :\!\ddot{F}\!:$ (b) $\ddot{F}\overset{\displaystyle\cdot\cdot\;\;\cdot\cdot}{\underset{\displaystyle :\!\ddot{F}\!:}{\ddot{F}\!-\!P\!-\!\ddot{F}}}$ (c) $\ddot{O}\!=\!C\!=\!\ddot{O}$

(d) $\left[:\!\ddot{O}\!=\!N\!=\!\ddot{O}\!:\right]^{+}$ (e) $\left[\begin{array}{c} :\!\ddot{Cl}\!: \\ | \\ :\!\ddot{Cl}\!-\!Al\!-\!\ddot{Cl}\!: \\ | \\ :\!\ddot{Cl}\!: \end{array}\right]^{-}$

65. Find the formal charges on the N and B atoms in (a) NH_3, (b) BF_3, (c) NH_4^+, (d) BF_4^-, (e) NH_2^-, (f) $H_3B\!-\!NH_3$.

66. Find the formal charge of each atom, other than H and halogen atoms, in the following:
(a) $Cl_3P\!-\!O$

(b) $\left[\begin{array}{c} CH_3O\!-\!H \\ | \\ H \end{array}\right]^{+}$ (c) $\left[\begin{array}{c} O \\ || \\ O\!-\!S\!-\!O \\ | \\ O \end{array}\right]^{2-}$

*67. With the aid of formal charges, explain which Lewis formula is more likely to be correct for each given molecule.

(a) For Cl_2O, $:\!\ddot{Cl}\!-\!\ddot{O}\!-\!\ddot{Cl}\!:$ or $:\!\ddot{Cl}\!-\!\ddot{Cl}\!-\!\ddot{O}\!:$

(b) For HN_3, $H\!-\!\ddot{N}\!=\!N\!=\!\ddot{N}$ or $H\!-\!\ddot{N}\!\equiv\!N\!-\!\ddot{N}\!:$

(c) For N_2O, $\ddot{N}\!=\!O\!=\!\ddot{N}$ or $:\!N\!\equiv\!N\!-\!\ddot{O}\!:$

68. Write Lewis formulas for three different atomic arrangements with the molecular formula HCNO. Indicate all formal charges. Predict which arrangement is likely to be the least stable and justify your selection.

Ionic versus Covalent Character and Bond Polarities

69. Distinguish between polar and nonpolar covalent bonds.

70. Why is an HCl molecule polar while a Cl_2 molecule is nonpolar?

71. Why do we show only partial charges, and not full charges, on the atoms of a polar molecule?

72. (a) Which two of the following pairs of elements are most likely to form ionic bonds? Te and H, C and F, Ba and F, N and F, K and O. (b) Of the remaining three pairs, which one forms the least polar, and which the most polar, covalent bond?

73. (a) List three reasonable nonpolar covalent bonds between dissimilar atoms. (b) List three pairs of elements whose compounds should exhibit extreme ionic character.

74. The elements X, Y, and Z are in the same period of the periodic table and are in groups IIA, VIA, and VIIA, respectively. (a) Write the Lewis formula for the compound most likely to be formed between X and Z. Will this compound most likely be ionic or covalent? (b) Write the Lewis formula for the most probable compound of Y and Z. Will this compound be ionic or covalent?

75. Classify the bonding between the following pairs of atoms as ionic, polar covalent, or nonpolar covalent. (a) Si and O, (b) N and O, (c) Sr and F, (d) As and As.

76. Classify the bonding between the following pairs of atoms as ionic, polar covalent, or nonpolar covalent. (a) Li and O, (b) Br and I, (c) Ca and H, (d) O and O, (e) H and O.

77. State whether the structure of each substance is covalent or ionic, and give a reason for each answer. (a) HBr, mp $-85.5°C$, bp $-67°C$; (b) $MgCl_2$, mp $708°C$, bp $1418°C$ (the molten substance conducts electricity).

78. Look up the properties of NaCl and PCl_3 in a handbook of chemistry. Why is NaCl classified as an ionic compound and PCl_3 as a covalent compound?

79. The following properties can be found in a handbook of chemistry:

   camphor, $C_{10}H_{16}O$—colorless crystals; specific gravity 0.990 at 25°C; sublimes 204°C; insoluble in water; very soluble in alcohol and ether.

   praseodymium chloride, $PrCl_3$—blue-green needle crystals; specific gravity 4.02; melting point 786°C; boiling point 1700°C; solubility in cold water, 103.9 g/100 mL H_2O; very soluble in hot water.

   Would you classify each of these as ionic or covalent? Why?

80. (a) Write Lewis formulas for atoms of strontium, chlorine, and silicon. (b) Use the appropriate pairs of these atoms to show formation of (i) an ionic compound and (ii) a covalent molecule. (c) Indicate the polarity of the bond in the covalent molecule.

Oxidation Numbers (review)

81. What are oxidation numbers? How can they be useful?

82. What oxidation numbers are the following elements expected to exhibit in simple binary ionic compounds? Cl, O, Sr, K, Al, Se.

83. What oxidation numbers are the following elements expected to exhibit in simple binary ionic compounds? Ba, Li, Na, S, I, P.

84. Evaluate the oxidation number of N in NO, NO_2, N_2O_3, N_2O_5, N_2H_4, and H_2NOH.

85. Evaluate the oxidation number of (a) C in CH_4, CH_3OH, H_2CO, CO, and CO_2; (b) Cl in Cl_2, HClO, $HClO_2$, $HClO_3$, and $HClO_4$.

86. Evaluate the oxidation number of the underlined element in (a) $\underline{C}O_2$, (b) $\underline{S}OCl_2$, (c) $Na_2\underline{S}$, (d) $K_3\underline{P}O_4$, (e) $K_2\underline{P}HO_3$, (f) $K\underline{P}H_2O_2$, (g) $\underline{P}OCl_3$, (h) $K_2\underline{Mn}O_4$, and (i) H$\underline{C}$OOH. Assume that the other elements have their common oxidation numbers.

Naming Inorganic Compounds

87. Name the following monatomic cations, using the IUPAC system of nomenclature. (a) Li^+, (b) Cd^{2+}, (c) Fe^{2+}, (d) Mn^{2+}, (e) Al^{3+}.

88. Name the following monatomic cations, using the IUPAC system of nomenclature. (a) Au^+, (b) Au^{3+}, (c) Ba^{2+}, (d) Zn^{2+}, (e) Ag^+.

89. Write the chemical symbol for each of the following: (a) sodium ion, (b) zinc ion, (c) silver ion, (d) mercury(II) ion, (e) iron(III) ion.

90. Write the chemical symbol for each of the following: (a) lithium ion, (b) bismuth(III) ion, (c) iron(II) ion, (d) chromium(III) ion, (e) potassium ion.

91. Name the following ions: (a) N^{3-}, (b) O^{2-}, (c) Se^{2-}, (d) F^-, (e) Br^-.

92. Write the chemical symbol for each of the following: (a) chloride ion, (b) sulfide ion, (c) telluride ion, (d) iodide ion, (e) phosphide ion.

93. Name the following ionic compounds: (a) Li_2S, (b) SnO_2, (c) RbI, (d) Li_2O, (e) Ba_3N_2.

94. Name the following ionic compounds: (a) NaI, (b) Hg_2S, (c) Li_3N, (d) $MnCl_2$, (e) CuF_2, (f) FeO.

95. Write the chemical formula for each of the following compounds: (a) sodium fluoride, (b) zinc oxide, (c) barium oxide, (d) magnesium bromide, (e) hydrogen iodide, (f) copper(I) chloride.

96. Write the chemical formula for each of the following compounds: (a) sodium oxide, (b) calcium phosphide, (c) iron(II) oxide, (d) iron(III) oxide, (e) manganese(IV) oxide, (f) silver fluoride.

97. Name the following compounds: (a) NH_4CN, (b) $K_2Cr_2O_7$, (c) $Ca_3(PO_4)_2$, (d) $CaCO_3$, (e) $NaNO_3$.

98. Name the following compounds: (a) $(NH_4)_2SO_4$, (b) $Al(NO_3)_3$, (c) $Fe(ClO_4)_2$, (d) Li_2CO_3, (e) BaO_2.

99. Write the chemical formula for each of the following compounds: (a) potassium sulfite, (b) calcium permanganate, (c) sodium peroxide, (d) ammonium dichromate, (e) ammonium acetate.

100. Write the chemical formula for each of the following compounds: (a) iron(II) chlorate, (b) potassium nitrite, (c) barium phosphate, (d) copper(I) sulfate, (e) sodium carbonate.

101. Name the following common acids: (a) HCl, (b) H_3PO_4, (c) $HClO_4$, (d) HNO_3, (e) H_2SO_3.

102. Write the chemical formula for each of the following acids and bases: (a) nitrous acid, (b) sulfuric acid, (c) bromic acid, (d) sodium hydroxide, (e) calcium hydroxide.

103. What is the name of the acid with the formula H_2CO_3? Write the formulas of the two anions derived from it and name these ions.

104. What is the name of the acid with the formula H_3PO_3? What is the name of the HPO_3^{2-} ion?

105. Name the following binary molecular compounds: (a) CO, (b) CO_2, (c) SF_6, (d) $SiCl_4$, (e) IF.

106. Name the following binary molecular compounds: (a) AsF_3, (b) Br_2O, (c) BrF_5, (d) CSe_2, (e) Cl_2O_7.

107. Write the chemical formula for each of the following compounds: (a) iodine bromide, (b) silicon dioxide, (c) phosphorus trichloride, (d) tetrasulfur dinitride, (e) bromine trifluoride, (f) hydrogen telluride, (g) xenon tetrafluoride.

108. Write the chemical formula for each of the following compounds: (a) diboron trioxide, (b) dinitrogen pentasulfide, (c) phosphorus triiodide, (d) sulfur tetrachloride, (e) silicon sulfide, (f) hydrogen sulfide, (g) diphosphorus trioxide.

109. Write formulas for the compounds that are expected to be formed by the following pairs of ions:

	A. F^-	B. OH^-	C. SO_3^{2-}	D. PO_4^{3-}	E. NO_3^-
1. NH_4^+		Omit – see note			
2. K^+					
3. Mg^{2+}					
4. Cu^{2+}					
5. Fe^{3+}					
6. Ag^+					

NOTE: The compound NH_4OH does not exist. The solution commonly labeled "NH_4OH" is aqueous ammonia, $NH_3(aq)$.

110. Write the names for the compounds of Exercise 109.

111. Write the formula for each of the following substances and balance each equation.
(a) aluminum + iron(III) oxide →
aluminum oxide + iron
(b) potassium hydroxide + zinc chlorate →
zinc hydroxide + potassium chlorate
(c) silver nitrate + hydrogen sulfide →
silver sulfide + nitric acid
(d) sodium carbonate + hydrogen chloride →
sodium chloride + carbon dioxide + water

Mixed Exercises

112. Write a name for each formula or a formula for each name: (a) $Al(OH)_3$, (b) nitrogen trichloride, (c) tin(IV) oxide, (d) chromium(VI) oxide, (e) PbS, (f) $NaNO_2$.

113. Find the charge of the ion formed in each of the following ionization reactions, and name each of the resulting ions:
(a) $Fe \rightarrow Fe$ ion $+ 2\ e^-$ (c) $P + 3\ e^- \rightarrow P$ ion
(b) $S + 2\ e^- \rightarrow S$ ion   (d) $Al \rightarrow Al$ ion $+ 3\ e^-$

114. The following properties can be found in a handbook of chemistry:
barium chloride, $BaCl_3$—colorless cubic crystals; specific gravity 3.92; melting point 963°C; boiling point 1560°C; very soluble in water; very slightly soluble in alcohol.
Would you classify this substance as ionic or covalent? Why?

115. Describe the types of bonding in sodium sulfite, Na_2SO_3.

$$2[Na]^+ \quad \left[\ddot{\underset{\cdot\cdot}{O}}{-}\underset{\underset{\displaystyle :\ddot{O}:}{|}}{S}{-}\ddot{\underset{\cdot\cdot}{O}}: \right]^{2-}$$

116. Write the formula for the compound that forms between (a) calcium and nitrogen, (b) aluminum and oxygen, (c) potassium and selenium, (d) strontium and bromine. Name each compound.

117. Write Lewis formulas for the covalent molecules (a) silicon tetrahydride, (b) iodine chloride, (c) phosphorus pentafluoride, (d) carbon disulfide.

118. Draw the Lewis formulas for the nitric acid molecule (HNO_3) that are consistent with the following bond length data: 1.405 Å for the bond between the nitrogen atom and the oxygen atom that is attached to the hydrogen atom; 1.206 Å for the bonds between the nitrogen atom and each of the other oxygen atoms.

119. Which of the following species contain(s) at least one atom that violates the octet rule?

(a) $\underset{\ddot{\underset{\cdot\cdot}{Cl}}\cdot}{\overset{\ddot{Cl}\cdot}{\diagdown}} C{=}\ddot{O}$ (b) $:\ddot{F}{-}\underset{\underset{:\ddot{F}\cdot\ \cdot\ddot{F}:}{|}}{P}{-}\ddot{F}:$ (c) $\left[\underset{H}{\overset{H}{\underset{|}{\overset{|}{H{-}O{-}H}}}} \right]^+$

120. Name the following compounds: (a) K_2CrO_4, (b) Na_2SO_3, (c) $FeCO_3$, (d) $Fe_2(SO_4)_3$, (e) $FeSO_4$.
121. Write the chemical formula for each of the following compounds: (a) silver nitrate, (b) uranium(IV) sulfate, (c) aluminum acetate, (d) manganese(II) phosphate.
122. Write the formula of each substance and balance each equation:
    (a) ammonium sulfate + sodium hydroxide $\rightarrow$
        ammonia + sodium sulfate + water

(b) sulfur tetrafluoride + water $\rightarrow$
        hydrogen fluoride + sulfur dioxide
(c) phosphorus trichloride + chlorine $\rightarrow$
        phosphorus pentachloride
(d) hydrogen chloride + manganese(IV) oxide $\rightarrow$
        manganese(II) chloride + chlorine + water

8 Molecular Structure and Covalent Bonding Theories

The primary genetic material of all cells is deoxyribonucleic acid, DNA. A DNA molecule is a double helix of nucleotides, each of which contains one molecule of a base, one molecule of deoxyribose (a sugar), and one molecule of phosphoric acid. A model of a small portion of this helix is shown here. Genetic information is encoded in DNA in the order in which pairs of bases (shown here in red and yellow) are stacked.

Objectives

As you study this chapter, you should learn

☐ The basic ideas of the valence shell electron pair repulsion (VSEPR) theory

☐ To use the VSEPR theory to predict electronic geometry of polyatomic molecules and ions

☐ To use the VSEPR theory to predict molecular geometry of polyatomic molecules and ions

☐ The relationships between molecular shapes and molecular polarities

☐ To predict whether a molecule is polar or nonpolar

☐ The basic ideas of the valence bond (VB) theory

☐ To analyze the hybrid orbitals used in bonding in polyatomic molecules and ions

☐ To use hybrid orbitals to describe the bonding in double and triple bonds

We know a great deal about the molecular structures of many thousands of compounds, all based on reliable experiments. In our discussion of theories of covalent bonding, we must keep in mind that the theories represent *an attempt to explain and organize experimental observations*. For bonding theories to be valid, they must be consistent with the large body of experimental observations about molecular structure. In this chapter, we shall study two theories of covalent bonding, which allow us to predict structures and properties that are usually accurate. Like any simplified theories, they are not entirely satisfactory in describing *every* known structure; however, their successful application to many thousands of structures justifies their continued use.

8-1 An Overview of the Chapter

The electrons in the outer shell, or **valence shell**, of an atom are the electrons involved in bonding. In most of our discussion of covalent bonding, we will focus attention on these electrons. Valence shell electrons may be thought of as those that were not present in the *preceding* noble gas, ignoring *filled* sets of d and f orbitals. Lewis dot formulas show the number of valence shell electrons in a polyatomic molecule or ion (Sections 7-7 through 7-10). We shall draw Lewis dot formulas for each molecule or polyatomic ion we discuss. The theories introduced in this chapter apply equally well to polyatomic molecules and to ions.

Two theories go hand in hand in a discussion of covalent bonding. The *valence shell electron pair repulsion (VSEPR) theory* helps us to understand and predict the spatial arrangement of atoms in a polyatomic molecule or ion. However, it does not explain *how* bonding occurs, just *where* it occurs, as well as where lone pairs of valence shell electrons are directed. The *valence bond (VB) theory* describes *how* the bonding takes place, in terms of *overlapping atomic orbitals*. In this theory, the atomic orbitals discussed in Chapter 5 are often "mixed," or *hybridized,* to form new orbitals with different spatial orientations. These two simple ideas, used together, enable us to understand the bonding, molecular shapes, and properties of a wide variety of polyatomic molecules and ions.

We shall first discuss the basic ideas and application of these two theories. Then we shall learn how an important molecular property, *polarity,* depends on molecular shape. The major part of this chapter will then be devoted to studying how these ideas are applied to various types of polyatomic molecules and ions.

In Chapter 7 we used valence bond terminology to discuss the bonding in H_2 although we did not name the theory there.

Important Note

Different instructors prefer to cover these two theories in different ways. We believe that one of these two alternative approaches is most effective. Your instructor will tell you the order in which you should study the material in this chapter. However you study this chapter, Tables 8-1, 8-2, 8-3, and 8-4 are important summaries, and you should refer to them often.

A. One approach is to discuss both the VSEPR theory and the VB theory together, emphasizing how they complement one another. If your instructor prefers this parallel approach, you should study the chapter in the order in which it is presented.

B. An alternative approach is to first master the VSEPR theory and the related topic of molecular polarity for many different structures, and then learn how the VB theory describes the overlap of bonding orbitals in these structures. If your instructor takes this approach, you should study this chapter in the following order:

1. Read the summary material under the main heading "Molecular Shapes and Bonding" preceding Section 8-5.

2. *VSEPR theory, molecular polarity.* Study Sections 8-2 and 8-3; then in Sections 8-5 through 8-12, study only the subsections marked A and B.

3. *VB theory.* Study Section 8-4; then in Sections 8-5 through 8-12, study the valence bond subsections, marked C; then study Sections 8-13 and 8-14.

No matter which order your instructor prefers, the following procedure will help you to analyze the structure and bonding in any compound.

Learn this procedure and use it as a mental "checklist." Trying to do this reasoning in a different order often leads to confusion or wrong answers.

1. Draw the Lewis formula for the molecule or polyatomic ion, and identify a *central atom*—an atom that is bonded to more than one other atom (Section 8-2).

2. Count the *number of regions of high electron density* on the central atom (Section 8-2).

3. Apply the VSEPR theory to determine the arrangement of the *regions of high electron density* (the *electronic geometry*) about the central atom (Section 8-2; Tables 8-1 and 8-4).

4. Using the Lewis formula as a guide, determine the arrangement of the *bonded atoms* (the *molecular geometry*) about the central atom, as well as the location of the unshared valence electron pairs on that atom (parts B of Sections 8-5 through 8-12; Tables 8-3 and 8-4). This description includes predicted bond angles.

Never try to skip to Step 5 until you have done Step 4. The electronic geometry and the molecular geometry may or may not be the same; knowing the electronic geometry first will enable you to find the correct molecular geometry.

5. If there are lone pairs of electrons on the central atom, consider how their presence might modify somewhat the *ideal* molecular geometry and bond angles deduced in Step 4 (Section 8-2; parts B of Sections 8-8 through 8-12).

6. Use the VB theory to determine the *hybrid orbitals* utilized by the central atom; describe the overlap of these orbitals to form bonds and the orbitals that contain unshared valence shell electron pairs on the central atom (parts C of Sections 8-5 through 8-12; Sections 8-13; 8-14; Tables 8-2 and 8-4).

7. If more than one atom can be identified as a central atom, repeat Steps 2 through 6 for each central atom, to build up a picture of the geometry and bonding in the entire molecule.

8. When all central atoms have been accounted for, use the entire molecular geometry, electronegativity differences, and the presence of lone pairs of valence shell electrons on the central atom to predict *molecular polarity* (Section 8-3; parts B of Sections 8-5 through 8-12).

The following diagram summarizes this procedure.

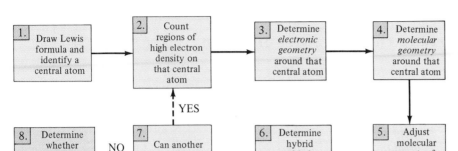

In Section 7-7 we showed that Lewis dot formulas of polyatomic ions can be constructed in the same way as dot formulas of neutral molecules. We must take into account the "extra" electrons on anions and the "missing" electrons of cations. Once the Lewis formula of an ion is known, we use the VSEPR and VB theories to deduce its electronic geometry, shape, and hybridization, just as for neutral molecules.

8-2 Valence Shell Electron Pair Repulsion (VSEPR) Theory

The basic ideas of the **valence shell electron pair repulsion (VSEPR) theory** follow:

Each set of valence shell electrons on a central atom is significant. The sets of valence shell electrons on the *central atom* repel one another. They are arranged about the *central atom* so that repulsions among them are as small as possible.

This results in maximum separation among regions of high electron density about the central atom.

A **central atom** is any atom that is bonded to more than one other atom. In some molecules, more than one central atom may be present. In such cases, we determine the arrangement around each in turn, to build up a picture of the overall shape of the molecule or ion. We first count the number of **regions of high electron density** around the *central atom,* as follows:

1. Each bonded atom is counted as *one* region of high electron density, *whether the bonding is single, double, or triple.*
2. Each unshared pair of valence electrons on the central atom is counted as *one* region of high electron density.

As examples of this way of counting, consider the following molecules and polyatomic ions.

Formula:	CH_4	NH_3	CO_2	SO_4^{2-}

Lewis dot formula:

$$H-\overset{\overset{\displaystyle H}{|}}{\underset{\underset{\displaystyle H}{|}}{C}}-H \qquad :\overset{\overset{\displaystyle H}{|}}{\underset{\underset{\displaystyle H}{|}}{N}}-H \qquad \overset{..}{\overset{}{O}}::C::\overset{..}{O}{\overset{}{:}} \qquad \left[\overset{\overset{\displaystyle :\ddot{O}:}{|}}{\underset{\underset{\displaystyle :\ddot{O}:}{|}}{:\ddot{O}-S-\ddot{O}:}} \right]^{2-}$$

Central atom:	C	N	C	S
Number of atoms bonded to *central atom:*	4	3	2	4
Number of lone pairs on *central atom:*	0	1	0	0
Total number of regions of high electron density on *central atom:*	4	4	2	4

According to VSEPR theory, the structure is as stable as possible when the regions of high electron density on the central atom are as far apart as possible. For instance, two regions of high electron density are most stable on opposite sides of the central atom (the linear arrangement). Three regions are most stable when they are arranged at the corners of an equilateral triangle (the trigonal planar arrangement). The arrangement of these *regions of high electron density* around the central atom is referred to as the **electronic geometry** of the central atom.

Table 8-1 shows the relationship between the common numbers of regions of high electron density and the corresponding electronic geometries. After we know the electronic geometry (and *only then*), we consider how many of these regions of high electron density connect (bond) the central atom to other atoms. This lets us deduce the arrangement of *atoms* around the central atom, called the **molecular geometry**. If necessary, we repeat this procedure for each central atom in the molecule or ion. These procedures are illustrated in parts B of Sections 8-5 through 8-12.

> Although the terminology is not as precise as we might wish, we use "molecular geometry" to describe the arrangement of atoms in *polyatomic ions* as well as in molecules.

8-3 Polar Molecules—The Influence of Molecular Geometry

In Chapter 7 we saw that the unequal sharing of electrons between two atoms with different electronegativities, $\Delta EN > 0$, results in a *polar bond*. For heteronuclear diatomic molecules such as HF, this bond polarity results in a *polar molecule*. Then the entire molecule acts as a dipole, and we would measure the *dipole moment* of such a molecule to be nonzero. When a molecule consists of more than two atoms joined by polar bonds, we must also take into account the *arrangement* of the resulting bond dipoles. For such a case, we first use VSEPR theory to deduce the atomic arrangement (molecular geometry), as described in the preceding section and exemplified in parts A and B of Sections 8-5 through 8-12. Then we determine whether

Table 8-1
Number of Regions of High Electron Density about a Central Atom

Number of Regions of High Electron Density	Electronic Geometry*	Angles†
2	linear	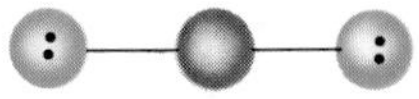 180°
3	trigonal planar	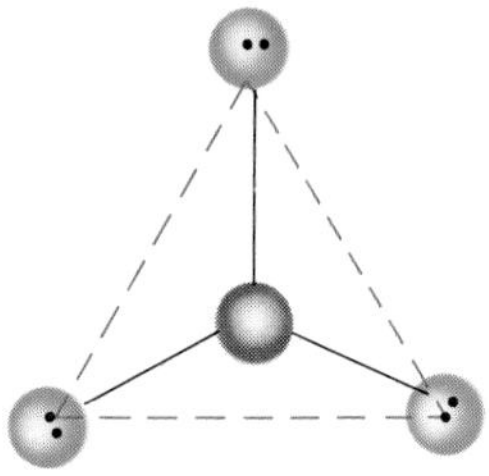 120°
4	tetrahedral	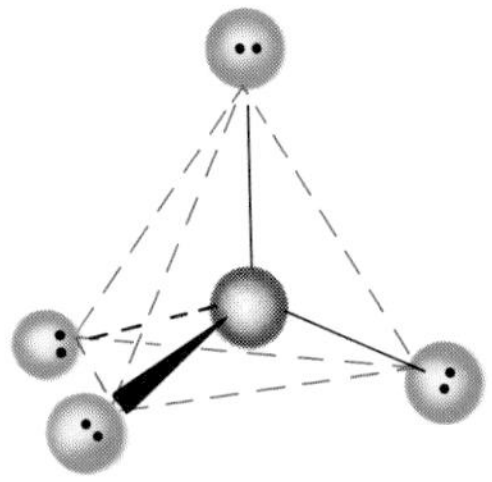 109.5°
5	trigonal bipyramidal	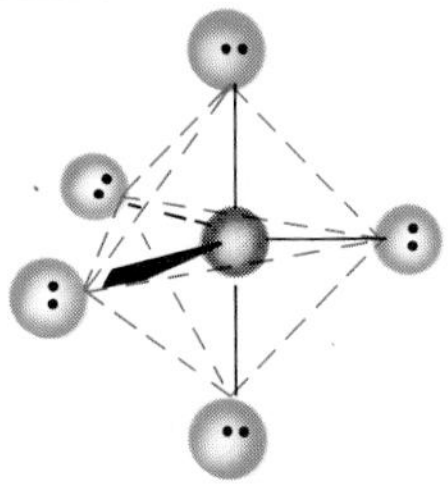 90°, 120°, 180°
6	octahedral	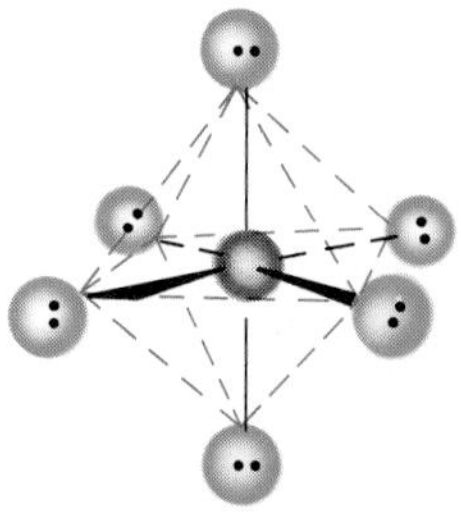 90°, 180°

* Electronic geometries are illustrated here using only single pairs of electrons as regions of high electron density. The symbol ⊙ represents the regions of high electron density about the central atom ●. By convention, a line in the plane of the drawing is represented by a solid line _____, a line behind this plane is shown as a dashed line ---, and a line in front of this plane is shown as a wedge ◀ with the fat end of the wedge nearest the viewer. Each shape is outlined in blue dashed lines to help you visualize it.

† Angles made by imaginary lines through the nucleus and the centers of regions of high electron density.

the bond dipoles are arranged in such a way that they cancel (so that the resulting molecule is *nonpolar*) or do not cancel (so that the resulting molecule is *polar*).

In this section, we shall discuss the ideas of cancellation of dipoles in general terms, using general atomic symbols X and Y. These ideas will be applied to specific molecular geometries and molecular polarities in parts B of Sections 8-5 through 8-12.

Let us consider a heteronuclear molecule with the formula XY_2 (X is the central atom). Such a molecule must have one of the following two molecular geometries:

The angular form could have different angles, but either the molecule is linear or it is not.

$$Y—X—Y \qquad \text{or} \qquad \begin{array}{c} Y—X \\ \qquad \searrow \\ \qquad Y \end{array}$$

linear   angular

Suppose that atom Y has a higher electronegativity than atom X. Then each X—Y bond is polar, with the negative end of the bond dipole pointing toward Y. Each bond dipole can be viewed as an *electronic vector,* with a *magnitude* and a *direction.* In the linear XY_2 arrangement, the two bond dipoles are *identical* in magnitude and *opposite* in direction. Therefore, they cancel to give a nonpolar molecule (dipole moment equal to zero).

$$\overset{\longleftarrow + \quad + \longrightarrow}{Y—X—Y}$$

Net dipole = 0
(nonpolar molecule)

In the case of the angular arrangement, the two equal dipoles *do not cancel,* but add to give a nonzero result. The angular molecular arrangement represents a polar molecule (dipole moment greater than zero).

$$\overset{\longleftarrow +}{Y—X}$$

Net dipole > 0
(polar molecule)

If the electronegativity differences were reversed in this Y—X—Y molecule—that is, if X were more electronegative than Y—the directions of all

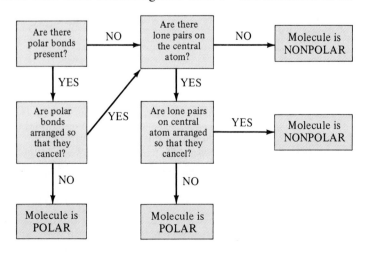

Figure 8-1
A guide to determining whether a polyatomic molecule is polar or nonpolar. Study the more detailed presentation in the text.

bond polarities would be reversed. But the bond polarities would still cancel in the linear arrangement, to give a nonpolar molecule. In the angular arrangement, bond polarities would still add to give a polar molecule, but with the net dipole pointing in the opposite direction from that described above.

Similar arguments based on addition of bond dipoles can be made for other arrangements. As we shall see in Section 8-8, lone pairs on the central atom also affect the direction and the magnitude of the net molecular dipole, so the presence of lone pairs on the central atom must always be taken into account.

> For a molecule to be polar, *both* of the following conditions must be met:
>
> **1.** There must be at least one polar bond or one lone pair on the central atom.
> *and*
> **2.** a. The polar bonds, if there are more than one, must not be so symmetrically arranged that their bond polarities cancel.
> *or*
>   b. If there are two or more lone pairs on the central atom, they must not be so symmetrically arranged that their polarities cancel.

Put another way, if there are no polar bonds or lone pairs of electrons on the central atom, the molecule *cannot* be polar. Even if polar bonds or lone pairs are present, they might be arranged so that their polarities cancel one another, resulting in a nonpolar molecule.

For instance, carbon dioxide, CO_2, is a three-atom molecule in which each carbon–oxygen bond is *polar* because of the electronegativity difference between C and O. But the molecule *as a whole* is shown by experiment to be nonpolar. This tells us that the polar bonds are arranged in such a way that the bond polarities cancel. Water, H_2O, on the other hand, is a very polar molecule; this tells us that the H—O bond polarities do not cancel one another. Molecular shapes clearly play a crucial role in determining molecular dipole moments. We must develop a better understanding of molecular shapes in order to understand molecular polarities.

The logic used in deducing whether a molecule is polar or nonpolar is outlined in Figure 8-1. The approach described in this section will be applied to various electronic and molecular geometries in parts B of Sections 8-5 through 8-12.

linear molecule;
bond dipoles cancel;
molecule is nonpolar

angular molecule;
bond dipoles do not cancel;
molecule is polar

8-4 Valence Bond (VB) Theory

In Chapter 7 we described covalent bonding as electron pair sharing that results from the overlap of orbitals from two atoms. This is the basic idea of the **valence bond (VB) theory**—it describes *how* bonding occurs. In many examples throughout this chapter, we shall first use the VSEPR theory to describe the *orientations* of the regions of high electron density. Then we shall use the VB theory to describe the atomic orbitals that overlap to produce the bonding with that geometry. We shall also assume that each

VSEPR theory describes the locations of bonded atoms around the central atom, as well as where its lone pairs of valence shell electrons are directed.

Table 8-2
Relation between Electronic Geometries and Hybridization

Regions of High Electron Density	Electronic Geometry	Hybridization	Atomic Orbitals Mixed from Valence Shell of Central Atom
2	linear	sp	one s, one p
3	trigonal planar	sp^2	one s, two p's
4	tetrahedral	sp^3	one s, three p's
5	trigonal bipyramidal	sp^3d	one s, three p's, one d
6	octahedral	sp^3d^2	one s, three p's, two d's

lone pair occupies a separate orbital. Thus, the two theories work together to give a fuller description of the bonding.

The atomic orbitals that we described in Chapter 5 refer to isolated atoms. Usually these "pure" atomic orbitals do not have the correct energies or the correct orientations to describe where the electrons are when an atom is bonded to other atoms. To explain the experimentally observed geometry, we usually need to invoke the concept of **hybrid orbitals**, which are formed by the combination, or **hybridization**, of the atom's valence shell atomic orbitals. The number of regions of high electron density about a central atom in a molecule or polyatomic ion (Table 8-1) suggests the kind of hybridization of that atom's valence atomic orbitals that occurs. The designation (label) given to a set of hybridized orbitals reflects the *number and kind* of atomic orbitals that hybridize to produce the set (Table 8-2). Further details about hybridization and hybrid orbitals appear in the following sections. Throughout the text, hybrid orbitals and, when appropriate, the atomic orbitals to which they are related are shaded in green.

Molecular Shapes and Bonding

We are now ready to study the structures of some simple molecules. In the sections that follow, we use generalized chemical formulas in which "A" represents the central atom, "B" represents an atom bonded to A, and "U" represents an unshared valence shell electron pair (lone pair) on the central atom A. For instance, AB_3U (Section 8-8) would represent any molecule with three B atoms bonded to a central atom A, with one unshared valence pair on A. For each type of bonding arrangement, we shall follow the eight steps of analysis outlined in Section 8-1. We shall first give the known (experimentally determined) facts about polarity and shape, and draw the Lewis dot formula (part A of each section). Then we shall explain these facts in terms of the VSEPR and VB theories. The simpler VSEPR theory will be used to explain (or predict) first the *electronic geometry* and then the *molecular geometry* in the molecule (part B). We shall then show how the molecular polarity of each molecule is a result of bond polarities, lone pairs, and molecular geometry. Finally, we shall use the VB theory to describe the bonding in molecules in more detail, usually using hybrid orbitals (part C). As you study each section, refer frequently to the summaries that appear in Table 8-3 on pages 324–325.

See the "Important Note" in Section 8-1 and consult your instructor for guidance on the order in which you should study Sections 8-5 through 8-12.

8-5 Linear Electronic Geometry—AB$_2$ Species (No Unshared Pairs of Electrons on A)

A. Experimental Facts and Lewis Formulas

Several linear molecules consist of a central atom plus two atoms of another element, abbreviated as AB$_2$. These compounds include $BeCl_2$, $BeBr_2$, and BeI_2, as well as CdX_2 and HgX_2, where X = Cl, Br, or I. All of these are known to be linear (bond angle = 180°), nonpolar, covalent compounds, although the individual bonds are polar.

Let's focus on *gaseous* $BeCl_2$ molecules (mp 405°C). The electronic structures of Be and Cl *atoms* in their ground states are

The high melting point of $BeCl_2$ is due to its polymeric nature in the solid state.

1s	2s			1s	2s	2p	3s	3p

$$\text{Be } \uparrow\downarrow \ \uparrow\downarrow \quad \text{and} \quad \text{Cl } \uparrow\downarrow \ \uparrow\downarrow \ \uparrow\downarrow\uparrow\downarrow\uparrow\downarrow \ \uparrow\downarrow \ \uparrow\downarrow\uparrow\downarrow\uparrow$$

We drew the Lewis dot formula for $BeCl_2$ in Example 7-5. It shows two single covalent bonds, with Be and Cl each contributing one electron to each bond:

$$:\!\ddot{Cl}\!:Be:\!\ddot{Cl}\!: \quad \text{or} \quad :\!\ddot{Cl}\!-\!Be\!-\!\ddot{Cl}\!:$$

B. VSEPR Theory

Valence Shell Electron Pair Repulsion theory places the two electron pairs on Be 180° apart, i.e., with *linear electronic geometry*. Both electron pairs are bonding pairs, so VSEPR predicts a linear atomic arrangement, or *linear molecular geometry*, for $BeCl_2$.

VSEPR theory assumes that regions of high electron density (electron pairs) on the central atom will be as far from one another as possible.

$$:\!\ddot{Cl}\!\overset{180°}{\frown}\!Be\!-\!\ddot{Cl}\!:$$

If we examine the bond dipoles, we see that the electronegativity difference (see Table 6-3) is large (1.5 units) and the bonds are quite polar:

$$\begin{array}{ccc} & Cl-Be-Cl \\ EN = & 3.0 \ \ 1.5 \ \ 3.0 \\ \Delta(EN) = & 1.5 \ \ \ \ 1.5 \end{array}$$

$$:\!\ddot{Cl}\!\overset{\longleftarrow +}{}\!Be\!\overset{+\longrightarrow}{}\!\ddot{Cl}\!:$$

Net dipole = 0

A model of a linear AB$_2$ molecule, e.g., $BeCl_2$.

The two bond dipoles are *identical* in magnitude and *opposite* in direction. Therefore, they cancel to give nonpolar molecules.

It is important to distinguish between nonpolar bonds and nonpolar molecules.

The difference in electronegativity between Be and Cl is so large that we might expect ionic bonding. However, the radius of Be^{2+} is so small (0.31 Å) and its **charge density** (ratio of charge to size) is so high that most simple beryllium compounds are covalent rather than ionic. The high charge density of Be^{2+} causes it to attract and distort the electron cloud of monatomic anions of all but the most electronegative elements. As a result, electrons are shared rather than being localized on ions. Two exceptions are BeF_2 and BeO. They are ionic compounds that contain very electronegative elements.

We say that the Be^{2+} ion polarizes the anions, Cl^-.

C. Valence Bond Theory

Consider the ground state electronic configuration for Be. There are two electrons in the $1s$ orbital, but these nonvalence (inner) electrons are *not* involved in bonding. There are two more electrons *paired* in the $2s$ orbital. How, then, will two Cl atoms bond to Be? The Be atom must somehow make available one orbital for each bonding Cl electron (the unpaired p electrons). The following *ground state* electron configuration for Be is the configuration for an isolated Be atom. Another configuration may be more stable in a bonding environment. Suppose that the Be atom "promoted" one of the paired $2s$ electrons to one of the $2p$ orbitals, the next higher energy orbitals.

Then there would be two Be orbitals available for bonding, but we find a discrepancy between this description and experimental fact. The Be $2s$ and $2p$ orbitals would not overlap a Cl $3p$ orbital with equal effectiveness. So this "promoted pure atomic" arrangement would predict two *nonequivalent* Be—Cl bonds. Yet we observe experimentally that the Be—Cl bonds are *identical* in bond length and bond strength. So we reject the idea of simple "promotion" as an explanation.

For these two orbitals on Be to become equivalent, they must hybridize to give two orbitals intermediate between the s and p orbitals. These are called *sp* **hybrid orbitals**. Consistent with Hund's Rule, the two valence electrons of Be would occupy each of these orbitals individually:

Hund's Rule is discussed in Section 5-16.

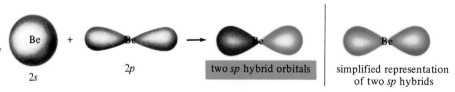

The *sp* hybrid orbitals are described as *linear orbitals*, and we say that Be has *linear electronic geometry*.

As we did for pure atomic orbitals, we often draw hybrid orbitals more slender than they actually are. Such drawings are intended to remind us of the orientations and general shapes of orbitals.

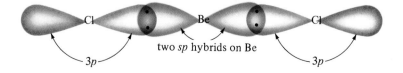

two sp hybrid orbitals   simplified representation of two sp hybrids

Recall that each Cl atom has a $3p$ orbital that contains only one electron, and so overlap with the sp hybrids of Be is possible. We picture the bonding in $BeCl_2$ in the following diagram, in which only the bonding electrons are represented:

Lone pairs of e^- on Cl atoms are not shown. The hybrid orbitals on the central atom are shown in green in this and subsequent drawings.

two sp hybrids on Be

The Be and two Cl nuclei lie on a straight line. *This is consistent with the experimental observation that the molecule is linear.*

One additional idea about hybridization is worth special emphasis:

> The number of hybrid orbitals is always equal to the number of atomic orbitals that hybridize.

Hybrid orbitals are named by indicating the *number and kind* of atomic orbitals hybridized. Hybridization of *one s* orbital and *one p* orbital gives *two sp* hybrid orbitals. We shall see presently that hybridization of *one s* and *two p* orbitals gives *three sp²* hybrid orbitals; hybridization of *one s* orbital and *three p* orbitals gives *four sp³* hybrids, and so on (Table 8-2).

Hybridization usually involves orbitals from the same main shell (same *n*).

The structures of beryllium bromide, $BeBr_2$, and beryllium iodide, BeI_2, are similar to that of $BeCl_2$. The chlorides, bromides, and iodides of cadmium, CdX_2, and mercury, HgX_2, are also linear, covalent molecules (where X = Cl, Br, or I). A cadmium ion has two electrons in its $5s$ orbitals, and its $5p$ orbitals are vacant. Similarly, a mercury atom has two electrons in its $6s$ orbital, and its $6p$ orbitals are vacant. Thus, the possibility of sp hybridization exists in both metals. CdX_2 and HgX_2 are additional examples of this kind of covalent bonding.

The two X's within one structure are identical.

> sp hybridization occurs at the central atom whenever there are two regions of high electron density around the central atom. AB_2 molecules and ions with no lone pairs on the central atom have linear electronic geometry, linear molecular geometry, and sp hybridization on the central atom.

8-6 Trigonal Planar Electronic Geometry—AB₃ Species (No Unshared Pairs of Electrons on A)

A. Experimental Facts and Lewis Formulas

Boron is a Group IIIA element that forms many covalent compounds by bonding to three other atoms. Typical examples include boron trifluoride, BF_3 (mp −127°C); boron trichloride, BCl_3 (mp −107°C); boron tribromide, BBr_3 (mp −46°C); and boron triiodide, BI_3 (mp 50°C). All are trigonal planar nonpolar molecules.

A trigonal planar molecule is a flat molecule in which all three bond angles are 120°.

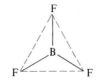

The solid lines represent bonds between B and F atoms. The dashed blue lines emphasize the shape of the molecule.

The Lewis dot formula for BF_3 is derived from the following: (a) each B atom has three electrons in its valence shell and (b) each B atom is bonded to three F (or Cl, Br, I) atoms. In Example 7-6 we drew the Lewis dot formula for BCl_3. F and Cl are both members of Group VIIA, and so the dot formula for BF_3 should be similar.

A model of a trigonal planar AB_3 molecule, e.g., BF_3.

The B^{3+} ion is so small (radius = 0.20 Å) that boron does not form simple ionic compounds.

We see that BF_3 and other similar molecules have central elements that do *not* attain a noble gas configuration by sharing electrons. Boron shares only six electrons.

B. VSEPR Theory

Boron, the central atom, has three regions of high electron density (three bonded atoms, no unshared pairs on B). VSEPR theory predicts **trigonal planar** *electronic geometry* for molecules such as BF_3 because this structure gives maximum separation among the three regions of high electron density. There are no lone pairs of electrons associated with the boron atom, so a fluorine atom is at each corner of the equilateral triangle, and the *molecular geometry* is also trigonal planar. The maximum separation of any three items (electron pairs) around a fourth item (B atom) is at 120° angles in a single plane. All four atoms are in the same plane. The three F atoms are at the corners of an equilateral triangle, with the B atom in the center. The structures of BCl_3, BBr_3, and BI_3 are similar.

Examination of the bond dipoles of BF_3 shows that the electronegativity difference (Table 6-3) is very large (2.0 units) and that the bonds are very polar.

$$\text{EN} = \underbrace{\overset{\text{B—F}}{2.0 \quad 4.0}}$$
$$\Delta(\text{EN}) = 2.0$$

Net molecular dipole = 0

However, the three bond dipoles are symmetrical and cancel to give nonpolar molecules.

C. Valence Bond Theory

To be consistent with experimental findings and the predictions of VSEPR theory, the VB theory must explain three *equivalent* B—F bonds. Again we invoke hybridization. Now the $2s$ orbital and two of the $2p$ orbitals of B hybridize to form a set of three degenerate sp^2 **hybrid orbitals**.

"Degenerate" refers to orbitals of the same energy.

B [He] $\quad \frac{\uparrow\downarrow}{2s} \quad \frac{\uparrow \ \_ \ \_}{2p} \quad \xrightarrow{\text{hybridize}} \quad$ B [He] $\quad \frac{\uparrow \ \uparrow \ \uparrow}{sp^2} \quad \overline{\ \ 2p\ \ }$

Three sp^2 hybrid orbitals point toward the corners of an equilateral triangle:

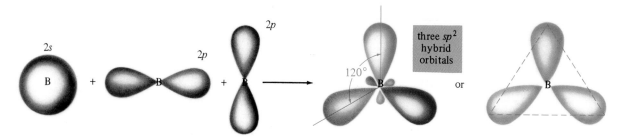

Each of the three F atoms has a $2p$ orbital with one unpaired electron. The $2p$ orbitals can overlap the three sp^2 hybrid orbitals on B. Three electron pairs are shared among one B and three F atoms:

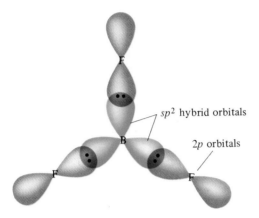

sp^2 hybrid orbitals

$2p$ orbitals

Lone pairs of e^- are not shown for the F atoms.

> sp^2 hybridization occurs at the central atom whenever there are three regions of high electron density around the central atom. AB₃ molecules and ions with no lone pairs on the central atom have trigonal planar electronic geometry, trigonal planar molecular geometry, and sp^2 hybridization on the central atom.

A molecule that has fewer than eight electrons in the valence shell of the central atom frequently reacts by accepting a share in an electron pair from another species. A substance that behaves in this way is called a **Lewis acid** (to be discussed more fully in Section 10-10). Both beryllium chloride, $BeCl_2$, and boron trichloride, BCl_3, react as Lewis acids. The fact that both compounds so readily take a share of additional pairs of electrons tells us that Be and B atoms do not have octets of electrons in these gaseous compounds.

8-7 Tetrahedral Electronic Geometry—AB₄ Species (No Unshared Pairs of Electrons on A)

A. Experimental Facts and Lewis Formulas

Each Group IVA element has four electrons in its highest occupied energy level. The Group IVA elements form many covalent compounds by sharing those four electrons with four other atoms. Typical examples include CH_4 (mp $-182°C$), CF_4 (mp $-184°C$), CCl_4 (mp $-23°C$), SiH_4 (mp $-185°C$), and SiF_4 (mp $-90°C$). All are tetrahedral, nonpolar molecules (bond angles = 109.5°). In each of them, the IVA atom is located in the center of a regular tetrahedron. The other four atoms are located at the four corners of the tetrahedron.

The Group IVA element contributes four electrons in a tetrahedral AB₄ molecule, and the other four atoms contribute one electron each. The Lewis dot formulas for methane, CH_4, and carbon tetrafluoride, CF_4, are typical:

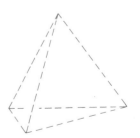

The names of many solid figures are based on the numbers of plane faces they have. A *regular* tetrahedron is a three-dimensional figure with four equal-sized equilateral triangular faces (the prefix *tetra-* means "four").

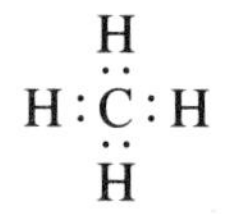

$$
\begin{array}{c}
\text{H} \\
\cdot\cdot \\
\text{H} : \text{C} : \text{H} \\
\cdot\cdot \\
\text{H}
\end{array}
$$

CH₄, methane

$$
\begin{array}{c}
: \ddot{\text{F}} : \\
: \ddot{\text{F}} : \text{C} : \ddot{\text{F}} : \\
: \ddot{\text{F}} :
\end{array}
$$

CF₄, carbon tetrafluoride

Ammonium ion, NH_4^+, and sulfate ion, SO_4^{2-}, are familiar examples of polyatomic ions of this type. In each of these ions, the central atom is located at the center of a regular tetrahedron with the other atoms at the corners (H—N—H and O—S—O bond angles = 109.5°).

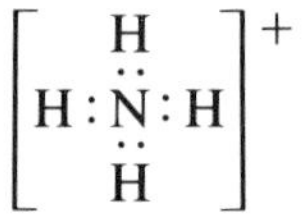

$$
\left[
\begin{array}{c}
\text{H} \\
\cdot\cdot \\
\text{H} : \text{N} : \text{H} \\
\cdot\cdot \\
\text{H}
\end{array}
\right]^+
$$

NH₄⁺, ammonium ion

$$
\left[
\begin{array}{c}
: \ddot{\text{O}} : \\
| \\
: \ddot{\text{O}} - \text{S} - \ddot{\text{O}} : \\
| \\
: \ddot{\text{O}} :
\end{array}
\right]^{2-}
$$

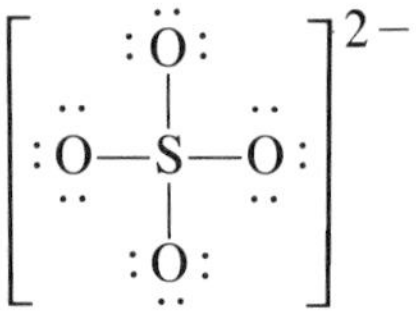

SO₄²⁻, sulfate ion

B. VSEPR Theory

VSEPR theory predicts that the four electron pairs are directed toward the corners of a regular tetrahedron. That shape gives the maximum separation for four electron pairs around one atom. Thus, VSEPR theory predicts a *tetrahedral electronic structure* for an AB₄ molecule that has no unshared electrons on A. Because there are no lone pairs of electrons on the central atom, another atom is at each corner of the tetrahedron. VSEPR theory predicts a *tetrahedral molecular geometry* of each of these molecules. Again, let us describe CH₄ and CF₄ molecules.

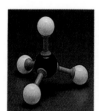

Models of two tetrahedral AB₄ molecules: CH₄ (left) and CF₄ (right).

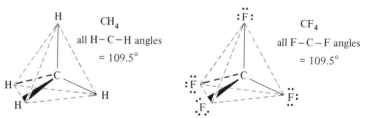

Examination of bond dipoles shows that in CH₄ the individual bonds are only slightly polar, whereas in CF₄ the bonds are quite polar. In CH₄ the bond dipoles are directed toward carbon, but in CF₄ they are directed away from carbon. Both molecules are quite symmetrical, so the bond dipoles cancel, and both molecules are nonpolar. This is true for *tetrahedral* AB₄ molecules in which there are *no unshared electron pairs on the central element* and all four B atoms are identical.

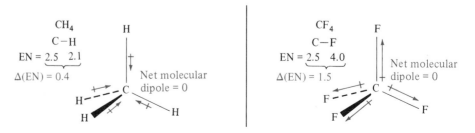

In some tetrahedral molecules, the atoms bonded to the central atom are not all the same. Such molecules may be polar, depending on the relative sizes of the bond dipoles present. In CH$_3$F or CH$_2$F$_2$, for example, the addition of unequal dipoles makes the molecule polar.

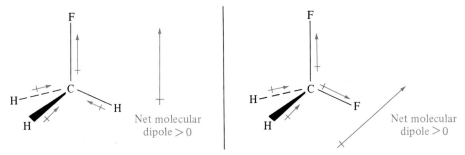

Net molecular dipole > 0

Net molecular dipole > 0

VSEPR theory predicts that NH$_4^+$ and SO$_4^{2-}$ ions each have tetrahedral electronic geometry. Each region of high electron density bonds the central atom to another atom (H in NH$_4^+$, O in SO$_4^{2-}$) at the corner of the tetrahedral arrangement. We describe the molecular geometry of each of these ions as tetrahedral.

You may wonder whether square planar AB$_4$ molecules exist. They do, in compounds of some of the transition metals. All *simple* square planar AB$_4$ molecules have unshared electron pairs on A. The bond angles in square planar molecules are only 90°. Nearly all AB$_4$ molecules are tetrahedral, however, with larger bond angles (109.5°) and greater separation of valence electron pairs around A.

C. Valence Bond Theory

According to VB theory, each Group IVA atom (C in our example) must make four equivalent orbitals available for bonding. To do this, C forms four **sp^3 hybrid orbitals** by mixing the s and all three p orbitals in its outer shell. This results in four unpaired electrons:

C [He]   2s ↑↓   2p ↑ ↑ ___   →(hybridize)   C [He]   sp^3 ↑ ↑ ↑ ↑

These sp^3 hybrid orbitals are directed toward the corners of a regular tetrahedron, which has a 109.5° angle from any corner to center to any other corner.

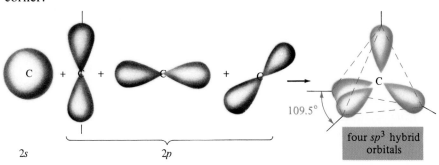

$2s$ $2p$ 109.5°   four sp^3 hybrid orbitals

Each of the four atoms that bond to C possesses a half-filled atomic orbital that can overlap the half-filled sp^3 hybrids, as illustrated for CH_4 and CF_4.

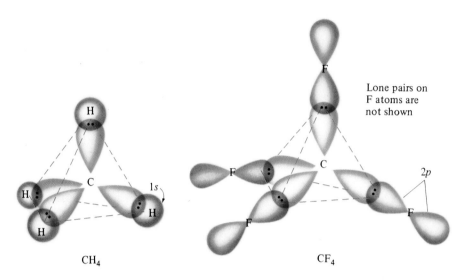

Lone pairs on F atoms are not shown

CH_4 CF_4

We can give the same VB description for the hybridization of the central atoms in polyatomic ions. In NH_4^+ and SO_4^{2-}, the N and S atoms, respectively, form four sp^3 hybrid orbitals directed toward the corners of a regular tetrahedron. Each of these sp^3 hybrid orbitals overlaps with an orbital on a neighboring atom (H in NH_4^+, O in SO_4^{2-}) to form a bond.

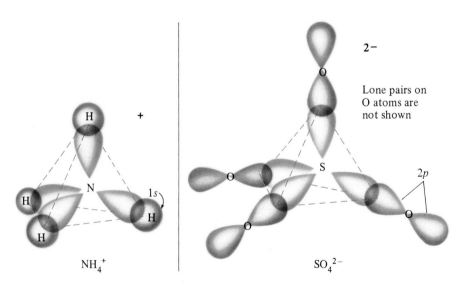

Lone pairs on O atoms are not shown

NH_4^+ SO_4^{2-}

sp^3 hybridization occurs at the central atom whenever there are four regions of high electron density around the central atom. AB_4 molecules and ions with no lone pairs on the central atom have tetrahedral electronic geometry, tetrahedral molecular geometry, and sp^3 hybridization on the central atom.

The symbol "U" in the generalized chemical formula stands for an unshared valence electron pair on atom A.

8-8 Tetrahedral Electronic Geometry—AB₃U Species (One Unshared Pair of Electrons on A)

A. Experimental Facts and Lewis Formulas

Each Group VA element has five electrons in its valence shell. The Group VA elements form some covalent compounds by sharing three of those electrons with three other atoms. Let us describe two examples: ammonia, NH_3, and nitrogen trifluoride, NF_3. Each is a pyramidal, polar molecule with an unshared pair on the nitrogen atom. The Lewis dot formulas for NH_3 and NF_3 are

Some Group VA elements also form covalent compounds by sharing all five valence electrons (Sections 7-10 and 8-11).

$$NH_3 \quad H:\overset{..}{N}:H \qquad\qquad NF_3 \quad :\overset{..}{F}:\overset{..}{N}:\overset{..}{F}:$$
$$\underset{H}{\vphantom{H}} \qquad\qquad\qquad\qquad :\overset{..}{F}:$$

Sulfite ion, $SO_3{}^{2-}$, is an example of a polyatomic ion of the AB₃U type. It is a pyramidal ion with an unshared pair on the sulfur atom.

$$\left[\,:\overset{..}{\underset{..}{O}}-\overset{..}{S}-\overset{..}{\underset{..}{O}}:\,\right]^{2-}$$
$$\left.\quad\quad\; |\quad\quad\right.$$
$$:\overset{}{\underset{..}{O}}:$$

B. VSEPR Theory

As in Section 8-7, VSEPR theory predicts that the *four* regions of high electron density around a central atom will be directed toward the corners of a tetrahedron, because this gives maximum separation. Thus, in both molecules N has tetrahedral electronic geometry.

At this point let us reemphasize the distinction between electronic geometry and molecular geometry. *Electronic geometry* refers to the geometric arrangement of *regions of electron density* around the central atom. *Molecular geometry* refers to the arrangement of *atoms* (that is, nuclei), not just pairs of electrons, around the central atom. For example, CH_4, CF_4, NH_3, and NF_3 all have tetrahedral electronic geometry. But CH_4 and CF_4 have tetrahedral molecular geometry, whereas NH_3 and NF_3 have pyramidal molecular geometry.

As we have seen, the term "lone pair" refers to a pair of valence electrons that is associated with only one nucleus. The known geometries of many molecules and polyatomic ions, based on measurements of bond angles, show that *lone pairs of electrons occupy more space than bonding pairs.* This is due to the fact that a lone pair has only one atom exerting strong attractive forces on it, so it resides closer to the nucleus than do bonding electrons. Additional observations indicate that the relative magnitudes of the repulsive forces between pairs of electrons on an atom are

$$lp/lp \gg lp/bp > bp/bp$$

where *lp* refers to lone pairs and *bp* refers to bonding pairs of valence shell electrons. We are most concerned with the repulsions involving the electrons in the valence shell of the *central atom* of a molecule or polyatomic ion. The angles at which repulsive forces among valence shell electron pairs are

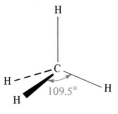

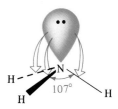

exactly balanced are the angles at which the bonding pairs and lone pairs (and therefore nuclei) are found in covalently bonded molecules and polyatomic ions. Thus, the bond angles in NH_3 and NF_3 are *less* than the angles of 109.5° we observed in CH_4 and CF_4 molecules.

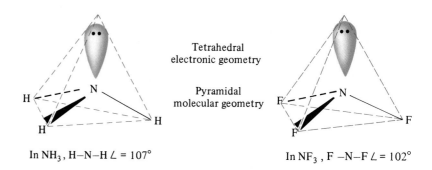

Tetrahedral electronic geometry

Pyramidal molecular geometry

In NH_3, H–N–H ∠ = 107°   In NF_3, F –N–F ∠ = 102°

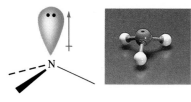

A drawing and a model of a pyramidal molecule (AB₃U).

The formulas are frequently written as :NH_3 and :NF_3 to emphasize the lone pairs of electrons. The lone pairs must be considered as the polarities of these molecules are examined; they are extremely important! The contribution of each lone pair to polarity can be depicted as shown in the margin.

The electronegativity differences in NH_3 and NF_3 are nearly equal, *but* the resulting nearly equal bond polarities are in opposite directions.

$$
\begin{array}{ll}
\text{N—H} & \overset{\longleftarrow +}{\text{N—H}} \\
\text{EN} = \underbrace{3.0 \quad 2.1} & \\
\Delta(\text{EN}) = \quad 0.9 &
\end{array}
\qquad
\begin{array}{ll}
\text{N—F} & \overset{+\longrightarrow}{\text{N—F}} \\
\text{EN} = \underbrace{3.0 \quad 4.0} & \\
\Delta(\text{EN}) = \quad 1.0 &
\end{array}
$$

Thus, we have

In NH_3 the bond dipoles *reinforce* the effect of the lone pair, so NH_3 is very polar ($\mu = 1.5$ D). In NF_3 the bond dipoles *oppose* the effect of the lone pair, so NF_3 is only slightly polar ($\mu = 0.2$ D).

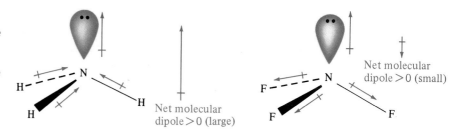

Net molecular dipole > 0 (large)

Net molecular dipole > 0 (small)

We can now use this information to explain the bond angles observed in NF_3 and NH_3. Because of the direction of the bond dipoles in NH_3, the electron-rich end of each N—H bond is at the central atom, N. In NF_3, on the other hand, the fluorine end of each bond is the electron-rich end. As a result, the lone pair can more closely approach the N in NF_3 than in NH_3. Therefore, in NF_3 the lone pair exerts greater repulsion toward the bonded pairs than in NH_3. In addition, the longer N—F bond length makes the *bp–bp* distance greater in NF_3 than in NH_3, so that the *bp–bp* repulsion in NF_3 is less than that in NH_3. The net effect is that the bond angles are reduced more in NF_3. We can represent this situation as follows:

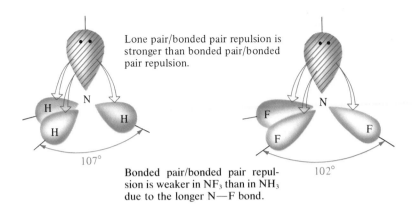

Lone pair/bonded pair repulsion is stronger than bonded pair/bonded pair repulsion.

107°

Bonded pair/bonded pair repulsion is weaker in NF₃ than in NH₃ due to the longer N—F bond.

102°

We might expect the larger F atoms ($r = 0.64$ Å) to repel each other more strongly than H atoms ($r = 0.37$ Å), leading to larger bond angles in NF₃ than in NH₃. This is not the case, however, because the N—F bond is longer than the N—H bond. The N—F bond density is farther from the N than the N—H bond density.

With the same kind of reasoning, VSEPR theory predicts that sulfite ion, SO_3^{2-}, has tetrahedral electronic geometry. One of these tetrahedral locations is occupied by the sulfur lone pair, and oxygen atoms are at the other three locations. The molecular geometry of this ion is trigonal pyramidal, the same as for other AB₃U species.

C. Valence Bond Theory

Experimental results suggest four nearly equivalent orbitals (three involved in bonding, a fourth to accommodate the lone pair), so we again need four sp^3 hybrid orbitals:

In both NH₃ and NF₃ the unshared pair of electrons occupies one of the sp^3 hybrid orbitals. Each of the other three sp^3 orbitals participates in bonding by sharing electrons with another atom. They overlap with half-filled H 1s orbitals and F 2p orbitals in NH₃ and NF₃, respectively.

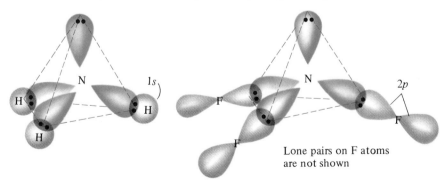

Lone pairs on F atoms are not shown

As we shall see in Section 10-10, many compounds that have an unshared electron pair on the central element are called Lewis bases. **Lewis bases** react by making available an electron pair that can be shared by other species.

The structure of nitrogen triiodide, NI₃, is analogous to that of NF₃, described in the text. When dry, this dark brown compound is so sensitive that it explodes when touched by a small tree branch. The photo at the bottom was taken just as the compound exploded. *Caution! Do not attempt this experiment yourself.*

We must remember that *theory* (and its application) depends on fact, not the other way around. Sometimes the experimental facts are not consistent with the existence of hybrid orbitals. In such cases, we just use the "pure" atomic orbitals rather than hybrid orbitals. There appears to be no need to use hybridization to describe bonding in PH_3 and AsH_3. Each H—P—H bond angle is 93°, and each H—As—H bond angle is 92°. These angles very nearly correspond to three *p* orbitals at 90° to each other.

The sulfur atom in the sulfite ion, SO_3^{2-}, can be described as sp^3 hybridized. One of these hybrid orbitals contains the sulfur lone pair, and the remaining three overlap with oxygen orbitals to form bonds.

> AB_3U molecules and ions, each having four regions of high electron density around the central atom, *usually* have tetrahedral electronic geometry, trigonal pyramidal molecular geometry, and sp^3 hybridization on the central atom.

8-9 Tetrahedral Electronic Geometry—AB_2U_2 Species (Two Unshared Pairs of Electrons on A)

A. Experimental Facts and Lewis Formulas

A model of H_2O, an angular molecule (AB_2U_2).

Each Group VIA element has six electrons in its highest energy level. The Group VIA elements form many covalent compounds by acquiring a share in two additional electrons from two other atoms. Typical examples are H_2O, H_2S, and Cl_2O. All are angular, polar molecules. The Lewis dot formulas for these molecules are

$$H_2O \quad H:\overset{..}{\underset{H}{O}}: \qquad H_2S \quad H:\overset{..}{\underset{H}{S}}: \qquad Cl_2O \quad \overset{..}{\underset{..}{:}}\overset{.}{Cl}:\overset{..}{\underset{..}{O}}: \\ \qquad\qquad\qquad\qquad\qquad\qquad\qquad\qquad :\overset{..}{\underset{..}{Cl}}:$$

Let us consider the structure of water in detail. The bond angle in water is 104.5°, and the molecule is very polar.

B. VSEPR Theory

Six electrons come from oxygen and two from the hydrogens.

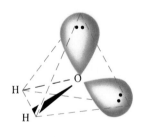

VSEPR theory predicts that the four electron pairs around the oxygen atom in H_2O should be 109.5° apart in a tetrahedral arrangement. When we take into account increased repulsions between unshared pairs and bonding electron pairs, this theory satisfactorily explains the angular structure of water molecules and the observed bond angle of only 104.5°.

The electronegativity difference is large (1.4 units) and so the bonds are quite polar. Additionally, the bond dipoles *reinforce* the effect of the two lone pairs, so the H_2O molecule is very polar. Its dipole moment is 1.7 D. Water has unusual properties, which can be explained in large part by its high polarity.

$$\begin{array}{c} \quad\; O\text{—}H \\ EN = \underline{3.5 \quad 2.1} \\ \Delta(EN) = \quad 1.4 \end{array}$$

Molecular dipole; includes effect of two unshared electron pairs.

C. Valence Bond Theory

The bond angle in H_2O (104.5°) is near the tetrahedral value (109.5°). Valence bond theory postulates four sp^3 hybrid orbitals centered on the O atom: two to participate in bonding and two to hold the two lone pairs.

We can easily explain the observed bond angle of 104.5°. The expected bond angle for sp^3 hybridization (tetrahedral electronic geometry) is 109.5°. However, the two lone pairs strongly repel each other and the bonding pairs of electrons. These repulsions force the bonding pairs closer together and result in the decreased bond angle. The decrease in the H—O—H bond angle (from 109.5° to 104.5°) is greater than the corresponding decrease in the H—N—H bond angles in ammonia (from 109.5° to 107°).

Hydrogen sulfide, H_2S, is also an angular molecule, but the H—S—H bond angle is 92°. This is very close to the 90° angles between two unhybridized $3p$ orbitals of S. Therefore, we *do not* postulate hybrid orbitals. The two H atoms are able to exist at approximately right angles to each other when they are bonded to the larger S atom. Likewise, the bond angles in H_2Se and H_2Te are 91° and 89.5°, respectively.

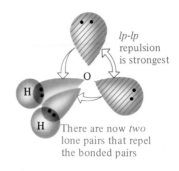

lp-lp repulsion is strongest

There are now *two* lone pairs that repel the bonded pairs

Sulfur is located directly below oxygen in Group VIA.

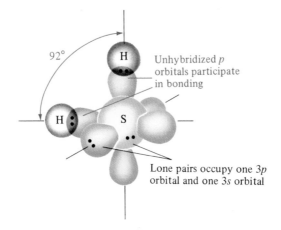

92°

Unhybridized *p* orbitals participate in bonding

Lone pairs occupy one $3p$ orbital and one $3s$ orbital

AB_2U_2 molecules and ions, each having four regions of high electron density around the central atom, *usually* have tetrahedral electronic geometry, angular molecular geometry, and sp^3 hybridization on the central atom.

8-10 Tetrahedral Electronic Geometry—ABU₃ Species (Three Unshared Pairs of Electrons on A)

Each Group VIIA element has seven electrons in its highest occupied energy level. The Group VIIA elements form covalent molecules such as H—F, H—Cl, Cl—Cl, and I—I by sharing one of those electrons with another atom. The other atom contributes one electron to bonding. Lewis dot formulas for these molecules are shown in the margin. All diatomic molecules are of necessity linear. Neither VSEPR theory nor VB theory adds anything to what we already know about the molecular geometry of such molecules.

We represent the halogen as X.

HX, H : Ẍ :   X₂, : Ẍ : Ẍ :

In the latter case, either halogen may be considered the "A" atom of AB.

8-11 Trigonal Bipyramidal Electronic Geometry—AB₅, AB₄U, AB₃U₂, and AB₂U₃

A. Experimental Facts and Lewis Formulas

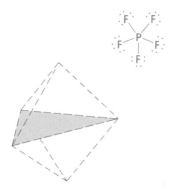

In Section 8-8 we saw that the Group VA elements have five electrons in their outermost occupied shells and form some covalent molecules by sharing only three of these electrons with other atoms (for example, NH_3 and NF_3). Other Group VA elements (P, As, and Sb) form some covalent compounds by sharing all five of their valence electrons with five other atoms (Section 7-10). Phosphorus pentafluoride, PF_5 (mp −83°C), is such a compound. Each P atom has five valence electrons to share with five F atoms. The Lewis formula for PF_5 (Example 7-7) is shown in the margin. PF_5 molecules are *trigonal bipyramidal* nonpolar molecules. A **trigonal bipyramid** is a six-sided polyhedron consisting of two pyramids joined at a common triangular (trigonal) base.

A trigonal bipyramid. The triangular base common to the two pyramids is shaded.

B. VSEPR Theory

VSEPR theory predicts that the five regions of high electron density around the phosphorus atom in PF_5 should be as far apart as possible. Maximum separation of five items around a sixtn item is achieved when the five items (bonding pairs) are placed at the corners and the sixth item (P atom) is placed in the center of a trigonal bipyramid. This is in agreement with experimental observation.

A model of a trigonal bipyramidal AB_5 molecule, e.g., PF_5.

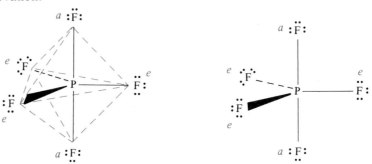

The three F atoms marked *e* are at the corners of the common base, in the same plane as the P atom. These are called *equatorial* F atoms (*e*). The other two F atoms, one above and one below the plane, are called *axial* F atoms (*a*). The axial P—F bonds are longer than the equatorial P—F bonds. The F—P—F bond angles are 90° (axial to equatorial), 120° (equatorial to equatorial), and 180° (axial to axial).

As an exercise in geometry, in how many different ways can five fluorine atoms be arranged *symmetrically* around a phosphorus atom? Compare the hypothetical bond angles in such arrangements with those in a trigonal bipyramidal arrangement.

The large electronegativity difference between P and F (1.9) indicates very polar bonds. Let's consider the bond dipoles in two groups, because there are two different kinds of P—F bonds in PF_5 molecules, axial and equatorial.

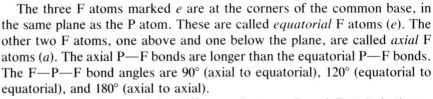

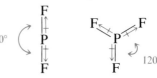

$$\text{P—F}$$
$$EN = \underbrace{2.1 \quad 4.0}$$
$$\Delta(EN) = \quad 1.9$$

The two axial bond dipoles cancel each other, and the three equatorial bond dipoles cancel, so PF_5 molecules are nonpolar.

C. Valence Bond Theory

Because phosphorus is the central element in PF_5 molecules, it must have available five half-filled orbitals to form bonds with five F atoms. Hybridization is again the explanation. Now it involves one d orbital from the vacant set of $3d$ orbitals and the $3s$ and $3p$ orbitals of the P atom.

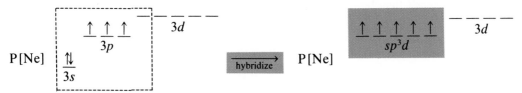

The five sp^3d **hybrid orbitals** point toward the corners of a trigonal bipyramid. Each is overlapped by a singly occupied $2p$ orbital of an F atom. The resulting pairing of P and F electrons forms five covalent bonds.

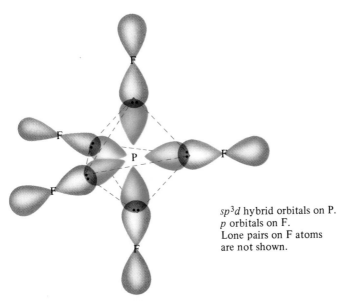

sp^3d hybrid orbitals on P.
p orbitals on F.
Lone pairs on F atoms
are not shown.

> sp^3d hybridization occurs at the central atom whenever there are five regions of high electron density around the central atom. AB_5 molecules and ions with no lone pairs on the central atom have trigonal bipyramidal electronic geometry, trigonal bipyramidal molecular geometry, and sp^3d hybridization on the central atom.

We see that sp^3d hybridization uses an available d orbital in the outermost occupied shell of the central atom, P. The heavier Group VA elements—P, As, and Sb—can form *five-coordinate compounds* using this hybridization. But nitrogen, also in Group VA, cannot form such five-coordinate compounds. This is because the valence shell of N has only one s and three p orbitals (and no d orbitals). The set of s and p orbitals in a given energy level (and therefore any set of hybrids composed only of s and p orbitals)

The P atom is said to have an expanded valence shell (Section 7-10).

can accommodate a *maximum* of eight electrons and participate in a *maximum* of four covalent bonds. The same is true of all elements of the second period, because they have only *s* and *p* orbitals in their valence shells. No atoms before the third period can exhibit expanded valence.

D. Unshared Valence Electron Pairs in Trigonal Bipyramidal Electronic Geometry

Relative magnitudes of repulsive forces:

$lp/lp \gg lp/bp > bp/bp.$

As we saw in Sections 8-8 and 8-9, lone pairs of electrons occupy more space than bonding pairs, resulting in increased repulsions from lone pairs. What happens when one or more of the five regions of high electron density on the central atom are unshared electron pairs? Let us first consider a molecule such as SF_4, for which the Lewis formula is

$$\text{:}\ddot{F} \quad \text{:}\ddot{F}\text{:}$$
$$\searrow \underset{\cdot\cdot}{S}\diagdown$$
$$\text{:}\ddot{F}\text{:} \qquad \diagdown\ddot{F}\text{:}$$

The central atom, S, is bonded to four atoms and has one unshared valence electron pair. This is an example of the general formula AB_4U. Sulfur has five regions of high electron density, so we know that the electronic geometry is trigonal bipyramidal and that the bonding orbitals are sp^3d hybrids. But now a new question arises: Is the lone pair more stable in an axial (*a*) or in an equatorial (*e*) position? If it were in an axial position, it would be 90° from the *three* closest other pairs (the pairs bonding three F atoms in equatorial positions) and 180° from the other axial pair. If it were in an equatorial position, only the *two* axial pairs would be at 90°, while the other two equatorial pairs would be less crowded at 120° apart.

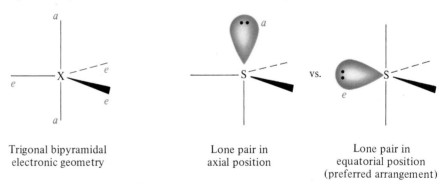

Trigonal bipyramidal electronic geometry

Lone pair in axial position

vs.

Lone pair in equatorial position (preferred arrangement)

We conclude that the lone pair would be less crowded in an *equatorial* position. The four F atoms then occupy the remaining four positions. We describe the resulting arrangement of *atoms* as a **seesaw arrangement**.

Imagine rotating the arrangement so that the line joining the two axial positions is the board on which the two seesaw riders sit, and the two bonded equatorial positions are the pivot of the seesaw.

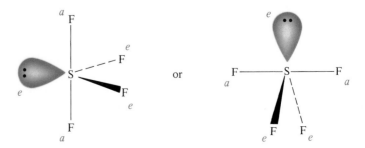

or

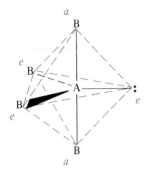

AB$_4$U	4 bonded atoms (B) 1 lone pair (U) in equatorial position

Seesaw molecular geometry
Example: SF$_4$

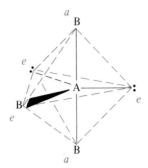

AB$_3$U$_2$	3 bonded atoms (B) 2 lone pairs (U) in equatorial positions

T-shaped molecular geometry
Examples: ICl$_3$, ClF$_3$

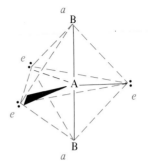

AB$_2$U$_3$	2 bonded atoms (B) 3 lone pairs (U) in equatorial positions

Linear molecular geometry
Examples: XeF$_2$, I$_3^-$

Figure 8-2
Arrangements of bonded atoms and lone pairs (five regions of high electron density—trigonal bipyramidal electronic geometry).

As we saw in Sections 8-8 and 8-9, the differing magnitudes of repulsions involving lone pairs and bonding pairs often result in actual bond angles that deviate slightly from idealized values. For instance, *bp/lp* repulsion in the seesaw molecule SF$_4$ causes distortion of the axial S—F bonds away from the lone pair, to an angle of 177°; the two equatorial S—F bonds, ideally at 120°, also move closer together to an angle of 104°.

By the same reasoning, we understand why additional lone pairs also take equatorial positions (AB$_3$U$_2$ with both lone pairs equatorial or AB$_2$U$_3$ with all three lone pairs equatorial). These arrangements are summarized in Figure 8-2.

8-12 Octahedral Electronic Geometry—AB$_6$, AB$_5$U, and AB$_4$U$_2$

A. Experimental Facts and Lewis Formulas

The heavier Group VIA elements form some covalent compounds of the AB$_6$ type by sharing their six valence electrons with six other atoms. Sulfur hexafluoride SF$_6$ (mp −51°C), an unreactive gas, is an example. Sulfur hexafluoride molecules are nonpolar octahedral molecules. Hexafluorophosphate ion, PF$_6^-$, is an example of a polyatomic ion of the type AB$_6$.

B. VSEPR Theory

In an SF$_6$ molecule we have six valence electron pairs and six F atoms surrounding one S atom. Because there are no lone pairs in the valence shell of sulfur, the electronic and molecular geometries are identical. The maximum separation possible for six electron pairs around one S atom is achieved when the electron pairs are at the corners and the S atom is at the center of a regular octahedron. Thus, VSEPR theory is consistent with the observation that SF$_6$ molecules are octahedral.

In a regular octahedron, each of the eight faces is an equilateral triangle.

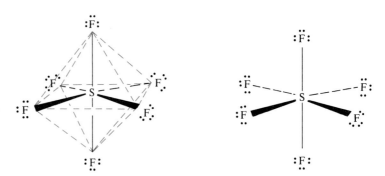

In this octahedral molecule the F—S—F bond angles are 90° and 180°. Each S—F bond is quite polar, but each bond dipole is cancelled by an equal dipole at 180° from it. So the large bond dipoles cancel and the SF_6 molecule is nonpolar.

By similar reasoning, VSEPR theory predicts octahedral electronic geometry and octahedral molecular geometry for the PF_6^- ion, which has six valence electron pairs and six F atoms surrounding one P atom.

C. Valence Bond Theory

Sulfur atoms can use one $3s$, three $3p$, and two $3d$ orbitals to form six hybrid orbitals that accommodate six electron pairs:

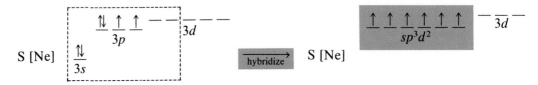

The six **sp^3d^2 hybrid orbitals** are directed toward the corners of a regular octahedron. Each sp^3d^2 hybrid orbital is overlapped by a half-filled $2p$ orbital from fluorine, to form a total of six covalent bonds.

Se and Te, in the same group, form analogous compounds. O cannot do so, for the reasons discussed earlier for N (Section 8-11).

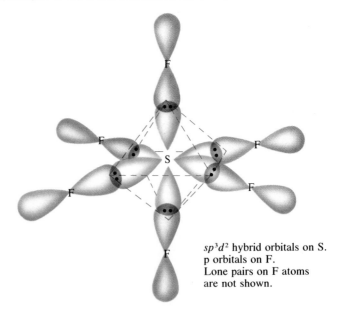

sp^3d^2 hybrid orbitals on S.
p orbitals on F.
Lone pairs on F atoms
are not shown.

(Left) A model of an octahedral AB_6 molecule, e.g., SF_6.
(Right) Some familiar octahedral toys.

An analogous picture could be drawn for the PF_6^- ion.

> sp^3d^2 hybridization occurs at the central atom whenever there are six regions of high electron density around the central atom. AB_6 molecules and ions with no lone pairs on the central atom have octahedral electronic geometry, octahedral molecular geometry, and sp^3d^2 hybridization on the central atom.

D. Unshared Valence Electron Pairs in Octahedral Electronic Geometry

We can reason along the lines of part D in Section 8-10 to predict the preferred locations of unshared electron pairs on the central atom in octahedral electronic geometry. Because of the high symmetry of the octahedral arrangement, all six positions are equivalent, so it does not matter in which position in the drawing we put the first lone pair. AB_5U molecules and ions are described as having square pyramidal molecular geometry. When a second lone pair is present, the two lone pairs (which repel one another more strongly than bonding pairs are repelled) are most stable when they are in two octahedral positions at 180° from one another. This leads to a square planar molecular geometry for AB_4U_2 species. These arrangements are shown in Figure 8-3. Table 8-3 summarizes a great deal of information—study this table carefully.

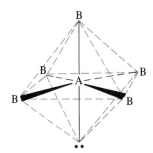

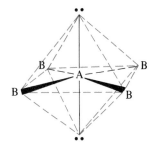

AB_5U	5 bonded atoms (B)
	1 lone pair (U)

Square pyramidal molecular geometry
Examples: IF_5, BrF_5

AB_4U_2	4 bonded atoms (B)
	2 lone pairs (U)

Square planar molecular geometry
Examples: XeF_4, IF_4^-

Figure 8-3
Arrangements of bonded atoms and lone pairs (six regions of high electron density—octahedral electronic geometry).

Table 8-3
Molecular Geometry of Species with Lone Pairs (U) on the Central Atom

General Formula	Regions of High Electron Density	Electronic Geometry	Hybridization at Central Atom	Lone Pairs	Molecular Geometry	Examples
AB_2U	3	trigonal planar	sp^2	1	Angular	O_3, NO_2^-
AB_3U	4	tetrahedral	sp^3	1	Pyramidal	NH_3, SO_3^{2-}
AB_2U_2	4	tetrahedral	sp^3	2	Angular	H_2O
AB_4U	5	trigonal bipyramidal	sp^3d	1	Seesaw	SF_4

Table 8-3
(*continued*)

General Formula	Regions of High Electron Density	Electronic Geometry	Hybridization at Central Atom	Lone Pairs	Molecular Geometry	Examples
AB$_3$U$_2$	5	trigonal bipyramidal	sp^3d	2	T-shaped	ICl$_3$, ClF$_3$
AB$_2$U$_3$	5	trigonal bipyramidal	sp^3d	3	Linear	XeF$_2$, I$_3{}^-$
AB$_5$U	6	octahedral	sp^3d^2	1	Square pyramidal	IF$_5$, BrF$_5$
AB$_4$U$_2$	6	octahedral	sp^3d^2	2	Square planar	XeF$_4$, IF$_4{}^-$

8-13 Compounds Containing Double Bonds

In Chapter 7 we constructed dot formulas for some molecules and polyatomic ions that contain double and triple bonds. We have not yet considered bonding and shapes for such species. Let us consider ethene (ethylene), C_2H_4, as a specific example. Its dot formula is

$$S = N - A$$
$$= 24 - 12 = \underline{12e^- \text{ shared}}$$

Here each C atom is considered a central atom.

There are three regions of high electron density around each C atom. VSEPR theory tells us that each C atom is at the center of a trigonal plane.

Valence bond theory pictures each doubly bonded carbon atom as sp^2 hybridized with one electron in each sp^2 hybrid orbital and one electron in the unhybridized $2p$ orbital. This $2p$ orbital is perpendicular to the plane of the three sp^2 hybrid orbitals:

Relative energies of pure atomic orbitals and hybridized orbitals are not indicated here.

Recall that sp^2 hybrid orbitals are directed toward the corners of an equilateral triangle. Figure 8-4 shows top and side views of these hybrid orbitals.

The two C atoms interact by head-on (end-to-end) overlap of sp^2 hybrids pointing toward each other to form a *sigma* (σ) *bond* and by side-on overlap of the unhybridized $2p$ orbitals to form a *pi* (π) *bond*. A **sigma bond** is a bond resulting from head-on overlap of atomic orbitals. *The region of electron sharing is along and cylindrically around the imaginary line connecting the bonded atoms.* All single bonds are sigma bonds. We have seen that many kinds of pure atomic orbitals and hybridized orbitals can be involved in sigma bond formation. A **pi bond** is a bond resulting from side-on overlap of atomic orbitals. *The regions of electron sharing are on opposite sides of the imaginary line connecting the bonded atoms and parallel to this line.* A pi bond can form *only* if there is *also* a sigma bond between the same two atoms. The sigma and pi bonds together make a double bond (Figure 8-5).

Figure 8-4
(a) A top view of three sp^2 hybrid orbitals (green). The remaining unhybridized p orbital (not shown in this view) is perpendicular to the plane of the drawing. (b) A side view of a carbon atom in a trigonal planar (sp^2-hybridized) environment, showing the remaining p orbital (purple). This p orbital is perpendicular to the plane of the three sp^2 hybrid orbitals.

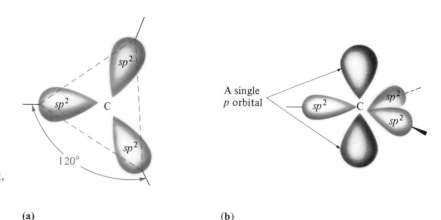

(a)   (b)

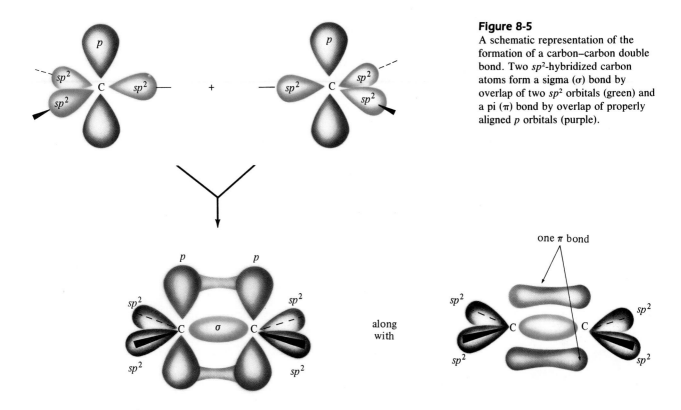

Figure 8-5
A schematic representation of the formation of a carbon–carbon double bond. Two sp^2-hybridized carbon atoms form a sigma (σ) bond by overlap of two sp^2 orbitals (green) and a pi (π) bond by overlap of properly aligned p orbitals (purple).

The 1s orbitals (with one e^- each) of four hydrogen atoms overlap the remaining four sp^2 orbitals (with one e^- each) on the carbon atoms to form four C—H sigma bonds (Figure 8-6).

A double bond consists of one sigma bond and one pi bond.

As a consequence of the sp^2 hybridization of C atoms in carbon–carbon double bonds, each carbon atom is at the center of a trigonal plane. The p orbitals that overlap to form the π bond must be parallel to each other for effective overlap to occur. This adds the further restriction that these trigonal planes (sharing a common corner) must also be *coplanar*. Thus, all four atoms attached to the doubly bonded C atoms lie in the same plane (Figure 8-6). Many other important organic compounds contain carbon–carbon double bonds. Several are described in Chapter 31.

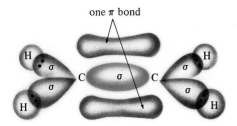

Figure 8-6
Four C—H σ bonds (gray), one C—C σ bond (green), and one C—C π bond (purple) in the planar C_2H_4 molecule.

8-14 Compounds Containing Triple Bonds

One compound that contains a triple bond is ethyne (acetylene), C_2H_2. This has the dot formula

$$S = N - A \qquad\qquad H:C::C:H \qquad H—C\equiv C—H$$
$$= 20 - 10 = \underline{10e^- \text{shared}}$$

VSEPR theory predicts that the two regions of high electron density around each carbon atom are 180° apart.

Valence bond theory postulates that each triply bonded carbon atom is sp-hybridized (see Section 8-5) because each has two regions of high electron density. Let us designate the p_x orbitals as the ones involved in hybridization. Carbon has one electron in each sp hybrid orbital and one electron in each of the $2p_y$ and $2p_z$ orbitals (before bonding is considered). See Figure 8-7.

> The three p orbitals in a set are indistinguishable. We label the one involved in hybridization as "p_x" to help us visualize the orientations of the two unhybridized p orbitals on carbon.

Each carbon atom forms one sigma bond with the other C atom and a sigma bond with one H atom.

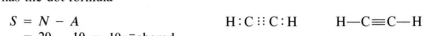

The unhybridized atomic $2p_y$ and $2p_z$ orbitals are perpendicular to each other and to a line through the centers of the two sp hybrid orbitals (Figure 8-8). The sp hybrids are 180° apart. The sp hybrids on the two C atoms overlap head-on. Thus, the entire molecule must be linear.

A triple bond consists of one sigma bond and two pi bonds.

Other molecules containing triply bonded atoms are nitrogen, $:N\equiv N:$, propyne, $CH_3—C\equiv C—H$, and hydrogen cyanide, $H—C\equiv N:$. In each case, both atoms involved in the triple bonds are sp-hybridized. In the triple bond, each atom participates in one sigma and two pi bonds. The C atom in carbon dioxide, $:O=C=O:$, must participate in two pi bonds (to two

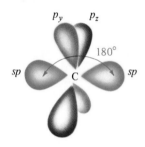

Figure 8-7
Diagram of the two linear hybridized sp orbitals (green) of an atom. These lie in a straight line, and the two unhybridized p orbitals p_y (orange) and p_z (purple) lie in the perpendicular plane.

> In propyne, the C atom in the CH_3 group is sp^3-hybridized and at the center of a tetrahedral arrangement.

Acetylene (ethyne) is used in welding.

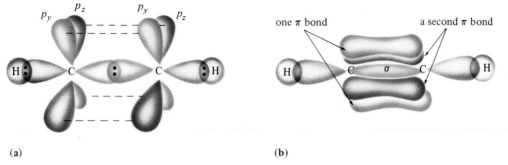

Figure 8-8
The acetylene molecule, C_2H_2. (a) The overlap diagram of two *sp*-hybridized carbon atoms and two *s* orbitals from two hydrogen atoms. One hybridized *sp* orbital on each C is shown in green, the other *sp* orbital on each atom is shown in gray, and the unhybridized *p* orbitals are shown in purple and orange. The dashed lines, each connecting two lobes, indicate the side-by-side overlap of the four unhybridized *p* orbitals to form two π bonds. There are two C—H σ bonds (gray), one C—C σ bond (green), and two C—C π bonds (purple and orange). This makes the net carbon–carbon bond a triple bond. (b) The π bonding orbitals (purple and orange) are positioned with one above and below the line of the σ bonds (green) and the other behind and in front of the line of the σ bonds.

different O atoms). It also participates in two sigma bonds, so it is also *sp*-hybridized and the molecule is linear.

8-15 A Summary of Electronic and Molecular Geometries

We have discussed several common types of polyatomic molecules and ions, and provided a reasonable explanation for the observed structures and polarities of these species. Table 8-4 provides a summation of the points developed in this chapter.

Our discussion of covalent bonding illustrates two important points:

1. Molecules and polyatomic ions have definite shapes.
2. The properties of molecules and polyatomic ions are determined to a great extent by their shapes. Incompletely filled electron shells and unshared pairs of electrons on the central element are very important.

Our ideas about chemical bonding have developed over many years. As experimental techniques for determining the *structures* of molecules have improved, our understanding of chemical bonding has improved also. Experimental observations on molecular geometry support our ideas about chemical bonding. The ultimate test for any theory is this: Can it correctly predict the results of experiments before they are performed? When the answer is *yes*, we have confidence in the theory. When the answer is *no*, the theory must be modified. Current theories of chemical bonding enable us to make predictions that are usually accurate.

Table 8-4
A Summary of Electronic and Molecular Geometries of Polyatomic Molecules and Ions

Regions of High Electron Density[a]	Electronic Geometry	Hybridization at Central Atoms (A)	Hybridized Orbital Orientation	Examples	Molecular Geometry
2	linear	sp (180°)		$BeCl_2$ $HgBr_2$ CdI_2 CO_2[b] C_2H_2[c]	linear linear linear linear linear
3	trigonal planar	sp^2 (120°)		BF_3^- BCl_3^- NO_3^{-e} $SO_2^{d,e}$ $NO_2^{-d,e}$ $C_2H_4^f$	trigonal planar trigonal planar trigonal planar angular (AB_2U) angular (AB_2U) planar (trig. planar at each C)
4	tetrahedral	sp^3 (109.5°)		CH_4 CCl_4 NH_4^+ SO_4^{2-} $CHCl_3$ NH_3^d SO_3^{2-d} H_3O^{+d} H_2O^d	tetrahedral tetrahedral tetrahedral tetrahedral distorted tet. pyramidal (AB_3U) pyramidal (AB_3U) pyramidal (AB_3U) angular (AB_2U_2)
5	trigonal bipyramidal	sp^3d or dsp^3 (90°, 120°, 180°)		PF_5 $AsCl_5$ SF_4^d ICl_3^d XeF_2^d I_3^{-d}	trigonal bipyramidal trigonal bipyramidal seesaw (AB_4U) T-shaped (AB_3U_2) linear (AB_2U_3) linear (AB_2U_3)
6	octahedral	sp^3d^2 or d^2sp^3 (90°, 180°)		SF_6 SeF_6 PF_6^- BrF_5^d XeF_4^d	octahedral octahedral octahedral square pyramidal (AB_5U) square planar (AB_4U_2)

[a] The number of locations of high electron density around the central atom. A region of high electron density may be a single bond, a double bond, a triple bond, or a lone pair. These determine the electronic geometry, and thus hybridization at the central element.
[b] Contains two double bonds.
[c] Contains a triple bond.
[d] Central atom in molecule or ion has lone pair(s) of electrons.
[e] Contains a resonant double bond.
[f] Contains one double bond.

Key Terms

Central atom   An atom in a molecule or polyatomic ion that is bonded to more than one other atom.

Electronic geometry   The geometric arrangement of orbitals containing the shared and unshared electron pairs surrounding the central atom of a molecule or polyatomic ion.

Hybridization   The mixing of a set of atomic orbitals to form a new set of atomic orbitals with the same total electron capacity and with properties and energies intermediate between those of the original unhybridized orbitals.

Ionic geometry   The arrangement of atoms (not lone pairs of electrons) about the central atom of a polyatomic ion.

Lewis acid   A substance that accepts a share in a pair of electrons from another species.

Lewis base   A substance that makes available a share in an electron pair.

Lewis dot formula   A method of representing a molecule or formula unit by showing atoms and only outer shell electrons; does not show shape.

Molecular geometry   The arrangement of atoms (*not* lone pairs of electrons) around a central atom of a molecule or polyatomic ion.

Octahedral   A term used to describe a molecule or polyatomic ion that has one atom in the center and six atoms at the corners of a regular octahedron.

Octahedron   A polyhedron with eight equal-sized, equilateral triangular faces and six apices (corners).

Overlap of orbitals   The interaction of orbitals on different atoms in the same region of space.

Pi (π) bond   A bond resulting from the side-on overlap of atomic orbitals, in which the regions of electron sharing are on opposite sides of and parallel to the imaginary line connecting the bonded atoms.

Sigma (σ) bond   A bond resulting from the head-on overlap of atomic orbitals, in which the region of electron sharing is along and (cylindrically) symmetrical to the imaginary line connecting the bonded atoms.

Square planar   A term used to describe molecules and polyatomic ions that have one atom in the center and four atoms at the corners of a square.

Tetrahedral   A term used to describe a molecule or polyatomic ion that has one atom in the center and four atoms at the corners of a regular tetrahedron.

Tetrahedron   A polyhedron with four equal-sized, equilateral triangular faces and four apices (corners).

Trigonal bipyramid   A six-sided polyhedron with five apices (corners), consisting of two pyramids sharing a common triangular base.

Trigonal bipyramidal   A term used to describe a molecule or polyatomic ion that has one atom in the center and five atoms at the corners of a trigonal bipyramid.

Trigonal planar   A term used to describe a molecule or polyatomic ion that has one atom in the center and three atoms at the corners of an equilateral triangle.

Valence bond (VB) theory   Assumes that covalent bonds are formed when atomic orbitals on different atoms overlap and electrons are shared.

Valence shell electron pair repulsion (VSEPR) theory   Assumes that electron pairs are arranged around the central element of a molecule or polyatomic ion so that there is maximum separation (and minimum repulsion) among regions of high electron density.

Exercises

VSEPR Theory—General Concepts

1. State in your own words the basic idea of the VSEPR theory.
2. (a) Distinguish between "lone pairs" and "bonding pairs" of electrons. (b) Which has the greater spatial requirement? How do we know this? (c) Indicate the order of increasing repulsions among lone pairs and bonding pairs of electrons.
3. Distinguish between electronic geometry and molecular geometry.
4. Under what conditions is molecular (or ionic) geometry identical to electronic geometry about a central atom?
5. What two shapes can a triatomic species have? How would the electronic geometries for the two shapes differ?
6. When using VSEPR theory to predict molecular geometry, how are double and triple bonds treated? How is a single nonbonding electron treated?
7. How does the presence of lone pairs of electrons on an atom influence the bond angles around that atom?
8. Sketch the three different possible arrangements of the two B atoms around the central atom A for the molecule AB_2U_3. Which of these structures correctly describes the molecular geometry? Why?
9. Sketch the three different possible arrangements of the three B atoms around the central atom A for the molecule AB_3U_2. Which of these structures correctly describes the molecular geometry? Why? What are the predicted ideal bond angles? How would the actual bond angles deviate from these values?

Valence Bond Theory—General Concepts

10. Describe the orbital overlap model of covalent bonding.

11. What are hybridized atomic orbitals? How is the theory of hybridized orbitals useful?

12. Prepare sketches of the overlaps of the following atomic orbitals: (a) s with s, (b) s with p along the bond axis, (c) p with p along the bond axis (head-on overlap), (d) p with p perpendicular to the bond axis (side-on overlap).

13. Prepare a sketch of the cross-section (through the atomic centers) taken between two atoms that have formed (a) a single σ bond, (b) a double bond consisting of a σ bond and a π bond, and (c) a triple bond consisting of a σ bond and two π bonds.

14. Prepare sketches of the orbitals around atoms that are (a) sp, (b) sp^2, (c) sp^3, (d) sp^3d, and (e) sp^3d^2 hybridized. Show in the sketches any unhybridized p orbitals that might participate in multiple bonding.

15. What form of hybridization is associated with these electronic geometries: trigonal planar, linear, tetrahedral, octahedral, trigonal bipyramidal?

16. What angles are associated with orbitals in the following hybridized sets of orbitals? (a) sp, (b) sp^2, (c) sp^3, (d) sp^3d, (e) sp^3d^2

17. What types of hybridization would you predict for molecules having the following general formulas? (a) AB_3, (b) AB_2U_2, (c) AB_3U, (d) ABU_4, (e) ABU_3

18. Repeat Exercise 17 for (a) ABU_5, (b) AB_2U_4, (c) AB_4, (d) AB_3U_2, (e) AB_5

19. What are the primary factors upon which we base a decision on whether the bonding in a molecule is better described in terms of simple orbital overlap or overlap involving hybridized atomic orbitals?

Electronic and Molecular Geometry

20. Draw a Lewis dot formula for each of the following species. Indicate the number of regions of high electron density and describe the electronic and molecular geometries. (a) H_2Be, (b) molecular $AlCl_3$, (c) SiH_4, (d) SF_6, (e) IO_4^-, (f) NCl_3

21. Draw a Lewis dot formula for each of the following species. Indicate the number of regions of high electron density and describe the electronic and molecular or ionic geometries. (a) IF_4^-, (b) CO_2, (c) AlH_4^-, (d) NH_4^+, (e) $AsCl_3$, (f) ClO_3^-

22. Draw a Lewis dot formula for each of the following species. Indicate the number of regions of high electron density and describe the electronic and molecular or ionic geometries. (a) SeF_6, (b) $ONCl$ (N is the central atom), (c) Cl_2CO, (d) PCl_5, (e) BCl_3, (f) ClO_4^-

23. Draw a Lewis dot formula for each of the following species. Indicate the number of regions of high electron density and describe the electronic and molecular geometries. (a) molecular $HgCl_2$, (b) ClO_2, (c) XeF_2, (d) CCl_4, (e) $CdCl_2$, (f) $AsCl_5$

24. Draw a Lewis dot formula for each of the following species. Indicate the number of regions of high electron density and the electronic and molecular or ionic geometries. (a) H_2O, (b) $SnCl_4$, (c) SeF_6, (d) SbF_6^-

25. Draw a Lewis dot formula for each of the following species. Indicate the number of regions of high electron density and the electronic and molecular or ionic geometries. (a) BF_3, (b) NF_3, (c) BrO_3^-, (d) $SiCl_4$

26. (a) What would be the ideal bond angles in the species in Exercise 24, ignoring lone pair effects? (b) How do these differ, if at all, from the actual values? Why?

27. (a) What would be the ideal bond angles in the molecules or ions in Exercise 25, ignoring lone pair effects? (b) Are these values greater than, less than, or equal to the actual values? Why?

28. Carbon forms two common oxides, CO and CO_2, both of which are linear. It forms a third (very uncommon) oxide, carbon suboxide, C_3O_2, which is also linear. The structure has terminal oxygen atoms on both ends. Draw Lewis dot and dash formulas for C_3O_2. How many regions of high electron density are there about each of the three carbon atoms?

29. Pick the member of each pair that you would expect to have the smaller bond angles, if different, and explain why. (a) SF_2 and SO_2, (b) BF_3 and BCl_3, (c) CF_4 and SF_4, (d) NH_3 and H_2O.

30. Draw a Lewis dot formula, sketch the three-dimensional shape, and name the electronic and ionic geometries for the following polyatomic ions. (a) $AsCl_4^-$, (b) PCl_6^-, (c) PCl_4^-, (d) $AsCl_4^+$

31. As the name implies, the interhalogens are compounds that contain two halogens. Draw Lewis dot formulas and three-dimensional structures for the following. Name the electronic and molecular geometries of each. (a) IF_5, (b) IBr, (c) ICl_5

32. A number of ions derived from the interhalogens are known. Draw Lewis dot formulas and three-dimensional structures for the following ions. Name the electronic and ionic geometries of each. (a) IF_4^+, (b) ICl_2^-, (c) BrF_4^-.

*33. (a) Draw a Lewis dot formula for each of the following molecules: BF_3, NF_3, BrF_3. (b) Contrast the molecular geometries of these three molecules. Account for differences in terms of the VSEPR theory.

*34. (a) Draw a Lewis dot formula for each of the following molecules: GeF_4, SF_4, XeF_4. (b) Contrast the molecular geometries of these three molecules. Account for differences in terms of the VSEPR theory.

35. Draw the Lewis structures and predict the shapes of these very reactive carbon-containing species: H_3C^+ (a carbocation), $H_3C:^-$ (a carbanion), and $:CH_2$ (a carbene whose unshared electrons are paired).

36. Draw the Lewis structures and predict the shapes of (a) ICl_4^-, (b) $TeCl_4$, (c) XeO_3, (d) $BrNO$ (N is the central atom), (e) $ClNO_2$ (N is the central atom), (f) Cl_2SO (S is the central atom).

37. Describe the shapes of these polyatomic ions: (a) BO_3^{3-}, (b) PO_4^{3-}, (c) SO_3^{2-}, (d) NO_2^-.
38. Describe the shapes of these polyatomic ions: (a) H_3O^+, (b) GeF_3^-, (c) ClF_3^{2-}, (d) $IO_2F_2^-$.
39. Which of the following molecules are polar? Why? (a) CH_4, (b) CH_3Br, (c) CH_2Br_2, (d) $CHBr_3$, (e) CBr_4
40. Which of the following molecules are polar? Why? (a) CdI_2, (b) BCl_3, (c) PCl_3, (d) H_2O, (e) SF_6
41. Which of the following molecules have dipole moments of zero? Justify your answer. (a) SO_3, (b) IF, (c) Cl_2O, (d) $AsCl_3$
42. The PF_3Cl_2 molecule has a dipole moment of zero. Use this information to sketch its three-dimensional shape. Justify your choice.
*43. In what two major ways does the presence of lone pairs of valence electrons affect the polarity of a molecule? Describe two molecules for which the presence of lone pairs on the central atom helps to make the molecules polar. Can you think of a bonding arrangement that has lone pairs of valence electrons on the central atom but that is nonpolar?

Valence Bond Theory

44. What is the hybridization on the central atom in each of the following? (a) H_2Be, (b) molecular $AlCl_3$, (c) SiH_4, (d) SF_6, (e) IO_4^-, (f) NCl_3
45. What is the hybridization on the central atom in each of the following? (a) ICl_4^-, (b) CO_2, (c) AlH_4^-, (d) NH_4^+, (e) $AsCl_3$, (f) ClO_3^-
46. What is the hybridization on the central atom in each of the following? (a) SeF_6, (b) $ONCl$ (N is the central atom), (c) Cl_2CO, (d) PCl_5, (e) BCl_3, (f) ClO_4^-
47. What is the hybridization on the central atom in each of the following? (a) molecular $MgCl_2$, (b) ClO_2, (c) XeF_2, (d) CCl_4, (e) $CdCl_2$, (f) $AsCl_5$
48. (a) Describe the hybridization of the central atom in each of these covalent species. (1) $CHCl_3$, (2) BeH_2, (3) NCl_3, (4) ClO_3^-, (5) IF_6^+, (6) SiF_6^{2-}. (b) Give the shape of each species.
49. Describe the hybridization of the underlined atoms in $\underline{C}_2Cl_4$, $\underline{C}_2Cl_2$, $\underline{N}_2F_4$, and $(\underline{H_2N})_2\underline{C}O$.
50. (a) Describe the hybridization of N in NO_2^+ and NO_2^-. (b) Predict the bond angle in each case.
*51. After comparing experimental and calculated dipole moments, Charles A. Coulson suggested that the Cl atom in HCl is *sp*-hybridized. (a) Give the orbital electronic structure for an *sp*-hybridized Cl atom. (b) Which HCl molecule would have a larger dipole moment—one in which the chlorine uses pure *p* orbitals for bonding with the H atom or one in which *sp* hybrid orbitals are used?
*52. Predict the hybridization at each carbon atom in each of the following molecules.
    (a) ethanol (ethyl alcohol or grain alcohol):

(b) alanine (an amino acid):

(c) tetracyanoethylene:

(d) chloroprene (used to make neoprene, a synthetic rubber):

(e) 3-penten-1-yne:

*53. Predict the hybridization at the numbered atoms (①, ②, and so on) in the following molecules and predict the approximate bond angles at those atoms.
    (a) diethyl ether, an anesthetic:

(b) caffeine, a stimulant in coffee and in many over-the-counter medicinals:*

---

* In this kind of structural drawing, each intersection of lines represents a C atom; sometimes H atoms are not shown at all these intersections.

(c) acetylsalicylic acid (aspirin):

(d) nicotine:

(e) ephedrine (used as a nasal decongestant):

54. How many sigma and how many pi bonds are there in each of the following molecules?

*55. How many sigma and how many pi bonds are there in each of the following molecules?

(d) $CH_2CHCHCHCH_2CH_3$

56. Describe the bonding in the N_2 molecule with a three-dimensional VB structure. Show the orbital overlap and label the orbitals.

57. Draw the Lewis structures for molecular oxygen and ozone. Assuming that all of the oxygen atoms are hybridized, what will be the hybridization of the oxygen atoms in each substance? Prepare sketches of the molecules.

*58. A water solution of cadmium bromide, $CdBr_2$, contains not only Cd^{2+} and Br^- ions, but also $CdBr^+$, $CdBr_2$, $CdBr_3^-$, and $CdBr_4^{2-}$. Describe the type of hybrid orbital used by Cd in each polyatomic species and describe the shape of the species.

*59. In their crystalline states, PCl_5 exists as $(PCl_4)^+(PCl_6)^-$, and PBr_5 exists as $PBr_4^+Br^-$. (a) Predict the shapes of all the polyatomic ions. (b) Indicate the hybrid orbital structure for the P atom in each of its different types of ions.

*60. Draw a dash formula and a three-dimensional structure for each of the following polycentered molecules. Indicate hybridizations and bond angles at each carbon atom. (a) Butane, C_4H_{10}; (b) 1-butene, $H_2C=CHCH_2CH_3$; (c) 1-butyne, $HC\equiv CCH_2CH_3$; (d) acetaldehyde, CH_3CHO

61. How many σ bonds and how many π bonds are there in each of the molecules of Exercise 60?

62. (a) Describe the hybridization of N in each of these species? (1) $:NH_3$, (2) NH_4^+, (3) $HN=NH$, (4) $HC\equiv N:$.
(b) Give an orbital description for each species, specifying the location of any unshared pairs and the orbitals used for the multiple bonds.

63. Draw the Lewis structures and predict the orbital types and the shapes of these polyatomic ions and covalent molecules: (a) $HgCl_2$, (b) BF_3, (c) BF_4^-, (d) SeF_2, (e) $AsCl_5$, (f) SbF_6^-.

64. (a) What is the hybridized state of each C in these molecules? (1) $Cl_2C=O$, (2) $HC\equiv N$, (3) CH_3CH_3, (4) ketene, $H_2C=C=O$. (b) Describe the shape of each molecule.

*65. The following fluorides of xenon have been well characterized: XeF_2, XeF_4, and XeF_6. (a) Draw Lewis structures for these substances and decide what type of hybridization of the Xe atomic orbitals has taken place. (b) Draw all of the possible atomic arrangements of XeF_2 and discuss your choice of molecular geometry. (c) What shape do you predict for XeF_4?

*66. Iodine and fluorine form a series of interhalogen molecules and ions. Among these are IF (minute quantities observed spectroscopically), IF_3, IF_4^-, IF_5, IF_6^-, and IF_7. (a) Draw Lewis structures for each of these species. (b) Identify the type of hybridization that the orbitals of the iodine atom have undergone in each substance. (c) Identify the shape of the molecule or ion.

Mixed Exercises

67. In the pyrophosphate ion, $P_2O_7^{4-}$, one oxygen atom is bonded to both phosphorus atoms. Draw a Lewis dot formula and sketch the three-dimensional shape of the ion. Describe the ionic geometry with respect to the central O atom and with respect to each P atom.

68. Briefly discuss the bond angles in the hydroxylamine molecule in terms of the ideal geometry and the small changes caused by electron pair repulsions.

$$H-\overset{\cdot\cdot}{\underset{|}{N}}-\overset{\cdot\cdot}{\underset{\cdot\cdot}{O}}-H$$
$$\overset{|}{H}$$

69. Repeat Exercise 68 for nitric acid.

$$H-\overset{\cdot\cdot}{\underset{\cdot\cdot}{O}}-\overset{\overset{\cdot}{O}\overset{\cdot}{}}{\underset{\|}{N}}-\overset{\cdot\cdot}{\underset{\cdot\cdot}{O}}: \longleftrightarrow H-\overset{\cdot\cdot}{\underset{\cdot\cdot}{O}}-\overset{:\overset{\cdot\cdot}{O}:}{\underset{|}{N}}=\overset{\cdot\cdot}{O}$$

*70. The methyl free radical $\cdot CH_3$ has bond angles of about 120°, whereas the methyl carbanion $:CH_3^-$ has bond angles of about 109°. What can you infer from these facts about the repulsive force exerted by an unpaired, unshared electron as compared to that exerted by an unshared pair of electrons?

*71. Two Lewis structures can be written for the square-planar molecule $PtCl_2Br_2$:

$$\begin{matrix} Br & & Cl \\ & \diagdown \diagup & \\ & Pt & \\ & \diagup \diagdown & \\ Br & & Cl \end{matrix} \quad and \quad \begin{matrix} Br & & Cl \\ & \diagdown \diagup & \\ & Pt & \\ & \diagup \diagdown & \\ Cl & & Br \end{matrix}$$

Show how a difference in dipole moments can distinguish between these two possible structures.

72. Prepare a sketch of the molecule $CH_3CH=CH_2$ showing orbital overlaps. Identify the type of hybridization of atomic orbitals for each carbon atom.

*73. The skeleton for the nitrous acid molecule, HNO_2, is shown. What are the hybridizations at the middle O and N atoms? Draw the dash formula. Is this consistent with your predictions?

$$\begin{matrix} H & & \\ \diagdown & 104° & \\ 0.98\ \text{Å} & \overset{}{O} \xrightarrow{1.45\ \text{Å}} N & \\ & & \diagdown \\ & 116° & \\ & & O \end{matrix}$$

74. Describe the change in hybridization that occurs at the central atom of the reactant at the left in each of the following reactions.

(a) $PF_5 + F^- \longrightarrow PF_6^-$
(b) $2\ CO + O_2 \longrightarrow 2\ CO_2$
(c) $AlI_3 + I^- \longrightarrow AlI_4^-$
(d) What change in hybridization occurs in the following reaction?

$$:NH_3 + BF_3 \longrightarrow H_3N:BF_3$$

*75. Consider the following proposed Lewis structures for ozone (O_3):

(i) $:\overset{\cdot\cdot}{O}-\overset{\cdot\cdot}{O}=\overset{\cdot\cdot}{O}: \longleftrightarrow :\overset{\cdot\cdot}{O}=\overset{\cdot\cdot}{O}-\overset{\cdot\cdot}{O}:$ (ii) $:\overset{:\overset{\cdot\cdot}{O}:}{\underset{\diagup\diagdown}{O}}\overset{\cdot\cdot}{\underset{}{O}}:$

(iii) $:\overset{\cdot\cdot}{O}-\overset{\cdot\cdot}{O}-\overset{\cdot\cdot}{O}:$

(a) Which of these structures correspond to a polar molecule?
(b) Which of these structures predict covalent bonds of equal lengths and strengths?
(c) Which of these structures predict a diamagnetic molecule?
(d) The properties listed in (a), (b), and (c) are those observed for ozone. Which structure correctly predicts all three?
(e) Which of these structures contains a considerable amount of "strain"?

76. What hybridizations are predicted for the central atoms in molecules having the formulas AB_2U_2 and AB_3U? What are the predicted bond angles for these molecules? The actual bond angles for representative substances are

H_2O	104.45°	NH_3	106.67°
H_2S	92.2°	PH_3	93.67°
H_2Se	91.0°	AsH_3	91.8°
He_2Te	89.5°	SbH_3	91.3°

What would be the predicted bond angle if no hybridization occurred? What conclusion can you draw concerning the importance of hybridization for molecules of compounds involving elements with higher atomic numbers?

77. Describe the orbitals (s, p, sp^2, and so on) of the central atom used for bond orbitals and unshared pairs in (a) H_2S (bond angle 91°), (b) CH_3OCH_3 (C—O—C angle 110°), (c) S_8 (S—S—S angle 105°), (d) $(CH_3)_2Sn^{2+}$ (C—Sn—C angle 180°), (e) $(CH_3)_3Sn^+$ (planar with respect to the Sn and 3 C atoms), (f) $SnCl_2$ (angle 95°).

Objectives

As you study this chapter, you should learn to

☐ Understand the basic ideas of molecular orbital theory

☐ Relate the shapes and overlap of atomic orbitals to the shapes and energies of the resulting molecular orbitals

☐ Distinguish between bonding and antibonding orbitals

☐ Apply the Aufbau Principle to find the molecular orbital descriptions for homonuclear diatomic molecules and their ions

☐ Apply the Aufbau Principle to find the molecular orbital descriptions for heteronuclear diatomic molecules and their ions

☐ Find the bond order in diatomic molecules

☐ Relate bond order to bond stability

☐ Use the MO concept of delocalization for molecules in which VB theory would postulate resonance

An early triumph of molecular orbital theory was its ability to account for the observed paramagnetism of oxygen, O_2. According to earlier theories, O_2 was expected to be diamagnetic, that is, to have only paired electrons.

W e have described bonding and molecular geometry in terms of valence bond theory. In valence bond theory, we postulate that bonds result from the sharing of electrons in overlapping orbitals of different atoms. These orbitals may be *pure atomic orbitals* or *hybridized atomic orbitals* of *individual* atoms. We describe electrons in overlapping orbitals of different atoms as being localized in the bonds between the two atoms involved, rather than delocalized over the entire molecule. We then invoke hybridization when it helps to account for the geometry of a molecule.

In **molecular orbital theory**, we postulate

the combination of atomic orbitals on different atoms forms **molecular orbitals** (MOs), so that electrons in them belong to the molecule as a whole.

The valence bond and molecular orbital approaches have strengths and weaknesses that are complementary. They are alternative descriptions of chemical bonding. Valence bond theory is descriptively attractive and it lends itself well to visualization. Molecular orbital (MO) theory gives better descriptions of electron cloud distributions, bond energies, and magnetic properties, but its results are not as easy to visualize.

In some polyatomic molecules, a molecular orbital may extend over only a fraction of the molecule. Molecular orbitals also exist for polyatomic ions such as CO_3^{2-}, SO_4^{2-}, and NH_4^+.

The valence bond picture of bonding in the O_2 molecule, involves a double bond and no unpaired electrons. It predicts sp^2 hybridization at each oxygen because there are three sets of valence electrons on each O atom.

$$\ddot{\text{O}}::\ddot{\text{O}}$$

However, experiments show that O_2 is paramagnetic; therefore, it has unpaired electrons. Thus, the VB structure is inconsistent with experiment and cannot be accepted as a suitable description of the bonding. Molecular orbital theory *predicts* that O_2 has two unpaired electrons. This ability of MO theory to explain the paramagnetism of O_2 brought it to the forefront as a major theory in bonding. In the following sections, we shall develop some of the notions of MO theory. Then we shall apply them to some molecules and ions.

9-1 Molecular Orbitals

We learned in Chapter 5 that each solution to the Schrödinger equation, called a wave function, represents an atomic orbital. The mathematical pictures of hybrid orbitals in valence bond theory can be generated by combining the wave functions that describe two or more atomic orbitals on a *single* atom. Similarly, combining wave functions that describe atomic orbitals on *separate* atoms generates mathematical descriptions of molecular orbitals.

An orbital has physical meaning only when we square its wave function to describe the electron density. Thus, the overall sign on the wave function that describes an atomic orbital is arbitrary. But when we *combine* two orbitals, the signs relative to one another do matter. When waves are combined, they may interact either constructively or destructively (Figure 9-1). Likewise, when two atomic orbitals overlap, they can be in phase (added) or out of phase (subtracted). When they overlap in phase, constructive interference occurs in the region between the nuclei, and a **bonding orbital** is produced. The energy of the bonding orbital is always lower (more stable) than the energies of the combining orbitals. When they overlap out of phase, destructive interference reduces the probability of finding electrons in the region between the nuclei, and an **antibonding orbital** is produced. This is higher in energy (less stable) than the original atomic orbitals. The overlap of two atomic orbitals always produces two MOs—one bonding and one antibonding.

We can illustrate this basic principle by considering the combination of the 1s atomic orbitals *on two different atoms* (Figure 9-2). When these orbitals are occupied by electrons, the shapes of the orbitals are really plots of electron density. These plots show the regions in molecules where the probabilities of finding electrons are greatest.

In the bonding orbital, the two 1s orbitals have reinforced one another in the region between the two nuclei by in-phase overlap, or addition of their electron waves. In the antibonding orbital, they have canceled one another in this region by out-of-phase overlap, or subtraction of their electron waves. We designate both molecular orbitals as *sigma* (σ) *orbitals* (which indicates that they are cylindrically symmetrical about the internuclear axis). We

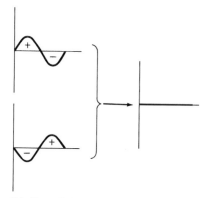

(a) In-phase overlap (add)

(b) Out-of-phase overlap (subtract)

Figure 9-1

An illustration of constructive and destructive interaction of waves. (a) If the two identical waves shown at the left are added, they interfere constructively to produce the more intense wave at the right. (b) Conversely, if they are subtracted, it is as if the phases (signs) of one wave were reversed and added to the first wave. This causes destructive interference, resulting in the wave at the right with zero amplitude.

Although this structure is correct by the standards of valence bond theory, it is inconsistent with the observed paramagnetism of O_2.

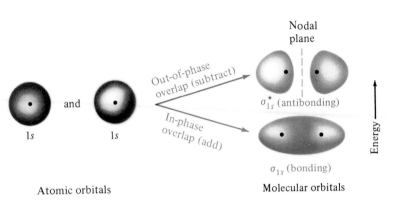

Atomic orbitals

Molecular orbitals

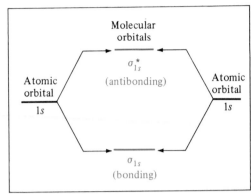

Figure 9-2
Molecular orbital (MO) diagram for the combination of the 1s atomic orbitals on two identical atoms (at the left) to form two MOs. One is a *bonding* orbital, σ_{1s} (blue), resulting from addition of the wave functions of the 1s orbitals. The other is an *antibonding* orbital, $\sigma_{1s}^{\star}$ (red), at higher energy resulting from subtraction of the waves that describe the combining 1s orbitals. In all σ-type MOs, the electron density is symmetrical about an imaginary line connecting the two nuclei. The terms "subtraction of waves," "out of phase," and "destructive interference in the region between the nuclei" all refer to the formation of an antibonding MO. Nuclei are represented by dots.

indicate with subscripts the atomic orbitals that have been combined. The star denotes an antibonding orbital. Thus, two 1s orbitals produce a σ_{1s} (read "sigma-1s") bonding orbital and a $\sigma_{1s}^{\star}$ (read "sigma-1s-star") antibonding orbital. The right-hand side of Figure 9-2 shows the relative energy levels of these orbitals. All sigma antibonding orbitals have nodal planes bisecting the internuclear axis. A **node** or **nodal plane** is a region in which the probability of finding electrons is zero.

Another way of viewing the relative stabilities of these orbitals follows. In a bonding molecular orbital, there is high electron density *between* the two atoms, where it stabilizes the arrangement by exerting a strong attraction for both nuclei. By contrast, an antibonding orbital has a node (a region of zero electron density) between the nuclei; this allows a strong net repulsion between the nuclei, destabilizing the arrangement. Electrons are *more* stable (have lower energy) in bonding molecular orbitals than in the individual atoms. Placing electrons in antibonding orbitals, on the other hand, requires raising their energy, which makes them *less* stable than in the individual atoms.

For any two sets of *p* orbitals on two different atoms, corresponding orbitals such as p_x orbitals can overlap *head-on*. This gives σ_p and $\sigma_p^{\star}$ orbitals, as shown in Figure 9-3 for the head-on overlap of $2p_x$ orbitals on two atoms. If the remaining *p* orbitals overlap (p_y with p_y and p_z with p_z), they must do so sideways, or *side-on*, forming *pi (π) molecular orbitals*. De-

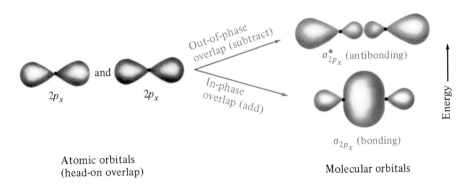

Atomic orbitals
(head-on overlap)

Molecular orbitals

How we name the axes is arbitrary. We shall designate the internuclear axis as the *x* direction.

Figure 9-3
Production of σ_{2p_x} and $\sigma_{2p_x}^{\star}$ molecular orbitals by overlap of $2p_x$ orbitals on two atoms.

If we had chosen the z-axis as the axis of head-on overlap of the 2p orbitals in Figure 9-3, side-on overlap of the $2p_x$–$2p_x$ and $2p_y$–$2p_y$ orbitals would form the π-type molecular orbitals.

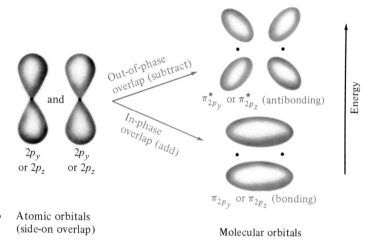

Atomic orbitals
(side-on overlap)

Molecular orbitals

Figure 9-4
The π_{2p} and $\pi^\star_{2p}$ molecular orbitals from overlap of one pair of 2p atomic orbitals (for instance, $2p_y$ orbitals). There can be an identical pair of molecular orbitals at right angles to these, formed by another pair of p orbitals on the same two atoms (in this case, $2p_z$ orbitals).

This would involve rotating Figures 9-2, 9-3, and 9-4 by 90° so that the internuclear axes are perpendicular to the plane of the pages.

pending on whether all p orbitals overlap, there can be as many as two π_p and two $\pi^\star_p$ orbitals. Figure 9-4 illustrates the overlap of two corresponding 2p orbitals on two atoms to form π_{2p} and $\pi^\star_{2p}$ molecular orbitals. There is a nodal plane along the internuclear axis for all pi molecular orbitals. If one views a sigma molecular orbital along the internuclear axis, it appears to be symmetrical around the axis like a pure s atomic orbital. A similar cross-sectional view of a pi molecular orbital looks like a pure p atomic orbital, with a node along the internuclear axis.

> The number of molecular orbitals (MOs) formed is equal to the number of atomic orbitals that are combined. When two atomic orbitals are combined, one of the resulting MOs is at a *lower* energy than the original atomic orbitals; this is a *bonding* orbital. The other MO is at a *higher* energy than the original atomic orbitals; this is an *antibonding* orbital.

9-2 Molecular Orbital Energy-Level Diagrams

In the same way that atomic orbitals can be arranged by increasing energy into an energy-level diagram (Figure 5-28), we can draw molecular orbital energy-level diagrams for simple molecules. The simplest examples, shown in Figure 9-5, apply to homonuclear diatomic molecules of elements in the first and second periods. Each diagram is an extension of the right-hand diagram in Figure 9-2, to which we have added the molecular orbitals formed from 2s and 2p atomic orbitals.

For the cases shown in Figure 9-5a, the two π_{2p} orbitals are lower in energy than the σ_{2p} orbital. However, molecular orbital calculations indicate that for O_2, F_2, and (hypothetical) Ne_2 molecules, the σ_{2p} orbital is lower in energy than the π_{2p} orbitals (Figure 9-5b).

Spectroscopic data support these orders.

Diagrams such as these are used to describe the bonding in a molecule in MO terms. We simply apply the Aufbau Principle and follow the same rules that we did for atomic energy-level diagrams (Section 5-16).

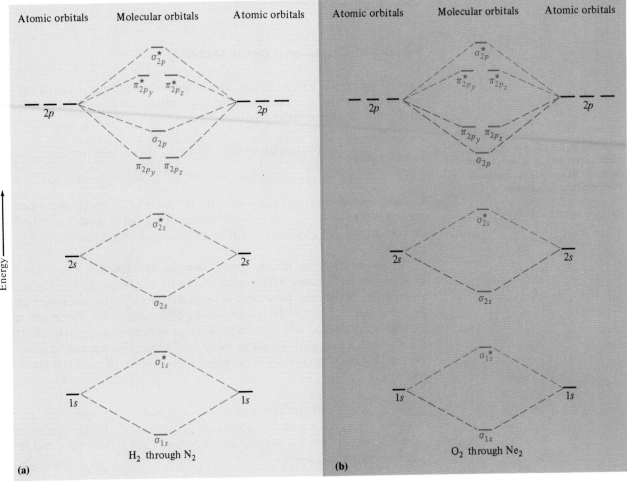

Atomic orbitals   Molecular orbitals   Atomic orbitals

$\sigma^{\star}_{2p}$

$\pi^{\star}_{2p_y}$ $\pi^{\star}_{2p_z}$

2p

2p

σ_{2p}

π_{2p_y} π_{2p_z}

$\sigma^{\star}_{2s}$

2s

2s

σ_{2s}

$\sigma^{\star}_{1s}$

1s

1s

σ_{1s}

H_2 through N_2

(a)

Atomic orbitals   Molecular orbitals   Atomic orbitals

$\sigma^{\star}_{2p}$

$\pi^{\star}_{2p_y}$ $\pi^{\star}_{2p_z}$

2p

2p

π_{2p_y} π_{2p_z}

σ_{2p}

$\sigma^{\star}_{2s}$

2s

2s

σ_{2s}

$\sigma^{\star}_{1s}$

1s

1s

σ_{1s}

O_2 through Ne_2

(b)

Energy →

Figure 9-5
Energy-level diagrams for first- and second-period homonuclear diatomic molecules and
ions. The solid lines represent the relative energies of the indicated atomic and molecular
orbitals. (a) The diagram for H_2, He_2, Li_2, Be_2, B_2, C_2, and N_2 molecules and their ions.
(b) The diagram for O_2, F_2, and Ne_2 molecules and their ions.

1. Draw (or select) the appropriate molecular orbital energy-level diagram.
2. Determine the *total* number of electrons in the molecule. Note that
   in applying MO theory, we usually account for *all* electrons. This
   includes both the core electrons and the valence electrons.
3. Add these electrons to the energy-level diagram, putting each electron into the lowest energy level available.
   a. A maximum of *two* electrons can occupy any given molecular
      orbital, and then only if they have opposite spin (Pauli Exclusion
      Principle).
   b. Electrons must occupy all the orbitals of the same energy singly
      before pairing begins. These unpaired electrons have parallel
      spins (Hund's Rule).

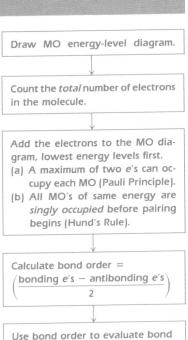

Draw MO energy-level diagram.

↓

Count the *total* number of electrons
in the molecule.

↓

Add the electrons to the MO diagram, lowest energy levels first.
(a) A maximum of two e's can oc-
    cupy each MO (Pauli Principle).
(b) All MO's of same energy are
    singly occupied before pairing
    begins (Hund's Rule).

↓

Calculate bond order =
$\left(\dfrac{\text{bonding } e\text{'s} - \text{antibonding } e\text{'s}}{2}\right)$

↓

Use bond order to evaluate bond
stability.

9-3 Bond Order and Bond Stability

Now all we need is a way to judge the stability of a molecule, once its energy-level diagram has been filled with the appropriate number of electrons. This criterion is the **bond order** (bo):

Electrons in bonding orbitals are often called **bonding electrons**; electrons in antibonding orbitals, **antibonding electrons**.

$$\text{bond order} = \frac{(\text{no. of bonding electrons}) - (\text{no. of antibonding electrons})}{2}$$

Usually the bond order corresponds to the number of bonds described by the valence bond theory. Fractional bond orders exist in species that contain an odd number of electrons, such as the nitrogen oxide molecule, NO (15 electrons).

A bond order *equal to zero* means that the molecule has equal numbers of electrons in bonding MOs (more stable than in separate atoms) and in antibonding MOs (less stable than in separate atoms); such a molecule would be *unstable*. A bond order *greater than zero* means that there are more electrons in bonding MOs (stabilizing) than in antibonding MOs (destabilizing). Such a molecule would be more stable than the separate atoms, and we predict that its existence is possible; such a molecule could still be quite reactive.

> The greater the bond order of a diatomic molecule or ion, the more stable we predict it to be. Likewise, for a bond between two given atoms, the greater the bond order, the shorter the bond length and the greater the bond energy.

The **bond energy** is the amount of energy necessary to break a mole of bonds (Section 15-9); it is therefore a measure of bond strength.

9-4 Homonuclear Diatomic Molecules

"Homonuclear" means consisting only of atoms of the same element. "Diatomic" means consisting of two atoms.

The electron distributions for the homonuclear diatomic molecules of the first and second periods are shown in Table 9-1 together with their bond orders, bond lengths, and bond energies. We shall now look at some of these as examples.

The Hydrogen Molecule, H₂

The overlap of the $1s$ orbitals of two hydrogen atoms produces σ_{1s} and $\sigma_{1s}^{\star}$ molecular orbitals. The two electrons of the molecule occupy the lower-energy σ_{1s} orbital (Figure 9-6).

Because the two electrons in an H_2 molecule are in a bonding orbital, the bond order is one. We conclude that the H_2 molecule would be stable, and it is. The energy associated with two electrons in the H_2 molecule is less than that associated with the same two electrons in $1s$ atomic orbitals. The lower the energy of a system, the more stable it is.

H_2 bo $= \dfrac{2 - 0}{2} = 1$

Table 9-1
Molecular Orbitals for First and Second Row Diatomic Molecules[a]

Increasing energy (not to scale)		H₂	He₂[c]	Li₂[b]	Be₂[b]	B₂[b]	C₂[b]	N₂		O₂	F₂	Ne₂[c]
σ^*_{2p}		—	—	—	—	—	—	—		—	—	↑↓
$\pi^*_{2p_y}, \pi^*_{2p_z}$		— —	— —	— —	— —	— —	— —	— —		↑ ↑	↑↓ ↑↓	↑↓ ↑↓
σ_{2p}		—	—	—	—	—	—	↑↓	π_{2p_y}, π_{2p_z}	↑↓ ↑↓	↑↓ ↑↓	↑↓ ↑↓
π_{2p_y}, π_{2p_z}		— —	— —	— —	— —	↑ ↑	↑↓ ↑↓	↑↓ ↑↓	σ_{2p}	↑↓	↑↓	↑↓
σ^*_{2s}		—	—	—	↑↓	↑↓	↑↓	↑↓		↑↓	↑↓	↑↓
σ_{2s}		—	—	↑↓	↑↓	↑↓	↑↓	↑↓		↑↓	↑↓	↑↓
σ^*_{1s}		—	↑↓	↑↓	↑↓	↑↓	↑↓	↑↓		↑↓	↑↓	↑↓
σ_{1s}		↑↓	↑↓	↑↓	↑↓	↑↓	↑↓	↑↓		↑↓	↑↓	↑↓
Paramagnetic?		no	no	no	no	yes	no	no		yes	no	no
Bond order		1	0	1	0	1	2	3		2	1	0
Observed bond length (Å)		0.74	—	2.67	—	1.59	1.31	1.09		1.21	1.43	—
Observed bond energy (kJ/mol)		435	—	110	9	~270	602	946		498	159	—

[a] Electron distribution in molecular orbitals, bond order, bond length, and bond energy of homonuclear diatomic molecules of the first- and second-row elements. Note that nitrogen molecules, N₂, have the highest bond energies listed; they have a bond order of three. The species C₂ and O₂, with a bond order of two, have the next highest bond energies.
[b] Exists only in the vapor state at elevated temperatures.
[c] Unknown species.

The Helium Molecule (hypothetical), He₂

The energy-level diagram for He₂ is similar to that for H₂ except that it has two more electrons. These occupy the antibonding σ^*_{1s} orbital (Figures 9-5a and 9-6b and Table 9-1), giving it a bond order of zero. That is, two electrons in He₂ would be *more stable* than in the separate atoms, but the other two would be *less stable* by the same amount. Thus, the molecule would be no more stable than the separate atoms and would not exist. In fact, He₂ is not known.

$$\text{He}_2 \text{ bo} = \frac{2-2}{2} = 0$$

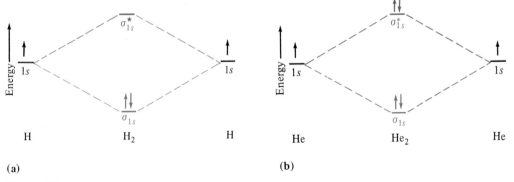

Figure 9-6
Molecular orbital diagrams for (a) H₂ and (b) He₂.

The Boron Molecule, B_2

The boron atom has the configuration $1s^2 2s^2 2p^1$. Now the p electrons do participate in bonding. Figure 9-5a and Table 9-1 show that the π_{p_y} and π_{p_z} molecular orbitals are lower in energy than the σ_{2p} for B_2. Thus, the electron configuration is $\sigma_{1s}^2\,\sigma_{1s}^{\star 2}\,\sigma_{2s}^2\,\sigma_{2s}^{\star 2}\,\pi_{2p_y}^1\,\pi_{2p_z}^1$. The unpaired electrons are consistent with the observed paramagnetism of B_2. Here we illustrate Hund's Rule in molecular orbital theory. The π_{2p_y} and π_{2p_z} orbitals are equal in energy and contain a total of two electrons. As a result, one electron occupies each orbital. The bond order is one. Experiments verify that the molecule exists in the vapor state.

$$B_2\ bo = \frac{6-4}{2} = 1$$

Orbitals of equal energy are called *degenerate* orbitals. Hund's Rule for filling degenerate orbitals was discussed in Section 5-16.

The Nitrogen Molecule, N_2

Experimental thermodynamic data show that the N_2 molecule is stable, is diamagnetic, and has a very high bond energy, 946 kJ/mol. This is consistent with molecular orbital theory. Each nitrogen atom has seven electrons, so the diamagnetic N_2 molecule has 14 electrons, distributed as follows:

$$\sigma_{1s}^2 \quad \sigma_{1s}^{\star 2} \quad \sigma_{2s}^2 \quad \sigma_{2s}^{\star 2} \quad \pi_{2p_y}^2 \quad \pi_{2p_z}^2 \quad \sigma_{2p}^2$$

There are six more electrons in bonding orbitals than in antibonding orbitals, so the bond order is three. We see (Table 9-1) that N_2 has a very short bond length, only 1.09 Å, the shortest of any diatomic species except H_2.

$$N_2\ bo = \frac{10-4}{2} = 3$$

In the valence bond representation, N_2 is shown as N≡N, with a triple bond.

The Oxygen Molecule, O_2

Among the homonuclear diatomic molecules, only N_2 and the very small H_2 have shorter bond lengths than O_2, 1.21 Å. Recall that VB theory predicts that O_2 is diamagnetic. However, experiments show that it is paramagnetic, with two unpaired electrons. MO theory predicts a structure consistent with this observation. For O_2, the σ_{2p} orbital is lower in energy than the π_{2p_y} and π_{2p_z} orbitals. Each oxygen atom has eight electrons, so the O_2 molecule has 16, distributed as follows:

$$\sigma_{1s}^2 \quad \sigma_{1s}^{\star 2} \quad \sigma_{2s}^2 \quad \sigma_{2s}^{\star 2} \quad \sigma_{2p}^2 \quad \pi_{2p_y}^2 \quad \pi_{2p_z}^2 \quad \pi_{2p_y}^{\star 1} \quad \pi_{2p_z}^{\star 1}$$

The two unpaired electrons reside in the *degenerate* antibonding orbitals, $\pi_{2p_y}^{\star}$ and $\pi_{2p_z}^{\star}$. Because there are four more electrons in bonding orbitals than in antibonding orbitals, the bond order is two (Figure 9-5b and Table 9-1). We understand why the molecule is much more stable than two free O atoms.

$$O_2\ bo = \frac{10-6}{2} = 2$$

Molecular orbital theory can also be used to predict the structures and stabilities of ions, as Example 9-1 shows.

Example 9-1

Predict the stabilities and bond orders of the ions (a) O_2^+ and (b) O_2^-.

Plan

(a) The O_2^+ ion is obtained by removing one electron from the O_2 molecule, given above. The electrons that are withdrawn most easily are those in the highest energy orbitals. (b) The superoxide ion, O_2^-, results from adding an electron to the O_2 molecule.

Solution

(a) We remove one of the $\pi_{2p}^{\star}$ electrons of O_2 to find the configuration of O_2^+:

$$\sigma_{1s}^2 \quad \sigma_{1s}^{\star 2} \quad \sigma_{2s}^2 \quad \sigma_{2s}^{\star 2} \quad \sigma_{2p}^2 \quad \pi_{2p_y}^2 \quad \pi_{2p_z}^2 \quad \pi_{2p_y}^{\star 1}$$

There are five more electrons in bonding orbitals than in antibonding orbitals, so the bond order is 2.5. We conclude that the ion would be reasonably stable relative to other diatomic ions, and it does exist.

In fact, the unusual ionic compound $[O_2^+][PtF_6^-]$ played an important role in the discovery of the first noble gas compound, $XePtF_6$ (Section 24-3).

(b) We add one electron to the appropriate orbital of O_2 to find the configuration of O_2^-. Following Hund's Rule, we add this electron into the $\pi_{2p_y}^{\star}$ orbital to form a pair:

$$\sigma_{1s}^2 \quad \sigma_{1s}^{\star 2} \quad \sigma_{2s}^2 \quad \sigma_{2s}^{\star 2} \quad \sigma_{2p}^2 \quad \pi_{2p_y}^2 \quad \pi_{2p_z}^2 \quad \pi_{2p_y}^{\star 2} \quad \pi_{2p_z}^{\star 1}$$

The bond order is 1.5 because there are three more bonding electrons than antibonding electrons. Thus, we can conclude that the ion should exist but be less stable than O_2.

The known superoxides of the heavier Group IA elements—KO_2, RbO_2, and CsO_2—contain the superoxide ion, O_2^-. These compounds are formed by combination of the free metals with oxygen (Section 6-8, part 2).

EOC 32

The Fluorine Molecule, F_2

Each fluorine atom has the $1s^2 2s^2 2p^5$ configuration. The 18 electrons of F_2 are distributed as follows:

$$\sigma_{1s}^2 \quad \sigma_{1s}^{\star 2} \quad \sigma_{2s}^2 \quad \sigma_{2s}^{\star 2} \quad \sigma_{2p}^2 \quad \pi_{2p_y}^2 \quad \pi_{2p_z}^2 \quad \pi_{2p_y}^{\star 2} \quad \pi_{2p_z}^{\star 2}$$

The bond order is one. Experiments show that F_2 exists. The F—F bond distance is longer (1.43 Å) than the bond distances in O_2 (1.21 Å), N_2 (1.09 Å), and even C_2 (1.31 Å) molecules. The bond energy of the F_2 molecules is quite low (159 kJ/mol), so F_2 molecules are very reactive.

$$F_2 \text{ bo} = \frac{10 - 8}{2} = 1$$

Heavier Homonuclear Diatomic Molecules

It might appear reasonable to use the same types of molecular orbital diagrams to predict the stability or existence of homonuclear diatomic molecules of the third and subsequent periods. However, the heavier halogens, Cl_2, Br_2, and I_2, which contain only sigma (single) bonds, are the only well-characterized examples at room temperature. We would predict from both molecular orbital theory and valence bond theory that the other (nonhalogen) homonuclear diatomic molecules from below the second period would exhibit pi bonding and therefore multiple bonding.

Some heavier elements exist as diatomic species, such as S_2, in the vapor phase at elevated temperatures. These are neither prominent nor very stable.

The instability appears to be directly related to the inability of the heavier elements to form strong pi bonds with each other. For larger atoms, the sigma bond length is too great to allow the atomic p orbitals on different atoms to overlap very effectively. Therefore, the strength of pi bonding decreases rapidly with increasing atomic size. For example, N_2 is *much* more stable than P_2. This is because the $3p$ orbitals on one P atom do not overlap side by side in a pi-bonding manner with corresponding $3p$ orbitals on another P atom nearly as effectively as do the corresponding $2p$ orbitals on the smaller N atoms. MO theory does not predict multiple bonding for Cl_2, Br_2, or I_2, each of which has a bond order of one.

9-5 Heteronuclear Diatomic Molecules

Heteronuclear Diatomic Molecules of Second-Period Elements

The corresponding atomic orbitals of two different elements, such as the $2s$ orbitals of carbon and oxygen atoms, have different energies because their differently charged nuclei have different attractions for the electrons. The atomic orbitals of the *more electronegative element* are *lower* in energy than the corresponding orbitals of the less electronegative element. As a result, a molecular orbital diagram such as Figure 9-5 is inappropriate for *heter-*

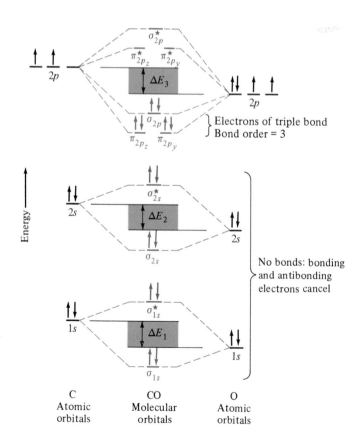

Figure 9-7

MO energy-level diagram for carbon monoxide, CO, a slightly polar heteronuclear diatomic molecule ($\mu = 0.11$ D). The atomic orbitals of oxygen, the more electronegative element, are a little lower in energy than the corresponding atomic orbitals of carbon, the less electronegative element. For this molecule, the energy differences ΔE_1, ΔE_2, and ΔE_3 are not very large; the molecule is not very polar.

onuclear diatomic molecules. If the two elements are similar (as in CO, NO, or CN molecules, for example), we can modify the diagram of Figure 9-5 by skewing it slightly. Figure 9-7 shows the energy-level diagram and electron configuration for carbon monoxide, CO.

The closer the energy of a molecular orbital is to the energy of one of the atomic orbitals of which it is composed, the more of the character of that atomic orbital it shows. Thus, in the CO molecule, the bonding MOs have more oxygen-like atomic orbital character, and the antibonding orbitals have more carbon-like atomic orbital character.

In general, the energy differences ΔE_1, ΔE_2, and ΔE_3 (shown in green in Figure 9-7) depend on the difference in electronegativities between the two atoms. The greater these energy differences, the more polar is the bond joining the atoms (the greater is its ionic character). On the other hand, the energy differences reflect the degree of overlap between atomic orbitals; the smaller these differences, the more the orbitals can overlap, and the greater is the covalent character of the bond.

We see that CO has a total of 14 electrons, making it isoelectronic with the stable N_2 molecule. Therefore, the distribution of electrons is the same in CO as in N_2, although we expect the energy levels of the MOs to be different. In accord with our predictions, carbon monoxide is a very stable molecule. It has a bond order of three, a short carbon–oxygen bond length of 1.13 Å, a low dipole moment of 0.11 D, and a very high bond energy of 1071 kJ/mol.

Note: CN is a reactive molecule, not the cyanide ion, CN⁻.

The Hydrogen Fluoride Molecule, HF

The electronegativity difference between hydrogen (EN = 2.1) and fluorine (EN = 4.0) is very large (ΔEN = 1.9). The hydrogen fluoride molecule contains a very polar bond (μ = 1.9 D). The bond in HF involves the $1s$ electron of H and an unpaired electron from a F $2p$ orbital. Figure 9-8 shows the overlap of the $1s$ orbital of H with a $2p$ orbital of F to form σ_{sp} and $\sigma_{sp}^{\star}$ molecular orbitals. The remaining two F $2p$ orbitals have no net overlap with H orbitals. They are called **nonbonding** orbitals. The same is true for the F $2s$ and $1s$ orbitals. These nonbonding orbitals retain the characteristics of the F atomic orbitals from which they are formed. The MO diagram of HF is shown in Figure 9-9.

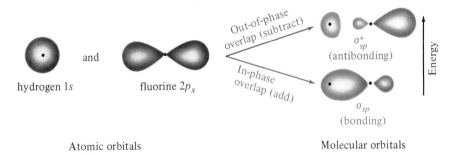

Figure 9-8
Formation of σ_{sp} and $\sigma_{sp}^{\star}$ molecular orbitals in HF by overlap of the $1s$ orbital of H with a $2p$ orbital of F.

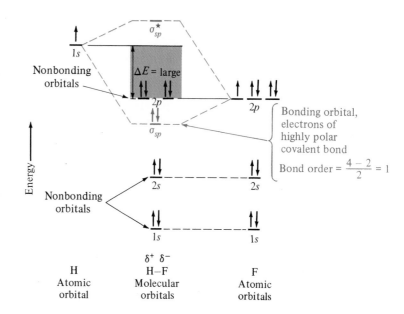

Figure 9-9
MO energy-level diagram for hydrogen fluoride, HF, a very polar molecule diagram (μ = 1.9 D). Because of the large electronegativity difference, ΔE is large.

9-6 Delocalization and the Shapes of Molecular Orbitals

In Section 7-9 we described resonance formulas for molecules and polyatomic ions. Resonance is said to exist when two or more equivalent Lewis dot formulas can be drawn for the same species and a single such formula does not account for the properties of a substance. In molecular orbital terminology, a more appropriate description involves *delocalization* of electrons. The shapes of molecular orbitals for species in which electron delocalization occurs can be determined by combining all the contributing atomic orbitals.

The Carbonate Ion, $CO_3{}^{2-}$

The average carbon–oxygen bond order in the $CO_3{}^{2-}$ ion is $1\frac{1}{3}$.

Consider the trigonal planar carbonate ion, $CO_3{}^{2-}$, as an example. All the carbon–oxygen bonds in the ion have equal bond length and equal bond energy, intermediate between those of typical C—O and C═O bonds. Valence bond theory describes the ion in terms of three contributing resonance structures (Figure 9-10a). No one of the three resonance forms adequately describes the bonding.

According to valence bond theory, the C atom is described as sp^2-hybridized, and it forms one sigma bond with each of the three O atoms. This leaves one unhybridized $2p$ atomic orbital on the C atom, say the $2p_z$ orbital. This orbital is capable of overlapping and mixing with the $2p_z$ orbital of any of the three O atoms. The sharing of two electrons in the resulting localized pi orbital would form a pi bond. Thus, three equivalent resonance structures can be drawn in valence bond terms (Figure 9-10b). We must emphasize that there is *no evidence* for the existence of these separate resonance structures.

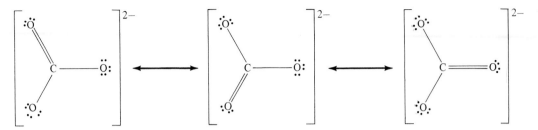

(a) Lewis formulas for valence bond resonance structures

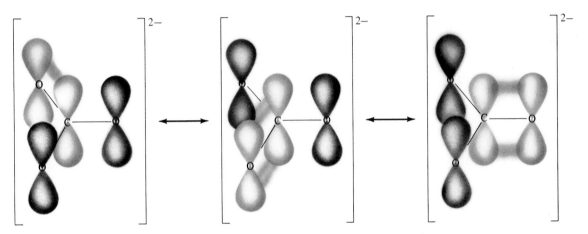

(b) *p*-orbital overlap in valence bond resonance structures

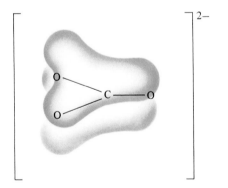

(c) Delocalized MO representation

Figure 9-10
Alternative representations of the bonding in the carbonate ion, CO_3^{2-}. (a) Lewis formulas of the three valence bond resonance structures. (b) Representation of the *p* orbital overlap in the valence bond resonance structures. In each resonance form, the *p* orbitals on two atoms would overlap to form the π components of the hypothetical double bonds. Each O atom has two additional *sp²* orbitals (not shown) in the plane of the nuclei. Each of these additional *sp²* orbitals contains an oxygen lone pair. (c) In the MO description, the electrons in the π-bonded region are spread out, or *delocalized*, over all four atoms of the CO_3^{2-} ion. This MO description is more consistent with the experimental observation of equal bond lengths and energies than are the valence bond pictures in (a) and (b).

The MO description of the pi bonding involves the simultaneous overlap and mixing of the carbon $2p_z$ orbital with the $2p_z$ orbitals of all three oxygen atoms. This forms a delocalized bonding pi molecular orbital system extending above and below the plane of the sigma system, as well as an antibonding pi orbital system. Electrons are said to occupy the entire set of bonding pi MOs, as depicted in Figure 9-10c. The shape is obtained by averaging the three contributing valence bond resonance structures. The bonding in such species as nitrate ion, NO_3^-, and sulfur dioxide, SO_2, can be described in a similar manner.

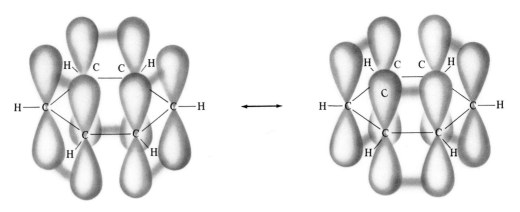

(a) Lewis formulas for valence bond resonance structures

(b) *p*-orbital overlap in valence bond resonance structures

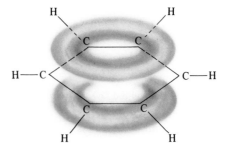

(c) Delocalized MO representation

Figure 9-11

Representations of the bonding in the benzene molecule, C_6H_6. (a) Lewis formulas of the two valence bond resonance structures. (b) The six *p* orbitals of the benzene ring, shown overlapping to form the (hypothetical) double bonds of the two resonance forms of valence bond theory. (c) In the MO description, the six electrons in the pi-bonded region are *delocalized*, occupying an extended pi-bonding region above and below the plane of the six C atoms.

The Benzene Molecule, C_6H_6

Now let us consider the benzene molecule, C_6H_6, whose two valence bond resonance forms are shown in Figure 9-11a. The valence bond description involves sp^2 hybridization at each C atom. Each C atom is at the center of a trigonal plane, and the entire molecule is known to be planar. There are sigma bonds from each C atom to the two adjacent C atoms and to one H atom. This leaves one unhybridized $2p_z$ orbital on each C atom and one remaining valence electron for each. According to valence bond theory, adjacent pairs of $2p_z$ orbitals and the six remaining electrons occupy the regions of overlap to form a total of three pi bonds in either of the two ways shown in Figure 9-11b.

There is no evidence for the existence of either of these forms of benzene. The MO description of benzene is far superior to the VB description.

Experimental studies of the C_6H_6 structure prove that it does *not* contain alternating single and double carbon–carbon bonds. The usual C—C single bond length is 1.54 Å, and the usual C=C double bond length is 1.34 Å. All six of the carbon–carbon bonds in benzene are the same length, 1.39 Å, intermediate between those of single and double bonds.

This is well explained by the MO theory, which predicts that the six $2p_z$ orbitals of the C atoms overlap and mix to form three pi bonding and three pi antibonding molecular orbitals. For instance, the most strongly bonding pi molecular orbital in the benzene pi–MO system is that in Figure 9-11c. The six pi electrons occupy three bonding MOs of this extended (delocalized) system. Thus, they are distributed throughout the molecule as a whole, above and below the plane of the sigma-bonded framework. This results in identical character for all carbon–carbon bonds in benzene. Each carbon–carbon bond has a bond order of $1\frac{1}{2}$. The MO representation of the extended pi system is the same as that obtained by averaging the two contributing valence bond resonance structures.

Key Terms

Antibonding orbital A molecular orbital higher in energy than any of the atomic orbitals from which it is derived; when populated with electrons, lends instability to a molecule or ion. Denoted with a star ($\star$) superscript on its symbol.

Bond energy The amount of energy necessary to break one mole of bonds of a given kind (in the gas phase).

Bond order Half the number of electrons in bonding orbitals minus half the number of electrons in antibonding orbitals.

Bonding orbital A molecular orbital lower in energy than any of the atomic orbitals from which it is derived; when populated with electrons, lends stability to a molecule or ion.

Delocalization The formation of a set of molecular orbitals that extend over more than two atoms; important in species that valence bond theory describes in terms of *resonance*.

Heteronuclear Consisting of different elements.

Homonuclear Consisting of only one element.

Molecular orbital (MO) An orbital resulting from overlap and mixing of atomic orbitals on different atoms. An MO belongs to the molecule as a whole.

Molecular orbital theory A theory of chemical bonding based upon the postulated existence of molecular orbitals.

Nodal plane (node) A region in which the probability of finding an electron is zero.

Nonbonding orbital A molecular orbital derived only from an atomic orbital of one atom; lends neither stability nor instability to a molecule or ion when populated with electrons.

Pi (π) bond A bond resulting from electron occupation of a pi molecular orbital.

Pi (π) orbital A molecular orbital resulting from side-on overlap of atomic orbitals.

Sigma (σ) bond A bond resulting from electron occupation of a sigma molecular orbital.

Sigma (σ) orbital A molecular orbital resulting from head-on overlap of two atomic orbitals.

Exercises

MO Theory—General Concepts

1. Describe the main differences between the valence bond theory and the molecular orbital theory.

2. What is a molecular orbital? What two types of information can be obtained from molecular orbital calculations? How do we use such information to describe the bonding within a molecule?

3. What is the relationship between the maximum number of electrons that can be accommodated by a set of molecular orbitals and the maximum number that can be accommodated by the atomic orbitals from which the MOs are formed? What is the maximum number of electrons that one MO can hold?

4. Answer Exercise 3 after replacing "molecular orbitals" with "hybridized atomic orbitals."

5. What differences and similarities exist among (a) atomic orbitals, (b) localized hybridized atomic orbitals according to valence bond theory, and (c) molecular orbitals?

6. What is the relationship between the energy of a bonding molecular orbital and the energies of the original atomic orbitals? What is the relationship between the energy of an antibonding molecular orbital and the energies of the original atomic orbitals?

7. Draw an energy-level diagram for the formation of molecular orbitals from two atomic orbitals of equal energy. Identify the bonding and antibonding molecular orbitals.

8. In terms of overlapping AO's, describe three ways in which a σ MO can be formed.

9. Complete the following energy-level diagram for the formation of molecular orbitals:

$$\overline{2p_x}\ \overline{2p_y}\ \overline{2p_z} \qquad\qquad \overline{2p_x}\ \overline{2p_y}\ \overline{2p_z}$$

atomic orbitals   molecular orbitals   atomic orbitals

Assume the bonding axis to be the x axis. Identify the bonding and antibonding molecular orbitals that are formed.

10. State the three rules for placing electrons in molecular orbitals.

11. What is meant by the term "bond order"? How is the value of the bond order calculated?

12. Compare and illustrate the differences between (a) atomic orbitals and molecular orbitals, (b) bonding and antibonding molecular orbitals, (c) σ bonds and π bonds, and (d) localized and delocalized molecular orbitals.

13. How are the electron occupancies of bonding, nonbonding, and antibonding orbitals related to the stability of a molecule or ion?

14. From memory, draw the energy-level diagram of molecular orbitals produced by the overlap of orbitals of two identical atoms from the second period. (Show the π_{2p_y} and π_{2p_z} orbitals below the σ_{2p} in energy.)

15. Is it possible for a molecule or complex ion in its ground state to have a negative bond order? Why?

Homonuclear Diatomic Species

16. What do we mean when we say that a molecule or ion is (a) *homonuclear,* (b) *heteronuclear,* or (c) *diatomic?*

17. Use the appropriate molecular orbital energy diagram to write the electron configuration for each of the following molecules or ions, calculate the bond order of each, and predict which would exist. (a) H_2^+, (b) H_2, (c) H_2^-, (d) H_2^{2-}

18. Repeat Exercise 17 for (a) He_2^+ and (b) He_2.

19. Repeat Exercise 17 for (a) N_2, (b) Ne_2, and (c) C_2.

20. Repeat Exercise 17 for (a) Li_2, (b) Li_2^+, and (c) F_2.

21. Determine the electron configurations of the following molecules and ions: (a) Be_2, Be_2^+, Be_2^-; (b) B_2, B_2^+, B_2^-; (c) C_2^+.

22. What is the bond order of each of the species in Exercise 21?

23. Which of the species in Exercise 21 are diamagnetic (D) and which are paramagnetic (P)?

24. Apply MO theory to predict relative stabilities of the species in Exercise 21. Comment on the validity of these predictions. What else *must* be considered in addition to electron occupancy of MOs?

*25. Which homonuclear diatomic molecules or ions of the second period have the following electron distributions in MOs? In other words, identify X in each of the following cases.

(a) X_2 $\sigma_{1s}^2\ \sigma_{1s}^{\star 2}\ \sigma_{2s}^2\ \sigma_{2s}^{\star 2}\ \pi_{2p_y}^2\ \pi_{2p_z}^2$

(b) X_2^+ $\sigma_{1s}^2\ \sigma_{1s}^{\star 2}\ \sigma_{2s}^2\ \sigma_{2s}^{\star 2}\ \sigma_{2p}^2\ \pi_{2p_y}^2\ \pi_{2p_z}^2\ \pi_{2p_y}^{\star 2}\ \pi_{2p_z}^{\star 1}$

(c) X_2^- $\sigma_{1s}^2\ \sigma_{1s}^{\star 2}\ \sigma_{2s}^2\ \sigma_{2s}^{\star 2}\ \pi_{2p_y}^2\ \pi_{2p_z}^2\ \sigma_{2p}^2\ \pi_{2p_y}^{\star 1}$

26. What is the bond order of each of the species in Exercise 25?

27. (a) Give the MO designations for O_2, O_2^-, O_2^{2-}, O_2^+, and O_2^{2+}. (b) Give the bond order in each case. (c) Match these species with the following observed bond lengths: 1.04 Å, 1.12 Å, 1.21 Å, 1.33 Å, and 1.49 Å.

28. (a) Give the MO designations for N_2, N_2^-, and N_2^+. (b) Give the bond order in each case. (c) Rank these three species by increasing predicted bond length.

29. Assuming that the σ_{2p} MO is lower in energy than the π_{2p_y} and π_{2p_z} MOs for the following species, write out electronic configurations for all of them. (a) F_2, F_2^+, F_2^-; (b) Ne_2, Ne_2^+.

30. (a) What is the bond order of each of the species in Exercise 29? (b) Are they diamagnetic or paramagnetic? (c) What would MO theory predict about the stabilities of these species?

Heteronuclear Diatomic Species

The following is the molecular orbital energy-level diagram for a heteronuclear diatomic molecule, XY, in which both X and Y are from Period 2 and Y is the more electronegative element. This diagram may be useful in answering questions in this section.

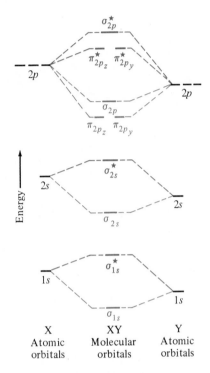

	X Atomic orbitals	XY Molecular orbitals	Y Atomic orbitals

31. Use the preceding diagram to fill in an MO diagram for nitrogen monoxide, NO. What is the bond order of NO? Is it paramagnetic? How would you assess its stability?

32. Assuming that the preceding MO diagram is valid for CN, CN$^+$, and CN$^-$, write the MO descriptions for these species. Which would be most stable? Why?

33. For each of the two species OF and OF$^+$: (a) Draw MO energy-level diagrams. (b) Write out electronic configurations. (c) Determine bond orders and predict relative stabilities. (d) Predict diamagnetism or paramagnetism. Refer to the preceding diagram. Assume that the σ_{2p} MO is lower in energy than the π_{2p_y} and π_{2p_z} MOs.

34. For each of the two species NF and NF$^-$: (a) Draw MO energy-level diagrams. (b) Write out electronic configurations. (c) Determine bond orders and predict relative stabilities. (d) Predict diamagnetism or paramagnetism. Refer to the preceding diagram. Assume

that the σ_{2p} MO is lower in energy than the π_{2p_y} and π_{2p_z} MOs.

35. Considering the shapes of MO energy-level diagrams for nonpolar covalent and polar covalent molecules, what would you predict about MO diagrams, and therefore about overlap of atomic orbitals, for ionic compounds?

36. To increase the strength of the bonding in the hypothetical compound BO, would you add or subtract an electron? Explain your answer with the aid of an MO electron structure.

Delocalization

37. Draw Lewis dot formulas depicting the resonance structures of the following species from the valence bond point of view, and then draw MOs for the delocalized π systems. (a) SO$_2$, sulfur dioxide; (b) HCO$_3^-$, hydrogen carbonate ion (H is bonded to O); (c) NO$_3^-$, nitrate ion.

38. Draw Lewis dot formulas depicting the resonance structures of the following species from the valence bond point of view, and then draw MOs for the delocalized π systems: (a) SO$_3$, sulfur trioxide; (b) O$_3$, ozone; (c) HCO$_2^-$, formate ion (H is bonded to C).

Mixed Exercises

39. Draw and label the complete MO energy-level diagrams for the following species. For each, determine the bond order, predict the stability of the species, and predict whether the species will be paramagnetic. (a) He$_2^+$; (b) CN; (c) HeH$^+$.

40. Draw and label the complete MO energy-level diagrams for the following species. For each, determine the bond order, predict the stability of the species, and predict whether the species will be paramagnetic. (a) O$_2^{2+}$; (b) HO$^-$; (c) HCl.

41. Which of these species would you expect to be paramagnetic? (a) He$_2^+$; (b) NO; (c) NO$^+$; (d) N$_2^{2+}$; (e) CO^{2-}; (f) F$_2^+$.

42. Rationalize the following observations in terms of the stabilities of σ and π bonds: (a) The most common form of nitrogen is N$_2$, whereas the most common form of phosphorus is P$_4$ (see the structure in Figure 2-3). (b) The most common forms of oxygen are O$_2$ and (less common) O$_3$, whereas the most common form of sulfur is S$_8$.

43. (a) Give the MO designations for (1) Be$_2$, (2) F$_2$, (3) HeH$^+$, (4) OF, (5) Ne$_2$. (b) Which of these species is/are unlikely to exist? Explain.

Outline

Objectives

As you study this chapter, you should learn

☐ To understand the Arrhenius theory
☐ About hydrated hydrogen ions
☐ To understand the Brønsted–Lowry theory
☐ The properties of aqueous solutions of acids
☐ The properties of aqueous solutions of bases
☐ How to predict the strengths of binary acids

☐ About acid–base reactions
☐ About acidic and basic salts
☐ To understand the strengths of ternary acids
☐ About amphoterism
☐ How to prepare acids
☐ To understand the Lewis theory

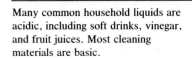

Many common household liquids are acidic, including soft drinks, vinegar, and fruit juices. Most cleaning materials are basic.

I n highly developed societies, acids, bases, and salts are indispensable compounds. Table 4-10 lists the 18 such compounds that were included in the top 50 chemicals produced in the United States in 1990. The production of H_2SO_4 (number 1) was more than twice as great as the production of NH_3 (number 3). Sixty-five percent of the H_2SO_4 is used in the production of fertilizers.

Many acids, bases, and salts occur in nature and serve a wide variety of purposes. For instance, your "digestive juice" contains approximately 0.10 mole of hydrochloric acid per liter. The liquid in your automobile battery is approximately 40% H_2SO_4 by mass. Sodium hydroxide, a base, is used in the manufacture of soaps, paper, and many other chemicals. "Drāno" is solid NaOH that contains some aluminum turnings. Sodium chloride is used to season food and as a food preservative. Calcium chloride is used to melt ice on highways and in the emergency treatment of cardiac arrest. Several ammonium salts are used as fertilizers.

You will encounter many of these in your laboratory work.

Many organic acids (carboxylic acids) and their derivatives occur in nature. Amino acids are the building blocks of proteins, which are important materials in the bodies of animals, including humans. Amino acids are carboxylic acids that also contain basic groups derived from ammonia. The pleasant odors and flavors of ripe fruit are due in large part to the presence of esters (Section 31-13), which are formed from the acids that are present in unripe fruit.

10-1 The Arrhenius Theory

In 1680 Robert Boyle noted that acids (1) dissolve many substances, (2) change the colors of some natural dyes (indicators), and (3) lose their characteristic properties when mixed with alkalis (bases). By 1814 J. Gay-Lussac concluded that acids *neutralize* bases and that the two classes of substances can be defined only in terms of their reactions with each other.

In 1884 Svante Arrhenius presented this theory of electrolytic dissociation, which resulted in the Arrhenius theory of acid–base reactions. In his view, an **acid** is a substance that contains hydrogen and produces H^+ in aqueous solution. A **base** is a substance that contains the OH group and produces hydroxide ions, OH^-, in aqueous solution. **Neutralization** is defined as the combination of H^+ ions with OH^- ions to form H_2O molecules:

$$H^+(aq) + OH^+(aq) \longrightarrow H_2O(\ell) \qquad \text{(neutralization)}$$

The Arrhenius theory of acid–base behavior satisfactorily explained reactions of **protonic acids** (those containing acidic hydrogen atoms) with metal hydroxides (hydroxy bases). It was a significant contribution to chemical thought and theory in the latter part of the nineteenth century. We used this theory in introducing acids and bases and discussing some of their reactions. The Arrhenius model of acids and bases, although limited in scope, led to the development of other general theories of acid–base behavior. They will be considered in later sections.

You may wish to review Sections 4-2, 4-5, 6-7, and 6-8.

10-2 The Hydrated Hydrogen Ion

Although Arrhenius described H^+ ions in water as bare ions (protons), we now know that they are hydrated in aqueous solution and exist as $H^+(H_2O)_n$ in which n is some small integer. This is due to the attraction of the H^+ ions, or protons, for the oxygen end ($\delta-$) of water molecules. While we do not know the extent of hydration of H^+ in most solutions, we often represent the hydrated hydrogen ion as the hydronium ion, H_3O^+, or $H^+(H_2O)_n$ in which $n = 1$. The hydrated hydrogen ion is the species that gives aqueous solutions of acids their characteristic acidic properties. Whether we use the designation $H^+(aq)$ or H_3O^+, we are always referring to the hydrated hydrogen ion.

$$H^+ + \;:\!\overset{\cdots}{\underset{\overset{\displaystyle|}{H}}{O}}\!:H \longrightarrow H\!:\!\overset{\cdots}{\underset{\overset{\displaystyle|}{H}}{O}}\!:H^+$$

10-3 The Brønsted–Lowry Theory

In 1923 J. N. Brønsted and T. M. Lowry independently presented logical extensions of the Arrhenius theory. Brønsted's contribution was more thorough than Lowry's, and the result is known as the **Brønsted theory** or the **Brønsted–Lowry theory**.

An **acid** is defined as a *proton donor,* H^+, and a **base** is defined as a *proton acceptor*. The definitions are sufficiently broad that any hydrogen-containing molecule or ion capable of releasing a proton, H^+, is an acid, while any molecule or ion that can accept a proton is a base. According to the Brønsted–Lowry theory,

An acid–base reaction is the transfer of a proton from an acid to a base.

Thus, the complete ionization of hydrogen chloride, HCl, a *strong* acid, in water is an acid–base reaction in which water acts as a base, a proton acceptor.

Step 1:	$HCl(aq)$	$\longrightarrow H^+(aq) + Cl^-(aq)$	(Arrhenius description)
Step 2:	$H^+(aq) + H_2O(\ell) \longrightarrow H_3O^+$		
Overall:	$HCl(aq) + H_2O(\ell) \longrightarrow H_3O^+ + Cl^-(aq)$		(Brønsted–Lowry description)

We use red to indicate acids and blue to indicate bases. We use rectangles to indicate one conjugate acid–base pair and ovals to indicate the other pair.

The ionization of hydrogen fluoride, a *weak* acid, is similar, but it occurs to only a slight extent. So we use double arrows to indicate that it is reversible.

Various measurements (electrical conductivity, freezing point depression, and so on) indicate that HF is only *slightly* ionized in water.

We can describe Brønsted–Lowry acid–base reactions in terms of **conjugate acid–base pairs**. These are species that differ by a proton. In the preceding equation, HF ($acid_1$) and F^- ($base_1$) are one conjugate acid–base pair, and H_2O ($base_2$) and H_3O^+ ($acid_2$) are the other pair. The members of each conjugate pair are designated by the same numerical subscript. In the forward reaction, HF and H_2O act as acid and base, respectively. In the reverse reaction, H_3O^+ acts as the acid, or proton donor, and F^- acts as the base, or proton acceptor.

It makes no difference which conjugate acid–base pair, HF and F^- or H_3O^+ and H_2O, is assigned the subscripts 1 and 2.

When HF is dissolved in water, the HF molecules give up some H^+ ions that can be accepted by either of two bases, F^- and H_2O. The fact that HF is only slightly ionized tells us that F^- is a stronger base than H_2O. When HCl (a *strong* acid) is dissolved in water, the HCl molecules give up H^+

ions that can be accepted by either of two bases, Cl^- and H_2O. The fact that HCl is completely ionized in dilute aqueous solution tells us that Cl^- is a weaker base than H_2O. Thus, the stronger acid, HCl, has the weaker conjugate base, Cl^-. The weaker acid, HF, has the stronger conjugate base, F^-. We can generalize:

> The weaker an acid is, the greater is the base strength of its conjugate base. Likewise, the weaker a base is, the stronger is its conjugate acid.

"Strong" and "weak," like many other adjectives, are used in a relative sense. We do not mean to imply that the fluoride ion, F^-, is a strong base compared with species such as the hydroxide ion, OH^-. We mean that *relative to the anions of strong acids,* which are very weak bases, F^- is a much stronger base.

Ammonia acts as a weak Brønsted–Lowry base, and water acts as an acid in the ionization of aqueous ammonia:

Ammonia is very soluble in water (~15 mol/L at 25°C). In 0.10 M solution, NH_3 is only 1.3% ionized and 98.7% nonionized.

$$NH_3(aq) \; + \; H_2O(\ell) \; \rightleftharpoons \; NH_4^+(aq) \; + \; OH^-(aq)$$

$$\text{base}_1 \qquad\quad \text{acid}_2 \qquad\qquad \text{acid}_1 \qquad\quad \text{base}_2$$

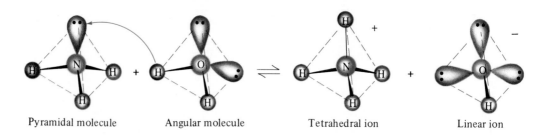

As we see in the reverse reaction, ammonium ion, NH_4^+, is the conjugate acid of NH_3. The hydroxide ion, OH^-, is the conjugate base of water. In three dimensions, the molecular structures are

| Pyramidal molecule | Angular molecule | Tetrahedral ion | Linear ion |

Water acts as an acid (H^+ donor) in its reaction with NH_3, whereas it acts as a base (H^+ acceptor) in its reaction with HCl and HF.

> Whether water acts as an acid or as a base depends on the other species present.

Careful measurements show that pure water ionizes ever so slightly to produce equal numbers of hydrated hydrogen ions and hydroxide ions:

$$H_2O + H_2O \rightleftharpoons H_3O^+ + OH^-$$

Base$_1$ Acid$_2$ Acid$_1$ Base$_2$

In simplified notation,

$$H_2O \rightleftharpoons H^+(aq) + OH^-(aq)$$

The **autoionization** (self-ionization) of water is an acid–base reaction according to the Brønsted–Lowry theory. One H_2O molecule (the acid) donates a proton to another H_2O molecule (the base). The H_2O molecule that donates a proton becomes an OH^- ion, the conjugate base of water. The H_2O molecule that accepts a proton becomes an H_3O^+ ion. Examination of the reverse reaction (right to left) shows that H_3O^+ (an acid) donates a proton to OH^- (a base) to form two H_2O molecules. One H_2O molecule behaves as an acid while the other acts as a base in the autoionization of water. Water is said to be **amphiprotic**; that is, H_2O molecules can accept and donate protons.

The prefix "amphi-" means "of both kinds." "Amphiprotism" refers to amphoterism by accepting and donating a proton in different reactions.

As we saw in Section 4-5, H_3O^+ and OH^- ions combine to form nonionized water molecules when strong acids and strong soluble bases react to form soluble salts and water. The reverse reaction, the autoionization of water, occurs only slightly, as expected.

10-4 Properties of Aqueous Solutions of Acids and Bases

Aqueous solutions of most protonic acids exhibit certain properties, which are properties of hydrated hydrogen ions.

1. They have a sour taste. Pickles are usually preserved in vinegar, a 5% solution of acetic acid. Many pickled condiments contain large amounts of sugar so that the taste of acetic acid is not so pronounced. Lemons contain citric acid, which is responsible for their characteristic sour taste.
2. They change the colors of many indicators (highly colored dyes). Acids turn blue litmus red, and bromthymol blue changes from blue to yellow in acids.
3. Nonoxidizing acids react with metals above hydrogen in the activity series (Section 4-6, part 2) to liberate hydrogen gas, H_2.

The indicator bromthymol blue is yellow in acidic solution and blue in basic solution.

4. They react with (neutralize) metal oxides and metal hydroxides to form salts and water (Section 4-5).
5. They react with salts of weaker or more volatile acids to form the weaker or more volatile acid and a new salt.
6. Aqueous solutions of protonic acids conduct an electric current because they are wholly or partly ionized.

Aqueous solutions of most bases also exhibit certain properties. These are due to the hydrated hydroxide ions that are present in aqueous solutions of bases:

1. They have a bitter taste.
2. They have a slippery feeling. Soaps are common examples; they are mildly basic. A solution of household bleach feels very slippery because it is strongly basic.
3. They change the colors of many indicators: litmus changes from red to blue, and bromthymol blue changes from yellow to blue, in bases.
4. They react with (neutralize) protonic acids to form salts and water.
5. Their aqueous solutions conduct an electric current because they are ionized or dissociated.

10-5 Strengths of Binary Acids

A *weak* acid may be very reactive. For example, HF dissolves sand and glass. The equation for its reaction with sand is

$$SiO_2(s) + 4HF(g) \longrightarrow$$
$$SiF_4(g) + 2H_2O(\ell)$$

The reaction with glass and other silicates is similar. These reactions are not related to acid strength.

The ease of ionization of binary protonic acids depends on both (1) the ease of breaking H—X bonds and (2) the stability of the resulting ions in solution. Let us consider the relative strengths of the Group VIIA hydrohalic acids. Hydrogen fluoride ionizes only slightly in dilute aqueous solutions:

$$HF(aq) \overset{H_2O}{\rightleftharpoons} H^+(aq) + F^-(aq)$$

However, HCl, HBr, and HI ionize completely or nearly completely in dilute aqueous solutions because the H—X bonds are much weaker.

$$HX(aq) \overset{H_2O}{\longrightarrow} H^+(aq) + X^-(aq) \qquad X = Cl, Br, I$$

The order of *bond strengths* for the hydrogen halides is

$$HF \gg HCl > HBr > HI$$

Bond strength is shown by the bond energies introduced in Chapter 7 and tabulated in Section 15-9. The strength of the H—F bond is due largely to the very small size of the F atom.

To understand why HF is so much weaker an acid than the other hydrogen halides, let us consider the following factors:

1. In HF the electronegativity difference is 1.9, compared with 0.9 in HCl, 0.7 in HBr, and 0.4 in HI. We expect the very polar H—F bond in HF to ionize easily. The fact that HF is the *weakest* of these acids suggests that this effect must be of minor importance.
2. The bond strength is considerably greater in HF than in the other three molecules. This tells us that the H—F bond is harder to break than the H—Cl, H—Br, and H—I bonds.
3. The small, highly charged F^- ion, formed when HF ionizes, causes increased ordering of the water molecules. This increase is unfavorable to the process of ionization.

Table 10-1
Relative Strengths of Conjugate Acid–Base Pairs

	Acid				Base	

Acid strength increases →

$HClO_4$
HI } 100% ionized in
HBr } dilute aq. soln.
HCl } No molecules of
HNO_3 } nonionized acid.

Negligible base
strength in water.

ClO_4^-
I^-
Br^-
Cl^-
NO_3^-

$$\xrightleftharpoons[+H^+]{-H^+}$$

H_3O^+
HF
CH_3COOH } Equilibrium mixture
HCN } of nonionized
NH_4^+ } molecules of acid,
H_2O } conjugate base, and
NH_3 } H^+ (aq).

H_2O
F^-
CH_3COO^-
CN^-
NH_3
OH^-

Reacts completely with H_2O;
cannot exist in aqueous solution. } NH_2^-

Base strength increases

The net result of all factors is that the order of *acid strengths* is

$$HF \ll HCl < HBr < HI$$

In dilute aqueous solutions, hydrochloric, hydrobromic, and hydroiodic acids are completely ionized and all show the same apparent acid strength. Water is sufficiently basic that it does not distinguish among the acid strengths of HCl, HBr, and HI, and therefore it is referred to as a **leveling solvent**. It is not possible to determine the order of the strengths of these three acids in water because they are so nearly completely ionized.

When these compounds are dissolved in anhydrous acetic acid or other solvents less basic than water, however, significant differences in their acid strengths are observed.

$$HCl < HBr < HI \quad \text{(strongest acid)}$$

One more observation is appropriate to describe the leveling effect:

The hydrated hydrogen ion is the strongest acid that can exist in aqueous solution.

Acids stronger than H^+(aq) react with water to produce H^+(aq) and their conjugate bases. For example, $HClO_4$ (see Table 10-1) reacts with H_2O completely to form H^+(aq) and ClO_4^-(aq).

Similar observations can be made for aqueous solutions of strong soluble bases such as NaOH and KOH. Both are completely dissociated in dilute aqueous solutions:

$$Na^+OH^-(aq) \xrightarrow{H_2O} Na^+(aq) + OH^-(aq)$$

The strengths of *ternary* acids are discussed in Section 10-8.

> The hydroxide ion is the strongest base that can exist in aqueous solution.

The amide ion, NH_2^-, is a stronger base than OH^-.

Bases stronger than OH^- react with H_2O to produce OH^- and their conjugate acids. When metal amides such as sodium amide, $NaNH_2$, are placed in H_2O, the amide ion, NH_2^-, reacts with H_2O completely, as shown by the following equation:

$$NH_2^- + H_2O \longrightarrow NH_3(aq) + OH^-(aq)$$

Thus, we see that H_2O is a leveling solvent for all bases stronger than OH^-.

Acid strengths for other *vertical* series of binary acids vary in the same way as those of the VIIA elements. The order of bond strengths for the VIA hydrides is

The trends in binary acid strengths *across* a period (e.g., $CH_4 < NH_3 < H_2O < HF$) are *not* those predicted from trends in bond energies and electronegativity differences. The correlations used for *vertical* trends cannot be used for *horizontal* trends. This is because a "horizontal" series of compounds has different stoichiometries and different numbers of lone pairs of electrons on its central atoms.

$$H_2O \gg H_2S > H_2Se > H_2Te$$

H—O bonds are much stronger than the other H—El bonds. As we might expect, the order of acid strengths for these hydrides is just the reverse of the order of bond strengths.

$$H_2O \ll H_2S < H_2Se < H_2Te \qquad \text{(strongest acid)}$$

Table 10-1 displays relative acid and base strengths of a number of conjugate acid–base pairs.

10-6 Reactions of Acids and Bases

Common Strong Acids	
Binary	**Ternary**
HCl	$HClO_4$
HBr	$HClO_3$
HI	HNO_3
	H_2SO_4

Strong Soluble Bases	
LiOH	
NaOH	
KOH	$Ca(OH)_2$
RbOH	$Sr(OH)_2$
CsOH	$Ba(OH)_2$

In Section 4-5 we introduced classical acid–base reactions. We defined neutralization as the reaction of an acid with a base to form a salt and (in most cases) water. Most *salts* are ionic compounds that contain a cation other than H^+ and an anion other than OH^- or O^{2-}. The *strong acids* and *strong soluble bases* are listed in the margin. Recall that other common acids may be assumed to be weak. The other common metal hydroxides (bases) are insoluble in water.

Arrhenius and Brønsted–Lowry acid–base neutralization reactions all have one thing in common. They involve the reaction of an acid with a base to form a salt that contains the cation characteristic of the base and the anion characteristic of the acid. Water is also usually formed. This is indicated in the formula unit (molecular) equation. However, the general form of the net ionic equation and the essence of the reaction are different for different acid–base reactions. They depend upon the solubility and extent of ionization or dissociation of each reactant and product.

In writing ionic equations, we always write the formulas of the predominant forms of the compounds in, or in contact with, aqueous solution. Writing ionic equations from formula unit equations requires a knowledge of the lists of strong acids and strong soluble bases, as well as of the generalizations on solubilities of inorganic compounds. Please review carefully all of Sections 4-2 and 4-3. Study Tables 4-9 and 4-10 carefully because they summarize much information that you are about to use again.

In Section 4-5 we examined some reactions of strong acids with strong soluble bases to form soluble salts. Let us illustrate one additional example.

Perchloric acid, $HClO_4$, reacts with sodium hydroxide to produce sodium perchlorate, $NaClO_4$, a soluble ionic salt.

$$HClO_4(aq) + NaOH(aq) \longrightarrow NaClO_4(aq) + H_2O(\ell)$$

The total ionic equation for this reaction is

$$[H^+(aq) + ClO_4^-(aq)] + [Na^+(aq) + OH^-(aq)] \longrightarrow$$
$$[Na^+(aq) + ClO_4^-(aq)] + H_2O(\ell)$$

Eliminating the spectator ions, Na^+ and ClO_4^-, gives the net ionic equation

$$H^+(aq) + OH^-(aq) \longrightarrow H_2O(\ell)$$

This is the same as
$H_3O^+ + OH^- \rightarrow 2H_2O$.

This is the net ionic equation for the reaction of all strong acids with strong soluble bases to form soluble salts and water.

Many weak acids react with strong soluble bases to form soluble salts and water. For example, acetic acid, CH_3COOH, reacts with sodium hydroxide, $NaOH$, to produce sodium acetate, $NaCH_3COO$.

$$CH_3COOH(aq) + NaOH(aq) \longrightarrow NaCH_3COO(aq) + H_2O(\ell)$$

The total ionic equation for this reaction is

$$CH_3COOH(aq) + [Na^+(aq) + OH^-(aq)] \longrightarrow$$
$$[Na^+(aq) + CH_3COO^-(aq)] + H_2O(\ell)$$

Elimination of Na^+ from both sides gives the net ionic equation

$$CH_3COOH(aq) + OH^-(aq) \longrightarrow CH_3COO^-(aq) + H_2O(\ell)$$

In general terms, the reaction of a *weak monoprotic acid* with a *strong soluble base* to form a *soluble salt* may be represented as

$$HA(aq) + OH^-(aq) \longrightarrow A^-(aq) + H_2O(aq) \qquad \text{(net ionic equation)}$$

Monoprotic acids contain one, diprotic acids contain two, and triprotic acids contain three acidic (ionizable) hydrogen atoms per formula unit. Polyprotic acids (those that contain more than one ionizable hydrogen atom) are discussed in detail in Chapter 18.

Example 10-1

Write (a) formula unit, (b) total ionic, and (c) net ionic equations for the neutralization of aqueous ammonia with nitric acid.

Plan

(a) The salt contains the cation of the base, NH_4^+, and the anion of the acid, NO_3^-. The salt is NH_4NO_3.
(b) HNO_3 is a strong acid—we write it in ionic form. Ammonia is a weak base. NH_4NO_3 is a soluble salt that is completely dissociated—we write it in ionic form.
(c) We cancel the spectator ions, NO_3^-, and obtain the net ionic equation.

Solution

(a) $\quad HNO_3(aq) + NH_3(aq) \longrightarrow NH_4NO_3(aq)$

(b) $\quad [H^+(aq) + NO_3^-(aq)] + NH_3(aq) \longrightarrow [NH_4^+(aq) + NO_3^-(aq)]$

(c) $\quad H^+(aq) + NH_3(aq) \longrightarrow NH_4^+(aq)$

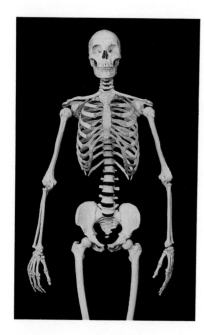

Our bones are mostly calcium phosphate, $Ca_3(PO_4)_2$, an insoluble compound. They are formed by reactions of calcium ions, Ca^{2+}, with phosphate ions, $PO_4{}^{3-}$. Calcium and phosphorus are essential elements in human nutrition. Do you see why children require calcium and phosphorus for normal growth?

Example 10-2

Write (a) formula unit, (b) total ionic, and (c) net ionic equations for the complete neutralization of phosphoric acid, H_3PO_4, with calcium hydroxide, $Ca(OH)_2$.

Plan

(a)   The salt contains the cation of the base, Ca^{2+}, and the anion of the acid, $PO_4{}^{3-}$. The salt is $Ca_3(PO_4)_2$.

(b) H_3PO_4 is a weak acid—it is not written in ionic form. $Ca(OH)_2$ is a strong soluble base, and so it is written in ionic form. $Ca_3(PO_4)_2$ is an *insoluble* salt, and so it is not written in ionic form.

(c)   There are no species common to both sides of the equation, and so there are no spectator ions. The net ionic equation is the same as the total ionic equation except that there are no brackets.

Solution

(a) $H_3PO_4(aq) + 3Ca(OH)_2(aq) \longrightarrow Ca_3(PO_4)_2(s) + 6H_2O(\ell)$

(b) $2H_3PO_4(aq) + 3[Ca^{2+}(aq) + 2OH^-(aq)] \longrightarrow Ca_3(PO_4)_2(s) + 6H_2O(\ell)$

(c) $2H_3PO_4(aq) + 3Ca^{2+}(aq) + 6OH^-(aq) \longrightarrow Ca_3(PO_4)_2(s) + 6H_2O(\ell)$

EOC 50, 52

There are other kinds of reactions between acids and bases. The preceding examples illustrate how chemical equations for them are written.

10-7 Acidic Salts and Basic Salts

To this point we have examined acid–base reactions in which stoichiometric amounts of acids and bases were mixed to form *normal salts*. As the name implies, **normal salts** contain no unreacted H^+ or OH^- ions.

If less than stoichiometric amounts of bases react with *polyprotic* acids, the resulting salts are known as **acidic salts** because they are still capable of neutralizing bases. The reaction of phosphoric acid, H_3PO_4, a weak acid, with strong bases can produce three different salts, depending on the relative amounts of acid and base used.

Sodium hydrogen carbonate, baking soda, is the most familiar example of an acidic salt. It can neutralize strong bases, but its aqueous solutions are slightly basic, as the color of the indicator shows.

$$H_3PO_4(aq) + NaOH(aq) \longrightarrow NaH_2PO_4(aq) + H_2O(\ell)$$
$$\text{1 mole} \qquad \text{1 mole} \qquad \text{sodium dihydrogen phosphate,}$$
$$\text{an acidic salt}$$

$$H_3PO_4(aq) + 2NaOH(aq) \longrightarrow Na_2HPO_4(aq) + 2H_2O(\ell)$$
$$\text{1 mole} \qquad \text{2 moles} \qquad \text{sodium hydrogen phosphate,}$$
$$\text{an acidic salt}$$

$$H_3PO_4(aq) + 3NaOH(aq) \longrightarrow Na_3PO_4(aq) + 3H_2O(\ell)$$
$$\text{1 mole} \qquad \text{3 moles} \qquad \text{sodium phosphate,}$$
$$\text{a normal salt}$$

There are many additional examples of acidic salts. Sodium hydrogen carbonate, $NaHCO_3$, commonly called sodium bicarbonate, is classified as an acidic salt. However, it is the acidic salt of an extremely weak acid—carbonic acid, H_2CO_3—and solutions of sodium bicarbonate are slightly basic.

Polyhydroxy bases (bases that contain more than one OH per formula unit) react with less than stoichiometric amounts of acids to form **basic salts**, i.e., salts that contain unreacted OH groups. For example, the reaction of aluminum hydroxide with hydrochloric acid can produce three different salts.

$$Al(OH)_3(s) \ + \ HCl(aq) \ \longrightarrow \ Al(OH)_2Cl(s) \ + \ H_2O(\ell)$$

     1 mole   1 mole   aluminum dihydroxide chloride, a basic salt

These basic aluminum salts are called "aluminum chlorohydrate." They are components of many deodorants.

$$Al(OH)_3(s) \ + \ 2HCl(aq) \ \longrightarrow \ Al(OH)Cl_2(s) \ + \ 2H_2O(\ell)$$

     1 mole   2 moles   aluminum hydroxide dichloride, a basic salt

$$Al(OH)_3(s) \ + \ 3HCl(aq) \ \longrightarrow \ AlCl_3(aq) \ + \ 3H_2O(\ell)$$

     1 mole   3 moles   aluminum chloride, a normal salt

Aqueous solutions of basic salts are not necessarily basic, but they can neutralize acids. Most basic salts are rather insoluble in water.

10-8 Strengths of Ternary Acids and Amphoterism

Ternary acids are *hydroxides of nonmetals* that ionize to produce $H^+(aq)$. The formula for nitric acid is commonly written HNO_3 to emphasize the presence of an acidic hydrogen atom, but it could also be written as $NO_2(OH)$, as its structure shows (see margin).

We usually reserve the term "hydroxide" for substances that produce basic solutions, and call the other "hydroxides" acids because they ionize to produce $H^+(aq)$. In ternary acids the hydroxyl oxygen is bonded to a fairly electronegative element (usually a nonmetal). In nitric acid the nitrogen draws the electrons of the N—O (hydroxyl) bond closer to itself than would a less electronegative element such as sodium. The oxygen pulls the electrons of the O—H bond close enough so that the hydrogen atom ionizes as H^+, leaving NO_3^-:

Bond that breaks to form H^+ and NO_3^-

Hydroxyl group

$$HNO_3(aq) \longrightarrow H^+(aq) + NO_3^-(aq)$$

In contrast, let us consider hydroxides of metals. Oxygen is much more electronegative than most metals, such as sodium. It draws the electrons of the sodium–oxygen bond in NaOH (a strong soluble base) so close to itself that the bonding is ionic. Therefore, NaOH exists as Na^+ and OH^- ions, even in the solid state, and dissociates into Na^+ and OH^- ions when it dissolves in H_2O.

$$Na^+OH^-(s) \xrightarrow{H_2O} Na^+(aq) + OH^-(aq)$$

We usually write the formula for sulfuric acid as H_2SO_4 to emphasize the fact that it is an acid. However, the formula can also be written as $SO_2(OH)_2$, because the structure of sulfuric acid (see margin) shows clearly that H_2SO_4 contains two —O—H groups bound to a sulfur atom. Because the O—H bonds are easier to break than the S—O bonds, sulfuric acid ionizes as an acid.

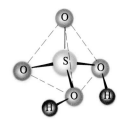

Sulfuric acid is called a polyprotic acid because it has more than one ionizable hydrogen atom per molecule. It is the only common polyprotic acid that is also a strong acid.

The autoionization of water (Section 10-3) was described in terms of Brønsted–Lowry theory. In Lewis theory terminology, this is also an acid–base reaction. The acceptance of a proton, H^+, by a base involves the formation of a *coordinate covalent bond*.

$$H:\overset{..}{\underset{..}{O}}:\ +\ H:\overset{..}{\underset{..}{O}}: \ \rightleftharpoons \ H:\overset{..}{\underset{H}{O}}: \overset{H^+}{} +\ H:\overset{..}{\underset{..}{O}}:^-$$

base   acid

Theoretically, any species that contains an unshared electron pair could act as a base. In fact, most ions and molecules that contain unshared electron pairs do undergo some reactions by sharing their electron pairs. Conversely, many Lewis acids contain only six electrons in the highest occupied energy level of the central element. They react by accepting a share in an additional pair of electrons. These species are said to have an **open sextet**. Many compounds of the Group IIIA elements are Lewis acids, as illustrated by the reaction of boron trichloride with ammonia, presented earlier.

Anhydrous aluminum chloride is a common Lewis acid that is used to catalyze many organic reactions. The dissolution of $AlCl_3$ in hydrochloric acid gives a solution that contains $AlCl_4^-$ ions.

$$AlCl_3(s) + Cl^-(aq) \longrightarrow AlCl_4^-(aq)$$

acid   base   product

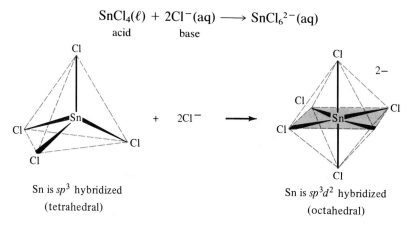

Other ions and molecules behave as Lewis acids by expansion of the valence shell of the central element. Anhydrous tin(IV) chloride, often called stannic chloride, is a colorless liquid that also is frequently used as a Lewis acid catalyst. The tin atom (Group IVA) can expand its valence shell by utilizing vacant d orbitals. It can accept shares in two additional electron pairs, as its reaction with hydrochloric acid illustrates.

$$SnCl_4(\ell) + 2Cl^-(aq) \longrightarrow SnCl_6^{2-}(aq)$$

acid   base

Sn is sp^3 hybridized
(tetrahedral)

Sn is sp^3d^2 hybridized
(octahedral)

Experienced chemists find the Lewis theory to be very useful because so many chemical reactions are covered by it. The less experienced sometimes find the theory less useful, but as their knowledge expands so does its utility.

Key Terms

Acid A substance that produces $H^+(aq)$ ions in aqueous solution. Strong acids ionize completely or almost completely in dilute aqueous solution; weak acids ionize only slightly.

Acid anhydride The oxide of a nonmetal that reacts with water to form an acid.

Acidic salt A salt containing an ionizable hydrogen atom; does not necessarily produce acidic solutions.

Amphiprotism The ability of a substance to exhibit amphoterism by accepting or donating protons.

Amphoterism Ability of a substance to act as either an acid or a base.

Anhydrous Without water.

Autoionization An ionization reaction between identical molecules.

Base A substance that produces $OH^-(aq)$ ions in aqueous solution. Strong soluble bases are soluble in water and are completely *dissociated*. Weak bases ionize only slightly.

Basic anhydride The oxide of a metal that reacts with water to form a base.

Basic salt A salt containing an ionizable OH group.

Brønsted–Lowry acid A proton donor.

Brønsted–Lowry base A proton acceptor.

Conjugate acid–base pair In Brønsted–Lowry terminology, a reactant and product that differ by a proton, H^+.

Coordinate covalent bond A covalent bond in which both shared electrons are furnished by the same species; the bond between a Lewis acid and a Lewis base.

Dissociation In aqueous solution, the process in which a *solid ionic compound* separates into its ions.

Electrolyte A substance whose aqueous solutions conduct electricity.

Formula unit (molecular) equation A chemical equation in which all compounds are represented by complete formulas.

Hydration The process by which water molecules bind to ions or molecules in the solid state or in solution.

Hydride A binary compound of hydrogen.

Hydrolysis Reaction of a substance with water.

Ionization In aqueous solution, the process in which a *molecular* compound reacts with water and forms ions.

Leveling effect The effect by which all acids stronger than the acid that is characteristic of the solvent react with the solvent to produce that acid; a similar statement applies to bases. The strongest acid (base) that can exist in a given solvent is the acid (base) characteristic of that solvent.

Lewis acid Any species that can accept a share in an electron pair.

Lewis base Any species that can make available a share in an electron pair.

Net ionic equation The equation that results from canceling spectator ions and eliminating brackets from a total ionic equation.

Neutralization The reaction of an acid with a base to form a salt and water; usually, the reaction of hydrogen ions with hydroxide ions to form water molecules.

Nonelectrolyte A substance whose aqueous solutions do not conduct electricity.

Normal oxide A metal oxide containing the oxide ion, O^{2-} (oxygen in the −2 oxidation state).

Normal salt A salt containing no ionizable H atoms or OH groups.

Open sextet Refers to species that have only six electrons in the highest energy level of the central element (many Lewis acids).

Polyprotic acid An acid that contains more than one ionizable hydrogen atom per formula unit.

Protonic acid An Arrhenius (classical) acid, or a Brønsted–Lowry acid.

Salt A compound that contains a cation other than H^+ and an anion other than OH^- or O^{2-}.

Spectator ions Ions in solution that do not participate in a chemical reaction.

Strong electrolyte A substance that conducts electricity well in dilute aqueous solution.

Ternary acid An acid that contains three elements—usually H, O, and another nonmetal.

Ternary compound A compound that contains three elements.

Total ionic equation The equation for a chemical reaction written to show the predominant form of all species in aqueous solution or in contact with water.

Weak electrolyte A substance that conducts electricity poorly in dilute aqueous solution.

Exercises

Basic Ideas

1. Robert Boyle observed that acids have certain properties. Which ones did he observe?
2. Gay-Lussac reached an important conclusion about acids and bases. What was it?
3. What is the significance of the idea that acids and bases can be defined only in terms of their reactions with each other?

The Arrhenius Theory

4. Outline Arrhenius' ideas about acids and bases. (a) How did he define the following terms: acid, base, neutralization? (b) Give an example that illustrates each term.
5. Define and illustrate the following terms clearly and concisely. Give an example of each.
   (a) strong electrolyte   (e) strong soluble base
   (b) weak electrolyte   (f) weak acid
   (c) nonelectrolyte   (g) weak base
   (d) strong acid
6. Distinguish between the following pairs of terms and provide a specific example of each.
   (a) strong acid and weak acid
   (b) strong soluble base and weak base
   (c) strong soluble base and insoluble base
7. Write formulas and names for
   (a) the common strong acids
   (b) three weak acids
   (c) the common strong soluble bases
   (d) the most common weak base
   (e) three soluble ionic salts
   (f) three insoluble salts
8. List at least three characteristic properties of acids and three of bases.
9. Summarize the electrical properties of strong electrolytes, weak electrolytes, and nonelectrolytes.
10. Describe an experiment for classifying each of these compounds as a strong electrolyte, a weak electrolyte, or a nonelectrolyte. Classify each. K_2CO_3, HCN, C_2H_5COOH, CH_3OH, H_2S, H_2SO_4, H_2CO, NH_3
11. Limestone, $CaCO_3$, is a water-insoluble material, whereas $Ca(HCO_3)_2$ is soluble. Caves are formed when rainwater containing dissolved CO_2 passes over limestone for long periods of time. Write a chemical equation for the acid–base reaction.

The Hydrated Hydrogen Ion

12. Describe the hydrated hydrogen ion in words and with formulas.
13. Why is the hydrated hydrogen ion important?
14. Criticize the following statement: "The hydrated hydrogen ion should always be represented as H_3O^+."

Brønsted–Lowry Theory

15. State the basic ideas of the Brønsted–Lowry theory.
16. Use Brønsted–Lowry terminology to define the following terms. Illustrate each of the following with a specific example.
   (a) acid
   (b) conjugate base
   (c) base
   (d) conjugate acid
   (e) conjugate acid–base pair
17. Write balanced equations that describe the ionization of the following acids in dilute aqueous solution. Use a single arrow ($\rightarrow$) to represent complete, or nearly complete, ionization and a double arrow ($\rightleftharpoons$) to represent a small extent of ionization. (a) HNO_3, (b) CH_3COOH, (c) HBr, (d) HCN, (e) HF, (f) $HClO_4$
18. Use words and equations to describe how ammonia can act as a base in (a) aqueous solution and (b) the pure state, i.e., as gaseous ammonia molecules when it reacts with gaseous hydrogen chloride or a similar anhydrous acid.
19. What does autoionization mean? How can the autoionization of water be described as an acid–base reaction?
20. Autoionization occurs when an ion other than an H^+ is transferred, as exemplified by the transfer of a Cl^- ion from one PCl_5 molecule to another. Write the equation for this reaction. What are the shapes of the two ions that are formed?
21. What structural features must a compound have to be able to undergo autoionization?
22. What do we mean when we say that water is amphiprotic? (a) Can we also describe water as amphoteric? Why? (b) Illustrate the amphiprotic nature of water by writing two equations for reactions in which water exhibits this property.
23. In terms of Brønsted–Lowry theory, state the differences between (a) a strong and a weak base and (b) a strong and a weak acid.
24. Illustrate, with appropriate equations, the fact that these species are bases in water: NH_3, HS^-, CH_3COO^-, O^{2-}.
25. Illustrate, with appropriate equations, the fact that the following species are acids in water: HBr, HNO_2, $HBrO_4$.
26. Give the products in the following acid–base reactions. Identify the conjugate acid–base pairs.
   (a) $NH_4^+ + CN^-$
   (b) $HS^- + HSO_4^-$
   (c) $HClO_4 + [H_2NNH_3]^+$
   (d) $H^- + H_2O$
27. List the conjugate acids of H_2O, OH^-, Cl^-, HCl, AsO_4^{3-}, NH_2^-, HPO_4^{2-}, and SO_4^{2-}.
28. List the conjugate bases of H_2O, HS^-, HCl, PH_4^+, and $HOCH_3$.

29. Identify the Brønsted–Lowry acids and bases in these reactions and group them into conjugate acid–base pairs.
(a) $NH_3 + HBr \rightleftharpoons NH_4^+ + Br^-$
(b) $NH_4^+ + OH^- \rightleftharpoons NH_3 + H_2O$
(c) $H_3O^+ + PO_4^{3-} \rightleftharpoons HPO_4^{2-} + H_2O$
(d) $HSO_3^- + CN^- \rightleftharpoons HCN + SO_3^{2-}$

Properties of Aqueous Solutions of Acids and Bases

30. Write equations and designate conjugate pairs for the stepwise reactions in water of (a) sulfuric acid, H_2SO_4, and (b) H_3AsO_4.
31. List six properties of aqueous solutions of protonic acids.
32. List five properties of bases in aqueous solution. Does aqueous ammonia exhibit these properties? Why or why not?
33. We say that strong acids, weak acids, and weak bases *ionize* in water, while strong soluble bases *dissociate* in water. What is the difference?
34. Distinguish between solubility in water and extent of ionization in water. Provide specific examples that illustrate the meanings of both terms.
35. Write three general statements that describe the extents to which acids, bases, and salts are ionized in dilute aqueous solutions.

Strengths of Binary Acids

36. Classify each of the hydrides LiH, BeH_2, BH_3, CH_4, NH_3, H_2O, and HF as a Brønsted–Lowry base, a Brønsted–Lowry acid, or neither.
37. What does "acid strength" mean?
38. What does "base strength" mean?
39. Which of the following substances are (a) strong soluble bases, (b) insoluble bases, (c) strong acids, or (d) weak acids? LiOH, HCl, $Ca(OH)_2$, $Fe(OH)_2$, H_2S, H_2CO_3, H_2SO_4, $Zn(OH)_2$.
40. (a) What are binary protonic acids? (b) Write names and formulas for four binary protonic acids.
41. (a) How can the order of increasing acid strength in a series of similar binary protonic acids be explained?
(b) Illustrate your answer for the series HF, HCl, HBr, and HI.
(c) What is the order of increasing base strength of the conjugate bases of the acids in (b)? Why?
(d) Is your explanation applicable to the series H_2O, H_2S, H_2Se, and H_2Te? Why?
42. What does the term "leveling effect" mean? Illustrate your answer with three specific examples.
43. (a) Which is the stronger acid of each pair? (1) NH_4^+, NH_3; (2) H_2O, H_3O^+; (3) HS^-, H_2S. (b) How are acidity and charge related?
44. Arrange the members of each group in order of decreasing acidity: (a) H_2O, H_2Se, H_2S; (b) HI, HCl, HF, HBr; (c) H_2S, S^{2-}, HS^-; (d) SiH_4, HCl, PH_3, H_2S.
45. Illustrate the leveling effect of water by writing reactions for HCl and HNO_3.

Reactions of Acids and Bases

46. Why are acid–base reactions described as neutralization reactions?
47. Distinguish among (a) formula unit equations, (b) total ionic equations, and (c) net ionic equations. What are the advantages and limitations of each?
48. Classify each substance as either an electrolyte or a nonelectrolyte: NH_4Cl, HI, C_6H_6, RaF_2, $Zn(CH_3COO)_2$, $Cu(NO_3)_2$, CH_3COOH, $C_{12}H_{22}O_{11}$ (table sugar), LiOH, $KHCO_3$, CCl_4, $La_2(SO_4)_3$, I_2.
49. Classify each substance as either a strong or a weak electrolyte, and then list (a) the strong acids, (b) the strong bases, (c) the weak acids, and (d) the weak bases. NaCl, $MgSO_4$, HCl, $(COOH)_2$, $Ba(NO_3)_2$, H_3PO_4, $Sr(OH)_2$, HNO_3, HI, $Ba(OH)_2$, LiOH, C_3H_5COOH, NH_3, CH_3NH_2, KOH, $MgMoO_4$, HCN, $HClO_4$.

For Exercises 50–52, write balanced (1) formula unit, (2) total ionic, and (3) net ionic equations for reactions between the following acid–base pairs. Name all compounds except water. Assume complete neutralization.

50. (a) $HNO_3 + KOH \longrightarrow$
(b) $H_2SO_4 + NaOH \longrightarrow$
(c) $HCl + Ca(OH)_2 \longrightarrow$
(d) $CH_3COOH + KOH \longrightarrow$
51. (a) $H_2CO_3 + Sr(OH)_2 \longrightarrow$
(b) $H_2SO_4 + Ba(OH)_2 \longrightarrow$
(c) $H_3PO_4 + Ca(OH)_2 \longrightarrow$
(d) $H_2S + KOH \longrightarrow$
(e) $H_3AsO_4 + KOH \longrightarrow$
52. (a) $HClO_4 + Ba(OH)_2 \longrightarrow$
(b) $HCl + NH_3 \longrightarrow$
(c) $HNO_3 + NH_3 \longrightarrow$
(d) $H_2SO_4 + Fe(OH)_3 \longrightarrow$
(e) $H_3PO_4 + Ba(OH)_2 \longrightarrow$
53. Complete these equations by writing the formulas of the omitted compounds.
(a) $Ba(OH)_2 + ? \rightarrow BaSO_4(s) + H_2O$
(b) $FeO(s) + ? \rightarrow Fe(NO_3)_2(aq) + H_2O$
(c) $HCl(aq) + ? \rightarrow AlCl_3(aq) + ?$
(d) $Na_2O + ? \rightarrow 2NaOH(aq)$
(e) $NaOH + ? \rightarrow Na_2HPO_4(aq) + ?$
(two possible answers)
54. Although many salts may be formed by a variety of reactions, salts are usually thought of as being derived from the reaction of an acid with a base. For each of the salts listed below, choose the acid and base that react with each other to form the salt. Write the (1) formula unit, (2) total ionic, and (3) net ionic equations for the formation of each salt. (a) $Pb(NO_3)_2$, (b) $FeCl_3$, (c) $(NH_4)_2CO_3$, (d) $Ca(ClO_4)_2$, (e) $Al_2(SO_4)_3$

Acidic and Basic Salts

55. What are polyprotic acids? Write names and formulas for five polyprotic acids.

56. What are acidic salts? Write balanced equations to show how the following acidic salts can be prepared from the appropriate acid and base: $NaHSO_3$, $NaHCO_3$, NaH_2PO_4, Na_2HPO_4.

57. Indicate the molar ratio of acid and base required in each case in Exercise 56.

58. The following salts are components of fertilizers. They are made by reacting gaseous NH_3 with concentrated solutions of acids. The heat produced by the reactions evaporates most of the water. Write balanced molecular equations that show the formation of each. (a) NH_4NO_3, (b) $NH_4H_2PO_4$, (c) $(NH_4)_2HPO_4$, (d) $(NH_4)_3PO_4$

59. Some of the acid formed in tissues is excreted through the kidneys. One of the bases removing the acid is HPO_4^{2-}. Write the equation for the reaction. Could Cl^- serve this function?

*60. Acids react with metal carbonates and hydrogen carbonates to form carbon dioxide and water.
(a) Write the balanced equation for the reaction that occurs when baking soda, $NaHCO_3$, and vinegar, 5% acetic acid, are mixed. What causes the "fizz"?
(b) Lactic acid, $CH_3CH(OH)COOH$, is found in sour milk and in buttermilk. Many of its reactions are very similar to those of acetic acid. Write the balanced equation for the reaction of baking soda, $NaHCO_3$, with lactic acid. Explain why bread "rises" during the baking process.

61. What are polyhydroxy bases? Write names and formulas for five polyhydroxy bases.

62. What are basic salts?
(a) Write balanced equations to show how the following basic salts can be prepared from the appropriate acid and base: $Ba(OH)Cl$, $Al(OH)_2Cl$, $Al(OH)Cl_2$.
(b) Indicate the molar ratio of acid and base required in each case.

Strengths of Ternary Acids and Amphoterism

63. What are ternary acids? Write names and formulas for four of them.

64. Why can we describe nitric and sulfuric acids as "hydroxides" of nonmetals?

65. Explain the order of increasing acid strength for the following groups of acids and the order of increasing base strength for their conjugate bases. (a) H_2SO_3, H_2SO_4; (b) HNO_2, HNO_3; (c) H_3PO_3, H_3PO_4; (d) $HClO$, $HClO_2$, $HClO_3$, $HClO_4$.

66. (a) Write a generalization that describes the order of acid strengths for a series of ternary acids that contain different elements in the same oxidation state from the same group in the periodic table.
(b) Indicate the order of acid strengths for the following: (1) HNO_3, H_3PO_4; (2) H_3PO_4, H_3AsO_4; (3) H_2SO_4, H_2SeO_4; (4) $HClO_3$, $HBrO_3$.

67. List the following acids in order of increasing strength: (a) sulfuric, phosphoric, and perchloric; (b) HIO_3, HIO_2, HIO, and HIO_4; (c) selenous, sulfurous, and tellurous; (d) hydrosulfuric, hydroselenic, and hydrotelluric; (e) H_2CrO_4, H_2CrO_2, $HCrO_3$, and H_3CrO_3; (f) $H_4P_2O_7$, $HP_2O_7^{3-}$, $H_3P_2O_7^-$, and $H_2P_2O_7^{2-}$.

68. NaOH behaves as a base in water, while ClOH behaves as an acid. Clearly explain this behavior and the general principles involved for any El—O—H compound.

69. What are amphoteric metal hydroxides? (a) Are they bases? (b) Write the names and formulas for five amphoteric metal hydroxides.

70. Chromium(III) hydroxide and lead(II) hydroxide are typical amphoteric hydroxides.
(a) Write the formula unit, total ionic, and net ionic equations for the complete reaction of each hydroxide with nitric acid.
(b) Write the same kinds of equations for the reaction of each hydroxide with an excess of potassium hydroxide solution. Reference to Table 10-2 may be helpful.

Preparation of Acids

71. Volatile acids such as nitric acid, HNO_3, and acetic acid, CH_3COOH, can be prepared by adding concentrated H_2SO_4 to salts of the acids.
(a) Write chemical equations for the reaction of H_2SO_4 with (1) sodium acetate and (2) sodium nitrate (called chile saltpeter.)
(b) Why can't a dilute aqueous solution of H_2SO_4 be used?

72. Outline a method of preparing each of the following acids and write appropriate balanced equations for each preparation: (a) H_2S, (b) HCl, (c) HNO_3.

73. Repeat Exercise 72 for (a) carbonic acid, (b) perchloric acid, (c) permanganic acid, and (d) phosphoric acid (two methods).

74. Give the formula for an example chosen from the representative elements for (a) an acidic oxide, (b) an amphoteric oxide, and (c) a basic oxide.

The Lewis Theory

75. Define and illustrate the following terms clearly and concisely. Write an equation to illustrate the meaning of each term. (a) Lewis acid, (b) Lewis base, (c) neutralization (according to Lewis theory)

76. What are the advantages and limitations of the Brønsted–Lowry theory?

77. What are the advantages and limitations of the Lewis theory?

78. Draw a Lewis formula for each species in the following equations. Label the acids and bases using Lewis theory terminology.
(a) $H_2O + H_2O \rightleftharpoons H_3O^+ + OH^-$
(b) $HCl(g) + H_2O \longrightarrow H_3O^+ + Cl^-$
(c) $NH_3(g) + H_2O \rightleftharpoons NH_4^+ + OH^-$
(d) $NH_3(g) + HF(g) \longrightarrow NH_4F(s)$

79. What is the term for a single covalent bond in which both electrons in the shared pair come from the same atom? Identify the Lewis acid and base and the donor and acceptor atoms in the following.

$$H-N: \quad + \quad B-F \longrightarrow H-N-B-F:$$

80. Identify the Lewis acid and base and the donor and acceptor atoms in each of the following.

(a) $H-O: + H^+ \longrightarrow \left[H-O-H \right]^+$

(b) $6 \left[:Cl: \right]^- + Pt^{4+} \longrightarrow \left[:Cl-Pt-Cl: \right]^{2-}$

81. Iodine, I_2, is much more soluble in a water solution of potassium iodide, KI, than it is in H_2O. The anion found in the solution is I_3^-. Write an equation for this reaction, indicating the Lewis acid and the Lewis base.

82. A group of very strong acids are the fluoroacids, H_mXF_n. Two such acids are formed by Lewis acid–base reactions.

(a) Identify the Lewis acid and the Lewis base:

$HF + SbF_5 \longrightarrow H(SbF_6)$ (called a "super" acid, hexafluoroantimonic acid)

$HF + BF_3 \longrightarrow H(BF_4)$ (tetrafluoroboric acid)

(b) To which atom is the H of the product bonded? How is the H bonded?

Mixed Exercises

83. Sort the following list of chemicals into (i) acidic, (ii) basic, and (iii) amphoteric species. Assume all oxides are dissolved in water. Do not be intimidated by the way in which the formula of the compound is written. (a) Cs_2O, (b) N_2O_5, (c) HCl, (d) $SO_2(OH)_2$, (e) HNO_2, (f) Al_2O_3, (g) BaO, (h) H_2O, (i) CO_2

84. Indicate which of the following substances—(a) HCl, (b) $H_2PO_3^-$, (c) H_2CaO_2, (d) $ClO_3(OH)$, (e) $Sb(OH)_3$— can act as (i) an acid, (ii) a base, or (iii) both according to the (α) Arrhenius (classical) theory and/or the (β) Brønsted–Lowry theory. Do not be confused by the way in which the formulas are written.

85. (a) Write equations for the reactions (1) $HCO_3^- + H_3O^+$ and (2) $HCO_3^- + OH^-$, and indicate the conjugate acid–base pairs in each case.

(b) A substance such as HCO_3^- that reacts with both H_3O^+ and OH^- is said to be _____. (Fill in the missing word.)

86. (a) List the conjugate bases of (1) H_3PO_4, (2) NH_4^+, and (3) OH^- and the conjugate acids of (4) HSO_4^-, (5) PH_3, and (6) PO_4^{3-}.

(b) Given that NO_2^- is a stronger base than NO_3^-, which is the stronger acid—nitric acid, HNO_3, or nitrous acid, HNO_2?

87. To determine the relative basicities of bases (a) stronger than OH^-, use an acid that is _____ (*stronger than, the same strength as,* or *weaker than*) H_2O; (b) weaker than H_2O, use an acid that is _____ (*stronger than, the same strength as,* or *weaker than*) H_3O^+.

88. Write net ionic equations for the reactions of the amphoteric hydroxide $Sn(OH)_2$ with (a) HCl and (b) NaOH.

*****89.** A 0.1 M solution of copper(II) chloride, $CuCl_2$, causes the light bulb in Figure 4-2 to glow brightly. When hydrogen sulfide, H_2S, a very weak acid, is added to the solution, a black precipitate of copper(II) sulfide, CuS, forms and the bulb still glows brightly. The experiment is repeated with a 0.1 M solution of copper(II) acetate, $Cu(C_2H_3O_2)_2$, which also causes the bulb to glow brightly. Again, CuS forms, but this time the bulb glows dimly. With the aid of ionic equations, explain the difference in behavior of the $CuCl_2$ and $Cu(C_2H_3O_2)_2$ solutions.

90. Referring again to Figure 4-2, explain the following results of a conductivity experiment (use ionic equations).
(a) Individual solutions of NaOH and HCl cause the bulb to glow brightly. When the solutions are mixed, the bulb still glows brightly.
(b) Individual solutions of NH$_3$ and CH$_3$COOH cause the bulb to glow dimly. When the solutions are mixed, the bulb glows brightly.

91. Which statements are true? Rewrite any false statement so that it is correct.

(a) Strong acids and bases are virtually 100% ionized or dissociated in dilute aqueous solutions.
(b) The leveling effect is the seemingly identical strengths of all acids and bases in aqueous solutions.
(c) A conjugate acid is a molecule or ion formed by the addition of a proton to a base.
(d) Amphoterism and amphiprotism are the same in aqueous solution.

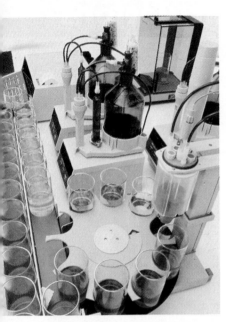

Automatic titrators are used in modern analytical laboratories. Such titrators rely on electrical properties of the solutions. Methyl red indicator changes from yellow to red at the end point of this titration.

Objectives

As you study this chapter, you should
☐ Review molarity calculations and expand your understanding of molarity
☐ Learn about standardization and the use of standard solutions of acids and bases using
  The mole method and molarity

Equivalent weights and normality
☐ Review redox reactions and balancing redox equations
☐ Learn about redox titrations using
  The mole method and molarity
  Equivalent weights and normality

ydrochloric acid, HCl, is called "stomach acid" because it is the main acid ($\sim 0.10\ M$) in our digestive juices. When the concentration of HCl is too high in humans, problems result. These range from "heartburn" to ulcers, which can eat through the lining of the stomach wall. Snakes have very high concentrations of HCl in their digestive juices so that they can digest whole small animals and birds.

Automobile batteries contain 40% H_2SO_4 by mass. When the battery has "run down," the concentration of H_2SO_4 is significantly lower than 40%. A technician checks an automobile battery by drawing some battery acid into a hydrometer, which indicates the density of the solution. This density is related to the concentration of H_2SO_4.

There are many practical applications of acid–base chemistry in which we must know the concentration of a solution of an acid or a base.

Concentrations and Aqueous Acid–Base Reactions

11-1 Calculations Involving Molarity

In Sections 3-6 through 3-9, we introduced methods for expressing concentrations of solutions and discussed some related calculations. Review of those sections will be helpful as we learn more about acid–base reactions in solutions.

In *some cases,* one mole of an acid reacts with one mole of a base:

$$HCl + NaOH \longrightarrow NaCl + H_2O$$

$$HNO_3 + KOH \longrightarrow KNO_3 + H_2O$$

Because one mole of each acid reacts with one mole of each base in these examples, *one liter of a one-molar solution of either of these acids* reacts with *one liter of a one-molar solution of either of these bases.* These acids have only one acidic hydrogen per formula unit, and these bases have one hydroxide ion per formula unit, so one formula unit of base reacts with one hydrogen ion.

The *reaction ratio* is the relative numbers of moles of reactants and products shown in the balanced equation.

Example 11-1

If 100.0 mL of 1.00 M HCl solution and 100.0 mL of 1.00 M NaOH are mixed, what is the molarity of the salt in the resulting solution? You may assume that the volumes are additive.

Plan

We first write the balanced equation for the acid–base reaction, and then construct the reaction summary that shows the amounts (millimoles) of NaOH and HCl. We determine the amount of salt formed from the reaction summary. The final (total) volume is the sum of the volumes mixed. Then we can calculate the molarity of the salt.

Solution

The following tabulation shows that equal numbers of millimoles or moles of HCl and NaOH are mixed, and therefore the resulting solution contains only NaCl, the salt formed by the reaction, and water:

	NaOH	+	HCl	$\longrightarrow$ NaCl	+	H_2O
Rxn ratio:	1 mmol		1 mmol	1 mmol		1 mmol
Start:	$\left[100 \text{ mL} \left(\dfrac{1.00 \text{ mmol}}{\text{mL}}\right)\right]$		$\left[100 \text{ mL} \left(\dfrac{1.00 \text{ mmol}}{\text{mL}}\right)\right]$	0 mmol		
	= 100 mmol NaOH		= 100 mmol HCl			
Change:	−100 mmol		−100 mmol	+100 mmol		
After rxn:	0 mmol		0 mmol	100 mmol		

The HCl and NaOH neutralize each other exactly, and the resulting solution contains 100 mmol of NaCl in 200 mL of solution. Its molarity is

$$? \frac{\text{mmol NaCl}}{\text{mL}} = \frac{100 \text{ mmol}}{200 \text{ mL}} = \boxed{0.500 \; M \text{ NaCl}}$$

Experiments have shown that volumes of dilute aqueous solutions are very nearly additive. No significant error is introduced by making this assumption. 100 mL of NaOH solution mixed with 100 mL of HCl solution gives 200 mL of solution.

The millimole (mmol) was introduced at the end of Section 2-6. Please review Example 2-11. Recall that
1 mol = 1000 mmol
1 L = 1000 mL

Example 11-2

If 100 mL of 1.00 M HCl and 100 mL of 0.75 M NaOH solutions are mixed, what is the molarity of the resulting solution?

Plan

We proceed as we did in Example 11-1. This reaction summary shows that NaOH is the limiting reactant and that we have excess HCl.

Solution

$$HCl \quad + \quad NaOH \quad \longrightarrow \quad NaCl \quad + \quad H_2O$$

Rxn ratio:	1 mmol	1 mmol	1 mmol	1 mmol
Start:	100 mmol	75 mmol	0 mmol	
Change:	−75 mmol	−75 mmol	+75 mmol	
After rxn:	25 mmol	0 mmol	75 mmol	

Because two solutes are present in the solution after reaction, we must calculate the concentrations of both:

$$\underline{?}\frac{\text{mmol HCl}}{\text{mL}} = \frac{25 \text{ mmol HCl}}{200 \text{ mL}} = \boxed{0.12 \; M \text{ HCl}}$$

$$\underline{?}\frac{\text{mmol NaCl}}{\text{mL}} = \frac{75 \text{ mmol NaCl}}{200 \text{ mL}} = \boxed{0.38 \; M \text{ NaCl}}$$

Both HCl and NaCl are strong electrolytes, so the solution is 0.12 M in $H^+(aq)$, $(0.12 + 0.38)$ $M = 0.50$ M in Cl^-, and 0.38 M in Na^+ ions.

EOC 10, 11

In many cases more than one mole of a base will be required to neutralize completely one mole of an acid, or more than one mole of an acid will be required to neutralize completely one mole of a base.

$$H_2SO_4 + 2NaOH \longrightarrow Na_2SO_4 + 2H_2O$$
$$\text{1 mol} \qquad \text{2 mol} \qquad \qquad \text{1 mol}$$

$$2HCl + Ca(OH)_2 \longrightarrow CaCl_2 + 2H_2O$$
$$\text{2 mol} \qquad \text{1 mol} \qquad \quad \text{1 mol}$$

The first equation shows that one mole of H_2SO_4 reacts with two moles of NaOH. Thus, *two* liters of 1 M NaOH solution are required to neutralize one liter of 1 M H_2SO_4 solution. The second equation shows that two moles of HCl react with one mole of $Ca(OH)_2$. Thus, *two* liters of HCl solution are required to neutralize one liter of $Ca(OH)_2$ solution of equal molarity.

Example 11-3

What volume of 0.00300 M HCl solution would just neutralize 30.0 mL of 0.00100 M $Ca(OH)_2$ solution?

Plan

We write the balanced equation for the reaction to determine the reaction ratio. Then we convert (1) milliliters of $Ca(OH)_2$ solution to moles of $Ca(OH)_2$ using molarity as a unit factor, 0.00100 mol $Ca(OH)_2$/1000 mL $Ca(OH)_2$ solution; (2) moles of $Ca(OH)_2$ to moles of HCl using the unit factor, 2 mol HCl/1 mol $Ca(OH)_2$ (from the balanced equation); and (3) moles of HCl to milliliters of HCl solution using the unit factor, 1000 mL HCl/0.00300 mol HCl.

mL Ca(OH)$_2$ soln	$\longrightarrow$	mol Ca(OH)$_2$ present	$\longrightarrow$	mol HCl needed	$\longrightarrow$	mL HCl(aq) needed

Solution

The balanced equation for the reaction is

$$2HCl + Ca(OH)_2 \longrightarrow CaCl_2 + 2H_2O$$

$$\text{2 mol}\text{1 mol}\text{1 mol}\text{2 mol}$$

$$\underset{?}{?} \text{ mL HCl} = 30.0 \text{ mL Ca(OH)}_2 \times \frac{0.00100 \text{ mol Ca(OH)}_2}{1000 \text{ mL Ca(OH)}_2} \times \frac{2 \text{ mol HCl}}{1 \text{ mol Ca(OH)}_2} \times \frac{1000 \text{ mL HCl}}{0.00300 \text{ mol HCl}}$$

$$= \boxed{20.0 \text{ mL HCl}}$$

EOC 7

In the preceding example we used the unit factor, 2 mol HCl/1 mol $Ca(OH)_2$, to convert moles of $Ca(OH)_2$ to moles of HCl because the balanced equation for the reaction shows that two moles of HCl are required to neutralize one mole of $Ca(OH)_2$. We must always write the balanced equation for the reaction and determine the *reaction ratio*.

Example 11-4

If 100.0 mL of 1.00 M H_2SO_4 solution is mixed with 200.0 mL of 1.00 M KOH, what salt is produced, and what is its molarity?

Plan

We proceed as we did in Example 11-1. We note that the reaction ratio is 1 mmol of H_2SO_4 to 2 mmol of KOH to 1 mmol of K_2SO_4.

Solution

	H_2SO_4	+	2KOH	$\longrightarrow$	K_2SO_4	+ 2H$_2$O
Rxn ratio:	1 mmol		2 mmol		1 mmol	
Start:	$\left[100.0 \text{ mL} \left(\frac{1.00 \text{ mmol}}{\text{mL}}\right)\right]$		$\left[200.0 \text{ mL} \left(\frac{1.00 \text{ mmol}}{\text{mL}}\right)\right]$			
	= 100 mmol		= 200 mmol		0 mmol	
Change:	−100 mmol		−200 mmol		+100 mmol	
After rxn:	0 mmol		0 mmol		100 mmol	

The reaction produces 100 mmol of potassium sulfate. This is contained in 300 mL of solution, and so the concentration is

$$\underset{?}{?} \frac{\text{mmol } K_2SO_4}{\text{mL}} = \frac{100 \text{ mmol } K_2SO_4}{300 \text{ mL}} = \boxed{0.333 \text{ } M \text{ } K_2SO_4}$$

Because K_2SO_4 is a strong electrolyte, this corresponds to 0.666 M K^+ and 0.333 M SO_4^{2-}.

EOC 12

Standardization of Solutions of Acids and Bases

In Section 3-9 we discussed *titrations* of solutions of acids and bases and introduced the terminology used to describe titrations. Please review Section 3-9 thoroughly and study Figure 3-4.

Standardization is the process by which one determines the concentration of a solution by measuring accurately the volume of the solution required to react with an exactly known amount of a primary standard. The standardized solution is then known as a secondary standard and is used in the analysis of unknowns.

The properties of an ideal *primary standard* include the following:

CO_2, H_2O, and O_2 are present in the atmosphere. They react with many substances.

1. It must not react with or absorb the components of the atmosphere, such as water vapor, oxygen, and carbon dioxide.
2. It must react according to one invariable reaction.
3. It must have a high percentage purity.
4. It should have a high formula weight to minimize the effect of error in weighing.
5. It must be soluble in the solvent of interest.
6. It should be nontoxic.

The first five of these characteristics minimize the errors involved in analysis. An additional factor, low cost, is desirable but not necessary. Because primary standards are often costly and difficult to prepare, secondary standards are often used in day-to-day work.

11-2 Acid–Base Titrations: The Mole Method and Molarity

Refer to the Brønsted–Lowry theory (Section 10-3).

Let us now describe the use of a few primary standards for acids and bases. One primary standard for solutions of acids is sodium carbonate, Na_2CO_3, a solid compound.

$$H_2SO_4 + Na_2CO_3 \longrightarrow Na_2SO_4 + CO_2 + H_2O$$
$$\text{1 mol} \qquad \text{1 mol} \qquad \text{1 mol} \quad \text{1 mol} \quad \text{1 mol}$$

1 mol Na_2CO_3 = 106.0 g   and   1 mmol Na_2CO_3 = 0.1060 g

Sodium carbonate is a salt. However, because a base can be broadly defined as a substance that reacts with hydrogen ions, in *this* reaction Na_2CO_3 can be thought of as a base.

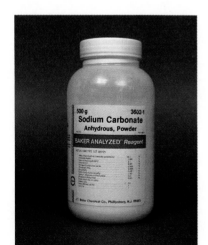

Sodium carbonate is often used as a primary standard for acids.

Example 11-5

Calculate the molarity of a solution of H_2SO_4 if 40.0 mL of the solution neutralize 0.364 gram of Na_2CO_3.

Plan

We know from the balanced equation that 1 mol of H_2SO_4 reacts with 1 mol of Na_2CO_3, 106.0 g. This provides the unit factors that convert 0.364 g of Na_2CO_3 to the corresponding number of moles of H_2SO_4, from which we can calculate molarity.

| g Na_2CO_3 available | $\longrightarrow$ | mol Na_2CO_3 present | $\longrightarrow$ | mol H_2SO_4 used | $\longrightarrow$ | molarity of H_2SO_4 |

Solution

$$\underline{?}\text{ mol } H_2SO_4 = 0.364 \text{ g } Na_2CO_3 \times \frac{1 \text{ mol } Na_2CO_3}{106.0 \text{ g } Na_2CO_3} \times \frac{1 \text{ mol } H_2SO_4}{1 \text{ mol } Na_2CO_3}$$

$$= 0.00343 \text{ mol } H_2SO_4 \qquad \text{(present in 40.0 mL of solution)}$$

Now we calculate the molarity of the H_2SO_4 solution:

$$\frac{?\ \text{mol}\ H_2SO_4}{L} = \frac{0.00343\ \text{mol}\ H_2SO_4}{0.0400\ L} = \boxed{0.0858\ M\ H_2SO_4}$$

EOC 31

Most inorganic bases are metal hydroxides, all of which are solids. However, even in the solid state, most inorganic bases react rapidly with CO_2 (an acid anhydride) from the atmosphere. Most metal hydroxides also absorb H_2O from the air. These properties make it *very* difficult to accurately weigh out samples of pure metal hydroxides. Chemists obtain solutions of bases of accurately known concentration by standardizing the solutions against an acidic salt, potassium hydrogen phthalate, $KC_6H_4(COO)(COOH)$. This is produced by neutralization of one of the two ionizable hydrogens of an organic acid, phthalic acid:

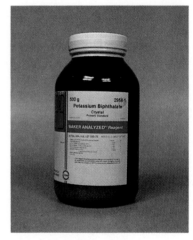

Very pure KHP is available.

$$C_6H_4(COOH)_2 \quad \text{phthalic acid} \;+\; K\,OH \longrightarrow H_2O \;+\; KC_6H_4(COO)(COOH) \quad \text{potassium hydrogen phthalate (KHP)}$$

This acidic salt, known as KHP, has one acidic hydrogen (highlighted) that reacts with bases. KHP is easily obtained in a high state of purity, and is soluble in water. It is used as a primary standard for bases.

Example 11-6

A 20.00-mL sample of a solution of NaOH reacts with 0.3641 gram of KHP. Calculate the molarity of the basic solution.

Plan

We first write the balanced equation for the reaction between NaOH and KHP. We then calculate the number of moles of NaOH in 20.00 mL of solution from the amount of KHP that reacts with it. Then we can calculate the molarity of the NaOH solution.

$$\boxed{\begin{array}{c}\text{g KHP}\\ \text{available}\end{array}} \longrightarrow \boxed{\begin{array}{c}\text{mol KHP}\\ \text{available}\end{array}} \longrightarrow \boxed{\begin{array}{c}\text{mol NaOH}\\ \text{required}\end{array}} \longrightarrow \boxed{\begin{array}{c}\text{molarity}\\ \text{of NaOH}\end{array}}$$

Solution

$$\text{NaOH} + \text{KHP} \longrightarrow \text{NaKP} + H_2O$$
$$\text{1 mol} \quad\;\; \text{1 mol} \quad\quad\; \text{1 mol} \quad\; \text{1 mol}$$

The P in KHP stands for the phthalate ion, $C_6H_4(COO)_2^{2-}$, *not* phosphorus.

We see that NaOH and KHP react in a 1:1 mole ratio. One mole of KHP is 204.2 g.

$$? \text{ mol NaOH} = 0.3641 \text{ g KHP} \times \frac{1 \text{ mol KHP}}{204.2 \text{ g KHP}} \times \frac{1 \text{ mol NaOH}}{1 \text{ mol KHP}}$$

$$= 0.001783 \text{ mol NaOH}$$

Then we calculate the molarity of the NaOH solution.

$$\frac{? \text{ mol NaOH}}{\text{L}} = \frac{0.001783 \text{ mol NaOH}}{0.02000 \text{ L}} = \boxed{0.08915 \text{ } M \text{ NaOH}}$$

EOC 28, 31

Impure samples of acids can be titrated with standard solutions of bases. The results can be used to determine percentage purity of the samples.

Example 11-7

Oxalic acid is used to remove iron stains and some ink stains from fabrics. A 0.1743-gram sample of *impure* oxalic acid, $(COOH)_2$, required 39.82 mL of 0.08915 M NaOH solution for complete neutralization. No acidic impurities were present. Calculate the percentage purity of the $(COOH)_2$.

Plan

We write the balanced equation for the reaction and calculate the number of moles of NaOH in the standard solution. Then we calculate the mass of $(COOH)_2$ in the sample, which gives us the information we need to calculate percentage purity.

Solution

The equation for the reaction of NaOH with $(COOH)_2$ is

$$\begin{array}{ccccc} 2\text{NaOH} & + & (COOH)_2 & \longrightarrow & \text{Na}_2(COO)_2 + 2\text{H}_2\text{O} \\ 2 \text{ mol} & & 1 \text{ mol} & & 1 \text{ mol} \qquad 2 \text{ mol} \end{array}$$

> Each molecule of $(COOH)_2$ contains two acidic H's.
>
> $$\begin{array}{cc} & O \quad O \\ & \| \quad \| \\ H-O-C-C-O-H \end{array}$$
>
> 1 mol = 90.04 g

Two moles of NaOH neutralize completely one mole of $(COOH)_2$. The number of moles of NaOH that react is the volume times the molarity of the solution:

$$? \text{ mol NaOH} = 0.03982 \text{ L} \times \frac{0.08915 \text{ mol NaOH}}{\text{L}} = 0.003550 \text{ mol NaOH}$$

Now we calculate the mass of $(COOH)_2$ that reacts with 0.003550 mol NaOH.

$$? \text{ g } (COOH)_2 = 0.003550 \text{ mol NaOH} \times \frac{1 \text{ mol } (COOH)_2}{2 \text{ mol NaOH}} \times \frac{90.04 \text{ g } (COOH)_2}{1 \text{ mol } (COOH)_2}$$

$$= 0.1598 \text{ g } (COOH)_2$$

The sample contained 0.1598 g of $(COOH)_2$, and its percentage purity was

$$\% \text{ purity} = \frac{0.1598 \text{ g } (COOH)_2}{0.1743 \text{ g sample}} \times 100\% = \boxed{91.68\% \text{ pure } (COOH)_2}$$

EOC 32, 34

> Any calculation that can be carried out with equivalent weights and normality can also be done by the mole method using molarity. However, the methods of this section are widely used in health related fields and in many industrial laboratories.

11-3 Acid–Base Titrations: Equivalent Weights and Normality

Because one mole of an acid does not necessarily neutralize one mole of a base, many chemists prefer a method of expressing concentration other than molarity to retain a one-to-one relationship. Concentrations of solutions of acids and bases are frequently expressed as *normality* (*N*). The **normality** of a solution is defined as the number of equivalent weights, or simply equivalents (eq), of solute per liter of solution. (The term "equivalent mass" is not widely used.) Normality may be represented symbolically as

$$\text{normality} = \frac{\text{number of equivalent weights of solute}}{\text{liter of solution}} = \frac{\text{no. eq}}{\text{L}}$$

An **equivalent weight** is often referred to simply as an **equivalent** (eq).

By definition there are 1000 milliequivalent weights (meq) in one equivalent weight of an acid or base. Normality may also be represented as

$$\text{normality} = \frac{\text{number of milliequivalent weights of solute}}{\text{milliliter of solution}} = \frac{\text{no. meq}}{\text{mL}}$$

A **milliequivalent weight** is often referred to simply as a **milliequivalent** (meq).

In acid–base reactions, one **equivalent weight**, or **equivalent (eq), of an acid** is defined as the mass of the acid (expressed in grams) that will furnish 6.022×10^{23} hydrogen ions (1 mol) or that will react with 6.022×10^{23} hydroxide ions (1 mol). One mole of an acid contains 6.022×10^{23} formula units of the acid. Consider hydrochloric acid as a typical monoprotic acid:

$$\text{HCl} \xrightarrow{\text{H}_2\text{O}} \text{H}^+(aq) \quad + \quad \text{Cl}^-(aq)$$

1 mol	1 mol	1 mol
36.46 g	1.008 g	35.45 g
6.022×10^{23} FU	6.022×10^{23} FU	6.022×10^{23} FU

We see that one mole of HCl can produce 6.022×10^{23} H$^+$, and so *one mole of HCl is one equivalent*. The same is true for all monoprotic acids.

Sulfuric acid is a diprotic acid. One molecule of H_2SO_4 can furnish 2H$^+$ ions.

$$\text{H}_2\text{SO}_4 \xrightarrow{\text{H}_2\text{O}} 2\text{H}^+(aq) \quad + \quad \text{SO}_4{}^{2-}(aq)$$

1 mol	2 mol	1 mol
98.08 g	2(1.008 g)	96.06 g
6.022×10^{23} FU	$2(6.022 \times 10^{23})$ FU	6.022×10^{23} FU

This equation shows that one mole of H_2SO_4 can produce $2(6.022 \times 10^{23})$ H$^+$; therefore, one mole of H_2SO_4 is *two* equivalent weights in all reactions in which *both* acidic hydrogen atoms react.

One **equivalent weight of a base** is defined as the mass of the base (expressed in grams) that will furnish 6.022×10^{23} hydroxide ions or the mass of the base that will react with 6.022×10^{23} hydrogen ions.

The equivalent weight of an *acid* is obtained by dividing its formula weight in grams either by the number of acidic hydrogens furnished by one formula unit of the acid *or* by the number of hydroxide ions with which one formula unit of the acid reacts. The equivalent weight of a *base* is obtained by dividing its formula weight in grams either by the number of hydroxide ions furnished by one formula unit *or* by the number of hydrogen ions with which one formula unit of the base reacts. Equivalent weights of some common acids and bases are given in Table 11-1.

From the definitions of one equivalent of an acid and of a base, we see that *one equivalent of any acid reacts with one equivalent of any base*. It is *not* true that one mole of any acid reacts with one mole of any base in a specific chemical reaction. As a consequence of the definition of equivalents,

Table 11-1
Equivalent Weights* of Some Acids and Bases

Acids			Bases		
Symbolic Representation		One eq	Symbolic Representation		One eq
$\dfrac{HNO_3}{1}$	$= \dfrac{63.02\ g}{1}$	$= 63.02\ g\ HNO_3$	$\dfrac{NaOH}{1}$	$= \dfrac{40.00\ g}{1}$	$= 40.00\ g\ NaOH$
$\dfrac{CH_3COO\underline{H}}{1}$	$= \dfrac{60.03\ g}{1}$	$= 60.03\ g\ CH_3COO\underline{H}$	$\dfrac{NH_3}{1}$	$= \dfrac{17.04\ g}{1}$	$= 17.04\ g\ NH_3$
$\dfrac{KC_6H_4(COO)(COO\underline{H})}{1}$	$= \dfrac{204.2\ g}{1}$	$= 204.2\ g\ KC_6H_4(COO)(COO\underline{H})$	$\dfrac{Ca(OH)_2}{2}$	$= \dfrac{74.10\ g}{2}$	$= 37.05\ g\ Ca(OH)_2$
$\dfrac{H_2SO_4}{2}$	$= \dfrac{98.08\ g}{2}$	$= 49.04\ g\ H_2SO_4$	$\dfrac{Ba(OH)_2}{2}$	$= \dfrac{171.36\ g}{2}$	$= 85.68\ g\ Ba(OH)_2$

* Complete neutralization is assumed.

The notation $\simeq$ is read "is equivalent to."

1 eq acid $\simeq$ 1 eq base. In general, we may write the following for *all* acid–base reactions that go to completion:

> no. eq acid = no. eq base   *or*   no. meq acid = no. meq base

The product of the volume of a solution, in liters, and its normality is equal to the number of equivalents of solute contained in the solution. For a solution of an acid,

Remember that the product of volume and concentration equals the amount of solute.

$$L_{acid} \times N_{acid} = L_{acid} \times \frac{eq\ acid}{L_{acid}} = eq\ acid$$

Alternatively,

$$mL_{acid} \times N_{acid} = mL_{acid} \times \frac{meq\ acid}{mL_{acid}} = meq\ acid$$

Similar relationships can be written for a solution of a base. Because 1 eq of acid *always* reacts with 1 eq of base, we may write

> no. eq acid = no. eq base   so

$$L_{acid} \times N_{acid} = L_{base} \times N_{base} \quad or \quad mL_{acid} \times N_{acid} = mL_{base} \times N_{base}$$

Example 11-8
What volume of 0.100 *N* HNO$_3$ solution is required to neutralize completely 50.0 mL of a 0.150 *N* solution of Ba(OH)$_2$?

Plan
We know three of the four variables in the relationship
$mL_{acid} \times N_{acid} = mL_{base} \times N_{base}$, and so we solve for mL_{acid}.

Solution

$$? \, mL_{acid} = \frac{mL_{base} \times N_{base}}{N_{acid}} = \frac{50.0 \, mL \times 0.150 \, N}{0.100 \, N}$$

$$= \boxed{75.0 \, mL \, of \, HNO_3 \, solution}$$

EOC 42, 43

Example 11-9
What is the normality of a solution of 4.202 grams of HNO_3 in 600 mL of solution?

Plan

We convert grams of HNO_3 to moles of HNO_3 and then to equivalents of HNO_3, which lets us calculate the normality.

$$\boxed{\frac{g \, HNO_3}{L}} \longrightarrow \boxed{\frac{mol \, HNO_3}{L}} \longrightarrow \boxed{\frac{eq \, HNO_3}{L}} = \boxed{N \, HNO_3}$$

Solution

$$N = \frac{no. \, eq \, HNO_3}{L}$$

$$? \, \frac{eq \, HNO_3}{L} = \underbrace{\frac{4.202 \, g \, HNO_3}{0.600 \, L} \times \frac{1 \, mol \, HNO_3}{63.02 \, g \, HNO_3}}_{M_{HNO_3}} \times \frac{1 \, eq \, HNO_3}{mol \, HNO_3} = \boxed{0.111 \, N \, HNO_3}$$

Because normality is equal to molarity times the number of equivalents per mole of solute, a solution's normality is always equal to or greater than its molarity.

$$normality = molarity \times \frac{no. \, eq}{mol} \quad or \quad N = M \times \frac{no. \, eq}{mol}$$

Example 11-10
What is (a) the molarity and (b) the normality of a solution that contains 9.50 grams of barium hydroxide in 2000 mL of solution?

Plan

(a) We use the same kind of logic we used in Example 11-9.
(b) Because each mole of $Ba(OH)_2$ produces 2 moles of OH^- ions, 1 mole of $Ba(OH)_2$ is 2 equivalents. Thus

$$N = M \times \frac{2 \, eq}{mol} \quad or \quad M = \frac{N}{2 \, eq/mol}$$

Solution

(a) $$? \, \frac{mol \, Ba(OH)_2}{L} = \frac{9.50 \, g \, Ba(OH)_2}{2.00 \, L} \times \frac{1 \, mol \, Ba(OH)_2}{171.4 \, g \, Ba(OH)_2}$$

$$= \boxed{0.0277 \, M \, Ba(OH)_2}$$

Because each formula unit of $Ba(OH)_2$ contains two OH^- ions,

1 mol $Ba(OH)_2$ = 2 eq $Ba(OH)_2$

Thus, molarity is one half of normality for $Ba(OH)_2$ solutions.

(b) $\underset{?}{}\dfrac{eq\ Ba(OH)_2}{L} = \dfrac{0.0277\ mol\ Ba(OH)_2}{L} \times \dfrac{2\ eq\ Ba(OH)_2}{1\ mol\ Ba(OH)_2}$

$$= \boxed{0.0554\ N\ Ba(OH)_2}$$

EOC 38, 39

In Example 11-11, let us again solve Example 11-5, this time using normality rather than molarity. The balanced equation for the reaction of H_2SO_4 with Na_2CO_3, interpreted in terms of equivalent weights, is

> By definition, there must be equal numbers of equivalents of all reactants and products in a balanced chemical equation.

$$H_2SO_4 + Na_2CO_3 \longrightarrow Na_2SO_4 + CO_2 + H_2O$$

1 mol	1 mol	1 mol	1 mol	1 mol
2 eq	2 eq	2 eq	2 eq	2 eq
98.08 g	106.0 g			

$$1\ eq\ Na_2CO_3 = 53.0\ g \qquad and \qquad 1\ meq\ Na_2CO_3 = 0.0530\ g$$

Example 11-11

Calculate the normality of a solution of H_2SO_4 if 40.0 mL of the solution reacts completely with 0.364 gram of Na_2CO_3.

Plan

We refer to the balanced equation. We are given the mass of Na_2CO_3, so we convert grams of Na_2CO_3 to milliequivalents of Na_2CO_3, then to milliequivalents of H_2SO_4, which lets us calculate the normality of the H_2SO_4 solution.

$$\boxed{\begin{array}{c} g\ Na_2CO_3 \\ present \end{array}} \longrightarrow \boxed{\begin{array}{c} meq\ Na_2CO_3 \\ present \end{array}} \longrightarrow \boxed{\begin{array}{c} meq\ H_2SO_4 \\ needed \end{array}} \longrightarrow \boxed{\begin{array}{c} meq\ H_2SO_4 \\ \hline mL \end{array}}$$

Solution

First we calculate the number of milliequivalents of Na_2CO_3 in the sample:

$$no.\ meq\ Na_2CO_3 = 0.364\ g\ Na_2CO_3 \times \dfrac{1\ meq\ Na_2CO_3}{0.0530\ g\ Na_2CO_3} = 6.87\ meq\ Na_2CO_3$$

Because no. meq H_2SO_4 = no. meq Na_2CO_3, we may write

$$mL_{H_2SO_4} \times N_{H_2SO_4} = 6.87\ meq\ H_2SO_4$$

> The normality of this H_2SO_4 solution is twice the molarity obtained in Example 11-5 because 1 mol of H_2SO_4 is 2 eq.

$$N_{H_2SO_4} = \dfrac{6.87\ meq\ H_2SO_4}{mL_{H_2SO_4}} = \dfrac{6.87\ meq\ H_2SO_4}{40.0\ mL} = \boxed{0.172\ N\ H_2SO_4}$$

Example 11-12

A 0.1743-gram sample of impure oxalic acid, $(COOH)_2$, required 39.82 mL of 0.08915 N NaOH solution for complete neutralization. No acidic impurities were present. Calculate the percentage purity of the $(COOH)_2$.

> This is Example 11-7 using normality rather than molarity.

Plan

We write the balanced equation for the reaction. We calculate the number of equivalents of NaOH in the standard solution, which tells us the number of equivalents of $(COOH)_2$ in the sample. Then we convert equivalents of $(COOH)_2$ to grams of $(COOH)_2$, which allows us to calculate the percentage purity.

$$N \text{ NaOH soln} \longrightarrow \text{eq NaOH} = \text{eq (COOH)}_2 \longrightarrow \text{g (COOH)}_2 \longrightarrow \% \text{ (COOH)}_2$$

Solution

$$2NaOH + (COOH)_2 \longrightarrow Na_2(COO)_2 + 2H_2O$$

2 mol	1 mol	
2 eq	2 eq	$= 90.04$ g; therefore 1 eq $= 45.02$ g (COOH)$_2$

The equation for the reaction of NaOH with $(COOH)_2$ shows that *one mole* of $(COOH)_2$ is *two* equivalents. Therefore, one equivalent weight of $(COOH)_2$ is 90.04 g/2 $= 45.02$ g. The number of equivalents of NaOH that react is the volume times the normality of the solution.

$$\underline{?} \text{ eq NaOH} = 0.03982 \text{ L} \times \frac{0.08915 \text{ eq NaOH}}{L} = 0.003550 \text{ eq NaOH}$$

Now we calculate the mass of $(COOH)_2$ that reacts with 0.003550 eq NaOH:

$$\underline{?} \text{ g (COOH)}_2 = 0.003550 \text{ eq NaOH} \times \frac{1 \text{ eq (COOH)}_2}{1 \text{ eq NaOH}} \times \frac{45.02 \text{ g (COOH)}_2}{1 \text{ eq (COOH)}_2}$$

$$= 0.1598 \text{ g (COOH)}_2$$

The sample contained 0.1598 g of $(COOH)_2$, and its percent purity is

$$\% \text{ purity} = \frac{0.1598 \text{ g (COOH)}_2}{0.1743 \text{ g sample}} \times 100\% = \boxed{91.68\% \text{ pure (COOH)}_2}$$

EOC 45, 46

Redox Reactions

One method of analyzing samples quantitatively for the presence of *oxidizable* or *reducible* substances is by **redox titration**. In such analyses, the concentration of a solution is determined by allowing it to react with a carefully measured amount of a *standard* solution of an oxidizing or reducing agent.

As in acid–base titrations, amounts of solutes can be described in terms of either moles or equivalent weights. Concentrations of solutions involved in redox titrations can be expressed in terms of either molarity or normality. We shall illustrate redox titrations, separately, by both the mole–molarity method and the equivalent weight–normality method.

11-4 Redox Titrations: The Mole Method and Molarity

As in other kinds of chemical reactions, we must pay particular attention to the mole ratio in which oxidizing agents and reducing agents react. Please review Sections 4-9 through 4-11 carefully.

Potassium permanganate, $KMnO_4$, is a strong oxidizing agent. Through the years it has been the "workhorse" of redox titrations. For example, in acidic solution, $KMnO_4$ reacts with iron(II) sulfate, $FeSO_4$, according to the balanced equations below. A strong acid, such as H_2SO_4, is used in such titrations. See Figure 11-1.

$$2KMnO_4 + 10FeSO_4 + 8H_2SO_4 \longrightarrow 2MnSO_4 + 5Fe_2(SO_4)_3 + K_2SO_4 + 8H_2O$$

$$MnO_4^-(aq) + 5Fe^{2+}(aq) + 8H^+(aq) \longrightarrow Mn^{2+}(aq) + 5Fe^{3+}(aq) + 4H_2O(\ell)$$

Because it has an intense purple color, $KMnO_4$ acts as its own indicator. One drop of 0.020 M $KMnO_4$ solution imparts a pink color to a liter of pure water. When $KMnO_4$ solution is added to a solution of a reducing agent, the end point in the titration is taken as the point at which a pale pink color appears in the solution being titrated and persists for at least 30 seconds.

A word about terminology. The preceding reaction involves MnO_4^- ions and Fe^{2+} ions in acidic solution. The source of MnO_4^- ions is the soluble ionic compound $KMnO_4$. We often refer to "permanganate solutions." Clearly such solutions also contain cations—in this case, K^+. Likewise, we often refer to "iron(II) solutions" without specifying what the anion is.

Example 11-13

What volume of 0.0200 M $KMnO_4$ solution is required to oxidize 40.0 mL of 0.100 M $FeSO_4$ in sulfuric acid solution (Figure 11-1)?

Plan

We refer to the balanced equation in the preceding discussion to find the reaction ratio, 1 mol MnO_4^-/5 mol Fe^{2+}. Then we calculate the number of moles of Fe^{2+} to be titrated, which lets us find the number of moles of MnO_4^- required *and* the volume in which this number of moles of $KMnO_4$ is contained.

Solution

The reaction ratio is

One mole of $KMnO_4$ contains one mole of MnO_4^- ions. Therefore, the number of moles of $KMnO_4$ is *always* equal to the number of moles of MnO_4^- ions required in a reaction. Similarly, one mole of $FeSO_4$ contains 1 mole of Fe^{2+} ions.

$$MnO_4^-(aq) + 8H^+(aq) + 5Fe^{2+}(aq) \longrightarrow 5Fe^{3+}(aq) + Mn^{2+}(aq) + 4H_2O$$

rxn ratio: 1 mol   5 mol

The number of moles of Fe^{2+} to be titrated is

$$\underset{?}{} \text{ mol } Fe^{2+} = 40.0 \text{ mL} \times \frac{0.100 \text{ mol } Fe^{2+}}{1000 \text{ mL}} = 4.00 \times 10^{-3} \text{ mol } Fe^{2+}$$

We use the balanced equation to find the number of moles of MnO_4^- required:

$$\underset{?}{} \text{ mol } MnO_4^- = 4.00 \times 10^{-3} \text{ mol } Fe^{2+} \times \frac{1 \text{ mol } MnO_4^-}{5 \text{ mol } Fe^{2+}}$$

$$= 8.00 \times 10^{-4} \text{ mol } MnO_4$$

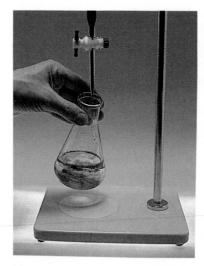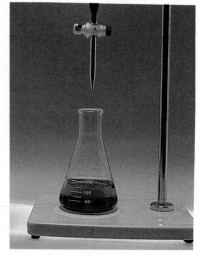

Figure 11-1
(a) Nearly colorless $FeSO_4$ solution is titrated with deep-purple $KMnO_4$.
(b) The end point is the point at which the solution becomes pink, owing to a *very small* excess of $KMnO_4$. Here a considerable excess of $KMnO_4$ was added so that the pink color could be reproduced photographically.

(a)   (b)

Each formula unit of $KMnO_4$ contains one MnO_4^- ion, and so

$$1 \text{ mol } KMnO_4 \mathrel{\hat{=}} 1 \text{ mol } MnO_4^-$$

The volume of $0.0200\ M$ $KMnO_4$ solution that contains 8.00×10^{-4} mol of $KMnO_4$ is

$$\underline{?} \text{ mL } KMnO_4 \text{ soln} = 8.00 \times 10^{-4} \text{ mol } KMnO_4 \times \frac{1000 \text{ mL } KMnO_4 \text{ soln}}{0.0200 \text{ mol } KMnO_4}$$

$$= 40.0 \text{ mL } KMnO_4 \text{ soln}$$

EOC 48

Potassium dichromate, $K_2Cr_2O_7$, is another frequently used oxidizing agent. However, an indicator must be used when reducing agents are titrated with dichromate solutions. $K_2Cr_2O_7$ is orange, and its reduction product, Cr^{3+}, is green.

Consider the oxidation of sulfite ions, SO_3^{2-}, to sulfate ions, SO_4^{2-}, by $Cr_2O_7^{2-}$ ions in the presence of a strong acid such as sulfuric acid. We shall balance the equation by the ion–electron method.

$K_2Cr_2O_7$ is orange in acidic solution. $Cr_2(SO_4)_3$ is green in acidic solution.

$$Cr_2O_7^{2-} \longrightarrow Cr^{3+} \qquad \text{(red, half-rxn)}$$

$$Cr_2O_7^{2-} \longrightarrow 2Cr^{3+}$$

$$14H^+ + Cr_2O_7^{2-} \longrightarrow 2Cr^{3+} + 7H_2O$$

$$6e^- + 14H^+ + Cr_2O_7^{2-} \longrightarrow 2Cr^{3+} + 7H_2O \qquad \text{(balanced red. half-rxn)}$$

$$SO_3^{2-} \longrightarrow SO_4^{2-} \qquad \text{(ox. half-rxn)}$$

$$SO_3^{2-} + H_2O \longrightarrow SO_4^{2-} + 2H^+$$

$$SO_3^{2-} + H_2O \longrightarrow SO_4^{2-} + 2H^+ + 2e^- \qquad \text{(balanced ox. half-rxn)}$$

We now equalize the electron transfer, add the balanced half-reactions, and eliminate common terms:

$$(6e^- + 14H^+ + Cr_2O_7^{2-} \longrightarrow 2Cr^{3+} + 7H_2O) \qquad \text{(reduction)}$$

$$3(SO_3^{2-} + H_2O \longrightarrow SO_4^{2-} + 2H^+ + 2e^-) \qquad \text{(oxidation)}$$

$$\overline{8H^+(aq) + Cr_2O_7^{2-}(aq) + 3SO_3^{2-}(aq) \longrightarrow 2Cr^{3+}(aq) + 3SO_4^{2-}(aq) + 4H_2O(\ell)}$$

The balanced equation tells us that the reaction ratio is 3 mol SO_3^{2-}/mol $Cr_2O_7^{2-}$ or 1 mol $Cr_2O_7^{2-}$/3 mol SO_3^{2-}. Potassium dichromate is the usual source of $Cr_2O_7^{2-}$ ions, and Na_2SO_3 is the usual source of SO_3^{2-} ions. Thus, the preceding reaction ratio could also be expressed as 1 mol $K_2Cr_2O_7$/3 mol Na_2SO_3.

Example 11-14

A 20.00-mL sample of Na_2SO_3 was titrated with 36.30 mL of 0.05130 M $K_2Cr_2O_7$ solution in the presence of H_2SO_4. Calculate the molarity of the Na_2SO_3 solution.

Plan

We can calculate the number of moles of $Cr_2O_7{}^{2-}$ in the standard solution. Then we refer to the balanced equation in the preceding discussion, which gives us the reaction ratio, 3 mol $SO_3{}^{2-}$/1 mol $Cr_2O_7{}^{2-}$. The reaction ratio lets us calculate the number of moles of $SO_3{}^{2-}$(Na_2SO_3) that reacted and the molarity of the solution.

$$\boxed{\text{L } Cr_2O_7{}^{2-}\text{ soln}} \longrightarrow \boxed{\text{mol } Cr_2O_7{}^{2-}} \longrightarrow \boxed{\text{mol } SO_3{}^{2-}} \longrightarrow \boxed{M\text{ } SO_3{}^{2-}\text{ soln}}$$

Solution

From the preceding discussion we know the balanced equation and the reaction ratio:

$$\underset{\text{3 mol}}{3SO_3{}^{2-}} + \underset{\text{1 mol}}{Cr_2O_7{}^{2-}} + 8H^+ \longrightarrow 3SO_4{}^{2-} + 2Cr^{3+} + 4H_2O$$

The number of moles of $Cr_2O_7{}^{2-}$ used is

$$\underline{?}\text{ mol } Cr_2O_7{}^{2-} = 0.03630 \text{ L} \times \frac{0.05130 \text{ mol } Cr_2O_7{}^{2-}}{\text{L}} = 0.001862 \text{ mol } Cr_2O_7{}^{2-}$$

The number of moles of $SO_3{}^{2-}$ that reacted with 0.001862 mol of $Cr_2O_7{}^{2-}$ is

$$\underline{?}\text{ mol } SO_3{}^{2-} = 0.001862 \text{ mol } Cr_2O_7{}^{2-} \times \frac{3 \text{ mol } SO_3{}^{2-}}{1 \text{ mol } Cr_2O_7{}^{2-}} = 0.005586 \text{ mol } SO_3{}^{2-}$$

The Na_2SO_3 solution contained 0.005586 mol of $SO_3{}^{2-}$ (and 0.005586 mol of Na_2SO_3). Its molarity is

$$\underline{?}\frac{\text{mol } Na_2SO_3}{\text{L}} = \frac{0.005586 \text{ mol } Na_2SO_3}{0.02000 \text{ L}} = \boxed{0.2793 \text{ } M \text{ } Na_2SO_3}$$

EOC 49

11-5 Redox Titrations: Equivalent Weights and Normality

The equivalent weight of an oxidizing agent or reducing agent depends upon the specific reaction it undergoes.

When we studied acid–base reactions, we defined the term "equivalent weight" so that one equivalent weight, or one equivalent (eq), of any acid reacts with one equivalent of any base. We now define the term as it applies to *redox reactions* so that one equivalent of any oxidizing agent reacts with one equivalent of any reducing agent. In redox reactions, one **equivalent weight**, or **equivalent**, of a substance is the mass of the oxidizing or reducing substance that gains or loses 6.022×10^{23} electrons.

For all redox reactions we may write

$$\text{no. eq oxidizing agent} = \text{no. eq reducing agent}$$

or

$$\text{no. meq oxidizing agent} = \text{no. meq reducing agent}$$

$$L_O \times N_O = L_R \times N_R \qquad or \qquad mL_O \times N_O = mL_R \times N_R$$

The only difference between calculations for acid–base reactions and those for redox reactions is the *definition* of the equivalent.

Example 11-15

What volume of 0.1000 N KMnO$_4$ would oxidize 40.0 mL of 0.100 N FeSO$_4$ in acidic solution?

This is Example 11-13 using normality rather than molarity.

Plan

From the preceding discussion we know the balanced equation and that $mL_O \times N_O = mL_R \times N_R$. So we solve for mL_R.

Solution

The balanced equation is

$$MnO_4^-(aq) + 8H^+(aq) + 5Fe^{2+}(aq) \longrightarrow 5Fe^{3+}(aq) + Mn^{2+}(aq) + 4H_2O(\ell)$$

Because $mL_{MnO_4^-} \times N_{MnO_4^-} = mL_{Fe^{2+}} \times N_{Fe^{2+}}$, we can solve for the number of milliliters of KMnO$_4$ solution.

$$\underline{?}\ mL_{MnO_4^-} = \frac{mL_{Fe^{2+}} \times N_{Fe^{2+}}}{N_{MnO_4^-}} = \frac{40.0\ mL \times 0.100\ N}{0.1000\ N}$$

$$= \boxed{40.0\ mL\ KMnO_4\ solution}$$

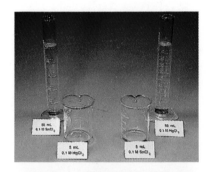

We saw earlier that the balanced net ionic equation and half-reactions for the reaction of KMnO$_4$ with FeSO$_4$ in acidic solution are

$$MnO_4^-(aq) + 8H^+(aq) + 5Fe^{2+}(aq) \longrightarrow Mn^{2+}(aq) + 5Fe^{3+}(aq) + 4H_2O(\ell)$$

$$Fe^{2+}(aq) \longrightarrow Fe^{3+}(aq) + 1e^- \qquad \text{(oxidation)}$$

$$MnO_4^-(aq) + 8H^+(aq) + 5e^- \longrightarrow Mn^{2+}(aq) + 4H_2O(\ell) \quad \text{(reduction)}$$

Each Fe^{2+} ion loses one electron, and so one mole of FeSO$_4$ loses 6.022×10^{23} electrons. One mole of FeSO$_4$ (151.9 g) is one equivalent *in this reaction*.

Each MnO$_4^-$ ion gains five electrons, and so each mole of MnO$_4^-$ gains $5(6.022 \times 10^{23}$ electrons). *In this reaction*, one mole of KMnO$_4$ (158.0 g) is five equivalents.

When they are mixed together, SnCl$_2$ is a reducing agent and HgCl$_2$ is an oxidizing agent. When excess HgCl$_2$ is present, SnCl$_2$ reduces it to white insoluble Hg$_2$Cl$_2$ (right).

$$SnCl_2(aq) + 2HgCl_2(aq) \longrightarrow$$
$$SnCl_4(aq) + Hg_2Cl_2(s)$$

An excess of SnCl$_2$ reduces HgCl$_2$ to elemental mercury (black) in a two-step reaction (left).

$$SnCl_2(aq) + 2HgCl_2(aq) \longrightarrow$$
$$SnCl_4(aq) + Hg_2Cl_2(s)$$

$$SnCl_2(aq) + Hg_2Cl_2(s) \longrightarrow$$
$$SnCl_4(aq) + 2Hg(\ell)$$

$$SnCl_2(aq) + HgCl_2(aq) \longrightarrow$$
$$SnCl_4(aq) + Hg(\ell)$$

The equivalent weight of HgCl$_2$ in these reactions depends upon whether it is the limiting reactant.

	e^- Transferred per Formula Unit	One Mole	One eq
Oxidizing agent: KMnO$_4$	5	158.0 g	$\dfrac{158.0\,g}{5} = 31.60\,g$
Reducing agent: FeSO$_4$	1	151.9 g	$\dfrac{151.9\,g}{1} = 151.9\,g$

In *this* reaction, 31.60 grams of KMnO$_4$ reacts with 151.9 grams of FeSO$_4$.

Example 11-16

How many grams of KMnO$_4$ are contained in 35.0 mL of 0.0500 N KMnO$_4$ used in the following reaction in acidic solution?

$$MnO_4^-(aq) + 5Fe^{2+}(aq) + 8H^+(aq) \longrightarrow Mn^{2+}(aq) + 5Fe^{3+}(aq) + 4H_2O(\ell)$$

Plan

We are given the balanced equation, and we have just seen that one mole of $KMnO_4$ is five equivalents *in this reaction*. We relate milliliters of $KMnO_4$ solution to equivalents of $KMnO_4$, then to moles of $KMnO_4$, and finally to grams of $KMnO_4$.

$$\text{mL } KMnO_4 \text{ soln} \longrightarrow \text{eq } KMnO_4 \longrightarrow \text{mol } KMnO_4 \longrightarrow \text{g } KMnO_4$$

Solution

One mole of $KMnO_4$ is five equivalents in this reaction.

$$\underline{?} \text{ g } KMnO_4 = 35.0 \text{ mL} \times \frac{0.0500 \text{ eq } KMnO_4}{1000 \text{ mL}} \times \frac{1 \text{ mol}}{5 \text{ eq}} \times \frac{158.0 \text{ g}}{1 \text{ mol}}$$

$$= 0.0553 \text{ g } KMnO_4$$

EOC 58, 59

The equivalent weight of an oxidizing or reducing agent depends upon the specific reaction it undergoes. In the following reaction, the equivalent weight of $KMnO_4$ is different from the value we calculated before, because the MnO_4^- ion undergoes a three-electron change rather than a five-electron change. Dissolving metallic zinc in mildly basic $KMnO_4$ solution produces solid manganese(IV) oxide and zinc hydroxide. In this reaction the half-reactions and balanced equation are as follows:

$$2(MnO_4^- + 2H_2O + 3e^- \longrightarrow MnO_2 + 4OH^-) \qquad \text{(reduction)}$$

$$\underline{3(Zn + 2OH^- \longrightarrow Zn(OH)_2 + 2e^-) \qquad \text{(oxidation)}}$$

$$2MnO_4^-(aq) + 3Zn(s) + 4H_2O(\ell) \longrightarrow 2MnO_2(s) + 3Zn(OH)_2(s) + 2OH^-(aq)$$

One mole of $KMnO_4$ is three equivalents *in this reaction*. One mole of zinc is two equivalents, as the following tabulation shows.

Compare the equivalent weight of $KMnO_4$ with that in the previous example.

	Substance	e^- Transferred per Formula Unit	One Mole	One eq
Oxidizing agent:	$KMnO_4(aq)$	3	158.0 g	$\dfrac{158.0 \text{ g}}{3} = 52.67 \text{ g}$
Reducing agent:	$Zn(s)$	2	65.39 g	$\dfrac{65.39 \text{ g}}{2} = 32.69 \text{ g}$

Here one equivalent of $KMnO_4$ is 52.67 grams, *not* 31.60 grams as before.

Example 11-17

How many grams of $KMnO_4$ are contained in 35.0 mL of 0.0500 N $KMnO_4$ used in the following reaction in basic solution?

$$2MnO_4^-(aq) + 3Zn(s) + 4H_2O(\ell) \longrightarrow 2MnO_2(s) + 3Zn(OH)_2(s) + 2OH^-(aq)$$

Plan

We use the same kind of logic we used in Example 11-16. Note that 1 mole of $KMnO_4$ is 3 equivalents in *this* reaction.

Solution

$$\underline{?} \, g \, KMnO_4 = 35.0 \, mL \times \frac{0.0500 \, eq \, KMnO_4}{1000 \, mL} \times \frac{1 \, mol}{3 \, eq} \times \frac{158.0 \, g}{1 \, mol} = \boxed{0.0922 \, g \, KMnO_4}$$

EOC 60

To further illustrate how normality is used in oxidation–reduction titrations, let us solve Example 11-14 using normality rather than molarity.

Example 11-18

A 20.00-mL sample of Na_2SO_3 was titrated with 36.30 mL of 0.3078 N $K_2Cr_2O_7$ solution in the presence of H_2SO_4. (a) Calculate the normality of the Na_2SO_3 solution. (b) What mass of Na_2SO_3 was present in the sample? (c) What mass of $K_2Cr_2O_7$ was used to prepare 500 mL of the $K_2Cr_2O_7$ solution?

Plan

We write the balanced equation for the reaction as well as the balanced half-reactions. We apply the logic of earlier examples.

Solution

The balanced equation for the oxidation of SO_3^{2-} ions to SO_4^{2-} ions by $Cr_2O_7^{2-}$ ions in acidic solution and the balanced half-reactions are

$$8H^+ + Cr_2O_7^{2-} + 3SO_3^{2-} \longrightarrow 2Cr^{3+} + 3SO_4^{2-} + 4H_2O$$

$$6e^- + 14H^+ + Cr_2O_7^{2-} \longrightarrow 2Cr^{3+} + 7H_2O \qquad \text{(red. half-rxn)}$$

$$SO_3^{2-} + H_2O \longrightarrow SO_4^{2-} + 2H^+ + 2e^- \qquad \text{(ox. half-rxn)}$$

(a) We know the volume and normality of the $K_2Cr_2O_7$ solution as well as the volume of the Na_2SO_3 solution:

$$mL_{Cr_2O_7^{2-}} \times N_{Cr_2O_7^{2-}} = mL_{SO_3^{2-}} \times N_{SO_3^{2-}}$$

$$N_{SO_3^{2-}} = \frac{mL_{Cr_2O_7^{2-}} \times N_{Cr_2O_7^{2-}}}{mL_{SO_3^{2-}}} = \frac{36.30 \, mL \times 0.3078 \, N}{20.00 \, mL} = \boxed{0.5587 \, N \, Na_2SO_3}$$

This tells us that 1.000 L of Na_2SO_3 solution contains 0.5587 eq of Na_2SO_3.

The half-reactions (above) give us the information needed to answer (b) and (c):

$$1 \, mol \, K_2Cr_2O_7 = 294.2 \, g \, K_2Cr_2O_7 = 6 \, eq \, K_2Cr_2O_7$$

$$1 \, mol \, Na_2SO_3 = 126.0 \, g \, Na_2SO_3 = 2 \, eq \, Na_2SO_3$$

(b) $\underline{?} \, g \, Na_2SO_3 = 20.00 \, mL \times \dfrac{0.5587 \, eq \, Na_2SO_3}{1000 \, mL} \times \dfrac{126.0 \, g \, Na_2SO_3}{2 \, eq \, Na_2SO_3}$

$$= \boxed{0.7040 \, g \, Na_2SO_3}$$

(c) $\underline{?} \, g \, K_2Cr_2O_7 = 500 \, mL \times \dfrac{0.3078 \, eq \, K_2Cr_2O_7}{1000 \, mL} \times \dfrac{294.2 \, g \, K_2Cr_2O_7}{6 \, eq \, K_2Cr_2O_7}$

$$= \boxed{7.546 \, g \, K_2Cr_2O_7}$$

EOC 61, 63

The answer to Example 11-14 is 0.2793 M. Is 0.5587 N the same concentration?

Key Terms

Buret A piece of volumetric glassware, usually graduated in 0.1-mL intervals, that is used to deliver solutions to be used in titrations in a quantitative (dropwise) manner.

Disproportionation reaction A redox reaction in which the oxidizing agent and the reducing agent are the same species.

End point The point at which an indicator changes color and a titration is stopped.

Equivalence point The point at which chemically equivalent amounts of reactants have reacted.

Equivalent weight in acid–base reactions The mass of an acid or base that furnishes or reacts with 6.022×10^{23} H_3O^+ or OH^- ions.

Equivalent weight of oxidizing or reducing agent The mass that gains (oxidizing agents) or loses (reducing agents) 6.022×10^{23} electrons in a redox reaction.

Half-reaction Either the oxidation part or the reduction part of a redox reaction.

Indicator For acid–base titrations, an organic compound that exhibits different colors in solutions of different acidities; used to determine the point at which the reaction between two solutes is complete.

Normality The number of equivalent weights (equivalents) of solute per liter of solution.

Oxidation An algebraic increase in oxidation number; may correspond to a loss of electrons.

Oxidation–reduction reaction A reaction in which oxidation and reduction occur; also called redox reaction.

Oxidizing agent The substance that oxidizes another substance and is reduced.

Primary standard A substance of a known high degree of purity that undergoes one invariable reaction with the other reactant of interest.

Redox reaction An oxidation–reduction reaction.

Redox titration The quantitative analysis of the amount or concentration of an oxidizing or reducing agent in a sample by observing its reaction with a known amount or concentration of a reducing or oxidizing agent.

Reducing agent The substance that reduces another substance and is oxidized.

Reduction An algebraic decrease in oxidation number; may correspond to a gain of electrons.

Secondary standard A solution that has been titrated against a primary standard. A standard solution is a secondary standard.

Standard solution A solution of accurately known concentration.

Standardization The process by which the concentration of a solution is accurately determined by titrating it against an accurately known amount of a primary standard.

Titration The process by which the volume of a standard solution required to react with a specific amount of a substance is determined.

Exercises

Molarity

1. Why can we describe molarity as a "method of convenience" for expressing concentrations of solutions?

2. Why can molarity be expressed in mol/L *and* in mmol/mL?

3. Calculate the molarities of solutions that contain the following masses of solute in the indicated volumes:
   (a) 75 g of H_3AsO_4 in 500 mL of solution
   (b) 8.3 g of $(COOH)_2$ in 600 mL of solution
   (c) 13.0 g of $(COOH)_2 \cdot 2H_2O$ in 750 mL of solution

4. Calculate the molarities of solutions that contain the following:
   (a) 17.5 g of K_2SO_4 in 300 mL of solution
   (b) 143 g of $Al_2(SO_4)_3$ in 3.00 L of solution
   (c) 143 g of $Al_2(SO_4)_3 \cdot 18H_2O$ in 3.00 L of solution

5. Calculate the mass of NaOH required to prepare 2.50 L of 3.15 *M* NaOH solution.

6. What mass of $(NH_4)_2SO_4$ is required to prepare 3.75 L of 0.288 *M* $(NH_4)_2SO_4$ solution?

7. What volume of 0.300 *M* potassium hydroxide solution would just neutralize 30.0 mL of 0.100 *M* H_2SO_4 solution?

8. Calculate the molarity of a solution that is 39.77% H_2SO_4 by mass. The specific gravity of the solution is 1.305.

9. What is the molarity of a solution that is 19.0% HNO_3 by mass? The specific gravity of the solution is 1.11.

10. If 200 mL of 4.32 *M* HCl solution is added to 400 mL of 2.16 *M* NaOH solution, the resulting solution will be _____ molar in NaCl.

11. If 400 mL of 0.400 *M* HCl solution is added to 800 mL of 0.0800 *M* $Ba(OH)_2$ solution, the resulting solution will be _____ molar in $BaCl_2$ and _____ molar in _____ .

12. If 225 mL of 3.68 *M* H_3PO_4 solution is added to 775 mL of 3.68 *M* NaOH solution, the resulting solution will be _____ molar in Na_3PO_4 and _____ molar in _____ .

13. What volumes of 1.00 *M* KOH and 0.750 *M* HNO_3 solution would be required to produce 8.40 g of KNO_3?

14. What volumes of 1.00 M NaOH and 1.50 M H_3PO_4 solutions would be required to form 1.00 mol of Na_3PO_4?

15. A household ammonia solution is 5.03% ammonia. Its density is 0.979 g/mL. What is the molarity of this ammonia solution?

16. A vinegar solution is 5.11% acetic acid. Its density is 1.007 g/mL. What is its molarity?

Before you work Examples 17 and 18, *think* about the description of each solution. Note that the percentage by mass of solute is *nearly* the same for the solutions. Would you expect them to contain *approximately* the same number of moles of solute? *Hint:* Think molecular weights.

17. Refer to Exercises 15 and 16. What volume of the vinegar solution would be required to neutralize 1.00 L of the household ammonia solution?

18. Refer to Exercises 15 and 16. What volume of the household ammonia solution would be required to neutralize 1.00 L of the vinegar?

Standardization and Acid–Base Titrations—Mole Method

19. Define and illustrate the following terms clearly and concisely: (a) standard solution, (b) titration, (c) primary standard, (d) secondary standard.

20. Describe the preparation of a standard solution of NaOH, a compound that absorbs both CO_2 and H_2O from the air.

21. Distinguish between the *equivalence point* and the *end point* of a titration.

22. (a) What are the properties of an ideal primary standard? (b) What is the importance of each property?

23. Why can sodium carbonate be used as a primary standard for solutions of acids?

(a) What is potassium hydrogen phthalate? (b) For what is it used?

25. What volume of 0.275 M hydrochloric acid solution neutralizes 36.4 mL of 0.150 M sodium hydroxide solution?

26. What volume of 0.112 M sodium hydroxide solution would be required to neutralize completely 25.3 mL of 0.400 M sulfuric acid solution?

27. Calculate the molarity of a KOH solution if 27.63 mL of the KOH solution reacted with 0.4084 g of potassium hydrogen phthalate.

28. A solution of sodium hydroxide is standardized against potassium hydrogen phthalate. From the following data, calculate the molarity of the NaOH solution:

mass of KHP	0.8407 g
buret reading before titration	0.23 mL
buret reading after titration	46.16 mL

29. The secondary standard solution of NaOH of Exercise 28 was used to titrate a solution of unknown concentration of HCl. A 30.00-mL sample of the HCl solution required 24.21 mL of the NaOH solution for complete neutralization. What is the molarity of the HCl solution?

30. A 34.53-mL sample of a solution of sulfuric acid, H_2SO_4, is neutralized by 27.86 mL of the NaOH solution of Exercise 28. Calculate the molarity of the sulfuric acid solution.

31. If 37.38 mL of a sulfuric acid solution reacts with 0.2888 g of Na_2CO_3, what is the molarity of the sulfuric acid solution.

*32. An impure sample of $(COOH)_2 \cdot 2H_2O$ that had a mass of 2.00 g was dissolved in water and titrated with standard NaOH solution. The titration required 40.32 mL of 0.198 M NaOH solution. Calculate the percentage of $(COOH)_2 \cdot 2H_2O$ in the sample.

*33. A 50.0-mL sample of 0.0500 M $Ca(OH)_2$ is added to 10.0 mL of 0.200 M HNO_3. (a) Is the resulting solution acidic or basic? (b) How many moles of excess acid or base are present? (c) How many mL of 0.0500 M $Ca(OH)_2$ or 0.0500 M HNO_3 would be required to neutralize the solution?

*34. An antacid tablet containing calcium carbonate as an active ingredient requires 22.6 mL of 0.0932 M HCl for complete neutralization. What mass of $CaCO_3$ did the tablet contain?

*35. Butyric acid, whose empirical formula is C_2H_4O, is the acid responsible for the odor of rancid butter. The acid has one ionizable hydrogen per molecule. A 1.000-g sample of butyric acid is neutralized by 54.42 mL of 0.2088 M NaOH solution. What are (a) the molecular weight and (b) the molecular formula of butyric acid?

*36. The typical concentration of HCl in stomach acid (gastric juice) is a concentration of about 8.0×10^{-2} M. One experiences "acid stomach" when the stomach contents reach about 1.0×10^{-1} M HCl. One Rolaids® tablet (an antacid) contains 334 mg of active ingredient, $NaAl(OH)_2CO_3$. Assume that you have acid stomach and that your stomach contains 800 mL of 1.0×10^{-1} M HCl. Calculate the number of mmol of HCl in the stomach and the number of mmol of HCl that the tablet *can* neutralize. Which is greater? The neutralization reaction produces NaCl, $AlCl_3$, CO_2, and H_2O.

Standardization and Acid–Base Titrations—Equivalent Weight Method

In answering Exercises 37–41, assume that the acids and bases will be completely neutralized.

37. Calculate the normality of a solution that contains 4.93 g of H_2SO_4 in 125 mL of solution.

38. What is the normality of a solution that contains 7.08 g of H_3PO_4 in 185 mL of solution?

39. Calculate the molarity and the normality of a solution that was prepared by dissolving 34.2 g of barium hydroxide in enough water to make 4000 mL of solution.

40. Calculate the molarity and the normality of a solution that was prepared by dissolving 19.6 g of arsenic acid, H_3AsO_4, in enough water to make 600 mL of solution.

41. What are the molarity and normality of a sulfuric acid solution that is 19.6% H_2SO_4 by mass? The density of the solution is 1.14 g/mL.

42. A 25.0-mL sample of 0.206 normal nitric acid solution required 39.3 mL of barium hydroxide solution for neutralization. Calculate the molarity of the barium hydroxide solution.

43. Vinegar is an aqueous solution of acetic acid, CH_3COOH. Suppose you titrate a 25.00-mL sample of vinegar with 17.62 mL of a standardized 0.1060 N solution of NaOH. (a) What is the normality of acetic acid in this vinegar? (b) What is the mass of acetic acid contained in 1.000 L of vinegar?

44. A 44.4-mL sample of sodium hydroxide solution was titrated with 41.8 mL of 0.100 N sulfuric acid solution. A 36.0-mL sample of hydrochloric acid required 47.2 mL of the sodium hydroxide solution for titration. What is the normality of the hydrochloric acid solution?

45. Calculate the normality and molarity of an H_2SO_4 solution if 20.0 mL of the solution reacts with 0.212 g of Na_2CO_3.

$$H_2SO_4 + Na_2CO_3 \longrightarrow Na_2SO_4 + CO_2 + H_2O$$

46. Calculate the normality and molarity of an HCl solution if 33.1 mL of the solution reacts with 0.318 g of Na_2CO_3.

$$2HCl + Na_2CO_3 \longrightarrow 2NaCl + CO_2 + H_2O$$

Redox Titrations—Mole Method and Molarity

47. In a redox titration, we must have a(n) _____ species and a(n) _____ species.

48. What volume of 0.10 M $KMnO_4$ would be required to oxidize 20 mL of 0.10 M $FeSO_4$ in acidic solution? Refer to Example 11-13.

49. What volume of 0.10 M $K_2Cr_2O_7$ would be required to oxidize 60 mL of 0.10 M Na_2SO_3 in acidic solution? The products include Cr^{3+} and SO_4^{2-} ions. Refer to Example 11-14.

50. What volume of 0.10 M $KMnO_4$ would be required to oxidize 50 mL of 0.10 M KI in acidic solution? Products include Mn^{2+} and I_2.

51. What volume of 0.10 M $K_2Cr_2O_7$ would be required to oxidize 50 mL of 0.10 M KI in acidic solution? Products include Cr^{3+} and I_2.

52. (a) A solution of sodium thiosulfate, $Na_2S_2O_3$, is 0.1455 M. 25.00 mL of this solution reacts with 26.36 mL of I_2 solution. What is the molarity of the I_2 solution?

$$2Na_2S_2O_3 + I_2 \longrightarrow Na_2S_4O_6 + 2NaI$$

(b) 25.32 mL of the I_2 solution is required to titrate a sample containing As_2O_3. Calculate the mass of As_2O_3 (197.8 g/mol) in the sample.

$$As_2O_3 + 5H_2O + 2I_2 \longrightarrow 2H_3AsO_4 + 4HI$$

53. Copper(II) ion, Cu^{2+}, can be determined by the net reaction

$$2Cu^{2+} + 2I^- + 2S_2O_3^{2-} \longrightarrow 2CuI(s) + S_4O_6^{2-}$$

A 2.075-g sample containing $CuSO_4$ and excess KI is titrated with 41.75 mL of 0.1214 M solution of $Na_2S_2O_3$. What is the percentage of $CuSO_4$ (159.6 g/mol) in the sample?

54. Find the volume of 0.150 M HI solution required to titrate
   (a) 25.0 mL of 0.100 M NaOH
   (b) 5.03 g of $AgNO_3$ ($Ag^+ + I^- \longrightarrow AgI(s)$)
   (c) 0.621 g $CuSO_4$ ($2Cu^{2+} + 4I^- \longrightarrow 2CuI(s) + I_2(s)$)

*55. The iron in a sample containing some Fe_2O_3 is reduced to Fe^{2+}. The Fe^{2+} is titrated with 12.02 mL of 0.1167 M $K_2Cr_2O_7$ in an acid solution.

$$6Fe^{2+} + Cr_2O_7^{2-} + 14H^+ \longrightarrow$$
$$6Fe^{3+} + 2Cr^{3+} + 7H_2O$$

Find (a) the mass of Fe and (b) the percentage of Fe in a 5.675-g sample.

*56. Limonite is an ore of iron that contains $Fe_2O_3 \cdot 1\frac{1}{2}H_2O$ (or $2Fe_2O_3 \cdot 3H_2O$). A 0.5166-g sample of limonite is dissolved in acid and treated so that all the iron is converted to ferrous ion, Fe^{2+}. This sample requires 42.96 mL of 0.02130 M sodium dichromate solution, $Na_2Cr_2O_7$, for titration. Fe^{2+} is oxidized to Fe^{3+} and $Cr_2O_7^{2-}$ is reduced to Cr^{3+}. What is the percentage of iron in the limonite?

*57. A 0.683-g sample of an ore of iron is dissolved in acid and converted to the ferrous form. The sample is oxidized by 38.50 mL of 0.161 M ceric sulfate, $Ce(SO_4)_2$, solution during which the ceric ion, Ce^{4+}, is reduced to Ce^{3+} ion.
   (a) Write a balanced equation for the reaction.
   (b) What is the percentage of iron in the ore?

Redox Titrations—Equivalent Weights and Normality
In Exercises 58 and 59, calculate the mass of the compound in the indicated volume of solution.

58. (a) Grams of $KMnO_4$ in 475 mL of 0.137 N $KMnO_4$ solution used in the reaction in Exercise 48.
   (b) Grams of NaI in 325 mL of 0.267 N NaI solution used in the reaction in Exercise 50. NaI and KI undergo similar reactions.
   (c) Grams of $K_2Cr_2O_7$ in 198 mL of 0.183 N $K_2Cr_2O_7$ solution used in the reaction in Exercise 51.

59. (a) Grams of $KMnO_4$ in 475 mL of 0.137 N $KMnO_4$ solution used in a reaction in which a product contains MnO_2.
   (b) Grams of $NaNO_2$ in 1.65 L of 0.325 N $NaNO_2$ solution used in a reaction in which a product contains NO_3^-.
   (c) Grams of H_2O_2 in 1.50 L of 0.789 N H_2O_2 used in a reaction that produces O_2.

60. Calculate the molarity and normality of a solution that contains 15.8 g of $KMnO_4$ in 500 mL of solution to be used in the reaction that produces MnO_4^{2-} ions.

61. Calculate the molarity and normality of a solution that contains 2.94 g of $K_2Cr_2O_7$ in 100 mL of solution to be used in the reaction in Exercise 51.

62. Calculate the molarity and normality of a solution that contains 16.2 g of $FeSO_4$ in 200 mL of solution to be used in the reaction in Exercise 48.

63. Calculate the molarity and normality of a solution that contains 12.6 g of Na_2SO_3 in 1.00 L of solution to be used in the reaction in Exercise 49.

Mixed Exercises

64. Calculate the molarity of a hydrochloric acid solution if 27.63 mL of it reacts with 0.2013 g of sodium carbonate.

65. Calculate the molarity and the normality of a sulfuric acid solution if 27.63 mL of it reacts with 0.2013 g of sodium carbonate.

66. Is water a very weak electrolyte? Why?

67. Are slightly soluble ionic compounds considered to be strong or weak electrolytes? Why?

68. What volume of 0.1123 M HCl is needed to neutralize 1.29 g of $Ca(OH)_2$?

69. What mass of NaOH is needed to neutralize 50.0 mL of 0.1036 M HCl? If the NaOH is available as a 0.1021 M aqueous solution, what volume will be required?

70. What volume of 0.203 M H_2SO_4 solution would be required to neutralize completely 37.4 mL of 0.302 M KOH solution?

71. What volume of 0.203 N H_2SO_4 solution would be required to neutralize completely 37.4 mL of 0.302 N KOH solution?

72. What volume of 0.2045 normal sodium hydroxide would be required to neutralize completely 41.39 mL of 0.1023 normal H_2SO_4 solution?

73. Benzoic acid, C_6H_5COOH, is sometimes used as a primary standard for the standardization of solutions of bases. A 1.862-g sample of the acid is neutralized by 31.62 mL of an NaOH solution. What is the molarity of the base solution?

$$C_6H_5COOH(s) + NaOH(aq) \longrightarrow$$
$$C_6H_5COONa(aq) + H_2O(\ell)$$

12 Gases and the Kinetic–Molecular Theory

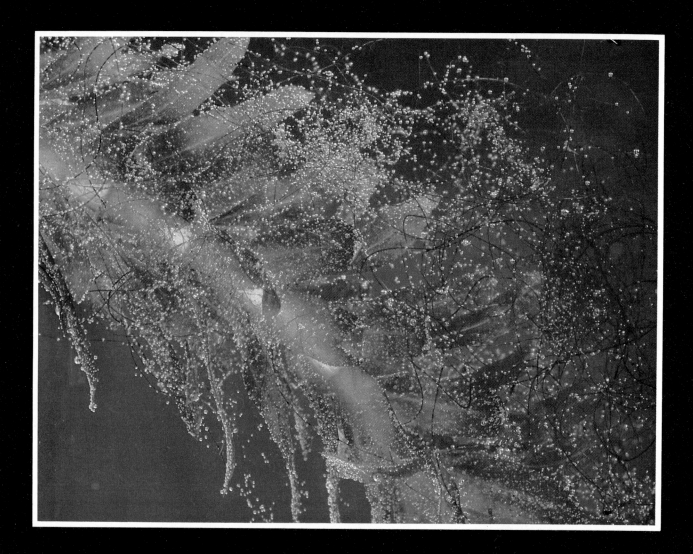

Outline

Immersed in water, this green plant oxidizes the water to form bubbles of gaseous oxygen, O_2.

Objectives

As you study this chapter, you should learn

☐ About the properties of gases and how they differ from liquids and solids

☐ How we measure pressure

☐ To understand the absolute (Kelvin) temperature scale

☐ About the relationships among pressure, volume, temperature, and amount of gas (Boyle's Law, Charles' Law, Avogadro's Law, and the Combined Gas Law, and the limitations of each one)

☐ To use Boyle's Law, Charles' Law, Avogadro's Law, and the Combined Gas Law, as appropriate, to calculate *changes* in pressure, volume, temperature, and amount of gas

☐ To do calculations about gas densities and the standard molar volume

☐ About the ideal gas equation, and how to use it to do calculations about a sample of gas

☐ To determine molecular weights and formulas of gaseous substances from measured properties of gases

☐ To describe how mixtures of gases behave and to predict their properties (Dalton's Law of Partial Pressures)

☐ About the kinetic–molecular theory of gases, and how this theory is consistent with the observed gas laws

☐ About molecular motion, diffusion, and effusion of gases

☐ What molecular features are responsible for nonideal behavior of real gases, and when this nonideal behavior is important

☐ To carry out calculations about the gases involved in chemical reactions

12-1 Comparison of Solids, Liquids, and Gases

Some compounds decompose before melting or boiling.

Matter exists in three physical states: solids, liquids, and gases. In the solid state water is known as ice, in the liquid state it is called water, and in the gaseous state it is known as steam or water vapor. Most, but not all, substances can exist in all three states. Most solids change to liquids and most liquids change to gases as they are heated. Liquids and gases are known as **fluids** because they flow freely. Solids and liquids are referred to as **condensed states** because they have much higher densities than gases. Table 12-1 displays the densities of a few common substances in different states.

Ice is less dense than liquid water. This behavior is quite unusual; most substances are more dense in the solid state.

As the data in Table 12-1 indicate, solids and liquids are many times more dense than gases. The molecules must be very far apart in gases and much closer together in liquids and solids. For example, the volume of one mole of liquid water is about 18 milliliters, whereas one mole of steam occupies about 30,600 milliliters at 100°C and at atmospheric pressure. Gases are easily compressed, and they completely fill any container in which they are present. This tells us that the molecules in a gas are far apart and that interactions among them are weak. The possibilities for interaction among gaseous molecules would be minimal (because they are so far apart) were it not for their rapid motion.

Volatile liquids evaporate readily. They have low boiling points and high vapor pressures (Sections 13-7 and 13-8).

All substances that are gases at room temperature may be liquefied by cooling and compressing them. Volatile liquids are easily converted to gases at room temperature or slightly above. The term "**vapor**" refers to a gas that is formed by evaporation of a liquid or sublimation of a solid; it is commonly used when some of the liquid or solid remains in contact with the gas.

12-2 Composition of the Atmosphere and Some Common Properties of Gases

Many important chemical substances are gases. The earth's atmosphere is a mixture of gases and particles of liquids and solids (Table 12-2). The major gaseous components are N_2 (bp −195.79°C) and O_2 (bp −182.98°C), with smaller concentrations of other gases. All gases are *miscible*; that is, they mix completely *unless* they react with each other.

Several scientists, notably Torricelli (1643), Boyle (1660), Charles (1787), and Graham (1831), laid an experimental foundation upon which our present

Table 12-1
Comparison of Densities (g/mL) and Molar Volumes (mL/mol) of Substances in Different States at Atmospheric Pressure*

Substance	Solid		Liquid (20°C)		Gas (100°C)	
water (H_2O)	0.917 (0°C)	19.6	0.998	18.0	0.000588	30,600
benzene (C_6H_6)	0.899 (0°C)	86.7	0.879	88.7	0.00255	30,600
carbon tetrachloride (CCl_4)	1.7 (−25°C)	90.6	1.59	96.8	0.00503	30,600

* The *molar volume* of a substance is the volume occupied by one mole of that substance. Densities are given in black, molar volumes in red.

Table 12-2
Volume Percentage Composition of Dry Air*

Gas	% by Volume*
N_2	78.09
O_2	20.94
Ar	0.93
CO_2	0.03†
He, Ne, Kr, Xe	0.002
CH_4	0.00015†
H_2	0.00005
All others combined‡	< 0.00004

* Also equal to mole percentage (Avogadro's Law, Section 12-8).
† Variable.
‡ Atmospheric moisture varies.

understanding of gases is based. For example, their investigations showed that

1. Gases can be compressed into smaller volumes; that is, their densities can be increased by applying increased pressure.
2. Gases exert pressure on their surroundings; in turn, pressure must be exerted to confine gases.
3. Gases expand without limits, and so gas samples completely and uniformly occupy the volume of any container.
4. Gases diffuse into each other, and so samples of gas placed in the same container mix completely. Conversely, different gases in a mixture do not separate on standing.
5. The amounts and properties of gases are described in terms of temperature, pressure, the volume occupied, and the number of molecules present. For example, a sample of gas occupies a greater volume hot than it does when cold at the same pressure, but the number of molecules does not change.

Diffusion of bromine vapor in air. A drop of liquid bromine was placed in each cylinder. As the liquid evaporated, the resulting reddish-brown gas diffused. Here diffusion has occurred for 2 minutes in the left cylinder and for 20 minutes in the right cylinder.

Chemistry in Use. . .
The Greenhouse Effect

During the last century, the great increase in our use of fossil fuels has caused a significant rise in the concentration of carbon dioxide, CO_2, in the atmosphere. Scientists believe that the concentration of atmospheric CO_2 could double by early in the 21st century, compared with its level just before the Industrial Revolution. During the last 100 to 200 years, the CO_2 concentration has increased by 25%; nearly a fifth of this rise took place just in the decade from 1975 to 1985. The curve in Figure (a) shows the recent steady rise in atmospheric CO_2 concentration.

Energy from the sun reaches the earth in the form of light. Neither CO_2 nor H_2O vapor absorbs the visible light in sunlight, so they do not prevent it from reaching the surface of the earth. However, the energy given off by the earth in the form of lower-energy infrared (heat) radiation is readily absorbed by both CO_2 and H_2O (as it is by the glass or plastic of greenhouses). Thus, some of the heat the earth must lose to maintain thermal equilibrium can

become trapped in the atmosphere, causing the temperature to rise (Figure b). This phenomenon, called the **greenhouse effect**, has been the subject of much discussion among scientists and of many articles in the popular press. The anticipated rise in average global temperature by the year 2050 due to increased CO_2 concentration is predicted to be 2 to 5°C. Indeed, eight of the ten warmest years on record in the Northern Hemisphere were between 1980 and 1989.

An increase of 2 to 5°C may not seem like much. However, this is thought to be enough to cause a dramatic change in climate, transforming now arable land into desert and altering the habitats of many animals and plants beyond their ability to adapt. Another drastic consequence of even this small temperature rise would be the partial melting of the polar ice caps. The resulting rise in sea level, though only a few feet, would mean that water would inundate coastal cities such as Los Angeles, New York, and Houston, and low-lying coastal

areas such as southern Florida and Louisiana. On a global scale, the effects would be devastating.

The earth's forests and jungles play a crucial role in maintaining the balance of gases in the atmosphere, removing CO_2 and supplying O_2. The massive destruction, for economic reasons, of heavily forested areas such as the Amazon rain forest in South America is cited as another long-term contributor to global environmental problems. Worldwide, more than three million square miles of once-forested land are now barren for some reason; at least 60% of this land is now unused. Environmental scientists estimate that if even one quarter of this land could be reforested, the vegetation would absorb 1.1 billion tons of CO_2 annually.

Some scientists are more skeptical about the role of human-produced CO_2 in climate changes and, indeed, about whether global warming is a significant phenomenon or simply another of the recognized warm–cold cycles that have occurred throughout the earth's his-

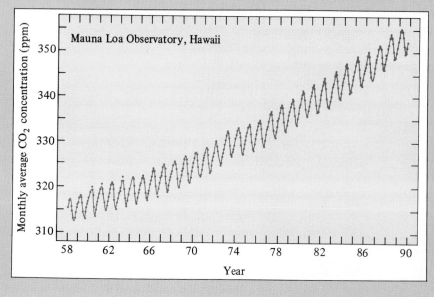

(a) A plot of the monthly average CO_2 concentration in parts per million (ppm), measured at Mauna Loa Observatory, Hawaii, far from significant sources of CO_2 from human activities. Annual fluctuations occur because plants in the Northern Hemisphere absorb CO_2 in the spring and release it as they decay in the fall.

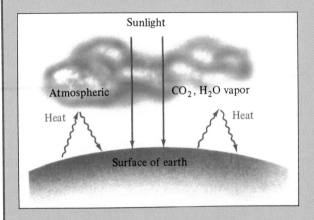

(b) The greenhouse effect. Visible light passes through atmospheric H_2O and CO_2, but heat radiated from the surface of the earth is absorbed by these gases.

tory. Such skeptics point out an unexplained increase in atmospheric CO_2 during an extended period in the 17th century, and an even higher and more prolonged peak about 130,000 years ago. However, even the most skeptical observers seem to agree that responsible stewardship of the planet requires that we do something in a reasoned fashion to reduce production of greenhouse gases, primarily CO_2, and that this will involve decreasing our dependence on energy from fossil fuels. Despite the technical and political problems of waste disposal, an essentially all-electric economy based on nuclear power is viewed by some as the most reasonable course in the long run.

Much CO_2 is eventually absorbed by the vast amount of water in the oceans, where the carbonate–bicarbonate buffer system almost entirely counteracts any adverse effects of ocean water acidity. Ironically, there is also evidence to suggest that other types of air pollution in the form of particulate matter may partially counteract the greenhouse effect. The particles reflect visible (sun) radiation rather than absorbing it, blocking some light from entering the atmosphere. However, it seems foolish to depend on one form of pollution to help rescue us from the effects of another! Real solutions to current environmental problems such as the greenhouse effect are not subject to quick fixes, but depend on long-term, cooperative international efforts that are based on the firm knowledge resulting from scientific research.

Tropical rain forests are a major factor in maintaining the balance of CO_2 and O_2 in the earth's atmosphere. In recent years, a portion of the South American forests (by far the world's largest) larger than France has been destroyed, either by flooding caused by hydroelectric dams or by clearing of forest land for agricultural or ranching use. Such destruction continues at a rate of more than 20,000 square kilometers per year. If current trends continue, many of the world's rain forests will be severely reduced or even obliterated by the year 2000. The fundamental question—"What are the overall consequences of the destruction of tropical rain forests?"—remains unanswered.

12-3 Pressure

Pressure is defined as force per unit area—for example, lb/in², commonly known as *psi*. Pressure may be expressed in many different units, as we shall see. The mercury **barometer** is a simple device for measuring atmospheric pressures. Figure 12-1a illustrates the "heart" of the mercury barometer. A glass tube (about 800 mm long) is sealed at one end, filled with mercury, and then carefully inverted into a dish of mercury without air being allowed to enter. The mercury in the tube falls to the level at which the pressure of the air on the surface of the mercury in the dish equals the gravitational pull downward on the mercury in the tube. The air pressure is measured in terms of the height of the mercury column, i.e., the vertical distance between the surface of the mercury in the open dish and that inside the closed tube. The pressure exerted by the atmosphere is equal to the pressure exerted by the column of mercury.

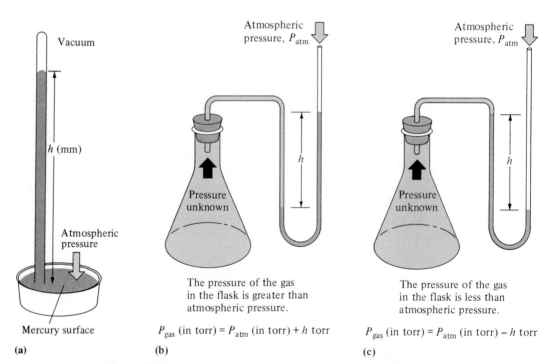

The pressure of the gas in the flask is greater than atmospheric pressure.

P_{gas} (in torr) $= P_{atm}$ (in torr) $+ h$ torr

The pressure of the gas in the flask is less than atmospheric pressure.

P_{gas} (in torr) $= P_{atm}$ (in torr) $- h$ torr

(a)   **(b)**   **(c)**

Figure 12-1
Some laboratory devices for measuring pressure. (a) Schematic diagram of a closed-end barometer. At the level of the lower mercury surface, the pressure both inside and outside the tube must be equal to that of the atmosphere. Inside the tube, the pressure is exerted only by the mercury column h mm high. Hence, the atmospheric pressure must equal h mm Hg or h torr. (b) The two-arm mercury barometer is called a manometer. In this sample, the pressure of the gas is *greater than* the external atmospheric pressure. At the level of the lower mercury surface, the total pressure on the mercury in the left arm must equal the total pressure on the mercury in the right arm. The pressure exerted by the gas is equal to the external pressure *plus* the pressure exerted by the mercury column of height h mm, or P_{gas} (in torr) $= P_{atm}$ (in torr) $+ h$ torr. (c) When the gas pressure measured by the manometer is *less than* the external atmospheric pressure, the pressure exerted by the atmosphere is equal to the gas pressure *plus* the pressure exerted by the mercury column, or $P_{atm} = P_{gas} + h$. We can rearrange this to write P_{gas} (in torr) $= P_{atm}$ (in torr) $- h$ torr.

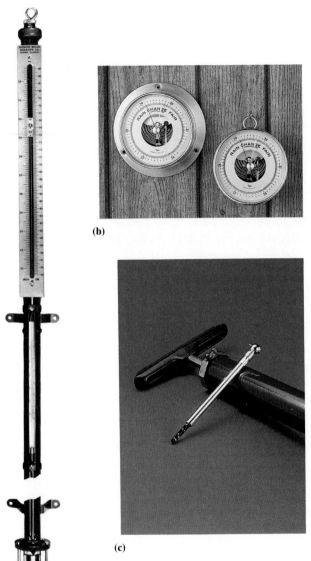

(b)

(c)

Figure 12-2
Some commercial pressure-measuring devices. (a) A commercial mercury barometer. (b) Portable barometers. This type is called an *aneroid* (''not wet'') barometer. Some of the air has been removed from the airtight box, which is made of thin, flexible metal. When the pressure of the atmosphere changes, the remaining air in the box expands or contracts (Boyle's Law), moving the flexible box surface and an attached pointer along a scale. (c) A tire gauge. This kind of gauge registers ''relative'' pressure, i.e., the *difference* between internal pressure and the external atmospheric pressure. For instance, when the gauge reads 30 psi (pounds per square inch), the total gas pressure in the tire is 30 psi + 1 atm, or about 45 psi. In engineering terminology, this is termed ''psig'' (g = gauge).

(a)

Mercury barometers are simple and well known, so gas pressures are frequently expressed in terms of millimeters of mercury (mm Hg, or just mm). In recent years the unit **torr** has been used to indicate pressure; it is defined as 1 torr = 1 mm Hg.

A mercury **manometer** consists of a glass U-tube partially filled with mercury. One arm is open to the atmosphere and the other is connected to a container of gas (Figure 12-1b,c).

Atmospheric pressure varies with atmospheric conditions and distance above sea level. It is lower at high elevations because the air at low levels is compressed by the air above it. Approximately one half of the matter in the atmosphere is less than 20,000 feet above sea level. Thus, atmospheric pressure is only about one-half as great at 20,000 feet as it is at sea level. Mountain climbers and pilots use portable barometers to determine their altitudes (Figure 12-2).

The unit *torr* was named for Evangelista Toricelli (1608–1647), who invented the mercury barometer.

407

At sea level, at a latitude of 45°, the average atmospheric pressure supports a column of mercury 760 mm high in a simple mercury barometer when the mercury is at 0°C. This average sea-level pressure of 760 mm Hg is called **one atmosphere of pressure**:

1 atmosphere (atm) = 760 mm Hg at 0°C = 760 torr

The SI unit of pressure is the **pascal** (Pa), defined as the pressure exerted by a force of one newton acting on an area of one square meter. By definition, one **newton** (N) is the force required to give a mass of one kilogram an acceleration of one meter/second per second. Symbolically we represent one newton as

Acceleration is the change in velocity (m/s) per unit time (s), m/s².

$$1 \text{ N} = \frac{1 \text{ kg} \cdot \text{m}}{\text{s}^2} \qquad \text{so} \qquad 1 \text{ Pa} = \frac{1 \text{ N}}{\text{m}^2} = \frac{1 \text{ kg}}{\text{m} \cdot \text{s}^2}$$

One atmosphere of pressure = 1.013×10^5 Pa, or 101.3 kPa.

12-4 Boyle's Law: The Relation of Volume to Pressure

Early experiments on the behavior of gases were carried out by Robert Boyle in the 17th century. In a typical experiment (Figure 12-3), a sample of a gas was trapped in a U-tube and allowed to come to constant temperature. Then its volume and the difference in the heights of the two mercury columns were recorded. This difference in height plus the pressure of the atmosphere represents the pressure on the gas. Addition of more mercury to the tube increases the pressure by changing the height of the mercury column. As a result, the gas volume decreases. The results of several such experiments are tabulated in Figure 12-3b.

Figure 12-3
(a) A representation of Boyle's experiment. A sample of air is trapped in a tube in such a way that the pressure on the air can be changed and the volume of the air measured. P_{atm} is the atmospheric pressure, measured with a barometer. $P_1 = h_1 + P_{atm}$, $P_2 = h_2 + P_{atm}$. (b) Some typical data from such an experiment. Measured values of P and V are presented in the first two columns, on an arbitrary scale.

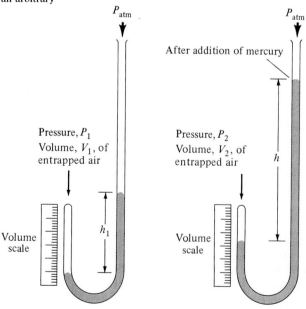

P	V	$P \times V$	$1/P$
5.0	40.0	200	0.20
10.0	20.0	200	0.10
15.0	13.3	200	0.067
17.0	11.8	201	0.059
20.0	10.0	200	0.050
22.0	9.10	200	0.045
30.0	6.70	201	0.033
40.0	5.00	200	0.025

(a)

(b)

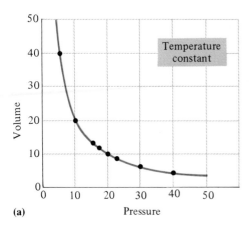

(a)

Pressure

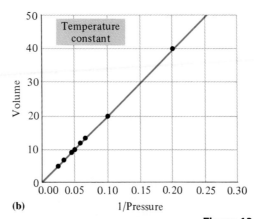

(b)

1/Pressure

Figure 12-4
Graphical representations of Boyle's Law, using the data of Figure 12-2. (a) V versus P. (b) V versus $1/P$.

Boyle noticed that for a given sample of gas at constant temperature, the product of pressure and volume, $P \times V$, was always the same number.

At a given temperature, the product of pressure and volume of a definite mass of gas is constant.

$$PV = k \qquad \text{(constant } n, T)$$

This relationship is **Boyle's Law**. The value of k depends on the amount (number of moles, n) of gas present and on the temperature, T. The units of k are determined by the units used to express the volume (V) and pressure (P).

When the volume of a gas is plotted against its pressure at constant temperature, the resulting curve is one branch of a hyperbola. Figure 12-4a is a graphic illustration of this inverse relationship. When volume is plotted versus the reciprocal of the pressure, $1/P$, a straight line results (Figure 12-4b). In 1662 Boyle summarized the results of his experiments on various samples of gases in an alternative statement of Boyle's Law:

Because neither V nor P of a sample of a gas can be less than zero, the other branch of the hyperbola (third quadrant) has no physical significance.

At constant temperature the volume, V, occupied by a definite mass of a gas is inversely proportional to the applied pressure, P.

$$V \propto \frac{1}{P} \qquad \text{or} \qquad V = k\left(\frac{1}{P}\right) \qquad \text{(constant } n, T)$$

The symbol $\propto$ reads "is proportional to." A proportionality is converted into an equality by introducing a proportionality constant, k.

At normal temperatures and pressure, most gases obey Boyle's Law rather well. We call this *ideal behavior*. Deviations from ideality are discussed in Section 12-14.

Let us think about a fixed mass of gas at constant temperature, but at two different conditions of pressure and volume (Figure 12-3). For the first condition we can write

$$P_1 V_1 = k \qquad \text{(constant } n, T)$$

and for the second condition we can write

$$P_2 V_2 = k \qquad \text{(constant } n, T)$$

Because the right-hand sides of these two equations are the same, the left-hand sides must be equal, or

$$P_1V_1 = P_2V_2 \quad \text{(for a given amount of a gas at constant temperature)}$$

This form of Boyle's Law is quite useful for calculations involving pressure and volume changes, as the following examples demonstrate.

Example 12-1

A sample of gas occupies 12 liters under a pressure of 1.2 atm. What would its volume be if the pressure were increased to 2.4 atm? Assume that the temperature of the gas sample does not change.

Plan

We know the volume at one pressure and wish to find the volume at another pressure (constant temperature). We can solve Boyle's Law for the second volume and substitute. Alternatively, we can multiply the original volume by a "Boyle's Law factor" to change the volume in the direction required by the stated pressure change.

Solution

It is often helpful to tabulate what is given and what is asked for in a problem.

We have $P_1 = 1.2$ atm; $V_1 = 12$ L; $P_2 = 2.4$ atm; $V_2 = ?$. Solving Boyle's Law, $P_1V_1 = P_2V_2$, for V_2 and substituting yields

$$V_2 = \frac{P_1V_1}{P_2} = \frac{(1.2 \text{ atm})(12 \text{ L})}{2.4 \text{ atm}} = \boxed{6.0 \text{ L}}$$

Pressure and volume are inversely proportional; doubling the pressure halves the volume of a sample of gas at constant temperature. To find a Boyle's Law factor, we reason that the pressure increases from 1.2 atm to 2.4 atm, so the volume *decreases* by the factor (1.2 atm/2.4 atm). The solution then becomes

$$? \text{ L} = 12 \text{ L} \times \text{(Boyle's Law factor that describes a decrease in volume)}$$

This is a *correction factor*, not a unit factor.

$$= 12 \text{ L} \times \left(\frac{1.2 \text{ atm}}{2.4 \text{ atm}}\right) = \boxed{6.0 \text{ L}}$$

Example 12-2

A sample of oxygen occupies 10.0 liters under a pressure of 790 torr (105 kPa). At what pressure would it occupy 13.4 liters if the temperature did not change?

Plan

We know the pressure at one volume and wish to find the pressure at another volume (at constant temperature). We can solve Boyle's Law for the second pressure and substitute. Alternatively, we can multiply the original pressure by a Boyle's Law factor to change the pressure in the direction required by the stated volume change.

Solution

We have $P_1 = 790$ torr; $V_1 = 10.0$ L; $P_2 = ?$; $V_2 = 13.4$ L. Solving Boyle's Law, $P_1V_1 = P_2V_2$, for P_2 and substituting yields

$$P_2 = \frac{P_1V_1}{V_2} = \frac{(790 \text{ torr})(10.0 \text{ L})}{13.4 \text{ L}} = \boxed{590 \text{ torr}} \quad \left(\times \frac{101.3 \text{ kPa}}{760 \text{ torr}} = \boxed{78.6 \text{ kPa}} \right)$$

The problem tells us that V *increases* from 10.0 L to 13.4 L (at constant temperature). Therefore P must *decrease*. We can also solve the problem by multiplying the original pressure by the volume factor less than unity—that is, by 10.0 L/13.4 L.

$$? \text{ torr} = 790 \text{ torr} \times \frac{10.0 \text{ L}}{13.4 \text{ L}} = \boxed{590 \text{ torr}} \qquad (78.6 \text{ kPa})$$

EOC 14, 16

Jacques Charles' first ascent in a hydrogen balloon at the Tuileries, Paris, December 1, 1783.

12-5 Charles' Law: The Relation of Volume to Temperature; The Absolute Temperature Scale

In his pressure–volume studies on gases, Robert Boyle noticed that heating a sample of gas caused some volume change, but he did not follow up on this observation. About 1800, two French scientists—Jacques Charles and Joseph Gay-Lussac, pioneer balloonists at the time—began studying the expansion of gases with increasing temperature. Their studies showed that the rate of expansion with increased temperature was constant and was the same for all the gases they studied as long as pressure remained constant. The implications of their discovery were not fully recognized until nearly a century later. Then scientists used this behavior of gases as the basis of a new temperature scale, the absolute temperature scale.

The change of volume with temperature, at constant pressure, is illustrated in Figure 12-5. From the table of typical data in Figure 12-5b, we see that volume (V, mL) increases as temperature (t, °C) increases, but the exact nature of the relationship is not yet obvious. These data are plotted in Figure 12-5c (line A), together with similar data for the same gas sample at different pressures (lines B and C).

Lord Kelvin, a British physicist, noticed that an extension of the different temperature–volume lines back to zero volume (dashed line) yields a common intercept at −273.15°C on the temperature axis. Kelvin named this temperature **absolute zero**. The degrees are the same size over the entire scale, so 0°C becomes 273.15 degrees above absolute zero. In honor of Lord Kelvin's work, this scale is called the Kelvin temperature scale. As pointed out in Section 1-12, the relationship between the Celsius and Kelvin temperature scales is K = °C + 273.15°.

If we convert temperatures (°C) to absolute temperatures (K), the green scale in Figure 12-5c, the volume–temperature relationship becomes obvious. This relationship is known as **Charles' Law**:

At constant pressure, the volume occupied by a definite mass of a gas is directly proportional to its absolute temperature.

We can express Charles' Law in mathematical terms as

$$V \propto T \quad \text{or} \quad V = kT \qquad (\text{constant } n, P)$$

Rearranging the expression gives $V/T = k$, a concise statement of Charles' Law. As the temperature increases, the volume must increase proportionally. If we let subscripts 1 and 2 represent values for the same sample of gas at two different temperatures, we obtain

Lord Kelvin (1842–1907) was born William Thompson. At the age of ten he was admitted to Glasgow University. Because its new appliance was based on Kelvin's theories, a refrigerator company named its product the Kelvinator.

Recall that temperatures on the Kelvin scale are expressed in kelvins (not degrees Kelvin) and represented by K, not °K.

Absolute zero may be thought of as the limit of thermal contraction for an ideal gas.

$$\frac{V_1}{T_1} = \frac{V_2}{T_2} \qquad \text{(for a definite mass of gas at constant pressure)}$$

which is the more useful form of Charles' Law. This relationship is valid *only* when temperature, T, is expressed on an absolute (usually the Kelvin) scale.

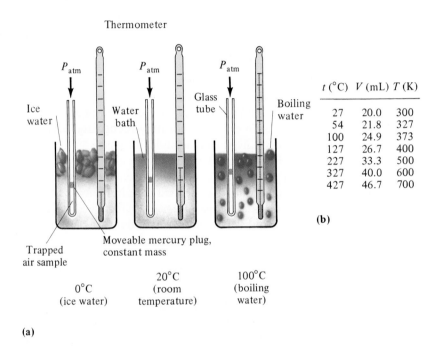

t (°C)	V (mL)	T (K)
27	20.0	300
54	21.8	327
100	24.9	373
127	26.7	400
227	33.3	500
327	40.0	600
427	46.7	700

(b)

(a)

Figure 12-5
An experiment showing that the volume of an ideal gas increases as the temperature is increased at constant pressure. (a) A mercury plug of constant weight, plus atmospheric pressure, maintains a constant pressure on the trapped air. (b) Some representative volume–temperature data at constant pressure. The relationship becomes clear when t (°C) is converted to T (K) by adding 273°. (c) A graph in which volume is plotted versus temperature on two different scales. Lines A, B, and C represent the same mass of the same ideal gas at different pressures. Line A represents the data tabulated in part (b). Graph D shows the behavior of a gas that condenses to form a liquid (in this case, at 50°C) as it is cooled.

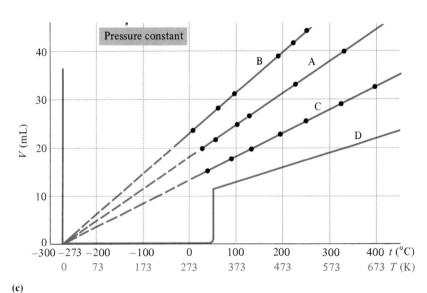

(c)

Example 12-3

A sample of nitrogen occupies 117 mL at 100°C. At what temperature would it occupy 234 milliliters if the pressure did not change?

Plan

We know the volume of the sample at one temperature, and wish to know its temperature corresponding to a second volume (constant pressure). We can solve Charles' Law for the second temperature. Alternatively, we can multiply the original temperature by a Charles' Law factor to account for the desired volume change. We must remember to carry out calculations with all temperatures expressed on the Kelvin scale, converting to or from Celsius as necessary.

Solution

$$V_1 = 117 \text{ mL} \qquad\qquad V_2 = 234 \text{ mL}$$

$$T_1 = 100°C + 273° = 373K \qquad T_2 = \underline{?}$$

$$\frac{V_1}{T_1} = \frac{V_2}{T_2} \quad \text{and} \quad T_2 = \frac{V_2 T_1}{V_1} = \frac{(234 \text{ mL})(373 \text{ K})}{(117 \text{ mL})} = \boxed{746 \text{ K}}$$

$$°C = 746 \text{ K} - 273° = \boxed{473°C}$$

The temperature doubles on the Kelvin scale, from 373 K to 746 K, so the volume doubles. Alternatively we can use a Charles' Law factor. The volume *increases* from 117 mL to 234 mL, which corresponds to an *increase* in the Kelvin temperature by the factor 234 mL/117 mL:

$$\underline{?} \text{ K} = 373 \text{ K} \times \frac{234 \text{ mL}}{117 \text{ mL}} = \boxed{746 \text{ K } (473°C)}$$

EOC 23, 24

When balloons filled with air are cooled in liquid nitrogen (bp −196°C), each shrinks to a small fraction of its original volume. Because the boiling points of the other components of air, except He and Ne, are lower than −196°C, they condense to form liquids. When the balloons are removed from the liquid nitrogen, the liquids vaporize to form gases again. As the air warms to room temperature, the balloons expand to their original volume (Charles' Law).

12-6 Standard Temperature and Pressure

We have seen that both temperature and pressure affect the volumes (and therefore the densities) of gases. It is often convenient to choose some "standard" temperature and pressure as a reference point for discussing gases. **Standard conditions of temperature and pressure (STP or SC)** are, by international agreement, 0°C (273.15 K) and one atmosphere of pressure (760 torr).

12-7 The Combined Gas Law Equation

Boyle's Law relates the pressure and volume of a sample of gas at constant temperature, $P_1 V_1 = P_2 V_2$. Charles' Law relates the volume and temperature at constant pressure, $V_1/T_1 = V_2/T_2$. Combination of the essence of Boyle's Law and Charles' Law into a single expression gives the **combined gas law equation**:

$$\frac{P_1 V_1}{T_1} = \frac{P_2 V_2}{T_2} \qquad \text{(constant } n\text{)}$$

Notice that the combined gas law equation becomes
1. $P_1V_1 = P_2V_2$ (Boyle's Law) when T is constant;
2. $\dfrac{V_1}{T_1} = \dfrac{V_2}{T_2}$ (Charles' Law) when P is constant; and
3. $\dfrac{P_1}{T_1} = \dfrac{P_2}{T_2}$ when V is constant.

When any five of the variables in the equation are known, the sixth variable can be calculated.

Example 12-4

A sample of neon occupies 105 liters at 27°C under a pressure of 985 torr. What volume would it occupy at standard conditions?

Plan

A sample of gas is changing in all three quantities P, V, and T. This suggests that we use the combined gas law equation. We tabulate what is known and what is asked for, solve the combined gas law equation for the unknown quantity, V_2, and substitute known values.

Solution

$$V_1 = 105 \text{ L} \qquad P_1 = 985 \text{ torr} \qquad T_1 = 27°C + 273° = 300 \text{ K}$$

$$V_2 = \underline{?} \qquad P_2 = 760 \text{ torr} \qquad T_2 = 273 \text{ K}$$

Solving for V_2,

$$\frac{P_1V_1}{T_1} = \frac{P_2V_2}{T_2} \quad \text{so} \quad V_2 = \frac{P_1V_1T_2}{P_2T_1} = \frac{(985 \text{ torr})(105 \text{ L})(273 \text{ K})}{(760 \text{ torr})(300 \text{ K})} = \boxed{124 \text{ L}}$$

Alternatively, we can multiply the original volume by a Boyle's Law factor and a Charles' Law factor. As the pressure decreases from 985 torr to 760 torr, the volume increases, so the Boyle's Law factor is 985 torr/760 torr. As the temperature decreases from 300 K to 273 K, the volume decreases, so the Charles' Law factor is 273 K/300 K. Multiplication of the original volume by these factors gives the same result.

$$\underline{?} \text{ L} = 105 \text{ L} \times \frac{985 \text{ torr}}{760 \text{ torr}} \times \frac{273 \text{ K}}{300 \text{ K}} = \boxed{124 \text{ L}}$$

The temperature decrease (from 300 K to 273 K) alone would give only a small *decrease* in the volume of neon. The pressure decrease (from 985 torr to 760 torr) alone would result in a greater *increase* in the volume. The result of the two changes is that the volume increases from 105 liters to 124 liters.

Example 12-5

A sample of gas occupies 10.0 liters at 240°C under a pressure of 80.0 kPa. At what temperature would the gas occupy 20.0 liters if we increased the pressure to 107 kPa?

As a reference, one atm is 101.3 kPa.

Plan

The approach is the same as for Example 12-4 except that the unknown quantity is the temperature, T_2.

Solution

$$V_1 = 10.0 \text{ L} \qquad P_1 = 80.0 \text{ kPa} \qquad T_1 = 240°C + 273° = 513 \text{ K}$$

$$V_2 = 20.0 \text{ L} \qquad P_2 = 107 \text{ kPa} \qquad T_2 = \underline{?}$$

We solve the combined gas law equation for T_2:

$$\frac{P_1V_1}{T_1} = \frac{P_2V_2}{T_2} \quad \text{so} \quad T_2 = \frac{P_2V_2T_1}{P_1V_1} = \frac{(107 \text{ kPa})(20.0 \text{ L})(513 \text{ K})}{(80.0 \text{ kPa})(10.0 \text{ L})} = \boxed{1.37 \times 10^3 \text{ K}}$$

$$\text{K} = {}^\circ\text{C} + 273^\circ \quad \text{so} \quad {}^\circ\text{C} = 1.37 \times 10^3 \text{ K} - 273^\circ = \boxed{1.10 \times 10^3 \text{ }^\circ\text{C}}$$

EOC 32, 34

12-8 Avogadro's Law and the Standard Molar Volume

In 1811 Amedeo Avogadro postulated:

At the same temperature and pressure, equal volumes of all gases contain the same number of molecules.

Many experiments have demonstrated that Avogadro's hypothesis is correct to within about ±2%, and the statement is now known as **Avogadro's Law**. Avogadro's Law can also be stated as follows:

At constant temperature and pressure, the volume occupied by a gas sample is directly proportional to the number of moles of gas.

$$V \propto n \quad \text{or} \quad V = kn \quad \text{or} \quad \frac{V}{n} = k \quad \text{(constant } P, T\text{)}$$

For two different samples of gas at the same temperature and pressure, the relation between volumes and numbers of moles can be represented as

$$\frac{V_1}{n_1} = \frac{V_2}{n_2} \quad \text{(same } T, P\text{)}$$

The volume occupied by a mole of gas at *standard temperature and pressure,* STP, is referred to as the standard molar volume. It is nearly constant for all gases (Table 12-3).

Deviations in volume indicate that the gases are not behaving ideally. Such behavior is discussed in molecular terms in Section 12-14.

Table 12-3
Standard Molar Volumes and Densities of Some Gases

Gas	Formula	(g/mol)	Standard Molar Volume (L/mol)	STP (g/L)
hydrogen	H_2	2.02	22.428	0.090
helium	He	4.003	22.426	0.178
neon	Ne	20.18	22.425	0.900
nitrogen	N_2	28.01	22.404	1.250
oxygen	O_2	32.00	22.394	1.429
argon	Ar	39.95	22.393	1.784
carbon dioxide	CO_2	44.01	22.256	1.977
ammonia	NH_3	17.03	22.094	0.771
chlorine	Cl_2	70.91	22.063	3.214

> The **standard molar volume** of an ideal gas is taken to be 22.4 liters per mole at STP.

Gas densities depend on pressure and temperature. However, the number of moles of gas in a sample does not change with temperature or pressure. Recall that density is defined as mass per unit volume. Pressure changes affect volumes of gases according to Boyle's Law, while temperature changes affect volumes of gases according to Charles' Law. We can use these laws to convert gas densities measured at various temperatures and pressures to *standard temperature and pressure*. Table 12-3 gives the densities of several gases at standard conditions.

Example 12-6

One (1.00) mole of a gas occupies 27.0 liters, and its density is 1.41 g/L at a particular temperature and pressure. What is its molecular weight? What is the density of the gas at STP?

Plan

We can use dimensional analysis to convert the density, 1.41 g/L, to molecular weight, g/mol. To calculate the density at STP, we recall that the volume occupied by one mole would be 22.4 L.

Solution

We multiply the density under the original conditions by the unit factor 27.0 L/1.00 mol to generate the appropriate units, g/mol:

$$\frac{?\text{ g}}{\text{mol}} = \frac{1.41\text{ g}}{\text{L}} \times \frac{27.0\text{ L}}{\text{mol}} = \boxed{38.1\text{ g/mol}}$$

At STP, 1.00 mol of the gas, 38.1 g, would occupy 22.4 L, and its density would be

$$\text{density} = \frac{38.1\text{ g}}{22.4\text{ L}} = \boxed{1.70\text{ g/L}} \quad \text{at STP}$$

EOC 38, 40

12-9 Summary of the Gas Laws—The Ideal Gas Equation

Let us summarize what we have learned about gases. Any sample of gas can be described in terms of its pressure, temperature (Kelvin), volume, and the number of moles, n, present. Any three of these variables determine the fourth. The gas laws we have studied give several relationships among these variables. An **ideal gas** is one that obeys these gas laws exactly. Many real gases show slight deviations from ideality, but at normal temperatures and pressures the deviations are usually small enough to be ignored. We shall do so for the present and discuss deviations later.

We can summarize the behavior of ideal gases as follows:

Boyle's Law $V \propto \dfrac{1}{P}$ (at constant T and n)

Charles' Law $V \propto T$ (at constant P and n)

Avogadro's Law $V \propto n$ (at constant T and P)

Summarizing $V \propto \dfrac{nT}{P}$ (no restrictions)

As before, a proportionality can be written as an equality by introducing a proportionality constant, for which we'll use the symbol R. This gives

$$V = R\left(\frac{nT}{P}\right) \quad \text{or, rearranging,} \quad \boxed{PV = nRT}$$

This relationship is called the **ideal gas equation** or the *ideal gas law*. The numerical value of R, the **universal gas constant**, depends on the choices of the units for P, V, and T. Recall that 1.00 mole of an ideal gas occupies 22.4 liters at 1.00 atmosphere and 273 K (STP). Solving the ideal gas law for R gives

> This equation takes into account the values of n, T, P, and V. Therefore, restrictions that apply to the individual gas laws are not needed for the ideal gas equation.

$$R = \frac{PV}{nT} = \frac{(1.00 \text{ atm})(22.4 \text{ L})}{(1.00 \text{ mol})(273 \text{ K})} = 0.0821 \; \frac{\text{L} \cdot \text{atm}}{\text{mol} \cdot \text{K}}$$

The more exact value is 0.08206 L · atm/mol · K. R may be expressed in other units.

Example 12-7

R can have any *energy* units per mole per kelvin. Calculate R in terms of joules per mole per kelvin and in SI units of kPa · dm³/mol · K.

Plan

We apply dimensional analysis to convert to the required units.

Solution

Appendix C shows that 1 L · atm = 101.32 joules.

$$R = \frac{0.08206 \text{ L} \cdot \text{atm}}{\text{mol} \cdot \text{K}} \times \frac{101.32 \text{ J}}{1 \text{ L} \cdot \text{atm}} = \boxed{8.314 \text{ J/mol} \cdot \text{K}}$$

Let us now evaluate R in SI units. One atmosphere pressure is 101.3 kilopascals, and the molar volume at STP is 22.4 dm³.

> Recall that 1 dm³ = 1 L.

$$R = \frac{PV}{nT} = \frac{101.3 \text{ kPa} \times 22.4 \text{ dm}^3}{1 \text{ mol} \times 273 \text{ K}} = \boxed{8.31 \; \frac{\text{kPa} \cdot \text{dm}^3}{\text{mol} \cdot \text{K}}}$$

EOC 45

We can now express R, the universal gas constant, in three different sets of units:

$$R = 0.08206 \; \frac{\text{L} \cdot \text{atm}}{\text{mol} \cdot \text{K}} = \frac{8.314 \text{ J}}{\text{mol} \cdot \text{K}} = 8.314 \; \frac{\text{kPa} \cdot \text{dm}^3}{\text{mol} \cdot \text{K}}$$

The usefulness of the ideal gas equation is that it relates the four variables, P, V, n, and T, that describe a sample of gas at *one set of conditions*. If any three of these variables are known, the fourth can be calculated.

Example 12-8

What is the volume of a gas balloon filled with 4.00 moles of He when the atmospheric pressure is 748 torr and the temperature is 30°C?

Plan

We first list the variables with the proper units. Then we solve the ideal gas equation for V and substitute values.

Solution

Let's first list the variables with the proper units:

$$P = 748 \text{ torr} \times \frac{1 \text{ atm}}{760 \text{ torr}} = 0.984 \text{ atm} \qquad n = 4.00 \text{ mol}$$

$$T = 30°C + 273° = 303 \text{ K} \qquad V = \underline{?}$$

Solving $PV = nRT$ for V and substituting gives

$$PV = nRT; \qquad V = \frac{nRT}{P} = \frac{(4.00 \text{ mol}) \left(0.0821 \dfrac{\text{L} \cdot \text{atm}}{\text{mol} \cdot \text{K}}\right) (303 \text{ K})}{0.984 \text{ atm}} = \boxed{101 \text{ L}}$$

EOC 46, 48

You may wonder why pressures are given in torr or mm Hg and temperatures in °C. This is because pressures are often measured with mercury barometers, and temperatures are measured with Celsius thermometers.

Example 12-9

A helium-filled weather balloon with a diameter of 24.0 feet has a volume of 7240 cubic feet. How many grams of helium would be required to inflate this balloon to a pressure of 745 torr at 21°C? ($1 \text{ ft}^3 = 28.3 \text{ L}$)

Plan

We shall use the ideal gas equation to find n, the number of moles required, and then convert to grams. We must remember to convert each quantity to one of the units stated for R. ($R = 0.0821 \text{ L} \cdot \text{atm/mol} \cdot \text{K}$)

Solution

$$P = 745 \text{ torr} \times \frac{1 \text{ atm}}{760 \text{ torr}} = 0.980 \text{ atm} \qquad T = 21°C + 273 = 294 \text{ K}$$

$$V = 7240 \text{ ft}^3 \times \frac{28.3 \text{ L}}{1 \text{ ft}^3} = 2.05 \times 10^5 \text{ L} \qquad n = \underline{?}$$

Solving $PV = nRT$ for n and substituting gives

$$n = \frac{PV}{RT} = \frac{(0.980 \text{ atm})(2.05 \times 10^5 \text{ L})}{\left(0.0821 \dfrac{\text{L} \cdot \text{atm}}{\text{mol} \cdot \text{K}}\right)(294 \text{ K})} = 8.32 \times 10^3 \text{ mol He}$$

A helium-filled weather balloon.

$$? \text{ g He} = (8.32 \times 10^3 \text{ mol He})\left(4.00 \ \frac{g}{\text{mol}}\right) = \boxed{3.33 \times 10^4 \text{ g He}}$$

Example 12-10
What pressure, in kPa, is exerted by 54.0 grams of Xe in a 1.00-liter flask at 20°C?

Plan
We list the variables with the proper units. Then we solve the ideal gas equation for P and substitute values.

Solution

$$V = 1.00 \text{ L} = 1.00 \text{ dm}^3 \qquad n = 54.0 \text{ g Xe} \times \frac{1 \text{ mol}}{131.3 \text{ g Xe}} = 0.411 \text{ mol}$$

$$T = 20°C + 273° = 293 \text{ K} \qquad P = ?$$

Solving $PV = nRT$ for P and substituting gives

$$P = \frac{nRT}{V} = \frac{(0.411 \text{ mol})\left(\dfrac{8.31 \text{ kPa} \cdot \text{dm}^3}{\text{mol} \cdot \text{K}}\right)(293 \text{ K})}{1.00 \text{ dm}^3} = \boxed{1.00 \times 10^3 \text{ kPa}} \quad (9.87 \text{ atm})$$

Summary of the Ideal Gas Laws

1. The individual gas laws are usually used to calculate the *changes* in conditions for a sample of gas (subscripts can be thought of as "before" and "after"):

Boyle's Law	$P_1V_1 = P_2V_2$	(for a given amount of a gas at constant temperature)
Charles' Law	$\dfrac{V_1}{T_1} = \dfrac{V_2}{T_2}$	(for a given amount of a gas at constant pressure)
Combined gas law	$\dfrac{P_1V_1}{T_1} = \dfrac{P_2V_2}{T_2}$	(for a given amount of a gas)
Avogadro's Law	$\dfrac{V_1}{n_1} = \dfrac{V_2}{n_2}$	(for gas samples at the same temperature and pressure)

2. The ideal gas equation is used to calculate one of the four variables P, V, n, and T, that describe a sample of gas at *any single set of conditions*.

$$PV = nRT$$

The ideal gas equation can also be used to calculate the densities of gases.

Example 12-11
Nitric acid, a very important industrial chemical, is made by dissolving the gas nitrogen dioxide, NO_2, in water. Calculate the density of NO_2 gas, in g/L, at 1.24 atm and 50°C.

Plan

We use the ideal gas equation to find the number of moles, n, in any volume, V, at the specified pressure and temperature. Then we convert moles to grams. Because we want to express density in g/L, we choose a volume of one liter.

Solution

$$V = 1.00 \text{ L} \qquad\qquad n = \underline{?}$$

$$T = 50°\text{C} + 273° = 323 \text{ K} \qquad P = 1.24 \text{ atm}$$

Solving $PV = nRT$ for n and substituting gives

$$n = \frac{PV}{RT} = \frac{(1.24 \text{ atm})(1.00 \text{ L})}{\left(0.0821 \dfrac{\text{L} \cdot \text{atm}}{\text{mol} \cdot \text{K}}\right)(323 \text{ K})} = 0.0468 \text{ mol}$$

So there is 0.0468 mol NO_2/L at the specified P and T. Converting this to grams of NO_2,

$$\text{density} = \frac{\underline{?} \text{ g}}{\text{L}} = \frac{0.0468 \text{ mol } NO_2}{\text{L}} \times \frac{46.0 \text{ g } NO_2}{\text{mol } NO_2} = \boxed{2.15 \text{ g/L}}$$

EOC 48

12-10 Determination of Molecular Weights and Molecular Formulas of Gaseous Substances

In Section 2-9 we distinguished between simplest and molecular formulas of compounds. We showed how simplest formulas can be calculated from percent compositions of compounds. The molecular weight must be known in order to determine the molecular formula for a compound. For compounds that are gases at reasonable temperatures and pressures, the ideal gas law provides a basis for determining molecular weights.

Example 12-12

A 0.109-gram sample of a pure gaseous compound occupies 112 mL at 100°C and 750 torr. What is the molecular weight of the compound?

Plan

We first use the ideal gas law, $PV = nRT$, to find the number of moles of gas. Then, knowing the mass of that number of moles of gas, we calculate the mass of one mole, the molecular weight.

Solution

$$V = 0.112 \text{ L}; \ T = 100°\text{C} + 273 = 373 \text{ K}; \ P = 750 \text{ torr} \times \frac{1 \text{ atm}}{760 \text{ torr}} = 0.987 \text{ atm}$$

$$n = \frac{PV}{RT} = \frac{(0.987 \text{ atm})(0.112 \text{ L})}{\left(0.0821 \dfrac{\text{L} \cdot \text{atm}}{\text{mol} \cdot \text{K}}\right)(373 \text{ K})} = 0.00361 \text{ mol}$$

The mass of 0.00361 mole is 0.109 g, so the mass of one mole is

$$\text{molecular weight} = \frac{?\ g}{\text{mol}} = \frac{0.109\ g}{0.00361\ \text{mol}} = \boxed{30.2\ g/\text{mol}}$$

EOC 54

The molecular weight of the gas is 30.2 amu. The gas could be ethane, C_2H_6, MW = 30.1 amu. Can you think of other possibilities?

Example 12-13

Analysis of a sample of a gaseous compound shows that it contains 85.7% carbon and 14.3% hydrogen by mass. At standard conditions, 100 mL of the compound has a mass of 0.188 gram. What is its molecular formula?

Plan

We first find the simplest formula for the compound as we did in Section 2-8 (Examples 2-13 and 2-14). We use the gas volume and mass data to find the molecular weight as in Example 12-12. To find the molecular formula, we reason as in Example 2-17. We use the experimentally known molecular weight to find the ratio

$$n = \frac{\text{molecular weight}}{\text{simplest formula weight}}$$

The molecular weight is n times the simplest formula weight, so the molecular formula is n times the simplest formula.

Solution

We first determine the simplest formula.

Element	Relative Mass of Element	Relative Number of Atoms (divide mass by AW)	Divide by Smallest Number	Smallest Whole-Number Ratio of Atoms
C	85.7	$\frac{85.7}{12.0} = 7.14$	$\frac{7.14}{7.14} = 1.00$	1
H	14.3	$\frac{14.3}{1.01} = 14.2$	$\frac{14.2}{7.14} = 1.99$	2

The ratio gives CH_2.

The simplest formula is CH_2 (simplest formula weight = 14 amu), so the molecular formula of the compound must be some multiple of CH_2. We use the ideal gas law, $PV = nRT$, to find the number of moles in the sample.

$$V = 0.100\ \text{L} \qquad P = 1.00\ \text{atm} \qquad T = 273\ \text{K}$$

$$n = \frac{PV}{RT} = \frac{(1.00\ \text{atm})(0.100\ \text{L})}{\left(0.0821\ \dfrac{\text{L} \cdot \text{atm}}{\text{mol} \cdot \text{K}}\right)(273\ \text{K})} = 0.00446\ \text{mol}$$

This result tells us that 0.00446 mole has a mass of 0.188 g, which enables us to calculate the molecular weight:

$$\frac{?\ g}{\text{mol}} = \frac{0.188\ g}{0.00446\ \text{mol}} = \boxed{42.2\ g/\text{mol}}$$

We know that one mole of the compound has a mass of 42.2 grams. To determine its molecular formula, we divide the molecular weight of the compound (which corresponds to its molecular formula) by the simplest formula weight to obtain the nearest integer:

$$\frac{\text{molecular weight}}{\text{simplest formula weight}} = \frac{42.2\ \text{amu}}{14\ \text{amu}} = 3$$

propene

cyclopropane

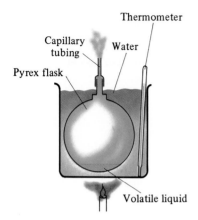

Thermometer

Capillary tubing

Water

Pyrex flask

Volatile liquid

Figure 12-6
Determination of molecular weight by the Dumas method. A volatile liquid is vaporized in the boiling-water bath, and excess vapor is driven from the flask. The remaining vapor, at known *P*, *T*, and *V*, is condensed and weighed.

Because the resulting vapor is only slightly above its liquefaction point (the boiling point of the liquid), it does not behave very ideally. Hence, errors of a few percent are common in results of the Dumas method. Such errors are acceptable in determining molecular formulas (Example 12-15).

This tells us that the molecular formula is three times the simplest formula. Therefore, the molecular formula is $(CH_2)_3 = \boxed{C_3H_6}$. The gas could be either propene or cyclopropane, both of which have the formula C_3H_6.

EOC 53, 57

This kind of calculation of molecular weights can be extended to volatile liquids. In the **Dumas method**, a sample of volatile liquid is placed in a previously weighed flask, and the flask is placed in a boiling-water bath. The liquid is allowed to evaporate, and the excess vapor escapes into the air (Figure 12-6). When the last bit of liquid evaporates, the container is filled completely with vapor of the volatile liquid at 100°C. The container is then removed from the boiling-water bath, cooled quickly, and weighed again. Rapid cooling condenses the vapor that filled the flask at 100° C, and air reenters the flask. Recall that the volume of a liquid is extremely small in comparison with the volume occupied by the same mass of the gaseous compound at atmospheric pressure. The difference between the mass of the flask containing the condensed liquid and the mass of the empty flask is the mass of the condensed liquid. It is also equal to the mass of the vapor that filled the flask at 100°C and the atmospheric pressure. The "empty" flask contains very nearly the same amount of air both times it is weighed, so the mass of air does not affect the result.

Although the compound is a liquid at room temperature, it is a gas at 100°C, and therefore we can apply the gas laws. If the water surrounding the flask is replaced by a liquid with a higher boiling point, the Dumas method can be applied to liquids that vaporize at or below the boiling point of the liquid surrounding the flask. Example 12-14 illustrates the determination of the molecular weight of a volatile liquid by the Dumas method.

Example 12-14
The molecular weight of a volatile liquid was determined by the Dumas method. A 120-mL flask contained 0.345 gram of vapor at 100°C and 1.00 atm pressure. What is the molecular weight of the compound?

Plan
We use the ideal gas law, $PV = nRT$, to determine the number of moles of vapor that filled the flask. Then, knowing the mass of this number of moles, we can calculate the mass of one mole.

Solution

$$V = 0.120 \text{ L} \qquad P = 1.00 \text{ atm} \qquad T = 100°C + 273 = 373 \text{ K}$$

$$n = \frac{PV}{RT} = \frac{(1.00 \text{ atm})(0.120 \text{ L})}{\left(0.0821 \dfrac{\text{L} \cdot \text{atm}}{\text{mol} \cdot \text{K}}\right)(373 \text{ K})} = 0.00392 \text{ mol}$$

The mass of 0.00392 mol of vapor is 0.345 g, so the mass of one mole is

$$\frac{?\text{ g}}{\text{mol}} = \frac{0.345 \text{ g}}{0.00392 \text{ mol}} = \boxed{88.0 \text{ g/mol}}$$

The molecular weight of the vapor, and therefore of the volatile liquid, is 88.0 amu.

EOC 55, 56

Let's carry the calculation one step further in the next example.

Example 12-15

Analysis of the volatile liquid in Example 12-14 showed that it contained 54.5% carbon, 9.10% hydrogen, and 36.4% oxygen by mass. What is its molecular formula?

Plan

We reason just as we did in Example 12-13. We first determine the simplest formula for the compound. Then we use the molecular weight that we determined in Example 12-14 to find the molecular formula.

Solution

Element	Relative Mass of Element	Relative Number of Atoms (divide mass by AW)	Divide by Smallest Number	Smallest Whole-Number Ratio of Atoms
C	54.5	$\dfrac{54.5}{12.0} = 4.54$	$\dfrac{4.54}{2.28} = 1.99$	2
H	9.10	$\dfrac{9.10}{1.01} = 9.01$	$\dfrac{9.01}{2.28} = 3.95$	4 ———— C_2H_4O
O	36.4	$\dfrac{36.4}{16.0} = 2.28$	$\dfrac{2.28}{2.28} = 1.00$	1

The simplest formula is C_2H_4O; the simplest formula weight is 44 amu.

Division of the molecular weight by the simplest formula weight gives

$$\frac{\text{molecular weight}}{\text{simplest formula weight}} = \frac{88 \text{ amu}}{44 \text{ amu}} = 2$$

Therefore, the molecular formula is $(C_2H_4O)_2 = $ $\boxed{C_4H_8O_2}$.

One compound that has this molecular formula is ethyl acetate, a common solvent used in nail polishes and polish removers.

12-11 Dalton's Law of Partial Pressures

Many gas samples, including our atmosphere, are mixtures that consist of different kinds of gases. The total number of moles in a mixture of gases is

$$n_{\text{total}} = n_A + n_B + n_C + \cdots$$

where n_A, n_B, and so on represent the number of moles of each kind of gas present. Rearranging the ideal gas equation, $P_{\text{total}}V = n_{\text{total}}RT$, for the total pressure, P_{total}, and then substituting for n_{total} gives

$$P_{\text{total}} = \frac{n_{\text{total}}RT}{V} = \frac{(n_A + n_B + n_C + \cdots)RT}{V}$$

Multiplying out the right-hand side gives

$$P_{total} = \frac{n_A RT}{V} + \frac{n_B RT}{V} + \frac{n_C RT}{V} + \cdots$$

Now $n_A RT/V$ is the partial pressure P_A that the n_A moles of gas A alone would exert in the container at temperature T; similarly, $n_B RT/V = P_B$, and so on. Substituting these into the equation for P_{total}, we obtain **Dalton's Law of Partial Pressures** (Figure 12-7):

$$P_{total} = P_A + P_B + P_C + \cdots \qquad \text{(constant } V, T)$$

The total pressure exerted by a mixture of ideal gases is the sum of the partial pressures of those gases.

John Dalton was the first to notice this effect. He did so in 1807 while studying the compositions of moist and dry air. The pressure that each gas exerts in a mixture is called its *partial pressure*. No way has been devised to measure the pressure of an individual gas in a mixture; it must be calculated from other quantities.

Dalton's Law is useful in describing real gaseous mixtures at moderate pressures because it allows us to relate total measured pressures to the composition of mixtures.

Example 12-16

A 10.0-liter flask contains 0.200 mole of methane, 0.300 mole of hydrogen, and 0.400 mole of nitrogen at 25°C. (a) What is the pressure, in atmospheres, inside the flask? (b) What is the partial pressure of each component of the mixture of gases?

Plan

(a) We are given the number of moles of each component. The ideal gas law is then used to calculate the total pressure from the total number of moles. (b) The partial pressure of each gas in the mixture can be calculated by substituting the number of moles of each gas into $PV = nRT$ individually.

Solution

(a) $n = 0.200$ mol $CH_4 + 0.300$ mol $H_2 + 0.400$ mol $N_2 = 0.900$ mol of gas

     $V = 10.0$ L $T = 25°C + 273° = 298$ K

Solving $PV = nRT$ for P gives $P = nRT/V$. Substitution gives

$$P = \frac{(0.900 \text{ mol}) \left(0.0821 \dfrac{\text{L} \cdot \text{atm}}{\text{mol} \cdot \text{K}}\right)(298 \text{ K})}{10.0 \text{ L}} = \boxed{2.20 \text{ atm}}$$

Figure 12-7
An illustration of Dalton's Law. When the two gases A and B are mixed in the same container at the same temperature, they exert a total pressure equal to the sum of their partial pressures.

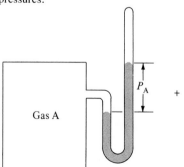
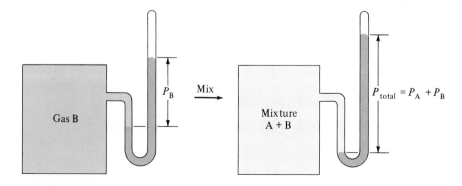

(b) Now we find the partial pressures. For CH_4, $n = 0.200$ mol, and the values for V and T are the same as above.

$$P_{CH_4} = \frac{(n_{CH_4})RT}{V} = \frac{(0.200 \text{ mol}) \left(0.0821 \dfrac{L \cdot atm}{mol \cdot K} \right) (298 \text{ K})}{10.0 \text{ L}} = \boxed{0.489 \text{ atm}}$$

Similar calculations for the partial pressures of hydrogen and nitrogen give

$$P_{H_2} = \boxed{0.734 \text{ atm}} \qquad \text{and} \qquad P_{N_2} = \boxed{0.979 \text{ atm}}$$

As a check, we use Dalton's Law: $P_{total} = P_A + P_B + P_C + \cdots$. Addition of the partial pressures in this mixture gives the total pressure:

$$P_{total} = P_{CH_4} + P_{H_2} + P_{N_2} = 0.489 \text{ atm} + 0.734 \text{ atm} + 0.979 \text{ atm} = \boxed{2.20 \text{ atm}}$$

The total pressure exerted by the mixture of gases is 2.20 atmospheres.

EOC 60

We can describe the composition of any mixture in terms of the mole fraction of each component. The **mole fraction**, X_A, of component A in a mixture is defined as

$$X_A = \frac{\text{no. mol } A}{\text{total no. mol of all components}}$$

Like any other fraction, mole fraction is a dimensionless quantity. For each component in a mixture, the mole fraction is

$$X_A = \frac{\text{no. mol } A}{\text{no. mol } A + \text{no. mol } B + \cdots},$$

$$X_B = \frac{\text{no. mol } B}{\text{no. mol } A + \text{no. mol } B + \cdots}, \qquad \text{and so on}$$

The sum of all mole fractions in a mixture is equal to 1.

$X_A + X_B + \cdots = 1$ for any mixture

We can use this relationship to check mole fraction calculations or to find a remaining mole fraction if we know all the others.

For a gaseous mixture, we can relate the mole fraction of each component to its partial pressure as follows. From the ideal gas equation, the number of moles of each component can be written as

$$n_A = P_A V/RT, \qquad n_B = P_B V/RT, \qquad \text{and so on}$$

and the total number of moles is

$$n_{total} = P_{total} V/RT$$

Substituting into the definition of X_A,

$$X_A = \frac{n_A}{n_A + n_B + \cdots} = \frac{P_A V/RT}{P_{total} V/RT}$$

The quantities V, R, and T cancel to give

$$X_A = \frac{P_A}{P_{total}}; \qquad \text{similarly,} \quad X_B = \frac{P_B}{P_{total}}; \qquad \text{and so on.}$$

We can rearrange these equations to give another statement of Dalton's Law of Partial Pressures:

$$P_A = X_A \times P_{\text{total}}; \qquad P_B = X_B \times P_{\text{total}}; \qquad \text{and so on}$$

The partial pressure of each gas is equal to its mole fraction in the gaseous mixture times the total pressure of the mixture.

Example 12-17
Calculate the mole fractions of the three gases in Example 12-16.

Plan
One way to solve this problem is to use the numbers of moles given in the problem. Alternatively, we could use the partial pressures and the total pressure from Example 12-16.

Solution
Using the moles given in the Example 12-16,

$$X_{\text{CH}_4} = \frac{n_{\text{CH}_4}}{n_{\text{total}}} = \frac{0.200 \text{ mol}}{0.900 \text{ mol}} = \boxed{0.222}$$

$$X_{\text{H}_2} = \frac{n_{\text{H}_2}}{n_{\text{total}}} = \frac{0.300 \text{ mol}}{0.900 \text{ mol}} = \boxed{0.333}$$

$$X_{\text{N}_2} = \frac{n_{\text{N}_2}}{n_{\text{total}}} = \frac{0.400 \text{ mol}}{0.900 \text{ mol}} = \boxed{0.444}$$

Using the partial and total pressures calculated in Example 12-16,

$$X_{\text{CH}_4} = \frac{P_{\text{CH}_4}}{P_{\text{total}}} = \frac{0.489 \text{ atm}}{2.20 \text{ atm}} = \boxed{0.222}$$

$$X_{\text{H}_2} = \frac{P_{\text{H}_2}}{P_{\text{total}}} = \frac{0.734 \text{ atm}}{2.20 \text{ atm}} = \boxed{0.334}$$

The difference between the two calculated results is due to rounding.

$$X_{\text{N}_2} = \frac{P_{\text{N}_2}}{P_{\text{total}}} = \frac{0.979 \text{ atm}}{2.20 \text{ atm}} = \boxed{0.445}$$

EOC 62

Example 12-18
The mole fraction of oxygen in the atmosphere is 0.2094. Calculate the partial pressure of O_2 in air when the atmospheric pressure is 760. torr.

Plan
The partial pressure of each gas in a mixture is equal to its mole fraction in the mixture times the total pressure of the mixture.

Solution

$$P_{\text{O}_2} = X_{\text{O}_2} \times P_{\text{total}}$$

$$P_{\text{O}_2} = 0.2094 \times 760. \text{ torr} = \boxed{159 \text{ torr}}$$

Dalton's Law can be used in combination with other gas laws, as the following example shows.

Example 12-19

Two tanks are connected by a closed valve. Each tank is filled with gas as shown, and both tanks are held at the same temperature. We open the valve and allow the gases to mix.

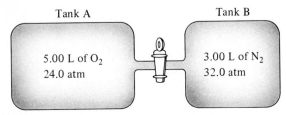

(a) After the gases mix, what is the partial pressure of each gas, and what is the total pressure? (b) What is the mole fraction of each gas in the mixture?

Plan

(a) Each gas expands to fill the entire 8.00 liters. We can use Boyle's Law to calculate the partial pressure that each gas would exert after expansion. The total pressure is equal to the sum of the partial pressures of the two gases. (b) The mole fractions can be calculated from the ratio of the partial pressure of each gas to the total pressure.

Solution

(a)

For O_2: $P_1V_1 = P_2V_2$ or $P_{2,O_2} = \dfrac{P_1V_1}{V_2} = \dfrac{24.0 \text{ atm} \times 5.00 \text{ L}}{8.00 \text{ L}} = \boxed{15.0 \text{ atm}}$

For N_2: $P_1V_1 = P_2V_2$ or $P_{2,N_2} = \dfrac{P_1V_1}{V_2} = \dfrac{32.0 \text{ atm} \times 3.00 \text{ L}}{8.00 \text{ L}} = \boxed{12.0 \text{ atm}}$

The total pressure is the sum of the partial pressures.

$$P_{\text{total}} = P_{2,O_2} + P_{2,N_2} = 15.0 \text{ atm} + 12.0 \text{ atm} = \boxed{27.0 \text{ atm}}$$

(b) $$X_{O_2} = \dfrac{P_{2,O_2}}{P_{\text{total}}} = \dfrac{15.0 \text{ atm}}{27.0 \text{ atm}} = \boxed{0.556}$$

$$X_{N_2} = \dfrac{P_{2,N_2}}{P_{\text{total}}} = \dfrac{12.0 \text{ atm}}{27.0 \text{ atm}} = \boxed{0.444}$$

As a check, the sum of the mole fractions is 1. Notice that this problem has been solved without calculating the number of moles of either gas.

EOC 64, 68

Some gases can be collected over water. Figure 12-8 illustrates the collection of a sample of hydrogen over water. A gas produced in a reaction displaces the more dense water from the inverted water-filled jars. The

Gases that are soluble in water or that react with water cannot be collected by this method. Other liquids can be used.

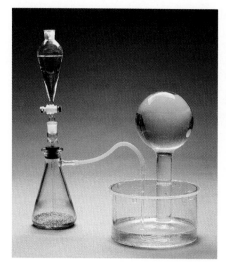

Figure 12-8

Apparatus for preparing hydrogen from zinc and sulfuric acid.

$$Zn(s) + 2H^+(aq) \longrightarrow$$
$$Zn^{2+}(aq) + H_2(g)$$

The hydrogen is collected over water.

The partial pressure exerted by the vapor above a liquid is called the vapor pressure of that liquid. A more extensive table of the vapor pressure of water appears in Appendix E.

pressure on the gas inside the collection vessel can be made equal to atmospheric pressure by raising or lowering the vessel until the water level inside is the same as that outside.

One complication arises, however. A gas in contact with water soon becomes saturated with water vapor. The pressure inside the vessel is the sum of the partial pressure of the gas itself *plus* the partial pressure exerted by the water vapor in the gas mixture (the **vapor pressure** of water). Every liquid shows a characteristic vapor pressure that varies only with temperature, and *not* with the volume of vapor present, so long as both liquid and vapor are present. Table 12-4 displays the vapor pressure of water near room temperature.

The relevant point here is that a gas collected over water is "moist"; that is, it is saturated with water vapor. Measuring the atmospheric pressure at which the gas is collected, we can write

$$P_{atm} = P_{gas} + P_{H_2O} \qquad \text{or} \qquad P_{gas} = P_{atm} - P_{H_2O}$$

Example 12-20 provides a detailed illustration.

Table 12-4
Vapor Pressure of Water Near Room Temperature

Temperature (°C)	Vapor Pressure of Water (torr)	Temperature (°C)	Vapor Pressure of Water (torr)
19	16.48	24	22.38
20	17.54	25	23.76
21	18.65	26	25.21
22	19.83	27	26.74
23	21.07	28	28.35

Example 12-20

A 300-mL sample of hydrogen was collected over water at 21°C on a day when the atmospheric pressure was 748 torr. (a) How many moles of H_2 were present? (b) What is the mole fraction of hydrogen in the moist gas mixture? (c) What would be the mass of the hydrogen sample if it were dry?

Plan

(a) The vapor pressure of H_2O, P_{H_2O} at 21°C, is obtained from Table 12-4. Applying Dalton's Law, $P_{H_2} = P_{atm} - P_{H_2O}$. We then use the partial pressure of H_2 in the ideal gas equation to find the number of moles of H_2 present. (b) The mole fraction of H_2 is the ratio of its partial pressure to the total pressure. (c) The number of moles found in (a) can be converted to mass of H_2.

Solution

(a) $\quad P_{H_2} = P_{atm} - P_{H_2O} = (748 - 19)\ \text{torr} = 729\ \text{torr} \times \dfrac{1\ \text{atm}}{760\ \text{torr}} = 0.959\ \text{atm}$

We also know

$$V = 300\ \text{mL} = 0.300\ \text{L} \qquad \text{and} \qquad T = 21°\text{C} + 273 = 294\ \text{K}$$

Solving the ideal gas equation for n_{H_2} gives

$$n_{H_2} = \frac{P_{H_2}V}{RT} = \frac{(0.959\ \text{atm})(0.300\ \text{L})}{\left(0.0821\ \dfrac{\text{L} \cdot \text{atm}}{\text{mol} \cdot \text{K}}\right)(294\ \text{K})} = \boxed{1.19 \times 10^{-2}\ \text{mol}\ H_2}$$

(b) $\quad X_{H_2} = \dfrac{P_{H_2}}{P_{total}} = \dfrac{729\ \text{torr}}{748\ \text{torr}} = \boxed{0.974}$

(c) $\quad ?\ \text{g}\ H_2 = 1.19 \times 10^{-2}\ \text{mol} \times \dfrac{2.02\ \text{g}}{1\ \text{mol}} = \boxed{2.40 \times 10^{-2}\ \text{g}\ H_2}$

EOC 66

> Remember that *each* gas occupies the *total* volume of the container.

> At STP, this dry hydrogen would occupy 266 mL. Can you work this out?

12-12 The Kinetic–Molecular Theory

As early as 1738, Daniel Bernoulli envisioned gaseous molecules in ceaseless motion striking the walls of their container and thereby exerting pressure. In 1857 Rudolf Clausius published a theory that attempted to explain various experimental observations that had been summarized by Boyle's, Dalton's, Charles', and Avogadro's laws. The basic assumptions of the **kinetic–molecular theory** for an ideal gas follow.

1. Gases consist of discrete molecules. The individual molecules are very small and are very far apart compared to their own sizes.
2. The gas molecules are in continuous, random, straight-line motion with varying velocities (see Figure 12-9).
3. The collisions between gas molecules and with the walls of the container are elastic; the total energy is conserved during a collision; i.e., there is no net energy gain or loss.

1. The observation that gases can be easily compressed indicates that the molecules are far apart. At ordinary temperatures and pressures, the gas molecules themselves occupy an insignificant fraction of the total volume of the container.

2. Near temperatures and pressures at which a gas liquefies, the gas does not behave ideally (Section 12-14), and attractions or repulsions among gas molecules *are* significant.

3. At any given instant, only a small fraction of the molecules are involved in collisions.

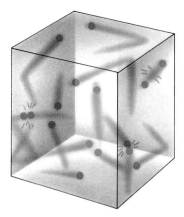

Figure 12-9

A representation of molecular motion. Gaseous molecules, in constant motion, undergo collisions with one another and with the walls of the container.

4. Between collisions, the molecules exert no attractive or repulsive forces on one another; instead, each molecule travels in a straight line with a constant velocity.

Kinetic energy is the energy a body possesses by virtue of its motion. It is $\frac{1}{2}mu^2$, where m, the body's mass, can be expressed in grams and u, its velocity, can be expressed in meters per second, m/s. The assumptions of the kinetic–molecular theory can be used to find the relationship between temperature and molecular kinetic energy (see the Enrichment Box, pages 432–434).

The average kinetic energy of gaseous molecules is directly proportional to the absolute temperature of the sample. The average kinetic energies of molecules of different gases are equal at a given temperature.

For instance, in samples of H_2, He, CO_2, and SO_2 at the same temperature, all the molecules have the same average kinetic energies. But the lighter molecules, H_2 and He, have much higher average velocities than do the heavier molecules, CO_2 and SO_2, at the same temperature.

We can summarize this very important result from the kinetic–molecular theory:

or
$$\text{average molecular } KE \propto T$$

$$\text{average molecular speed} \propto \sqrt{\frac{T}{\text{molecular weight}}}$$

Molecular kinetic energies of gases increase with increasing temperature and decrease with decreasing temperature. We have referred only to the *average* kinetic energy; in a given sample, some molecules may be moving quite rapidly while others are moving more slowly. Figure 12-10 shows the distribution of velocities of gaseous molecules at two temperatures.

Figure 12-10

The Maxwellian distribution function for molecular speeds. This graph shows the relative numbers of O_2 molecules having a given speed at 25°C and at 1000°C. At 25°C, most O_2 molecules have speeds between 200 and 600 m/s. The graph approaches the horizontal axis, but does not reach it; a finite fraction of the molecules have very high speeds.

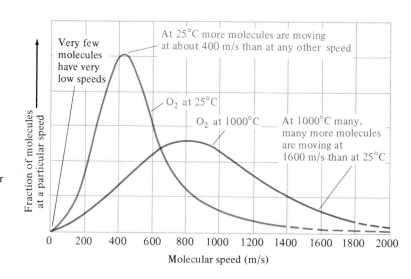

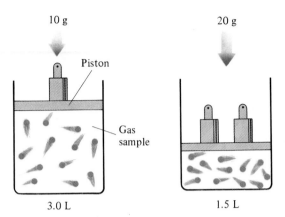

10 g   20 g

Piston

Gas
sample

3.0 L   1.5 L

Figure 12-11
A molecular interpretation of
Boyle's Law—the change in
pressure of a gas with changes in
volume (at constant
temperature). The entire
apparatus is enclosed in a
vacuum. In the smaller volume,
more molecules strike the walls
per unit time. This gives a higher
pressure.

The kinetic–molecular theory satisfactorily explains most of the observed
behavior of gases in terms of molecular behavior. Let's look at the gas laws
in light of the kinetic–molecular theory.

Boyle's Law

The pressure exerted by a gas upon the walls of its container is caused by
gas molecules striking the walls. Clearly, pressure depends upon two factors:
(1) the number of molecules striking the walls per unit time and (2) how
vigorously the molecules strike the walls. If the temperature is held constant,
the mean speed and the force of the collisions remain the same. But halving
the volume of a sample of gas doubles the pressure because twice as many
molecules strike a given area on the walls per unit time. Likewise, doubling
the volume of a sample of gas halves the pressure because only half as many
gas molecules strike a given area on the walls per unit time (Figure 12-11).

Dalton's Law

In a gas sample, the molecules are very far apart and do not attract one
another significantly. Each kind of gas molecule acts independently of the
presence of the other kind. The molecules of each gas thus collide with the
walls with a frequency and vigor that do not change even if other molecules
are present (Figure 12-12). As a result, each gas exerts a partial pressure
that is independent of the presence of the other gas, and the total pressure
is due to the sum of all the molecule–wall collisions.

Charles' Law

Recall that average kinetic energy is directly proportional to the absolute
temperature. Doubling the *absolute* temperature of a sample of gas doubles
the average kinetic energy of the gaseous molecules, and the increased force
of the collisions of molecules with the walls doubles the volume at constant
pressure. Similarly, halving the absolute temperature decreases kinetic energy to one-half its original value; at constant pressure, the volume decreases

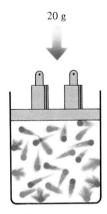

20 g

Figure 12-12
A molecular interpretation of Dalton's
Law. The molecules act independently,
so each gas exerts its own partial
pressure due to its molecular collisions
with the walls.

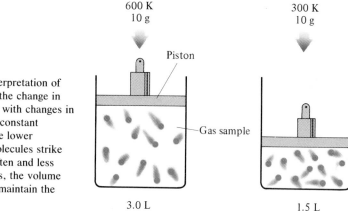

Figure 12-13
A molecular interpretation of Charles' Law—the change in volume of a gas with changes in temperature (at constant pressure). At the lower temperature, molecules strike the walls less often and less vigorously. Thus, the volume must be less to maintain the same pressure.

by one half because of reduced vigor of collision with the container walls (Figure 12-13).

Enrichment

Kinetic–Molecular Theory, the Ideal Gas Equation, and Molecular Speeds

In 1738, Daniel Bernoulli derived Boyle's Law from Newton's laws of motion applied to gas molecules. This derivation was the basis for an extensive mathematical development of the kinetic–molecular theory more than a century later by Clausius, Maxwell, Boltzmann, and others. Although we do not need to study the detailed mathematical presentation of this theory, we can gain some insight into its concepts from the reasoning behind Bernoulli's derivation. Here we present that reasoning based on proportionality arguments.

In the kinetic–molecular theory, pressure is viewed as the result of collisions of gas molecules with the walls of the container. As each molecule strikes a wall, it exerts a small impulse. The pressure is the total force thus exerted on the walls divided by the area of the walls. The total force on the wall (and thus the pressure) is proportional to two factors: (1) the impulse exerted by each collision and (2) the rate of collisions (number of collisions in a given time interval).

$$P \propto (\text{impulse per collision}) \times (\text{rate of collisions})$$

Let us represent the mass of an individual molecule by m and its speed by u. The heavier the molecule is (greater m) and the faster it is moving (greater u), the harder it pushes on the wall when it collides. The impulse due to each molecule is proportional to its *momentum, mu*.

> Recall that momentum is mass × speed.

$$\text{impulse per collision} \propto mu$$

The rate of collisions, in turn, is proportional to two factors. First, the rate of collision must be proportional to the molecular speed; the faster the molecules move, the more often they reach the wall to collide. Second, this collision rate must be proportional to the number of molecules per unit volume, N/V. The greater the number of molecules, N, in a given volume, the more molecules collide in a given time interval.

rate of collisions $\propto$ (molecules per unit volume) $\times$ (molecular speed)

or

$$\text{rate of collisions} \propto \left(\frac{N}{V}\right) \times (u)$$

We can introduce these proportionalities into the one describing pressure, to conclude that

$$P \propto (mu) \times u \times \frac{N}{V} \quad \text{or} \quad P \propto \frac{Nmu^2}{V} \quad \text{or} \quad PV \propto Nmu^2$$

At any instant not all molecules are moving at the same speed, u. We should reason in terms of the *average* behavior of the molecules, and express the quantity u^2 in average terms as $\overline{u^2}$, the **mean-square speed**.

$$PV \propto Nm\overline{u^2}$$

Not all molecules collide with the walls at right angles, so we must average (using calculus) over all the trajectories. This gives a proportionality constant of 1/3, and

$$PV = \tfrac{1}{3}Nm\overline{u^2}$$

This describes the quantity PV (pressure $\times$ volume) in terms of *molecular quantities*—number of molecules, molecular masses, and molecular speeds. The number of molecules, N, is given by the number of moles, n, times Avogadro's number, N_{Av}, or $N = nN_{Av}$. Making this substitution, we obtain

$$PV = \tfrac{1}{3}nN_{Av}m\overline{u^2}$$

The ideal gas equation already describes (pressure $\times$ volume) in terms of *measurable quantities*—number of moles and absolute temperature:

$$PV = nRT$$

So we see that the ideas of the kinetic–molecular theory lead to an equation of the same form as the macroscopic ideal gas equation. Thus, the molecular picture of the theory is consistent with the ideal gas equation and gives support to the theory. Equating the right-hand sides of these last two equations and cancelling n gives

$$\tfrac{1}{3}N_{Av}m\overline{u^2} = RT$$

This equation can also be written as

$$\tfrac{1}{3}N_{Av} \times (2 \times \tfrac{1}{2}m\overline{u^2}) = RT$$

From physics we know that the *kinetic energy* of a particle of mass m moving at speed u is $\tfrac{1}{2}mu^2$. So we can write

$$\tfrac{2}{3}N_{Av} \times (\text{avg } KE \text{ per molecule}) = RT$$

or

$$N_{Av} \times (\text{avg } KE \text{ per molecule}) = \tfrac{3}{2}RT$$

A bar over a quantity denotes an *average* of that quantity. $\overline{u^2}$ is the average of the squares of the various speeds. It is proportional to the square of the average speed, but the two quantities are not equal.

This equation shows that the absolute temperature is directly proportional to the average molecular kinetic energy, as postulated by the kinetic–molecular theory. Because there are N_{Av} molecules in a mole, the left-hand side of this equation is equal to the total kinetic energy of a mole of molecules.

total kinetic energy per mole of gas $= \frac{3}{2}RT$

With this interpretation, the total molecular kinetic energy of a mole of gas depends *only* on the temperature, and not on the mass of the molecules or the gas density.

We can also obtain some useful equations for molecular speeds from the above reasoning. Solving the equation

$$\frac{1}{3}N_{Av}m\overline{u^2} = RT$$

for root-mean-square speed, $u_{rms} = \sqrt{\overline{u^2}}$, we obtain

$$u_{rms} = \sqrt{\frac{3RT}{N_{Av}m}}$$

Recalling that m is the mass of a single molecule, $N_{Av}m$ is the mass of Avogadro's number of molecules, or one mole of substance; this is equal to the *molecular weight*, M, of the gas.

$$u_{rms} = \sqrt{\frac{3RT}{M}}$$

Example 12-21

Calculate the root-mean-square speed of H_2 molecules at 20°C in meters per second. Recall that $1\ J = 1\ \dfrac{kg \cdot m^2}{s^2}$.

Plan

We substitute the appropriate values into the equation relating u_{rms} to temperature and molecular weight. Remember that R must be expressed in the appropriate units:

$$R = 8.314\ \frac{J}{mol \cdot K} = 8.314\ \frac{kg \cdot m^2}{mol \cdot K \cdot s^2}$$

Solution

$$u_{rms} = \sqrt{\frac{3RT}{M}} = \sqrt{\frac{3 \times 8.314\ \dfrac{kg \cdot m^2}{mol \cdot K \cdot s^2} \times 293\ K}{2.016\ \dfrac{g}{mol} \times \dfrac{1\ kg}{1000\ g}}}$$

$$u_{rms} = \sqrt{3.62 \times 10^6\ m^2/s^2} = \boxed{1.90 \times 10^3\ m/s} \qquad \text{(about 4250 mi/hr)}$$

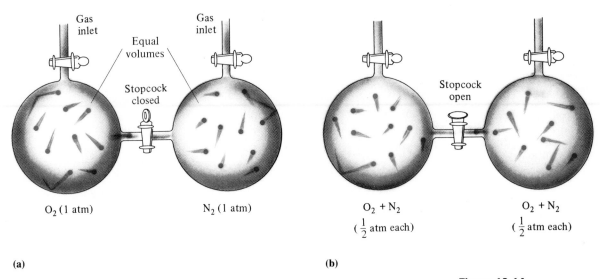

O₂ (1 atm) N₂ (1 atm)

O_2 (1 atm) N_2 (1 atm)

$O_2 + N_2$ ($\frac{1}{2}$ atm each) $O_2 + N_2$ ($\frac{1}{2}$ atm each)

(a) (b)

12-13 Graham's Law: Diffusion and Effusion of Gases

Because gas molecules are in constant, rapid, random motion, they diffuse quickly throughout any container (Figure 12-14). For example, if hydrogen sulfide (the essence of rotten eggs) is released in a large room, the odor can eventually be detected throughout the room. If a mixture of gases is placed in a container with porous walls, the molecules effuse through the walls. Because they move faster, lighter gas molecules always effuse through the tiny openings of porous materials faster than heavier molecules (Figure 12-15).

Figure 12-14
A representation of diffusion of gases. The space between the molecules allows for ease of mixing of one gas into another. Collisions of molecules with the walls of the container are responsible for the pressure of the gas.

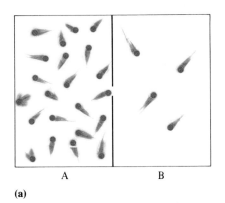

A B

(a)

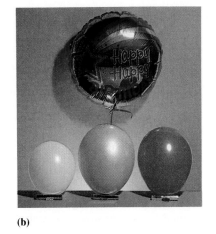

(b)

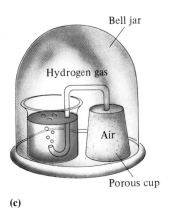

(c)

Figure 12-15
Effusion of gases. (a) A molecular interpretation of effusion. Molecules are in constant motion; occasionally they strike the opening and escape. (b) Rubber balloons were filled with the same volumes of He (yellow), N_2 (blue), and O_2 (red). Lighter molecules, such as He, effuse through the tiny pores of these balloons more rapidly than does N_2 or O_2. The silver party balloon is made of a metal-coated polymer with pores that are too small to allow rapid He effusion. (c) If a bell jar full of hydrogen is brought down over a porous cup full of air, rapidly moving hydrogen diffuses into the cup faster than the oxygen and nitrogen in the air can effuse out of the cup. This causes an increase in pressure in the cup sufficient to produce bubbles in the water in the beaker.

Scientists use the word "effusion" to describe the escape of a gas through a tiny hole, and the word "diffusion" to describe movement of a gas into a space or the mixing of one gas with another. The distinction made by chemists is somewhat sharper than that found in the dictionary.

In 1832 Thomas Graham showed that

> the rates of effusion of gases are inversely proportional to the square roots of their molecular weights (or densities).

This statement is known as **Graham's Law**, which can be represented as

$$\frac{\text{rate of effusion of gas A}}{\text{rate of effusion of gas B}} = \sqrt{\frac{\text{molecular weight of gas B}}{\text{molecular weight of gas A}}}$$

Or, to use simpler notation,

$$\frac{\text{rate}_A}{\text{rate}_B} = \sqrt{\frac{M_B}{M_A}}$$

where "rate" refers to rates of effusion and M refers to molecular weights

The following examples illustrate the use of Graham's Law for known and unknown gases.

Example 12-22

Calculate the ratio of the rate of effusion of methane to that of sulfur dioxide, that is, $\text{rate}_{CH_4}/\text{rate}_{SO_2}$.

Plan

We substitute the known molecular weights in the Graham's Law equation.

Solution

Substitution of molecular weights gives

$$\frac{\text{rate}_{CH_4}}{\text{rate}_{SO_2}} = \sqrt{\frac{M_{SO_2}}{M_{CH_4}}} = \sqrt{\frac{64 \text{ amu}}{16 \text{ amu}}} = \sqrt{\frac{4}{1}} = \frac{2}{1}$$

Because $\text{rate}_{CH_4}/\text{rate}_{SO_2} = 2/1$, we can write $\text{rate}_{CH_4} = 2 \times \text{rate}_{SO_2}$, which tells us

At a given temperature, the average speed of CH_4 molecules is twice that of SO_2 molecules.

that   the rate of effusion of CH_4 is twice that of SO_2.

EOC 82

Example 12-23

A 100-mL sample of hydrogen effuses through a porous container four times as rapidly as an unknown gas. Calculate the molecular weight of the unknown gas.

Plan

We use the Graham's Law equation to relate the given ratio of rates to the ratio of molecular weights. The molecular weight of H_2 is known, so the equation can be solved for the unknown molecular weight.

Solution

$$\frac{\text{rate}_{H_2}}{\text{rate}_{Unk}} = \sqrt{\frac{M_{Unk}}{M_{H_2}}}$$

Let's square both sides of the equation, solve the resulting expression for M_{Unk}, and substitute the known values into the expression

$$\left(\frac{\text{rate}_{H_2}}{\text{rate}_{Unk}}\right)^2 = \frac{M_{Unk}}{M_{H_2}} \quad\text{and}\quad M_{Unk} = \left(\frac{\text{rate}_{H_2}}{\text{rate}_{Unk}}\right)^2 M_{H_2}$$

H_2 effuses four times as rapidly as the unknown gas, so its rate must be four times as great:

$$\text{rate}_{H_2} = 4 \times \text{rate}_{Unk} \quad\text{or}\quad \frac{\text{rate}_{H_2}}{\text{rate}_{Unk}} = 4$$

Substituting,

$$M_{Unk} = (4)^2(2.0 \text{ amu}) = (16)(2.0 \text{ amu}) = \boxed{32 \text{ amu}}$$

The unknown gas may be oxygen.

EOC 79

Interestingly, Graham's Law can be derived from the kinetic–molecular theory, which says that *at a given temperature the average kinetic energies of molecules of different gases are equal.* If we represent the average kinetic energy of gas molecules A as $KE_A = \frac{1}{2}m_A\overline{u_A}^2$ and the average kinetic energy of gas molecules B as $KE_B = \frac{1}{2}m_B\overline{u_B}^2$, we can write

$$KE_A = KE_B \quad\text{or}\quad \tfrac{1}{2}m_A\overline{u_A}^2 = \tfrac{1}{2}m_B\overline{u_B}^2$$

Multiplying the latter expression by 2 gives $m_A\overline{u_A}^2 = m_B\overline{u_B}^2$. Rearranging this expression, we obtain

$$\frac{\overline{u_A}^2}{\overline{u_B}^2} = \frac{m_B}{m_A}$$

Taking the square root of both sides gives

$$\frac{u_{rms} \text{ for gas A}}{u_{rms} \text{ for gas B}} = \sqrt{\frac{m_B}{m_A}}$$

This tells us that the ratio of the speed of molecule A to that of molecule B is the square root of the ratio of the mass of molecule B to the mass of molecule A. Rates of effusion are directly proportional to speeds, and molecular weights are numerically equal to masses of molecules. Therefore, we may write

$$\frac{\text{rate}_A}{\text{rate}_B} = \sqrt{\frac{M_B}{M_A}}$$

which is the usual form of Graham's Law. We can also show that the preceding equation is equivalent to

$$\frac{\text{rate}_A}{\text{rate}_B} = \sqrt{\frac{D_B}{D_A}} \quad\text{where } D \text{ refers to gas densities at the same } P \text{ and } T.$$

This discussion involves the ratio of two *root-mean-square speeds*, u_{rms}. The root-mean-square speed is given by

$$u_{rms} = \sqrt{\overline{u^2}}$$

It is the square root of the mean-squared speeds of the individual molecules. This is slightly different from the average speed, u; however, the *ratio* of two average speeds is the same as the ratio of u_{rms} values. Graham's Law can also be derived by taking the ratio of the expressions for $u_{rms} = \sqrt{\frac{3RT}{M}}$ for the two gases A and B. Can you carry out that derivation?

Can *you* show this?

Although they are the most abundant elements in the universe, hydrogen and helium occur as gases only in trace amounts in our atmosphere. This is due to the high average molecular speeds resulting from their low molecular weights. At temperatures in our atmosphere, these molecules reach speeds exceeding the escape velocity required for them to break out of the earth's gravitational pull and diffuse into interplanetary space. Thus, most of the gaseous atmospheric hydrogen and helium that were probably present in large concentrations in the earth's early atmosphere have long since diffused away. The same is true of other small planets in our solar system, especially those with higher average temperatures than ours (Mercury and Venus). The Mariner 10 spacecraft in 1974 revealed measurable amounts of He in the atmosphere of Mercury; the source of this helium is unknown. Massive bodies such as stars (including our own sun) are mainly H and He.

12-14 Real Gases—Deviations from Ideality

Our discussions up to now have dealt with *ideal* behavior of gases. By this we mean that the identity of a gas does not affect how it behaves, and the same equations should work equally well for all gases. Under ordinary conditions most *real* gases do behave ideally; their P and V are properly predicted by the ideal gas laws, so they do obey the postulates of the kinetic–molecular theory. According to the kinetic–molecular model, (1) all but a negligible volume of a gas sample is empty space, and (2) the molecules of *ideal* gases do not attract each other because they are so far apart.

However, under some conditions most gases can have pressures and/or volumes that are *not* accurately predicted by the ideal gas laws. This tells us that they are not behaving entirely as postulated by the kinetic–molecular theory.

> Nonideal gas behavior (deviations from the predictions of the ideal gas laws) is most significant at *high pressures* and/or *low temperatures,* i.e., near the conditions under which the gas liquefies.

Johannes van der Waals studied deviations of real gases from ideal behavior. In 1867 he empirically adjusted the ideal gas equation

$$P_{\text{ideal}}V_{\text{ideal}} = nRT$$

to take into account two complicating factors:

1. According to the kinetic–molecular theory, the molecules are so small, compared to the total volume of the gas, that each molecule has virtually the entire *measured volume* of the container, V_{measured}, available in which to move (Figure 12-16a). But under high pressures, a gas is compressed so that the volume of the molecules themselves becomes a significant fraction of the total volume occupied by the gas. As a result, the *available volume*, $V_{\text{available}}$, for any molecule to move in is less than the *measured volume* by an amount that depends on the volume excluded by the presence of the other molecules (Figure 12-16b). To account for this, we subtract a correction factor, nb.

$$V_{\text{ideally available}} = V_{\text{measured}} - nb$$

The factor nb corrects for the volume occupied by the molecules themselves. Larger molecules have greater values of b, and the more molecules (higher n), the larger the correction. When the volume is large, the correction term becomes negligibly small.

2. The kinetic–molecular theory describes pressure as resulting from molecular collisions with the walls of the container, and assumes that there

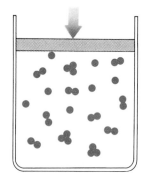

(a) Low temperature

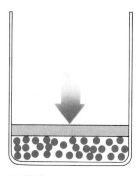

(b) High pressure

Figure 12-16
A molecular interpretation of deviations from ideal behavior. (a) A sample of gas at a low temperature. Each sphere represents a molecule. Because of their low kinetic energies, attractive forces between molecules can now cause molecules to stick together. (b) A sample of gas under high pressure. The molecules are quite close together. The free volume is now much smaller.

are no significant attractive forces between molecules. For any real gas, the molecules can attract one another, but at higher temperatures the intermolecular attraction is negligibly small due to the molecules' rapid motion and to the great distances between them. But when the temperature is quite low, the molecules move so slowly that even small attractive forces between molecules *do* become important. This perturbation becomes even more important when the molecules are very close together (at high pressure). As a result, the molecules deviate from their straight-line paths and take longer to reach the walls, so fewer collisions take place in a given time interval. Further, for a molecule about to collide with the wall, the attraction by its neighbors causes the collision to be less energetic than it would otherwise be. As a consequence, the pressure that the gas exerts, $P_{measured}$, is less than the pressure it would exert if attractions were truly negligible, $P_{ideally\ exerted}$. To correct for this, we subtract a correction factor, n^2a/V^2, from the ideal pressure:

$$P_{measured} = P_{ideally\ exerted} - \frac{n^2a}{V^2_{measured}} \qquad \text{or}$$

$$P_{ideally\ exerted} = P_{measured} + \frac{n^2a}{V^2_{measured}}$$

In this correction term, large values of a indicate strong attractive forces. When more molecules are present (greater n) and when the molecules are close together (smaller V^2 in the denominator), the correction term becomes larger. Again, when the volume is large, the correction term becomes negligibly small.

When we substitute these two expressions for corrections into the ideal gas equation, we obtain the equation

$$\left(P_{measured} + \frac{n^2a}{V^2_{measured}}\right)(V_{measured} - nb) = nRT$$

or

$$\left(P + \frac{n^2a}{V^2}\right)(V - nb) = nRT$$

This is the **van der Waals equation**. In this equation, P, V, T, and n represent the *measured* values of pressure, volume, temperature (expressed on the absolute scale), and number of moles, respectively, just as in the ideal gas equation. The quantities a and b are experimentally derived constants that differ for different gases (Table 12-5). We see that when a and b are both zero, the van der Waals equation reduces to the ideal gas equation.

We can understand the relative values of a and b in Table 12-5 in terms of molecular properties. Note that a for helium is very small. This is the case for all noble gases and many other nonpolar molecules, because only very weak attractive forces, called London forces, exist between them. **London forces** result from short-lived electrical dipoles produced by the attraction of one atom's nucleus for an adjacent atom's electrons. These forces exist for all molecules but are especially important for nonpolar molecules, which would never liquefy if the London forces did not exist. Polar molecules such as ammonia, NH_3, have permanent charge separations (dipoles) and therefore exhibit greater forces of attraction for each other. This

Table 12-5
van der Waals Constants

Gas	a (L^2 · atm/mol^2)	b (L/mol)
H_2	0.244	0.0266
He	0.034	0.0237
N_2	1.39	0.0391
NH_3	4.17	0.0371
CO_2	3.59	0.0427
CH_4	2.25	0.0428

The van der Waals equation, like the ideal gas equation, is known as an *equation of state*, i.e., an equation that describes a state of matter.

explains the high value of a for ammonia. London forces and permanent dipole forces of attraction are discussed in more detail in Chapter 13.

Larger molecules have greater values of b. For instance, H_2, a first-row diatomic molecule, has a greater b value than the first-row monatomic He. The b value for CO_2, which contains three second-row atoms, is greater than that for N_2, which contains only two second-row atoms.

The following example illustrates the deviation of methane, CH_4, from ideal gas behavior under high pressure.

Example 12-24

Calculate the pressure exerted by 1.00 mole of methane, CH_4, in a 500-mL vessel at 25.0°C assuming (a) ideal behavior and (b) nonideal behavior.

Plan

(a) Ideal gases obey the ideal gas equation. We can solve this equation for P.
(b) To describe methane as a nonideal gas, we use the van der Waals equation and solve for P.

Solution

(a) Using the ideal gas equation to describe ideal gas behavior,

$$PV = nRT$$

$$P = \frac{nRT}{V} = \frac{(1.00 \text{ mol}) \left(0.0821 \frac{L \cdot atm}{mol \cdot K} \right) (298 \text{ K})}{0.500 \text{ L}} = \boxed{48.9 \text{ atm}}$$

(b) Using the van der Waals equation to describe nonideal gas behavior,

$$\left(P + \frac{n^2 a}{V^2} \right) (V - nb) = nRT$$

For CH_4, $a = 2.25 \text{ L}^2 \cdot atm/mol^2$ and $b = 0.0428 \text{ L/mol}$ (Table 12-5).

$$\left[P + \frac{(1.00 \text{ mol})^2 (2.25 \text{ L}^2 \cdot atm/mol^2)}{(0.500 \text{ L})^2} \right] \left[0.500 \text{ L} - (1.00 \text{ mol}) \left(0.0428 \frac{L}{mol} \right) \right]$$

$$= (1.00 \text{ mol}) \left(0.0821 \frac{L \cdot atm}{mol \cdot K} \right) (298 \text{ K})$$

Combining terms and canceling units, we get

$$[P + 9.00 \text{ atm}][0.457 \text{ L}] = 24.5 \text{ L} \cdot atm$$

$$P + 9.00 \text{ atm} = 53.6 \text{ atm}$$

$$P = \boxed{44.6 \text{ atm}}$$

EOC 90, 91

The pressure is 4.3 atm (8.8%) less than that calculated from the ideal gas law. We see that a significant error would be introduced by assuming that methane behaved as an ideal gas under these conditions. Repeating the calculations of Example 12-24 with $V = 10.0$ L gives pressures (ideal and nonideal, respectively) of 2.45 and 2.44 atm, a difference of only 0.4%.

Many other equations have been developed to describe the behavior of real gases. Each of these contains quantities that must be empirically derived for each gas. One equation that is often used in engineering is the *virial equation*:

$$PV = nRT \left(1 + \frac{Bn}{V} + \frac{Cn^2}{V^2} + \frac{Dn^3}{V^3} + \cdots \right)$$

In this equation, the constants B, C, and so on depend not only on the identity of the gas but also on temperature. This equation, unlike most other empirical gas equations, can be adjusted for mixtures of gases.

Stoichiometry in Reactions Involving Gases

12-15 Gay-Lussac's Law of Combining Volumes

Gases react in simple, definite proportions by volume. For example, *one* volume of hydrogen always combines (reacts) with *one* volume of chlorine to form *two* volumes of hydrogen chloride:

$$H_2(g) + Cl_2(g) \longrightarrow 2HCl(g)$$

| 1 volume | volume | 2 volumes |

> It is understood that all volumes are measured at the same temperature and pressure. Volumes may be expressed in any units as long as the same units are used for all.

Gay-Lussac (1788–1850) summarized experimental observations on combining volumes of gases in **Gay-Lussac's Law** (or the **Law of Combining Volumes**):

> At constant temperature and pressure, the volumes of reacting gases can be expressed as a ratio of small whole numbers.

This ratio is equal to the ratio of the coefficients in the balanced equation for the reaction. Clearly, the law applies only to *gaseous* substances at the same temperature and pressure. No generalizations can be made about the volume of solids and liquids as they undergo chemical reactions. Consider the following examples:

1. One volume of nitrogen reacts with three volumes of hydrogen to form two volumes of ammonia:

$$N_2(g) + 3H_2(g) \longrightarrow 2NH_3(g)$$

1 volume   3 volumes   2 volumes

2. Sulfur (a solid) reacts with one volume of oxygen to form one volume of sulfur dioxide:

$$S(s) + O_2(g) \longrightarrow SO_2(g)$$

1 volume   1 volume

3. Four volumes of ammonia burn in five volumes of oxygen to produce four volumes of nitric oxide and six volumes of steam:

$$4NH_3(g) + 5O_2(g) \longrightarrow 4NO(g) + 6H_2O(g)$$

4 volumes   5 volumes   4 volumes   6 volumes

Avogadro's Law (Section 12-8) provides an explanation of Gay-Lussac's Law. Consider the reaction of fluorine and hydrogen to produce hydrogen fluoride. Experiments show that

NH_3 gas (*left*) and HCl gas (*right*) escape from concentrated aqueous solutions. The white smoke (solid NH_4Cl) shows where the gases mix and react.

$$NH_3(g) + HCl(g) \longrightarrow NH_4Cl(s)$$

See Exercise 80 at the end of the chapter.

$$S(s) + O_2(g) \rightarrow SO_2(g)$$

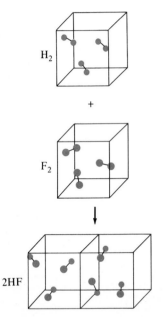

Figure 12-17
A representation of the reaction of fluorine and hydrogen to form hydrogen fluoride.

$$H_2(g) + F_2(g) \longrightarrow 2HF(g)$$

We can use any units we choose to describe the volumes of H_2, O_2, and H_2O, as long as we use the same units for all.

1 volume fluorine + 1 volume hydrogen $\longrightarrow$ 2 volumes hydrogen fluoride

Figure 12-17 illustrates the volumetric relationships in the combination of fluorine and hydrogen to form hydrogen fluoride. Avogadro's Law says that equal volumes of fluorine, hydrogen, and hydrogen fluoride contain equal numbers of molecules.

The deductions based on these experimentally observed facts follow.

1. Two molecules of hydrogen fluoride are formed from one molecule of fluorine and one molecule of hydrogen.
2. Each hydrogen fluoride molecule must contain *at least* one fluorine atom and one hydrogen atom.
3. Therefore, two hydrogen fluoride molecules contain *at least* two fluorine atoms and two hydrogen atoms, which must have been present in one fluorine molecule and one hydrogen molecule.

$$1 \text{ volume of } F_2 + 1 \text{ volume of } H_2 \longrightarrow 2 \text{ volumes of HF}$$

$$1 \text{ } F_2 \text{ molecule} + 1 \text{ } H_2 \text{ molecule} \longrightarrow 2 \text{ HF molecules}$$

There are no known reactions in which one molecule of fluorine or one molecule of hydrogen contains enough atoms to form more than two molecules of product. Therefore, we may safely assume that each fluorine molecule and each hydrogen molecule contains *exactly* two atoms, i.e., that the formulas for the molecules are F_2 and H_2.

Similar experiments and reasoning showed that some other common elements are diatomic: oxygen, O_2; nitrogen, N_2; chlorine, Cl_2; bromine, Br_2; and iodine, I_2.

Other reactions involving gases can be explained by Avogadro's and Gay-Lussac's laws. In the first decade of the 19th century, Dalton considered, but then quite properly rejected, the notion that equal volumes of gases contain the same number of *atoms*. The idea of the existence of diatomic (and more complex) molecules had not yet evolved, and it didn't occur to Dalton. Gay-Lussac's Law predicts the volumes of gases involved in reactions.

Example 12-25
What volume of hydrogen would combine with 42 liters of oxygen to form steam at 750°C and atmospheric pressure?

Plan
By Gay-Lussac's Law, the ratio of the volumes of gaseous reactants and products is the same as the ratio of molecules. This ratio is given by the coefficients in the balanced equation for the reaction.

Solution
The balanced equation for the reaction

$$2H_2(g) + O_2(g) \longrightarrow 2H_2O(g)$$

tells us that two hydrogen molecules combine with one oxygen molecule to form two water molecules. Or, applying Gay-Lussac's Law, we can write

$$2H_2 + O_2 \longrightarrow 2H_2O$$

2 volumes   1 volume   2 volumes

and we have

$$? \text{ L H}_2 = 42 \text{ L O}_2 \times \frac{2 \text{ volumes H}_2}{1 \text{ volume O}_2} = \boxed{84 \text{ L H}_2}$$

EOC 93

12-16 Mass–Volume Relationships in Reactions Involving Gases

Many chemical reactions produce gases. For instance, the combustion of a hydrocarbon in excess oxygen at high temperatures produces both carbon dioxide and water as gases, as illustrated for octane:

$$2C_8H_{18}(g) + 25O_2(g) \longrightarrow 16CO_2(g) + 18H_2O(g)$$

The N_2 gas produced by the very rapid decomposition of sodium azide, $NaN_3(s)$, inflates air bags used as safety devices in automobiles.

We know that one mole of gas, measured at STP, occupies 22.4 liters; we can use the ideal gas equation to find the volume of a mole of gas at any other conditions. This information can be utilized in stoichiometry calculations (Section 3-2).

Small amounts of oxygen can be produced in the laboratory by heating solid potassium chlorate, $KClO_3$, in the presence of a catalyst, manganese dioxide, MnO_2. Solid potassium chloride, KCl, is also produced (CAUTION: Heating $KClO_3$ can be dangerous.)

$$2KClO_3(s) \xrightarrow[\Delta]{MnO_2} 2KCl(s) + 3O_2(g)$$

$$\begin{array}{ccc} 2 \text{ mol} & 2 \text{ mol} & 3 \text{ mol} \\ 2(122.6 \text{ g}) & 2(74.6 \text{ g}) & 3(22.4 \text{ L}_{STP}) \end{array}$$

Unit factors can be constructed using any two of these quantities.

The nitrogen gas formed in the rapid reaction

$$2NaN_3(s) \longrightarrow 2Na(s) + 3N_2(g)$$

fills an automobile air bag during a collision. The air bag fills within 1/20th of a second after a front collision.

Example 12-26
What volume of O_2 (STP) can be produced by heating 112 grams of $KClO_3$?

Plan
The preceding equation shows that two moles of $KClO_3$ produce three moles of O_2. We construct appropriate unit factors from the balanced equation and the standard molar volume of a gas to solve the problem.

Solution

$$? \text{ L}_{STP} \text{ O}_2 = 112 \text{ g KClO}_3 \times \frac{1 \text{ mol KClO}_3}{122.6 \text{ g KClO}_3} \times \frac{3 \text{ mol O}_2}{2 \text{ mol KClO}_3} \times \frac{22.4 \text{ L}_{STP} \text{ O}_2}{1 \text{ mol O}_2}$$

$$= \boxed{30.7 \text{ L}_{STP} \text{ O}_2}$$

Alternatively, we could have solved this problem using a single unit factor, $3(22.4 \text{ L}_{STP}) \text{ O}_2/2(122.6 \text{ g}) \text{ KClO}_3$, read directly from the equation:

$$? \text{ L}_{STP} \text{ O}_2 = 112 \text{ g KClO}_3 \times \frac{3(22.4 \text{ L}_{STP}) \text{ O}_2}{2(122.6 \text{ g}) \text{ KClO}_3} = \boxed{30.7 \text{ L}_{STP} \text{ O}_2}$$

Production of a gas by a reaction.

$$2NaOH(g) + 2Al(s) + 6H_2O(\ell) \longrightarrow 2Na[Al(OH)_4](aq) + 3H_2(g)$$

This reaction is used in some solid drain cleaners.

This calculation shows that the thermal decomposition of 112 grams of $KClO_3$ produces 30.7 liters of oxygen measured at standard conditions.

EOC 98

Example 12-27

A 1.80-gram mixture of potassium chlorate, $KClO_3$, and potassium chloride, KCl, was heated until all of the $KClO_3$ had decomposed. The liberated oxygen, after drying, occupied 405 mL at 25°C when the barometric pressure was 745 torr. (a) How many moles of O_2 were produced? (b) What percentage of the mixture was $KClO_3$? KCl?

Plan

(a) The number of moles of O_2 produced can be calculated from the ideal gas equation. (b) Then we use the balanced chemical equation to relate the known number of moles of O_2 formed and the mass of $KClO_3$ that decomposed to produce it.

Solution

(a) $V = 405 \text{ mL} = 0.405 \text{ L}; P = 745 \text{ torr} \times \dfrac{1 \text{ atm}}{760 \text{ torr}} = 0.980 \text{ atm}$

   $T = 25°C + 273 = 298 \text{ K}$

Solving the ideal gas equation for n and evaluating gives

$$n = \frac{PV}{RT} = \frac{(0.980 \text{ atm})(0.405 \text{ L})}{\left(0.0821 \dfrac{\text{L} \cdot \text{atm}}{\text{mol} \cdot \text{K}} (298 \text{ K})\right)} = \boxed{0.0162 \text{ mol } O_2}$$

(b) $? \text{ g } KClO_3 = 0.0162 \text{ mol } O_2 \times \dfrac{2 \text{ mol } KClO_3}{3 \text{ mol } O_2} \times \dfrac{122.6 \text{ g } KClO_3}{1 \text{ mol } KClO_3}$

   $= 1.32 \text{ g } KClO_3$

The sample contained 1.32 grams of $KClO_3$. The percent of $KClO_3$ in the sample is

$$\% \ KClO_3 = \frac{\text{g } KClO_3}{\text{g sample}} \times 100\% = \frac{1.32 \text{ g}}{1.80 \text{ g}} \times 100\% = \boxed{73.3\% \ KClO_3}$$

The sample contains 73.3% $KClO_3$ and $(100.0 - 73.3)\% = \boxed{26.7\% \text{ KCl.}}$

EOC 104, 105

Key Terms

Absolute zero   The zero point on the absolute temperature scale; −273.15°C or 0 K; theoretically, the temperature at which molecular motion ceases.

Atmosphere   A unit of pressure; the pressure that will support a column of mercury 760 mm high at 0°C.

Avogadro's Law   At the same temperature and pressure, equal volumes of all gases contain the same number of molecules.

Barometer   A device for measuring atmospheric pressure. See Figures 12-1 and 12-2.

Boyle's Law   At constant temperature, the volume occupied by a definite mass of a gas is inversely proportional to the applied pressure.

Charles' Law   At constant pressure, the volume occupied by a definite mass of a gas is directly proportional to its absolute temperature.

Condensed states   The solid and liquid states.

Dalton's Law   See *Law of Partial Pressures*.

Dumas method   A method used to determine the molecular weights of volatile liquids. See Figure 12-5.

Equation of state   An equation that describes the behavior of matter in a given state; e.g., the van der Waals equation describes the behavior of the gaseous state.

Fluids   Substances that flow freely; gases and liquids.

Gay-Lussac's Law   See *Law of Combining Volumes*.

Graham's Law   The rates of effusion of gases are inversely proportional to the square roots of their molecular weights (or densities).

Ideal gas   A hypothetical gas that obeys exactly all postulates of the kinetic–molecular theory.

Ideal Gas Law   The product of the pressure and volume of an ideal gas is directly proportional to the number of moles of the gas and the absolute temperature.

Kinetic–molecular theory   A theory, originally published by Clausius in 1857, that attempts to explain macroscopic observations on gases in microscopic or molecular terms.

Law of Combining Volumes   At constant temperature and pressure, the volumes of reacting gases (and any gaseous products) can be expressed as ratios of small whole numbers; also known as Gay-Lussac's Law.

Law of Partial Pressures   The total pressure exerted by a mixture of gases is the sum of the partial pressures of the individual gases; also called Dalton's Law.

Manometer   A two-armed barometer. See Figure 12-1.

Mole fraction   The number of moles of a component of a mixture divided by the total number of moles in the mixture.

Partial pressure   The pressure exerted by one gas in a mixture of gases.

Pressure   Force per unit area.

Real gases   Gases that deviate from ideal gas behavior.

Root-mean-square speed, u_{rms}   The square root of the mean-squared speed, $\sqrt{\overline{u^2}}$. This is equal to $\sqrt{\dfrac{3RT}{M}}$ for an ideal gas. The root-mean-square speed is slightly different from the average speed, but the two quantities are proportional.

Standard conditions (STP or SC)   Standard temperature, 0°C, and standard pressure, one atmosphere, are standard conditions for gases.

Standard molar volume   The volume occupied by one mole of an ideal gas under standard conditions; 22.4 liters.

Universal gas constant R, the proportionality constant in the ideal gas equation, $PV = nRT$.

van der Waals equation   An equation of state that extends the ideal gas law to real gases by inclusion of two empirically determined parameters, which are different for different gases.

Vapor   A gas formed by boiling or evaporation of a liquid or sublimation of a solid; a term commonly used when some of the liquid or solid remains in contact with the gas.

Vapor pressure   The pressure exerted by a vapor in equilibrium with its mother liquid.

Exercises

You may assume *ideal gas behavior* unless otherwise indicated.

Basic Ideas

1. What are the three states of matter? Compare and contrast them.

2. State whether each property is characteristic of all gases, some gases, or no gas: (a) transparent to light, (b) colorless, (c) unable to pass through filter paper, (d) more difficult to compress than liquid water, (e) odorless, (f) settles on standing.

3. Suppose you were asked to supply a particular mass of a specified gas in a container of fixed volume at a specified pressure and temperature. Is it likely that you could fulfill the request? Explain.

4. State whether each of the following samples of matter is a gas. If the information is insufficient for you to decide, write "insufficient information."

    (a) A material is in a steel tank at 100 atm pressure. When the tank is opened to the atmosphere, the material immediately expands, increasing its volume many-fold.

    (b) A material, on being emitted from an industrial smokestack, rises about 10 m into the air. Viewed against a clear sky, it has a white appearance.

    (c) 1.0 mL of material weighs 8.2 g.

    (d) When a material is released from a point 30 ft below the level of a lake at sea level (equivalent in pressure to about 76 cm of mercury), it rises rapidly to the surface, at the same time doubling its volume.

    (e) A material is transparent and pale green in color.

    (f) One cubic meter of a material contains as many molecules as 1 m³ of air at the same temperature and pressure.

Pressure

5. (a) What is pressure? (b) Describe the mercury barometer. How does it work?

6. What is a manometer? How does it work?

7. Express a pressure of 650 torr in the following units: (a) cm Hg, (b) atm, (c) Pa, (d) kPa.

8. A typical laboratory atmospheric pressure reading is 745 torr. Convert this value to (a) psi, (b) mm Hg, (c) inches Hg, (d) Pa, (e) atm, and (f) ft H_2O.

9. Complete the following table.

	atm	torr	Pa	kPa
Standard atmosphere	1			
Partial pressure of nitrogen in the atmosphere		593		
A tank of compressed hydrogen			1.21×10^5	
Atmospheric pressure at the summit of Mt. Everest				33.7

*10. Consider a container of mercury open to the atmosphere. Calculate the total pressure, in torr and atmospheres, within the mercury at depths of (a) 100 mm and (b) 5.04 cm. The barometric pressure is 758 torr.

*11. The densities of mercury and corn oil are 13.5 g/mL and 0.92 g/mL, respectively. If corn oil were used in a barometer, what would be the height of the column, in meters, at standard atmospheric pressure? (The vapor pressure of the oil is negligible.)

Boyle's Law: The Pressure–Volume Relationship

12. (a) On what kinds of observations (measurements) is Boyle's Law based? State the law.
(b) Use the statement of Boyle's Law to derive a simple mathematical expression for Boyle's Law.

13. Could the words "a fixed number of moles" be substituted for "a definite mass" in the statement of Boyle's Law? Why?

14. What pressure is needed to confine an ideal gas to 75 L after it has expanded from 25 L and 1.00 atm at constant temperature?

15. A sample of krypton gas occupies 40.0 mL at 0.400 atm. If the temperature remained constant, what volume would the krypton occupy at (a) 4.00 atm, (b) 0.00400 atm, (c) 765 torr, (d) 4.00 torr, and (e) 3.5×10^{-2} torr?

16. A 50-L sample of gas of the upper atmosphere at a pressure of 6.5 torr is compressed into a 150-mL container at the same temperature. (a) What is the new pressure, in atmospheres? (b) To what volume would the original sample have had to be compressed to reach a pressure of 10.0 atm?

17. Assume that, for some set of conditions, the value of k in $PV = k$ is 12.
(a) Plot the graph of P (x axis) versus V (y axis) for the values $V = 1, 2, 3, 4, 6,$ and 12. What is the shape of the curve?
(b) Plot the graph of $1/P$ (x axis) versus V (y axis) for the same values of V. What plot is obtained?

*18. A cylinder containing 44 L of helium gas at a pressure of 170 atm is to be used to fill toy balloons to a pressure of 1.1 atm. Each inflated balloon has a volume of 2.0 L. What is the maximum number of balloons that can be inflated? (Remember that 44 L of helium at 1.1 atm will remain in the "exhausted" cylinder.)

The Absolute Temperature Scale

19. (a) Can an absolute temperature scale based on Fahrenheit rather than Celsius degrees be evolved? Why?
(b) Can an absolute temperature scale that is based on a "degree" twice as large as a Celsius degree be developed? Why?

20. (a) What does "absolute temperature scale" mean?
(b) Describe the experiments that led to the evolution of the absolute temperature scale. What is the relationship between the Celsius and Kelvin temperature scales?
(c) What does "absolute zero" mean?

21. Complete the table by making the required temperature conversions. Pay attention to significant figures.

	Temperature	
	K	°C
Normal boiling point of water		100
Standard for thermodynamic data	298.15	
Dry ice becomes a gas at atmospheric pressure		−78.5
The center of the sun (more or less)	1.5×10^7	

Charles' Law: The Volume–Temperature Relationship

22. (a) Why is a plot of volume versus temperature at constant pressure a straight line (Figure 12-4)?
(b) On what kind of observations (measurements) is Charles' Law based? State the law.

23. A gas occupies a volume of 30.3 L at 17.0°C. If the gas temperature rises to 34.0°C at constant pressure, (a) would you expect the volume to double to 60.6 L? Explain. Calculate the new volume (b) at 34.0°C, (c) at 400 K, and (d) at −34.0°C.

24. A sample of neon occupies 75.0 mL at 15°C. To what temperature, in °C, must it be cooled, at constant pressure, to reduce its volume to 25.0 mL?

25. Which of the following statements are true? Which are false? Why is each true or false? *Assume constant pressure* in each case.
(a) If a sample of gas is heated from 100°C to 200°C, the volume will double.
(b) If a sample of gas is heated from 0°C to 273°C, the volume will double.
(c) If a sample of gas is cooled from 1273°C to 500°C, the volume will decrease by a factor of two.
(d) If a sample of gas is cooled from 1000°C to 200°C, the volume will decrease by a factor of five.
(e) If a sample of gas is heated from 473°C to 1219°C, the volume will increase by a factor of two.

*26. The device shown below is a **gas thermometer**.
(a) At the ice point, the gas volume is 1.400 L. What would be the new volume if the gas temperature were raised from the ice point to 6.0°C?
(b) Assume the cross-sectional area of the graduated arm is 1.0 cm². What would be the difference in height if the gas temperature went up from 0°C to 6.0°C?
(c) What modifications could be made to increase the sensitivity of the thermometer?

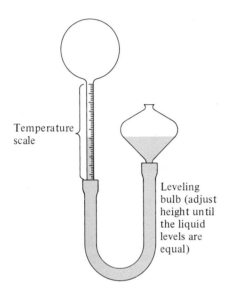

Temperature scale

Leveling bulb (adjust height until the liquid levels are equal)

27. What volume of gas should we use at 25°C and 1 atm if we wish to have the sample occupy 145 L at 225°C and 1 atm?

28. Calculate the volume of an ideal gas at the temperatures of Dry Ice (−78.5°C), liquid N_2 (−195.8°C), and liquid He (−268.9°C) if it occupies 10.00 L at 25.0°C. Assume constant pressure. Plot your results and extrapolate to zero volume. At what temperature would zero volume be reached?

The Combined Gas Law

29. Classify the relationship between the variables (a) P and V, (b) V and T, and (c) P and T as either (i) directly proportional or (ii) inversely proportional.

30. Prepare sketches of plots of (a) P vs. V, (b) P vs. $1/V$, (c) V vs. T, and (d) P vs. T for an ideal gas.

31. A sample of gas occupies 500 mL at STP. Under what pressure would this sample occupy 250 mL if the temperature were increased to 819°C?

32. A sample of hydrogen occupies 375 mL at STP. If the temperature were increased to 819°C, what final pressure would be necessary to keep the volume constant at 375 mL?

33. A 326-mL sample of a gas exerts a pressure of 1.67 atm at 12°C. What volume would it occupy at 100°C and 1.00 atm?

34. A 350-ml sample of oxygen exerts a pressure of 830 torr at 22°C. At what temperature would it exert a pressure of 600 torr in a volume of 500 mL?

35. What temperature would be necessary to double the volume of an ideal gas initially at STP if the pressure decreased by 25%?

Standard Conditions, Standard Molar Volume, and Gas Densities

36. (a) What is Avogadro's Law? What does it mean?
(b) What does "standard molar volume" mean?

37. How many molecules of an ideal gas are contained in a 1.00-L flask at STP?

38. The limit of sensitivity for the analysis of carbon monoxide, CO, in air is 1 ppb (ppb = parts per billion) by volume. What is the smallest number of CO molecules that can be detected in 10 L of air at STP?

39. Sodium vapor has been detected recently as a major component of the thin atmosphere of Mercury using a ground-based telescope and a spectrometer. Its concentration is estimated to be about 10^5 atoms per cm³.
(a) Express this in moles per liter.
(b) The maximum temperature of the atmosphere was measured by Mariner 10 to be about 970°C. What is the approximate partial pressure of sodium vapor at that temperature?

40. Ethylene dibromide (EDB; formerly used as a fumigant for fruits and grains, but now banned because of its potential carcinogenicity) is a liquid that boils at 109°C. Its molecular weight is 188 g/mol. Calculate the density of its vapor at 200°C and 0.50 atm.

41. A student's lab instructor asks her to calculate the number of moles of gas contained in a 250-mL bulb in the laboratory. The student has just determined that the pressure and temperature of the gas are 672 torr and 11.4°C. Her calculations follow.

$$\underline{?}\ mol = 0.250\ L \times \frac{1\ mol}{22.4\ L} = 0.0112\ mol$$

Is she correct? Why?

42. A laboratory technician forgot what the color coding on some commercial cylinders of gas meant, but remembered that each of two specific tanks contained one of the following gases: He, Ne, Ar, or Kr. Density measurements at STP were made on samples of the gases from the two cylinders and were found to be 0.178 g/L and 0.900 g/L. Which of these gases was present in each tank?

*43. A 503-mL flask contains 0.0179 mol of an ideal gas at a given set of temperature and pressure conditions. Another flask contains 0.0256 mol of the gas at the same temperature and pressure conditions. What is the volume of the second flask?

The Ideal Gas Equation

44. (a) What is an ideal gas?
(b) What is the ideal gas equation?
(c) Outline the logic used to obtain the ideal gas equation.
(d) What is R? How is it obtained?

45. Calculate R in L · atm/mol · K, in kPa · dm³/mol · K, in J/mol · K, and in kJ/mol · K.

46. Calculate the pressure needed to contain 5.29 mol of an ideal gas at 45°C in a volume of 3.45 L.

47. (a) A chemist is preparing to carry out a reaction at high pressure that requires 45 mol of hydrogen gas. The chemist pumps the hydrogen into a 10.5-L rigid steel vessel at 25°C. To what pressure (in atmospheres) must the hydrogen be compressed? (b) What would be the density of the high-pressure hydrogen?

48. A 6.00-mol sample of helium is confined in a 4.5-L vessel. (a) What is the temperature if the pressure is 3.0 atm? (b) What is the density of the sample?

49. How many gaseous molecules are in a 1.00-L container if the pressure is 1.60×10^{-9} torr and the temperature is 1475 K?

*50. A barge containing 640 tons of liquid chlorine was involved in an accident. What volume would this amount of chlorine occupy if it were all converted to a gas at 740 torr and 15°C? Assume that the chlorine is confined to a width of 0.5 mile and an average depth of 50 ft. How long would this chlorine "cloud" be?

Molecular Weights and Formulas for Gaseous Compounds

51. A sample of a liquid with a boiling point of 56.5°C was vaporized in a Dumas apparatus (Figure 12-6) as the 300-mL bulb was immersed into boiling water. The

barometric pressure was 733 torr. The boiling point of water at this pressure is 99°C. Filled only with air, the bulb had a mass of 156.872 g; filled with the vapor, it had a mass of 157.421 g. Determine the molecular weight of the liquid.

*52. A student was given a container of ethane, C_2H_6, that had been closed at STP. By making appropriate measurements, he found that the mass of the sample of ethane was 0.218 g and the volume of the container was 165 mL. Use his data to calculate the molecular weight of ethane. What percent error is obtained? Suggest some possible sources of the error.

53. Analysis of a volatile liquid shows that it contains 37.23% carbon, 7.81% hydrogen, and 54.96% chlorine by mass. At 150°C and 1.00 atm, 500 mL of the vapor has a mass of 0.922 g.
(a) What is the molecular weight of the compound?
(b) What is its molecular formula?

54. What is the molecular weight of an ideal gas if 0.52 g of the gas occupies 610 mL at 385 torr and 45°C?

55. The Dumas method was used to determine the molecular weight of a liquid. The vapor occupied a 103-mL volume at 99°C and 721 torr. The condensed vapor has a mass of 0.800 g. Calculate the molecular weight of the liquid.

*56. A highly volatile liquid was allowed to vaporize completely into a 250-mL flask immersed in boiling water. From the following data, calculate the molecular weight of the liquid. Mass of empty flask = 65.347 g, mass of flask filled with water at room temperature = 327.4 g, mass of flask and condensed liquid = 65.739 g, atmospheric pressure = 743.3 torr, temperature of boiling water = 99.8°C, density of water at room temperature = 0.997 g/mL.

57. A pure gas contains 81.71% carbon and 18.29% hydrogen by mass. Its density is 1.97 g/L at STP. What is its molecular formula?

Gas Mixtures and Dalton's Law

58. (a) What are partial pressures of gases? (b) State Dalton's Law. Express it symbolically.

59. A sample of oxygen of mass 24.0 g is confined in a vessel at 0°C and 1000 torr. If 6.00 g of hydrogen is now pumped into the vessel at constant temperature, what will be the final pressure in the vessel (assuming only mixing with no reaction)?

60. A gaseous mixture contains 4.18 g of chloroform, $CHCl_3$, and 1.95 g of ethane, C_2H_6. What pressure is exerted by the mixture inside a 50.0-mL metal bomb at 375°C? What pressure is contributed by the $CHCl_3$?

61. A cyclopropane–oxygen mixture can be used as an anesthetic. If the partial pressures of cyclopropane and oxygen are 150 torr and 550 torr, respectively, what is the ratio of the number of moles of cyclopropane to the number of moles of oxygen in this mixture?

62. What is the mole fraction of each gas in a mixture that contains 0.267 atm of He, 0.369 atm of Ar, and 0.394 atm of Xe?

*63. Assume that unpolluted air has the composition shown in Table 12-2. (a) Calculate the number of molecules of N_2, of O_2, and of Ar in 1.00 L of air at 21°C and 1.00 atm. (b) Calculate the mole fractions of N_2, O_2, and Ar in the air.

64. Individual samples of O_2, N_2, and He are present in three 3.50-L vessels. Each exerts a pressure of 1.50 atm.
(a) If all three gases are forced into the same 1.00-L container with no change in temperature, what will be the resulting pressure?
(b) What is the partial pressure of O_2 in the mixture?
(c) What are the partial pressures of N_2 and He?

65. A sample of hydrogen, collected over water at 20°C and 765 torr, occupies 28.8 mL. What volume would the dry gas occupy at STP?

66. A sample of dry nitrogen occupies 331 mL at STP. What would its volume be if it were collected over water at 26°C and 740 torr?

*67. A study of climbers who reached the summit of Mt. Everest without supplemental oxygen revealed that the partial pressures of oxygen and CO_2 in their lungs were 35 torr and 7.5 torr, respectively. The barometric pressure at the summit was 253 torr. Assume that the lung gases are saturated with moisture at a body temperature of 37°C. Calculate the partial pressure of inert gas (mostly nitrogen) in the climbers' lungs.

68. A 5.00-L flask containing He at 5.00 atm is connected to a 4.00-L flask containing N_2 at 4.00 atm and the gases are allowed to mix.
(a) Find the partial pressures of each gas after they are allowed to mix.
(b) Find the total pressure of the mixture.
(c) What is the mole fraction of helium?

The Kinetic–Molecular Theory and Molecular Speeds

69. Outline the kinetic–molecular theory.

70. How do average speeds of gaseous molecules vary with temperature?

71. How does the kinetic–molecular theory explain (a) Boyle's Law? (b) Dalton's Law? (c) Charles' Law?

72. SiH_4 molecules are heavier than CH_4 molecules; yet, according to kinetic–molecular theory, their average kinetic energies at the same temperature are equal. How can this be?

*73. At 25°C, Cl_2 molecules have some rms speed (which we need not calculate). At what temperature would the rms speed of F_2 molecules be the same?

*74. Calculate the ratio of the rms speed of N_2 molecules at 100°C to the rms speed of the same molecules at 0°C.

Effusion and Diffusion of Gases; Graham's Law

75. State Graham's Law. What does it mean?

*76. Show that the following equations are equivalent for ideal gases, A and B, under the same conditions:

$$\frac{R_A}{R_B} = \sqrt{\frac{M_B}{M_A}} \qquad \frac{R_A}{R_B} = \sqrt{\frac{D_B}{D_A}}$$

77. How much faster would neon effuse than argon?

78. (a) Explain how the hydrogen bubbler (Figure 12-15c) works. (b) Could gaseous sulfur hexafluoride, SF_6, be used in this bubbler? Why?

79. A sample of unknown gas flows through the wall of a porous cup in 39.9 min. An equal volume of hydrogen, measured at the same temperature and pressure, flows through in 9.75 min. What is the molecular weight of the unknown gas?

80. Gaseous molecules of NH_3 and HCl escape from aqueous ammonia and hydrochloric acid solutions. When they diffuse together, a white cloud (smoke) of solid NH_4Cl is formed. Suppose we put a cotton plug saturated with HCl solution into one end of a 1.00-m tube and simultaneously insert a plug saturated with aqueous NH_3 into the other end. How many centimeters from the HCl end of the tube will the white smoke first form? See figure below.

81. The uranium used as a fuel in nuclear reactors is fissionable ^{235}U (Sections 30-13 and 30-15). It must first be separated from the much more abundant nonfissionable ^{238}U. This is accomplished by separating $^{235}UF_6$ from $^{238}UF_6$ by gaseous effusion. This "enrichment" process involves approximately 2000 passes of gaseous UF_6, which consists of both components, through porous barriers at higher temperatures. Calculate the ratio of the rate of effusion of $^{235}UF_6$ to $^{238}UF_6$. There

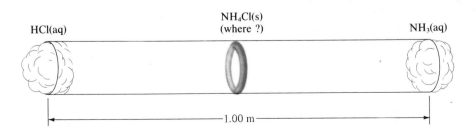

HCl(aq)   NH₄Cl(s) (where ?)   NH₃(aq)

|← 1.00 m →|

is only one naturally occurring form of fluorine, ^{19}F. Isotopic masses are ^{235}U, 235.0439 amu; ^{238}U, 238.0508 amu; and ^{19}F, 18.99840 amu.

82. What would be the relative rates of effusion of gaseous H_2, HD, and D_2? (D is a chemical symbol used to represent deuterium, an isotope of hydrogen that has an atomic mass of 2.0140 amu as compared to 1.0078 amu for H.)

Real Gases and Deviations from Ideality

83. (a) How do "real" and "ideal" gases differ?
    (b) Under what kinds of conditions are deviations from ideality most important? Why?

84. Which of the following gases would be expected to behave most nearly ideally under the same conditions? H_2, F_2, HF. Which one would deviate the most from ideal behavior? Explain both answers.

85. Does the effect of intermolecular attraction on the properties of a gas become more significant or less significant if (a) the gas is compressed to a smaller volume at constant temperature? (b) more gas is forced into the same volume at the same temperature? (c) the temperature of the gas is raised at constant pressure?

86. Does the effect of molecular volume on the properties of a gas become more significant or less significant if (a) the gas is compressed to a smaller volume at constant temperature? (b) more gas is forced into the same volume at the same temperature? (c) the temperature of the gas is raised at constant pressure?

87. A sample of gas has a molar volume of 10.3 L at a pressure of 745 torr and a temperature of $-138°C$. Is the gas behaving ideally?

88. Calculate the compressibility factor, $(P_{real})(V_{real})/RT$, for a 1.00-mol sample of NH_3 under the following conditions: in a 500-mL vessel at $-10.0°C$ it exerts a pressure of 30.0 atm. What would be the *ideal* pressure for 1.00 mol of NH_3 at $-10.0°C$ in a 500-mL vessel? Compare this with the real pressure and account for the difference.

89. What is the van der Waals equation? How does it differ from the ideal gas equation?

90. Find the pressure of a sample of carbon tetrachloride, CCl_4, if 1.00 mol occupies 30.0 L at 77.0°C (slightly above its normal boiling point). Assume that CCl_4 obeys (a) the ideal gas law; (b) the van der Waals equation. The van der Waals constants for CCl_4 are $a = 20.39$ $L^2 \cdot atm/mol^2$ and $b = 0.1383$ L/mol.

91. Repeat the calculations of Exercise 90 using a 3.25-mol gas sample confined to 6.25 L at 115°C.

Stoichiometry in Reactions Involving Gases

92. (a) What is Gay-Lussac's Law? (b) What does it mean? (c) What is the Law of Combining Volumes?

93. What volume of chlorine under the same temperature and pressure conditions will react with 2.00 L of each of the following gases: H_2, C_2H_4, CO, C_2H_2? The equations are
    (a) $H_2(g) + Cl_2(g) \longrightarrow 2HCl(g)$
    (b) $C_2H_4(g) + Cl_2(g) \longrightarrow C_2H_4Cl_2(g)$
    (c) $CO(g) + Cl_2(g) \longrightarrow COCl_2(g)$
    (d) $C_2H_2(g) + 2Cl_2(g) \longrightarrow C_2H_2Cl_4(g)$

94. Which reaction requires the smallest volume of gaseous oxygen for the reaction of 1 volume of the other gas?
    (a) $CH_4(g) + 2O_2(g) \longrightarrow CO_2(g) + 2H_2O(\ell)$
    (b) $2CH_3OH(g) + 3O_2(g) \longrightarrow 2CO_2(g) + 4H_2O(\ell)$
    (c) $2C_2H_2(g) + 5O_2(g) \longrightarrow 4CO_2(g) + 2H_2O(\ell)$
    (d) $CH_3CH_2OH(g) + 3O_2(g) \longrightarrow 2CO_2(g) + 3H_2O(\ell)$

95. Assuming the volumes of all gases in the reaction are measured at the same temperature and pressure conditions, calculate the volume of water vapor obtainable by the explosion of a mixture of 350 mL of hydrogen gas and 175 mL of oxygen gas.

*96. One liter of sulfur vapor at 600°C and 1 atm is burned in pure oxygen to give 8.00 L of sulfur dioxide gas, SO_2, measured at the same temperature and pressure. How many atoms are there in a molecule of sulfur in the gaseous state?

97. What volume of N_2 is required to convert 25.0 L of H_2 to NH_3? Assume that all gases are at the same temperature and pressure, and that the reaction is complete.

$$N_2(g) + 3H_2(g) \longrightarrow 2NH_3(g)$$

98. "Air" bags for automobiles are inflated during a collision by the explosion of sodium azide, NaN_3. The equation for the decomposition is

$$2NaN_3 \longrightarrow 2Na + 3N_2$$

What mass of sodium azide would be needed to inflate a 25.0-L bag to a pressure of 1.3 atm at 25°C?

99. Calculate the volume of methane, CH_4, measured at 300 K and 800 torr, that can be produced by the bacterial breakdown of 1.00 kg of a simple sugar according to the equation

$$C_6H_{12}O_6 \longrightarrow 3CH_4 + 3CO_2$$

*100. A common laboratory preparation of oxygen is

$$2KClO_3(s) \xrightarrow[\Delta]{MnO_2} 2KCl(s) + 3O_2(g)$$

If you were designing an experiment to generate four bottles (each containing 250 mL) of O_2 at 25°C and 741 torr and allowing for 25% waste, what mass of potassium chlorate would be required?

101. Many campers use small propane stoves to cook meals. What volume of air (assumed to be 20% O_2 by volume) will be required to burn 10.0 L of propane, C_3H_8? Assume all gas volumes are measured at the same temperature and pressure. The equation is

$$C_3H_8(g) + 5O_2(g) \longrightarrow 3CO_2(g) + 4H_2O(g)$$

102. If 1.67 L of nitrogen gas and 4.42 L of hydrogen gas were allowed to react, what volume of $NH_3(g)$ can form? Assume all gases are at the same temperature and pressure, and that the reaction is complete.

$$N_2(g) + 3H_2(g) \longrightarrow 2NH_3(g)$$

103. We burn 12.44 L of ammonia gas in 16.33 L of oxygen at 500°C. What volume of nitric oxide, NO, gas can form? What volume of steam, $H_2O(g)$, is formed? Assume all gases are at the same temperature and pressure, and that the reaction is complete.

$$4NH_3(g) + 5O_2(g) \longrightarrow 4NO(g) + 6H_2O(g)$$

104. What mass of KNO_3 would have to be decomposed to produce 18.4 L of oxygen gas measured at STP?

$$2KNO_3(s) \xrightarrow{\Delta} 2KNO_2(s) + O_2(g)$$

105. Refer to Exercise 104. An impure sample of KNO_3 that had a mass of 48.2 g was heated until all of the KNO_3 had decomposed. The liberated oxygen occupied 4.22 L at STP. What percentage of the sample was KNO_3?

*106. Heating a 6.862-g sample of an ore containing a metal sulfide, in the presence of excess oxygen, produces 1053 mL of dry SO_2, measured at 66°C and 739 torr. Calculate the percentage by mass of sulfur in the ore.

*107. The following reactions occur in a gas mask (self-contained breathing apparatus) sometimes used by underground miners. The H_2O and CO_2 come from exhaled air, and O_2 is inhaled as it is produced. KO_2 is potassium superoxide. The CO_2 is converted to the solid salt $KHCO_3$, potassium hydrogen carbonate, so that CO_2 is not inhaled in significant amounts.

$$4KO_2(s) + 2H_2O(\ell) \longrightarrow 4KOH(s) + 3O_2(g)$$

$$CO_2(g) + KOH(s) \longrightarrow KHCO_3(s)$$

(a) What volume of O_2, measured at STP, is produced by the complete reaction of 1.00 g of KO_2?

(b) What is this volume at body temperature, 37°C, and 1.00 atm?
(c) What mass of KOH is produced in (a)?
(d) What volume of CO_2, measured at STP, will react with the mass of KOH of (c)?
(e) What is the volume of CO_2 of (d) measured at 37°C and 1.00 atm?

*108. Let us represent gasoline as octane, C_8H_{18}. When such hydrocarbon fuels burn in the presence of sufficient oxygen, CO_2 is formed.

$$\text{Reaction A:}\quad 2C_8H_{18} + 25O_2 \longrightarrow 16CO_2 + 18H_2O$$

But when the supply of oxygen is limited, the poisonous gas carbon monoxide, CO, is formed.

$$\text{Reaction B:}\quad 2C_8H_{18} + 17O_2 \longrightarrow 16CO + 18H_2O$$

Any automobile engine, no matter how well tuned, burns its fuel by some combination of these two reactions. Suppose an automobile engine is running at idle speed in a closed garage with air volume 85.5 m³. This engine burns 95.0% of its fuel by reaction A, and the remainder by reaction B.

(a) How many liters of octane, density 0.702 g/mL, must be burned for the CO to reach a concentration of 2.00 g/m³?
(b) If the engine running at idle speed burns fuel at the rate of 1.00 gal/h (0.0631 L/min), how long does it take to reach the CO concentration in (a)?

Mixed Exercises

109. A tilting McLeod gauge is used to measure very low pressures of gases in glass vacuum lines in the laboratory. It operates by compressing a large volume of gas at low pressure to a much smaller volume so that the pressure is more easily measured. What is the pressure of a gas in a vacuum line if a 48.6-mL volume of the gas, when compressed to 0.118 mL, supports a 19.7-mm column of mercury?

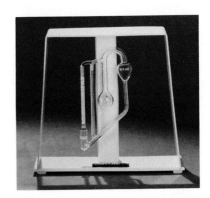

110. Imagine that you live in a cabin with an interior volume of 150 m³. On a cold morning your indoor air temperature is 10°C, but by the afternoon the sun has warmed

the cabin air to 18°C. The cabin is not sealed; therefore, the pressure inside is the same as it is outdoors. Assume that the pressure remains constant during the day. How many cubic meters of air would have been forced out of the cabin by the sun's warming? How many liters?

111. A particular tank can safely hold gas up to a pressure of 36.2 atm. When the tank contains 36.0 g of N_2 at 25°C, the gas exerts a pressure of 12.7 atm. What is the highest temperature to which the gas sample can be heated safely?

112. The molecular weights of xenon and argon are 131 and 39.9, respectively. Which gas effuses faster through a narrow opening? By what factor?

113. A flask of unknown volume was filled with air to a pressure of 3.6 atm. This flask was then attached to an evacuated flask with a known volume of 4.9 L, and the air was allowed to expand into the flask. The final pressure of the air (in both flasks) was 2.5 atm. Calculate the volume of the first flask.

114. Find the molecular weight of Freon-12 (a chlorofluoromethane); 8.29 L of vapor at 200°C and 790 torr has a mass of 26.8 g.

115. A 350-mL flask contains 0.0131 mol of neon gas at a pressure of 744 torr. Are these data sufficient to allow you to calculate the temperature of the gas? If not, what is missing? If so, what is the temperature in °C?

*116. Relative humidity is the ratio of the pressure of water vapor in the air to the pressure of water vapor in air that is saturated with water vapor at the same temperature.

relative humidity =

$$\frac{\text{actual partial pressure of } H_2O \text{ vapor}}{\text{partial pressure of } H_2O \text{ vapor if sat'd}}$$

Often this quantity is multiplied by 100 to give the percent relative humidity. Suppose the percent rela-

tive humidity is 80.0% at 92.0°F (33.0°C) in a house with volume 245 m³. Then an air conditioner is turned on. Due to the condensation of water vapor on the cold coils of the air conditioner, water vapor is also removed from the air as it cools. After the air temperature has reached 77.0°F (25.0°C), the percent relative humidity is measured to be 15.0%.

(a) What mass of water has been removed from the air in the house? (Reminder: take into account the difference in saturated water vapor pressure at the two temperatures.)

(b) What volume would this liquid water occupy at 25°C? (Density of liquid water at 25.0°C = 0.997 g/cm³.)

*117. The average speed of hydrogen molecules at 25°C is about 31 km/min. Pentane vapor is about 36 times as dense as hydrogen at the same temperature and pressure. What is the average speed of the pentane molecules at 25°C?

118. Use both the ideal gas law and the van der Waals equation to calculate the pressure exerted by a 10.0-mol sample of ammonia in a 60.0-L container at 100°C. By what percentage do the two results differ?

119. What volume of hydrogen fluoride at 743 torr and 24°C will be released by the reaction of 74.2 g of xenon difluoride with a stoichiometric amount of water? The unbalanced equation is

$$XeF_2(s) + H_2O(\ell) \longrightarrow Xe(g) + O_2(g) + HF(g)$$

What volumes of oxygen and xenon will be released under these conditions?

120. Cyanogen is 46.2% carbon and 53.8% nitrogen by mass. At a temperature of 25°C and a pressure of 750 torr, 1.00 g of cyanogen gas occupies 0.476 L. Determine the empirical formula and the molecular formula of cyanogen.

Soap bubbles are due to surface tension, an important physical property of liquids. White light striking the bubbles gives brightly colored interference patterns.

Objectives

As you study this chapter, you should learn

☐ About the properties of liquids and solids and how they differ from gases

☐ To understand the kinetic–molecular description of liquids and solids, and how this description differs from that for gases

☐ To use the terminology of phase changes

☐ To understand various kinds of intermolecular attractions and how they are related to physical properties such as vapor pressure, viscosity, melting point, boiling point, and so on

☐ To describe evaporation, condensation, and boiling in molecular terms

☐ To do calculations about the heat transfer involved in warming or cooling without change of phase

☐ To do calculations about the heat transfer involved in phase changes

☐ To describe melting, solidification, sublimation, and deposition in molecular terms

☐ To interpret P vs. T phase diagrams

☐ About the regular structure of crystalline solids

☐ About the various types of solids

☐ To relate the properties of different types of solids to the bonding or interactions among particles in these solids

☐ To visualize some common simple arrangements of atoms in solids

☐ To carry out calculations relating atomic arrangement, density, unit cell size, and ionic or atomic radii in some simple crystalline arrangements

☐ About the bonding in metals

☐ Why some substances are conductors, some are insulators, and others are semiconductors

he molecules of most gases are so widely separated at ordinary temperatures and pressures that they do not interact with each other significantly. Consequently, the physical properties of gases are reasonably well described by the relatively simple relationships described in Chapter 12. In liquids and solids, the so-called **condensed phases**, the particles are closely spaced and interact strongly. Although the properties of liquids and solids can be described, they cannot be adequately explained by simple mathematical relationships. Figure 13-1 and Table 13-1 summarize some of the characteristics of gases, liquids, and solids.

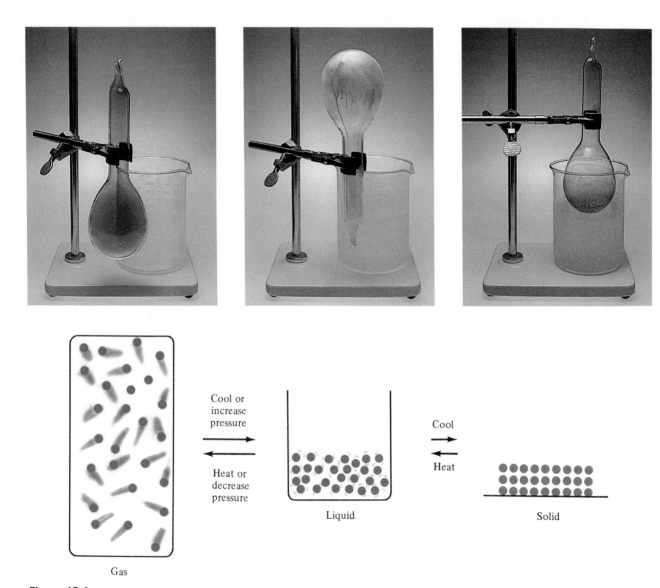

Figure 13-1
The three phases of matter. The photos show the three phases of NO_2. The brown gas (left) is frozen in liquid nitrogen to a solid (right). The blue solid melts to form liquid NO_2 (center) when warmed. The drawings represent the kinetic–molecular interpretation of the three phases.

**Table 13-1
Some Characteristics of Gases, Liquids, and Solids**

Gases	Liquids	Solids
1. Have no definite shape (fill containers completely)	1. Have no definite shape (assume shapes of containers)	1. Have definite shape (resist deformation)
2. Are compressible	2. Have definite volume (are only very slightly compressible)	2. Are nearly incompressible
3. Have low density	3. Have high density	3. Usually have higher density than liquids
4. Are fluid	4. Are fluid	4. Are not fluid
5. Diffuse rapidly	5. Diffuse through other liquids	5. Diffuse only very slowly through solids
6. Consist of extremely disordered particles and much empty space; particles have rapid, random motion in three dimensions	6. Consist of disordered clusters of particles that are quite close together; particles have random motion in three dimensions	6. Have an ordered arrangement of particles that are very close together; particles have vibrational motion only

13-1 Kinetic–Molecular Description of Liquids and Solids

The properties listed in Table 13-1 can be qualitatively explained in terms of the kinetic–molecular theory of Chapter 12. We saw in Section 12-12 that the average kinetic energy of a collection of gas molecules decreases as the temperature is lowered. As a sample of gas is cooled and compressed, the rapid, random motion of gaseous molecules decreases. The molecules approach each other and the intermolecular attractions increase. Eventually these increasing intermolecular attractions overcome the decreasing kinetic energies. At this point condensation (liquefaction) occurs. Because different kinds of molecules have different attractive forces, the temperatures and pressures required for condensation vary from gas to gas.

In the liquid state, the forces of attraction among particles are great enough that disordered clustering occurs. The particles are so close together that very little of the volume occupied by a liquid is empty space. As a result, it is very hard to compress a liquid. Particles in liquids have sufficient energy of motion to overcome partially the attractive forces among them. They are able to slide past each other so that liquids assume the shapes of their containers up to the volume of the liquid. Liquids diffuse into other liquids with which they are *miscible*. For example, a drop of red food coloring added to a glass of water causes the water to become red throughout after diffusion is complete. The natural diffusion rate is slow. Because the average separations among particles in liquids are far less than those in gases, the densities of liquids are much higher than the densities of gases (Table 12-1).

Cooling a liquid lowers its molecular kinetic energy and causes its molecules to slow down even more. If the temperature is lowered sufficiently, at ordinary pressures, stronger but shorter-range attractive interactions overcome the decreasing kinetic energies of the molecules to cause *solidification*. The temperature required for *crystallization* at a given pressure depends on

Intermolecular attractions are those between different molecules or ions. Intramolecular attractions are those between atoms within a single molecule or ion.

The *miscibility* of two liquids refers to their ability to mix and produce a homogeneous solution.

Solidification and *crystallization* refer to the process in which a liquid changes to a solid.

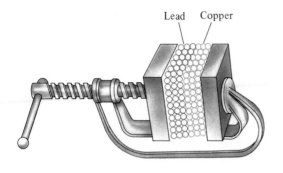

Figure 13-2

A representation of diffusion in solids. When blocks of two different metals are clamped together for a long time, a few atoms of each metal diffuse into the other metal.

the nature of short-range interactions among the particles and is characteristic of each substance.

Most solids have ordered arrangements of particles with a very restricted range of motion. Particles in the solid state cannot move freely past one another and only vibrate about fixed positions. Consequently, solids have definite shapes and volumes. Because the particles are so close together, solids are nearly incompressible and are very dense relative to gases. Solid particles do not diffuse readily into other solids. However, analysis of two blocks of different solids, such as copper and lead, that have been pressed together for a period of years shows that each block contains some atoms of the other element. This demonstrates that solids do diffuse, but very slowly (Figure 13-2).

13-2 Intermolecular Attractions and Phase Changes

We have already seen (Section 12-14) how the presence of strong attractive forces between gas molecules can cause gas behavior to become quite non-ideal when the molecules get close together. In liquids and solids, the molecules are already much closer together than in gases. Therefore, the liquid properties we shall consider, such as boiling point, vapor pressure, viscosity, and heat of vaporization, depend markedly on the strengths of the intermolecular attractive forces. These forces are also directly related to the properties of solids, such as melting point (Section 13-10) and heat of fusion (Section 13-11). Let us preface our study of these condensed phases with a consideration of the types of attractive forces that can exist between molecules and ions.

*Inter*molecular forces refer to the forces *between* individual particles (atoms, molecules, ions) of a substance. These forces are quite weak relative to *intra*molecular forces, i.e., covalent and ionic bonds *within* compounds. For example, 920 kJ of energy is required to decompose one mole of water vapor into H and O atoms. This reflects the strength of intramolecular forces (chemical bonds). But only 40.7 kJ is required to convert one mole of liquid water into steam at 100°C. This reflects the strength of the intermolecular forces of attraction between the water molecules, mainly *hydrogen bonding*.

If it were not for the existence of intermolecular attractions, condensed phases (liquids and solids) could not exist. These are the forces that hold the particles together in liquids and solids. As we shall see, the effects of these attractions on melting points of solids parallel those on boiling points of liquids. High boiling points are associated with compounds exhibiting strong intermolecular attractions. Let us consider the general types of forces that exist among ionic, covalent, and monatomic species and their effects on boiling points.

Ion–Ion Interactions

According to Coulomb's Law, the *force of attraction* between two oppositely charged ions is directly proportional to the charges on the ions, q^+ and q^-, and inversely proportional to the square of the distance between them, d:

$$F \propto \frac{q^+ q^-}{d^2}$$

Energy has the units of force × distance, $F \times d$, so the *energy of attraction* between two oppositely charged ions is directly proportional to the charges on the ions and inversely proportional to the distance of separation:

$$E \propto \frac{q^+ q^-}{d}$$

Ionic compounds such as NaCl, CaBr$_2$, and K$_2$SO$_4$ exist as extended arrays of discrete ions in the solid state. As we shall see in Section 13-16, the oppositely charged ions in these arrays are quite close together. As a result of these small distances, d (in the denominator of the equation for energy), the energies of attraction in these solids are substantial. Most ionic bonding is strong, and as a result most ionic compounds have high melting points (Table 13-2). At high enough temperatures, ionic solids melt as the added heat energy overcomes the potential energy associated with the attraction of oppositely charged ions. The ions in the resulting *molten* samples are free to move about, which accounts for the excellent electrical conductivity of molten ionic compounds.

> Ionic bonding may be thought of as both *inter-* and *intramolecular* bonding.

For most substances, the liquid is less dense than the solid (H$_2$O is one of the rare exceptions). Therefore, melting a solid nearly always produces greater average separations among the ions. This means that the forces (and energies) of attractions among the ions in the liquid are less than in the solid state because average d is greater in the melt. However, these energies of attraction are still much greater in magnitude than the energies of attraction among neutral species (molecules or atoms).

The product $q^+ q^-$ increases as the charges on ions increase. Ionic substances containing multiply charged ions such as Al^{3+}, Mg^{2+}, O^{2-}, and S^{2-} ions usually have higher melting and boiling points than ionic compounds

Table 13-2
Melting Points of Some Ionic Compounds

Compound	mp (°C)	Compound	mp (°C)	Compound	mp (°C)
NaF	993	CaF$_2$	1423	MgO	2800
NaCl	801	Na$_2$S	1180	CaO	2580
NaBr	747	K$_2$S	840	BaO	1923
KCl	770				

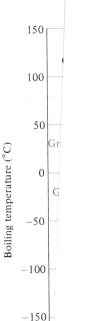

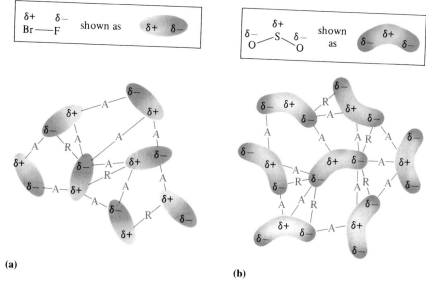

(a)

(b)

Figure 13-3
Dipole–dipole interactions among polar molecules. Each polar molecule is shaded with regions of highest negative charge ($\delta-$) darkest and regions of highest positive charge ($\delta+$) lightest. Attractive forces are shown as —A—, and repulsive forces are shown as —R—. Molecules tend to arrange themselves to maximize attractions by bringing regions of opposite charge together while minimizing repulsions by separating regions of like charge. (a) Bromine fluoride, BrF. (b) Sulfur dioxide, SO_2.

containing only singly charged ions such as Na^+, K^+, F^-, and Cl^- (Table 13-2). For series of ions of similar charges, the closer approach of smaller ions results in higher interionic attractive forces and higher melting points (compare NaF, NaCl, and NaBr in Table 13-2).

Dipole–Dipole Interactions

Permanent dipole–dipole interactions occur between polar covalent molecules because of the attraction of the $\delta+$ atoms of one molecule to the $\delta-$ atoms of another molecule (Section 7-6).

Electrostatic forces between two ions decrease by the factor $1/d^2$ as their separation, d, increases. But dipole–dipole forces vary as $1/d^4$. Because of the higher power of d in the denominator, $1/d^4$ diminishes with increasing d much more rapidly than does $1/d^2$. Therefore, dipole forces are effective only over very short distances. In addition, such forces are weaker than in the ion–ion cases because q^+ and q^- represent only "partial charges." Average dipole–dipole interaction energies are approximately 4 kJ per mole of bonds. They are much weaker than ionic and covalent bonds, which have typical energies of about 400 kJ per mole of bonds. Substances in which permanent dipole–dipole interactions affect physical properties include bromine fluoride, BrF, and sulfur dioxide, SO_2. Dipole–dipole interactions are illustrated in Figure 13-3. All dipole–dipole interactions, including hydrogen bonding (discussed in the following section), are somewhat directional. An increase in temperature causes an increase in translational, rotational, and vibrational motion of molecules. This produces more randomness of orientation of molecules relative to each other. Consequently, the strength of dipole–dipole interactions decreases as temperature increases. All these factors make compounds having dipole–dipole interactions more volatile than ionic compounds.

Figure 13-6

An illustration of how a temporary dipole can be induced in an atom. (a) An isolated argon atom, with spherical charge distribution (no dipole). (b) When a cation approaches the argon atom, the outer portion of the electron cloud is weakly attracted by the ion's positive charge. This induces a weak *temporary* dipole in the argon atom. (c) A temporary dipole can also be induced if the argon atom is approached by an anion. (d) The approach of a molecule with a permanent dipole (for instance, HF) could also temporarily polarize the argon atom. (e) Even in pure argon, the close approach of one argon atom to another results in temporary dipole formation in both atoms as each atom's electron cloud is attracted by the nucleus of the other atom or is repelled by the other atom's electron cloud. The resulting temporary dipoles cause weak attractions among the argon atoms. Molecules are even more easily polarized than isolated atoms.

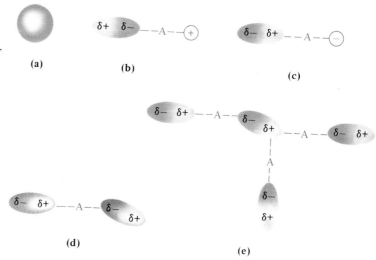

in the case of some polar covalent molecules. The increasing effectiveness of London forces, for example, accounts for the increase in boiling points in the sequences HCl < HBr < HI and H_2S < H_2Se < H_2Te, which involve nonhydrogen-bonded polar covalent molecules. The differences in electronegativities between hydrogen and other nonmetals *decrease* in these sequences, and the increasing London forces override the decreasing permanent dipole–dipole forces. Therefore, the *permanent* dipole–dipole interactions have very little effect on the boiling points of these compounds.

Let us compare the magnitudes of the various contributions to the total energy of interactions in a group of simple molecules. Table 13-3 shows the permanent dipole moments and the energy contributions for five simple molecules. The contribution from London forces is substantial in all cases. The permanent dipole–dipole energy is greatest for the substances in which hydrogen bonding occurs. The variations of these total energies of interaction are closely related to molar heats of vaporization. As we shall see in Section 13-9, the heat of vaporization measures the amount of energy required to overcome the attractive forces that hold the molecules together in a liquid.

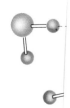

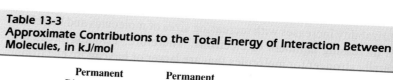

Honey is a very viscous liquid.

Table 13-3
Approximate Contributions to the Total Energy of Interaction Between Molecules, in kJ/mol

Molecule	Permanent Dipole Moment (D)	Permanent Dipole–Dipole Energy	London Energy	Total Energy	Molar Heat of Vaporization (kJ/mol)
Ar	0	0	8.5	8.5	6.7
CO	0.1	~0	8.7	8.7	8.0
HCl	1.03	3.3	17.8	21	16.2
NH_3	1.5	13*	16.3	29	27.4
H_2O	1.8	36*	10.9	47	40.7

* Hydrogen-bonded.

containing only singly charged ions such as Na^+, K^+, F^-, and Cl^- (Table 13-2). For series of ions of similar charges, the closer approach of smaller ions results in higher interionic attractive forces and higher melting points (compare NaF, NaCl, and NaBr in Table 13-2).

Dipole–Dipole Interactions

> Permanent dipole–dipole interactions occur between polar covalent molecules because of the attraction of the δ+ atoms of one molecule to the δ− atoms of another molecule (Section 7-6).

Electrostatic forces between two ions decrease by the factor $1/d^2$ as their separation, d, increases. But dipole–dipole forces vary as $1/d^4$. Because of the higher power of d in the denominator, $1/d^4$ diminishes with increasing d much more rapidly than does $1/d^2$. Therefore, dipole forces are effective only over very short distances. In addition, such forces are weaker than in the ion–ion cases because q^+ and q^- represent only "partial charges." Average dipole–dipole interaction energies are approximately 4 kJ per mole of bonds. They are much weaker than ionic and covalent bonds, which have typical energies of about 400 kJ per mole of bonds. Substances in which permanent dipole–dipole interactions affect physical properties include bromine fluoride, BrF, and sulfur dioxide, SO_2. Dipole–dipole interactions are illustrated in Figure 13-3. All dipole–dipole interactions, including hydrogen bonding (discussed in the following section), are somewhat directional. An increase in temperature causes an increase in translational, rotational, and vibrational motion of molecules. This produces more randomness of orientation of molecules relative to each other. Consequently, the strength of dipole–dipole interactions decreases as temperature increases. All these factors make compounds having dipole–dipole interactions more volatile than ionic compounds.

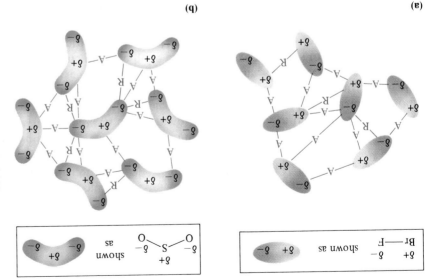

Figure 13-3

Dipole–dipole interactions among polar molecules. Each polar molecule is shaded with regions of highest negative charge (δ−) darkest and regions of highest positive charge (δ+) lightest. Attractive forces are shown as —A—, and repulsive forces are shown as —R—. Molecules tend to arrange themselves to maximize attractions by bringing regions of opposite charge together while minimizing repulsions by separating regions of like charge. (a) Bromine fluoride, BrF. (b) Sulfur dioxide, SO_2.

(a)

Br—F
shown as
δ+ δ−

(b)

$$\underset{\delta-}{O}\overset{\delta+}{\underset{S}{\diagdown}}\underset{\delta-}{O}$$
shown as
δ+ δ−

Figure 13-6

An illustration of how a temporary dipole can be induced in an atom. (a) An isolated argon atom, with spherical charge distribution (no dipole). (b) When a cation approaches the argon atom, the outer portion of the electron cloud is weakly attracted by the ion's positive charge. This induces a weak *temporary dipole* in the argon atom. (c) A temporary dipole can also be induced if the argon atom is approached by an anion. (d) The approach of a molecule with a permanent dipole (for instance, HF) could also temporarily polarize the argon atom. (e) Even in pure argon, the close approach of one argon atom to another results in temporary dipole formation in both atoms as each atom's electron cloud is attracted by the nucleus of the other atom or is repelled by the other atom's electron cloud. The resulting temporary dipoles cause weak attractions among the argon atoms. Molecules are even more easily polarized than isolated atoms.

Honey is a very viscous liquid.

in the case of some polar covalent molecules. The increasing effectiveness of London forces, for example, accounts for the increase in boiling points in the sequences HCl < HBr < HI and $H_2S < H_2Se < H_2Te$, which involve nonhydrogen-bonded polar covalent molecules. The differences in electronegativities between hydrogen and other nonmetals *decrease* in these sequences, and the increasing London forces override the decreasing permanent dipole–dipole forces. Therefore, the *permanent* dipole–dipole interactions have very little effect on the boiling points of these compounds.

Let us compare the magnitudes of the various contributions to the total energy of interactions in a group of simple molecules. Table 13-3 shows the permanent dipole moments and the energy contributions for five simple molecules. The contribution from London forces is substantial in all cases. The permanent dipole–dipole energy is greatest for the substances in which hydrogen bonding occurs. The variations of these total energies of interaction are closely related to molar heats of vaporization. As we shall see in Section 13-9, the heat of vaporization measures the amount of energy required to overcome the attractive forces that hold the molecules together in a liquid.

Table 13-3
Approximate Contributions to the Total Energy of Interaction Between Molecules, in kJ/mol

Molecule	Permanent Dipole Moment (D)	Permanent Dipole–Dipole Energy	London Energy	Total Energy	Molar Heat of Vaporization (kJ/mol)
Ar	0	0	8.5	8.5	6.7
CO	0.1	~0	8.7	8.7	8.0
HCl	1.03	3.3	17.8	21	16.2
NH_3	1.5	13*	16.3	29	27.4
H_2O	1.8	36*	10.9	47	40.7

* Hydrogen-bonded.

The Liquid State

We shall briefly describe several properties of the liquid state. These properties vary markedly among various liquids, depending on the nature and strength of the attractive forces among the particles (atoms, molecules, ions) making up the liquid.

13-3 Viscosity

Viscosity is the resistance to flow of a liquid. Honey has a high viscosity at room temperature, and freely flowing gasoline has a low viscosity. The viscosity of a liquid can be measured with a viscometer such as the one in Figure 13-7.

For a liquid to flow, the molecules must be able to slide past one another. In general, the stronger the intermolecular forces of attraction, the more viscous the liquid is. Substances that have a great ability to form hydrogen bonds, especially involving several hydrogen-bonding sites per molecule, such as glycerine (see margin), usually have high viscosities. Increasing the size and surface area of molecules generally results in increased viscosity, due to the increased London forces. For instance, the shorter-chain hydrocarbon *n*-pentane (a free-flowing liquid at room temperature) is less viscous than *n*-dodecane (an oily liquid at room temperature). The longer the molecules are, the more they can get "tangled up" in the liquid, and the harder it is for them to flow.

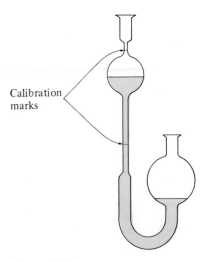

Figure 13-7
The Ostwald viscometer, a device used to measure viscosity of liquids. The time it takes for a known volume of a liquid to flow through a small neck of known size is measured. Liquids with low viscosities flow rapidly.

n-pentane, C_5H_{12}
viscosity = 0.24 centipoise

$$H-\overset{\overset{\displaystyle H}{|}}{\underset{\underset{\displaystyle H}{|}}{C}}-\overset{\overset{\displaystyle H}{|}}{\underset{\underset{\displaystyle H}{|}}{C}}-\overset{\overset{\displaystyle H}{|}}{\underset{\underset{\displaystyle H}{|}}{C}}-\overset{\overset{\displaystyle H}{|}}{\underset{\underset{\displaystyle H}{|}}{C}}-\overset{\overset{\displaystyle H}{|}}{\underset{\underset{\displaystyle H}{|}}{C}}-H$$

n-dodecane, $C_{12}H_{26}$
viscosity = 1.35 centipoise

$$H-C-C-C-C-C-C-C-C-C-C-C-C-H$$

glycerine

As temperature increases and the molecules move more rapidly, their kinetic energies are better able to overcome intermolecular attractions. Thus, viscosity decreases with increasing temperature, as long as no changes in composition occur.

The *poise* is the unit used to express viscosity. The viscosity of water at 25°C is 0.89 centipose.

13-4 Surface Tension

Molecules below the surface of a liquid are influenced by intermolecular attractions from all directions. Those on the surface are attracted only toward the interior, as shown in Figure 13-8. The attractions pull the surface layer toward the center. The most stable situation is one in which the surface area is minimal. For a given volume, a sphere has the least possible surface area, so drops of liquid tend to assume spherical shapes. **Surface tension** is a

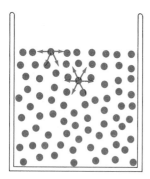

Figure 13-8
A molecular-level view of the attractive forces experienced by molecules at and below the surface of a liquid.

Coating glass with a silicone polymer greatly reduces the adhesion of water to the glass. The left side of each glass has been treated with Rain-X, which contains a silicone polymer. Water on the treated side forms droplets that are easily swept away.

measure of the inward forces that must be overcome to expand the surface area of a liquid.

13-5 Capillary Action

The surface tension of water supports this water strider. The nonpolar surfaces of its feet also help to repel the water.

All forces holding a liquid together are called **cohesive forces**. The forces of attraction between a liquid and another surface are **adhesive forces**. The positively charged H atoms of water hydrogen-bond strongly to the partial negative charges on the oxygen atoms at the surface of the glass. As a result, water *adheres* to glass, or is said to *wet* the glass. As the water creeps up the side of the glass tube, its favorable area of contact with the glass increases. The surface of the water, its **meniscus**, has a concave shape. (Figure 13-9). On the other hand, mercury does not wet glass because its cohesive forces are much stronger than its attraction to glass. Thus, its meniscus is convex. **Capillary action** occurs when one end of a capillary tube, a glass tube with small bore (inside diameter), is immersed in a liquid. If adhesive forces exceed cohesive forces, the liquid continually creeps up the sides of the tube until a balance is reached between adhesive forces and the weight of liquid. The smaller the bore, the higher the liquid climbs. Capillary action helps plant roots take up water and dissolved nutrients from the soil and transmit them up the stems. The roots, like glass, exhibit strong adhesive forces for water. Osmotic pressure (Section 14-15) also plays a major role in this process.

Droplets of mercury lying on a glass surface. The small droplets are almost spherical, whereas the larger droplets are flattened. This shows that surface tension has more influence on the shape of the small (lighter) droplets.

13-6 Evaporation

Kinetic energies of molecules in liquids depend on temperature in the same way as do those in gases. The distribution of kinetic energies among liquid molecules at two different temperatures is shown in Figure 13-10. **Evaporation**, or **vaporization**, is the process by which molecules on the surface of the liquid break away and go into the gas phase (Figure 13-11). To break away, the molecules must possess at least some minimum kinetic energy. Figure 13-10 shows that at a higher temperature, a greater fraction of molecules possess at least that minimum energy. The rate of evaporation increases as temperature increases.

Further, it is only the higher-energy molecules that can escape from the liquid phase. The average molecular kinetic energy of the molecules remaining in the liquid state is thereby lowered, resulting in a lower temperature in the liquid. Because the liquid would then be cooler than its surroundings, it absorbs heat from its surroundings. The cooling of your body by evaporation of perspiration is a familiar example of the cooling effect of evaporation on the surroundings of a liquid.

A molecule in the vapor may later strike the liquid surface and be captured there. This process, the reverse of evaporation, is called **condensation**. As evaporation occurs in a closed container, the volume of liquid decreases and the number of gas molecules above the surface increases. Because more gas-phase molecules can collide with the surface, the rate of condensation increases. The system composed of the liquid and gas molecules of the same substance eventually achieves a **dynamic equilibrium** in which the rate of evaporation equals the rate of condensation in the closed container:

$$\text{liquid} \xrightleftharpoons[\text{condensation}]{\text{evaporation}} \text{vapor}$$

Figure 13-9
The meniscus, as observed in glass tubes with water and with mercury.

The two opposing rates are not zero, but are equal to one another—hence we call this ''dynamic,'' rather than ''static,'' equilibrium. Even though evaporation and condensation are both continuously occurring, *no net change occurs* because the rates are equal.

However, if the vessel were left open to the air, this equilibrium could not be established. Molecules would diffuse away and slight air currents would also sweep some gas molecules away from the liquid surface. This would allow more evaporation to occur to replace the lost vapor molecules. Consequently, a liquid can eventually evaporate entirely if it is left uncovered. This situation illustrates **LeChatelier's Principle**, which states:

As an analogy, suppose that 45 students per minute leave a classroom, moving into the closed hallway outside, and 45 students per minute enter it. The total number of students in the room would remain constant, as would the total number of students outside the room.

> A system at equilibrium, or changing toward equilibrium, responds in the way that tends to relieve or ''undo'' any stress placed upon it.

This is one of the guiding principles that allows us to understand chemical equilibrium. It is discussed further in Chapter 17.

In this example, the stress is the removal of molecules in the vapor phase. The response is the continued evaporation of the liquid.

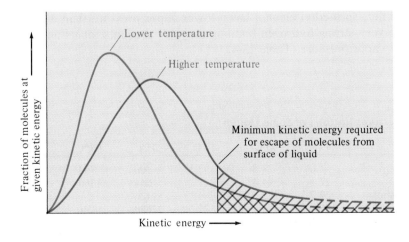

Figure 13-10
Distribution of kinetic energies of molecules in a liquid at different temperatures. At the lower temperature, fewer molecules have the energy required to escape from the liquid, so evaporation is slower and the equilibrium vapor pressure (Section 13-7) is lower.

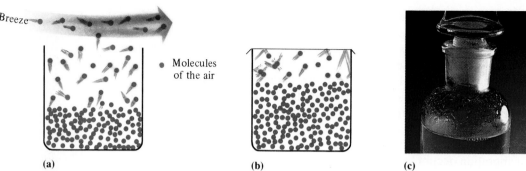

Figure 13-11
(a) Liquid continuously evaporates from an open vessel. (b) Equilibrium between liquid and vapor is established in a closed container in which molecules return to the liquid at the same rate as they leave it. (c) A bottle in which liquid–vapor equilibrium has been established. Note that droplets have condensed.

As long as some liquid remains in contact with the vapor, the pressure does not depend on the volume or surface area of the liquid.

13-7 Vapor Pressure

Vapor molecules cannot escape when vaporization of a liquid occurs in a closed container. As more molecules leave the liquid, more gaseous molecules collide with the walls of the container, with each other, and with the liquid surface, so more condensation occurs. This is responsible for the formation of liquid droplets that adhere to the sides of the vessel above a liquid surface and for the eventual establishment of equilibrium between liquid and vapor (Figure 13-11b and c).

> The partial pressure of vapor molecules above the surface of a liquid at equilibrium at a given temperature is the **vapor pressure (vp)** of the liquid at that temperature. Because the rate of evaporation increases and the rate of condensation decreases with increasing temperature, vapor pressures of liquids *always* increase as temperature increases.

Easily vaporized liquids are called **volatile** liquids, and they have relatively high vapor pressures. (The most volatile liquid in Table 13-4 is diethyl ether. Water is the least volatile.) Vapor pressures can be measured with manometers (Figure 13-12).

Higher cohesive forces tend to hold molecules in the liquid state. London forces generally increase with increasing molecular size, so substances composed of larger molecules have lower vapor pressures. Methyl alcohol molecules are strongly linked by hydrogen bonding, whereas diethyl ether molecules are not, so methyl alcohol has a lower vapor pressure than diethyl ether. The very strong hydrogen bonding in water accounts for its anomalously low vapor pressure (Table 13-4).

Table 13-4
Vapor Pressures (in torr) of Some Liquids

	0°C	25°C	50°C	75°C	100°C	125°C
water	4.6	23.8	92.5	300	760	1741
benzene	27.1	94.4	271	644	1360	
methyl alcohol	29.7	122	404	1126		
diethyl ether	185	470	1325	2680	4859	

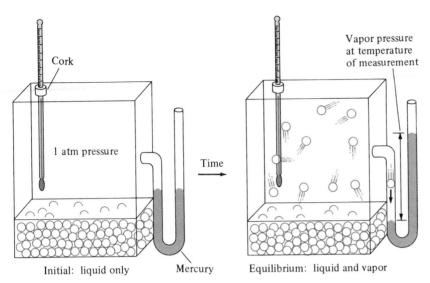

Figure 13-12
A simplified representation of the measurement of vapor pressure of a liquid at a given temperature. At the instant the liquid is added to the container, the space above the liquid is occupied by air only. Some of the liquid then vaporizes until equilibrium is established. The increase in height of the mercury column is a measure of the vapor pressure of the liquid at that temperature.

At a given temperature the vapor pressures of different liquids differ (Table 13-4 and Figure 13-13) because their cohesive forces are different.

13-8 Boiling Points and Distillation

When heat energy is added to a liquid, it increases the kinetic energy of the molecules, and the temperature of the liquid increases. Heating a liquid always increases its vapor pressure. When a liquid is heated to a sufficiently high temperature under a given applied (usually atmospheric) pressure, bubbles of vapor begin to form below the surface. They rise to the surface and burst, releasing the vapor into the air. This process is called *boiling* and is distinctly different from evaporation. If the vapor pressure inside the bubbles is less than the applied pressure on the surface of the liquid, the bubbles collapse as soon as they form and no boiling occurs. The **boiling point** of a liquid under a given pressure is the temperature at which its vapor pressure

As water is being heated, but before it boils, small bubbles may appear in the container. This is not boiling, but rather the formation of bubbles of dissolved gases such as CO_2 and O_2, whose solubilities in water decrease with increasing temperature.

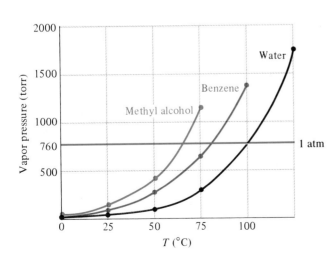

Figure 13-13
Plots of the vapor pressures of some of the liquids in Table 13-4. The *normal* boiling point of a liquid is the temperature at which its vapor pressure is equal to one atmosphere. Normal boiling points (°C) are: water, 100; benzene, 80.1; and methyl alcohol, 65.0. Notice that the increase in vapor pressure is *not* linear with temperature.

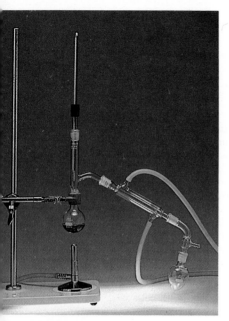

Figure 13-14
A laboratory setup for distillation. During distillation of an impure liquid, nonvolatile substances remain in the distilling flask. The liquid is vaporized and condensed before being collected in the receiving flask. A diagram of this kind of distillation apparatus is shown in Figure 1-12a.

is just equal to the applied pressure. The **normal boiling point** is the temperature at which the vapor pressure of a liquid is equal to exactly one atmosphere (760 torr). The vapor pressure of water is 760 torr at 100°C, its normal boiling point. As heat energy is added to a pure liquid *at its boiling point*, the temperature remains constant, because the energy goes to overcome the cohesive forces in the liquid to form vapor.

If the applied pressure is lower than 760 torr, say on the top of a mountain, water boils below 100°C. The chemical reactions involved in cooking food occur more slowly at the lower temperature, so it takes longer to cook food in boiling water at high altitudes than at sea level. A pressure cooker cooks food rapidly because water boils at higher temperatures under increased pressures. The higher temperature of the boiling water increases the rate of cooking.

Different liquids have different cohesive forces, so they have different vapor pressures and boil at different temperatures. A mixture of liquids with different enough boiling points can often be separated into its components by **distillation** (Section 1-6). In this process the mixture is heated slowly until the temperature reaches the point at which the most volatile liquid boils off. If this component is a liquid under ordinary conditions, it is subsequently recondensed in a water-cooled condensing column (Figures 13-14 and 1-12a) and collected as a distillate. After enough heat has been added to vaporize all of the most volatile liquid, the temperature again rises slowly until the boiling point of the next substance is reached, and the process continues. Any nonvolatile substances dissolved in the liquid do not boil, but remain in the distilling flask. Impure water can be purified and separated from its dissolved salts by distillation. Compounds with similar boiling points, especially those that interact very strongly with each other, are not well separated by simple distillation but require a modification called fractional distillation (see Section 14-10).

13-9 Heat Transfer Involving Liquids

Heat must be added to a liquid to raise its temperature (Section 1-13). The **specific heat** (J/g · °C) or **molar heat capacity** (J/mol · °C) of a liquid is the amount of heat that must be added to the stated mass of liquid to raise its temperature by one degree Celsius. If heat is added at a constant rate to a liquid under constant pressure, the temperature rises at a constant rate until its boiling point is reached. At that point the temperature remains constant until enough heat has been added to boil away all the liquid. The **molar heat of vaporization** (ΔH_{vap}) of a liquid is the amount of heat that must be added to one mole of the liquid at its boiling point to convert it to vapor with no change in temperature. Heats of vaporization can also be expressed in joules per gram. For example, the heat of vaporization for water at its boiling point is 40.7 kJ/mol, or 2.26×10^3 J/g:

The specific heat and heat capacity of a substance change somewhat with its temperature. For most substances, this variation is small enough to ignore.

Molar heats of vaporization are often expressed in kilojoules rather than joules. They are greater than molar heat capacities. The units of heat of vaporization do *not* include temperature. This is because boiling occurs with *no change in temperature*.

$$\frac{?\ J}{g} = \frac{40.7\ kJ}{mol} \times \frac{1000\ J}{kJ} \times \frac{1\ mol}{18.0\ g} = 2.26 \times 10^3\ J/g$$

Like many other properties of liquids, their heats of vaporization reflect the strengths of intermolecular forces. Heats of vaporization generally in-

Table 13-5
Heats of Vaporization, Boiling Points, and Vapor Pressures of Some Common Liquids

Liquid	Vapor Pressure (torr at 20°C)	Boiling Point at 1 atm (°C)	Heat of Vaporization at Boiling Point	
			J/g	kJ/mol
water (MW 18.0)	17.5	100	2260	40.7
ethyl alcohol (MW 46.1)	43.9	78.3	858	39.3
benzene (MW 78.1)	74.6	80	395	30.8
diethyl ether (MW 74.1)	442	34.6	351	26.0
carbon tetrachloride (MW 153.8)	85.6	76.8	213	32.8
ethylene glycol (MW 67.1)	0.1	197.3	984	58.9

The heat of vaporization of water is higher at 37°C (normal body temperature) than at 100°C (2.41 kJ/g as opposed to 2.26 kJ/g).

crease as boiling points and intermolecular forces increase and as vapor pressures decrease. Table 13-5 illustrates this.

The very high heat of vaporization of water and ethylene glycol and, to a lesser extent, that of ethyl alcohol are due mainly to the strong hydrogen-bonding interactions in these liquids (Section 13-2). The high value for water makes it very effective as a coolant and, in the form of steam, as a source of heat.

Even below their boiling points, liquids can evaporate. The water in perspiration is an effective coolant for our bodies. As it evaporates, each gram absorbs 2.41 kJ of heat from the body and carries it into the air in the form of water vapor. We feel even cooler in a breeze because perspiration evaporates faster (Section 13-6), so heat is removed more rapidly.

Condensation is the reverse of evaporation. The amount of heat that must be removed from a vapor to condense it (without change in temperature) is called the **heat of condensation**.

Heat of vaporization, like other heats of transition, is an *extensive property* of a substance, because the amount of heat absorbed depends on the amount of liquid that vaporizes.

$$\text{liquid} + \text{heat} \underset{\text{condensation}}{\overset{\text{evaporation}}{\rightleftharpoons}} \text{vapor}$$

The heat of condensation of a liquid is equal in magnitude, but opposite in sign, to the heat of vaporization. It is released by the vapor during condensation.

Because 2.26 kJ must be absorbed to vaporize one gram of water at 100°C, that same amount of heat must be released to the environment when one gram of steam at 100°C condenses to form liquid water at 100°C. In steam-heated radiators, steam condenses and releases 2.26 kJ of heat per gram as its molecules collide with the cooler radiator walls and condense there. The metallic walls conduct heat well. They transfer the heat to the air in contact with the outside walls of the radiator. The heats of condensation and vaporization of other substances, such as benzene, have smaller magnitudes (see Table 13-5). Therefore, they are much less effective as heating and cooling agents.

Because of the large amount of heat released by steam as it condenses, burns caused by steam are much more severe than burns caused by liquid water at 100°C.

Example 13-1
Calculate the amount of heat, in joules, required to convert 180 grams of water at 10.0°C to steam at 105.0°C.

Steps 1 and 3 of this example involve warming with *no* phase change. These calculations were introduced in Section 1-13.

Plan

The total amount of heat absorbed is the sum of the amounts required to (1) warm the liquid water from 10.0°C to 100.0°C, (2) convert the liquid water to steam at 100°C, and (3) warm the steam from 100.0°C to 105.0°C. Steps 1 and 3 involve the specific heats of water and steam, 4.18 J/g · °C and 2.03 J/g · °C, respectively (see Appendix E), whereas Step 2 involves the heat of vaporization of water (2.26 × 10³ J/g).

Solution

1. $\underline{?}J = 180 \text{ g} \times \dfrac{4.18 \text{ J}}{\text{g} \cdot °\text{C}} \times (100.0°\text{C} - 10.0°\text{C}) = 6.77 \times 10^4 \text{ J} \quad = 0.677 \times 10^5 \text{ J}$

2. $\underline{?}J = 180 \text{ g} \times \dfrac{2.26 \times 10^3 \text{ J}}{\text{g}} \qquad\qquad\qquad\qquad = 4.07 \times 10^5 \text{ J}$

3. $\underline{?}J = 180 \text{ g} \times \dfrac{2.03 \text{ J}}{\text{g} \cdot °\text{C}} \times (105.0°\text{C} - 100.0°\text{C}) = 1.83 \times 10^3 \text{ J} = 0.0183 \times 10^5 \text{ J}$

Total amount of heat absorbed = 4.76×10^5 J

EOC 49

Example 13-2

Compare the amount of "cooling" experienced by an individual who drinks 400 mL of ice water (0°C) with the amount of "cooling" experienced by an individual who "sweats out" 400 mL of water. Assume that all the sweat evaporates. The density of water is very nearly 1.00 g/mL at both 0°C and 37.0°C, average body temperature. The heat of vaporization of water is 2.41 kJ/g at 37.0°C.

Plan

In the case of drinking ice water, the body is cooled by the amount of heat required to raise the temperature of 400 mL (400 g) of water from 0.0°C to 37.0°C. The amount of heat lost by perspiration is equal to the amount of heat required to vaporize 400 g of water at 37.0°C.

Solution

Raising the temperature of 400 g of water from 0.0°C to 37.0°C requires

$$\underline{?}J = (400 \text{ g})(4.18 \text{ J/g} \cdot °\text{C})(37.0°\text{C}) = 6.19 \times 10^4 \text{ J, or} \quad 61.9 \text{ kJ}$$

Evaporating (i.e., "sweating out") 400 mL of water at 37°C requires

$$\underline{?}J = (400 \text{ g})(2.41 \times 10^3 \text{ J/g}) = 9.64 \times 10^5 \text{ J, or} \quad 964 \text{ kJ}$$

For health reasons, it is important to replace the water lost by perspiration.

Thus, we see that "sweating out" 400 mL of water removes 964 kJ of heat from one's body, whereas drinking 400 mL of ice water cools it by only 61.9 kJ. Stated differently, sweating removes (964/61.9) = 15.6 times more heat than drinking ice water!

EOC 57

Enrichment

The Clausius–Clapeyron Equation

We have seen (Figure 13-13) that vapor pressure increases with increasing temperature. Let us now discuss the quantitative expression of this relationship.

When the temperature of a liquid is changed from T_1 to T_2, the vapor pressure of the liquid changes from P_1 to P_2. These changes are related to the molar heat of vaporization, ΔH_{vap}, for the liquid by the **Clausius–Clapeyron equation**:

$$\ln\left(\frac{P_2}{P_1}\right) = \frac{\Delta H_{vap}}{R}\left(\frac{1}{T_1} - \frac{1}{T_2}\right) \quad \text{or} \quad \log\left(\frac{P_2}{P_1}\right) = \frac{\Delta H_{vap}}{2.303R}\left(\frac{1}{T_1} - \frac{1}{T_2}\right)$$

Although ΔH_{vap} changes somewhat with temperature, it is usually adequate to use the value tabulated at the normal boiling point of the liquid (Appendix E) unless more precise values are available. The units of R must be consistent with those of ΔH_{vap}.

The Clausius–Clapeyron equation is used for three types of calculations: (1) to predict the vapor pressure of a liquid at a specified temperature, as in Example 13-3; (2) to determine the temperature at which a liquid has a specified vapor pressure; and (3) to calculate ΔH_{vap} from measurement of vapor pressures at different temperatures.

Example 13-3

The normal boiling point of ethanol, C_2H_5OH, is 78.3°C and its molar heat of vaporization is 39.3 kJ/mol (Appendix E). What would be the vapor pressure, in torr, of ethanol at 50.0°C?

Plan

The normal boiling point of a liquid is the temperature at which its vapor pressure is 760 torr, so we designate this as one of the conditions (subscript 1). We wish to find the vapor pressure at another temperature (subscript 2), and we know the molar heat of vaporization. We use the Clausius–Clapeyron equation to solve for P_2.

The normal boiling point of acetone is 56.2°C, and its ΔH_{vap} is 32.0 kJ/mol. It boils at lower temperatures under reduced pressure. Can you use the Clausius–Clapeyron equation to calculate the pressure at which acetone boils at 14°C, as in this photograph?

Solution

$$P_1 = 760 \text{ torr} \quad \text{at} \quad T_1 = 78.3°C + 273.2 = 351.5 \text{ K}$$
$$P_2 = ? \quad \text{at} \quad T_2 = 50.0°C + 273.2 = 323.2 \text{ K}$$
$$\Delta H_{vap} = 39.3 \text{ kJ/mol} \quad \text{or} \quad 3.93 \times 10^4 \text{ J/mol}$$

We solve for P_2:

$$\log\left(\frac{P_2}{760 \text{ torr}}\right) = \frac{3.93 \times 10^4 \text{ J/mol}}{2.303\left(8.314\ \dfrac{\text{J}}{\text{mol}\cdot\text{K}}\right)}\left(\frac{1}{351.5 \text{ K}} - \frac{1}{323.2 \text{ K}}\right)$$

$$\log\left(\frac{P_2}{760 \text{ torr}}\right) = -0.511 \quad \text{so} \quad \left(\frac{P_2}{760 \text{ torr}}\right) = 10^{-0.511} = 0.308$$

$$P_2 = 0.308(760 \text{ torr}) = \boxed{234 \text{ torr}} \quad \text{(lower vapor pressure at lower temperature)}$$

EOC 40, 41

We have described many properties of liquids and discussed how they depend on intermolecular forces of attraction. The general effects of these attractions on the physical properties of liquids are summarized in Table 13-6. "High" and "low" are relative terms. Table 13-6 is intended to show only very general trends. Example 13-4 illustrates the use of intermolecular attractions to predict boiling points.

Table 13-6
General Effects of Intermolecular Attractions on Physical Properties of Liquids

Property	Volatile Liquids (weak intermolecular attractions)	Nonvolatile Liquids (strong intermolecular attractions)
Cohesive forces	Low	High
Viscosity	Low	High
Surface tension	Low	High
Specific heat	Low	High
Vapor pressure	High	Low
Rate of evaporation	High	Low
Boiling point	Low	High
Heat of vaporization	Low	High

Example 13-4

Predict the order of increasing boiling points for the following: H_2S, H_2O, CH_4, H_2, KBr.

Plan

We analyze the polarity and size of each substance to determine what kinds of intermolecular forces are present. In general, the stronger the intermolecular forces, the higher the boiling point of the substance.

Solution

KBr is ionic, so it boils at the highest temperature. Water exhibits hydrogen bonding and boils at the next highest temperature. Hydrogen sulfide is the only other polar covalent substance in the list, so it boils below H_2O but above the other two substances. Both CH_4 and H_2 are nonpolar. The larger CH_4 molecule is more easily polarized than the very small H_2, so London forces are stronger in CH_4. Thus, CH_4 boils at a higher temperature than H_2.

$$H_2 < CH_4 < H_2S < H_2O < KBr$$
$$\text{increasing boiling points} \longrightarrow$$

EOC 20

The Solid State

13-10 Melting Point

The **melting point (freezing point)** of a substance is the temperature at which its solid and liquid phases exist in equilibrium.

$$\text{liquid} \underset{\text{melting}}{\overset{\text{freezing}}{\rightleftarrows}} \text{solid}$$

The *melting point* of a solid is the same as the *freezing point* of its liquid. It is the temperature at which the rate of melting of a solid is the same as the rate of freezing of its liquid under a given applied pressure.

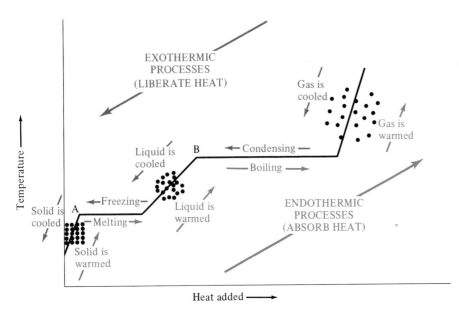

Figure 13-15
A typical heating curve at constant pressure. When heat energy is added to a solid below its melting point, the temperature of the solid rises until its melting point is reached (point A). In this region of the plot, the slope is rather steep because of the low specific heats of solids [e.g., 2.09 J/g · °C for $H_2O(s)$]. If the solid is heated at its melting point (A), its temperature remains constant until the solid has melted, because the melting process requires energy. The length of this horizontal line is proportional to the heat of fusion of the substance—the higher the heat of fusion, the longer the line. When all of the solid has melted, heating the liquid raises its temperature until its boiling point is reached (point B). The slope of this line is less steep than that for warming the solid, because the specific heat of the liquid phase [e.g., 4.18 J/g · °C for $H_2O(\ell)$] is usually greater than that of the corresponding solid. If heat is added to the liquid at its boiling point (B), the added heat energy is absorbed as the liquid boils. This horizontal line is longer than the previous one, because the heat of vaporization of a substance is always higher than its heat of fusion. When all of the liquid has been converted to a gas (vapor), the addition of more heat raises the temperature of the gas. This segment of the plot has a steep slope because of the relatively low specific heat of the gas phase [e.g., 2.03 J/g · °C for $H_2O(g)$]. Each step in the process can be reversed by removing the same amount of heat.

The **normal melting point** of a substance is its melting point at one atmosphere pressure. Variations in melting and boiling points of substances are generally parallel because intermolecular forces are similar.

13-11 Heat Transfer Involving Solids

When heat is added to a solid below its melting point, its temperature rises. After enough heat has been added to bring the solid to its melting point, additional heat is required to convert the solid to liquid. During this melting process, the temperature remains constant at the melting point until all of the substance has melted. After melting is complete, the continued addition of heat results in an increase in the temperature of the liquid, until the boiling point is reached. This is illustrated graphically in the first three segments of the heating curve in Figure 13-15.

The amount of heat required to melt one gram of a solid at its melting point is its **heat of fusion**. The **molar heat of fusion** (ΔH_{fus}; kJ/mol) is the amount of heat required to melt one mole of a solid at its melting point. The heat of fusion depends on the *inter*molecular forces of attraction in the solid state. These forces "hold the molecules together" as a solid. Heats of fusion are *usually* higher for substances with higher melting points. Values for some common compounds are shown in Table 13-7.

The **heat of solidification**, ΔH_{sol}, of a liquid is equal in magnitude, but opposite in sign, to the heat of fusion, ΔH_{fus}:

$$\Delta H_{solidification} = -\Delta H_{fusion}$$

It represents removal of a sufficient amount of heat from a given amount (1 g or 1 mol) of liquid to solidify the liquid at its freezing point. For water,

$$\text{water} \underset{\substack{+6.01 \text{ kJ/mol} \\ \text{or } +334 \text{ J/g}}}{\overset{\substack{-6.01 \text{ kJ/mol} \\ \text{or } -334 \text{ J/g}}}{\rightleftharpoons}} \text{ice} \qquad \text{(at 0°C)}$$

Melting is always endothermic. The term "fusion" means "melting."

Table 13-7
Melting Points and Heats of Fusion of Some Common Compounds

Substance	Melting Point (°C)	Heat of Fusion	
		J/g	kJ/mol
methane	−182	58.6	0.92
ethyl alcohol	−117	109	5.02
water	0	334	6.02
naphthalene	80.2	147	18.8
silver nitrate	209	67.8	11.5
aluminum	658	395	10.6
sodium chloride	801	519	30.3

Example 13-5

The molar heat of fusion, ΔH_{fus}, of Na at its melting point, 97.5°C, is 2.6 kJ/mol. How much heat must be absorbed by 5.0 grams of solid Na at 97.5°C to convert it to molten Na?

Plan

The molar heat of fusion tells us that every mole of Na, 23 grams, will absorb 2.6 kJ of heat at 97.5°C during the melting process. We want to know the amount of heat that 5.0 grams would absorb. We use the appropriate unit factors, constructed from the atomic weight and ΔH_{fus}, to calculate the amount of heat absorbed.

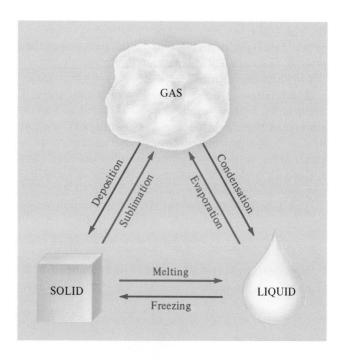

Transitions among the three states of matter. The transitions shown in blue are endothermic (absorb heat); those shown in red are exothermic (release heat).

Solution

$$?kJ = 5.0 \text{ g Na} \times \frac{1 \text{ mol Na}}{23 \text{ g Na}} \times \frac{2.6 \text{ kJ}}{1 \text{ mol Na}} = \boxed{0.57 \text{ kJ}}$$

EOC 46

These two unit factors could be combined into a single unit factor,
$$\frac{2.6 \text{ kJ}}{23 \text{ g Na}}.$$

Example 13-6

Calculate the amount of heat that must be absorbed by 50.0 grams of ice at $-12.0°C$ to convert it to water at $20.0°C$.

Plan

We must determine the amount of heat absorbed during three steps: (1) warming 50.0 g of ice from $-12.0°C$ to its melting point, $0.0°C$ (we use the specific heat of ice, 2.09 J/g · °C); (2) melting the ice with no change in temperature (we use the heat of fusion of ice at $0.0°C$, 334 J/g); and (3) warming the resulting water from $0.0°C$ to $20.0°C$ (we use the specific heat of water, 4.18 J/g · °C).

Ice is very efficient for cooling because considerable heat is required to melt a given mass of it. However, ΔH_{vap} is generally much greater than ΔH_{fusion}, so evaporative cooling is preferable when possible.

Solution

1. $50.0 \text{ g} \times \dfrac{2.09 \text{ J}}{\text{g} \cdot °C} \times [0.0 - (-12.0)]°C = 1.25 \times 10^3 \text{ J} = 0.125 \times 10^4 \text{ J}$

2. $50.0 \text{ g} \times \dfrac{334 \text{ J}}{\text{g}} \qquad\qquad\qquad\qquad = 1.67 \times 10^4 \text{ J}$

3. $50.0 \text{ g} \times \dfrac{4.18 \text{ J}}{\text{g} \cdot °C} \times (20.0 - 0.0)°C = 4.18 \times 10^3 \text{ J} \quad = 0.418 \times 10^4 \text{ J}$

Total amount of heat absorbed = $\boxed{2.21 \times 10^4 \text{ J} = 22.1 \text{ kJ}}$

Note that most of the heat absorbed was required in Step 2, melting the ice.

EOC 58

13-12 Sublimation and the Vapor Pressure of Solids

Some solids, such as iodine and carbon dioxide, vaporize at atmospheric pressure without passing through the liquid state. We say that they **sublime**. Solids exhibit vapor pressures just as liquids do, but they generally have much lower vapor pressures. Solids with high vapor pressures sublime easily. The characteristic odors of the common household solids naphthalene (mothballs) and *para*-dichlorobenzene (bathroom deodorizer) are due to sublimation. The reverse process, by which a vapor solidifies without passing through the liquid phase, is called **deposition**:

$$\text{solid} \underset{\text{deposition}}{\overset{\text{sublimation}}{\rightleftharpoons}} \text{gas}$$

Some impure solids can be purified by sublimation and subsequent deposition of the vapor (as a solid) onto the surface of a cooler object. Purification by sublimation is illustrated in Figure 13-16. Iodine is commonly purified by sublimation.

Figure 13-16
Sublimation can be used to purify volatile solids. The high vapor pressure of the solid substance causes it to sublime when heated. Crystals of purified substance are formed when the vapor is deposited as solid on the cooler (inner) portion of the apparatus. Iodine, I_2, sublimes readily.

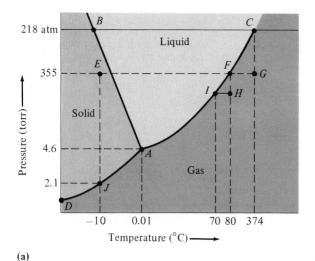

(a)

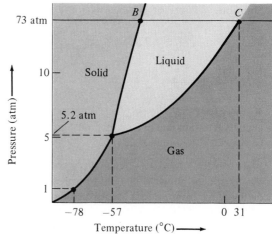

(b)

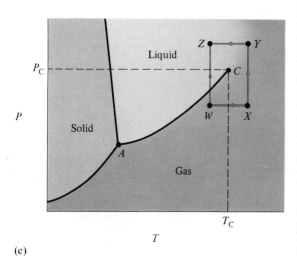

(c)

Figure 13-17

Phase diagrams (not to scale). (a) Diagram for water. For water and a few other substances for which the solid is less dense than the liquid, the solid–liquid equilibrium line (*AB*) has negative slope, i.e., up and to the left. (b) Diagram for carbon dioxide, a substance for which the solid is more dense than the liquid. Note that the solid–liquid equilibrium line has positive slope, i.e., up and to the right. This is true for most substances. (c) Two paths by which a gas can be liquefied. (1) Below the critical temperature. Compressing the sample at *constant* temperature is represented by the vertical line *WZ*. Where this line crosses the vapor pressure curve *AC*, the gas liquefies; at that set of conditions, *two distinct phases,* gas and liquid, are present in equilibrium with one another. These two phases have different properties—e.g., different densities. Raising the pressure further results in a completely liquid sample at point *Z*. (2) Above the critical temperature. Suppose that we instead first warm the gas at constant pressure from *W* to *X*, a temperature above its critical temperature. Then, holding the temperature constant, we increase the pressure to point *Y*. Along this path, the sample increases *smoothly* in density, with no sharp transition between phases. From *Y*, we then decrease the temperature to reach final point *Z*, where the sample is clearly a liquid.

13-13 Phase Diagrams (*P* vs. *T*)

Now that we have discussed the general properties of the three phases of matter, we can describe **phase diagrams**. They show the equilibrium pressure–temperature relationships among the different phases of a given substance in a closed system. Our discussion of phase diagrams applies only to *closed systems* (e.g., a sample in a sealed container), in which matter does not escape into the surroundings. This limitation is especially important when the vapor phase is involved. Figure 13-17 shows a portion of the phase diagrams for water and carbon dioxide. The curves are not drawn to scale. The distortion allows us to describe the changes of state accompanying different pressure or temperature changes using one diagram rather than several.

The curved line from *A* to *C* in Figure 13-17a is a vapor pressure curve obtained experimentally by measuring the vapor pressures of water at var-

Table 13-8
Points on the Vapor Pressure Curve for Water

temperature (°C)	−10	0	20	30	50	70	90	95	100	101
vapor pressure (torr)	2.1	4.6	17.5	31.8	92.5	234	526	634	760	788

Camphor, which is used in inhalers, has a high vapor pressure. When stored in a bottle, camphor sublimes and then deposits elsewhere in the bottle.

ious temperatures (Table 13-8). Points along this curve represent the temperature–pressure combinations for which liquid and gas (vapor) coexist in equilibrium. At points above *AC*, the stable form of water is liquid; below the curve, it is vapor.

Line *AB* represents the liquid–solid equilibrium conditions. Note that it has a negative slope. Water is one of the very few substances for which this is the case. The slope up and to the left indicates that increasing the pressure sufficiently on the surface of ice causes it to melt. This is because ice is *less dense* than liquid water in the vicinity of the liquid–solid equilibrium. The network of hydrogen bonding in ice is more extensive than that in liquid water and requires a greater separation of H_2O molecules. This causes ice to float in liquid water. Almost all other solids are more compact (more dense) than their corresponding liquids; they would have positive slopes associated with line *AB*.

The stable form of water at points to the left of *AB* is solid (ice). Thus, *AB* is called a *melting curve*. At pressures and temperatures along *AD*, the *sublimation curve*, solid and vapor are in equilibrium. There is only one point, *A*, at which all three phases—ice, liquid water, and water vapor—can coexist at equilibrium. This is called the **triple point**. For water it occurs at 4.6 torr and 0.01°C.

At pressures below the triple-point pressure, the liquid phase does not exist; rather, the substance goes directly from solid to gas (sublimes) or the reverse happens (crystals deposit from the gas).

Consider CO_2 (Figure 13-17b). The triple point is at 5.2 atmospheres and −57°C. Because this pressure is *above* normal atmospheric pressure, liquid

The CO_2 in common fire extinguishers is liquid. As you can see from Figure 13-17b, the liquid must be at some pressure greater than 10 atm for temperatures above 0°C. It is ordinarily at about 65 atm (more than 900 lb/in²), so these cylinders must be *handled with care*.

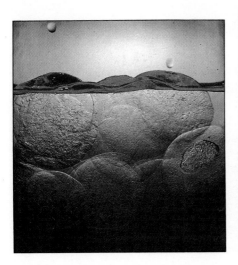

Benzene is *more* dense as a solid than as a liquid, so the solid sinks in the liquid (right). This is the behavior shown by nearly all known substances except water (left).

CO_2 cannot exist at atmospheric pressure. Dry Ice (solid CO_2) sublimes and does not melt at atmospheric pressure.

The **critical temperature** is the temperature above which a gas cannot be liquefied, i.e., the temperature above which the liquid and gas do not exist as distinct phases. A substance at a temperature above its critical temperature is called a *supercritical fluid*. The **critical pressure** is the pressure required to liquefy a gas (vapor) *at* its critical temperature. The combination of critical temperature and critical pressure is called the **critical point** (*C* in Figure 13-17). For H_2O, the critical point is 374°C and 218 atmospheres; for CO_2, 31°C and 73 atmospheres.

To illustrate the use of a phase diagram in determining the physical state or states of a system under different sets of pressures and temperatures, let's consider a sample of water at point *E* in Figure 13-17a (355 torr and −10°C). At this point all the water is in the form of ice. Suppose that we hold the pressure constant and gradually increase the temperature—in other words, trace a path from left to right along *EG*. At the temperature at which *EG* intersects *AB*, the melting curve, some of the ice melts. If we stopped here, equilibrium between ice and liquid water would eventually be established, and both phases would be present. If we added more heat, all the ice would melt with no temperature change. Remember that all phase changes of pure substances occur at constant temperature.

Once the ice is completely melted, additional heat causes the temperature to rise. Eventually, at point *F* (355 torr and 80°C), some of the liquid begins to boil; liquid and vapor are in equilibrium. Adding more heat at constant pressure vaporizes the rest of the water with no temperature change. Adding still more heat warms the gas from *F* to *G*. Complete vaporization would also occur if, at point *F* and before all the liquid had vaporized, the temperature were held constant and the pressure were decreased to, say, 234 torr at point *H*. If we wished to hold the pressure at 234 torr and condense some of the vapor, it would be necessary to cool the vapor to 70°C, point *I*, which lies on the vapor pressure curve, *AC*. To state this in another way, the vapor pressure of water at 70°C is 234 torr.

Compare the description here with that accompanying Figure 13-15. Each horizontal line on a phase diagram contains the information in one heating curve obtained at a particular pressure. Each horizontal line in the warming curve (Figure 13-15) corresponds to crossing one of the phase equilibrium lines in the phase diagram (Figure 13-17) at a particular pressure. Phase diagrams are obtained by combining the results of heating curves measured experimentally at different pressures.

A weighted wire cuts through a block of ice. The ice melts under the high pressure of the wire, and then refreezes behind the wire.

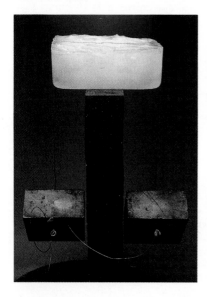

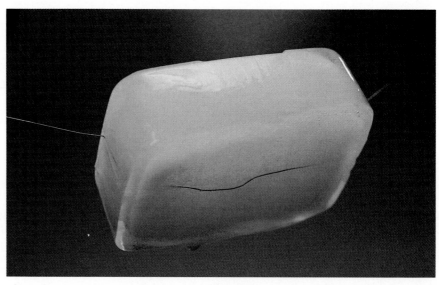

Suppose we move back to ice at point E (355 torr and $-10°C$). If we now hold the temperature at $-10°C$ and reduce the pressure, we move vertically down along EJ. At a pressure of 2.1 torr we reach the sublimation curve, at which point some of the solid passes directly into the gas phase (sublimes). Further decreasing the pressure vaporizes all the ice. An important application of this phenomenon is in the freeze-drying of foods. In this process a water-containing food is cooled below the freezing point of water to form ice, which is then removed as a vapor by decreasing the pressure.

Let us clarify the nature of the fluid phases (liquid and gas) and of the critical point by describing two different ways by which a gas can be liquefied. A sample at point W in the phase diagram of Figure 13-17c is in the vapor (gas) phase, below its critical temperature. Suppose we compress the sample at constant T from point W to point Z. We can identify a definite pressure (the interaction of line WZ with the vapor pressure curve AC) where the transition from gas to liquid takes place. However, if we go *around* the critical point by the path $WXYZ$, no such clear-cut transition takes place. By this second path, the density and other properties of the sample vary in a continuous manner; there is no definite point at which we can say that the sample changes from gas to liquid. So we use the term *fluid* to describe either a gas or a liquid.

A fluid *below* its critical temperature may properly be identified as a liquid or as a gas. *Above* the critical temperature, we should use the term "fluid." The somewhat unusual properties of supercritical fluids lead to some novel uses. For instance, supercritical CO_2 is a useful solvent that can be removed much more completely than more conventional solvents.

13-14 Amorphous Solids and Crystalline Solids

We have already seen that solids have definite shapes and volumes, are not very compressible, are dense, and diffuse only very slowly into other solids. They are generally characterized by compact, ordered arrangements of particles that vibrate about fixed positions in their structures.

However, some noncrystalline solids, called **amorphous solids**, have no well-defined, ordered structure. Examples include rubber, some kinds of plastics, and amorphous sulfur.

Glasses are sometimes called amorphous solids and sometimes called **supercooled liquids**. The justification for calling them supercooled liquids is that, like liquids, they flow, although *very* slowly. The irregular structures of glasses are intermediate between those of freely flowing liquids and those of crystalline solids; there is only short-range order. Unlike crystalline solids, glasses and other amorphous solids do not exhibit sharp melting points, but soften over a temperature range. Crystalline solids such as ice and sodium chloride have well-defined, sharp melting temperatures. Because particles in amorphous solids are irregularly arranged, intermolecular forces among particles vary in strength within a sample. Melting occurs at different temperatures for various portions of the same sample as the intermolecular forces are overcome.

The shattering of a crystalline solid produces fragments having the same (or related) interfacial angles and structural characteristics as the original sample. The shattering of a large cube of rock salt produces several small cubes of rock salt. This cleaving of crystals occurs preferentially along crystal lattice planes between which the interionic or intermolecular forces of attraction are weakest. Amorphous solids with irregular structures, such as glasses, shatter irregularly to yield pieces with jagged edges and irregular angles.

The regular external shape of a crystal is the result of regular internal arrangements of atoms, molecules, or ions. Crystals of the ionic solid sodium chloride, NaCl, from a kitchen saltshaker have the same shape as the large crystal shown here.

One test for the purity of a crystalline solid is the sharpness of its melting point. Impurities disrupt the intermolecular forces and cause melting to occur over a considerable temperature range.

The lattice planes are planes within the crystal containing ordered arrangements of particles.

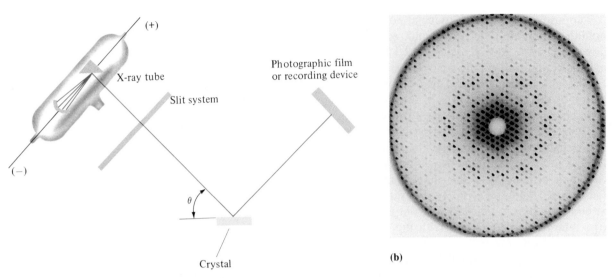

(a)

Figure 13-18

(a) X-ray diffraction by crystals (schematic). (b) A photograph of the X-ray diffraction pattern from a crystal of the enzyme histidine decarboxylase (MW ~37,000 amu). The crystal was rotated so that many different lattice planes with different spacings were moved in succession into diffracting position (Figure 13-19).

<table>
<tr><td>Enrichment</td><td>

X-Ray Diffraction

Atoms, molecules, and ions are much too small to be seen with the eye. The arrangements of particles in crystalline solids are determined indirectly by X-ray diffraction (scattering). In 1912 the German physicist Max von Laue showed that any crystal could serve as a three-dimensional diffraction grating for incident electromagnetic radiation, with wavelengths approximating the internuclear separations of atoms in the crystal. Such radiation is in the X-ray region of the electromagnetic spectrum. Using an apparatus similar to that shown in Figure 13-18, a monochromatic (single-wavelength) X-ray beam is defined by a system of slits and directed onto a crystal. The crystal is rotated to vary the angle of incidence θ. At various angles, strong beams of deflected X-rays hit a photographic plate. Upon development, the plate shows a set of symmetrically arranged spots due to deflected X-rays. Different kinds of crystals produce different arrangements of spots.

In 1913 the English scientists William and Lawrence Bragg found that diffraction photographs are more easily interpreted by treating the crystal as a reflection grating rather than a diffraction grating. The analysis of the spots is somewhat complicated, but an experienced crystallographer can determine the separations between atoms within identical layers and the distances between layers of atoms. It is also possible to determine the identities of individual atoms. The more electrons an atom has, the more strongly it scatters X-rays.

Figure 13-19 illustrates the determination of spacings between layers of atoms. The X-ray beam strikes parallel layers of atoms in the crystal at an angle θ. Those rays colliding with atoms in the first layer are reflected at the same angle θ. Those passing through the first layer may be reflected

</td></tr>
</table>

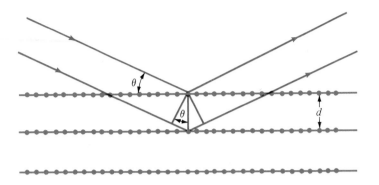

Figure 13-19
Reflection of a monochromatic beam of X-rays by two lattice planes (layers of atoms) of a crystal.

from the second layer, third layer, and so forth. A reflected beam results only if all rays are in phase.

In order for the waves to be in phase (interact constructively), the difference in path length must be equal to the wavelength, λ, times an integer, n. This leads to the condition known as the **Bragg equation**:

$$n\lambda = 2d \sin \theta \quad \text{or} \quad \sin \theta = \frac{n\lambda}{2d}$$

It tells us that for X-rays of a given wavelength λ, atoms in planes separated by distances d give rise to reflections at angles of incidence θ. The reflection angles increase with increasing order, $n = 1, 2, 3 \ldots$.

13-15 Structures of Crystals

All crystals contain regularly repeating arrays of atoms, molecules, or ions. They are analogous (but in three dimensions) to a wallpaper pattern (Figure 13-20). Once we discover the pattern of a wallpaper, we can repeat it in two dimensions to cover a wall. To describe such a repeating pattern we must specify two things: (1) the size and shape of the repeating unit and (2) the contents of this unit. In the wallpaper pattern of Figure 13-20a, several choices of the repeating unit are outlined. Repeating unit A contains one complete cat; unit B, with the same area, contains parts of several different

Figure 13-20
Patterns that repeat in two dimensions. Such patterns might be used to make wallpaper. We must imagine that the pattern extends indefinitely (to the end of the wall). In each pattern, two of the many possible choices of unit cells are outlined. Once we identify a unit cell and its contents, repetition by translating this unit generates the entire pattern. In (a), the unit cell contains only one cat. In (b), each cell contains two cats related to one another by a 180° rotation. Any crystal is an analogous pattern in which the contents of the three-dimensional unit cell consist of atoms, molecules, or ions. The pattern extends in *three* dimensions to the boundaries of the crystal, usually including many thousands of unit cells.

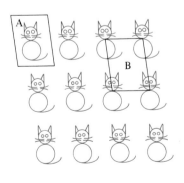

(a)

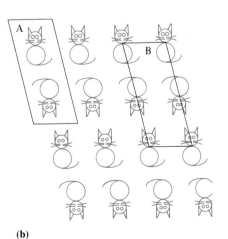

(b)

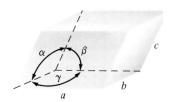

Figure 13-21
A representation of a unit cell.

cats, but these still add up to one complete cat. Whichever unit we choose, we can then obtain the entire pattern by repeatedly translating the contents of that unit in two dimensions.

In a crystal, the repeating unit is three-dimensional; its contents consist of atoms, molecules, or ions. The smallest unit of volume of a crystal that shows all the characteristics of the crystal's pattern is a **unit cell**. We note that the unit cell is just the fundamental *box* that describes the arrangement. The unit cell is described by the lengths of its edges—a, b, c (which are related to the spacings between layers, d)—and the angles between the edges—α, β, γ (Figure 13-21). Unit cells are stacked in three dimensions to build a lattice, the three-dimensional *arrangement* corresponding to the crystal. It can be proven that unit cells must fit into one of the seven crystal systems (Table 13-9). Each crystal system is distinguished by the relations between the unit cell lengths and angles *and* by the symmetry of the resulting three-dimensional patterns. Crystals have the same symmetry as their constituent unit cells because all crystals are repetitive multiples of such cells.

Let us replace each repeat unit in the crystal by a point (called a *lattice point*) placed at the same place in the unit. All such points have the same environment and are indistinguishable from each other. The resulting three-dimensional array of points is called a **lattice**. It is a simple but complete description of the way in which a crystal structure is built up.

The unit cells shown in red in Figure 13-22 are the simple, or primitive, unit cells corresponding to the seven crystal systems listed in Table 13-9. Each of these unit cells corresponds to *one* lattice point. As a two-dimensional representation of the reasoning behind this statement, look at the unit cell marked "B" in Figure 13-20a. Each corner of the unit cell is a lattice point, and can be imagined to represent one cat. The cat at each corner is shared among four unit cells (remember—we are working in two dimensions here). The unit cell has four corners, and in the corners of the unit cell are enough pieces to make one complete cat. Thus, unit cell B contains one cat, the same as the alternative unit cell choice marked "A." Now imagine that each lattice point in a three-dimensional crystal represents an object (a molecule, an atom, and so on). Such an object at a corner (Figure 13-23a) is shared by the eight unit cells that meet at that corner. Because each unit cell has eight corners, it contains eight "pieces" of the object, so it contains $8(\frac{1}{8}) = 1$ object. Similarly, an object on an edge, but not at a corner, is shared

Table 13-9
The Unit Cell Relationships for the Seven Crystal Systems*

System	Unit Cell Lengths	Unit Cell Angles	Example (common name)
cubic	$a = b = c$	$\alpha = \beta = \gamma = 90°$	NaCl (rock salt)
tetragonal	$a = b \neq c$	$\alpha = \beta = \gamma = 90°$	TiO_2 (rutile)
orthorhombic	$a \neq b \neq c$	$\alpha = \beta = \gamma = 90°$	$MgSO_4 \cdot 7H_2O$ (epsomite)
monoclinic	$a \neq b \neq c$	$\alpha = \gamma = 90°; \beta \neq 90°$	$CaSO_4 \cdot 2H_2O$ (gypsum)
triclinic	$a \neq b \neq c$	$\alpha \neq \beta \neq \gamma \neq 90°$	$K_2Cr_2O_7$ (potassium dichromate)
hexagonal	$a = b \neq c$	$\alpha = \beta = 90°; \gamma = 120°$	SiO_2 (silica)
rhombohedral	$a = b = c$	$\alpha = \beta = \gamma \neq 90°$	$CaCO_3$ (calcite)

* In these definitions, the sign $\neq$ means "is not *necessarily* equal to."

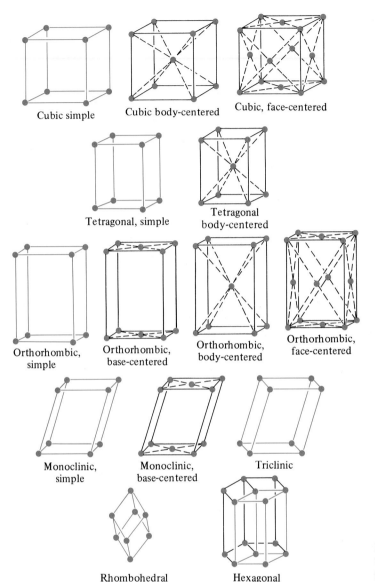

Cubic simple

Cubic body-centered

Cubic, face-centered

Tetragonal, simple

Tetragonal body-centered

Orthorhombic, simple

Orthorhombic, base-centered

Orthorhombic, body-centered

Orthorhombic, face-centered

Monoclinic, simple

Monoclinic, base-centered

Triclinic

Rhombohedral

Hexagonal

Figure 13-22
Unit cells of the 14 crystal lattices. The primitive cells of the seven crystal systems are shown in red.

by four unit cells (Figure 13-23b), and an object on a face is shared by two unit cells (Figure 13-23c).

Each unit cell contains atoms, molecules, or ions in a definite arrangement. Often the unit cell contents are related by some additional symmetry. For instance, the unit cell in Figure 13-20b contains *two* cats, related to one another by a rotation of 180°. Different substances that crystallize in the same type of lattice with the same atomic arrangement are said to be **isomorphous**. A single substance that can crystallize in more than one arrangement is said to be **polymorphous**.

Points equivalent to the unit cell corners may also appear at certain other positions. This results in additional lattices besides the simple ones, giving a total of 14 crystal lattices whose unit cells are shown in Figure 13-22. In a *simple*, or *primitive*, lattice, only the eight corners of the cubic unit cell

A crystal of one form of manganese metal has Mn atoms at the corners of a simple cubic unit cell that is 6.30 Å on edge (Example 13-7).

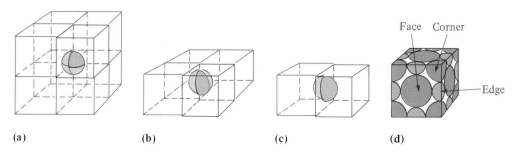

(a)   (b)   (c)   (d)

Figure 13-23

Representation of the sharing of an object (an atom, ion, or molecule) among unit cells. The fraction of each sphere that "belongs" to a single unit cell is shown in red. (a) The sharing of an object in a face by two unit cells. (b) The sharing of an object on an edge by four unit cells. (c) The sharing of an object at a corner by eight unit cells. (d) A representation of a unit cell that illustrates the portions of atoms presented in more detail in Figure 13-27. The green ion at each corner is shared by eight unit cells, as in (a). The gray ion at each edge is shared by four unit cells, as in (b). The green ion in each face is shared by two unit cells, as in (c).

Each point in a face is shared between two unit cells, so it is counted $\frac{1}{2}$ in each; there are six faces in each unit cell.

are equivalent. In other types of crystals, particles like those forming the outline of the unit cell may occupy extra positions within the unit cell. A *body-centered* lattice has equivalent points at the eight unit cell corners *and* at the center of the unit cell (top row center, Figure 13-22). Iron, chromium, and many other metals crystallize in a body-centered cubic (bcc) arrangement. The unit cell of such a metal contains $8(\frac{1}{8}) = 1$ atom at the corners of the cell *plus* one atom at the center of the cell (and therefore entirely in this cell); this makes a total of *two* atoms per unit cell. A *face-centered* structure involves the eight points at the corners and six more equivalent points, one in the middle of each of the six square faces of the cell. A metal (calcium and silver are cubic examples) that crystallizes in this arrangement has $8(\frac{1}{8}) = 1$ atom at the corners *plus* $6(\frac{1}{2}) = 3$ more in the faces, for a total of *four* atoms per unit cell. In more complicated crystals, each lattice site may represent several atoms or an entire molecule.

We have discussed a few simple structures that are easy to visualize. Many more complex compounds crystallize in structures with unit cells that can be more difficult to describe. Experimental determination of the crystal structures of such solids is usually correspondingly more complex. However, modern computer-controlled instrumentation is used to collect and analyze the large amounts of X-ray diffraction data used in such studies. This now allows analysis of structures ranging from simple metals to complex biological molecules such as proteins and nucleic acids. Most of our knowledge about the three-dimensional arrangements of atoms is based on such crystal structure studies.

13-16 Bonding in Solids

We classify crystalline solids into categories that depend on the types of particles in the crystal and the bonding or interactions among them. The four categories are (1) metallic solids, (2) ionic solids, (3) molecular solids, and (4) covalent solids. Table 13-10 summarizes these categories of solids and their typical properties.

Metallic Solids

Metals crystallize as solids in which metal ions may be thought to occupy the lattice sites and are embedded in a cloud of delocalized valence electrons. Practically all metals crystallize in one of three types of lattices: (1) body-centered cubic (bcc), (2) face-centered cubic (fcc; also called cubic close-

Table 13-10
Characteristics of Types of Solids

	Metallic	Ionic	Molecular	Covalent
Particles of unit cell	Metal ions in "electron gas"	Anions, cations	Molecules (or atoms)	Atoms
Strongest interparticle forces	Metallic bonds (attraction between cations and e^-'s)	Electrostatic	London, dipole–dipole, and/or hydrogen bonds	Covalent bonds
Properties	Soft to very hard; good thermal and electrical conductors; wide range of melting points (-39 to $3400°C$)	Hard; brittle; poor thermal and electrical conductors; high melting points (400 to $3000°C$)	Soft; poor thermal and electrical conductors; low melting points (-272 to $400°C$)	Very hard; poor thermal and electrical conductors;* high melting points (1200 to $4000°C$)
Examples	Li, K, Ca, Cu, Cr, Ni (metals)	$NaCl$, $CaBr_2$, K_2SO_4 (typical salts)	CH_4 (methane), P_4, O_2, Ar, CO_2, H_2O, S_8	C (diamond), SiO_2 (quartz)

* Exceptions: diamond is a good conductor of heat; graphite is soft and conducts electricity well.

packed), and (3) hexagonal close-packed. The latter two types are called close-packed structures because the particles (in this case metal atoms) are packed together as closely as possible. The differences between the two close-packed structures are illustrated in Figure 13-24 and 13-25. We shall let spheres of equal size represent the identical metal atoms, or any other particles, that form close-packed structures. Consider a layer of spheres packed in a plane, *A*, as closely as possible (Figure 13-25a). An identical plane of spheres, *B*, is placed in the depressions of plane *A*. If the third plane is placed with its spheres directly above those in plane *A*, the *ABA* arrangement results. This is the hexagonal close-packed structure (Figure 13-25a). The extended pattern of arrangement of planes is *ABABAB* .... If the third layer is placed in the alternate set of depressions in the second layer so that spheres in the first and third layers are *not* directly above and below each other, the cubic close-packed structures, *ABCABCABC* ..., results (Figure 13-25b). In close-packed structures, each sphere has a co-ordination number of 12, i.e., 12 nearest neighbors. In ideal close-packed structures, 74% of a given volume is due to spheres and 26% is empty space. The body-centered cubic structure is less efficient in packing; each sphere has only eight nearest neighbors, and there is more empty space.

The term "coordination number" is used in crystallography in a somewhat different sense from that in coordination chemistry (Section 29-31). Here it refers to the number of near neighbors.

 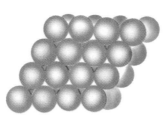

(a) **(b)**

Figure 13-24
(a) Spheres in the same plane, packed as closely as possible. Each sphere touches six others. (b) Spheres in two planes, packed as closely as possible. Real crystals have far more than two planes. Each sphere touches six others in its own layer, three in the layer below it, and three in the layer above it; i.e., it contacts a total of 12 other spheres (has a coordination number of 12).

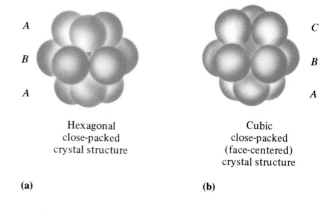

Hexagonal
close-packed
crystal structure

(a)

Cubic
close-packed
(face-centered)
crystal structure

(b)

Figure 13-25
There are two crystal structures in which atoms are packed together
as compactly as possible. The diagrams show the structures
expanded to clarify the difference between them. (a) In the hexagonal
close-packed structure, the first and third layers are oriented in the
same direction, so that each atom in the third layer (*A*) lies directly
above an atom in the first layer (*A*). (b) In the cubic close-packed
structure, the first and third layers are oriented in opposite
directions, so that no atom in the third layer (*C*) is directly above an
atom in either of the first two layers (*A* and *B*). In both cases, every
atom is surrounded by 12 other atoms if the structure is extended
indefinitely, so each atom has a coordination number of 12. Although
it is not obvious from this figure, the cubic close-packed structure is
face-centered cubic. To see this, we would have to include additional
atoms and tilt the resulting cluster of atoms.

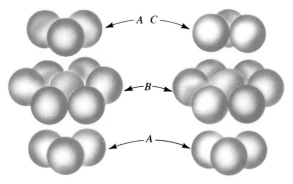

Expanded view

Example 13-7
In the simple cubic form of manganese metal, there are Mn atoms at the corners
of a simple cubic unit cell that is 6.30 Å on edge. (a) What is the shortest distance
between neighboring Mn atoms? (b) How many nearest neighbors does each atom
have?

Plan
We visualize the simple cubic cell:

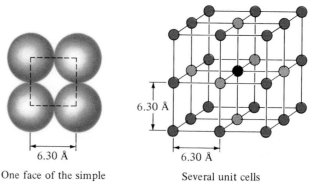

One face of the simple
cubic unit cell

Several unit cells
(atoms shown smaller for clarity)

Solution

(a) One face of the cubic unit cell is shown in the left-hand drawing, with the atoms touching. The nearest-neighbor atoms are separated by one unit cell edge, at the distance 6.30 Å.

(b) A three-dimensional representation of eight unit cells is also shown. In that drawing, the atoms are shown smaller for clarity. Some atoms are shaded to aid in visualizing the arrangement, but *all atoms are identical.* Consider the atom shown in black at the center (at the intersection of the eight unit cells). Its nearest neighbors in all of the unit cell directions are shown as shaded atoms. As we can see, there are six nearest neighbors. The same would be true of any other atom in the structure.

Example 13-8

Gold crystals are face-centered cubic, with a cell edge of 4.10 Å. (a) What is the distance between centers of the two closest Au atoms? (b) How many nearest neighbors does each atom have?

Plan

We reason as in Example 13-7, except that now the two atoms closest to one another are those along the face diagonal.

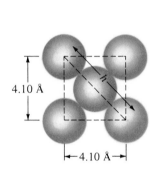

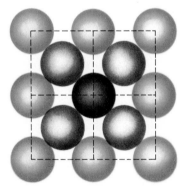

One face of the face-centered   One face of four adjacent unit cells
cubic unit cell   (x-y plane)

Solution

(a) One face of the face-centered cubic unit cell is shown in the left-hand drawing, with the atoms touching. The nearest-neighbor atoms are the ones along the diagonal of the face of the cube. We may visualize the face as consisting of two right isosceles triangles sharing a common hypotenuse, h, and having sides of length $a = 4.10$ Å. The hypotenuse is equal to *twice* the center-to-center distance. The hypotenuse can be calculated from the Pythagorean theorem, $h^2 = a^2 + a^2$. The length of the hypotenuse equals the square root of the sum of the squares of the sides:

$$h = \sqrt{a^2 + a^2} = \sqrt{2a^2} = \sqrt{2(4.10 \text{ Å})^2} = 5.80 \text{ Å}$$

The distance between centers of adjacent gold atoms is one half of h, so

$$\text{distance} = \frac{5.80 \text{ Å}}{2} = \boxed{2.90 \text{ Å}}$$

(b) To see the number of nearest neighbors, let us expand the left-hand drawing to include several unit cells, as shown in the right-hand drawing. Suppose that

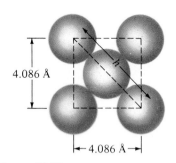

Figure 13-26
One face of the face-centered cubic unit cell of metallic silver (Example 13-9).

this is the *x–y* plane. The atom shown in black has four nearest neighbors in this plane. There are four more such neighbors in the *x–z* plane (perpendicular to the *x–y* plane), and four additional neighbors in the *y–z* plane (also perpendicular to the *x–y* plane). This gives a total of   12 nearest neighbors.

EOC 88, 92

Example 13-9

The unit cell of metallic silver is face-centered cubic (or cubic close-packed) with $a = b = c = 4.086$ Å. Calculate (a) the radius of an Ag atom; (b) the volume of an Ag atom in cm³, given that, for a sphere, $V = (\frac{4}{3})\pi r^3$; (c) the percentage of the volume of a unit cell that is occupied by Ag atoms; and (d) the percentage that is empty space.

Plan

(a) One face of the fcc unit cell can be visualized (Figure 13-26) as consisting of (parts of) five Ag atoms describing two right triangles that share a hypotenuse. The hypotenuse, *h*, is four times the radius of the silver atom. We can use the Pythagorean theorem to evaluate r_{Ag}. (b) The volume of a sphere is $(\frac{4}{3})\pi r^3$. (c) We can use the length of the edge of the unit cell to find its total volume. The volume occupied by the Ag atoms is equal to the number of atoms per unit cell times the volume of each atom. These two values can be used to calculate the required percentage. (d) The percentage that is empty space is 100% minus the percentage that is occupied.

Solution

(a) $h = \sqrt{a^2 + a^2} = \sqrt{2a^2} = \sqrt{2(4.086\text{ Å})^2} = 5.778\text{ Å} = 4r_{Ag}$

$$r_{Ag} = \frac{5.778\text{ Å}}{4} = \boxed{1.444\text{ Å}}$$

(b)   The volume of an Ag atom is $(\frac{4}{3})\pi r^3_{Ag}$:

$$V = (\tfrac{4}{3})\pi(1.444\text{ Å})^3 = 12.61\text{ Å}^3$$

We now convert Å³ to cm³:

$$?\text{ cm}^3 = 12.61\text{ Å}^3 \times \left(\frac{10^{-8}\text{ cm}}{\text{Å}}\right)^3 = \boxed{1.261 \times 10^{-23}\text{ cm}^3}$$

(c)   The entire fcc unit cell contains $8(\frac{1}{8}) + 6(\frac{1}{2}) = 4$ Ag atoms. Each Ag atom has a volume of 1.261×10^{-23} cm³. Thus, the volume of four Ag atoms is

$$V_{Ag\ atoms} = 4\text{ Ag atoms} \times \frac{1.261 \times 10^{-23}\text{ cm}^3}{\text{Ag atom}} = 5.044 \times 10^{-23}\text{ cm}^3$$

Because the unit cell is cubic, its volume is a^3:

$$V_{unit\ cell} = (4.086\text{ Å})^3 = 68.22\text{ Å}^3 \times \left(\frac{10^{-8}\text{ cm}}{\text{Å}}\right)^3 = 6.822 \times 10^{-23}\text{ cm}^3$$

The percentage of the volume of the unit cell occupied by the silver atoms is

$$\%\ Ag_{volume} = \frac{V_{Ag\ atoms}}{V_{unit\ cell}} \times 100\% = \frac{5.044 \times 10^{-23}\text{ cm}^3}{6.822 \times 10^{-23}\text{ cm}^3} \times 100\% = \boxed{73.9\%}$$

This result agrees well with our earlier observation that ideal close-packed structures are about 26% empty space.

(d)   % empty space = $(100.0 - 73.9)\% = \boxed{26.1\%}$

EOC 114

Example 13-10

From data in Example 13-9, calculate the density of metallic silver.

Plan

We first determine the mass of a unit cell, i.e., the mass of four atoms of silver. The density of the unit cell, and therefore of silver, is its mass divided by its volume.

Solution

$$\underline{?}\text{ g Ag per unit cell} = \frac{4\text{ Ag atoms}}{\text{unit cell}} \times \frac{1\text{ mol Ag}}{6.022 \times 10^{23}\text{ Ag atoms}} \times \frac{107.87\text{ g Ag}}{1\text{ mol Ag}}$$

$$= 7.165 \times 10^{-22}\text{ g Ag/unit cell}$$

$$\text{density} = \frac{7.165 \times 10^{-22}\text{ g Ag/unit cell}}{6.822 \times 10^{-23}\text{ cm}^3/\text{unit cell}} = \boxed{10.50\text{ g/cm}^3}$$

A handbook gives 10.5 g/cm³ as the density of silver at 20°C.

EOC 90

Data obtained from crystal structures and observed densities give us information from which we can calculate the value of Avogadro's number. The next example illustrates these calculations.

Example 13-11

Titanium crystallizes in a body-centered cubic unit cell with an edge length of 3.306 Å. The density of titanium metal is measured to be 4.401 g/cm³. Use these data to calculate Avogadro's number.

Plan

We relate the density and the volume of the unit cell to find the total mass contained in one unit cell. Knowing the number of atoms per unit cell, we can then find the mass of one atom. Comparing this to the known atomic weight, which is the mass of one mole (Avogadro's number) of atoms, we can evaluate Avogadro's number.

Solution

We first determine the volume of the unit cell:

$$V_{\text{cell}} = (3.306\text{ Å})^3 = 36.13\text{ Å}^3$$

We now convert Å³ to cm³:

$$\underline{?}\text{ cm}^3 = 36.13\text{ Å}^3 \times \left(\frac{10^{-8}\text{ cm}}{\text{Å}}\right)^3 = 3.613 \times 10^{-23}\text{ cm}^3$$

The mass of the unit is its volume times the measured density:

$$\text{mass of unit cell} = 3.613 \times 10^{-23}\text{ cm}^3 \times \frac{4.401\text{ g}}{\text{cm}^3} = 1.590 \times 10^{-22}\text{ g}$$

The bcc unit cell contains $8(\frac{1}{8}) + 1 = 2$ Ti atoms, so this represents the mass of two Ti atoms. The mass of a single Ti atom is

$$\text{mass of atom} = \frac{1.590 \times 10^{-22}\text{ g}}{2\text{ atoms}} = 7.950 \times 10^{-23}\text{ g/atom}$$

From the known atomic weight of Ti (47.88), we know that the mass of one mole of Ti is 47.88 g/mol. Avogadro's number represents the number of atoms per mole, and can be calculated as

$$N_{Av} = \frac{47.88 \text{ g}}{\text{mol}} \times \frac{1 \text{ atom}}{7.950 \times 10^{-23} \text{ g}} = \boxed{6.023 \times 10^{23} \text{ atoms/mol}}$$

Ionic Solids

Most salts crystallize as ionic solids with ions occupying the unit cell. Sodium chloride (Figure 13-27) is an example. Many other salts crystallize in the sodium chloride (face-centered cubic) arrangement. Examples are the halides of Li^+, K^+, and Rb^+, and $M^{2+}X^{2-}$ oxides and sulfides such as MgO, CaO, CaS, and MnO. Two other common ionic structures are those of cesium chloride, CsCl (simple cubic lattice), and zincblende, ZnS (face-centered cubic lattice), shown in Figure 13-28. Salts that are isomorphous with the CsCl structure include CsBr, CsI, NH_4Cl, TlCl, TlBr, and TlI. The sulfides of Be^{2+}, Cd^{2+}, and Hg^{2+}, together with CuBr, CuI, AgI, and ZnO, are isomorphous with the zincblende structure.

Because the ions in an ionic solid can vibrate only about their fixed positions, ionic solids are poor electrical and thermal conductors. However, molten ionic compounds are excellent conductors because their ions are freely mobile.

In certain types of solids, including ionic crystals, particles *different* from those at the corners of the unit cell may occupy extra positions within the unit cell. For example, the face-centered cubic unit cell of sodium chloride can be visualized as having chloride ions at the corners and middles of the faces; sodium ions are on the edges between the chloride ions and in the center (Figure 13-27). Thus, a unit cell of NaCl contains the following:

$$Cl^-: \quad \begin{matrix} \text{eight at} \\ \text{corners} \\ 8(\frac{1}{8}) \end{matrix} \quad + \quad \begin{matrix} \text{six in middles} \\ \text{of faces} \\ 6(\frac{1}{2}) \end{matrix} \quad = 1 + 3 = 4 \text{ Cl}^- \text{ ions/unit cell}$$

$$Na^+: \quad \begin{matrix} \text{twelve on} \\ \text{edges} \\ 12(\frac{1}{4}) \end{matrix} \quad + \quad \begin{matrix} \text{one in center} \\ \\ + 1 \end{matrix} \quad = 3 + 1 = 4 \text{ Na}^+ \text{ ions/unit cell}$$

The unit cell contains equal numbers of Na^+ and Cl^- ions, as required by its chemical formula. Alternatively, we could translate the unit cell by half its length in any axial direction within the lattice, and visualize the unit cell in which sodium and chloride ions have exchanged positions. Such an exchange is not always possible. You should confirm that this alternative description also gives four chloride ions and four sodium ions per unit cell.

Figure 13-27

Some representations of the crystal structure of sodium chloride, NaCl. Sodium ions are shown in gray and chloride ions are shown in green.
(a) One unit cell of the crystal structure of sodium chloride. (b) A cross section of the space lattice of NaCl, showing the repeating pattern of its unit cell at the right. The dashed lines outline an alternative choice of the unit cell. The entire pattern is generated by repeating either unit cell (and its contents) in all three directions. Several such choices of unit cells are usually possible.
(c) The three-dimensional representation of part (b). One unit cell is outlined in red. (d) A representation of the unit cell of sodium chloride that indicates the relative sizes of the Na^+ and Cl^- ions as well as how ions are shared between unit cells. Particles at the corners, edges, and faces of unit cells are shared by other unit cells. Remember that there is an additional Na^+ ion at the center of the cube.

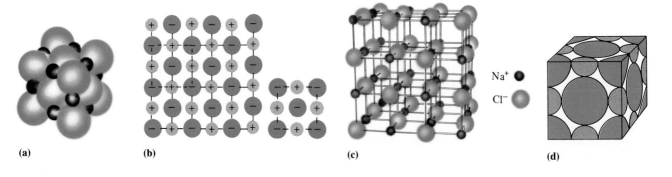

Na$^+$ ●
Cl$^-$ ◯

(a)   (b)   (c)   (d)

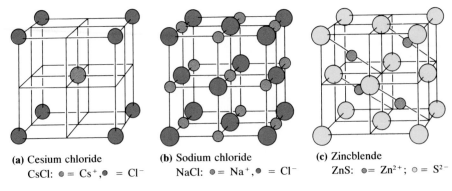

(a) Cesium chloride
CsCl: ● = Cs^+, ● = Cl^-

(b) Sodium chloride
NaCl: ● = Na^+, ● = Cl^-

(c) Zincblende
ZnS: ● = Zn^{2+}; ○ = S^{2-}

Figure 13-28
Crystal structures of some ionic compounds of the MX type. The gray circles represent cations. One unit cell of each structure is shown. (a) The structure of cesium chloride, CsCl, is simple cubic. It is *not* body-centered, because the point at the center of the cell (Cs^+) is not the same as the point at a corner of the cell (Cl^-). (b) Sodium chloride, NaCl is face-centered cubic. (c) Zincblende, ZnS, is face-centered cubic, with four Zn^{2+} and four S^{2-} ions per unit cell. The Zn^{2+} ions are related by the same translations as the S^{2-} ions.

Ionic radii such as those in Figure 6-1 and Table 14-1 are obtained from X-ray crystallographic determinations of unit cell dimensions, assuming that adjacent ions are in contact with each other.

Example 13-12

Lithium bromide, LiBr, crystallizes in the NaCl face-centered cubic structure with a unit cell edge length of $a = b = c = 5.501$ Å. Assume that the Br^- ions at the corners of the unit cell are in contact with those at the centers of the faces. Determine the ionic radius of the Br^- ion. One face of the unit cell is depicted in Figure 13-29.

Plan

We may visualize the face as consisting of two right isosceles triangles sharing a common hypotenuse, h, and having sides of length $a = 5.501$ Å. The hypotenuse is equal to four times the radius of the bromide ion, $h = 4r_{Br^-}$.

Solution

The hypotenuse can be calculated from the Pythagorean theorem, $h^2 = a^2 + a^2$. The length of the hypotenuse equals the square root of the sum of the squares of the sides.

$$h = \sqrt{a^2 + a^2} = \sqrt{2a^2} = \sqrt{2(5.501 \text{ Å})^2} = 7.780 \text{ Å}$$

The radius of the bromide ion is one fourth of h, so

$$r_{Br^-} = \frac{7.780 \text{ Å}}{4} = \boxed{1.945 \text{ Å}}$$

EOC 82, 83

This value is fairly close to the ionic radius of the Br^- ion listed in Figure 6-1. However, the tabulated value is the *average* value obtained from a number of crystal structures of compounds containing Br^- ions. Calculations of ionic radii assume that anion–anion contact exists. This is not always true. Therefore, calculated radii vary from structure to structure. We should not place too much emphasis on a value of radius obtained from any *single* structure determination.

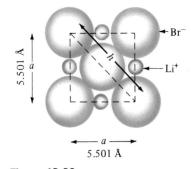

Br⁻
Li⁺
5.501 Å
h
a
a
5.501 Å

Figure 13-29
One face of the face-centered cubic unit cell of lithium bromide (Example 13-12).

Example 13-13

Refer to Example 13-12. Calculate the ionic radius of Li^+ in LiBr, assuming anion–cation contact along an edge of the unit cell.

Plan

The edge length, $a = 5.501$ Å, is twice the radius of the Br^- ion plus twice the radius of the Li^+ ion. We know from Example 13-12 that the radius for the Br^- ion is 1.945 Å.

Solution

$$5.501 \text{ Å} = 2\,r_{Br^-} + 2\,r_{Li^+}$$

$$2\,r_{Li^+} = 5.501 \text{ Å} - 2(1.945 \text{ Å}) = 1.611 \text{ Å}$$

$$r_{Li^+} = \boxed{0.806 \text{ Å}}$$

The radius of Li^+ is calculated to be 0.806 Å, but Figure 6-1 lists it as 0.60 Å. The discrepancy results from the assumption that the Li^+ ion is in contact with both Br^- ions simultaneously. It is too small and is free to vibrate about a fixed-center position between two large Br^- ions. We now see that there is some difficulty in determining precise values of ionic radii. Similar difficulties can arise in the determination of atomic radii from molecular and covalent solids or of metallic radii from solid metals.

Molecular Solids

The lattice positions that describe unit cells of molecular solids represent molecules of monatomic elements (sometimes referred to as monatomic molecules). Figure 13-30 shows the unit cells of two simple molecular crystals. Although the bonds *within* molecules are covalent and strong, the forces of attraction *between* molecules are much weaker. They range from hydrogen bonds and weaker dipole–dipole interactions in polar molecules such as H_2O

Figure 13-30

The packing arrangement in a molecular crystal depends on the shape of the molecule as well as on the electrostatic interactions of any regions of excess positive and negative charge in the molecules. The arrangements in some molecular crystals are shown here, with one unit cell outlined: (a) carbon dioxide, CO_2; (b) benzene, C_6H_6.

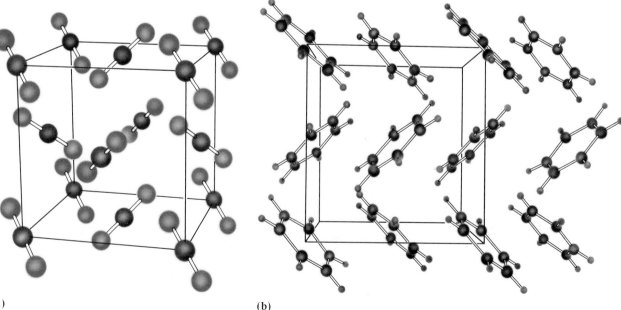

(a)   (b)

and SO_2 to very weak London forces in symmetrical, nonpolar molecules such as CH_4, CO_2, and O_2 and monatomic elements, e.g., the noble gases. Because of the relatively weak intermolecular forces of attraction, molecules can be easily displaced. Thus, molecular solids are usually soft substances with low melting points. Because electrons do not move from one molecule to another under ordinary conditions, molecular solids are poor electrical conductors and good insulators.

London forces are also present among polar molecules.

Covalent Solids

Covalent solids (or "network solids") can be considered giant molecules that consist of covalently bonded atoms in an extended, rigid crystalline network. Diamond (one crystalline form of carbon; Section 27-1) and quartz (Section 27-9) are examples of covalent solids (Figure 13-31). Because of their rigid, strongly bonded structures, *most* covalent solids are very hard and melt at high temperatures. Because electrons are localized in covalent bonds, they are not freely mobile. As a result, covalent solids are *usually* poor thermal and electrical conductors at ordinary temperatures. (However,

(text continued on p. 495)

Figure 13-31
Portions of the atomic arrangements in three covalent solids. (a) Diamond. Each C is bonded tetrahedrally to four others through sp^3-sp^3 σ-bonds (1.52 Å). (b) Quartz (SiO_2). Each Si atom (gray) is bonded tetrahedrally to four O atoms (red). (c) Graphite. C atoms are linked (1.42 Å) in planes by sp^2-sp^2 σ-bonds. Electrons move freely through the π-bonding network in these planes, but they do not jump between planes easily. The crystal is soft, owing to the weakness of the attractions between planes (3.40 Å).

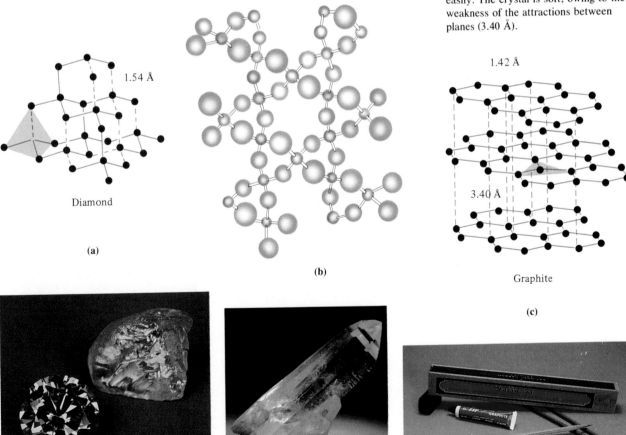

1.54 Å

Diamond

(a)

(b)

1.42 Å

3.40 Å

Graphite

(c)

Diamond

A natural quartz crystal

Graphite

Chemistry in Use. . .
Superconductivity

Superconductivity was discovered by a Dutch scientist named Heike K. Onnes. In 1908 he was the first to accomplish the liquefaction of helium. Using liquid helium as a coolant, he then began to study the low-temperature conductive properties of metals. In 1911 he observed that the resistivity of mercury displays a remarkable behavior as it is cooled to a temperature approaching absolute zero. At about 4 K, all electrical resistance of mercury is suddenly lost within a temperature range of 0.01 K. In 1913 Onnes concluded that "mercury has passed into a new state that, on account of its remarkable electrical properties, may be called the superconducting state." The temperature at which a material enters the superconducting state is now called its *superconducting transition temperature*, T_c.

Soon after its initial discovery, superconductivity was observed in lead at 7.2 K and in tin at 3.7 K. Work done in the past 80 years has shown that superconductivity is a widespread phenomenon. Today about half the metallic elements have been shown to be superconductors. The highest superconducting transition temperature for a pure metal is that observed for niobium, $T_c = 9.5$ K. A number of compounds, as well as a large number of alloys, also exhibit superconductivity. Many of them display transition temperatures higher than those displayed by the pure metals.

Among the more familiar superconductors are those described as A15 superconductors ("A15" is a crystallographic symbol for the β-tungsten structure). This class includes materials such as Nb_3Al, Nb_3Ge, and Nb_3Sn, which exhibit T_c values up to 23.3 K. Many other superconductors have cubic β-1

(NaCl) structures; these have T_c values up to 18 K. Examples include NbN, MoN, and PdH.

Superconductivity has also been observed in a series of materials called *intercalation compounds*. Such compounds are formed by the insertion of atomic or molecular layers of a guest species, an *intercalant*, into the host material. The electrical, magnetic, optical, and conductive properties of the material are affected by the presence of the intercalant. Graphite and some metal sulfides, selenides, and tellurides are examples of intercalation hosts.

Surprisingly, the phenomenon of superconductivity has been observed in unusual cases in materials such as polymers and organic crystals. In 1975, superconductivity at 0.3 K was discovered in a material called polythiazyl, $(SN)_x$ (the subscript x indicates a large number of variable size). This observation provided the first, and still the only, example of superconductivity in a polymeric system, as well as the first example of superconductivity at ambient pressure in a material containing no metallic elements. Superconducting materials often contain an organic species or a metal in more than one oxidation state (sometimes called a mixed-valence species). Superconductivity ($T_c = 0.9$ K) in such an organic material (containing no metallic elements) was discovered in 1980 in the selenium-based salt $(TMTSF)_2PF_6$. TMTSF is tetramethyltetraselenafulvalene.

TMTSF

Since this initial discovery, a number of additional (TMTSF)X derivatives have been synthesized and found to be superconductors. In

these, X is TaF_6^-, AsF_6^-, ReO_4^-, FSO_3^-, or ClO_4^-. More recently, superconductivity has been discovered in similar salts of the sulfur-based system $(BEDT-TTF)_2X$, where BEDT-TTF is bis-ethylenedithiotetrathiafulvalene.

BEDT-TTF

Until recently, superconductivity was considered a low-temperature phenomenon. In 1986, however, "high-temperature" superconductivity (transition temperature ≈ 35 K) was discovered in a copper oxide material, $La_{1.8}Ba_{0.2}CuO_4$, by two Swiss scientists, George Bednorz and Karl Muller. Before their discovery, superconductivity was thought to be limited to tempera-

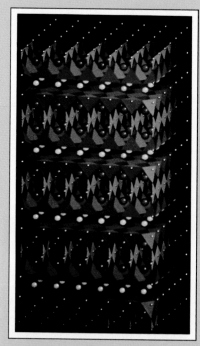

A representation of the structure of the $YBa_2Cu_3O_7$ superconductor.

tures below 23.3 K. The discovery of superconductivity in the ceramic copper oxide materials stimulated tremendous interest in the scientific community. Moreover, within four months of the publication of the Swiss work, Paul Chu and Maw-Kuen Wu found that a similar material, $YBa_2Cu_3O_7$, is superconducting at temperatures up to 92 K. While 92 K is not warm by most standards, it is well above the boiling point of liquid nitrogen (77 K). Liquid nitrogen is an inexpensive coolant that costs about 25 cents per liter. Liquid helium, on the other hand, which was required to cool the previous generations of low-temperature superconductors, is an inefficient refrigerant and costs about \$4 per liter. Thus, there could be a tremendous economic advantage to using the new copper oxide superconductors in practical applications.

Development of liquid-nitrogen-cooled superconductors could have dramatic effects on research, industry, electronics, transportation, medicine, and utilities. Currently, nearly one third of the electric power transmitted over long distances is lost because of resistive heating in the power lines. Superconductive transmission lines could carry current for hundreds of miles without such dissipative losses;

Until recently, only small, coin-size magnets could be floated above the new high-temperature superconductors, due to limited particle size in the ceramic superconductor. Attempts to increase the particle size caused the brittle ceramic to crack. A team of workers at the International Superconductivity Technical Center in Tokyo added a small amount of silver during the preparation of a superconductor. The result was a great increase in the weight that could be raised. Here, a goldfish in a 4.5-pound tank resting on a ring magnet is levitated.

hence, power plants such as nuclear generators could be located far from population centers. Smaller, more powerful computers could be made with superconductive microchips. The strong magnetic fields that are generated with superconducting coils could lead to many applications. High-speed trains could be levitated above the tracks by superconductor-generated magnetic fields, thereby eliminating friction at the track surface. Magnetic resonance imaging (MRI) scanners, used in diagnostic medicine to make images of body tissue, could be built more cheaply and simply, making them accessible to many more hospitals and clinics. Furthermore, the strong magnetic fields produced by superconductors could aid high-energy physics research. Currently, the plasma produced in fusion reactions cannot be contained by any known material, but with the use of superconducting magnets the plasma might be contained by strong magnetic fields.

Much work must still be done to translate laboratory results into such revolutionary technological applications. Intensive research continues to develop such applications, to search for other superconductive materials, and to illuminate the underlying mechanism that is responsible for high-temperature superconductivity.

Professor John T. McDevitt
University of Texas at Austin

(text continued from p. 493)

diamond is a good conductor of heat; jewelers use this property to distinguish diamonds from imitations.)

An important exception to these generalizations about properties is *graphite*, an allotropic form of carbon. It has the layer structure shown in Figure 13-31c. The overlap of an extended π-electron network in each plane makes graphite an excellent conductor. The very weak attraction between layers allows these layers to slip easily. Graphite is used as a lubricant, as an additive for motor oil, and in pencil "lead" (combined with clay and other fillers to control hardness).

It is interesting to note that the allotropes of carbon are one very hard substance and one very soft substance. They differ only in the arrangement and bonding of the C atoms.

Because of their malleability and ductility, metals can be formed into many shapes.

13-17 Band Theory of Metals

As described in the previous section, most metals crystallize in close-packed structures. The ability of metals to conduct electricity and heat must result from strong electronic interactions among the 8 to 12 nearest neighbors. This is somewhat difficult to rationalize if we recall that each Group IA and Group IIA metal atom has only one or two valence electrons available for bonding. This is too few to participate in bonds localized between it and each of its nearest neighbors.

Bonding in metals is called **metallic bonding**. It results from the electrical attractions among positively charged metal ions and mobile, delocalized electrons belonging to the crystal as a whole. The properties associated with metals—metallic luster, high thermal and electrical conductivity, and so on—are well explained by the **band theory** of metals, which we shall now describe.

The interaction of two atomic orbitals, say the $3s$ orbitals of two sodium atoms, produces two molecular orbitals, one bonding orbital and one antibonding orbital (Chapter 9). If N atomic orbitals interact, N molecular orbitals are formed. In a single metallic crystal containing one mole of sodium atoms, for example, the interaction of 6.022×10^{23} $3s$ atomic orbitals produces 6.022×10^{23} molecular orbitals. Atoms interact more strongly with those nearby than with those farther away. As discussed in Chapter 9, the energy separating bonding and antibonding molecular orbitals resulting from two given atomic orbitals decreases as the interaction (overlap) between the atomic orbitals decreases. When we consider all possible interactions among the mole of Na atoms, there results a series of very closely spaced molecular orbitals (formally σ_{3s} and $\sigma_{3s}^{\star}$). These comprise a nearly continuous **band** of orbitals belonging to the crystal as a whole. One mole of Na atoms contributes 6.022×10^{23} valence electrons (see Figure 13-32a), so the 6.022×10^{23} orbitals in the band are half-filled.

The empty $3p$ atomic orbitals of the Na atoms also interact to form a wide band of $3 \times 6.022 \times 10^{23}$ orbitals. The $3s$ and $3p$ atomic orbitals are quite close in energy, so the fanned-out bands of molecular orbitals overlap, as shown in Figure 13-32b. The two overlapping bands contain $4 \times 6.022 \times 10^{23}$ orbitals and only 6.022×10^{23} electrons. Because each orbital can hold two electrons, the resulting combination of bands is only one-eighth full.

Figure 13-32
(a) The band of orbitals resulting from interaction of the $3s$ orbitals in a crystal of sodium. (b) Overlapping of a half-filled "$3s$" band (black) with an empty "$3p$" band (red) of Na_N crystal.

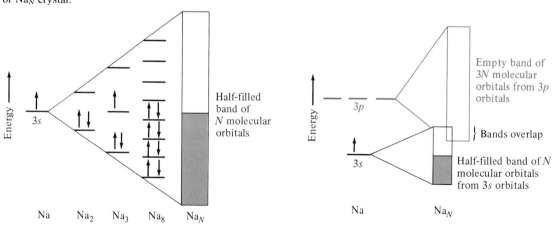

The ability of metallic Na to conduct electricity is due to the ability of any of the highest-energy electrons in the "3s" band to jump to a slightly higher-energy vacant orbital in the same band when an electric field is applied. The resulting net flow of electrons through the crystal is in the direction of the applied field.

The fact that the "3s" and "3p" bands overlap is not necessary to explain the ability of Na or of any other alkali metal to conduct electricity. It could do so utilizing only the half-filled "3s" band. In the alkaline earth metals, however, such overlap is important. Consider a crystal of magnesium as an example. The 3s atomic orbital of an isolated Mg atom is filled with two electrons. Thus, without this overlap, the "3s" band in a crystal of Mg is also filled. Mg is a good conductor at room temperature because the highest-energy electrons are able to move readily into vacant orbitals in the "3p" band (Figure 13-33).

According to band theory, the highest-energy electrons of metallic crystals occupy either a partially filled band or a filled band that overlaps an empty band. A band within which (or into which) electrons must move to allow electrical conduction is called a **conduction band**. The electrical conductivity of a metal decreases as temperature increases. The increase in temperature causes thermal agitation of the metal ions. This impedes the flow of electrons when an electrical field is applied.

Crystalline nonmetals, such as diamond and phosphorus, are **insulators**—they do not conduct electricity. The reason is that their highest-energy electrons occupy filled bands of molecular orbitals that are separated from the lowest empty band (conduction band) by a large energy gap called a **forbidden zone**. This is an energy difference that is too large for electrons to jump to get to the conduction band (see Figure 13-34).

Elements that are **semiconductors** have filled bands that are only slightly below, but do not overlap with, empty bands. They do not conduct electricity at low temperatures, but a small increase in temperature is sufficient to excite some of the highest-energy electrons into the empty conduction band.

The alkali metals are those of Group IA; the alkaline earth metals are those of Group IIA.

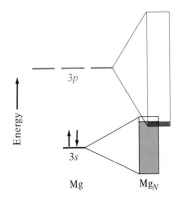

Figure 13-33
Overlapping of a filled "3s" band (blue) with an empty "3p" band of Mg$_N$ crystal. The higher energy electrons are able to move into the "3p" band (red) as a result of this overlap.

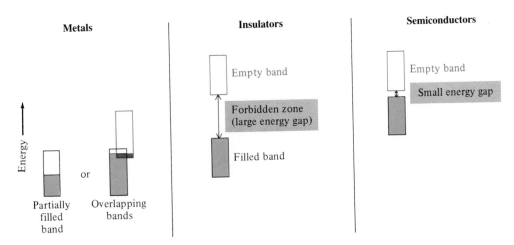

Figure 13-34
Distinction among metals, insulators, and semiconductors. In each case, an unshaded area represents a conduction band.

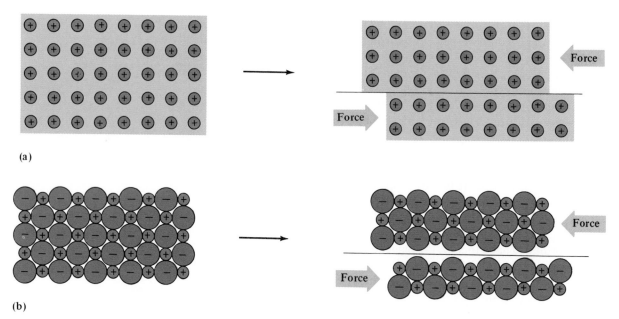

(a)

(b)

Figure 13-35

(a) In a metal, the positively charged metal ions are immersed in a "sea of electrons." When the metal is distorted (e.g., rolled into sheets or drawn into wires), the environment around the metal atoms is essentially unchanged, and no new repulsive forces occur. This explains why metal sheets and wires remain intact. (b) By contrast, when an ionic crystal is subjected to a force that causes it to slip along a plane, the increased repulsive forces between like-charged ions cause the crystal to break.

A **malleable** substance can be rolled or pounded into sheets. A **ductile** substance can be drawn into wires.

Let us now explain some of the physical properties of metals in terms of the band theory of metallic bonding.

1. We have just accounted for the *ability of metals to conduct electricity.*
2. Metals are also *conductors of heat.* They can absorb heat as electrons become thermally excited to low-lying vacant orbitals in a conduction band. The reverse process accompanies the release of heat.
3. Metals have a *lustrous appearance* because the mobile electrons can absorb a wide range of wavelengths of radiant energy as they jump to higher energy levels. They then immediately emit photons of visible light and fall back to lower levels within the conduction band.
4. Metals are *malleable and/or ductile.* A crystal of a metal is easily deformed when a mechanical stress is applied to it. All of the metal ions are identical, and they are imbedded in a "sea of electrons." As bonds are broken, new ones are readily formed with adjacent metal ions. The features of the lattice remain unchanged, and the environment of each metal ion is the same as before the deformation occurred (Figure 13-35). The breakage of bonds involves the promotion of electrons to higher energy levels. The formation of bonds is accompanied by the return of the electrons to the original energy levels.

A 6-inch wafer of ultrapure silicon, a semiconductor (left). Gallium phosphide, GaP, is a semiconducting compound (right). The analogous semiconducting compound gallium arsenide, GaAs, is used in many solid-state electronic devices.

Chemistry in Use. . .
Semiconductors

A **semiconductor** is an element or a compound with filled bands that are only slightly below, but do not overlap with, empty bands. The difference between an insulator and a semiconductor is only the size of the energy gap, and there is no sharp distinction between them. An **intrinsic** semiconductor (i.e., a semiconductor in its pure form) is a much poorer conductor of electricity than a metal because, for conduction to occur in a semiconductor, electrons must be excited from bonding orbitals in the filled *valence band* into the empty *conduction band*. Figure (a) shows how this happens. An electron that is given an excitation energy greater than or equal to the **band gap** (E_g) enters the conduction band and leaves behind a positively charged **hole** (h^+, the absence of a bonding electron) in the valence band. Both the electron and the hole reside in *delocalized* orbitals, and both can move in an electric field, much as electrons move in a metal. (Holes can migrate because an electron in a nearby orbital can move to fill in the hole, thereby creating a new hole in the nearby orbital.) Electrons and holes move in opposite directions in an electric field.

Silicon, a semiconductor of great importance in electronics, has a band gap of 1.94×10^{-22} kJ, or 1.21 *electron volts* (eV). This is the energy needed to create one electron and one hole or, put another way, the energy needed to break one Si—Si bond. This energy can be supplied either thermally or by using light with a photon energy greater than the band gap. To excite one *mole* of electrons from the valence band to the conduction band, an energy of

$$(6.022 \times 10^{23} \text{ electrons/mol}) \times (1.94 \times 10^{-22} \text{ kJ/electron}) = 117 \text{ kJ/mol}$$

is required. Because this is a large amount of energy, there are very few mobile electrons and holes (about one electron in a trillion— i.e., 1 in 10^{12}—is excited thermally at room temperature); the conductivity of pure silicon is therefore about 10^{11} times lower than that of highly conductive metals such as silver. The number of electrons excited thermally is proportional to $e^{-E_g/RT}$. Increasing the temperature or decreasing the band gap energy leads to higher conductivity for an intrinsic semiconductor. Insulators such as diamond and silicon dioxide (quartz), which have large values of E_g, have conductivities 10^{15} to 10^{20} times lower than most metals.

The electrical conductivity of a semiconductor can be greatly increased by **doping** with impurities. For example, silicon, a Group IVA element, can be doped by adding small amounts of a Group VA element, such as phosphorus, or a Group IIIA element, such as boron. Figure (b) shows the effect of substituting phosphorus for silicon in the crystal structure (silicon has the same structure as diamond, Figure 13-31a). There are exactly enough valence band orbitals to accommodate four of the valence electrons from the phosphorus atom. However, the phosphorus atom has one more electron (and one more proton in its nucleus) than does silicon. The fifth electron enters a higher energy orbital that is localized in the lattice near the phosphorus atom; the energy of this orbital, called a **donor level**, is just below the conduction band, within the energy gap. An electron in this orbital can easily become *delocalized* when a small amount of thermal energy promotes it into the conduction band. Because the phosphorus-doped silicon contains mobile, *negatively* charged carriers (electrons), it is said to be doped **n-type**. Doping the silicon crystal with boron produces a related, but opposite, effect. Each boron atom contributes only three valence electrons to bonding orbitals in the valence band, and therefore a *hole* is localized near each boron atom. Thermal energy is enough to separate the

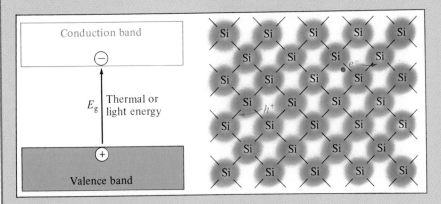

(a) Generation of an electron–hole pair in silicon, an intrinsic semiconductor. The electron (e^-) and hole (h^+) have opposite charges, and so move in opposite directions in an electric field.

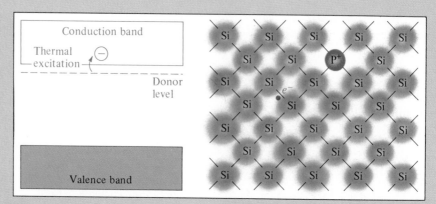

(b) *n*-Type doping of silicon by phosphorus. The extra valence electron from a phosphorus atom is thermally excited into the conduction band, leaving a fixed positive charge on the phosphorus atom.

negatively charged boron atom from the hole, delocalizing the latter. In this case the charge carriers are *positive* holes, and the crystal is doped **p-type**. In both *p*- and *n*-type doping, an extremely small concentration of dopants (as little as one part per billion) is enough to cause a significant increase in conductivity. For this reason, great pains are taken to purify the semiconductors used in electronic devices.

Even in a doped semiconductor, mobile electrons and holes are both present, although one carrier type is predominant. For example, in a sample of silicon doped with arsenic (*n*-type doping), the concentrations of mobile electrons are slightly less than the concentration of arsenic atoms (usually expressed in terms of atoms/cm³), and the concentrations of mobile holes are extremely low. Interestingly, the concentrations of electrons and holes always follow an equilibrium expression that is entirely analogous to that for the autodissociation of water into H^+ and OH^- ions (Chapter 18); that is,

$$[e^-][h^+] = K_{eq}$$

where the equilibrium constant K_{eq} depends only on the identity of the semiconductor and the absolute temperature. For silicon at room temperature, $K_{eq} = 4.9 \times 10^{19}$ carriers²/cm⁶.

Doped semiconductors are extremely important in electronic applications. A **p–n junction** is formed by joining *p*- and *n*-type semiconductors. At the junction, free electrons and holes combine, annihilating each other and leaving positively and negatively charged dopant atoms on opposite sides. The unequal charge distribution on the two sides of the junction causes an electric field to develop and gives rise to current rectification (electrons can flow, with a small applied voltage, only from the *n* side to the *p* side of the junction; holes flow only in the reverse direction). Devices such as **diodes** and bipolar **transistors**, which form the bases of most analog and digital electronic circuits, are composed of *p–n* junctions.

Professor Thomas A. Mallouk
University of Texas at Austin

The colors of semiconductors are determined by the band gap energy E_g. Only photons with energy greater than E_g can be absorbed. From the Planck radiation formula ($E = h\nu$) and $\lambda\nu = c$, we calculate that the wavelength, λ, of an absorbed photon must be less than hc/E_g. Gallium arsenide (GaAs; $E_g = 1.4$ eV) absorbs photons of wavelengths shorter than 890 nm, which is in the near infrared region. Because it absorbs all wavelengths of visible light, gallium arsenide appears black to the eye. Iron oxide (Fe_2O_3; $E_g = 2.2$ eV) absorbs light of wavelengths shorter than 570 nm; it absorbs both yellow and blue light, and therefore appears red. Cadmium sulfide (CdS; $E_g = 2.6$ eV), which absorbs blue light ($\lambda \leq 470$ nm), appears yellow. Strontium titanate ($SrTiO_3$; $E_g = 3.2$ eV) absorbs only in the ultraviolet ($\lambda \leq 390$ nm) and therefore appears white to the eye.

Key Terms

Adhesive force   Force of attraction between a liquid and another surface.

Allotropes   Different forms of the same element in the same physical state.

Amorphous solid   A noncrystalline solid with no well-defined, ordered structure.

Band   A series of very closely spaced, nearly continuous molecular orbitals that belong to the crystal as a whole.

Band theory of metals   A theory that accounts for the bonding and properties of metallic solids.

Boiling point   The temperature at which the vapor pressure of a liquid is equal to the applied pressure; also the condensation point.

Capillary action   The drawing of a liquid up the inside of a small-bore tube when adhesive forces exceed cohesive forces, or the depression of the surface of the liquid when cohesive forces exceed adhesive forces.

Cohesive forces   All the forces of attraction among particles of a liquid.

Condensation   Liquefaction of vapor.

Condensed phases   The liquid and solid phases; phases in which particles interact strongly.

Conduction band   A partially filled band or a band of vacant energy levels just higher in energy than a filled band; a band within which, or into which, electrons must be promoted to allow electrical conduction to occur in a solid.

Coordination number   In describing crystals, the number of nearest neighbors of an atom or ion.

Critical point   The combination of critical temperature and critical pressure of a substance.

Critical pressure   The pressure required to liquefy a gas (vapor) at its critical temperature.

Critical temperature   The temperature above which a gas cannot be liquefied; the temperature above which a substance cannot exhibit distinct gas and liquid phases.

Crystal lattice   The pattern of arrangement of particles in a crystal.

Crystalline solid   A solid characterized by a regular, ordered arrangement of particles.

Deposition   The direct solidification of a vapor by cooling; the reverse of sublimation.

Dipole–dipole interactions   Attractive interactions between polar molecules, i.e., between molecules with permanent dipoles.

Dipole-induced dipole interaction   See *London forces*.

Distillation   The separation of a liquid mixture into its components on the basis of differences in boiling points.

Dynamic equilibrium   A situation in which two (or more) processes occur at the same rate so that no net change occurs.

Evaporation   Vaporization of a liquid below its boiling point.

Forbidden zone   A relatively large energy separation be-

tween an insulator's highest filled electron energy band and the next higher energy vacant band.

Heat of condensation   The amount of heat that must be removed from one gram of a vapor at its condensation point to condense the vapor with no change in temperature.

Heat of crystallization   The amount of heat that must be removed from one gram of a liquid at its freezing point to freeze it with no change in temperature.

Heat of fusion   The amount of heat required to melt one gram of a solid at its melting point with no change in temperature; usually expressed in J/g. The *molar heat of fusion* is the amount of heat required to melt one mole of a solid at its melting point with no change in temperature and is usually expressed in kJ/mol.

Heat of vaporization   The amount of heat required to vaporize one gram of a liquid at its boiling point with no change in temperature; usually expressed in J/g. The *molar heat of vaporization* is the amount of heat required to vaporize one mole of a liquid at its boiling point with no change in temperature and is usually expressed in kJ/mol.

Hydrogen bond   A fairly strong dipole–dipole interaction (but still considerably weaker than covalent or ionic bonds) between molecules containing hydrogen directly bonded to a small, highly electronegative atom, such as N, O, or F.

Insulator   A poor conductor of electricity and heat.

Intermolecular forces   Forces *between* individual particles (atoms, molecules, ions) of a substance.

Intramolecular forces   Forces between atoms (or ions) *within* molecules (or formula units).

Isomorphous   Refers to crystals having the same atomic arrangement.

LeChatelier's Principle   A system at equilibrium, or striving to attain equilibrium, responds in such a way as to counteract any stress placed upon it.

London forces   Very weak and very short-range attractive forces between short-lived temporary (induced) dipoles; also called dispersion forces.

Melting point   The temperature at which liquid and solid coexist in equilibrium; also the freezing point.

Meniscus   The upper surface of a liquid in a cylindrical container.

Metallic bonding   Bonding within metals due to the electrical attraction of positively charged metal ions for mobile electrons that belong to the crystal as a whole.

Normal boiling point   The temperature at which the vapor pressure of a liquid is equal to one atmosphere pressure.

Normal melting point   The melting (freezing) point at one atmosphere pressure.

Phase diagram   A diagram that shows equilibrium temperature–pressure relationships for different phases of a substance.

Polymorphous   Refers to substances that crystallize in more than one crystalline arrangement.

Semiconductor   A substance that does not conduct electricity well at low temperatures but that does at higher temperatures.

Sublimation   The direct vaporization of a solid by heating without passing through the liquid state.

Supercooled liquids   Liquids that, when cooled, apparently solidify but actually continue to flow very slowly under the influence of gravity.

Supercritical fluid   A substance at a temperature above its critical temperature.

Surface tension   A measure of the inward intermolecular forces of attraction among liquid particles that must be overcome to expand the surface area.

Triple point   The point on a phase diagram that corresponds to the only pressure and temperature at which the solid, liquid, and gas phases of a substance can coexist at equilibrium.

Unit cell   The smallest repeating unit showing all the structural characteristics of a crystal.

Vapor pressure   The partial pressure of a vapor at the surface of its parent liquid.

Viscosity   The tendency of a liquid to resist flow; the inverse of its fluidity.

Volatility   The ease with which a liquid vaporizes.

Exercises

General Concepts

1. Explain why liquids are nearly incompressible and gases are very compressible.

2. Why does a gas completely fill its container, a liquid spread to take the shape of its container, and a solid retain its shape?

3. What causes London forces? What factors determine the strengths of London forces between molecules?

4. What is a hydrogen bond? Under what conditions can strong hydrogen bonds be formed?

5. Which of the following substances have permanent dipole–dipole forces? (a) SiH_4, (b) molecular $MgCl_2$, (c) NCl_3, (d) F_2O.

6. Which of the following substances have permanent dipole–dipole forces? (a) Molecular $AlCl_3$, (b) SF_6, (c) NO, (d) SeF_4.

7. For which of the substances in Exercise 5 are London forces the only important forces in determining boiling points?

8. For which of the substances in Exercise 6 are London forces the only important forces in determining boiling points?

9. For each of the following pairs of compounds, predict which compound would exhibit stronger hydrogen bonding. Justify your prediction. It may help to draw a Lewis structure for each compound. (a) Water, H_2O, or hydrogen sulfide, H_2S; (b) difluoromethane, CH_2F_2, or fluoroamine, NH_2F; (c) acetone, C_3H_6O (contains a C=O double bond) or ethyl alcohol, C_2H_6O (contains one C—O single bond); (d) nitric acid, HNO_3 (H bonded to O) or nitrogen trifluoride, NF_3.

10. For each of the following pairs of compounds, predict which would exhibit stronger hydrogen bonding. Justify your prediction. It may help to draw a Lewis structure for each compound. (a) Ammonia, NH_3, or phosphine, PH_3; (b) ethylene, C_2H_4, or hydrazine, N_2H_4; (c) hydrogen fluoride, HF, or hydrogen chloride, HCl;

(d) ethyl alcohol, C_2H_6O (contains one C—O single bond) or dimethyl ether, C_2H_6O (contains two C—O single bonds).

11. Classify the intermolecular forces that would most strongly influence the properties of (a) bromine pentafluoride, BrF_5; (b) acetone, C_3H_6O (contains a central C=O double bond); (c) carbonyl fluoride, F_2CO.

12. Classify the intermolecular forces that would most strongly influence the properties of (a) ethyl alcohol, C_2H_6O (contains one C—O single bond); (b) phosphine, PH_3; (c) sulfur hexafluoride, SF_6.

13. Give the correct names for these changes in state: (a) Crystals of *para*-dichlorobenzene, used as a moth repellent, gradually become vapor without passing through the liquid phase. (b) As you enter a warm room from the outdoors on a cold winter day, your eyeglasses become fogged with a film of moisture. (c) On the same winter day, a pan of water is left outdoors. Some of the water turns to vapor, the rest to ice.

14. The normal boiling point of trichlorofluoromethane, CCl_3F, is 24°C, and its freezing point is −111°C. Complete these sentences by supplying the proper terms. (a) At standard temperature and pressure, CCl_3F is a _____. (b) In an arctic winter at −40°C and 1 atm pressure, CCl_3F is a _____. If it is cooled further to −196°C and the molecules arrange themselves in an orderly lattice, the CCl_3F will _____ and become a _____. (c) If crystalline CCl_3F is held at a temperature of −120°C while a stream of helium gas is blown over it, the crystals will gradually disappear by the process of _____. If liquid CCl_3F is boiled at atmospheric pressure, it is converted to a _____ at a temperature of _____.

15. Why does HF have a lower boiling point and lower heat of vaporization than H_2O, even though their molecular weights are nearly the same and the hydrogen bonds between molecules of HF are stronger?

*16. Many carboxylic acids, which contain the group —COOH, form dimers in which two molecules are "stuck together." These dimers result from the formation of *two* hydrogen bonds between the two individual molecules. Use acetic acid to draw a likely structure for this kind of hydrogen-bonded dimer.

$$CH_3-\overset{\overset{\displaystyle \cdot\cdot\hspace{2pt}\cdot\cdot}{O}}{\underset{\displaystyle \|}{C}}-\overset{\cdot\cdot}{\underset{\cdot\cdot}{O}}-H$$

The Liquid State

17. Describe the behavior of liquids with changing temperature in terms of kinetic–molecular theory. Why are liquids more dense than gases?

18. Distinguish between evaporation and boiling. Explain the dependence of rate of evaporation on temperature in terms of kinetic–molecular theory.

19. Support or criticize the statement that liquids with high normal boiling points have low vapor pressures. Give examples of three liquids that have relatively high vapor pressures at 25°C and three that have low vapor pressures at 25°C.

20. Within each group, assign each of the boiling points to the respective substances on the basis of intermolecular forces. (a) Ne, Ar, Kr: −246°C, −186°C, −152°C; (b) NH_3, H_2O, HF: −33°C, 20°C, 100°C.

21. Within each group, assign each of the boiling points to the respective substances on the basis of intermolecular forces. (a) N_2, HCN, C_2H_6: −196°C, −89°C, 26°C; (b) H_2, HCl, Cl_2: −35°C, −259°C, −85°C.

22. Why is it necessary to specify the atmospheric pressure over a liquid when measuring a boiling point? What is the definition of the normal boiling point?

23. What factors determine how viscous a liquid is? How does viscosity change with an increase in temperature?

24. What is the surface tension of a liquid? What causes this property? How does surface tension change with an increase in temperature?

25. What happens inside a capillary tube if a liquid "wets" the tube? What happens if a liquid does not "wet" the tube?

26. Are the following statements true or false? If a statement is false, indicate why. (a) The equilibrium vapor pressure of a liquid is independent of the volume occupied by the vapor above the liquid. (b) The normal boiling point of a liquid changes with changing atmospheric pressure. (c) The vapor pressure of a liquid will increase if the mass of liquid is increased.

27. Are the following statements true or false? If a statement is false, indicate why. (a) The vapor pressure of a liquid will decrease if the volume of liquid decreases. (b) The normal boiling point of a liquid is the temperature at which the external pressure equals the vapor pressure of the liquid. (c) The vapor pressures of liq-

uids in a similar series tend to increase with increasing molecular weight.

28. Choose from each pair the substance that, in the liquid state, would have the greater vapor pressure at a given temperature. Base your choice on predicted strengths of intermolecular forces. (a) $BiBr_3$ or $BiCl_3$, (b) CO or CO_2, (c) N_2 or NO, (d) CH_3COOH or $HCOOCH_3$.

29. Repeat Exercise 28 for (a) C_6H_6 or C_6Cl_6, (b) $H_2C=O$ or CH_3OH, (c) He or H_2.

30. The temperatures at which the vapor pressures of the following liquids are all 100 torr are given. Predict the order of increasing boiling points of the liquids. Normal butane, C_4H_{10}, −44.2°C; 1-butanol, $C_4H_{10}O$, 70.1°C; diethyl ether, $C_4H_{10}O$, −11.5°C.

31. A closed flask contains water at 75.0°C. The total pressure of the air-and-water-vapor mixture is 633.5 torr. The vapor pressure of water at this temperature is given in Appendix E as 289.1 torr. What is the partial pressure of the air in the flask?

*32. The total pressure in a flask containing dry air and silicon tetrachloride, $SiCl_4$, was 988 torr at 25°C. The volume of the flask was halved and the pressure changed to 1742 torr while the temperature was held constant. What is the vapor pressure of $SiCl_4$ at this temperature? Assume that a small amount of liquid $SiCl_4$ is present in the container at all times.

33. The vapor pressure of liquid bromine at room temperature is 168 torr. Suppose that bromine is introduced drop by drop into a closed system containing air at 745 torr and room temperature. (The liquid bromine volume is negligible compared to the gas volume.) If the bromine is added until no more vaporizes and a few drops of liquid are present in the flask, what will be the total pressure in the flask? What will be the total pressure if the volume of this closed system is decreased to one-half its original value?

*34. The heat of vaporization of water at 100°C is 2.26 kJ/g; at 37°C (body temperature) it is 2.41 kJ/g.
(a) Convert the latter value to standard molar heat of vaporization, ΔH^0_{vap}, at 37°C.
(b) Why is the heat of vaporization greater at 37°C than at 100°C?

*35. For any substance, $\Delta H_{vaporization}$ is almost always greater than ΔH_{fusion}, yet the *nature* of interactions that must be overcome in the vaporization and fusion processes are similar. Why should $\Delta H_{vaporization}$ be greater?

*36. A fellow student comes to you with this problem: "I looked up the vapor pressure of water in a table; it is 26.7 torr at 300 K and 92,826 torr at 600 K. That means that when the absolute temperature doubles, the vapor pressure is multiplied by 3477. But I thought the pressure was proportional to the absolute temperature, $P = nRT/V$. Why doesn't the pressure just double?" How would you help the student?

37. Plot a vapor pressure curve for $C_2Cl_2F_4$ from the following vapor pressures. Determine the boiling point

of $C_2Cl_2F_4$ under a pressure of 300 torr from the plot:

t (°C)	−95.4	−72.3	−53.7	−39.1	−12.0	3.5
vp (torr)	1	10	40	100	400	760

38. Plot a vapor pressure curve for $C_2H_4F_2$ from the following vapor pressures. From the plot, determine the boiling point of $C_2H_4F_2$ under a pressure of 200 torr.

t (°C)	−77.2	−51.2	−31.1	−15.0	14.8	31.7
vp (torr)	1	10	40	100	400	760

Clausius–Clapeyron Equation

39. Toluene, $C_6H_5CH_3$, is a liquid used in the manufacture of TNT. The normal boiling point of toluene is 111.0°C, and its molar heat of vaporization is 35.9 kJ/mol. What would be the vapor pressure (torr) of toluene at 75.00°C?

40. At the normal boiling point, the heat of vaporization of water (100.00°C) is 40,656 J/mol, and that of heavy water (101.41°C) is 41,606 J/mol. Use these data to calculate the vapor pressure of each liquid at 80.00°C.

41. (a) Use the Clausius–Clapeyron equation to calculate the temperature (°C) at which pure water would boil at a pressure of 400.0 torr. (b) Compare this result with the temperature read from Figure 13-13. (c) Compare the results of (a) and (b) with a value obtained from Appendix E.

*42. Show that the Clausius–Clapeyron equation can be written as

$$\log P = \frac{-\Delta H_{vap}}{2.303\ RT} + B$$

where B is a constant that has different values for different substances. This is an equation for a straight line. (a) What is the expression for the slope of this line? (b) Using the following vapor pressure data, plot $\log P$ vs. $1/T$ for ethyl acetate, $CH_3COOC_2H_5$, a common organic solvent used in nail polish removers.

t (°C)	−43.4	−23.5	−13.5	−3.0	+9.1
vp (torr)	1	5	10	20	40

t (°C)	16.6	27.0	42.0	59.3
vp (torr)	60	100	200	400

(c) From the plot, estimate ΔH_{vap} for ethyl acetate. (d) From the plot, estimate the normal boiling point of ethyl acetate.

*43. Repeat Exercise 42 for mercury, using the following data for liquid mercury. Then compare this value with the one in Appendix E.

t (°C)	126.2	184.0	228.8	261.7	323.0
vp (torr)	1	10	40	100	400

Phase Changes and Associated Heat Transfer

The following values will be useful in working some of the exercises in this section:

Specific heat of ice	2.09 J/g · °C
Heat of fusion of ice at 0°C	334 J/g
Specific heat of liquid H_2O	4.18 J/g · °C
Heat of vaporization of liquid H_2O at 100°C	2.26×10^3 J/g
Specific heat of steam	2.03 J/g · °C

44. Is the equilibrium that is established between two physical states of matter an example of static or dynamic equilibrium? Explain your answer.

45. Which of the following changes of state are exothermic? (a) Fusion, (b) liquefaction, (c) sublimation, (d) deposition.

46. Suppose that heat is added to a 21.8-g sample of solid zinc at the rate of 8.2 J/s. After the temperature reached the normal melting point of zinc, 420°C, it remained constant for 4.5 minutes. Calculate ΔH^0_{fusion} at 420°C, in J/mol, for zinc.

47. The specific heat of silver is 0.237 J/g · °C. Its melting point is 961°C. Its heat of fusion is 11 J/g. How much heat is needed to change 7.50 g of silver from solid at 25°C to liquid at 961°C?

48. The heat of fusion of thallium is 21 J/g, and its heat of vaporization is 795 J/g. The melting and boiling points are 304°C and 1457°C. The specific heat of liquid thallium is 0.13 J/g · °C. How much heat is needed to change 10.00 g of solid thallium at 304°C to vapor at 1457°C and 1 atm?

49. Use data in Appendix E to calculate the amount of heat required to warm 100 g of mercury from 25°C to its boiling point and then to vaporize it.

50. Aluminum is produced as a liquid. The specific heat of liquid aluminum is 1.090 J/g · °C. Use this value and data in Appendix E to calculate how much heat must be removed to cool one metric ton (10^6 g) of liquid aluminum at 1000°C to solid aluminum at 25°C?

51. Calculate the amount of heat required to raise the temperature of 30.0 g of water from 25.0°C to 80.0°C?

52. If 275 g of liquid water at 100°C and 525 g of water at 30.0°C are mixed in an insulated container, what is the final temperature?

53. Calculate the amount of heat required to convert 25.0 g of ice at 0.0°C to liquid water at 100°C.

54. Calculate the amount of heat required to convert 25.0 g of ice at −15.0°C to steam at 125.0°C.

55. If 25.0 g of ice at −10.0°C and 25.0 g of liquid water at 100°C are mixed in an insulated container, what will the final temperature be?

56. If 140 g of liquid water at 0.0°C and 14.0 g of steam at 110°C are mixed in an insulated container, what will the final temperature be?

57. Water can be cooled in hot climates by the evaporation of water from the surfaces of canvas bags. What mass of water can be cooled from 35.0°C to 20.0°C by the evaporation of one gram of water? Assume that ΔH_{vap} does not change with temperature.

58. (a) How much heat must be removed to prepare 10.00 g of ice at 0°C from water at 25.0°C? (b) Suppose this heat is to be absorbed by vaporization of Freon-12 (a common household refrigerant, CCl_2F_2). What mass of refrigerant must be vaporized? The heat of vaporization of Freon-12 is 165.1 J/g.

59. We slowly bubble 25.0 g of steam at 110.0°C into 100.0 g of liquid water at 0.0°C in an insulated container. Will all of the steam be condensed?

60. If heat is supplied at an identical, constant rate in each case, which will take longer, heating 100 g of water from 0°C to 100°C, or vaporizing 100 g of water at 100°C?

Phase Diagrams

61. How many phases exist at a triple point? Describe what would happen if a small amount of heat were added under constant-volume conditions to a sample of water at the triple point. Assume a negligible volume change during fusion.

62. What is the critical point? Will a substance always be a liquid below the critical temperature? Why or why not?

Refer to the phase diagram of CO_2 in Figure 13-17b to answer Exercises 63–66.

63. What phase of CO_2 exists at 2 atm pressure and a temperature of −90°C? −60°C? 0°C?

64. What phases of CO_2 are present (a) at a temperature of −78°C and a pressure of 1 atm? (b) at −57°C and a pressure of 5.2 atm?

65. List the phases that would be observed if a sample of CO_2 at 8 atm pressure were heated from −80°C to 40°C.

66. How does the melting point of CO_2 change with pressure? What does this indicate about the relative density of solid CO_2 versus liquid CO_2?

67. You are given the following data for ethanol, C_2H_5OH:

Normal melting point	−114°C
Normal boiling point	78.5°C
Critical temperature	243°C
Critical pressure	63.0 atm

Assume that the triple point is slightly lower in temperature than the melting point and that the vapor pressure at the triple point is about 10^{-5} torr. (a) Sketch a phase diagram for ethanol. (b) Ethanol at 1 atm and 140°C is compressed to 70 atm. Are two phases present at any time during this process? (c) Ethanol at 1 atm and 270°C is compressed to 70 atm. Are two phases present at any time during this process?

68. You are given the following data for butane, C_4H_{10}:

Normal melting point	−138°C
Normal boiling point	0°C
Critical temperature	152°C
Critical pressure	38 atm

Assume that the triple point is slightly lower in temperature than the melting point and that the vapor pressure at the triple point is 3×10^{-5} torr. (a) Sketch a phase diagram for butane. (b) Butane at 1 atm and 140°C is compressed to 40 atm. Are two phases present at any time during this process? (c) Butane at 1 atm and 200°C is compressed to 40 atm. Are two phases present at any time during this process?

Exercises 69 and 70 refer to the phase diagram for sulfur (below). (The vertical axis is on a logarithmic scale.) Sulfur has two *solid* forms, monoclinic and rhombic.

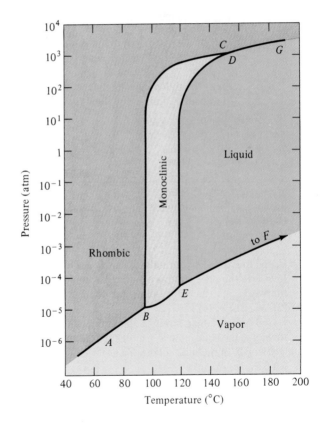

*69. (a) How many triple points are there for sulfur? (b) Indicate the approximate pressure and temperature at each triple point. (c) What phases are in equilibrium at each triple point?

*70. Which physical states should be present at equilibrium under the following conditions? (a) 10^{-1} atm and 140°C, (b) 10^{-5} atm and 80°C, (c) 5×10^3 atm and 160°C, (d) 10^{-1} atm and 110°C, (e) 10^{-5} atm and 140°C, (f) 1 atm and 140°C.

The Solid State

71. Comment on the following statement: "The only perfectly ordered state of matter is the crystalline state."

72. Ice floats in water. Why? Would you expect solid mercury to float in liquid mercury at its freezing point? Explain why or why not.

73. Distinguish among and compare the characteristics of molecular, covalent, ionic, and metallic solids. Give two examples of each kind of solid.

74. Classify each of the following substances, in the solid state, as a molecular, ionic, covalent (network), or metallic solid:

	Melting Point (°C)	Boiling Point (°C)	Electrical Conductor	
			Solid	Liquid
SiO_2 (crystobalite)	1713	2230	no	no
Na_2S	1180	—	no	yes
$Cr(CO)_6$	110	210 (dec.)	no	no
Ti	1660	3287	yes	yes

75. Classify each of the following substances, in the solid state, as a molecular, ionic, covalent (network), or metallic solid:

	Melting Point (°C)	Boiling Point (°C)	Electrical Conductor	
			Solid	Liquid
NH_4NO_3	167	210	no	yes
Mg	649	1090	yes	yes
GeO_2	1115	—	no	no
S_8 (rhombic)	113	445	no	no

76. Based only on their formulas, classify each of the following in the solid state as a molecular, ionic, covalent (network), or metallic solid: (a) SO_2F, (b) MgF_2, (c) W, (d) Pb, (e) PF_5.

77. Based only on their formulas, classify each of the following in the solid state as a molecular, ionic, covalent (network), or metallic solid: (a) Au, (b) NO_2, (c) CaF_2, (d) SF_4, (e) $C_{diamond}$.

78. Arrange the following solids in order of increasing melting points and account for the order: NaF, MgF_2, AlF_3.

79. Arrange the following solids in order of increasing melting points and account for the order: MgO, CaO, SrO, BaO.

80. Distinguish among and sketch simple cubic, body-centered cubic (bcc), and face-centered cubic (fcc) lattices. Use CsCl, sodium, and nickel as examples of solids existing in simple cubic, bcc, and fcc lattices, respectively.

81. What is a unit cell? Sketch a unit cell and label the dimensions a, b, and c and the angles α, β, and γ. How do the relations among unit cell lengths and angles differ in cubic, tetragonal, and hexagonal crystal systems?

82. Consider a unit cell that consists of a cube in which there is a cation at each corner and an anion at the center of each face. (a) Sketch the unit cell. How many (b) cations and (c) anions are present? (d) The simplest formula of the compound is of the type _____.

83. Consider a unit cell that consists of a cube in which there is an anion at each corner and one at the center of the unit cell, and a cation at the center of each face. How many cations and anions make up the unit cell? The simplest formula of this compound is of the type _____.

84. Choose two different unit cells in the two-dimensional lattice shown. Make one as simple as possible and the other orthogonal (containing right angles). Which of these is simpler to use to calculate area and so on?

Unit Cell Data; Atomic and Ionic Sizes

85. Refer to Figure 13-28a. (a) If the unit cell edge is represented as a, what is the distance (center to center) from Cs^+ to its nearest neighbor? (b) How many equidistant nearest neighbors does each Cs^+ ion have? What are the identities of these nearest neighbors? (c) What is the distance (center to center), in terms of a, from a Cs^+ ion to the nearest Cs^+ ion? (d) How many equidistant nearest neighbors does each Cl^- ion have? What are their identities?

86. Refer to Figure 13-28b. (a) If the unit cell edge is represented as a, what is the distance (center to center) from Na^+ to its nearest neighbor? (b) How many equidistant nearest neighbors does each Na^+ ion have? What are the identities of these nearest neighbors? (c) What is the distance (center to center), in terms of a, from an Na^+ ion to the nearest Na^+ ion? (d) How many equidistant nearest neighbors does each Cl^- ion have? What are their identities?

87. Polonium crystallizes in a simple cubic unit cell with an edge length of 3.36 Å. (a) What is the mass of the unit cell? (b) What is the volume of the unit cell? (c) What is the theoretical density of Po?

88. Calculate the density of Na metal. The length of the body-centered cubic unit cell is 4.24 Å.

89. Tungsten has a density of 19.3 g/cm³ and crystallizes in a cubic lattice whose unit cell edge length is 3.16 Å. Which type of cubic unit cell is it?

90. A Group IVA element with a density of 11.35 g/cm³ crystallizes in a face-centered cubic lattice whose unit cell edge length is 4.95 Å. Calculate its atomic weight. What is the element?

91. The crystal structure of CO_2 is cubic, with a cell edge length of 5.540 Å. A diagram of the cell is shown in Figure 13-30a. (a) What is the number of molecules of CO_2 per unit cell?
(b) Is this structure face-centered cubic? How can you tell?
(c) What is the density of solid CO_2 at this temperature?

92. Magnesium crystallizes in the hexagonal closest-packed unit cell with $a = 3.203$ Å and $c = 5.196$ Å. The volume of the unit cell is given by $a^2c \cdot \sin 60°$. (a) What mass of Mg is contained in the unit cell? (b) Calculate the volume of the unit cell. (c) Calculate the theoretical density of Mg.

93. The crystal structure of ice has a hexagonal unit cell with $a = 4.534$ Å and $c = 7.410$ Å. The number of molecules in the unit cell is four, and the volume of the unit cell is given by $a^2c \cdot \sin 60°$. (a) What mass of H_2O is contained in the unit cell? (b) Calculate the volume of the unit cell. (c) Calculate the theoretical density of ice.

94. The structure of diamond is shown below, with each sphere representing a carbon atom. (a) How many carbon atoms are there per unit cell in the diamond structure? (b) Verify, by extending the drawing if necessary, that each carbon atom has four nearest neighbors. What is the arrangement of these nearest neighbors? (c) What is the distance (center to center) from any carbon atom to its nearest neighbor, expressed in terms of a, the unit cell edge? (d) The observed unit cell edge length in diamond is 3.567 Å. What is the C—C single bond length in diamond? (e) Calculate the density of diamond.

95. Crystalline silicon has the same structure as diamond, with a unit cell edge length of 5.430 Å. (a) What is the Si—Si distance in this crytal? (b) Calculate the density of crystalline silicon.

96. (a) What types of electromagnetic radiation are suitable for diffraction studies of crystals? (b) Describe the X-ray diffraction experiment. (c) What must be the relationship between the wavelength of incident radiation and the spacing of the particles in a crystal for diffraction to occur?

97. (a) Write the Bragg equation. Identify each symbol. (b) X-rays from a palladium source ($\lambda = 0.576$ Å) were reflected by a sample of copper at an angle of 9.40°. This reflection corresponds to the unit cell length ($d = a$) with $n = 2$ in the Bragg equation. Calculate the length of the copper unit cell.

98. The spacing between successive planes of platinum atoms parallel to the cubic unit cell face is 2.256 Å. When X-radiation emitted by copper strikes a crystal of platinum metal, the minimum diffraction angle of X-rays is 19.98°. What is the wavelength of the Cu radiation?

99. Gold crystallizes in an fcc structure. When X-radiation of 0.70926 Å wavelength from molybdenum is used to determine the structure of metallic gold, the minimum diffraction angle of X-rays by the gold is 8.683°. Calculate the spacing between parallel layers of gold atoms.

Metallic Bonding and Semiconductors

100. Describe metallic bonding. Nonmetals do not form metallic bonds. Why not?

101. Compare the temperature dependence of electrical conductivity of a metal with that of a typical metalloid. Explain the difference.

102. In general, metallic solids are ductile and malleable, whereas ionic salts are brittle and shatter readily (although they are hard). Explain this observation.

103. What single factor accounts for the ability of metals to conduct both heat and electricity in the solid state? Why are ionic solids poor conductors of heat and electricity even though they are composed of charged particles?

Mixed Exercises

104. The three major components of air are N_2 (bp $-196°C$), O_2 (bp $-183°C$), and Ar (bp $-186°C$). Suppose we have a sample of liquid air at $-200°C$. In what order will these gases evaporate as the temperature is raised?

*105. A 10.0-g sample of liquid ethanol, C_2H_5OH, absorbs 3.42×10^3 J of heat at its normal boiling point, 78.5°C. The molar enthalpy of vaporization of ethanol, ΔH_{vap}, is 39.3 kJ/mol.
(a) What volume of C_2H_5OH vapor is produced? The volume is measured at 78.5°C and 1.00 atm pressure.
(b) What mass of C_2H_5OH remains in the liquid state?

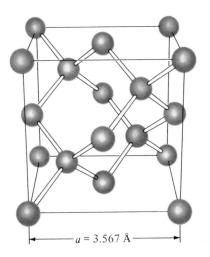

$\longleftarrow a = 3.567$ Å $\longrightarrow$

*106. What is the pressure predicted by the ideal gas law for one mole of steam in 31.0 L at 100.00°C? What is the pressure predicted by the van der Waals equation (Section 12-14) given that $a = 5.464$ $L^2 \cdot atm/mol^2$ and $b = 0.03049$ L/mol? What is the percent difference between these values? Does steam deviate from ideality significantly at 100°C? Why?

*107. The boiling points of HCl, HBr, and HI increase with increasing molecular weight. Yet the melting and boiling points of the sodium halides, NaCl, NaBr, and NaI, decrease with increasing formula weight. Explain why the trends are opposite.

108. The structures for three molecules having the formula $C_2H_2Cl_2$ are

Describe the intermolecular forces present in each of these compounds and predict which has the lowest boiling point.

109. The vapor pressure of $CH_3CH_2CH_2Cl$ at room temperature is 385 torr. The total pressure of $CH_3CH_2CH_2Cl$ and air in a container is 745 torr. What will happen to the pressure if the volume of the container is doubled at constant temperature? Assume that a small amount of liquid $CH_3CH_2CH_2Cl$ is present in the container at all times.

110. Tantalum (181 g/mol, density 16.7 g/cm^3) crystallizes in a cubic lattice whose unit cell edge length is 3.32 Å. How many atoms are in one unit cell? What is the type of cubic crystal lattice?

111. Isopropyl alcohol, C_3H_8O, is marketed as "rubbing alcohol." Its vapor pressure is 100 torr at 39.5°C and 400 torr at 67.8°C.
(a) Estimate the molar heat of vaporization of isopropyl alcohol.
(b) Predict the normal boiling point of isopropyl alcohol.

112. The heat of vaporization of cyclohexane, C_6H_{12}, is 390 J/g at its boiling point, 80.7°C. For how many minutes would heat have to be supplied at a rate of 10.0 J/s to a 1.00 mol sample of liquid cyclohexane at 80.7°C to vaporize all of it?

113. The van der Waals constants (Section 12-14) are $a = 19.01$ $L^2 \cdot atm/mol^2$, $b = 0.1460$ L/mol for n-pentane, and $a = 18.05$ $L^2 \cdot atm/mol^2$, $b = 0.1417$ L/mol for isopentane.

n-pentane   *isopentane*

(a) Basing your reasoning on intermolecular forces, why would you expect a for n-pentane to be greater?
(b) Basing your reasoning on molecular size, why would you expect b for n-pentane to be greater?

114. The density of solid lead is 11.288 g/cm^3 at 20°C; that of liquid lead is 10.43 g/cm^3 at 500°C; and that of gaseous lead is 1.110 g/L at 2000°C and 1 atm pressure. (a) Calculate the volume occupied by one mole of lead in each state. The radius of a lead atom is 1.75 Å. Calculate (b) the volume actually occupied by one mole of lead atoms and (c) the fraction of volume of each state actually occupied by the atoms.

115. Refer to the sulfur phase diagram that accompanies Exercises 69 and 70. (a) Can rhombic sulfur be sublimed? If so, under what conditions? (b) Can monoclinic sulfur be sublimed? If so, under what conditions? (c) Describe what happens if rhombic sulfur is slowly heated from 80°C to 140°C at constant 1-atm pressure. (d) What happens if rhombic sulfur is heated from 80°C to 140°C under constant pressure of 5 × 10^{-6} atm?

116. The normal boiling point of ammonia, NH_3, is −33°C, and its freezing point is −78°C. Fill in the blanks. (a) At STP (0°C, 1 atm pressure), NH_3 is a _____. (b) If the temperature drops to −40°C, the ammonia will _____ and become a _____. (c) If the temperature drops further to −80°C and the molecules arrange themselves in an orderly pattern, the ammonia will _____ and become a _____. (d) If crystals of ammonia are left on the planet Mars at a temperature of −100°C, they will gradually disappear by the process of _____ and form a _____.

14 Solutions

The very polar sugar molecules are strongly attracted to polar molecules of water and are pulled away from the surface of the crystal. As the more concentrated sugar solution falls, regions of different concentration in the solution bend light rays differently.

Outline

Objectives

As you study this chapter, you should learn

☐ To understand the factors that favor the dissolution process
☐ About the dissolution of solids in liquids, liquids in liquids, and gases in liquids
☐ How temperature and pressure affect dissolution and rates of saturation
☐ To express concentrations of solutions in terms of molality and mole fractions
☐ About colligative properties of solutions: lowering of vapor pressure (Raoult's Law), boiling point elevation, and freezing point depression

☐ To use colligative properties to determine molecular weights of compounds
☐ About dissociation and ionization of compounds, and their effects on colligative properties
☐ About membrane osmotic pressure and some of its applications
☐ About colloids: the Tyndall effect, the adsorption phenomenon, hydrophilic and hydrophobic colloids

Solutions are common in nature and are extremely important in all life processes, in all scientific areas, and in many industrial processes. The body fluids of all forms of life are solutions. Variations in their concentrations, especially those of blood and urine, give physicians valuable clues about a person's health.

Solutions include many different combinations in which a solid, liquid, or gas acts as either solvent or solute. Most commonly the solvent is a liquid. For instance, seawater is an aqueous solution of many salts and some gases (such as carbon dioxide and oxygen). Carbonated water is a saturated solution of carbon dioxide in water. Examples of solutions in which the solvent is not a liquid also are quite common. Air is a solution of gases (with variable composition). Dental fillings are solid amalgams, or solutions of liquid mercury dissolved in solid metals. Alloys are solid solutions of solid metals dissolved in one another.

A solution is defined as a *homogeneous mixture* of pure substances in which no settling occurs. A true solution consists of a solvent and one or more solutes, whose proportions vary from one solution to another. By contrast, a pure substance has fixed composition. The *solvent* is the medium in which the *solutes* are dissolved. The fundamental units of solutes are usually ions or molecules.

It is usually obvious which of the components of a solution is the solvent and which is (are) the solute(s): The solvent is usually the most abundant species present. In a cup of instant coffee, the coffee and any added sugar are considered solutes, and the hot water is the solvent. If we mix 10 grams of alcohol with 90 grams of water, the alcohol is the solute. If we mix 10 grams of water with 90 grams of alcohol, the water is the solute. But which is the solute and which the solvent in a solution of 50 grams of water and 50 grams of alcohol? In such cases, the terminology is arbitrary and, in fact, unimportant.

Many naturally occurring fluids contain particulate matter suspended in a solution. For example, blood contains a solution (plasma) with suspended blood cells, and seawater contains dissolved substances as well as suspended solids.

The Dissolution Process

14-1 Spontaneity of the Dissolution Process

In Section 4-2, part 5, we listed the solubility rules for aqueous solutions. Now we shall investigate the major factors that influence solubility in general. A substance may dissolve with or without reaction with the solvent. Metallic sodium "dissolves" in water with the evolution of bubbles of hydrogen and a great deal of heat. A chemical change occurs in which H_2 and soluble ionic sodium hydroxide, NaOH, are produced. The total ionic equation is

$$2Na(s) + 2H_2O \longrightarrow 2[Na^+(aq) + OH^-(aq)] + H_2(g)$$

If the resulting solution is evaporated to dryness, solid sodium hydroxide, NaOH, is obtained rather than metallic sodium. This, along with the production of bubbles of hydrogen, is evidence of a reaction with the solvent.

Solid sodium chloride, NaCl, on the other hand, dissolves in water with no evidence of chemical reaction:

$$NaCl(s) \xrightarrow{H_2O} Na^+(aq) + Cl^-(aq)$$

Evaporation of the sodium chloride solution yields the original NaCl. We shall consider dissolution of the latter type, in which no irreversible reaction occurs between components.

The ease of the dissolution process depends upon two factors: (1) the change in energy (exothermicity or endothermicity) and (2) the change in

Strictly speaking, even solutes that do not "react" with the solvent undergo solvation. This is a kind of reaction in which molecules of solvent are attached in oriented clusters to the solute particles.

disorder (called entropy change) accompanying the process. In Chapter 15 we shall study both these factors in detail for many kinds of physical and chemical changes. For now, we point out that a process is *favored* by (1) a *decrease in the energy* of the system, which corresponds to an *exothermic process*, and (2) an *increase in the disorder*, or randomness, of the system.

Let us look at the first of these factors. If the solution gets hotter as the substance dissolves, this means that energy is being released in the form of heat. This change is called the **heat of solution, $\Delta H_{solution}$.** It depends mainly on how strongly solute and solvent particles interact. A negative value of $\Delta H_{solution}$ designates the release of heat. More negative (less positive) values of $\Delta H_{solution}$ favor the dissolution process.

In a pure liquid, the intermolecular forces are all between like molecules; when the liquid and a solute are mixed, each molecule then experiences forces from molecules (or ions) unlike it as well as from like molecules. The relative strengths of these interactions help to determine the extent of solubility of a solute in a solvent. The main interactions that affect the dissolution of a solute in a solvent follow.

> **1.** Strong solvent–solute attractions favor solubility.
> **2.** Weak solvent–solvent attractions favor solubility.
> **3.** Weak solute–solute attractions favor solubility.

Figure 14-1 illustrates the interplay of these factors. The intermolecular or interionic attractions among solute particles in the pure solute must be overcome (step a) to dissolve the solute. This part of the process requires an *input* of energy. Separating the solvent molecules from each other (step b) to "make room" for the solute particles also requires the *input* of energy. However, energy is *released* as the solute particles and solvent molecules interact in the solution (step c). The dissolution process is exothermic (and favored) if the amount of heat absorbed in hypothetical steps a and b is less than the amount of heat released in step c. The process is endothermic (and disfavored) if the amount of heat absorbed in (hypothetical) steps a and b is greater than the amount of heat released in step c.

However, many solids do dissolve in liquids by *endothermic* processes. The reason such processes can occur is that the endothermicity is outweighed

We can consider the energy changes separately, even though the actual process cannot be carried out in these separate steps.

Figure 14-1

A diagram representing the changes in heat content associated with the hypothetical three-step sequence in a dissolution process—in this case, for a solid solute dissolving in a liquid solvent. (Similar considerations would apply to other combinations.) An *exothermic* process is depicted here. The amount of heat absorbed in steps a and b is *less* than the amount of heat released in step c, so the heat of the solution is favorable. In an *endothermic* process (not shown), the heat content of the solution would be *higher* than that of the original solvent plus solute. Thus, the amount of heat absorbed in steps a and b would be *greater* than the amount of heat released in step c, so heat of solution would be unfavorable.

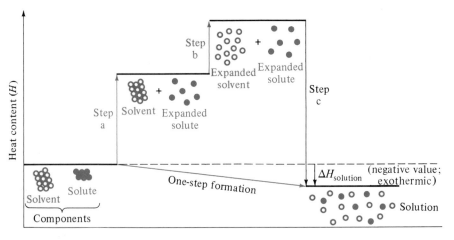

by a large increase in disorder of the solute during the dissolution process. The solute particles are highly ordered in a solid crystal, but are free to move about randomly in liquid solutions. Likewise, the degree of disorder in the solvent increases as the solution is formed, because solvent molecules are then in a more random environment. They are surrounded by a mixture of solvent and solute particles (Figure 14-1).

Nearly all dissolution processes are accompanied by an increase in the disorder of both solvent and solute. Thus, the disorder factor is usually *favorable* to solubility. The determining factor, then, is whether the heat of solution also favors dissolution or, if it does not, whether it is small enough to be outweighed by the favorable effects of the increasing disorder. In gases, for instance, the molecules are so far apart that intermolecular forces are quite weak. Thus, when gases are mixed, changes in the intermolecular forces are very slight. So the very favorable increase in disorder that accompanies mixing is always more important than possible changes in intermolecular attractions (energy). Hence, gases can always be mixed with each other in any proportion. (This statement does not apply to gases that react with each other chemically.)

The most common types of solutions are those in which the solvent is a liquid. In the next several sections, we shall consider these in more detail.

One of very few exceptions is the dissolution of NaF. The water molecules become more ordered around the small F^- ions. This is due to the strong hydrogen bonding between H_2O molecules and F^- ions.

$$O—H\text{---}F^-\text{---}H—O$$
$$|\phantom{H\text{---}F^-\text{---}H—}|$$
$$H\phantom{\text{---}F^-\text{---}H—O}H$$

However, the heat released on mixing outweighs the disadvantage of this ordering.

14-2 Dissolution of Solids in Liquids

The ability of a solid to go into solution depends most strongly on its crystal lattice energy, or the strength of attractions among the particles making up the solid. The **crystal lattice energy** is defined as the energy change accompanying the formation of one mole of formula units in the crystalline state from constituent particles in the gaseous state. This process is always exothermic; i.e., crystal lattice energies are always *negative*. For an ionic solid, the process is written as

$$M^+(g) + X^-(g) \longrightarrow MX(s) + energy$$

The amount of energy involved in this process depends on the attraction between ions in the solid. When these attractions are strong, a large amount of energy is released as the solid forms, and so the solid is very stable.

The reverse of the crystal formation reaction, the separation of the crystal into ions,

$$MX(s) + energy \longrightarrow M^+(g) + X^-(g)$$

can be considered the hypothetical first step (step a in Figure 14-1) in forming a solution of a solid in liquid. It is always endothermic. The smaller the magnitude of the crystal lattice energy (a measure of the solute–solute interactions), the more readily dissolution occurs. Less energy must be supplied to start the dissolution process.

If the solvent is water, the energy that must be supplied to expand the solvent (step b in Figure 14-1) includes that required to break up some of the hydrogen bonding between water molecules.

The third major factor contributing to heat of solution is the extent to which the solvent molecules interact with particles of the solid. The process in which solvent molecules surround and interact with solute ions or molecules is called **solvation**. When the solvent is water, the more specific term is **hydration**. **Hydration energy** (equal to the sum of steps b and c in Figure 14-1) is defined as the energy change involved in the (exothermic) hydration of one mole of gaseous ions:

$$M^{n+}(g) + xH_2O \longrightarrow M(OH_2)_x^{n+} + \text{energy (for cation)}$$

$$X^{y-}(g) + rH_2O \longrightarrow X(H_2O)_r^{y-} + \text{energy (for anion)}$$

Hydration is generally highly exothermic for ionic or polar covalent compounds, because the polar water molecules interact very strongly with ions and polar molecules (as we saw in Section 13-2). In fact, the only solutes that are appreciably soluble in water either undergo dissociation or ionization or are able to form hydrogen bonds with water.

The overall heat of solution for a solid dissolving in a liquid is equal to the heat of solvation minus the crystal lattice energy:

$$\Delta H_{\text{solution}} = \text{heat of solvation} - \text{crystal lattice energy}$$

(Remember that both terms on the right are always negative.)

Nonpolar solids such as naphthalene, $C_{10}H_8$, do not dissolve appreciably in polar solvents such as water because the two substances do not attract each other significantly. This is true despite the fact that crystal lattice energies of solids consisting of nonpolar molecules are much less negative than those of ionic solids. Naphthalene dissolves readily in nonpolar solvents such as benzene because there are no strong attractive forces between solute molecules or between solvent molecules. These facts help explain the observation that "like dissolves like."

In such cases, the increase in disorder controls the process.

Let us consider what happens when a piece of sodium chloride, a typical ionic solid, is placed in water. The $\delta+$ ends of water molecules attract the negative chloride ions on the surface of the solid NaCl, as shown in Figure 14-2. Likewise, the $\delta-$ ends of H_2O molecules (O atoms) orient themselves toward the Na^+ ions and solvate them. These attractions help to overcome the forces holding the ions in the crystal.

$$NaCl(s) \xrightarrow{H_2O} Na^+(aq) + Cl^-(aq)$$

For simplicity, we frequently omit the (aq) designations from dissolved ions. But we must remember that all ions are hydrated in aqueous solution, whether this is indicated or not.

When we write $Na^+(aq)$ and $Cl^-(aq)$, we refer to hydrated ions. The number of H_2O molecules attached to an ion differs with different ions. Sodium ions are thought to be hexahydrated; that is, $Na^+(aq)$ probably represents $[Na(OH_2)_6]^+$. Most cations in aqueous solution are surrounded by four to nine H_2O molecules, with six being the most common. Generally, larger cations can accommodate more H_2O molecules than smaller cations.

Many solids that are appreciably soluble in water are ionic in nature. Magnitudes of crystal lattice energies generally increase with increasing charge and decreasing size of ions. That is, the size of the lattice energy increases as the ionic charge densities increase and, therefore, as the strength of electrostatic attractions within the crystal increases. Hydration energies vary in the same order (Table 14-1). As we indicated earlier, crystal lattice energies and hydration energies are generally much smaller in magnitude for molecular solids than for ionic solids.

Review the sizes of ions in Table 6-1 carefully.

Hydration and the effects of attractions in a crystal oppose each other in the dissolution process. Hydration energies and lattice energies are usually of about the same magnitude for low-charge species, so they often nearly

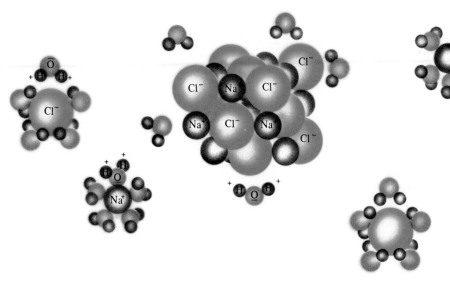

Figure 14-2
Electrostatic attraction in the dissolution of NaCl in water. The $\delta+$ H of the polar H_2O molecule helps to attract Cl^- away from the crystal. Likewise, Na^+ is attracted by the $\delta-$ O. Once they are separated from the crystal, both kinds of ions are surrounded by water molecules, to complete the hydration process.

cancel each other. As a result, the dissolution process is slightly endothermic for many ionic substances. Ammonium nitrate, NH_4NO_3, is an example of a salt that dissolves endothermically. This property is used in the "instant ice packs" used to treat sprains and other minor injuries. Ammonium nitrate and water are packaged in a plastic bag in which they are kept separate by a partition that is easily broken when squeezed. As the NH_4NO_3 dissolves in the H_2O, the mixture absorbs heat from its surroundings and the bag becomes cold to the touch.

Some ionic solids dissolve with the release of heat. Examples are anhydrous sodium sulfate, Na_2SO_4; calcium acetate, $Ca(CH_3COO)_2$; calcium chloride, $CaCl_2$; and lithium sulfate monohydrate, $Li_2SO_4 \cdot H_2O$.

As the charge-to-size ratio (charge density) increases for ions in ionic solids, the magnitude of the crystal lattice energy usually increases more than the hydration energy. This makes dissolution of solids that contain highly charged ions—such as aluminum fluoride, AlF_3; magnesium oxide, MgO; and chromium(III) oxide, Cr_2O_3—very endothermic. This means that these compounds are not very soluble in water.

Solid ammonium nitrate, NH_4NO_3, dissolves in water in a very endothermic process, absorbing heat from its surroundings. It is used in instant cold packs for early treatment of injuries such as sprains and bruises, to minimize swelling.

The charge/radius ratio is the ionic charge divided by the ionic radius in angstroms. This is a measure of the *charge density* around the ion. Negative values for heat of hydration indicate that heat is *released* during hydration.

Table 14-1
Ionic Radii, Charge/Radius Ratios, and Heats of Hydration for Some Cations

Ion	Ionic Radius (Å)	Charge/Radius Ratio	Heat of Hydration (kJ/mol)
K^+	1.33	0.75	−351
Na^+	0.95	1.05	−435
Li^+	0.60	1.67	−544
Ca^{2+}	0.99	2.02	−1650
Fe^{2+}	0.76	2.63	−1980
Zn^{2+}	0.74	2.70	−2100
Cu^{2+}	0.72	2.78	−2160
Fe^{3+}	0.64	4.69	−4340
Cr^{3+}	0.62	4.84	−4370
Al^{3+}	0.50	6.00	−4750

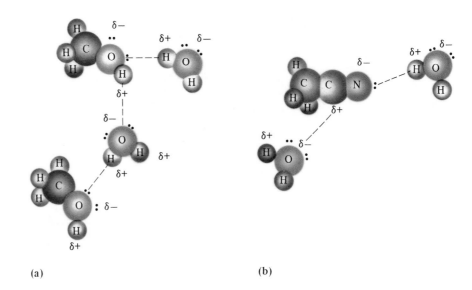

Figure 14-3
(a) Hydrogen bonding in methanol–water solution. (b) Dipolar interaction in acetonitrile–water solution.

(a)   (b)

14-3 Dissolution of Liquids in Liquids (Miscibility)

Miscibility is the ability of one liquid to dissolve in another. The three kinds of attractive interactions (solvent–solute, solvent–solvent, and solute–solute) must be considered for liquid–liquid solutions just as they were for solid–liquid solutions. Because solute–solute attractions are usually much lower for liquid solutes than for solids, this factor is less important and so the mixing process is often exothermic for miscible liquids. Polar liquids tend to interact strongly with and dissolve readily in other polar liquids. Methanol, CH_3OH; ethanol, CH_3CH_2OH; acetonitrile, CH_3CN; and sulfuric acid, H_2SO_4, are all polar liquids that are soluble in most polar solvents (such as water). The hydrogen bonding between methanol and water molecules and the dipolar interaction between acetonitrile and water molecules are depicted in Figure 14-3.

Because hydrogen bonding is so strong between sulfuric acid, H_2SO_4, and water, large amounts of heat are released when concentrated H_2SO_4 is diluted with water (Figure 14-4). This can cause the solution to boil and spatter. If the major component of the mixture is water, this heat can be absorbed with

Hydrogen bonding and dipolar interactions were discussed in Section 13-2.

Figure 14-4
The heat released by pouring 50 mL of sulfuric acid into 50 mL of water increases the temperature by 100°C (from 21°C to 121°C)!

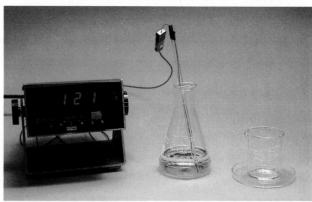

less increase in temperature because of the unusually high specific heat of H_2O. For this reason, *sulfuric acid (as well as other mineral acids) is always diluted by adding the acid slowly and carefully to water. Water is never added to the acid.* If spattering does occur when the acid is added to water, it is mainly water that spatters, not the corrosive concentrated acid.

Nonpolar liquids that do not react with the solvent generally are not very soluble in polar liquids because of the mismatch of forces of interaction. Nonpolar liquids are, however, usually quite soluble in other nonpolar liquids. Between nonpolar molecules (whether alike or different) there are only London forces, which are weak and easily overcome. As a result, when two nonpolar liquids are mixed, their molecules just "slide between" each other.

14-4 Dissolution of Gases in Liquids

Based on Section 13-2 and the foregoing discussion, we should expect that polar gases are most soluble in polar solvents and nonpolar gases are most soluble in nonpolar liquids. Although carbon dioxide and oxygen are nonpolar gases, they do dissolve slightly in water. CO_2 is somewhat more soluble because it reacts with water to some extent to form carbonic acid, H_2CO_3. This in turn ionizes slightly in two steps to give hydrogen ions, bicarbonate ions, and carbonate ions:

$$CO_2(g) + H_2O(\ell) \rightleftharpoons H_2CO_3(aq) \qquad \text{carbonic acid (exists only in solution)}$$

$$H_2CO_3(aq) \rightleftharpoons H^+(aq) + HCO_3^-(aq)$$

$$HCO_3^-(aq) \rightleftharpoons H^+(aq) + CO_3^{2-}(aq)$$

Approximately 1.45 grams of CO_2 (0.033 mole) dissolves in a liter of water at 25°C and one atmosphere pressure.

Oxygen, O_2, is less soluble in water than CO_2, but it does dissolve to a noticeable extent due to London forces (induced dipoles, Section 13-2). Only about 0.041 gram of O_2 (1.3×10^{-3} mole) dissolves in a liter of water at 25°C and one atmosphere pressure. Yet this is sufficient to support aquatic life.

The hydrogen halides, HF, HCl, HBr, and HI, are all polar covalent gases. In the gas phase the interactions among the widely separated molecules are not very strong, so solute–solute attractions are minimal, and the dissolution processes in water are exothermic. The resulting solutions, called hydrohalic acids, contain predominantly ionized HX (X = Cl, Br, I). The ionization involves *protonation* of a water molecule by HX to form a hydrated hydrogen ion and halide ion X^- (which is also hydrated). HCl is used as an example.

$$H:\ddot{C}l: + H:\ddot{O}: \longrightarrow :\ddot{C}l:^- + \left[H:\ddot{O}: \atop H \right]^+ \atop H$$

HF is only slightly ionized in aqueous solution because of the strong covalent bond between highly electronegative fluorine and hydrogen atoms. In addition, the more polar bond between H and the small F atoms in HF

When water is added to concentrated acid, the danger is due more to the spattering of the acid itself than to the steam from boiling water.

The nonpolar molecules in oil do not attract polar water molecules, so oil and water are immiscible. The polar water molecules attract each other strongly—they "squeeze out" the nonpolar molecules in the oil. Oil is less dense than water, so it floats on water.

Carbon dioxide is called an acid anhydride, i.e., an "acid without water." As noted in Section 6-8, part 2, many other oxides of nonmetals, such as N_2O_5, SO_3, and P_4O_{10}, are also acid anhydrides. Most dissolve in water by reacting with it.

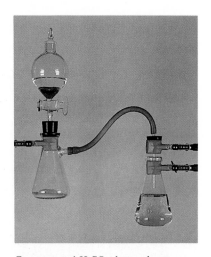

Concentrated H_2SO_4 dropped onto solid NaCl in the flask on the left produces gaseous HCl. The flask at the right contains water to which the indicator methyl orange has been added. As the HCl gas dissolves in (reacts with) the water, the indicator turns red, its color in acidic solutions.

causes very strong hydrogen bonding between H_2O and the largely intact HF molecules.

$$\overset{\delta+}{H}-\overset{\delta-}{\underset{\underset{\displaystyle H}{|}}{\ddot{O}}}\!:\!\cdots\overset{\delta+}{H}-\overset{\delta-}{\ddot{\underset{\cdot\cdot}{F}}}\!:\qquad \text{as well as} \qquad \overset{\delta+}{H}-\overset{\delta-}{\ddot{\underset{\cdot\cdot}{F}}}\!:\!\cdots\overset{\delta+}{H}-\overset{\delta-}{\underset{\underset{\displaystyle H}{|}}{\ddot{O}}}\!:$$

> The only gases that dissolve appreciably in water are those that are capable of hydrogen bonding (such as HF), those that ionize (such as HCl, HBr, and HI), and those that react with water (such as CO_2).

14-5 Rates of Dissolution and Saturation

At a given temperature, the rate of dissolution of a solid increases if large crystals are ground to a powder (Figure 14-5). Grinding increases the surface area, which in turn increases the number of solute ions or molecules in contact with the solvent. When a solid is placed in water, some of its particles solvate and dissolve. The rate of this process slows as time passes because the surface area of each crystallite gets smaller and smaller. At the same time, the number of particles in solution increases, so they collide with the solid more frequently. Some of these collisions result in recrystallization. The rates of the two opposing processes become equal after some time. The solid and dissolved ions are then in equilibrium with each other.

$$\text{solid} \underset{\text{crystallization}}{\overset{\text{dissolution}}{\rightleftharpoons}} \text{dissolved ions}$$

Dynamic equilibria occur in all saturated solutions; for instance, there is a continuous exchange of oxygen molecules across the surface of water in an open container. This is fortunate for fish, which "breathe" dissolved oxygen.

Such a solution is said to be **saturated**. Saturation occurs at very low concentrations of dissolved species for slightly soluble substances and at high concentrations for very soluble substances. When imperfect crystals are placed in saturated solutions of their ions, surface defects on the crystals are slowly "patched" with no net increase in mass of solid. Often, after some time has passed, we see fewer but larger crystals. These observations provide evidence of the dynamic nature of the solubility equilibrium. After

Oxygen gas is sufficiently soluble in water to support a wide variety of aquatic life.

Figure 14-5
A mortar and pestle are used for grinding solids.

A saturated solution of copper(II) sulfate, $CuSO_4$, in water. As it evaporates, crystals of blue $CuSO_4 \cdot 5H_2O$ form. They are in dynamic equilibrium with the saturated solution.

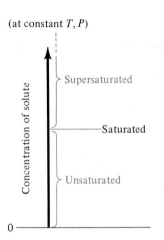

(at constant T, P)

Concentration of solute

Supersaturated

Saturated

Unsaturated

0

A solution that contains less than the amount of solute necessary for saturation is said to be unsaturated.

A tiny crystal of sodium acetate, $NaCH_3COO$, was added to a clear, colorless, supersaturated solution of $NaCH_3COO$. This photo shows solid $NaCH_3COO$ just beginning to crystallize in a very rapid process.

equilibrium is established, no more solid dissolves without the simultaneous crystallization of an equal mass of dissolved ions.

The solubilities of many solids increase at higher temperatures. **Supersaturated solutions** actually contain higher-than-saturation concentrations of solute. They can sometimes be prepared by saturating a solution at a high temperature. The saturated solution is cooled slowly, without agitation, to a temperature at which the solute is less soluble. At this point, the resulting supersaturated solution is metastable. A **metastable** state may be thought of as a state of pseudoequilibrium in which the system is at a higher energy than that required for its most stable state. In such a case, the solute has not yet become sufficiently organized for crystallization to begin. A super-

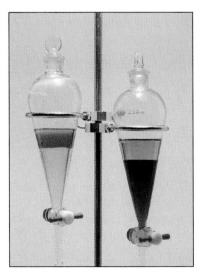

Solid iodine, I_2, dissolves to a limited extent in water to give an orange solution. This aqueous solution does not mix with nonpolar carbon tetrachloride, CCl_4 (left). Iodine is much more soluble in the nonpolar carbon tetrachloride. After the funnel is shaken and the liquids are allowed to separate (right), the upper aqueous phase is lighter orange and the lower CCl_4 layer is much more highly colored. This is because iodine is much more soluble in the nonpolar carbon tetrachloride than in water; much of the iodine dissolves preferentially in the lower (CCl_4) phase. The design of the separatory funnel allows the lower (more dense) layer to be drained off. Fresh CCl_4 could be added and the process repeated. This method of separation is called *extraction*. It takes advantage of the different solubilities of a solute in two immiscible liquids.

Figure 14-6
Another method of seeding a supersaturated solution is by pouring it very slowly onto a seed crystal. A supersaturated sodium acetate solution was used in this figure.

saturated solution produces crystals rapidly if it is slightly disturbed or if it is "seeded" with a dust particle or a crystallite. Under such conditions enough solid crystallizes to leave a solution that is just saturated (Figure 14-6).

14-6 Effect of Temperature on Solubility

In Section 13-6 we introduced LeChatelier's Principle, which states that *when a stress is applied to a system at equilibrium, the system responds in a way that best relieves the stress.* Recall that exothermic processes release heat and endothermic processes absorb heat.

$$\text{Exothermic:} \qquad \text{reactants} \longrightarrow \text{products} + \text{heat}$$

$$\text{Endothermic:} \qquad \text{reactants} + \text{heat} \longrightarrow \text{products}$$

Many solids dissolve by endothermic processes. Their solubilities in water usually *increase* as heat is added and the temperature increases. For example, KCl dissolves endothermically:

$$KCl(s) + 17.2 \text{ kJ} \xrightarrow{\text{H}_2\text{O}} K^+(aq) + Cl^-(aq)$$

Figure 14-7 shows that the solubility of KCl increases as the temperature increases because more heat is available to drive the dissolving process. Raising the temperature (adding *heat*) causes a stress on a saturated solution. This stress favors the process that *consumes* heat. In this case, more KCl dissolves.

Calcium acetate, $Ca(CH_3COO)_2$, is more soluble in cold water than in hot water. When a cold, concentrated solution of calcium acetate is heated, solid calcium acetate precipitates.

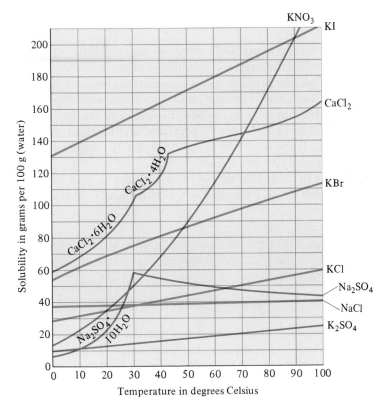

Figure 14-7
A graph that illustrates the effect of temperature on the solubilities of some salts. Some compounds exist either as nonhydrated crystalline substances or as hydrated crystals. Hydrated and nonhydrated crystal forms of the same compounds often have different solubilities, because of the different total forces of attraction in the solids. The discontinuities in the solubility curves for $CaCl_2$ and Na_2SO_4 are due to transition between hydrated and nonhydrated crystal forms.

Some solids, such as anhydrous Na_2SO_4, and many liquids and gases dissolve by exothermic processes. Their solubilities, therefore, usually decrease as temperature increases. The solubility of O_2 in water decreases (by 22%) from 0.041 gram per liter of water at 25°C to 0.03 gram per liter at 50°C. Raising the temperature of rivers and lakes by dumping heated waste water from industrial plants and nuclear power plants is called **thermal pollution**. A slight increase in the temperature of the water causes a small but significant decrease in the concentration of dissolved oxygen. As a result, the water can no longer support the marine life it ordinarily could.

The dissolution of anhydrous calcium chloride, $CaCl_2$, in water is quite exothermic. This dissolution process is utilized in commercial instant hot packs for quick treatment of injuries requiring heat.

14-7 Effect of Pressure on Solubility

Changing the pressure has no appreciable effect on the solubilities of either solids or liquids in liquids. However, the solubilities of gases in all solvents increase as the partial pressures of the gases increase. Carbonated water is a saturated solution of carbon dioxide in water under pressure. When a can or bottle of a carbonated beverage is opened, the pressure on the surface of the beverage is reduced to atmospheric pressure, and much of the CO_2 bubbles out of solution. If the container is left open, the beverage becomes "flat" because the released CO_2 escapes.

Henry's Law applies to gases that do not react with the solvent in which they dissolve (or, in some cases, gases that react incompletely). It is usually stated as follows:

> The pressure of a gas above the surface of a solution is proportional to the concentration of the gas in the solution. Henry's Law can be represented symbolically as
>
> $$P_{gas} = k\, C_{gas}$$

in which P_{gas} is the pressure of the gas above the solution and k is a constant for a particular gas and solvent at a particular temperature. C_{gas} represents the concentration of dissolved gas; it is usually expressed either as molarity (Section 3-6) or as mole fraction (Section 14-8). The relationship is valid at low concentrations and low pressures (Figure 14-8).

The solubility of CO_2 in water at 25°C (approximately room temperature) and one atmosphere pressure is only 1.45 grams per liter (Section 14-4).

14-8 Molality and Mole Fraction

We saw in Section 3-6 that concentrations of solutions are often expressed as percent by mass of solute or as molarity. Discussion of many physical properties of solutions is often made easier by expressing concentrations either in molality units or as mole fractions.

Molality

> The **molality**, m, of a solute in solution is the number of moles of solute *per kilogram of solvent* (not solution).
>
> $$\text{molality} = \frac{\text{number of moles solute}}{\text{number of kilograms solvent}}$$

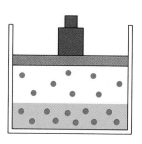
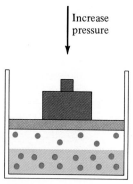

Increase
pressure

Figure 14-8
An illustration of Henry's Law. The solubility of a gas (that does not react completely with the solvent) increases with increasing pressure of the gas above the solution.

Example 14-1
What is the molality of a solution that contains 128 grams of CH_3OH in 108 grams of water?

Plan
We convert the amount of solute (CH_3OH) to moles, express the amount of solvent (water) in kilograms, and apply the definition of molality.

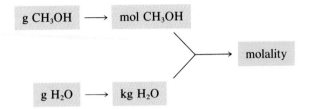

Solution

$$\frac{?\ mol\ CH_3OH}{kg\ H_2O} = \frac{128\ g\ CH_3OH}{0.108\ kg\ H_2O} \times \frac{1\ mol\ CH_3OH}{32.0\ g\ CH_3OH} = \frac{37.0\ mol\ CH_3OH}{kg\ H_2O}$$

$$= \boxed{37.0\ m\ CH_3OH}$$

EOC 28

___

Example 14-2

How many grams of H_2O must be used to dissolve 50.0 grams of sucrose to prepare a 1.25 m solution of sucrose, $C_{12}H_{22}O_{11}$?

Plan

We convert the amount of solute ($C_{12}H_{22}O_{11}$) to moles, solve the expression for molality for kilograms of solvent (water), and then express the result in grams.

Solution

$$?\ mol\ C_{12}H_{22}O_{11} = 50.0\ g\ C_{12}H_{22}O_{11} \times \frac{1\ mol\ C_{12}H_{22}O_{11}}{342\ g\ C_{12}H_{22}O_{11}} = 0.146\ mol\ C_{12}H_{22}O_{11}$$

$$molality\ of\ solution = \frac{number\ of\ mol\ C_{12}H_{22}O_{11}}{number\ of\ kg\ H_2O}$$

Rearranging gives

$$number\ of\ kg\ H_2O = \frac{number\ of\ mol\ C_{12}H_{22}O_{11}}{molality\ of\ solution}$$

$$= \frac{0.146\ mol\ C_{12}H_{22}O_{11}}{1.25\ m} = 0.117\ kg\ H_2O = \boxed{117\ g\ H_2O}$$

Each beaker holds the amount of a crystalline ionic compound that will dissolve in 100 grams of water at 100°C. The compounds are (top row, left to right) 39 grams of sodium chloride (NaCl, white), 102 grams of potassium dichromate ($K_2Cr_2O_7$, red-orange), 341 grams of nickel sulfate hexahydrate ($NiSO_4 \cdot 6H_2O$, green); (bottom row; left to right) 79 grams of potassium chromate (K_2CrO_4, yellow), 191 grams of cobalt(II) chloride hexahydrate ($CoCl_2 \cdot 6H_2O$, dark red), and 203 grams of copper sulfate pentahydrate ($CuSO_4 \cdot 5H_2O$, blue).

In other examples later in this chapter, we shall calculate several properties of the solution in Example 14-2.

Mole Fraction

Recall that in Chapter 12 the **mole fractions**, X_A and X_B, of each component in a mixture containing components A and B were defined as

$$X_A = \frac{no.\ mol\ A}{no.\ mol\ A + no.\ mol\ B} \quad and \quad X_B = \frac{no.\ mol\ B}{no.\ mol\ A + no.\ mol\ B}$$

Note that mole fraction is a dimensionless quantity.

Simple solutions, such as salt in water, are readily understood. More complex solutions, such as multi-component mixtures in water, are more complicated. At an extreme, the collection of cells, bodily fluids, and organs that make up a human being are a sort of solution—one that is exceedingly complex and imperfectly understood. The use of nuclear magnetic resonance (NMR), an application of quantum mechanical principles of living subjects, and serendipity combine to provide exciting insight into the structures revealed by the internal chemistry of humans.

Before discussing the impact of NMR on biology and medicine, we should examine some of its scientific background. Felix Bloch, one of the preeminent physicists and teachers of the 20th century, characterized himself as an "applied" physicist. That is, he applied quantum mechanics to investigations in physics. Nowhere is this purpose better exemplified than in Bloch's translation of abstract quantum mechanical concepts to concrete physical science through the agency of nuclear magnetic resonance (NMR). Quantum mechanics sometimes appears to be complicated, hard to use, arcane, and without relevance to the "real" world. A chemistry student may well ask, "How can I understand the relevance of quantum mechanics without becoming a specialist?" The answer is summarized in this statement about Bloch by one of his colleagues: "Bloch, a virtuoso of quantum mechanics, deliberately used classical language in his voyage of discovery towards NMR, in the same way as he did when describing his findings." The

Bloch description of NMR has guided several generations of chemists in their studies of solutions.

The development of NMR started nearly 60 years ago, building on a set of observations that defied, then and now, the ability of scientists to explain them entirely using models from classical physics and chemistry. Our descriptions of nuclear particles include a spin quantum number, just as those of electrons do (Section 5-14). NMR observations of molecules in solution always involve nuclei with nonzero spin quantum numbers. Consider a hydrogen atom, which contains one proton. For a proton (^1H) the spin is $\pm\frac{1}{2}$. The interpretation of the behavior of this particle in a magnetic field is that the spin (a magnetic vector) aligns either with the field or opposed to the field, but it cannot assume any other orientation. We say that two states are allowed—no more, no fewer. This either–or property is quantum mechanical; any *classical* magnet, even the nuclear one, could assume any orientation in space, no matter what the orientation of the molecule in solution. Indirect evidence reveals that the spin of any individual nucleus spontaneously changes from one state to the other every few seconds. One rule of quantum mechanics is that these spontaneous conversions cannot be directly observed. A second rule is that a conversion between the two spin states can be caused by the application of energy of the proper strength. In a magnetic field of 1.5 tesla (a field strong enough to lift screwdrivers out of your pocket from a distance of several feet), the energy required for this conversion is supplied by a radio frequency (rf) field. The required radio frequency is 64 MHz when measured at 1.5 tesla. Application of 64 MHz radio frequency causes the spin states to convert, a

resonance condition. For effective observation, the 64 MHz frequency must be tuned to within about 0.6 Hz of the resonance. The resolution of the NMR signal is about 1 part in 10^9 of the fundamental resonance frequency. Numbers of this kind are hard to imagine. If an optical telescope had such resolving power, an earthbound astronomer could distinguish the two ears of a cat on the moon. Some NMR measurements have precisions of 1 part in 10^{11}; one can surmise how much more could now be observed about the lunar cat. This phenomenon, the excitation and observation of the conversion between nuclear spin states, is NMR.

One principal use of the NMR technique is to study the pattern of resonances associated with an individual substance. Interpretation using a number of paradigms, based in quantum mechanics but stated in everyday language, permits the investigator to confirm a guessed molecular structure and even to propose a new structure. The NMR method of chemical structure determination has become very sophisticated and is now pervasive in chemistry.

One feature of NMR is that the resonance frequency changes with the strength of the applied magnetic field. The relationship is linear—if we double the magnetic field strength, the resonance frequency is doubled. This informs us that a change in field strength of a few parts per million will alter the resonance frequency by a few parts per million. The sharpness of the resonance assures that we can detect this change accurately and precisely. In the last two decades, a remarkable scientific and commercial discipline has been built around a novel concept in NMR. If a magnetic field can be generated in which the field strength varies from

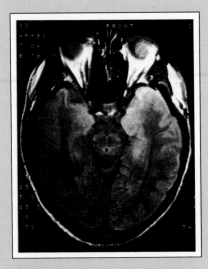

(a) A transverse view of the brain, in a plane containing the eyes and sinus cavities.

point to point in space, a new observation can be made. The frequency of the resonance becomes a mapping parameter so that we can make a Cartesian plot of NMR response versus spatial location. *We can make an NMR image.* In this manner, the research community has constructed a new type of camera. When this technique is applied to a biological system, the physical basis of the image is the quantum mechanical resonance of the water molecule. The output of the measurement is a picture that can be printed on photographic media, so it appears that we have created a camera capable of examining very complex solutions.

How good is this camera? What types of material can be examined? How precise is the spatial mapping? What will we really do with it? Partial answers are provided by medical and biological science. When a human was first placed in a large magnet and the NMR image was formed, there were a number of surprises. (First we needed to name

the technology. The word "nuclear" has alarming connotations in our society, so the medical community renamed NMR imaging "magnetic resonance imaging" or MRI.) Applied to the observation of human beings, MRI can present a thin section in exquisite two-dimensional detail. The orientation of the thin slice is under the complete control of the operator of the MRI scanner. An image of a transverse section of the brain of a normal (or at least a college professor!) volunteer is shown in Figure (a). Comparison of the detailed anatomical features in the image with corresponding cross-sections from the anatomy lab are striking; the MR image looks like the laboratory result. One significant exception is that the MR image was obtained from a live subject, by a noninvasive method, and no known harm was done.

The operation of an MR imager is analogous to the operation of a high-priced camera. One need not know about optics, film manufacture, processing chemicals, filtering of the light prior to printing, and so on, to be a good photographer. On the other hand, the very best photographers are knowledgeable about all of these features. In the same sense, a good MR imager can find more anatomical details when the device is operated at its technical limit. A magnified cross-section of the middle of the cerebellum in Figures (b) and (c) illustrates the state of the art (1991). This cross-section, obtained from the anatomical midline (a sagittal view), shows at least as much detail as gross anatomy studies do. The result is a revelation to the medical student and even to the practicing physician.

Although these images may seem to be of interest only to the medical community, three features are worth noting. First, the methodology applies to ordinary chemical so-

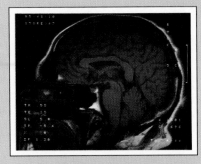

(b) A view of the brain through the anatomical midline (a sagittal view).

lutions. Second, the technology and intellectual basis are similar to those of X-ray crystallography, so MRI is part of a cohesive scientific package. Finally, MRI is a tribute to pure mathematics and to the application of quantum mechanics to useful systems. MRI is a mere infant. In the next decade, this technique will provide surprising advances in materials science, chemistry, physics, and medical science.

Professor M. R. Willcott III
Vanderbilt Medical School
Vanderbilt University

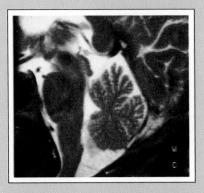

(c) A magnified view of the cerebellum at the midline, enhanced to show more detail.

Example 14-3

What are the mole fractions of CH_3OH and H_2O in the solution described in Example 14-1? It contains 128 grams of CH_3OH and 108 grams of H_2O.

Plan

We express the amounts of both components in moles, and then apply the definition of mole fraction.

Solution

$$? \text{ mol } CH_3OH = 128 \text{ g } CH_3OH \times \frac{1 \text{ mol } CH_3OH}{32.0 \text{ g } CH_3OH} = 4.00 \text{ mol } CH_3OH$$

$$? \text{ mol } H_2O = 108 \text{ g } H_2O \times \frac{1 \text{ mol } H_2O}{18.0 \text{ g } H_2O} = 6.00 \text{ mol } H_2O$$

Now we calculate the mole fraction of each component.

$$X_{CH_3OH} = \frac{\text{no. mol } CH_3OH}{\text{no. mol } CH_3OH + \text{no. mol } H_2O} = \frac{4.00 \text{ mol}}{(4.00 + 6.00) \text{ mol}} = \boxed{0.400}$$

$$X_{H_2O} = \frac{\text{no. mol } H_2O}{\text{no. mol } CH_3OH + \text{no. mol } H_2O} = \frac{6.00 \text{ mol}}{(4.00 + 6.00) \text{ mol}} = \boxed{0.600}$$

In any mixture the sum of the mole fractions must be 1:

$$0.400 + 0.600 = 1$$

EOC 40

Colligative Properties of Solutions

Colligative means "tied together."

Physical properties of solutions that depend upon the *number*, but not the *kind*, of solute particles in a given amount of solvent are called **colligative properties**. The four colligative properties of a solution affect the solvent by (1) lowering its vapor pressure, (2) raising its boiling point, (3) lowering its melting point, and (4) generating an osmotic pressure. These properties of a solution depend on the *total concentration of all solute particles*, regardless of their ionic or molecular nature, charge, or size. For most of this chapter, we shall consider *nonelectrolyte* solutes (Section 4-2, part 1); these substances dissolve to give one mole of dissolved particles for each mole of solute. In Section 14-14, we shall learn to modify our predictions of colligative properties to account for ion formation in electrolyte solutions.

14-9 Lowering of Vapor Pressure and Raoult's Law

Many experiments have shown that a solution containing a *nonvolatile* liquid or a solid as a solute always has a lower vapor pressure than the pure solvent (Figure 14-9). The vapor pressure of a liquid depends on the ease with which the molecules are able to escape from the surface of the liquid. When a solute is dissolved in a liquid, some of the total volume of the solution is occupied by solute molecules, and so there are fewer solvent molecules *per unit area* at the surface. As a result, solvent molecules vaporize at a slower rate than if no solute were present. The increase in disorder that accompanies evaporation is also a significant factor. Because a solution is already more disordered ("mixed up") than a pure solvent, the evaporation of the pure solvent involves a larger increase in disorder, and is thus more favorable. Hence, the pure solvent exhibits a higher vapor pressure than does the solution. The lowering of the vapor pressure of the solution is a colligative property. It is a function of the number, and not the kind, of solute particles in solution. We emphasize that solutions of gases or low-boiling (volatile) liquids have *higher* total vapor pressures than the pure solvents, so this discussion does not apply to them.

The lowering of vapor pressure associated with *nonvolatile, nonionizing* solutes is summarized by **Raoult's Law**:

> The vapor pressure of a solvent in an ideal solution decreases as its mole fraction decreases.

The relationship can be expressed mathematically as

$$P_{solvent} = X_{solvent} P^0_{solvent}$$

in which $X_{solvent}$ represents the mole fraction of the solvent in a solution, $P^0_{solvent}$ is the vapor pressure of the *pure* solvent, and $P_{solvent}$ is the vapor pressure of the solvent *in the solution*. (See Figure 14-10.)

The *lowering* of vapor pressure, $\Delta P_{solvent}$, is defined as

$$\Delta P_{solvent} = P^0_{solvent} - P_{solvent}$$

Thus,

$$\Delta P_{solvent} = P^0_{solvent} - (X_{solvent} P^0_{solvent}) = (1 - X_{solvent})P^0_{solvent}$$

Now $X_{solvent} + X_{solute} = 1$, so $1 - X_{solvent} = X_{solute}$. We can express the *lowering* of vapor pressure in terms of the mole fraction of solute:

$$\Delta P_{solvent} = X_{solute} P^0_{solvent}$$

Solutions that obey this relationship exactly are called **ideal solutions**. The vapor pressures of many solutions do not behave ideally.

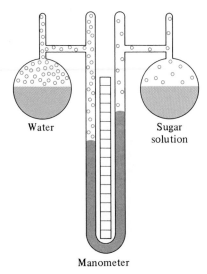

Figure 14-9

Lowering of vapor pressure. If no air is present in the apparatus, the pressure above each liquid is due to water vapor. This pressure is less over the solution of sugar and water.

Example 14-4

Sucrose is a nonvolatile, nonionizing solute in water. Determine the vapor pressure lowering, at 25°C, of the 1.25 *m* sucrose solution in Example 14-2. Assume that the solution behaves ideally. From Appendix E, the vapor pressure of pure water at 25°C is 23.8 torr.

Plan

We found in Example 14-2 that the solution was made by dissolving 50.0 grams of sucrose (0.146 mol sucrose) in 117 grams of water. We convert the known amount of water to number of moles and then calculate the mole fraction of solute in the solution. We apply Raoult's Law to find the vapor pressure lowering, $\Delta P_{solvent}$.

Solution

$$\underline{?}\text{ mol H}_2\text{O} = 117\text{ g H}_2\text{O} \times \frac{1\text{ mol H}_2\text{O}}{18.0\text{ g H}_2\text{O}} = 6.50\text{ mol H}_2\text{O}$$

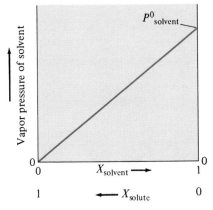

Figure 14-10

Raoult's Law for an ideal solution of a solute in a volatile liquid. The vapor pressure exerted by the liquid is proportional to *its* mole fraction in the solution.

The vapor pressure of water in the solution is 23.8 torr − 0.524 torr = 23.3 torr. We could calculate this vapor pressure directly from the mole fraction of the solvent (water) in the solution, using the relationship $P_{solvent} = (X_{solvent})(P^0_{solvent})$.

Then $X_{sucrose} = \dfrac{0.146 \text{ mol}}{0.146 \text{ mol} + 6.50 \text{ mol}} = 0.0220$

Applying Raoult's Law in terms of the vapor pressure lowering,

$$\Delta P_{solvent} = (X_{solute})(P^0_{solvent}) = (0.0220)(23.8 \text{ torr}) = \boxed{0.524 \text{ torr}}$$

EOC 44, 46

When a solution consists of two components that are very similar, each component behaves essentially as it would if it were pure. For example, the two liquids *n*-heptane, C_7H_{16}, and *n*-octane, C_8H_{18}, are so similar that each heptane molecule experiences nearly the same intermolecular forces whether it is near another heptane molecule or near an *n*-octane molecule, and similarly for each *n*-octane molecule. The properties of such a solution can be predicted from a knowledge of its composition and the properties of each component. Such a solution is very nearly ideal.

Consider an ideal solution of two volatile components, A and B. The vapor pressure of each component above the solution is proportional to its mole fraction in the solution.

$$P_A = X_A P^0_A \qquad \text{and} \qquad P_B = X_B P^0_B$$

The total vapor pressure of the solution is, by Dalton's Law of Partial Pressures (Section 12-11), equal to the sum of the vapor pressures:

$$P_{total} = P_A + P_B \qquad \text{or} \qquad P_{total} = X_A P^0_A + X_B P^0_B$$

This is shown graphically in Figure 14-11. We can use these relationships to predict the vapor pressures of an ideal solution, as Example 14-5 illustrates.

Example 14-5

At 40°C, the vapor pressure of pure *n*-heptane is 92.0 torr and the vapor pressure of pure *n*-octane is 31.0 torr. Consider a solution that contains 1.00 mole of *n*-heptane and 4.00 moles of *n*-octane. Calculate the vapor pressure of each component and the total vapor pressure above the solution.

Plan

We first calculate the mole fraction of each component in the liquid solution. Then we apply Raoult's Law to each of the two volatile components. By Dalton's Law of Partial Pressures, the total vapor pressure is the sum of the vapor pressures of the components.

Solution

We first calculate the mole fraction of each component in the liquid solution.

$$X_{heptane} = \frac{1.00 \text{ mol heptane}}{(1.00 \text{ mol heptane}) + (4.00 \text{ mol octane})} = 0.200$$

$$X_{octane} = 1 - X_{heptane} = 0.800$$

Then, applying Raoult's Law for volatile components,

$$P_{heptane} = X_{heptane} P^0_{heptane} = (0.200)(92.0 \text{ torr}) = \boxed{18.4 \text{ torr}}$$

$$P_{octane} = X_{octane}P^0_{octane} = (0.800)(31.0 \text{ torr}) = \boxed{24.8 \text{ torr}}$$

$$P_{total} = P_{heptane} + P_{octane} = 18.4 \text{ torr} + 24.8 \text{ torr} = \boxed{43.2 \text{ torr}}$$

EOC 47

The vapor in equilibrium with a solution of two or more volatile components is *richer* in the more volatile component *than the liquid solution*. The next example illustrates this.

Example 14-6
Calculate the mole fractions of *n*-heptane and *n*-octane in the vapor that is in equilibrium with the solution in Example 14-5.

Plan
We learned in Section 12-11 that the mole fraction of a component in a gaseous mixture equals the ratio of its partial pressure to the total pressure. In Example 14-5, we calculated the partial pressure of each component in the vapor and the total vapor pressure.

Solution
In the *vapor*,

$$X_{n\text{-heptane}} = \frac{P_{n\text{-heptane}}}{P_{total}} = \frac{18.4 \text{ torr}}{43.2 \text{ torr}} = \boxed{0.426}$$

$$X_{n\text{-octane}} = \frac{P_{n\text{-octane}}}{P_{total}} = \frac{24.8 \text{ torr}}{43.2 \text{ torr}} = \boxed{0.574}$$

EOC 48

n-Heptane (pure vapor pressure = 92.0 torr at 40°C) is a more volatile liquid than *n*-octane (pure vapor pressure = 31.0 torr at 40°C). Its mole fraction in the vapor, 0.426, is higher than its mole fraction in the liquid, 0.200.

Most solutions behave ideally, provided they are sufficiently dilute. Some solutions do not behave ideally over the entire concentration range. For

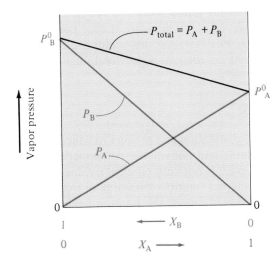

Figure 14-11
Raoult's Law for an ideal solution of two volatile components. Note that the left-hand side of the plot corresponds to pure B ($X_A = 0$, $X_B = 1$), and the right-hand side corresponds to pure A ($X_A = 1$, $X_B = 0$). Of these hypothetical liquids, pure B is more volatile than pure A ($P^0_B > P^0_A$).

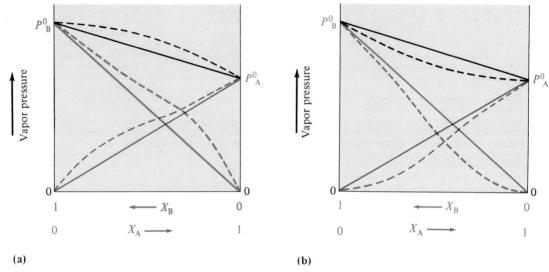

(a)   (b)

Figure 14-12
Deviations from Raoult's Law for two
volatile components. (a) Positive
deviation. (b) Negative deviation. (See
text for explanation.)

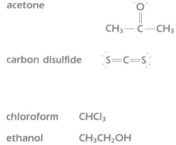

some solutions, the observed vapor pressure is greater than that predicted
by Raoult's Law (Figure 14-12a). This kind of deviation, known as a *positive
deviation*, is due to differences in polarity of the two components. On the
molecular level, the two substances do not mix entirely randomly, so there
is self-association of each component with local regions enriched in one type
of molecule or the other. In a region enriched in A molecules, substance A
acts as though its mole fraction were greater than it is in the solution as a
whole, and the vapor pressure due to A is greater than if the solution were
ideal. A similar description applies to component B. The total vapor pressure
is then greater than it would be if the solution were behaving ideally. For
instance, a solution of acetone and carbon disulfide shows a positive deviation from Raoult's Law.

Another, more common type of deviation occurs when the total vapor
pressure is less than that predicted (Figure 14-12b). This is called a *negative
deviation*. Such an effect is due to unusually strong attractions (such as
hydrogen bonding) between *unlike* molecules. As a result, unlike species
hold one another especially tightly in the liquid phase, so fewer molecules
escape to the vapor phase. The observed vapor pressure of each component
is thus less than ideally predicted. For instance, an acetone–chloroform
solution behaves in this way, as does an ethanol–water solution.

14-10 Fractional Distillation

In Section 13-8 we described *simple* distillation as a process in which a liquid
solution can be separated into volatile and nonvolatile components. Separation of volatile components, however, is not very efficient by this method.
Consider the simple distillation of a liquid solution consisting of two volatile
components. If the temperature is slowly raised, the solution begins to boil
when the sum of the vapor pressures of the components reaches the applied
pressure on the surface of the solution. Both components exert vapor pres-

The applied pressure is often atmospheric pressure.

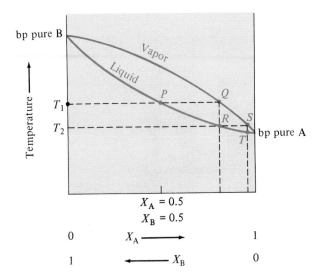

Figure 14-13
A boiling point diagram for a solution of two volatile liquids, A and B. The lower curve represents the boiling point of a liquid mixture with the indicated composition. The upper curve represents the composition of the *vapor* in equilibrium with the boiling liquid mixture at the indicated temperature. Pure liquid A boils at a lower temperature than pure liquid B; hence, A is the more volatile liquid in this illustration. Suppose we begin with an ideal equimolar mixture ($X_A = X_B = 0.5$) of liquids A and B. The point P represents the temperature at which this solution boils, T_1. The vapor that is present at this equilibrium is indicated by point Q ($X_A \approx 0.8$). Condensation of that vapor at temperature T_2 gives a liquid of the same composition (point R). At this point we have described one step of simple distillation. The boiling liquid at point R is in equilibrium with vapor of composition indicated by point S ($X_A > 0.95$), and so on.

sures, so both are carried away as a vapor. The resulting distillate is richer than the original liquid in the more volatile component (Example 14-6).

As a mixture of volatile liquids is distilled, the compositions of both the liquid and the vapor, as well as the boiling point of the solution, change continuously. *At constant pressure*, these quantities are conveniently represented in a **boiling point diagram**, Figure 14-13. In such a diagram, the lower curve represents the boiling point of a liquid mixture with the indicated composition. The upper curve represents the composition of the *vapor* in equilibrium with the boiling liquid mixture at the indicated temperature. The intercepts at the two vertical axes represent the boiling points of the two pure liquids. The distillation of the two liquids is described in the legend to Figure 14-13.

From the boiling point diagram in Figure 14-13, we see that two or more volatile liquids cannot be completely separated from each other by a single distillation step. The vapor collected at any boiling temperature is always enriched in the more volatile component (A); however, at any temperature it is still some mixture of the two vapors. A *series* of simple distillations would provide distillates increasingly richer in the more volatile component, but the repeated distillations would be very tedious.

Repeated distillations may be avoided by using **fractional distillation**. A *fractionating column* is inserted above the solution and attached to the condenser, as shown in Figures 1-12b and 14-14a. The column is constructed so that it has a large surface area or is packed with many small glass beads. These provide surfaces upon which condensation can occur. Contact between the vapor and the packing favors condensation of the less volatile component. The column is cooler at the top than at the bottom. By the time the vapor reaches the top of the column, practically all of the less volatile component has condensed and fallen back down the column. The more volatile component goes into the condenser, where it is liquefied and delivered as a highly enriched distillate into the collection flask. The longer the column or the more packing, the better the separation.

Distillation under vacuum lowers the applied pressure. This allows boiling at lower temperatures than under atmospheric pressure. This technique allows distillation of some substances that would decompose at higher temperatures.

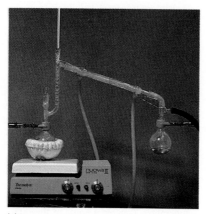

(a)

Figure 14-14
(a) A fractional distillation apparatus. The vapor phase rising in the column is in equilibrium with the liquid phase that has condensed out and is flowing slowly back down the column.
(b) Fractional distillation is used for separations in many industrial processes. In this Pennsylvania plant, atmospheric air is liquefied by cooling and compression and then is separated by distillation in towers. This plant produces more than 1000 tons per day of gases from air (nitrogen, oxygen, and argon).

(b)

14-11 Boiling Point Elevation

Recall that the boiling point of a liquid is the temperature at which its vapor pressure just equals the applied pressure on its surface. For liquids in open containers, this is atmospheric pressure. We have seen that the vapor pressure of a solvent at a given temperature is lowered by the presence in it of a *nonvolatile* solute. Such a solution must be heated to a higher temperature than the pure solvent to cause the vapor pressure of the solvent to equal atmospheric pressure (Figure 14-15). Therefore, in accord with Raoult's Law, the elevation of the boiling point of a solvent caused by the presence of a nonvolatile, nonionized solute is proportional to the number of moles of solute dissolved in a given mass of solvent. Mathematically, this is usually expressed as

> When the solute is nonvolatile, only the *solvent* distills from the solution.

$$\Delta T_b = K_b m$$

> $\Delta T_b = T_{b(soln)} - T_{b(solvent)}$. Note that ΔT_b is always positive for solutions that contain nonvolatile solutes, because the boiling points of such solutions are always higher than the boiling points of the pure solvents.

The term ΔT_b represents the elevation of boiling point of the solvent, i.e., the boiling point of the solution minus the boiling point of the pure solvent.

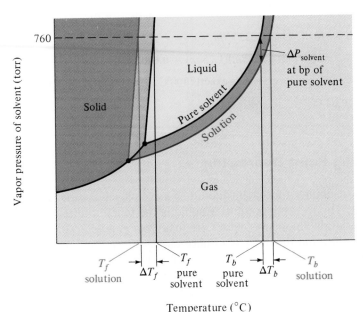

Figure 14-15
Because a *nonvolatile* solute lowers the vapor pressure of a solvent, the boiling point of a solution is higher and the freezing point lower than the corresponding points for the pure solvent. The magnitude of boiling point elevation, ΔT_b, is less than the magnitude of freezing point depression, ΔT_f.

The m is the molality of the solute, and K_b is a proportionality constant called the **molal boiling point elevation constant**. This constant is different for different solvents and does not depend on the solute (Table 14-2).

K_b corresponds to the change in boiling point produced by a one-molal *ideal* solution of a nonvolatile nonelectrolyte.

Elevations of boiling points and depressions of freezing points, which will be discussed later, are usually quite small for solutions of typical concentrations. As a result, they are often measured with specially constructed (and expensive) differential thermometers that measure small temperature changes accurately to the nearest 0.001°C.

Example 14-7
What is the normal boiling point of the 1.25 m sucrose solution of Example 14-2?

Plan
We first find the *increase* in boiling point from the relationship $\Delta T_b = K_b m$. The boiling point is *higher* by this amount than the normal boiling point of pure water.

Table 14-2
Some Properties of Common Solvents

Solvent	bp (pure)	K_b (°C/m)	fp (pure)	K_f (°C/m)
water	100	0.512	0	1.86
benzene	80.1	2.53	5.48	5.12
acetic acid	118.1	3.07	16.6	3.90
nitrobenzene	210.88	5.24	5.7	7.00
phenol	182	3.56	43	7.40
camphor	207.42	5.61	178.40	40.0

Solution
From Table 14-2, K_b for H_2O = 0.512°C/m, so

$$\Delta T_b = (0.512°C/m)(1.25\ m) = 0.640°C$$

The normal boiling point of pure water is exactly 100°C, so at 1.00 atm this solution

boils at 100°C + 0.640°C = $\boxed{100.640°C.}$

14-12 Freezing Point Depression

Molecules of most liquids approach each other more closely as the temperature is lowered. The freezing point of a liquid is the temperature at which the forces of attraction among molecules are just great enough to overcome their kinetic energies and thus cause a phase change from the liquid to the solid state. Strictly speaking, the freezing (melting) point of a substance is the temperature at which the liquid and solid phases are in equilibrium. When a solution freezes, it is the *solvent* that begins to solidify first, leaving the solute in a more concentrated solution. Solvent molecules in a solution are somewhat more separated from each other (because of solute particles) than they are in the pure solvent. Consequently, the temperature of a solution must be lowered below the freezing point of the pure solvent to freeze it.

The freezing point depressions of solutions of nonelectrolytes have been found to be equal to the molality of the solute times a proportionality constant called the **molal freezing point depression constant, K_f**:

ΔT_f is the *depression* of freezing point. It is defined as

$$\Delta T_f = T_{f(solvent)} - T_{f(soln)}$$

so it is always *positive*.

$$\Delta T_f = K_f m$$

The values of K_f for a few solvents are given in Table 14-2. Each is numerically equal to the freezing point depression of a one-molal *ideal* solution of a nonvolatile nonelectrolyte in that solvent.

Lime, CaO, is added to molten iron ore during the manufacture of pig iron. It lowers the melting point of the mixture. The metallurgy of iron is discussed in more detail in Chapter 22.

Example 14-8
When 15.0 grams of ethyl alcohol, C_2H_5OH, is dissolved in 750 grams of formic acid, the freezing point of the solution is 7.20°C. The freezing point of pure formic acid is 8.40°C. Evaluate K_f for formic acid.

Plan
We solve the equation $\Delta T_f = K_f m$ for K_f and substitute values for ΔT_f and m:

$$K_f = \frac{\Delta T_f}{m}$$

The molality and the depression of the freezing point are calculated first.

Solution

$$\frac{?\ mol\ C_2H_5OH}{kg\ formic\ acid} = \frac{15.0\ g\ C_2H_5OH}{0.750\ kg\ formic\ acid} \times \frac{1\ mol\ C_2H_5OH}{46.0\ g\ C_2H_5OH} = 0.435\ m$$

$$\Delta T_f = (T_{f[formic\ acid]}) - (T_{f[solution]}) = 8.40°C - 7.20°C = 1.20°C\ \text{(depression)}$$

Then $K_f = \dfrac{\Delta T_f}{m} = \dfrac{1.20°C}{0.435\ m} = \boxed{2.76°C/m}$ for formic acid.

Example 14-9

What is the freezing point of the 1.25 m sucrose solution of Example 14-2?

Plan

We first find the *decrease* in freezing point from the relationship $\Delta T_f = K_f m$. The temperature at which the solution freezes is *lower* by this amount than the freezing point of pure water.

Solution

From Table 14-2, K_f for $H_2O = 1.86°C/m$, so

$$\Delta T_f = (1.86°C/m)(1.25\ m) = 2.32°C$$

The temperature at which the solution freezes is 2.32°C *below* the freezing point of pure water, or

$$T_{f(solution)} = 0.00°C - 2.32°C = \boxed{-2.32°C}$$

EOC 55, 58

You may be familiar with several examples of the effects we have studied. Seawater does not freeze on some days when fresh water does, because seawater contains higher concentrations of solutes, largely ionic solutes. Spreading soluble salts such as sodium chloride, NaCl, or calcium chloride, $CaCl_2$, on an icy road lowers the freezing point of the ice, causing the ice to melt.

A familiar application is the addition of "permanent" antifreeze, mostly ethylene glycol, $HOCH_2CH_2OH$, to the water in an automobile radiator. Because the boiling point of the solution is elevated, addition of a solute as a winter antifreeze also helps to protect against loss of the coolant by summer "boil-over." The amounts by which the freezing and boiling points change depend on the concentration of the ethylene glycol solution. However, the addition of too much ethylene glycol is counterproductive. The freezing point of pure ethylene glycol is about $-12°C$. A solution that is mostly ethylene glycol would have a somewhat lower freezing point due to the presence of water as a solute. Suppose you graph the freezing point depression of water below 0°C as ethylene glycol is added, and also graph the freezing point depression of ethylene glycol below $-12°C$ as water is added. Obviously, these two curves would intersect at some temperature, indicating the limit of lowering that can occur. (At these high concentrations, the solutions do not behave ideally, so the temperatures could not be accurately predicted by the equations we have introduced in this chapter, but the main ideas still apply.) Most antifreeze labels recommend a 50:50 mixture by volume (fp $-34°F$, bp 265°F with a 15-pound pressure cap on the radiator), and cite the limit of possible protection with a 70:30 mixture by volume of anti-freeze:water (fp $-84°F$, bp 276°F with a 15-pound pressure cap).

The total concentration of all dissolved solute species determines the colligative properties. Thus, as we shall emphasize in Section 14-14, we must take into account the extent of ion formation in solutions of ionic solutes.

Ethylene glycol, $HOCH_2CH_2OH$, is the major component of "permanent" antifreeze. It depresses the freezing point of water in an automobile radiator and also raises its boiling point. The solution remains in the liquid phase over a wider temperature range than does pure water. This protects against both freezing and boil-over.

14-13 Determination of Molecular Weight by Freezing Point Depression or Boiling Point Elevation

The colligative properties of freezing point depression and, to a lesser extent, boiling point elevation are useful in the determination of molecular weights of solutes. The solutes *must* be nonvolatile in the temperature range of the investigation if boiling point elevations are to be determined. They *must* also be nonelectrolytes.

Example 14-10

A 1.20-gram sample of an unknown covalent compound is dissolved in 50.0 grams of benzene. The solution freezes at 4.92°C. Determine the molecular weight of the compound.

Plan

To calculate the molecular weight of the unknown compound, we find the number of moles that is represented by the 1.20 grams of unknown compound. We first use the freezing point data to find the molality of the solution. The molality relates the number of moles of solute and the mass of solvent (known), so this allows us to calculate the number of moles of unknown.

Solution

From Table 14-2, the freezing point of pure benzene is 5.48°C and K_f is 5.12°C/m.

$$\Delta T_f = 5.48°C - 4.92°C = 0.56°C$$

$$m = \frac{\Delta T_f}{K_f} = \frac{0.56°C}{5.12°C/m} = 0.11\ m$$

The molality is the number of moles of solute per kilogram of benzene, so the number of moles of solute in 50.0 g (0.0500 kg) of benzene can be calculated:

$$0.11\ m = \frac{?\ \text{mol solute}}{0.0500\ \text{kg benzene}}$$

$$\underline{?}\ \text{mol solute} = (0.11\ m)(0.0500\ \text{kg}) = 0.0055\ \text{mol solute}$$

$$\text{mass of 1.0 mol} = \frac{\text{no. of g solute}}{\text{no. of mol solute}} = \frac{1.20\ \text{g solute}}{0.0055\ \text{mol solute}} = 2.2 \times 10^2\ \text{g/mol}$$

$$\text{molecular weight} = \boxed{2.2 \times 10^2\ \text{amu}}$$

EOC 62

Example 14-11

Either camphor ($C_{10}H_{16}O$, molecular weight = 152.24 g/mol) or naphthalene ($C_{10}H_8$, molecular weight 128.19 g/mol) can be used to make mothballs. A 5.0-gram sample of mothballs was dissolved in 100.0 grams of ethyl alcohol, and the resulting solution had a boiling point of 78.90°C. Determine whether the mothballs were made of camphor or naphthalene. Pure ethyl alcohol has a boiling point of 78.41°C, and $K_b = 1.22°C/m$ for this solvent.

Plan

We can distinguish between the two possibilities by determining the molecular weight of the unknown solute. We do this by the method shown in Example 14-10, except that now we use the observed boiling point data.

Solution

The observed boiling point elevation is

$$\Delta T_b = T_{b(\text{solution})} - T_{b(\text{solvent})} = (78.90 - 78.41)°C = 0.49°C$$

Using $\Delta T_b = 0.49°C$ and $K_b = 1.22°C/m$, we can find the molality of the solution:

$$\text{molality} = \frac{\Delta T_b}{K_b} = \frac{0.49°C}{1.22°C/m} = 0.40\ m$$

The number of moles of solute in the 100.0 g (0.1000 kg) of solvent used is

$$\left(0.40 \ \frac{\text{mol solute}}{\text{kg solvent}}\right) (0.1000 \ \text{kg solvent}) = 0.040 \ \text{mol solute}$$

The molecular weight of the solute is its mass divided by the number of moles.

$$\frac{? \ \text{g}}{\text{mol}} = \frac{5.0 \ \text{g}}{0.040 \ \text{mol}} = \sim 125 \ \text{g/mol}$$

The value ~125 g/mol for the molecular weight indicates that   naphthalene   was used to make these mothballs.

EOC 61

14-14 Dissociation of Electrolytes and Colligative Properties

As we have emphasized, colligative properties depend on the *number* of solute particles in a given mass of solvent. A 0.100 molal *aqueous* solution of a covalent compound that does not ionize gives a freezing point depression of 0.186°C. If dissociation were complete, 0.100 *m* KBr would have an *effective* molality of 0.200 *m* (that is, 0.100 *m* K$^+$ + 0.100 *m* Br$^-$). So we might predict that a 0.100 molal solution of this 1:1 strong electrolyte would have a freezing point depression of 2 × 0.186°C, or 0.372°C. In fact, the *observed* depression is only 0.349°C.

In an ionic solution, the solute particles are not randomly distributed. Rather, each positive ion has more negative than positive ions near it. The resulting electrical interactions cause the solution to behave nonideally. Some of the ions undergo **association** in solution (Figure 14-16). At any given instant, some K$^+$ and Br$^-$ ions collide and "stick together." During the brief time that they are in contact, they behave as a single particle. This tends to reduce the effective molality. Therefore, the freezing point depression is reduced (as well as the boiling point elevation and the lowering of vapor pressure).

A (more concentrated) 1.00 *m* solution of KBr might be expected to have a freezing point depression of 2 × 1.86°C = 3.72°C, but the observed depression is 3.29°C. There is a greater deviation from the depression predicted (neglecting ionic association) in the more concentrated solution. This is because the ions are closer together and collide more often in the more concentrated solution. Consequently, the ionic association is greater.

One measure of the extent of dissociation (or ionization) of an electrolyte in water is the **van't Hoff factor**, *i*, for the solution. This is the ratio of the *actual* colligative property to the value that *would* be observed *if no dissociation occurred*.

$$i = \frac{\Delta T_{f(\text{actual})}}{\Delta T_f \ (\text{if nonelectrolyte})} = \frac{K_f m_{\text{effective}}}{K_f m_{\text{stated}}} = \frac{m_{\text{effective}}}{m_{\text{stated}}}$$

The ideal, or limiting, value of *i* for a solution of KBr would be 2, and the value for a 2:1 electrolyte such as Na$_2$SO$_4$ would be 3; these values would

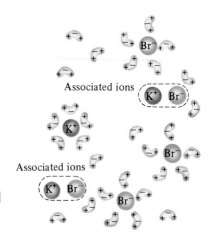

Figure 14-16
Diagrammatic representation of the various species thought to be present in a solution of KBr in water. This would explain unexpected values for its colligative properties, such as freezing point depression.

$\Delta T_f = K_f m = (1.86°C/m)(0.100 \ m)$

$= 0.186°C$

Ionic solutions are elegantly described by the Debye–Hückel theory, which is beyond the scope of this text.

Table 14-3
Actual and Ideal van't Hoff factors, i, for Aqueous Solutions of Nonelectrolytes and Strong Electrolytes

Compound	i for 1.0 m Solution	i for 0.10 m Solution
sucrose, $C_{12}H_{22}O_{11}$	1.00	1.00
nonelectrolytes	1.00	1.00
KBr	1.77	1.88
NaCl	1.83	1.87
If 2 ions/formula unit	2.00	2.00
K_2CO_3	2.39	2.45
K_2CrO_4	1.95	2.39
If 3 ions/formula unit	3.00	3.00
$K_3[Fe(CN)_6]$		2.85
If 4 ions/formula unit	4.00	4.00

apply to infinitely dilute solutions in which no ion association occurs. For 0.10 m and 1.0 m solutions of KBr, i is *less than* 2.

For 0.10 m: $i = \dfrac{0.349°C}{0.186°C} = 1.88$ For 1.0 m: $i = \dfrac{3.29°C}{01.86°C} = 1.77$

Table 14-3 lists actual and ideal values of i for solutions of some strong electrolytes, based on measurements of freezing point depressions.

Many weak electrolytes are quite soluble in water, but they ionize only slightly. The percent ionization and i value for a weak electrolyte in solution can be determined from freezing point depression data (Example 14-12).

Weak acids and weak bases (Section 4-2) are weak electrolytes.

Example 14-12

Lactic acid, $C_2H_4(OH)(COOH)$, is found in sour milk. It is also formed in muscles during intense physical activity and is responsible for the pain felt during strenuous exercise. It is a weak monoprotic acid and therefore a weak electrolyte. The freezing point of a 0.0100 m aqueous solution of lactic acid is $-0.0206°C$. Calculate (a) the i value and (b) the percent ionization in the solution.

Plan

(a) To evaluate the van't Hoff factor, i, we first calculate $m_{effective}$ from the observed freezing point depression and K_f for water; we then compare $m_{effective}$ and m_{stated} to find i. (b) The percent ionization is given by

$$\% \text{ ionization} = \frac{m_{ionized}}{m_{original}} \times 100\% \qquad (\text{where } m_{original} = m_{stated} = 0.0100 \ m)$$

The freezing point depression is caused by the $m_{effective}$, the total *concentration of all dissolved species*—in this case, the sum of the concentrations of HA, H^+, and A^-. We know the value of $m_{effective}$ from part (a). Thus, we need to construct an expression for the effective molality in terms of the amount of lactic acid that ionizes. We represent the molality of lactic acid that ionizes as an unknown, x, and write the concentrations of all species in terms of this unknown.

Solution

(a) $m_{effective} = \dfrac{\Delta T_f}{K_f} = \dfrac{0.0206°C}{1.86°C/m} = 0.0111\ m$

$i = \dfrac{m_{effective}}{m_{stated}} = \dfrac{0.0111\ m}{0.0100\ m} = \boxed{1.11}$

(b) In many calculations, it is helpful to write down (1) the values, or symbols for the values, of initial concentrations; (2) changes in concentrations due to reaction; and (3) final concentrations, as shown below. The coefficients of the equation are all ones, so the reaction ratio must be 1:1:1:1.

Let x = molality of lactic acid that reacts; then

x = molality of H^+ and lactate ion that have been formed

	HA	$\longrightarrow$	H^+	+	A^-
Start	$0.0100\ m$		0		0
Change	$-xm$		$+xm$		$+xm$
Final	$(0.0100 - x)m$		xm		xm

To simplify the notation, we denote the weak acid as HA and its anion as A^-.

The $m_{effective}$ is equal to the sum of the molalities of all the solute particles.

$m_{effective} = m_{HA} \qquad\quad + m_{H^+} + m_{A^-}$

$\qquad\quad = (0.0100 - x)m + xm\ + xm = (0.0100 + x)m$

This must equal the value for $m_{effective}$ calculated earlier, $0.0111\ m$.

$0.0111\ m = (0.0100 + x)m$

$x = 0.0011\ m$ = molality of the acid that ionizes

We can now calculate the percent ionization:

% ionization $= \dfrac{m_{ionized}}{m_{original}} \times 100\% = \dfrac{0.0011\ m}{0.0100\ m} \times 100\% = \boxed{11\%}$

This experiment shows that in $0.0100\ m$ solution, only 11% of the lactic acid has been converted into H^+ and $C_2H_4(OH)COO^-$ ions. The remainder, 89%, exists as nonionized molecules.

EOC 74, 76

14-15 Membrane Osmotic Pressure

Osmosis is the spontaneous process by which the solvent molecules pass through a semipermeable membrane from a solution of lower concentration of solute into a solution of higher concentration of solute. A **semipermeable membrane** (such as cellophane) separates two solutions. Solvent molecules may pass through the membrane in either direction, but the rate at which they pass into the more concentrated solution is found to be greater than the rate in the reverse direction. The initial difference between the two rates is directly proportional to the difference in concentration between the two solutions. Solvent particles continue to pass through the membrane (Figure 14-17a). The column of liquid continues to rise until the hydrostatic pressure due to the weight of the water in the column is sufficient to force solvent

Osmosis is one of the main ways in which water molecules move into and out of living cells. The membranes and cell wells in living organisms allow solvent to pass through. Some of these also selectively permit passage of ions and other small solute particles.

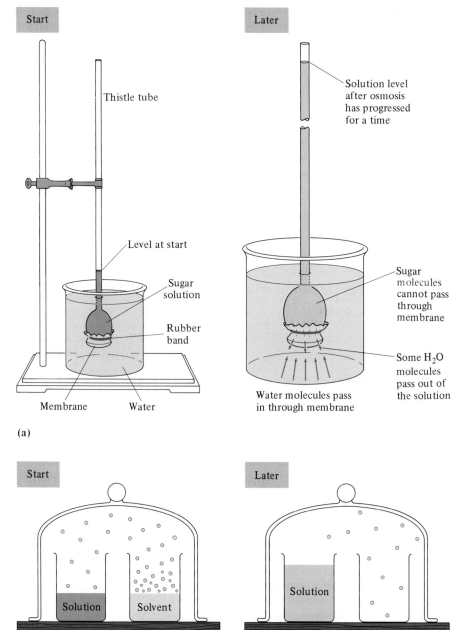

Figure 14-17
(a) Laboratory apparatus for demonstrating osmosis. The picture at the right gives some details of the process, which is analogous to the transfer of solvent into a solution through the space above them (b). In (b), the solute particles cannot pass through the vapor phase because they are nonvolatile. In (a), the solute particles cannot pass through the semipermeable membrane.

molecules back through the membrane at the same rate at which they enter from the dilute side. The pressure exerted under this condition is called the **osmotic pressure** of the solution.

Osmotic pressure depends on the number, and not the kind, of solute particles in solution; it is therefore a colligative property.

The osmotic pressure of a given aqueous solution can be measured with an apparatus such as that depicted in Figure 14-17a. The solution of interest is placed inside an inverted thistle tube that has a membrane firmly fastened across the bottom. This part of the thistle tube and its membrane are then immersed in a container of pure water. As time passes, the height of the solution in the neck rises until the pressure it exerts just counterbalances the osmotic pressure.

Alternatively, we can view osmotic pressure as the external pressure exactly sufficient to prevent osmosis. The pressure required (Figure 14-18) is equal to the osmotic pressure of the solution.

Like molecules of an ideal gas, solute particles are widely separated in very dilute solutions and do not interact significantly with each other. For very dilute solutions, osmotic pressure, π, is found to follow the equation

$$\pi = \frac{nRT}{V}$$

In this equation, n is the number of moles of solute in volume V (in liters) of the solution. The other quantities have the same meaning as in the ideal gas law. The term n/V is a concentration term. In terms of molarity (M),

$$\pi = MRT$$

Osmotic pressures increases with T because T affects the number of solvent–membrane collisions per unit time. It also increases with M because M affects the difference in the numbers of solvent molecules hitting the membrane from the two sides, and because higher M leads to a stronger drive to equalize the concentration difference by dilution and to increase disorder in the solution. For *dilute aqueous solutions*, the molarity is approximately equal to the molality (because the density is nearly 1 kg/L), so

$$\pi = mRT \quad \text{(dilute aqueous solutions)}$$

Osmotic pressures represent very significant forces. For example, a 1.0 molal solution of a nonelectrolyte in water at 0°C produces an equilibrium osmotic pressure of approximately 22.4 atmospheres ($\sim$330 lb/in^2).

The greater the number of solute particles, the greater the height to which the column rises, and the greater the osmotic pressure.

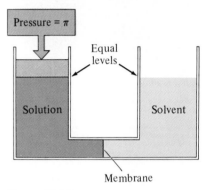

Figure 14-18
The pressure that is just sufficient to prevent solvent flow from the pure solvent side through the semipermeable membrane to the solution side is a measure of the osmotic pressure of the solution.

For a solution of an electrolyte, $\pi = m_{effective}RT$.

Example 14-13

What osmotic pressure would the 1.25 m sucrose solution of Example 14-2 exhibit at 25°C? The density of this solution is 1.34 g/mL.

Plan

We note that the approximation $M \approx m$ is not very good for this solution, because the density of this solution is quite different from 1 g/mL or kg/L. Thus, we must first find the molarity of sucrose, and then use the relationship $\pi = MRT$.

Solution

Recall that 167 g of solution contains 50.0 g of sucrose (0.146 mol) in 117 g of H_2O. The volume of this solution is

$$\text{vol solution} = 167 \text{ g} \times \frac{1 \text{ mL}}{1.34 \text{ g}} = 125 \text{ mL, or } 0.125 \text{ L}$$

Thus, the molarity of sucrose in the solution is

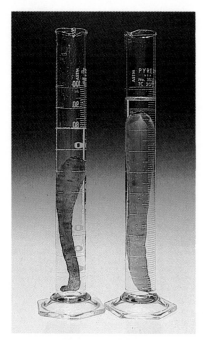

An illustration of osmosis. When a carrot is soaked in a concentrated salt solution, water flows out of the plant cells by osmosis. A carrot soaked overnight in salt solution (left) has lost much water and become limp. A carrot soaked overnight in pure water (right) is little affected.

$$M_{sucrose} = \frac{0.146 \text{ mol}}{0.125 \text{ L}} = 1.17 \text{ mol/L}$$

Now we can calculate the osmotic pressure:

$$\pi = MRT = (1.17 \text{ mol/L}) \left(0.0821 \frac{\text{L} \cdot \text{atm}}{\text{mol} \cdot \text{K}} \right) (298 \text{ K}) = \boxed{28.6 \text{ atm}}$$

EOC 82

Now let us compare the magnitudes of the four colligative properties for this 1.25 m sucrose solution:

vapor pressure lowering	= 0.524 torr	(Ex. 14-4)
boiling point elevation	= 0.640°C	(Ex. 14-7)
freezing point depression	= 2.32°C	(Ex. 14-9)
osmotic pressure	= 28.6 atm	(Ex. 14-13)

The first of these is so small that it would be hard to measure precisely. Even this small lowering of the vapor pressure is sufficient to raise the boiling point by an amount that could be measured, although with difficulty. The freezing point depression is greater yet, but still could not be measured very precisely without special apparatus. The osmotic pressure, on the other hand, is so large that it could be measured much more precisely. Thus, osmotic pressure is often the most easily measured of the four colligative properties, especially when very dilute solutions are used.

The use of measurements of osmotic pressure for the determination of molecular weights has several advantages. Even very dilute solutions give easily measurable osmotic pressures. This method therefore is useful in determination of the molecular weights of (1) very expensive substances, (2) substances that can be prepared only in very small amounts, and (3) substances of very high molecular weight. Because many high-molecular-weight materials are difficult, and in some cases impossible, to obtain in a high state of purity, determinations of their molecular weights are not as accurate as we might like. Nonetheless, osmotic pressures provide a very useful method of estimating molecular weights.

Example 14-14

Pepsin is an enzyme present in the human digestive tract. An enzyme is a protein that acts as a biological catalyst. Pepsin catalyzes the metabolic cleavage of amino acid chains (called peptide chains) in other proteins. A solution of a 0.500-gram sample of purified pepsin in 30.0 mL of benzene solution exhibits an osmotic pressure of 8.92 torr at 27.0°C. Estimate the molecular weight of pepsin.

Plan

As we did in earlier molecular weight determinations (Section 14-13), we must first find n, the number of moles that 0.500 grams of pepsin represents. Because this solution is not aqueous, we use the relationship $\pi = MRT$. The molarity of pepsin is equal to the number of moles of pepsin per liter of solution, n/V. We substitute this for M and solve for n.

Solution

$$\pi = MRT = \left(\frac{n}{V}\right)RT \quad \text{or} \quad n = \frac{\pi V}{RT}$$

We convert 8.92 torr to atmospheres to be consistent with the units of R.

$$n = \frac{\pi V}{RT} = \frac{\left(8.92 \text{ torr} \times \dfrac{1 \text{ atm}}{760 \text{ torr}}\right)(0.0300 \text{ L})}{\left(0.0821 \dfrac{\text{L} \cdot \text{atm}}{\text{mol} \cdot \text{K}}\right)(300 \text{ K})} = 1.43 \times 10^{-5} \text{ mol pepsin}$$

Thus, 0.500 g of pepsin is 1.43×10^{-5} mol. We now estimate its molecular weight.

$$\underline{?} \text{ g/mol} = \frac{0.500 \text{ g}}{1.43 \times 10^{-5} \text{ mol}} = \boxed{3.50 \times 10^4 \text{ g/mol}}$$

The molecular weight of pepsin is approximately 35,000 amu. This is typical for medium-size proteins.

EOC 89, 90

The freezing point depression of this very dilute solution would be only about 0.003°C, which would be difficult to measure accurately. The osmotic pressure of 8.92 torr, on the other hand, is easily measured.

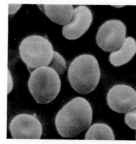

Cells expand in solution of lower solute concentration

↑

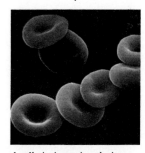

Normal cells in isotonic solution

↓

Cells shrink in solution of greater solute concentration

Living cells contain solutions. When living cells are put in contact with solutions having different total solute concentrations, the resulting osmotic pressures can cause solvent to flow into (top) or out of (bottom) the cells.

Colloids

At the beginning of the chapter, a solution was defined as a homogeneous mixture in which no settling occurs and in which solute particles are at the molecular or ionic state of subdivision. This represents one extreme of mix-

Table 14-4
Types of Colloids

Dispersed (solute-like) Phase		Dispersing (solvent-like) Medium	Common Name	Examples
Solid	in	Solid	Solid sol	Many alloys (such as steel and duralumin), some colored gems, reinforced rubber, porcelain, pigmented plastics
Liquid	in	Solid	Solid emulsion	Cheese, butter, jellies
Gas	in	Solid	Solid foam	Sponge, rubber, pumice, Styrofoam
Solid	in	Liquid	Sols and gels	Milk of magnesia, paints, mud, puddings
Liquid	in	Liquid	Emulsion	Milk, face cream, salad dressings, mayonnaise
Gas	in	Liquid	Foam	Shaving cream, whipped cream, foam on beer
Solid	in	Gas	Solid aerosol	Smoke, airborne viruses and particulate matter, auto exhaust
Liquid	in	Gas	Liquid aerosol	Fog, mist, aerosol spray, clouds

Table 14-5
Approximate Sizes of Dispersed Particles

Mixture	Example	Particle Size
Suspension	Sand in water	Larger than 10,000 Å
Colloidal dispersion	Starch in water	10–10,000 Å
Solution	Sugar in water	1–10 Å

Freshly made wines are often cloudy because of colloidal particles. Removing these colloidal particles clarifies the wine.

tures. The other extreme is a suspension, a clearly heterogeneous mixture in which solute-like particles immediately settle out after mixing with a solvent-like phase. Such a situation results when a handful of sand is stirred into water. **Colloids** (**colloidal suspensions** or **colloidal dispersions**) represent an intermediate kind of mixture in which the solute-like particles, or **dispersed phase**, are suspended in the solvent-like phase, or **dispersing medium**. The particles of the dispersed phase are small enough that settling is negligible. However, they are large enough to make the mixture appear cloudy (and, in many cases, opaque) because light is scattered as it passes through the colloid.

Table 14-4 indicates that all combinations of solids, liquids, and gases can form colloids except mixtures of nonreacting gases (all of which are homogeneous and, therefore, true solutions). Whether a given mixture forms a solution, a colloidal dispersion, or a suspension depends upon the size of the solute-like particles (Table 14-5), as well as solubility and miscibility.

14-16 The Tyndall Effect

Figure 14-19
The dispersion of a beam of light by colloidal particles is called the Tyndall effect. The presence of colloidal particles is easily detected with the aid of a light beam.

The scattering of light by colloidal particles is called the **Tyndall effect** (Figure 14-19). A particle cannot scatter light if it is too small. Solute particles in solutions are below this limit. The maximum dimension of colloidal particles is about 10,000 Å.

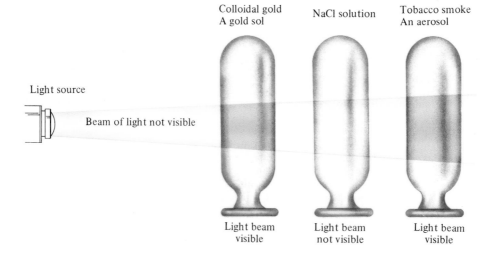

Light source

Beam of light not visible

Colloidal gold
A gold sol

NaCl solution

Tobacco smoke
An aerosol

Light beam
visible

Light beam
not visible

Light beam
visible

The scattering of light from automobile headlights by fogs and mists is an example of the Tyndall effect, as is the scattering of a light beam in a laser show by dust particles in the air in a darkened room.

14-17 The Adsorption Phenomenon

Much of the chemistry of everyday life is the chemistry of colloids, as one can tell from the examples in Table 14-4. Because colloidal particles are so finely divided, they have tremendously high total surface area in relation to their volume. It is not surprising, therefore, that an understanding of colloidal behavior requires an understanding of surface phenomena.

Atoms on the surface of a colloidal particle are bonded only to other atoms of the particle on and below the surface. Because these atoms interact in all directions, they have a tendency to interact with whatever comes in contact with the surface. Colloidal particles often adsorb ions or other charged particles, as well as gases and liquids. The process of **adsorption** involves adhesion of any such species onto the surfaces of particles. For example, a bright red **sol** (solid dispersed in liquid) is formed by mixing hot water with a concentrated aqueous solution of iron(III) chloride:

$$2x[Fe^{3+}(aq) + 3Cl^-(aq)] + x(3 + y)H_2O \longrightarrow$$

yellow solution

$$[Fe_2O_3 \cdot yH_2O]_x(s) + 6x[H^+ + Cl^-]$$

bright red sol

Each colloidal particle of this sol is a cluster of many formula units of hydrated Fe_2O_3. Each attracts positively charged Fe^{3+} ions to its surface. Because each particle is then surrounded by a shell of positively charged ions, the particles repel each other and cannot combine to the extent necessary to cause precipitation (see Figure 14-20).

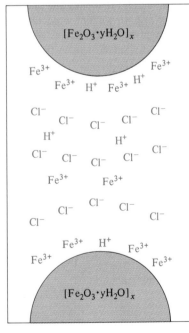

Figure 14-20
Stabilization of a colloid (Fe_2O_3 sol) by electrostatic forces. Each colloidal particle of this red sol is a cluster of many formula units of hydrated Fe_2O_3. Each attracts positively charged Fe^{3+} ions to its surface. (Fe^{3+} ions fit readily into the crystal structure, so they are preferentially adsorbed rather than Cl^-.) Because each particle is then surrounded by a shell of positively charged ions, the particles repel one another and cannot combine to the extent necessary to cause actual precipitation. The suspended particles scatter light, making the path of the light beam through the suspension visible.

Semipermeable membranes play important roles in the normal functioning of many living systems. In addition, they are used in a wide variety of industrial and medical applications. One of these is the purification of water by reverse osmosis.

Suppose we place a semipermeable membrane between a saline (salt) solution and pure water. If the saline solution is pressurized under a greater pressure than its osmotic pressure, the direction of osmosis can be reversed. That is, the net flow of water molecules will be from the saline solution through the membrane into the pure water. This process, called **reverse osmosis**, is depicted in Figure (a). The membrane usually consists of cellulose acetate or hollow fibers of a material structurally similar to Nylon. This method has been used for the purification of brackish (mildly saline) water. It has the economic advantages of low cost, ease of apparatus construction, and simplicity of operation. Because this method of water purification requires no heat, it has a great advantage over distillation.

The city of Sarasota, Florida, has built a large reverse osmosis plant to purify drinking water. It processes more than 4 million gallons of water per day from local wells. Total dissolved solids are reduced in concentration from 1744 parts per million (0.1744% by mass) to 90 parts per million (ppm). This water is mixed with additional well water purified by an ion exchange system. The final product is more than 10 million gallons of water per day containing less than 500 ppm of total dissolved solids, the standard for drinking water set by the World Health Organization. The Kuwaiti and Saudi water purification plants that were of strategic concern in the Persian Gulf war use reverse osmosis in one of their primary stages.

The human kidney carries out a wide variety of important functions. One of the most crucial is the removal of metabolic waste products (such as creatinine, urea, and uric acid) from the blood without removal of substances needed by the body (such as glucose, electrolytes, and amino acids). The process by which this is accomplished in the kidney involves *dialysis*, a phenomenon in which the membrane allows transfer of both solvent molecules *and* certain solute molecules and ions, usually small ones. A patient

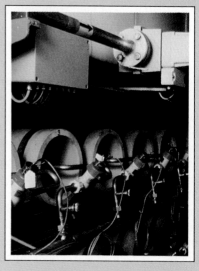

A reverse osmosis unit used to provide all the fresh water (82,500 gallons per day) for the steamship *Norway*.

whose kidneys have failed can often have this dialysis performed by an artificial kidney machine. In this mechanical procedure, called *hemodialysis*, the blood is withdrawn from the body and passed through or by a semipermeable membrane (Figure b). The small pore size of the membrane prevents passage of blood cells out of the blood.

The membrane separates the blood from a dialyzing solution, or

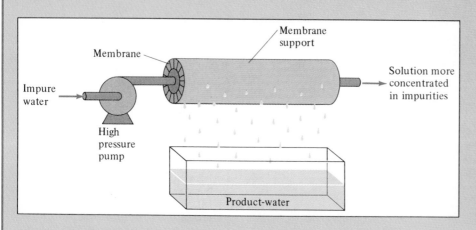

(a) The reverse osmosis method of water purification.

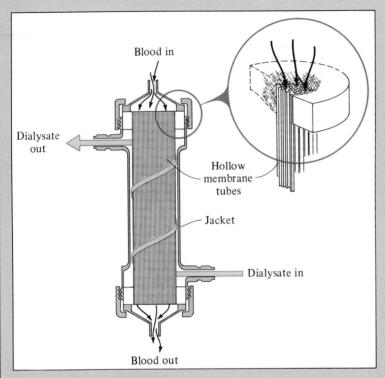

Blood in

Dialysate out

Hollow membrane tubes

Jacket

Dialysate in

Blood out

(b) A schematic diagram of the hollow fiber (or capillary) dialyzer, the most commonly used artificial kidney. The blood flows through many small tubes constructed of semipermeable membrane; these tubes are bathed in the dialyzing solution.

dialysate, that is similar to blood plasma in its concentration of needed substances (such as electrolytes and amino acids) but contains none of the waste products. Because the concentrations of undesirable substances are thus higher in the blood than in the dialysate, they flow preferentially out of the blood and are washed away. The concentrations of *needed* substances are the same on both sides of the membrane, so these substances are maintained at the proper concentrations in the blood. A patient with total kidney failure may require up to four hemodialysis sessions per week, at 3 to 4 hours per session. To help hold down the cost of such treatment, the dialysate solution is later purified by a combination of filtration, distillation, and reverse osmosis and is then re-used.

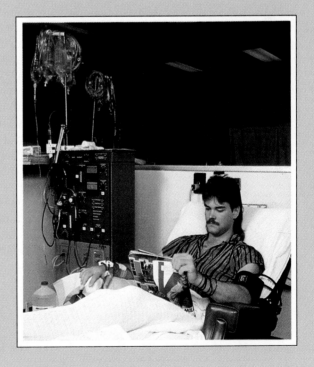

Fe(OH)₃ is a gelatinous precipitate (a gel).

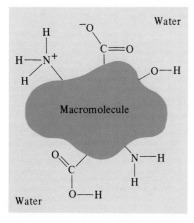

Figure 14-21
Examples of hydrophilic groups at the surface of a giant molecule (macromolecule) that help keep the macromolecule suspended in water.

"Dry" cleaning, on the other hand, does not involve water. The solvents that are used in dry cleaning dissolve grease to form true solutions.

14-18 Hydrophilic and Hydrophobic Colloids

Colloids are classified as **hydrophilic** ("water loving") or **hydrophobic** ("water hating") based on the surface characteristics of the dispersed particles.

Hydrophilic Colloids

Proteins such as the oxygen-carrier hemoglobin form hydrophilic sols when they are suspended in saline aqueous body fluids, e.g., blood plasma. Such proteins are macromolecules (giant molecules) that fold and twist in an aqueous environment so that polar groups are exposed to the fluid whereas nonpolar groups are encased (see Figure 14-21). Protoplasm and human cells are examples of **gels**, which are special types of sols in which the solid particles (in this case mainly proteins and carbohydrates) join together in a semirigid network structure that encloses the dispersing medium. Other examples of gels are gelatin, jellies, and gelatinous precipitates such as $Al(OH)_3$ and $Fe(OH)_3$.

Hydrophobic Colloids

Hydrophobic colloids cannot exist in polar solvents without the presence of **emulsifying agents**, or **emulsifiers**. These agents coat the particles of the dispersed phase to prevent their coagulation into a separate phase. Milk and mayonnaise are examples of hydrophobic colloids (milk fat in milk, vegetable oil in mayonnaise) that stay suspended with the aid of emulsifying agents (casein in milk and egg yolk in mayonnaise).

Consider the mixture resulting from vigorous shaking of salad oil (nonpolar) and vinegar (polar). Droplets of hydrophobic oil are temporarily suspended in the water. However, in a short time, the very polar water molecules, which attract each other strongly, squeeze out the nonpolar oil molecules. The oil then coalesces and floats to the top. If we add an emulsifying agent, such as egg yolk, and shake or beat the mixture, a stable emulsion (mayonnaise) results.

Oil and grease are mostly long-chain hydrocarbons that are nonpolar. Our most common solvent is water, a polar substance that does not dissolve nonpolar substances. In order to use water to wash soiled fabrics, greasy dishes, or our bodies, we must enable the water to suspend and remove nonpolar substances. Soaps and detergents are emulsifying agents that accomplish this. Their function is controlled by the intermolecular interactions that result from their structures.

Solid soaps are usually sodium salts of long-chain organic acids called fatty acids. They have a polar "head" and a nonpolar "hydrocarbon tail." Sodium stearate, a typical soap, is shown below.

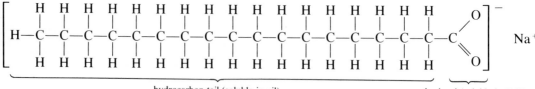

hydrocarbon tail (soluble in oil)   polar head (soluble in H_2O)

sodium stearate (a soap)

Powdered sulfur floats on pure water because of the high surface tension of water. When a drop of detergent solution is added to the water, its surface tension is lowered and sulfur sinks. This lowering of surface tension enhances the cleaning action of detergent solutions.

The stearate ion is typical of the anions in soaps. It has a polar carboxylate

Sodium stearate is also a major component of many stick deodorants.

head, $-\overset{\overset{\displaystyle O}{\|}}{C}-O^-$, and a long nonpolar tail, $CH_3(CH_2)_{16}-$. The head of the stearate ion is compatible with ("soluble in") water, whereas the hydrocarbon tail is compatible with ("soluble in") oil and grease. Groups of such ions can be dispersed in water because they form *micelles* (Figure 14-22a). They have their "water-insoluble" tails in the interior of a micelle and their polar heads on the outside where they can interact with the polar water molecules. When sodium stearate is stirred into water, the result is not a true solution. Instead it contains negatively charged micelles of stearate ions, surrounded by the positively charged Na^+ ions. The result is a suspension of micelles in water. These micelles are large enough to scatter light, and so a soap–water mixture appears cloudy. Oil and grease "dissolve" in soapy water because the nonpolar oil and grease are taken into the nonpolar interior of micelles (Figure 14-22b). Micelles form a true emulsion in water, so the oil and grease can be washed away. Sodium stearate is called a **surfactant** (meaning "surface-active agent") or wetting agent because it has the ability to suspend and wash away oil and grease. Other soaps and detergents behave similarly.

Some edible colloids.

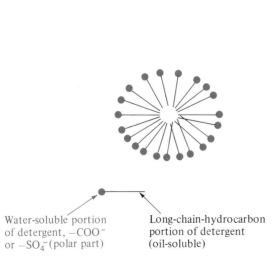

(a)

Water-soluble portion of detergent, $-COO^-$ or $-SO_4^-$ (polar part)

Long-chain-hydrocarbon portion of detergent (oil-soluble)

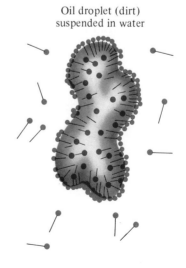

Oil droplet (dirt) suspended in water

(b)

Figure 14-22
(a) A representation of a micelle. The nonpolar tails "dissolve" in one another in the center of the cluster and the polar heads on the outside interact favorably with the polar water molecules. (b) Attachment of soap or detergent molecules to a droplet of oily dirt to suspend it in water.

The 14th-century Saint Martin's Bridge over the Tajo River in Toledo, Spain. The presence of nonbiodegradable detergents is responsible for the foam.

"Hard" water contains Fe^{3+}, Ca^{2+}, and/or Mg^{2+} ions, all of which displace Na^+ from soaps to form precipitates. This removes the soap from the water and puts an undesirable coating on the bathtub or on the fabric being laundered. **Synthetic detergents** are soap-like emulsifiers that contain sulfonate, $—SO_3^-$, or sulfate, $—OSO_3^-$, instead of carboxylate groups, $—COO^-$. They do not precipitate the ions of hard water, so they can be used in hard water as soap substitutes without forming undesirable scum.

Phosphates have been added to commercial detergents for various purposes (Section 26-21). They complex the metal ions that contribute to water hardness and keep them dissolved, control acidity, and influence micelle formation. The use of detergents containing phosphates is now discouraged, however, because they cause **eutrophication** in rivers and streams that receive sewage. This is a condition (not related to colloids) in which there is an overgrowth of vegetation caused by the high concentration of phosphorus, a plant nutrient. This overgrowth and the subsequent decay of the dead plants lead to decreased dissolved O_2 in the water, which causes the gradual elimination of marine life. There is also a foaming problem associated with alkylbenzenesulfonate (ABS) detergents in streams and in pipes, tanks, and pumps of sewage treatment plants. Such detergents are not **biodegradable**; that is, they cannot be broken down by bacteria.

a sodium alkylbenzenesulfonate (ABS)—a nonbiodegradable detergent

Currently used linear-chain alkylsulfonate (LAS) detergents are biodegradable and do not cause such foaming.

sodium lauryl benzenesulfonate
a linear alkylbenzenesulfonate (LAS)—a biodegradable detergent

550

Key Terms

The following terms were defined at the end of Chapter 3: **concentration, dilution, molarity, percent by mass, solute, solution,** and **solvent.**

Adsorption   Adhesion of species onto surfaces of particles.

Associated ions   Short-lived species formed by the collision of dissolved ions of opposite charge.

Biodegradability   The ability of a substance to be broken down into simpler substances by bacteria.

Boiling point elevation   The increase in the boiling point of a solvent caused by dissolution of a nonvolatile solute.

Colligative properties   Physical properties of solutions that depend on the number but not the kind of solute particles present.

Colloid   A heterogeneous mixture in which solute-like particles do not settle out.

Crystal lattice energy   The energy change when one mole of formula units of a crystalline solid is formed from its ions, atoms, or molecules in the gas phase; always negative.

Detergent   A soap-like emulsifier that contains a sulfonate, $-SO_3^-$, or sulfate, $-OSO_3^-$, group instead of a carboxylate, $-COO^-$, group.

Differential thermometer   A thermometer used for accurate measurement of very small changes in temperature.

Dispersed phase   The solute-like species in a colloid.

Dispersing medium   The solvent-like phase in a colloid.

Distillation   The process in which components of a mixture are separated by boiling away the more volatile liquid.

Effective molality   The sum of the molalities of all solute particles in solution.

Emulsifier   See *Emulsifying agent.*

Emulsifying agent   A substance that coats the particles of a dispersed phase and prevents coagulation of colloidal particles; an emulsifier.

Emulsion   A colloidal suspension of a liquid in a liquid.

Eutrophication   The undesirable overgrowth of a vegetation caused by high concentrations of plant nutrients in bodies of water.

Foam   A colloidal suspension of a gas in a liquid.

Fractional distillation   The process in which a fractionating column is used in a distillation apparatus to separate components of a liquid mixture that have different boiling points.

Freezing point depression   The decrease in the freezing point of a solvent caused by the presence of a solute.

Gel   A colloidal suspension of a solid dispersed in a liquid; a semirigid sol.

Hard water   Water containing Fe^{3+}, Ca^{2+}, and/or Mg^{2+} ions, which form precipitates with soaps.

Heat of solution (molar)   The amount of heat absorbed in the formation of a solution that contains one mole of solute; the value is positive if heat is absorbed (endothermic) and negative if heat is released (exothermic).

Henry's Law   The pressure of the gas above a solution is proportional to the concentration of the gas in the solution.

Hydration   The interaction (surrounding) of a solute particle with water molecules.

Hydration energy (molar) of an ion   The energy change accompanying the hydration of a mole of gaseous ions.

Hydrophilic colloids   Colloidal particles that attract water molecules.

Hydrophobic colloids   Colloidal particles that repel water molecules.

Ideal solution   A solution that obeys Raoult's Law exactly.

Liquid aerosol   A colloidal suspension of a liquid in gas.

Micelle   A cluster of a large number of soap or detergent molecules or ions, assembled with their hydrophobic tails directed toward the center and their hydrophilic ends directed outward.

Miscibility   The ability of one liquid to mix with (dissolve in) another liquid.

Molality (m)   Concentration expressed as number of moles of solute per kilogram of solvent.

Mole fraction of a component in solution   The number of moles of the component divided by the total number of moles of all components.

Osmosis   The process by which solvent molecules pass through a semipermeable membrane from a dilute solution into a more concentrated solution.

Osmotic pressure   The hydrostatic pressure produced on the surface of a semipermeable membrane by osmosis.

Percentage ionization of weak electrolytes   The percentage of the weak electrolyte that ionizes in a solution of given concentration.

Raoult's Law   The vapor pressure of a solvent in an ideal solution decreases as its mole fraction decreases.

Reverse osmosis   The forced flow of solvent molecules through a semipermeable membrane from a concentrated solution into a dilute solution. This is accomplished by application of hydrostatic pressure on the concentrated side greater than the osmotic pressure that is opposing it.

Saturated solution   A solution in which no more solute will dissolve.

Semipermeable membrane   A thin partition between two solutions through which certain molecules can pass but others cannot.

Soap   An emulsifier that can disperse nonpolar substances in water; the sodium salt of a long chain organic acid; consists of a long hydrocarbon chain attached to a carboxylate group, $-CO_2^-Na^+$.

Sol   A colloidal suspension of a solid dispersed in a liquid.

Solid aerosol   A colloidal suspension of a solid in a gas.

Solid emulsion   A colloidal suspension of a liquid dispersed in a solid.

Solid foam   A colloidal suspension of a gas dispersed in a solid.

Solid sol   A colloidal suspension of a solid dispersed in a solid.

Solvation   The process by which solvent molecules surround and interact with solute ions or molecules.

Supersaturated solution   A (metastable) solution that contains a higher-than-saturation concentration of solute; slight disturbance or seeding causes crystallization of excess solute.

Surfactant   A "surface-active agent"; a substance that has the ability to emulsify and wash away oil and grease in an aqueous suspension.

Suspension   A heterogeneous mixture in which solute-like particles settle out of a solvent-like phase some time after their introduction.

Thermal pollution   Introduction of heated waste water into natural waters.

Tyndall effect   The scattering of light by colloidal particles.

Exercises

General Concepts—The Dissolving Process

1. Support or criticize the statement "Solutions and mixtures are the same thing."

2. Give an example of a solution that contains each of the following: (a) a solid dissolved in a liquid, (b) a gas dissolved in a gas, (c) a gas dissolved in a liquid, (d) a liquid dissolved in a liquid, (e) a solid dissolved in a solid. Identify the substances that are the solvent and the solute in each case.

3. There are no *true* solutions in which the solvent is gaseous and the solute is either liquid or solid. Why?

4. Explain why (a) solute–solute, (b) solvent–solvent, and (c) solute–solvent interactions are important in determining the extent to which a solute dissolves in a solvent.

5. What is the relative importance of each factor listed in Exercise 4 when (a) solids, (b) liquids, and (c) gases dissolve in water?

6. Why do many solids dissolve in water in endothermic processes, whereas most miscible liquids mix with each other in exothermic processes?

7. Define and distinguish between solvation and hydration.

8. The amount of heat released or absorbed in the dissolution process is important in determining whether the dissolution process is spontaneous, i.e., whether it can occur. What is the other important factor? How does it influence solubility?

9. An old saying is that "oil and water don't mix." Why is this true?

10. Two liquids, A and B, do not react chemically and are completely miscible. What would be observed as one is poured into the other? What would be observed in the case of two completely immiscible liquids and in the case of two partially miscible liquids?

11. Consider the following solutions. In each case, predict whether the solubility of the solute should be high or low. Justify your answers. (a) $CaCl_2$ in hexane, C_6H_{14};
(b) $CaCl_2$ in H_2O; (c) C_6H_{14} in H_2O; (d) CCl_4 in C_6H_{14}; (e) C_6H_{14} in CCl_4.

12. Consider the following solutions. In each case, predict whether the solubility of the solute should be high or low. Justify your answers. (a) HCl in H_2O, (b) HF in H_2O, (c) Al_2O_3 in H_2O, (d) S_8 in H_2O.

13. For those solutions in Exercise 11 that can be prepared in "reasonable" concentrations, classify the solutes as nonelectrolytes, weak electrolytes, or strong electrolytes.

14. For those solutions in Exercise 12 that can be prepared in "reasonable" concentrations, classify the solutes as nonelectrolytes, weak electrolytes, or strong electrolytes.

15. Both methanol, CH_3OH, and ethanol, CH_3CH_2OH, are completely miscible with water at room temperature because of strong solvent–solute intermolecular forces. Predict the trend in solubility in water for 1-propanol, $CH_3CH_2CH_2OH$; 1-butanol, $CH_3CH_2CH_2CH_2OH$; and 1-pentanol, $CH_3CH_2CH_2CH_2CH_2OH$.

16. (a) Does the solubility of a solid in a liquid exhibit an appreciable dependence on pressure? (b) Is the same true for the solubility of a liquid in a liquid?

17. Describe the effect of increasing pressure on the solubilities of gases in liquids.

18. What is the effect of raising temperature on the solubilities of most gases in water?

*19. A handbook lists the value of the Henry's Law constant as 3.02×10^4 atm for ethane, C_2H_6, dissolved in water at 25°C. The absence of concentration units on k means that the constant is meant to be used with the concentration expressed as a mole fraction. Calculate the mole fraction of ethane in water at an ethane pressure of 0.15 atm.

*20. The mole fraction of methane, CH_4, dissolved in water can be calculated from the Henry's Law constants of 4.13×10^4 atm at 25°C and 5.77×10^4 atm at 50°C.

Calculate the solubility of methane at these temperatures for a methane pressure of 15 atm above the solution. Does the solubility increase or decrease with increasing temperature? (See Exercise 19 for interpretation of units.)

21. Choose the ionic compound from each pair that should have the largest crystal lattice energy. Justify your choice. (a) LiF or LiBr, (b) KF or CaF_2, (c) FeF_2 or FeF_3, (d) LiF or KF.

22. Choose the ion from each pair that should be more strongly hydrated in aqueous solution. Justify your choice.
   (a) Na^+ or Rb^+, (b) F^- or Cl^-, (c) Fe^{3+} or Fe^{2+}, (d) Na^+ or Mg^{2+}.

*23. The crystal lattice energy, ΔH_{xtal}, for LiBr(s) is -818.6 kJ/mol at 25°C. The hydration energy of the ions of LiBr is -867.4 kJ/mol at 25°C (for infinite dilution). (a) What is the heat of solution of LiBr(s) at 25°C (for infinite dilution)? (b) The heat of hydration of Li^+ (g) is -544 kJ/mol at 25°C. What is ΔH_{hyd} for Br^-(g) at 25°C?

24. Describe and explain the effect of changing the temperature on the solubility of (a) a solid that dissolves by an exothermic process, and (b) a solid that dissolves by an endothermic process.

Concentrations of Solutions

25. Under what conditions are the molarity and molality of a solution nearly the same? Which concentration unit is more useful when measuring volume with burets, pipets, and volumetric flasks in the laboratory? Why?

26. Many handbooks list solubilities in units of (g solute/100 g H_2O). How would you convert from this unit to mass percent?

27. A 60.0-mL sample of ethyl ether, $(C_2H_5)_2O$, is dissolved in enough methyl alcohol, CH_3OH, to make 300.0 mL of solution. The density of the ether is 0.714 g/mL. What is the molarity of this solution?

28. Calculate the molality of a solution that contains 80.0 g of benzoic acid, C_6H_5COOH, in 400 mL of ethanol, C_2H_5OH. The density of ethanol is 0.789 g/mL.

29. A solution contained 20.0 g N_2H_4CO, 5.0 g $C_6H_{12}O_6$, and 75.0 g H_2O. Calculate the mole fraction of water.

30. What masses of NaCl and H_2O are present in 180 g of a 15.0% aqueous solution of NaCl?

31. Sodium fluoride has a solubility of 4.22 g in 100.0 g of water at 18°C. Express the concentration (a) in mass percent and (b) in mole fraction.

32. The solubility of K_2ZrF_6 at 100°C in 100 g of H_2O is 25 g. Express this concentration in terms of (a) molality and (b) mole fraction.

*33. A piece of jewelry is marked "14 carat gold," meaning that on a mass basis the jewelry is 14/24 pure gold.

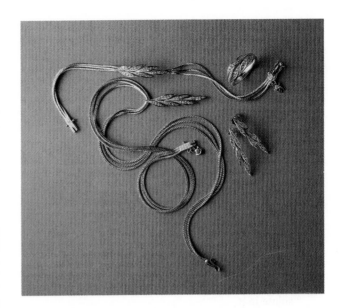

What is the molality of this alloy—assuming the other metal is the solvent?

*34. Describe how to prepare 1.000 L of 0.250 m NaCl. The density of this solution is 1.011 g/mL.

35. Urea, N_2H_4CO, is a product of metabolism of proteins. An aqueous solution is 32.0% urea by mass and has a density of 1.087 g/mL. Calculate the molality of urea in the solution.

*36. A solution that is 24.0% fructose, $C_6H_{12}O_6$, in water has a density of 1.10 g/mL at 20°C. (a) What is the molality of fructose in this solution? (b) At a higher temperature, the density would be lower. Would the molality be less than, greater than, or the same as the molality at 20°C? Explain.

37. The density of a sulfuric acid solution taken from a car battery is 1.225 g/cm³. This corresponds to a 3.75 M solution. Express the concentration of this solution in molality, mole fraction of H_2SO_4, and % of water by mass.

38. The density of an aqueous solution containing 10.00 g K_2SO_4 in 100.00 g solution is 1.0825 g/mL. Calculate the concentration of this solution in molarity, molality, percent of K_2SO_4, and mole fraction of solvent.

39. What are the mole fractions of ethanol, C_2H_5OH, and water in a solution prepared by mixing 80.0 mL of ethanol with 40.0 mL of water at 25°C? The density of ethanol is 0.789 g/mL, and that of water is 1.00 g/mL.

40. What are the mole fractions of ethanol, C_2H_5OH, and water in a solution prepared by mixing 80.0 g of ethanol with 40.0 g of water?

*41. (a) What is the mass % ethanol in an aqueous solution in which the mole fraction of each component is 0.500? (b) What is the molality of ethanol in such a solution?

Raoult's Law and Vapor Pressure

42. In your own words, explain briefly *why* the vapor pressure of a solvent is lowered by dissolving a nonvolatile solute in it.

43. (a) Calculate the lowering of vapor pressure associated with dissolving 40.0 g of table sugar, $C_{12}H_{22}O_{11}$, in 400 g of water at 25.0°C. (b) What is the vapor pressure of the solution? Assume that the solution is ideal. The vapor pressure of pure water at 25°C is 23.76 torr. (c) What is the vapor pressure of the solution at 100°C?

44. Calculate (a) the lowering of vapor pressure and (b) the vapor pressure of a solution prepared by dissolving 50.0 g of naphthalene, $C_{10}H_8$ (a nonvolatile nonelectrolyte), in 150.0 g of benzene, C_6H_6, at 20°C. Assume that the solution is ideal. The vapor pressure of pure benzene is 74.6 torr at 20°C.

45. What is the vapor pressure at 25°C above a solution containing 150.0 g of water and 15.0 g of urea, $CO(NH_2)_2$, a nonvolatile solute? The vapor pressure of pure water at 25°C is 23.76 torr.

46. What mass of a nonvolatile solute having a molecular weight of 325 g/mol would be required to decrease the vapor pressure of 1.00 kg of water by 1.00 torr at 100°C?

47. Using Raoult's Law, predict the partial pressures over a solution containing 0.200 mol acetone ($P^0 = 345$ torr) and 0.300 mol chloroform ($P^0 = 295$ torr). What is the total pressure over this solution?

48. At -100°C the vapor pressure of pure ethane, CH_3CH_3, is 394 torr and that of pure propane, $CH_3CH_2CH_3$, is 22 torr. What is the vapor pressure over a solution containing equal molar amounts of these substances? What is the composition of the vapor?

49. Use the following diagram to estimate (a) the partial pressure of chloroform, (b) the partial pressure of acetone, and (c) the total vapor pressure of a solution in which the mole fraction of $CHCl_3$ is 0.4, assuming *ideal* behavior.

50. Answer Exercise 49 for the *real* solution of acetone and chloroform.

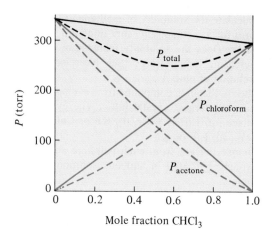

51. A solution is prepared by mixing 50.0 g of dichloromethane, CH_2Cl_2, and 30.0 g of dibromomethane, CH_2Br_2, at 0°C. The vapor pressure at 0°C of pure CH_2Cl_2 is 0.175 atm, and that of CH_2Br_2 is 0.015 atm. (a) Assuming ideal behavior, calculate the total vapor pressure of the solution. (b) Calculate the mole fractions of CH_2Cl_2 and CH_2Br_2 in the *vapor* above the liquid. Assume that both the vapor and the solution behave ideally.

Boiling Point Elevation and Freezing Point Depression—Solutions of Nonelectrolytes

52. Refer to Table 14-2. Suppose you had a 0.100 *m* solution of a nonvolatile nonelectrolyte in each of the solvents listed there. Which one would have (a) the greatest freezing point depression, (b) the lowest freezing point, (c) the greatest boiling point elevation, and (d) the highest boiling point?

53. Explain qualitatively why boiling points are elevated, whereas freezing points are depressed, by dissolving nonvolatile solutes in solvents.

54. What is the significance of (a) the molal freezing point depression constant, K_f, and (b) the molal boiling point elevation constant, K_b? How could their values for given solvents be determined in the laboratory?

55. What are the boiling and freezing points of a 2.15 *m* aqueous solution of ethylene glycol, a nonelectrolyte?

56. The molal freezing point constant for copper is 23°C/m. If pure copper melts at 1083°C, what will be the melting point of a brass made of 10 mass % Zn and 90 mass % Cu?

57. A solution is prepared by dissolving 6.41 g of ordinary sugar (sucrose, $C_{12}H_{22}O_{11}$) in 32.0 g of water. Calculate the freezing and boiling points of the solution. Sucrose is a nonvolatile nonelectrolyte.

58. Calculate the freezing and boiling points of a solution that contains 30.0 g of urea, N_2H_4CO, in 250 g of water. Urea is a nonvolatile nonelectrolyte.

59. How many grams of the nonelectrolyte sucrose, $C_{12}H_{22}O_{11}$, should be dissolved in 600.0 g of water to produce a solution that freezes at -2.50°C?

60. What mass of naphthalene, $C_{10}H_8$, should be dissolved in 300 g of nitrobenzene, $C_6H_5NO_2$, to produce a solution that boils at 214.20°C? See Table 14-2.

61. A solution was made by dissolving 3.75 g of a nonvolatile solute in 104.0 g of acetone. The solution boiled at 56.58°C. The boiling point of pure acetone is 55.95°C, and $K_b = 1.71$°C/m. Calculate the molar mass of the solute.

62. The molecular weight of an organic compound was determined by measuring the freezing point depression of a benzene solution. A 0.500-g sample was dissolved in 50.0 g of benzene, and the resulting depression was 0.42°C. What is the approximate molecular weight? The compound gave the following elemental analysis: 40.0 % C, 6.67 % H, 53.3 % O by mass. Determine the formula and exact molecular weight of the substance.

63. When 0.163 g of sulfur is finely ground and melted with 4.38 g of camphor, the freezing point of the camphor is lowered by 5.47°C. What is the molecular weight of sulfur? What is its molecular formula?

*64. (a) Suppose we dissolve a 6.00-g sample of a mixture of naphthalene, $C_{10}H_8$, and anthracene, $C_{14}H_{10}$, in 360 g of benzene. The solution is observed to freeze at 4.85°C. Find the percent composition (by mass) of the sample. (b) At what temperature would the solution boil? Naphthalene and anthracene are nonvolatile nonelectrolytes.

Boiling Point Elevation and Freezing Point Depression— Solutions of Electrolytes

65. What is ion association in solution? Can you suggest why the term "ion pairing" is sometimes used to describe this phenomenon?

66. You have separate 0.10 M aqueous solutions of the following salts: $LiNO_3$, $Ca(NO_3)_2$, and $Al(NO_3)_3$. In which one would you expect the solute to be the most completely dissociated? Which solution would you expect to conduct electricity most strongly? Explain your reasoning.

67. What is the significance of the van't Hoff factor, i?

68. What is the value of the van't Hoff factor, i, for the following strong electrolytes at infinite dilution? (a) Na_2SO_4, (b) KOH, (c) $Al_2(SO_4)_3$, (d) $Ba(OH)_2$.

69. Compare the number of solute particles that are present in solutions of equal concentrations of strong electrolytes, weak electrolytes, and nonelectrolytes.

70. Four beakers contain 0.010 m aqueous solutions of CH_3OH, $NaOH$, $CaCl_2$, and CH_3COOH, respectively. Which of these solutions has the lowest freezing point?

*71. A 0.050 m aqueous solution of $K_3[Fe(CN)_6]$ has a freezing point of -0.2800°C. Calculate the total concentration of solute particles in this solution and interpret your results.

*72. One gram each of NaCl, NaBr, and NaI was dissolved in 100.0 g water. What is the vapor pressure above the solution at 100°C? Assume complete dissociation of the three salts.

73. Formic acid, HCOOH, dissolves in water according to

$$HCOOH \rightleftharpoons H^+ + HCOO^-$$

A 0.0100 m formic acid solution freezes at -0.02092°C. Calculate the percentage ionization of HCOOH in this solution.

74. A 0.100 m acetic acid solution in water freezes at -0.1884°C. Calculate the percentage ionization of CH_3COOH in this solution.

75. A weak acid, HZ, ionizes in water according to

$$HZ \rightleftharpoons H^+ + Z^-$$

In a 0.100 m solution, HZ is 8.09% ionized. Calculate the freezing point of this solution.

76. CsCl dissolves in water according to

$$CsCl \longrightarrow Cs^+ + Cl^-$$

A 0.121 m solution of CsCl freezes at -0.403°C. Calculate i and the apparent percentage dissociation of CsCl in this solution.

*77. In a home ice cream freezer, we lower the freezing point of the water bath surrounding the ice cream can by dissolving NaCl in water to make a brine solution. A 15.0% brine solution is observed to freeze at -10.888°C. What is the van't Hoff factor, i, for this solution?

*78. (a) Solutions of benzoic acid, C_6H_5COOH, in nonpolar solvents such as benzene, C_6H_6, yield i values that approach $\frac{1}{2}$ in freezing point depression experiments. How can you account for this? (b) What is the apparent molecular weight of benzoic acid that would be determined by measuring the freezing point depression of a solution of benzoic acid in benzene?

Osmotic Pressure

79. What are osmosis and osmotic pressure?

*80. Show numerically that the molality and molarity of 1.00×10^{-4} M aqueous sodium chloride are nearly equal. Why is this true? Would this be true if another solvent, say acetonitrile, CH_3CN, replaced water? Why or why not? The density of CH_3CN is 0.786 g/mL at 20°C.

81. Show how the expression $\pi = MRT$, where π is osmotic pressure, is similar to the ideal gas law. Rationalize qualitatively why this should be so.

82. What is the osmotic pressure associated with a 0.0100 M aqueous solution of a nonvolatile nonelectrolyte solute at 75°C?

83. The osmotic pressure of an aqueous solution of a nonvolatile nonelectrolyte solute is 12.9 atm at 0°C. What is the molarity of the solution?

*84. If a sugar maple tree grows to a height of 45 feet, what must be the concentration of sugar in its sap so that osmotic pressure forces the sap to the top of the tree at 0°C? The density of mercury is 13.6 g/cm³, and the density of the sap can be considered to be 1.00 g/cm³. The cells containing the sap are in contact with water.

85. Estimate the osmotic pressure associated with 24.0 g of an enzyme of molecular weight 4.21×10^6 dissolved in 1740 mL of benzene solution at 38.0°C.

86. Calculate the freezing point depression and boiling point elevation associated with the solution in Exercise 83.

87. Estimate the osmotic pressure at 25°C of 0.10 m K_2CrO_4 in water, assuming no ion association.

88. Calculate the osmotic pressure at 25°C of 0.10 m K_2CrO_4 in water, taking ion association into account. Refer to Table 14-3.

89. Estimate the molecular weight of a biological macromolecule if a 0.183-g sample dissolved in 86.3 mL of benzene solution has an osmotic pressure of 13.67 torr at 25.0°C. The density of this benzene solution is 0.879 g/mL.

*90. Many biological compounds are isolated and purified in short supply. We dissolve 10.0 mg of a biological macromolecule with molecular weight of 2.00×10^4 in 10.0 g of water. (a) Calculate the freezing point of the solution. (b) Calculate the osmotic pressure of the solution at 25°C. (c) Suppose we are trying to use freezing point measurements to *determine* the molecular weight of this substance and that we make an error of only 0.001°C in the temperature measurement. What

percentage error would this cause in the calculated molecular weight? (d) Suppose we could measure the osmotic pressure with an error of only 0.1 torr (not a very difficult experiment). What percentage error would this cause in the calculated molecular weight?

Colloids

91. How does a colloidal dispersion differ from a true solution?

92. Distinguish among (a) sol, (b) gel, (c) emulsion, (d) foam, (e) solid sol, (f) solid emulsion, (g) solid foam, (h) solid aerosol, and (i) liquid aerosol. Try to give an example of each that is not listed in Table 14-4.

93. What is the Tyndall effect, and how it is caused?

94. Distinguish between hydrophilic and hydrophobic colloids.

95. What is an emulsifier?

96. Distinguish between soaps and detergents. How do they interact with hard water? Write an equation to show the interaction between a soap and hard water that contains Ca^{2+} ions.

97. What is the disadvantage of alkylbenzenesulfonate (ABS) detergents compared to linear alkyl sulfonate (LAS) detergents?

Mixed Exercises

*98. The heat of solution (for infinite dilution) of KF(s) is -17.7 kJ/mol at 25°C. The crystal lattice energy, ΔH_{xtal}, is -825.9 kJ/mol at 25°C. What is the hydration energy of KF for infinite dilution at 25°C? [Here we refer to the sum of the hydration energies of $K^+(g)$ and $F^-(g)$.]

99. Dry air contains 20.94% O_2 by volume. The solubility of O_2 in water at 25°C is 0.041 gram O_2 per liter of water. How many liters of water would dissolve the O_2 in one liter of dry air at 25°C and 1.00 atm?

100. (a) The freezing point of a 1.00 % aqueous solution of acetic acid, CH_3COOH, is -0.310°C. What is the approximate formula weight of acetic acid in water? (b) A 1.00 % solution of acetic acid in benzene has a freezing point depression of 0.441°C. What is the formula weight of acetic acid in this solvent? Explain the difference.

101. An aqueous ammonium chloride solution contains 6.50 mass % NH_4Cl. The density of the solution is 1.0201 g/mL. Express the concentration of this solution in molarity, molality, and mole fraction of solute.

102. Starch contains C—C, C—H, C—O, and O—H bonds. Hydrocarbons contain only C—C and C—H bonds. Both starch and hydrocarbon oils can form colloidal dispersions in water. (a) Which dispersion is classified as hydrophobic? (b) Which is hydrophilic? (c) Which dispersion would be easier to make and maintain?

*103. Suppose we put some one-celled microorganisms in various aqueous NaCl solutions. We observe that the cells remain unperturbed in 0.7% NaCl, whereas they shrink in more concentrated solutions and expand in more dilute solutions. Assume that 0.7% NaCl behaves as an *ideal* 1 : 1 electrolyte. Calculate the osmotic pressure of the fluid within the cells at 25°C.

*104. A sample of a drug ($C_{21}H_{23}O_5N$, molecular weight = 369 g/mol) mixed with lactose (a sugar, $C_{12}H_{22}O_{11}$, molecular weight = 342 g/mol) was analyzed by osmotic pressure to determine the amount of sugar present. If 100.00 mL of solution containing 1.00 g of the drug–sugar mixture has an osmotic pressure of 527 torr at 25°C, what is the percent sugar present?

105. A solution containing 3.81 g of a nonelectrolyte polymer per liter of benzene solution has an osmotic pressure of 0.646 torr at 20.0°C. (a) Calculate the molecular weight of the polymer. (b) Assume that the density of the dilute solution is the same as that of benzene, 0.879 g/mL. What would be the freezing point depression for this solution? (c) Why are boiling point elevations and freezing point depressions difficult to use to measure molecular weights of polymers?

106. On what basis would you choose the components to prepare an ideal solution of a molecular solute? Which of the following combinations would you expect to act most nearly ideally? (a) $CH_4(\ell)$ and $CH_3OH(\ell)$, (b) $CH_3OH(\ell)$ and $NaCl(s)$, (c) $CH_4(\ell)$ and $CH_3CH_3(\ell)$.

*107. At what temperature would a 1.00 *M* aqueous solution of sugar have an osmotic pressure of 1.00 atm? Is this answer reasonable?

*108. In the Signer method for estimating molecular weights, separate solutions of two compounds are placed at constant temperature in an evacuated system similar to that in Figure 14-17b. The same solvent is used in each solution. In one solution is dissolved a known mass of a compound of known molecular weight, while the other solution contains a known mass of the substance of unknown molecular weight. The solvent evaporates preferentially from the more dilute solution and condenses preferentially in the most concentrated solution. Eventually the mole fractions of solute in both solutions become equal, at which point the vapor pressures of the two solutions are equal (Raoult's Law); at this point, equilibrium is reached. If both solutions are sufficiently dilute, two approximations can be made: (1) The volume due to each solute is negligible compared to the volume due to the solvent, and (2) the number of moles of each solute is negligible compared to the number of moles of solvent. The volumes of the two solutions are measured, and the molecular weight of the unknown compound is calculated. Suppose we dissolve 40.6 mg of an unknown compound to form one solution and 45.1 mg of azobenzene (molecular weight 168.2) to form the other solution. After four days at 50°C, when equilibrium has been reached, the volume of the azobenzene solution is 1.80 mL and the volume of the "unknown" solution is 1.41 mL. Calculate the molecular weight of the unknown compound.

Outline

Methane, CH_4, is the main component of natural gas. When methane is burned, chemical energy is converted into heat energy and light energy.

Objectives

As you study this chapter, you should learn to

☐ Pay close attention to the terminology of thermodynamics, especially the significance of the signs of changes

☐ Use the concept of state functions

☐ Use the First Law of Thermodynamics to relate heat, work, and energy changes

☐ Relate the work done on or by a system to changes in its volume

☐ Carry out calculations of calorimetry to determine changes in energy and enthalpy

☐ Use Hess' Law to find the enthalpy change, ΔH, for a reaction by combining thermochemical equations with known ΔH values

☐ Use Hess' Law to find the enthalpy change, ΔH, for a reaction by using tabulated values of standard molar enthalpies of formation

☐ Use bond energies to estimate heats of reaction for gas-phase reactions; use ΔH values for gas-phase reactions to find bond energies

☐ Use the Born–Haber cycle to find the crystal lattice energy of an ionic solid

☐ Understand what is meant by the spontaneity of a process

☐ Understand the relationship of entropy to the order/disorder of a system

☐ Understand how the spontaneity of a process is related to entropy changes—the Second Law of Thermodynamics

☐ Use tabulated values of absolute entropies to calculate the entropy change, ΔS, for a process

☐ Calculate changes in Gibbs free energy, ΔG, (a) from values of ΔH and ΔS and (b) from tabulated values of standard molar free energies of formation; know when to use each type of calculation

☐ Use ΔG to predict the spontaneity of a process at constant T and P

☐ Understand how changes in temperature can affect the spontaneity of a process

E nergy is very important in every aspect of our daily lives. The food we eat supplies the energy to sustain life with all of its activities and concerns. The availability of relatively inexpensive energy is the basis for our technological society. This is seen in the costs of fuel, heating and cooling our homes and workplaces, and the electricity to power our lights, appliances, and computers. It is also seen in the costs of the goods and services we purchase, because a substantial part of the cost of production is for energy in one form or another. We must understand the storage and use of energy on a scientific basis to learn how to decrease our dependence on consumable oil and natural gas as our main energy sources. Such understanding has profound ramifications, ranging from our daily lifestyles to international relations.

The concept of energy is at the very heart of science. All physical and chemical processes are accompanied by the transfer of energy. Because energy cannot be created or destroyed, we must understand how to do the "accounting" of energy transfers from one body or one substance to another.

In the science of **thermodynamics** we study the energy changes that accompany physical and chemical processes. Usually these energy changes involve *heat*—hence the "thermo-" part of the term. In this chapter we shall study the two main aspects of thermodynamics. The first is **thermochemistry**. This practical subject is concerned with how we *observe, measure*, and *predict* energy changes for both physical changes and chemical reactions. The second part of the chapter addresses a more fundamental aspect of thermodynamics. Here we will learn to use energy changes to tell whether or not a given process can occur under specified conditions, and how to make a process more (or less) favorable.

Some instructors prefer to discuss thermochemistry earlier in the course. With this in mind, this chapter is presented so that the first ten sections (or selected portions of them) could be discussed earlier if desired.

Heat and Energy Changes: Thermochemistry

15-1 The First Law of Thermodynamics

Energy is a somewhat abstract, but very useful, concept.

Energy is the capacity to do work or to transfer heat.

We classify energy into two general types—kinetic and potential. **Kinetic energy** is the energy of motion. The kinetic energy of an object is equal to one-half its mass, m, times the square of its velocity, v.

$$E_{\text{kinetic}} = \tfrac{1}{2}mv^2$$

The heavier a hammer is and the more rapidly it moves, the more work it can accomplish.

Potential energy is the energy that a system possesses by virtue of its position or composition. The work that we do to lift an object is stored in the object as energy; we describe this as potential energy. If we drop a hammer, its potential energy is converted into kinetic energy as it falls, and it could do work on something it hits—for example, drive a nail or break a piece of glass. Similarly, an electron in an atom has potential energy because of the electrostatic force between it and the positively charged nucleus. Energy can manifest itself in many other forms: electrical energy, light (radiant energy), nuclear energy, and chemical energy. At the atomic or molecular level, however, we can think of each of these as kinetic or potential energy.

The chemical energy in a fuel or food can be viewed as potential energy stored in the electrons and nuclei due to their arrangements in the molecules. This stored chemical energy can be released when the compounds are burned or metabolized. Reactions that release energy in the form of heat are called **exothermic** reactions.

The combustion reactions of fossil fuels are familiar examples. Hydrocarbons—including methane, the main component of natural gas, and *n*-octane, a minor component of gasoline—undergo combustion with an excess of O_2 to yield CO_2 and H_2O. These reactions release heat energy. The amounts of heat energy released at constant pressure are shown for the reactions of one mole of methane and of two moles of octane:

> A hydrocarbon is a binary compound of hydrogen and carbon. Hydrocarbons may be gaseous, liquid, or solid. All burn.

$$CH_4(g) + 2O_2(g) \longrightarrow CO_2(g) + 2H_2O(\ell) + 890 \text{ kJ}$$

$$2C_8H_{18}(\ell) + 25O_2(g) \longrightarrow 16CO_2(g) + 18H_2O(\ell) + 1.090 \times 10^4 \text{ kJ}$$

In such reactions, the total energy of the products is lower than that of the reactants by the amount of energy released, most of which is heat. Some initial activation by heat is needed to get these reactions started. This is shown for CH_4 in Figure 15-1. However, this initiation energy *plus* 890 kJ is released as one mole of CO_2 and two moles of H_2O are formed. A process

> The amount of heat shown in such an equation always refers to the reaction for the number of moles of reactants and products specified by the coefficients. We call this *one mole of reaction*. It is important to specify the physical states of all substances, because different physical states have different energy contents.

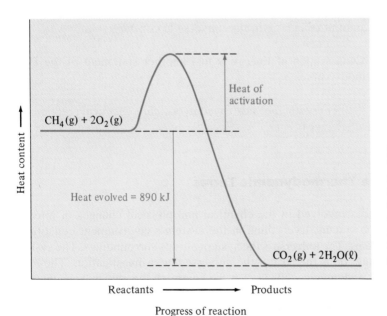

Figure 15-1

The difference between the heat content of the reactants—one mole of $CH_4(g)$ and two moles of $O_2(g)$—and that of the products—one mole of $CO_2(g)$ and two moles of $H_2O(\ell)$—is the amount of heat evolved in this *exothermic* reaction at constant pressure. For this reaction, it is 890 kJ/mol of reaction. Some initial activation by heat is needed to get the reaction started. In the absence of such initiating energy, a mixture of CH_4 and O_2 can be kept at room temperature for a long time without reacting. For an *endothermic* reaction, the final level is higher than the initial level.

(a)

(b)

Figure 15-2
An endothermic process. (a) When solid hydrated barium hydroxide [Ba(OH)₂ · 8H₂O] and *excess* solid ammonium nitrate [NH₄NO₃] are mixed, a reaction occurs.

$$Ba(OH)_2 \cdot 8H_2O(s) + 2NH_4NO_3(s) \longrightarrow$$
$$Ba(NO_3)_2(s) + 2NH_3(g) + 10H_2O(\ell)$$

The excess ammonium nitrate dissolves in the water produced in the reaction. (b) The dissolution process is very endothermic. If the flask is placed on a wet wooden block, the water freezes and attaches the block to the flask.

In Chapter 1 we pointed out the equivalence of matter and energy. The word "energy" is understood to include the energy equivalent of all matter in the universe. Stated differently, the total amount of mass and energy in the universe is constant.

that absorbs energy from its surroundings is called **endothermic**. One such process is shown in Figure 15-2.

Energy changes accompany physical changes, too (Chapter 13). For example, the melting of one mole of ice at 0°C at constant pressure must be accompanied by the absorption of 6.02 kJ of energy.

$$H_2O(s) + 6.02 \text{ kJ} \longrightarrow H_2O(\ell)$$

This tells us that the total energy of the water is raised by 6.02 kJ in the form of heat during the phase change.

Some important ideas about energy are summarized in the **First Law of Thermodynamics**:

> The total amount of energy in the universe is constant.

The **Law of Conservation of Energy** is just another statement of the First Law of Thermodynamics:

> Energy is neither created nor destroyed in ordinary chemical reactions and physical changes.

15-2 Some Thermodynamic Terms

The substances involved in the chemical and physical changes of interest are called the **system**. Everything in the system's environment constitutes its **surroundings**. The **universe** is the system plus its surroundings. The system may be thought of as the part of the universe under investigation. The First Law of Thermodynamics tells us that the energy of the universe is constant; i.e., energy is neither created nor destroyed, but is only transferred between the system and its surroundings.

The **thermodynamic state of a system** is defined by a set of conditions that completely specifies all the properties of the system. This set commonly includes the temperature, pressure, composition (identity and number of moles of each component), and physical state (gas, liquid, or solid) of each part of the system. Once the state has been specified, all other properties—both physical and chemical—are fixed.

The properties of a system—e.g., P, V, T—are called **state functions**. The *value* of a state function depends *only* on the state of the system and not on the way in which the system came to be in that state. A *change* in a state function describes a *difference* between the two states. It is independent of the process or pathway by which the change occurs.

For instance, consider a sample of one mole of pure liquid water at 30°C and 1 atm pressure. If at some later time the temperature of the sample is 22°C at the same pressure and with no change of physical state or composition, then it is in a different thermodynamic state. We can tell that the *net* temperature change is −8°C. It does not matter whether (1) the cooling took place directly (either slowly or rapidly) from 30°C to 22°C, or (2) the sample was first heated to 36°C, then cooled to 10°C, and finally warmed to 22°C, or (3) any other conceivable path was followed from the initial state to the final state. The change in other properties (e.g., the pressure) of the sample is likewise independent of path.

You can consider a state function as analogous to a bank account. With a bank account, at any time you can measure the amount of money in your account (your balance) in convenient terms—dollars and cents. Changes in this balance can occur for several reasons, such as deposit of your paycheck, writing of checks, or service charges assessed by the bank. In our analogy, the changes are *not* state functions, but they do cause *changes in* the state function (the balance in the account). You can think of the bank balance on a vertical scale; a deposit of $150 changes the balance by +150 dollars, no matter what it was at the start, just as a withdrawal of $150 would change the balance by −150 dollars. Similarly, we shall see that the energy of a system is a state function that can be changed—for instance, by an energy "deposit" of heat absorbed or work done on the system, or by an energy "withdrawal" of heat given off or work done by the system.

The most important use of state functions in thermodynamics is to describe *changes*. We describe the difference in any quantity, X, as

$$\Delta X = X_{final} - X_{initial}$$

When X increases, the final value is greater than the initial value, so ΔX is *positive*; a decrease in X makes ΔX a *negative* value.

We can describe *differences* between levels of a state function, regardless of where the zero level happens to be located. In the case of a bank balance, the "natural" zero level is obviously the point at which we open the account, before any deposits or withdrawals. In contrast, the zero levels on most temperature scales are set arbitrarily. When we say that the temperature of an ice–water mixture is "zero degrees Celsius," we are not saying that the mixture contains no temperature! We have simply chosen to describe this point on the temperature scale by the number *zero*; conditions of higher temperature are described by positive temperature values, and those of lower temperature have negative values, "below zero." The phrase "15 degrees

State functions are represented by capital letters. Here P refers to pressure, V to volume, and T to absolute temperature.

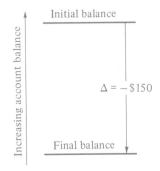

Suppose your bank balance last Thursday was $150 *higher* than it is today. We express the *change* in your bank balance as $\Delta \$ = \$_{final} - \$_{initial}$. Your final balance (today) is *less* than your initial balance (last Thursday), so the result is *negative*, indicating a *decrease*. There are many ways to get this same net change— one large withdrawal or some combination of deposits, withdrawals, interest earned, and service charges. All of the Δ values we shall see in this chapter can be thought of in this way.

cooler'' has the same meaning at any point on the scale. Many of the scales that we use in thermodynamics are thus arbitrarily defined. Arbitrary scales are useful when we are interested only in *changes* in the quantity being described.

Any property of a system that depends only on the values of its state functions is also a state function. For instance, the volume of a sample of matter, which depends only on temperature, pressure, composition, and physical state, is a state function. We shall encounter other thermodynamic state functions.

15-3 Changes in Internal Energy, ΔE

Internal energy is a state function.

The **internal energy**, E, of a specific amount of a substance represents all the energy contained within the substance. It includes such forms as kinetic energies of the molecules; energies of attraction and repulsion among sub-atomic particles, atoms, ions, or molecules; and other forms of energy. The internal energy of a collection of molecules is a state function. The difference between the internal energy of the products and the internal energy of the reactants of a chemical reaction or physical change, ΔE, is given by the equation

For many years this equation was written in the form $\Delta E = q - w$, using the reverse convention for the sign of w. In this text we use the convention of physical chemistry. Heat added to the system or work done on the system would increase its energy.

$$\Delta E = E_{final} - E_{initial} = E_{products} - E_{reactants} = q + w$$

The terms q and w represent heat and work, respectively. These are two ways in which energy can flow into or out of a system. **Work** involves a change of energy in which a body is moved through a distance, d, against some force, f; that is, $w = fd$.

$\Delta E = $ (amount of heat added to system) + (amount of work done on system)

The following conventions apply to the signs of q and w:

q is positive: Heat is *absorbed* by the system from the surroundings.
q is negative: Heat is *released* by the system to the surroundings.
w is positive: Work is done *on* the system by the surroundings.
w is negative: Work is done *by* the system on the surroundings.

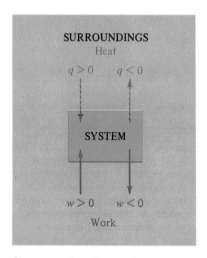

Sign conventions for q and w.

In other words, whenever energy is added to a system, either as heat or as work, the energy of the system increases.

When energy is released by a reacting system, ΔE is negative; energy can be written as a product in the equation for the reaction. When the system absorbs energy from the surroundings, ΔE is positive; energy can be written as a reactant in the equation.

For example, the complete combustion of CH_4 at constant volume at 25°C *releases* energy:

At 25°C the change in internal energy for the combustion of methane is −887 kJ/mol CH_4. The change in heat content is −890 kJ/mol CH_4 (Section 15-1). The small difference is due to work done on the system as it is compressed by the atmosphere.

$$CH_4(g) + 2O_2(g) \longrightarrow CO_2(g) + 2H_2O(\ell) + 887 \text{ kJ}$$

indicates release of energy

The coefficients in such a description *must* be interpreted as *number of moles*. Thus, the amount of energy indicated, 887 kJ, is released when *one* mole of $CH_4(g)$ reacts with *two* moles of $O_2(g)$ to form *one* mole of $CO_2(g)$ and *two* moles of $H_2O(\ell)$ at constant volume at 25°C. We can refer to this amount of reaction as one **mole of reaction**, which we shall abbreviate "mol rxn." This interpretation will allow us to write various unit factors as desired, such as

$$\frac{1 \text{ mol } CH_4}{1 \text{ mol rxn}}, \qquad \frac{2 \text{ mol } O_2}{1 \text{ mol rxn}}, \qquad \text{and so on}$$

In these terms, we can write the *change in energy* that accompanies this reaction as

$$CH_4(g) + 2O_2(g) \longrightarrow CO_2(g) + 2H_2O(\ell) \qquad \Delta E = -887 \text{ kJ/mol rxn}$$

As discussed in Section 15-2, the negative sign indicates a *decrease* in energy of the system, or a *release* of energy by the system.

The reverse of this reaction *absorbs* energy. It could be written as

$$CO_2(g) + 2H_2O(\ell) + 887 \text{ kJ} \longrightarrow CH_4(g) + 2O_2(g)$$

indicates absorption of energy

or

$$CO_2(g) + 2H_2O(\ell) \longrightarrow CH_4(g) + 2O_2(g) \qquad \Delta E = +887 \text{ kJ/mol rxn}$$

If the latter reaction could be forced to occur, the system would have to absorb 887 kJ of energy per mole of reaction from its surroundings.

The only type of work involved in most chemical and physical changes is pressure–volume work. From dimensional analysis we can see that the product of pressure and volume is work. Pressure is the force exerted per unit area, where area is distance squared, d^2; volume is distance cubed, d^3. Thus, the product of pressure and volume is force times distance, which is work. When a gas is produced against constant external pressure, such as in an open vessel at atmospheric pressure, the gas does work as it expands against the pressure of the atmosphere. If no heat is absorbed, this results in a decrease in the internal energy of the system. On the other hand, when a gas is consumed in a reaction, the atmosphere does work on the reacting system.

Let us illustrate the latter case. Consider the complete reaction of a $2:1$ mole ratio of H_2 and O_2 to produce steam at some constant temperature above 100°C and at one atmosphere pressure (Figure 15-3). Assume that the constant-temperature bath surrounding the reaction vessel completely absorbs all the evolved heat. The temperature of the gases does not change. The volume of the system decreases by one third (3 mol gaseous reactants → 2 mol gaseous products). The surroundings exert a constant pressure of one atmosphere and do work on the system by compressing it. The internal energy of the system increases by an amount equal to the amount of work done on it.

The work done on or by a system depends on the *external* pressure and the volume. When the external pressure is constant during a change, the work is equal to this pressure times the change in volume. The work done *on* a system equals $-P\Delta V$ or $-P(V_2 - V_1)$. When a gas expands, $(V_2 - V_1) > 0$ and $-P\Delta V$ is a negative quantity; the system does work on its surroundings, so w is negative. When a gas is compressed, $(V_2 - V_1) < 0$ and $-P\Delta V$ is a positive quantity; the surroundings do work on the system, so

$$\frac{F}{d^2} \times d^3 = Fd = w$$
$$\uparrow \qquad \uparrow$$
$$P \qquad V$$

V_2 is the final volume and V_1 is the initial volume.

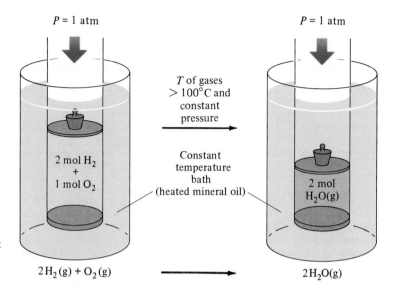

$P = 1$ atm

T of gases
$> 100°C$ and
constant
pressure

2 mol H_2
+
1 mol O_2

Constant
temperature
bath
(heated mineral oil)

$P = 1$ atm

2 mol
H_2O(g)

Figure 15-3
An illustration of the one-third decrease in volume that accompanies the reaction of H_2 with O_2 at constant temperature. The temperature is above 100°C.

$$2 H_2(g) + O_2(g) \longrightarrow 2 H_2O(g)$$

w is positive. We substitute $-P\Delta V$ for w in the equation $\Delta E = q + w$ to obtain

$$\Delta E = q - P\Delta V$$

Do not make the error of setting work equal to $V\Delta P$.

In constant-volume reactions, no $P\Delta V$ work is done. Although the pressure varies, the absence of a change in volume means that nothing "moves through a distance," so $d = 0$ and $fd = 0$. The change in internal energy of the system is just the amount of heat absorbed or released at constant volume, q_v:

$$\Delta E = q_v$$

A subscript v indicates a constant-volume process, a subscript p indicates a constant-pressure process, and so on.

Solids and liquids do not expand or contract significantly as pressure changes. Their production or consumption involves only negligible work ($\Delta V \approx 0$). In reactions in which equal numbers of moles of gases are produced and consumed at constant temperature and pressure, essentially no work is done. By the ideal gas law, $P\Delta V = (\Delta n)RT$ and $\Delta n = 0$, where Δn equals the number of moles of gaseous products minus the number of moles of gaseous reactants. Thus, the work term w has a significant value only when there are different numbers of moles of gaseous products and reactants so that the volume of the system changes.

Example 15-1
For each of the following chemical reactions carried out at constant temperature and constant pressure, predict the sign of w, and tell whether work is done *on* or *by* the system. Consider the reaction mixture to be the system.
(a) Ammonium nitrate, commonly used as a fertilizer, decomposes explosively:

$$2NH_4NO_3(s) \longrightarrow 2N_2(g) + 4H_2O(g) + O_2(g)$$

This reaction was responsible for an explosion in 1947 that destroyed nearly the entire port of Texas City, Texas, and killed 576 people.

(b) The combination of hydrogen and chlorine forms hydrogen chloride gas:

$$H_2(g) + Cl_2(g) \longrightarrow 2HCl(g)$$

(c) The oxidation of sulfur dioxide to sulfur trioxide is one step in the production of sulfuric acid.

$$2SO_2(g) + O_2(g) \longrightarrow 2SO_3(g)$$

Plan

For a process at constant pressure, $w = -P\Delta V = -(\Delta n)RT$. For each reaction, we evaluate Δn, the change in the number of moles of *gaseous* substances in the reaction.

Δn = no. of moles of gaseous products − no. of moles of gaseous reactants
(in balanced equation)

Because both R and T (on the Kelvin scale) are positive quantities, the sign of w is opposite from that of Δn, and it tells us whether the work is done *on* ($w = +$) or *by* ($w = -$) the system.

Solution

(a) $\Delta n = [2 \text{ mol } N_2(g) + 4 \text{ mol } H_2O(g) + 1 \text{ mol } O_2(g)] - 0 \text{ mol}$
(no *gaseous* reactants)

$$= 7 \text{ mol} - 0 \text{ mol} = +7 \text{ mol}$$

Δn is positive, so w is negative. This tells us that work is done

by the system as the reaction proceeds. The large amount of gas formed by the reaction pushes against the surroundings (this happened with devastating effect in the Texas City case).

(b) $\Delta n = [2 \text{ mol } HCl(g)] - [1 \text{ mol } H_2(g) + 1 \text{ mol } Cl_2(g)]$
$$= 2 \text{ mol} - 2 \text{ mol} = 0 \text{ mol}$$

Thus, $w = 0$, and no work is done as the reaction proceeds. We can see from the balanced equation that for every two moles (total) of gas that react, two moles of gas are formed, so the volume neither expands nor contracts as the reaction occurs.

(c) $\Delta n = [2 \text{ mol } SO_3(g)] - [2 \text{ mol } SO_2(g) + 1 \text{ mol } O_2(g)]$
$$= 2 \text{ mol} - 3 \text{ mol} = -1 \text{ mol}$$

Δn is negative, so w is positive. This tells us that work is done *on the*

system as the reaction proceeds. The surroundings push against the diminishing volume of gas.

EOC 13, 14

15-4 Calorimetry

We can determine the energy change associated with a chemical or physical process by using an experimental technique called **calorimetry**. This technique is based on observing the temperature change when a system absorbs

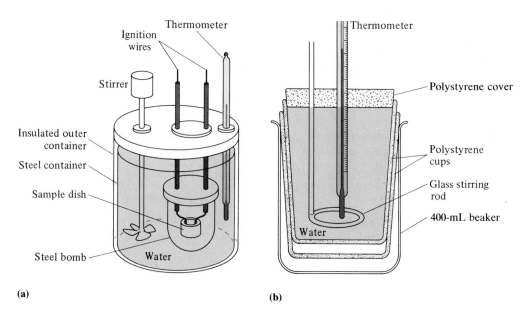

(a)   **(b)**

Figure 15-4
Calorimeters. (a) A bomb calorimeter measures q_v, the amount of heat evolved or absorbed by a reaction occurring at constant *volume*. The amount of energy introduced via the ignition wires is measured and taken into account. (b) A coffee-cup calorimeter. The stirring rod is moved up and down to ensure thorough mixing and uniform heating of the solution during reaction. The polystyrene walls and top provide insulation so that very little heat escapes. This kind of calorimeter measures q_p, the heat transfer due to a reaction occurring at constant *pressure*.

or releases energy in the form of heat. The experiment is carried out in a device called a **calorimeter**, in which the temperature change of a known amount of substance (often water) of known specific heat is measured. The temperature change is caused by the absorption or release of heat by the chemical or physical process under study. A review of calculations involved with heat transfer (Sections 1-13, 13-9, and 13-11) would be helpful for understanding this section.

A bomb calorimeter is a device that measures the amount of heat evolved or absorbed by a reaction occurring at constant volume (Figure 15-4a). A strong steel vessel (the bomb) is immersed in a large volume of water. As heat is produced or absorbed by a reaction inside the steel vessel, the heat is transferred to or from the large volume of water. Thus, only rather small temperature changes occur. For all practical purposes, the energy changes associated with the reactions are measured at constant volume and constant temperature. No work is done when a reaction is carried out in a bomb calorimeter, even if gases are involved, because $\Delta V = 0$. Therefore,

$$\Delta E = q_v \qquad \text{(constant volume)}$$

The concepts and methods of Sections 1-13, 13-9, and 13-11 are applied to calorimetric data to relate changes in the temperature of water to amounts of heat transferred and, therefore, to ΔE. For exothermic reactions, we may write

$$\begin{pmatrix} \text{amount of heat} \\ \text{lost by reaction} \end{pmatrix} = \begin{pmatrix} \text{amount of heat gained} \\ \text{by calorimeter bomb} \end{pmatrix} + \begin{pmatrix} \text{amount of heat} \\ \text{gained by water} \end{pmatrix}$$

The heat capacity of a calorimeter is determined by adding a known amount of heat and measuring the rise in temperature of the calorimeter and of the water it contains. This heat capacity of a calorimeter is sometimes called its *calorimeter constant*. It depends on the materials of which the calorimeter is constructed.

The amount of heat absorbed by a calorimeter is sometimes expressed as the *heat capacity* of the calorimeter, in joules per degree.

Example 15-2

To calibrate a calorimeter, we add 3000 grams of water and then burn a sufficient amount of a particular compound to produce 9.598 kJ of heat. We observe that the water temperature rises by 0.629°C. What is the heat capacity of the calorimeter?

Plan

We first calculate the amount of heat absorbed by the water as its temperature increases. The remainder of the 9.598 kJ must be absorbed by the calorimeter apparatus as it warms by 0.629°C; we scale this to one Celsius degree to describe the heat capacity of the calorimeter.

Solution

$$\begin{matrix} \text{heat gained} \\ \text{by water} \end{matrix} = \text{mass of water} \times \text{specific heat of water} \times \text{temperature change}$$

$$= 3000 \text{ g} \times \frac{4.184 \text{ J}}{\text{g} \cdot °\text{C}} \times (0.629°\text{C}) = 7895 \text{ J}$$

$$\text{heat gained by calorimeter} = (\text{heat lost by sample}) - (\text{heat gained by water})$$

$$= 9598 \text{ J} - 7895 \text{ J} = 1703 \text{ J, or } 1.703 \text{ kJ}$$

We see that 1.703 kJ of heat went to warm the calorimeter by 0.629°C. We must scale this to find the amount of heat required to raise the calorimeter temperature by one Celsius degree.

$$\begin{matrix} \text{heat capacity of calorimeter} \\ \text{(also called calorimeter constant)} \end{matrix} = \frac{1.703 \text{ kJ}}{0.629°\text{C}} = \boxed{2.71 \text{ kJ/°C}}$$

EOC 20, 21

Benzoic acid, C_6H_5COOH, is often used to determine the heat capacity of a calorimeter. it is a solid that can be compressed into pellets. Its heat of combustion is accurately known: 3227 kJ/mol benzoic acid, or 26.46 kJ/g benzoic acid. Another way to measure the heat capacity of a calorimeter is to add a known amount of heat electrically.

Example 15-3

A 1.000-gram sample of ethanol, C_2H_5OH, was burned in the sealed bomb calorimeter calibrated in Example 15-2, with 3000 grams of water in the calorimeter. The temperature of the water rose from 24.284°C to 26.255°C. Determine ΔE for the reaction in joules per gram of ethanol, then in kilojoules per mole of ethanol. The specific heat of water is 4.184 J/g · °C. The combustion reaction is

$$C_2H_5OH(\ell) + 3O_2(g) \longrightarrow 2CO_2(g) + 3H_2O(\ell)$$

Plan

The amount of heat given off by the sealed compartment (the system) raises the temperature of the calorimeter and its water. The amount of heat absorbed by the water can be calculated using the specific heat of water; similarly, we use the heat capacity of the calorimeter to find the amount of heat absorbed by the calorimeter. The sum of these two amounts of heat is the total amount of heat released by the combustion of 1.000 gram of ethanol. We must then scale that result to correspond to one mole of ethanol.

Solution

The increase in temperature is

$$\underline{?} °\text{C} = 26.225°\text{C} - 24.284°\text{C} = 1.941°\text{C rise}$$

The number of kilojoules of heat responsible for this increase in temperature of 3000 grams of water is

$$\text{heat to warm water} = 1.941°\text{C} \times \frac{4.184 \text{ J}}{\text{g} \cdot °\text{C}} \times 3000 \text{ g} = 2.436 \times 10^4 \text{ J} = 24.36 \text{ kJ}$$

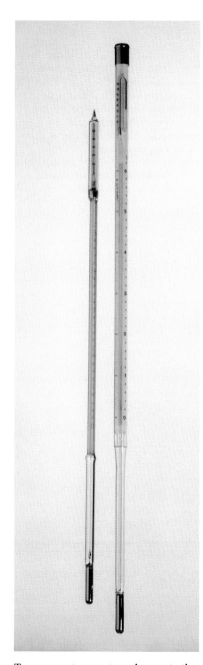

To measure temperature changes to the nearest thousandth of a degree, very sensitive and expensive differential thermometers are used. Modern solid state devices provide another way to measure temperature changes quite precisely.

The number of kilojoules of heat responsible for the warming of the calorimeter is

$$\text{heat to warm calorimeter} = 1.941°C \times \frac{2.71 \text{ kJ}}{°C} = 5.26 \text{ kJ}$$

The total amount of heat absorbed by the calorimeter *and* the water is

$$\text{total amount of heat} = 24.36 \text{ kJ} + 5.26 \text{ kJ} = 29.62 \text{ kJ}$$

Combustion of one gram of C_2H_5OH thus liberates 29.62 kJ of energy in the form of heat; that is,

$$\Delta E = q_v = -29.62 \text{ kJ/g ethanol}$$

The negative sign indicates that energy is released by the system to the surroundings.

Now we may evaluate ΔE in kJ/mol of ethanol by converting grams of C_2H_5OH to moles.

$$\frac{? \text{ kJ}}{\text{mol ethanol}} = \frac{-29.62 \text{ kJ}}{\text{g}} \times \frac{46.07 \text{ g } C_2H_5OH}{1 \text{ mol } C_2H_5OH} = -1365 \text{ kJ/mol ethanol}$$

$$\Delta E = -1365 \text{ kJ/mol}$$

This calculation shows that for the combustion of ethanol at constant temperature and constant volume, the change in internal energy is −1365 kJ/mol ethanol.

EOC 24

The balanced chemical equation involves one mole of ethanol, so we can write the unit factor $\frac{1 \text{ mol ethanol}}{1 \text{ mol rxn}}$. Then we express the result of Example 15-3 as

$$\Delta E = \frac{-1365 \text{ kJ}}{\text{mol ethanol}} \times \frac{1 \text{ mol ethanol}}{1 \text{ mol rxn}} = -1365 \text{ kJ/mol rxn}$$

As we shall see in the next section, q_p, the amount of heat evolved or absorbed by a reaction at *constant pressure*, is equal to the change in another thermodynamic state function. A "coffee-cup" calorimeter (Figure 15-4b) is often used in teaching laboratories to measure "heats of reaction" at constant pressure, q_p, in aqueous solutions. Reactions are chosen so that there are no gaseous reactants or products. Thus, all reactants and products remain in the vessel throughout the experiment. Such a calorimeter could be used to measure the amount of heat evolved with the neutralization of known quantities of hydrochloric acid and potassium hydroxide. The formula unit and net ionic equations are

$$HCl(aq) + KOH(aq) \longrightarrow KCl(aq) + H_2O(\ell) + \text{heat}$$

$$H^+(aq) + OH^-(aq) \longrightarrow H_2O(\ell) + \text{heat}$$

The amount of heat evolved is related to the rise in the temperature of the known mass of solution and to the previously determined amount of heat the calorimeter itself absorbs. The following example illustrates such a calculation.

Example 15-4

A 50.0-mL sample of 0.400 M copper(II) sulfate solution at 23.35°C is mixed with 50.0 mL of 0.600 M sodium hydroxide solution, also at 23.35°C, in a coffee-cup calorimeter whose heat capacity has already been determined to be 24.0 J/°C. After the reaction occurs, the temperature of the resulting mixture is measured to be 26.65°C. The density of the final solution is 1.02 g/mL. Calculate the amount of heat evolved. Assume that the specific heat of the solution is the same as that of pure water, 4.184 J/g · C.

$$CuSO_4(aq) + 2NaOH(aq) \longrightarrow Cu(OH)_2(s) + Na_2SO_4(aq)$$

Plan

The amount of heat released by the reaction is absorbed by the solution *and* the calorimeter (assuming negligible loss to the surroundings). To find the amount of heat absorbed by the solution, we must know the mass of solution; to find that, we assume that the volume of the reaction mixture is the sum of volumes of the initial solutions.

When *dilute aqueous solutions* are mixed, their volumes are very nearly additive. This is *not* true for most solutions.

Solution

The mass of solution is

$$\underline{?} \text{ g soln} = (50.0 + 50.0) \text{ mL} \times \frac{1.02 \text{ g soln}}{\text{mL}} = 102 \text{ g soln}$$

The amount of heat absorbed by the solution *plus* the calorimeter is

$$\underline{?} \text{ J} = \overbrace{102 \text{ g} \times \frac{4.18 \text{ J}}{\text{g} \cdot °\text{C}} \times (26.65 - 23.35)°\text{C}}^{\substack{\text{amount of heat} \\ \text{absorbed by solution}}} + \overbrace{\frac{24.0 \text{ J}}{°\text{C}} \times (26.65 - 23.35)°\text{C}}^{\substack{\text{amount of heat} \\ \text{absorbed by calorimeter}}}$$

$$= 1.41 \times 10^3 \text{ J} + 79 \text{ J} = 1.49 \times 10^3 \text{ J absorbed by solution plus calorimeter}$$

Thus, the reaction must have liberated 1.49×10^3 J, or 1.49 kJ, of heat.

EOC 22a

15-5 Enthalpy Change, ΔH, and Thermochemical Equations

Measuring the amount of heat transferred at constant volume may not be easy or even desirable, because the reaction must be carried out in a closed, rigid container with constant volume. Reactions in a living system or in an open beaker in the laboratory take place at constant pressure, not at constant volume. It is usually much easier to measure the amount of heat released or absorbed by a reaction at constant *pressure*, e.g., under laboratory conditions. The heat change for a reaction at constant pressure, q_p, is also called the **heat of reaction** or **enthalpy change**, ΔH, for the reaction. For a process at constant temperature and pressure, ΔH is defined as follows.

$$\Delta H = \Delta E + P\Delta V \qquad \text{(constant } T \text{ and } P\text{)}$$

From Section 15-3 we know that $\Delta E = q + w$, so

$$\Delta H = q + w + P\Delta V \qquad \text{(constant } T \text{ and } P\text{)}$$

At constant pressure, $w = -P\Delta V$, so

$$\Delta H = q + (-P\Delta V) + P\Delta V$$

$$\Delta H = q_p \quad \text{(constant } T \text{ and } P\text{)}$$

The difference between ΔE and ΔH is the amount of $P\Delta V$ work (expansion work) that the system can do. Unless there is a change in the number of moles of gas present, this difference is extremely small and can usually be neglected. For an ideal gas, $PV = nRT$. At constant temperature and constant pressure, $P\Delta V = (\Delta n)RT$, a work term. Substituting gives

In this equation, Δn refers to the number of moles of *gaseous products* minus the number of moles of *gaseous reactants*.

$$\Delta H = \Delta E + (\Delta n)RT \quad \text{or} \quad \Delta E = \Delta H - (\Delta n)RT \quad \text{(constant } T \text{ and } P\text{)}$$

In Example 15-3 we found that the change in internal energy, ΔE, for the combustion of ethanol is -1365 kJ/mol ethanol at 298 K. Combustion of one mole of ethanol at 298 K and constant pressure releases 1367 kJ of heat. Therefore, we can say

$$\Delta H = -1367 \frac{\text{kJ}}{\text{mol ethanol}}$$

The difference between ΔH and ΔE is due to the work term, $-P\Delta V$ or $-(\Delta n)RT$. In this reaction there are fewer moles of gaseous products than of gaseous reactants: $\Delta n = 2 - 3 = -1$.

$$C_2H_5OH(\ell) + 3O_2(g) \longrightarrow 2CO_2(g) + 3H_2O(\ell)$$

Thus, the atmosphere does work on the system (compresses it). Let us find the work done on the system per mole of reaction.

$$w = -P\Delta V = -(\Delta n)RT$$

$$w = -(\Delta n)RT = -(-1 \text{ mol})\left(\frac{8.314 \text{ J}}{\text{mol} \cdot \text{K}}\right)(298 \text{ K}) = +2.48 \times 10^3 \text{ J}$$

$$w = +2.48 \text{ kJ} \quad \text{or} \quad (\Delta n)RT = -2.48 \text{ kJ}$$

The positive sign is consistent with the fact that work is done on the system. The balanced equation involves one mole of ethanol, so this is the amount of work done when one mole of ethanol undergoes combustion.

We can now calculate ΔE for the reaction from ΔH and $(\Delta n)RT$ values:

$$\Delta E = \Delta H - (\Delta n)RT = [-1367 - (-2.48)] = -1365 \text{ kJ/mol rxn}$$

This value agrees with the result that we obtained in Example 15-3. The size of the work term ($+2.48$ kJ) is very small compared with ΔH (-1367 kJ/mol rxn). This is true for many reactions. Of course, if $\Delta n = 0$, then $\Delta H = \Delta E$.

A balanced chemical equation together with a designation of its value of ΔH is called a **thermochemical equation**. For example,

Historically, the term *thermochemical equation* has been used as described here. It would be logical to extend the same term to include descriptions of changes in other thermodynamic quantities.

$$C_2H_5OH(\ell) + 3O_2(g) \longrightarrow 2CO_2(g) + 3H_2O(\ell) \qquad \Delta H = -1367 \text{ kJ}$$

is a thermochemical equation that describes the combustion (burning) of liquid ethanol, $C_2H_5OH(\ell)$ at constant temperature and pressure. The negative sign indicates that this is an *exothermic* reaction (it *gives off* heat).

> We can interpret ΔH as (enthalpy change)/(mole of reaction), where the denominator means "for the number of moles of each substance shown in the balanced equation."

We can then use several unit factors to interpret this thermochemical equation:

$$\frac{1367 \text{ kJ given off}}{\text{mol of reaction}} = \frac{1367 \text{ kJ given off}}{\text{mol } C_2H_5OH(\ell) \text{ consumed}} = \frac{1367 \text{ kJ given off}}{3 \text{ mol } O_2(g) \text{ consumed}}$$

$$= \frac{1367 \text{ kJ given off}}{2 \text{ mol } CO_2(g) \text{ formed}} = \frac{1367 \text{ kJ given off}}{3 \text{ mol } H_2O(\ell) \text{ formed}}$$

The reverse reaction would require the absorption of 1367 kJ under the same conditions; i.e., it is *endothermic* with $\Delta H = +1367$ kJ.

$$2CO_2(g) + 3H_2O(\ell) \longrightarrow C_2H_5OH(\ell) + 3O_2(g) \qquad \Delta H = +1367 \text{ kJ/mol rxn}$$

It is important to remember the following conventions regarding thermochemical equations:

1. The coefficients in a balanced thermochemical equation refer to the numbers of *moles* of reactants and products involved. In the thermodynamic interpretation of equations, we *never* interpret the coefficients as *numbers of molecules*. Thus, it is acceptable to write coefficients as fractions rather than as integers, when necessary.
2. The numerical value of ΔH (or any other thermodynamic change) refers to the *number of moles* of substances specified by the equation. This amount of change of substances is *one mole of reaction*, so we can express ΔH in units of energy/mol rxn. For brevity, the units of ΔH are sometimes written kJ/mol or even just kJ. No matter what units are used, be sure that you interpret the thermodynamic change *per mole of reaction for the balanced chemical equation to which it refers*. If a different amount of material is involved in the reaction, then the ΔH (or other change) must be scaled accordingly.
3. The physical states of all species are important and must be specified.
4. The value of ΔH usually does not change significantly with changing temperature.

Example 15-5

When aluminum metal is exposed to atmospheric oxygen (as in aluminum doors and windows), it is oxidized to form aluminum oxide. How much heat is released by the complete oxidation of 24.2 grams of aluminum at 25°C and 1 atm? The thermochemical equation is

$$4Al(s) + 3O_2(g) \longrightarrow 2Al_2O_3(s) \qquad \Delta H = -3352 \text{ kJ/mol rxn}$$

Plan

The thermochemical equation tells us that 3352 kJ of heat is released for every mole of reaction, that is, for every 4 moles of Al that reacts. We convert 24.2 g of Al to moles, and then calculate the number of kilojoules corresponding to that number of moles of Al, using the unit factors

$$\frac{-3352 \text{ kJ}}{\text{mol rxn}} \quad \text{and} \quad \frac{1 \text{ mol rxn}}{4 \text{ mol Al}}$$

The *sign* tells us that heat was released, but it would be grammatical
nonsense to say in words that "−751
kJ of heat was released." As an analogy, suppose you give your friend
$5. Your Δ$S$ is −$5, but in describing
your action you would not say "I
gave her minus five dollars," but
rather "I gave her five dollars."

Solution

For 24.2 g Al,

$$? \text{ kJ} = 24.2 \text{ g Al} \times \frac{1 \text{ mol Al}}{27.0 \text{ g Al}} \times \frac{1 \text{ mol rxn}}{4 \text{ mol Al}} \times \frac{-3352 \text{ kJ}}{\text{mol rxn}} = -751 \text{ kJ}$$

This tells us that $\boxed{751 \text{ kJ of heat is released to the surroundings}}$ during the ox-

idation of 24.2 grams of aluminum.

EOC 30, 31

Example 15-6

Write the thermochemical equation for the reaction in Example 15-4.

Plan

We must determine *how much* reaction occurred—that is, how many moles of
reactants were consumed. We first multiply the volume, in liters, of each solution
by its concentration in mol/L (molarity), to determine the number of moles of
each reactant mixed. Then we scale the amount of heat released in the experiment
to correspond to the number of moles of reactant shown in the balanced equation.

Solution

Using the data from Example 15-4:

$$? \text{ mol CuSO}_4 = 0.0500 \text{ L} \times \frac{0.400 \text{ mol CuSO}_4}{1.00 \text{ L}} = 0.0200 \text{ mol CuSO}_4$$

$$? \text{ mol NaOH} = 0.0500 \text{ L} \times \frac{0.600 \text{ mol NaOH}}{1.00 \text{ L}} = 0.0300 \text{ mol NaOH}$$

We determine which is the limiting reactant (Section 3-3):

Required Ratio	Available Ratio
$\dfrac{1 \text{ mol CuSO}_4}{2 \text{ mol NaOH}} = \dfrac{0.50 \text{ mol CuSO}_4}{1.00 \text{ mol NaOH}}$	$\dfrac{0.0200 \text{ mol CuSO}_4}{0.0300 \text{ mol NaOH}} = \dfrac{0.667 \text{ mol CuSO}_4}{1.00 \text{ mol NaOH}}$

NaOH is the limiting reactant.

More $CuSO_4$ is available than is required to react with the NaOH. Thus, 1.49 kJ
of heat was released during the consumption of 0.0300 mol of NaOH. The amount
of heat released per "mole of reaction" is

$$\frac{? \text{ kJ released}}{\text{mol of rxn}} = \frac{1.49 \text{ kJ released}}{0.0300 \text{ mol NaOH}} \times \frac{2 \text{ mol NaOH}}{\text{mol of rxn}} = \frac{99.3 \text{ kJ released}}{\text{mol of rxn}}$$

Thus, when the reaction occurs *to the extent indicated by the balanced chemical
equation*, 99.3 kJ is released. Remembering that exothermic reactions have negative values of ΔH, we write

$$CuSO_4(aq) + 2NaOH(aq) \longrightarrow Cu(OH)_2(s) + Na_2SO_4(aq)$$
$$\Delta H = -99.3 \text{ kJ/mol rxn}$$

EOC 22(b)

15-6 Standard States and Standard Changes in Energy and Enthalpy

A temperature of 25°C is 77°F. This is
slightly above room temperature,
which is about 20°C, or 68°F.

The **thermochemical standard state** of a substance is its most stable state
under standard pressure (one atmosphere) and at some specific temperature
(25°C unless otherwise specified). Examples of elements in their standard

states at 25°C are hydrogen, gaseous diatomic molecules, $H_2(g)$; mercury, a silver-colored liquid metal, $Hg(\ell)$; sodium, a silvery-white solid metal, $Na(s)$; and carbon, a grayish-black solid called graphite, $C(graphite)$. We use $C(graphite)$ instead of $C(s)$ to distinguish it from another form of carbon, $C(diamond)$. Examples of standard states of compounds include ethanol (ethyl alcohol or grain alcohol), a liquid, $C_2H_5OH(\ell)$; water, a liquid, $H_2O(\ell)$; calcium carbonate, a solid, $CaCO_3(s)$; and carbon dioxide, a gas, $CO_2(g)$. Keep in mind the following conventions for thermochemical standard states:

1. For a *pure* substance in the liquid or solid phase, the standard state is the pure liquid or solid.
2. For a gas, the standard state is the gas at a pressure of *one atmosphere*; in a mixture of gases, its partial pressure must be one atmosphere.
3. For a substance in solution, the standard state refers to *one-molar* concentration.

For ease of comparison and tabulation, we often refer to thermochemical or thermodynamic changes "at standard states" or, more simply, to a *standard change*. To indicate a change at standard pressure, we add a superscript zero. If some temperature other than standard temperature of 25°C (298 K) is specified, we indicate it with a subscript; if no subscript appears, a temperature of 298 K is implied. For instance, the **standard enthalpy change, ΔH^0**, for a process

$$\text{reactants} \longrightarrow \text{products}$$

This is sometimes referred to as the *standard heat of reaction*. It is sometimes represented as ΔH_{rxn}^0.

refers to the ΔH when the specified number of moles of reactants, all at standard states, are converted *completely* to the specified number of moles of products, all at standard states. We allow the reaction to take place, with changes in temperature or pressure if necessary; when the reaction is complete, we return the products to the same conditions of temperature and pressure that we started with, *keeping track of energy or enthalpy changes* as we do so. We sometimes describe this as taking place "at constant T and P," but we merely mean that the initial and final conditions are the same. Because we are dealing with changes in state functions, the net change is the same as the change we would have obtained hypothetically with T and P actually held constant.

15-7 Standard Molar Enthalpies of Formation, ΔH_f^0

It is not possible to determine the total enthalpy content of a substance on an absolute scale. However, we need to describe only *changes* in this state function, so we can define an *arbitrary scale* as follows.

The **standard molar enthalpy of formation, ΔH_f^0**, of a substance is the amount of heat absorbed in a reaction in which *one mole* of the substance in a specified state is formed from its elements in their standard states.

We can think of ΔH_f^0 as the enthalpy content of each substance, in its standard state, relative to the enthalpy content of the elements, in their standard states.

Standard molar enthalpy of formation is often called **standard molar heat of formation** or, more simply, **heat of formation**. The superscript zero in

Table 15-1
Selected Standard Molar Enthalpies of Formation at 298 K

Substance	ΔH_f^0 (kJ/mol)	Substance	ΔH_f^0 (kJ/mol)
$Br_2(\ell)$	0	HgS(s) red	-58.2
$Br_2(g)$	30.91	$H_2(g)$	0
C(diamond)	1.897	HBr(g)	-36.4
C(graphite)	0	$H_2O(\ell)$	-285.8
$CH_4(g)$	-74.81	$H_2O(g)$	-241.8
$C_2H_4(g)$	52.26	NO(g)	90.25
$C_6H_6(\ell)$	49.03	Na(s)	0.
$C_2H_5OH(\ell)$	-277.7	NaCl(s)	-411.0
CO(g)	-110.52	$O_2(g)$	0
$CO_2(g)$	-393.51	$SO_2(g)$	-296.8
CaO(s)	-635.5	$SiH_4(g)$	34
$CaCO_3(s)$	-1207	$SiCl_4(g)$	-657.0
$Cl_2(g)$	0	$SiO_2(s)$	-910.9

The ΔH_f^0 values of $Br_2(g)$ and C(diamond) are *not equal to 0* at 298 K. The standard states of these elements are $Br_2(\ell)$ and C(graphite), respectively.

ΔH_f^0 signifies standard pressure, 1 atmosphere. Negative values for ΔH_f^0 describe exothermic formation reactions, whereas positive values for ΔH_f^0 describe endothermic formation reactions.

The enthalpy change for a balanced equation may not give directly a molar enthalpy of formation for the compound formed. Consider the following exothermic reaction at standard state conditions:

$$H_2(g) + Br_2(\ell) \longrightarrow 2HBr(g) + 72.8 \text{ kJ} \quad \text{or} \quad \Delta H_{rxn}^0 = -72.8 \text{ kJ/mol rxn}$$

We see that *two* moles of HBr(g) are formed in the reaction as written. Half as much energy, 36.4 kJ, is liberated when *one mole* of HBr(g) is produced from its constituent elements in their standard states. For HBr(g), $\Delta H_f^0 = -36.4$ kJ/mol. This can be shown by dividing all coefficients in the balanced equation by 2:

$$\tfrac{1}{2}H_2(g) + \tfrac{1}{2}Br_2(\ell) \longrightarrow HBr(g) \qquad \Delta H_{rxn}^0 = -36.4 \text{ kJ/mol rxn}$$

$$\Delta H_{f\,HBr(g)}^0 = -36.4 \text{ kJ/mol HBr}$$

The coefficients $\tfrac{1}{2}$ preceding $H_2(g)$ and $Br_2(\ell)$ do *not* imply half a molecule of each. In thermochemical equations, the coefficients always refer to the number of *moles* under consideration.

Standard heats of formation of some common substances are tabulated in Table 15-1. Appendix K contains a larger listing.

When referring to a thermodynamic quantity for a *substance*, we often omit the description of the substance from the units. Units for tabulated ΔH_f^0 values are given as "kJ/mol"; we must interpret this as "per mole of the substance in the specified state." For instance, for HBr(g) the tabulated ΔH_f^0 value of -36.4 kJ/mol should be interpreted as $\dfrac{-36.4 \text{ kJ}}{\text{mol HBr(g)}}$.

By the definition of ΔH_f^0, the formation of an element in its standard state from that element in its standard state is "no reaction." Thus,

The ΔH_f^0 value for any *element in its standard state* is zero.

Example 15-7
The standard molar enthalpy of formation of ethanol, $C_2H_5OH(\ell)$, is -277.7 kJ/mol. Write the equation for the reaction for which $\Delta H_{rxn}^0 = -277.7$ kJ/mol rxn.

Plan

The usual balanced chemical equation shows the formation of two moles of C_2H_5OH:

$$4C(graphite) + 6H_2(g) + O_2(g) \longrightarrow 2C_2H_5OH(\ell)$$

Because ΔH_f^0 values apply to the formation of *one* mole of the substance, we divide the coefficients in the chemical equation by 2:

Solution

$$2C(graphite) + 3H_2(g) + \tfrac{1}{2}O_2(g) \longrightarrow C_2H_5OH(\ell) \qquad \Delta H = -277.7 \text{ kJ/mol rxn}$$

EOC 38

15-8 Hess' Law

In 1840, G. H. Hess published his **law of heat summation**, which he derived on the basis of numerous thermochemical observations:

> The enthalpy change for a reaction is the same whether it occurs by one step or by any series of steps.

As an analogy, consider traveling from Kansas City (elevation 884 ft above sea level) to Denver (elevation 5280 ft). The change in elevation is (5280 − 884) ft = 4396 ft, regardless of the route taken.

This is consistent with the fact that enthalpy is a state function. Therefore, its *change* is independent of the pathway by which a reaction occurs. We do not need to know whether the reaction *does*, or even *can*, occur by the series of steps used in the calculation. The steps must (if only "on paper") result in the overall reaction. Hess' Law lets us calculate enthalpy changes for reactions for which the changes could be measured only with difficulty, if at all. In general terms, Hess' Law of heat summation may be represented as

$$\Delta H_{rxn}^0 = \Delta H_a^0 + \Delta H_b^0 + \Delta H_c^0 + \cdots$$

Here a, b, c, ... refer to balanced thermochemical equations that can be summed to give the equation for the desired reaction.

Consider the following reaction:

$$C(graphite) + \tfrac{1}{2}O_2(g) \longrightarrow CO(g) \qquad \Delta H^0 = \underline{?}$$

The enthalpy change for this reaction cannot be measured directly. Even though $CO(g)$ is the predominant product of the reaction of graphite with a *limited* amount of $O_2(g)$, some $CO_2(g)$ is always produced as well. However, the following reactions do go to completion with excess $O_2(g)$. Therefore, ΔH^0 values have been measured experimentally for them.

$$C(graphite) + O_2(g) \longrightarrow CO_2(g) \qquad \Delta H^0 = -393.5 \text{ kJ/mol rxn} \qquad (1)$$

$$CO(g) + \tfrac{1}{2}O_2(g) \longrightarrow CO_2(g) \qquad \Delta H^0 = -283.0 \text{ kJ/mol rxn} \qquad (2)$$

We can "work backwards" to find out how to combine these two known reactions to obtain the desired reaction. We want one mole of CO on the right, so we reverse equation (2); heat is then absorbed instead of released, so we must change the sign of its ΔH^0 value. Then we add it to equation (1), canceling equal numbers of moles of the same species on each side. This

You are familiar with the addition and subtraction of algebraic equations. This method of combining thermochemical equations is exactly analogous.

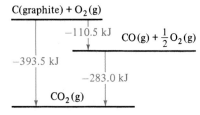

A schematic representation of the enthalpy changes for the reaction C(graphite) + $\frac{1}{2}$O$_2$(g) → CO(g). The ΔH value for each step refers to the number of moles of each substance indicated.

gives the equation for the reaction we want. Adding the corresponding enthalpy changes gives the enthalpy change we seek.

$$
\begin{array}{lll}
 & & \Delta H^0 \\
\text{C(graphite)} + \text{O}_2(g) \longrightarrow \text{CO}_2(g) & -393.5 \text{ kJ/mol rxn} & (1) \\
\text{CO}_2(g) \longrightarrow \text{CO}(g) + \tfrac{1}{2}\text{O}_2(g) & -(-283.0 \text{ kJ/mol rxn}) & (-2) \\
\hline
\text{C(graphite)} + \tfrac{1}{2}\text{O}_2(g) \longrightarrow \text{CO}(g) & \Delta H^0_{rxn} = -110.5 \text{ kJ/mol rxn} &
\end{array}
$$

Example 15-8

Use the thermochemical equations given below to determine ΔH^0_{rxn} at 25°C for the following reaction.

$$\text{C(graphite)} + 2\text{H}_2(g) \longrightarrow \text{CH}_4(g)$$

$$
\begin{array}{lll}
 & \Delta H^0 & \\
\text{C(graphite)} + \text{O}_2(g) \longrightarrow \text{CO}_2(g) & -393.5 \text{ kJ/mol rxn} & (1) \\
\text{H}_2(g) + \tfrac{1}{2}\text{O}_2(g) \longrightarrow \text{H}_2\text{O}(\ell) & -285.8 \text{ kJ/mol rxn} & (2) \\
\text{CO}_2(g) + 2\text{H}_2\text{O}(\ell) \longrightarrow \text{CH}_4(g) + 2\text{O}_2(g) & +890.3 \text{ kJ/mol rxn} & (3)
\end{array}
$$

Plan

(i) We want one mole of C(graphite) as reactant, so we write down equation (1).
(ii) We want two moles of H$_2$(g) as reactants, so we multiply equation (2) by 2 [designated below as 2 × (2)].
(iii) We want one mole of CH$_4$(g) as product, so we write down equation (3).
(iv) We do the same operations on each ΔH^0 value.
(v) Then we add these equations term by term. The result is the desired thermochemical equation, with all unwanted substances cancelling. The sum of the ΔH^0 values is the ΔH^0 for the desired reaction.

We have used a series of reactions for which ΔH^0 values can be easily measured to calculate ΔH^0 for a reaction that cannot be carried out.

Solution

$$
\begin{array}{lll}
 & \Delta H^0 & \\
\text{C(graphite)} + \cancel{\text{O}_2(g)} \longrightarrow \cancel{\text{CO}_2(g)} & -393.5 \text{ kJ/mol rxn} & (1) \\
2\text{H}_2(g) + \cancel{\text{O}_2(g)} \longrightarrow \cancel{2\text{H}_2\text{O}(\ell)} & 2(-285.8 \text{ kJ/mol rxn}) & 2 \times (2) \\
\cancel{\text{CO}_2(g)} + \cancel{2\text{H}_2\text{O}(\ell)} \longrightarrow \text{CH}_4(g) + \cancel{2\text{O}_2(g)} & +890.3 \text{ kJ/mol rxn} & (3)
\end{array}
$$

$$\text{C(graphite)} + 2\text{H}_2(g) \longrightarrow \text{CH}_4(g) \qquad \boxed{\Delta H^0_{rxn} = -74.8 \text{ kJ}}$$

Example 15-9

Given the following thermochemical equations, calculate the heat of reaction at 298 K for the reaction of ethylene with water to form ethanol:

$$\text{C}_2\text{H}_4(g) + \text{H}_2\text{O}(\ell) \longrightarrow \text{C}_2\text{H}_5\text{OH}(\ell)$$

$$
\begin{array}{lll}
 & \Delta H^0 & \\
\text{C}_2\text{H}_5\text{OH}(\ell) + 3\text{O}_2(g) \longrightarrow 2\text{CO}_2(g) + 3\text{H}_2\text{O}(\ell) & -1367 \text{ kJ/mol rxn} & (1) \\
\text{C}_2\text{H}_4(g) + 3\text{O}_2(g) \longrightarrow 2\text{CO}_2(g) + 2\text{H}_2\text{O}(\ell) & -1411 \text{ kJ/mol rxn} & (2)
\end{array}
$$

Plan

We reverse equation (1) to give (−1); when the equation is reversed, the sign of ΔH^0 is changed because the reverse of an exothermic reaction is endothermic. Then we add it to equation (2).

Solution

$$\begin{array}{lr}
& \Delta H° \\
2\cancel{CO_2}(g) + 3H_2O(\ell) \longrightarrow C_2H_5OH(\ell) + \cancel{3O_2(g)} & +1367 \text{ kJ/mol rxn} \quad (-1) \\
C_2H_4(g) + \cancel{3O_2(g)} \longrightarrow \cancel{2CO_2(g)} + 2H_2O(\ell) & -1411 \text{ kJ/mol rxn} \quad (2) \\
\hline
C_2H_4(g) + H_2O(\ell) \longrightarrow C_2H_5OH(\ell) & \boxed{\Delta H^0_{rxn} = -44 \text{ kJ/mol rxn}}
\end{array}$$

EOC 42, 44

The ΔH^0 for the reaction in Example 15-9 is -44 kJ for each mole of $C_2H_5OH(\ell)$ formed. However, this reaction does not involve formation of $C_2H_5OH(\ell)$ from its constituent elements. ΔH^0_{rxn} is *not* ΔH^0_f for $C_2H_5OH(\ell)$.

Another interpretation of Hess' Law lets us use tables of ΔH^0_f values to calculate the enthalpy change accompanying a reaction. Let us illustrate this approach by reconsidering the reaction of Example 15-9:

$$C_2H_4(g) + H_2O(\ell) \longrightarrow C_2H_5OH(\ell)$$

A table of ΔH^0_f values (Appendix K) gives $\Delta H^0_{f\,C_2H_5OH(\ell)} = -277.7$ kJ/mol, $\Delta H^0_{f\,C_5H_4(g)} = 52.3$ kJ/mol, and $\Delta H^0_{f\,H_2O(\ell)} = -285.8$ kJ/mol. We may express this information in the form of the following thermochemical equations:

$$\begin{array}{lll}
& & \Delta H° = \Delta H^0_f \\
C_2H_5OH(\ell): & 2C(\text{graphite}) + 3H_2(g) + \tfrac{1}{2}O_2(g) \longrightarrow C_2H_5OH(\ell) & -277.7 \text{ kJ/mol rxn} \quad (1) \\
C_2H_4(g): & 2C(\text{graphite}) + 2H_2(g) \longrightarrow C_2H_4(g) & 52.3 \text{ kJ/mol rxn} \quad (2) \\
H_2O(\ell): & H_2(g) + \tfrac{1}{2}O_2(g) \longrightarrow H_2O(\ell) & -285.8 \text{ kJ/mol rxn} \quad (3)
\end{array}$$

We may generate the equation for the desired net reaction by adding equation (1) to the reverse of equations (2) and (3). The value of ΔH^0 for the desired reaction is then simply the sum of the corresponding ΔH^0 values:

$$\begin{array}{lr}
& \Delta H° \\
2C(\text{graphite}) + 3H_2(g) + \tfrac{1}{2}O_2(g) \longrightarrow C_2H_5OH(\ell) & -277.7 \text{ kJ/mol rxn} \quad (1) \\
C_2H_4(g) \longrightarrow 2C(\text{graphite}) + 2H_2(g) & -52.3 \text{ kJ/mol rxn} \quad (-2) \\
H_2O(\ell) \longrightarrow H_2(g) + \tfrac{1}{2}O_2(g) & +285.8 \text{ kJ/mol rxn} \quad (-3) \\
\hline
\text{net rxn:} \quad C_2H_4(g) + H_2O(\ell) \longrightarrow C_2H_5OH(\ell) & \Delta H^0_{rxn} = -44.2 \text{ kJ/mol rxn}
\end{array}$$

We see that ΔH^0 for this reaction is given by

$$\Delta H^0_{rxn} = \Delta H^0_{(1)} + \Delta H^0_{(-2)} + \Delta H^0_{(-3)}$$

Or by

$$\Delta H^0_{rxn} = \underset{\text{product}}{\Delta H^0_{f\,C_2H_5OH(\ell)}} - [\underset{\text{reactants}}{\Delta H^0_{f\,C_2H_4(g)} + \Delta H^0_{f\,H_2O(\ell)}}]$$

In general terms, this is a very useful form of Hess' Law:

$$\Delta H^0_{rxn} = \Sigma\, n\Delta H^0_{f\,\text{products}} - \Sigma\, n\Delta H^0_{f\,\text{reactants}}$$

The standard enthalpy change of a reaction is equal to the sum of the standard molar enthalpies of formation of the products, each multiplied by its coefficient, n, in *the balanced equation*, minus the corresponding sum of the standard molar enthalpies of formation of the reactants.

The capital Greek letter sigma (Σ) is read "the sum of." The $\Sigma\, n$ means that the ΔH^0_f value of each product and reactant must be multiplied by its coefficient, n, in the balanced equation. The resulting values are then added.

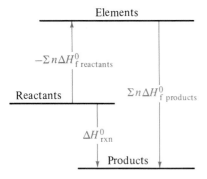

Figure 15-5
A schematic representation of Hess' Law. The red arrow represents the *direct* path from reactants to products. The series of blue arrows is a path (hypothetical) in which reactants are converted to elements, and they in turn are converted to products—all in their standard states.

$O_2(g)$ is an element in its standard state, so its ΔH_f^0 is zero.

In effect, this form of Hess' Law supposes that the reaction occurs by converting reactants to the elements in their standard states, then converting these to products (see Figure 15-5). Almost no reactions actually occur by such an extreme pathway. Nevertheless, the ΔH^0 for this *hypothetical* path for *reactants* → *products* would be the same as that for any other pathway—including the one by which the reaction actually occurs.

Example 15-10

Calculate ΔH_{rxn}^0 for the following reaction at 298 K.

$$SiH_4(g) + 2O_2(g) \longrightarrow SiO_2(s) + 2H_2O(\ell)$$

Plan

We apply Hess' Law in the form $\Delta H_{rxn}^0 = \Sigma\, n\Delta H_{f\,products}^0 - \Sigma\, n\Delta H_{f\,reactants}^0$, using the ΔH_f^0 values tabulated in Appendix K.

Solution

$$\Delta H_{rxn}^0 = \Sigma\, n\Delta H_{f\,products}^0 - \Sigma\, n\Delta H_{f\,reactants}^0$$

$$\Delta H_{rxn}^0 = [\Delta H_{f\,SiO_2(s)}^0 + 2\Delta H_{f\,H_2O(\ell)}^0] - [\Delta H_{f\,SiH_4(g)}^0 + 2\Delta H_{f\,O_2(g)}^0]$$

$$\Delta H_{rxn}^0 = \left[\frac{1\text{ mol } SiO_2(s)}{\text{mol rxn}} \times \frac{-910.9\text{ kJ}}{\text{mol } SiO_2(s)} + \frac{2\text{ mol } H_2O(\ell)}{\text{mol rxn}} \times \frac{-285.8\text{ kJ}}{\text{mol } H_2O(\ell)}\right]$$
$$- \left[\frac{1\text{ mol } SiH_4(g)}{\text{mol rxn}} \times \frac{+34\text{ kJ}}{\text{mol } SiH_4(g)} + \frac{2\text{ mol } O_2(g)}{\text{mol rxn}} \times \frac{0\text{ kJ}}{\text{mol } O_2(g)}\right]$$

$$\Delta H_{rxn}^0 = \boxed{-1516\text{ kJ/mol rxn}}$$

EOC 46, 47

Each term in the sums on the right-hand side of the solution in Example 15-10 has the units

$$\frac{\text{mol substance}}{\text{mol rxn}} \times \frac{\text{kJ}}{\text{mol substance}} \quad \text{or} \quad \frac{\text{kJ}}{\text{mol rxn}}$$

> For brevity, we shall omit units in the intermediate steps of subsequent calculations of this type, and simply assign the proper units to the answer. Be sure that you understand how these units arise.

Suppose we measure ΔH_{rxn}^0 at 298 K and know all but one of the ΔH_f^0 values for reactants and products. We can then calculate the unknown ΔH_f^0 value.

We will consult Appendix K only after working the problem, to check the answer.

Example 15-11

Use the following information to determine ΔH_f^0 for PbO(s, yellow).

$$PbO(s, yellow) + CO(g) \longrightarrow Pb(s) + CO_2(g) \qquad \Delta H_{rxn}^0 = -65.69\text{ kJ}$$

$$\Delta H_f^0 \text{ for } CO_2(g) = -393.5\text{ kJ/mol} \quad \text{and} \quad \Delta H_f^0 \text{ for } CO(g) = -110.5\text{ kJ/mol}$$

Plan

We again use Hess' Law in the form $\Delta H^0_{rxn} = \Sigma\, n\Delta H^0_{f\,products} - \Sigma\, n\Delta H^0_{f\,reactants}$. Now we are given ΔH^0_{rxn} and the ΔH^0_f values for all substances *except* PbO(s, yellow). We can solve for this unknown.

Solution

$$\Delta H^0_{rxn} = \Sigma\, n\Delta H^0_{f\,products} \qquad\qquad - \Sigma\, n\Delta H^0_{f\,reactants}$$

$$\Delta H^0_{rxn} = \Delta H^0_{f\,Pb(s)} + \Delta H^0_{f\,CO_2(g)} - [\Delta H^0_{f\,PbO(s,\,yellow)} + \Delta H^0_{f\,CO(g)}]$$

$$-65.69 = 0 + (-393.5) \qquad - [\Delta H^0_{f\,PbO(s,\,yellow)} + (-110.5)]$$

Rearranging to solve for $\Delta H^0_{f\,PbO(s,\,yellow)}$, we have

$$\Delta H^0_{f\,PbO(s,\,yellow)} = 65.69 - 393.5 + 110.5 = \boxed{-217.3 \text{ kJ/mol of PbO}}$$

EOC 52, 53

In the next two sections we shall study two important applications of Hess' Law.

15-9 Bond Energies

Chemical reactions involve the breaking and making of chemical bonds. Energy is always required to break a chemical bond.

> The **bond energy (B.E.)** is the amount of energy necessary to break *one mole* of bonds in a gaseous covalent substance to form products in the gaseous state.

Consider the following reaction:

$$H_2(g) \longrightarrow 2H(g) \qquad \Delta H^0_{rxn} = \Delta H_{H-H} = +435 \text{ kJ/mol H—H bonds}$$

The bond energy of the hydrogen–hydrogen bond is 435 kJ/mol of bonds. This endothermic reaction (ΔH^0_{rxn} is positive) could be written

$$H_2(g) + 435 \text{ kJ} \longrightarrow 2H(g)$$

Some average bond energies are listed in Tables 15-2 and 15-3.

For all practical purposes, the bond energy is the same as bond enthalpy. Tabulated values of average bond energies are actually average bond enthalpies. We use the term "bond *energy*" rather than "bond *enthalpy*" because it is common practice to do so.

Table 15-2
Some Average Single Bond Energies in kJ/mol of Bonds

H	C	N	O	F	Si	P	S	Cl	Br	I	
435	414	389	464	569	293	318	339	431	368	297	H
	347	293	351	439	289	264	259	330	276	238	C
		159	201	272	—	209	—	201	243?	—	N
			138	184	368	351	—	205	—	201	O
				159	540	490	327	255	197?	—	F
					176	213	226	360	289	213	Si
						213	230	331	272	213	P
							213	251	213	—	S
								243	218	209	Cl
									192	180	Br
										151	I

Bond energies for double and triple bonds are *not* simply two or three times those for the corresponding single bonds. A single bond is a σ bond, whereas double and triple bonds involve a combination of σ and π bonding. The bond energy measures the difficulty of overcoming the orbital overlap, and we should not expect the stability of a π bond to be the same as that of a σ bond between the same two atoms.

Table 15-3
Some Average Multiple Bond Energies in kJ/mol of Bonds

N=N	418	C=C	611	O=O	498
N≡N	946	C≡C	837		
C=N	615	C=O	741		
C≡N	891	C≡O	1070		

Let us consider more complex reactions. For example,

$$CH_4(g) \longrightarrow C(g) + 4H(g) \qquad \Delta H^0_{rxn} = 1.66 \times 10^3 \text{ kJ/mol rxn}$$

The four hydrogen atoms are identical, so all the C—H bonds are identical in bond length and energy *in methane molecules*. However, the energies required to break the individual C—H bonds differ for successively broken bonds, as shown below.

The hydrogen atoms are indistinguishable, so it makes no difference which one is removed first.

$CH_4(g)$	$\longrightarrow$	$CH_3(g)$	+	$H(g)$	$\Delta H^0 = +\ 427$ kJ/mol rxn
$CH_3(g)$	$\longrightarrow$	$CH_2(g)$	+	$H(g)$	$\Delta H^0 = +\ 439$ kJ/mol rxn
$CH_2(g)$	$\longrightarrow$	$CH(g)$	+	$H(g)$	$\Delta H^0 = +\ 452$ kJ/mol rxn
$CH(g)$	$\longrightarrow$	$C(g)$	+	$H(g)$	$\Delta H^0 = +\ 347$ kJ/mol rxn
$CH_4(g)$	$\longrightarrow$	$C(g)$	+	$4H(g)$	$\Delta H^0 = +\ 1665$ kJ/mol rxn

$$\text{"average" C—H bond energy in } CH_4 = \Delta H^0/4 \text{ mol bonds}$$
$$= 416 \text{ kJ/mol bonds}$$

We see that the *average C—H bond energy in methane* is 416 kJ/mol of bonds. No mole of single C—H bonds is actually broken by absorption of exactly that amount of energy. Average C—H bond energies differ slightly from compound to compound, as in CH_4, CH_3Cl, CH_3NO_2, and so on. Nevertheless, they are sufficiently constant to be useful, as illustrated below.

A special case of Hess' Law involves the use of bond energies to *estimate* heats of reaction. Consider the enthalpy diagrams in Figure 15-6. In general terms, ΔH^0_{rxn} is related to the bond energies of the reactants and products *in gas phase reactions* by the following version of Hess' Law:

Note that this equation involves bond energies of *reactants* minus bond energies of *products*.

$$\Delta H^0_{rxn} = \Sigma \text{ B.E.}_{reactants} - \Sigma \text{ B.E.}_{products} \qquad \text{in gas phase reactions only}$$

This relationship may allow us to estimate ΔH^0 values for gas phase reactions when the required ΔH^0_f values are not available.

The net enthalpy change of a reaction is the amount of energy required to break all the bonds in reactant molecules *minus* the amount of energy required to break all the bonds in product molecules. Stated in another way, the amount of energy released when a bond is formed is equal to the amount absorbed when the same bond is broken. The heat of reaction for a gas phase reaction can be described as the amount of energy released in forming all the bonds in the products minus the amount of energy released in forming all the bonds in the reactants (Figure 15-6).

The definition of bond energies is limited to the bond-breaking process *only*, and does not include any provision for changes of state. Thus, it is valid only in the gaseous state. Therefore, the calculations of this section apply *only* when all substances in the reaction are gases. If liquids or solids

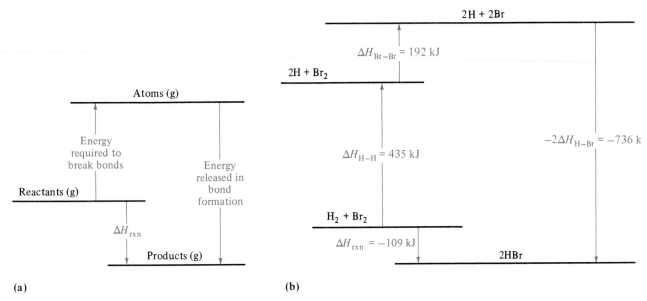

Figure 15-6
A schematic representation of the relationship between bond energies and ΔH_{rxn} for gas phase reactions. (a) For a general reaction (exothermic). (b) For the gas phase reaction

$$H_2(g) + Br_2(g) \longrightarrow 2HBr(g)$$

As usual for such diagrams, the value shown for each change refers to the number of moles of substances or bonds indicated.

were involved, then additional information such as heats of vaporization, fusion, and so on would be needed to account for phase changes.

Example 15-12

Use the bond energies listed in Table 15-2 to estimate the heat of reaction at 298 K for the following reaction. All bonds are single bonds.

$$Br_2(g) + 3F_2(g) \longrightarrow 2BrF_3(g)$$

Plan

Each BrF_3 molecule contains three Br—F bonds, so two moles of BrF_3 contain six moles of Br—F bonds. Three moles of F_2 contain a total of three moles of F—F bonds, and one mole of Br_2 contains one mole of Br—Br bonds. Using the bond energy form of Hess' Law,

Solution

$$\Delta H^0_{rxn} = [\Delta H_{Br—Br} + 3\Delta H_{F—F}] - [6\Delta H_{Br—F}]$$

$$= 192 + 3(159) - 6(197) = \boxed{-513 \text{ kJ/mol rxn}}$$

For each term in the sum, the units are

$$\frac{\text{mol bonds}}{\text{mol rxn}} \times \frac{\text{kJ}}{\text{mol bonds}}$$

EOC 58, 59

Example 15-13

Use the bond energies listed in Table 15-2 to estimate the heat of reaction at 298 K for the following reaction.

$$C_3H_8(g) \quad + \ Cl_2(g) \ \longrightarrow \quad C_3H_7Cl(g) \quad + \ HCl(g)$$

We would get the same value for ΔH^0_{rxn} if we used the full bond energy form of Hess' Law and assumed that *all* bonds in reactants were broken and then *all* bonds in products were formed. In such a calculation, the bond energies for the unchanged bonds would cancel. Why? Try it!

Plan

Two moles of C—C bonds and seven moles of C—H bonds are the same before and after reaction, so we do not need to include them in the bond energy calculation. The only reactant bonds that are broken are one mole of C—H bonds and one mole of Cl—Cl bonds. On the product side, the only new bonds formed are one mole of C—Cl bonds and one mole of H—Cl bonds. We need to take into account only the bonds that are different on the two sides of the equation. As before, we add and subtract the appropriate bond energies, using values from Table 15-2.

Solution

$$\Delta H^0_{rxn} = [\Delta H_{C-H} + \Delta H_{Cl-Cl}] - [\Delta H_{C-Cl} + \Delta H_{H-Cl}]$$

$$= [414 + 243] - [330 + 431] = \boxed{-104 \text{ kJ/mol rxn}}$$

EOC 60, 61

If we measure ΔH^0_{rxn} for a gas phase reaction and know all but one of the bond energies, we can calculate the missing bond energy.

Example 15-14

Given the following equation and average bond energies,

$$C_3H_8(g) + 5O_2(g) \longrightarrow 3CO_2(g) + 4H_2O(g) \qquad \Delta H^0_{rxn} = -2.05 \times 10^3 \text{ kJ/mol rxn}$$

Bond	Energy	Bond	Energy
C—C	347 kJ/mol of bonds	C=O	741 kJ/mol of bonds
C—H	414 kJ/mol of bonds	O—H	464 kJ/mol of bonds

Without consulting Table 15-3, estimate the energy of the oxygen–oxygen bond in O_2 molecules. Compare this with the value listed in Table 15-3.

Plan

There are eight moles of C—H bonds and two moles of C—C bonds per mole of C_3H_8. As before, we add and subtract the appropriate bond energies, and then rearrange for the single unknown, $\Delta H_{O=O}$.

Solution

$$\Delta H^0_{rxn} = [2\Delta H_{C-C} + 8\Delta H_{C-H} + 5\Delta H_{O=O}] - [6\Delta H_{C=O} + 8\Delta H_{O-H}]$$

$$-2.05 \times 10^3 = [2(347) + 8(414) + 5\Delta H_{O=O}] - [6(741) + 8(464)]$$

Rearranging, we obtain

$$-5\Delta H_{O=O} = 694 + 3.31 \times 10^3 - 4.45 \times 10^3 - 3.71 \times 10^3 + 2.05 \times 10^3$$

$$5\Delta H_{O=O} = 2.11 \times 10^3$$

$$\Delta H_{O=O} = \boxed{422 \text{ kJ per mol of O=O bonds}}$$

Table 15-3 gives 498 kJ. The calculated value is in error by about 15%. It is a reasonable estimate of the O=O bond energy, but it does point out the limitations of using *average* bond energies.

EOC 64, 66

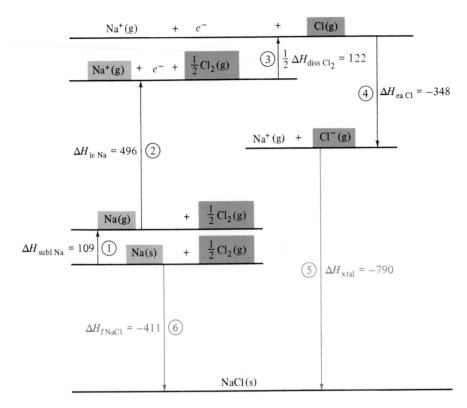

Figure 15-7
The Born–Haber cycle for NaCl(s). Each ΔH value is given in kilojoules for the number of moles of reaction involved in that step. The red arrow represents the crystal lattice energy, the quantity for which we usually solve in the Born–Haber cycle. The blue arrow represents the standard enthalpy of formation of the solid ionic crystal from the elements. The circled numbers correspond to the numbered steps in the discussion in the text.

15-10 The Born–Haber Cycle

Covalent substances exist as discrete molecules, but solid ionic substances exist as arrays of alternating positive and negative ions. The amount of energy holding a crystal together is called the **crystal lattice energy, ΔH_{xtal}**. For an ionic substance, it is essentially the enthalpy change for the reaction in which a mole of the ionic substance is formed from its gaseous ions. For example, for sodium chloride,

$$Na^+(g) + Cl^-(g) \longrightarrow NaCl(s) \qquad \Delta H_{rxn} = \Delta H_{xtal}$$

Because energy is always released in such processes, crystal lattice energies have negative values.

Crystal lattice energies *cannot* be measured directly. However, we can evaluate ΔH_{xtal} by applying Hess' Law to a series of reactions, beginning with the elements in their standard states and ending with the ionic compound in its standard state. Such a sequence of reactions for ionic substances is called a **Born–Haber cycle** after Max Born and Fritz Haber, who developed the general procedure. The Born–Haber cycle for NaCl is shown in Figure 15-7.

The enthalpy of formation for NaCl is known (Appendix K).

$$Na(s) + \tfrac{1}{2}Cl_2(g) \longrightarrow NaCl(s) \qquad \Delta H_{rxn} = \Delta H^0_{f\,NaCl(s)} = -411 \text{ kJ/mol}$$

The following equations and experimental data can be summed to give the same equation. The ΔH for each change refers to the number of moles of substances indicated.

We can use this approach to calculate any one of the ΔH values, provided we know the others.

Sublimation is the process by which a solid is converted directly to a gas (Section 13-12).

1. *Sublimation of Na metal.* The enthalpy change is ΔH_f^0 for Na(g) = ΔH_{subl}.

$$Na(s) \longrightarrow Na(g) \qquad \Delta H_{subl} = 109 \text{ kJ/mol}$$

2. *Ionization of one mole of Na atoms.* The enthalpy change is essentially the first ionization energy for Na, expressed in kJ/mol.

$$Na(g) \longrightarrow Na^+(g) + e^- \qquad \Delta H_{ie} = 496 \text{ kJ/mol}$$

3. *Dissociation of one-half mole of $Cl_2(g)$ into gaseous Cl atoms.* The enthalpy change is ΔH_f^0 for Cl(g) = $\frac{1}{2}\Delta H_{diss}$ [also half the bond energy for $Cl_2(g)$].

$$\tfrac{1}{2}Cl_2(g) \longrightarrow Cl(g) \qquad \tfrac{1}{2}\Delta H_{diss} = 122 \text{ kJ/mol}$$

4. *Addition of one mole of electrons to one mole of gaseous chlorine atoms.* This enthalpy change is essentially the electron affinity of Cl, expressed in kilojoules.

$$Cl(g) + e^- \longrightarrow Cl^-(g) \qquad \Delta H_{ea} = -348 \text{ kJ/mol}$$

5. *Condensation of the gaseous ions to form a mole of solid NaCl.* This enthalpy change cannot be measured directly.

$$Na^+(g) + Cl^-(g) \longrightarrow NaCl(s) \qquad \Delta H_{xtal} = ?$$

Summing these five reactions and their ΔH values gives $\Delta H_{f \, NaCl(s)}$:

We want to solve for the ΔH for Step 5.

1. Na(s)	$\longrightarrow$	Na(g)	ΔH_{subl}	$=$	109 kJ/mol
2. Na(g)	$\longrightarrow$	Na$^+$(g) + e^-	ΔH_{ie}	$=$	496 kJ/mol
3. $\tfrac{1}{2}Cl_2(g)$	$\longrightarrow$	Cl(g)	$\tfrac{1}{2}\Delta H_{diss}$	$=$	122 kJ/mol
4. Cl(g) + e^-	$\longrightarrow$	Cl$^-$(g)	ΔH_{ea}	$=$	-348 kJ/mol
5. Na$^+$(g) + Cl$^-$(g)	$\longrightarrow$	NaCl(s)	ΔH_{xtal}	$=$	?
6. Na(s) + $\tfrac{1}{2}Cl_2(g)$	$\longrightarrow$	NaCl(s)	$\Delta H_{f \, NaCl(s)}$	$=$	-411 kJ

$$\Delta H_{f \, NaCl(s)} = \Delta H_{subl} + \Delta H_{ie} + \tfrac{1}{2}\Delta H_{diss} + \Delta H_{ea} + \Delta H_{xtal}$$

$$-411 = 109 \quad + 496 \quad + 122 \quad + (-348) + \Delta H_{xtal}$$

Solving for ΔH_{xtal}, we get

$$\Delta H_{xtal} = (-411 - 109 - 496 - 122 + 348)$$

$$= -790 \text{ kJ per mole of NaCl(s)}$$

(Review Figure 15-7.)

The crystal lattice energy is a measure of the stability of an ionic solid. The more negative its value, the more energy would be required to separate a mole of the ionic solid into its isolated gaseous ions. The large negative crystal lattice energy, -790 kJ per mol, indicates that NaCl is a very stable solid.

Remember that crystal lattice energies cannot be measured directly. They can be calculated from other known enthalpy values.

Spontaneity of Physical and Chemical Changes

Another major concern of thermodynamics is predicting *whether* a particular process can occur under specified conditions. We may summarize this concern in the question "Which would be more stable at the given conditions—

the collection of reactants or the collection of products?" A change for which the collection of products is thermodynamically *more stable* than the collection of reactants under the given conditions is said to be **spontaneous** under those conditions. A change for which the products are thermodynamically *less stable* than the reactants under the given conditions is described as **nonspontaneous** under those conditions. Some changes are spontaneous under all conditions; others are nonspontaneous under all conditions. The great majority of changes, however, are spontaneous under some conditions but not under others. One goal of thermodynamics is to predict conditions for which the latter type of reactions can occur.

The concept of spontaneity has a very specific interpretation in thermodynamics. A spontaneous chemical reaction or physical change is one that can happen without any continuing outside influence. Any spontaneous change has a natural direction, like the rusting of a piece of iron, the burning of a piece of paper, or the melting of ice at room temperature. We can think of a spontaneous process as one for which products are favored over reactants at the specified conditions. Although a spontaneous reaction *might* occur rapidly, thermodynamic spontaneity is unrelated to speed. The fact that a process is spontaneous does not even mean that it will occur at an observable rate. It may occur rapidly, at a moderate rate, or very slowly. The rate at which a spontaneous reaction occurs is addressed by kinetics (Chapter 16). In the remainder of this chapter, we shall study the factors that influence spontaneity of a physical or chemical change.

15-11 The Two Parts of Spontaneity

Many spontaneous reactions are exothermic. For instance, the combustion (burning) reactions of hydrocarbons such as methane and octane are all exothermic and spontaneous. The enthalpy contents of the products are lower than those of the reactants. However, not all exothermic changes are spontaneous, nor are all spontaneous changes exothermic. As an example, consider the freezing of water, which is an exothermic process (heat is released). This process is spontaneous at temperatures below 0°C, but it certainly is not spontaneous at temperatures above 0°C. Likewise, we can find conditions at which the melting of ice, an endothermic process, is spontaneous. When heat is released during a chemical reaction or a physical change, spontaneity is *favored* but not required.

Another factor, related to the disorder of reactants and products, also plays a role in determining spontaneity. The dissolution of ammonium nitrate, NH_4NO_3, in water is spontaneous. Yet a beaker in which this process occurs becomes colder (see Figure 15-2). In this case, the system (consisting of the water, the solid NH_4NO_3, and the resulting hydrated NH_4^+ and NO_3^- ions) absorbs heat from the surroundings as the endothermic process occurs. Nevertheless, the process is spontaneous because the system increases in disorder as the regularly arranged ions of crystalline ammonium nitrate become more randomly distributed hydrated ions in solution (Figure 15-8). An increase in disorder in the system favors the spontaneity of a reaction (Section 15-13). In this particular case, the increase in disorder overrides the effect of endothermicity. The balance of these two effects is considered in Section 15-14.

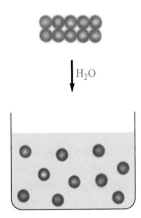

Figure 15-8
As particles leave a crystal to go into solution, they become more disordered. This increase in disorder favors the dissolution of the crystal.

Suppose we shake a beaker containing marbles. A disordered arrangement (left) is more likely than an ordered arrangement (right) in which all marbles of the same color remain together.

15-12 The Second Law of Thermodynamics

We have stated that two factors determine whether a reaction is spontaneous under a given set of conditions. The effect of the first factor, the enthalpy change, is that spontaneity is favored (but not required) by exothermicity, and nonspontaneity is favored by endothermicity. The effect of the other factor is summarized in the **Second Law of Thermodynamics**:

> In spontaneous changes the universe tends toward a state of greater disorder.

The Second Law is based on our experiences. Some examples illustrate this law in the macroscopic world. When a mirror is dropped, it can shatter. When a drop of food coloring is added to a glass of water, it diffuses until a homogeneously colored solution results. When a truck is driven down the street, it consumes fuel and oxygen, producing carbon dioxide, water vapor, and other emitted substances.

The reverse of any spontaneous change is nonspontaneous, because if it did occur, the universe would tend toward a state of greater order. This is contrary to our experience. We would be very surprised if we dropped some pieces of silvered glass on the floor and a mirror spontaneously assembled. A truck cannot be driven along the street, even in reverse gear, so that it sucks up CO_2, water vapor, and other substances and produces fuel and oxygen.

15-13 Entropy, S

The thermodynamic state function **entropy**, S, is a measure of the disorder of the system. The greater the disorder of a system, the higher its entropy. For any particular substance, the particles in the solid state are more highly ordered than those in the liquid state. These, in turn, are more highly ordered than those in the gaseous state. Thus, the entropy of a given substance increases as it goes from solid to liquid to gas (Figure 15-9).

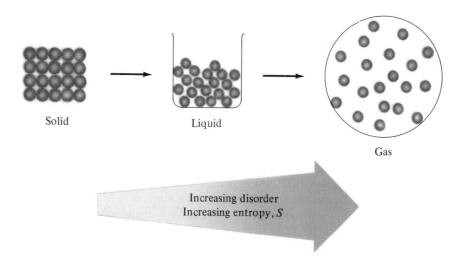

Solid   Liquid

Gas

Increasing disorder
Increasing entropy, S

Figure 15-9
As a sample changes from solid to liquid to gas, its particles become increasingly less ordered, and its entropy increases.

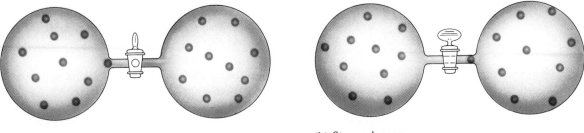

(a) Stopcock closed   (b) Stopcock open

Figure 15-10
(a) A sample of gas in which all molecules of one gas are in one bulb and all molecules of the other gas are in the other bulb. (b) A sample of gas that contains the same number of each kind of molecule as in (a), but with the two kinds randomly mixed in the two bulbs. Sample (b) has greater disorder (higher entropy), and is thus more probable.

If the entropy of a system increases during a process, the spontaneity of the process is favored (but not required). The second law says that the entropy of the *universe* (not the system) increases during a spontaneous process. Thus, for a spontaneous process,

$$\Delta S_{universe} = \Delta S_{system} + \Delta S_{surroundings} > 0$$

Of the two ideal gas samples in Figure 15-10, the more ordered arrangement (Figure 15-10a) has lower entropy than the randomly mixed arrangement with the same volume (Figure 15-10b). Because these ideal gas samples mix without absorbing or releasing heat and without a change in total volume, they do not interact with the surroundings, so the entropy of the surroundings does not change. In this case,

$$\Delta S_{universe} = \Delta S_{system}$$

If we open the stopcock between the two bulbs in Figure 15-10a, we expect the gases to mix spontaneously, with an increase in the disorder of the system, that is, ΔS_{system} is positive.

unmixed gases $\longrightarrow$ mixed gases $\Delta S_{universe} = \Delta S_{system} > 0$

We do not expect the more homogeneous sample in Figure 15-10b to spontaneously "unmix" to give the arrangement in Figure 15-10a (which would correspond to a decrease in ΔS_{system}).

mixed gases $\longrightarrow$ unmixed gases $\Delta S_{universe} = \Delta S_{system} < 0$

The ideas of entropy, order, and disorder are related to probability. The more ways an event can happen, the more probable it is. In Figure 15-10b, each individual red molecule is equally likely to be in either container, as is each individual blue molecule. As a result, there are many ways in which the mixed arrangement of Figure 15-10b can occur, so the probability of its occurrence is high, and so its entropy is high. In contrast, there is only one way the unmixed arrangement in Figure 15-10a can occur. The resulting probability is extremely low, and the entropy of this arrangement is low.

The entropy of a system can decrease during a spontaneous process or increase during a nonspontaneous process, depending on the accompanying $\Delta S_{surroundings}$. If ΔS_{sys} is negative (decrease in disorder), then ΔS_{univ} may still be positive (overall increase in disorder) if ΔS_{surr} is more positive than ΔS_{sys} is negative. A refrigerator provides a good illustration. It removes heat from inside the box (the system) and ejects that heat, *plus* the heat generated by the compressor, into the room (the surroundings). The entropy of the system decreases because the air molecules inside the box move more slowly. How-

We shall abbreviate these subscripts as follows: system = sys, surroundings = surr, and universe = univ.

ever, the increase in the entropy of the surroundings more than makes up for that, so the entropy of the universe (refrigerator + room) increases.

Similarly, if ΔS_{sys} is positive and ΔS_{surr} is even more negative, then ΔS_{univ} is still negative. Such a process will be nonspontaneous.

Similar arguments apply for condensing a gas at its condensation point (boiling point).

Let's consider the entropy changes that occur when a liquid solidifies at a temperature *below* its freezing (melting) point (Figure 15-11). ΔS_{sys} is negative because a solid forms from its liquid, yet we know that this is a spontaneous process. A liquid releases heat to its surroundings (atmosphere) as it crystallizes. The released heat increases the motion (disorder) of the molecules of the surroundings, so ΔS_{surr} is positive. As the temperature decreases, the ΔS_{surr} contribution becomes more important. When the temperature is low enough (below the freezing point), the positive ΔS_{surr} outweighs the negative ΔS_{sys}. At that point, ΔS_{univ} becomes positive and the freezing process becomes spontaneous.

The situation is reversed when a liquid is boiled or a solid is melted (Figure 15-11b). For example, at temperatures above its melting point, a solid spontaneously melts, and ΔS_{sys} is positive. The heat absorbed when the solid (system) melts comes from its surroundings. This decreases the motion of the molecules of the surroundings. Thus, ΔS_{surr} is negative (the surroundings become less disordered). However, the positive ΔS_{sys} is greater in magnitude that the negative ΔS_{surr}, so ΔS_{univ} is positive and the process is spontaneous.

Can you develop a comparable table for boiling (liquid → gas) and condensation (gas → liquid)? (See Table 15-4).

Above the melting point, ΔS_{univ} is positive for melting. Below the melting point, ΔS_{univ} is positive for freezing. At the melting point, ΔS_{surr} is equal in magnitude and opposite in sign to ΔS_{sys}. Then ΔS_{univ} is zero for both melting and freezing; the system is at *equilibrium*. Table 15-4 lists the entropy effects for these changes of physical state.

We have said that ΔS_{univ} is positive for all spontaneous processes. Unfortunately, it is not possible to make direct measurements of ΔS_{univ}. Con-

(a) Freezing below mp	(b) Melting above mp
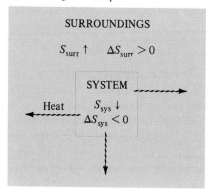	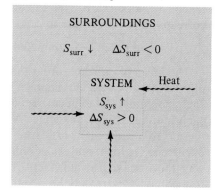

Below mp, S_{surr} increases more than S_{sys} decreases, so $\Delta S_{univ} > 0$, and the process is spontaneous.

Above mp, S_{sys} increases more than S_{surr} decreases, so $\Delta S_{univ} > 0$, and the process is spontaneous.

Figure 15-11
A schematic representation of heat flow and entropy changes for (a) freezing and (b) melting of a pure substance.

Above mp, S_{surr} is already high, so it cannot increase enough to compensate for the decrease in S_{sys}, and the process is not spontaneous.

Below mp, too little heat flows into the system, so S_{sys} cannot increase enough to make the process spontaneous.

Table 15-4
Entropy Effects Associated with Melting and Freezing

Change	Temperature	Sign of ΔS_{sys}	Sign of ΔS_{surr}	(Magnitude of ΔS_{sys}) Compared with (Magnitude of ΔS_{surr})	$\Delta S_{univ} = \Delta S_{sys} + \Delta S_{surr}$	Spontaneity
1. Melting (solid → liquid)	(a) >mp	+	−	>	>0	Spontaneous
	(b) =mp	+	−	=	=0	Equilibrium
	(c) <mp	+	−	<	<0	Nonspontaneous
2. Freezing (liquid → solid)	(a) >mp	−	+	>	<0	Nonspontaneous
	(b) =mp	−	+	=	=0	Equilibrium
	(c) <mp	−	+	<	>0	Spontaneous

sequently, entropy changes accompanying physical and chemical changes are reported in terms of ΔS_{sys}. The subscript "system" is usually omitted. The symbol ΔS refers to the change in the entropy of the reacting system, just as ΔH refers to the change in enthalpy of the reacting system.

The **Third Law of Thermodynamics** establishes the zero of the entropy scale:

> The entropy of a pure, perfect crystalline substance (perfectly ordered) is zero at absolute zero (0 K).

As the temperature of a substance increases, the particles vibrate more vigorously, so the entropy increases (Figure 15-12). Further heat input causes either increased temperature (still higher entropy) or phase transitions (melting, sublimation, or boiling) that also result in higher entropy. The entropy of a substance at any condition is its **absolute entropy**, also called standard molar entropy. Consider as examples the absolute entropies at 298 K listed

Enthalpies are measured only as *differences* with respect to an arbitrary standard state. Entropies, in contrast, are defined relative to an absolute zero level. In either case, the *per mole* designation means *per mole of substance in the specified state.*

(a)   (b)

Figure 15-12
(a) A simplified representation of a side view of a "perfect" crystal of a polar substance at 0 K. Note the perfect alignment of the dipoles in all molecules in a perfect crystal. This causes its entropy to be zero at 0 K. However, there are no perfect crystals, because even the purest substances that scientists have prepared are contaminated by traces of impurities that occupy a few of the positions in the crystal structure. Additionally, there are some vacancies in the crystal structures of even very highly purified substances such as those used in semiconductors (Section 13-17). (b) A simplified representation of the same "perfect" crystal at a temperature above 0 K. Vibrations of the individual molecules within the crystal cause some dipoles to be oriented in directions other than those in a perfect arrangement. The entropy of such a crystalline solid is greater than zero, because there is disorder in the crystal.

in Table 15-5. The absolute entropies, S_{298}^0, of various substances under standard conditions are tabulated in Appendix K. At 298 K, *any* substance is more disordered than if it were in a perfect crystalline state at absolute zero, so tabulated S_{298}^0 values are *always positive*. Notice especially that S_{298}^0 of an element, unlike its ΔH_f^0, is *not* equal to zero. The reference state for absolute entropy specified by the Third Law is different from that for ΔH_f^0 (Section 15-7).

The **standard entropy change**, ΔS^0, of a reaction can be determined from the absolute entropies of reactants and products. The relationship is analogous to Hess' Law:

The Σ n means that each S^0 value must be multiplied by the appropriate coefficient, n, from the balanced equation. These values are then added.

$$\Delta S_{rxn}^0 = \Sigma \; nS_{products}^0 - \Sigma \; nS_{reactants}^0$$

S^0 values are tabulated in units of $J/mol \cdot K$ rather than the larger units involving kilojoules that are used for enthalpy changes. The "mol" term in the units for a *substance* refers to a mole of the substance, whereas for a *reaction* it refers to a mole of reaction. Each term in the sums on the right-hand side of the equation has the units

$$\frac{\text{mol substance}}{\text{mole rxn}} \times \frac{J}{(\text{mol substance}) \cdot K} = \frac{J}{(\text{mol rxn}) \cdot K}$$

The result is usually abbreviated as $J/mol \cdot K$, or sometimes even as J/K. As before, we shall usually omit units in intermediate steps and simply apply appropriate units to the result.

Table 15-5
Absolute Entropies at 298 K for a Few Common Substances

	S^0 (J/mol · K)
C(diamond)	2.38
C(g)	158.0
$H_2O(\ell)$	69.91
$H_2O(g)$	188.7
$I_2(s)$	116.1
$I_2(g)$	260.6

Example 15-15

Use the values of standard molar entropies in Appendix K to calculate the entropy change at 25°C and one atmosphere pressure for the reaction of hydrazine with hydrogen peroxide. This explosive reaction has been used for rocket propulsion. Do you think the reaction is spontaneous? The balanced equation for the reaction is

$$N_2H_4(\ell) + 2H_2O_2(\ell) \longrightarrow N_2(g) + 4H_2O(g) + 642.2 \text{ kJ}$$

Plan

We use the equation for standard entropy change to calculate ΔS_{rxn}^0 from the tabulated values of standard molar entropies, S_{298}^0, for the substances in the reaction.

Small booster rockets adjust the course of a satellite in orbit. Some of these small rockets are powered by the N_2H_4/H_2O_2 reaction.

Solution

$$\Delta S^0_{rxn} = [S^0_{N_2(g)} + 4S^0_{H_2O(g)}] - [S^0_{N_2H_4(\ell)} + 2S^0_{H_2O_2(\ell)}]$$

$$= [1\,(191.5) + 4\,(188.7)] - [1\,(121.2) + 2\,(109.6)]$$

$$\Delta S^0_{rxn} = +605.9 \text{ J/mol} \cdot \text{K}$$

The "mol" designation for ΔS^0_{rxn} refers to a mole of reaction, that is, one mole of $N_2H_4(\ell)$, two moles of $H_2O_2(\ell)$, and so on. Although it may not appear to be, $+605.9$ J/mol · K is a relatively large value of ΔS^0 (for the system). The positive entropy change favors spontaneity. In this case, the reaction is also exothermic: $\Delta H^0 = -642.2$ kJ/mol. As we shall see in the next section, this reaction *must* be spontaneous, because ΔS^0 is positive and ΔH^0 is negative.

EOC 88

Realizing that changes in the thermodynamic quantity *entropy* may be understood in terms of changes in *molecular disorder*, we can often predict the sign of ΔS_{sys}. The following illustrations emphasize several common types of processes that result in predictable entropy changes.

(a) *Phase changes*. For instance, when melting occurs, the molecules or ions are taken from their quite ordered crystalline arrangement to a more disordered one in which they are able to move past one another in the liquid. Thus, a melting process is always accompanied by an entropy increase ($\Delta S_{sys} > 0$). Likewise, vaporization and sublimation both take place with large increases in disorder, and hence with increases in entropy. For the reverse processes of freezing, condensation, and deposition, entropy decreases because order increases.

(b) *Temperature changes*—for example, warming a gas from 25°C to 50°C. As any sample is warmed, the molecules undergo more (random) motion; hence entropy increases ($\Delta S_{sys} > 0$) as temperature increases. Likewise, as the temperature of a solid increases, the particles vibrate more vigorously about their positions in the crystal, so that at any instant there is a larger average displacement from their mean positions; this results in an increase in entropy.

(c) *Volume changes*. When the volume of a sample of gas increases, the molecules can occupy more positions, and hence are more randomly arranged than if they were closer together in a smaller volume. Hence, an expansion is accompanied by an increase in entropy ($\Delta S_{sys} > 0$). Conversely, as a sample is compressed, the molecules are more restricted in their locations, and a situation of greater order (lower entropy) results.

(d) *Mixing of substances* without chemical reaction. Situations in which the molecules are more "mixed up" are more disordered, and hence are at higher entropy. We pointed out that the mixed gases of Figure 15-10b were more disordered than the separated gases of Figure 15-10a, and that the former was a situation of higher entropy. We see, then, that mixing of gases by diffusion is a process for which $\Delta S_{sys} > 0$; we know from experience that it is always spontaneous. We have already pointed out (Section 14-2) that the increase in disorder (entropy increase) that accompanies mixing often provides the driving force for solubility of one substance in another. For example, when one mole of solid NaCl dissolves in water, NaCl(s) → NaCl(aq), the entropy (Appendix K) increases from 72.4 J/mol · K to 115.5 J/mol · K,

Table 15-6
Entropy Changes for Some Processes $X_2 \rightarrow 2X$

Reaction	ΔS^0 (J/mol · K)
$H_2(g) \rightarrow 2H(g)$	98.0
$N_2(g) \rightarrow 2N(g)$	114.9
$F_2(g) \rightarrow 2F(g)$	114.5
$Cl_2(g) \rightarrow 2Cl(g)$	107.2
$Br_2(\ell) \rightarrow 2Br(g)$	197.5
$I_2(s) \rightarrow 2I(g)$	245.3

Do you think that the process

$2H_2(g) + O_2(g) \longrightarrow 2H_2O(\ell)$

would have a higher or lower value of ΔS^0 than when the water is in the gas phase? Confirm by calculation.

or $\Delta S^0 = +43.1$ J/mol · K. The term "mixing" can be applied rather liberally. For example, the process $H_2(g) + Cl_2(g) \rightarrow 2HCl(g)$ has $\Delta S^0 > 0$; in the reactants, each atom is paired with like atoms, a less "mixed-up" situation than in the products, where unlike atoms are bonded together.

(e) *Increase in the number of particles*, as in the dissociation of a diatomic gas such as $F_2(g) \rightarrow 2F(g)$. Any process in which the number of particles increases results in an increase in entropy, $\Delta S_{sys} > 0$. Values of ΔS^0 calculated for several reactions of this type are given in Table 15-6. As you can see, the ΔS^0 values for the dissociation process $X_2 \rightarrow 2X$ are all similar for X = H, F, Cl, and N. Why is the value given in Table 15-6 so much larger for X = Br? The answer lies in the fact that the value is for the process starting with *liquid* Br_2. The total process $Br_2(\ell) \rightarrow 2Br(g)$, for which $\Delta S^0 = 197.5$ J/mol · K, can be treated as the result of *two* processes. The first of these is *vaporization*, $Br_2(\ell) \rightarrow Br_2(g)$, for which $\Delta S^0 = 93.1$ J/mol · K. The second step is the dissociation of gaseous bromine, $Br_2(g) \rightarrow 2Br(g)$, for which $\Delta S^0 = 104.4$ J/mol · K; this entropy increase is about the same as for the other processes that involve *only* dissociation of a gaseous diatomic species. Can you rationalize the even higher value given in the table for $I_2(s) \rightarrow 2I(g)$?

(f) *Changes in the number of moles of gaseous substances*. Processes that result in an increase in the number of moles of gaseous substances have $\Delta S_{sys} > 0$. Example 15-15 illustrates this. The reactants of the balanced chemical equation include no gaseous substances, whereas the products include five moles of gas. Conversely, the process $2H_2(g) + O_2(g) \rightarrow 2H_2O(g)$ has a negative ΔS^0 value; in the balanced equation, three moles of gas are consumed while only two moles are produced, for a net decrease in the number of moles in the gas phase. You should be able to calculate the value of ΔS^0 for this reaction from the values in Appendix K.

15-14 Free Energy Change, ΔG, and Spontaneity

Energy is the capacity to do work. If heat is released in a chemical reaction (ΔH is negative), *some* of the heat may be converted into useful work. Some of it may be expended to increase the order of the system (if ΔS is negative). If a system becomes more disordered ($\Delta S > 0$), more useful energy becomes available than indicated by ΔH alone. A prominent 19th-century American professor of mathematics and physics, J. Willard Gibbs, formulated the relationship between enthalpy and entropy changes for a process in terms of another state function change, ΔG, called the **Gibbs free energy change**. The relationship, which holds at constant temperature and pressure, is

This is often called simply the Gibbs energy change or the free energy change.

$$\Delta G = \Delta H - T\Delta S \quad \text{(constant } T \text{ and } P\text{)}$$

The amount by which Gibbs free energy decreases is the *maximum useful energy* obtainable in the form of work from a given process at constant temperature and pressure. It is also the *indicator of spontaneity of a reaction or physical change* at constant T and P. If there is a net release of useful energy, ΔG is negative and the process is spontaneous. We see from the equation that ΔG becomes more negative as (1) ΔH becomes more negative (exothermic) and (2) ΔS becomes more positive (increase in disorder). If there is a net absorption of free energy by the system during a process, ΔG is positive and the process is nonspontaneous. This means that the reverse process is spontaneous under the given conditions. When $\Delta G = 0$, there is no net transfer of free energy; both the forward and reverse processes are equally favorable. Thus, $\Delta G = 0$ describes a system at *equilibrium*.

The relationship between ΔG and spontaneity may be summarized as follows:

ΔG	Spontaneity of Reaction (constant T and P)
ΔG is positive	Reaction is nonspontaneous
ΔG is zero	System is at equilibrium
ΔG is negative	Reaction is spontaneous

The free energy content of a system depends on temperature and pressure (and, for mixtures, on concentrations). The value of ΔG for a process depends on the states and the concentrations of the various substances involved. Just as for other thermodynamic variables, we choose some set of conditions as a standard state reference. The standard state for free energy is the same as for enthalpy—1 atm and the specified temperature, usually 25°C (298 K). Values of standard molar free energy of formation, ΔG_f^0, for many substances are tabulated in Appendix K. For *elements* in their standard states, $\Delta G_f^0 = 0$. The values of ΔG_f^0 may be used to calculate the standard free energy change of a reaction *at 298 K* by using the following relationship:

$$\Delta G_{rxn}^0 = \Sigma \, n\Delta G_{f \text{ products}}^0 - \Sigma \, n\Delta G_{f \text{ reactants}}^0 \qquad \text{(1 atm and 298 K } only)$$

The value of ΔG_{rxn}^0 allows us to predict the spontaneity of a very special hypothetical reaction that we call the *standard reaction*.

In the **standard reaction**, the numbers of moles of reactants shown in the balanced equation, all in their standard states, are *completely* converted to the numbers of moles of products shown in the balanced equation, also all in their standard states.

In other words, are the *reactants* or the *products* more stable *in their standard states*?

But we must remember that it is ΔG, and not ΔG^0, that is the general criterion for spontaneity. ΔG depends on concentrations of reactants and products in the mixture. For most reactions, there is an *equilibrium mixture* that is more stable than either all reactants or all products. In Chapter 17 we shall see how to find ΔG for mixtures and study the concept of equilibrium.

Example 15-16

Diatomic nitrogen and oxygen molecules make up about 99% of all the molecules in reasonably "unpolluted" air. Evaluate ΔG^0 for the following reaction at 298 K, using ΔG_f^0 values from Appendix K. Is the standard reaction spontaneous?

$$N_2(g) + O_2(g) \longrightarrow 2NO(g) \qquad \text{(nitrogen oxide)}$$

Plan

The reaction conditions are 1 atm and 298 K, so we can use the tabulated values of ΔG_f^0 for each substance in Appendix K to evaluate ΔG_{rxn}^0 in the equation given above. The treatment of units for calculation of ΔG^0 is the same as that for ΔH^0 in Example 15-10.

Solution

$$\Delta G^0_{rxn} = 2\Delta G^0_{f\,NO(g)} \quad - [\Delta G^0_{f\,N_2(g)} + \Delta G^0_{f\,O_2(g)}]$$

$$= (2(86.57) \quad - [0 \quad + 0])$$

$$\Delta G^0_{rxn} = +173.1 \text{ kJ/mol rxn} \qquad \text{for the reaction as written}$$

For the reverse reaction at 298 K, $\Delta G^0_{rxn} = -173.1$ kJ/mol. It is spontaneous, but very slow at room temperature. The NO formed in automobile engines is oxidized to even more harmful NO_2 much more rapidly than it spontaneously decomposes to N_2 and O_2. The oxides of nitrogen in the atmosphere represent a major environmental problem.

Because ΔG^0 is positive, the reaction is nonspontaneous at 298 K under standard state conditions. This is fortunate, for otherwise we *might* have to breathe nitrogen oxide, which is poisonous. (Remember that thermodynamic spontaneity would not guarantee that a process would occur at an observable rate.)

EOC 95

ΔG^0 can also be calculated by the equation

$$\Delta G^0 = \Delta H^0 - T\Delta S^0 \qquad \text{(constant } T \text{ and } P\text{)}$$

Strictly, this last equation applies to standard conditions. However, ΔH^0 and ΔS^0 often do not vary much with temperature, so it can often be used to estimate free energy changes at other temperatures.

Example 15-17

Make the same determination as in Example 15-16, using heats of formation and absolute entropies rather than free energies of formation.

Plan

First we calculate ΔH^0_{rxn} and ΔS^0_{rxn}. We use the relationship $\Delta G^0 = \Delta H^0 - T\Delta S^0$ to evaluate the free energy change under standard state conditions at 298 K.

Solution

$$\Delta H^0_{rxn} = \Sigma \, n\Delta H^0_{f\,products} - \Sigma \, n\Delta H^0_{f\,reactants}$$

$$= 2\Delta H^0_{f\,NO(g)} \quad - [\Delta H^0_{f\,N_2(g)} + \Delta H^0_{f\,O_2(g)}]$$

$$= (2[90.25] \quad - [0 + 0]) = 180.5$$

$$\Delta S^0_{rxn} = \Sigma \, nS^0_{products} - \Sigma \, nS^0_{reactants}$$

$$= 2S^0_{NO(g)} \quad - [S^0_{N_2(g)} + S^0_{O_2(g)}]$$

$$= (2[210.7] \quad - [191.5 + 205.0]) = 24.9 \text{ J/mol} \cdot \text{K} = 0.0249 \text{ kJ/mol} \cdot \text{K}$$

Now we use the relationship $\Delta G^0 = \Delta H^0 - T\Delta S^0$, with $T = 298$ K, to evaluate the free energy change under standard state conditions at 298 K.

$$\Delta G^0_{rxn} = \Delta H^0_{rxn} \quad - T\Delta S^0_{rxn}$$

$$= 180.5 \text{ kJ/mol} \quad - (298 \text{ K})(0.0249 \text{ kJ/mol} \cdot \text{K})$$

$$= 180.5 \text{ kJ/mol} \quad - 7.42 \text{ kJ/mol}$$

$$\Delta G^0_{rxn} = +173.1 \text{ kJ/mol rxn}, \quad \text{the same value obtained in Example 15-16.}$$

EOC 96

15-15 The Temperature Dependence of Spontaneity

The methods developed in Section 15-14 can also be used to estimate the temperature at which a process is in equilibrium.

Example 15-18

Use the thermodynamic data in Appendix K to estimate the normal boiling point of bromine, Br_2. Assume that ΔH and ΔS do not change with temperature.

Plan

The process we must consider is

$$Br_2(\ell) \longrightarrow Br_2(g)$$

By definition, the normal boiling point of a liquid is the temperature at which pure liquid and pure gas exist in equilibrium at 1 atm. Therefore, $\Delta G = 0$. We assume that $\Delta H_{rxn} = \Delta H^0_{rxn}$ and $\Delta S_{rxn} = \Delta S^0_{rxn}$. We can evaluate these two quantities, substitute them in the relationship $\Delta G = \Delta H - T\Delta S$, and then solve for T.

Solution

$$\Delta H_{rxn} = \Delta H^0_{f\,Br_2(g)} - \Delta H^0_{f\,Br_2(\ell)}$$

$$= 30.91 \quad - 0 = 30.91 \text{ kJ/mol}$$

$$\Delta S_{rxn} = S^0_{Br_2(g)} \quad - S^0_{Br_2(\ell)}$$

$$= (245.4 \quad - 152.2) = 93.2 \text{ J/mol} \cdot \text{K} = 0.0932 \text{ kJ/mol} \cdot \text{K}$$

We can now solve for the temperature at which the system is in equilibrium, i.e., the boiling point of Br_2:

$$\Delta G_{rxn} = \Delta H_{rxn} - T\Delta S_{rxn} = 0 \quad \text{so} \quad \Delta H_{rxn} = T\Delta S_{rxn}$$

$$T = \frac{\Delta H_{rxn}}{\Delta S_{rxn}} = \frac{30.91 \text{ kJ/mol}}{0.0932 \text{ kJ/mol} \cdot \text{K}} = \boxed{332 \text{ K } (59°C)}$$

This is the temperature at which the system is in equilibrium, i.e., the boiling point of Br_2. The value listed in a handbook of chemistry and physics is 58.78°C.

EOC 103, 104, 106

Actually, both ΔH^0_{rxn} and ΔS^0_{rxn} vary with temperature, but usually not enough to introduce significant errors for modest temperature changes. The value of ΔG^0_{rxn} on the other hand, is strongly dependent on the temperature.

The free energy change and spontaneity of a reaction depend upon both enthalpy and entropy changes. Both ΔH and ΔS may be either positive or negative, so we can classify reactions into four categories with respect to spontaneity (see Figure 15-13).

$\Delta G = \Delta H - T\Delta S$	*(constant temperature and pressure)*
1. $\Delta H = -$, $\Delta S = +$	Rxns are spontaneous at all temperatures.
2. $\Delta H = -$, $\Delta S = -$	Rxns become spontaneous at lower temperatures.
3. $\Delta H = +$, $\Delta S = +$	Rxns become spontaneous at higher temperatures.
4. $\Delta H = +$, $\Delta S = -$	Rxns are nonspontaneous at all temperatures.

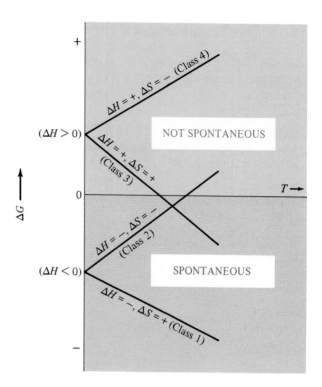

Figure 15-13
A graphical representation of the dependence of ΔG and spontaneity on temperature for each of the four classes of reactions listed in the text and in Table 15-7.

Table 15-7 gives examples of reactions in each category, as well as temperatures at which the changeover from spontaneous to nonspontaneous occurs, where appropriate. The numbering system corresponds to that above.

We can estimate the temperature range over which a chemical reaction is spontaneous by evaluating ΔH^0_{rxn} and ΔS^0_{rxn} from tabulated data.

Example 15-19

Mercury(II) sulfide is a dark red mineral called cinnabar. Metallic mercury is obtained by roasting the sulfide in a limited amount of air (Section 22-3). Estimate the temperature range in which the *standard* reaction is spontaneous.

$$HgS(s) + O_2(g) \longrightarrow Hg(\ell) + SO_2(g)$$

Plan

We evaluate ΔH^0_{rxn} and ΔS^0_{rxn} and assume that their values are independent of temperature. When we do this, we find that both factors are favorable to spontaneity.

Solution

$$\Delta H^0_{rxn} = \Delta H^0_{f\ Hg(\ell)} + \Delta H^0_{f\ SO_2(g)} - [\Delta H^0_{f\ HgS(s)} + \Delta H^0_{f\ O_2(g)}]$$

$$= (0 - 296.8 + 58.2 - 0) = -238.6 \text{ kJ/mol}$$

$$\Delta S^0_{rxn} = S^0_{Hg(\ell)} + S^0_{SO_2(g)} - [S^0_{HgS(s)} + S^0_{O_2(g)}]$$

$$= (76.02 + 248.1 - 82.4 - 205.0) = +36.7 \text{ J/mol} \cdot \text{K}$$

Heating red HgS in air produces liquid Hg. The gaseous SO_2 escapes. Cinnabar, an important ore of mercury, contains HgS.

Table 15-7
Thermodynamic Classes of Reactions

Class	Examples	ΔH (kJ/mol)	ΔS (J/mol · K)	Temperature Range of Spontaneity
1	$2H_2O_2(\ell) \longrightarrow 2H_2O(\ell) + O_2(g)$ $H_2(g) + Br_2(\ell) \longrightarrow 2HBr(g)$	-196 -72.8	$+126$ $+114$	All temperatures All temperatures
2	$NH_3(g) + HCl(g) \longrightarrow NH_4Cl(s)$ $2H_2S(g) + SO_2(g) \longrightarrow 3S(s) + 2H_2O(\ell)$	-176 -233	-285 -424	Lower temperatures (<619 K) Lower temperatures (<550 K)
3	$NH_4Cl(s) \longrightarrow NH_3(g) + HCl(g)$ $CCl_4(\ell) \longrightarrow C(graphite) + 2Cl_2(g)$	$+176$ $+136$	$+285$ $+235$	Higher temperatures (>619 K) Higher temperatures (>517 K)
4	$2H_2O(\ell) + O_2(g) \longrightarrow 2H_2O_2(\ell)$ $3O_2(g) \longrightarrow 2O_3(g)$	$+196$ $+285$	-126 -137	Nonspontaneous, all temperatures Nonspontaneous, all temperatures

ΔH^0_{rxn} is negative and ΔS^0_{rxn} is positive, so the reaction is spontaneous at all temperatures. The reverse reaction is, therefore, nonspontaneous at all temperatures.

The fact that the preceding reaction is spontaneous at all temperatures does not mean that the reaction occurs fast enough to be useful at all temperatures. As a matter of fact, $Hg(\ell)$ can be obtained from $HgS(s)$ only at high temperatures.

Example 15-20

Estimate the temperature range for which the following standard reaction is spontaneous.

$$SiO_2(s) + 2C(graphite) + 2Cl_2(g) \longrightarrow SiCl_4(g) + 2CO(g)$$

Plan

When we proceed as in Example 15-19, we find that ΔS^0_{rxn} is favorable to spontaneity, whereas ΔH^0_{rxn} is unfavorable. We can set ΔG^0 equal to zero in the equation $\Delta G^0 = \Delta H^0 - T\Delta S^0$ and solve for the temperature at which the system is *at equilibrium*. This will represent the temperature at which the reaction changes from nonspontaneous to spontaneous.

Solution

$$\Delta H^0_{rxn} = [\Delta H^0_{f\ SiCl_4(g)} + 2\Delta H^0_{f\ CO(g)}] - [\Delta H^0_{f\ SiO_2(s)} + 2\Delta H^0_{f\ C(graphite)} + 2\Delta H^0_{f\ Cl_2(g)}]$$

$$= [(-657.0) + 2(-110.5)] - [(-910.9) + 2(0) + 2(0)]$$

$$= +32.9 \text{ kJ/mol}$$

$$\Delta S^0_{rxn} = S^0_{SiCl_4(g)} + 2S^0_{CO(g)} - [S^0_{SiO_2(s)} + 2S^0_{C(graphite)} + 2S^0_{Cl_2(g)}]$$

$$= [330.6 + 2(197.6)] - [41.84 + 2(5.740) + 2(223.0)]$$

$$= 226.5 \text{ J/mol} \cdot \text{K} = 0.2265 \text{ kJ/mol} \cdot \text{K}$$

When $\Delta G^0 = 0$, neither the forward nor the reverse reaction is spontaneous. Let's find the temperature at which $\Delta G^0 = 0$ and the system is at equilibrium.

$$\Delta G^0 = \Delta H^0 - T\Delta S^0 = 0$$

$$\Delta H^0 = T\Delta S^0$$

$$T = \frac{\Delta H^0}{\Delta S^0} = \frac{+32.9 \text{ kJ/mol}}{+0.2265 \text{ kJ/mol} \cdot \text{K}}$$

$$\underline{T = 145 \text{ K}}$$

At temperatures above 145 K, the $T\Delta S^0$ would be greater ($-T\Delta S^0$ would be more negative) than the ΔH^0 term, which would make ΔG^0 negative; so the reaction would be spontaneous above 145 K. At temperatures below 145 K, the $T\Delta S^0$ term would be smaller than the ΔH^0 term, which would make ΔG^0 positive; so the reaction would be nonspontaneous below 145 K.

However, 145 K ($-128°C$) is a very low temperature. For all practical purposes, the reaction is spontaneous at all but very low temperatures. In practice, it is carried out at 800 to 1000°C because of the greater reaction rate at these higher temperatures. This gives a useful and economical rate of production of $SiCl_4$, an important industrial chemical.

EOC 101, 102

Key Terms

Absolute entropy (of a substance)   The increase in the entropy of a substance as it goes from a perfectly ordered crystalline form at 0 K (where its entropy is zero) to the temperature in question.

Bomb calorimeter   A device used to measure the heat transfer between system and surroundings at constant volume.

Bond energy   The amount of energy necessary to break one mole of bonds in a substance, dissociating the substance in the gaseous state into atoms of its elements in the gaseous state.

Born–Haber cycle   A series of reactions (and accompanying enthalpy changes) that, when summed, represents the hypothetical one-step reaction by which elements in their standard states are converted into crystals of ionic compounds (and the accompanying enthalpy changes).

Calorimeter   A device used to measure the heat transfer between system and surroundings.

Calorimeter constant   The heat capacity of a calorimeter; determined by adding a known amount of heat to the calorimeter and measuring the observed temperature change.

Crystal lattice energy   The amount of energy that holds a crystal together; the energy change when a mole of solid is formed from its constituent molecules or ions (for ionic compounds) in the gaseous state.

Endothermicity   The absorption of heat by a system as a process occurs.

Enthalpy, H   The heat content of a specific amount of a substance; defined as $E - PV$.

Entropy, S   A thermodynamic state property that measures the degree of disorder or randomness of a system; $T\Delta S$ represents a form of energy.

Equilibrium   A state of dynamic balance in which the rates of forward and reverse processes (reactions) are equal; the state of a system when neither the forward nor the reverse process is thermodynamically favored.

Exothermicity   The release of heat by a system as a process occurs.

First Law of Thermodynamics   The total amount of energy in the universe is constant (also known as the Law of Conservation of Energy); energy is neither created nor destroyed in ordinary chemical reactions and physical changes.

Free energy, G   The thermodynamic state function of a system that indicates the amount of energy available for the system to do useful work at constant T and P. Also known as Gibbs free energy.

Free energy change, ΔG   The indicator of spontaneity of a process at constant T and P. If ΔG is negative, the process is spontaneous.

Gibbs free energy   See *Free energy, G*.

Hess' Law of heat summation   The enthalpy change for a reaction is the same whether it occurs in one step or a series of steps.

Internal energy, E   All forms of energy associated with a specific amount of a substance.

Mole of reaction (mol rxn)   The amount of reaction corresponding to the number of moles of each substance shown in the balanced equation.

Pressure–volume work Work done by a gas when it expands against an external pressure or work done on a system as gases are compressed or consumed in the presence of an external pressure.

Second Law of Thermodynamics The universe tends toward a state of greater disorder in spontaneous processes.

Spontaneity Of a process, its property of being energetically favorable and therefore capable of proceeding in the forward direction (but not necessarily at an observable rate).

Standard entropy, S^0 (of a substance) The absolute entropy of a substance in its standard state at 298 K.

Standard molar enthalpy of formation, ΔH_f^0 (of a substance) The amount of heat absorbed in the formation of one mole of a substance in a specified state from its elements in their standard states.

Standard reaction A reaction in which the numbers of moles of reactants shown in the balanced equation, all in their standard states, are *completely* converted to the numbers of moles of products shown in the balanced equation, also all at their standard states.

State function A variable that defines the state of a system; a function that is independent of the pathway by which a process occurs.

Surroundings Everything in the environment of the system.

System The substances of interest in a process; the part of the universe under investigation.

Thermochemical equation A balanced chemical equation together with a designation of the corresponding value of ΔH_{rxn}. Sometimes used with changes in other thermodynamic quantities.

Thermodynamics The study of the energy transfers accompanying physical and chemical processes.

Thermodynamic state of a system A set of conditions that completely specifies all of the properties of the system.

Thermodynamic standard state of a substance The most stable state of the substance under one atmosphere pressure and at some specific temperature (25°C unless otherwise specified).

Third Law of Thermodynamics The entropy of a hypothetical pure, perfect, crystalline substance at absolute zero temperature is zero.

Universe The system plus the surroundings.

Work The application of a force through a distance; for physical changes or chemical reactions at constant external pressure, the work done on the system is $-P\Delta V$.

Exercises

General Concepts

1. State precisely the meaning of each of the following terms. You may need to review Chapter 1 to refresh your memory concerning terms introduced there. (a) Energy, (b) kinetic energy, (c) potential energy, (d) calorie, (e) joule.

2. State precisely the meaning of each of the following terms. You may need to review Chapter 1 to refresh your memory concerning terms introduced there. (a) Heat, (b) temperature, (c) system, (d) surroundings, (e) thermodynamic state of system, (f) work.

3. (a) Give an example of the conversion of heat into work.
   (b) Give an example of the conversion of work into heat.

4. (a) Give an example of the conversion of potential energy into kinetic energy.
   (b) Give an example of the conversion of kinetic energy into potential energy.

5. What is a state function? Would Hess' Law be a law if enthalpy were not a state function?

6. Distinguish between endothermic and exothermic processes. If we know that a reaction is endothermic in one direction, what can be said about the reaction in the reverse direction?

7. According to the First Law of Thermodynamics, the total amount of energy in the universe is constant.

Why, then, do we say that we are experiencing a declining supply of energy?

Internal Energy and Changes in Internal Energy

8. (a) What are the sign conventions for q, the amount of heat added to or removed from a system?
   (b) What are the sign conventions for w, the amount of work done on or by a system?

9. What happens to ΔE for a system during a process in which (a) $q < 0$ and $w < 0$, (b) $q = 0$ and $w > 0$, (c) $q > 0$ and $w < 0$?

10. What happens to ΔE for a system during a process in which (a) $q > 0$ and $w > 0$, (b) $q = w = 0$, (c) $q < 0$ and $w > 0$?

11. A system performs 720 L · atm of pressure–volume work (1 L · atm = 101.325 J) on its surroundings and absorbs 6300 J of heat from its surroundings. What is the change in internal energy of the system?

12. A system receives 73 J of electrical work, delivers 227 J of pressure–volume work, and releases 212 J of heat. What is the change in internal energy of the system?

13. For each of the following chemical and physical changes carried out at constant pressure, state whether work is done by the system (the substances undergoing the change) on the surroundings or by the surroundings on the system, or whether the amount of work is negligible.

(a) $C_6H_6(\ell) \longrightarrow C_6H_6(s)$
(b) $\frac{1}{2}N_2(g) + \frac{3}{2}H_2(g) \longrightarrow NH_3(g)$
(c) $3H_2S(g) + 2HNO_3(g) \longrightarrow$
$$2NO(g) + 4H_2O(\ell) + 3S(s)$$

14. Repeat Exercise 13 for
(a) $4HNO_3(\ell) + P_4O_{10}(s) \longrightarrow 2N_2O_5(s) + 4HPO_3(s)$
(b) $2MnO_4^- + 5H_2O_2(aq) + 6H^+ \longrightarrow$
$$2Mn^{2+} + 5O_2(g) + 8H_2O(\ell)$$
(c) $CO_2(g) + H_2O(\ell) + CaCO_3(s) \longrightarrow Ca^{2+} + 2HCO_3^-$

15. Assuming that the gases are ideal, calculate the amount of work done (in joules) in each of the following reactions. In each case, is the work done *on* or *by* the system?
(a) A reaction in the Mond process for purifying nickel that involves formation of the volatile gas, nickel(0) tetracarbonyl, at 50 to 100°C. Assume one mole of nickel is used and a constant temperature of 75°C is maintained.

$$Ni(s) + 4CO(g) \longrightarrow Ni(CO)_4(g)$$

(b) The conversion of one mole of brown nitrogen dioxide into colorless dinitrogen tetroxide at 10.0°C:

$$2NO_2(g) \longrightarrow N_2O_4(g)$$

16. Assuming that the gases are ideal, calculate the amount of work done (in joules) in each of the following reactions. In each case, is the work done *on* or *by* the system?
(a) The oxidation of one mole of HCl(g) at 200°C:

$$4HCl(g) + O_2(g) \longrightarrow 2Cl_2(g) + 2H_2O(g)$$

(b) The decomposition of one mole of nitric oxide (an air pollutant) at 300°C:

$$2NO(g) \longrightarrow N_2(g) + O_2(g)$$

*17. When an ideal gas expands at *constant temperature*, there is no change in molecular kinetic energy (kinetic energy is proportional to temperature), and there is no change in potential energy due to intermolecular attractions (these are zero for an ideal gas). Thus, for the isothermal (constant temperature) expansion of an ideal gas, $\Delta E = 0$. Suppose we allow an ideal gas to expand isothermally from 2.00 L to 5.00 L in two steps: (i) against a constant external pressure of 3.00 atm until equilibrium is reached, then (ii) against a constant external pressure of 2.00 atm until equilibrium is reached. Calculate q and w for this two-step expansion.

Calorimetry

18. What is a bomb calorimeter? How do bomb calorimeters give us useful data?

19. What is a coffee-cup calorimeter? How do coffee-cup calorimeters give us useful information?

20. A calorimeter contained 75.0 g of water at 16.95°C. A 75.2-g sample of iron at 63.14°C was placed in it, giving a final temperature of 19.68°C for the system. Calculate the heat capacity of the calorimeter. The specific heats are 4.184 J/g · °C for H_2O and 0.444 J/g · °C for Fe.

21. A student wishes to determine the heat capacity of a coffee-cup calorimeter. After she mixes 100.0 g of water at 61.0°C with 100.0 g of water, already in the calorimeter, at 24.2°C, the final temperature of the mixed water is 41.5°C. (a) Calculate the heat capacity of the calorimeter in J/°C. Use 4.18 J/g · °C as the specific heat of water. (b) Why is it more useful to express this value in J/°C rather than units of J/(g calorimeter · °C)?

22. A coffee-cup calorimeter having a heat capacity of 472 J/°C is used to measure the heat evolved when the following aqueous solutions, both initially at 22.6°C, are mixed: 100 g of solution containing 6.62 g of lead(II) nitrate, $Pb(NO_3)_2$, and 100 g of solution containing 6.00 g of sodium iodide, NaI. The final temperature is 24.2°C. Assume that the specific heat of the mixture is the same as that for water, 4.18 J/g · °C. The reaction that occurs is

$$Pb(NO_3)_2(aq) + 2NaI(aq) \longrightarrow$$
$$PbI_2(s) + 2NaNO_3(aq)$$

(a) Calculate the heat evolved in the reaction.
(b) Calculate the ΔH for the reaction [per mole of $Pb(NO_3)_2(aq)$ consumed] under the conditions of the experiment.

23. A coffee-cup calorimeter is used to determine the heat of reaction for the acid–base neutralization

$$CH_3COOH(aq) + NaOH(aq) \longrightarrow$$
$$NaCH_3COO(aq) + H_2O(\ell)$$

When we add 20.00 mL of 0.625 M NaOH at 23.000°C to 30.00 mL of 0.500 M CH_3COOH already in the calorimeter at the same temperature, the resulting temperature is observed to be 25.947°C. The heat capacity of the calorimeter has previously been determined to be 27.8 J/°C. Assume that the specific heat of the mixture is the same as that of water, 4.18 J/g · °C and that density of the mixture is 1.02 g/mL. (a) Calculate the amount of heat given off in the reaction. (b) Determine ΔH for the reaction under the conditions of the experiment.

24. The combustion of 1.048 g of benzene, $C_6H_6(\ell)$, in a bomb calorimeter compartment surrounded by 945 g of water raised the temperature of the water from 23.640°C to 32.692°C. The heat capacity of the calorimeter is 891 J/°C. (a) Write the balanced equation for the combustion reaction, assuming that $CO_2(g)$ and $H_2O(\ell)$ are the only products. (b) Use the calorimetric data to calculate ΔE for the combustion of benzene in kJ/g and in kJ/mol.

25. A 2.00-g sample of hydrazine, N_2H_4, is burned in a bomb calorimeter that contains 6.40×10^3 g of H_2O, and the temperature increases from 25.00°C to 26.17°C. The heat capacity of the calorimeter is 3.76 kJ/°C.

Calculate ΔE for the combustion of N_2H_4 in kJ/g and in kJ/mol.

Enthalpy and Changes in Enthalpy

26. (a) Distinguish between ΔH and ΔH^0 for a reaction.
    (b) Distinguish between ΔH^0 and ΔH_f^0.

27. How do enthalpy and internal energy differ?

28. For each of these reactions, (a) does the enthalpy increase or decrease; (b) is $H_{reactant} > H_{product}$ or is $H_{product} > H_{reactant}$; (c) is ΔH positive or negative?
    (1) $Al_2O_3(s) \longrightarrow 2Al(s) + 1\frac{1}{2}O_2(g)$ (endothermic)
    (2) $Sn(s) + Cl_2(g) \longrightarrow SnCl_2(s)$ (exothermic)

29. (a) The combustion of 0.0222 g of octane vapor, C_8H_{18}, at constant pressure raises the temperature of a calorimeter 0.400°C. The heat capacity of the calorimeter is 2.48 kJ/°C. Find the molar heat of combustion of C_8H_{18}:

    $$C_8H_{18}(g) + 12\frac{1}{2}O_2(g) \longrightarrow 8CO_2(g) + 9H_2O(\ell)$$

    (b) How many grams of $C_8H_{18}(g)$ must be burned to obtain 105 kJ of heat energy?

30. Methanol, CH_3OH, is an efficient fuel with a high octane rating that can be produced from coal and hydrogen:

    $$CH_3OH(g) + 1\frac{1}{2}O_2(g) \longrightarrow CO_2(g) + 2H_2O(\ell)$$
    $$\Delta H = -76.2 \text{ kJ/mol rxn}$$

    Find (a) the heat evolved when 30.0 g $CH_3OH(g)$ burns and (b) the mass of O_2 consumed when 950 kJ of heat is given out.

31. Methylhydrazine is burned with dinitrogen tetroxide in the altitude-control engines of the space shuttle:

    $$CH_6N_2(\ell) + \tfrac{5}{4}N_2O_4(\ell) \longrightarrow$$
    $$CO_2(g) + 3H_2O(\ell) + \tfrac{9}{4}N_2(g)$$

    The two substances ignite instantly on contact, producing a flame temperature of 3000 K. The energy liberated per 0.100 g of CH_6N_2 at constant atmospheric pressure after the products are cooled back to 25°C is 750 J. (a) Find ΔH for the reaction as written. (b) How many kilojoules are liberated (i) when 32.0 g of N_2 is produced and (ii) when 13.5 g of NO_2 is converted to N_2O_4 and consumed?

32. Which is more exothermic, the combustion of one mole of methane to form $CO_2(g)$ and liquid water or the combustion of one mole of methane to form $CO_2(g)$ and steam? Why? (No calculations are necessary.)

33. Which is more exothermic, the combustion of one mole of gaseous benzene or the combustion of one mole of liquid benzene? Why? (No calculations are necessary.)

Thermochemical Equations, ΔH_f^0, and Hess' Law

34. Explain the meaning of the term "thermochemical standard state of a substance."

35. Explain the significance of each word in the term "standard molar enthalpy of formation."

36. From the data in Appendix K, determine the form that represents the standard state for each of the following elements: (a) fluorine, (b) iron, (c) carbon, (d) iodine, (e) oxygen.

37. From the data in Appendix K, determine the form that represents the standard state for each of the following elements: (a) sulfur, (b) hydrogen, (c) phosphorus, (d) chromium, (e) titanium.

38. Write the balanced chemical equation whose ΔH_{rxn}^0 value is equal to ΔH_f^0 for each of the following substances: (a) silicon tetrachloride, $SiCl_4(g)$; (b) ethane, $C_2H_6(g)$; (c) sodium bromide, $NaBr(s)$; (d) calcium fluoride, $CaF_2(s)$; (e) phosgene, $COCl_2(g)$; (f) propane, $C_3H_8(g)$; (g) atomic nitrogen, $N(g)$.

39. Write the balanced chemical equation whose ΔH_{rxn}^0 value is equal to ΔH_f^0 for each of the following substances: (a) hydrogen bromide, $HBr(g)$; (b) magnesium bromide, $MgBr_2(s)$; (c) atomic sodium, $Na(g)$; (d) benzoic acid, $C_6H_5COOH(s)$; (e) hydrogen peroxide, $H_2O_2(\ell)$; (f) dinitrogen pentoxide, $N_2O_5(s)$; (g) ruthenium(III) hydroxide, $Ru(OH)_3(s)$.

*40. We burn 17.5 g of lithium in excess oxygen at constant atmospheric pressure to form Li_2O. Then we bring the reaction mixture back to 25°C. In this process, 735 kJ of heat is given off. What is the standard molar enthalpy of formation of Li_2O?

*41. We burn 17.5 g of magnesium in excess nitrogen at constant atmospheric pressure to form Mg_3N_2. Then we bring the reaction mixture back to 25°C. In this process, 166.1 kJ of heat is given off. What is the standard molar enthalpy of formation of Mg_3N_2?

42. Find ΔH for the reaction

    $$2HCl(g) + F_2(g) \longrightarrow 2HF(\ell) + Cl_2(g)$$

    from the following enthalpies of reaction:

    $$4HCl(g) + O_2(g) \longrightarrow 2H_2O(\ell) + 2Cl_2(g)$$
    $$\Delta H = -148.4 \text{ kJ/mol rxn}$$

    $$HF(\ell) \longrightarrow \tfrac{1}{2}H_2(g) + \tfrac{1}{2}F_2(g)$$
    $$\Delta H = +600.0 \text{ kJ/mol rxn}$$

    $$H_2(g) + \tfrac{1}{2}O_2(g) \longrightarrow H_2O(\ell)$$
    $$\Delta H = -285.8 \text{ kJ/mol rxn}$$

43. Calculate ΔH for

$$Ca^{2+}(aq) + 2OH^-(aq) + CO_2(g) \longrightarrow$$
$$CaCO_3(s) + H_2O(\ell)$$

from the following heats of reaction:

$$CaCO_3(s) \longrightarrow CaO(s) + CO_2(g)$$
$$\Delta H = +178.1 \text{ kJ/mol rxn}$$

$$CaO(s) + H_2O(\ell) \longrightarrow Ca(OH)_2(s)$$
$$\Delta H = -64.8 \text{ kJ/mol rxn}$$

$$Ca(OH)_2(s) \longrightarrow Ca^{2+}(aq) + 2OH^-(aq)$$
$$\Delta H = +11.7 \text{ kJ/mol rxn}$$

44. Use the following thermochemical equations to find ΔH_f^0 for $CuCl_2(s)$.

$$2Cu(s) + Cl_2(g) \longrightarrow 2CuCl(s)$$
$$\Delta H^0 = -274.4 \text{ kJ/mol}$$

$$2CuCl(s) + Cl_2(g) \longrightarrow 2CuCl_2(s)$$
$$\Delta H^0 = -165.8 \text{ kJ/mol}$$

45. Find the heat of formation of liquid hydrogen peroxide at 25°C from the following thermochemical equations.

$$H_2(g) + \tfrac{1}{2}O_2(g) \longrightarrow H_2O(g) \quad \Delta H^0 = -241.818 \text{ kJ/mol}$$

$$2H(g) + O(g) \longrightarrow H_2O(g) \quad \Delta H^0 = -926.919 \text{ kJ/mol}$$

$$2H(g) + 2O(g) \longrightarrow H_2O_2(g) \quad \Delta H^0 = -1070.60 \text{ kJ/mol}$$

$$2O(g) \longrightarrow O_2(g) \quad \Delta H^0 = -498.340 \text{ kJ/mol}$$

$$H_2O_2(\ell) \longrightarrow H_2O_2(g) \quad \Delta H^0 = 51.46 \text{ kJ/mol}$$

46. Use data in Appendix K to find the enthalpy of reaction for
(a) $NH_4NO_3(s) \longrightarrow N_2O(g) + 2H_2O(g)$
(b) $2Fe_3O_4(s) + \tfrac{1}{2}O_2(g) \longrightarrow 3Fe_2O_3(s)$
(c) $KCl(s) + Na(s) \longrightarrow K(s) + NaCl(s)$

47. Repeat Exercise 46 for
(a) $C(s, \text{graphite}) + CO_2(g) \longrightarrow 2CO(g)$
(b) $2HI(g) + F_2(g) \longrightarrow 2HF(g) + I_2(s)$
(c) $2SO_2(g) + O_2(g) \longrightarrow 2SO_3(g)$

48. What is the enthalpy change during the formation of ozone as (a) 1.00 mol O_2 reacts, (b) 1.00 mol O_3 is formed, (c) 1.00 g O_2 reacts, (d) 1.00 g O_3 is formed?

$$3O_2(g) \longrightarrow 2O_3(g) \quad \Delta H^0 = 285.4 \text{ kJ/mol rxn}$$

49. Compare the enthalpy change per mole of iron formed when the oxides Fe_3O_4 and Fe_2O_3 react with aluminum.

$$3Fe_3O_4(s) + 8Al(s) \longrightarrow 4Al_2O_3(s) + 9Fe(s)$$
$$\Delta H^0 = -3347.6 \text{ kJ/mol rxn}$$

$$Fe_2O_3(s) + 2Al(s) \longrightarrow Al_2O_3(s) + 2Fe(s)$$
$$\Delta H^0 = -851.4 \text{ kJ/mol rxn}$$

50. The thermite reaction, used for welding iron, is the reaction of Fe_3O_4 with Al:

$$8Al(s) + 3Fe_3O_4(s) \longrightarrow 4Al_2O_3(s) + 9Fe(s)$$
$$\Delta H^0 = -3347.6 \text{ kJ/mol rxn}$$

Because this large amount of heat cannot be rapidly dissipated to the surroundings, the reacting mass may reach temperatures near 3000°C. How much heat is released by the reaction of 8.0 g of Al with 20.0 g of Fe_3O_4?

51. When a welder uses an acetylene torch, the combustion of acetylene liberates the intense heat needed for welding metals together. The equation for this process is

$$2C_2H_2(g) + 5O_2(g) \longrightarrow 4CO_2(g) + 2H_2O(g)$$

The heat of combustion of acetylene is -1300 kJ/mol of C_2H_2. How much heat is liberated when 0.550 kg of C_2H_2 is burned?

52. Silicon carbide, or carborundum, SiC, is one of the hardest substances known and is used as an abrasive. It has the structure of diamond with half of the carbons replaced by silicon. It is prepared industrially by reduction of sand (SiO_2) with coke (C) in the electric furnace:

$$SiO_2(s) + 3C(s) \longrightarrow SiC(s) + 2CO(g)$$

ΔH^0 for this reaction is 624.6 kJ, and the ΔH_f^0 for $SiO_2(s)$ and $CO(g)$ are -910.9 kJ/mol and -110.5 kJ/mol, respectively. Calculate ΔH_f^0 for silicon carbide.

53. The tungsten used in filaments for light bulbs can be prepared from tungsten(VI) oxide by reduction with hydrogen at 1200°C:

$$114.9 \text{ kJ} + WO_3(s) + 3H_2(g) \longrightarrow W(s) + 3H_2O(g)$$

The ΔH_f^0 value for $H_2O(g)$ is -241.8 kJ/mol. What is ΔH_f^0 for $WO_3(s)$?

*54. Natural gas is mainly methane, $CH_4(g)$. Assume that gasoline is octane, $C_8H_{18}(\ell)$, and that kerosene is $C_{10}H_{22}(\ell)$.
(a) Write the balanced equations for the complete combustion (in excess O_2) of each of these three hydrocarbons. Assume that the products are $CO_2(g)$ and $H_2O(\ell)$.
(b) Calculate ΔH_{rxn}^0 at 25°C for each of the combustion reactions. ΔH_f^0 for $C_{10}H_{22}$ is -249.6 kJ/mol.
(c) When burned at standard conditions, which of these three fuels would produce more heat per mole?
(d) When burned at standard conditions, which of the three would produce more heat per gram?

*55. Methane, $CH_4(g)$, is the main constituent of natural gas. In excess oxygen, methane burns to $CO_2(g)$ and $H_2O(\ell)$, whereas in limited oxygen, the products are $CO(g)$ and $H_2O(\ell)$. Which would result in a higher temperature—a gas–air flame or a gas–oxygen flame? How can you tell?

Bond Energies

56. (a) How is the heat released or absorbed in a *gas phase reaction* related to bond energies of products and reactants?
(b) Hess' Law states that

$$\Delta H_{rxn}^0 = \Sigma \, n\Delta H_{f \text{ products}}^0 - \Sigma \, n\Delta H_{f \text{ reactants}}^0$$

The relationship between ΔH_{rxn}^0 and bond energies for a *gas phase reaction* is

$$\Delta H_{rxn}^0 = \Sigma \text{ bond energies}_{\text{reactants}}$$
$$- \Sigma \text{ bond energies}_{\text{products}}$$

It is *not* true, in general, that ΔH_f^0 for a substance is equal to the negative of the sum of the bond energies of the substance. Why?

57. (a) Suggest a reason for the fact that different amounts of energy are required for the successive removal of the three hydrogen atoms of an ammonia molecule, even though all N—H bonds in ammonia are equivalent.
(b) Suggest why the N—H bonds in different compounds such as ammonia, NH_3; methylamine, CH_3NH_2; and ethylamine, $C_2H_5NH_2$, have slightly different bond energies.

58. Use the bond energies in Table 15-2 to estimate the heat of reaction for

$$2HI(g) + F_2(g) \longrightarrow 2HF(g) + I_2(g)$$

59. Use the bond energies in Table 15-2 to estimate the heat of reaction for

$$NF_3(g) + H_2(g) \longrightarrow NF_2H(g) + HF(g)$$

60. Use the bond energies in Table 15-2 to estimate the heat of reaction for

$$\begin{array}{c} \overset{\displaystyle Cl}{\underset{\displaystyle F}{\mid}} \\ Cl\!-\!\overset{\mid}{C}\!-\!F(g) + F\!-\!F(g) \longrightarrow \\ \underset{\displaystyle |}{|} \end{array}$$

$$\begin{array}{c} \overset{\displaystyle F}{\underset{\displaystyle F}{\mid}} \\ F\!-\!\overset{\mid}{C}\!-\!F(g) + Cl\!-\!Cl(g) \\ \underset{\displaystyle |}{|} \end{array}$$

61. Estimate ΔH for the burning of one mole of butane, using the bond energies in Tables 15-2 and 15-3.

$$\begin{array}{cccc} H & H & H & H \\ | & | & | & | \\ H\!-\!C\!-\!C\!-\!C\!-\!C\!-\!H(g) + \tfrac{13}{2}O\!=\!O(g) \longrightarrow \\ | & | & | & | \\ H & H & H & H \end{array}$$

$$4O\!=\!C\!=\!O(g) + 5H\!-\!O\!-\!H(g)$$

62. Using data in Appendix K, calculate the average O—F bond energy in $OF_2(g)$.

63. Using data in Appendix K, calculate the average H—S bond energy in $H_2S(g)$.

64. Using data in Appendix K, calculate the average S—F bond energy in $SF_6(g)$.

*65. (a) Calculate the carbon–oxygen bond energies in CO_2 and CO. For $CO_2(g)$, $\Delta H_f^0 = -393.5$ kJ/mol; for $CO(g)$, $\Delta H_f^0 = -110.5$ kJ/mol. For $O(g)$ and $C(g)$, $\Delta H_f^0 = 249.2$ kJ/mol and 716.7 kJ/mol, respectively.
(b) Compare these bond energies with those listed in Tables 15-2 and 15-3. Would you classify the bonds as single, double, or triple covalent bonds?

*66. Ethylamine undergoes an endothermic gas phase dissociation to produce ethylene (or ethene) and ammonia:

$$54.68 \text{ kJ} + \text{H}-\overset{\overset{\displaystyle H}{|}}{\underset{\underset{\displaystyle H}{|}}{C}}-\overset{\overset{\displaystyle H}{|}}{\underset{\underset{\displaystyle H}{|}}{C}}-\overset{\displaystyle H}{\underset{\displaystyle H}{N}}: \quad \overset{\Delta}{\longrightarrow}$$

$$\underset{\text{H}}{\overset{\text{H}}{\diagdown}}C=C\underset{\text{H}}{\overset{\text{H}}{\diagup}} \quad + \quad \underset{\text{H}\ \ \text{H}}{\overset{..}{N}}\text{H}$$

The following average bond energies per mole of bonds are given: C—H = 414 kJ, C—C = 347 kJ, C=C = 611 kJ, N—H = 389 kJ. Calculate the C—N bond energy in ethylamine. Compare this with the value in Table 15-2.

*67. Methane undergoes several different exothermic reactions with gaseous chlorine. One of these forms chloroform, $CHCl_3(g)$:

$$CH_4(g) + 3Cl_2(g) \longrightarrow CHCl_3(g) + 3HCl(g)$$
$$\Delta H^0 = -277 \text{ kJ/mol rxn}$$

Given the following average bond energies per mole of bonds: C—H = 414 kJ, Cl—Cl = 243 kJ, H—Cl = 431 kJ. Calculate the average C—Cl bond energy in chloroform. Compare this with the value in Table 15-2.

Born–Haber Cycles and Crystal Lattice Energy

68. The value of ΔH_{ie1} is *always* positive. Yet the ionization of a metal atom to form a metal ion with a noble gas electron configuration, $M(g) \longrightarrow M^{n+}(g) + ne^-$, occurs readily in the formation of many ionic solids from their elements. What is the primary factor that makes the overall process favorable?

69. Sketch the Born–Haber cycle for the formation of $M_2O(s)$ from the alkali metal M(s) and $O_2(g)$. Write the expression for ΔH_f^0 in terms of the various enthalpy changes involved.

*70. Sketch the Born–Haber cycle for the formation of $MBr_2(s)$ from the alkaline earth metal M(s) and $Br_2(\ell)$. Write the expression for ΔH_f^0 in terms of the various enthalpy changes involved.

71. Determine the crystal lattice energy for LiF(s), given the following: sublimation energy for Li = 155 kJ/mol, ΔH_f^0 for F(g) = 78.99 kJ/mol, the first ionization energy of Li = 520 kJ/mol, the electron affinity of fluorine = −322 kJ/mol, and the standard molar enthalpy of formation of LiF(s) = −589.5 kJ/mol.

72. From the following information, determine the crystal lattice energy of KI. Sublimation energy for potassium = 90 kJ/mol, first ionization energy for potassium = 419 kJ/mol, sublimation energy for iodine = 62 kJ/mol, dissociation energy for iodine = 151 kJ/mol, and electron affinity for iodine = −295 kJ/mol.

73. Calculate the electron affinity for chlorine from the following information: ΔH_f^0 for $MgCl_2(s)$ = −642 kJ/mol, ΔH_{subl} for Mg = 151 kJ/mol, the first and second ionization energies for Mg are 738 and 1451 kJ/mol, ΔH_{diss} for Cl_2 = 243 kJ/mol, and the crystal lattice energy for $MgCl_2(s)$ = −2529 kJ/mol.

74. Calculate the crystal lattice energy for $CaBr_2$ from the following information: ΔH_{subl} for calcium = 193 kJ/mol, ΔH_{ie1} for calcium = 590 kJ/mol, ΔH_{ie2} for calcium = 1145 kJ/mol, ΔH_{vap} for Br_2 = 315 kJ/mol [$Br_2(\ell)$ is standard state for bromine], ΔH_{diss} for $Br_2(g)$ = 193 kJ/mol, ΔH_{ea} for bromine = −324 kJ/mol, and ΔH_f^0 for $CaBr_2(g)$ = −675 kJ/mol.

Entropy and Entropy Changes

75. State the Second Law of Thermodynamics. We cannot use ΔS_{univ} directly as a measure of the spontaneity of a reaction. Why?

76. State the Third Law of Thermodynamics. What does it mean?

77. What do we call the quantitative measure of the randomness or disorder in a system? Place the following systems in order of increasing randomness: (a) 1 mol of gas A, (b) 1 mol of solid A, (c) 1 mol of liquid A.

78. Why would you expect a decrease in entropy as a gas condenses? Would this change be as great a decrease as when a liquid sample of the same substance crystallizes?

79. When solid sodium chloride is cooled from 25°C to 0°C, the entropy change is −4.4 J/mol · K. Is this an increase or decrease in randomness? Explain this entropy change in terms of what is happening in the solid at the molecular level.

80. When a one-mole sample of argon gas at 0°C is compressed to one-half its original volume, the entropy change is −5.76 J/mol · K. Is this an increase or a decrease in randomness? Explain this entropy change in terms of what is happening in the gas at the molecular level.

81. Which of the following processes are accompanied by an increase in entropy of the system? (No calculation is necessary.)
(a) A building is constructed from bricks, mortar, lumber, and nails.
(b) A building collapses into bricks, mortar, lumber, and nails.
(c) Iodine sublimes, $I_2(s) \rightarrow I_2(g)$.
(d) White silver sulfate, Ag_2SO_4, precipitates from a solution containing silver ions and sulfate ions.
(e) A marching band is gathered into formation.
(f) A partition is removed to allow two gases to mix.

82. Which of the following processes are accompanied by an increase in entropy of the system? (No calculation is necessary.)
(a) Thirty-five pennies are removed from a bag and placed heads up on a table.
(b) The pennies of (a) are swept off the table and back into the bag.

(c) Water freezes.

(d) Carbon tetrachloride, CCl_4, evaporates.

(e) The reaction $PCl_5(g) \rightarrow PCl_3(g) + Cl_2(g)$ occurs.

(f) The reaction $PCl_3(g) + Cl_2(g) \rightarrow PCl_5(g)$ occurs.

83. For each of the following processes, tell whether the entropy of the *universe* increases, decreases, or remains constant.

(a) melting one mole of ice to water at 0°C

(b) freezing one mole of water to ice at 0°C

(c) freezing one mole of water to ice at −10°C

(d) freezing one mole of water to ice at 0°C and then cooling it to −10°C

84. For each of the following processes, tell whether the entropy of the *system* increases, decreases, or remains constant.

(a) melting one mole of ice to water at 0°C

(b) freezing one mole of water to ice at 0°C

(c) freezing one mole of water to ice at −10°C

(d) freezing one mole of water to ice at 0°C and then cooling it to −10°C

*85. Compare the entropy change for melting one mole of water at 273 K ($\Delta H^0_{273} = 6010$ J/mol for fusion) with that for vaporizing a mole of water at 373 K ($\Delta H^0_{373} = 40{,}660$ J/mol for vaporization). Why is the second value so much larger?

*86. Consider the boiling of a pure liquid at constant pressure. (a) Is ΔS_{sys} greater than, less than, or equal to zero? (b) Is ΔH_{sys} greater than, less than, or equal to zero? (c) Is ΔT_{sys} greater than, less than, or equal to zero?

87. Use S^0 data from Appendix K to calculate the value of ΔS^0_{298} for each of the following reactions. Compare the signs and magnitudes for these ΔS^0_{298} values and explain your observations.

(a) $2NO(g) + H_2(g) \longrightarrow N_2O(g) + H_2O(g)$

(b) $2N_2O_5(s) \longrightarrow 4NO_2(g) + O_2(g)$

(c) $N_2O_5(s) \longrightarrow 2NO_2(g) + \frac{1}{2}O_2(g)$

88. Use S^0 data from Appendix K to calculate the value of ΔS^0_{298} for each of the following reactions. Compare the signs and magnitudes for these ΔS^0_{298} values and explain your observations.

(a) $CaCO_3(s) \longrightarrow CaO(s) + CO_2(g)$

(b) $2SO_2(g) + O_2 \longrightarrow 2SO_3(g)$

(c) $2NO(g) \longrightarrow N_2(g) + O_2(g)$

Free Energy Changes and Spontaneity

89. (a) What are the two factors that favor spontaneity of a process?

(b) What is free energy? What is change in free energy?

(c) Most spontaneous reactions are exothermic, but some are not. Explain.

(d) Explain how the signs and magnitudes of ΔH and ΔS are related to the spontaneity of a process and how they affect it.

90. Which of the following conditions would predict a process that is (a) always spontaneous, (b) always nonspontaneous, or (c) spontaneous or nonspontaneous depending on the temperature and magnitudes of ΔH and ΔS? (i) $\Delta H > 0$, $\Delta S > 0$; (ii) $\Delta H > 0$, $\Delta S < 0$; (iii) $\Delta H < 0$, $\Delta S > 0$; (iv) $\Delta H < 0$, $\Delta S < 0$.

91. For the decomposition of $O_3(g)$ to $O_2(g)$

$$2O_3(g) \longrightarrow 3O_2(g)$$

$\Delta H^0 = -285.4$ kJ/mol and $\Delta S^0 = 137.55$ J/mol · K at 25°C. Calculate ΔG^0 for the reaction. Is the reaction spontaneous? Is either or both the driving forces (ΔH^0 and ΔS^0) for the reaction favorable?

92. Calculate ΔG^0 at 25°C for the reaction

$$2NO_2(g) \longrightarrow N_2O_4(g)$$

given $\Delta H^0 = -57.20$ kJ/mol and $\Delta S^0 = -175.83$ J/mol · K. Is this reaction spontaneous? What is the driving force for spontaneity?

93. The standard state free energy of formation is −286.06 kJ/mol for $NaI(s)$, −261.905 kJ/mol for $Na^+(aq)$, and − 51.57 kJ/mol for $I^-(aq)$ at 25°C. Calculate ΔG^0 for the reaction represented by the equation

$$NaI(s) \xrightarrow{H_2O} Na^+(aq) + I^-(aq)$$

*94. Use the following equations to find the ΔG^0 of formation of $HBr(g)$ at 25°C.

$Br_2(\ell) \longrightarrow Br_2(g)$	$\Delta G^0 = 3.110$ kJ/mol
$HBr(g) \longrightarrow H(g) + Br(g)$	$\Delta G^0 = 339.09$ kJ/mol
$Br_2(g) \longrightarrow 2Br(g)$	$\Delta G^0 = 164.792$ kJ/mol
$H_2(g) \longrightarrow 2H(g)$	$\Delta G^0 = 406.494$ kJ/mol

95. Using values of standard free energy of formation, ΔG^0_f, from Appendix K, calculate the standard free energy change for each of the following reactions at 25°C and 1 atm.

(a) $3NO_2(g) + H_2O(\ell) \longrightarrow 2HNO_3(\ell) + NO(g)$

(b) $SnO_2(s) + 2CO(g) \longrightarrow 2CO_2(g) + Sn(s)$

(c) $2Na(s) + 2H_2O(\ell) \longrightarrow 2NaOH(aq) + H_2(g)$

96. Make the same calculations as in Exercise 95, using values of standard enthalpy of formation and absolute entropy instead of values of ΔG^0_f.

97. Calculate ΔG^0 for the reduction of the oxides of iron and copper by carbon at 700 K represented by the equations

$$2Fe_2O_3(s) + 3C(graphite) \longrightarrow 4Fe(s) + 3CO_2(g)$$

$$2CuO(s) + C(graphite) \longrightarrow 2Cu(s) + CO_2(g)$$

given that the standard free energy of formation is −92 kJ/mol for $CuO(s)$, −637 kJ/mol for $Fe_2O_3(s)$, and −395 kJ/mol for $CO_2(g)$. Which oxide can be reduced using carbon in a wood fire (which has a temperature of about 700 K), assuming standard state conditions?

Temperature Range of Spontaneity

98. Are the following statements true or false? Justify your answer.
(a) An exothermic reaction is spontaneous.
(b) If ΔH and ΔS are both positive, then ΔG will decrease when the temperature increases.
(c) A reaction for which ΔS_{sys} is positive is spontaneous.

99. For the reaction

$$C(s) + O_2(g) \longrightarrow CO_2(g)$$

$\Delta H^0 = -393.509$ kJ/mol and $\Delta S^0 = 2.86$ J/mol · K at 25°C. Does this reaction become more or less favorable as the temperature increases? For the reaction

$$C(s) + \tfrac{1}{2}O_2(g) \longrightarrow CO(g)$$

$\Delta H^0 = -110.525$ kJ/mol and $\Delta S^0 = 89.365$ J/mol · K at 25°C. Does this reaction become more or less favorable as the temperature increases? Compare the temperature dependencies of these reactions.

100. How does the value of ΔG^0 change for the reaction

$$CCl_4(\ell) + H_2(g) \longrightarrow HCl(g) + CHCl_3(\ell)$$

if the reaction is carried out at 65°C rather than at 25°C? At 25°C, $\Delta G^0 = -103.75$ kJ/mol and $\Delta H^0 = -91.34$ kJ/mol for the reaction.

101. The enthalpy of reaction under standard state conditions at 25°C for the combustion of CO,

$$CO(g) + \tfrac{1}{2}O_2(g) \longrightarrow CO_2(g)$$

is -282.984 kJ/mol, and ΔG^0 for the reaction at 25°C is -257.191 kJ/mol. At what temperature will this reaction no longer be spontaneous under standard state conditions?

102. (a) Calculate ΔH^0, ΔG^0, and ΔS^0 for the reaction

$$2H_2O_2(\ell) \longrightarrow 2H_2O(\ell) + O_2(g)$$

at 25°C. (b) Is there any temperature at which $H_2O_2(\ell)$ is stable at 1 atm?

103. Estimate the normal boiling point (the temperature at which liquid and vapor are in equilibrium with each other) of tin(IV) chloride, $SnCl_4$, at 1 atm pressure, using Appendix K.

104. (a) Estimate the normal boiling point of water, at 1 atm pressure, using Appendix K.
(b) Compare the temperature obtained with the known boiling point of water. Can you explain the discrepancy?

105. Sublimation and subsequent deposition onto a cold surface are a common method of purification of I_2 and other solids that sublime readily. Estimate the sublimation temperature (solid to vapor) of the dark-violet solid iodine, I_2, at 1 atm pressure, using the data of Appendix K.

***106.** (a) Is the reaction C(diamond) → C(graphite) spontaneous at 25°C and 1 atm? (b) Now are you worried about your diamonds turning to graphite? Why or why not? (c) Is there a temperature at which diamond and graphite are in equilibrium? If so, what is this temperature? (d) How do you account for the formation of diamonds in the first place? (*Hint*: Diamond has a higher density than graphite.)

***107.** An ice calorimeter, shown below, can be used to measure the amount of heat released or absorbed by a reaction that is carried out at a constant temperature of 0°C. If heat is transferred from the system to the bath, some of the ice melts. A given mass of liquid water has a smaller volume than the same mass of ice, so the total volume of the ice/water mixture decreases. Measuring the volume decrease using the scale at the left indicates the amount of heat released by the reacting system. As long as some ice remains in the bath, the temperature remains at 0°C. In Example 15-4, we saw that the reaction

$$CuSO_4(aq) + 2NaOH(aq) \longrightarrow$$
$$Cu(OH)_2(s) + Na_2SO_4(aq)$$

releases 1.49 kJ of heat at constant temperature and pressure when 50.0 mL of 0.400 M $CuSO_4$ solution and 50.0 mL of 0.600 M NaOH solution are allowed to react. (Because no gases are involved in the reaction, the volume change of the reaction mixture is negligible.) Calculate the change in volume of the ice/water mixture that would be observed if we carried out the same experiment in an ice calorimeter. The density of $H_2O(\ell)$ at 0°C is 0.99987 g/mL and that of ice is 0.917 g/mL. The heat of fusion of ice at 0°C is 334 J/g.

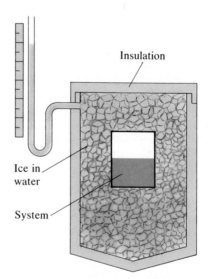

Insulation

Ice in water

System

Mixed Exercises

***108.** Energy to power muscular work is produced from stored carbohydrate (glycogen) or fat (triglycerides). Metabolic consumption and production of energy are described with the nutritional "Calorie," which is equal to 1 kilocalorie. Average energy output per minute for various activities is given: sitting (1.7 kcal); walking, level, 3.5 mph (5.5 kcal); cycling, level, 13 mph (10 kcal); swimming (8.4 kcal); running 10 mph (19 kcal). Approximate energy values of some common foods are also given: large apple (100 kcal); 8-oz cola drink (105 kcal); malted milkshake (8 oz milk, 500 kcal); $\frac{3}{4}$ cup pasta with tomato sauce and cheese (195 kcal); hamburger on bun with burger sauce (350 kcal); 10-oz sirloin steak, including fat (1000 kcal). To maintain body weight, fuel intake should balance energy output. Prepare a table showing (a) each given food, (b) its fuel value, and (c) the minutes of each activity that would balance the kcal of each food.

109. Write the chemical equation for the combination reaction between ethene, C_2H_4, and hydrogen (a hydrogenation reaction) to form ethane, C_2H_6. Find the enthalpy of hydrogenation of ethene using the following thermochemical equations.

$$C_2H_4(g) + 3O_2(g) \longrightarrow 2CO_2(g) + 2H_2O(\ell)$$
$$\Delta H^0 = -1410.9 \text{ kJ/mol}$$

$$C_2H_6(g) + \tfrac{7}{2}O_2(g) \longrightarrow 2CO_2(g) + 3H_2O(\ell)$$
$$\Delta H^0 = -1559.8 \text{ kJ/mol}$$

$$H_2(g) + \tfrac{1}{2}O_2(g) \longrightarrow H_2O(\ell)$$
$$\Delta H^0 = -285.8 \text{ kJ/mol}$$

110. It is difficult to prepare many compounds directly from the elements, so ΔH_f^0 values for these compounds cannot be measured directly. For many organic compounds, it is easier to measure the standard enthalpy of combustion by reaction of the compound with excess $O_2(g)$ to form $CO_2(g)$ and $H_2O(\ell)$. From the following standard enthalpies of combustion at 25°C, determine ΔH_f^0 for the compound.
(a) cyclohexane, $C_6H_{12}(\ell)$, a useful organic solvent: $\Delta H_{combustion}^0 = -3920 \text{ kJ/mol}$
(b) phenol, $C_6H_5OH(s)$, used as a disinfectant and in the production of thermo-setting plastics: $\Delta H_{combustion}^0 = -3053 \text{ kJ/mol}$

*111. Standard entropy changes cannot be measured directly in the laboratory. They are calculated from experimentally obtained values of ΔG^0 and ΔH^0. From the data given below, calculate ΔS^0 at 298 K for each of the following reactions.
(a) $OF_2(g) + H_2O(g) \longrightarrow O_2(g) + 2HF(g)$
   oxygen difluoride

$$\Delta H^0 = -323 \text{ kJ/mol} \qquad \Delta G^0 = -358.4 \text{ kJ/mol}$$

(b) $CaC_2(s) + 2H_2O(\ell) \longrightarrow Ca(OH)_2(s) + C_2H_2(g)$
   calcium carbide   acetylene

$$\Delta H^0 = -125.4 \text{ kJ/mol} \qquad \Delta G^0 = -145.4 \text{ kJ/mol}$$

(c) $2PbS(s) + 3O_2(g) \longrightarrow 2PbO(s) \text{ (red)} + 2SO_2(g)$

$$\Delta H^0 = -830.8 \text{ kJ/mol} \qquad \Delta G^0 = -780.9 \text{ kJ/mol}$$

*112. (a) A student heated a sample of a metal weighing 32.6 g to 99.83°C and put it into 100.0 g of water at 23.62°C in a calorimeter. The final temperature was 24.41°C. The student calculated the specific heat of the metal, but neglected to use the heat capacity of the calorimeter. The specific heat of water is 4.184 J/g · °C. What was his answer? The metal was known to be either chromium, molybdenum, or tungsten. By comparing the value of the specific heat to those of the metals (Cr, 0.460; Mo, 0.250; W, 0.135 J/g · °C), the student identified the metal. What was the metal?
(b) A student at the next laboratory bench did the same experiment, obtained the same data, and used the heat capacity of the calorimeter in his calculations. The heat capacity of the calorimeter was 410 J/°C. Was his identification of the metal different?

113. Find the heat of formation of hydrogen peroxide, $H_2O_2(\ell)$, from $2H_2O_2(\ell) \rightarrow 2H_2O(\ell) + O_2(g)$; $\Delta H = -196.0 \text{ kJ/mol}$ and $\Delta H_f^0 \ H_2O(\ell) = -285.8 \text{ kJ/mol}$.

*114. Calculate q, w, and ΔE for the vaporization of 1.00 g of liquid ethanol (C_2H_5OH) at 1.00 atm at 78.0°C, to form gaseous ethanol at 1.00 atm at 78.0°C. Make the following simplifying assumptions: (i) the density of liquid ethanol at 78.0°C is 0.789 g/mL, and (ii) gaseous ethanol is adequately described by the ideal gas equation. The heat of vaporization of ethanol is 854 J/g.

115. Diagram the Born–Haber cycle for the production of calcium oxide, CaO(s), from Ca(s) and $O_2(g)$. It is not necessary to include values of enthalpy changes for the various steps.

116. (a) The accurately known molar heat of combustion of naphthalene, $C_{10}H_8(s)$, $\Delta H = -5156.8 \text{ kJ/mol } C_{10}H_8$, is used to calibrate calorimeters. The complete combustion of 0.01520 g of $C_{10}H_8$ at constant pressure raises the temperature of a calorimeter by 0.212°C. Find the heat capacity of the calorimeter. (b) The initial temperature of the calorimeter (part a) is 22.102°C; 0.1040 g of $C_8H_{18}(\ell)$, octane (molar heat of combustion $\Delta H = -1303 \text{ kJ/mol } C_8H_{18}$), is completely burned in the calorimeter. Find the final temperature of the calorimeter.

117. When a gas expands suddenly, it may not have time to absorb a significant amount of heat: $q = 0$. Assume that 1.00 mol N_2 expands suddenly, doing 3000 J of work.
(a) What is ΔE for the process? (b) The heat capacity of N_2 is 20.9 J/mol · °C. How much does its temperature fall during this expansion? (This is the principle of most snow-making machines, which use compressed air mixed with water vapor.)

Objectives

As you study this chapter, you should learn

☐ How to express the rate of a chemical reaction in terms of changes in concentrations of reactants and products with time
☐ About the experimental factors that affect the rates of chemical reactions
☐ How to use the rate-law expression for a reaction—the relationship between concentration and rate
☐ How to use the concept of order of a reaction
☐ To apply the method of initial rates to find the rate-law expression for a reaction
☐ How to use the integrated rate-law expression for a reaction—the relationship between concentration and time
☐ How to analyze concentration-versus-time data to determine the order of a reaction

☐ About the collision theory of reaction rates
☐ The main aspects of transition state theory and the role of activation energy in determining the rate of a reaction
☐ How the mechanism of a reaction is related to its rate-law expression
☐ How temperature affects rates of reactions
☐ How to use the Arrhenius equation to relate the activation energy for a reaction to changes in its rate constant with changing temperature
☐ How a catalyst changes the rate of a reaction
☐ About homogeneous catalysis and heterogeneous catalysis

A burning building is an example of a rapid, highly exothermic reaction. Firefighters use basic principles of chemical kinetics. When water is sprayed onto a fire, its evaporation absorbs a large amount of energy; this lowers the temperature and slows the reaction. Other common methods for extinguishing fires include covering them with CO_2 (as with most household extinguishers), which decreases the supply of oxygen, and backburning (for grass and forest fires), which removes combustible material. In both cases, the removal of a reactant slows (or stops) the reaction.

W e are all familiar with processes in which some quantity changes with time—an automobile travels at 40 miles/hour, a faucet delivers water at 3 gallons/minute, or a factory produces 32,000 tires/day. Each of these ratios is called a rate. The **rate of a reaction** describes how fast reactants are used up and products are formed. **Chemical kinetics** is the study of *rates* of chemical reactions and the *mechanisms* (the series of steps) by which they occur.

611

Our experience tells us that different chemical reactions occur at very different rates. For instance, combustion reactions—such as the burning of the methane, CH_4, in natural gas and the combustion of *iso*-octane, C_8H_{18}, in gasoline—proceed very rapidly, sometimes even explosively.

$$CH_4(g) + 2O_2(g) \longrightarrow CO_2(g) + 2H_2O(g)$$

$$2C_8H_{18}(g) + 25O_2(g) \longrightarrow 16CO_2(g) + 18H_2O(g)$$

On the other hand, a chemical change such as the rusting of iron takes place only very slowly.

In our study of thermodynamics we learned to assess whether a particular reaction was favorable. The question of whether substantial reaction would occur in a certain period is addressed by kinetics. If a reaction is not thermodynamically favored, it will not occur appreciably under the given conditions. If a reaction is thermodynamically favored, it can occur but not necessarily at a measurable rate.

The reactions of strong acids with strong bases are thermodynamically favored *and* occur at very rapid rates. Consider, for example, the reaction of one-molar hydrochloric acid solution with solid magnesium hydroxide. It is thermodynamically spontaneous, as indicated by the negative ΔG^0 value under standard state conditions. It also occurs rapidly.

$$2HCl(aq) + Mg(OH)_2(s) \longrightarrow MgCl_2(aq) + 2H_2O(\ell) \qquad \Delta G^0 = -97 \text{ kJ/mol}$$

The reaction of diamond with oxygen is also spontaneous:

$$C(\text{diamond}) + O_2(g) \longrightarrow CO_2(g) \qquad \Delta G^0 = -396 \text{ kJ/mol}$$

However, we know from experience that diamonds exposed to air, even over long periods, do not react to form carbon dioxide. The reaction does not occur at an observable rate at room temperature. This observation is explained by kinetics, not thermodynamics.

16-1 The Rate of a Reaction

To describe the rate of a reaction, we must determine the concentrations of some reactants or products at various times as the reaction proceeds. Devising effective methods for this is a continuing challenge for chemists who study kinetics of reactions. If a reaction is slow enough, we can take samples from the reaction mixture after successive time intervals and then analyze them. For instance, if one reaction product is an acid, its concentration after each time interval can be determined by titration (Section 11-2). The hydrolysis (reaction with water) of ethyl acetate in the presence of a small amount of strong acid produces acetic acid. The extent of the reaction at any desired time can be determined by titration of the acetic acid.

$$\underset{\text{ethyl acetate}}{CH_3\overset{\displaystyle O}{\overset{\|}{C}}-OCH_2CH_3(aq)} + H_2O \xrightarrow{H^+} \underset{\text{acetic acid}}{CH_3\overset{\displaystyle O}{\overset{\|}{C}}-OH(aq)} + \underset{\text{ethanol}}{CH_3CH_2OH(aq)}$$

This approach is suitable only if the reaction is sufficiently slow that the time elapsed during withdrawal and analysis of the sample is negligible. Sometimes the sample that is withdrawn is quickly cooled ("quenched").

The rusting of iron is a complicated process. It can be represented in simplified form as

$$4Fe(s) + 3O_2(g) \rightarrow 2Fe_2O_3(s).$$

This is one of the reactions that occur in a human digestive system when an antacid containing relatively insoluble magnesium hydroxide neutralizes excess stomach acid.

This slows the reaction (Section 16-10) so much that the desired concentration does not change significantly while the analysis is performed.

It is more convenient, especially when a reaction is rapid, to use a technique that measures the change in some physical property of the system. If one of the reactants or products is colored, the increase (or decrease) in intensity of its color might be used to measure a decrease or increase in its concentration. Such an experiment is a special case of *spectroscopic* methods. These methods involve passing light (visible, infrared, or ultraviolet) through the sample. The light should have a wavelength that is absorbed by some substance whose concentration is changing (Figure 16-1). An appropriate light-sensing apparatus then provides a signal that depends on the concentration of the absorbing substance. Modern techniques that use computer-controlled pulsing and sensing of lasers have enabled scientists to sample concentrations at intervals on the order of picoseconds (1 picosecond = 10^{-12} second) or even femtoseconds (1 femtosecond = 10^{-15} second).

If the progress of a reaction causes a change in the total number of moles of gas present, the change in pressure of the reaction mixture (held at constant temperature and constant volume) lets us measure how far the reaction has gone. For instance, the decomposition of dinitrogen pentoxide, $N_2O_5(g)$, has been studied by this method.

$$2N_2O_5(g) \longrightarrow 4NO_2(g) + O_2(g)$$

For every 2 moles of N_2O_5 gas that react, a total of 5 moles of gas is formed (4 moles of NO_2 and 1 mole of O_2). The resulting increase in pressure can be related by the ideal gas equation to the total number of moles of gas present. This indicates the extent to which the reaction has proceeded.

Once we have measured the changes in concentrations of reactants or products with time, how do we describe the rate of a reaction? Consider a hypothetical reaction:

$$aA + bB \longrightarrow cC + dD$$

The amount of each substance present can be given by its concentration, usually expressed as molarity (mol/L) and designated by brackets. The rate at which the reaction proceeds can be described in terms of the rate at which one of the reactants disappears, $-\Delta[A]/\Delta t$ or $-\Delta[B]/\Delta t$, or the rate at which one of the products appears, $\Delta[C]/\Delta t$ or $\Delta[D]/\Delta t$. The reaction rate must be positive because it describes the forward (left-to-right) reaction, which consumes A and B. The concentrations of reactants A and B decrease in the time interval Δt. Thus, $\Delta[A]/\Delta t$ and $\Delta[B]/\Delta t$ would be *negative* quantities. The effect of the negative sign in the definition of rate of reaction is to make the rate a positive quantity.

In reactions involving gases, rates of reactions may be related to rates of change of partial pressures. Pressures of gases and concentrations of gases are directly proportional.

Figure 16-1
A spectroscopic method for determining reaction rates. Light of a wavelength that is absorbed by some substance whose concentration is changing is passed through a reaction chamber. Recording the change in light intensity gives a measure of the changing concentration of a reactant or product as the reaction progresses.

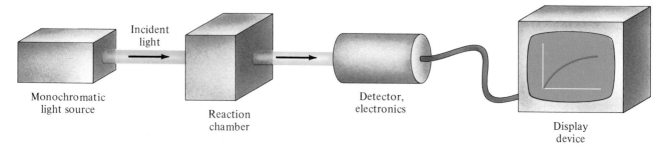

Monochromatic light source

Incident light

Reaction chamber

Detector, electronics

Display device

Rates of reactions are usually expressed as moles per liter per unit time. If we know the chemical equation for a reaction, its rate can be determined by following the change in concentration of any product or reactant that can be detected quantitatively.

If no other reaction takes place, the changes in concentration are related to each other. For every a mol/L that [A] decreases, [B] must decrease by b mol/L, [C] must increase by c mol/L, and so on. We wish to describe the rate of reaction on a basis that is the same regardless of which reactant or product we choose to measure. Therefore, we divide each change by its coefficient in the balanced equation. This allows us to write the rate of reaction based on the rate of change of concentration of each species:

$$\text{Rate of reaction} = \overbrace{\frac{-\Delta[A]}{a\Delta t} = \frac{-\Delta[B]}{b\Delta t}}^{\substack{\text{in terms of}\\\text{reactants}}} = \overbrace{\frac{\Delta[C]}{c\Delta t} = \frac{\Delta[D]}{d\Delta t}}^{\substack{\text{in terms of}\\\text{products}}}$$

This representation gives several equalities, any one of which can be used to relate changes in observed concentrations to the rate of reaction.

As an analogy, suppose we are making sardine sandwiches by the following procedure:

$$2 \text{ bread slices} + 3 \text{ sardines} + 1 \text{ pickle} \longrightarrow 1 \text{ sandwich}$$

The number of sandwiches increases, so Δ(sandwiches) is positive; the rate of the process is given by Δ(sandwiches)/Δ(time). Alternatively, we could

---

Table 16-1
Concentration and Rate Data for Reaction of 2.000 M ICl and 1.000 M H$_2$ at 230°C

[ICl] (mol/L)	[H$_2$] (mol/L)	Average Rate During One Time Interval $= \dfrac{-\Delta[H_2]}{\Delta t}$ (mol · L^{-1} · s^{-1})	Time (t) (seconds)
2.000	1.000		0
		0.326	
1.348	0.674		1
		0.148	
1.052	0.526		2
		0.090	
0.872	0.436		3
		0.062	
0.748	0.374		4
		0.046	
0.656	0.328		5
		0.035	
0.586	0.293		6
		0.028	
0.530	0.265		7
		0.023	
0.484	0.242		8
		0.020	
0.444	0.222		9
		0.016	
0.412	0.206		10

For example, the *average* rate over the interval from 1 to 2 seconds can be calculated as

$$-\frac{\Delta[H_2]}{\Delta t} = -\frac{(0.526 - 0.674)\text{ mol} \cdot L^{-1}}{(2 - 1)\text{ s}}$$
$$= 0.148 \text{ mol} \cdot L^{-1} \cdot s^{-1}$$
$$= 0.148 \ M^{-1} \cdot s$$

This does *not* mean that the reaction proceeds at this rate during the entire interval.

count the decreasing number of pickles at various times. Because Δ(pickles) is negative, we must multiply by (-1) to make the rate positive; Rate = $-\Delta$(pickles)/Δ(time). If we choose to measure the rate by counting slices of bread, we must also take into account that bread slices are consumed *twice as fast* as sandwiches are produced, so rate = $-\frac{1}{2}\Delta$(bread)/Δ(time). Four different ways of describing the rate all have the same numerical value:

$$\text{Rate} = \frac{\Delta(\text{sandwiches})}{\Delta t} = -\frac{\Delta(\text{bread})}{2\Delta t} = -\frac{\Delta(\text{sardines})}{3\Delta t} = -\frac{\Delta(\text{pickles})}{\Delta t}$$

Consider as a specific chemical example the gas-phase reaction of 1.000 mole of hydrogen and 2.000 moles of iodine chloride at 230°C in a closed 1.000-liter container:

$$H_2(g) + 2ICl(g) \longrightarrow I_2(g) + 2HCl(g)$$

The coefficients tell us that 1 mole of H_2 disappears for every 2 moles of ICl that disappear and for every 1 mole of I_2 and 2 moles of HCl that are formed. In other terms, the rate of disappearance of moles of H_2 is one-half the rate of disappearance of moles of ICl, and so on. So we write the rate of reaction as

$$\text{Rate of reaction} = -\left(\begin{array}{c}\text{rate of}\\\text{decrease}\\\text{in }[H_2]\end{array}\right) = -\frac{1}{2}\left(\begin{array}{c}\text{rate of}\\\text{decrease}\\\text{in }[ICl]\end{array}\right) = \left(\begin{array}{c}\text{rate of}\\\text{increase}\\\text{in }[I_2]\end{array}\right) = \frac{1}{2}\left(\begin{array}{c}\text{rate of}\\\text{increase}\\\text{in }[HCl]\end{array}\right)$$

$$\text{Rate of reaction} = \frac{-\Delta[H_2]}{\Delta t} = \frac{-\Delta[ICl]}{2\Delta t} = \frac{\Delta[I_2]}{\Delta t} = \frac{\Delta[HCl]}{2\Delta t}$$

As we see from this discussion, the units of the rate of a reaction are $\dfrac{\text{mol rxn}}{\text{L}\cdot\text{time}}$. We often abbreviate this as mol/L · time or $M^{-1}\cdot\text{time}^{-1}$.

Table 16-1 lists the concentrations of reactants remaining at 1-second intervals, beginning with the time of mixing ($t = 0$ seconds). The *average* rate of reaction over each 1-second interval is indicated in terms of the rate of decrease in concentration of hydrogen. Verify for yourself that the rate of loss of ICl is twice that of H_2. Therefore, the rate of reaction could also be expressed as Rate = $-\Delta[ICl]/2\Delta t$. Increases in concentrations of products could be used instead. Figure 16-2 shows graphically the rates of change of concentrations of all reactants and products.

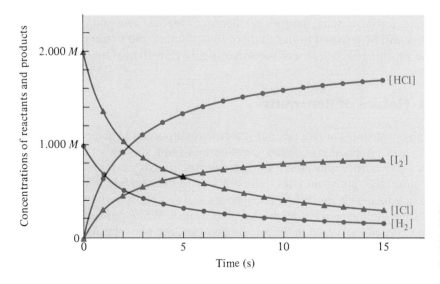

Figure 16-2
Plot of concentrations of all reactants and products versus time in the reaction of 2.000 *M* ICl with 1.000 *M* H_2 at 230°C, from data in Table 16-1 (and a few more points).

Figure 16-3
Plot of H_2 concentration versus time for the reaction of 2.000 M ICl with 1.000 M H_2. The instantaneous rate of reaction at any time, t, equals the negative of the slope of the tangent to this curve at time t. The initial rate of the reaction is equal to the negative of the initial slope ($t = 0$). The determination of the instantaneous rate at $t = 2$ seconds is illustrated. (If you do not recall how to find the slope of a straight line, refer to Figure 16-5.)

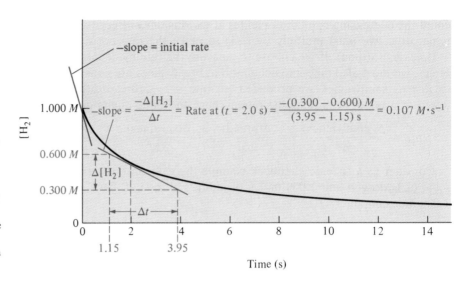

Suppose a driver goes 40 miles in an hour; we describe his average speed (rate) as 40 mi/h. This does not necessarily mean that he traveled at a steady speed. He might have stopped at a few traffic signals, made a fuel stop, driven sometimes faster, sometimes slower—his *instantaneous rate* (the rate at which he was traveling at any instant) was quite changeable.

Figure 16-3 is a plot of the hydrogen concentration versus time, using data of Table 16-1. The initial rate, or the rate at the instant of mixing the reactants, is the negative of the slope at $t = 0$. The *instantaneous* rate of reaction at time t (2.0 seconds, for example) is the negative of the slope of the tangent to the curve at time t. We see that the rate decreases with time; lower concentrations of H_2 and ICl result in slower reaction. Had we plotted concentration of a product versus time, the rate would have been related to the *positive* slope of the tangent.

Factors That Affect Reaction Rates

Four factors have marked effects on the rates of chemical reactions. They are (1) nature of reactants, (2) concentrations of reactants, (3) temperature, and (4) presence of a catalyst. Understanding their effects can help us control the rates of reactions. The study of these factors gives important insight into the molecular processes by which a reaction occurs. This kind of study is the basis for developing theories of chemical kinetics. The remainder of this chapter will be devoted to the study of these factors and to the presentation of the resulting theories—collision theory and transition state theory.

16-2 Nature of Reactants

The physical states of reacting substances are important in determining their reactivities. A puddle of liquid gasoline can burn smoothly, but gasoline vapors can burn explosively. Two immiscible liquids may react slowly at their interface, but if they are intimately mixed to provide better contact, the reaction speeds up. White phosphorus and red phosphorus are different solid forms (allotropes) of elemental phosphorus. White phosphorus ignites when exposed to oxygen in the air. By contrast, red phosphorus can be kept in open containers for long periods of time.

The extent of subdivision of solids or liquids can be crucial in determining reaction rates. Large chunks of most metals do not burn. But many powdered metals, with larger surface areas and hence more atoms exposed to the oxygen of the air, burn easily. One pound of fine iron wire rusts much more rapidly than a solid one-pound chunk of iron. Violent explosions sometimes occur in grain elevators, coal mines, and chemical plants in which large amounts of powdered substances are produced. These explosions are examples of the effect of large surface areas on rates of reaction.

Chemical identities of elements and compounds affect reaction rates. Metallic sodium, with its low ionization energy, reacts rapidly with water at room temperature; metallic calcium has a higher ionization energy and reacts only slowly with water at room temperature. Solutions of a strong acid and a strong base react rapidly when they are mixed because the interactions involve mainly electrostatic attractions between ions in solution. Reactions that involve the breaking of covalent bonds are usually slower.

Samples of dry solid potassium sulfate, K_2SO_4, and dry solid barium nitrate, $Ba(NO_3)_2$, can be mixed with no appreciable reaction occurring for several years. But if aqueous solutions of the two are mixed, a reaction occurs rapidly, forming a white precipitate, barium sulfate.

$$Ba^{2+}(aq) + SO_4^{2-}(aq) \longrightarrow BaSO_4(s)$$

Two allotropes of phosphorus. White phosphorus (above) ignites and burns rapidly when exposed to oxygen in the air, so it is stored under water. Red phosphorus (below) reacts with air much more slowly, and can be stored in contact with air.

16-3 Concentrations of Reactants: The Rate-Law Expression

As the concentrations of reactants change at constant temperature, the rate of reaction changes. We write the **rate-law expression** (often called simply the **rate law**) for a reaction to describe how its rate depends on concentrations; this rate law is experimentally deduced for each reaction from a study of how its rate varies with concentration.

> The rate-law expression for a reaction in which A, B, . . . are reactants has the general form
>
> $$\text{Rate} = k[A]^x[B]^y \cdots$$

The constant k is called the **specific rate constant** (or just the **rate constant**) for the reaction at a particular temperature. The values of the exponents, x and y, and of the rate constant, k, can be determined *only experimentally*. They bear no necessary relationship to the coefficients in the *balanced chemical equation* for the reaction.

The powers to which the concentrations are raised, x and y, are usually integers or zero but are occasionally fractional. A power of *one* means that the rate is directly proportional to the concentration of that reactant. A power of *two* means that the rate increases as the *square* of that concentration. A power of *zero* means that the rate does not depend on the concentration of that reactant, so long as some of the reactant is present. The value of x is said to be the **order** of the reaction with respect to A, and y is the order of the reaction with respect to B. The *overall order* of the reaction is $x + y$. Examples of observed rate laws for three reactions follow.

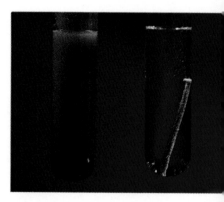

Powdered iron reacts rapidly with dilute sulfuric acid because it has a large total surface area. An iron nail reacts much more slowly.

There is *no way* to predict reaction orders from the balanced overall chemical equation. They must be determined experimentally.

1. $3NO(g) \longrightarrow N_2O(g) + NO_2(g)$

   Rate = $k[NO]^2$ second order in NO; second order overall

2. $2NO_2(g) + F_2(g) \longrightarrow 2NO_2F(g)$

   Rate = $k[NO_2][F_2]$ first order in NO_2 and F_2; second order overall

Any number raised to the zero power is one.

3. $H_2O_2(aq) + 3I^-(aq) + 2H^+(aq) \longrightarrow 2H_2O(\ell) + I_3^-(aq)$

   Rate = $k[H_2O_2][I^-]$ first order in H_2O_2 and I^-; zero order in H^+; second order overall

It is important to remember the following points about the specific rate constant, k:

More details about values and units of k will be discussed in later sections.

1. Its value is for a *specific reaction*, represented by a balanced equation.
2. Its units depend on the *overall order* of the reaction.
3. Its value does not change with concentrations of either reactants or products.
4. Its value does not change with time (Section 16-4).
5. Its value refers to the reaction *at a particular temperature* and changes if we change the temperature (Section 16-10).
6. Its value depends on whether a *catalyst* is present (Section 16-11).
7. Its value must be determined experimentally for the reaction at appropriate conditions.

Usually we know the concentrations of all reactants at the beginning of the reaction. We can then measure the *initial rate* of the reaction, corresponding to these initial concentrations. Let us see how we can use the **method of initial rates** to deduce the rate law from experimentally measured rate data. The following tabulated data refer to the hypothetical reaction

$$A + 2B \longrightarrow AB_2$$

at a specific temperature. The brackets indicate the concentrations of the reacting species *at the beginning* of each experimental run listed in the first column—i.e., the initial concentrations for each experiment.

In such an experiment, we often keep some initial concentrations the same and vary others by simple factors, such as 2 or 3. This makes it easier to assess the effect of each change on the rate.

Experiment	Initial [A]	Initial [B]	Initial Rate of Formation of AB$_2$
1	$1.0 \times 10^{-2} M$	$1.0 \times 10^{-2} M$	$1.5 \times 10^{-4} M \cdot s^{-1}$
2	$1.0 \times 10^{-2} M$	$2.0 \times 10^{-2} M$	$1.5 \times 10^{-4} M \cdot s^{-1}$
3	$2.0 \times 10^{-2} M$	$3.0 \times 10^{-2} M$	$6.0 \times 10^{-4} M \cdot s^{-1}$

Because we are describing the same reaction in each experiment, each is governed by the same rate-law expression. This expression has the form

$$\text{Rate} = k[A]^x[B]^y$$

Let's compare the initial rates of formation of product (reaction rates) for different experimental runs to see how changes in concentrations of reactants affect the rate of reaction. This lets us evaluate x and y, and then k.

We see that the initial concentration of A is the same in experiments 1 and 2; for these trials, any change in reaction rate would be due to different initial concentrations of B. However, the initial rate of formation of product

AB_2 is the same in these two experiments. So we conclude that the rate is independent of [B], or the exponent $y = 0$. Thus far, we know that the rate expression is

$$\text{Rate} = k[A]^x[B]^0 \quad \text{or} \quad \text{Rate} = k[A]^x$$

To evaluate x, we observe that the concentrations of [A] are different in experiments 1 and 3. The [B] concentrations also change, but we have already deduced that the rate does not depend on [B]. So we know that any change in observed rate between experiments 1 and 3 must be due *only* to the changed [A] concentration. Comparing these two experiments, we see that

[A] has been *multiplied* by a factor of
$$(2.0 \times 10^{-2})/(1.0 \times 10^{-2}) = 2.0 = [A] \text{ ratio}$$

Rate changes by a factor of $(6.0 \times 10^{-4})/(1.5 \times 10^{-4}) = 4.0 = $ Rate ratio

The exponent x can be deduced from

$$\text{Rate ratio} = ([A] \text{ ratio})^x$$

$$4.0 = (2.0)^x \quad \text{so} \quad x = 2$$

The reaction is second order in [A]. We can now write the rate-law expression for this reaction:

$$\text{Rate} = k[A]^2$$

The specific rate constant, k, can be evaluated by substituting any of the three sets of data into the rate-law expression. Using the data from experiment 1 gives

$$\text{Rate}_1 = k[A]_1^2 \quad \text{or} \quad k = \frac{\text{Rate}_1}{[A]_1^2}$$

$$k = \frac{1.5 \times 10^{-4} \, M \cdot s^{-1}}{(1.0 \times 10^{-2} \, M)^2} = 1.5 \, M^{-1} \cdot s^{-1}$$

At the temperature at which the measurements were made, the rate-law expression for this reaction is

$$\text{Rate} = k[A]^2 \quad \text{or} \quad \text{Rate} = 1.5 \, M^{-1} \cdot s^{-1} \, [A]^2$$

We could check our result by evaluating k from one of the other sets of data.

An Alternative Method   We can also use a simple algebraic approach to find the exponents in a rate-law expression. Consider the set of rate data given earlier for the hypothetical reaction

$$A + 2B \longrightarrow AB_2$$

Experiment	Initial [A]	Initial [B]	Initial Rate of Formation of AB_2 $(M \cdot s^{-1})$
1	$1.0 \times 10^{-2} \, M$	$1.0 \times 10^{-2} \, M$	1.5×10^{-4}
2	$1.0 \times 10^{-2} \, M$	$2.0 \times 10^{-2} \, M$	1.5×10^{-4}
3	$2.0 \times 10^{-2} \, M$	$3.0 \times 10^{-2} \, M$	6.0×10^{-4}

Margin notes:

$n^0 = 1$ for any value of n. Here, $[B]^0 = 1$.

Alternatively, we could have used data from experiments 2 and 3 to deduce the value of x.

Remember that the specific rate constant k does *not* change with concentration. Only a temperature change or the introduction of a catalyst can change the value of k.

The units of k depend on the overall order of the reaction, consistent with converting the product of concentrations on the right to concentration/time on the left. For any reaction that is second order overall, the units of k are $M^{-1} \cdot \text{time}^{-1}$ or $L/(\text{mol} \cdot \text{time})$.

When heated in air, steel wool glows but does not burn rapidly, due to the low O_2 concentration in air (about 21%). When the glowing steel wool is put into pure oxygen, it burns vigorously because of the much greater accessibility of O_2 reactant molecules.

Because we are describing the same reaction in each experiment, all the experiments are governed by the same rate-law expression,

$$\text{Rate} = k[A]^x[B]^y$$

The initial concentration of A is the same in experiments 1 and 2, so any change in the initial rates for these experiments would be due to different initial concentrations of B. To evaluate y, we solve the ratio of the rate-law expressions of these two experiments for y. We divide the first rate-law expression by the corresponding terms in the second rate-law expression.

$$\frac{\text{Rate}_{(1)}}{\text{Rate}_{(2)}} = \frac{k[A]^x_{(1)}[B]^y_{(1)}}{k[A]^x_{(2)}[B]^y_{(2)}}$$

The value of k always cancels from such a ratio because it is constant at a particular temperature. The initial concentrations of A are equal, so they too cancel. Thus, the expression simplifies to

$$\frac{\text{Rate}_{(1)}}{\text{Rate}_{(2)}} = \left(\frac{[B]_{(1)}}{[B]_{(2)}}\right)^y$$

The only unknown in this equation is y. We substitute data from experiments 1 and 2 into the equation, which gives us

$$\frac{1.5 \times 10^{-4}\ M \cdot s^{-1}}{1.5 \times 10^{-4}\ M \cdot s^{-1}} = \left(\frac{1.0 \times 10^{-2}\ M}{2.0 \times 10^{-2}\ M}\right)^y$$

$$1.0 = (0.5)^y \qquad \text{so} \qquad y = 0$$

Because the units of $\text{Rate}_{(1)}$ and $\text{Rate}_{(2)}$ are identical, they cancel. The units of $[B]_{(1)}$ and $[B]_{(2)}$ are identical, and they too cancel. Thus far, we know that the rate-law expression is

$$\text{Rate} = k[A]^x[B]^0 \qquad \text{or} \qquad \text{Rate} = k[A]^x$$

Next we evaluate x. In experiments 1 and 3, the initial concentration of A is doubled and the rate increases by a factor of four. We neglect the initial concentration of B because we have just shown that it does *not* affect the rate of the reaction.

$$\frac{\text{Rate}_{(3)}}{\text{Rate}_{(1)}} = \frac{k[A]^x_{(3)}}{k[A]^x_{(1)}} = \left(\frac{[A]_{(3)}}{[A]_{(1)}}\right)^x$$

$$\frac{6.0 \times 10^{-4}\ M \cdot s^{-1}}{1.5 \times 10^{-4}\ M \cdot s^{-1}} = \left(\frac{2.0 \times 10^{-2}\ M}{1.0 \times 10^{-2}\ M}\right)^x$$

$$4.0 = (2.0)^x \qquad \text{so} \qquad x = 2$$

The power to which [A] is raised in the rate-law expression is 2, so the rate-law expression for this reaction is the same as that obtained earlier:

$$\text{Rate} = k[A]^2[B]^0 \qquad \text{or} \qquad \text{Rate} = k[A]^2$$

Example 16-1

Given the following data, determine the rate-law expression for the reaction

$$2A + B_2 + C \longrightarrow A_2B + BC$$

Experiment	Initial [A]	Initial [B₂]	Initial [C]	Initial Rate of Formation of BC
1	0.20 M	0.20 M	0.20 M	2.4×10^{-6} $M \cdot min^{-1}$
2	0.40 M	0.30 M	0.20 M	9.6×10^{-6} $M \cdot min^{-1}$
3	0.20 M	0.30 M	0.20 M	2.4×10^{-6} $M \cdot min^{-1}$
4	0.20 M	0.40 M	0.60 M	7.2×10^{-6} $M \cdot min^{-1}$

Plan

The coefficient of BC is one, so the rate of the *reaction* is the same as the rate of formation of BC. The rate law is of the form Rate = $k[A]^x[B_2]^y[C]^z$. We must evaluate x, y, z, and k. We can use the reasoning outlined earlier; in this presentation the first method is used.

The alternative method outlined above could also be used.

Solution

Dependence on [B₂]: In experiments 1 and 3, the initial concentrations of A and C are the same. Thus, any change in the rate would be due to the change in concentration of B_2. But we see that the rate is the same in experiments 1 and 3.

Thus, the reaction rate is independent of [B₂], so $y = 0$. We can neglect [B₂] changes in the subsequent reasoning. The rate law must be

$$Rate = k[A]^x[C]^z$$

Dependence on [C]: Experiments 1 and 4 involve the same initial concentration of A; thus, the observed change in rate must be due entirely to the changed [C]. So we compare experiments 1 and 4 to find z:

[C] has been *multiplied* by a factor of $(0.60)/(0.20) = 3.0 = $ [C] ratio

The effect on rate is that

Rate changes by a factor of $(7.2 \times 10^{-6})/(2.4 \times 10^{-6}) = 3.0 = $ rate ratio

The exponent z can be deduced from

Rate ratio = ([C] ratio)^z

$3.0 = (3.0)^z$ so $z = 1$ The reaction is first order in [C].

Now we know that the rate law is of the form

$$Rate = k[A]^x[C]$$

Dependence on [A]: We can use experiments 1 and 2 to evaluate x, because [A] is changed, [B₂] does not matter, and [C] is unaltered. The observed rate change is due *only* to the changed [A]:

[A] has been *multiplied* by a factor of $(0.40)/(0.20) = 2.0 = $ [A] ratio

The effect on rate is that

Rate changes by a factor of $(9.6 \times 10^{-6})/(2.4 \times 10^{-6}) = 4.0 = $ Rate ratio

The exponent x can be deduced from

Rate ratio = ([A] ratio)^x

$4.0 = (2.0)^x$ so $x = 2$ The reaction is second order in [A].

From these results we can write the complete rate-law expression:

$$Rate = k[A]^2[B_2]^0[C]^1 \quad or \quad Rate = k[A]^2[C]$$

We can evaluate the specific rate constant, k, by substituting any of the four sets of data into the rate-law expression we have just derived. Data from experiment 2 give

$$Rate_{(2)} = k[A]_{(2)}^2[C]_{(2)}$$

$$k = \frac{Rate}{[A]^2[C]} = \frac{9.6 \times 10^{-6} \ M \cdot min^{-1}}{(0.40 \ M)^2(0.20 \ M)} = 3.0 \times 10^{-4} \ M^{-2} \cdot min^{-1}$$

The rate-law expression, Rate $= k[A]^2[C]$, can now be written

$$Rate = 3.0 \times 10^{-4} \ M^{-2} \cdot min^{-1} \ [A]^2[C]$$

This expression allows us to calculate the rate at which this reaction occurs with any known concentrations of A and C (so long as some B_2 is present). As we shall see presently, changes in temperature change reaction rates. This value of k is valid *only* at the temperature at which the data were collected.

Example 16-2

Use the following initial rate data to determine the form of the rate-law expression for the reaction $3A + 2B \rightarrow 2C + D$.

If the rate data provided consisted of the initial rate of formation of C, we would need to divide each value by 2 to find the initial rate of the *reaction*.

Experiment	Initial [A]	Initial [B]	Initial Rate of Formation of D
1	$1.00 \times 10^{-2} \ M$	$1.00 \times 10^{-2} \ M$	$6.00 \times 10^{-3} \ M \cdot min^{-1}$
2	$2.00 \times 10^{-2} \ M$	$3.00 \times 10^{-2} \ M$	$1.44 \times 10^{-1} \ M \cdot min^{-1}$
3	$1.00 \times 10^{-2} \ M$	$2.00 \times 10^{-2} \ M$	$1.20 \times 10^{-2} \ M \cdot min^{-1}$

Plan

The rate law is of the form Rate $= k[A]^x[B]^y$. Let us use the alternative method presented earlier to evaluate x, y, and k.

Solution

The initial concentration of A is the same in experiments 1 and 3. We divide the third rate-law expression by the corresponding terms in the first one.

$$\frac{Rate_{(3)}}{Rate_{(1)}} = \frac{k[A]_{(3)}^x[B]_{(3)}^y}{k[A]_{(1)}^x[B]_{(1)}^y}$$

The initial concentrations of A are equal, so they cancel, as does k. Simplifying and then substituting known values of rates and [B],

$$\frac{Rate_{(3)}}{Rate_{(1)}} = \frac{[B]_{(3)}^y}{[B]_{(1)}^y} \quad \text{or} \quad \frac{1.20 \times 10^{-2} \ M \cdot min}{6.00 \times 10^{-3} \ M \cdot min} = \left(\frac{2.00 \times 10^{-2} \ M}{1.00 \times 10^{-2} \ M}\right)^y$$

$$2.0 = (2.0)^y \quad \text{or} \quad y = 1 \quad \text{The reaction is first order in [B].}$$

No two of the experimental runs have the same concentrations of B, so we must proceed somewhat differently. Let us compare experiments 1 and 2. The observed change in rate must be due to the *combination* of the changes in [A] and [B]. We can divide the second rate-law expression by the corresponding terms in the first one, cancel the equal k values, and collect terms:

$$\frac{Rate_{(2)}}{Rate_{(1)}} = \frac{k[A]_{(2)}^x[B]_{(2)}^y}{k[A]_{(1)}^x[B]_{(1)}^y} = \left(\frac{[A]_{(2)}}{[A]_{(1)}}\right)^x\left(\frac{[B]_{(2)}}{[B]_{(1)}}\right)^y$$

Now let's insert the known values for rates and concentrations and the known [B] exponent of 1:

$$\frac{1.44 \times 10^{-1}\ M \cdot min^{-1}}{6.00 \times 10^{-3}\ M \cdot min^{-1}} = \left(\frac{2.00 \times 10^{-2}\ M}{1.00 \times 10^{-2}\ M}\right)^{x}\left(\frac{3.00 \times 10^{-2}\ M}{1.00 \times 10^{-2}\ M}\right)^{1}$$

$$24.0 = (2.00)^{x}(3.00)$$

$$8.00 = (2.00)^{x} \quad \text{or} \quad \boxed{x = 3} \quad \text{The reaction is third order in [A].}$$

The rate-law expression has the form $\boxed{\text{Rate} = k[A]^{3}[B].}$

EOC 14, 16, 20

16-4 Concentration versus Time: The Integrated Rate Equation

Often we wish to know the concentration of a reactant that would remain after some specified time, or how long it would take for some amount of the reactants to be used up.

> The equation that relates *concentration* and *time* is the **integrated rate equation**. We can also use it to calculate the **half life**, $t_{1/2}$, of a reactant—the time it takes for half of that reactant to be converted into product. The integrated rate equation and the half-life are different for reactions of different order.

We shall look at relationships for some simple cases. (Students who know calculus will be interested in the derivation of the integrated rate equations. This development is presented in the Enrichment Box at the end of this section.)

First-Order Reactions

For reactions involving $aA \rightarrow$ products that are *first order in* A and *first order overall*, the integrated rate equation is

a represents the coefficient of reactant A in the balanced overall equation.

$$\ln\left(\frac{[A]_0}{[A]}\right) = akt \quad \text{or} \quad \log\left(\frac{[A]_0}{[A]}\right) = \frac{akt}{2.303} \quad \text{(first order)}$$

$[A]_0$ is the initial concentration of A, and $[A]$ is its concentration at some time, t, after the reaction begins. Solving this relationship for t gives

$$t = \frac{1}{ak}\ln\left(\frac{[A]_0}{[A]}\right) \quad \text{or} \quad t = \frac{2.303}{ak}\log\left(\frac{[A]_0}{[A]}\right)$$

By definition, $[A] = \frac{1}{2}[A]_0$ at $t = t_{1/2}$. Thus (using the ln form),

$$t_{1/2} = \frac{1}{ak}\ln\frac{[A]_0}{\frac{1}{2}[A]_0} = \frac{1}{ak}\ln(2)$$

The same expression for $t_{1/2}$ would be obtained using the base-10 log form above.

$$t_{1/2} = \frac{\ln 2}{ak} = \frac{0.693}{ak} \qquad \text{(first order)}$$

This relates the half-life of a reactant in a *first-order reaction* and its rate constant, k. In such reactions, the half-life *does not depend* on the initial concentration of A. This is not true for reactions having overall orders other than first order.

Nuclear decay (Chapter 30) is a very important first-order process. Exercises at the end of this chapter involve calculations of nuclear decay rates.

Example 16-3

Compound A decomposes to form B and C in a reaction that is first order with respect to A and first order overall. At 25°C, the specific rate constant for the reaction is 0.0450 s^{-1}. What is the half-life of A at 25°C?

$$A \longrightarrow B + C$$

Plan

We use the equation given above for $t_{1/2}$ for a first-order reaction. The value of k is given in the problem; the coefficient of reactant A is $a = 1$.

Solution

$$t_{1/2} = \frac{\ln 2}{ak} = \frac{0.693}{1(0.0450 \text{ s}^{-1})} = \boxed{15.4 \text{ s}}$$

After 15.4 seconds of reaction, half of the original reactant remains, so that $[A] = \frac{1}{2}[A]_0$.

Example 16-4

The reaction $2N_2O_5(g) \rightarrow 2N_2O_4(g) + O_2(g)$ obeys the rate law Rate $= k[N_2O_5]$, in which the specific rate constant is 0.0840 s^{-1} at a certain temperature. If 2.50 moles of N_2O_5 were placed in a 5.00-liter container at that temperature, how many moles of N_2O_5 would remain after 1.00 minute?

Plan

We shall apply the first-order integrated rate equation:

$$\log\left(\frac{[N_2O_5]_0}{[N_2O_5]}\right) = \frac{akt}{2.303}$$

First we must determine $[N_2O_5]_0$, the original molar concentration of N_2O_5. Then we solve for $[N_2O_5]$, the molar concentration after 1.00 minute. We must remember to express k and t using the same time units. Finally, we shall need to convert molar concentration of N_2O_5 to moles remaining.

Alternatively, we could use the natural log (ln) form of the first-order integrated rate equation.

Solution

The original concentration of N_2O_5 is

$$[N_2O_5]_0 = \frac{2.50 \text{ mol}}{5.00 \text{ L}} = 0.500 \text{ } M$$

Tabulating the other quantities,

$$a = 2 \qquad k = 0.0840 \text{ s}^{-1} \qquad t = 1.00 \text{ min} = 60.0 \text{ s} \qquad [N_2O_5] = \underline{?}$$

The only unknown in the integrated rate equation is $[N_2O_5]$ after 1.00 minute. Let us solve for the unknown. Because $\log x/y = \log x - \log y$,

$$\log\frac{[N_2O_5]_0}{[N_2O_5]} = \log [N_2O_5]_0 - \log [N_2O_5] = \frac{akt}{2.303}$$

$$\log [N_2O_5] = \log [N_2O_5]_0 - \frac{akt}{2.303}$$

$$= \log (0.500) - \frac{(2)(0.0840 \text{ s}^{-1})(60.0 \text{ s})}{2.303} = -0.301 - 0.438 \qquad \text{1.00 minute} = 60.0 \text{ seconds}$$

$$\log [N_2O_5] = -0.739$$

Taking the inverse logarithm of both sides gives

$$[N_2O_5] = 10^{-0.739} = 1.82 \times 10^{-1} \; M$$

Thus, after 1.00 minute of reaction, the concentration of N_2O_5 is 0.182 M. The number of moles of N_2O_5 left in the 5.00-L container is

$$\underline{?} \text{ mol } N_2O_5 = 5.00 \text{ L} \times \frac{0.182 \text{ mol}}{L} = \boxed{0.910 \text{ mol } N_2O_5}$$

EOC 33

Second-Order Reactions

For reactions involving $aA \rightarrow$ products that are *second order with respect to A and second order overall*, the integrated rate equation is

$$\frac{1}{[A]} - \frac{1}{[A]_0} = akt \qquad \left(\begin{array}{l}\text{second order in A,} \\ \text{second order overall}\end{array}\right)$$

For $t = t_{1/2}$, we have $[A] = \frac{1}{2}[A]_0$, so

$$\frac{1}{\frac{1}{2}[A]_0} - \frac{1}{[A]_0} = akt_{1/2}$$

Simplifying and solving for $t_{1/2}$, we obtain the relationship between the rate constant and $t_{1/2}$:

Can you carry out the algebraic steps to solve for $t_{1/2}$?

$$t_{1/2} = \frac{1}{ak[A]_0} \qquad \left(\begin{array}{l}\text{second order in A,} \\ \text{second order overall}\end{array}\right)$$

In this case, $t_{1/2}$ *depends on the initial concentration of* A. This equation may also be used for reactions that are second order overall and first order with respect to each of two reactants initially present in equal concentrations. Figure 16-4 illustrates the different behavior of half-life for first- and second-order reactions.

Example 16-5

Compounds A and B react to form C and D in a reaction that was found to be second order overall and second order in A. The rate constant at 30°C is 0.622 liter per mole per minute. What is the half-life of A if $4.10 \times 10^{-2} \; M$ A is mixed with excess B?

$$A + B \longrightarrow C + D$$

Figure 16-4
(a) Plot of concentration versus time for a first-order reaction. The first half-life, 1.73 seconds, is the time required for the concentration to fall from 1.00 M to 0.50 M. An additional 1.73 seconds is required for the concentration to fall by half again, from 0.50 M to 0.25 M, and so on. For a first-order reaction, $t_{1/2} = \dfrac{\ln 2}{ak} = \dfrac{0.693}{ak}$; $t_{1/2}$ does not depend on the concentration at the beginning of that time period.
(b) Plot of concentration versus time for a second-order reaction. The same values are used for a, $[A]_0$, and k as in part (a). The first half-life, 2.50 seconds, is required for the concentration to fall from 1.00 M to 0.50 M. The concentration falls by half again from 2.50 to 7.50 seconds, so the second half-life is 5.00 seconds. The half-life beginning at 0.25 M is 10.00 seconds. For a second-order reaction, $t_{1/2} = \dfrac{1}{ak[A]_0}$; $t_{1/2}$ is inversely proportional to the concentration at the beginning of that time period.

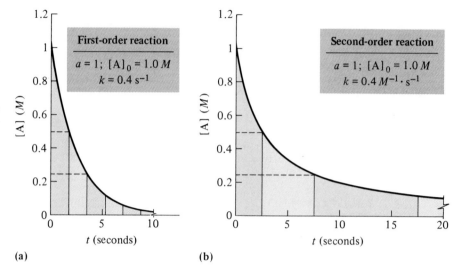

(a)   (b)

Plan

As long as some B is present, only the concentration of A affects the rate. The reaction is second order in [A] and second order overall, so we use the appropriate equation for the half-life.

Solution

$$t_{1/2} = \frac{1}{ak[A]_0} = \frac{1}{(1)(0.622\ M^{-1}\cdot min^{-1})(4.10\times 10^{-2}\ M)} = \boxed{39.2\ min}$$

Example 16-6

The gas-phase decomposition of NOBr is second order in [NOBr], with $k = 0.810$ $M^{-1}\cdot s^{-1}$ at 10°C. We start with $4.00\times 10^{-3}\ M$ NOBr in a flask at 10°C. How many seconds does it take to use up $1.50\times 10^{-3}\ M$ of this NOBr?

$$2NOBr(g) \longrightarrow 2NO(g) + Br_2(g) \qquad Rate = k[NOBr]^2$$

Plan

We must first determine what concentration of NOBr remains after 1.50×10^{-3} M is used up. Then we use the second-order integrated rate equation to determine the time required to reach that concentration.

Solution

$$\underline{?}\ M\ NOBr\ remaining = (0.00400 - 0.00150)\ M = 0.00250\ M = [NOBr]$$

We solve the integrated rate equation $\dfrac{1}{[NOBr]} - \dfrac{1}{[NOBr]_0} = akt$ for t.

The coefficient of NOBr is $a = 2$.

$$t = \frac{1}{ak}\left(\frac{1}{[NOBr]} - \frac{1}{[NOBr]_0}\right) = \frac{1}{(2)(0.810\ M^{-1}\cdot s^{-1})}\left(\frac{1}{0.00250\ M} - \frac{1}{0.00400\ M}\right)$$

$$= \frac{1}{1.62\ M^{-1}\cdot s^{-1}}(400\ M^{-1} - 250\ M^{-1})$$

$$= \boxed{92.6\ s\ (1.54\ min)}$$

Example 16-7

Consider the reaction of Example 16-6 at 10°C. What concentration of NOBr will remain after 5.00 minutes of reaction, starting with $2.40\times 10^{-3}\ M$ of NOBr?

Plan

We use the second-order integrated rate equation to solve for the concentration remaining at $t = 5.00$ minutes.

Solution

Again, we start with the expression $\dfrac{1}{[NOBr]} - \dfrac{1}{[NOBr]_0} = akt$. Let us put in the known values and solve for [NOBr]:

$$\frac{1}{[NOBr]} - \frac{1}{2.40 \times 10^{-3}\ M} = (2)(0.810\ M^{-1} \cdot s^{-1})(5.00\ min)\left(\frac{60\ s}{1\ min}\right)$$

$$\frac{1}{[NOBr]} - 4.17 \times 10^2\ M^{-1} = 486\ M^{-1}$$

$$\frac{1}{[NOBr]} = 486\ M^{-1} + 417\ M^{-1} = 903\ M^{-1}$$

$$[NOBr] = \boxed{1.11 \times 10^{-3}\ M} \qquad (46\% \text{ remains unreacted})$$

Thus, about 54% of the original concentration of NOBr reacts within the first 5 minutes. This is reasonable because, as you may easily verify, the reaction has an initial half-life of 257 seconds, or 4.29 minutes.

EOC 32, 34

You may prefer to rearrange the original equation to solve for [NOBr] before substituting values.

Zero-Order Reaction

For a reaction $aA \rightarrow$ products that is zero order, we can write the rate-law expression as

$$-\frac{\Delta[A]}{a\Delta t} = k$$

The corresponding integrated rate equation is

$$[A] = [A]_0 - akt \qquad \text{(zero order)}$$

and the half-life is

$$t_{1/2} = \frac{[A]_0}{2ak} \qquad \text{(zero order)}$$

Calculus Derivation of Integrated Rate Equations

Enrichment

The derivation of the integrated rate equation is an example of the use of calculus in chemistry. The following derivation is for a reaction that is assumed to be first order in A and first order overall. If you do not know calculus, you can still use the results of this derivation, as we have already shown in this section. For the reaction

$$aA \rightarrow \text{products}$$

the rate is expressed as

$$\text{Rate} = -\frac{\Delta[A]}{a\Delta t}$$

For a first-order reaction, the rate is proportional to the first power of [A]:

$$-\frac{\Delta[A]}{a\Delta t} = k[A]$$

In calculus terms, we express the change during an infinitesimally short time as the derivative of [A] with respect to time:

$$-\frac{1}{a}\frac{d[A]}{dt} = k[A]$$

Separating variables, we obtain

$$-\frac{d[A]}{[A]} = (ak)dt$$

We integrate this equation to get

$$-\ln[A] = akt + C$$

We evaluate C, the constant of integration, by setting $t = 0$. This refers to the beginning of the reaction. At that time, the concentration of A is $[A]_0$, its initial value, so $C = -\ln[A]_0$.

$$-\ln[A] = akt - \ln[A]_0$$

We now rearrange the equation, remembering that $\ln(x) - \ln(y) = \ln(x/y)$:

$$\ln[A]_0 - \ln[A] = akt$$

$$\ln\frac{[A]_0}{[A]} = akt \qquad \text{(first order)}$$

This is the integrated rate equation for a reaction that is first order in a reactant A and first order overall.

Integrated rate equations can be derived similarly from other simple rate laws. For a reaction $aA \rightarrow$ products that is second order in a reactant A and second order overall, we can write the rate equation as

$$-\frac{d[A]}{adt} = k[A]^2$$

Again, using the methods of calculus, we can separate variables, integrate, and rearrange to obtain the corresponding second-order integrated rate equation:

$$\frac{1}{[A]} - \frac{1}{[A]_0} = akt \qquad \text{(second order)}$$

For a reaction $aA \rightarrow$ products that is zero order overall, we can write the rate equation as

$$-\frac{d[A]}{a\,dt} = k$$

In this case, the calculus derivation already described leads to the zero-order integrated rate equation

$$[A] = [A]_0 - akt \qquad \text{(zero order)}$$

16-5 Using Integrated Rate Equations to Determine Reaction Order

The integrated rate equation can help us to analyze concentration-versus-time data to determine reaction order. A graphical approach is often used. We can rearrange the first-order integrated rate equation

$$\ln\frac{[A]_0}{[A]} = akt$$

as follows. Remembering that the logarithm of a quotient, $\ln (x/y)$, is equal to the difference of the logarithms, $\ln x - \ln y$, we can write

$$\ln [A]_0 - \ln [A] = akt \qquad \text{or} \qquad \ln [A] = -akt + \ln [A]_0$$

Recall that the equation for a straight line may be written

$$y = mx + b$$

where y is the variable plotted along the ordinate (vertical axis), x is the variable plotted along the abscissa (horizontal axis), m is the slope of the line, and b is the intercept of the line with the y axis (Figure 16-5). If we compare the last two equations, we find that $\ln [A]$ can be interpreted as y, and t as x.

$$\underbrace{\ln [A]}_{\displaystyle y} = \underbrace{-ak}_{\displaystyle m}\,\underbrace{t}_{\displaystyle x} + \underbrace{\ln [A]_0}_{\displaystyle b}$$

The quantity $-ak$ is a constant as the reaction proceeds, so it can be interpreted as m. The initial concentration of A is fixed, so $\ln [A]_0$ is a constant

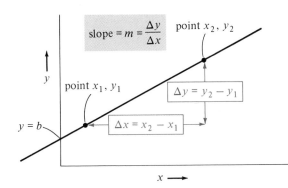

Figure 16-5
Plot of the equation $y = mx + b$, where m and b are constant. The slope of the line (positive in this case) is equal to m; the intercept on the y axis is equal to b.

Figure 16-6

Plot of ln [A] versus time for a reaction $aA \rightarrow$ products that follows first-order kinetics. The observation that such a plot gives a straight line would confirm that the reaction is first order in [A] and first order overall, i.e., Rate = k[A]. The slope is equal to $-ak$. Because a and k are positive numbers, the slope of the line is always negative. Logarithms are dimensionless, so the slope has the units (time)$^{-1}$. A plot of (base-10) log [A] for a first-order reaction would give a straight line with a slope equal to $-ak/2.303$. The logarithm of a quantity less than 1 is negative, so data points for concentrations less than 1 molar would appear below the time-axis.

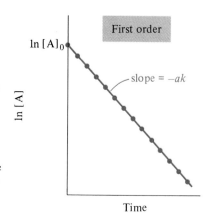

for each experiment, and ln [A]$_0$ can be interpreted as b. Thus, a plot of ln [A] versus time for a first-order reaction would be expected to give a straight line (Figure 16-6) with the slope of the line equal to $-ak$ and the intercept equal to ln [A]$_0$. The same analysis of the integrated rate equation in terms of base-10 logs shows that a plot of log[A] versus time for a first-order reaction would be expected to give a straight line with the slope of the line equal to $-ak/2.303$ and the intercept equal to log [A]$_0$.

In similar fashion we can work with the integrated rate equation for a reaction that is second order in A and second order overall. We rearrange

$$\frac{1}{[A]} - \frac{1}{[A]_0} = akt \qquad \text{to read} \qquad \frac{1}{[A]} = akt + \frac{1}{[A]_0}$$

Again comparing this with the equation for a straight line, we see that a plot of 1/[A] versus time would be expected to give a straight line (Figure 16-7). The line would have a slope equal to ak and an intercept equal to 1/[A]$_0$.

For a zero-order reaction, we can rearrange the integrated rate equation

$$[A]_0 - [A] = akt \qquad \text{to} \qquad [A] = -akt + [A]_0$$

Comparing this with the equation for a straight line, we see that a straight-line plot would be obtained by plotting concentration versus time, [A] versus t. The slope of this line is $-ak$, and the intercept is [A]$_0$.

This discussion suggests another way to deduce an unknown rate-law expression from experimental concentration data. The following approach is particularly useful for any decomposition reaction, one that involves only one reactant:

$$aA \longrightarrow \text{products}$$

We plot the data in various ways as suggested above. *If* the reaction followed zero-order kinetics, *then* a plot of [A] versus t would give a straight line. But *if* the reaction followed first-order kinetics, *then* a plot of ln [A] versus t (or log [A] versus t) would give a straight line whose slope could be

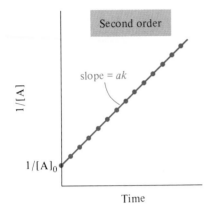

Figure 16-7
Plot of $1/[A]$ versus time for a reaction $aA \rightarrow$ products that follows second-order kinetics. The observation that such a plot gives a straight line would confirm that the reaction is second order in $[A]$ and second order overall, i.e., Rate = $k[A]^2$. The slope is equal to ak. Because a and k are positive numbers, the slope of the line is always positive. Because concentrations cannot be negative, $1/[A]$ is always positive, and the line is always above the time-axis.

interpreted to derive a value of k. *If* the reaction were second order in A and second order overall, *then* neither of these plots would give a straight line, but a plot of $1/[A]$ versus t would. If none of these plots gave a straight line (within expected scatter due to experimental error), we would know that none of these is the correct order (rate law) for the reaction. Plots to test for other orders can be devised, as can graphical tests for rate-law expressions involving more than one reactant, but those are subjects for more advanced texts. The graphical approach that we have described is illustrated in the following example.

It is not possible for *all* of the plots suggested here to yield straight lines for a given reaction. However, the nonlinearity of the plots may not become obvious if the reaction times used are too short.

Example 16-8

We carry out the reaction $A \rightarrow B + C$ at a particular temperature. As the reaction proceeds, we measure the molarity of the reactant, $[A]$, at various times, observing the following results:

Time (min)	0.00	1.00	2.00	3.00	4.00	5.00	6.00	7.00	8.00	9.00	10.00
[A] (mol/L)	2.000	1.488	1.107	0.823	0.612	0.455	0.339	0.252	0.187	0.139	0.104

(a) Plot $[A]$ versus time. (b) Plot log $[A]$ versus time. (c) Plot $1/[A]$ versus time. (d) What is the order of the reaction? (e) Write the rate-law expression for the reaction. (f) What is the value of k at this temperature?

Plan
For parts (a)–(c), we use the observed data to make the required plots, calculating related values as necessary. (d) We can determine the order of the reaction by observing which of these plots gives a straight line. (e) Knowing the order of the reaction, we can write the rate-law expression. (f) The value of k can be determined from the slope of the straight-line plot.

Solution
(a) The plot of $[A]$ versus time is given in Figure 16-8a.
(b) We first use the given data to calculate the log $[A]$ column in Figure 16-8b. These data are then used to plot log $[A]$ versus time, as shown in Figure 16-8c.

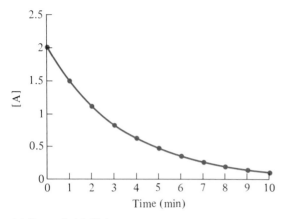

(a) Example 16–8(a).

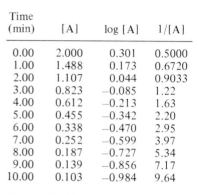

Time (min)	[A]	log [A]	1/[A]
0.00	2.000	0.301	0.5000
1.00	1.488	0.173	0.6720
2.00	1.107	0.044	0.9033
3.00	0.823	−0.085	1.22
4.00	0.612	−0.213	1.63
5.00	0.455	−0.342	2.20
6.00	0.338	−0.470	2.95
7.00	0.252	−0.599	3.97
8.00	0.187	−0.727	5.34
9.00	0.139	−0.856	7.17
10.00	0.103	−0.984	9.64

(b) Data for Example 16–8.

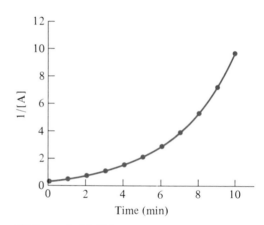

(c) Example 16–8(b).

(d) Example 16–8(c)

Figure 16-8
Plots and data conversion for Example 16-8. (a) Plot of [A] versus time using the data given. (b) The data are used to calculate the two columns log [A] and 1/[A]. (c) Plot of log [A] versus time. The observation that this plot gives a straight line indicates that the reaction follows first-order kinetics. (d) Plot of 1/[A] versus time. If the reaction had followed second-order kinetics, this plot would have resulted in a straight line, and the plot in Figure 16-8c would not.

(c) The given data are used to calculate the 1/[A] column in Figure 16-8b. Then we plot 1/[A] versus time, as shown in Figure 16-8d.

(d) It is clear from the answer to part (b) that the plot of log [A] versus time gives

a straight line. This tells us that   the reaction is first order in [A].

(e) Putting our answer to part (d) in the form of a rate-law expression, we write

Rate = k[A].

(f) We can use the straight-line plot in Figure 16-8c to determine the value of the rate constant for this first-order reaction from the relationship

$$\text{slope} = -\frac{ak}{2.303} \quad \text{or} \quad k = -\frac{2.303(\text{slope})}{a}$$

To determine the slope of the straight line, we can pick any two points, such as P and Q, on the line. From their coordinates, we calculate

$$\text{slope} = \frac{\text{change in ordinate}}{\text{change in abscissa}} = \frac{(-0.90) - (0.00)}{(9.30 - 2.29)\ \text{min}} = -0.128\ \text{min}^{-1}$$

$$k = -\frac{2.303(\text{slope})}{a} = -\frac{(2.303)(-0.128\ \text{min}^{-1})}{1} = \boxed{0.295\ \text{min}^{-1}}$$

You should confirm that a straight line would also be observed by plotting ln [A] versus time, and that proper interpretation of its slope would give the same value for k.

EOC 38, 39

> Remember that the ordinate is the vertical axis and the abscissa is the horizontal one. If you are not careful to keep the points in the same order in the numerator and denominator, you will get the wrong sign for the slope.

Another method of assessing reaction order is based on comparing successive half-lives. As we saw in Section 16-4, $t_{1/2}$ for a first-order reaction does not depend on initial concentration. We can measure the time required for different concentrations of a reactant to fall to half of their original values. If this time remains constant, it is an indication that the reaction is first order for that reactant and first order overall (Figure 16-4a). By contrast, for other orders of reaction, $t_{1/2}$ would change depending on initial concentration. For a second-order reaction, successively measured $t_{1/2}$ values would increase as $[A]_0$ (measured at the *beginning of a particular measurement period*) decreases (Figure 16-4b).

> You should test this method using the concentration-versus-time data of Example 16-8, plotted in Figure 16-8a.

16-6 Summary of Concentration Effects

Table 16-2 summarizes the relationships that were presented in Sections 16-3, 16-4, and 16-5.

16-7 Collision Theory of Reaction Rates

The fundamental notion of the **collision theory of reaction rates** is that

> for reaction to occur, molecules, atoms, or ions must first collide.

Increased concentrations of reacting species result in greater numbers of collisions per unit time. However, not all collisions result in reaction; i.e., not all collisions are **effective collisions**. For a collision to be effective, the reacting species must (1) possess at least a certain minimum energy necessary to rearrange outer electrons in breaking bonds and forming new ones and (2) have the proper orientations toward each other at the time of collision. That is, collisions must occur in order for a chemical reaction to proceed, but they do not guarantee that a reaction will occur.

A collision between atoms, molecules, or ions is not like one between two hard billiard balls. Whether or not chemical species "collide" depends on the distance at which they can interact with one another. For instance, the gas-phase ion–molecule reaction $CH_4^+ + CH_4 \rightarrow CH_5^+ + CH_3$ can

Table 16-2
Summary of Relationships for Various Orders of the Reaction $aA \rightarrow$ Products

	Order		
	Zero	First	Second
Rate-law expression	Rate $= k$	Rate $= k[A]$	Rate $= k[A]^2$
Integrated rate equation	$[A] = [A]_0 - akt$	$\ln\dfrac{[A]_0}{[A]} = akt$ or $\log\dfrac{[A]_0}{[A]} = \dfrac{akt}{2.303}$	$\dfrac{1}{[A]} - \dfrac{1}{[A]_0} = akt$
Half-life, $t_{1/2}$	$\dfrac{[A]_0}{2ak}$	$\dfrac{\ln 2}{ak} = \dfrac{0.693}{ak}$	$\dfrac{1}{ak[A]_0}$
Plot that gives straight line	$[A]$ vs. t	$\ln[A]$ vs. t or $\log[A]$ vs. t	$\dfrac{1}{[A]}$ vs. t
Direction of straight-line slope	down with time	down with time	up with time
Interpretation of slope	$-ak$	$-ak$ (for ln plot) or $-\dfrac{ak}{2.303}$ (for log plot)	ak
Interpretation of intercept	$[A]_0$	$\ln[A]_0$ or $\log[A]_0$	$\dfrac{1}{[A]_0}$

occur with a fairly long-range contact. This is because the interactions between ions and induced dipoles are effective over a long distance. By contrast, the reacting species in the gas reaction $CH_3 + CH_3 \rightarrow C_2H_6$ are both neutral. They interact appreciably only through very short-range forces between induced dipoles, so they must approach one another very closely before we would say that they "collide."

$Zn(s) + 2H^+(aq) \longrightarrow Zn^{2+}(aq) + H_2(g)$

Dilute sulfuric acid reacts slowly with zinc metal (left), whereas more concentrated acid reacts rapidly (right). The $H^+(aq)$ concentration is higher in the more concentrated acid, and so more $H^+(aq)$ ions collide with Zn per unit time.

Recall from Chapter 12 that the average kinetic energy of a collection of molecules is proportional to the absolute temperature. At higher temperatures, more of the molecules possess sufficient energy to react (Section 16-10).

For colliding molecules to react, they must have the proper orientations. If they have improper orientations, they do not react even though they may possess sufficient energy. Figure 16-9 depicts collisions between molecules of NO and N_2O. We assume that each possesses sufficient energy to react according to the equation

$$NO + N_2O \longrightarrow NO_2 + N_2$$

For many reactions, a heterogeneous catalyst (Section 16-11) can increase the fraction of colliding molecules that have the proper orientations.

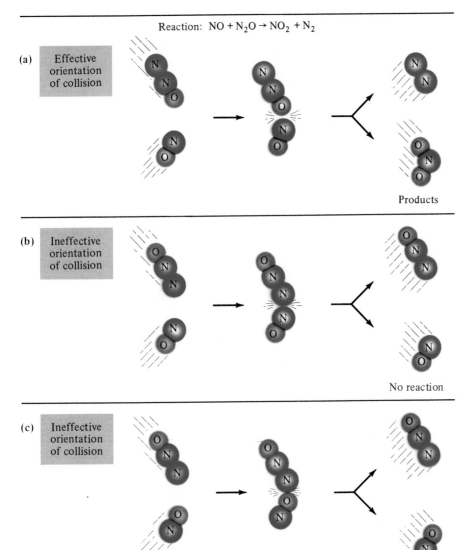

Reaction: $NO + N_2O \rightarrow NO_2 + N_2$

(a) Effective orientation of collision

Products

(b) Ineffective orientation of collision

No reaction

(c) Ineffective orientation of collision

No reaction

Figure 16-9
Some possible collisions between N_2O and NO molecules in the gas phase. (a) A collision that could be effective in producing the reaction. (b, c) Collision would be ineffective. The molecules must have the proper orientations relative to each other *and* have sufficient energy to react.

16-8 Transition State Theory

Chemical reactions involve the making and breaking of chemical bonds. The energy associated with a chemical bond is a form of potential energy. Reactions are accompanied by changes in potential energy. Consider the following hypothetical, one-step *exothermic* reaction at a certain temperature:

$$A + B_2 \longrightarrow AB + B + \text{heat}$$

The "reaction coordinate" represents the *progress along the pathway* leading from reactants to products. This coordinate is sometimes labeled "progress of reaction."

Figure 16-10 shows a plot of potential energy versus reaction coordinate. In (a) the ground state energy of the reactants, A and B_2, is higher than the ground state energy of the products, AB and B. The energy released in the reaction is the difference between these two energies, ΔE. It is related to the change in enthalpy or heat content.

Quite often, for reaction to occur, some covalent bonds must be broken so that others can be formed. This can occur only if the molecules collide *with enough kinetic energy* to overcome the potential energy stabilization of the bonds. According to the **transition state theory**, the reactants pass through a short-lived, high-energy intermediate state, called a **transition state**, before the products are formed.

$$A + B\text{—}B \longrightarrow A\text{---}B\text{---}B \longrightarrow A\text{—}B + B$$

reactants	transition state	products
$A + B_2$	AB_2	$AB + B$

The **activation energy**, E_a, is the additional energy that must be absorbed by the reactants in their ground states to allow them to reach the transition state. If A and B_2 do not possess the necessary amount of energy, E_a, above their ground states when they collide, no reaction will occur. If they do possess sufficient energy to "climb the energy barrier" to reach the transition state, the reaction can proceed. When the atoms go from the transition state arrangement to the product molecules, energy is *released*. If the reaction results in a net *release* of energy (Figure 16-10a), *more* energy than the activation energy is returned to the surroundings. If the reaction results in a *net absorption* of energy (Figure 16-10b), an amount less than E_a is given

Figure 16-10

A potential energy diagram. (a) A reaction that releases energy (exothermic). An example of such a reaction is

$$H + I_2 \longrightarrow HI + I$$

(b) A reaction that absorbs energy (endothermic). An example of such a reaction is

$$I + H_2 \longrightarrow HI + H$$

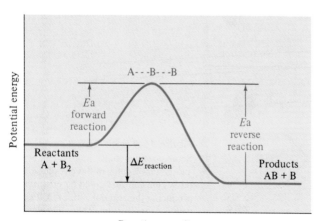

(a)

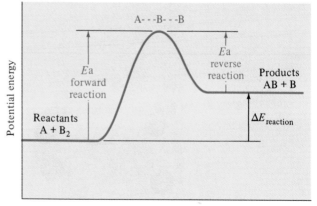

(b)

off when the transition state is converted to products. Thus, the activation energy must be supplied to the system from its environment, but some of that energy is subsequently released to the surroundings. The *net* release of energy is ΔE.

When the reverse reaction occurs, an increase in energy equal to the reverse activation energy, $E_{a\ reverse}$, is required to convert the AB product molecules to the transition state. As you can see from the potential energy diagrams in Figure 16-10,

Remember that the ΔE relates product energy to reactant energy, regardless of the pathway.

$$E_{a\ forward} - E_{a\ reverse} = \Delta E_{reaction}$$

As we shall see, increasing the temperature changes the rate by altering the fraction of molecules that can get over a given energy barrier (Section 16-10). Introducing a catalyst changes the rate by lowering the barrier (Section 16-11).

As a specific example of the ideas of collision theory and transition state theory, consider the reaction of iodide ions with chloroform.

$$I^- + CH_3Cl \longrightarrow CH_3I + Cl^-$$

Extensive studies have established that this reaction proceeds as shown in Figure 16-11a. The I^- ion must approach the CH_3Cl from the "back side" of the C—Cl bond, through the middle of the three hydrogen atoms. A collision of the I^- ion with the CH_3Cl molecules from any other angle does not lead to reaction. But a collision in the appropriate orientation could allow

The CH_3Cl and CH_3I molecules are (distorted) tetrahedra.

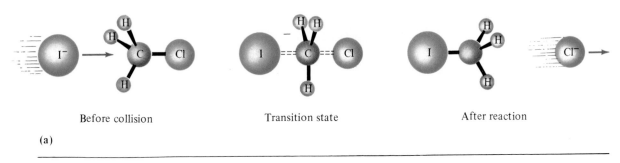

| Before collision | Transition state | After reaction |

(a)

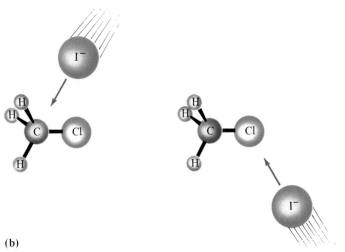

(b)

Figure 16-11
(a) A collision that could lead to reaction of $I^- + CH_3Cl$ to give $CH_3I + Cl^-$. The I^- must approach along the "back side" of the C—Cl bond. (b) Two collisons that are not in the "correct" orientation.

the new I—C bond to form at the same time that the C—Cl bond is breaking. This collection of atoms, which we might write as

$$I\text{---}\underset{\underset{H}{|}}{\overset{\overset{H\quad H}{\diagdown\diagup}}{C}}\text{---}Cl$$

is what we mean by the transition state of this reaction (Figure 16-11b). From this state, either of two things could happen: (1) the I—C bond could finish forming and the C—Cl bond could finish breaking with Cl^- leaving, leading to products, or (2) the I—C bond could fall apart with I^- leaving, and the C—Cl bond could re-form, leading back to reactants.

16-9 Reaction Mechanisms and the Rate-Law Expression

The step-by-step pathway by which a reaction occurs is called its **mechanism**. Some reactions take place in a single step, but most reactions occur in a series of **elementary steps**. The reaction orders *for any single elementary step* are equal to the coefficients for that step. In many mechanisms, however, one step is much slower than the others. This slow step is called the **rate-determining step**. The speed at which the slow step occurs limits the rate at which the overall reaction occurs. That is, a reaction can never occur faster than its slowest step.

> For the general overall reaction
>
> $$aA + bB \longrightarrow cC + dD$$
>
> the experimentally determined rate-law expression has the form
>
> $$\text{Rate} = k[A]^x[B]^y$$
>
> The values of x and y are related to the coefficients of the reactants in the slowest step, influenced in some cases by earlier steps.

The balanced equation for the overall reaction is equal to the sum of *all* the individual steps, including any steps that might follow the rate-determining step. Once again, we emphasize that the rate-law exponents *do not necessarily match* the coefficients of the overall balanced equation.

Using a combination of experimental data and chemical intuition, we can *postulate* a mechanism by which a reaction could occur. There is no way to prove absolutely that a proposed mechanism is correct. All we can do is postulate a mechanism that is *consistent* with experimental data. If the results of further investigations are inconsistent with a postulated mechanism, it must be modified to conform to all observations. For instance, we might identify reaction-intermediate species that are not part of the proposed mechanism.

As an example, the reaction of nitrogen dioxide and carbon monoxide has been found to be second order with respect to NO_2 and zero order with respect to CO below 225°C.

$$NO_2(g) + CO(g) \longrightarrow NO(g) + CO_2(g) \qquad Rate = k[NO_2]^2$$

The balanced equation for the overall reaction shows the stoichiometry but *does not necessarily mean* that the reaction simply occurs by one molecule of NO_2 colliding with one molecule of CO. If the reaction really took place in *that* one step, the observed rate would be first order in NO_2 and first order in CO, or Rate $= k[NO_2][CO]$. The fact that the observed orders do not match the coefficients in the overall balanced equation tells us that *the reaction does not take place in one step.*

The following proposed two-step mechanism is consistent with the observed rate-law expression:

(1) $NO_2 + NO_2 \longrightarrow N_2O_4$ (slow)

(2) $N_2O_4 + CO \longrightarrow NO + CO_2 + NO_2$ (fast)

      $NO_2 + CO \longrightarrow NO + CO_2$ overall

The rate-determining step of this mechanism involves a *bimolecular* collision between two NO_2 molecules. This is consistent with the rate expression involving $[NO_2]^2$. Because the CO is involved only after the slow step has occurred, the reaction rate does not depend on [CO] (that is, the reaction would be zero order in CO) if this were the actual mechanism. In this proposed mechanism, N_2O_4 is formed in one step and is completely consumed in a later step. Such a species is called a **reaction intermediate**.

However, in other studies of this reaction, nitrogen trioxide, NO_3, has been detected as a transient (short-lived) intermediate in this reaction. The mechanism now thought to be correct is

Some reaction intermediates are so unstable that it is very difficult to prove experimentally that they exist.

(1) $NO_2 + NO_2 \longrightarrow NO_3 + NO$ (slow)

(2) $NO_3 + CO \longrightarrow NO_2 + CO_2$ (fast)

      $NO_2 + CO \longrightarrow NO + CO_2$ overall

In this proposed mechanism, two molecules of NO_2 collide to produce one molecule each of NO_3 and NO. The reaction intermediate NO_3 then collides with one molecule of CO and reacts very rapidly to produce one molecule each of NO_2 and CO_2. Even though two NO_2 molecules are consumed in the first step, one is produced in the second step. The net result is that only one NO_2 molecule is consumed in the overall reaction.

Each of these proposed mechanisms meets both criteria for a plausible mechanism: (1) The steps add to give the equation for the overall reaction, and (2) the mechanism is consistent with the experimentally determined rate-law expression (in that two NO_2 molecules and no CO molecules are reactants in the slow step). The NO_3 that has been detected is evidence in favor of the second mechanism, but this does not unequivocally *prove* that mechanism; it might be possible to think of other mechanisms that would involve NO_3 as an intermediate and would also be consistent with the observed rate law.

The gas-phase reaction of nitrogen oxide and bromine is known to be second order in NO and first order in Br_2.

$$2NO(g) + Br_2(g) \longrightarrow 2NOBr(g) \qquad Rate = k[NO]^2[Br_2]$$

A one-step collision involving two NO molecules and one Br_2 molecule would be consistent with the experimentally determined rate-law expression. However, the likelihood of all *three* molecules colliding simultaneously is far less

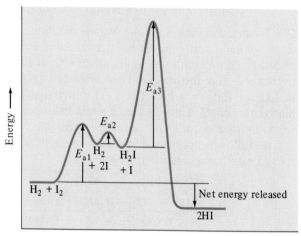

Figure 16-12
A graphical representation of the relative energies of activation for a postulated mechanism for the gas-phase reaction $H_2 + I_2 \longrightarrow 2HI$.

Think how unlikely it is for three moving billiard balls to collide simultaneously.

than the likelihood of two colliding. *Routes involving only bimolecular collisions or unimolecular decompositions are thought to be more favorable in reaction mechanisms.* The mechanism is believed to be

$$
\begin{array}{lll}
(1) & NO + Br_2 \rightleftharpoons NOBr_2 & \text{(fast, equilibrium)} \\
(2) & NOBr_2 + NO \longrightarrow 2NOBr & \text{(slow)} \\
\hline
& 2NO + Br_2 \longrightarrow 2NOBr & \text{overall}
\end{array}
$$

The first step involves the collision of one NO and one Br_2 to produce the intermediate species $NOBr_2$. However, $NOBr_2$ can react rapidly to re-form NO and Br_2. We say that this is an *equilibrium step*. One of the $NOBr_2$ molecules can then react relatively slowly with one NO molecule to produce two NOBr molecules.

Denoting the rate constant for Step 2 as k_2, we could express the rate of this step as

The rate-law expression of Step 2 (the rate-determining step) determines the rate law for the overall reaction.

$$
\text{Rate} = k_2[NOBr_2][NO]
$$

However, $NOBr_2$ is a reaction intermediate, so its concentration at the beginning of the second step cannot be measured directly. Because $NOBr_2$ is formed in a fast equilibrium step, we can relate its concentration to the concentrations of the original reactants. When a reaction or reaction step is at *equilibrium*, its forward (f) and reverse (r) rates are equal:

$$
\text{Rate}_{1f} = \text{Rate}_{1r}
$$

Because this is an elementary step, we can write the rate expression for both directions from the equation for the elementary step

$$
k_{1f}[NO][Br_2] = k_{1r}[NOBr_2]
$$

and then rearrange for $[NOBr_2]$:

$$
[NOBr_2] = \frac{k_{1f}}{k_{1r}}[NO][Br_2]
$$

When we substitute the right side of this equation for $[NOBr_2]$ in the rate expression for the rate-determining step, Rate $= k_2[NOBr_2][NO]$, we arrive at the experimentally determined rate-law expression.

$$\text{Rate} = k_2\left(\frac{k_{1f}}{k_{1r}}[NO][Br_2]\right)[NO] \quad \text{or} \quad \text{Rate} = k[NO]^2[Br_2]$$

The product and quotient of constants k_2, k_{1f}, and k_{1r} is another constant, k.

Similar interpretations apply to most other overall third- or higher-order reactions, as well as many lower-order reactions. However, when several steps are about equally slow, the analysis of experimental data is much more complex. Fractional or negative reaction orders result from complex multistep mechanisms.

One of the earliest kinetic studies involved the gas-phase reaction of hydrogen and iodine to form hydrogen iodide. The reaction was found to be first order in both hydrogen and iodine:

$$H_2(g) + I_2(g) \longrightarrow 2HI(g) \qquad \text{Rate} = k[H_2][I_2]$$

The mechanism that was accepted for many years involved collision of single molecules of H_2 and I_2 in a simple one-step reaction. However, current evidence indicates a more complex process. Most kineticists now accept the following mechanism:

$$
\begin{array}{llll}
(1) & I_2 & \rightleftharpoons 2I & \text{(fast, equilibrium)} \\
(2) & I + H_2 \rightleftharpoons H_2I & & \text{(fast, equilibrium)} \\
(3) & H_2I + I \longrightarrow 2HI & & \text{(slow)} \\
\hline
& H_2 + I_2 \longrightarrow 2HI & & \text{overall}
\end{array}
$$

In this case, neither original reactant appears in the rate-determining step, but both appear in the rate-law expression. Each step is a reaction in itself. It follows (according to transition state theory) that each step has its own activation energy. Because Step 3 is the slowest, we know that its activation energy is the highest, as shown in Figure 16-12.

In summary:

Apply the algebraic approach described above to show that this mechanism is consistent with the observed rate-law expression.

> The experimentally determined reaction orders of reactants indicate the number of molecules of those reactants involved in (1) the slow step only or (2) the slow step *and* any fast equilibrium steps preceding the slow step.

16-10 Temperature: The Arrhenius Equation

The average kinetic energy of a collection of molecules is proportional to the absolute temperature. At any particular temperature, T_1, a definite fraction of the reactant molecules have sufficient energy, E_a, to react to form product molecules upon collision. At a higher temperature, T_2, a greater fraction of the molecules possess the necessary activation energy, and the reaction proceeds at a greater rate. This situation is depicted in Figure 16-13.

From experimental observations, Svante Arrhenius developed the mathematical relationship among activation energy, absolute temperature, and the specific rate constant of a reaction, k, at that temperature. The relationship, called the Arrhenius Equation, is

$$k = Ae^{-E_a/RT}$$

Which egg will cook faster—the one in ice water (left) or the one in boiling water (right)? Why—in terms of what you have learned in this chapter?

$e = 2.718$ is the base of *natural* logarithms (ln). The number 2.303 (which is the natural logarithm of 10) is a conversion factor from natural to base-10 logarithms (log).

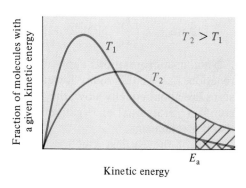

The area between the distribution curve and the horizontal axis in Figure 16-13 is proportional to the total number of molecules present. The total area is the same at T_1 and T_2. The shaded areas represent the number of particles that exceed the energy of activation, E_a.

Figure 16-13
The effect of temperature on the number of molecules that have kinetic energies greater than E_a. At T_2, a higher fraction of molecules possess at least E_a, the activation energy.

or, in logarithmic form,

$$\ln k = \ln A - \frac{E_a}{RT} \qquad \text{or} \qquad \log k = \log A - \frac{E_a}{2.303\, RT}$$

In this expression, A is a constant having the same units as the rate constant. It is proportional to the frequency of collisions between reacting molecules. R is the universal gas constant, expressed with the same energy units in its numerator as are used for E_a. For instance, when E_a is known in J/mol, the value $R = 8.314$ J/mol · K is appropriate. Here the unit "mol" is interpreted as "mole of reaction," as described in Chapter 15. One important point is the following: The greater the value of E_a, the smaller the value of k and the lower the reaction rate (other factors being equal). This is because fewer collisions take place with sufficient energy to get over a high energy barrier.

The Arrhenius equation predicts that increasing T results in faster reaction for the same E_a and concentrations.

| **If T increases** | $\Rightarrow$ | E_a/RT decreases | $\Rightarrow$ | $-E_a/RT$ increases | $\Rightarrow$ | $e^{-E_a/RT}$ increases | $\Rightarrow$ | k increases | $\Rightarrow$ | **Reaction speeds up** |

Let's look at how the rate constant varies with temperature for a given single reaction. Assume that the activation energy and the factor A do not

Antimony powder reacts with bromine more rapidly at 75°C (left) than at 25°C (right).

depend on temperature. Then we can write the Arrhenius equation for two different temperatures. We subtract one equation from the other, and rearrange the result to obtain, in natural logarithm (ln) form,

$$\ln \frac{k_2}{k_1} = \frac{E_a}{R}\left(\frac{1}{T_1} - \frac{1}{T_2}\right) \quad \text{or} \quad \ln \frac{k_2}{k_1} = \frac{E_a}{R}\left(\frac{T_2 - T_1}{T_1 T_2}\right)$$

In base-10 logarithm (log) form, this equation is written as

$$\log \frac{k_2}{k_1} = \frac{E_a}{2.303R}\left(\frac{1}{T_1} - \frac{1}{T_2}\right) \quad \text{or} \quad \log \frac{k_2}{k_1} = \frac{E_a}{2.303R}\left(\frac{T_2 - T_1}{T_1 T_2}\right)$$

Let's substitute some typical values into this equation. The activation energy for many reactions that occur near room temperature is about 50 kJ/mol (or 12 kcal/mol). For such a reaction, a temperature increase from 300 K to 310 K would result in

$$\log \frac{k_2}{k_1} = \frac{50,000 \text{ J/mol}}{(2.303)(8.314 \text{ J/mol} \cdot \text{K})}\left(\frac{310 \text{ K} - 300 \text{ K}}{(300 \text{ K})(310 \text{ K})}\right) = 0.281$$

$$\frac{k_2}{k_1} = 1.91 \approx 2$$

Chemists sometimes use the rule of thumb that near room temperature the rate of a reaction approximately doubles with a 10°C rise in temperature. Such a "rule" must be used with care, however, because it obviously depends on the activation energy.

Example 16-9

The specific rate constant, k, for the following first-order reaction is 9.16×10^{-3} s^{-1} at 0.0°C. The activation energy of this reaction is 88.0 kJ/mol. Determine the value of k at 2.0°C.

$$N_2O_5 \longrightarrow NO_2 + NO_3$$

Plan

First we tabulate the values, remembering to convert temperature to the Kelvin scale:

$E_a = 88,000 \text{ J/mol}$ $R = 8.314 \text{ J/mol} \cdot \text{K}$

$k_1 = 9.16 \times 10^{-3} \text{ s}^{-1}$ at $T_1 = 0.0°C + 273 = 273 \text{ K}$

$k_2 = ?$ at $T_2 = 2.0°C + 273 = 275 \text{ K}$

We use these values in the "two-temperature" form of the Arrhenius equation.

Solution

$$\ln \frac{k_2}{k_1} = \frac{E_a}{R}\left(\frac{T_2 - T_1}{T_1 T_2}\right)$$

$$\ln \left(\frac{k_2}{9.16 \times 10^{-3} \text{ s}^{-1}}\right) = \frac{88,000 \text{ J/mol}}{8.314 \dfrac{\text{J}}{\text{mol} \cdot \text{K}}}\left(\frac{275 \text{ K} - 273 \text{ K}}{(273 \text{ K})(275 \text{ K})}\right) = 0.282$$

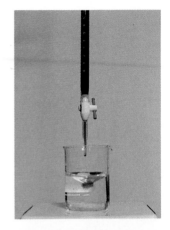

Oxalic acid, $(COOH)_2$, reduces intensely colored permanganate ions, MnO_4^-, to nearly colorless Mn^{2+} ions in acidic solution (top). The reaction is catalyzed by Mn^{2+} ions. If no Mn^{2+} ions are present, the reaction occurs *very* slowly, and MnO_4^- ions are reduced to brown solid MnO_2 (bottom). When we titrate $(COOH)_2$ with $KMnO_4$, we should always add the first few drops of $KMnO_4$ solution *very slowly* to produce some Mn^{2+} ions that can then catalyze the desired reaction.

For illustration, we use the *ln* version of the Arrhenius equation; if you prefer, you should establish that you get the same result with the *log* version.

Taking inverse (natural) logarithms of both sides,

$$\frac{k_2}{9.16 \times 10^{-3}\ \text{s}^{-1}} = 1.32$$

$$k_2 = 1.32\,(9.16 \times 10^{-3}\ \text{s}^{-1}) = \boxed{1.21 \times 10^{-2}\ \text{s}^{-1}}$$

We see that a very small temperature difference, only 2°C, causes an increase in the rate constant (and hence in the reaction rate for the same concentrations) of about 32%. Such sensitivity of rate to temperature change makes the control and measurement of temperature extremely important in kinetics experiments.

EOC 45

Example 16-10

Ethyl iodide decomposes to give ethylene and hydrogen iodide in the first-order gas phase reaction

$$C_2H_5I \longrightarrow C_2H_4 + HI$$

At 600 K, the value of k is determined to be $1.60 \times 10^{-6}\ \text{s}^{-1}$. When the temperature is raised to 700 K, the value of k increases to $6.36 \times 10^{-3}\ \text{s}^{-1}$. What is the activation energy for this reaction?

Plan

We know k at two different temperatures. We solve the two-temperature form of the Arrhenius equation for E_a and evaluate.

Solution

$$k_1 = 1.60 \times 10^{-5}\ \text{s}^{-1} \text{ at } T_1 = 600\ \text{K} \qquad k_2 = 6.36 \times 10^{-3}\ \text{s}^{-1} \text{ at } T_2 = 700\ \text{K}$$

$$R = 8.314\ \text{J/mol} \cdot \text{K} \qquad\qquad E_a = \underline{?}$$

We rearrange the Arrhenius equation for E_a:

$$\log \frac{k_2}{k_1} = \frac{E_a}{2.303\,R}\left(\frac{T_2 - T_1}{T_1 T_2}\right)$$

$$E_a = (2.303R)\left(\frac{T_1 T_2}{T_2 - T_1}\right)\log \frac{k_2}{k_1}$$

$$= (2.303R)\left(\frac{(600\ \text{K})(700\ \text{K})}{700\ \text{K} - 600\ \text{K}}\right)\log \frac{6.36 \times 10^{-3}\ \text{s}^{-1}}{1.60 \times 10^{-5}\ \text{s}^{-1}}$$

$$= (2.303)\left(8.314\ \frac{\text{J}}{\text{mol} \cdot \text{K}}\right)(4.20 \times 10^3\ \text{K})(2.60)$$

$$= \boxed{2.09 \times 10^5\ \text{J/mol} \qquad \text{or} \qquad 209\ \text{kJ/mol}}$$

EOC 46

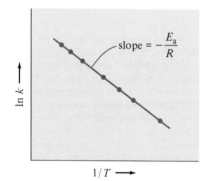

Figure 16-14
A graphical method for determining activation energy, E_a. At each of several different temperatures, the rate constant, k, is determined by methods such as those in Sections 16-3 through 16-6. A plot of ln k versus $1/T$ gives a straight line with negative slope. The slope of this straight line is $-E_a/R$. Alternatively, a plot of log k versus $1/T$ gives a straight line whose slope is $-E_a/2.303R$. Use of this graphical method is often desirable, because it partially compensates for experimental errors in individual k and T values.

The determination of E_a in the manner illustrated in Example 16-10 may be subject to considerable error, because it depends on the measurement of k at only two temperatures. Any error in either of these k values would greatly affect the resulting value of E_a. A more reliable method that uses many measured values for the same reaction is based on a graphical approach. Let us rearrange the single-temperature logarithmic form of the Arrhenius equation and compare it with the equation for a straight line.

$$\underbrace{\ln k}_{} = -\underbrace{\left(\frac{E_a}{R}\right)}_{}\underbrace{\left(\frac{1}{T}\right)}_{} + \underbrace{\ln A}_{}$$

$$y \;\;=\;\; m \quad x \;\;+\;\; b$$

Compare the approach to that described in Section 16-5 for determining k. We could plot base-10 log k versus $1/T$; then the slope would be equal to $-E_a/2.303R$.

The value of the collision frequency factor, A, is very nearly constant over moderate temperature changes. Thus, $\ln A$ can be interpreted as the constant term in the equation (the intercept). The slope of the straight line obtained by plotting $\ln k$ versus $1/T$ can be interpreted as $-E_a/R$. This allows us to determine the value of the activation energy from the slope (Figure 16-14). End-of-chapter Exercises 51 and 52 use this method.

16-11 Catalysts

Catalysts are substances that can be added to reacting systems to change the rate of reaction. They allow reactions to occur via alternative pathways, and these affect reaction rates by changing activation energies.

The activation energy is lowered in most catalyzed reactions, as depicted in Figures 16-15 and 16-16. (In the rare case of inhibitory catalysis, the activation energy is raised as the reaction is forced to occur by a less favorable route.) Although a catalyst enters into a reaction, it does not appear in the balanced equation for the reaction. If a catalyst does react, all of it is regenerated in subsequent steps. If a species is not regenerated, it is not a catalyst but a reactant.

For constant T and the same concentrations,

| **If E_a** **decreases** | $\Rightarrow$ | E_a/RT decreases | $\Rightarrow$ | $-E_a/RT$ increases | $\Rightarrow$ | $e^{-E_a/RT}$ increases | $\Rightarrow$ | k increases | $\Rightarrow$ | **Reaction** **speeds up** |

We can divide catalysts into two categories: (1) homogeneous catalysts and (2) heterogeneous catalysts, or contact catalysts. A **homogeneous catalyst**

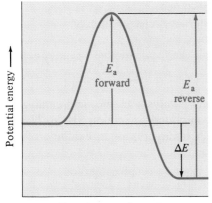

Reaction coordinate for uncatalyzed reaction

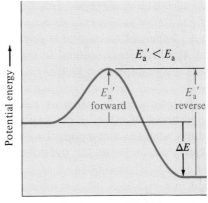

Reaction coordinate for catalyzed reaction

Figure 16-15
Potential energy diagrams showing the effect of a catalyst. The catalyst provides a different, lower-energy mechanism for the formation of the products. A catalyzed reaction typically occurs in several steps, each with its own barrier, but the over-all energy barrier is lower than the uncatalyzed reaction. ΔE has the same value for each path. The value of ΔE depends only on the states of the reactants and products.

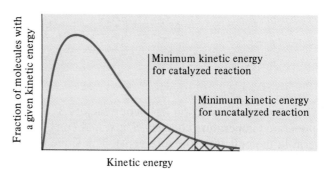

Figure 16-16
When a catalyst is present, the energy barrier is lowered. Thus, more molecules possess the minimum kinetic energy necessary for reaction. This is analogous to allowing more students to pass a course by lowering the requirements.

exists in the same phase as the reactants. Ceric ion, Ce^{4+}, is an important laboratory oxidizing agent that is used in many redox titrations (Sections 11-4 and 11-5). For example, Ce^{4+} oxidizes thallium(I) ions in solution; this reaction is catalyzed by the addition of a very small amount of a soluble salt containing manganous ions, Mn^{2+}. The Mn^{2+} acts as a homogeneous catalyst.

$$2Ce^{4+} + Tl^+ \xrightarrow{Mn^{2+}} 2Ce^{3+} + Tl^{3+}$$

This reaction is thought to proceed by the following sequence of elementary steps:

$$
\begin{array}{ll}
Ce^{4+} + Mn^{2+} \longrightarrow Ce^{3+} + Mn^{3+} & \text{Step 1} \\
Ce^{4+} + Mn^{3+} \longrightarrow Ce^{3+} + Mn^{4+} & \text{Step 2} \\
\underline{Mn^{4+} + Tl^+ \longrightarrow Mn^{2+} + Tl^{3+}} & \text{Step 3} \\
2Ce^{4+} + Tl^+ \longrightarrow 2Ce^{3+} + Tl^{3+} & \text{Overall}
\end{array}
$$

Some Mn^{2+}, the catalyst, reacts in Step 1, but in equal amount is regenerated in Step 3 and is thus available to react again. The two ions shown in blue, Mn^{3+} and Mn^{4+}, are *reaction intermediates*. Mn^{3+} ions are formed in Step 1 and consumed in an equal amount in Step 2; similarly, Mn^{4+} ions are formed in Step 2 and consumed in an equal amount in Step 3.

Be sure you can distinguish among the various species that appear in a reaction mechanism:

1. *Reactant*: more is consumed than is formed.
2. *Product*: more is formed than is consumed.
3. *Reaction intermediate*: formed in earlier steps, then consumed in an equal amount in later steps.
4. *Catalyst*: consumed in earlier steps, then regenerated in an equal amount in later steps.

Strong acids function as homogeneous catalysts in the acid-catalyzed hydrolysis of esters (a class of organic compounds—Section 31-13). Using ethyl acetate (a component of nail polish removers) as an example of an ester, we can write the overall reaction as follows:

"Hydrolysis" means reaction with water.

$$\underset{\text{ethyl acetate}}{CH_3\overset{O}{\overset{\|}{C}}-OCH_2CH_3(aq)} + H_2O \xrightarrow{H^+} \underset{\text{acetic acid}}{CH_3\overset{O}{\overset{\|}{C}}-OH(aq)} + \underset{\text{ethanol}}{CH_3CH_2OH(aq)}$$

This is a thermodynamically favored reaction. Because of its high energy of activation, it occurs only very, very slowly when no catalyst is present. In the presence of strong acids, however, the reaction occurs more rapidly. In this acid-catalyzed hydrolysis, different intermediates with lower activation energies are formed. The *postulated* sequence of steps follows.

Groups of atoms that are involved in the change in each step are shown in blue.

$$CH_3-\overset{\overset{O}{\|}}{C}-OCH_2CH_3 + H^+ \longrightarrow \left[CH_3-\overset{\overset{+}{OH}}{\underset{\|}{C}}-OCH_2CH_3\right] \qquad \text{Step 1}$$

$$CH_3-\overset{\overset{+}{OH}}{\underset{|}{C}}-OCH_2CH_3 + H_2O \longrightarrow \left[CH_3-\overset{OH}{\underset{\underset{\underset{H \quad H}{}}{\overset{+}{O}}}{\underset{|}{C}}}-OCH_2CH_3\right] \qquad \text{Step 2}$$

$$CH_3-\overset{OH}{\underset{\underset{\underset{H \quad H}{}}{\overset{+}{O}}}{\underset{|}{C}}}-OCH_2CH_3 \longrightarrow \left[CH_3-\overset{OH}{\underset{\underset{HO \quad H}{}}{\underset{|}{C}}}-\overset{+}{O}-CH_2CH_3\right] \qquad \text{Step 3}$$

$$CH_3-\overset{OH}{\underset{\underset{HO \quad H}{}}{\underset{|}{C}}}-\overset{+}{O}-CH_2CH_3 \longrightarrow \left[CH_3-\overset{\overset{+}{OH}}{\underset{\|}{C}}-OH\right] + HOC_2H_5 \qquad \text{Step 4}$$
ethanol

$$CH_3-\overset{\overset{+}{OH}}{\underset{\|}{C}}-OH \longrightarrow CH_3-\overset{\overset{O}{\|}}{C}-OH + H^+ \qquad \text{Step 5}$$
acetic acid

$$CH_3-\overset{\overset{O}{\|}}{C}-OCH_2CH_3 + H_2O \overset{H^+}{\longrightarrow} CH_3-\overset{\overset{O}{\|}}{C}-OH + CH_3CH_2OH \quad \text{Overall}$$
ethyl acetate   acetic acid   ethanol

Note that H$^+$ is a reactant in Step 1, but it is completely regenerated in Step 5. Therefore, H$^+$ is a catalyst. The species shown in brackets in Steps 1 through 4 are *reaction intermediates*. All intermediates in this sequence of elementary steps are charged species. Ethyl acetate and water are the reactants, and acetic acid and ethanol are the products of the overall catalyzed reaction.

Heterogeneous catalysts exist in a different phase than the reactants. They are usually solids, and they lower activation energies by providing surfaces on which reactions can occur. The first step in the catalytic process is usually *adsorption*, in which one or more of the reactants become attached to the solid surface. Some reactant molecules may be held in particular orientations; in other molecules, some bonds may be broken to form atoms or smaller molecular fragments. This causes *activation* of the reactants. As a result, *reaction* occurs more readily than would otherwise be possible. In a

The petroleum industry uses numerous heterogeneous catalysts. Many of them contain highly colored compounds of transition metal ions. Several are shown here.

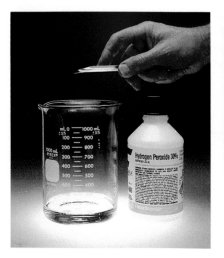  

The decomposition of a 30% hydrogen peroxide solution is catalyzed by very small amounts of transition metal oxides. This catalyzed reaction is rapid, so the exothermic reaction quickly heats the solution to the boiling point of water, forming steam. The temperature increase further accelerates the decomposition. One should never use a metal syringe tip to withdraw a sample from a 30% hydrogen peroxide solution.

final step, *desorption*, the product molecules leave the surface, freeing reaction sites to be used again. Most contact catalysts are more effective as small particles, providing large surface areas.

The catalytic converters built into automobile exhaust systems contain two types of heterogeneous catalysts, powdered noble metals and powdered transition metal oxides. They catalyze the oxidation of unburned fuel and of partial combustion products such as carbon monoxide (see Figure 16-17).

$$2C_8H_{18}(g) + 25O_2(g) \xrightarrow[\text{NiO}]{\text{Pt}} 16CO_2(g) + 18H_2O(g)$$

iso-octane (a component of gasoline)

$$2CO(g) + O_2(g) \xrightarrow[\text{NiO}]{\text{Pt}} 2CO_2(g)$$

It is desirable to carry out these reactions in automobile exhaust systems. Carbon monoxide is very poisonous. The latter reaction is so slow that a mixture of CO and O_2 gas at the exhaust temperature would remain unreacted for thousands of years in the absence of a catalyst! Yet the addition of only a small amount of a solid, finely divided transition metal catalyst promotes

Figure 16-17
(a) The arrangement of a catalytic converter in an automobile. (b) A cutaway view of a catalytic converter, showing the pellets of catalyst.

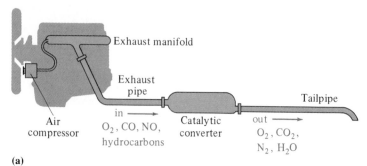

Exhaust manifold

Exhaust pipe

Tailpipe

Air compressor

in →

O_2, CO, NO, hydrocarbons

Catalytic converter

out →

O_2, CO_2, N_2, H_2O

(a)

(b)

the production of up to a mole of CO_2 per minute. Because this reaction is a very simple but important one, it has been studied extensively by surface chemists. It is one of the best understood of the heterogeneously catalyzed reactions. The major features of the catalytic process are shown in Figure 16-18.

The same catalysts also catalyze the decomposition of nitrogen oxide, NO, formed during the combustion of any fuel in air, into harmless N_2 and O_2:

$$2NO(g) \xrightarrow[\text{NiO}]{\text{Pt}} N_2(g) + O_2(g)$$

Nitrogen oxide is a serious air pollutant because it is oxidized to nitrogen dioxide, NO_2. This reacts with water to form nitric acid and with alcohols to form nitrites (eye irritants).

Figure 16-18
A simplified representation of the catalysis of the reaction
$2CO(g) + O_2(g) \rightarrow 2CO_2(g)$
on a metal surface.

(a) *Adsorption*: CO and O_2 reactant molecules become bound to the surface:

$$CO(g) \longrightarrow CO(\text{surface}) \quad \text{and} \quad O_2(g) \longrightarrow O_2(\text{surface})$$

The CO molecules are linked through their C atoms to one or more metal atoms on the surface. The O_2 molecules are more weakly bound.

(b) *Activation*: The O_2 molecules dissociate into O atoms, which are held in place more tightly:

$$O_2(\text{surface}) \longrightarrow 2O(\text{surface})$$

The CO molecules stick to the surface, but they migrate easily across the surface.

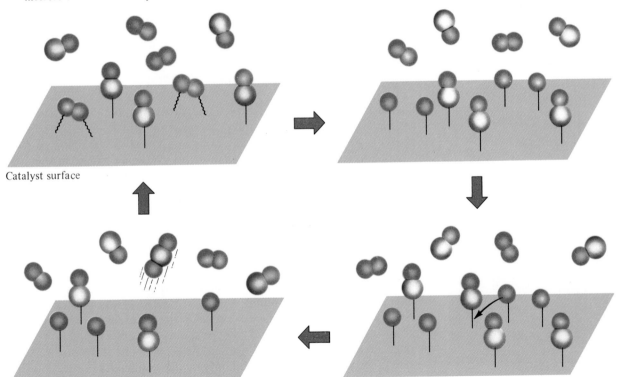

Catalyst surface

(d) *Desorption*: The CO_2 product molecules leave the surface:

$$CO_2(\text{surface}) \longrightarrow CO_2(g)$$

Fresh reactant molecules can then replace them to start the cycle again [back to step (a)].

(c) *Reaction*: O atoms react with bound CO molecules, to form CO_2 molecules:

$$CO(\text{surface}) + O(\text{surface}) \longrightarrow CO_2(\text{surface})$$

The resulting CO_2 molecules bind to the surface *very poorly*.

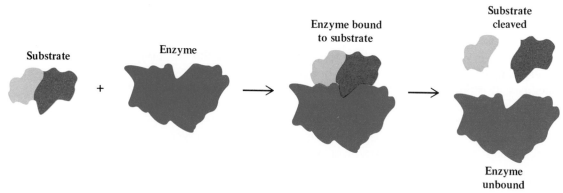

Substrate

Enzyme

Enzyme bound
to substrate

Substrate
cleaved

Enzyme
unbound

Figure 16-19

A schematic representation of a simplified mechanism (lock-and-key) for enzyme reaction. The substrates (reactants) fit the active sites of the enzyme molecule much as keys fit locks. When the reaction is complete, the products do not fit the active sites as well as the reactants did. They separate from the enzyme, leaving it free to catalyze the reaction of additional reactant molecules. The enzyme is not permanently changed by the process. The illustration here is for a process in which a complex reactant molecule is split to form two simpler product molecules. The formation of simple sugars from complex carbohydrates is a similar reaction. Some enzymes catalyze the combination of simple molecules to form more complex ones.

Maintaining the continued efficiency of all three reactions in a "three-way" catalytic converter is a delicate matter. It requires control of such factors as the O_2 supply pressure and the order in which the reactants reach the catalyst. Some modern automobile engines use microcomputer chips, based on an O_2 sensor in the exhaust stream, to control air valves.

These three reactions, catalyzed in catalytic converters, are all exothermic and thermodynamically favored. Unfortunately, other energetically favored reactions are also accelerated by the mixed catalysts. All fossil fuels contain sulfur compounds, which are oxidized to sulfur dioxide during combustion. Sulfur dioxide, itself an air pollutant, undergoes further oxidation to form sulfur trioxide as it passes through the catalytic bed:

$$2SO_2(g) + O_2(g) \xrightarrow[\text{NiO}]{\text{Pt}} 2SO_3(g)$$

Sulfur trioxide is probably a worse pollutant than sulfur dioxide, because SO_3 is the acid anhydride of strong, corrosive sulfuric acid. Sulfur trioxide reacts with water vapor in the air, as well as in auto exhausts, to form sulfuric acid droplets. This problem must be overcome if the current type of catalytic converter is to see continued use. These same catalysts also suffer from the problem of being "poisoned"—i.e., made inactive—by lead. Leaded fuels contain tetraethyl lead, $Pb(C_2H_5)_4$, and tetramethyl lead, $Pb(CH_3)_4$. Such fuels are not suitable for automobiles equipped with catalytic converters and are excluded by law from use in such cars.

Reactions that occur in the presence of a solid catalyst, as on a metal surface (heterogeneous catalysis) often follow zero-order kinetics. For instance, the rate of decomposition of $NO_2(g)$ at higher pressures on a platinum metal surface does not change if we add more NO_2. This is because only the NO_2 molecules on the surface can react. If the metal surface is completely covered with NO_2 molecules, no additional molecules can be *adsorbed* until the ones already there have reacted and the products have *desorbed*. Thus, the rate of the reaction is controlled only by the availability of reaction sites on the Pt surface, and not by the total number of NO_2 molecules available.

Enzymes are proteins that act as catalysts in living systems for specific biochemical reactions. The reactants in enzyme-catalyzed reactions are called **substrates**. Thousands of vital processes in our bodies are catalyzed by as many distinct enzymes. For instance, the enzyme carbonic anhydrase catalyzes the combination of CO_2 and water (the substrates), facilitating most

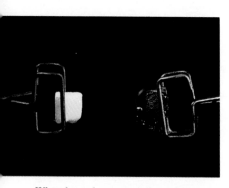

When heated, a sugar cube (sucrose, melting point 185°C) melts but does not burn. A sugar cube rubbed in cigarette ash burns before it melts. The cigarette ash contains trace amounts of metal compounds that catalyze the combustion of sugar.

of the transport of carbon dioxide in the blood. This combination reaction, ordinarily uselessly slow, proceeds rapidly in the presence of carbonic anhydrase; a single molecule of this enzyme can promote the conversion of more than 1 million molecules of carbon dioxide each second. Each enzyme is extremely specific, catalyzing only a few closely related reactions—or, in many cases, only one particular reaction—for only certain substrates. Modern theories of enzyme action attribute this to the requirement of very specific matching of shapes (molecular geometries) for a particular substrate to bind to a particular enzyme (Figure 16-19).

Enzyme-catalyzed reactions are important examples of zero-order reactions; that is, the rate of each of them is independent of the concentration of the substrate (provided *some* substrate is present).

$$\text{Rate} = k$$

The active site on an enzyme can bind to only one substrate molecule at a time (or one pair, if the reaction is one that links two reactant molecules), no matter how many other substrate molecules are available in the vicinity.

Ammonia is a very important industrial chemical. The Haber process for its preparation involves the use of iron as a *catalyst* at 450°C to 500°C and high pressures. The reaction is thermodynamically spontaneous, but very slow:

$$N_2(g) + 3H_2(g) \xrightarrow{Fe} 2NH_3(g) \qquad \Delta G^0 = -194.7 \text{ kJ/mol (at 500°C)}$$

The ammonia formation reaction is discussed in detail in Section 17-6.

In the absence of a catalyst, the reaction does not occur at an observable rate at room temperature. Even with a catalyst, it is not efficient at atmospheric pressure.

Most of the essential nutrients for both plants and animals contain nitrogen. The reaction between N_2 and H_2 to form NH_3 is catalyzed at room temperature and atmospheric pressure by a class of enzymes, called nitrogenases, that are present in some bacteria. Legumes are plants that support these bacteria; they are able to obtain nitrogen as N_2 from the atmosphere and convert it to ammonia.

The process is called nitrogen fixation. The ammonia can be used in the synthesis of many nitrogen-containing biological compounds such as proteins and nucleic acids.

In comparison with manufactured catalysts, most enzymes are tremendously efficient under very mild conditions. If chemists and biochemists could develop catalysts with even a small fraction of the efficiency of enzymes, they could be a great boon to the world's health and economy. One of the most active areas of current chemical research involves attempts to discover or synthesize catalysts that can mimic the efficiency of naturally occurring enzymes such as nitrogenases. Such a development would be important in industry. It would eliminate the costs of the high temperature and high pressure that are necessary in the Haber process. This could decrease the cost of food grown with the aid of ammonia-based fertilizers. Ultimately this would help greatly to feed the world's growing population.

In the presence of a hot copper catalyst, acetone (CH_3COCH_3) is converted into ketene (C_2H_2O) and methane (CH_4) in an exothermic reaction.

$$CH_3COCH_3(g) \xrightarrow[\Delta]{Cu} C_2H_2O(g) + CH_4(g) + \text{heat}$$

Hot copper pennies are suspended over acetone in the flask. Acetone vapor reacts on the hot copper surface. The heat liberated by the reaction causes the pennies to glow brightly.

Key Terms

Activation energy   The amount of energy that must be absorbed by reactants in their ground states in order for them to reach the transition state so that a reaction can occur.

Arrhenius equation   An equation that relates the specific rate constant to activation energy and temperature.

Catalyst   A substance that alters (usually increases) the rate at which a reaction occurs.

Chemical kinetics   The study of rates and mechanisms of chemical reactions and of the factors on which they depend.

Collision theory   A theory of reaction rates that states that effective collisions between reactant molecules must occur for reaction to occur.

Contact catalyst   See *Heterogeneous catalyst*.

Effective collision   A collision between molecules that results in reaction; one in which molecules collide with proper relative orientations and with sufficient energy to react.

Elementary step   An individual step in the mechanism by which a reaction occurs. For each elementary step, the reaction orders *do* match the reactant coefficients in that step.

Enzyme   A protein that acts as a catalyst in biological systems.

Half-life of a reactant   The time required for half of that reactant to be converted into product(s).

Heterogeneous catalyst   A catalyst that exists in a different phase (solid, liquid, or gas) from the reactants; a contact catalyst.

Homogeneous catalyst   A catalyst that exists in the same phase (solid, liquid, or gas) as the reactants.

Inhibitory catalyst   An inhibitor; a catalyst that decreases the rate of a reaction.

Integrated rate equation   An equation that gives the concentration of a reactant remaining after a specified time; has different mathematical forms for different orders of reaction.

Mechanism   The sequence of steps by which reactants are converted into products.

Method of initial rates   A method of determining the rate-

law expression by carrying out a reaction with different initial concentrations and analyzing the resulting changes in initial rates.

Order of a reactant   The power to which the reactant's concentration is raised in the rate-law expression.

Order of a reaction   The sum of the powers to which all concentrations are raised in the rate-law expression (also called overall order of a reaction).

Rate-determining step   The slowest step in a mechanism; the step that determines the overall rate of reaction.

Rate-law expression (also called **rate law**)   An equation that relates the rate of a reaction to the concentrations of the reactants and the specific rate constant; Rate $= k[A]^x[B]^y$. The exponents of reactant concentrations *do not necessarily* match the coefficients in the overall balanced chemical equation. The rate-law expression must be determined from experimental data.

Rate of reaction   The change in concentration of a reactant or product per unit time.

Reaction coordinate   The progress along the pathway from reactants to products; sometimes called "progress of reaction."

Reaction intermediate   A species that is produced and then entirely consumed during a reaction; usually short-lived.

Specific rate constant (also called **rate constant**)   An experimentally determined proportionality constant that is different for different reactions and that changes for a given reaction only with temperature or the presence of a catalyst; k in the rate-law expression Rate $= k[A]^x[B]^y$.

Substrate   A reactant in an enzyme-catalyzed reaction.

Thermodynamically favorable (spontaneous) reaction   A reaction that occurs with a net release of free energy, G; a reaction for which ΔG is negative (see Section 15-14).

Transition state   A relatively high-energy state in which bonds in reactant molecules are partially broken and new ones are partially formed.

Transition state theory   A theory of reaction rates that states that reactants pass through high-energy transition states before forming products.

Exercises

General Concepts

1. Briefly summarize the effects of each of the four factors that affect rates of reactions.
2. Describe the basic features of collision theory and transition state theory.
3. What is a rate-law expression? Describe how it is determined for a particular reaction.
4. Distinguish between reactions that are thermodynamically favorable and reactions that are kinetically favorable. What can be said about relationships between the two?
5. What is meant by the order of a reaction?
6. What, if anything, can be said about the relationship between the coefficients of the balanced *overall* equa-

tion for a reaction and the powers to which concentrations are raised in the rate-law expression? To what are these powers related?

7. Express the rate of reaction in terms of the rate of change of each reactant and each product in the following reactions:
   (a) $S^{2-}(aq) + H_2O(\ell) \rightarrow HS^-(aq) + OH^-(aq)$
   (b) $2H_2O(g) \xrightarrow{\Delta} 2H_2(g) + O_2(g)$
   (c) $CH_3COOH(aq) + OH^-(aq) \rightarrow CH_3COO^-(aq) + H_2O(\ell)$

8. Express the rate of reaction in terms of the rate of change of each reactant and each product in the following reactions:
   (a) $2H_2O_2(aq) \rightarrow 2H_2O(\ell) + O_2(g)$
   (b) $I^-(aq) + ClO^-(aq) \rightarrow IO^-(aq) + Cl^-(aq)$
   (c) $C_2H_4(g) + Br_2(g) \rightarrow C_2H_4Br_2(g)$

9. At the instant when N_2 is reacting at a rate of 0.25 M/min, what is the rate at which H_2 is disappearing, and what is the rate at which NH_3 is forming?

$$N_2 + 3H_2 \longrightarrow 2NH_3$$

10. At the instant when NH_3 is reacting at a rate of 0.80 M/min, what is the rate at which the other reactant is disappearing, and what is the rate at which each product is being formed?

$$4NH_3 + 5O_2 \longrightarrow 4NO + 6H_2O$$

Rate-Law Expression

11. If doubling the initial concentration of a reactant doubles the initial rate of reaction, what is the order of the reaction with respect to the reactant? If the rate increases by a factor of four, what is the order? If the rate remains the same, what is the order?

12. Use times expressed in seconds to give the units of the rate constant for reactions that are overall (a) first order; (b) second order; (c) third order; (d) of order $1\frac{1}{2}$.

13. Rate data were obtained at 25°C for the following reaction. What is the rate-law expression for this reaction?

$$A + 2B \longrightarrow C + 2D$$

Expt.	[A] [mol/L]	[B] [mol/L]	Initial Rate of Formation of C
1	0.10	0.10	2.0×10^{-4} M/min
2	0.30	0.30	6.0×10^{-4} M/min
3	0.30	0.10	2.0×10^{-4} M/min
4	0.40	0.20	4.0×10^{-4} M/min

14. Rate data were obtained for the following reaction at 25°C. What is the rate-law expression for the reaction?

$$2A + B + 2C \longrightarrow D + 2E$$

Expt.	Initial [A]	Initial [B]	Initial [C]	Initial Rate of Formation of D
1	0.10 M	0.20 M	0.10 M	5.0×10^{-4} M/min
2	0.20 M	0.20 M	0.30 M	1.5×10^{-3} M/min
3	0.30 M	0.20 M	0.10 M	5.0×10^{-4} M/min
4	0.40 M	0.60 M	0.30 M	4.5×10^{-3} M/min

15. The reaction $2NO + 2H_2 \rightarrow N_2 + 2H_2O$ gives the following initial rates at 904°C.

Expt.	$[NO]_0$ (mol/L)	$[H_2]_0$ (mol/L)	Initial Rate of Formation of N_2
1	0.420	0.122	0.136 M/s
2	0.210	0.122	0.0339 M/s
3	0.210	0.244	0.0678 M/s
4	0.105	0.488	0.0339 M/s

(a) Write the rate-law expression for this reaction.
(b) Find the rate of formation of N_2 at the instant when $[NO] = 0.550\ M$ and $[H_2] = 0.199\ M$.

*16. The reaction $2NO + O_2 \rightarrow 2NO_2$ gives the following initial rates.

Expt.	$[NO]_0$ (mol/L)	$[O_2]_0$ (mol/L)	Initial Rate of Disappearance of NO
1	0.020	0.010	1.0×10^{-4} M/s
2	0.040	0.010	4.0×10^{-4} M/s
3	0.020	0.040	4.0×10^{-4} M/s

(*Reminder*: Express the rate of *reaction* in terms of the rate of change of concentration of each reactant and each product.)
(a) Write the rate-law expression for this reaction.
(b) Find the rate of reaction at the instant when [NO] = 0.045 M and $[O_2]$ = 0.025 M.
(c) At the time described in part (b), what will be the rate of disappearance of NO? the rate of disappearance of O_2? the rate of formation of NO_2?

17. (a) A certain reaction is zero order in reactant A and second order in reactant B. If the concentrations of both reactants are doubled, what happens to the reaction rate?
(b) What would happen to the reaction rate if the reaction in part (a) were first order in A and first order in B?

18. The rate expression for the following reaction is Rate = $k[A]^2[B_2]$. If, during a reaction, the concentrations of both A and B_2 were suddenly halved, the rate of the reaction would _____ by a factor of _____.

$$A + B_2 \longrightarrow products$$

19. Rate data were collected for the following reaction at a particular temperature.

$$A + B \longrightarrow products$$

Expt.	[A]₀ (mol/L)	[B]₀ (mol/L)	Initial Rate of Reaction
1	0.10	0.10	0.0090 M/s
2	0.20	0.10	0.036 M/s
3	0.10	0.20	0.018 M/s
4	0.10	0.30	0.027 M/s

(a) What is the rate-law expression for this reaction?
(b) Describe the order of the reaction with respect to each reactant and to the overall order.

20. Rate data were collected for the following reaction at a particular temperature.

$$2ClO_2(aq) + 2OH^-(aq) \longrightarrow$$
$$ClO_3^-(aq) + ClO_2^-(aq) + H_2O(\ell)$$

Expt.	[ClO₂]₀ (mol/L)	[OH⁻]₀ (mol/L)	Initial Rate of Reaction
1	0.015	0.015	3.88×10^{-4} M/s
2	0.030	0.015	1.55×10^{-3} M/s
3	0.015	0.030	7.76×10^{-4} M/s
4	0.030	0.030	3.11×10^{-3} M/s

(a) What is the rate-law expression for this reaction?
(b) Describe the order of the reaction with respect to each reactant and to the overall order.

21. The reaction $(C_2H_5)_2(NH)_2 + I_2 \rightarrow (C_2H_5)_2N_2 + 2HI$ gives the following initial rates.

Expt.	[(C₂H₅)₂(NH)₂]₀ (mol/L)	[I₂]₀ (mol/L)	Initial Rate of Formation of (C₂H₅)₂N₂
1	0.010	0.010	2.0 M/s
2	0.010	0.020	4.0 M/s
3	0.030	0.020	12.0 M/s

Write the rate-law expression.

22. Given the following data for the reaction $A + B \rightarrow C$, write the rate-law expression.

Expt.	Initial [A]	Initial [B]	Initial Rate of Formation of C
1	0.20 M	0.10 M	5.0×10^{-6} M/s
2	0.30 M	0.10 M	7.5×10^{-6} M/s
3	0.40 M	0.20 M	4.0×10^{-5} M/s

23. Given the following data for the reaction $A + B \rightarrow C$, write the rate-law expression.

Expt.	Initial [A]	Initial [B]	Initial Rate of Formation of C
1	0.25	0.15	8.0×10^{-5} M/s
2	0.25	0.30	3.2×10^{-4} M/s
3	0.50	0.30	1.28×10^{-3} M/s

24. Given the following data for the reaction $A + B \rightarrow C$, write the rate-law expression.

Expt.	Initial [A]	Initial [B]	Initial Rate of Formation of C
1	0.10 M	0.10 M	2.0×10^{-4} M/s
2	0.20 M	0.10 M	8.0×10^{-4} M/s
3	0.40 M	0.20 M	2.56×10^{-2} M/s

*25. Consider a chemical reaction between compounds A and B that is first order in A and first order in B. From the information given below, fill in the blanks.

Expt.	Rate ($M \cdot s^{-1}$)	[A]	[B]
1	0.10	0.20 M	0.050 M
2	0.20	___ M	0.050 M
3	0.80	0.40 M	___ M

*26. Consider a chemical reaction of compounds A and B that was found to be first order in A and second order in B. From the information given below, fill in the blanks.

Expt.	Rate ($M \cdot s^{-1}$)	[A]	[B]
1	0.050	1.0 M	0.20 M
2	___	2.0 M	0.20 M
3	___	2.0 M	0.40 M

*27. The decomposition of NO_2 by the following reaction at some temperature proceeds at a rate 5.4×10^{-5} mol NO_2/L · s when $[NO_2] = 0.0100$ mol/L.

$$2NO_2(g) \longrightarrow 2NO(g) + O_2(g)$$

(a) Assume that the rate law is Rate = $k[NO_2]$. What rate of disappearance of NO_2 would be predicted when $[NO_2] = 0.00500$ mol/L? (b) Now assume that the rate law is Rate = $k[NO_2]^2$. What rate of disappearance of NO_2 would be predicted when $[NO_2] = 0.00500$ mol/L? (c) The observed rate when $[NO_2] = 0.00500$ mol/L is observed to be 1.4×10^{-5} mol NO_2/L · s. Which rate law is correct? (d) Calculate the rate constant. (*Reminder*: Express the rate of reaction in terms of rate of disappearance of NO_2.)

Integrated Rate Equations and Half-Life
28. What is meant by the half-life of a reactant?
29. The rate law for the reaction of sucrose in water,

$$C_{12}H_{22}O_{11} + H_2O \longrightarrow 2C_6H_{12}O_6$$

is Rate = $k[C_{12}H_{22}O_{11}]$. After 2.57 hours at 25°C, 5.00 g/L of $C_{12}H_{22}O_{11}$ has decreased to 4.50 g/L. Evaluate k for this reaction at 25°C.

30. The rate constant for the decomposition of nitrogen dioxide, $NO_2 \rightarrow NO + \frac{1}{2}O_2$, with a laser beam is 3.40 $M^{-1} \cdot min^{-1}$. Find the time, in seconds, needed to decrease 2.00 mol/L of NO_2 to 1.50 mol/L.

31. The rate constant for the decomposition of gaseous azomethane, $CH_3N{=}NCH_3 \rightarrow N_2 + C_2H_6$, is 40.8 min⁻¹ at 425°C. Find the number of moles of $CH_3N{=}NCH_3$

and N_2 in a flask 0.0500 min after 2.00 g CH_3N=NCH_3 is introduced.

32. The second-order rate constant for the following gas-phase reaction is 0.0488 $M^{-1} \cdot s^{-1}$. We start with 0.120 mol C_2F_4 in a 2.00-liter container, with no C_4F_8 initially present.

$$2C_2F_4 \longrightarrow C_4F_8$$

(a) What will be the concentration of C_2F_4 after 1.00 hour? (b) What will be the concentration of C_4F_8 after 1.00 hour? (c) What is the half-life of the reaction for the initial C_2F_4 concentration given in part (a)? (d) How long will it take for half of the C_2F_4 that remains after 1.00 hour to disappear?

33. The first-order rate constant for the conversion of cyclobutane to ethylene at 1000°C is 87 s^{-1}.

cyclobutane   ethylene

(a) What is the half-life of this reaction at 1000°C?
(b) If one started with 1.00 g of cyclobutane, how long would it take to consume 0.060 g of it?
(c) How much of an initial 1.00-g sample of cyclobutane would remain after 1.00 s?

*34. For the reaction

$$2NO_2 \longrightarrow 2NO + O_2$$

the rate equation is

Rate = $1.4 \times 10^{-10} M^{-1} \cdot s^{-1}[NO_2]^2$ at 25°C

(a) If 4.00 mol of NO_2 is initially present in a sealed 2.00-L vessel at 25°C, what is the half-life of the reaction?
(b) Refer to (a). What concentration and how many grams of NO_2 remain after 125 years?
(c) Refer to (b). What concentration of NO would have been produced during the same period of time?

35. The first-order rate constant for the radioactive decay of radium-224 is 0.189 day^{-1}. What is the half-life of radium-224?

36. Cyclopropane rearranges to form propene

cyclopropane   propene

in a reaction that follows first-order kinetics. At 800 K, the specific rate constant for this reaction is 2.74×10^{-3} s^{-1}. Suppose we start with a cyclopropane concentration of 0.250 M. How long will it take for 99.0% of the cyclopropane to disappear according to this reaction?

37. The thermal decomposition of ammonia, at high temperatures was studied in the presence of inert gases. Data at 2000 K are given for a single experiment after various reaction times have elapsed.

$$NH_3 \longrightarrow NH_2 + H$$

Time (hours)	[NH₃] (mol/L)
0	8.000×10^{-7}
25	6.75×10^{-7}
50	5.84×10^{-7}
75	5.15×10^{-7}

Plot the appropriate concentration expressions against time to find the order of the reaction. Find the rate constant of the reaction from the slope of the line. Use the given data and the appropriate integrated rate equation to check your answer.

38. The following data were obtained from a study of the decomposition of a sample of HI on the surface of a gold wire. (a) Plot the data to find the order of the reaction, the rate constant, and the rate equation. (b) Calculate the HI concentration in mmol/L at 900 sec.

t (seconds)	[HI] (mol/L)
0	5.46
250	4.10
500	2.73
750	1.37

39. The decomposition of SO_2Cl_2 in the gas phase,

$$SO_2Cl_2 \longrightarrow SO_2 + Cl_2$$

can be studied by measuring the concentration of Cl_2 gas as the reaction proceeds. We begin with $[SO_2Cl_2]_0$ = 0.250 M. Holding the temperature constant at 320°C, we monitor the Cl_2 concentration at intervals of 2 hours, with the following results:

Time (hours)	[Cl₂] (mol/L)
0.00	0.000
2.00	0.037
4.00	0.068
6.00	0.095
8.00	0.117
10.00	0.137
12.00	0.153
14.00	0.168
16.00	0.180
18.00	0.190
20.00	0.199

(a) Plot $[Cl_2]$ versus t.
(b) Plot $[SO_2Cl_2]$ versus t.
(c) Determine the rate law for this reaction.
(d) What is the value, with units, for the specific rate constant at 320°C?
(e) How long would it take for 90% of the original SO_2Cl_2 to react?

*40. At some temperature, the rate constant for the decomposition of HI on a gold surface is $0.080 \ M \cdot s^{-1}$.

$$2HI(g) \longrightarrow H_2(g) + I_2(g)$$

(a) What is the order of the reaction? (b) How long will it take for the concentration of HI to drop from 1.00 M to 0.10 M?

Activation Energy, Temperature, and Catalysts

41. Draw typical reaction-energy diagrams for one-step reactions that release energy and that absorb energy. Distinguish between the net energy change for each kind of reaction and the activation energy. Indicate potential energies of products and reactants for both kinds of reactions.

42. Describe and illustrate with graphs how the presence of a catalyst can affect the rate of a reaction.

43. How do homogeneous catalysts and heterogeneous catalysts differ? What is an inhibitor?

44. What is the effect of a catalyst on the energy of the reactants? of the transition state? of the products?

45. For the gas-phase reaction $2N_2O_5 \rightarrow 2N_2O_4 + O_2$, $E_a = 103$ kJ/mol and the rate constant is 0.0900 min^{-1} at 328 K. Find the rate constant at 338 K.

46. The rate constant of a reaction is tripled when the temperature is increased from 298 K to 308 K. Find E_a.

47. The activation energy for the reaction

$$2HI(g) \longrightarrow H_2(g) + I_2(g) \qquad \Delta E^0 = 9.478 \text{ kJ/mol}$$

is 179 kJ/mol. Construct a diagram similar to Figure 16-10 for this reaction. (*Hint*: How does ΔH^0 compare to ΔE^0 for this reaction?)

*48. The activation energy for the reaction between O_3 and NO is 9.6 kJ/mol.

$$O_3(g) + NO(g) \longrightarrow NO_2(g) + O_2(g)$$

(a) Use the thermodynamic quantities in Appendix K to calculate ΔH^0 for this reaction.
(b) Prepare an activation energy plot similar to Figure 16-10 for this reaction. (*Hint*: How does ΔH^0 compare to ΔE^0 for this reaction?)

49. The rate constant for the decomposition of N_2O

$$2N_2O(g) \longrightarrow 2N_2(g) + O_2(g)$$

is 2.6×10^{-11} s^{-1} at 300°C and 2.1×10^{-10} s^{-1} at 330°C. What is the activation energy for this reaction? Prepare a reaction coordinate diagram like Figure 16-10 using -164.1 kJ/mol as the energy of reaction.

50. For the reaction

$$N_2O_5(g) \longrightarrow 2NO_2(g) + \tfrac{1}{2}O_2(g) \qquad \Delta E^0 = 51.51 \text{ kJ/mol}$$

$k = 8.0 \times 10^{-7}$ s^{-1} at 0.0°C and 8.9×10^{-4} s^{-1} at 50.0°C. Prepare a reaction coordinate diagram like Figure 16-10 for this reaction.

*51. You are given the rate constant as a function of temperature for the exchange reaction

$$Mn(CO)_5(CH_3CN)^+ + NC_5H_5 \longrightarrow$$
$$Mn(CO)_5(NC_5H_5)^+ + CH_3CN$$

T (K)	k (min^{-1})
298	0.0409
308	0.0818
318	0.157

(a) Calculate E_a from a plot of log k versus $1/T$. (b) Use the graph to predict the value of k at 311 K. (c) What is the numerical value of the collision frequency factor, A, in the Arrhenius equation?

*52. The rearrangement of cyclopropane to propene described in Exercise 36 has been studied at various temperatures. The following values for the specific rate constant have been determined experimentally:

T (K)	k (s^{-1})
600	3.30×10^{-9}
650	2.19×10^{-7}
700	7.96×10^{-6}
750	1.80×10^{-4}
800	2.74×10^{-3}
850	3.04×10^{-2}
900	2.58×10^{-1}

(a) From the appropriate plot of these data, determine the value of the activation energy for this reaction.
(b) Use the graph to estimate the value of k at 400 K.
(c) Use the graph to estimate the temperature at which the value of k would be equal to 1.00×10^{-5} s^{-1}.

53. Biological reactions nearly always occur in the presence of enzymes as catalysts. The enzyme catalase, which acts on peroxides, reduces the E_a for the reaction from 72 kJ/mol (uncatalyzed) to 28 kJ/mol (catalyzed). By what factor does the reaction rate increase at normal body temperature, 37.0°C, for the same reactant (peroxide) concentration? Assume that the collision factor, A, remains constant.

*54. The enzyme carbonic anhydrase catalyzes the hydration of CO_2, $CO_2 + H_2O \rightarrow H_2CO_3$, a critical reaction involved in the transfer of CO_2 from tissues to the lung via the bloodstream. One enzyme molecule hydrates 10^6 molecules of CO_2 per second. How many kilograms of CO_2 are hydrated in 1 hour in 1 L by 5×10^{-6} M enzyme?

55. The following gas-phase reaction follows first-order kinetics:

$$FClO_2 \longrightarrow FClO + O$$

The activation energy of this reaction is measured to be 186 kJ/mol. The value of k at 322°C is determined to be 6.76×10^{-4} s^{-1}. (a) What would be the value of k for this reaction at 25°C? (b) At what temperature would this reaction have a k value of 6.00×10^{-2} s^{-1}?

56. The following gas-phase reaction is first order:

$$N_2O_5 \longrightarrow NO_2 + NO_3$$

The activation energy of this reaction is measured to be 88 kJ/mol. The value of k at 0°C is determined to be 9.16×10^{-3} s^{-1}. (a) What would be the value of k for this reaction at room temperature, 25°C? (b) At what temperature would this reaction have a k value of 6.00×10^{-2} s^{-1}?

Reaction Mechanisms

57. Define reaction mechanism. Why do we believe that only bimolecular collisions and unimolecular decompositions are important in most reaction mechanisms?

58. The rate equation for the reaction

$$Cl_2(aq) + H_2S(aq) \longrightarrow S(s) + 2HCl(aq)$$

is found to be Rate = $k[Cl_2][H_2S]$. Which of the following mechanisms are consistent with the rate law?

(a)
$Cl_2 \longrightarrow Cl^+ + Cl^-$	(slow)
$Cl^- + H_2S \longrightarrow HCl + HS^-$	(fast)
$Cl^+ + HS^- \longrightarrow HCl + S$	(fast)
$Cl_2 + H_2S \longrightarrow S + 2HCl$	overall

(b)
$Cl_2 + H_2S \longrightarrow HCl + Cl^+ + HS^-$	(slow)
$Cl^+ + HS^- \longrightarrow HCl + S$	(fast)
$Cl_2 + H_2S \longrightarrow S + 2HCl$	overall

(c)
$Cl_2 \rightleftharpoons Cl + Cl$	(fast, equilibrium)
$Cl + H_2S \rightleftharpoons HCl + HS$	(fast, equilibrium)
$HS + Cl \longrightarrow HCl + S$	(slow)
$Cl_2 + H_2S \longrightarrow S + 2HCl$	overall

59. The ozone, O_3, of the stratosphere can be decomposed by reaction with nitrogen oxide (commonly called nitric oxide), NO, from high-flying jet aircraft.

$$O_3(g) + NO(g) \longrightarrow NO_2(g) + O_2(g)$$

The rate expression is Rate = $k[O_3][NO]$. Which of the following mechanisms are consistent with the observed rate expression?

(a)
$NO + O_3 \longrightarrow NO_3 + O$	(slow)
$NO_3 + O \longrightarrow NO_2 + O_2$	(fast)
$O_3 + NO \longrightarrow NO_2 + O_2$	overall

(b)
$NO + O_3 \longrightarrow NO_2 + O_2$	(slow)
	(one step)

(c)
$O_3 \longrightarrow O_2 + O$	(slow)
$O + NO \longrightarrow NO_2$	(fast)
$O_3 + NO \longrightarrow NO_2 + O_2$	overall

(d)
$NO \longrightarrow N + O$	(slow)
$O + O_3 \longrightarrow 2O_2$	(fast)
$O_2 + N \longrightarrow NO_2$	(fast)
$O_3 + NO \longrightarrow NO_2 + O_2$	overall

(e)
$NO \rightleftharpoons N + O$	(fast, equilibrium)
$O + O_3 \longrightarrow 2O_2$	(slow)
$O_2 + N \longrightarrow NO_2$	(fast)
$O_3 + NO \longrightarrow NO_2 + O_2$	overall

60. The energy of activation for a hypothetical reaction A + B → C is 345 kJ/mol for the uncatalyzed reaction and 165 kJ/mol for the reaction catalyzed on a metal surface. ΔE for this reaction is 90 kJ/mol of reaction. Draw and label a reaction coordinate diagram similar to Figure 16-15 for this reaction.

61. A mechanism for the gas-phase reaction $H_2 + I_2 \rightarrow 2HI$ was discussed in the chapter.
(a) Does the following mechanism also predict the correct rate law, Rate = $k[H_2][I_2]$?

$I_2 \rightleftharpoons 2I$	(fast, equilibrium)
$I + H_2 \rightleftharpoons IH_2$	(fast, equilibrium)
$IH_2 + I \longrightarrow 2HI$	(slow)

(b) Identify any reaction intermediates in this proposed mechanism.

62. The combination of Cl atoms is catalyzed by $N_2(g)$. The following mechanism is suggested:

$N_2 + Cl \rightleftharpoons ClN_2$	(fast, equilibrium)
$ClN_2 + Cl \longrightarrow Cl_2 + N_2$	(slow)

(a) Identify any reaction intermediates in this proposed mechanism.
(b) Is this mechanism consistent with the experimental rate law, Rate = $k[N_2][Cl]^2$?

63. The reaction between NO and Br_2 was discussed in Section 16-9. The following mechanism has also been proposed.

$2NO \rightleftharpoons N_2O_2$	(fast, equilibrium)
$N_2O_2 + Br_2 \longrightarrow 2NOBr$	(slow)

Is this mechanism consistent with the observation that the reaction is second order in NO and first order in Br_2?

*64. The following mechanism for the reaction between H_2 and CO has been proposed.

$H_2 \rightleftharpoons 2H$	(fast, equilibrium)
$H + CO \longrightarrow HCO$	(slow)
$H + HCO \longrightarrow HCHO$	(fast)

(a) Write the balanced equation for the overall reaction.

(b) The observed rate dependence is found to be one-half order in H_2 and first order in CO. Is this proposed reaction mechanism consistent with the observed rate dependence?

Mixed Exercises

65. The following explanation of the operation of a pressure cooker appears in a cookbook: "Boiling water in the presence of air can never produce a temperature higher than 212°F, no matter how high the heat source. But in a pressure cooker, the air is withdrawn first, so the boiling water can be maintained at higher temperatures." Support or criticize this explanation.

*66. A cookbook gives the following general guideline for use of a pressure cooker: "For steaming vegetables, cooking time at a gauge pressure of 15 pounds per square inch (psi) is $\frac{1}{3}$ that at atmospheric pressure." Remember that gauge pressure is measured relative to the external atmospheric pressure, which is 15 psi at sea level. From this information, estimate the activation energy for the process of steaming vegetables. (*Hint:* Clausius and Clapeyron may be able to help you.)

67. The reaction between ozone and nitrogen dioxide, $2NO_2 + O_3 \rightarrow N_2O_5 + O_2$, has been studied at 231 K. The experimental rate equation is Rate $= k[NO_2][O_3]$.

(a) What is the order of the reaction?

(b) Is either of the following proposed mechanisms consistent with the given kinetic data? Show how you arrived at your answer.

(a) $NO_2 + NO_2 \rightleftharpoons N_2O_4$ (fast, equilibrium)
$N_2O_4 + O_3 \longrightarrow N_2O_5 + O_2$ (slow)

(b) $NO_2 + O_3 \longrightarrow NO_3 + O_2$ (slow)
$NO_3 + NO_2 \longrightarrow N_2O_5$ (fast)

68. Data are given for the decomposition of N_2O_5 in CCl_4 solution at 45°C:

$$2N_2O_5(sol) \longrightarrow 4NO_2(sol) + O_2(g)$$

The rate is determined by measuring the volume of O_2 produced.

$[N_2O_5]_0$ (*M*)	Initial Rate (mol O_2/L · s)
0.600	3.7×10^{-4}
0.300	1.9×10^{-4}
0.100	6.2×10^{-5}

Write the rate-law expression.

69. Refer to the reaction and data in Exercise 55. We begin with 4.50 mol of $FClO_2$ in a 2.00-L container.

(a) How many moles of $FClO_2$ would remain after 1.00 min at 25°C?

(b) How much time would be required for 99.9% of the $FClO_2$ to decompose at 25°C?

70. Refer to the reaction and data in Exercise 56. We begin with 4.50 mol of N_2O_5 in a 2.00-L container.

(a) How many moles of N_2O_5 would remain after 1.00 min at 25°C?

(b) How much time would be required for 99.9% of the N_2O_5 decompose at 25°C?

71. The decomposition of gaseous dimethyl ether, $CH_3OCH_3 \rightarrow CH_4 + CO + H_2$, follows first-order kinetics. Its half-life is 25.0 min at 500°C. (a) Starting with 8.00 g of dimethyl ether at 500°C, how many grams would remain after 125 min? (b) In part (a), how many grams would remain after 145 min? (c) In part (b), what fraction remains and what fraction reacts? (d) Calculate the time, in minutes, required to decrease 7.60 ng of dimethyl ether to 2.25 ng.

72. The rate of the hemoglobin (Hb)–carbon monoxide reaction, $4Hb + 3CO \rightarrow Hb_4(CO)_3$, has been studied at 20°C. Concentrations are expressed in micromoles per liter (μmol/L):

Concentration (μmol/L)		Rate of Disappearance of Hb (μmol/L · s)
[Hb]	[CO]	
3.36	1.00	0.941
6.72	1.00	1.88
6.72	3.00	5.64

(a) Write the rate equation for the reaction. (b) Calculate the rate constant for the reaction. (c) Calculate the rate, at the instant when [Hb] = 1.50 and [CO] = 0.600 μmol/L.

*73. In the chapter it was calculated that the rate constant, k, for a reaction whose activation energy is 50 kJ/mol approximately doubles (factor of 1.9) when the temperature is increased from 300 K to 310 K. Over this same temperature range, determine the factor by which k increases for a reaction whose activation energy is 80 kJ/mol. Can you rationalize this result?

17 Chemical Equilibrium

A night-time photo of a large plant for the commercial production of ammonia, NH_3. Such an installation can produce up to 7000 metric tons of ammonia per day. There are nearly a hundred such plants in the world.

Objectives

As you study this chapter, you should learn

☐ The basic ideas of chemical equilibrium
☐ What an equilibrium constant is and what it tells us
☐ What a reaction quotient is and what it tells us
☐ To use equilibrium constants to describe systems at equilibrium
☐ To recognize the factors that affect equilibria and predict the resulting effects

☐ About equilibrium constants expressed in terms of partial pressures (K_P's) and how these are related to K_c's
☐ About heterogeneous equilibria
☐ About relationships between thermodynamics and equilibrium
☐ How to estimate equilibrium constants at different temperatures

17-1 Basic Concepts

Most chemical reactions do not go to completion. That is, when reactants are mixed in stoichiometric quantities, they are not completely converted to products. Reactions that do not go to completion *and* that can occur in either direction are called *reversible reactions*.

Reversible reactions can be represented in general terms as follows, where the capital letters represent formulas and the lowercase letters represent the stoichiometric coefficients in the balanced equation.

$$a\mathrm{A} + b\mathrm{B} \rightleftharpoons c\mathrm{C} + d\mathrm{D}$$

The double arrow ($\rightleftharpoons$) indicates that the reaction is reversible—i.e., both the forward and reverse reactions occur simultaneously. When A and B react to form C and D at the same rate at which C and D react to form A and B, the system is at *equilibrium*.

Chemical equilibrium exists when two opposing reactions occur simultaneously at the same rate.

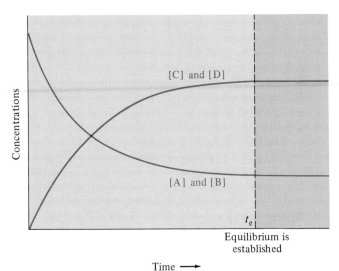

[C] and [D]

[A] and [B]

t_e

Equilibrium is established

Time →

Concentrations

Brackets, [], represent the concentration, in moles per liter, of the species enclosed within them.

Figure 17-1
Variation in the concentrations of species present in the $A + B \rightleftharpoons C + D$ system as equilibrium is approached, beginning with equal concentrations of A and B only. For this reaction, production of products is favored. As we shall see, this corresponds to a value of the equilibrium constant greater than 1.

Chemical equilibria are **dynamic equilibria**; that is, individual molecules are continually reacting, even though the overall composition of the reaction mixture does not change. In a system at equilibrium, the equilibrium is said to lie toward the right if more C and D are present than A and B, and to lie toward the left if more A and B are present.

Consider a case in which the coefficients in the equation for a reaction are all 1. When substances A and B react, the rate of the forward reaction decreases as time passes because the concentrations of A and B decrease.

$$A + B \longrightarrow C + D \qquad (1)$$

As the concentrations of C and D build up, they start to form A and B.

$$C + D \longrightarrow A + B \qquad (2)$$

As more C and D molecules are formed, more can react, and so the rate of reaction between C and D increases with time. Eventually, the two reactions occur at the same rate, and the system is at equilibrium (see Figure 17-1).

$$A + B \rightleftharpoons C + D$$

If a reaction begins with only C and D present, the rate of reaction (2) decreases with time. The rate of reaction (1) increases with time until the two rates are equal.

The dynamic nature of chemical equilibrium can be proved experimentally by "tagging" a small percentage of molecules with radioactive atoms and following them through the reaction. Even when the initial mixture is at equilibrium, radioactive atoms eventually appear in both reactant and product molecules.

The SO₂–O₂–SO₃ System

Consider the reversible reaction of sulfur dioxide with oxygen to form sulfur trioxide at 1500 K.

$$2SO_2(g) + O_2(g) \rightleftharpoons 2SO_3(g)$$

Suppose 0.400 mole of SO_2 and 0.200 mole of O_2 are injected into a closed 1.00-liter container. When equilibrium is established (at time t_e, Figure 17-2a), we find that 0.056 mole of SO_3 has formed and that 0.344 mole of SO_2 and 0.172 mole of O_2 remain unreacted. The reaction does not go to completion.

The numbers in this discussion were determined experimentally.

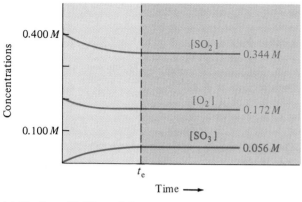

(a) Starting with SO_2 and O_2.

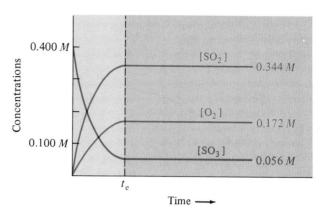

(b) Starting with SO_3.

Figure 17-2

Establishment of equilibrium in the $2SO_2 + O_2 \rightleftharpoons 2SO_3$ system. (a) Beginning with stoichiometric amounts of SO_2 and O_2 and no SO_3. (b) Beginning with only SO_3 and no SO_2 or O_2. Greater changes in concentrations occur to establish equilibrium when starting with SO_3 than when starting with SO_2 and O_2. The equilibrium favors SO_2 and O_2.

These changes are summarized below, using molarity units rather than moles. (They are numerically identical here because the volume of the reaction vessel is 1.00 liter.) The *net reaction* is represented by the *changes* in concentrations.

	$2SO_2(g)$	$+$	$O_2(g)$	$\rightleftharpoons$	$2SO_3(g)$
initial conc'n	0.400 *M*		0.200 *M*		0
change due to rxn	−0.056 *M*		−0.028 *M*		+0.056 *M*
equilibrium conc'n	0.344 *M*		0.172 *M*		0.056 *M*

The ratio in the "change due to rxn" line is determined by the coefficients in the balanced equation.

In another experiment, 0.400 mole of SO_3 is introduced alone into a closed 1.00-liter container. The same total numbers of sulfur and oxygen atoms are present as in the previous experiment. When equilibrium is established (time t_e in Figure 17-2b), 0.056 mole of SO_3, 0.172 mole O_2, and 0.344 mole of SO_2 are present. (These are the same amounts found at equilibrium in the previous case.) This time the net reaction proceeds from *right to left* as the equation is written. The changes in concentration are in the same 2:1:2 ratio as in the previous case, as required by the coefficients of the balanced equation. The time required to reach equilibrium may be longer or shorter.

A setup such as this is called a "reaction summary."

	$2SO_2(g)$	$+$	$O_2(g)$	$\rightleftharpoons$	$2SO_3(g)$
initial conc'n	0		0		0.400 *M*
change due to rxn	+0.344 *M*		+0.172 *M*		−0.344 *M*
equilibrium conc'n	0.344 *M*		0.172 *M*		0.056 *M*

The results of these experiments are summarized below and in Figure 17-2.

	Initial Concentrations			Equilibrium Concentrations		
	[SO_2]	[O_2]	[SO_3]	[SO_2]	[O_2]	[SO_3]
Experiment 1	0.400 *M*	0.200 *M*	0 *M*	0.344 *M*	0.172 *M*	0.056 *M*
Experiment 2	0 *M*	0 *M*	0.400 *M*	0.344 *M*	0.172 *M*	0.056 *M*

17-2 The Equilibrium Constant

Suppose a reversible reaction occurs by a *one-step mechanism*:

$$2A + B \rightleftharpoons A_2B$$

The rate of the forward reaction is $\text{Rate}_f = k_f[A]^2[B]$; the rate of the reverse reaction is $\text{Rate}_r = k_r[A_2B]$. In these expressions, k_f and k_r are the *specific rate constants* of the forward and reverse reactions, respectively. By definition, the two rates are equal *at equilibrium* ($\text{Rate}_f = \text{Rate}_r$). So we may write

$$k_f[A]^2[B] = k_r[A_2B] \qquad \text{(at equilibrium)}$$

Dividing both sides of this equation by k_r and by $[A]^2[B]$ gives

$$\frac{k_f}{k_r} = \frac{[A_2B]}{[A]^2[B]}$$

At any specific temperature, both k_f and k_r are constants, so k_f/k_r is a constant.

This ratio is given a special name and symbol—the **equilibrium constant**, K_c or simply K.

$$K_c = \frac{[A_2B]}{[A]^2[B]} \qquad \text{(at equilibrium)}$$

The subscript c refers to concentrations. The brackets, [], in this expression indicate *equilibrium* concentrations in moles per liter.

We have described a one-step reaction. Suppose, instead, that this reaction involves a *two-step mechanism* with the following rate constants.

$$\text{(step 1)} \quad 2A \underset{k_{1r}}{\overset{k_{1f}}{\rightleftharpoons}} A_2 \quad \text{followed by} \quad \text{(step 2)} \quad A_2 + B \underset{k_{2r}}{\overset{k_{2f}}{\rightleftharpoons}} A_2B$$

We add (1) and (2) and see that the overall reaction is $2A + B \rightleftharpoons A_2B$. An equilibrium constant expression can be written for each step:

$$K_1 = \frac{k_{1f}}{k_{1r}} = \frac{[A_2]}{[A]^2} \quad \text{and} \quad K_2 = \frac{k_{2f}}{k_{2r}} = \frac{[A_2B]}{[A_2][B]}$$

If we multiply K_1 by K_2, the $[A_2]$ term can be eliminated. Because K_1 and K_2 are both constants, their product is also a constant, and we obtain

$$K_1 \times K_2 = \frac{[A_2]}{[A]^2} \times \frac{[A_2B]}{[A_2][B]} = \frac{[A_2B]}{[A]^2[B]} = K_c$$

Regardless of the mechanism by which this reaction occurs, the equilibrium constant expression has the same form. For a reaction in general terms, the equilibrium constant can always be written as follows:

$$\text{For} \quad \underbrace{aA + bB}_{\text{reactants}} \rightleftharpoons \underbrace{cC + dD}_{\text{products}}, \quad K_c = \frac{[C]^c[D]^d}{[A]^a[B]^b} \quad \begin{matrix} \leftarrow \text{products} \\ \\ \leftarrow \text{reactants} \end{matrix}$$

The equilibrium constant, K_c, is defined as the product of the *equilibrium concentrations* (moles per liter) of the products, each raised to the power that corresponds to its coefficient in the balanced equation, divided by the product of the *equilibrium concentrations* of reactants, each raised to the power that corresponds to its coefficient in the balanced equation.

In general, numerical values for K_c can come only from experiments. Some equilibrium constant expressions and their numerical values at 25°C are

$$N_2(g) + O_2(g) \rightleftharpoons 2NO(g) \qquad K_c = \frac{[NO]^2}{[N_2][O_2]} = 4.5 \times 10^{-31}$$

$$CH_4(g) + Cl_2(g) \rightleftharpoons CH_3Cl(g) + HCl(g) \qquad K_c = \frac{[CH_3Cl][HCl]}{[CH_4][Cl_2]} = 1.2 \times 10^{18}$$

$$N_2(g) + 3H_2(g) \rightleftharpoons 2NH_3(g) \qquad K_c = \frac{[NH_3]^2}{[N_2][H_2]^3} = 3.6 \times 10^8$$

Activities were discussed in Section 14-14.

The thermodynamic definition of the equilibrium constant involves activities rather than concentrations. The **activity** of a component of an ideal mixture is the ratio of its concentration or partial pressure to a standard concentration (1 *M*) or pressure (1 atm). For now, we can consider the activity of each species to be a dimensionless quantity whose numerical value can be determined as follows. (a) For pure liquids and solids, the activity is taken as 1. (b) For components of ideal solutions, the activity of each component is taken to be equal to its molar concentration. (c) For gases in an ideal mixture, the activity of each component is taken to be equal to its partial pressure in atmospheres. Because of the use of activities, *the equilibrium constant has no units; the values we put into K_c are numerically equal to molar concentrations, but are dimensionless*, i.e., have no units.

> The magnitude of K_c is a measure of the extent to which reaction occurs. For any reaction, the value of K_c (1) varies only with temperature, (2) is constant at a given temperature, and (3) is independent of the initial concentrations.

A value of K_c *much* greater than 1 indicates that at equilibrium most of the reactants would be converted into products. On the other hand, if K_c is quite small, equilibrium is established when most of the reactants remain unreacted and only small amounts of products are formed.

For a given chemical reaction at a specific temperature, the product of the concentrations of the products formed by the reaction, each raised to the appropriate power, divided by the product of the concentrations of the reactants, each raised to the appropriate power, always has the same value. This does *not* mean that the individual equilibrium concentrations for a given reaction are always the same, but it does mean that this particular numerical combination of their values is constant.

Consider again the SO_2–O_2–SO_3 equilibrium, at 1500 K, described earlier in this chapter. The equilibrium concentrations were the same in the two experiments. We can use these equilibrium concentrations to calculate the value of the equilibrium constant for this reaction.

$$2SO_2(g) + O_2(g) \rightleftharpoons 2SO_3(g)$$
$$\text{equil conc'n} \quad 0.344\ M \quad 0.172\ M \quad 0.056\ M$$

Substituting the numerical values (without units) into the equilibrium expression gives the value of the equilibrium constant.

$$K_c = \frac{[SO_3]^2}{[SO_2]^2[O_2]} = \frac{(0.056)^2}{(0.344)^2(0.172)} = 1.5 \times 10^{-1}$$

For the reaction written *as it is*, K_c is 0.15 at 1500 K.

Example 17-1

We place some nitrogen and hydrogen in an empty 5.00-liter vessel at 500°C. When equilibrium is established, 3.01 mol of N_2, 2.10 mol of H_2, and 0.565 mol of NH_3 are present. Evaluate K_c for the following reaction at 500°C.

$$N_2(g) + 3H_2(g) \rightleftharpoons 2NH_3(g)$$

Plan

The *equilibrium concentrations* are obtained by dividing the number of moles of each reactant and product by the volume, 5.00 liters. Then we substitute the equilibrium values into the equilibrium constant expression.

Solution

The equilibrium concentrations are

$$[N_2] = 3.01 \text{ mol}/5.00 \text{ L} = 0.602 \ M$$

$$[H_2] = 2.10 \text{ mol}/5.00 \text{ L} = 0.420 \ M$$

$$[NH_3] = 0.565 \text{ mol}/5.00 \text{ L} = 0.113 \ M$$

We substitute these numerical values into the expression for K_c:

$$K_c = \frac{[NH_3]^2}{[N_2][H_2]^3} = \frac{(0.113)^2}{(0.602)(0.420)^3} = 0.286$$

Thus, for the reaction of H_2 and N_2 to form NH_3 at 500°C, we can write

$$K_c = \frac{[NH_3]^2}{[N_2][H_2]^3} = \boxed{0.286}$$

The small value of K_c indicates that the equilibrium lies to the left.

EOC 14, 15

> The values (activities) that we put into the K_c expression are equal to the molar concentrations, but with *units omitted*. We use values of *concentrations*, not numbers of moles, in calculations of equilibrium constants.

17-3 Variation of K_c with the Form of the Balanced Equation

The value of K_c depends on the form of the balanced equation for the reaction. We wrote the equation for the reaction of SO_2 and O_2 to form SO_3 and its equilibrium constant expression as

$$2SO_2(g) + O_2(g) \rightleftharpoons 2SO_3(g) \qquad \text{and} \qquad K_c = \frac{[SO_3]^2}{[SO_2]^2[O_2]} = 0.15$$

Suppose we write the equation for the same reaction in reverse. The equilibrium constant K_c' for the reaction, written this way, is

$$2SO_3(g) \rightleftharpoons 2SO_2(g) + O_2(g) \qquad \text{and} \qquad K_c' = \frac{[SO_2]^2[O_2]}{[SO_3]^2} = \frac{1}{K_c} = 6.5$$

> Reversing an equation is the same as multiplying all coefficients by -1. This reverses the roles of "reactants" and "products."

We see that K_c', the equilibrium constant for the reaction written in reverse, is *the reciprocal of* K_c, the equilibrium constant for the original reaction.

If the equation for the reaction were written as

We see that K_c'' is the square root of K_c. $K_c^{1/2}$ means the square root of K_c.

$$SO_2(g) + \tfrac{1}{2}O_2(g) \rightleftharpoons SO_3(g) \qquad K_c'' = \frac{[SO_3]}{[SO_2][O_2]^{1/2}} = K_c^{1/2} = 0.39$$

If an *equation* for a reaction is *multiplied* by a positive or negative number, n, then the *original value of K_c* is raised to the nth power. Thus, we must always write the balanced chemical equation, as well as the value of K_c, for a chemical reaction.

Example 17-2

You are given the following reaction and its equilibrium constant at a given temperature:

$$2HBr(g) + Cl_2(g) \rightleftharpoons 2HCl(g) + Br_2(g) \qquad K_c = 4.0 \times 10^4$$

Write the expression for, and calculate the numerical value of, the equilibrium constant for each of the following at the same temperature.

A coefficient of $\tfrac{1}{2}$ refers to $\tfrac{1}{2}$ of a mole, not $\tfrac{1}{2}$ of a molecule.

(a) $\qquad\qquad 4HBr(g) + 2Cl_2(g) \rightleftharpoons 4HCl(g) + 2Br_2(g)$

(b) $\qquad\qquad HBr(g) + \tfrac{1}{2}Cl_2(g) \rightleftharpoons HCl(g) + \tfrac{1}{2}Br_2(g)$

Plan

We recall the definition of the equilibrium constant. For the original equation,

$$K_c = \frac{[HCl]^2[Br_2]}{[HBr]^2[Cl_2]} = 4.0 \times 10^4$$

Solution

(a) The original equation has been multiplied by 2, so K_c must be squared.

$$K_c' = \frac{[HCl]^4[Br_2]^2}{[HBr]^4[Cl_2]^2} \qquad K_c' = (K_c)^2 = (4.0 \times 10^4)^2 = \boxed{1.6 \times 10^9}$$

(b) The original equation has been multiplied by $\tfrac{1}{2}$ (divided by 2), so K_c must be raised to the $\tfrac{1}{2}$ power. This is the same as extracting the square root of K_c.

$$K_c'' = \frac{[HCl][Br_2]^{1/2}}{[HBr][Cl_2]^{1/2}} = \sqrt{K_c} = \sqrt{4.0 \times 10^4} = \boxed{2.0 \times 10^2}$$

EOC 18

17-4 The Reaction Quotient

The **mass action expression** or **reaction quotient**, Q, for the general reaction is given as follows:

$$\text{For } aA + bB \rightleftharpoons cC + dD, \qquad Q = \frac{[C]^c[D]^d}{[A]^a[B]^b} \qquad \begin{array}{l} \textit{not necessarily} \\ \textit{equilibrium} \\ \textit{concentrations} \end{array}$$

The reaction quotient has the same *form* as the equilibrium constant, but it involves specific values that are not *necessarily* equilibrium concentrations. However, when they *are* equilibrium concentrations, then $Q = K_c$. The concept of reaction quotient is very useful. We can compare the magnitude of Q with that of K for a reaction under given conditions to decide whether net forward or reverse reaction must occur to establish equilibrium.

When the forward reaction occurs to a greater extent than the reverse reaction, we say that a *net* forward reaction has occurred.

If at any time $Q < K$, the forward reaction must occur to a greater extent than the reverse reaction for equilibrium to be established. This is because when $Q < K$, the numerator of Q is too small and the denominator is too large. Reducing the denominator and increasing the numerator requires that A and B react to produce C and D. Conversely, if $Q > K$, the reverse reaction must occur to a greater extent than the forward reaction for equilibrium to be reached. When $Q = K$, the system is at equilibrium, so no further *net* reaction occurs.

$Q < K$ Forward reaction predominates until equilibrium is established.

$Q = K$ System is at equilibrium.

$Q > K$ Reverse reaction predominates until equilibrium is established.

Example 17-3

At a very high temperature, $K_c = 1.0 \times 10^{-13}$ for the following reaction:

$$2HF(g) \rightleftharpoons H_2(g) + F_2(g)$$

At a certain time the following concentrations were detected. Is the system at equilibrium? If not, what must occur for equilibrium to be established?

$$[HF] = 0.500\ M, \quad [H_2] = 1.00 \times 10^{-3}\ M, \quad \text{and}\ [F_2] = 4.00 \times 10^{-3}\ M$$

These concentrations could occur if we started with a mixture of HF, H_2, and F_2.

Plan

We substitute these concentrations into the mass action expression to calculate Q. Then we compare Q with K_c to see whether the system is at equilibrium.

Solution

$$Q = \frac{[H_2][F_2]}{[HF]^2} = \frac{(1.00 \times 10^{-3})(4.00 \times 10^{-3})}{(0.500)^2} = 1.60 \times 10^{-5}$$

But $K_c = 1.0 \times 10^{-13}$, so $Q > K_c$. The system is *not* at equilibrium. For equilibrium to be established, the value of Q must become smaller until it equals K_c. This can occur only if the numerator decreases and the denominator increases. Thus, the right-to-left reaction must occur to a greater extent than the forward reaction; i.e., H_2 and F_2 must react to form more HF.

EOC 31, 32

In Example 17-3 we calculated the value for Q and compared it with the *known* value of K_c to predict the direction of the net reaction that leads to equilibrium.

17-5 Uses of the Equilibrium Constant, K_c

The equilibrium constant is a "constant" only if the temperature does not change.

We have seen (Section 17-2) how to calculate the value of K_c from one set of equilibrium concentrations. Once that value has been obtained, the process can be turned around to calculate equilibrium *concentrations* from the equilibrium *constant*.

Example 17-4

The equation for the following reaction and the value of K_c at a given temperature are given. In a system at equilibrium in a 1.00-liter container, we find 0.25 mol of PCl_5 and 0.16 mol of PCl_3. What equilibrium concentration of Cl_2 must be present?

$$PCl_3(g) + Cl_2(g) \rightleftharpoons PCl_5(g) \qquad K_c = 1.9$$

Plan

We write the equilibrium constant expression and its value. Only one term, $[Cl_2]$, is unknown. We solve for it.

Solution

The equilibrium constant expression and its numeric value are

$$K_c = \frac{[PCl_5]}{[PCl_3][Cl_2]} = 1.9$$

$$[Cl_2] = \frac{[PCl_5]}{K_c[PCl_3]} = \frac{(0.25)}{(1.9)(0.16)} = 0.82\ M$$

EOC 34, 35

Often we know the starting concentrations and want to know how much of each reactant and each product would be present at equilibrium.

Example 17-5

We have represented chemical formulas by single letters to simplify the notation in these calculations.

For the following reaction, the equilibrium constant is 49.0 at a certain temperature. If 0.400 mol each of A and B are placed in a 2.00-liter container at that temperature, what concentrations of all species are present at equilibrium?

$$A + B \rightleftharpoons C + D$$

Plan

First we find the initial concentrations. Then we write the reaction summary and represent the equilibrium concentrations. Finally we substitute the representations into the K_c expression and find the equilibrium concentrations.

Solution

The initial concentrations are

$$[A] = \frac{0.400\ \text{mol}}{2.00\ \text{L}} = 0.200\ M \qquad [C] = 0\ M$$

$$[B] = \frac{0.400\ \text{mol}}{2.00\ \text{L}} = 0.200\ M \qquad [D] = 0\ M$$

We know that the reaction can only proceed to the right because only "reactants" are present. The reaction summary includes the values, or symbols for the values, of (1) initial concentrations, (2) changes in concentrations, and (3) concentrations at equilibrium.

Let x = moles per liter of A that react; then x = moles per liter of B that react and x = moles per liter of C and D that are formed.

The coefficients in the equation are all 1's, so the reaction ratio must be 1:1:1:1.

Reaction Summary

	A	+	B	⇌	C	+	D
initial	0.200 M		0.200 M		0 M		0 M
change due to rxn	$-x\ M$		$-x\ M$		$+x\ M$		$+x\ M$
at equilibrium	$(0.200 - x)\ M$		$(0.200 - x)\ M$		$x\ M$		$x\ M$

Now K_c is known but concentrations are not. However, the equilibrium concentrations have all been expressed in terms of the single variable x. We substitute the equilibrium concentrations (*not* the initial ones) into the K_c expression and solve for x.

$$K_c = \frac{[C][D]}{[A][B]} = 49.0$$

$$\frac{(x)(x)}{(0.200 - x)(0.200 - x)} = \frac{x^2}{(0.200 - x)^2} = 49.0$$

This quadratic equation has a perfect square on both sides. We solve it by taking the square roots of both sides of the equation and then rearranging for x.

$$\frac{x}{0.200 - x} = 7.0$$

$$x = 1.40 - 7.0\,x \qquad 8.0x = 1.40 \qquad x = \frac{1.40}{8.0} = 0.175$$

Now we know the value of x, so the equilibrium concentrations are

[A] $= (0.200 - x)\ M =$ 0.025 M ; [C] $= x\ M =$ 0.175 M

[B] $= (0.200 - x)\ M =$ 0.025 M ; [D] $= x\ M =$ 0.175 M

We see that the equilibrium concentrations of products are much greater than those of reactants. K_c is much greater than 1.

We can easily check the calculation by calculating Q *with the equilibrium values* (Q_e) and verifying that the value is equal to K_c. Recall that K_c = 49.0.

$$Q_e = \frac{[C][D]}{[A][B]} = \frac{(0.175)(0.175)}{(0.025)(0.025)} = 49.0$$

The ideas developed in Example 17-5 can be applied to cases in which the reactants are mixed in nonstoichiometric amounts.

Example 17-6

Consider the same system as in Example 17-5 at the same temperature. If 0.600 mol of A and 0.200 mol of B are mixed in a 2.00-liter container and allowed to reach equilibrium, what are the equilibrium concentrations of all species?

Plan

We proceed as we did in Example 17-5. The only difference is that we have *nonstoichiometric* amounts of reactants.

Solution

Let x = mol/L of A that react; then x = mol/L of B that react, and x = mol/L of C and D formed.

	A	+	B	$\rightleftharpoons$	C	+	D
initial	0.300 M		0.100 M		0 M		0 M
change due to rxn	$-x$ M		$-x$ M		$+x$ M		$+x$ M
equilibrium	$(0.300 - x)$ M		$(0.100 - x)$ M		x M		x M

The initial concentrations are governed by the amounts of reactants mixed together. But *changes in concentrations* due to reaction must occur in a 1:1:1:1 ratio, as required by the coefficients in the balanced equation.

The left side of this equation is *not* a perfect square.

$$K_c = \frac{[C][D]}{[A][B]} = 49.0 \quad \text{so} \quad \frac{(x)(x)}{(0.300 - x)(0.100 - x)} = 49.0$$

We can arrange this quadratic equation into the standard form:

$$\frac{x^2}{0.0300 - 0.400x + x^2} = 49.0$$

$$x^2 = 1.47 - 19.6x + 49.0x^2$$

$$48.0x^2 - 19.6x + 1.47 = 0$$

All quadratic equations of the form

$$ax^2 + bx + c = 0$$

can be solved by use of the quadratic formula, which is

$$x = \frac{-b \pm \sqrt{b^2 - 4ac}}{2a} \quad \text{(see Appendix A)}$$

In this case, $a = 48.0$, $b = -19.6$, and $c = 1.47$. Substituting these values gives

$$x = \frac{-(-19.6) \pm \sqrt{(-19.6)^2 - 4(48.0)(1.47)}}{2(48.0)} = \frac{19.6 \pm \sqrt{384 - 282}}{96.0}$$

$$= \frac{19.6 \pm \sqrt{102}}{96.0} = \frac{19.6 \pm 10.1}{96.0} = 0.309 \quad \text{or} \quad 0.099$$

Solving a quadratic equation always yields two roots. One root (the answer) has physical meaning. The other root, while mathematically correct, is extraneous; i.e., it has no physical meaning. The value of x is defined as the number of moles of A per liter that react and the number of moles of B per liter that react. No more B can be consumed than was initially present (0.100 M), as would be the case if $x = 0.309$. Thus, $x = 0.099$ is the root in which we are interested, and the extraneous root is 0.309. The equilibrium concentrations are

[A] = $(0.300 - x)$ M = 0.201 M ;   [B] = $(0.100 - x)$ M = 0.001 M ;

[C] = [D] = x M = 0.099 M

EOC 37, 38

The following table summarizes Examples 17-5 and 17-6.

	Initial Concentrations (M)				Equilibrium Concentrations (M)			
	[A]	[B]	[C]	[D]	[A]	[B]	[C]	[D]
Example 17-5	0.200	0.200	0	0	0.025	0.025	0.175	0.175
Example 17-6	0.300	0.100	0	0	0.201	0.001	0.099	0.099

The data from the table can be substituted into the reaction quotient expression, Q, as a check. Even though the reaction is initiated by different relative amounts of reactants in the two cases, the ratios of equilibrium concentrations of products to reactants (each raised to the first power) agree within roundoff error.

Check Example 17-5:

$$Q = \frac{[C][D]}{[A][B]} = \frac{(0.175)(0.175)}{(0.025)(0.025)}$$

$$Q = 49 = K_c$$

Check Example 17-6:

$$Q = \frac{(0.099)(0.099)}{(0.201)(0.001)}$$

$$Q = 49 = K_c$$

17-6 Factors That Affect Equilibria

Once a reacting system has reached equilibrium, it remains at equilibrium until it is disturbed by some change of conditions. Let us consider some different types of changes. The guiding principle is known as **LeChatelier's Principle** (Section 13-6):

> If a change of conditions (stress) is applied to a system at equilibrium, the system responds in the way that best tends to reduce the stress in reaching a new state of equilibrium.

Remember that the *value* of an equilibrium constant changes only with temperature.

The reaction quotient, Q, helps us predict the direction of this response. There are four types of changes to consider:

1. Concentration changes
2. Pressure changes (volume changes for gas-phase reactions)
3. Temperature changes
4. Introduction of catalysts

For reactions involving gases at constant temperature, changes in volume cause changes in pressure, and vice versa.

We shall now look at the effects of these types of stresses from a qualitative, or descriptive, point of view. In Section 17-7 we shall illustrate the validity of our discussion with several quantitative examples.

Changes in Concentration

Consider the following system *starting at equilibrium*:

$$A + B \rightleftharpoons C + D \qquad K_c = \frac{[C][D]}{[A][B]}$$

If more of any reactant or product is *added* to the system, the stress is relieved by shifting the equilibrium in the direction that consumes some of the added substance. Let us compare the mass action expressions for Q and K. If more A or B is added, then $Q < K$, and the forward reaction occurs to a greater extent than the reverse reaction until equilibrium is reestablished. If more C or D is added, $Q > K$, and the reverse reaction occurs to a greater extent until equilibrium is reestablished.

We can understand LeChatelier's Principle in the kinetic terms we used to introduce equilibrium. The rate of the forward reaction is proportional to the reactant concentrations raised to some powers,

The terminology used here is not as precise as we might like, but it is widely used. When we say that the equilibrium is "shifted to the left," we mean that the reaction to the left occurs to a greater extent than the reaction to the right.

$$\text{rate}_f = k_f[\text{A}]^x[\text{B}]^y$$

When we add more A to an equilibrium mixture, this rate increases so that it no longer matches the rate of the reverse reaction. As the reaction proceeds to the right, consuming some A and B and forming more C and D, the forward rate diminishes and the reverse rate increases until they are again equal. At that point, a new equilibrium condition has been reached, with more C and D than were present at the original equilibrium.

If a reactant or product is *removed* from a system at equilibrium, the reaction that produces *that* substance occurs to a greater extent than its reverse. For example, if some C is removed, then $Q < K$, and the forward reaction is favored until equilibrium is reestablished. If A is removed, the reverse reaction is favored.

Stress	Q	Direction of Shift of $A + B \rightleftharpoons C + D$
Increase concentration of A or B	$Q < K$	$\rightarrow$ right
Increase concentration of C or D	$Q > K$	left $\leftarrow$
Decrease concentration of A or B	$Q > K$	left $\leftarrow$
Decrease concentration of C or D	$Q < K$	$\rightarrow$ right

> When a "new equilibrium" is established, (1) the rates of the forward and reverse reactions are equal again, and (2) K_c is again satisfied by the concentrations of reactants and products.

Practical applications of changes of this type are of great economic importance. Removing a product of a reversible reaction forces the reaction to produce more product than could be obtained if the reaction were simply allowed to reach equilibrium.

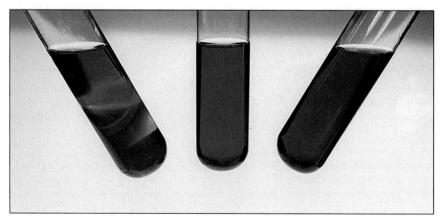

Effects of changes in concentration on the equilibrium

$$[\text{Co(OH}_2)_6]^{2+} + 4\text{Cl}^- \rightleftharpoons [\text{CoCl}_4]^{2-} + 6\text{H}_2\text{O}$$

A solution of $CoCl_2 \cdot 6H_2O$ in isopropyl alcohol is blue due to the $[CoCl_4]^{2-}$ ion. Addition of H_2O favors the reaction to the left to form a mixture of the pink and blue complexes, which is purple (middle test tube). When we add concentrated HCl, the excess Cl^- shifts the reaction to the right (blue, right). Adding $AgNO_3(aq)$ removes some Cl^- by precipitation of $AgCl(s)$ and favors the reaction to the left. The resulting solution is pink (left).

Changes in Volume and Pressure

Changes in pressure have little effect on the concentrations of solids or liquids because they are only slightly compressible. However, changes in pressure do cause significant changes in concentrations of gases. Therefore, such changes affect the value of Q for reactions in which the number of moles of gaseous reactants differs from the number of moles of gaseous products. For an ideal gas,

$$PV = nRT \qquad \text{or} \qquad P = (n/V)(RT)$$

The term (n/V) represents concentration. At constant temperature n, R, and T are constants. Thus, if the volume occupied by a gas decreases, its partial pressure increases and its concentration (n/V) increases. If the volume of a gas increases, both its partial pressure and its concentration decrease.

Consider the following gaseous system at equilibrium:

$$A(g) \rightleftharpoons 2D(g) \qquad K = \frac{[D]^2}{[A]}$$

At constant temperature, a decrease in volume (increase in pressure) increases the concentrations of both A and D. In the mass action expression, the concentration of D is squared and the concentration of A is raised to the first power. As a result, the numerator of Q increases more than the denominator as pressure increases. Thus, $Q > K$, and this equilibrium shifts to the left. Conversely, an increase in volume (decrease in pressure) shifts this reaction to the right until equilibrium is reestablished, because $Q < K$. We can summarize the effect of pressure (volume) changes on *this* gas-phase system at equilibrium:

Stress	Q^*	Direction of Shift of $A(g) \rightleftharpoons 2D(g)$
Volume decrease, pressure increase	$Q > K$	Toward smaller number of moles of gas (left for *this* reaction)
Volume increase, pressure decrease	$Q < K$	Toward larger number of moles of gas (right for *this* reaction)

* Refers to Q for *this* reaction, in which there are more moles of gaseous product than gaseous reactant.

A decrease in volume (increase in pressure) shifts the reaction in the direction that produces the smaller number of moles of gas.

An increase in volume (decrease in pressure) shifts the reaction in the direction that produces the larger number of moles of gas.

If there is no change in the number of moles of gases in a reaction, a volume (pressure) change does not affect the position of equilibrium.

One practical application of these ideas is illustrated later in this section by the Haber process.

The foregoing argument applies only when pressure changes are due to volume changes. It *does not apply* if the total pressure of a gaseous system is raised by merely pumping in a gas that does not react. In such a situation, the *partial* pressure of each reacting gas remains constant, so the system remains at equilibrium.

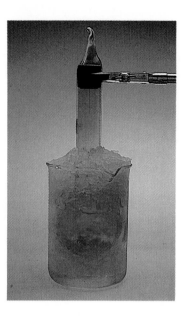

The gas-phase equilibrium for the *exothermic* reaction

$$2NO_2(g) \rightleftharpoons N_2O_4(g)$$

The two flasks contain the same *total* amounts of gas. NO_2 is brown, whereas N_2O_4 is colorless. The flask on the left, at the temperature of ice water, contains very little brown NO_2 gas. The higher temperature (50°C) of the flask on the right favors the reverse reaction; this mixture is more highly colored because it contains more NO_2.

Changes in Temperature

Consider the following system at equilibrium:

$$A + B \rightleftharpoons C + D + heat \qquad (\Delta H \text{ is negative})$$

Heat is produced by the forward (exothermic) reaction. Suppose we increase the temperature at constant pressure by adding heat to the system. This favors the reaction to the left, removing some of the extra heat. Lowering the temperature favors the reaction to the right as the system replaces some of the heat that was removed.

By contrast, for the endothermic reaction at equilibrium,

$$W + X + heat \rightleftharpoons Y + Z \qquad (\Delta H \text{ is positive})$$

an increase in temperature at constant pressure favors the reaction to the right. A decrease in temperature favors the reaction to the left.

> An increase in temperature favors endothermic reactions.
>
> A decrease in temperature favors exothermic reactions.

In fact, the *values* of equilibrium constants change as temperature changes.

The K's of exothermic reactions decrease with increasing T, and the K's of endothermic reactions increase with increasing T, as we shall see in Section 17-12. No other stresses affect the value of K.

Introduction of a Catalyst

Can you use the Arrhenius equation (Section 16-10) to show that lowering the activation energy barrier increases forward and reverse rates by the same factor?

Adding a catalyst to a system changes the rate of the reaction (Section 16-11), but this *cannot* shift the equilibrium in favor of either products or reactants. Because a catalyst affects the activation energy of *both* forward and reverse reactions, it changes both rates equally. Equilibrium is established more quickly in the presence of a catalyst.

Effect of temperature changes on the equilibrium

$$[Co(OH_2)_6]^{2+} + 4Cl^- + heat \rightleftharpoons [CoCl_4]^{2-} + 6H_2O$$

We begin with a purple equilibrium mixture of the pink and blue complexes (middle test tube). In hot water the forward reaction is favored, so the solution is blue (right). At 0°C the reverse reaction is favored, so the solution is pink (left).

Not all reactions attain equilibrium; they may occur too slowly, or else products or reactants may be continually added or removed. Such is the case with most reactions in biological systems. On the other hand, some reactions, such as typical acid–base neutralizations, achieve equilibrium very rapidly.

Example 17-7

Given the following reaction at equilibrium in a closed container at 500°C, how would the equilibrium be influenced by the following: (a) increasing the temperature, (b) lowering the temperature, (c) increasing the pressure by decreasing the volume, (d) introducing some platinum catalyst, (e) forcing more H_2 into the system, and (f) removing some NH_3 from the system?

$$N_2(g) + 3H_2(g) \rightleftharpoons 2NH_3(g) \qquad \Delta H^0 = -92 \text{ kJ/mol rxn}$$

Plan

We apply LeChatelier's Principle to each part of the question individually.

Solution

(a) The negative value for ΔH tells us that the forward reaction is exothermic.

Increasing the temperature favors the endothermic reaction (reverse in this case).

(b) Lowering the temperature favors the exothermic reaction (forward in this case).

(c)   Increasing the pressure favors the reaction that produces the smaller number of moles of gas (forward in this case).

(d)   A catalyst does not favor either reaction.

(e)   Adding a substance favors the reaction that uses up that substance (forward in this case).

(f)   Removing a substance favors the reaction that produces that substance (forward in this case).

EOC 45, 46, 51

Now we shall illustrate the commercial importance of these changes.

The Haber Process

Nitrogen, N_2, is very unreactive. The Haber process is the economically important industrial process by which atmospheric N_2 is converted to ammonia, NH_3, a soluble, reactive compound. Innumerable dyes, plastics, explosives, fertilizers, and synthetic fibers are made from ammonia. The Haber process provides insight into kinetic and thermodynamic factors that influence reaction rates and the positions of equilibria. In this process the exothermic, reversible reaction between N_2 and H_2 to produce NH_3 is never allowed to reach a state of equilibrium, but moves toward it.

$$N_2(g) + 3H_2(g) \rightleftharpoons 2NH_3(g) + 92.22 \text{ kJ}$$

$$K_c = \frac{[NH_3]^2}{[N_2][H_2]^3} = 3.6 \times 10^8 \qquad \text{(at 25°C)}$$

> Approximately 130 pounds of NH_3 are required for each person per year in the United States. Haber developed the process to provide a cheaper and more reliable source of explosives as Germany prepared for World War I. (Britain controlled the seas and access to the natural nitrates in India and Chile.) The current use of the process is more humanitarian: most NH_3 is used to produce fertilizers.

The process is diagrammed in Figure 17-3. The reaction is run at about 450°C under pressures ranging from 200 to 1000 atmospheres. Hydrogen is obtained from coal gas or petroleum refining and nitrogen from liquefied air.

The value of K_c is 3.6×10^8 at 25°C. This very large value of K_c indicates that *at equilibrium* virtually all of the N_2 and H_2 (mixed in a 1:3 mole ratio) would be converted into NH_3. However, at 25°C the reaction occurs so slowly that no measurable amount of NH_3 is produced within a reasonable time. Thus, the large equilibrium constant (a thermodynamic factor) indicates that the reaction proceeds toward the right almost completely. However, it tells us *nothing* about how fast the reaction occurs (a kinetic factor).

There are four moles of gases on the left side of the equation and only two moles of gas on the right, so increasing the pressure favors the production of NH_3. Therefore, the Haber process is carried out at very high pressures, as high as the equipment will safely stand.

> In practice, the mixed reactants are compressed by special pumps and injected into the heated reaction vessel.

The reaction is exothermic ($\Delta H_{rxn}^0 = -92.22$ kJ/mol), so increasing the temperature favors the *decomposition* of NH_3 (the reverse reaction). But, the rates of both forward and reverse reactions increase as temperature increases.

The addition of a catalyst of finely divided iron and small amounts of selected oxides also speeds up both the forward and reverse reactions. This

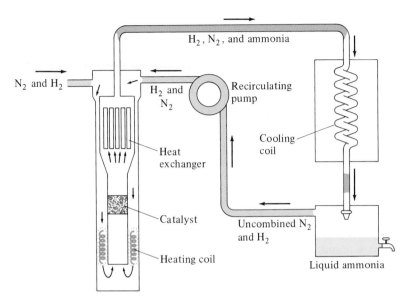

Figure 17-3
A simplified representation of the Haber process for synthesizing ammonia.

means that NH_3 is produced faster and at a lower temperature, which favors yield of NH_3 and extends the life of the equipment.

Table 17-1 shows the effects of increases in temperature and pressure on the equilibrium yield of NH_3, starting with 1:3 mole ratios of $N_2 : H_2$. K_c decreases by more than ten orders of magnitude, from 3.6×10^8 at 25°C to only 1.4×10^{-2} at 758°C. This tells us that the reaction proceeds *very far to the left* at high temperatures. Casual examination of the data might suggest that the reaction should be run at lower temperatures, because a

Ten orders of magnitude is 10^{10}, i.e., 10 billion.

$$1 \times 10^{10} = 10{,}000{,}000{,}000$$

A commercial plant for the production of ammonia.

Table 17-1
Effect of *T* and *P* on Yield of Ammonia

°C	K_c	Mole % NH₃ in Equilibrium Mixture		
		10 atm	100 atm	1000 atm
209	650	51	82	98
467	0.5	4	25	80
758	0.014	0.5	5	13

higher percentage of the N_2 and H_2 is converted into NH_3. However, the reaction occurs so slowly, even in the presence of a catalyst, that it cannot be run economically at temperatures below about 450°C.

The emerging reaction mixture is cooled down, and NH_3 (bp = −33.43°C) is removed as a liquid. This favors the forward reaction. The unreacted N_2 and H_2 are recycled. Excess N_2 is used to favor the reaction to the right.

17-7 Application of Stress to a System at Equilibrium

We can use equilibrium constants to determine new equilibrium concentrations that result from adding one or more species to a system at equilibrium.

Example 17-8
Some hydrogen and iodine are mixed at 229°C in a 1.00-liter container. When equilibrium is established, the following concentrations are present: [HI] = 0.490 *M*, [H_2] = 0.080 *M*, and [I_2] = 0.060 *M*. If an additional 0.300 mol of HI is then added, what concentrations will be present when the new equilibrium is established?

$$H_2(g) + I_2(g) \rightleftharpoons 2HI(g)$$

Plan
We use the initial equilibrium concentrations to calculate the value of K_c. Then we determine the new concentrations after some HI has been added and calculate Q. The value of Q tells us which reaction is favored. Now we can represent the new equilibrium concentrations. We substitute them into the K_c expression to find the new equilibrium concentrations.

Solution
We first calculate the value of K_c from the first set of equilibrium concentrations:

$$K_c = \frac{[HI]^2}{[H_2][I_2]} = \frac{(0.490)^2}{(0.080)(0.060)} = 50$$

When we add 0.300 mol of HI to the 1.00-liter container, the [HI] instantaneously increases by 0.300 *M*.

	$H_2(g)$	+	$I_2(g)$	$\rightleftharpoons$	$2HI(g)$
equilibrium	0.080 *M*		0.060 *M*		0.490 *M*
mol/L added	0 *M*		0 *M*		+0.300 *M*
new initial conc'n	0.080 *M*		0.060 *M*		0.790 *M*

Substitution of these concentrations into the reaction quotient gives

$$Q = \frac{[HI]^2}{[H_2][I_2]} = \frac{(0.790)^2}{(0.080)(0.060)} = 130$$

Because $Q > K$, the reaction proceeds to the left to establish a new equilibrium. The new equilibrium concentrations can be determined as follows. Let $x = $ mol/L of H_2 formed; so $x = $ mol/L of I_2 formed, and $2x = $ mol/L of HI consumed.

	$H_2(g)$	$+$	$I_2(g)$	$\rightleftharpoons$	$2HI(g)$
new initial	0.080 M		0.060M		0.790 M
change due to rxn	$+x\ M$		$+x\ M$		$-2x\ M$
new equilibrium	(0.080 + x) M		(0.060 + x) M		(0.790 − 2x) M

Here it is obvious that adding some HI favors the reaction to the left. In many cases we cannot tell which reaction will be favored. Calculating Q always lets us make the decision.

Equal concentrations of H_2 and I_2 must be formed by the change.

Substitution of these values into K_c allows us to evaluate x:

$$K_c = 50 = \frac{(0.790 - 2x)^2}{(0.080 + x)(0.060 + x)} = \frac{0.624 - 3.16x + 4x^2}{0.0048 + 0.14x + x^2}$$

$$0.24 + 7.0x + 50x^2 = 0.624 - 3.16x + 4x^2$$

$$46x^2 + 10.2x - 0.38 = 0$$

Solution by the quadratic formula gives $x = 0.032$ or -0.25.

Clearly, $x = -0.25$ is the extraneous root, because x cannot be less than zero. This reaction does not consume a negative quantity of HI, because the reaction is proceeding toward the left. Thus, $x = 0.032$ is the root with physical meaning, so the new equilibrium concentrations are

To "consume a negative quantity of HI" would be to form HI. The value $x = -0.25$ would lead to $[H_2] = (0.080 + x)\ M = (0.080 - 0.25)\ M = -0.17\ M$. A negative concentration is impossible, so $x = -0.25$ is the extraneous root.

$$[H_2] = (0.080 + x)\ M = (0.080 + 0.032)\ M = \boxed{0.112\ M}$$

$$[I_2] = (0.060 + x)\ M = (0.060 + 0.032)\ M = \boxed{0.092\ M}$$

$$[HI] = (0.790 - 2x)\ M = (0.790 - 0.064)\ M = \boxed{0.726\ M}$$

In summary,

Original Equilibrium	Stress Applied	New Equilibrium
$[H_2] = 0.080\ M$		$[H_2] = 0.112\ M$
$[I_2] = 0.060\ M$	Add 0.300 M HI	$[I_2] = 0.092\ M$
$[HI] = 0.490\ M$		$[HI] = 0.726\ M$

We see that some of the additional HI is consumed, but not all of it. More HI remains after the new equilibrium is established than was present before the stress was imposed. However, the new equilibrium $[H_2]$ and $[I_2]$ are substantially greater than the original equilibrium concentrations.

EOC 56

We can also use the equilibrium constant to calculate new equilibrium concentrations that result from decreasing the volume (increasing the pressure) of a gaseous system that was initially at equilibrium.

Example 17-9

At 22°C the equilibrium constant, K_c, for the following reaction is 4.66×10^{-3}. (a) If 0.800 mol of N_2O_4 were injected into a closed 1.00-liter container at 22°C, what would be the equilibrium concentrations of the two gases? (b) What would the concentrations be at equilibrium if the volume were suddenly halved at constant temperature?

$$N_2O_4(g) \rightleftharpoons 2NO_2(g) \qquad K_c = \frac{[NO_2]^2}{[N_2O_4]} = 4.66 \times 10^{-3}$$

Plan

(a) We are given the value for K_c and the initial concentration of N_2O_4. We write the reaction summary, which gives the representation of the equilibrium concentrations. Then we substitute these into the K_c expression and solve.
(b) We repeat the process we used in part (a).

Solution

(a) We let x = mol/L of N_2O_4 consumed and $2x$ = mol/L of NO_2 formed.

	$N_2O_4(g)$	$\rightleftharpoons$	$2NO_2(g)$
initial	0.800 M		0 M
change due to rxn	$-x\ M$		$+2x\ M$
equilibrium	$(0.800 - x)\ M$		$2x\ M$

$$K_c = \frac{[NO_2]^2}{[N_2O_4]} = 4.66 \times 10^{-3} = \frac{(2x)^2}{0.800 - x} = \frac{4x^2}{0.800 - x}$$

$$3.73 \times 10^{-3} - 4.66 \times 10^{-3}\,x = 4x^2$$

$$4x^2 + 4.66 \times 10^{-3}\,x - 3.73 \times 10^{-3} = 0$$

> The value of x is the number of moles per liter of N_2O_4 that react. So $0 < x < 0.800\ M$.

Solving by the quadratic formula gives $x = 3.00 \times 10^{-2}$ and $x = -3.11 \times 10^{-2}$. We use $x = 3.00 \times 10^{-2}$.

The original equilibrium concentrations are

$$[NO_2]\ = 2x\ M = \boxed{6.00 \times 10^{-2}\ M}$$

$$N_2O_4\ = (0.800 - x)\ M = (0.800 - 3.00 \times 10^{-2})\ M = \boxed{0.770\ M}$$

(b) When the volume of the reaction vessel is halved, the concentrations are doubled, so the new *initial* concentrations of N_2O_4 and NO_2 are $2(0.770\ M) = 1.54\ M$ and $2(6.00 \times 10^{-2}\ M) = 0.120\ M$, respectively.

> LeChatelier's Principle tells us that a decrease in volume (increase in pressure) favors the production of N_2O_4.
>
> $$Q = \frac{(0.120)^2}{1.54} = 9.35 \times 10^{-3}$$
>
> $$Q > K$$
>
> $$\therefore \xleftarrow{\text{shift left}}$$

	$N_2O_4(g)$	$\rightleftharpoons$	$2NO_2(g)$
new initial	1.54 M		0.120 M
change due to rxn	$+x\ M$		$-2x\ M$
new equilibrium	$(1.54 + x)\ M$		$(0.120 - 2x)\ M$

$$K_c = \frac{[NO_2]^2}{[N_2O_4]} = 4.66 \times 10^{-3} = \frac{(0.120 - 2x)^2}{1.54 + x}$$

Rearranging into the standard form of a quadratic equation gives

$$x^2 - 0.121x + 1.81 \times 10^{-3} = 0$$

Solving as before gives $x = 0.104$ and $x = 0.017$.

> The root $x = 0.104$ would give a *negative* concentration for NO_2, which is impossible.

The maximum value of x is 0.060 M, because $2x$ may not exceed the concentration of NO_2 that was present after the volume was halved. Thus, $x = 0.017\ M$ is the root with physical significance. The new equilibrium concentrations in the 0.500-liter container are

$$[NO_2]\ = (0.120 - 2x)\ M = (0.120 - 0.034)\ M = \boxed{0.086\ M}$$

$$[N_2O_4]\ = (1.54 + x)\ M\ = (1.54 + 0.017)\ M\ = \boxed{1.56\ M}$$

In summary,

First Equilibrium	Stress	New Equilibrium
$[N_2O_4] = 0.770\ M$	Decrease volume from	$[N_2O_4] = 1.56\ M$
$[NO_2] = 0.0600\ M$	1.00 L to 0.500 L	$[NO_2] = 0.086\ M$

Both $[N_2O_4]$ and $[NO_2]$ *increase* because of the large decrease in volume. You may wish to verify that the *number of moles* of N_2O_4 increases, while the *number of moles* of NO_2 decreases. We predict this from LeChatelier's Principle.

EOC 58

17-8 Partial Pressures and the Equilibrium Constant

It is often more convenient to measure pressures rather than concentrations of gases. Solving the ideal gas equation, $PV = nRT$, for pressure gives

$$P = \frac{n}{V}(RT)$$

The pressure of a gas is directly proportional to its concentration (n/V). For reactions in which all substances that appear in the equilibrium constant expression are gases, we sometimes prefer to express the equilibrium constant in terms of partial pressures *in atmospheres* (K_P) rather than in terms of concentrations (K_c).

In general, for a reaction involving gases,

$$a\text{A(g)} + b\text{B(g)} \rightleftharpoons c\text{C(g)} + d\text{D(g)} \qquad K_P = \frac{(P_C)^c(P_D)^d}{(P_A)^a(P_B)^b}$$

K_P has no units for the same reasons that were explained for K_c in Section 17-2.

For instance, for the following reversible reaction,

$$N_2\text{(g)} + 3H_2\text{(g)} \rightleftharpoons 2NH_3\text{(g)} \qquad K_P = \frac{(P_{NH_3})^2}{(P_{N_2})(P_{H_2})^3}$$

Example 17-10

In an equilibrium mixture at 500°C, we find $P_{NH_3} = 0.147$ atm, $P_{N_2} = 1.41$ atm, and $P_{H_2} = 6.00$ atm. Evaluate K_P at 500°C for the following reaction:

$$N_2\text{(g)} + 3H_2\text{(g)} \rightleftharpoons 2NH_3\text{(g)}$$

Plan

We are given equilibrium partial pressures of all reactants and products. So we write the expression for K_P and substitute partial pressures in atmospheres into it.

Solution

$$K_P = \frac{(P_{NH_3})^2}{(P_{N_2})(P_{H_2})^3} = \frac{(0.147)^2}{(1.41)(6.00)^3} = 7.10 \times 10^{-5}$$

EOC 67, 68

17-9 Relationship between K_P and K_c

If the ideal gas equation is rearranged, the molar concentration of a gas is

$\left(\dfrac{n}{V}\right)$ is just $\left(\dfrac{\text{no. mol}}{L}\right)$.

$$\left(\frac{n}{V}\right) = \frac{P}{RT} \quad \text{or} \quad [\text{conc'n}] = \frac{P}{RT}$$

Substituting P/RT for n/V in the K_c expression for the N_2–H_2–NH_3 equilibrium gives the relationship between K_c and K_P for *this* reaction:

$$K_c = \frac{[NH_3]^2}{[N_2][H_2]^3} = \frac{\left(\dfrac{P_{NH_3}}{RT}\right)^2}{\left(\dfrac{P_{N_2}}{RT}\right)\left(\dfrac{P_{H_2}}{RT}\right)^3} = \frac{(P_{NH_3})^2}{(P_{N_2})(P_{H_2})^3} \times \frac{\left(\dfrac{1}{RT}\right)^2}{\left(\dfrac{1}{RT}\right)^4}$$

$$K_c = K_P\,(RT)^2 \quad \text{and} \quad K_P = K_c\,(RT)^{-2}$$

In general, the relationship between K_c and K_P is

Δn refers to the numbers of moles of gaseous substances in the balanced equation, *not* in the reaction vessel.

$$K_P = K_c(RT)^{\Delta n} \quad \text{or} \quad K_c = K_P(RT)^{-\Delta n} \qquad \Delta n = (n_{\text{gas prod}}) - (n_{\text{gas react}})$$

Because K_c refers to mol/L and K_P refers to atmospheres, R must be expressed in L · atm/mol · K. For reactions in which there are equal numbers of moles of gases on both sides of the equation, $\Delta n = 0$ and $K_P = K_c$.

For the ammonia equilibrium,

$$N_2(g) + 3H_2(g) \rightleftharpoons 2NH_3(g) \qquad \Delta n = 2 - 4 = -2$$

$K_P = 7.10 \times 10^{-5}$ at 500°C (773 K). So we have

$$K_c = K_P(RT)^{-\Delta n} = (7.10 \times 10^{-5})[(0.0821)(773)]^{-(-2)} = 0.286$$

This agrees with the value from Example 17-1.

For *gas-phase reactions*, we can calculate the amounts of substances present at equilibrium using either K_P or K_c. The results are the same by either method (when they are expressed in the same terms). To illustrate, let us solve the following problem by both methods.

Example 17-11

We place 10.0 grams of $SbCl_5$ in a 5.00-liter container at 448°C, and allow the reaction to attain equilibrium. How many grams of $SbCl_5$ are present at equilibrium? Solve this problem

a. Using K_c and molar concentrations.
b. Using K_P and partial pressures.

$$SbCl_5(g) \rightleftharpoons SbCl_3(g) + Cl_2(g) \qquad K_c = 2.51 \times 10^{-2} \text{ at } 448°C$$

$$K_P = 1.48 \text{ at } 448°C$$

a. Plan (using K_c)

We calculate the initial concentration of $SbCl_5$, write the reaction summary and represent the equilibrium concentrations, and then substitute into the K_c expression to obtain the equilibrium concentrations.

a. Solution (using K_c)

Because we are given K_c, we use concentrations. The initial concentration of $SbCl_5$ is

$$[SbCl_5] = \frac{10.0 \text{ g } SbCl_5}{5.00 \text{ L}} \times \frac{1 \text{ mol}}{299 \text{ g}} = 0.00669 \; M \; SbCl_5$$

Let x = mol/L of $SbCl_5$ that react. In terms of molar concentrations, the reaction summary is

	$SbCl_5$	$\rightleftharpoons$	$SbCl_3$	$+$	Cl_2
initial	$0.00669 \; M$		0		0
change due to rxn	$-x \; M$		$+x \; M$		$+x \; M$
equilibrium	$(0.00669 - x) \; M$		$x \; M$		$x \; M$

$$K_c = \frac{[SbCl_3][Cl_2]}{[SbCl_5]}$$

$$2.51 \times 10^{-2} = \frac{(x)(x)}{0.00669 - x}$$

$$x^2 = 1.68 \times 10^{-4} - 2.51 \times 10^{-2} x$$

$$x^2 + 2.51 \times 10^{-2} x - 1.68 \times 10^{-4} = 0$$

Solving by the quadratic formula gives

$$x = 5.49 \times 10^{-3} \quad \text{and} \quad -3.06 \times 10^{-2} \text{ (extraneous root)}$$

$$[SbCl_5] = (0.00669 - x) \; M = (0.00669 - 0.00549) \; M = 1.20 \times 10^{-3} \; M$$

$$\underline{?} \text{ g } SbCl_5 = 5.00 \text{ L} \times \frac{1.20 \times 10^{-3} \text{ mol}}{L} \times \frac{299 \text{ g}}{\text{mol}} = \boxed{1.79 \text{ g } SbCl_5}$$

Let us now solve the same problem *using K_P and partial pressures.*

b. Plan (using K_P)

We calculate the initial partial pressure of $SbCl_5$ and write the reaction summary. Substitution of the representation of the equilibrium partial pressures into K_P gives their values.

b. Solution (using K_P)

We calculate the initial *pressure* of $SbCl_5$ in atmospheres, using $PV = nRT$.

$$P_{SbCl_5} = \frac{nRT}{V} = \frac{(10.0 \text{ g})\left(\dfrac{1 \text{ mol}}{299 \text{ g}}\right)\left(0.0821 \dfrac{L \cdot atm}{mol \cdot K}\right)(721 \text{ K})}{5.00 \text{ L}} = 0.396 \text{ atm}$$

Clearly, $P_{SbCl_3} = 0$ and $P_{Cl_2} = 0$ because only PCl_5 is present initially. The *partial pressure* of each substance is proportional to the *number of moles* of that substance. We write the reaction summary in terms of partial pressures in atmospheres, because K_P refers to pressures in atmospheres.

Let y = decrease in pressure (atm) of $SbCl_5$ due to reaction. In terms of partial pressures, the reaction summary is

	$SbCl_5$	$\rightleftharpoons$	$SbCl_3$	$+$	Cl_2
initial	0.396 atm		0		0
change due to rxn	$-y$ atm		$+y$ atm		$+y$ atm
equilibrium	$(0.396 - y)$ atm		y atm		y atm

$$K_P = \frac{(P_{SbCl_3})(P_{Cl_2})}{P_{SbCl_5}}$$

$$1.48 = \frac{(y)(y)}{0.396 - y}$$

$$0.586 - 1.48y = y^2$$

$$y^2 + 1.48y - 0.586 = 0$$

If we did not know the value of K_P, we could calculate it from the known value of K_c. $K_P = K_c(RT)^{\Delta n}$.

Solving by the quadratic formula gives

$$y = 0.325 \text{ or } -1.80 \text{ (extraneous root)}$$

$$P_{SbCl_5} = (0.396 - y) = (0.396 - 0.325) = 0.071 \text{ atm}$$

We use the ideal gas law, $PV = nRT$, to calculate the number of moles of $SbCl_5$.

$$n = \frac{PV}{RT} = \frac{(0.071 \text{ atm})(5.00 \text{ L})}{\left(0.0821 \dfrac{L \cdot atm}{mol \cdot K}\right)(721 \text{ K})} = 0.0060 \text{ mol } SbCl_5$$

$$? \text{ g } SbCl_5 = 0.0060 \text{ mol} \times \frac{299 \text{ g}}{mol} = \boxed{1.8 \text{ g } SbCl_5}$$

We see that within roundoff range the same result is obtained by both methods.

EOC 65, 70

17-10 Heterogeneous Equilibria

Thus far, we have considered only equilibria involving species in a single phase, i.e., homogeneous equilibria. **Heterogeneous equilibria** involve species in more than one phase. Consider the following reversible reaction at 25°C:

$$2HgO(s) \rightleftharpoons 2Hg(\ell) + O_2(g)$$

When equilibrium is established for this system, a solid, a liquid, and a gas are present. Neither solids nor liquids are significantly affected by changes in pressure. The fundamental definition of the equilibrium constant in thermodynamics is in terms of the activities of the substances involved.

> For pure solids or liquids, the activity is taken as 1 (Section 17-2), so terms for pure liquids and pure solids *do not* appear in the K expressions for heterogeneous equilibria.

Thus, for the reaction

$$2HgO(s) \rightleftharpoons 2Hg(\ell) + O_2(g) \qquad K_c = [O_2] \quad \text{and} \quad K_P = P_{O_2}$$

These equilibrium constant expressions indicate that equilibrium exists at a given temperature for *one and only one* concentration and one partial pressure of oxygen in contact with liquid mercury and solid mercury(II) oxide.

The reaction

$$2HgO(s) \rightleftharpoons 2Hg(\ell) + O_2(g)$$

The reaction is not at equilibrium here, because O_2 gas has been allowed to escape.

Example 17-12

Write both K_c and K_P expressions for the following reversible reactions:
(a) $2SO_2(g) + O_2(g) \rightleftharpoons 2SO_3(g)$
(b) $2NH_3(g) + H_2SO_4(\ell) \rightleftharpoons (NH_4)_2SO_4(s)$
(c) $S(s) + H_2SO_3(aq) \rightleftharpoons H_2S_2O_3(aq)$

Plan

We apply the definitions of K_c and K_P to each reaction.

Solution

(a) $\qquad K_c = \dfrac{[SO_3]^2}{[SO_2]^2[O_2]} \qquad\qquad\qquad K_P = \dfrac{(P_{SO_3})^2}{(P_{SO_2})^2(P_{O_2})}$

(b) $\quad K_c = \dfrac{1}{[NH_3]^2} = [NH_3]^{-2} \qquad K_P = \dfrac{1}{(P_{NH_3})^2} = (P_{NH_3})^{-2}$

(c) $\quad K_c = \dfrac{[H_2S_2O_3]}{[H_2SO_3]} \qquad\qquad K_P$ undefined; no gases involved

EOC 75

Example 17-13

The value of K_P is 27 for the thermal decomposition of potassium chlorate at a given high temperature. What is the partial pressure of oxygen in a closed container in which the following system is at equilibrium at the given temperature? (This can be a dangerous reaction.)

$$2KClO_3(s) \xrightarrow{\Delta} 2KCl(s) + 3O_2(g)$$

Plan

Because one gas, O_2, and two solids, $KClO_3$ and KCl, are involved, we see that K_P involves only partial pressures.

Solution

We are given

$$K_P = (P_{O_2})^3 = 27$$

We let x atm $= P_{O_2}$ at equilibrium. Then we have

$$(P_{O_2})^3 = 27 = x^3 \qquad \boxed{x = 3.0 \text{ atm}}$$

This tells us that the partial pressure of oxygen is 3.0 atm at equilibrium.

17-11 Relationship between ΔG^0 and the Equilibrium Constant

In thermodynamic terms, let us consider what may happen when two substances are mixed together at constant temperature and pressure. First, as a result of mixing, there is usually an increase in entropy due to the increase in disorder. If the two substances can react with each other, the chemical reaction begins, heat is released or absorbed, and the concentrations of the substances in the mixture change. An additional change in entropy, which depends upon changes in the nature of the reactants and products, also begins to occur. The evolution or absorption of heat energy, the changes in entropy, and the changes in concentrations all continue until equilibrium is established. Equilibrium may be reached with large amounts of products formed, with virtually all of reactants remaining, or at *any* intermediate combination of concentrations.

The standard free energy change for a reaction is ΔG^0. This is the free energy change that would accompany *complete* conversion of *all* reactants initially present in their standard states to *all* products in their standard states—the standard reaction (Section 15-14). The free energy change for any other concentrations or pressures is ΔG (no superscript zero). The two quantities are related by the equation

$$\Delta G = \Delta G^0 + RT \ln Q \qquad \text{or} \qquad \Delta G = \Delta G^0 + 2.303RT \log Q$$

Thermodynamic standard states are (1) pure solids or pure liquids at 1 atm, (2) solutions of one-molar concentrations, and (3) gases at partial pressures of 1 atm.

R is the universal gas constant, T is the absolute temperature, and Q is the reaction quotient (Section 17-4). When a system is *at equilibrium*, $\Delta G = 0$ (Section 15-14) and $Q = K$ (Section 17-4). Recall that the reaction quotient may represent nonequilibrium concentrations (or partial pressures) of products and reactants. As reaction occurs, the free energy of the mixture and the concentrations change until at equilibrium $\Delta G = 0$, and the concentrations of reactants and products satisfy the equilibrium constant. At that point, Q becomes equal to K (Section 17-4). Then

$$0 = \Delta G^0 + RT \ln K \qquad \text{or} \qquad 0 = \Delta G^0 + 2.303RT \log K \qquad \text{(at equilibrium)}$$

Rearranging gives

The energy units of *R must* match those of ΔG^0.

$$\Delta G^0 = -RT \ln K \qquad \text{or} \qquad \Delta G^0 = -2.303RT \log K$$

This equation shows the relationship between the standard free energy change and the **thermodynamic equilibrium constant**.

For the following generalized reaction, the thermodynamic equilibrium constant is defined in terms of the activities of the species involved.

$$a\text{A} + b\text{B} \rightleftharpoons c\text{C} + d\text{D} \qquad K = \frac{(a_\text{C})^c (a_\text{D})^d}{(a_\text{A})^a (a_\text{B})^b}$$

where a_A is the activity of substance A, and so on. The mass action expression to which it is related involves concentration terms for species in solution and partial pressures for gases.

> For equilibria that involve only gases, the thermodynamic equilibrium constant (related to ΔG^0) is K_P. For those that involve species in solution, it is equal to K_c.

Figure 17-4 displays the relationships between free energy and equilibrium. The *left* end of each curve represents the total free energy of the reactants and the *right* end of each curve represents the total free energy of the products at standard state conditions. The difference between them is ΔG^0; like K, ΔG^0 depends only on temperature and is a constant for any given reaction.

From the preceding equation that relates ΔG^0 and K, we see that when ΔG^0 is negative, K is greater than 1 (log K *must* be positive). This tells us that products are favored over reactants at equilibrium. This case is illustrated in Figure 17-4a. When ΔG^0 is positive, K is less than 1 (log K *must* be negative). This tells us that reactants are favored over products at equilibrium (Figure 17-4b). In the rare case of a chemical reaction for which $\Delta G^0 = 0$, then $K = 1$ and the numerator and the denominator must be equal in the equilibrium constant expression, (i.e., $[\text{C}]^c[\text{D}]^d \ldots = [\text{A}]^a[\text{B}]^b \ldots$). These relationships are summarized as follows.

ΔG^0	K	Product Formation
$\Delta G^0 < 0$	$K > 1$	Products favored over reactants at equilibrium
$\Delta G^0 = 0$	$K = 1$	At equilibrium when $[\text{C}]^c[\text{D}]^d \ldots = [\text{A}]^a[\text{B}]^b \ldots$ (very rare)
$\Delta G^0 > 0$	$K < 1$	Reactants favored over products at equilibrium

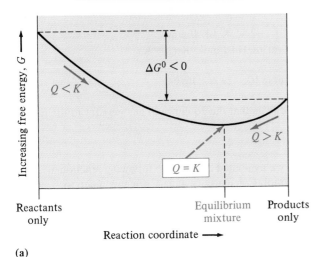

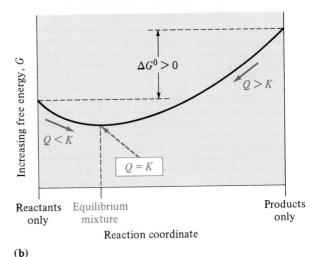

(a)

(b)

The direction of approach to equilibrium and the actual free energy change (ΔG) are *not* constants. They vary with the conditions and the initial concentrations. If the initial concentrations correspond to $Q < K$, equilibrium is approached from left to right on the curves in Figure 17-4, and the forward reaction is spontaneous. If $Q > K$, equilibrium is approached from right to left, and the reaction is spontaneous in the reverse direction.

The magnitude of ΔG^0 indicates the *extent* to which a chemical reaction occurs under standard state conditions, i.e., how far the reaction goes toward the formation of products before equilibrium is reached. The greater the magnitude of a negative ΔG^0 value, the larger the value of K and the more favorable the formation of products. We think of some reactions as going to "completion." These generally have large negative ΔG values. The greater the magnitude of a positive ΔG^0, the smaller the value of K and the smaller the amounts of product at equilibrium.

Figure 17-4
Variation in total free energy for a reversible reaction carried out at constant T. The *standard* free energy change, ΔG^0, represents the free energy change for the *standard reaction*—the *complete* conversion of reactants into products. In (a) this change is negative, indicating that the standard reaction is spontaneous; the collection of products would be more stable than the collection of reactants. However, the mixture of reactants and products corresponding to the minimum of the curve is even more stable, and represents the equilibrium mixture. Because ΔG^0 is negative, $K > 1$, and the equilibrium mixture contains more products than reactants. At any point on the curve, comparing Q and K indicates the direction in which the reaction must proceed to approach equilibrium, that is, which way is "downhill" in free energy. The plot in (b) is for positive ΔG^0 (the reverse standard reaction is thermodynamically spontaneous). In this case, $K < 1$, and the equilibrium mixture contains more reactants than products.

Example 17-14
Use the data in Appendix K to evaluate K_P for the following reaction at 25°C:

$$2C_2H_2(g) + 5O_2(g) \rightleftharpoons 4CO_2(g) + 2H_2O(g) \qquad (C_2H_2 \text{ is acetylene})$$

Plan
The temperature is 25°C, so we evaluate ΔG^0 for the reaction from ΔG_f^0 values in Appendix K. The reaction involves only gases so K is K_P. This means that $\Delta G^0 = -RT \ln K_P$ or $\Delta G^0 = -2.303RT \log K_P$. We solve for K_P.

Solution

$$\Delta G_{rxn}^0 = 4\Delta G_{f\ CO_2(g)}^0 + 2\Delta G_{f\ H_2O(g)}^0 - [2\Delta G_{f\ C_2H_2(g)}^0 + 5\Delta G_{f\ O_2(g)}^0]$$

$$= [4(-394.4) + 2(-228.6)] - [2(209.2) + 5(0)]$$

$$= -2.45 \times 10^3 \text{ kJ/mol} \qquad \text{or} \qquad -2.45 \times 10^6 \text{ J/mol}$$

This is a gas-phase reaction, so ΔG^0_{rxn} is related to K_P by

$$\Delta G^0_{rxn} = -2.303RT \log K_P$$

$$\log K_P = \frac{\Delta G^0_{rxn}}{-2.303RT}$$

Units cancel when we express ΔG^0 in joules per mole. We interpret this as meaning "per mole of reaction"— that is, for the number of moles of each substance shown in the balanced equation.

$$\log K_P = \frac{-2.45 \times 10^6 \text{ J/mol}}{-2.303(8.314 \text{ J/mol} \cdot \text{K})(298 \text{ K})} = 429$$

$$K_P = \text{inverse log } 429 = \boxed{10^{429}}$$

The very large value for K_P tells us that the equilibrium lies *very* far to the right.

Example 17-15

In Examples 15-16 and 15-17 we evaluated ΔG^0 for the following reaction at 25°C and found it to be +173.1 kJ/mol. Calculate K_P at 25°C for this reaction.

$$N_2(g) + O_2(g) \rightleftharpoons 2NO(g)$$

Plan

As in Example 17-14, $\Delta G^0 = -RT \ln K_P$.

Solution

Let us solve this example using natural logarithms, ln.

$$\Delta G^0 = -RT \ln K_P$$

$$\ln K_P = \frac{\Delta G^0}{-RT} = \frac{1.731 \times 10^5 \text{ J/mol}}{-(8.314 \text{ J/mol} \cdot \text{K})(298 \text{ K})}$$

$$\ln K_P = -69.87$$

On many calculators, we would evaluate e^x as follows: Enter the value of x, then press [INV] followed by [ln x].

$$K_P = e^{-69.87} = \boxed{4.5 \times 10^{-31}}$$

Clearly, this very small number indicates that at equilibrium almost no N_2 and O_2 are converted to NO at 25°C. For all practical purposes, the reaction does not occur at 25°C.

A very important application of the relationships in this section is the use of measured K values to calculate ΔG^0.

Example 17-16

The equilibrium constant, K_P, for the following reaction is 5.04×10^{17} at 25°C. Calculate ΔG^0_{298} for the hydrogenation of ethylene to form ethane.

$$C_2H_4(g) + H_2(g) \rightleftharpoons C_2H_6(g)$$

Plan

We use the relationship between ΔG^0 and K_P that we used in Example 17-14.

Solution

This time we use base-10 logarithms.

$$\Delta G^0_{298} = -2.303RT \log K_P$$
$$= -2.303(8.314)(298) \log (5.04 \times 10^{17})$$
$$= -5706(17.702)$$
$$= -1.010 \times 10^5 \text{ J/mol}$$

$$\Delta G^0_{298} = \boxed{-101 \text{ kJ/mol}}$$

EOC 85, 86

17-12 Evaluation of Equilibrium Constants at Different Temperatures

Chemists have determined equilibrium constants for thousands of reactions. It would be an impossibly huge task to catalog such constants at every temperature of interest for each reaction. Fortunately, there is no need to do this. If we determine the equilibrium constant, K_{T_1}, for a reaction at one temperature, T_1, and also its ΔH^0, we can then *estimate* the equilibrium constant at a second temperature, T_2, using the **van't Hoff equation**.

$$\ln \left(\frac{K_{T_2}}{K_{T_1}} \right) = \frac{\Delta H^0}{R} \left(\frac{1}{T_1} - \frac{1}{T_2} \right) \quad or \quad \ln \left(\frac{K_{T_2}}{K_{T_1}} \right) = \frac{\Delta H^0 (T_2 - T_1)}{RT_2T_1}$$

As a memory aid, compare the forms of these equations to the Arrhenius equation (Section 16-10) and to the Clausius–Clapeyron equation (Section 13-9).

Alternatively, we may write

$$\log \left(\frac{K_{T_2}}{K_{T_1}} \right) = \frac{\Delta H^0}{2.303R} \left(\frac{1}{T_1} - \frac{1}{T_2} \right) \quad or \quad \log \left(\frac{K_{T_2}}{K_{T_1}} \right) = \frac{\Delta H^0 (T_2 - T_1)}{2.303RT_2T_1}$$

Thus, if we know ΔH^0 for a reaction and K at a given temperature (say 298 K), we can use the van't Hoff equation to calculate the value of K at any other temperature.

Example 17-17

We found in Example 17-15 that $K_P = 4.6 \times 10^{-31}$ at 25°C (298 K) for the following reaction. $\Delta H^0 = 180.5$ kJ/mol for this reaction. Evaluate K_P at 2400 K, and then compare K_{2400} with K_{298}.

$$N_2(g) + O_2(g) \rightleftharpoons 2NO(g)$$

2400 K is a typical temperature inside the combustion chambers of automobile engines. Large quantities of N_2 and O_2 are present during gasoline combustion, because the gasoline is mixed with air.

Plan

We are given K_P at one temperature, 25°C, and the value for ΔH^0. We are given the second temperature, 2400 K. These data allow us to evaluate the right side of the van't Hoff equation, which gives us $\ln (K_{T_2}/K_{T_1})$. Because we know K_{T_1}, we can find the value for K_{T_2}.

Solution

Let $T_1 = 298$ K and $T_2 = 2400$ K. Then

$$\ln \left(\frac{K_{T_2}}{K_{T_1}} \right) = \frac{\Delta H^0 (T_2 - T_1)}{RT_2T_1}$$

Let us first evaluate the right side of the equation:

$$\ln \left(\frac{K_{T_2}}{K_{T_1}}\right) = \frac{(1.805 \times 10^5 \text{ J/mol})(2400 \text{ K} - 298 \text{ K})}{(8.314 \text{ J/mol} \cdot \text{K})(2400 \text{ K})(298 \text{ K})} = 63.81$$

Now we take the inverse logarithm of both sides:

$$\frac{K_{T_2}}{K_{T_1}} = e^{63.81} = 5.1 \times 10^{27}$$

Solving for K_{T_2} and substituting the known value of K_{T_1}, we obtain

$$K_{T_2} = (5.1 \times 10^{27})(K_{T_1}) = (5.1 \times 10^{27})(4.6 \times 10^{-31}) = \boxed{2.3 \times 10^{-3}} \text{ at 2400 K}$$

EOC 87, 88

The K_p value could be converted to K_c using the relationship $K_c = K_p (RT)^{-\Delta n}$ (Section 17-9).

In Example 17-17, we see that K_{T_2} (K_P at 2400 K) is quite small, which tells us that the equilibrium favors N_2 and O_2 rather than NO. However, K_{T_2} is very much larger than K_{T_1}, which is 4.6×10^{-31}. At 2400 K, significantly more NO is present at equilibrium, relative to N_2 and O_2, than at 298 K. So automobiles emit small amounts of NO into the atmosphere, which are sufficient to cause severe air pollution problems. Catalytic converters (Section 16-11) are designed to catalyze the breakdown of NO into N_2 and O_2.

$$2NO(g) \rightleftharpoons N_2(g) + O_2(g)$$

This reaction is spontaneous. Catalysts do not shift the position of equilibrium. They favor neither consumption nor production of NO. They merely allow the system to reach equilibrium more rapidly. The time factor is particularly important because the NO stays in the exhaust system for only a very short time.

Key Terms

Activity (of a component of an ideal mixture) A dimensionless quantity whose magnitude is equal to molar concentration in an ideal solution, equal to partial pressure (in atmospheres) in an ideal gas mixture, and defined as 1 for pure solids or liquids.

Chemical equilibrium A state of dynamic balance in which the rates of forward and reverse reactions are equal; there is no net change in concentrations of reactants or products while a system is at equilibrium.

Dynamic equilibrium An equilibrium in which processes occur continuously, with no *net* change.

Equilibrium constant, K A quantity that indicates the extent to which a reversible reaction occurs; its magnitude is equal to the mass action expression at equilibrium. K varies with temperature.

Heterogeneous equilibria Equilibria involving species in more than one phase.

Homogeneous equilibria Equilibria involving only species in a single phase, i.e., all gases, all liquids, or all solids.

LeChatelier's Principle If a stress (change of conditions) is applied to a system at equilibrium, the system shifts in the direction that reduces the stress.

Mass action expression For a reversible reaction,

$$aA + bB \rightleftharpoons cC + dD$$

the product of the concentrations of the products (species on the right), each raised to the power that corresponds to its coefficient in the balanced chemical equation, divided by the product of the concentrations of the reactants (species on the left), each raised to the power that corresponds to its coefficient in the balanced chemical equation. At equilibrium the mass action expression is equal to K; at other conditions, it is Q.

$$\frac{[C]^c[D]^d}{[A]^a[B]^b} = Q \text{ or, at equilibrium, } K$$

Reaction quotient, Q The mass action expression under any set of conditions (not necessarily equilibrium); its magnitude relative to K determines the direction in which reaction must occur to establish equilibrium.

Reversible reactions Reactions that do not go to completion and occur in both the forward and reverse directions.

van't Hoff equation The relationships between ΔH^0 for a reaction and its equilibrium constants at two different temperatures.

Exercises

Basic Concepts

1. Define and illustrate the following terms: (a) reversible reaction, (b) chemical equilibrium, (c) equilibrium constant, K.
2. What type of equilibrium is established in chemical reactions? When does a chemical reaction reach a state of equilibrium? Illustrate with a specific example.
3. Why are chemical equilibria called dynamic equilibria?
4. Consider the reaction described by the equation

$$H_2O(\ell) \rightleftharpoons H^+ + OH^-$$

What is the relationship between the rates of the forward and reverse reactions at equilibrium? Does this mean that there will be equal masses of reactants and products at equilibrium? Explain.

The Equilibrium Constant

5. Write the concentration equilibrium constant expression for each of the following reactions:
   (a) $SO_3(g) + H_2(g) \rightleftharpoons SO_2(g) + H_2O(g)$
   (b) $4NH_3(g) + 5O_2(g) \rightleftharpoons 4NO(g) + 6H_2O(g)$
   (c) $C_3H_8(g) + 5O_2(g) \rightleftharpoons 3CO_2(g) + 4H_2O(g)$
6. Repeat Exercise 5 for
   (a) $H_2O(g) \rightleftharpoons H(g) + OH(g)$
   (b) $H_2C{=}CH_2(g) + Cl_2(g) \rightleftharpoons H_2ClCCClH_2(g)$
   (c) $2O_3(g) \rightleftharpoons 3O_2(g)$
   (d) $IF(g) \rightleftharpoons \frac{1}{2}I_2(g) + \frac{1}{2}F_2(g)$
7. Write the equilibrium constant expression for the following chemical reaction:

$$H_2PO_4(aq) \rightleftharpoons H^+(aq) + H_2PO_4^-(aq)$$

Describe what happens at the molecular level as the H^+ ion from a concentrated strong acid such as hydrochloric acid is added to a system in which the H_3PO_4, H^+ ion, and $H_2PO_4^-$ ion have reached equilibrium.
8. Write the equilibrium constant expression for the following chemical reaction:

$$Cl_3(aq) + 2Br^-(aq) \rightleftharpoons 2Cl^-(aq) + Br_2(aq)$$

Describe what happens at the molecular level as the Br^- ion from solid NaBr, which is soluble, is added to a system in which the four species have reached equilibrium.
9. Explain the significance of (a) a very large value of K, (b) a very small value of K, and (c) a value of K of about 1.0.

10. What does the value of an equilibrium constant tell us about the time required for the reaction to reach equilibrium?
11. Why do we omit concentrations of pure solids and pure liquids from equilibrium constant expressions?

Calculation and Uses of K

12. On the basis of the equilibrium constant values, choose the reactions in which the *products* are favored:
    (a) $NH_3(aq) + H_2O(\ell) \rightleftharpoons NH_4^+(aq) + OH^-(aq)$
    $$K = 1.8 \times 10^{-5}$$
    (b) $Au^+(aq) + 2CN^-(aq) \rightleftharpoons [Au(CN)_2]^-(aq)$
    $$K = 2 \times 10^{38}$$
    (c) $PbC_2O_4(s) \rightleftharpoons Pb^{2+}(aq) + C_2O_4^{2-}(aq)$
    $$K = 4.8$$
    (d) $HS^-(aq) + H^+(aq) \rightleftharpoons H_2S(aq)$
    $$K = 1.0 \times 10^7$$
13. On the basis of the equilibrium constant values, choose the reactions in which the *reactants* are favored:
    (a) $H_2O(\ell) \rightleftharpoons H^+(aq) + OH^-(aq)$
    $$K = 1.0 \times 10^{-14}$$
    (b) $[AlF_6]^{3-}(aq) \rightleftharpoons Al^{3+}(aq) + 6F^-(aq)$
    $$K = 2 \times 10^{-24}$$
    (c) $Ca_3(PO_4)_2(s) \rightleftharpoons 3Ca^{2+}(aq) + 2PO_4^{3-}(aq)$
    $$K = 10^{-25}$$
    (d) $2Fe^{3+}(aq) + 3S^{2-}(aq) \rightleftharpoons Fe_2S_3(s)$
    $$K = 1 \times 10^{88}$$
14. The reaction between nitrogen and oxygen to form $NO(g)$ is represented by the chemical equation

$$N_2(g) + O_2(g) \rightleftharpoons 2NO(g)$$

The equilibrium concentrations of the gases at 1500 K are 1.7×10^{-3} mol/L for O_2, 6.4×10^{-3} mol/L for N_2, and 1.1×10^{-5} mol/L for NO. Calculate the value of K_c at 1500 K from these data.
15. At elevated temperatures, BrF_5 establishes the following equilibrium:

$$2BrF_5(g) \rightleftharpoons Br_2(g) + 5F_2(g)$$

The equilibrium concentrations of the gases at 1500 K are 0.0064 mol/L for BrF_5, 0.0018 mol/L for Br_2, and 0.0090 mol/L for F_2. Calculate the value of K_c.
16. The reaction described by the equation

$$PCl_3(g) + Cl_2(g) \rightleftharpoons PCl_5(g)$$

has come to equilibrium at a temperature at which the concentrations of PCl_3, Cl_2, and PCl_5 are 10, 9, and 12 mol/L, respectively. Calculate the value of K_c for this reaction at that temperature.

17. Nitrogen reacts with hydrogen as follows:

$$N_2(g) + 3H_2(g) \rightleftharpoons 2NH_3(g)$$

An equilibrium mixture at a given temperature is found to contain 0.31 mol/L N_2, 0.50 mol/L H_2, and 0.14 mol/L NH_3. Calculate the value of K_c at the given temperature.

18. The following equation is given. At 500 K, $K_c = 7.9 \times 10^{11}$.

$$H_2(g) + Br_2(g) \rightleftharpoons 2HBr(g)$$

(a) $\frac{1}{2}H_2(g) + \frac{1}{2}Br_2(g) \rightleftharpoons HBr(g)$ $K_c = ?$
(b) $2HBr(g) \rightleftharpoons H_2(g) + Br_2(g)$ $K_c = ?$
(c) $4HBr(g) \rightleftharpoons 2H_2(g) + 2Br_2(g)$ $K_c = ?$

19. $NO(g)$ and $O_2(g)$ are mixed in a container of fixed volume kept at 1000 K. Their initial concentrations are 0.0200 mol/L and 0.0300 mol/L, respectively. When the reaction

$$2NO(g) + O_2(g) \rightleftharpoons 2NO_2(g)$$

has come to equilibrium, the concentration of NO_2 is 2.2×10^{-3} mol/L. Calculate (a) the concentration of NO at equilibrium, (b) the concentration of O_2 at equilibrium, and (c) the equilibrium constant K_c for the reaction.

20. A sealed tube initially contains 9.84×10^{-4} mol H_2 and 1.38×10^{-3} mol I_2. It is kept at 350°C until the reaction

$$H_2(g) + I_2(g) \rightleftharpoons 2HI(g)$$

comes to equilibrium. At equilibrium, 4.73×10^{-4} mol I_2 is present. Calculate (a) the numbers of moles of H_2 and HI present at equilibrium; (b) the equilibrium constant of the reaction.

21. CO_2 and H_2 are admitted to a container of constant volume kept at 959 K. Their initial partial pressures, before any reaction, are 1.50 atm and 3.00 atm, respectively. The reaction

$$CO_2(g) + H_2(g) \rightleftharpoons CO(g) + H_2O(g)$$

then occurs. At equilibrium, the partial pressure of H_2O is 0.86 atm. Calculate the partial pressures of CO_2, H_2, and CO at equilibrium, and K_P for the reaction.

22. Arrange the following in order of increasing tendency of the forward reactions to proceed toward completion.
(a) $H_2O(g) \rightleftharpoons H_2O(\ell)$ $K_c = 782$
(b) $F_2(g) \rightleftharpoons 2F(g)$ $K_c = 4.9 \times 10^{-21}$
(c) $C(graphite) + O_2(g) \rightleftharpoons CO_2(g)$ $K_c = 1.3 \times 10^{69}$
(d) $H_2(g) + C_2H_4(g) \rightleftharpoons C_2H_6(g)$ $K_c = 9.8 \times 10^{18}$
(e) $N_2O_4(g) \rightleftharpoons 2NO_2(g)$ $K_c = 4.6 \times 10^{-3}$

23. Given the equilibrium constants for the following two reactions at a particular temperature,

$$NiO(s) + H_2(g) \rightleftharpoons Ni(s) + H_2O(g) \quad K_c = 40$$

$$NiO(s) + CO(g) \rightleftharpoons Ni(s) + CO_2(g) \quad K_c = 600$$

calculate the value for the equilibrium constant, K_c, for the reaction

$$CO_2(g) + H_2(g) \rightleftharpoons CO(g) + H_2O(g)$$

at the same temperature.

24. It is given that $A(g) + B(g) \rightleftharpoons C(g) + 2D(g)$. One mole of A and one mole of B are placed in a 0.400-L container. After equilibrium has been established, 0.20 mol of C is present in the container. Calculate the equilibrium constant for the reaction.

25. For the reaction

$$CO(g) + H_2O(g) \rightleftharpoons CO_2(g) + H_2(g)$$

the value of the equilibrium constant K_c is 1.845 at a given temperature. We place 0.500 mol CO and 0.500 mol H_2O in a 1.00-L container at this temperature, and allow the reaction to reach equilibrium. What will be the equilibrium concentrations of all substances present?

26. We place 0.500 mol CO, 0.500 mol H_2O, 0.500 mol CO_2, and 0.500 mol H_2 in a 1.00-L container under the conditions of Exercise 25 and allow the reaction to reach equilibrium. What will be the equilibrium concentrations of all substances present?

The Reaction Quotient, Q

27. Define the reaction quotient Q. Distinguish between Q and K.

28. Why is it useful to compare Q with K? What is the situation in the case of (a) $Q = K$? (b) $Q < K$? (c) $Q > K$?

29. How does the form of the reaction quotient compare with that of the equilibrium constant? What is the difference between these two expressions?

30. If the reaction quotient is larger than the equilibrium constant, what will happen to the reaction? What will happen if $Q < K$?

31. For the reaction

$$Cl_2(g) + F_2(g) \rightleftharpoons 2ClF(g)$$

$K_c = 19.9$. What will happen in a reaction mixture originally containing $[Cl_2] = 0.2$ mol/L, $[F_2] = 0.1$ mol/L, and $[ClF] = 3.65$ mol/L?

32. The concentration equilibrium constant for the gas-phase reaction

$$HCHO \rightleftharpoons H_2 + CO$$

has the numerical value 0.50 at a given temperature. A mixture of HCHO, H_2, and CO is introduced into a flask at this temperature. After a short time, analysis of a small sample of the reaction mixture shows the

concentrations to be [HCHO] = 0.50 M, [H$_2$] = 1.50 M, and [CO] = 0.25 M. Classify each of the following statements about this reaction mixture as true or false.
(a) The reaction mixture is at equilibrium.
(b) The reaction mixture is not at equilibrium, but no further reaction will occur.
(c) The reaction mixture is not at equilibrium, but will move toward equilibrium by using up more HCHO.
(d) The forward rate of this reaction is the same as the reverse rate.

33. The value of K_P at 25°C for

$$C(graphite) + CO_2(g) \rightleftharpoons 2CO(g)$$

is 9.0×10^{-22}. What is the value of K_c? Describe what will happen if 2 mol of CO and 1 mol of CO$_2$ are mixed in a 1-L container with a suitable catalyst to make the reaction "go" at this temperature.

Uses of the Equilibrium Constant, K_c

34. For the reaction described by the equation

$$N_2(g) + C_2H_2(g) \rightleftharpoons 2HCN(g)$$

$K_c = 2.3 \times 10^{-4}$ at 300°C. What is the equilibrium concentration of hydrogen cyanide if the initial concentrations of N$_2$ and acetylene (C$_2$H$_2$) were 5.0 mol/L and 2.0 mol/L, respectively?

35. The equilibrium constant for the reaction

$$Br_2(g) + F_2(g) \rightleftharpoons 2BrF(g)$$

is 54.7. What are the equilibrium concentrations of all these gases if the initial concentrations of bromine and fluorine were both 0.125 mol/L?

36. For the reaction

$$PCl_3(g) + Cl_2(g) \rightleftharpoons PCl_5(g)$$

$K_c = 96.2$ at 400 K. What is the concentration of Cl$_2$ at equilibrium if the initial concentrations were 0.10 mol/L for PCl$_3$ and 3.5 mol/L for Cl$_2$?

37. At 25°C, the equilibrium constant for the reaction

$$N_2O_4(g) \rightleftharpoons 2NO_2(g)$$

is $K_c = 5.85 \times 10^{-3}$. Twenty (20.0) grams of N$_2$O$_4$ is confined in a 5.00-L flask at 25°C. Calculate (a) the number of moles of NO$_2$ present at equilibrium and (b) the percentage of the original N$_2$O$_4$ that is dissociated.

38. Antimony pentachloride decomposes in a gas-phase reaction at 448°C as follows:

$$SbCl_5(g) \rightleftharpoons SbCl_3(g) + Cl_2(g)$$

An equilibrium mixture in a 5.00-L vessel is found to contain 3.84 g of SbCl$_5$, 9.14 g of SbCl$_3$, and 2.84 g of Cl$_2$. Evaluate K_c at 448°C.

39. If 10.0 g of SbCl$_5$ are placed in a 5.00-L container at 448°C and allowed to establish equilibrium as in Ex-

ercise 38, what will be the equilibrium concentrations of all species? For this reaction, $K_c = 2.51 \times 10^{-2}$.

40. If 5.00 g of SbCl$_5$ and 5.00 g of SbCl$_3$ are placed in a 5.00-L container and allowed to establish equilibrium as in Exercise 38, what will the equilibrium concentrations be? For this reaction, $K_c = 2.51 \times 10^{-2}$.

41. The following equilibrium is established in the presence of water and a particular enzyme (catalyst):

$$\underset{\text{fumaric acid}}{H_2C_4H_2O_4(aq)} + H_2O \rightleftharpoons \underset{\text{malic acid}}{H_2C_4H_4O_5(aq)}$$

One (1.00) liter of solution was prepared from water and 0.300 mol of pure malic acid. At equilibrium, the solution contained 0.067 mol of fumaric acid, in addition to some unchanged malic acid. Find the equilibrium constant for the reaction.

*42. At some temperature, the reaction

$$N_2(g) + 3H_2(g) \rightleftharpoons 2NH_3(g)$$

has an equilibrium constant K_c numerically equal to 1. For such a situation, state whether each of the following statements is true or false, and explain why.
(a) An equilibrium mixture must have the H$_2$ concentration three times that of N$_2$ and the NH$_3$ concentration twice that of H$_2$.
(b) An equilibrium mixture must have the H$_2$ concentration three times that of N$_2$.
(c) A mixture in which the H$_2$ concentration is three times that of N$_2$ *and* the NH$_3$ concentration is twice that of N$_2$ could be an equilibrium mixture.
(d) A mixture in which the concentration of each reactant and each product is 1 M is an equilibrium mixture.
(e) Any mixture in which the concentrations of all reactants and products are equal is an equilibrium mixture.
(f) An equilibrium mixture must have equal concentrations of all reactants and products.

*43. Bromine chloride, BrCl, a reddish covalent gas with properties similar to those of Cl$_2$, may eventually replace Cl$_2$ as a water disinfectant. One mole of chlorine and one mole of bromine are enclosed in a 5.00-L flask and allowed to reach equilibrium at a certain temperature.

$$Cl_2(g) + Br_2(g) \rightleftharpoons 2BrCl(g) \qquad K_c = 4.7 \times 10^{-2}$$

(a) What percentage of the chlorine has reacted at equilibrium?
(b) What mass of each species is present at equilibrium?
(c) How would a decrease in volume shift the position of equilibrium, if at all?

*44. At a certain temperature, K_c for the following reaction is 0.450. If 0.800 mol of POCl$_3$ is placed in a closed 2.00-L vessel at this temperature, what percentage of it will be dissociated when equilibrium is established?

$$POCl_3(g) \rightleftharpoons POCl(g) + Cl_2(g)$$

Factors That Influence Equilibrium

45. State LeChatelier's Principle. What factors do we usually consider to have an effect on a system at equilibrium? How does the presence of a catalyst or an inhibitor affect a system at chemical equilibrium? Explain your answer.

46. What will be the effect of increasing the total pressure on the equilibrium conditions for (a) a reaction that has more moles of gaseous products than gaseous reactants, (b) a reaction that has more moles of gaseous reactants than gaseous products, (c) a reaction that has the same number of moles of gaseous reactants and gaseous products, and (d) a reaction in which all reactants and products are either pure solids, pure liquids, or in solution?

47. What would be the effect on the equilibrium position of an equilibrium mixture of Br_2, F_2, and BrF_5 if the total pressure of the system were decreased?

$$2BrF_5(g) \rightleftharpoons Br_2(g) + 5F_2(g)$$

48. What would be the effect on the equilibrium position of an equilibrium mixture of carbon, oxygen, and carbon monoxide if the total pressure of the system were decreased?

$$2C(s) + O_2(g) \rightleftharpoons 2CO(g)$$

49. A weather indicator can be made with a hydrate of cobalt(II) chloride, which changes color as a result of the following reaction:

$$\underset{\text{pink}}{[Co(H_2O)_6]Cl_2(s)} \rightleftharpoons \underset{\text{blue}}{[Co(H_2O)_4]Cl_2(s)} + 2H_2O(g)$$

Does a pink color indicate "moist" or "dry" air? Explain.

50. Consider the reaction

$$CaCO_3(s) \rightleftharpoons CaO(s) + CO_2(g)$$

Will the mass of $CaCO_3$ at equilibrium (i) increase, (ii) decrease, or (iii) remain the same if (a) CO_2 is added to the equilibrium system? (b) the pressure is increased? (c) solid CaO is removed?

51. The reaction between NO and O_2 is exothermic.

$$2NO(g) + O_2(g) \rightleftharpoons 2NO_2(g) \; + \Delta H$$

Will the concentration of NO_2 at equilibrium (i) increase, (ii) decrease, or (iii) remain the same if (a) additional O_2 is introduced? (b) additional NO is introduced? (c) the total pressure is decreased? (d) the temperature is increased?

52. Predict whether the equilibrium for the photosynthesis reaction described by the equation

$$6CO_2(g) + 6H_2O(\ell) \rightleftharpoons C_6H_{12}O_6(s) + 6O_2(g)$$
$$\Delta H^0 = 2801.69 \text{ kJ/mol}$$

would (i) shift to the right, (ii) shift to the left, or (iii) remain unchanged if (a) $[CO_2]$ were increased; (b) P_{O_2}

were increased; (c) one half of the $C_6H_{12}O_6$ were removed; (d) the total pressure were decreased; (e) the temperature were decreased; (f) a catalyst were added.

53. What would be the effect of increasing the temperature on each of the following systems at equilibrium?
(a) $H_2(g) + I_2(g) \rightleftharpoons 2HI(g) + 9.45 \text{ kJ}$
(b) $PCl_5(g) + 92.5 \text{ kJ} \rightleftharpoons PCl_3(g) + Cl_2(g)$
(c) $2SO_2(g) + O_2(g) \rightleftharpoons 2SO_3(g); \Delta H^0 = -198 \text{ kJ/mol}$
(d) $2NOCl(g) \rightleftharpoons 2NO(g) + Cl_2(g); \Delta H^0 = 75 \text{ kJ/mol}$
(e) $C(s) + H_2O(g) + 131 \text{ kJ} \rightleftharpoons CO(g) + H_2(g)$

54. What would be the effect of increasing the pressure by decreasing the volume on each of the following systems at equilibrium?
(a) $2CO(g) + O_2(g) \rightleftharpoons 2CO_2(g)$
(b) $2NO(g) \rightleftharpoons N_2(g) + O_2(g)$
(c) $N_2O_4(g) \rightleftharpoons 2NO_2(g)$
(d) $Ni(s) + 4CO(g) \rightleftharpoons Ni(CO)_4(g)$
(e) $N_2(g) + 3H_2(g) \rightleftharpoons 2NH_3(g)$

55. The value of K_c is 0.020 at 2870°C for the reaction shown below. There are 0.800 mole of N_2, 0.500 mole of O_2, and 0.400 mole of NO in a 1.00-liter container at 2870°C. Is the system at equilibrium or must the forward or reverse reaction occur to a greater extent to bring the system to equilibrium?

$$N_2(g) + O_2(g) \rightleftharpoons 2NO(g)$$

56. Given: $A(g) + B(g) \rightleftharpoons C(g) + D(g)$
(a) At equilibrium a 1.00-liter container was found to contain 1.60 mole of C, 1.60 mole of D, 0.40 mole of A, and 0.40 mole of B. Calculate the equilibrium constant for this reaction.
(b) If 0.20 mole of A and 0.20 mole of B are added to this system, what will the new *equilibrium* concentration of A be?

57. Given: $A(g) + B(g) \rightleftharpoons C(g) + D(g)$
When one mole of A and one mole of B are mixed and allowed to reach equilibrium at room temperature, the mixture is found to contain $\frac{2}{3}$ mole of C.
(a) Calculate the equilibrium constant.
(b) If three moles of A were mixed with one mole of B and allowed to reach equilibrium, how many moles of C would be present at equilibrium?

58. Given: $A(g) \rightleftharpoons B(g) + C(g)$
(a) When the system is at equilibrium at 200°C, the concentrations are found to be: $[A] = 0.30 \, M$; $[B] = 0.20 \, M = [C]$. Calculate K_c.
(b) If the volume of the container in which the system is at equilibrium is suddenly doubled at 200°C, what will the new equilibrium concentrations be?
(c) Refer back to part (a). If the volume of the container is suddenly halved at 200°C, what will the new equilibrium concentrations be?

*59. The equilibrium constant, K_c, for the dissociation of phosphorus pentachloride is 4.2×10^{-2} at 250°C. How many moles and grams of PCl_5 must be added to a 3.0-liter flask to obtain a Cl_2 concentration of 0.15 M?

$$PCl_5(g) \rightleftharpoons PCl_3(g) + Cl_2(g)$$

*60. (a) What is K_P for the reaction of Exercise 59 at 250°C?
(b) What are the equilibrium partial pressures of each of the species?
(c) What is the total pressure of the system at equilibrium?
(d) What was the partial pressure of PCl_5 before any of it decomposed?

61. Consider the following system *at equilibrium*.

$$CO_2(g) + H_2(g) \rightleftharpoons H_2O(g) + CO(g)$$

A 2.00-liter vessel contains 0.48 mole of CO_2, 0.48 mole of H_2, 0.96 mole of H_2O, and 0.96 mole of CO.
(a) How many moles and how many grams of H_2 must be added to bring the concentration of CO to 0.60 M?
(b) How many moles and how many grams of CO_2 must be added to bring the CO concentration to 0.60 M?
(c) How many moles of H_2O must be removed to bring the CO concentration to 0.60 M?

62. At 25°C, K_c is 5.84×10^{-3} for the dissociation of dinitrogen tetroxide to nitrogen dioxide.

$$N_2O_4(g) \rightleftharpoons 2NO_2(g)$$

(a) Calculate the equilibrium concentrations of both gases when 2.50 grams of N_2O_4 are placed in a 2.00-liter flask at 25°C.
(b) What will be the new equilibrium concentrations if the volume of the system is suddenly increased to 4.00 liters at 25°C?
(c) What will be the new equilibrium concentrations if the volume is decreased to 1.00 liter at 25°C?

K in Terms of Partial Pressures

63. Write the K_P expression for each reaction in Exercise 5.

64. Under what conditions are K_c and K_P for a reaction numerically equal?

65. 0.0100 mol of NH_4Cl and 0.0100 mol of NH_3 are placed in a closed 2.00-L container and heated to 603 K. At this temperature, all the NH_4Cl vaporizes. When the reaction

$$NH_4Cl(g) \rightleftharpoons NH_3(g) + HCl(g)$$

has come to equilibrium, 5.8×10^{-3} mol of HCl is present. Calculate (a) K_c and (b) K_P for this reaction at 603 K.

66. CO_2 is passed over graphite at 500 K. The emerging gas stream contains 4.0×10^{-3} mol percent CO. The total pressure is 1.00 atm. Assume that equilibrium is attained. Find K_P for the reaction

$$C(graphite) + CO_2(g) \rightleftharpoons 2CO(g)$$

67. At 425°C, the equilibrium partial pressures of H_2, I_2, and HI are 0.06443 atm, 0.06540 atm, and 0.4821 atm, respectively. Calculate K_P for the following reaction at this temperature.

$$2HI(g) \rightleftharpoons H_2(g) + I_2(g)$$

68. At 27°C and 1.00 atm, N_2O_4 is 20.0% dissociated into NO_2. Calculate the value of K_P for

$$N_2O_4(g) \rightleftharpoons 2NO_2(g)$$

at this temperature.

69. For the reaction

$$H_2(g) + Cl_2(g) \rightleftharpoons 2HCl(g)$$

$K_c = 193$ at 2500 K. What is the value of K_P for this reaction?

70. For the reaction

$$Br_2(g) \rightleftharpoons 2Br(g)$$

$K_P = 2550$ at 4000 K. What is the value of K_c for this reaction?

*71. Chloromethane, CH_3Cl, is produced by the reaction of Cl_2 with methane, CH_4, as follows:

$$CH_4(g) + Cl_2(g) \rightleftharpoons CH_3Cl(g) + HCl(g)$$

Assuming that equal amounts of CH_4 and Cl_2 were placed in a reaction vessel so that their initial partial pressures were identical and equilibrium was allowed to be established, what is P_{CH_3Cl}/P_{CH_4} if $K_P = 7.4 \times 10^{17}$? (It is not necessary to find the individual pressures of the gases at equilibrium to calculate this ratio.)

72. For the reaction

$$2CO(g) \rightleftharpoons C(graphite) + CO_2(g)$$

$K_P = 6.0 \times 10^{-4}$ at 1473 K. CO(g) is initially at 1.00 atm. It is kept in contact with graphite at 1473 K until the reaction has come to equilibrium. What is the partial pressure of CO_2 at equilibrium?

73. A stream of gas containing H_2 at an initial partial pressure of 0.200 atm is passed through a tube in which CuO is kept at 500 K. The reaction

$$CuO(s) + H_2(g) \rightleftharpoons Cu(s) + H_2O(g)$$

comes to equilibrium. For this reaction, $K_P = 1.6 \times 10^9$. What is the partial pressure of H_2 in the gas leaving the tube? Assume that the total pressure of the stream is unchanged.

74. In the distant future, when hydrogen may be cheaper than coal, steel mills may make iron by the reaction

$$Fe_2O_3(s) + 3H_2(g) \rightleftharpoons 2Fe(s) + 3H_2O(g)$$

For this reaction, $\Delta H = 96$ kJ/mol and $K_c = 8.11$ at 1000 K. (a) What percentage of the H_2 remains unreacted after the reaction has come to equilibrium at 1000 K? (b) Is this percentage greater or less if the temperature is increased above 1000 K?

Heterogeneous Equilibrium

75. At $-10°C$, the solid compound $Cl_2(H_2O)_8$ is in equilibrium with gaseous chlorine, water vapor, and ice. The partial pressures of the two gases in equilibrium with a mixture of $Cl_2(H_2O)_8$ and ice are 0.20 atm for Cl_2 and 0.00262 atm for water vapor. Find the equilibrium constant K_P for each of these reactions:
(a) $Cl_2(H_2O)_8(s) \rightleftharpoons Cl_2(g) + 8H_2O(g)$
(b) $Cl_2(H_2O)_8(s) \rightleftharpoons Cl_2(g) + 8H_2O(s)$
Why are your two answers so different?

76. A flask contains $NH_4Cl(s)$ in equilibrium with its decomposition products:

$$NH_4Cl(s) \rightleftharpoons NH_3(g) + HCl(g)$$

For this reaction, $\Delta H = 176$ kJ/mol. How is the mass of NH_3 in the flask affected by each of the following disturbances? (a) The temperature is increased. (b) NH_3 is added. (c) HCl is added. (d) NH_4Cl is added, with no appreciable change in the gas volume. (e) A large amount of NH_4Cl is added, decreasing the volume available to the gases.

77. The equilibrium constant for the reaction

$$H_2(g) + Br_2(\ell) \rightleftharpoons 2HBr(g)$$

is $K_P = 4.5 \times 10^{18}$ at $25°C$. The vapor pressure of liquid Br_2 at this temperature is 0.28 atm. (a) Find K_P at $25°C$ for the reaction

$$H_2(g) + Br_2(g) \rightleftharpoons 2HBr(g)$$

(b) How will the equilibrium in part (a) be shifted by a decrease in the volume of the container if (1) liquid Br_2 is absent; (2) liquid Br_2 is present? Explain why the effect is different in these two cases.

78. A gaseous mixture of H_2 and H_2O is kept in contact with a solid mixture of Fe and Fe_3O_4 at $1150°C$ until the reaction

$$3Fe(s) + 4H_2O(g) \rightleftharpoons Fe_3O_4(s) + 4H_2(g)$$

has come to equilibrium. The following pressures are measured at equilibrium in three experiments:

	Total Pressure (torr)	Partial Pressure of H_2O (torr)
(1)	25.7	11.9
(2)	37.3	17.4
(3)	107.5	49.3

From the results of each experiment, find the average value of the equilibrium constant for the reaction.

Relationships Among K, ΔG^0, ΔH^0, and T

79. What kind of equilibrium constant can be calculated from a ΔG^0 value for a reaction involving only gases?

80. What must be true of the value of ΔG^0 for a reaction if (a) $K \gg 1$; (b) $K = 1$; (c) $K \ll 1$?

81. The equilibrium constant K_c of the reaction

$$H_2(g) + Br_2(g) \rightleftharpoons 2HBr(g)$$

is 1.6×10^5 at 1297 K and 3.5×10^4 at 1495 K. (a) Is ΔH^0 for this reaction positive or negative? (b) Find K_c for the reaction

$$\tfrac{1}{2}H_2(g) + \tfrac{1}{2}Br_2(g) \rightleftharpoons HBr(g)$$

at 1297 K. (c) Pure HBr is placed in a container of constant volume and heated to 1297 K. What percentage of the HBr is decomposed to H_2 and Br_2 at equilibrium?

82. The air pollutant sulfur dioxide can be partially removed from stack gases in industrial processes and converted to sulfur trioxide, the acid anhydride of commercially important sulfuric acid. Write the equation for the reaction, using the smallest whole-number coefficients. Calculate the value of the equilibrium constant for this reaction at $25°C$, from values of ΔG_f^0 in Appendix K.

83. The value of ΔH^0 for the reaction in Exercise 82 is -197.6 kJ/mol.
(a) Predict qualitatively (i.e., without calculation) whether the value of K_P for this reaction at $500°C$ would be greater than, the same as, or less than the value at room temperature ($25°C$).
(b) Now calculate the value of K_P at $500°C$.

84. Given that K_P is 4.6×10^{-14} at $25°C$ for the reaction

$$2Cl_2(g) + 2H_2O(g) \rightleftharpoons 4HCl(g) + O_2(g)$$
$$\Delta H^0 = +115 \text{ kJ/mol}$$

Calculate K_P and K_c for the reaction at $400°C$ and at $800°C$.

85. A mixture of 3.00 mol of Cl_2 and 3.00 mol of CO is enclosed in a 5.00-L flask at $600°C$. At equilibrium, 3.3% of the Cl_2 has been consumed.

$$CO(g) + Cl_2(g) \rightleftharpoons COCl_2(g)$$

(a) Calculate K_c for the reaction at $600°C$.
(b) Calculate ΔG^0 for the reaction at this temperature.

*86. The following is an example of an alkylation reaction that is important in the production of iso-octane (2,2,4-trimethylpentane) from two components of crude oil, isobutane and isobutene. Iso-octane is an antiknock additive for gasoline.

The thermodynamic equilibrium constant, K, for this reaction at 25°C is 4.3×10^6, and ΔH^0 is $- 78.58$ kJ/mol.
(a) Calculate ΔG^0 at 25°C.
(b) Calculate K at 800°C.
(c) Calculate ΔG^0 at 800°C.
(d) How does the spontaneity of the forward reaction at 800°C compare with that at 25°C?
(e) Why do you think the reaction mixture is heated in the industrial preparation of iso-octane?
(f) What is the purpose of the catalyst? Does it affect the forward reaction more than the reverse reaction?

87. At sufficiently high temperatures, chlorine gas dissociates, according to

$$Cl_2(g) \rightleftharpoons 2Cl(g)$$

At 800°C, K_P for this reaction is 5.63×10^{-7}.
(a) A sample originally contained Cl_2 at 1 atm and 800°C. Calculate the percentage dissociation of Cl_2 when this reaction has reached equilibrium.
(b) At what temperature would Cl_2 (originally at 1 atm pressure) be 1% dissociated into Cl atoms?

88. (a) Use the tabulated thermodynamic values of ΔH_f^0 and S^0 to calculate the value of K_P at 25°C for the gas-phase reaction

$$CO + H_2O \rightleftharpoons CO_2 + H_2$$

(b) Calculate the value of K_P for this reaction at 200°C, by the same method as in part (a).
(c) Repeat the calculation of part (a), using tabulated values of ΔG_f^0.

89. How does the form of the reaction quotient in terms of partial pressures compare with that of K_P? What is the difference between these two expressions?

90. If the reaction quotient in terms of partial pressures is larger than the equilibrium constant, what will happen to the reaction? What will happen if $Q_P < K_P$?

91. For the reaction

$$Cl_2(g) + F_2(g) \rightleftharpoons 2ClF(g)$$

$K_c = 19.9$. What will happen in a reaction mixture originally containing $[Cl_2] = 0.2$ mol/L, $[F_2] = 0.1$ mol/L, and $[ClF] = 0.365$ mol/L?

Mixed Exercises

92. The value of K_P at 25°C for

$$2CO(g) \rightleftharpoons C(graphite) + CO_2(g)$$

is 1.11×10^{21}. What is the value of K_c? Describe what will happen if 2 mol of CO and 1 mol of CO_2 are mixed in a 1-L container with a suitable catalyst to make the reaction "go" at this temperature.

93. $K_c = 0.21$ at 350°C for the reaction

$$2NO(g) + Br_2(g) \rightleftharpoons 2BrNO(g)$$

A 20.0-mL tube contains 1.0×10^{-4} mol NO, 2.0×10^{-4} mol Br_2, and 2.0×10^{-4} mol BrNO at 350°C. (a) Is the mixture in equilibrium? If it is not, which way can the reaction go? (Do not confuse number of moles and concentration.) (b) At equilibrium, the tube contains 2.8×10^{-4} mol NO and 1.6×10^{-5} mol BrNO. How many moles of Br_2 does it contain? (c) Do you need to know the volume of the tube to do this problem?

94. Given the reaction

$$H_2(g) + Br_2(g) \rightleftharpoons 2HBr(g)$$

(a) At 500 K, $K_c = 7.9 \times 10^{11}$. What is K_P?
(b) $\frac{1}{2}H_2(g) + \frac{1}{2}Br_2(g) \rightleftharpoons HBr(g)$ $K_P = ?$
(c) $2HBr(g) \rightleftharpoons H_2(g) + Br_2(g)$ $K_P = ?$
(d) $4HBr(g) \rightleftharpoons 2H_2(g) + 2Br_2(g)$ $K_P = ?$

95. What must be the pressure of hydrogen so that the reaction

$$WCl_6(g) + 3H_2(g) \rightleftharpoons W(s) + 6HCl(g)$$

will occur if $P_{WCl_6} = 0.012$ atm and $P_{HCl} = 0.10$ atm? $K_P = 1.37 \times 10^{21}$ at 900 K.

96. A mixture of CO, H_2, CH_4, and H_2O is kept at 1133 K until the reaction

$$CO(g) + 3H_2(g) \rightleftharpoons CH_4(g) + H_2O(g)$$

has come to equilibrium. The volume of the container is 0.100 L. The equilibrium mixture contains 1.21×10^{-4} mol CO, 2.47×10^{-4} mol H_2, 1.21×10^{-4} mol CH_4, and 5.63×10^{-8} mol H_2O. Calculate K_P for this reaction at 1133 K.

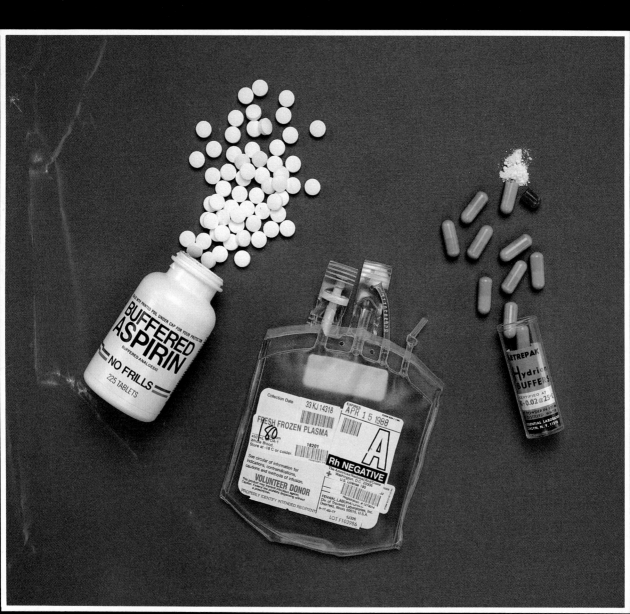

Objectives

As you study this chapter, you should learn

☐ To recognize strong electrolytes
  and calculate concentrations of
  their ions
☐ To understand the autoionization
  of water
☐ To understand the pH and pOH
  scales and how they are used
☐ About ionization constants for
  weak monoprotic acids and bases
☐ To use ionization constants
☐ What acid–base indicators are
  and how they function

☐ About the common ion effect
☐ About buffer solutions, how they
  are prepared, and how they
  function
☐ How to prepare buffer solutions
  of a given pH
☐ How polyprotic acids ionize in
  steps and how to calculate
  concentrations of all species in
  these solutions

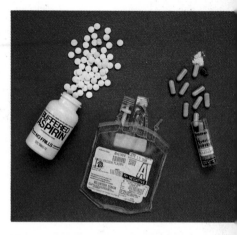

Three common examples of buffers.
Many medications are buffered to
minimize digestive upset. Most body
fluids, including blood plasma, contain
very efficient natural buffer systems.
Buffer capsules are used in laboratories
to prepare solutions of specified pH.

A queous solutions are very important. Approximately three fourths of the earth's surface is covered with water. Enormous numbers of chemical reactions occur in the oceans and smaller bodies of water. The body fluids of plants and animals are mostly water. The life processes (chemical reactions) of all plants and animals occur in aqueous solutions or in contact with water. All of us developed in sacs filled with aqueous solutions, which protected and nurtured us until we had developed to the point that we could live in the atmosphere.

Water-soluble compounds may be classified as either electrolytes or nonelectrolytes. **Electrolytes** are compounds that ionize (or dissociate into their constituent ions) to produce aqueous solutions that conduct an electric current. **Nonelectrolytes** exist as molecules in aqueous solution, and such solutions do not conduct an electric current.

18-1 A Brief Review of Strong Electrolytes

Strong electrolytes are ionized (or dissociated) completely, or very nearly completely, in dilute aqueous solutions. Strong electrolytes include strong acids, strong soluble bases, and most soluble salts. You should review the discussions of these substances in Sections 4-2 and 10-6. The common strong

Table 18-1
Common Strong Acids and Strong Soluble Bases

Strong Acids	
HCl	HNO_3
HBr	$HClO_4$
HI	$HClO_3$
	H_2SO_4

Strong Soluble Bases	
LiOH	
NaOH	
KOH	$Ca(OH)_2$
RbOH	$Sr(OH)_2$
CsOH	$Ba(OH)_2$

acids and strong soluble bases are listed again in Table 18-1. See Section 4-2, part 5, for the solubility rules for ionic compounds.

Concentrations of ions in aqueous solutions of strong electrolytes can be calculated directly from the molarity of the strong electrolyte, as the next two examples illustrate.

Example 18-1

Calculate the molar concentrations of Ba^{2+} and OH^- in 0.030 M barium hydroxide.

Plan

First we write the equation for the dissociation of $Ba(OH)_2$. Then we construct the reaction summary. $Ba(OH)_2$ is a strong soluble base that is completely dissociated.

Solution

From the equation for the dissociation of barium hydroxide, we see that *one* mole of $Ba(OH)_2$ produces *one* mole of Ba^{2+} ions and *two* moles of OH^- ions.

(*strong base*)	$Ba(OH)_2(s)$	$\longrightarrow$	$Ba^{2+}(aq)$	+	$2OH^-(aq)$
initial	0.030 M				
change due to rxn	−0.030 M		+0.030 M		+2(0.030) M
final	0 M		0.030 M		0.060 M

$$[Ba^{2+}] = 0.030\ M \quad \text{and} \quad [OH^-] = 0.060\ M$$

> Recall that we use a single arrow to indicate that a reaction goes to completion, or nearly to completion, in the indicated direction.

Example 18-2

Calculate the concentrations of Mg^{2+} and Br^- ions in a solution that contains 0.92 gram of $MgBr_2$ in 500 mL of solution.

Plan

First we calculate the concentration of $MgBr_2$ in moles per liter of solution. Then we write the equation for the dissociation of $MgBr_2$ and construct the reaction summary. The solubility rules tell us that $MgBr_2$ is a soluble ionic salt, so it is completely dissociated.

Solution

$$\frac{?\ mol\ MgBr_2}{L\ soln} = \frac{0.92\ g\ MgBr_2}{0.500\ L} \times \frac{1\ mol\ MgBr_2}{184\ g\ MgBr_2} = 0.010\ mol\ MgBr_2/L$$

The solution is 0.010 M in $MgBr_2$. The equation for the dissociation of $MgBr_2$ shows that *one* mole of $MgBr_2$ produces *one* mole of Mg^{2+} and *two* moles of Br^- ions:

> This does not imply the existence of "molecules" of $MgBr_2$ in solution.

(*soluble salt*)	$MgBr_2(s)$	$\longrightarrow$	$Mg^{2+}(aq)$	+	$2Br^-(aq)$
initial	0.010 M				
change due to rxn	−0.010 M		+0.010 M		+2(0.010) M
final	0 M		0.010 M		0.020 M

$$[Mg^{2+}] = 0.010\ M \text{ and } [Br^-] = 0.020\ M$$

EOC 2, 3

18-2 The Autoionization of Water

Careful experiments on its electrical conductivity have shown that pure water ionizes to a very slight extent:

$$H_2O(\ell) + H_2O(\ell) \rightleftharpoons H_3O^+ + OH^-(aq)$$

Because the H_2O is pure, its activity is 1, so we do not include its concentration in the equilibrium constant expression. This equilibrium constant is known as the **ion product** for water and is usually represented as K_w:

$$K_w = [H_3O^+][OH^-]$$

The formation of an H_3O^+ ion by the ionization of water is always accompanied by the formation of an OH^- ion. Thus, in *pure* water the concentration of H_3O^+ is *always* equal to the concentration of OH^-. Careful measurements show that, in pure water at 25°C,

$$[H_3O^+] = [OH^-] = 1.0 \times 10^{-7} \text{ mol/L}$$

Substituting these concentrations into the K_w expression gives

$$K_w = [H_3O^+][OH^-] = (1.0 \times 10^{-7})(1.0 \times 10^{-7})$$
$$= 1.0 \times 10^{-14} \quad \text{(at 25°C)}$$

Although the expression $K_w = [H_3O^+][OH^-] = 1.0 \times 10^{-14}$ was obtained for pure water, *it is valid for dilute aqueous solutions at 25°C*. This is one of the most useful relationships chemists have discovered. It gives a simple (inverse) relationship between H_3O^+ and OH^- concentrations in *all* dilute aqueous solutions.

Solutions in which the concentration of solute is less than about 1 mol/L are usually called dilute solutions.

Even though the *value* of K_w is different at different temperatures (Table 18-2), the *relationship* $K_w = [H_3O^+][OH^-]$ is still valid.

In this text, we shall assume a temperature of 25°C for all calculations involving aqueous solutions unless we specify another temperature.

Example 18-3
Calculate the concentrations of H_3O^+ and OH^- in a 0.050 *M* solution of nitric acid.

Plan
We write the equation for the ionization of HNO_3, a strong acid. Then we construct the reaction summary, which gives the concentration of H_3O^+ ions directly. Then we use the relationship $K_w = [H_3O^+][OH^-] = 1.0 \times 10^{-14}$ to find the concentration of OH^- ions.

Solution
The reaction summary for the ionization of HNO_3, a strong acid, is

(*strong acid*)	HNO_3	$+$	H_2O	$\longrightarrow$	H_3O^+	$+$	NO_3^-
initial	0.050 *M*				~0 *M*		0 *M*
change due to rxn	-0.050 *M*				$+0.050$ *M*		$+0.050$ *M*
at equil	0 *M*				0.050 *M*		0.050 *M*

Table 18-2
K_w at Some Temperatures

Temperature (°C)	K_w
0	1.13×10^{-15}
10	2.92×10^{-15}
25	1.00×10^{-14}
37*	2.38×10^{-14}
45	4.02×10^{-14}
60	9.61×10^{-14}

* Normal human body temperature.

So we see that the ionization of 0.050 mol/L of HNO_3 produces 0.050 mol/L of H_3O^+ and 0.050 mol/L of NO_3^-. Thus, $[H_3O^+] = [NO_3^-] = 0.050 \; M$.

The $[OH^-]$ is determined from the equation for the autoionization of water:

	$2H_2O$	$\rightleftharpoons$	$H_3O^+(aq)$	$+$	OH^-
initial			$0.050 \; M$		
change due to rxn	$-2x \; M$		$+x \; M$		$+x \; M$
at equil			$(0.050 + x) \; M$		$x \; M$

$$K_w = [H_3O^+][OH^-]$$

$$1.0 \times 10^{-14} = (0.050 + x)(x)$$

Because the product $(0.050 + x)(x)$ is such a small number, we know that x must be very small. Thus, it will not matter (much) whether we add x to 0.050; that is, we assume that $(0.050 + x) \approx 0.050$. We substitute this approximation into the equation and solve:

Recall that
$[OH^-]_{\text{from } H_2O} = [H_3O^+]_{\text{from } H_2O}$
in *all* aqueous solutions. So we know that $[H_3O^+]_{\text{from } H_2O}$ must be 2.0×10^{-13} *M* also.

$$1.0 \times 10^{-14} = (0.050)(x) \quad \text{or} \quad x = \frac{1.0 \times 10^{-14}}{0.050} = 2.0 \times 10^{-13} \; M = [OH^-]$$

We see that the assumption that x is much smaller than 0.050 was a good one.

EOC 13, 14

In solving Example 18-3, we assumed that *all* of the H_3O^+ (0.050 *M*) came from the ionization of HNO_3 and neglected the H_3O^+ formed by the ionization of H_2O. The ionization of H_2O produces only $2.0 \times 10^{-13} \; M \; H_3O^+$ and $2.0 \times 10^{-13} \; M \; OH^-$ *in this solution*. Thus, we were justified in assuming that the $[H_3O^+]$ is derived solely from the strong acid. A more concise way to carry out the calculation to find the $[OH^-]$ concentration is to write directly:

$$K_w = [H_3O^+][OH^-] = 1.0 \times 10^{-14} \quad \text{or} \quad [OH^-] = \frac{1.0 \times 10^{-14}}{[H_3O^+]}$$

Then we substitute to give

$$[OH^-] = \frac{1.0 \times 10^{-14}}{0.050} = 2.0 \times 10^{-13} \; M$$

From now on, we shall use this more direct approach for such calculations.

When nitric acid is added to water, large numbers of H_3O^+ ions are produced. The large increase in $[H_3O^+]$ shifts the water equilibrium far to the left (LeChatelier's Principle), and the $[OH^-]$ decreases:

$$H_2O(\ell) + H_2O(\ell) \rightleftharpoons H_3O^+(aq) + OH^-(aq)$$

In acidic solutions, the H_3O^+ concentration is always greater than the OH^- concentration. We should not conclude that acidic solutions contain no OH^- ions. Rather, the $[OH^-]$ is always less than $1.0 \times 10^{-7} \; M$ in such solutions. The reverse is true for basic solutions, in which the $[OH^-]$ is always greater than $1.0 \times 10^{-7} M$. By definition, "neutral" aqueous solutions at 25°C are solutions in which $[H_3O^+] = [OH^-] = 1.0 \times 10^{-7} \; M$.

Solution	General Condition		At 25°C	
acidic	$[H_3O^+] > [OH^-]$	$[H_3O^+] > 1.0 \times 10^{-7}$	$[OH^-] < 1.0 \times 10^{-7}$	
neutral	$[H_3O^+] = [OH^-]$	$[H_3O^+] = 1.0 \times 10^{-7}$	$[OH^-] = 1.0 \times 10^{-7}$	
basic	$[H_3O^+] < [OH^-]$	$[H_3O^+] < 1.0 \times 10^{-7}$	$[OH^-] > 1.0 \times 10^{-7}$	

18-3 The pH and pOH Scales

The pH scale provides a convenient way to express the acidity and basicity of dilute aqueous solutions. The **pH** of a solution is defined as

$$pH = \log \frac{1}{[H_3O^+]} \quad \text{or} \quad pH = -\log [H_3O^+]$$

Note that we use pH rather than pH_3O. At the time the pH concept was developed, H_3O^+ was represented as H^+. Various "p" terms are used. In general, a lowercase "**p**" before a symbol means "negative logarithm of the symbol." Thus, pH is the negative logarithm of H_3O^+ concentration, **pOH** is the negative logarithm of OH^- concentration, and **pK** refers to the negative logarithm of an equilibrium constant.

We always use the base-10 (common) logarithm, *not* the base-*e* (natural) logarithm, when dealing with pH. This is because pH is *defined* using base-10 logarithms.

$$pH = -\log[H_3O^+] \quad \text{or} \quad [H_3O^+] = 10^{-pH}$$

$$pOH = -\log[OH^-] \quad \text{or} \quad [OH^-] = 10^{-pOH}$$

Example 18-4
Calculate the pH of a solution in which the H_3O^+ concentration is 0.050 mol/L.

Plan
We are given the value for $[H_3O^+]$, and so we take the negative logarithm of this value.

Solution
We are given $[H_3O^+] = 0.050\ M = 5.0 \times 10^{-2}\ M$.

$$pH = -\log [H_3O^+] = -\log [5.0 \times 10^{-2}] = -(-1.30) = \boxed{1.30}$$

This answer contains only *two* significant figures. The "1" in 1.30 is *not* a significant figure; it comes from the power of ten.

Example 18-5
The pH of a solution is 3.301. What is the concentration of H_3O^+ in this solution?

Plan
By definition, $pH = -\log [H_3O^+]$. We are given pH, so we solve for $[H_3O^+]$.

Solution
From the definition of pH, we can write

$$-\log [H_3O^+] = 3.301$$

Multiplying through by -1 gives

$$\log [H_3O^+] = -3.301$$

The pH of some common substances is shown by a universal indicator. Refer to Figure 18-2 to interpret the indicator colors.

Taking the inverse logarithm (antilog) of both sides of the equation gives

$$[H_3O^+] = 10^{-3.301} \quad \text{so} \quad \boxed{[H_3O^+] = 5.00 \times 10^{-4} \; M}$$

EOC 24–26

A convenient relationship between pH and pOH in *all dilute solutions at 25°C* can be derived easily:

$$[H_3O^+][OH^-] = 1.0 \times 10^{-14}$$

Taking the logarithm of both sides of this equation gives

$$\log [H_3O^+] + \log [OH^-] = \log (1.0 \times 10^{-14})$$

Multiplying both sides of this equation by -1 gives

$$(-\log [H_3O^+]) + (-\log [OH^-]) = -\log (1.0 \times 10^{-14})$$

At any temperature, pH + pOH = pK_w.

or

$$pH + pOH = 14$$

We can now relate $[H_3O^+]$ and $[OH^-]$ as well as pH and pOH:

Remember these relationships!

$$[H_3O^+][OH^-] = 1.0 \times 10^{-14} \quad \text{and} \quad pH + pOH = 14 \quad \text{(at 25°C)}$$

From this relationship, we see that pH and pOH can *both* be positive only if *both* are less than 14. If either pH or pOH is greater than 14, the other is obviously negative.

Please study the following summary carefully. It will be helpful.

Solution	General Condition	At 25°C
acidic	$[H_3O^+] > [OH^-]$ pH $<$ pOH	$[H_3O^+] > 1.0 \times 10^{-7} \; M > [OH^-]$ pH $< 7.00 <$ pOH
neutral	$[H_3O^+] = [OH^-]$ pH $=$ pOH	$[H_3O^+] = 1.0 \times 10^{-7} \; M = [OH^-]$ pH $= 7.00 =$ pOH
basic	$[H_3O^+] < [OH^-]$ pH $>$ pOH	$[H_3O^+] < 1.0 \times 10^{-7} \; M < [OH^-]$ pH $> 7.00 >$ pOH

Example 18-6

Calculate $[H_3O^+]$, pH, $[OH^-]$, and pOH for 0.015 M HNO$_3$ solution.

Plan

We write the equation for the ionization of the strong acid HNO$_3$, which gives us $[H_3O^+]$. Then we calculate pH. We use the relationships pH + pOH = 14.00 and $[H_3O^+][OH^-] = 1.0 \times 10^{-14}$ to find pOH and $[OH^-]$.

Solution

All ions are hydrated in aqueous solution. We often omit the designations (ℓ), (g), (s), (aq), and so on.

$$HNO_3 + H_2O \longrightarrow H_3O^+ + NO_3^-$$

Because nitric acid is a strong acid (it ionizes completely), we know that

$$[H_3O^+] = \boxed{0.015 \; M}$$

$$pH = -\log [H_3O^+] = -\log (0.015) = -(-1.82) = \boxed{1.82}$$

We also know that pH + pOH = 14. Therefore,

$$pOH = 14 - pH = 14 - 1.82 = \boxed{12.18}$$

Because $[H_3O^+][OH^-] = 1.0 \times 10^{-14}$, $[OH^-]$ is easily calculated.

$$[OH^-] = \frac{1.0 \times 10^{-14}}{[H_3O^+]} = \frac{1.0 \times 10^{-14}}{0.015} = \boxed{6.7 \times 10^{-13} \ M}$$

EOC 28, 29

To develop familiarity with the pH and pOH scales, consider a series of solutions in which $[H_3O^+]$ varies from 10 M to 1.0×10^{-15} M. Obviously, $[OH^-]$ will vary from 1.0×10^{-15} M to 10 M in these solutions. Table 18-3 summarizes these scales.

The pH of a solution can be determined using a pH meter (Figure 18-1) or by the indicator method. Acid–base *indicators* are intensely colored complex organic compounds that have different colors in solutions of different pH (Section 18-5). Many are weak acids or weak bases that are useful over rather narrow ranges of pH values. *Universal indicators* are mixtures of several indicators; they show several color changes over a wide range of pH values.

In the indicator method, we prepare a series of solutions of known pH (standard solutions). We add a universal indicator to each; solutions with

An indicator is a compound that changes color as pH changes.

Table 18-3
Relationships Among [H₃O⁺], pH, pOH, and [OH⁻]

$[H_3O^+]$	pH		pOH	$[OH^-]$	
10^{-15}	15		-1	10^1	
10^{-14}	14		0	1	
10^{-13}	13		1	10^{-1}	
10^{-12}	12		2	10^{-2}	
10^{-11}	11	OH⁻ concentration	3	10^{-3}	
10^{-10}	10		4	10^{-4}	Increasing basicity
10^{-9}	9		5	10^{-5}	
10^{-8}	8		6	10^{-6}	
10^{-7}	7		7	10^{-7}	Neutral
10^{-6}	6		8	10^{-8}	
10^{-5}	5		9	10^{-9}	
10^{-4}	4	H₃O⁺ concentration	10	10^{-10}	Increasing acidity
10^{-3}	3		11	10^{-11}	
10^{-2}	2		12	10^{-12}	
10^{-1}	1		13	10^{-13}	
1	0		14	10^{-14}	
10^1	-1		15	10^{-15}	

Figure 18-1
A pH meter gives the pH of the solution directly. When the electrode is dipped into a solution, the meter displays the pH. The pH of this solution is 7.01 at 25.0°C. The pH meter is based on the glass electrode. This sensing device generates a voltage that is proportional to the pH of the solution in which the electrode is placed. The instrument has an electrical circuit to amplify the voltage from the electrode and a meter that relates the voltage to the pH of the solution. Before being used, a pH meter must be calibrated with a series of solutions of known pH.

different pH have different colors (Figure 18-2). We then add the same universal indicator to the unknown solution and compare its color to those of the standard solutions. Solutions with the same pH have the same color.

Universal indicator papers can also be used to determine pH. A drop of solution is placed on a piece of paper or a piece of the paper is dipped into a solution. The color of the paper is then compared with a color chart on the container to establish the pH of the solution.

Figure 18-2
Solutions containing a universal indicator. A universal indicator shows a wide range of colors as pH varies. The pH values are given by the black numbers. These solutions range from quite acidic (upper left) to quite basic (lower right).

pH Range for a Few Common Substances

Substance	pH Range
Gastric contents (human)	1.6–3.0
Soft drinks	2.0–4.0
Lemons	2.2–2.4
Vinegar	2.4–3.4
Tomatoes	4.0–4.4
Beer	4.0–5.0
Urine (human)	4.8–8.4
Milk (cow's)	6.3–6.6
Saliva (human)	6.5–7.5
Blood plasma (human)	7.3–7.5
Egg white	7.6–8.0
Milk of magnesia	10.5
Household ammonia	11–12

More acidic

More basic

18-4 Ionization Constants for Weak Monoprotic Acids and Bases

We have discussed strong acids and strong soluble bases. There are relatively few of these. Weak acids are much more numerous than strong acids. For this reason you were asked to learn the list of common strong acids (Table 18-1). You may assume that nearly all other acids you encounter in this text will be weak acids. Table 18-4 contains names, formulas, ionization constants, and pK_a values for a few common weak acids; Appendix F contains a longer list of K_a values.

Weak acids ionize only slightly in dilute aqueous solution. Our classification of acids as strong or weak is based only on the *extent to which they ionize in dilute aqueous solution.* You may know that hydrofluoric acid dissolves glass. But HF is *not* a strong acid. The reaction of glass with hydrofluoric acid occurs because silicates react with HF to produce silicon tetrafluoride, SiF_4, a very volatile compound. This reaction tells us nothing about the acid strength of hydrofluoric acid.

Several weak acids are familiar to us. Vinegar is a 5% solution of acetic acid, CH_3COOH. Carbonated beverages are saturated solutions of carbon dioxide in water, which produces carbonic acid ($CO_2 + H_2O \rightleftharpoons H_2CO_3$). Citrus fruits contain citric acid, $C_3H_5O(COOH)_3$. Some ointments and powders used for medicinal purposes contain boric acid, H_3BO_3 or $B(OH)_3$. These

How can you tell that carbonated beverages are *saturated* CO_2 solutions?

Table 18-4
Ionization Constants and pK_a Values for Some Weak Monoprotic Acids

Acid	Ionization Reaction		K_a at 25°C	pK_a
hydrofluoric acid	$HF + H_2O$	$\rightleftharpoons H_3O^+ + F^-$	7.2×10^{-4}	3.14
nitrous acid	$HNO_2 + H_2O$	$\rightleftharpoons H_3O^+ + NO_2^-$	4.5×10^{-4}	3.35
acetic acid	$CH_3COOH + H_2O$	$\rightleftharpoons H_3O^+ + CH_3COO^-$	1.8×10^{-5}	4.74
hypochlorous acid	$HClO + H_2O$	$\rightleftharpoons H_3O^+ + ClO^-$	3.5×10^{-8}	7.45
hydrocyanic acid	$HCN + H_2O$	$\rightleftharpoons H_3O^+ + CN^-$	4.0×10^{-10}	9.40

Would you think of using sulfuric or nitric acid for any of these purposes?

everyday uses of weak acids suggest that there is a significant difference between strong and weak acids. The difference is that *strong acids ionize completely in dilute aqueous solution, whereas weak acids ionize only slightly.*

Let us consider the reaction that occurs when a weak acid, such as acetic acid, is dissolved in water. The equation for the ionization of acetic acid is

$$CH_3COOH(aq) + H_2O(\ell) \rightleftharpoons H_3O^+(aq) + CH_3COO^-(aq)$$

The equilibrium constant for this reaction could be represented as

$$K_c = \frac{[H_3O^+][CH_3COO^-]}{[CH_3COOH][H_2O]}$$

This expression contains the concentration of water. If we restrict our discussion to *dilute* aqueous solutions, the concentration of water is very high. There are 55.5 moles of water in one liter of pure water. In dilute aqueous solutions, the concentration of water is *essentially* constant; if we assume that it *is* constant, we can rearrange the preceding expression to give

$$K_c[H_2O] = \frac{[H_3O^+][CH_3COO^-]}{[CH_3COOH]}$$

The thermodynamic approach is that the activity of the (nearly) pure H₂O is essentially 1. The activity of each dissolved species is numerically equal to its molar concentration.

Because both K_c and $[H_2O]$ are constant (at a specified temperature) the **ionization constant** of a weak acid is the product $K_c[H_2O]$; i.e., $K_a = K_c[H_2O]$. For acetic acid,

$$K_a = \frac{[H_3O^+][CH_3COO^-]}{[CH_3COOH]} = 1.8 \times 10^{-5}$$

This expression tells us that in dilute aqueous solutions of acetic acid, the concentration of H_3O^+ multiplied by the concentration of CH_3COO^- and then divided by the concentration of *nonionized* acetic acid is equal to 1.8 $\times$ 10^{-5}.

Ionization constants for weak acids (and bases) must be calculated from *experimentally determined data*. Conductivity, depression of freezing point, or measurements of pH provide data from which these constants can be calculated.

Example 18-7

Nicotinic acid, which is also called niacin (and is *not* physiologically related to nicotine), is a weak monoprotic organic acid that we can represent as HA.

$$HA + H_2O \rightleftharpoons H_3O^+ + A^-$$

A dilute solution of nicotinic acid was found to contain the following concentrations at equilibrum at 25°C. What is the value of K_a? [HA] = 0.049 M; [H⁺] = [A⁻] = 8.4 × 10⁻⁴ M.

Plan

We are given *equilibrium* concentrations, and so we substitute these into the expression for K_a.

Solution

$$HA + H_2O \rightleftharpoons H_3O^+ + A^- \qquad K_a = \frac{[H_3O^+][A^-]}{[HA]}$$

$$K_a = \frac{(8.4 \times 10^{-4})(8.4 \times 10^{-4})}{(0.049)} = 1.4 \times 10^{-5}$$

The structure of nicotinic acid is

The complete equilibrium constant expression is

$$K_a = \frac{[H_3O^+][A^-]}{[HA]} = 1.4 \times 10^{-5}$$

Example 18-8

In 0.0100 M solution, acetic acid is 4.2% ionized. Calculate its ionization constant.

Plan

We write the equation for the ionization of acid and its equilibrium constant expression. Next we use the percent ionization to complete the reaction summary and then substitute into the K_a expression.

Solution

The equations for the ionization of CH_3COOH and its ionization constant are

$$CH_3COOH + H_2O \rightleftharpoons H_3O^+ + CH_3COO^- \quad \text{and} \quad K_a = \frac{[H_3O^+][CH_3COO^-]}{[CH_3COOH]}$$

Because 4.2% of the CH_3COOH ionizes,

$$M_{CH_3COOH} \text{ that ionizes} = 0.042 \times 0.0100 \ M = 4.2 \times 10^{-4} \ M$$

Each mole of CH_3COOH that ionizes gives one mole of H_3O^+ and one mole of CH_3COO^-. We represent this in the reaction summary:

	CH_3COOH	+ H_2O $\rightleftharpoons$	H_3O^+	+	CH_3COO^-
initial	0.0100 M		~0 M		0 M
change	-4.2×10^{-4} M		$+4.2 \times 10^{-4}$ M		$+4.2 \times 10^{-4}$ M
at equil	9.58×10^{-3} M		4.2×10^{-4} M		4.2×10^{-4} M

Substitution of these values into the K_a expression gives the value for K_a.

$$K_a = \frac{[H_3O^+][CH_3COO^-]}{[CH_3COOH]} = \frac{(4.2 \times 10^{-4})(4.2 \times 10^{-4})}{9.58 \times 10^{-3}} = 1.8 \times 10^{-5}$$

Example 18-9

The pH of a 0.115 M solution of chloroacetic acid, $CH_2ClCOOH$, is measured to be 1.92. Calculate K_a for this weak monoprotic acid.

Plan

For simplicity, we represent $CH_2ClCOOH$ as HA. We write the ionization equation and the expression for K_a. Next we calculate $[H_3O^+]$ for the given pH and complete the reaction summary. Finally we substitute into the K_a expression.

Solution

The ionization of this weak monoprotic acid and its ionization constant expression may be represented as

$$HA + H_2O \rightleftharpoons H_3O^+ + A^- \quad \text{and} \quad K_a = \frac{[H_3O^+][A^-]}{[HA]}$$

We can calculate $[H_3O^+]$ from the definition of pH:

$$pH = -\log [H_3O^+]$$

$$[H_3O^+] = 10^{-pH} = 10^{-1.92} = 0.012 \ M$$

We can use the usual reaction summary as follows. At this point, we know the *original* [HA] and the *equilibrium* $[H_3O^+]$. From this information, we can fill out the "change" line and then deduce the other equilibrium values:

Some common household weak acids. A strip of paper impregnated with a universal indicator is convenient for estimating the pH of a solution.

We fill in the reaction summary in the order indicated by the numbered red arrows.

1. $[H_3O^+]_{equil} = 0.012\ M$, so we record this value
2. $[H_3O^+]_{initial} = 0$, so change in $[H_3O^+]$ due to rxn must be $+0.012\ M$
3. Formation of 0.012 M H_3O^+ consumes 0.012 M HA, so the change in [HA] = $-0.012\ M$
4. $[HA]_{equil} = [HA]_{orig} + [HA]_{chg}$

   $= 0.115\ M + (-0.012\ M)$

   $= 0.103\ M$
5. Formation of 0.012 M H_3O^+ also gives 0.012 M A^-
6. $[A^-]_{equil} = [A^-]_{orig} + [A^-]_{chg}$

   $= 0\qquad + 0.012\ M$

   $= 0.012\ M$

At equilibrium, $[H_3O^+] = 0.012\ M$ so

	HA	+ H_2O	$\rightleftharpoons$	H_3O^+	+	A^-
initial	0.115 M			~0 M		0 M
change due to rxn	−0.012 M			+0.012 M		0.012 M
at equil	0.103 M			0.012 M		0.012 M

Now that all concentrations are known, K_a can be calculated.

$$K_a = \frac{[H_3O^+][A^-]}{[HA]} = \frac{(0.012)(0.012)}{0.103} = \boxed{1.4 \times 10^{-3}}$$

EOC 38–40

Because ionization constants are equilibrium constants for ionization reactions, their values indicate the extents to which weak electrolytes ionize. Acids with larger ionization constants ionize to greater extents (and are stronger acids) than acids with smaller ionization constants. From Table 18-4 we see that the order of decreasing acid strength for these five weak acids is

$$HF > HNO_2 > CH_3COOH > HClO > HCN$$

Recall that in Brønsted–Lowry terminology, an acid forms its conjugate base by losing H^+.

Conversely, in Brønsted–Lowry terminology (Section 10-3), the order of increasing base strength of the anions of these acids is

$$F^- < NO_2^- < CH_3COO^- < ClO^- < CN^-$$

If we know the value of the ionization constant for a weak acid, we can calculate the concentrations of the species in solutions of known concentrations.

Example 18-10

Calculate the concentrations of the various species in 0.10 M hypochlorous acid, HOCl. For HOCl, $K_a = 3.5 \times 10^{-8}$.

We have written the formula for hypochlorous acid as HOCl rather than HClO to emphasize that its structure is H—O—Cl.

Plan

As usual, we write the equation for the ionization of the weak acid and its K_a expression. (We are given the value of K_a.) Then we represent the *equilibrium* concentrations algebraically and substitute into the K_a expression.

Solution

The equation for the ionization of HOCl and its K_a expression are

$$HOCl + H_2O \rightleftharpoons H_3O^+ + OCl^- \quad \text{and} \quad K_a = \frac{[H_3O^+][OCl^-]}{[HOCl]} = 3.5 \times 10^{-8}$$

We would like to know the concentrations of H_3O^+, OCl^-, and nonionized HOCl in solution. An algebraic representation of concentrations is required, because there is no other obvious way to obtain the concentrations.

Let x = mol/L of HOCl that ionizes, Then we can write the "change" line and complete the reaction summary.

We neglect the 1.0×10^{-7} mol/L of H_3O^+ produced by the ionization of *pure* water. Recall (Section 18-2) that the addition of an acid to water suppresses the ionization of H_2O, so $[H_3O^+]$ from H_2O is even less than $1.0 \times 10^{-7}\ M$.

	HOCl	+ H_2O	$\rightleftharpoons$	H_3O^+	+	OCl^-
initial	0.10 M			~0 M		0 M
change due to rxn	− x M			+x M		+x M
at equil	(0.10 − x) M			x M		x M

Substituting these algebraic representations into the K_a expression gives

$$K_a = \frac{[H_3O^+][OCl^-]}{[HOCl]} = \frac{(x)(x)}{(0.10 - x)} = 3.5 \times 10^{-8}$$

This is a quadratic equation, but it is not necessary to solve it by the quadratic formula. If we assume that $(0.10 - x)$ is very nearly equal to 0.10 (see the box, "Simplifying Quadratic Equations") the equation becomes

$$\frac{x^2}{0.10} \approx 3.5 \times 10^{-8} \qquad x^2 \approx 3.5 \times 10^{-9} \qquad \text{so} \qquad x \approx 5.9 \times 10^{-5}$$

In our algebraic representation, we let

$$[H_3O^+] = x \ M = \boxed{5.9 \times 10^{-5} \ M} \qquad [OCl^-] = x \ M = \boxed{5.9 \times 10^{-5} \ M}$$

$$[HOCl] = (0.10 - x) \ M = (0.10 - 0.000059) \ M = \boxed{0.10 \ M}$$

EOC 43, 44

Simplifying Quadratic Equations

We often encounter quadratic or higher-order equations in equilibrium calculations. With modern programmable calculators, solving such problems by iterative methods is often feasible. But frequently the problem can be made much simpler by using some mathematical common sense.

When the linear variable (x) in a *quadratic* equation is added to or subtracted from a much larger number, it can often be disregarded if it is sufficiently small. A reasonable rule of thumb for determining whether the variable can be disregarded in equilibrium calculations is this: If the exponent of 10 in the K value is -4 or less (-5, -6, -7, and so on), then the variable may be small enough to disregard when it is added to or subtracted from a number greater than 0.05. Solve the problem neglecting x; then compare the value of x with the number it would have been added to (or subtracted from). If x is more than 5% of that number, the assumption was *not* justified; solve the equation using the quadratic formula.

Let's examine the assumption as it applies to Example 18-10. Our quadratic equation is

$$\frac{(x)(x)}{(0.10 - x)} = 3.5 \times 10^{-8}$$

Because x is obviously a very small number, it is very small compared to 0.10, and so we can write $(0.10 - x) \approx 0.10$. The equation then becomes $\frac{x^2}{0.10} \approx 3.5 \times 10^{-8}$. To solve this, we rearrange and take the square roots of both sides. To check, we see that the result, $x = 5.9 \times 10^{-5}$, is only 0.059% of 0.10. This error is much less than 5%, so our assumption is justified. You may also wish to use the quadratic formula to verify that the answer obtained this way is correct to within roundoff error.

The preceding argument is purely algebraic. We could use our chemical intuition to reach the same conclusion. A small K_a value (10^{-4} or less) tells us that the extent of ionization is very small. Therefore, nearly all of the weak acid exists as nonionized molecules. The amount that ionizes is not significant compared to the concentration of nonionized weak acid.

From our calculations we can draw some conclusions. In a solution containing *only a weak monoprotic acid*, the concentration of H_3O^+ is equal to the concentration of the anion of the acid. Unless the solution is *very* dilute, say less than 0.050 M, the concentration of nonionized acid is approximately equal to the molarity of the solution. When the value of K_a for the weak acid is greater than $\sim 10^{-4}$, then the extent of ionization will be large enough to make a significant difference between the concentration of nonionized acid and the molarity of the solution. In such cases we *cannot* make the simplifying assumption.

Example 18-11
Calculate the percent ionization of a 0.10 M solution of acetic acid.

Plan
We write the ionization equation and the expression for K_a. Next we follow the procedure used in Example 18-10 to find the concentration of acid that ionized. Then we substitute the concentration of acid that ionized into the expression for percent ionization.

Solution
The equations for the ionization of CH_3COOH and its K_a are

$$CH_3COOH + H_2O \rightleftharpoons H_3O^+ + CH_3COO^- \qquad K_a = \frac{[H_3O^+][CH_3COO^-]}{[CH_3COOH]} = 1.8 \times 10^{-5}$$

Percentage is defined as (part/whole) $\times$ 100%, so the percent ionization is

$$\% \text{ ionization} = \frac{[CH_3COOH]_{ionized}}{[CH_3COOH]_{total}} \times 100\%$$

We proceed as we did in Example 18-10. Let $x = [CH_3COOH]_{ionized}$.

We could write the original $[H_3O^+]$ as 1.0×10^{-7} M. In very *dilute* solutions of weak acids, we might have to take this into account. In this acid solution, $(1.0 \times 10^{-7} + x) \approx x$.

	CH_3COOH	$+ \ H_2O \rightleftharpoons$	H_3O^+	$+ \ CH_3COO^-$
initial	0.10 M		~ 0 M	0 M
change due to rxn	$-x$ M		$+x$ M	$+x$ M
at equil	$(0.10 - x)$ M		x M	x M

Substituting into the ionization constant expression gives

$$K_a = \frac{[H_3O^+][CH_3COO^-]}{[CH_3COOH]} = \frac{(x)(x)}{(0.10 - x)} = 1.8 \times 10^{-5}$$

If we make the simplifying assumption that $(0.10 - x) \approx 0.10$, we have

$$\frac{x^2}{0.10} = 1.8 \times 10^{-5} \qquad x^2 = 1.8 \times 10^{-6} \qquad x = 1.3 \times 10^{-3}$$

This gives $[CH_3COOH]_{ionized} = x = 1.3 \times 10^{-3}$ M. Now we can calculate the percent ionization for 0.10 M CH_3COOH solution.

Note that we need not solve explicitly for the equilibrium concentrations $[H_3O^+]$ and $[CH_3COO^-]$ to answer the question. From the setup, we see that these are both 1.3×10^{-3} M. The pH of the solution is 2.89.

$$\% \text{ ionization} = \frac{[CH_3COOH]_{ionized}}{[CH_3COOH]_{total}} \times 100\% = \frac{1.3 \times 10^{-3} \ M}{0.10 \ M} \times 100\% = \boxed{1.3\%}$$

Our assumption that $(0.10 - x)$ is approximately 0.10 is reasonable because $(0.10 - x) = (0.10 - 0.0013)$. This is only about 1% different than 0.10. However, when K_a for a weak acid is significantly greater than 10^{-4}, this assumption would introduce considerable error.

EOC 47, 48

Table 18-5
Comparison of Extents of Ionization of Some Acids

Acid Solution	Ionization Constant	$[H_3O^+]$	pH	Percent Ionization
0.10 M HCl	very large	0.10 M	1.00	~100
0.10 M CH₃COOH	1.8×10^{-5}	0.0013 M	2.89	1.3
0.10 M HClO	3.5×10^{-8}	0.000059 M	4.23	0.059

An inert solid has been suspended in the liquid to improve the quality of this and following photographs of pH meters.

In dilute solutions, acetic acid exists primarily as nonionized molecules, as do all weak acids, and there are relatively few hydronium and acetate ions. In 0.10 M solution, CH_3COOH is 1.3% ionized; for each 1000 molecules of CH_3COOH originally placed in the solution, there are 13 H_3O^+ ions, 13 CH_3COO^- ions, and 987 nonionized CH_3COOH molecules. For weaker acids, the number of molecules of nonionized acid would be even larger.

By now we should have gained some "feel" for the strength of an acid by looking at its K_a value. Consider 0.10 M solutions of HCl (a strong acid), CH_3COOH (Example 18-11), and HOCl (Example 18-10). If we calculate the percent ionization for 0.10 M HOCl (as we did for 0.10 M CH_3COOH in Example 18-11), we find that it is 0.059% ionized. In 0.10 M solution, HCl is very nearly completely ionized. The data in Table 18-5 show that the $[H_3O^+]$ in 0.10 M HCl is approximately 77 times greater than that in 0.10 M CH_3COOH and approximately 1700 times greater than that in 0.10 M HOCl.

Many scientists prefer to use pK_a values rather than K_a values for weak acids. Recall that in general, "p" terms refer to negative logarithms. The pK_a value for a weak acid is just the negative logarithm of its K_a value.

Example 18-12

The K_a values for acetic acid and hydrofluoric acid are 1.8×10^{-5} and 7.2×10^{-4}, respectively. What are their pK_a values?

Plan

pK_a is defined as the negative logarithm of K_a—i.e., $pK_a = -\log K_a$—so we take the negative logarithm of each K_a.

Solution

For CH_3COOH,

$$pK_a = -\log K_a = -\log (1.8 \times 10^{-5}) = -(-4.74) = \boxed{4.74}$$

For HF,

$$pK_a = -\log K_a = -\log (7.2 \times 10^{-4}) = -(-3.14) = \boxed{3.14}$$

From Example 18-12, we see that the stronger acid has the larger K_a value (HF in this case) and the smaller pK_a value. Conversely, the weaker acid has the smaller K_a value (CH_3COOH in this case) and the larger pK_a value. The generalization is

A similar statement is true for weak bases; i.e., a stronger base has the greater K_b value and the smaller pK_b value.

The larger the value of K_a, the smaller the value of pK_a, and the stronger the acid.

Example 18-13

Given the following list of weak acids and their K_a values, arrange the acids in order of (a) increasing acid strength and (b) increasing pK_a values.

Acid	K_a
HOCl	3.5×10^{-8}
HCN	4.0×10^{-10}
HNO$_2$	4.5×10^{-4}

Plan

(a) We see that HNO$_2$ is the strongest acid in this group because it has the largest K_a value. HCN is the weakest because it has the smallest K_a value.

(b) We do not need to calculate pK_a values to answer the question. We recall that the weakest acid has the largest pK_a value and the strongest acid has the smallest pK_a value, so the order of increasing pK_a values is just the reverse of the order in (a).

Solution

(a)   Increasing acid strength: HCN < HOCl < HNO$_2$

(b)   Increasing pK_a values: HNO$_2$ < HOCl < HCN

EOC 61, 62

Thus far we have focused our attention on acids. Very few common weak bases are soluble in water. Aqueous ammonia is the most frequently encountered example. From our earlier discussion of bonding in covalent compounds (Section 8-8), we recall that there is one unshared pair of electrons on the nitrogen atom in NH$_3$. When ammonia dissolves in water, it accepts H$^+$ from a water molecule in a reversible reaction (Section 10-3). We say that NH$_3$ ionizes slightly when it undergoes this reaction. Aqueous solutions of NH$_3$ are basic because OH$^-$ ions are produced.

$$:NH_3 + H_2O \rightleftharpoons NH_4^+ + OH^-$$

Amines are derivatives of NH$_3$ in which one or more H atoms have been replaced by organic groups, as the following structures indicate.

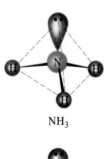

NH$_3$

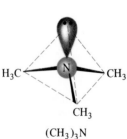

(CH$_3$)$_3$N

H $\mid$ H—N: $\mid$ H	H $\mid$ H$_3$C—N: $\mid$ H	H$_3$C $\mid$ H$_3$C—N: $\mid$ H	H$_3$C $\mid$ H$_3$C—N: $\mid$ H$_3$C
ammonia NH$_3$	methylamine CH$_3$NH$_2$	dimethylamine (CH$_3$)$_2$NH	trimethylamine (CH$_3$)$_3$N

Thousands of amines are known, and many are very important in biochemical processes. Low-molecular-weight amines are soluble weak bases. The ionization of trimethylamine, for example, forms trimethylammonium ions and OH$^-$ ions.

trimethylamine
$(CH_3)_3N$

trimethylammonium ion
$(CH_3)_3NH^+$

The structures of the ammonium and trimethylammonium ions are similar.

Now let us consider the behavior of ammonia in aqueous solutions. The reaction of ammonia with water and its ionization constant expression are

$$NH_3 + H_2O \rightleftharpoons NH_4^+ + OH^-$$

and

$$K_b = \frac{[NH_4^-][OH^-]}{[NH_3]} = 1.8 \times 10^{-5}$$

The subscript "b" indicates that the substance ionizes as a base.

The concentration of water is included in the ionization constant as it was for weak acids. The fact that K_b for aqueous NH_3 has the same value as K_a for CH_3COOH is pure coincidence. It does tell us that in aqueous solutions of the same concentration, CH_3COOH and NH_3 are ionized to the same extent. Table 18-6 lists K_b and pK_b values for a few common weak bases. Appendix G includes a longer list of K_b values.

We use K_b's for weak bases in the same way we used K_a's for weak acids. We can also use pK_b values for weak bases in the same way we used pK_a values for weak acids.

Example 18-14
Calculate the $[OH^-]$, pH, and percent ionization for 0.20 M aqueous NH_3.

Plan
We write the equation for the ionization of aqueous NH_3 and represent the equilibrium concentrations algebraically. Then we substitute into the K_b expression and solve for $[OH^-]$ and $[NH_3]_{ionized}$.

Solution
The equation for the ionization of aqueous ammonia and the algebraic representations of equilibrium concentrations follow. Let $x = [NH_3]_{ionized}$.

	NH_3	+	H_2O	$\rightleftharpoons$	NH_4^+	+	OH^-
initial	0.20 M				0 M		~0 M
change due to rxn	$-x$ M				$+x$ M		$+x$ M
at equil	$(0.20 - x)$ M				x M		x M

Table 18-6
Ionization Constants and pK_b Values for Some Weak Bases

Base	Ionization Reaction	K_b at 25°C	pK_b
ammonia	$NH_3 + H_2O \rightleftharpoons NH_4^+ + OH^-$	1.8×10^{-5}	4.74
methylamine	$(CH_3)NH_2 + H_2O \rightleftharpoons (CH_3)NH_3^+ + OH^-$	5.0×10^{-4}	3.30
dimethylamine	$(CH_3)_2NH + H_2O \rightleftharpoons (CH_3)_2NH_2^+ + OH^-$	7.4×10^{-4}	3.13
trimethylamine	$(CH_3)_3N + H_2O \rightleftharpoons (CH_3)_3NH^+ + OH^-$	7.4×10^{-5}	4.13
pyridine	$C_5H_5N + H_2O \rightleftharpoons C_5H_5NH^+ + OH^-$	1.5×10^{-9}	8.82

Substitution into the ionization constant expression gives

$$K_b = \frac{[NH_4^+][OH^-]}{[NH_3]} = 1.8 \times 10^{-5} = \frac{(x)(x)}{(0.20 - x)}$$

If we assume that $(0.20 - x) \approx 0.20$, we have

$$\frac{x^2}{0.20} = 1.8 \times 10^{-5} \qquad x^2 = 3.6 \times 10^{-6} \qquad x = 1.9 \times 10^{-3}\ M$$

Then $[OH^-] = x = \boxed{1.9 \times 10^{-3}\ M,}$ $pOH = 2.72$, and $pH = \boxed{11.28.}$

$[NH_3]_{ionized} = x$, so the percent ionization may be calculated.

$$\% \text{ ionization} = \frac{[NH_3]_{ionized}}{[NH_3]_{total}} \times 100\% = \frac{1.9 \times 10^{-3}}{0.20} \times 100\% = \boxed{0.95\% \text{ ionized}}$$

Example 18-15

The pH of a household ammonia solution is 11.50. What is its molarity?

Plan

We are given the pH of an aqueous NH_3 solution. We use $pH + pOH = 14.00$ to find pOH, which we can easily convert to $[OH^-]$. Then we complete the reaction summary and substitute the representations of equilibrum concentrations into the K_b expression.

Solution

Because $pH = 11.50$, we know that $pOH = 2.50$, so $[OH^-] = 10^{-2.50} = 3.2 \times 10^{-3}\ M$. This $[OH^-]$ results from the reaction, so we can write the change line. Then, letting x represent the *original* concentration of NH_3, we can complete the reaction summary:

At equilibrium $[OH^-] = 3.2 \times 10^{-3}\ M$, so ⟶

	NH_3	$+\ H_2O$ ⇌	NH_4^+	$+$	OH^- ①
initial	$x\ M$	④	$0\ M$ ③		$\sim 0\ M$
change	$-3.2 \times 10^{-3}\ M$ ◄-------		$+3.2 \times 10^{-3}\ M$ ◄---		$+3.2 \times 10^{-3}\ M$
at equil	$(x - 3.2 \times 10^{-3})M$ ⑥		⑤ $3.2 \times 10^{-3}\ M$	②	$3.2 \times 10^{-3}\ M$

Substituting these values into the K_b expression for aqueous NH_3 gives

$$K_b = \frac{[NH_4^+][OH^-]}{[NH_3]} = \frac{(3.2 \times 10^{-3})(3.2 \times 10^{-3})}{(x - 3.2 \times 10^{-3})} = 1.8 \times 10^{-5}$$

This suggests that $(x - 3.2 \times 10^{-3}) \approx x$. So we can approximate:

$$\frac{(3.2 \times 10^{-3})(3.2 \times 10^{-3})}{x} = 1.8 \times 10^{-5} \qquad \text{and} \qquad x = \boxed{0.57\ M\ NH_3}$$

The solution is $0.57\ M\ NH_3$. Our assumption that $(x - 3.2 \times 10^{-3}) \approx x$ was justified.

EOC 50–54

18-5 Acid–Base Indicators

In Section 3-9 we described acid–base titrations and the use of indicators to tell us when to stop a titration. Detection of the end point in an acid–base titration is only one of the important uses of indicators.

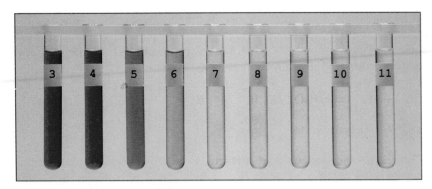

(a)

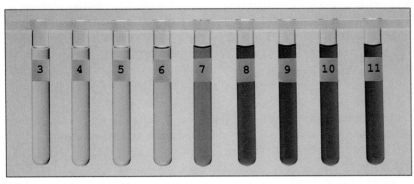

(b)

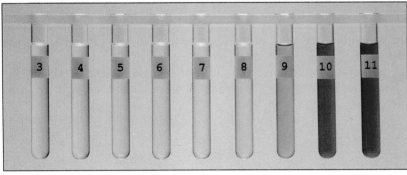

(c)

Figure 18-3
Three common indicators in solutions that cover the pH range 3 to 11 (the black numbers). (a) Methyl red is red at pH 4 and below; it is yellow at pH 7 and above. Between pH 4 and pH 7 it changes from red to red-orange, to orange, to yellow. (b) Bromthymol blue is yellow at pH 6 and below; it is blue at pH 8 and above. Between pH 6 and 8 it changes from yellow to yellow-green, to green, to blue-green, to blue. (c) Phenolphthalein is colorless below pH 8 and red above pH 10. It changes from colorless to pale pink, to pink, to red.

An indicator is an organic dye; its color depends on the concentration of H_3O^+ ions, or pH, in the solution. By the color an indicator displays, it "indicates" the acidity or basicity of a solution. Figure 18-3 displays solutions that contain three common indicators in solutions over the pH range 3 to 11. Study Figure 18-3 and its legend carefully.

The first indicators used were vegetable dyes. Litmus is a familiar example. Most of the indicators that we use in the laboratory today are synthetic compounds; i.e., they have been made in laboratories by chemists. Phenolphthalein is the most common acid–base indicator. It is colorless in solutions of pH less than 8 ($[H_3O^+] > 10^{-8}$ M) and turns red as pH approaches 10.

Phenolphthalein is also the active component of the laxative Ex-Lax. It is sometimes added to laboratory ethyl alcohol to discourage consumption.

Bromthymol blue indicator is yellow in acidic solutions and blue in basic solutions.

Recall that for bromthymol blue, $K_a = 7.9 \times 10^{-8}$.

Many acid–base indicators are weak organic acids, HIn, where "In" represents complex organic groups. Bromthymol blue is such an indicator. Its ionization constant is 7.9×10^{-8}. We can represent its ionization in dilute aqueous solution and its ionization constant expression as

$$HIn + H_2O \rightleftharpoons H_3O^+ + In^- \qquad K_a = \frac{[H_3O^+][In^-]}{[HIn]} = 7.9 \times 10^{-8}$$

color 1
yellow   ← for bromthymol blue →   color 2
blue

HIn represents nonionized acid molecules, and In^- represents the anion (conjugate base) of the weak acid. The essential characteristic of an acid–base indicator is that HIn and In^- *must* have quite different colors. The relative amounts of the two species determine the color of the solution. Adding an acid favors the reaction to the left and gives more HIn molecules (color 1). Adding a base favors the reaction to the right and gives more In^- ions (color 2). The ionization constant expression can be rearranged:

$$\frac{[H_3O^+][In^-]}{[HIn]} = K_a \qquad so \qquad \frac{[In^-]}{[HIn]} = \frac{K_a}{[H_3O^+]}$$

This shows clearly how the $[In^-]/[HIn]$ ratio depends on $[H_3O^+]$ (or on pH) and the K_a value for the indicator. As a rule-of-thumb, when $[In^-]/[HIn] \geq 10$, color 2 is observed; conversely, when $[In^-]/[HIn] \leq \frac{1}{10}$, color 1 is observed.

Universal indicators are mixtures of several indicators that display a continuous range of colors over a wide range of pH values. Figure 18-2 shows concentrated solutions of a universal indicator in flat dishes so that the colors are very intense. The juice of red (purple) cabbage is a universal indicator. Figure 18-4 shows the color of red cabbage juice in solutions within the pH range 1 to 13.

One important use of universal indicators is in commercial indicator papers, which are small strips of paper impregnated with solutions of universal

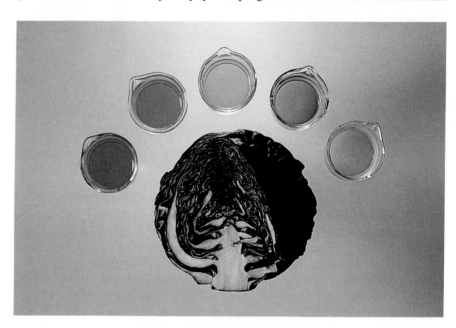

Figure 18-4
The juice of the red (purple) cabbage is a naturally occurring universal indicator. From left to right are solutions of pH 1, 4, 7, 10, and 13.

indicators. A strip of the paper is dipped into the solution of interest, and the color of the indicator on the paper indicates the pH of the solution. The photographs on page 703 (solutions of universal indicators) and page 709 (an indicator paper) illustrate the use of universal indicators to estimate pH. We shall describe the use of indicators in titrations more fully in Sections 19-6 and 19-7.

18-6 The Common Ion Effect and Buffer Solutions

In laboratory reactions, in industrial processes, and in the bodies of plants and animals, it is often necessary to keep the pH nearly constant despite the addition of acids or bases. The oxygen-carrying capacity of the hemo-globin in your blood and the activity of the enzymes in your cells depend very strongly on the pH of your body fluids. Our bodies use a combination of compounds known as a *buffer system* to keep the pH within a narrow range. The operation of a buffer solution depends on the *common ion effect*, a special case of LeChatelier's Principle.

Buffer systems resist changes in pH.

The Common Ion Effect

The term **common ion effect** is used to describe the behavior of a solution in which the same ion is produced by two different compounds. Many types of solutions exhibit this effect. Two of the most frequently encountered kinds are

1. A solution of a weak acid *plus* a soluble ionic salt of the weak acid
2. A solution of a weak base *plus* a soluble ionic salt of the weak base

Weak Acids Plus Salts of Weak Acids

Consider a solution that contains acetic acid *and* sodium acetate, a soluble ionic salt of CH_3COOH. The $NaCH_3COO$ is completely dissociated into its constituent ions, but CH_3COOH is only slightly ionized:

$$NaCH_3COO \xrightarrow{H_2O} Na^+ + \boxed{CH_3COO^-} \quad \text{(to completion)}$$
$$CH_3COOH + H_2O \rightleftharpoons \boxed{H_3O^+} + \boxed{CH_3COO^-} \quad \text{(reversible)}$$

Both CH_3COOH and $NaCH_3COO$ are sources of CH_3COO^- ions. The completely dissociated $NaCH_3COO$ provides a high $[CH_3COO^-]$. This shifts the ionization equilibrium of CH_3COOH far to the left as CH_3COO^- combines with H_3O^+ to form nonionized CH_3COOH and H_2O. The result is a drastic decrease in $[H_3O^+]$ in the solution.

LeChatelier's Principle (Section 17-6) is applicable to equilibria in aqueous solution.

> Solutions that contain a weak acid plus a salt of the weak acid are always less acidic than solutions that contain the same concentration of the weak acid alone.

Example 18-16
Calculate the concentration of H_3O^+ and the pH of a solution that is 0.10 M in CH_3COOH and 0.20 M in $NaCH_3COO$.

The two solutions of Table 18-7, in the presence of universal indicator. The CH_3COOH solution is on the left.

Plan

We write the appropriate equations for *both* $NaCH_3COO$ *and* CH_3COOH and the ionization constant expression for CH_3COOH. Then we represent the *equilibrium* concentrations algebraically and substitute them into the K_a expression.

Solution

The appropriate equations and ionization constant expression are

$$NaCH_3COO \longrightarrow Na^+ + CH_3COO^- \quad \text{(to completion)}$$

$$CH_3COOH + H_2O \rightleftharpoons H_3O^+ + CH_3COO^- \quad \text{(reversible)}$$

$$K_a = \frac{[H_3O^+][CH_3COO^-]}{[CH_3COOH]} = 1.8 \times 10^{-5}$$

This K_a expression is valid for *all solutions* that contain CH_3COOH. In solutions that contain both CH_3COOH and $NaCH_3COO$, CH_3COO^- ions come from two sources. The ionization constant is satisfied by the *total* CH_3COO^- concentration.

Because $NaCH_3COO$ is completely dissociated, the $[CH_3COO^-]$ *from* $NaCH_3COO$ will be 0.20 mol/L. Let $x = [CH_3COOH]$ that ionizes; then x is also equal to $[H_3O^+]$ *and* equal to $[CH_3COO^-]$ *from* CH_3COOH. The *total* concentration of CH_3COO^- is $(0.20 + x)\ M$. The concentration of nonionized CH_3COOH is $(0.10 - x)\ M$:

$$
\begin{array}{lcll}
NaCH_3COO & \longrightarrow & Na^+ + & \boxed{CH_3COO^-} \\
0.20\ M & \Longrightarrow & 0.20\ M & 0.20\ M \\
CH_3COOH + H_2O & \rightleftharpoons & H_3O^+ + & \boxed{CH_3COO^-} \\
(0.10 - x)\ M & & x\ M & x\ M
\end{array}
$$

$\longrightarrow$ Total $[CH_3COO^-] = (0.20 + x)\ M$

Substitution into the ionization constant expression for CH_3COOH gives

$$K_a = \frac{[H_3O^+][CH_3COO^-]}{[CH_3COOH]} = \frac{(x)(0.20 + x)}{(0.10 - x)} = 1.8 \times 10^{-5}$$

This equation suggests that x is very small. We make two assumptions:

$$(0.20 + x) \approx 0.20 \quad \text{and} \quad (0.10 - x) \approx 0.10$$

Introducing these assumptions gives

$$\frac{0.20\ x}{0.10} = 1.8 \times 10^{-5} \quad \text{and} \quad x = 9.0 \times 10^{-6}$$

You can verify the validity of the assumption by substituting the value for x, 9.0×10^{-6}, into the original equation.

$$x\ M = \boxed{[H_3O^+] = 9.0 \times 10^{-6}\ M} \quad \text{so} \quad \boxed{pH = 5.05}$$

EOC 74

To see how much the acidity of the $0.10\ M$ CH_3COOH solution is reduced by making it $0.20\ M$ in $NaCH_3COO$ also, refer back to Example 18-11. We found that in $0.10\ M$ CH_3COOH the H_3O^+ concentration is 1.3×10^{-3} mol/L (pH = 2.89).

Let us calculate the percent ionization in the solution of Example 18-16:

$$\% \text{ ionization} = \frac{[CH_3COOH]_{\text{ionized}}}{[CH_3COOH]_{\text{original}}} \times 100\%$$

$$= \frac{9.0 \times 10^{-6}\ M}{0.10\ M} \times 100\% = 0.009\% \text{ ionized}$$

Table 18-7
Comparison of [H₃O⁺] and pH in Acetic Acid and Sodium Acetate–Acetic Acid Solutions

Solution	% CH_3COOH Ionized	$[H_3O^+]$	pH	
0.10 M CH_3COOH	1.3%	$1.3 \times 10^{-3} M$	2.89	
0.10 M CH_3COOH and 0.20 M $NaCH_3COO$	0.009%	$9.0 \times 10^{-6} M$	5.05	$\Delta pH = 2.16$

This compares with 1.3% ionization in 0.10 M CH_3COOH (Example 18-11). Table 18-7 compares these solutions. The third column shows that $[H_3O^+]$ is about 140 times greater in 0.10 M CH_3COOH than in the solution to which 0.20 mol/L $NaCH_3COOH$ has been added (LeChatelier's Principle).

The calculation of $[H_3O^+]$ in solutions containing both a weak acid and a salt of the weak acid can be simplified greatly. Let us write the equation for the ionization of a *weak monoprotic acid* and its K_a in the following way.

$$HA + H_2O \rightleftharpoons H_3O^+ + A^- \quad \text{and} \quad \frac{[H_3O^+][A^-]}{[HA]} = K_a$$

HA and A⁻ represent the weak acid and its conjugate base, respectively.

Solving this expression for $[H_3O^+]$ gives

$$[H_3O^+] = \frac{[HA]}{[A^-]} \times K_a$$

We now impose two conditions: (1) The concentrations of both the weak acid and its salt are some reasonable values, say greater than 0.050 M, and (2) the salt, MA, contains a univalent cation. Under these conditions the concentration of the anion, $[A^-]$, in the solution can be assumed to be the same as the concentration of the salt. With these restrictions, the preceding expression for $[H_3O^+]$ becomes

These are the kinds of assumptions we made in Example 18-16.

$$[H_3O^+] = \frac{[\text{acid}]}{[\text{salt}]} \times K_a$$

[acid] is the concentration of nonionized weak acid (in most cases, this is the total acid concentration), and [salt] is the concentration of its salt.

If we take the logarithm of the preceding equation, we obtain

$$\log [H_3O^+] = \log \frac{[\text{acid}]}{[\text{salt}]} + \log K_a$$

Multiplying by -1 gives

$$-\log [H_3O^+] = -\log \frac{[\text{acid}]}{[\text{salt}]} - \log K_a$$

and rearrangement gives

The relationship is valid *only* for solutions that contain a weak *monoprotic* acid and a soluble, ionic salt of the weak acid with a *univalent* cation, both in reasonable concentrations.

$$pH = pK_a + \log \frac{[\text{salt}]}{[\text{acid}]} \quad \text{where} \quad pK_a = -\log K_a \quad \text{(acid/salt buffer)}$$

This equation is known as the **Henderson–Hasselbalch equation**. Workers in the biological sciences use it frequently. In general terms, we can also write

$$pH = pK_a + \log \frac{[\text{conjugate base}]}{[\text{acid}]}$$

Weak Bases Plus Salts of Weak Bases

Let us consider the second common kind of buffer solution, containing a weak base and its salt. A solution that contains aqueous NH_3 and ammonium chloride, NH_4Cl, a soluble ionic salt of NH_3, is typical. The NH_4Cl is completely dissociated, but aqueous NH_3 is only slightly ionized.

$$NH_4Cl \xrightarrow{H_2O} \boxed{NH_4^+} + Cl^- \qquad \text{(to completion)}$$
$$NH_3 + H_2O \rightleftharpoons \boxed{NH_4^+} + \boxed{OH^-} \qquad \text{(reversible)}$$

Both NH_4Cl and aqueous NH_3 produce NH_4^+ ions. The completely dissociated NH_4Cl provides a high $[NH_4^+]$. This shifts the ionization equilibrium of aqueous NH_3 far to the left, as NH_4^+ ions combine with OH^- ions to form nonionized NH_3 and H_2O. Thus, $[OH^-]$ is decreased significantly.

Solutions that contain a weak base plus a salt of the weak base are less basic than solutions that contain the same concentration of weak base alone.

Example 18-17

Calculate the concentration of OH^- and the pH of a solution that is 0.20 M in aqueous NH_3 *and* 0.10 M in NH_4Cl.

Plan

We write the appropriate equations for *both* NH_4Cl and NH_3 and the ionization constant expression for NH_3. Then we represent the *equilibrium* concentrations algebraically and substitute into the K_b expression.

Solution

The appropriate equations and algebraic representations of concentrations are

$$NH_4Cl \longrightarrow \boxed{NH_4^+} + Cl^-$$
$$0.10\ M \qquad\qquad 0.10\ M \quad\ 0.10\ M$$
$$NH_3 + H_2O \rightleftharpoons \boxed{NH_4^+} + OH^-$$
$$(0.20 - x)\ M \qquad\quad x\ M \qquad\ x\ M \longrightarrow \text{Total } [NH_4^+] = (0.10 + x)\ M$$

Substitution into the K_b expression for aqueous NH_3 gives

$$K_b = \frac{[NH_4^+][OH^-]}{[NH_3]} = 1.8 \times 10^{-5} = \frac{(0.10 + x)(x)}{(0.20 - x)}$$

Because K_b is small, we can assume that $(0.10 + x) \approx 0.10$ and $(0.20 - x) \approx 0.20$.

$$\frac{0.10x}{0.20} = 1.8 \times 10^{-5}\ M \qquad \text{and} \qquad x = 3.6 \times 10^{-5}\ M$$

$$x\,M = \boxed{[OH^-] = 3.6 \times 10^{-5}\,M} \qquad \text{so} \qquad pOH = 4.44 \quad \text{and} \quad \boxed{pH = 9.56}$$

EOC 75, 76

In Example 18-14 we calculated $[OH^-]$ and pH for 0.20 M aqueous NH_3. Let us compare that result with the results obtained here (Table 18-8). The concentration of OH^- is 53 times greater in the solution containing only 0.20 M aqueous NH_3 than in the solution to which 0.10 mol/L NH_4Cl has been added. This is another demonstration of LeChatelier's Principle.

We can derive a relationship for $[OH^-]$ in solutions containing weak bases *plus* salts of the weak bases, just as we did for weak acids. In general terms, the equation for the ionization of a monoprotic weak base and its K_b expression are

$$\text{Base} + H_2O \rightleftharpoons (\text{base})H^+ + OH^- \qquad \text{and} \qquad \frac{[(\text{base})H^+][OH^-]}{[\text{base}]} = K_b$$

base and (*base*)H^+ represent the weak base and its conjugate acid, respectively—e.g., NH_3 and NH_4^+.

Solving the K_b expression for $[OH^-]$ gives

$$[OH^-] = \frac{[\text{base}]}{[(\text{base})H^+]} \times K_b$$

Taking the logarithm of both sides of the equation gives

$$\log[OH^-] = \log\frac{[\text{base}]}{[(\text{base})H^+]} + \log K_b$$

For salts of weak bases that contain univalent anions, $[(\text{base})H^+] = [\text{salt}]$. Multiplication by -1 and rearrangement gives the *Henderson–Hasselbalch equation* for solutions containing a weak base plus a salt of the weak base.

$$pOH = pK_b + \log\frac{[\text{salt}]}{[\text{base}]} \qquad \text{where} \qquad pK_b = -\log K_b \quad \text{(base/salt buffer)}$$

The Henderson–Hasselbalch equation is valid for solutions of weak bases plus salts of weak bases with univalent anions in reasonable concentrations. In general terms, we can also write this equation as

$$pOH = pK_b + \log\frac{[\text{conjugate acid}]}{[\text{base}]}$$

For salts such as $(NH_4)_2SO_4$ that contain divalent anions, $[(\text{base})H^+] = 2[\text{salt}]$.

18-7 Buffering Action

The two common kinds of buffer solutions are the ones we have just discussed—namely, solutions containing (1) a weak acid plus a soluble ionic salt of the weak acid and (2) a weak base plus a soluble ionic salt of the weak base.

Table 18-8
Comparison of [OH⁻] and pH in Ammonia and Ammonium Chloride–Ammonia Solutions

Solution	%NH₃ Ionized	[OH⁻]		pH	
0.20 M aq NH₃	0.95%	$1.9 \times 10^{-3}\ M$	11.28		
0.20 M aq NH₃ and 0.10 M aq NH₄Cl	0.0018%	$3.6 \times 10^{-5}\ M$	9.56	ΔpH $= -1.72$	

The two solutions in Table 18-8, in the presence of universal indicator. The NH₃ solution is on the left. Can you calculate the percentage of NH₃ that is ionized in these two solutions?

A buffer solution contains a conjugate acid–base pair with both the acid and base in reasonable concentrations. The more acidic component reacts with added strong bases. The more basic component reacts with added acids.

> A buffer solution is able to react with either H_3O^+ and OH^- ions, whichever is added.

Thus, a buffer solution resists changes in pH. When we add a modest amount of a strong base or a strong acid to a buffer solution, the pH changes very little.

Solutions of a Weak Acid and a Salt of the Weak Acid

A solution containing acetic acid, CH_3COOH, and sodium acetate, $NaCH_3COO$, is an example of this kind of buffer solution. The more acidic component is CH_3COOH. The more basic component is $NaCH_3COO$ because the CH_3COO^- ion is the conjugate base of CH_3COOH. The operation of this buffer depends on the equilibrium

$$\textit{high conc}\ CH_3COOH + H_2O \rightleftharpoons H_3O^+ + CH_3COO^-\ \leftarrow\textit{high conc (from salt)}$$

If we add a strong acid such as HCl to this solution, it produces H_3O^+. As a result of the added H_3O^+, the reaction occurs to the *left*, to use up most of the added H_3O^+ and reestablish equilibrium. Because the $[CH_3COO^-]$ in the buffer solution is high, this can occur to a great extent. The net reaction is

$$H_3O^+ + CH_3COO^- \longrightarrow CH_3COOH + H_2O \qquad (\sim100\%)$$

or, as a formula unit equation,

$$\underset{\text{added acid}}{HCl} + \underset{\text{base}}{NaCH_3COO} \longrightarrow \underset{\text{weak acid}}{CH_3COOH} + \underset{\text{salt}}{NaCl} \qquad (\sim100\%)$$

This reaction goes nearly to completion because CH_3COOH is a *weak* acid; even when mixed from separate sources, its ions have a strong tendency to form nonionized CH_3COOH molecules rather than remain separate.

The net effect is to neutralize most of the H_3O^+ from HCl by forming nonionized CH_3COOH molecules. This slightly decreases the ratio $[CH_3COO^-]\,/\,[CH_3COOH]$, which governs the pH of the solution.

When a strong soluble base, such as NaOH, is added to the CH_3COOH–$NaCH_3COO$ buffer solution, it is consumed by the acidic component, CH_3COOH. This occurs in the following way. The additional OH^- causes the water autoionization reaction to proceed to the *left*:

$$2\ H_2O \rightleftharpoons H_3O^+ + OH^- \qquad (\text{shifts}\ \textit{left})$$

This uses up some H_3O^+, causing more CH_3COOH to ionize:

$$CH_3COOH + H_2O \rightleftharpoons CH_3COO^- + H_3O^+ \quad \text{(shifts } right\text{)}$$

Because the $[CH_3COOH]$ is high, this can occur to a great extent. The net result is the neutralization of OH^- by CH_3COOH:

$$OH^- + CH_3COOH \longrightarrow CH_3COO^- + H_2O \quad (\sim100\%)$$

or, as a formula unit equation,

$$\underset{\text{added base}}{NaOH} + \underset{\text{acid}}{CH_3COOH} \longrightarrow \underset{\text{salt}}{NaCH_3COO} + \underset{\text{water}}{H_2O} \quad (\sim100\%)$$

The net effect is to neutralize most of the OH^- from NaOH. This slightly increases the ratio $[CH_3COO^-]\,/\,[CH_3COOH]$, which governs the pH of the solution.

Example 18-18

If 0.010 mol of solid NaOH is added to 1.0 liter of a buffer solution that is 0.10 M in CH_3COOH and 0.10 M in $NaCH_3COO$, how much will $[H_3O^+]$ and pH change? Assume that there is no volume change due to the addition of solid NaOH.

Plan

First we calculate $[H_3O^+]$ and pH for the original buffer solution. Then we write the reaction summary that shows how much of the CH_3COOH is neutralized by NaOH. Then we calculate $[H_3O^+]$ and pH for the resulting buffer solution. Finally we calculate ΔpH.

Solution

For the 0.10 M CH_3COOH and 0.10 M $NaCH_3COO$ solution, we can write

$$[H_3O^+] = \frac{[\text{acid}]}{[\text{salt}]} \times K_a = \frac{0.10}{0.10} \times 1.8 \times 10^{-5} = 1.8 \times 10^{-5}\ M; \text{pH} = 4.74$$

When solid NaOH is added, it reacts with CH_3COOH to form more $NaCH_3COO$:

	NaOH	+	CH_3COOH	$\longrightarrow$	$NaCH_3COO$	+	H_2O
start	0.01 mol		0.10 mol		0.10 mol		
change due to rxn	−0.01 mol		−0.01 mol		+0.01 mol		
after rxn	0 mol		0.09 mol		0.11 mol		

The volume of the solution is 1.0 liter, so we now have a solution that is 0.09 M in CH_3COOH and 0.11 M in $NaCH_3COO$. In this solution.

$$[H_3O^+] = \frac{[\text{acid}]}{[\text{salt}]} \times K_a = \frac{0.09}{0.11} \times 1.8 \times 10^{-5} = 1.5 \times 10^{-5}\ M; \text{pH} = 4.82$$

The calculation shows that the addition of 0.010 mol of solid NaOH to 1.0 liter of this buffer solution decreases $[H_3O^+]$ from 1.8×10^{-5} M to 1.5×10^{-5} M and increases pH from 4.74 to 4.82, a change of 0.08 pH unit. This is a very slight change.

This is enough NaOH to neutralize 10% of the acid.

EOC 83

Addition of 0.010 mole of solid NaOH to one liter of 0.10 M CH_3COOH (pH = 2.89 from Table 18-7) gives a solution that is 0.09 M in CH_3COOH and 0.01 M in $NaCH_3COO$. The pH of this solution is 3.80, which is 0.91 pH unit higher than that of the 0.10 M CH_3COOH solution.

By contrast, adding 0.010 mole of NaOH to enough pure H_2O to give one liter produces a 0.010 M solution of NaOH: $[OH^-] = 1.0 \times 10^{-2}\ M$ and

**Table 18-9
Changes in pH Caused by Addition of Pure Acid or Base to One Liter of Solution**

We have 1.00 L of original solution	When we add 0.010 mol NaOH(s)		When we add 0.010 mol HCl(g)	
	pH increases by	$[H_3O^+]$ decreases by a factor of	pH decreases by	$[H_3O^+]$ increases by a factor of
buffer solution (0.010 M NaCH$_3$COO and 0.010 M CH$_3$COOH)	$+0.08$ pH unit	1.2	-0.08 pH unit	1.2
0.010 M CH$_3$COOH	$+0.91$	8.1	-0.89	7.8
pure H$_2$O	$+5.00$	100,000	-5.00	100,000

pH + pOH = 14

pOH = 2.00. The pH of this solution is 12.00, an increase of 5 pH units above that of pure H_2O. In summary, 0.010 mole of NaOH

added to the $CH_3COOH/NaCH_3COO$ buffer, pH 4.74 $\longrightarrow$ 4.82

added to 0.10 M CH_3COOH, pH 2.89 $\longrightarrow$ 3.80

added to pure H_2O, pH 7.00 $\longrightarrow$ 12.00

In similar fashion, we could calculate the effects of adding 0.010 mole of pure HCl(g) instead of pure NaOH to 1.00 liter of each of these three solutions. This would result in the following changes in pH:

added to the $CH_3COOH/NaCH_3COO$ buffer, pH 4.74 $\longrightarrow$ 4.66

added to 0.10 M CH_3COOH, pH 2.89 $\longrightarrow$ 2.00

added to pure H_2O, pH 7.00 $\longrightarrow$ 2.00

The results of adding NaOH or HCl to these solutions (Table 18-9) demonstrate the efficiency of the buffer solution. We recall that each change of 1 pH unit means that the $[H_3O^+]$ and $[OH^-]$ change by a *factor* of 10. In these terms, the effectiveness of the buffer solution in controlling pH is even more dramatic.

Solutions of a Weak Base and a Salt of the Weak Base
An example of this type of buffer solution is one that contains the weak base aqueous ammonia, NH_3, and its soluble ionic salt ammonium chloride, NH_4Cl. The reactions responsible for the operation of this buffer are

$$NH_4Cl \xrightarrow{H_2O} NH_4^+ + Cl^- \quad \text{(to completion)}$$

$$\underset{\textit{high } \text{conc}}{NH_3 + H_2O} \rightleftharpoons \underset{\substack{\textit{high } \text{conc} \\ \text{from salt}}}{NH_4^+} + OH^- \quad \text{(reversible)}$$

If a strong acid such as HCl is added to this buffer solution, the resulting H_3O^+ shifts the equilibrium reaction

$$2H_2O \rightleftharpoons H_3O^+ + OH^- \quad \text{(shifts \textit{left})}$$

strongly to the *left*. As a result of the diminished OH^- concentration, the reaction

$$NH_3 + H_2O \rightleftharpoons NH_4^+ + OH^- \qquad \text{(shifts } right\text{)}$$

shifts markedly to the *right*. Because the $[NH_3]$ in the buffer solution is high, this can occur to a great extent. The net reaction is

$$H_3O^+ + NH_3 \longrightarrow NH_4^+ + H_2O \qquad (\sim 100\%)$$

or, as a formula unit equation,

$$\underset{\text{added acid}}{HCl} + \underset{\text{base}}{NH_3} \longrightarrow \underset{\text{salt}}{NH_4Cl} \qquad (\sim 100\%)$$

> The net effect is to neutralize most of the H_3O^+ from HCl. This slightly increases the ratio $[NH_4^+]\,/\,[NH_3]$, which governs the pH of the solution.

When a strong soluble base such as NaOH is added to the *original* buffer solution, it is neutralized by the more acidic component, NH_4Cl, or NH_4^+, the conjugate acid of aqueous ammonia:

$$NH_3 + H_2O \rightleftharpoons NH_4^+ + OH^- \qquad \text{(shifts } left\text{)}$$

Because the $[NH_4^+]$ is high, this can occur to a great extent. The result is the neutralization of OH^- by NH_4^+:

$$OH^- + NH_4^+ \longrightarrow NH_3 + H_2O \qquad (\sim 100\%)$$

or, as a formula unit equation,

$$\underset{\text{added base}}{NaOH} + \underset{\text{acid}}{NH_4Cl} \longrightarrow \underset{\text{weak base}}{NH_3} + \underset{\text{water}}{H_2O} + NaCl \qquad (\sim 100\%)$$

> The net effect is to neutralize most of the OH^- from NaOH. This slightly decreases the ratio $[NH_4^+]\,/\,[NH_3]$, which governs the pH of the solution.

> **Summary**   Changes in pH are minimized in buffer solutions because the basic component can react with excess H_3O^+ ions and the acidic component can react with excess OH^- ions.

18-8 Preparation of Buffer Solutions

Buffer solutions can be prepared by mixing other solutions. When solutions are mixed, the volume in which each solute is contained increases, so solute concentrations change. These changes in concentration must be considered. If the solutions are dilute, we may assume that their volumes are additive.

Example 18-19

Calculate the concentration of H_3O^+ in a buffer solution prepared by mixing 200 mL of 0.10 M NaF and 100 mL of 0.050 M HF. $K_a = 7.2 \times 10^{-4}$ for HF.

Plan

We calculate the number of millimoles (or moles) of NaF and HF and then the molarity of each solute in the solution after mixing. We write the appropriate equations for both NaF and HF, represent the equilibrium concentrations algebraically, and substitute into the K_a expression for HF.

Solution

We assume that when two dilute solutions are mixed, their volumes are additive. The volume of the new solution will be 300 mL. Mixing a solution of a weak acid with a solution of its salt does not form any new species. So we have a straightforward buffer calculation. We calculate the number of millimoles (or moles) of each compound and the molarities in the new solution.

$$\underline{?} \text{ mmol NaF} = 200 \text{ mL} \times \frac{0.10 \text{ mmol NaF}}{\text{mL}} = 20 \text{ mmol NaF}$$

$$\underline{?} \text{ mmol HF} = 100 \text{ mL} \times \frac{0.050 \text{ mmol HF}}{\text{mL}} = 5.0 \text{ mmol HF}$$

in 300 mL

Recall that

$$M = \frac{\text{\# mol of solute}}{\text{L}}$$

or

$$M = \frac{\text{\# mmol of solute}}{\text{mL}}$$

The molarities of NaF and HF in the solution are

$$\frac{20 \text{ mmol NaF}}{300 \text{ mL}} = 0.067 \ M \text{ NaF} \quad \text{and} \quad \frac{5.0 \text{ mmol HF}}{300 \text{ mL}} = 0.017 \ M \text{ HF}$$

The appropriate equations and algebraic representations of concentrations are

$$
\begin{array}{ccccc}
\text{NaF} & \longrightarrow & \text{Na}^+ & + & \text{F}^- \\
0.067 \ M & \Longrightarrow & 0.067 \ M & & 0.067 \ M \\
\text{HF} \quad + \text{ H}_2\text{O} & \rightleftharpoons & \text{H}_3\text{O}^+ & + & \text{F}^- \\
(0.017 - x) \ M & & x \ M & & x \ M
\end{array}
$$

$\longrightarrow$ Total $[\text{F}^-] = (0.067 + x) \ M$

Substituting into the K_a expression for hydrofluoric acid gives

$$K_a = \frac{[\text{H}_3\text{O}^+][\text{F}^-]}{[\text{HF}]} = \frac{(x)(0.067 + x)}{(0.017 - x)} = 7.2 \times 10^{-4}$$

Can we assume that x is negligible compared with 0.067 and 0.017 in this expression? When in doubt, solve the equation using the simplifying assumption. Then decide if the assumption was valid. Assume that $(0.067 + x) \approx 0.067$ and $(0.017 - x) \approx 0.017$.

Our assumption is valid.

$$\frac{0.067x}{0.017} = 7.2 \times 10^{-4} \qquad x = \boxed{1.8 \times 10^{-4} \ M = [\text{H}_3\text{O}^+]}$$

EOC 94

Preparation of the buffer solution in Example 18-20. We add 3.5 grams of NH_4Cl to a 500-mL volumetric flask and dissolve it in a little of the 0.10 M NH_3 solution. We then dilute to 500 mL with the 0.10 M NH_3 solution.

We often need a buffer solution of a given pH. One method by which such solutions can be prepared involves adding a salt of a weak base (or weak acid) to a solution of the weak base (or weak acid).

Example 18-20

Calculate the numbers of moles and grams of NH_4Cl that must be used to prepare 500 mL of a buffer solution that is 0.10 M in aqueous NH_3 and has a pH of 9.15.

Plan

We convert the given pH to the desired $[OH^-]$ by the usual procedure. We write the appropriate equations for the reactions of NH_4Cl and NH_3 and represent the equilibrium concentrations. Then we substitute into the K_b expression and solve for the concentration of NH_4Cl required.

Solution

Because the desired pH = 9.15, pOH = 14.00 − 9.15 = 4.85.
So $[OH^-] = 10^{-pOH} = 10^{-4.85} = 1.4 \times 10^{-5}$ M desired.
Let x mol/L be the necessary molarity of NH_4Cl. Because $[OH^-] = 1.4 \times 10^{-5}$ M, this must be the $[OH^-]$ produced by ionization of NH_3. The equations and representations of equilibrium concentrations follow.

$$NH_4Cl \xrightarrow{100\%} \underbrace{NH_4^+}_{x\,M} + \underset{x\,M}{Cl^-}$$

$$\underset{(0.10 - 1.4 \times 10^{-5})\,M}{NH_3} + H_2O \rightleftharpoons \underset{1.4 \times 10^{-5}\,M}{NH_4^+} + \underset{1.4 \times 10^{-5}\,M}{OH^-}$$

$$\text{Total } [NH_4^+] = (x + 1.4 \times 10^{-5})\,M$$

Substitution into the K_b expression for aqueous ammonia gives

$$K_b = \frac{[NH_4^+][OH^-]}{[NH_3]} = 1.8 \times 10^{-5} = \frac{(x + 1.4 \times 10^{-5})(1.4 \times 10^{-5})}{0.10 - 1.4 \times 10^{-5}}$$

NH_4Cl is 100% dissociated, so $x \gg 1.4 \times 10^{-5}$. Then $(x + 1.4 \times 10^{-5}) \approx x$.

$$\frac{(x)(1.4 \times 10^{-5})}{0.10} = 1.8 \times 10^{-5} \qquad x = 0.13\,M = [NH_4^+] = M_{NH_4Cl}$$

> Here x does *not* represent a *change* in concentration, but rather the initial concentration of NH_4Cl. We do *not* assume that $x \ll 1.4 \times 10^{-5}$, but rather the reverse.

Now we calculate the number of moles of NH_4Cl that must be added to prepare 500 mL (0.500 L) of buffer solution.

$$\underline{?}\text{ mol } NH_4Cl = 0.500\text{ L} \times \frac{0.13\text{ mol } NH_4Cl}{L} = \boxed{0.065\text{ mol } NH_4Cl \quad (3.5\text{ g } NH_4Cl)}$$

EOC 97

18-9 Polyprotic Acids

Thus far we have considered only *monoprotic* weak acids. Acids that can furnish *two* or more hydronium ions per molecule are called **polyprotic acids**. The ionizations of polyprotic acids occur stepwise, i.e., one proton at a time. An ionization constant expression can be written for each step, as the following example illustrates. Consider phosphoric acid as a typical polyprotic acid. It contains three acidic hydrogen atoms and ionizes in three steps.

$$H_3PO_4 + H_2O \rightleftharpoons H_3O^+ + H_2PO_4^- \qquad K_1 = \frac{[H_3O^+][H_2PO_4^-]}{[H_3PO_4]} = 7.5 \times 10^{-3}$$

$$H_2PO_4^- + H_2O \rightleftharpoons H_3O^+ + HPO_4^{2-} \qquad K_2 = \frac{[H_3O^+][HPO_4^{2-}]}{[H_2PO_4^-]} = 6.2 \times 10^{-8}$$

$$HPO_4^{2-} + H_2O \rightleftharpoons H_3O^+ + PO_4^{3-} \qquad K_3 = \frac{[H_3O^+][PO_4^{3-}]}{[HPO_4^{2-}]} = 3.6 \times 10^{-13}$$

We see that K_1 is much greater than K_2 and that K_2 is much greater than K_3. This is generally true for polyprotic *inorganic* acids (refer to Appendix F). Successive ionization constants often decrease by a factor of approximately 10^4 to 10^6, although some differences are outside this range. Large decreases in the values of successive ionization constants mean that each step in the ionization of a polyprotic acid occurs to a much lesser extent than the previous step. Thus, the $[H_3O^+]$ produced in the first step is very large compared with the $[H_3O^+]$ produced in the second and third steps. As we shall see, in all except extremely dilute solutions of H_3PO_4, the concentration of H_3O^+ may be assumed to be that furnished by the first step in the ionization alone.

> Each K expression includes $[H_3O^+]$, so each K expression must be satisfied by the *total* concentration of H_3O^+ in the solution.

Example 18-21

Calculate the concentrations of all species present in 0.100 M H_3PO_4.

Plan

Because H_3PO_4 contains three acidic hydrogens per formula unit, we show its ionization in three steps. For each step we write the appropriate ionization equation, with its K_a expression and value. Then we represent the equilibrium concentrations from the *first*-step ionization and substitute into the $K_{a(1)}$ expression. We repeat the procedure for the second and third steps *in order*.

Solution

First we calculate the concentrations formed in the first-step ionization. Let $x = $ mol/L of H_3PO_4 that ionize; then $x = [H_3O^+]_{1st} = [H_2PO_4^-]$.

$$\underset{(0.100 - x)\, M}{H_3PO_4} + H_2O \rightleftharpoons \underset{x\, M}{H_3O^+} + \underset{x\, M}{H_2PO_4^-}$$

Substitution into the expression for K_1 gives

$$\frac{[H_3O^+][H_2PO_4^-]}{[H_3PO_4]} = \frac{(x)(x)}{(0.100 - x)} = 7.5 \times 10^{-3}$$

This equation must be solved by the quadratic formula because K is too large to neglect x relative to 0.100 M. Solving gives the *positive* root $x = 2.4 \times 10^{-2}$. Thus, from the first step in the ionization of H_3PO_4,

> $x = -3.1 \times 10^{-2}$ is the extraneous root of the quadratic equation.

$$x\, M = \boxed{[H_3O^+]_{1st} = [H_2PO_4^-] = 2.4 \times 10^{-2}\, M}$$

$$(0.100 - x)\, M = \boxed{[H_3PO_4] = 7.6 \times 10^{-2}\, M}$$

For the second step we use the $[H_3O^+]$ and $[H_2PO_4^-]$ from the first step. Let $y = $ mol/L of $H_2PO_4^{2-}$ that ionize; then $y = [H_3O^+]_{2nd} = [HPO_4^{2-}]$.

$$H_2PO_4^- \ + \ H_2O \ \rightleftharpoons \ H_3O^+ \ + \ HPO_4^{2-}$$
$$(2.4 \times 10^{-2} - y) \ M \qquad\qquad (2.4 \times 10^{-2} + y) \ M \quad y \ M$$

from 1st step   from 2nd step

Substitution into the expression for K_2 gives

$$\frac{[H_3O^+][HPO_4^{2-}]}{[H_2PO_4^-]} = \frac{(2.4 \times 10^{-2} + y)(y)}{(2.4 \times 10^{-2} - y)} = 6.2 \times 10^{-8}$$

Examination of this equation suggests that $y \ll 2.4 \times 10^{-2}$, so

$$\frac{(2.4 \times 10^{-2})(y)}{(2.4 \times 10^{-2})} = 6.2 \times 10^{-8} \qquad y = \boxed{6.2 \times 10^{-8} \ M = [HPO_4^{2-}]} = [H_3O^+]_{2nd}$$

We see that $[HPO_4^{2-}] = K_2$ and $[H_3O^+]_{2nd} \ll [H_3O^+]_{1st}$. In general, in solutions of reasonable concentration of weak polyprotic acids for which $K_1 \gg K_2$ and that contain no other electrolytes, *the concentration of the anion produced in the second-step ionization is always equal to K_2.*

The pH of most solutions of polyprotic acids is governed by the first-step ionization.

For the third step we use $[H_3O^+]$ from the *first* step and $[HPO_4^{2-}]$ from the *second* step. Let $z = $ mol/L of HPO_4^{2-} that ionize; then $z = [H_3O^+]_{3rd} = [PO_4^{3-}]$.

$$HPO_4^{2-} \ + \ H_2O \ \rightleftharpoons \ H_3O^+ \ + \ PO_4^{3-}$$
$$(6.2 \times 10^{-8} - z) \ M \qquad\qquad (2.4 \times 10^{-2} + z) \ M \quad z \ M$$

from 2nd step   from 3rd step   from 1st step   from 3rd step

$$\frac{[H_3O^+][PO_4^{3-}]}{[HPO_4^{2-}]} = \frac{(2.4 \times 10^{-2} + z)(z)}{(6.2 \times 10^{-8} - z)} = 3.6 \times 10^{-13}$$

$y = 6.2 \times 10^{-8}$ was disregarded in the second step and is also disregarded here.

We make the usual simplifying assumption, and find that

$$z \ M = \boxed{9.3 \times 10^{-19} \ M = [PO_4^{3-}]} = [H_3O^+]_{3rd}$$

EOC 98

We have calculated the concentrations of the species formed by the ionization of $0.100 \ M \ H_3PO_4$. These concentrations are compared in Table 18-10. The concentration of $[OH^-]$ in $0.100 \ M \ H_3PO_4$ is included. It was calculated from the known $[H_3O^+]$ using the ion product for water, $[H_3O^+][OH^-] = 1.0 \times 10^{-14}$.

Nonionized H_3PO_4 is present in greater concentration than any other species in $0.100 \ M \ H_3PO_4$ solution. The only other species present in significant concentrations are H_3O^+ and $H_2PO_4^-$. Similar statements can be made for other weak polyprotic acids for which the last K is very small.

Table 18-10
Concentrations of the Species in 0.10 M H₃PO₄

Species	Concentration (mol/L)
H_3PO_4	7.6×10^{-2} = 0.076
H_3O^+	2.4×10^{-2} = 0.024
$H_2PO_4^-$	2.4×10^{-2} = 0.024
HPO_4^{2-}	6.2×10^{-8} = 0.000000062
OH^-	4.2×10^{-13} = 0.00000000000042
PO_4^{3-}	9.3×10^{-19} = 0.00000000000000000093

Phosphoric acid is a typical *weak* polyprotic acid. Let us now describe solutions of sulfuric acid, a *very strong* polyprotic acid.

Example 18-22

Calculate concentrations of the species in 0.10 M H_2SO_4. $K_2 = 1.2 \times 10^{-2}$.

Plan

Because the first-step ionization of H_2SO_4 is complete, we read the concentrations from the first step of the balanced equation. The second-step ionization is *not* complete, and so we write the ionization equation, the $K_{a(2)}$ expression, and the algebraic representations of equilibrium concentrations. Then we substitute into K_2 for H_2SO_4.

Solution

The first-step ionization of H_2SO_4 is complete, as we pointed out.

$$H_2SO_4 + H_2O \xrightarrow{100\%} H_3O^+ + HSO_4^-$$
$$\text{0.10 } M \quad\quad \Rightarrow \quad \text{0.10 } M \quad \text{0.10 } M$$

However, the second-step ionization is not complete.

$$HSO_4^- + H_2O \rightleftharpoons H_3O^+ + SO_4^{2-} \quad\quad \text{and} \quad\quad K_2 = \frac{[H_3O^+][SO_4^{2-}]}{[HSO_4^-]} = 1.2 \times 10^{-2}$$

Let $x = [HSO_4^-]$ that ionizes. $[H_3O^+]$ is the sum of the concentrations produced in the first and second steps. So we represent the equilibrium concentrations as

$$HSO_4^- + H_2O \rightleftharpoons H_3O^+ + SO_4^{2-}$$
$$\text{(0.10} - x) M \quad\quad\quad \text{(0.10} + x) M \quad x M$$
$$\text{from 1st step} \quad \text{from 2nd step}$$

Substitution into the ionization constant expression for K_2 gives

$$K_2 = \frac{[H_3O^+][SO_4^{2-}]}{[HSO_4^-]} = \frac{(0.10 + x)(x)}{0.10 - x} = 1.2 \times 10^{-2}$$

Clearly, x cannot be disregarded because K is too large. This equation must be solved by the quadratic formula, which gives $x = 0.010$ and $x = -0.12$ (extraneous). So $[H_3O^+]_{2nd} = [SO_4^{2-}] = 0.010\ M$. The concentrations of species in 0.10 M H_2SO_4 are

$$[H_2SO_4] \approx 0\ M \quad\quad [HSO_4^-] = (0.10 - x)\ M = 0.09\ M \quad\quad [SO_4^{2-}] = 0.010\ M$$

$$[H_3O^+] = (0.10 + x)\ M = 0.11\ M$$

$$[OH^-] = \frac{K_w}{[H_3O^+]} = \frac{1.0 \times 10^{-14}}{0.11} = 9.1 \times 10^{-14}\ M$$

In 0.10 M H_2SO_4 solution, the extent of the second-step ionization is 10%.

Hydrosulfuric acid, H_2S, is a *very weak* diprotic acid. Saturated H_2S solutions (0.10 M) are used in qualitative analysis.

Example 18-23

Calculate the concentrations of the species in 0.10 M H_2S solution.
$K_1 = 1.0 \times 10^{-7}$ and $K_2 = 1.3 \times 10^{-13}$.

Plan

We use the same kind of logic as in Example 18-21.

Solution

Let x = mol/L of H_2S that ionize in the first step:

$$H_2S \; + H_2O \rightleftharpoons H_3O^+ + HS^-$$
$$\underset{(0.10 - x)\, M}{} \qquad\qquad \underset{x\, M}{} \;\; \underset{x\, M}{}$$

$$K_1 = \frac{[H_3O^+][HS^-]}{[H_2S]} = \frac{(x)(x)}{(0.10 - x)} = 1.0 \times 10^{-7}$$

Solving with the usual assumption gives $x = 1.0 \times 10^{-4}$.

$$[H_2S] = (0.10 - x)\, M = \boxed{0.10\ M}$$

$$[HS^-] = [H_3O^+] = x\, M = \boxed{1.0 \times 10^{-4}\ M} \qquad \text{(first step)}$$

The second step involves ionization of the anion produced in the first step. We represent equilibrium concentrations (where y = mol/L of HS^- that ionizes) as follows:

$$HS^- \quad + H_2O \rightleftharpoons \qquad H_3O^+ \quad + S^{2-}$$
$$\underset{(1.0 \times 10^{-4} - y)\, M}{} \qquad\qquad \underset{(1.0 \times 10^{-4} + y)\, M}{} \;\; \underset{y\, M}{\phantom{S^{2-}}}$$
$$\qquad\qquad \underset{\text{from 1st step}}{\nearrow} \quad \underset{\text{from 2nd step}}{\nearrow}$$

Substitution into K_2 gives

$$K_2 = \frac{[H_3O^+][S^{2-}]}{[HS^-]} = \frac{(1.0 \times 10^{-4} + y)(y)}{(1.0 \times 10^{-4} - y)} = 1.3 \times 10^{-13}$$

Assume that $(1.0 \times 10^{-4} + y) \approx 1.0 \times 10^{-4}$ and $(1.0 \times 10^{-4} - y) \approx 1.0 \times 10^{-4}$.

$$\frac{(1.0 \times 10^{-4})y}{1.0 \times 10^{-4}} = 1.3 \times 10^{-13} \qquad y = \boxed{1.3 \times 10^{-13}\ M = [HS^-]} = [H_3O^+]_{2nd}$$

Our assumption is valid. Very little HS^- ionizes ($1.3 \times 10^{-13}\ M$).

$$[S^{2-}] = y = \boxed{1.3 \times 10^{-13}\ M}$$

$$[HS^-] = (1.0 \times 10^{-4} - y) = (1.0 \times 10^{-4}) - (1.3 \times 10^{-13}) \approx \boxed{1.0 \times 10^{-4}\ M}$$

$$[H_3O^+] = (1.0 \times 10^{-4} + y) = (1.0 \times 10^{-4}) + (1.3 \times 10^{-13}) \approx \boxed{1.0 \times 10^{-4}\ M}$$

We have already calculated the $\boxed{\text{concentration of nonionized } H_2S\ (\sim 0.10\ M).}$

EOC 100

In Table 18-11 we compare 0.10 M solutions of these three polyprotic acids. Their acidities are very different. The $[H_3O^+]$ in 0.10 M H_2SO_4 is about 1100 times greater than that in 0.10 M H_2S!

Table 18-11
Comparison of Concentrations of H_3O^+ and Nonionized Acid in 0.10 M Solutions of Three Polyprotic Acids

	0.10 M H_2SO_4	0.10 M H_3PO_4	0.10 M H_2S
K_1	very large	7.5×10^{-3}	1.0×10^{-7}
K_2	1.2×10^{-2}	6.2×10^{-8}	1.3×10^{-13}
K_3		3.6×10^{-13}	
$[H_3O^+]$	0.11 M	2.4×10^{-2} M	1.0×10^{-4} M
[acid molecules]	~0 M	7.6×10^{-2} M	~0.10 M

The three solutions in Table 18-11.

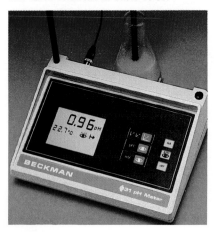

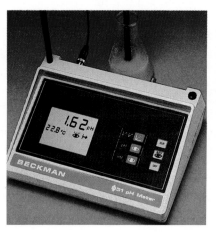

0.10 M H_2SO_4.
(Example 18-22)

0.10 M H_3PO_4.
(Example 18-21)

0.10 M H_2S.
(Example 18-23)

Key Terms

Amines   Derivatives of ammonia in which one or more hydrogen atoms have been replaced by organic groups.

Buffer solution   A solution that resists changes in pH when acids or bases are added. A buffer solution contains an acid and its conjugate base, so it can react with added base or acid. Common buffer solutions contain either (1) a weak acid and a soluble ionic salt of the weak acid *or* (2) a weak base and a soluble ionic salt of the weak base.

Common ion effect   Suppression of ionization of a weak electrolyte by the presence in the same solution of a strong electrolyte containing one of the same ions as the weak electrolyte.

Henderson–Hasselbalch equation   An equation that enables us to calculate the pH or pOH of a buffer solution directly.

For acid/salt buffer $pH = pK_a + \log \dfrac{[acid]}{[salt]}$

For base/salt buffer $pOH = pK_b + \log \dfrac{[base]}{[salt]}$

Indicator   An organic compound that exhibits different colors in solutions of different acidities. Many indicators are weak acids, HIn, in which HIn and In^- have different colors. Indicators may be used to estimate the pH of a solution.

Ionization constant   An equilibrium constant for the ionization of a weak electrolyte.

Ion product for water   An equilibrium constant for the ionization of water,

$$K_w = [H_3O^+][OH^-] = 1.00 \times 10^{-14} \text{ at } 25°C$$

Monoprotic acid   An acid that can form only one hydronium ion per molecule; may be strong or weak.

p[ ]   The negative logarithm of the concentration (mol/L) of the indicated species.

pH   The negative logarithm of the concentration (mol/L) of the $H_3O^+[H^+]$ ion; the commonly used scale ranges from 0 to 14.

pK_a The negative logarithm of K_a, the ionization constant for a weak acid.

pK_b The negative logarithm of K_b, the ionization constant for a weak base.

pOH The negative logarithm of the concentration (mol/L) of the OH^- ion; the commonly used scale ranges from 14 to 0.

Polyprotic acid An acid that can form two or more hydronium ions per molecule; often at least one step of the ionization is weak.

Exercises

Note: All exercises in this chapter assume a temperature of 25°C unless they specify otherwise. All logarithms are common (base 10).

A Brief Review of Strong Electrolytes

1. List names and formulas for
   (a) the common strong acids
   (b) six weak bases
   (c) the common strong soluble bases
   (d) ten soluble ionic salts
2. Calculate the concentrations of the constituent ions in solutions of the following compounds in the indicated concentrations.
   (a) 0.10 M HBr   (c) 0.0020 M $CaCl_2$
   (b) 0.040 M KOH
3. Calculate the concentrations of the constituent ions in solutions of the following compounds in the indicated concentrations.
   (a) 0.020 M $Sr(OH)_2$ (c) 0.0035 M K_2SO_4
   (b) 0.00030 M $HClO_3$
4. Calculate the concentrations of the constituent ions in the following solutions.
   (a) 3.0 g of KOH in 1.50 L of solution
   (b) 0.61 g of $Ba(OH)_2$ in 250 mL of solution
   (c) 2.64 g of $Ca(NO_3)_2$ in 100 mL of solution
5. Calculate the concentrations of the constituent ions in the following solutions.
   (a) 2.75 g of $Al_2(SO_4)_3$ in 400 mL of solution
   (b) 83.8 g of $CaCl_2 \cdot 6H_2O$ in 8.00 L of solution
   (c) 18.4 g of HBr in 450 mL of solution

The Autoionization of Water

6. (a) Write a chemical equation showing the ionization of water. (b) Write the equilibrium constant expression for this equation. (c) What is the special symbol used for this equilibrium constant? (d) What is the relationship between $[H^+]$ and $[OH^-]$ in pure water? (e) How can this relationship be used to define the terms "acidic" and "basic"?
7. Write mathematical definitions for pH and pOH. What is the relationship between pH and pOH? How can pH be used to define the terms "acidic" and "basic"?
8. Why is the ion product for water written in the form $[H_3O^+][OH^-] = 1.0 \times 10^{-14}$?

9. (a) Why is the concentration of H_3O^+ produced by the ionization of water neglected in calculating the concentration of H_3O^+ in a 0.10 M solution of HCl? (b) Demonstrate that it (H_3O^+ from H_2O) may be neglected.
10. (a) Why is the concentration of OH^- produced by the ionization of water neglected in calculating the concentration of OH^- in a 0.10 M solution of NaOH? (b) Demonstrate that it (OH^- from H_2O) may be neglected.
11. Calculate the concentrations of OH^- in the solutions described in Exercises 2(a), 3(b), and 5(c), and compare them with the OH^- concentration in pure water.
12. Calculate the concentrations of H_3O^+ in the solutions described in Exercises 2(b), 3(a), and 4(b), and compare them with the H_3O^+ concentration in pure water.
13. Calculate $[OH^-]$ that is in equilibrium with
    (a) $[H^+] = 1.3 \times 10^{-4}$ mol/L
    (b) $[H^+] = 7 \times 10^{-9}$ mol/L.
14. Calculate $[H^+]$ that is in equilibrium with

$$[OH^-] = 4.21 \times 10^{-6} \text{ mol/L}$$

15. The equilibrium constant of the reaction

$$2D_2O \rightleftharpoons D_3O^+ + OD^-$$

(where D is deuterium, ^2H) is 1.35×10^{-15} at 25°C. Calculate the pD of pure deuterium oxide (heavy water) at 25°C. What is the relationship between $[D_3O^+]$ and $[OD^-]$ in pure D_2O? Is pure D_2O acidic, basic, or neutral?

*16. In liquid ammonia at −50°C, the equilibrium constant for the reaction

$$2NH_3(\ell) \rightleftharpoons NH_4^+ + NH_2^-$$

is $K = 1.0 \times 10^{-33}$. (a) What is the concentration of NH_4^+ in pure liquid ammonia? (b) What is the concentration of NH_4^+ in a 1.0×10^{-3} M solution of $NaNH_2$ ($Na^+ + NH_2^-$) in liquid ammonia? $NaNH_2$ is a strong electrolyte in liquid NH_3.

The pH and pOH Scales

17. What are the logarithms of the following numbers?
    (a) 0.0027, (b) 0.000035, (c) 2.7×10^{-2}, (d) 6.6×10^{-6}.
18. What are the logarithms of the following numbers?
    (a) 0.00052, (b) 4.2, (c) 5.8×10^{-12}, (d) 4.9×10^{-7}.

19. What is the number that has each of the following as its logarithm? (a) 1.47, (b) 3.66, (c) −4.72, (d) −0.48.
20. What is the number that has each of the following as its logarithm? (a) 10.73, (b) −10.73, (c) −1.84, (d) 0.60.
21. Refer to Exercise 11 and calculate the pOH of each solution.
22. Refer to Exercise 12 and calculate the pH of each solution.
23. Calculate the pH of a 1.0×10^{-4} M solution of $HClO_4$, a strong acid, at 25°C.
24. Calculate the pH of a 2.0×10^{-4} M solution of NaOH at 25°C.
25. Calculate the pH of a 1.0×10^{-11} M solution of HCl at 25°C.
26. A solution of HNO_3 has pH 3.20. What is the molarity of the solution?
27. Complete the following table. Is there an obvious relationship between pH and pOH? What is it?

Solution	$[H_3O^+]$	$[OH^-]$	pH	pOH
0.10 M HI	——	——	——	——
0.040 M RbOH	——	——	——	——
0.020 M $Ba(OH)_2$	——	——	——	——
0.00030 M $HClO_4$	——	——	——	——

28. Calculate the following values for each solution:

Solution	$[H_3O^+]$	$[OH^-]$	pH	pOH
(a) 0.035 M NaOH	——	——	——	——
(b) 0.035 M HCl	——	——	——	——
(c) 0.035 M $Ca(OH)_2$	——	——	——	——

29. Complete the following table by appropriate calculations:

$[H_3O^+]$	pH	$[OH^-]$	pOH
(a) ——	4.84	——	——
(b) ——	12.61	——	——
(c) ——	——	——	1.34
(d) ——	——	——	9.47

30. Calculate (a) $[H^+]$ and (b) pH of pure water at 45°C.
31. Fill in the blanks in this table for given solutions a, b, c, and d.

Sol'n	Temp. (°C)	Concentration (mol/L)		pH
		$[H^+]$	$[OH^-]$	
(a)	25	1.0×10^{-6}	————	——
(b)	0	————	————	5.69
(c)	60	————	————	7.00
(d)	25	————	2.5×10^{-9}	——

32. (a) What is the pH of pure water at body temperature, 37°C? Refer to Table 18-2. (b) Is this acidic, basic, or neutral? Why?
33. Arrange the following common kitchen samples from most acidic to most basic:

carrot juice, pH 5.1   blackberry juice, pH 3.4
soap, pH 11.0   red wine, pH 3.7
egg white, pH 7.8   milk of magnesia, pH 10.5
sauerkraut, pH 3.5   lime juice, pH 2.0

Ionization Constants for Weak Monoprotic Acids and Bases

34. Write a chemical equation that represents the ionization of the acid HA. Write the equilibrium constant expression for this reaction. What is the special symbol used for this equilibrium constant?
35. What is the relationship between the strength of an acid and the numerical value of K_a? What is the relationship between the acid strength and the value of pK_a?
36. Write chemical equations that represent (a) the dissociation of the base MOH and (b) the equilibrium between water and the proton-accepting base B. Write the equilibrium constant expressions for these reactions. What is the special symbol used for these equilibrium constants?
37. What is the relationship between base strength and the value of K_b? What is the relationship between base strength and the value of pK_b?
38. In a solution of a weak acid, $HA + H_2O \rightleftharpoons H_3O^+ + A^-$, the following equilibrium concentrations are found: $[H_3O^+] = 0.0017$ M and $[HA] = 0.0983$ M. Calculate the ionization constant for the weak acid, HA.
39. A 0.040 M aqueous solution of a weak, monoprotic acid ($HA + H_2O \rightleftharpoons H_3O^+ + A^-$) is 0.85% ionized. Calculate the ionization constant for HA.
40. In 0.10 M solution a weak acid, $HX + H_2O \rightleftharpoons H_3O^+ + X^-$, is 2.7% ionized. Calculate the ionization constant for the weak acid, HX.
41. A 0.0100 molal solution of acetic acid ($CH_3COOH + H_2O \rightleftharpoons H_3O^+ + CH_3COO^-$) freezes at −0.01938°C. Use this information to calculate the ionization constant for acetic acid. A 0.0100 molal solution is sufficiently dilute that it may be assumed to be 0.0100 molar without introducing a significant error.
42. What is the concentration of CN^- in a solution having an equilibrium concentration of 0.0123 M HCN and a pH of 3.56?
43. Find the concentrations of the various species present in a 0.25 M solution of hydrofluoric acid, HF. What is the pH of the solution?
44. Calculate the concentrations of all the species present in a 0.35 M benzoic acid solution.
45. Electrical conductivity measurements show that 0.050 M acetic acid is 1.9% ionized at 25°C. Calculate K_a for the acid.
46. A 0.0202 M solution of $ClCH_2COOH(aq)$, chloroacetic acid, is 24.1% ionized at 25°C. Calculate K_a for this acid.
47. What is the extent of ionization of the molecules in (a) a 0.100 M CH_3COOH solution and (b) a 0.0100 M CH_3COOH solution?

48. What is the percent ionization in a 0.0500 M solution of formic acid, HCOOH?

49. What is the concentration of IO^- in equilibrium with $[H^+] = 0.035$ mol/L and $[HIO] = 0.427$ mol/L? $K_a = 2.3 \times 10^{-11}$ for HIO.

50. In a 0.0100 M aqueous solution of methylamine, CH_3NH_2, the equilibrium concentrations of the species are $[CH_3NH_2] = 0.0080$ mol/L and $[CH_3NH_3^+] = [OH^-] = 2.0 \times 10^{-3}$ mol/L. Calculate K_b for this weak base.

$$CH_3NH_2(aq) + H_2O(\ell) \rightleftharpoons CH_3NH_3^+ + OH^-$$

51. What is the concentration of NH_3 in equilibrium with $[NH_4^+] = 0.010$ mol/L and $[OH^-] = 1.2 \times 10^{-5}$ mol/L?

52. Calculate [OH], percent ionization, and pH for (a) 0.10 M aqueous ammonia, and (b) 0.10 M methylamine solution.

53. Calculate $[H_3O^+]$, [OH], pH, pOH, and percent ionization for 0.12 M aqueous ammonia solution.

54. Because K_b is larger for triethylamine

$$(C_2H_5)_3N(aq) + H_2O(\ell) \rightleftharpoons (C_2H_5)_3NH^+ + OH^-$$
$$K_b = 5.2 \times 10^{-4}$$

than for trimethylamine

$$(CH_3)_3N(aq) + H_2O(\ell) \rightleftharpoons (CH_3)_3NH^+ + OH^-$$
$$K_b = 7.4 \times 10^{-5}$$

an aqueous solution of triethylamine should have a larger concentration of OH^- ion than an aqueous solution of trimethylamine of the same concentration. Confirm this statement by calculating the $[OH^-]$ for 0.010 M solutions of both weak bases.

*55. The pH of a hydrofluoric acid solution is 2.75. Calculate the molarity of the solution. Can you make a simplifying assumption in this case?

*56. The buildup of lactic acid in muscles causes pain during extreme physical exertion. The K_a for lactic acid, C_2H_4OCOOH, is 8.4×10^{-4}. Calculate the pH of a 0.100 M solution of lactic acid. Can you make a simplifying assumption in this case?

57. A 0.20 M solution of a newly discovered monoprotic acid is found to have a pH of 3.22. What is the value of the ionization constant for the acid? Represent the new acid by the general formula HY.

58. Ascorbic acid, $C_5H_7O_4COOH$, known as vitamin C, is an essential vitamin for all mammals. Among mammals, only humans, monkeys, and guinea pigs cannot synthesize it in their bodies. K_a for ascorbic acid is 7.9×10^{-5}. Calculate $[H_3O^+]$ and pH in a 0.100 M solution of ascorbic acid.

59. The pain of bee and wasp stings is due to the injection of formic acid, HCOOH. Putting aqueous ammonia on the sting helps to relieve the pain. (a) Why is this so? (b) Write the equation for the neutralization of this monoprotic acid.

60. The emergency treatment for an overdose of an amphetamine (speed) is to administer ascorbic acid,

$C_5H_7O_4COOH$. As a result of the treatment, the drug is excreted in the urine. The formula of amphetamine is

$$C_6H_5CH-CH-\overset{..}{N}H_2$$
$$\overset{|}{CH_3}$$

(a) Can you tell why this treatment works? (b) Write the equation for the reaction.

61. Answer the following questions for 0.10 M solutions of the weak acids listed in Table 18-4. Which solution contains
(a) the highest concentration of H_3O^+?
(b) the highest concentration of OH^-?
(c) the lowest concentration of H_3O^+?
(d) the lowest concentration of OH^-?
(e) the highest concentration of nonionized acid molecules?
(f) the lowest concentration of nonionized acid molecules?

62. Answer the following questions for 0.10 M solutions of the weak bases listed in Table 18-6.
(a) In which solution is
  1. the pH highest?
  2. the pH lowest?
  3. the pOH highest?
  4. the pOH lowest?
(b) Which solution contains
  1. the highest concentration of the cation of the weak base?
  2. the lowest concentration of the cation of the weak base?

63. The smell of cooked fish is due to the presence of amines. The odor is lessened by adding lemon juice, which contains citric acid. Why does this work?

64. Niacin is essential to human nutrition. Its deficiency causes *pellagra*, a condition characterized by skin problems, loss of appetite, and nervous disorders. Niacin is nicotinic acid (which is *not* physiologically related to nicotine, found in tobacco), $K_a = 1.4 \times 10^{-5}$. What is the pH of a 0.050 M solution of nicotinic acid?

Acid–Base Indicators

65. K_a is 7.9×10^{-8} for bromthymol blue, an indicator that can be represented as HIn. HIn molecules are yellow, and In^- ions are blue. What color will bromthymol blue be in a solution in which (a) $[H_3O^+] = 1.0 \times 10^{-4}$ M and (b) pH = 10.30?

66. What are acid–base indicators? (b) What are the essential characteristics of acid–base indicators? (c) What determines the color of an acid–base indicator in an aqueous solution?

67. Demonstrate mathematically that neutral red is red in solutions of pH 3.00, whereas it is yellow in solutions of pH 10.00. HIn is red, and In^- is yellow. K_a is 2.0×10^{-7}.

*68. The indicator metacresol purple changes from yellow to purple at pH 8.2. At this point it exists in equal concentrations as the conjugate acid and the conjugate base. What are K_a and pK_a for metacresol purple, a weak acid represented as HIn?

69. Explain why universal indicators are useful.

The Common Ion Effect and Buffer Solutions

70. Consider solutions that contain the indicated concentrations of the following pairs of compounds. Which ones (i) are examples of the common ion effect and (ii) are buffer solutions? Why? (a) 0.15 M CH_3COOH and 0.30 M NaCl, (b) 0.10 M NH_3 and 0.15 M NaCl, (c) 0.10 M NH_3 and 0.10 M NH_4NO_3.

71. Consider solutions that contain the indicated concentrations of the following pairs of compounds. Which ones (i) are examples of the common ion effect and (ii) are buffer solutions? Why? (a) 0.10 M KOH and 0.10 M KCl, (b) 0.10 M HClO and 0.10 M KClO, (c) 0.30 M CH_3NH_2 and 0.20 M CH_3NH_3Cl.

72. A solution is 0.40 M in HCl and 1.00 M in formic acid, $HCHO_2$. Calculate the concentration of CHO_2^- ions in the solution.

73. Suppose that you have a solution that is 0.25 M in methylamine, CH_3NH_2, and 0.00050 M in the salt methylammonium chloride, CH_3NH_3Cl. Would you expect this to be an effective buffer solution? Why or why not?

74. Calculate pH for each of the following buffer solutions:
    (a) 0.10 M HF and 0.20 M KF
    (b) 0.050 M CH_3COOH and 0.025 M $Ba(CH_3COO)_2$

75. Calculate the concentration of OH^- and the pH for the following solutions:
    (a) 0.30 M $NH_3(aq)$ and 0.20 M NH_4NO_3
    (b) 0.10 M aniline and 0.20 M anilinium chloride, $C_6H_5NH_3Cl$

76. Calculate the concentration of OH^- and the pH for the following buffer solutions:
    (a) 0.20 M $NH_3(aq)$ and 0.30 M NH_4NO_3
    (b) 0.20 M $NH_3(aq)$ and 0.15 M $(NH_4)_2SO_4$

77. Calculate the ratio of concentrations $[NH_3]/[NH_4^+]$ that gives
    (a) solutions of pH = 10.00
    (b) solutions of pH = 9.25

*78. Buffer solutions are especially important in our body fluids and metabolism. Write net ionic equations to illustrate the buffering action of (a) the $H_2CO_3/NaHCO_3$ buffer system in blood and (b) the NaH_2PO_4/Na_2HPO_4 buffer system inside cells.

79. Suppose you are asked to calculate the $[H_3O^+]$ in two aqueous solutions: one is 0.010 M in HF ($K_a = 7.2 \times 10^{-4}$), and the other is 0.010 M in HF and 0.010 M in NaF. In the first calculation, it may *not* be assumed that the concentration of HF that ionizes (x M) is negligible compared with 0.010 molar; however, this as-

sumption *is* valid for the second solution. Explain why this is true.

Buffering Action

80. Consider the ionization of formic acid, HCOOH.

$$HCOOH + H_2O \rightleftharpoons HCOO^- + H_3O^+$$

What effect does the addition of sodium formate (NaHCOO) have on the fraction of formic acid molecules that undergo ionization in aqueous solution?

81. Briefly describe why the pH of a buffer solution remains nearly constant when small amounts of acid or base are added. Over what pH range do we observe the best buffering action (nearly constant pH)?

82. What is the pH of a solution that is 0.10 M in $HClO_4$ and 0.10 M $KClO_4$? Is this a buffer solution? $HClO_4$ is a strong acid.

83. (a) Find the pH of a solution 0.50 M in formic acid and 0.40 M in sodium formate. (b) Find the pH of the solution after 0.050 mol HCl/L has been added to it.

84. (a) Find the pH of a solution that is 1.00 M in NH_3 and 0.80 M in NH_4Cl. (b) Find the pH of the solution after 0.10 mol HCl/L has been added to it. (c) A solution of pH 9.34 is prepared by adding NaOH to pure water. Find the pH of this solution after 0.10 mol HCl/L has been added to it.

85. (a) Calculate the concentrations of CH_3COOH and CH_3COO^- in a solution in which the pH is 4.50 and the total concentration is 0.20 mol/L. (b) If 0.0100 mol of solid in NaOH is added to 1.00 L, how much does pH change?

86. A solution contains bromoacetic acid and sodium bromoacetate with a total concentration of 0.20 mol/L. If the pH is 3.00, what are the concentrations of the acid and the salt? $K_a = 2.0 \times 10^{-3}$ for $CH_2BrCOOH$.

87. Calculate the concentration of propionate ion, $CH_3CH_2COO^-$, in equilibrium with 0.010 M CH_3CH_2COOH (propionic acid) and 0.10 M H^+ from hydrochloric acid. $K_a = 1.3 \times 10^{-5}$ for CH_3CH_2COOH.

88. Calculate the concentration of $C_2H_5NH_3^+$ in equilibrium with 0.010 M $C_2H_5NH_2$ (ethylamine) and 0.0010 M OH^- ion from sodium hydroxide. $K_b = 4.7 \times 10^{-4}$ for ethylamine.

89. When chlorine gas is dissolved in water to make "chlorine water," HCl (a strong acid) and HClO (a weak acid) are produced in equal amounts:

$$Cl_2(g) + H_2O(\ell) \longrightarrow HCl(aq) + HClO(aq)$$

What is the concentration of ClO^- ion in a solution containing 0.010 mol of each acid in 1.00 L of solution?

Preparation of Buffer Solutions

90. A buffer solution of pH 5.30 is to be prepared from propionic acid and sodium propionate. The concentration of sodium propionate must be 0.50 mol/L. What

should be the concentration of the acid? K_a for CH_3CH_2COOH is 1.3×10^{-5}.

91. We need a buffer with pH 9.00. It can be prepared from NH_3 and NH_4Cl. What should be the $[NH_4^+]/[NH_3]$ ratio?

92. What volumes of 0.1000 M acetic acid and 0.1000 M NaOH solutions should be mixed to prepare 1.000 L of a buffer solution of pH 4.50 at 25°C?

93. One liter of a buffer solution is prepared by dissolving 0.100 mol of $NaNO_2$ and 0.050 mol of HCl in water. What is the pH of this solution? If the solution is diluted twofold with water, what is the pH?

94. One liter of a buffer solution is made by mixing exactly 500 mL of 1.00 M acetic acid and 500 mL of 0.500 M calcium acetate solutions. What is the concentration of each of the following in the buffer solution? (a) CH_3COOH, (b) Ca^{2+}, (c) CH_3COO^-, (d) H^+. (c) What is the pH?

95. What concentration of benzoate ion, $C_6H_5COO^-$, should be added to a 0.010 M benzoic acid, C_6H_5COOH, solution to prepare a buffer that has pH 5.00?

96. What concentration of chloroacetic acid, $CH_2ClCOOH$, should be added to a 0.010 M $NaCH_2ClCOO$ solution to prepare a buffer that has pH 3.00? $K_a = 1.4 \times 10^{-3}$ for $CH_2ClCOOH$.

97. What concentration of NH_4^+ should be added to a 0.050 M NH_3 solution to prepare a buffer that has pH 8.80?

Polyprotic Acids

98. Calculate the concentrations of the various species in 0.200 M H_3AsO_4 solution. Compare the concentrations with those of the analogous species in 0.100 M H_3PO_4 solution (Example 18-21 and Table 18-10 in the text).

99. Citric acid, the acid in lemons and other citrus fruits, has the structure

$$CH_2COOH$$
$$HO—C—COOH$$
$$CH_2COOH$$

which we may abbreviate as $C_3H_5O(COOH)_3$ or H_3A. It is a triprotic acid. Write the chemical equations for the three stages in the ionization of citric acid with the appropriate K_a values.

100. Calculate the concentrations of H_3O^+, OH^-, HCO_3^-, and CO_3^{2-} in 0.050 M H_2CO_3 solution.

101. Some kidney stones are crystalline deposits of calcium oxalate, a salt of oxalic acid, $(COOH)_2$. Calculate the concentrations of H_3O^+, OH^-, $H(COO)_2^-$, and $(COO^-)_2$ in 0.050 M $(COOH)_2$. Compare the concentrations with those obtained in Exercise 100. How can you explain the difference between the concentrations of HCO_3^- and $H(COO)_2^-$? between CO_3^{2-} and $(COO^-)_2$?

Mixed Exercises

102. A solution has $[OH^-] = 2.0 \times 10^{-9}$ mol/L at 25°C. Calculate $[H^+]$ and pH.

103. Calculate the pH at 25°C of these solutions: (a) 0.0050 M $Ca(OH)_2$; (b) 0.20 M chloroacetic acid, $CH_2ClCOOH$, $K_a = 1.4 \times 10^{-3}$; (c) 0.040 M pyridine, C_5H_5N (a monoprotic base).

104. A solution of oxalic acid, $(COOH)_2$, is 0.20 M. (a) Calculate the pH. (b) What is the concentration of the oxalate ion, $(COO)_2^{2-}$?

105. A solution is 0.050 M in HCl and 0.025 M in benzoic acid, $HC_7H_5O_2$. What is the concentration of benzoate ion, $C_7H_5O_2^-$?

*106. What is the pH of a solution that is a mixture of HClO and HIO, each at 0.10 M concentration? For HIO, $K_a = 2.3 \times 10^{-11}$.

Objectives

As you study this chapter, you should learn

☐ About the concepts of solvolysis and hydrolysis

☐ To apply these concepts to salts of strong bases and weak acids

☐ To apply these concepts to salts of weak bases and strong acids

☐ To apply these concepts to salts of weak bases and weak acids

☐ To apply these concepts to salts of small, highly charged cations

☐ About titration curves for (a) strong acids and strong bases and (b) weak acids and strong bases

Solutions of most detergents and other cleaning materials are basic because anions of weak acids hydrolyze to give basic solutions. A universal indicator shows the pH of each solution.

Solvolysis is the reaction of a substance with the solvent in which it is dissolved. The solvolysis reactions that we shall consider in this chapter occur in aqueous solutions, so they are called *hydrolysis* reactions. **Hydrolysis** is the reaction of a substance with water or its ions. One common kind of hydrolysis reaction involves the combination of the anion of a *weak acid* with H_3O^+ from water to form nonionized acid molecules. The removal of H_3O^+ upsets the H_3O^+/OH^- balance in water and produces basic solutions:

$$\underset{\substack{\text{anion of} \\ \text{weak acid}}}{A^-} + H_3O^+ \rightleftharpoons \underset{\text{weak acid}}{HA} + H_2O$$

This reaction is represented as

$$A^- + H_2O \rightleftharpoons HA + OH^- \qquad \text{(excess } OH^-\text{, so solution is basic)}$$

Recall that in

neutral solutions	$[H_3O^+] = [OH^-] = 1.0 \times 10^{-7}\,M$
basic solutions	$[H_3O^+] < [OH^-]$ or $[OH^-] > 1.0 \times 10^{-7}\,M$
acidic solutions	$[H_3O^+] > [OH^-]$ or $[H_3O^+] > 1.0 \times 10^{-7}\,M$

In Brønsted–Lowry terminology, anions of strong acids are extremely weak bases, whereas anions of weak acids are stronger bases (Section 10-3). Conversely, cations of strong bases are extremely weak acids and cations

Examples of conjugate acid–base pairs:

Acid	Conjugate Base
strong (HCl) $\longrightarrow$	weak (Cl$^-$)
weak (HCN) $\longrightarrow$	strong (CN$^-$)

Base	Conjugate Acid
strong (OH$^-$) $\longrightarrow$	weak (H$_2$O)
weak (NH$_3$) $\longrightarrow$	strong (NH$_4{}^+$)

of weak bases are stronger acids. To refresh your memory, consider the following examples.

Nitric acid, a typical strong acid, is essentially completely ionized in dilute aqueous solution. *Dilute* aqueous solutions of HNO$_3$ contain equal numbers of NO$_3{}^-$ and H$_3$O$^+$ ions. Nitrate ions show almost no tendency to combine with H$_3$O$^+$ ions to form nonionized HNO$_3$ in dilute aqueous solutions; thus NO$_3{}^-$ is a weak base.

$$HNO_3 + H_2O \xrightarrow{100\%} H_3O^+ + NO_3{}^-$$

On the other hand, acetic acid (a weak acid) is only slightly ionized in dilute aqueous solution. Acetate ions have a strong tendency to combine with H$_3$O$^+$ to form CH$_3$COOH molecules, which ionize only slightly.

$$CH_3COOH + H_2O \rightleftharpoons H_3O^+ + CH_3COO^-$$

Hence, the CH$_3$COO$^-$ ion is a much stronger base than the NO$_3{}^-$ ion.

In dilute solutions, strong acids and strong soluble bases are completely ionized or dissociated. Let us now consider dilute aqueous solutions of salts that contain no free acid or base. Based on our classification of acids and bases, we can identify four different kinds of salts:

1. Salts of strong soluble bases and strong acids
2. Salts of strong soluble bases and weak acids
3. Salts of weak bases and strong acids
4. Salts of weak bases and weak acids

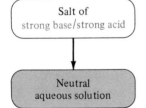

19-1 Salts of Strong Soluble Bases and Strong Acids

Salts derived from strong soluble bases and strong acids give *neutral* solutions because neither the cation nor the anion reacts with H$_2$O. Consider an aqueous solution of NaCl, the salt of NaOH and HCl. Sodium chloride is ionic even in the solid state. It dissociates into hydrated ions in H$_2$O. The H$_2$O ionizes slightly to produce equal numbers of H$_3$O$^+$ and OH$^-$ ions.

$$NaCl(solid) \xrightarrow[100\%]{H_2O} Na^+ + Cl^-$$

$$H_2O + H_2O \rightleftharpoons OH^- + H_3O^+$$

We see that aqueous solutions of NaCl contain four ions, Na$^+$, Cl$^-$, H$_3$O$^+$ and OH$^-$. The cation of the salt, Na$^+$, is such a weak acid that it does not react with the anion of water, OH$^-$. The anion of the salt, Cl$^-$, is such a weak base that it does not react with the cation of water, H$_3$O$^+$. Therefore, solutions of salts of strong bases and strong acids are *neutral* because neither ion of such a salt reacts to upset the H$_3$O$^+$/OH$^-$ balance in water.

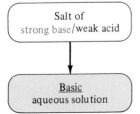

19-2 Salts of Strong Bases and Weak Acids

When salts derived from strong soluble bases and weak acids are dissolved in water, the resulting solutions are always basic. This is because anions of weak acids react with water to form hydroxide ions. Consider a solution of sodium acetate, NaCH$_3$COO, the salt of NaOH and CH$_3$COOH. It is soluble and dissociates completely in water.

$$NaCH_3COO(solid) \xrightarrow[100\%]{H_2O} Na^+ + \boxed{CH_3COO^-}$$

$$\underbrace{H_2O + H_2O \rightleftharpoons OH^- + \boxed{H_3O^+}}_{\text{Equilibrium is shifted}} \qquad (\text{result is excess } OH^-)$$

$$\Updownarrow$$

$$CH_3COOH + H_2O$$

Acetate ion is the conjugate base of a *weak* acid, CH_3COOH. Thus, it combines with H_3O^+ to form CH_3COOH. As H_3O^+ is removed from the solution, causing more H_2O to ionize, an excess of OH^- builds up. So the solution becomes basic. The preceding equations can be combined into a single equation.

$$CH_3COO^- + H_2O \rightleftharpoons CH_3COOH + OH^-$$

The equilibrium constant for this reaction is called a (base) hydrolysis constant, or K_b for CH_3COO^-.

$$K_b = \frac{[CH_3COOH][OH^-]}{[CH_3COO^-]} \qquad (K_b \text{ for } CH_3COO^-)$$

We can evaluate this equilibrium constant from other known expressions. If we multiply the preceding expression by $[H_3O^+]/[H_3O^+]$ (an algebraic manipulation that doesn't change the value of the expression) we have

$$K_b = \frac{[CH_3COOH][OH^-]}{[CH_3COO^-]} \times \frac{[H_3O^+]}{[H_3O^+]} = \frac{[CH_3COOH]}{[H_3O^+][CH_3COO^-]} \times \frac{[H_3O^+][OH^-]}{1}$$

We recognize that

$$K_b = \frac{1}{K_{a\,(CH_3COOH)}} \times \frac{K_w}{1} = \frac{K_w}{K_{a\,(CH_3COOH)}} = \frac{1.0 \times 10^{-14}}{1.8 \times 10^{-5}} \qquad \text{which gives}$$

$$K_b = \frac{[CH_3COOH][OH^-]}{[CH_3COO^-]} = 5.6 \times 10^{-10}$$

We have calculated K_b, the hydrolysis constant for the acetate ion, CH_3COO^-.

We can do the same kind of calculations for the anion of any weak monoprotic acid and find that $K_b = K_w/K_a$ where K_a refers to the ionization constant for the weak monoprotic acid from which the anion is derived. This equation can be rearranged to

> $K_w = K_aK_b$ (valid for *any conjugate acid–base pair* in aqueous solution)

If either K_a or K_b is known, the other can be calculated.

LeChatelier's Principle applies to equilibria in aqueous solution. It enables us to make accurate predictions.

Consider the similarity of this reaction to the ionization of aqueous ammonia.

$$NH_3 + H_2O \rightleftharpoons NH_4^+ + OH^-$$

Both bases remove an H^+ from an H_2O molecule and form an OH^- ion.

Hydrolysis constants, or K_b's, for anions of weak acids can be determined experimentally. The values obtained from experiments agree with the calculated values. Please note that this K_b refers to a reaction in which the anion of a weak acid acts as a base.

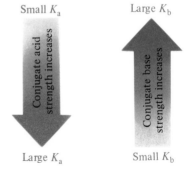

Example 19-1

(a) Write the equation for the reaction of the base CN^- with water.
(b) The value of the ionization constant for hydrocyanic acid, HCN, is 4.0×10^{-10}. What is the value of K_b for the cyanide ion, CN^-?

Plan
(a) The base CN^- accepts H^+ from H_2O to form the weak acid HCN and OH^- ions.
(b) We know that $K_aK_b = K_w$. So we solve for K_b and substitute into the equation.

An inert solid has been suspended in the liquids to improve the quality of photographs of pH meters.

Solution

(a) $CN^- + H_2O \rightleftharpoons HCN + OH^-$

(b) We are given $K_a = 4.0 \times 10^{-10}$ for HCN, and we know that $K_w = 1.0 \times 10^{-14}$.

$$K_b = \frac{K_w}{K_a} = \frac{1.0 \times 10^{-14}}{4.0 \times 10^{-10}} = \boxed{2.5 \times 10^{-5}}$$

EOC 12, 13

If we did not know K_b, we could use $K_{a(HCN)}$ to find $K_{b(CN^-)}$.

Example 19-2

Calculate $[OH^-]$, pH, and the percent hydrolysis for 0.10 M solutions of (a) $NaCH_3COO$, sodium acetate, and (b) NaCN, sodium cyanide. Both $NaCH_3COO$ and NaCN are soluble ionic salts that are completely dissociated in H_2O. From the text, K_b for $CH_3COO^- = 5.6 \times 10^{-10}$; from Example 19-1, K_b for $CN^- = 2.5 \times 10^{-5}$.

Plan

We recognize that both $NaCH_3COO$ and NaCN are salts of weak acids and strong bases. The anions in such salts hydrolyze to give basic solutions. As we did in Chapters 17 and 18, we write the appropriate chemical equation and equilibrium constant expression, we complete the reaction summary, and then we substitute the algebraic representations of equilibrium concentrations into the equilibrium constant expression and solve for the unknown concentration(s).

Solution

(a) The overall equation for the reaction of CH_3COO^- with H_2O and its equilibrium constant expression are

$$CH_3COO^- + H_2O \rightleftharpoons CH_3COOH + OH^-,$$

$$K_b = \frac{[CH_3COOH][OH^-]}{[CH_3COO^-]} = 5.6 \times 10^{-10}$$

Let x = mol/L of CH_3COO^- that hydrolyzes. Then $x = [CH_3COOH] = [OH^-]$.

	CH_3COO^-	$+ H_2O \longrightarrow$	CH_3COOH	$+ OH^-$
initial	0.10 M			
change due to rxn	$-x\ M$		$+x\ M$	$+x\ M$
at equil	$(0.10 - x)\ M$		$x\ M$	$x\ M$

Substitution into the equilibrium constant expression gives

$$\frac{[CH_3COOH][OH^-]}{[CH_3COO^-]} = 5.6 \times 10^{-10} = \frac{(x)(x)}{(0.10 - x)} \quad \text{so} \quad x = 7.5 \times 10^{-6}$$

$$x = \boxed{7.5 \times 10^{-6}\ M = [OH^-]} \qquad pOH = 5.12 \qquad \text{and} \qquad \boxed{pH = 8.88}$$

Table 19-1
Data for 0.10 M Solutions of $NaCH_3COO$, NaCN, and NH_3

	0.10 M $NaCH_3COO$	0.10 M NaCN	0.10 M aq NH_3
K_a for parent acid	1.8×10^{-5}	4.0×10^{-10}	
K_b for anion	5.6×10^{-10}	2.5×10^{-5}	K_b for $NH_3 = 1.8 \times 10^{-5}$
$[OH^-]$	$7.5 \times 10^{-6}\ M$	$1.6 \times 10^{-3}\ M$	$1.3 \times 10^{-3}\ M$
% hydrolysis	0.0075%	1.6%	1.3% ionized
pH	8.88	11.20	11.11

The 0.10 M NaCH$_3$COO solution is distinctly basic.

$$\% \text{ hydrolysis} = \frac{[\text{CH}_3\text{COO}^-]_{\text{hydrolyzed}}}{[\text{CH}_3\text{COO}^-]_{\text{total}}} \times 100\% = \frac{7.5 \times 10^{-6} \, M}{0.10 \, M} \times 100\%$$

$$= \boxed{0.0075\% \text{ hydrolysis}}$$

(b) We perform the same kind of calculation for 0.10 M NaCN. Let y = mol/L of CN$^-$ that hydrolyzes. Then y = [HCN] = [OH$^-$].

	CN$^-$	+	H$_2$O	$\rightleftharpoons$	HCN	+	OH$^-$
initial	0.10 M						
change due to rxn	$-y$ M				$+y$ M		$+y$ M
at equil	$(0.10 - y)$ M				y M		y M

$$K_b = \frac{[\text{HCN}][\text{OH}^-]}{[\text{CN}^-]} = 2.5 \times 10^{-5}$$

Substitution into this expression gives

$$\frac{(y)(y)}{(0.10 - y)} = 2.5 \times 10^{-5} \quad \text{so} \quad y = 1.6 \times 10^{-3} \, M$$

$$y = \boxed{[\text{OH}^-] = 1.6 \times 10^{-3} \, M} \quad \text{pOH} = 2.80 \quad \text{and} \quad \boxed{\text{pH} = 11.20}$$

The 0.10 M NaCN solution is even more basic than the 0.10 M NaCH$_3$COO solution in part (a).

$$\% \text{ hydrolysis} = \frac{[\text{CN}^-]_{\text{hydrolyzed}}}{[\text{CN}^-]_{\text{total}}} \times 100\% = \frac{1.6 \times 10^{-3} \, M}{0.10 \, M} \times 100\%$$

$$= \boxed{1.6\% \text{ hydrolysis}}$$

EOC 15, 16

The pH of 0.10 M NaCH$_3$COO is 8.88 (top). The pH of 0.10 M NaCN is 11.20 (bottom).

The 0.10 M solution of NaCN is much more basic than the 0.10 M solution of NaCH$_3$COO because CN$^-$ is a much stronger base than CH$_3$COO$^-$. This is expected because HCN is a much weaker acid than CH$_3$COOH.

We have just shown that the percent hydrolysis for 0.10 M NaCN (1.6%) is about 213 times greater than the percent hydrolysis for 0.10 M NaCH$_3$COO (0.0075%). Table 19-1 summarizes these calculations. Data for a 0.10 M solution of aqueous ammonia provide reference points.

19-3 Salts of Weak Bases and Strong Acids

The second common kind of hydrolysis reaction involves the reaction of the cation of a weak base with OH$^-$ from water to form nonionized molecules of the weak base. The removal of OH$^-$ upsets the H$_3$O$^+$/OH$^-$ balance in water and gives an excess of H$_3$O$^+$. This makes such solutions *acidic*. Consider a solution of ammonium chloride, the salt of aqueous NH$_3$ and HCl.

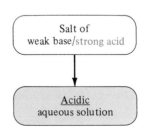

Salt of
weak base/strong acid

↓

Acidic
aqueous solution

Ammonium chloride is an ionic salt that is soluble in water.

$$NH_4Cl(solid) \xrightarrow[100\%]{H_2O} \left[\overline{NH_4{}^+} \right] + Cl^-$$

$$\underbrace{H_2O + H_2O \rightleftharpoons}_{\text{Equilibrium is shifted}} \left[OH^- \right] \quad H_3O^+ \qquad \text{(result is excess } H_3O^+\text{)}$$

$$\Updownarrow$$

$$NH_3 + H_2O$$

Ammonium ions from NH_4Cl react with OH^- to form nonionized NH_3 and H_2O molecules. This reaction removes OH^- from the system, so it causes more H_2O to ionize to produce an excess of H_3O^+. We combine the preceding equations into a single equation and write its equilibrium constant expression.

Similar equations can be written for cations derived from other weak bases.

$$NH_4{}^+ + H_2O \rightleftharpoons NH_3 + H_3O^+ \qquad K_a = \frac{[NH_3][H_3O^+]}{[NH_4{}^+]}$$

The expression $K_w = K_aK_b$ is valid for *any* conjugate acid–base pair in aqueous solution. We use it for the $NH_4{}^+/NH_3$ pair:

To derive $K_w = K_aK_b$ for this case, multiply the K_a expression by $[OH^-]/[OH^-]$ and simplify.

$$K_{a\,(NH_4{}^+)} = \frac{K_w}{K_{b\,(NH_3)}} = \frac{1.0 \times 10^{-14}}{1.8 \times 10^{-5}} = 5.6 \times 10^{-10} = \frac{[NH_3][H_3O^+]}{[NH_4{}^+]}$$

The fact that K_a for the ammonium ion, $NH_4{}^+$, is the same as K_b for the acetate ion, CH_3COO^-, should not be surprising. Recall that the ionization constants for CH_3COOH and aqueous NH_3 are equal (by coincidence). Thus, we expect CH_3COO^- to hydrolyze to the same extent as $NH_4{}^+$ does. We use hydrolysis constants just as we used other equilibrium constants.

Ammonium nitrate is widely used as a fertilizer. It contributes significantly to soil acidity.

Example 19-3

Calculate the pH of a 0.20 *M* solution of ammonium nitrate, NH_4NO_3. K_a for $NH_4{}^+ = 5.6 \times 10^{-10}$.

Plan

We recognize that NH_4NO_3 is the salt of a weak base, NH_3, and a strong acid, HNO_3, and that the cations of such salts hydrolyze to give acidic solutions. We proceed as we did in Example 19-2.

Solution

The cation of the weak base reacts with H_2O. We let $x = $ mol/L of $NH_4{}^+$ that hydrolyzes. Then $x = [NH_3] = [H_3O^+]$.

	$NH_4{}^+$	$+$	H_2O	$\rightleftharpoons$	NH_3	$+$	H_3O^+
initial	0.20 *M*						
change due to rxn	$-x$ *M*				$+x$ *M*		$+x$ *M*
at equil	$(0.20 - x)$ *M*				x *M*		x *M*

Substituting into the K_a expression gives

$$K_a = \frac{[NH_3][H_3O^+]}{[NH_4{}^+]} = \frac{(x)(x)}{(0.20 - x)} = 5.6 \times 10^{-10}$$

Making the usual simplifying assumption gives $x = 1.1 \times 10^{-5}$ *M* $= [H_3O^+]$

and pH = 4.96. The 0.20 *M* NH_4NO_3 solution is distinctly acidic.

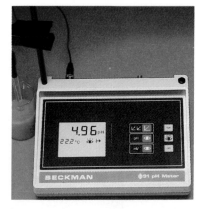

The pH of 0.20 *M* NH_4NO_3 solution is 4.96.

EOC 22

19-4 Salts of Weak Bases and Weak Acids

Salts of weak bases and weak acids are the fourth class of salts. Most are soluble. Salts of weak bases and weak acids contain cations that would give acidic solutions and anions that would give basic solutions. Will solutions of such salts be neutral, basic, or acidic? They may be any one of the three, depending on the relative strengths of the weak molecular acid and weak molecular base from which each salt is derived. Thus, salts of this class may be divided into three types, depending on the relative strengths of their parent weak bases and weak acids.

Salts of Weak Bases and Weak Acids for Which $K_b = K_a$

The common example of a salt of this type is ammonium acetate, NH_4CH_3COO, the salt of aqueous NH_3 and CH_3COOH. The ionization constants for both aqueous NH_3 and CH_3COOH are 1.8×10^{-5}. We know that ammonium ions react with water to produce H_3O^+.

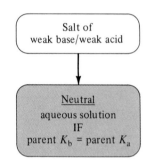

Salt of
weak base/weak acid

Neutral
aqueous solution
IF
parent K_b = parent K_a

$$NH_4^+ + H_2O \rightleftharpoons NH_3 + H_3O^+ \qquad K_a = \frac{[NH_3][H_3O^+]}{[NH_4^+]} = 5.6 \times 10^{-10}$$

We also recall that CH_3COO^- reacts with water to produce OH^- by the following reaction:

$$CH_3COO^- + H_2O \rightleftharpoons CH_3COOH + OH^- \qquad K_b = \frac{[CH_3COOH][OH^-]}{[CH_3COO^-]} = 5.6 \times 10^{-10}$$

Because these K values are equal, the NH_4^+ produces *exactly* as many H_3O^+ ions as the CH_3COO^- produces OH^- ions. Thus, we predict that solutions of NH_4CH_3COO are neutral.

Salts of Weak Bases and Weak Acids for Which $K_b > K_a$

Salts of weak bases and weak acids for which K_b is greater than K_a are always basic because the anion of the weaker acid hydrolyzes to a greater extent than the cation of the stronger base.

Consider NH_4CN, ammonium cyanide. Because K_a for HCN (4.0×10^{-10}) is much smaller than K_b for NH_3 (1.8×10^{-5}), then K_b for CN^- (2.5×10^{-5}) is much larger than K_a for NH_4^+ (5.6×10^{-10}). This tells us that the CN^- ion hydrolyzes to a much greater extent than the NH_4^+ ion, and so solutions of NH_4CN are distinctly basic. Stated differently, CN^- is much stronger as a base than NH_4^+ is as an acid.

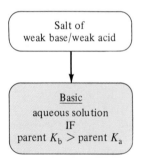

Salt of
weak base/weak acid

Basic
aqueous solution
IF
parent K_b > parent K_a

$$NH_4^+ + H_2O \rightleftharpoons NH_3 + H_3O^+$$
$$CN^- + H_2O \rightleftharpoons HCN + OH^-$$

$\rightarrow 2H_2O$

This reaction occurs to greater extent;
∴ solution is basic.

Salts of Weak Bases and Weak Acids for Which $K_b < K_a$

Salts of weak bases and weak acids for which K_b is less than K_a are acidic because the cation of the weaker base hydrolyzes to a greater extent than the anion of the stronger acid. Consider ammonium fluoride, NH_4F, the salt of aqueous ammonia and hydrofluoric acid.

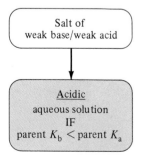

Salt of
weak base/weak acid

Acidic
aqueous solution
IF
parent K_b < parent K_a

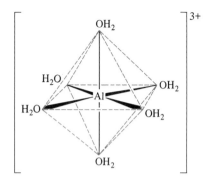

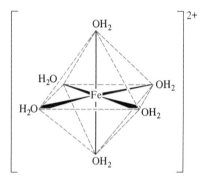

Figure 19-1
Structures of hydrated aluminum and iron(II) ions $[Al(OH_2)_6]^{3+}$ and $[Fe(OH_2)_6]^{2+}$.

Figure 19-2
Hydrolysis of hydrated aluminum ion to produce H_3O^+—that is, the removal of a proton from a coordinated H_2O molecule by a noncoordinated H_2O molecule.

Recall that K_a for aqueous NH_3 is 1.8×10^{-5} and that K_a for HF is 7.2×10^{-4}. The K_a value for NH_4^+ (5.6×10^{-10}) is slightly larger than the K_b value for F^- (1.4×10^{-11}). So the NH_4^+ ion hydrolyzes to a slightly greater extent than the F^- ion. In other words, NH_4^+ is slightly stronger as an acid than F^- is as a base. Solutions of NH_4F are slightly acidic.

$$NH_4^+ + H_2O \rightleftharpoons NH_3 + H_3O^+$$
$$F^- + H_2O \rightleftharpoons HF + OH^-$$

$\rightarrow 2H_2O$

This reaction occurs to greater extent; ∴ solution is acidic.

19-5 Salts That Contain Small, Highly Charged Cations

Solutions of certain common salts of strong acids are acidic. For this reason, many homeowners apply iron(II) sulfate, $FeSO_4 \cdot 7H_2O$, or aluminum sulfate, $Al_2(SO_4)_3 \cdot 18H_2O$, to the soil around "acid-loving" plants such as azaleas, camelias, and hollies. You are probably familiar with the sour, "acid" taste of alum, $KAl(SO_4)_2 \cdot 12H_2O$, a substance that is frequently added to pickles.

Each salt contains a small, highly charged cation and the anion of a strong acid. Solutions of such salts are acidic, because these cations hydrolyze to produce excess hydronium ions. Consider aluminum chloride, $AlCl_3$, as a typical example. When solid anhydrous $AlCl_3$ is added to water, the water becomes very warm. Ions are always hydrated in solution. In many cases, the interaction between positively charged ions and the negative ends of polar water molecules is so strong that salts crystallize from aqueous solution with definite numbers of water molecules. Salts containing Al^{3+}, Fe^{2+}, Fe^{3+}, and Cr^{3+} ions usually crystallize from aqueous solutions with six water molecules associated with each metal ion. The salts contain the hydrated cations $[Al(OH_2)_6]^{3+}$, $[Fe(OH_2)_6]^{2+}$, $[Fe(OH_2)_6]^{3+}$, and $[Cr(OH_2)_6]^{3+}$, respectively, in the solid state. Such species exist in aqueous solutions. Each of these species is octahedral; i.e., the metal ion (M^{n+}) is located at the center of a regular octahedron, and the O atoms in six H_2O molecules are located at the corners (Figure 19-1). In the metal–oxygen bonds of the hydrated cation, electron density is decreased around the O end of each H_2O molecule by the positively charged metal ion. This weakens the H—O bonds in coordinated H_2O molecules relative to the H—O bonds in noncoordinated H_2O molecules. Consequently, the coordinated H_2O molecules can donate H^+ to solvent H_2O molecules to form H_3O^+ ions. This produces acidic solutions (Figure 19-2).

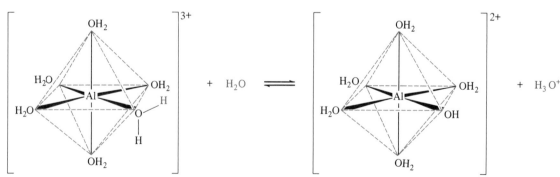

The equation for the hydrolysis of Al^{3+} is written as follows:

$$[Al(OH_2)_6]^{3+} + H_2O \rightleftharpoons [Al(OH)(OH_2)_5]^{2+} + H_3O^+$$

$$K_a = \frac{[[Al(OH)(OH_2)_5]^{2+}][H_3O^+]}{[[Al(OH_2)_6]^{3+}]} = 1.2 \times 10^{-5}$$

Removing an H^+ converts a coordinated water molecule to a coordinated hydroxide ion and decreases the positive charge on the hydrated species.

This can be written in more abbreviated form as

$$Al^{3+} + 2H_2O \rightleftharpoons Al(OH)^{2+} + H_3O^+$$

$$K_a = \frac{[Al(OH)^{2+}][H_3O^+]}{[Al^{3+}]} = 1.2 \times 10^{-5}$$

Hydrolysis of small, highly charged cations may occur beyond the first step. In many cases these reactions are quite complex. They may involve two or more cations reacting with each other to form large polymeric species. For most common cations, consideration of the first hydrolysis constant is adequate for our calculations.

Example 19-4

Calculate the pH and percent hydrolysis in 0.10 M $AlCl_3$ solution. For $[Al(OH_2)_6]^{3+}$ (often abbreviated Al^{3+}), $K_a = 1.2 \times 10^{-5}$.

Plan

We recognize that $AlCl_3$ contains a small, highly charged cation that hydrolyzes to give an acidic solution. We represent the equilibrium concentrations and proceed as we did in earlier examples.

Solution

The equation for the reaction and its hydrolysis constant are

$$Al^{3+} + 2H_2O \rightleftharpoons Al(OH)^{2+} + H_3O^+ \qquad K_a = \frac{[Al(OH)^{2+}][H_3O^+]}{[Al^{3+}]} = 1.2 \times 10^{-5}$$

Let $x = $ mol/L of Al^{3+} that hydrolyzes. Then $x = [Al(OH)^{2+}] = [H_3O^+]$, and $[Al^{3+}] = (0.10 - x)$.

$$\begin{array}{ccccc} Al^{3+} & + \ 2H_2O & \rightleftharpoons & Al(OH)^{2+} & + \ H_3O^+ \\ \text{equil} \ (0.10 - x) \ M & & & x \ M & x \ M \end{array}$$

$$\frac{(x)(x)}{(0.10 - x)} = 1.2 \times 10^{-5} \qquad \text{so} \qquad x = 1.1 \times 10^{-3}$$

$[H_3O^+] = 1.1 \times 10^{-3} \ M,$ **pH = 2.96,** and so the solution is quite acidic. We

let $x = [Al^{3+}]$ that hydrolyzes, so we can calculate the percent hydrolysis.

$$\% \text{ hydrolysis} = \frac{[Al^{3+}]_{\text{hydrolyzed}}}{[Al^{3+}]_{\text{total}}} \times 100\% = \frac{1.1 \times 10^{-3} \ M}{0.10 \ M} \times 100\%$$

$$= \boxed{1.1\% \text{ hydrolyzed}}$$

As a reference point, CH_3COOH is 1.3% ionized in 0.10 M solution (Example 18-11). In 0.10 M solution $AlCl_3$ is 1.1% hydrolyzed. The acidities of the two solutions are similar.

The pH of 0.10 M $AlCl_3$ is 2.96. The pH of 0.10 M CH_3COOH is 2.89.

EOC 30

Table 19-2
Ionic Radii and Hydrolysis Constants for Some Cations

Cation	Ionic Radius (Å)	Hydrated Cation	K_a
Li^+	0.60	$[Li(OH_2)_4]^+$	1×10^{-14}
Be^{2+}	0.31	$[Be(OH_2)_4]^{2+}$	1.0×10^{-5}
Na^+	0.95	$[Na(OH_2)_6]^+$ (?)	10^{-14}
Mg^{2+}	0.65	$[Mg(OH_2)_6]^{2+}$	3.0×10^{-12}
Al^{3+}	0.50	$[Al(OH_2)_6]^{3+}$	1.2×10^{-5}
Fe^{2+}	0.76	$[Fe(OH_2)_6]^{2+}$	3.0×10^{-10}
Fe^{3+}	0.64	$[Fe(OH_2)_6]^{3+}$	4.0×10^{-3}
Co^{2+}	0.74	$[Co(OH_2)_6]^{2+}$	5.0×10^{-10}
Co^{3+}	0.63	$[Co(OH_2)_6]^{3+}$	1.7×10^{-2}
Cu^{2+}	0.96	$[Cu(OH_2)_4]^{2+}$	1.0×10^{-8}
Zn^{2+}	0.74	$[Zn(OH_2)_6]^{2+}$	2.5×10^{-10}
Hg^{2+}	1.10	$[Hg(OH_2)_6]^{2+}$	8.3×10^{-7}
Bi^{3+}	0.74	$[Bi(OH_2)_6]^{3+}$	1.0×10^{-2}

Pepto-Bismol contains $BiO(HOC_4H_6COO)$, bismuth subsalicylate, a *hydrolyzed* bismuth salt. Such salts "coat" polar surfaces such as the lining of the stomach.

Smaller, more highly charged cations are stronger acids than larger, less highly charged cations (Table 19-2). This is because the smaller, more highly charged cations interact with coordinated water molecules more strongly.

For isoelectronic cations in the same period in the periodic table, the smaller, more highly charged cation is the stronger acid. (Compare K_a values for Li^+ and Be^{2+} and for Na^+, Mg^{2+}, and Al^{3+}.) For cations with the same charge from the same group in the periodic table, the smaller cation hydrolyzes to a greater extent. (Compare K_a values for Be^{2+} and Mg^{2+}.) If we compare cations of the same element in different oxidation states, the smaller, more highly charged cation is the stronger acid. (Compare K_a values for Fe^{2+} and Fe^{3+} and for Co^{2+} and Co^{3+}.)

19-6 Strong Acid–Strong Base Titration Curves

A **titration curve** is a plot of pH versus the amount (volume, usually) of acid or base added. It displays graphically the change in pH as acid or base is added to a solution and indicates clearly how pH changes near the equivalence point.

The point at which the color of an indicator changes in a titration is known as the **end point**. It is determined by the K_a value for the indicator (Section 18-5). Table 19-3 shows a few acid–base indicators and the pH ranges over which their colors change. Typically, color changes occur over a range of 1.5 to 2.0 pH units.

The **equivalence point** is the point at which chemically equivalent amounts of acid and base have reacted. Ideally, the end point and the equivalence point in a titration should coincide. In practice, we try to select an indicator whose range of color change includes the equivalence point. We use the same procedures in both standardization and analysis. This minimizes any error arising from a difference between end point and equivalence point.

Consider the titration of 100.0 mL of a 0.100 *M* solution of HCl with a 0.100 *M* solution of NaOH. As we know, NaOH and HCl react in a 1:1

Table 19-3
Range and Color Changes of Some Common Acid–Base Indicators

Indicators	pH Scale												
	1	2	3	4	5	6	7	8	9	10	11	12	13
methyl orange	←red→3.1—4.4←							yellow					→
methyl red	←red→4.4——6.2←						yellow						→
bromthymol blue	←yellow——6.2—7.6←						blue						→
neutral red	←red——→6.8—8.0←						yellow						→
phenolphthalein	←colorless——→8.0——10.0←							red→ colorless beyond 13.0					

ratio. We shall calculate the pH of the solution at several stages as NaOH is added.

Titrations are usually done with 50-mL or smaller burets. We have used 100 mL of solution in this example to simplify the arithmetic.

1. Before any NaOH is added to the 0.100 M HCl solution.

$$HCl + H_2O \xrightarrow{100\%} H_3O^+ + Cl^-$$

$$[H_3O^+] = 0.10 \ M \quad \text{so} \quad pH = 1.00$$

2. After 20.0 mL of 0.100 M NaOH has been added.

	HCl	+	NaOH	⟶	NaCl	+	H₂O
start	10.0 mmol		2.0 mmol		0 mmol		
change	−2.0 mmol		−2.0 mmol		+2.0 mmol		
after rxn	8.0 mmol		0 mmol		2.0 mmol		

We recall from Section 2-6 that we can use either moles or millimoles in calculations.

The concentration of unreacted HCl in the total volume of 120 mL is

$$M_{HCl} = \frac{8.00 \ \text{mmol HCl}}{120 \ \text{mL}} = 0.067 \ M \ \text{HCl}$$

$$[H_3O^+] = 6.7 \times 10^{-2} \ M \quad \text{so} \quad pH = 1.17$$

3. After 50.0 mL of 0.100 M NaOH has been added (midpoint of the titration):

	HCl	+	NaOH	⟶	NaCl	+	H₂O
start	10.0 mmol		5.0 mmol		0 mmol		
change	−5.0 mmol		−5.0 mmol		+5.0 mmol		
after rxn	5.0 mmol		0 mmol		5.0 mmol		

$$M_{HCl} = \frac{5.00 \ \text{mmol HCl}}{150 \ \text{mL}} = 0.033 \ M \ \text{HCl}$$

$$[H_3O^+] = 3.3 \times 10^{-2} \ M \quad \text{so} \quad pH = 1.48$$

4. After 100 mL of 0.100 M NaOH has been added.

	HCl	+	NaOH	⟶	NaCl	+	H₂O
start	10.0 mmol		10.0 mmol		0 mmol		
change	−10.0 mmol		−10.0 mmol		+10.0 mmol		
after rxn	0 mmol		0 mmol		10.0 mmol		

We have added enough NaOH to neutralize the HCl exactly, so this is the equivalence point. A strong acid and a strong base react to give a neutral salt solution, so pH = 7.00.

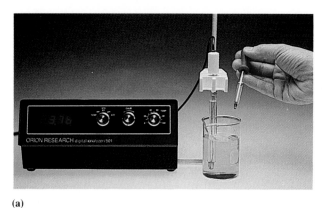

(a)

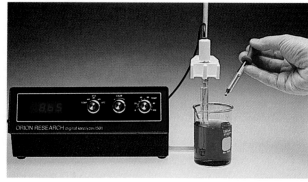

(b)

The end point of the titration of 0.100 M HCl with 0.100 M NaOH using another indicator, bromthymol blue.

5. After 110.0 mL of 0.100 M NaOH has been added, the pH is determined by the excess NaOH:

	HCl	+	NaOH	$\longrightarrow$	NaCl	+ H$_2$O
start	10.0 mmol		11.0 mmol		0 mmol	
change	-10.0 mmol		-10.0 mmol		$+10.0$ mmol	
after rxn	0 mmol		1.0 mmol		10.0 mmol	

$$M_{\text{NaOH}} = \frac{1.0 \text{ mmol NaOH}}{210 \text{ mL}} = 0.0048 \ M \text{ NaOH}$$

$[\text{OH}^-] = 4.8 \times 10^{-3} \ M$ so pOH = 2.32 and pH = 11.68

Table 19-4 displays the data for the titration of 100.0 mL of 0.100 M HCl by 0.100 M NaOH solution. A few additional points have been included to show the shape of the curve better. These data are plotted in Figure 19-3a.

This titration curve has a long "vertical section" over which pH changes very rapidly with the addition of very small amounts of base. The pH changes from 3.60 (99.5 mL NaOH added) to 10.40 (100.5 mL of NaOH added) in the vicinity of the equivalence point (100.0 mL NaOH added). The midpoint of the vertical section (pH = 7.00) is the equivalence point.

Table 19-4
Titration Data for 100.0 mL of 0.100 M HCl versus NaOH

mL of 0.100 M NaOH Added	mmol NaOH Added	mmol Excess Acid or Base	pH
0.0	0.00	10.0 H$_3$O$^+$	1.00
20.0	2.00	8.0	1.17
50.0	5.00	5.0	1.48
90.0	9.00	1.0	2.28
99.0	9.90	0.10	3.30
99.5	9.95	0.05	3.60
100.0	10.00	0.00 (eq. pt.)	7.00
100.5	10.05	0.05 OH$^-$	10.40
110.0	11.00	1.00	11.68
120.0	12.00	2.00	11.96

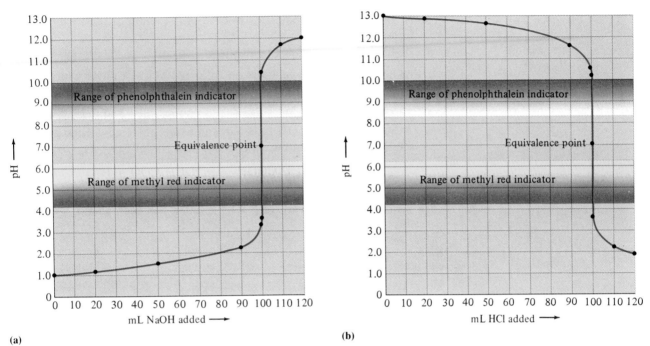

(a)

(b)

Figure 19-3
(a) The titration curve for 100 mL of 0.100 M HCl with 0.100 M NaOH. Note that the "vertical" section of the curve is quite long. The titration curves for other strong acids and bases are identical with this one *if* the same concentrations of acid and bases are used *and if* both are monoprotic.
(b) The titration curve for 100 mL of 0.100 M NaOH with 0.100 M HCl. This curve is similar to that in part (a), but inverted.

Ideally, the indicator color change should occur at pH = 7.00. For practical purposes, indicators with color changes in the range pH 4 to 10 can be used in the titration of strong acids and strong bases because the vertical portion of the titration curve is so long. Figure 19-3 shows the color ranges for methyl red and phenolphthalein, two widely used indicators. Both fall within the vertical section of the NaOH/HCl titration curve. When a strong acid is added to a solution of a strong base, the titration curve is inverted, but its essential characteristics are the same (Figure 19-3b).

In Figure 19-3a we see that the curve rises very slowly before the equivalence point because there is no buffering action. It then rises very rapidly near the equivalence point because there is no hydrolysis. The curve becomes almost flat beyond the equivalence point.

19-7 Weak Acid–Strong Base Titration Curves

In the last section we described titration curves for strong acids and strong bases. When a weak acid is titrated with a strong base, the curve is quite different. The solution is buffered *before* the equivalence point. It is basic *at* the equivalence point because salts of weak acids and strong bases hydrolyze to give basic solutions. So we can separate the calculations on this kind of titrations into four distinct types, which correspond to four regions of the titration curves.

1. Before any base is added, the pH depends on the weak acid alone.
2. After some base has been added, but before the equivalence point, a series of salt/weak acid buffer solutions determines the pH.

3. At the equivalence point, hydrolysis of the anion of the weak acid determines the pH.
4. Beyond the equivalence point, excess strong base determines the pH.

Consider the titration of 100.0 mL of 0.100 M CH_3COOH with 0.100 M NaOH solution. (The strong electrolyte is added to the weak electrolyte.)

1. Before any base is added, the pH is 2.89 (see Example 18-11).
2. As soon as some NaOH is added, and before the equivalence point, the solution is buffered because it contains both $NaCH_3COO$ and CH_3COOH.

$$NaOH + CH_3COOH \longrightarrow NaCH_3COO + H_2O$$
$$\text{lim amt} \qquad \text{excess}$$

For instance, after 20.0 mL of NaOH solution has been added, we have

	NaOH	+	CH_3COOH	$\longrightarrow$	$NaCH_3COO$	+ H_2O
start	2.00 mmol		10.00 mmol		0 mmol	
change	−2.00 mmol		−2.00 mmol		+2.00 mmol	
after rxn	0 mmol		8.00 mmol		2.00 mmol	

These amounts are present in 120 mL of solution, so the concentrations are

$$M_{CH_3COOH} = \frac{8.00 \text{ mmol } CH_3COOH}{120 \text{ mL}} = 0.0667 \ M \ CH_3COOH$$

$$M_{NaCH_3COO} = \frac{2.00 \text{ mmol } NaCH_3COO}{120 \text{ mL}} = 0.0167 \ M \ NaCH_3COO$$

The pH of this solution is calculated as in Example 18-16.

$$\frac{[H_3O^+][CH_3COO^-]}{[CH_3COOH]} = 1.8 \times 10^{-5}$$

$$[H_3O^+] = 1.8 \times 10^{-5} \times \frac{[CH_3COOH]}{[CH_3COO^-]} = 1.8 \times 10^{-5} \times \frac{0.0667}{0.0167}$$

$$= \underline{7.2 \times 10^{-5} \ M} \qquad \text{and} \qquad pH = \underline{4.14}$$

After some NaOH has been added, the solution contains both $NaCH_3COO$ and CH_3COOH, and so it is buffered until the equivalence point is reached. All points before the equivalence point are calculated in the same way.

3. At the equivalence point, the solution is 0.0500 M in $NaCH_3COO$.

	NaOH	+	CH_3COOH	$\longrightarrow$	$NaCH_3COO$	+ H_2O
start	10.0 mmol		10.0 mmol			
change	−10.0 mmol		−10.0 mmol		+10.0 mmol	
after rxn	0		0		10.0 mmol	

$$M_{NaCH_3COO} = \frac{10.0 \text{ mmol } NaCH_3COO}{200 \text{ mL}} = 0.0500 \ M \ NaCH_3COO$$

The pH of a 0.0500 M solution of $NaCH_3COO$ is 8.72 (See Example 19-2 for a similar calculation). The solution is distinctly basic at the equivalence point because of the hydrolysis of the acetate ion.

4. Beyond the equivalence point, the concentration of the excess NaOH determines the pH of the solution just as it did in the titration of a strong acid.

Just before the equivalence point, the solution contains relatively high concentrations of $NaCH_3COO$ and relatively low concentrations of CH_3COOH. Just after the equivalence point, the solution contains relatively high concentrations of $NaCH_3COO$ and relatively low concentrations of NaOH, both basic components. In both regions our calculations are only approximations. Exact calculations of pH in these regions are beyond the scope of this text.

Table 19-5
Titration Data for 100.0 mL of 0.100 M CH$_3$COOH with 0.100 M NaOH

mL 0.100 M NaOH Added		mmol Base Added	mmol Excess Acid or Base	pH
0.0 mL		0	10.0 CH$_3$COOH	2.89
20.0 mL		2.00	8.00	4.14
50.0 mL		5.00	5.00	4.74
75.0 mL	buffered region	7.50	2.50	5.22
90.0 mL		9.00	1.00	5.70
95.0 mL		9.50	0.50	6.02
99.0 mL		9.90	0.10	6.74
100.0 mL		10.0	0 (equivalence point)	8.72
101.0 mL		10.1	0.10 OH$^-$	10.70
110.0 mL		11.0	1.0	11.68
120.0 mL		12.0	2.0	11.96

Table 19-5 lists several points on the titration curve, and Figure 19-4 shows the titration curve for 100.0 mL of 0.100 M CH$_3$COOH titrated with a 0.100 M solution of NaOH. This titration curve has a short vertical section (pH ≈ 7 to 10), and the indicator range is limited. Phenolphthalein is the indicator commonly used to titrate weak acids with strong bases (see Table 19-3).

The titration curves for weak bases and strong acids are similar to those for weak acids and strong bases except that they are inverted (recall that strong is added to weak). Figure 19-5 displays the titration curve for 100.0 mL of 0.100 M aqueous ammonia titrated with 0.100 M HCl solution.

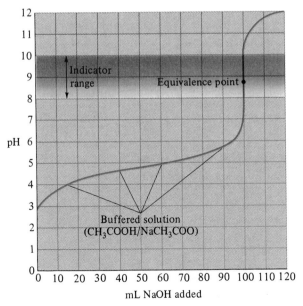

Figure 19-4
The titration curve for 100 mL of 0.100 M CH$_3$COOH with 0.100 M NaOH. The "vertical" section of this curve is much shorter than those in Figure 19-3 because the solution is buffered before the equivalence point.

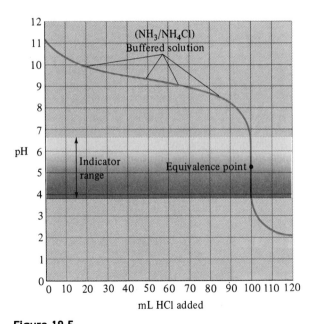

Figure 19-5
The titration curve for 100 mL of 0.100 M aqueous ammonia with 0.100 M HCl. The vertical section of the curve is relatively short because the solution is buffered before the equivalence point. The curve is very similar to that in Figure 19-4, but inverted.

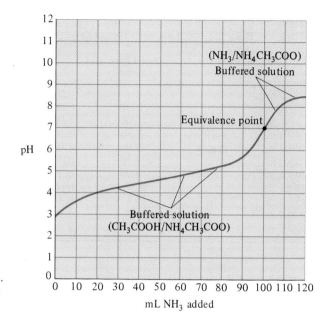

Figure 19-6
The titration curve for 100 mL of 0.100 M CH_3COOH with 0.100 M aqueous NH_3. Because the solution is buffered before and after the equivalence point, the vertical section of the curve is too short to be noticed. Therefore, color indicators cannot be used in such titrations. Physical methods such as conductivity measurements can be used to detect the end point.

In titration curves for weak acids and weak bases, change in pH is too gradual near the equivalence point for color indicators to be used. The solution is buffered both before and after the equivalence point. Figure 19-6 shows the titration curve for 100.0 mL of 0.100 M CH_3COOH solution titrated with 0.100 M aqueous NH_3.

Key Terms

Hydrolysis   The reaction of a substance with water or its ions.

Hydrolysis constant   An equilibrium constant for a hydrolysis reaction.

End point   The point at which an indicator changes color and a titration is stopped.

Equivalence point   The point at which chemically equivalent amounts of reactants have reacted.

Indicator (for acid–base titrations)   An organic compound that exhibits different colors in solutions of different acidities; used to indicate the point at which reaction between two solutes is complete.

Solvolysis   The reaction of a substance with the solvent in which it is dissolved.

Titration   A procedure in which one solution is added to another solution until the chemical reaction between the two solutes is complete; usually the concentration of one solution is known and that of the other is unknown.

Titration curve (for acid–base titration)   A plot of pH versus volume of acid or base solution added.

Exercises

Basic Ideas

1. Why can we say that 0.10 M solutions of HCl and HNO_3 contain essentially no molecules of nonionized acid?

2. Why can we say that 0.10 M solutions of HF and HNO_2 contain relatively few ions?

3. How can salts be classified conveniently into four classes? For each class, write the name and formula of a salt that fits into that category. Use examples other than those used in illustrations in this chapter.

4. Define and illustrate the following terms clearly and concisely: (a) solvolysis, (b) hydrolysis.

Salts of Strong Soluble Bases and Strong Acids

5. Some anions, when dissolved, undergo no significant reaction with water molecules. What is the relative base strength of such an anion compared to water? What effect will dissolution of such anions have on the pH of the solution?

6. Some cations in aqueous solution undergo no significant reactions with water molecules. What is the relative acid strength of such a cation compared to water? What effect will dissolution of such cations have on the pH of the solution?

7. Why do salts of strong soluble bases and strong acids give neutral aqueous solutions? Use KNO_3 to illustrate. Write names and formulas for three other salts of strong soluble bases and strong acids.

8. What determines whether the aqueous solution of a salt is acidic, neutral, or alkaline?

Salts of Strong Bases and Weak Acids

9. Some anions, when dissolved, react with water molecules. What is the relative base strength of such an anion compared to water? What effect will dissolution of such anions have on the pH of the solution?

10. Why do salts of strong soluble bases and weak acids give basic aqueous solutions? Use sodium hypochlorite, $NaClO$, to illustrate. (Recall that Clorox, Purex, and other "chlorine bleaches" are 5% $NaClO$.)

11. Write names and formulas for three salts of strong soluble bases and weak acids other than $NaClO$.

12. Calculate the equilibrium constant for the reaction of fluoride ions with water.

13. Calculate the equilibrium constant for the reaction of hypoiodite ion with water. $K_a = 2.3 \times 10^{-11}$ for HIO.

14. Calculate hydrolysis constants for the following anions of weak acids: (a) NO_2^-, (b) ClO^-, (c) $C_6H_5COO^-$. What is the relationship between K_a, the ionization constant for a weak acid, and K_b, the hydrolysis constant for the anion of the weak acid?

15. Calculate the pH of 0.15 M solutions of the following salts: (a) $NaNO_2$, (b) $NaClO$, (c) NaC_6H_5COO. (Refer to Exercise 14 for K values.)

16. What is the percent hydrolysis in each of the solutions in Exercise 15?

17. What is the pH of a 0.10 M solution of KIO? $K_b = 4.3 \times 10^{-4}$ for IO^-.

18. What is the pH of a 0.10 M solution of KF?

Salts of Weak Bases and Strong Acids

19. Why do salts of weak bases and strong acids give acidic aqueous solutions? Illustrate with NH_4NO_3, a common fertilizer.

20. Write names and formulas for four salts of weak bases and strong acids.

21. Calculate hydrolysis constants for the following cations of weak bases:
(a) NH_4^+
(b) $(CH_3)NH_3^+$, methylammonium ion
(c) $C_5H_5NH^+$, pyridinium ion

22. Calculate the pH of 0.15 M solutions of
(a) NH_4NO_3
(b) $(CH_3)NH_3NO_3$
(c) $C_5H_5NHNO_3$

23. Can you make a general statement relating parent base strength and extent of hydrolysis of the cations of Exercise 21 by using hydrolysis constants calculated in that exercise?

24. How do pH values for the following pairs of solutions compare? Why?
(a) 0.050 M NH_4Br, ammonium bromide, and 0.050 M NH_4NO_3, ammonium nitrate
(b) 0.010 M ammonium perchlorate, NH_4ClO_4, and 0.010 M ammonium sulfate, $(NH_4)_2SO_4$

Salts of Weak Bases and Weak Acids

25. Why are some aqueous solutions of salts of weak acids and weak bases neutral, whereas others are acidic and still others are basic?

26. Write the names and formulas for three salts of a weak acid and a weak base that give (a) neutral, (b) acidic, and (c) basic aqueous solutions.

27. If both the cation and anion of a salt react with water when dissolved, what determines whether the solution will be slightly acidic or slightly alkaline? Classify aqueous solutions of the following salts as (a) slightly acidic or (b) slightly alkaline: (i) $NH_4F(aq)$ and (ii) $NH_4IO(aq)$, $K_b = 1.8 \times 10^{-5}$ for NH_3, $K_a = 7.2 \times 10^{-4}$ for HF, and $K_a = 2.3 \times 10^{-11}$ for HIO.

Salts That Contain Small, Highly Charged Cations

28. Choose the cations that will react with water to form H^+: (a) K^+, (b) $[Be(OH_2)_4]^{2+}$, (c) $[Al(H_2O)_6]^{3+}$, (d) $[Fe(H_2O)_6]^{3+}$, (e) $[Sr(H_2O)_6]^{2+}$. Write chemical equations for the reactions.

29. Why do salts that contain cations related to insoluble bases (metal hydroxides) and anions related to strong acids give acidic aqueous solutions? Use $Fe(NO_3)_3$ to illustrate.

30. Calculate pH and percent hydrolysis for the following. (See Table 19-2.)
(a) 0.20 M $Al(NO_3)_3$, aluminum nitrate
(b) 0.085 M $Co(ClO_4)_2$, cobalt(II) perchlorate
(c) 0.15 M $MgCl_2$, magnesium chloride

*31. Given pH values for solutions of the following concentrations, calculate hydrolysis constants for the cations:
(a) 0.00050 M $CeCl_3$, cerium(III) chloride, pH = 5.99
(b) 0.10 M $Cu(NO_3)_2$, copper(II) nitrate, pH = 4.50
(c) 0.10 M $Sc(ClO_4)_3$, scandium perchlorate, pH = 3.44

Strong Acid–Strong Base Titration Curves

32. Make a rough sketch of the titration curve expected for the titration of a strong acid with a strong base. What determines the pH of the solution at the following points? (a) No base added, (b) half-equivalence point, (c) equivalence point, (d) excess base added. Compare your curve with Figure 19-3.

For Exercises 33, 36, 37, and 46, calculate and tabulate $[H_3O^+]$, $[OH^-]$, pH, and pOH at the indicated points (see Table 19-4). In each case assume that pure acid (or base) is added to exactly 1 L of a 0.0100 molar solution of the indicated base (or acid). This simplifies the arithmetic because we may assume that the volume of each solution is constant throughout the titration. Plot each titration curve with pH on the vertical axis and moles of base (or acid) added on the horizontal axis.

33. Solid NaOH is added to 1 L of 0.0100 M HCl solution. Number of moles of NaOH added: (a) none, (b) 0.00100, (c) 0.00300, (d) 0.00500 (50% titrated), (e) 0.00700, (f) 0.00900, (g) 0.00950, (h) 0.0100 (100% titrated), (i) 0.0105, (j) 0.0120, (k) 0.0150 (50% excess NaOH). Consult Table 19-3 and list the indicators that could be used in this titration.

34. A 25.0-mL sample of 0.250 M HNO_3 is titrated with 0.100 M NaOH. Calculate the pH of the solution (a) before the addition of NaOH and after the addition of (b) 10.0 mL, (c) 25.0 mL, (d) 50.0 mL, (e) 62.5 mL, (f) 75.0 mL of NaOH.

Weak Acid–Strong Base Titration Curves

35. Make a rough sketch of the titration curve expected for the titration of a weak monoprotic acid with a strong base. What determines the pH of the solution at the following points? (a) No base added, (b) half-equivalence point, (c) equivalence point, (d) excess base added. Compare your curve to Figure 19-4.

36. Solid NaOH is added to 1 L of 0.0100 M CH_3COOH solution. Number of moles NaOH added: (a) none, (b) 0.00200, (c) 0.00400, (d) 0.00500 (50% titrated), (e) 0.00700, (f) 0.00900, (g) 0.00950, (h) 0.0100 (100% titrated), (i) 0.0105, (j) 0.0120, (k) 0.0150 (50% excess NaOH). Consult Table 19-3 and list the indicators that could be used in this titration.

*37. Gaseous HCl is added to 1 L of 0.0100 M aqueous ammonia solution. Number of moles HCl added: (a) none, (b) 0.00100, (c) 0.00300, (d) 0.00500 (50% titrated), (e) 0.00700, (f) 0.00900, (g) 0.00950, (h) 0.0100 (100% titrated), (i) 0.0105, (j) 0.0120, (k) 0.0150 (50% excess HCl). Consult Table 19-3 and list the indicators that could be used in this titration.

38. A solution contains 2.000 mmol HF dissolved in 100 mL. This solution is titrated at 25°C with a 0.1000 M solution of NaOH. Calculate the pH when these volumes of the NaOH solution have been added: (a) 0 mL, (b) 10.00 mL, (c) 19.90 mL, (d) 20.00 mL, (e) 20.10 mL. (f) Plot these points on graph paper and sketch the titration curve. (g) Select a suitable indicator.

39. A solution contains an unknown weak monoprotic acid HA. It takes 46.24 mL of NaOH solution to titrate 50.00 mL of the HA solution. To another 50.00-mL sample

of the same HA solution, 23.12 mL of the same NaOH solution is added. The pH of the resulting solution in the second experiment is 5.14. What are pK_a and K_a of HA?

Mixed Exercises

40. Classify aqueous solutions of the following salts as (a) acidic, (b) alkaline, or (c) essentially neutral: (i) $(NH_4)HSO_4$; (ii) $(NH_4)_2SO_4$; (iii) KCl; (iv) LiCN; (v) $Al(NO_3)_3$.

41. Repeat Exercise 40 for (i) $NaClO_4$; (ii) $K_2C_2O_4$; (iii) $(NH_4)_2S$; (iv) NaH_2PO_4; (v) NH_4CN.

42. Some cations in aqueous solution react with water molecules. What is the relative acid strength of such a cation compared to water? What effect will dissolution of these cations have on the pH of the solution?

43. A 15.00-mL sample of 0.1063 M KOH is titrated with 0.1077 M HCl. Calculate the pH of the solution (a) before the addition of HCl and after the addition of (b) 10.00 mL, (c) 15.00 mL, (d) 20.00 mL of HCl.

44. 20.00 mL of 0.2000 M propionic acid (CH_3CH_2COOH, $K_a = 1.3 \times 10^{-5}$) is titrated with 20.00 mL of 0.2000 M KOH at 25°C. Calculate the pH at the equivalence point. Select a suitable indicator.

45. Calculate the pH at the equivalence point for the titration of a solution containing 150.0 mg of ethylamine, $C_2H_5NH_2$, with 0.1000 M HCl solution. The volume of the solution at the equivalence point is 200 mL. Select a suitable indicator.

*46. Gaseous NH_3 is added to 1 L of 0.0100 M CH_3COOH solution. Number of moles NH_3 added: (a) none, (b) 0.00100, (c) 0.00400, (d) 0.00500 (50% titrated), (e) 0.00900, (f) 0.00950, (g) 0.0100 (100% titrated), (h) 0.0105, (i) 0.0130. What is the major difference between the titration curve for the reaction of CH_3COOH and NH_3 and the other curves you have plotted? Consult Table 19-3. Can you suggest a satisfactory indicator for this titration?

47. Use Table 19-3 to choose one or more indicators that could be used to "signal" reaching a pH of (a) 2.4, (b) 7, (c) 10.3, (d) 5.1.

48. A solution of 0.01 M acetic acid is to be titrated with a 0.01 M NaOH solution. What is the approximate pH at the equivalence point? Choose an appropriate indicator for the titration.

49. Some plants require acidic soils for healthy growth. Which of the following could be added to the soil around such plants to increase the acidity of the soil? Write equations to justify your answers. (a) $FeSO_4$, (b) Na_2SO_4, (c) $Al_2(SO_4)_3$, (d) $Fe_2(SO_4)_3$, (e) $BaSO_4$. Arrange the salts that give acidic solutions in order of increasing acidity.

50. Some of the following salts are used in detergents and other cleaning materials because they produce basic solutions. Which of the following could *not* be used for this purpose? Write equations to justify your answers. (a) Na_2CO_3, (b) Na_2SO_4, (c) $(NH_4)_2SO_4$, (d) Na_3PO_4.

When the solubility product for silver chloride, AgCl, is exceeded, a white precipitate of AgCl forms.

Objectives

As you study this chapter, you should learn

☐ To recognize common slightly soluble compounds

☐ To write solubility product constant expressions

☐ How K_{sp}'s are determined

☐ To use K_{sp}'s

☐ About fractional precipitation and its importance

☐ How simultaneous equilibria are used to control solubility

☐ Some methods for dissolving precipitates

So far we have discussed mainly compounds that are quite soluble in water. Although most compounds dissolve in water to some extent, many are so slightly soluble that they are called "insoluble compounds." We shall now consider those that are only very slightly soluble. As a rough rule of thumb, compounds that dissolve in water to the extent of 0.020 mole/liter or more are classified as soluble. Refer to the solubility rules (Table 4-8) as necessary.

Slightly soluble compounds are important in many natural phenomena. Our bones and teeth are mostly calcium phosphate, $Ca_3(PO_4)_2$, a slightly soluble compound. (See also page 861.) Also, many natural deposits of $Ca_3(PO_4)_2$ rock are mined and converted into agricultural fertilizer. Limestone caves have been formed by acidic water slowly dissolving away calcium carbonate, $CaCO_3$. Sinkholes are created when acidic water dissolves away most of the underlying $CaCO_3$. The remaining limestone can no longer support the weight above it, so it collapses, and a sinkhole is formed.

Sinkholes are formed when the underlying limestone, mostly $CaCO_3$, is dissolved away by acidic water.

20-1 Solubility Product Constants

Suppose we add one gram of solid barium sulfate, $BaSO_4$, to 1.0 liter of water at 25°C and stir until the solution is *saturated*. Very little $BaSO_4$ dissolves. Only 0.0025 gram of $BaSO_4$ dissolves in 1.0 liter of water, no matter how much more $BaSO_4$ is added. The $BaSO_4$ that does dissolve is completely dissociated into its constituent ions.

$$BaSO_4(s) \rightleftharpoons Ba^{2+}(aq) + SO_4^{2-}(aq)$$

In equilibria that involve slightly soluble compounds in water, the equilibrium constant is called a **solubility product constant, K_{sp}.** The activity of the solid $BaSO_4$ is one. Hence, the concentration of the solid is not included in the equilibrium constant expression. For a saturated solution of $BaSO_4$ in contact with solid $BaSO_4$, we write

$$BaSO_4(s) \rightleftharpoons Ba^{2+}(aq) + SO_4^{2-}(aq) \quad \text{and} \quad K_{sp} = [Ba^{2+}][SO_4^{2-}]$$

The solubility product constant for $BaSO_4$ is the product of the concentrations of its constituent ions in a saturated solution.

> In general, the **solubility product expression** for a compound is the product of the concentrations of its constituent ions, each raised to the power that corresponds to the number of ions in one formula unit of the compound. The quantity is constant at constant temperature for a saturated solution of the compound. This statement is the **solubility product principle**.

The existence of a substance in the solid state is indicated several ways. For example, $BaSO_4(s)$, $\underline{BaSO_4}$, and $BaSO_4\downarrow$ are sometimes used to represent solid $BaSO_4$. We use the (s) notation for formulas of solid substances in equilibrium with their saturated aqueous solutions.

Consider dissolving slightly soluble calcium fluoride, CaF_2, in H_2O.

$$CaF_2(s) \rightleftharpoons Ca^{2+}(aq) + 2F^-(aq) \quad \text{and} \quad K_{sp} = [Ca^{2+}][F^-]^2$$

Dissolving solid bismuth sulfide, Bi_2S_3, in H_2O gives two bismuth ions and three sulfide ions per formula unit.

$$Bi_2S_3(s) \rightleftharpoons 2Bi^{3+}(aq) + 3S^{2-}(aq) \quad \text{and} \quad K_{sp} = [Bi^{3+}]^2[S^{2-}]^3$$

Generally, we may represent the dissolution of a slightly soluble compound and its K_{sp} expression as follows.

$$M_yX_z(s) \rightleftharpoons yM^{z+}(aq) + zX^{y-}(aq) \quad \text{and} \quad K_{sp} = [M^{z+}]^y[X^{y-}]^z$$

In some cases a compound contains more than two kinds of ions. The dissolution of the slightly soluble compound magnesium ammonium phosphate, $MgNH_4PO_4$, in water is represented as follows.

$$MgNH_4PO_4(s) \rightleftharpoons Mg^{2+}(aq) + NH_4^+(aq) + PO_4^{3-}(aq)$$

Its solubility product expression is

$$K_{sp} = [Mg^{2+}][NH_4^+][PO_4^{3-}]$$

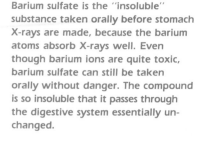

Barium sulfate is the "insoluble" substance taken orally before stomach X-rays are made, because the barium atoms absorb X-rays well. Even though barium ions are quite toxic, barium sulfate can still be taken orally without danger. The compound is so insoluble that it passes through the digestive system essentially unchanged.

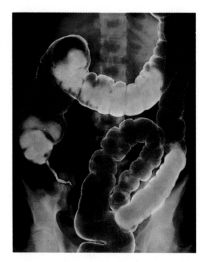

An X-ray photo of the gastrointestinal tract. The barium atoms in $BaSO_4$ absorb X-radiation well.

Barium sulfate, $BaSO_4$, occurs in the mineral barite (left). Calcium fluoride, CaF_2, occurs in the mineral fluorite (right). Both are white compounds. Minerals are often discolored by impurities.

We often shorten the term "solubility product constant" to "solubility product." Thus, the solubility products for barium sulfate, $BaSO_4$, and for calcium fluoride, CaF_2, are written

$$K_{sp} = [Ba^{2+}][SO_4^{2-}] = 1.1 \times 10^{-10} \quad \text{and} \quad K_{sp} = [Ca^{2+}][F^-]^2 = 3.9 \times 10^{-11}$$

The **molar solubility** of a compound is the number of moles that dissolve to give one liter of saturated solution.

Unless otherwise indicated, solubility product constants and solubility data are given for 25°C.

20-2 Determination of Solubility Product Constants

Solubility product constants can be determined in a number of ways. For example, careful measurements of conductivity show that one liter of a saturated solution of barium sulfate contains 0.0025 gram of dissolved $BaSO_4$. If the solubility of a compound is known, its solubility product can be calculated.

Example 20-1

We frequently use statements such as "The solution contains 0.0025 gram of dissolved $BaSO_4$." What we mean is that 0.0025 gram of solid $BaSO_4$ dissolves to give a solution that contains equal numbers of Ba^{2+} and SO_4^{2-} ions.

One (1.0) liter of saturated barium sulfate solution contains 0.0025 gram of dissolved $BaSO_4$. Calculate the solubility product constant for $BaSO_4$.

Plan

We write the chemical equation for the dissolution of $BaSO_4$ and the expression for its equilibrium constant, K_{sp}. From the stated solubility of $BaSO_4$ in H_2O at 25°C, we calculate its molar solubility and the concentrations of the ions. This enables us to calculate K_{sp}.

Solution

In a saturated solution, equilibrium exists between solid and dissolved solute.

$$BaSO_4(s) \rightleftharpoons Ba^{2+}(aq) + SO_4^{2-}(aq) \quad \text{and} \quad K_{sp} = [Ba^{2+}][SO_4^{2-}]$$

From the given solubility of $BaSO_4$ in H_2O we can calculate its *molar solubility*.

$$\frac{?\ mol\ BaSO_4}{L} = \frac{2.5 \times 10^{-3}\ g\ BaSO_4}{L} \times \frac{1\ mol\ BaSO_4}{233\ g\ BaSO_4}$$

$$= 1.1 \times 10^{-5}\ mol\ BaSO_4/L \quad \text{(dissolved)}$$

1.1×10^{-5} mole of solid $BaSO_4$ dissolves to give a liter of saturated solution.

We know the molar solubility of $BaSO_4$. The dissolution equation shows that each formula unit of $BaSO_4$ that dissolves produces one Ba^{2+} ion and one SO_4^{2-} ion.

$$BaSO_4(s) \quad \rightleftharpoons \quad Ba^{2+}(aq) \ + \ SO_4^{2-}(aq)$$
$$1.1 \times 10^{-5}\ mol/L \implies 1.1 \times 10^{-5}\ M \quad 1.1 \times 10^{-5}\ M$$
$$\text{(dissolved)}$$

In a saturated solution, $[Ba^{2+}] = [SO_4^{2-}] = 1.1 \times 10^{-5}\ M$. Substituting these values into the K_{sp} expression for $BaSO_4$ gives the value for K_{sp}.

The value calculated here is 1.2×10^{-10}, whereas the tabulated value is 1.1×10^{-10}. Roundoff error is responsible for the difference.

$$K_{sp} = [Ba^{2+}][SO_4^{2-}] = (1.1 \times 10^{-5})(1.1 \times 10^{-5}) = \boxed{1.2 \times 10^{-10}}$$

This expression is useful because it applies *to all saturated solutions of $BaSO_4$ at 25°C*. The *origin* of Ba^{2+} and SO_4^{2-} ions is not relevant. For example, suppose a solution of $BaCl_2$ and a solution of Na_2SO_4 are mixed at 25°C. (Both of these compounds are soluble and ionic.) As soon as the concentration of Ba^{2+} ions multiplied by the concentration of SO_4^{2-} ions exceeds 1.1×10^{-10}, solid $BaSO_4$ begins to precipitate and does so until the solution is just saturated. Or, if solid $BaSO_4$ is placed in water at 25°C, it dissolves until the $[Ba^{2+}]$ multiplied by the $[SO_4^{2-}]$ just equals 1.1×10^{-10}.

Example 20-2

One liter of a saturated solution of silver chromate at 25°C contains 0.0435 gram of Ag_2CrO_4. Calculate its solubility product constant.

Plan

We proceed as in Example 20-1.

Solution

The equation for the dissolution of silver chromate in water and its solubility product expression are

$$Ag_2CrO_4(s) \rightleftharpoons 2Ag^+(aq) + CrO_4^{2-}(aq) \quad \text{and} \quad K_{sp} = [Ag^+]^2[CrO_4^{2-}]$$

The molar solubility of silver chromate is calculated first.

$$\frac{? \text{ mol } Ag_2CrO_4}{L} = \frac{0.0435 \text{ g } Ag_2CrO_4}{L} \times \frac{1 \text{ mol } Ag_2CrO_4}{332 \text{ g } Ag_2CrO_4}$$

$$= 1.31 \times 10^{-4} \text{ mol/L} \quad \text{(dissolved)}$$

The equation for dissolution of Ag_2CrO_4 and its molar solubility give the concentrations of Ag^+ and CrO_4^{2-} ions in the saturated solution:

$$Ag_2CrO_4(s) \rightleftharpoons 2Ag^+(aq) + CrO_4^{2-}(aq)$$
$$1.31 \times 10^{-4} \text{ mol/L} \implies 2.62 \times 10^{-4} M \quad 1.31 \times 10^{-4} M$$
$$\text{(dissolved)}$$

> 1.31×10^{-4} mole of solid Ag_2CrO_4 dissolves to give a liter of saturated solution.

Substitution into the K_{sp} expression for Ag_2CrO_4 gives its value.

$$K_{sp} = [Ag^+]^2[CrO_4^{2-}] = (2.62 \times 10^{-4})^2(1.31 \times 10^{-4}) = \boxed{8.99 \times 10^{-12}}$$

The molar solubility of Ag_2CrO_4 is $1.31 \times 10^{-4} M$; its K_{sp} is 9.0×10^{-12}.

> Solubility products are usually given to only two significant digits.

EOC 8

The values for $BaSO_4$ and Ag_2CrO_4 are compared in Table 20-1. These data show that the molar solubility of Ag_2CrO_4 is greater than that of $BaSO_4$. However, K_{sp} for Ag_2CrO_4 is less than K_{sp} for $BaSO_4$ because the expression for Ag_2CrO_4 contains a *squared* term, $[Ag^+]^2$.

Table 20-1
Comparison of Solubilities of BaSO₄ and Ag₂CrO₄

Compound	Molar Solubility	K_{sp}
$BaSO_4$	1.1×10^{-5} mol/L	$[Ba^{2+}][SO_4^{2-}] = 1.1 \times 10^{-10}$
Ag_2CrO_4	1.3×10^{-4} mol/L	$[Ag^+]^2[CrO_4^{2-}] = 9.0 \times 10^{-12}$

If we compare K_{sp} values for two 1:1 compounds—e.g., AgCl and $BaSO_4$—the compound with the larger K_{sp} value has the higher molar solubility. The same is true for *any* two compounds that have the same stoichiometry, e.g., the 1:2 compounds CaF_2 and $Mg(OH)_2$.

Appendix H lists some K_{sp} values. Refer to it as needed.

> $K_{sp(AgCl)} = 1.8 \times 10^{-10}$
>
> $K_{sp(BaSO_4)} = 1.1 \times 10^{-10}$
>
> The molar solubility of AgCl is slightly higher than that of $BaSO_4$.

The Effects of Hydrolysis on Solubility

In Section 19-2 we discussed the hydrolysis of anions of weak acids. For example, we found that for CH_3COO^- and CN^- ions,

$$CH_3COO^- + H_2O \rightleftharpoons CH_3COOH + OH^-$$

$$K_b = \frac{[CH_3COOH][OH^-]}{[CH_3COO^-]} = 5.6 \times 10^{-10}$$

$$CN^- + H_2O \rightleftharpoons HCN + OH^- \qquad K_b = \frac{[HCN][OH^-]}{[CN^-]} = 2.5 \times 10^{-5}$$

We see that K_b for CN^-, the anion of a *very* weak acid, is much larger than K_b for CH_3COO^-, the anion of a much stronger acid. This tells us that in solutions of the same concentration, CN^- ions hydrolyze to a much greater extent than do CH_3COO^- ions. So we might expect that hydrolysis would have a much greater effect on the solubilities of cyanides such as AgCN than on the solubilities of acetates such as $AgCH_3COO$. It does.

Hydrolysis reduces the concentrations of anions of weak acids, so its effect must be taken into account when we do very precise solubility calculations. However, taking into account the effect of hydrolysis on solubilities of slightly soluble compounds is beyond the scope of this chapter.

20-3 Uses of Solubility Product Constants

When the solubility product for a compound is known, the solubility of the compound in H_2O at 25°C can be calculated, as Example 20-3 illustrates.

Example 20-3

Calculate the molar solubilities, concentrations of the constituent ions, and solubilities in grams per liter for (a) silver chloride, AgCl ($K_{sp} = 1.8 \times 10^{-10}$), and (b) zinc hydroxide, $Zn(OH)_2$ ($K_{sp} = 4.5 \times 10^{-17}$).

The values of K_{sp} are obtained from Appendix H.

Plan

We are given the value for each solubility product constant. In each case we write the appropriate equation, represent the equilibrium concentrations, and then substitute into the K_{sp} expression.

Solution

(a) The equation for the dissolution of silver chloride and its solubility product expression are

$$AgCl(s) \rightleftharpoons Ag^+(aq) + Cl^-(aq) \qquad \text{and} \qquad K_{sp} = [Ag^+][Cl^-] = 1.8 \times 10^{-10}$$

Each formula unit of AgCl that dissolves produces one Ag^+ and one Cl^-. We let $x = $ mol/L of AgCl that dissolve, i.e., the molar solubility.

$$AgCl(s) \rightleftharpoons Ag^+(aq) + Cl^-(aq)$$
$$x \text{ mol/L} \implies x\, M \qquad x\, M$$

Substitution into the solubility product expression gives

$$[Ag^+][Cl^-] = (x)(x) = 1.8 \times 10^{-10} \qquad x^2 = 1.8 \times 10^{-10} \qquad x = 1.3 \times 10^{-5}$$

molar solubility of AgCl $= x\, M = 1.3 \times 10^{-5}\, M$

20-4

On oc
others,
fractio
ions. T
periodi
some
ubility

Compou

AgCl
AgB
AgI

These
less s
izable
ion. S
easily

One liter of saturated AgCl contains 1.3×10^{-5} mole of AgCl at 25°C. We also know the concentrations of the constituent ions.

$$x = [Ag^+] = [Cl^-] = 1.3 \times 10^{-5} \text{ mol/L}$$

Now we can calculate the mass of AgCl in one liter of saturated solution.

$$\frac{?\text{ g AgCl}}{L} = \frac{1.3 \times 10^{-5}\text{ mol AgCl}}{L} \times \frac{143\text{ g AgCl}}{1\text{ mol AgCl}} = 1.9 \times 10^{-3} \text{ g AgCl/L}$$

A liter of saturated AgCl solution contains only 0.0019 g of AgCl.

(b) The dissolution of zinc hydroxide, $Zn(OH)_2$, in water and its solubility product expression are represented as

$$Zn(OH)_2(s) \rightleftharpoons Zn^{2+}(aq) + 2OH^-(aq) \qquad K_{sp} = [Zn^{2+}][OH^-]^2 = 4.5 \times 10^{-17}$$

We let x = molar solubility, so $[Zn^{2+}] = x$ and $[OH^-] = 2x$, and we have

$$Zn(OH)_2(s) \rightleftharpoons Zn^{2+}(aq) + 2OH^-(aq)$$
$$x \text{ mol/L} \Longrightarrow x\ M \qquad 2x\ M$$

Substitution into the solubility product expression gives

$$[Zn^{2+}][OH^-]^2 = (x)(2x)^2 = 4.5 \times 10^{-17}$$

$$4x^3 = 4.5 \times 10^{-17} \qquad x^3 = 11 \times 10^{-18} \qquad x = 2.2 \times 10^{-6}$$

$$\text{molar solubility} = x = 2.2 \times 10^{-6} \text{ mol } Zn(OH)_2/L$$

$$x = [Zn^{2+}] = 2.2 \times 10^{-6}\ M \qquad \text{and} \qquad 2x = [OH^-] = 4.4 \times 10^{-6}\ M$$

We can now calculate the mass of $Zn(OH)_2$ in one liter of saturated solution.

The $[OH^-]$ is twice the molar solubility of $Zn(OH)_2$ because each formula unit of $Zn(OH)_2$ produces two OH^-.

$$\frac{?\text{ g } Zn(OH)_2}{L} = \frac{2.2 \times 10^{-6}\text{ mol } Zn(OH)_2}{L} \times \frac{99\text{ g } Zn(OH)_2}{1\text{ mol } Zn(OH)_2}$$

$$= 2.2 \times 10^{-4} \text{ g } Zn(OH)_2/L$$

EOC 12

Exa
Solid
NaB
of th
poun
= 1.

Plan
We
ions
K_{sp}
salt
caus

Solu
We
hali

$[I^-]$
of /

In part (b) of Example 20-3 we found that $[OH^-] = 4.4 \times 10^{-6}\ M$ in a *saturated* $Zn(OH)_2$ solution. From this we see that pOH = 5.36 and pH = 8.64. A saturated $Zn(OH)_2$ solution is not very basic because $Zn(OH)_2$ is not very soluble in H_2O.

The Common Ion Effect in Solubility Calculations

The common ion effect applies to solubility equilibria just as it does to other ionic equilibria. The solubility of a compound is less in a solution that contains an ion common to the compound than in pure water (as long as no other reaction is caused by the presence of the common ion).

Th

so

Example 20-4

The molar solubility of magnesium fluoride, MgF_2, is 1.2×10^{-3} mol/L in pure water at 25°C. Calculate the molar solubility of MgF_2 in 0.10 M sodium fluoride, NaF, solution at 25°C. $K_{sp} = 6.4 \times 10^{-9}$ for MgF_2.

We find the *maximum* $[OH^-]$ *that does not cause precipitation* by substituting $[Mg^{2+}]$ into K_{sp} for $Mg(OH)_2$.

$$[Mg^{2+}][OH^-]^2 = 1.5 \times 10^{-11}$$

$$[OH^-]^2 = \frac{1.5 \times 10^{-11}}{[Mg^{2+}]} = \frac{1.5 \times 10^{-11}}{0.10} = 1.5 \times 10^{-10}$$

$$[OH^-] = 1.2 \times 10^{-5}\ M \longleftarrow \text{maximum } [OH^-] \text{ possible}$$

To prevent precipitation of $Mg(OH)_2$ in *this* solution, $[OH^-]$ must be equal to or less than $1.2 \times 10^{-5}\ M$. K_b for aqueous NH_3 is used to calculate the number of moles of NH_4Cl necessary to buffer $0.10\ M$ aqueous NH_3 so that $[OH^-] = 1.2 \times 10^{-5}\ M$. Let x = number of mol/L of NH_4Cl required.

You may wish to refer to Section 18-6 to refresh your memory on buffer solutions.

$$
\begin{array}{cccccc}
NH_4Cl(aq) & \longrightarrow & NH_4{}^+(aq) & + & Cl^-(aq) & \text{(to completion)} \\
x\ M & \Longrightarrow & x\ M & & x\ M & \\
NH_3(aq) & + H_2O \rightleftharpoons & NH_4{}^+(aq) & + & OH^-(aq) & \\
(0.10 - 1.2 \times 10^{-5})\ M & & 1.2 \times 10^{-5}\ M & & 1.2 \times 10^{-5}\ M &
\end{array}
$$

$$K_b = \frac{[NH_4{}^+][OH^-]}{[NH_3]} = 1.8 \times 10^{-5} = \frac{(x + 1.2 \times 10^{-5})(1.2 \times 10^{-5})}{(0.10 - 1.2 \times 10^{-5})}$$

We can assume that $(x + 1.2 \times 10^{-5}) \approx x$ and $(0.10 - 1.2 \times 10^{-5}) \approx 0.10$.

$$\frac{(x)(1.2 \times 10^{-5})}{0.10} = 1.8 \times 10^{-5}$$

$$x = 0.15 \text{ mol of } NH_4{}^+ \text{ per liter of solution}$$

Addition of 0.15 mol of NH_4Cl to 1.0 L of $0.10\ M$ aqueous NH_3 decreases $[OH^-]$ to $1.2 \times 10^{-5}\ M$, so that K_{sp} for $Mg(OH)_2$ is not exceeded in this solution.

EOC 36, 38, 43

All relevant equilibria must be satisfied when more than one equilibrium is required to describe a solution (Examples 20-10, 20-11, and 20-12).

20-6 Dissolving Precipitates

A precipitate dissolves when the concentrations of its ions are reduced so that K_{sp} is no longer exceeded, i.e., when $Q_{sp} < K_{sp}$. The precipitate then dissolves until $Q_{sp} = K_{sp}$. Precipitates can be dissolved by the following three types of reactions. All involve removing ions from solution.

Converting an Ion to a Weak Electrolyte

Solubility products, like other equilibrium constants, are thermodynamic quantities. They tell us nothing about how fast a given reaction occurs, only that it can, or cannot, occur under specified conditions.

Three specific illustrations follow.

1. Insoluble $Al(OH)_3$ dissolves in acids. H^+ ions react with OH^- ions (from the saturated $Al(OH)_3$ solution) to form the weak electrolyte H_2O. This can make $[Al^{3+}][OH^-]^3 < K_{sp}$, so that the dissolution equilibrium shifts to the right and $Al(OH)_3$ dissolves.

The recov
tions usec
photogra
just such
expensive
profitable
recoverec
constitute
water suj

Silver cl
chloride
containi

$$Al(OH)_3(s) \rightleftharpoons Al^{3+}(aq) + 3OH^-(aq)$$
$$3H^+(aq) + 3OH^-(aq) \longrightarrow 3H_2O(\ell)$$

overall rxn: $Al(OH)_3(s) + 3H^+(aq) \longrightarrow Al^{3+}(aq) + 3H_2O(\ell)$

2. Ammonium ions, from a salt such as NH_4Cl, dissolve insoluble $Mg(OH)_2$. The NH_4^+ ions combine with OH^- ions in the saturated $Mg(OH)_2$ solution. This forms the weak electrolytes NH_3 and H_2O. The result is $[Mg^{2+}][OH^-]^2 < K_{sp}$, and so the $Mg(OH)_2$ dissolves.

$$Mg(OH)_2(s) \rightleftharpoons Mg^{2+}(aq) + 2OH^-(aq)$$
$$2NH_4^+(aq) + 2OH^-(aq) \longrightarrow 2NH_3(aq) + 2H_2O(\ell)$$

overall rxn: $Mg(OH)_2(s) + 2NH_4^+(aq) \longrightarrow Mg^{2+}(aq) + 2NH_3(aq) + 2H_2O(\ell)$

This process, dissolution of $Mg(OH)_2$ in an NH_4Cl solution, is the reverse of the reaction we considered in Example 20-10. There, $Mg(OH)_2$ precipitated from a solution of aqueous NH_3.

3. Nonoxidizing acids dissolve some insoluble metal sulfides. For example, 6 M HCl dissolves MnS. The H^+ ions combine with S^{2-} ions to form H_2S, a gas that bubbles out of the solution. The result is $[Mn^{2+}][S^{2-}] < K_{sp}$, and so the MnS dissolves.

$$MnS(s) \rightleftharpoons Mn^{2+}(aq) + S^{2-}(aq)$$
$$2H^+(aq) + S^{2-}(aq) \longrightarrow H_2S(g)$$

overall rxn: $MnS(s) + 2H^+(aq) \longrightarrow Mn^{2+}(aq) + H_2S(g)$

Converting an Ion to Another Species by a Redox Reaction

Most insoluble metal sulfides dissolve in hot dilute HNO_3 because NO_3^- ions oxidize S^{2-} ions to elemental sulfur. This removes S^{2-} ions from the solution:

$$3S^{2-}(aq) + 2NO_3^-(aq) + 8H^+(aq) \longrightarrow 3S(s) + 2NO(g) + 4H_2O(\ell)$$

Consider copper(II) sulfide, CuS, in equilibrium with its ions. This equilibrium lies far to the left; $K_{sp} = 8.7 \times 10^{-36}$. Removal of the S^{2-} ions by oxidation to elemental sulfur favors the reaction to the right, and so CuS(s) dissolves in hot dilute HNO_3.

$$CuS(s) \rightleftharpoons Cu^{2+}(aq) + S^{2-}(aq)$$
$$3S^{2-}(aq) + 2NO_3^-(aq) + 8H^+(aq) \longrightarrow 3S(s) + 2NO(g) + 4H_2O(\ell)$$

We multiply the first equation by 3, add the two equations, and cancel like terms. This gives the net ionic equation for dissolving CuS(s) in hot dilute HNO_3:

$$3CuS(s) + 2NO_3^-(aq) + 8H^+(aq) \longrightarrow$$
$$3Cu^{2+}(aq) + 3S(s) + 2NO(g) + 4H_2O(\ell)$$

Complex Ion Formation

The cations in many slightly soluble compounds can form complex ions. This often results in dissolution of the slightly soluble compound. Some metal ions share electron pairs donated by molecules and ions such as NH_3,

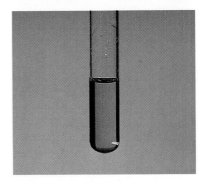

Manganese(II) sulfide, MnS, is salmon-colored. MnS dissolves in 6 M HCl. The resulting solution of $MnCl_2$ is pale pink.

Copper(II) sulfide, CuS, is black. As CuS dissolves in 6 M HNO_3, some NO is oxidized to brown NO_2 by O_2 in the air. The resulting solution of $Cu(NO_3)_2$ is blue.

CN^-, OH^-, F^-, Cl^-, Br^-, and I^-. Coordinate covalent bonds are formed as these ligands replace H_2O molecules from hydrated metal ions. The decrease in the concentration of the hydrated metal ion shifts the solubility equilibrium to the right.

"Ligand" is the name given to an atom or a group of atoms bonded to the central element in complex ions. Ligands are Lewis bases.

The stability of a complex ion can be described by K_d, the equilibrium constant for its dissociation into a metal ion plus ligands. For example, many copper(II) compounds react with excess aqueous NH_3 to form the deep-blue complex ion $[Cu(NH_3)_4]^{2+}(aq)$.

$$[Cu(NH_3)_4]^{2+}(aq) \rightleftharpoons Cu^{2+}(aq) + 4NH_3(aq)$$

$$K_d = \frac{[Cu^{2+}][NH_3]^4}{[[Cu(NH_3)_4]^{2+}]} = 8.5 \times 10^{-13}$$

As before, the outer brackets mean molar concentrations. The inner brackets are part of the formula of the complex ion.

Recall that $Cu^{2+}(aq)$ is really a hydrated ion, $[Cu(OH_2)_4]^{2+}$. The preceding reaction and its K_d expression are represented more accurately as

$$[Cu(NH_3)_4]^{2+} + 4H_2O \rightleftharpoons [Cu(OH_2)_4]^{2+} + 4NH_3$$

$$K_d = \frac{[[Cu(OH_2)_4]^{2+}][NH_3]^4}{[[Cu(NH_3)_4]^{2+}]} = 8.5 \times 10^{-13}$$

The more effectively a ligand competes with H_2O for a coordination site on the metal ions, the smaller K_d is. This tells us that, in a comparison of complexes with the same number of ligands, the smaller the K_d value, the more stable the complex ion. Some complex ions and their dissociation constants, K_d, are listed in Appendix I.

Example 20-13
What are the concentrations of hydrated Cu^{2+}, NH_3, and $[Cu(NH_3)_4]^{2+}$ in a 0.20 M solution of $[Cu(NH_3)_4]SO_4$?

Plan

We write the equation for the *complete dissociation* of the soluble compound that contains the complex ion. This gives us the concentration of the complex ion $[Cu(NH_3)_4]^{2+}$. Then we write the equation for the dissociation of the complex ion, represent the equilibrium concentrations algebraically, and substitute them into the equilibrium constant expression.

Solution

The soluble deep-blue complex salt $[Cu(NH_3)_4]SO_4$ dissociates completely to produce tetraamminecopper(II) ions and sulfate ions.

$$[Cu(NH_3)_4]SO_4(aq) \xrightarrow{100\%} [Cu(NH_3)_4]^{2+}(aq) + SO_4^{2-}(aq)$$

$$0.20 \, M \Longrightarrow 0.20 \, M \qquad 0.20 \, M$$

Some of the $[Cu(NH_3)_4]^{2+}$ ions then dissociate. Let x be the concentration of $[Cu(NH_3)_4]^{2+}$ that dissociates.

$$[Cu(NH_3)_4]^{2+} \rightleftharpoons Cu^{2+}(aq) + 4NH_3$$

$$(0.20 - x) \, M \qquad x \, M \qquad 4x \, M$$

$$K_d = \frac{[Cu^{2+}][NH_3]^4}{[[Cu(NH_3)_4]^{2+}]} = 8.5 \times 10^{-13} = \frac{x(4x)^4}{0.20 - x} = \frac{256 \, x^5}{0.20}$$

(assume that $(0.20 - x) \approx 0.20$)

$$x^5 = 6.6 \times 10^{-16}$$

Taking the fifth root of both sides of this equation gives $x = 9.2 \times 10^{-4}$.

$$[Cu^{2+}] = x \, M = 9.2 \times 10^{-4} \, M$$

$$[NH_3] = 4x \, M = 3.7 \times 10^{-3} \, M$$

$$[[Cu(NH_3)_4]^{2+}] = (0.20 - x) \, M \approx 0.20 \, M$$

$(0.20 - x) \approx 0.20$ is a valid assumption; 9.2×10^{-4} is much less than 0.20.

EOC 55

Concentrated aqueous NH_3 was added *slowly* to a solution of copper(II) sulfate, $CuSO_4$. Unreacted blue copper(II) sulfate solution remains in the bottom part of the test tube. The light-blue precipitate in the middle is copper(II) hydroxide. $Cu(OH)_2$. The top layer contains deep-blue $[Cu(NH_3)_4]^{2+}$ ions that were formed as some $Cu(OH)_2$ dissolved in excess aqueous NH_3.

Copper(II) hydroxide dissolves in an excess of aqueous NH_3 to form the deep-blue complex ion $[Cu(NH_3)_4]^{2+}$. This decreases the $[Cu^{2+}]$ so that $[Cu^{2+}][OH^-]^2 < K_{sp}$, and so the $Cu(OH)_2$ dissolves.

$$Cu(OH)_2(s) \rightleftharpoons Cu^{2+}(aq) + 2OH^-(aq)$$

$$Cu^{2+}(aq) + 4NH_3(aq) \rightleftharpoons [Cu(NH_3)_4]^{2+}(aq)$$

overall rxn: $\quad Cu(OH)_2(s) + 4NH_3(aq) \rightleftharpoons [Cu(NH_3)_4]^{2+}(aq) + 2OH^-(aq)$

On the other hand, zinc hydroxide is amphoteric (Section 10-8). This means that solid $Zn(OH)_2$ dissolves in excess NaOH solution to form the complex ion $[Zn(OH)_4]^{2-}$.

$$Zn(OH)_2(s) + 2OH^-(aq) \rightleftharpoons [Zn(OH)_4]^{2-}(aq)$$

Example 20-14

Some solid $Zn(OH)_2$ is suspended in a saturated solution of $Zn(OH)_2$. A solution of sodium hydroxide is added until all the $Zn(OH)_2$ just dissolves. The pH of the solution is 11.80. What are the concentrations of Zn^{2+} and $[Zn(OH)_4]^{2-}$ ions in the solution? K_{sp} for $Zn(OH)_2 = 4.5 \times 10^{-17}$, and K_d for $[Zn(OH)_4]^{2-} = 3.5 \times 10^{-16}$.

Plan

We write the appropriate chemical equations and equilibrium constants for the two reversible reactions. We see that $[OH^-]$ appears in both of these equilibrium constant expressions. We are given pH, from which we can calculate $[OH^-]$. Once we know $[OH^-]$, we can use K_{sp} for $Zn(OH)_2$ to calculate $[Zn^{2+}]$. Then, knowing both $[OH^-]$ and $[Zn^{2+}]$, we can solve for $[[Zn(OH)_4]^{2-}]$.

Solution

The important equilibria and their equilibrium constant expressions are

$$Zn(OH)_2(s) \rightleftharpoons Zn^{2+}(aq) + 2OH^-(aq) \qquad K_{sp} = [Zn^{2+}][OH^-]^2 = 4.5 \times 10^{-17}$$

$$[Zn(OH)_4]^{2-} \rightleftharpoons Zn^{2+}(aq) + 4OH^-(aq) \qquad K_d = \frac{[Zn^{2+}][OH^-]^4}{[[Zn(OH)_4]^{2-}]} = 3.5 \times 10^{-16}$$

We know the pH, and so we can calculate $[OH^-]$ from pH + pOH = 14.00:

$$pOH = 14.00 - pH = 14.00 - 11.80 = 2.20$$

$$[OH^-] = 10^{-pOH} = 10^{-2.20} = 6.3 \times 10^{-3} \, M = [OH^-]$$

We can use the solubility product expression for $Zn(OH)_2$ to calculate $[Zn^{2+}]$:

$$K_{sp} = [Zn^{2+}][OH^-]^2 = 4.5 \times 10^{-17}$$

$$[Zn^{2+}] = \frac{K_{sp}}{[OH^-]^2} = \frac{4.5 \times 10^{-17}}{(6.3 \times 10^{-3})^2} = \boxed{1.1 \times 10^{-12} \, M \; Zn^{2+}}$$

Both equilibria are established in the same solution, and so the same $[Zn^{2+}]$ and $[OH^-]$ also satisfy the complex ion dissociation equilibrium:

$$K_d = \frac{[Zn^{2+}][OH^-]^4}{[[Zn(OH)_4]^{2-}]} = 3.5 \times 10^{-16}$$

$$[[Zn(OH)_4]^{2-}] = \frac{[Zn^{2+}][OH^-]^4}{3.5 \times 10^{-16}} = \frac{(1.1 \times 10^{-12})(6.3 \times 10^{-3})^4}{3.5 \times 10^{-16}}$$

$$\boxed{[[Zn(OH)_4]^{2-}] = 5.0 \times 10^{-6} \, M}$$

EOC 62

Thus, we see that we are able to shift equilibria (in this case, dissolve $Zn(OH)_2$) by taking advantage of complex ion formation.

Key Terms

Complex ions   Ions resulting from the formation of coordinate covalent bonds between simple ions and other ions or molecules.

Dissociation constant   The equilibrium constant that applies to the dissociation of a complex ion into a simple ion and coordinating species (ligands).

Fractional precipitation   Removal of some ions from solution by precipitation while leaving other ions, with similar properties, in solution.

Insoluble compound   Actually a very slightly soluble compound.

Molar solubility   The number of moles of a solute that dissolve to produce a liter of saturated solution.

Precipitate   A solid formed by mixing in solution the constituent ions of a slightly soluble compound.

Solubility product constant   The equilibrium constant that applies to the dissolution of a slightly soluble compound.

Solubility product principle   The solubility product constant expression for a slightly soluble compound is the product of the concentrations of the constituent ions, each raised to the power that corresponds to the number of ions in one formula unit.

Exercises

Consult Appendix H for solubility product constant values and Appendix I for complex ion dissociation constants, as needed.

Solubility Product

1. (a) Are "insoluble" substances really insoluble? (b) What do we mean when we refer to insoluble substances?

2. State the solubility product principle. What is its significance?

3. (a) Why are solubility product constant expressions written as products of concentrations of ions raised to appropriate powers? (b) We do not include a term for the solid in a solubility product expression. Why?

4. What do we mean when we refer to the molar solubility of a compound?

5. Write a balanced chemical equation for the dissolution of each of the following slightly soluble compounds. Then write each solubility product constant expression. (a) Ag_2SO_4, (b) Fe_2S_3, (c) $AgBr$, (d) Ag_2CO_3, (e) $Mn_3(PO_4)_2$.

6. Write a balanced chemical equation for the dissolution of each of the following slightly soluble compounds. Then write each solubility product constant expression. (a) $SrCrO_4$, (b) $CdCO_3$, (c) Hg_2Cl_2 [contains mercury(I) ion, Hg_2^{2+}], (d) $Al(OH)_3$, (e) $Ba_3(PO_4)_2$.

Experimental Determination of K_{sp}

7. From the solubility data given for the following compounds, calculate their solubility product constants. Your calculated values may not agree exactly with the solubility products given in Appendix H, because roundoff errors are large in calculations to two significant figures.
   (a) CuI, copper(I) iodide, 4.4×10^{-4} g/L
   (b) AgI, silver iodide, 2.8×10^{-8} g/10 mL
   (c) $Pb_3(PO_4)_2$, lead(II) phosphate, 6.2×10^{-7} g/L
   (d) Ag_2SO_4, silver sulfate, 5.0 mg/mL

8. From the solubility data given for the following compounds, calculate their solubility product constants. Your calculated values may not agree exactly with the solubility products given in Appendix H, because roundoff errors are large in calculations to two significant figures.
   (a) $SrCrO_4$, strontium chromate, 1.2 mg/mL
   (b) BiI_3, bismuth iodide, 7.7×10^{-3} g/mL
   (c) $Fe(OH)_2$, iron(II) hydroxide, 1.1×10^{-3} g/L
   (d) $Zn(CN)_2$, zinc cyanide, 0.015 g/L

9. Construct a table like Table 20-1 for the compounds listed in Exercise 7. Which compound has (a) the highest molar solubility; (b) the lowest molar solubility; (c) the largest K_{sp}; (d) the smallest K_{sp}?

10. Construct a table like Table 20-1 for the compounds listed in Exercise 8. Which compound has (a) the highest molar solubility; (b) the lowest molar solubility; (c) the largest K_{sp}; (d) the smallest K_{sp}?

Uses of Solubility Product Constants

11. Calculate molar solubilities, concentrations of constituent ions, and solubilities in grams per liter for the following compounds at 25°C: (a) $AgCN$, silver cyanide; (b) MgF_2, magnesium fluoride; (c) $Pb_3(AsO_4)_2$, lead(II) arsenate; (d) Hg_2CO_3, mercury(I) carbonate [the formula for the mercury(I) ion is Hg_2^{2+}].

12. Calculate molar solubilities, concentrations of constituent ions, and solubilities in grams per liter for the following compounds at 25°C: (a) CuI, copper(I) iodide; (b) $Ba_3(PO_4)_2$, barium phosphate; (c) PbF_2, lead(II) fluoride; (d) Ag_3PO_4, silver phosphate.

13. Construct a table similar to Table 20-1 for the compounds listed in Exercise 11. Which compound has (a) the highest molar solubility; (b) the lowest molar solubility; (c) the highest solubility, expressed in grams per liter; (d) the lowest solubility, expressed in grams per liter?

14. Construct a table similar to Table 20-1 for the compounds listed in Exercise 12. Which compound has (a) the highest molar solubility; (b) the lowest molar solubility; (c) the highest solubility, expressed in grams per liter; (d) the lowest solubility, expressed in grams per liter?

15. Calculate the molar solubility of $AgBr$ in 0.10 M KBr solution.

16. Calculate the molar solubility of Ag_2SO_4 in 0.10 M K_2SO_4 solution.

17. Calculate the molar solubility of Ag_2CrO_4 (a) in pure water, (b) in 0.010 M $AgNO_3$, and (c) in 0.010 M K_2CrO_4.

18. Milk of magnesia is a suspension of the slightly soluble compound $Mg(OH)_2$ in water. (a) What is the molar solubility of $Mg(OH)_2$ in a 0.010 M NaOH solution? (b) What is the molar solubility of $Mg(OH)_2$ in a 0.010 M $MgCl_2$ solution?

19. A fluoridated water supply contains 1 mg/L of F^-. What is the maximum amount of Ca^{2+}, expressed in grams per liter, that can exist in this water supply?

20. Which is more soluble in 0.20 M K_2CrO_4 solution—$BaCrO_4$ or Ag_2CrO_4?

21. If 1.00 g of $AgNO_3$ is added to 50.0 mL of 0.050 M NaCl, will a precipitate form? If so, would you expect the precipitate to be visible?

22. Will a precipitate of $PbCl_2$ form when 5.0 g of solid $Pb(NO_3)_2$ is added to 1.00 L of 0.010 M NaCl? Assume that volume change is negligible.

23. Sodium bromide and lead nitrate are soluble in water. Will lead bromide precipitate when 1.03 g of NaBr and 0.332 g of $Pb(NO_3)_2$ are dissolved in sufficient water to make 1.00 L of solution?

24. Will a precipitate of $Cu(OH)_2$ form when 10.0 mL of 0.010 M NaOH is added to 1.00 L of 0.010 M $CuCl_2$?

25. Suppose you have three beakers that contain, respectively, 100 mL each of the following solutions: (i) 0.0015 M KOH, (ii) 0.0015 M K_2CO_3, (iii) 0.0015 M K_2S. (a) If solid lead nitrate, $Pb(NO_3)_2$, were added slowly to each beaker, what concentration of Pb^{2+} would be required to initiate precipitation? (b) If solid lead nitrate were added to each beaker until $[Pb^{2+}] = 0.0015$ M, what concentrations of OH^-, CO_3^{2-}, and S^{2-} would remain in solution, i.e., unprecipitated? Neglect any volume change when solute is added.

26. Suppose you have three beakers that contain, respectively, 100 mL of each of the following solutions: (i) 0.0015 M KOH, (ii) 0.0015 M K_2CO_3, (iii) 0.0015 M K_2S. (a) If solid zinc nitrate, $Zn(NO_3)_2$, were added slowly to each beaker, what concentration of Zn^{2+} would be required to initiate precipitation? (b) If solid zinc nitrate were added to each beaker until $[Zn^{2+}] = 0.0015$ M, what concentrations of OH^-, CO_3^{2-}, and S^{2-} would remain in solution, i.e., unprecipitated? Neglect any volume change when solid is added.

27. A solution is 0.100 M in Pb^{2+} ions. If 0.103 mol of solid Na_2SO_4 is added to 1.00 L of this solution (with negligible volume change), what percentage of the Pb^{2+} ions remain in solution?

28. A solution is 0.100 M in Pb^{2+} ions. If 0.103 mol of solid NaI is added to 1.00 L of this solution (with negligible volume change), what percentage of the Pb^{2+} ions remain in solution?

*29. A solution is 0.100 M in $Ba(NO_3)_2$. If 0.103 mole of solid Na_3PO_4 is added to 1.00 L of this solution (with negligible volume change), what percentage of the Ba^{2+} ions remain in solution?

Fractional Precipitation

30. What is fractional precipitation?

31. Solid Na_2SO_4 is added slowly to a solution that is 0.15 M in $Pb(NO_3)_2$ and 0.15 M in $Ba(NO_3)_2$. In what order will solid $PbSO_4$ and $BaSO_4$ form? Calculate the percentage of Ba^{2+} that precipitates just before $PbSO_4$ begins to precipitate.

32. To a solution that is 0.10 M in Cu^+, 0.10 M in Ag^+, and 0.10 M in Au^+, solid NaCl is added slowly. Assume that there is no volume change due to the addition of solid NaCl.
(a) Which compound will begin to precipitate first?
(b) Calculate $[Au^+]$ when AgCl just begins to precipitate. What percentage of the Au^+ has precipitated at this point?
(c) Calculate $[Au^+]$ and $[Ag^+]$ when CuCl just begins to precipitate.

33. A solution is 0.010 M in Pb^{2+} and 0.010 M in Ag^+. As Cl^- is introduced to the solution by the addition of solid NaCl, determine (a) which substance will precipitate first, AgCl or $PbCl_2$, and (b) the fraction of the metal ion in the first precipitate that remains in solution at the moment the precipitation of the second compound begins.

34. A solution is 0.10 M in K_2SO_4 and 0.10 M in K_2CrO_4. A solution of $Pb(NO_3)_2$ is added slowly without changing the volume appreciably. (a) Which salt, $PbSO_4$ or $PbCrO_4$, will precipitate first? (b) What is $[Pb^{2+}]$ when the salt in part (a) begins to precipitate? (c) What is $[Pb^{2+}]$ when the other lead salt begins to precipitate? (d) What are $[SO_4^{2-}]$ and $[CrO_4^{2-}]$ when the lead salt in part (c) begins to precipitate?

35. Solid $Pb(NO_3)_2$ is added slowly to a solution that is 0.020 M each in NaOH, K_2CO_3, and Na_2SO_4. (a) In what order will solid $Pb(OH)_2$, $PbCO_3$, and $PbSO_4$ begin to precipitate? (b) Calculate the percentages of OH^- and CO_3^{2-} that have precipitated when $PbSO_4$ begins to precipitate.

Simultaneous Equilibria

36. If a solution is made 0.080 M in $Mg(NO_3)_2$, 0.075 M in aqueous ammonia, and 3.5 M in NH_4NO_3, will $Mg(OH)_2$ precipitate? What is the pH of this solution?

37. If a solution is made 0.090 M in $Mg(NO_3)_2$, 0.090 M in aqueous ammonia, and 0.080 M in NH_4NO_3, will $Mg(OH)_2$ precipitate? What is the pH of this solution?

*38. Calculate the solubility of CaF_2 in a solution that is buffered at $[H^+] = 0.0100$ M.

*39. Calculate the solubility of AgCN in a solution that is buffered at $[H^+] = 0.000100$ M.

40. If a solution is 2.0×10^{-5} M in $Mn(NO_3)_2$ and 1.0×10^{-3} M in aqueous ammonia, will $Mn(OH)_2$ precipitate?

41. If a solution is 0.040 M in manganese(II) nitrate, $Mn(NO_3)_2$, and 0.080 M in aqueous ammonia, will manganese(II) hydroxide, $Mn(OH)_2$, precipitate?

*42. Find the minimum pH at which MnS will precipitate from a solution that is 0.100 M with respect to H_2S and 0.0100 M with respect to $MnCl_2$.

*43. What concentration of NH_4NO_3 is necessary to prevent precipitation of $Mn(OH)_2$ in the solution of Exercise 41?

44. (a) What is the pH of a saturated solution of $Mn(OH)_2$? (b) What is the solubility in g $Mn(OH)_2$/100 mL of solution?

45. (a) What is the pH of a saturated solution of $Mg(OH)_2$? (b) What is the solubility in g $Mg(OH)_2$/100 mL of solution?

Dissolution of Precipitates and Complex Ion Formation

46. Explain, by writing appropriate equations, how the following insoluble compounds can be dissolved by the addition of a solution of nitric acid. (Carbonates dissolve in strong acids to form carbon dioxide, which is evolved as a gas, and water.) What is the "driving force" for each reaction? (a) $Cu(OH)_2$, (b) $Al(OH)_3$, (c) $MnCO_3$, (d) $(PbOH)_2CO_3$.

47. Explain, by writing equations, how the following insoluble compounds can be dissolved by the addition of a solution of ammonium nitrate or ammonium chloride. (a) $Mg(OH)_2$, (b) $Mn(OH)_2$, (c) $Ni(OH)_2$.

48. The following insoluble sulfides can be dissolved in 3 M hydrochloric acid. Explain how this is possible and write the appropriate equations. (a) MnS, (b) FeS.

49. The following sulfides are less soluble than those listed in Exercise 48 and can be dissolved in hot 6 M nitric acid, an oxidizing acid. Explain how, and write the

appropriate balanced equations. (a) PbS, (b) CuS, (c) Bi_2S_3.

50. Why would MnS be expected to be more soluble in 0.10 M HCl solution than in water? Would the same be true for $Mn(NO_3)_2$?

*51. For each pair, choose the salt that would be expected to be more soluble in acidic solution than in pure water: (a) $Hg_2(CH_3COO)_2$ or Hg_2Br_2, (b) $Pb(OH)_2$ or PbI_2, (c) AgI or $AgNO_2$.

52. How can most water-insoluble metal hydroxides be dissolved? Write a chemical equation for the dissolution of $Fe(OH)_3$ in this way.

53. How does the presence of excess H_3O^+ aid in the dissolution of slightly soluble metal carbonates? Write the chemical equation for the dissolution of $MnCO_3$.

54. Find the concentration of Au^{3+} in a solution in which the other equilibrium concentrations are $[Cl^-] = 0.10$ M and $[AuCl_4^-] = 0.20$ M. K_d for $AuCl_4^- = 7.0 \times 10^{-26}$.

55. Calculate the concentrations of complex ion, metal ion, and ammonia in the following solutions. The complex ions are enclosed in brackets. All these compounds are soluble and ionic.

   (a) 0.100 M $[Ag(NH_3)_2]Cl$
   (b) 0.089 M $[Co(NH_3)_6]SO_4$
   (c) 0.114 M $[Co(NH_3)_6]_2(SO_4)_3$

*56. What is the concentration of Ag^+ in a solution that is 0.10 M in KSCN and 0.10 M in $[Ag(SCN)_4]^{3-}$? Will Ag_2SO_4 precipitate if $[SO_4^{2-}] = 0.10$ M? K_d for $[Ag(SCN)_4]^{3-} = 2.1 \times 10^{-10}$.

*57. (a) What is the concentration of Ag^+ in a solution that is 0.20 M in KSCN and 0.20 M in $[Ag(SCN)_4]^{3-}$? (b) Will Ag_2SO_4 precipitate if $[SO_4^{2-}] = 0.10$ M? K_d for $[Ag(SCN)_4]^{3-} = 2.1 \times 10^{-10}$.

Mixed Exercises

58. (a) Calculate the concentrations of CrO_4^{2-} ions in saturated solutions of Ag_2CrO_4 and of $BaCrO_4$. (b) Which salt has the greater molar solubility in water? Which salt has the larger solubility product? How do you account for any apparent disagreement?

59. We mix 20.0 mL of a 0.0030 M solution of $BaCl_2$ and 50.0 mL of a 0.050 M solution of NaF. (a) Find $[Ba^{2+}]$ and $[F^-]$ in the mixed solution at the instant of mixing (before any possible reaction occurs). (b) Would BaF_2 precipitate?

60. A solution is 0.010 M with respect to Cd^{2+} ions and also 0.010 M with respect to Pd^{2+} ions. Solid KBr is added to the solution (assume negligible change in volume) until $[Br^-] = 1.0$ M. Use the overall dissociation constants of the following complexes to calculate the concentrations of Cd^{2+} ions and Pd^{2+} ions once equilibrium is reached. K_d for $[CdBr_4]^{2-} = 2.0 \times 10^{-4}$ and for $[PdBr_4]^{2-} = 7.7 \times 10^{-14}$.

61. What concentration of Ag^+ ions remains in a solution that originally contained 0.10 M Ag^+ ions and 1.3 M NH_3?

*62. 0.010 mol of solid $Zn(OH)_2$ is suspended in a saturated $Zn(OH)_2$ solution. Some 6.0 M NaOH solution is added and the mixture is stirred vigorously. The volume of the solution is now 400 mL, and its pH is 13.15. (a) Does all the $Zn(OH)_2$ dissolve? (b) What is/was the minimum $[OH^-]$ necessary to dissolve $Zn(OH)_2$ completely?

*63. A concentrated, strong acid is added to a solid mixture of 0.010-mol samples of $Fe(OH)_2$ and $Cu(OH)_2$ placed in 1.0 L of water. At what values of pH will the dissolution of each hydroxide be complete? (Assume negligible volume change.)

64. Zinc ion forms a complex ion with $EDTA^{4-}$, where $EDTA^{4-}$ represents the ethylenediaminetetraacetate ion.

$$[Zn(EDTA)]^{2-} \rightleftharpoons Zn^{2+} + EDTA^{4-}$$
$$K_d = 2.6 \times 10^{-17}$$

Will ZnS form in a solution that was originally 0.10 M in $EDTA^{4-}$, 0.010 M in S^{2-}, and 0.010 M in Zn^{2+}?

65. A solution is 0.010 M in I^- ions and 0.010 M in Br^- ions. Ag^+ ions are introduced to the solution by the addition of solid $AgNO_3$. Determine (a) which compound will precipitate first, AgI or AgBr, and (b) the percentage of the halide ion in the first precipitate that is removed from solution before the precipitation of the second compound begins.

Outline

Objectives

As you study this chapter, you should learn

☐ How to use the terminology of electrochemistry (terms such as cell, electrode, cathode, anode)

☐ About the differences between electrolytic cells and voltaic (galvanic) cells

☐ To recognize oxidation and reduction half-reactions, and to know at which electrode each occurs

☐ To write half-reactions and overall cell reactions for electrolysis processes

☐ To use Faraday's Law of Electrolysis to calculate amounts of products formed, amounts of current passed, time elapsed, and oxidation state

☐ About the refining and plating of metals by electrolytic methods

☐ To describe the construction of simple voltaic cells from half-cells and a salt bridge, and to understand the function of each component

☐ To write half-reactions and overall cell reactions for voltaic cells

☐ To compare various voltaic cells to determine the relative strengths of oxidizing and reducing agents

The Statue of Liberty, 151 ft high and weighing 225 tons, was presented by the French people to the United States of America on July 4, 1884. It is made from copper plates attached to an iron framework. The outer surface of the statue, like other copper structures, has been protected by the blue-green patina that copper forms when exposed to weather. This patina is basic copper carbonate, $CuCO_3 \cdot Cu(OH)_2$, that has formed over the century that the statue has been in place. However, where the copper plates touch the iron framework, galvanic corrosion has taken its toll. The layers of shellac and asbestos originally used to insulate these metals had broken away over time, and rain water seeping in has caused the iron to rust. When the statue was recently refurbished, it was reequipped with stainless steel ribs and Teflon insulators.

Electrochemistry deals with the chemical changes produced by electric current and with the production of electricity by chemical reactions. Digital watches, automobile starters, pocket calculators, and heart pacemakers are just a few devices that depend on electrochemically produced power. Many metals are purified or plated onto jewelry by electrochemical methods. Corrosion of metals is an electrochemical process.

We have learned much about chemical reactions from the study of electrochemistry. The amount of electrical energy consumed or produced can be measured quite accurately. All electrochemical reactions involve the transfer of electrons and are therefore *oxidation–reduction* reactions. The sites of oxidation and reduction are separated physically so that oxidation occurs at one location while reduction occurs at the other. Electrochemical processes require some method of introducing a stream of electrons into a reacting chemical system and some means of withdrawing electrons. In most applications, the reacting system is contained in a **cell**, and an electric current enters or exits by **electrodes**.

We classify electrochemical cells into two types:

1. **Electrolytic cells** are those in which electrical energy from an external source causes *nonspontaneous* chemical reactions to occur.
2. **Voltaic cells** are those in which *spontaneous* chemical reactions produce electricity and supply it to an external circuit.

We shall discuss several electrochemical cells. From experimental observations we can deduce the electrode reactions and the overall reactions. We can then construct simplified diagrams of the cells.

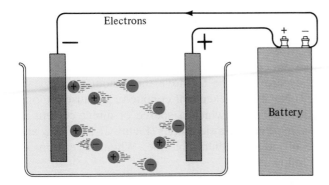

Figure 21-1
The motion of ions through a solution is an electric current. This accounts for ionic (electrolytic) conduction. Positively charged ions migrate toward the negative electrode, and negatively charged ions migrate toward the positive electrode. Here the rate of migration is greatly exaggerated for clarity. The ionic velocities are actually only slightly greater than random molecular speeds.

21-1 Electrical Conduction

Electric current represents transfer of charge. Charge can be conducted through pure liquid electrolytes or solutions containing electrolytes and through metals. The latter type of conduction is called **metallic conduction**. It involves the flow of electrons with no similar movement of the atoms of the metal and no obvious changes in the metal (Section 13-17). **Ionic** or **electrolytic conduction** is the conduction of electrical current by the motion of ions through a solution or a pure liquid. Positively charged ions migrate toward the negative electrode while negatively charged ions move toward the positive electrode. Both kinds of conduction, ionic and metallic, occur in electrochemical cells (Figure 21-1).

21-2 Electrodes

Electrodes are surfaces upon which oxidation or reduction half-reactions occur. They may or may not participate in the reactions. Those that do not react are called **inert electrodes**. Regardless of the kind of cell, electrolytic or voltaic, the electrodes are identified as follows.

The **cathode** is defined as the electrode at which *reduction* occurs as electrons are gained by some species. The **anode** is the electrode at which *oxidation* occurs as electrons are lost by some species.

Each of these can be either the positive or the negative electrode.

Electrolytic Cells

In some electrochemical cells, *nonspontaneous* chemical reactions are forced to occur by the input of electrical energy. This process is called **electrolysis**. An electrolytic cell consists of a container for the reaction material with electrodes immersed in the reaction material and connected to a source of direct current. Inert electrodes are often used so that they do not react.

Lysis means "splitting apart." In many electrolytic cells, compounds are split into their constituent elements.

21-3 The Electrolysis of Molten Sodium Chloride (the Downs Cell)

Solid sodium chloride does not conduct electricity. Although its ions vibrate about fixed positions, they are not free to move throughout the crystal. However, molten (melted) NaCl is an excellent conductor because its ions are freely mobile. Consider a cell in which a source of direct current is connected by wires to two inert graphite electrodes (Figure 21-2a). They are immersed in a container of molten sodium chloride. When the current is flowing, we observe the following:

Molten NaCl, melting point 801°C, is a clear, colorless liquid that looks like water.

The metal remains liquid because its melting point is only 97.8°C. It floats because it is less dense than the molten NaCl.

1. A pale green gas, which is chlorine, Cl_2, is liberated at one electrode.
2. Molten, silvery-white metallic sodium, Na, forms at the other electrode and floats on top of the molten sodium chloride.

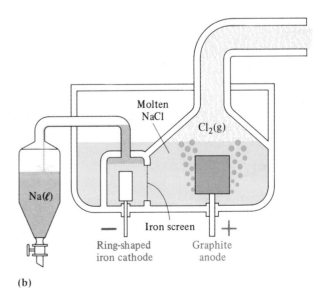

Figure 21-2
(a) Apparatus for electrolysis of molten sodium chloride. (b) The Downs cell, the apparatus in which molten sodium chloride is commercially electrolyzed to produce sodium metal and chlorine gas. The liquid Na floats on the more dense molten NaCl.

From these observations we can deduce the processes of the cell. Chlorine must be produced by oxidation of Cl^- ions, and the electrode at which this happens must be the anode. Metallic sodium is produced by reduction of Na^+ ions at the cathode, where electrons are being forced into the cell.

$$2Cl^- \longrightarrow Cl_2(g) + 2e^- \quad \text{(oxidation, anode half-reaction)}$$
$$\underline{2[Na^+ + e^- \longrightarrow Na(\ell)]} \quad \text{(reduction, cathode half-reaction)}$$
$$\underbrace{2Na^+ + 2Cl^-}_{2NaCl(\ell)} \longrightarrow 2Na(\ell) + Cl_2(g) \quad \text{(overall cell reaction)}$$

In this chapter, as in Chapters 4 and 11, we often use red type to emphasize reduction and blue type to emphasize oxidation.

The formation of metallic Na and gaseous Cl_2 from NaCl is *nonspontaneous* except at temperatures very much higher than 801°C. The direct current (dc) source must supply electrical energy to force this reaction to occur. Electrons are used in the cathode half-reaction (reduction) and produced in the anode half-reaction (oxidation). Therefore, they travel through the wire from *anode* to *cathode*. The dc source forces electrons to flow nonspontaneously from the positive electrode to the negative electrode. The anode is the positive electrode and the cathode the negative electrode *in all electrolytic cells*. Figure 21-2a is a simplified diagram of the cell.

The direction of *spontaneous* flow for negatively charged particles is from negative to positive.

Sodium and chlorine must not be allowed to come in contact with each other because they react spontaneously, rapidly, and explosively to form sodium chloride. Figure 21-2b shows the Downs cell that is used for the industrial electrolysis of sodium chloride. The Downs cell is expensive to run, mainly because of the cost of construction, the cost of the electricity, and the cost of heating the NaCl to melt it. However, electrolysis of a molten sodium salt is the most practical means by which metallic Na can be obtained, owing to its extremely high reactivity. Once liberated by the electrolysis, the liquid Na metal is drained off, cooled, and cast into blocks. These must be stored in an inert environment (e.g., in mineral oil) to prevent reaction with O_2 or other components of the atmosphere.

Electrolysis of molten compounds is also the common method of obtaining other Group IA metals, IIA metals (except barium), and aluminum (Chapter 22). The Cl_2 gas produced in the Downs cell is cooled, compressed, and marketed. This partially offsets the expense of producing metallic sodium. But most chlorine is produced by the cheaper electrolysis of aqueous NaCl.

21-4 The Electrolysis of Aqueous Sodium Chloride

Consider the electrolysis of a moderately concentrated solution of NaCl in water, using inert electrodes. The following experimental observations are made when a sufficiently high voltage is applied across the electrodes of a suitable cell.

1. H_2 gas is liberated at one electrode. The solution becomes basic in that vicinity.
2. Cl_2 gas is liberated at the other electrode.

Chloride ions are obviously being oxidized to Cl_2 in this cell, as they were in the electrolysis of molten NaCl. But Na^+ ions are not reduced to metallic Na. Instead, gaseous H_2 and aqueous OH^- ions are produced by reduction of H_2O molecules at the cathode. Water is more readily reduced than Na^+ ions. This is primarily because the reduction of Na^+ would produce the very

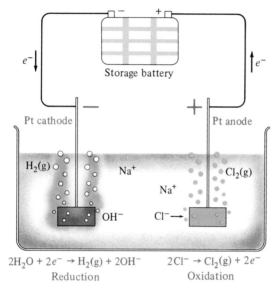

$2H_2O + 2e^- \rightarrow H_2(g) + 2OH^-$ $2Cl^- \rightarrow Cl_2(g) + 2e^-$
Reduction   Oxidation

Figure 21-3
Electrolysis of aqueous NaCl solution. Although several reactions occur at both the anode and the cathode, the net result is the production of $H_2(g)$ and NaOH at the cathode and $Cl_2(g)$ at the anode. A few drops of phenolphthalein indicator were added to the solution. The solution turns pink at the cathode, where OH^- ions are formed.

The electrolysis of the aqueous solution of KI, another Group IA–Group VIIA salt. At the cathode (left), water is reduced to $H_2(g)$ and OH^- ions, turning the phenolphthalein indicator pink. The characteristic reddish color of I_2 appears at the anode (right).

active metal Na, whereas the reduction of H_2O produces the more stable products $H_2(g)$ and $OH^-(aq)$. The active metals Li, K, Ca, and Na (Table 4-11) displace H_2 from aqueous solutions, so we do not expect these metals to be produced in aqueous solution. Later in this chapter (Section 21-15) we shall learn the quantitative basis for predicting which of several possible oxidations or reductions is favored. The half-reactions and overall cell reaction for this electrolysis are

We shall omit the notation that indicates states of substances—(s), (ℓ), (g), and (aq)—except where states are not obvious. This abbreviates writing equations.

$$2Cl^- \longrightarrow Cl_2 + 2e^- \qquad \text{(oxidation, anode)}$$
$$\underline{2H_2O + 2e^- \longrightarrow 2OH^- + H_2} \qquad \text{(reduction, cathode)}$$
$$2H_2O + 2Cl^- \longrightarrow 2OH^- + H_2 + Cl_2 \qquad \text{(overall cell reaction)}$$
$$\underbrace{+ 2Na^+} \longrightarrow \underbrace{+ 2Na^+} \qquad \text{(spectator ions)}$$
$$2NaCl \longrightarrow 2NaOH$$

The cell is illustrated in Figure 21-3. As before, the electrons flow from the anode (+) through the wire to the cathode (−).

Not surprisingly, the fluctuations in commercial prices of these widely used industrial products—H_2, Cl_2, and NaOH—nearly always parallel one another.

The overall cell reaction produces gaseous H_2 and Cl_2 and an aqueous solution of NaOH, called caustic soda. Solid NaOH is then obtained by evaporation of the residual solution. This is the most important commercial preparation of each of these substances. It is much less expensive than the electrolysis of molten NaCl, because it is not necessary to heat the solution.

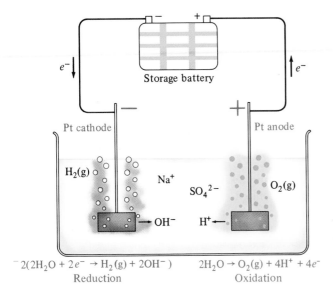

$-2(2H_2O + 2e^- \rightarrow H_2(g) + 2OH^-)$
Reduction

$2H_2O \rightarrow O_2(g) + 4H^+ + 4e^-$
Oxidation

Figure 21-4
The electrolysis of aqueous Na_2SO_4 produces $H_2(g)$ at the cathode and O_2 at the anode. Bromthymol blue indicator has been added to the solution. This indicator turns blue in the basic solution near the cathode (where OH^- is produced), and yellow in the acidic solution near the anode (where H^+ is formed).

21-5 The Electrolysis of Aqueous Sodium Sulfate

In the electrolysis of aqueous sodium sulfate using inert electrodes, we observe the following:

1. Gaseous H_2 is produced at one electrode. The solution becomes basic around that electrode.
2. Gaseous O_2 is produced at the other electrode. The solution becomes acidic around that electrode.

As in the previous example, water is reduced in preference to Na^+ at the cathode. Observation 2 suggests that water is also preferentially oxidized relative to the sulfate ion, SO_4^{2-}, at the anode (see Figure 21-4):

$$
\begin{array}{lll}
2(2H_2O + 2e^- & \longrightarrow & H_2 + 2OH^-) \quad \text{(reduction, cathode)} \\
2H_2O & \longrightarrow & O_2 + 4H^+ + 4e^- \quad \text{(oxidation, anode)} \\
\hline
6H_2O & \longrightarrow & 2H_2 + O_2 + \underbrace{4H^+ + 4OH^-}_{4H_2O} \quad \text{(overall cell reaction)} \\
2H_2O & \longrightarrow & 2H_2 + O_2 \quad \text{(net reaction)}
\end{array}
$$

The net result is the electrolysis of water. This happens because H_2O is more readily reduced than Na^+ and more readily oxidized than SO_4^{2-}. The ions of Na_2SO_4 conduct the current through the solution. They take no part in the reaction.

21-6 Faraday's Law of Electrolysis

In 1832–33, Michael Faraday's studies of electrolysis led to this conclusion:

> The amount of substance that undergoes oxidation or reduction at each electrode during electrolysis is directly proportional to the amount of electricity that passes through the cell.

This is **Faraday's Law of Electrolysis**. One quantitative unit of electricity is now called the faraday.

> One **faraday** is the amount of electricity that reduces one equivalent weight of a substance at the cathode and oxidizes one equivalent weight of a substance at the anode. This corresponds to the gain or loss, and therefore the passage, of 6.022×10^{23} electrons. Thus, *one equivalent weight* of any substance is the amount of that substance that supplies or consumes *one mole* of electrons.

One faraday of electricity corresponds to the passage of one mole of electrons. Review Section 11-5 on equivalent weights in redox reactions.

A smaller electrical unit commonly used in physics and electronics is the **coulomb (C)**. One coulomb is defined as the amount of charge that passes a given point when 1 ampere (A) of electrical current flows for 1 second. One ampere of current equals 1 coulomb per second. One faraday is equal to 96,487 coulombs of charge.

For comparison, a 100-watt household light bulb uses a current of about 0.8 ampere.

$$1 \text{ ampere} = 1\,\frac{\text{coulomb}}{\text{second}} \quad \text{or} \quad 1\text{ A} = 1\text{ C/s}$$

$$1 \text{ faraday} = 6.022 \times 10^{23}\ e^- = 96{,}487 \text{ C}$$

We may restate Faraday's Law in a very useful form.

> In an electrochemical process, 1 faraday of electricity (96,487 coulombs = 1 mole of electrons) reduces and oxidizes, respectively, one equivalent weight each of the oxidizing and reducing agents.

Table 21-1 shows the amounts of several elements produced during electrolysis by the passage of 1 faraday of electricity.

Example 21-1

Calculate the mass of copper metal produced during the passage of 2.50 amperes of current through a solution of copper(II) sulfate for 50.0 minutes.

Plan

The half-reaction that describes the reduction of copper(II) ions tells us the number of moles of electrons required to produce one mole of copper metal. Each mole of electrons corresponds to 1 faraday, or 96,500 coulombs, of charge. The product of current and time gives the number of coulombs.

$$\boxed{\begin{array}{c}\text{current,}\\\text{time}\end{array}} \longrightarrow \boxed{\begin{array}{c}\text{no. of}\\\text{coulombs}\end{array}} \longrightarrow \boxed{\begin{array}{c}\text{mol of } e^-\\\text{passed}\end{array}} \longrightarrow \boxed{\begin{array}{c}\text{mass}\\\text{of Cu}\end{array}}$$

Solution

The equation for the reduction of copper(II) ions to copper metal is

$$\underset{\substack{1\text{ mol}\\63.5\text{ g}}}{Cu^{2+}} + \underset{\substack{2(6.02\times10^{23})e^-\\2(96{,}500\text{ C})}}{2e^-} \longrightarrow \underset{\substack{1\text{ mol}\\63.5\text{ g}}}{Cu} \quad \text{(reduction, cathode)}$$

When the number of significant figures in the calculation warrants, the value 96,487 coulombs is usually rounded off to 96,500 coulombs.

We see that 63.5 grams of copper "plate out" for every 2 moles of electrons, or for every 2(96,500 coulombs) of charge. We first calculate the number of coulombs passing through the cell.

$$? \text{ C} = 50.0 \text{ min} \times \frac{60 \text{ s}}{1 \text{ min}} \times \frac{2.50 \text{ C}}{\text{s}} = 7.50 \times 10^3 \text{ C}$$

2.50 A = 2.50 C/s

We calculate the mass of copper produced by the passage of 7.50×10^3 coulombs.

$$? \text{ g Cu} = 7.50 \times 10^3 \text{ C} \times \frac{1 \text{ mol } e^-}{96{,}500 \text{ C}} \times \frac{63.5 \text{ g Cu}}{2 \text{ mol } e^-}$$

$$= \boxed{2.47 \text{ g Cu}} \qquad \text{(about the mass of a copper penny)}$$

Notice how little copper is deposited by this considerable current in 50 minutes.

Example 21-2

What volume of oxygen gas (measured at STP) is produced by the oxidation of water in the electrolysis of copper(II) sulfate in Example 21-1?

Plan

We use the same approach as in Example 21-1. Here we relate the amount of charge passed to the number of moles, and hence the volume at STP, of O_2 gas produced.

$$\boxed{\begin{array}{c} \text{current,} \\ \text{time} \end{array}} \longrightarrow \boxed{\begin{array}{c} \text{no. of} \\ \text{coulombs} \end{array}} \longrightarrow \boxed{\begin{array}{c} \text{mol of } e^- \\ \text{passed} \end{array}} \longrightarrow \boxed{\begin{array}{c} L_{STP} \\ \text{of } O_2 \end{array}}$$

Solution

The equation for the oxidation of water and the equivalence between the number of coulombs and the volume of oxygen produced at STP are

$$2H_2O \longrightarrow \underset{\substack{1 \text{ mol} \\ 22.4 \text{ L}_{STP}}}{O_2} + 4H^+ + \underset{\substack{4(6.02 \times 10^{23})e^- \\ 4(96{,}500 \text{ C})}}{4e^-} \qquad \text{(oxidation, anode)}$$

The number of coulombs passing through the cell is 7.50×10^3 C. For every 4(96,500 coulombs) passing through the cell, 22.4 liters of O_2 at STP is produced.

$$? \text{ L}_{STP} \text{ O}_2 = 7.50 \times 10^3 \text{ C} \times \frac{1 \text{ mol } e^-}{96{,}500 \text{ C}} \times \frac{22.4 \text{ L}_{STP} \text{ O}_2}{4 \text{ mol } e^-} = \boxed{0.435 \text{ L}_{STP} \text{ O}_2}$$

EOC 21, 26, 28

The amount of electricity in Examples 21-1 and 21-2 would be sufficient to light a 100-watt household light bulb for about 150 minutes, or 2.5 hours.

Table 21-1
Amounts of Elements Produced at One Electrode in Electrolysis by 1 Faraday of Electricity

Half-Reaction	Number of e^- in Half-Reaction	Product (electrode)	Amount (= 1 Eq Wt)
$Ag^+(aq) + e^- \longrightarrow Ag(s)$	1	Ag (cathode)	1 mol = 107.868 g
$2H^+(aq) + 2e^- \longrightarrow H_2(g)$	2	H_2 (cathode)	$\frac{1}{2}$ mol = 1.008 g
$Cu^{2+}(aq) + 2e^- \longrightarrow Cu(s)$	2	Cu (cathode)	$\frac{1}{2}$ mol = 31.773 g
$Au^{3+}(aq) + 3e^- \longrightarrow Au(s)$	3	Au (cathode)	$\frac{1}{3}$ mol = 65.6555 g
$2Cl^- \longrightarrow Cl_2(g) + 2e^-$	2	Cl_2 (anode)	$\frac{1}{2}$ mol = 35.4527 g = 11.2 L$_{STP}$
$2H_2O(\ell) \longrightarrow O_2(g) + 4H^+(aq) + 4e^-$	4	O_2 (anode)	$\frac{1}{4}$ mol = 7.9997 g = 5.60 L$_{STP}$

Notice how little product is formed by what seems to be a lot of electricity. This suggests why electrolytic production of gases and metals is so costly.

21-7 Determination of Oxidation State (Charge on an Ion) by Electrolysis

Faraday's Law can also be used to determine oxidation states. We electrolyze a solution containing an ion. We then relate the amount of an element produced to the number of moles of electrons passing through the cell.

Example 21-3

An aqueous solution of an unknown complex salt of chromium is electrolyzed by a current of 3.00 amperes for 1.00 hour. This produces 1.94 grams of chromium metal at the cathode. What is the oxidation state of chromium in this solution?

Plan

The magnitude of the oxidation state is equal to the number of moles of electrons required to oxidize or reduce one mole of any substance to the elemental state (i.e., to an oxidation state of zero).

$$\boxed{\text{current, time}} \longrightarrow \boxed{\text{no. of coulombs}} \longrightarrow \boxed{\text{mol of } e^- \text{ passed}} \quad \boxed{\text{mol Cr produced}} \longleftarrow \boxed{\text{g Cr produced}}$$

$$\boxed{\dfrac{\text{ox.}}{\text{state}} = \dfrac{\text{mol of } e^-}{\text{mol Cr}}}$$

Solution

As before, we first find the amount of charge passing through the cell.

$$\underline{?}\ C = 1.00\ \text{hr} \times \frac{60\ \text{min}}{1\ \text{hr}} \times \frac{60\ \text{s}}{1\ \text{min}} \times \frac{3.00\ \text{C}}{1\ \text{s}} = 1.08 \times 10^4\ \text{C}$$

We then find the number of moles of electrons passed.

$$\underline{?}\ \text{mol } e^- = 1.08 \times 10^4\ \text{C} \times \frac{1\ \text{mol } e^-}{96{,}500\ \text{C}} = 0.112\ \text{mol } e^-$$

We now find the number of moles of Cr formed.

$$\underline{?}\ \text{mol Cr formed} = 1.94\ \text{g Cr} \times \frac{1\ \text{mol Cr}}{52.0\ \text{g Cr}} = 0.0373\ \text{mol Cr}$$

$$\text{ox. state} = \frac{\text{mol of } e^-}{\text{mol Cr formed}} = \frac{0.112\ \text{mol } e^-}{0.0373\ \text{mol Cr formed}} = \frac{3\ \text{mol } e^-}{\text{mol Cr formed}}$$

Because three moles of electrons are required to reduce one mole of chromium ions to Cr metal, the chromium in solution must be   Cr(III).

EOC 31

21-8 Electrolytic Refining and Electroplating of Metals

Electrolytic reduction is the most practical means by which many active metals can be obtained (Section 22-3). Less active metals can be obtained from their ores by less expensive chemical reduction, and some metals occur in nature in the uncombined state. But these are frequently purified or refined by electrolysis. The electrolytic method for refining metals is also called *electroplating* when used to plate a metal onto a surface. For example, impure metallic copper obtained from the chemical reduction of Cu_2S and CuS (Section 22-8) is purified using an electrolytic cell like the one shown in Figure 21-5. Thin sheets of very pure copper are made to act as cathodes by connecting them to the negative terminal of a dc generator. Chunks of impure copper connected to the positive terminal function as anodes. The electrodes are immersed in a solution of copper(II) sulfate and sulfuric acid. When the cell operates, Cu from the impure anodes is oxidized and goes into solution as Cu^{2+} ions; Cu^{2+} ions from the solution are reduced and plate out as metallic Cu on the pure Cu cathodes. Other active metals from the impure bars also go into solution after oxidation. They do not plate out onto the cathode bars of pure Cu because of the far greater concentration of the more easily reduced Cu^{2+} ions that are already in solution. Overall, there is no net reaction, merely a simultaneous transfer of Cu from anode to solution and from solution to cathode:

A sludge called anode mud collects under the anodes. It contains such valuable and difficult-to-oxidize elements as Au, Pt, Ag, Se, and Te. The separation, purification, and sale of these elements reduce the cost of refined copper.

$$\begin{array}{lll}
\text{(impure) Cu} & \longrightarrow \; Cu^{2+} + 2e^- & \text{(oxidation, anode)} \\
Cu^{2+} + 2e^- & \longrightarrow \; \text{Cu (pure)} & \text{(reduction, cathode)} \\
\hline
\text{Cu (impure)} & \longrightarrow \; \text{Cu (pure)} & \text{(no net reaction)}
\end{array}$$

Nevertheless, the net effect is that small bars of very pure Cu and large bars of impure Cu are converted into large bars of very pure Cu and small bars of impure Cu. The energy provided by the electric generator forces a decrease in the entropy of the system by separating the Cu from its impurities in the impure bars. Copper can be plated onto other objects by the same mechanism (Figure 21-6).

Metal-plated articles are common in our society. Jewelry and tableware are often plated with silver. Gold is plated on jewelry and electrical contacts. Some automobiles have steel bumpers plated with thin films of chromium.

Figure 21-5
A schematic diagram of the electrolytic cell used for refining copper (a) before electrolysis and (b) after electrolysis. (c) Commercial electrolysis cells for refining copper.

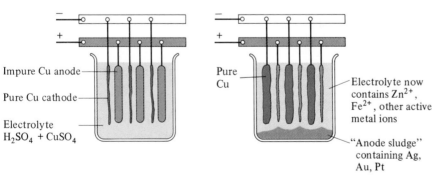

(a) Impure Cu anode — Pure Cu cathode — Electrolyte $H_2SO_4 + CuSO_4$

(b) Pure Cu — Electrolyte now contains Zn^{2+}, Fe^{2+}, other active metal ions — "Anode sludge" containing Ag, Au, Pt

(c)

Figure 21-6
Electroplating with copper. (a) The anode is made of pure copper, which dissolves during the electroplating process. This replenishes the Cu^{2+} ions that are removed from the solution as Cu plates out on the cathode. (b) A family memento that has been electroplated with copper. To aid in electroplating onto nonconductors such as shoes, the material is first soaked in a concentrated electrolyte solution to make it conductive.

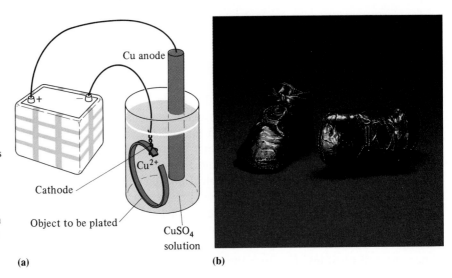

(a)   **(b)**

A typical bumper requires approximately 3 seconds of electroplating to produce a smooth, shiny surface only 0.0002 mm thick. When the atoms are deposited too rapidly, they are not able to form extended lattices. Rapid plating of metal results in rough, grainy, black surfaces. Slower plating produces smooth surfaces. "Tin cans" are steel cans plated electrolytically with tin; these are sometimes replaced by cans plated in $\frac{1}{3}$ second with an extremely thin chromium film.

Voltaic or Galvanic Cells

Voltaic, or **galvanic**, **cells** are electrochemical cells in which *spontaneous* oxidation–reduction reactions produce electrical energy. The two halves of the redox reaction are separated, requiring electron transfer to occur through an external circuit. In this way, useful electrical energy is obtained. Everyone is familiar with some voltaic cells. The dry cells commonly used in flashlights, transistor radios, photographic equipment, and many toys and appliances are voltaic cells. Automobile batteries consist of voltaic cells connected in series so that their voltages add. We shall first consider some simple laboratory cells used to measure the potential difference, or voltage, of a reaction under study. We shall then look at some common voltaic cells.

These are named for Allesandro Volta and Luigi Galvani, two Italian physicists of the 18th century.

21-9 The Construction of Simple Voltaic Cells

A **half-cell** contains the oxidized and reduced forms of an element, or other more complex species, in contact with each other. A common kind of half-cell consists of a piece of metal (the electrode) immersed in a solution of its ions. Consider two such half-cells in separate beakers (Figure 21-7). The electrodes are connected by a wire. A voltmeter can be inserted into the circuit to measure the potential difference between the two electrodes, or an ammeter can be inserted to measure the current flow. The electrical

Neither of these meters generates electrical energy.

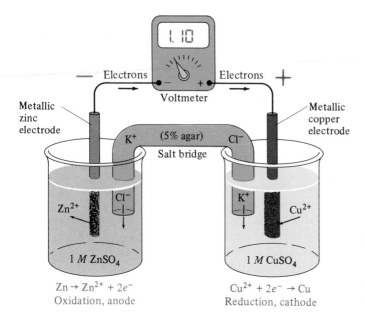

Figure 21-7

The zinc–copper voltaic cell utilizes the reaction

$$Zn(s) + Cu^{2+}(aq) \longrightarrow Zn^{2+}(aq) + Cu(s)$$

The standard potential of this cell is 1.10 volts. The cell can be represented as $Zn|Zn^{2+}(1.0\ M)\|Cu^{2+}(1.0\ M)|Cu$.

current is the result of the spontaneous redox reaction that occurs. We measure the potential of the cell.

The circuit between the two solutions is completed by a **salt bridge**. This can be any medium through which ions can slowly pass. A salt bridge can be made by bending a piece of glass tubing into the shape of a "U," filling it with a hot saturated salt/5% agar solution, and allowing it to cool. The cooled mixture "sets" to the consistency of firm gelatin. As a result, the solution does not run out when the tube is inverted (see Figure 21-7), but the ions in the gel are still able to move. A salt bridge serves three functions.

Agar is a gelatinous material obtained from algae.

1. It allows electrical contact between the two solutions.
2. It prevents mixing of the electrode solutions.
3. It maintains the electrical neutrality in each half-cell as ions flow into and out of the salt bridge.

A cell in which all reactants and products are in their thermodynamic standard states (1 M for dissolved species and 1 atm partial pressure for gases) is called a **standard cell**.

21-10 The Zinc–Copper Cell

Consider a standard cell made up of two half-cells, one a strip of metallic Cu immersed in 1.0 M copper(II) sulfate solution and the other a strip of Zn immersed in 1.0 M zinc sulfate solution (Figure 21-7). This cell is called the Daniell cell. The following experimental observations have been made about this cell:

1. The initial voltage is 1.10 volts.
2. The mass of the copper electrode increases. The concentration of Cu^{2+} decreases in the solution around this electrode as the cell operates.

3. The mass of the zinc electrode decreases. The concentration of Zn^{2+} increases in the solution around the zinc electrode as the cell operates.

We deduce that the half-reaction at the cathode is the reduction of copper(II) ions to Cu metal. This plates out on the Cu electrode. The Zn electrode is the anode. It loses mass because the Zn metal is oxidized to Zn^{2+} ions, which go into solution.

$$\begin{array}{rcll}
Zn & \longrightarrow & Zn^{2+} + 2e^- & \text{(oxidation, anode)} \\
Cu^{2+} + 2e^- & \longrightarrow & Cu & \text{(reduction, cathode)} \\
\hline
Cu^{2+} + Zn & \longrightarrow & Cu + Zn^{2+} & \text{(overall cell reaction)}
\end{array}$$

Electrons are released at the anode and consumed at the cathode. Therefore, they flow through the wire from anode to cathode, as in all electrochemical cells. In all *voltaic* cells, the electrons flow spontaneously from the negative electrode to the positive electrode. So, in contrast with electrolytic cells, the anode is negative and the cathode is positive. To maintain electroneutrality and complete the circuit, two Cl^- ions from the salt bridge migrate into the anode solution for every Zn^{2+} ion formed. Two K^+ ions migrate into the cathode solution to replace every Cu^{2+} ion reduced. Some Zn^{2+} ions from the anode vessel and some SO_4^{2-} ions from the cathode vessel also migrate into the salt bridge. Neither Cl^- nor K^+ ions are oxidized or reduced in preference to the zinc metal or Cu^{2+} ions.

As the reaction proceeds, the cell voltage decreases. When the cell voltage reaches zero, the reaction has reached equilibrium, and the reaction goes no further. At this point, however, the metal ion concentrations in the cell are *not* zero. This description applies to any voltaic cell.

Voltaic cells can be represented as follows for the zinc–copper cell:

salt bridge
↓

$$Zn|Zn^{2+}\ (1.0\ M)\|Cu^{2+}\ (1.0\ M)|Cu$$

species (and concentrations)
in contact with electrode surfaces

Compare the $-/+$, anode/cathode, and oxidation/reduction labels and the directions of electron flow in Figures 21-2a and 21-7.

(Left) A strip of zinc was placed in a blue solution of copper(II) sulfate, $CuSO_4$. The copper has been displaced from solution and has fallen to the bottom of the beaker. The resulting zinc sulfate solution is colorless. (Right) No reaction occurs when copper wire is placed in a colorless zinc sulfate solution.

(Left) A spiral of copper wire was placed in a colorless solution of silver nitrate, $AgNO_3$. The silver has been displaced from solution and adheres to the wire. The resulting copper nitrate solution is blue. (Right) No reaction occurs when silver wire is placed in a blue copper sulfate solution.

In this representation, a single line (|) represents an interface at which a potential develops, i.e., an electrode. It is conventional to write the anode half-cell on the left in this notation.

The same reaction occurs when a piece of Zn is dropped into a solution of $CuSO_4$. The Zn dissolves and the blue color of Cu^{2+} disappears. Copper forms on the Zn and then settles to the bottom of the container. But no electricity flows in an external circuit, because the two half-reactions are *not* physically separated.

21-11 The Copper–Silver Cell

Now consider a similar standard voltaic cell consisting of a strip of Cu immersed in 1.0 M $CuSO_4$ solution and a strip of Ag immersed in 1.0 M $AgNO_3$ solution. A wire and a salt bridge complete the circuit. The following observations have been made:

1. The initial voltage of the cell is 0.46 volt.
2. The mass of the copper electrode decreases. The Cu^{2+} ion concentration increases in the solution around the copper electrode.
3. The mass of the silver electrode increases. The Ag^+ ion concentration decreases in the solution around the silver electrode.

In this cell the Cu electrode is the anode because Cu metal is oxidized to Cu^{2+} ions. The Ag electrode is the cathode because Ag^+ ions are reduced to metallic Ag (Figure 21-8).

$$
\begin{array}{rcll}
Cu & \longrightarrow & Cu^{2+} + 2e^- & \text{(oxidation, anode)} \\
2(Ag^+ + e^- & \longrightarrow & Ag) & \text{(reduction, cathode)} \\
\hline
Cu + 2Ag^+ & \longrightarrow & Cu^{2+} + 2Ag & \text{(overall cell reaction)}
\end{array}
$$

As before, ions from the salt bridge migrate to maintain electroneutrality. Some NO_3^- ions (from the cathode vessel) and some Cu^{2+} ions (from the anode vessel) also migrate into the salt bridge.

Recall that in the zinc–copper cell the copper electrode is the *cathode*; now in the copper–silver cell the copper electrode is the *anode*.

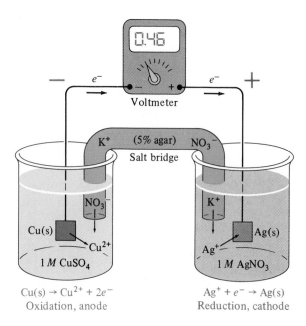

Figure 21-8

The copper–silver voltaic cell utilizes the reaction

$$Cu(s) + 2Ag^+(aq) \longrightarrow Cu^{2+}(aq) + 2Ag(s)$$

The standard potential of this cell is 0.46 volt. This cell can be represented as $Cu|Cu^{2+}(1.0\ M)\|Ag^+(1.0\ M)|Ag$.

Cu(s) → Cu^{2+} + 2e^-
Oxidation, anode

Ag^+ + e^- → Ag(s)
Reduction, cathode

> Whether a particular electrode acts as an anode or a cathode depends on what the other electrode of the cell is.

The two cells we have described show that the Cu^{2+} ion is more easily reduced (is a stronger oxidizing agent) than Zn^{2+}, so Cu^{2+} oxidizes metallic zinc to Zn^{2+}. By contrast, Ag^+ ion is more easily reduced (is a stronger oxidizing agent) than Cu^{2+} ion, so Ag^+ oxidizes Cu atoms to Cu^{2+}. Conversely, metallic Zn is a stronger reducing agent than metallic Cu, and metallic Cu is a stronger reducing agent than metallic Ag. We can now arrange the species we have studied in order of increasing strength as oxidizing agents and as reducing agents.

$$Zn^{2+} < Cu^{2+} < Ag^+ \qquad\qquad Zn > Cu > Ag$$

Increasing strength
as oxidizing agents

Increasing strength
as reducing agents

The standard $Cu|Cu^{2+}(1.0\ M)\|Ag^+\ (1.0\ M)|Ag$ cell.

Standard Electrode Potentials

The potentials of the standard zinc–copper and copper–silver voltaic cells are 1.10 volts and 0.46 volts, respectively. The magnitude of a cell's potential measures the spontaneity of its redox reaction. *Higher (more positive) cell potentials indicate greater driving force.* Under standard conditions, the oxidation of metallic Zn by Cu^{2+} ions has a greater tendency to go toward completion than does the oxidation of metallic Cu by Ag^+ ions. We would like to separate the total cell potential into the individual contributions of the two electrode half-reactions. Then we can determine the relative tendencies of particular oxidation or reduction half-reactions to occur. Such information gives us a quantitative basis for specifying strengths of oxidizing and reducing agents. In the next several sections we shall see how this is done for standard half-cells. Then we shall learn to correct for changes in temperature, concentration, and pressure (Section 21-20).

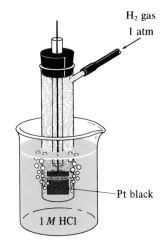

H$_2$ gas
1 atm

Pt black

1 M HCl

21-12 The Standard Hydrogen Electrode

Every oxidation must be accompanied by a reduction (that is, the electrons must have somewhere to go). So it is impossible to determine experimentally the potential of any *single* electrode. Therefore, we establish an arbitrary standard. The reference electrode is the **standard hydrogen electrode (SHE).** This electrode contains a piece of metal electrolytically coated with a grainy black surface of inert platinum metal, immersed in a 1.0 M H^+ solution. Hydrogen, H_2, is bubbled at 1 atm pressure through a glass envelope over the platinized electrode (Figure 21-9).

Figure 21-9
The standard hydrogen electrode (SHE).

By international agreement, the standard hydrogen electrode is arbitrarily assigned a potential of *exactly* 0.0000 . . . volt.

The superscript in E^0 indicates standard electrochemical conditions.

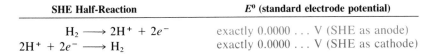

SHE Half-Reaction	E^0 (standard electrode potential)
$H_2 \longrightarrow 2H^+ + 2e^-$	exactly 0.0000 . . . V (SHE as anode)
$2H^+ + 2e^- \longrightarrow H_2$	exactly 0.0000 . . . V (SHE as cathode)

We then construct a standard cell consisting of a standard hydrogen electrode and some other standard electrode (half-cell). Because the defined electrode potential of the SHE contributes exactly 0 volt to the sum, the voltage of the overall cell then lets us determine the **standard electrode potential** of the other half-cell. This is its potential with respect to the standard hydrogen electrode, measured at 25°C when the concentration of each ion in the solution is 1 M and the pressure of any gas involved is 1 atm.

By agreement, we always present the standard cell potential for each half-cell as a *reduction* process.

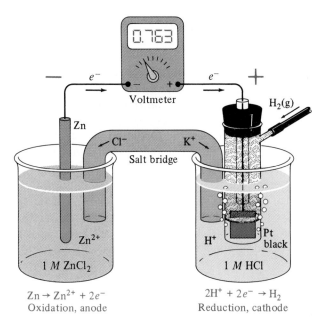

Figure 21-10
The $Zn|Zn^{2+}(1.0\ M)\|H^+(1.0\ M);\ H_2(1\ atm)|Pt$ cell, in which the following net reaction occurs:

$$Zn(s) + 2H^+(aq) \longrightarrow Zn^{2+}(aq) + H_2(g)$$

In this cell, the standard hydrogen electrode functions as the cathode.

$Zn \rightarrow Zn^{2+} + 2e^-$
Oxidation, anode

$2H^+ + 2e^- \rightarrow H_2$
Reduction, cathode

21-13 The Zinc–SHE Cell

This cell consists of an SHE in one beaker and a strip of zinc immersed in 1.0 M zinc sulfate solution in another beaker (Figure 21-10). A wire and a salt bridge complete the circuit. When the circuit is closed, the following observations can be made.

1. The initial potential of the cell is 0.763 volt.
2. As the cell operates, the mass of the zinc electrode decreases. The concentration of Zn^{2+} ions increases in the solution around the zinc electrode.
3. The H^+ concentration decreases in the SHE. Gaseous H_2 is produced.

We can conclude from these observations that the following half-reactions and cell reaction occur.

		E^0
(oxidation, anode)	$Zn \longrightarrow Zn^{2+} + 2e^-$	0.763 V
(reduction, cathode)	$2H^+ + 2e^- \longrightarrow H_2$	0.000 V (by definition)
(cell reaction)	$Zn + 2H^+ \longrightarrow Zn^{2+} + H_2$	$E^0_{cell} = 0.763$ V (measured)

The standard potential at the anode *plus* the standard potential at the cathode give the standard cell potential. The potential of the SHE is 0.000 volt, and the standard cell potential is found to be 0.763 volt. So the standard potential of the zinc anode must be 0.763 volt. The $Zn|Zn^{2+}(1.0\ M)\|H^+$ $(1.0\ M);\ H_2$ (1 atm)|Pt cell is depicted in Figure 21-10.

Note that in *this* cell the SHE is the *cathode*, and metallic zinc reduces H^+ to H_2. The zinc electrode is the *anode* in this cell.

Another way of thinking about this process is as follows. The negative reduction potential for the half-reaction

$$Zn^{2+} + 2e^- \longrightarrow Zn \qquad E^0 = -0.763\ V$$

says that this reaction is *less favorable* than the corresponding reduction to H_2,

$$2H^+ + 2e^- \longrightarrow H_2 \qquad E^0 = 0.000 \text{ V}$$

Before they are connected, each half-cell builds up a supply of electrons waiting to be released, thus generating an "electron pressure." Let us compare these electron pressures by reversing the two half-reactions to show production of electrons (and changing the signs of their E^0 values).

$$Zn \longrightarrow Zn^{2+} + 2e^- \qquad E^0_{oxidation} = +0.763 \text{ V}$$

$$H_2 \longrightarrow 2H^+ + 2e^- \qquad E^0_{oxidation} = 0.000 \text{ V}$$

The process with the more *positive* E^0 value is favored, so we reason that the electron pressure generated at the Zn electrode is greater than that at the H_2 electrode. As a result, when the cell is connected, electrons flow through the wire *from the Zn electrode to the H_2 electrode.* Oxidation occurs at the zinc electrode (anode), and reduction occurs at the hydrogen electrode (cathode).

21-14 The Copper–SHE Cell

Another cell consists of an SHE in one beaker and a strip of Cu metal immersed in 1.0 M copper(II) sulfate solution in another beaker. A wire and a salt bridge complete the circuit. For this cell, we observe the following (Figure 21-11):

1. The initial cell potential is 0.337 volt.
2. Gaseous hydrogen is used up. The H^+ concentration increases in the solution of the SHE.

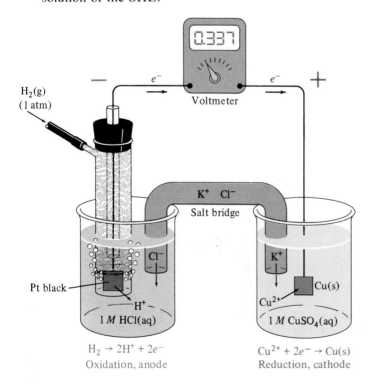

Figure 21-11
The standard copper–SHE cell, Pt|H^+(1.0 M); H_2(1 atm)‖Cu^{2+} (1.0 M)|Cu cell. In this cell, the standard hydrogen electrode functions as the anode. The net reaction is

$$H_2(g) + Cu^{2+}(aq) \longrightarrow 2H^+(aq) + Cu(s)$$

3. The mass of the copper electrode increases. The concentration of Cu^{2+} ions decreases in the solution around the copper electrode.

Thus, the following half-reactions and cell reaction occur:

		E^0
(oxidation, anode)	$H_2 \longrightarrow 2H^+ + 2e^-$	0.000 V (by definition)
(reduction, cathode)	$Cu^{2+} + 2e^- \longrightarrow Cu$	0.337 V
(cell reaction)	$H_2 + Cu^{2+} \longrightarrow 2H^+ + Cu$	$E^0_{cell} = 0.337$ V (measured)

In the Zn–SHE cell, the SHE is the cathode.

The SHE functions as the *anode* in this cell, and Cu^{2+} ions oxidize H_2 to H^+ ions. The standard electrode potential of the copper half-cell is 0.337 volt as a *cathode* in the Cu–SHE cell.

Again, we can think of $E^0_{oxidation}$ in the two half-cells as "electron pressures."

$$Cu \longrightarrow Cu^{2+} + 2e^- \qquad E^0_{oxidation} = -0.337 \text{ V}$$

$$H_2 \longrightarrow 2H^+ + 2e^- \qquad E^0_{oxidation} = 0.000 \text{ V}$$

Now the hydrogen electrode has the higher electron pressure. When the cell is connected, electrons flow through the wire from the hydrogen electrode to the copper electrode. H_2 is oxidized to $2H^+$ (anode), and Cu^{2+} is reduced to Cu (cathode).

21-15 The Electromotive Series (Activity Series) of the Elements

By measuring the potentials of other standard electrodes versus the SHE in the way we described for the standard Zn–SHE and standard Cu–SHE voltaic cells, we can develop a series of standard electrode potentials. When the electrodes involve metals or nonmetals in contact with their ions, the series is called the **electromotive series** or **activity series** of the elements. We saw (Section 21-13) that the standard Zn electrode behaves as the anode versus the SHE and that the standard *oxidation* potential for the Zn half-cell is 0.763 volt.

	$E^0_{oxidation}$
(as anode) $Zn \longrightarrow Zn^{2+} + 2e^-$	+0.763 V
reduced form $\longrightarrow$ oxidized form $+ ne^-$	(standard *oxidation* potential)

Therefore, the *reduction* potential for the standard zinc electrode (to act as a *cathode* relative to the SHE) is the negative of this, or −0.763 volt.

	$E^0_{reduction}$
(as cathode) $Zn^{2+} + 2e^- \longrightarrow Zn$	−0.763 V
oxidized form $+ ne^- \longrightarrow$ reduced form	(standard *reduction* potential)

Different conventions have been used for writing half-reactions and the signs for their potentials. To avoid confusion, use the convention presented here consistently.

By international convention, the standard potentials of electrodes are tabulated for *reduction half-reactions*. These indicate the tendencies of the electrodes to behave as cathodes toward the SHE. Electrodes with positive E^0 values for reduction half-reactions act as *cathodes* versus the SHE. Those with negative E^0 values for reduction half-reactions act as anodes versus the SHE.

Electrodes with *Positive* $E^0_{reduction}$	Electrodes with *Negative* $E^0_{reduction}$
Reduction occurs *more readily* than the reduction of $2H^+$ to H_2.	Reduction is *more difficult* than the reduction of $2H^+$ to H_2.
Electrode acts as a *cathode* versus the SHE.	Electrode acts as an *anode* versus the SHE.

The more positive the E^0 value for a half-reaction, the greater the tendency for the half-reaction to occur in the forward direction as written. Conversely, the more negative the E^0 value for a half-reaction, the greater the tendency for the half-reaction to occur in the reverse direction as written.

The electromotive series of the elements is shown in Table 21-2.

1. The species on the *left* side are all either cations of metals, hydrogen ions, or elemental nonmetals. These are all *oxidizing agents* (*oxidized forms* of the elements). Their strengths as oxidizing agents increase from top to bottom, i.e., as the $E^0_{reduction}$ values become more positive. Fluorine is the strongest oxidizing agent, and Li^+ is a very weak oxidizing agent.

2. The species on the *right* side are free metals, hydrogen, or anions of nonmetals. These are all *reducing agents* (*reduced forms* of the elements). Their strengths as reducing agents increase from bottom to top, i.e., as the $E^0_{reduction}$ values become more negative. Metallic Li is a very strong reducing agent, and F^- is a very weak reducing agent.

Table 21-2
Standard Aqueous Electrode Potentials at 25°C—The Electromotive Series

Element	Reduction Half-Reaction	Standard Reduction Potential E^0, volts	
Li	$Li^+ + e^- \longrightarrow Li$	−3.045	
K	$K^+ + e^- \longrightarrow K$	−2.925	
Ca	$Ca^{2+} + 2e^- \longrightarrow Ca$	−2.87	
Na	$Na^+ + e^- \longrightarrow Na$	−2.714	
Mg	$Mg^{2+} + 2e^- \longrightarrow Mg$	−2.37	
Al	$Al^{3+} + 3e^- \longrightarrow Al$	−1.66	
Zn	$Zn^{2+} + 2e^- \longrightarrow Zn$	−0.7628	
Cr	$Cr^{3+} + 3e^- \longrightarrow Cr$	−0.74	
Fe	$Fe^{2+} + 2e^- \longrightarrow Fe$	−0.44	
Cd	$Cd^{2+} + 2e^- \longrightarrow Cd$	−0.403	
Ni	$Ni^{2+} + 2e^- \longrightarrow Ni$	−0.25	
Sn	$Sn^{2+} + 2e^- \longrightarrow Sn$	−0.14	
Pb	$Pb^{2+} + 2e^- \longrightarrow Pb$	−0.126	
H_2	$2H^+ + 2e^- \longrightarrow H_2$	0.000	(reference electrode)
Cu	$Cu^{2+} + 2e^- \longrightarrow Cu$	+0.337	
I_2	$I_2 + 2e^- \longrightarrow 2I^-$	+0.535	
Hg	$Hg^{2+} + 2e^- \longrightarrow Hg$	+0.789	
Ag	$Ag^+ + e^- \longrightarrow Ag$	+0.7994	
Br_2	$Br_2 + 2e^- \longrightarrow 2Br^-$	+1.08	
Cl_2	$Cl_2 + 2e^- \longrightarrow 2Cl^-$	+1.360	
Au	$Au^{3+} + 3e^- \longrightarrow Au$	+1.50	
F_2	$F_2 + 2e^- \longrightarrow 2F^-$	+2.87	

Increasing strength as oxidizing agent

Increasing strength as reducing agent

The more positive the reduction potential, the stronger the species on the left is as an oxidizing agent and the weaker the species on the right is as a reducing agent. The elements with high ionization energies and highly negative electron affinities have the greatest tendencies to exist as anions. The elements with low ionization energies and positive or slightly negative electron affinities have the greatest tendencies to exist as cations.

21-16 Uses of the Electromotive Series

The most important application of the electromotive series is the prediction of the spontaneity of redox reactions. Standard electrode potentials can be used to determine the spontaneity of redox reactions in general, whether or not the reactions can take place in electrochemical cells.

Suppose we ask a question such as this: At standard conditions, will Cu^{2+} ions oxidize metallic Zn to Zn^{2+} ions, or will Zn^{2+} ions oxidize metallic copper to Cu^{2+}? One of the two possible reactions is spontaneous, and the reverse reaction is nonspontaneous. We must determine which one is spontaneous. We already know the answer to this question from experimental results (Section 21-10), but let us demonstrate the procedure for predicting the spontaneous reaction:

1. Choose the appropriate half-reactions from a table of standard reduction potentials.
2. Write the equation for the half-reaction with the more positive (or less negative) E^0 value *for reduction* first, along with its potential.
3. Then write the equation for the other half-reaction *as an oxidation* and write its *oxidation potential*; to do this, reverse the tabulated reduction half-reaction and change the sign of E^0. (Reversing a half-reaction or a complete reaction also changes the sign of its potential.)
4. Balance the electron transfer. *We do not multiply the potentials by the numbers used to balance the electron transfer!* The reason is that each potential represents the *tendency* for the process to occur; this does not depend on *how many times* it occurs. An electrical potential is an *intensive property*.
5. Add the reduction and oxidation half-reactions, and add the reduction and oxidation potentials. E^0_{cell} will be *positive* for the resulting overall cell reaction. *This indicates that the forward reaction is spontaneous.* (Any overall cell reaction for which E^0_{cell} is negative is nonspontaneous.)

For the cell described above, the Cu^{2+}/Cu couple has the more positive reduction potential, so we keep it as a reduction and reverse the other half-reaction. Following the steps outlined, we obtain the equation for the spontaneous reaction:

$$1(Cu^{2+} + 2e^- \longrightarrow Cu) \qquad\qquad +0.337 \text{ V} \longleftarrow \text{reduction potential}$$
$$\underline{1(Zn \longrightarrow Zn^{2+} + 2e^-) \qquad\qquad +0.763 \text{ V} \longleftarrow \text{oxidation potential}}$$
$$Cu^{2+} + Zn \longrightarrow Cu + Zn^{2+} \qquad E^0_{cell} = +1.100 \text{ V}$$

The fact that E^0_{cell} is positive tells us that the forward reaction is spontaneous at standard conditions. So we conclude that copper(II) ions will oxidize metallic zinc to Zn^{2+} ions as they are reduced to metallic copper.

(Section 21-10 shows that the potential of the standard zinc–copper voltaic cell is 1.10 volts. This is the spontaneous reaction that occurs.)

The reverse reaction has a negative E^0 and is nonspontaneous.

nonspontaneous
reaction: $Cu + Zn^{2+} \longrightarrow Cu^{2+} + Zn$ $E^0_{cell} = -1.10$ volts

To make it occur, we would have to supply electrical energy with a potential difference greater than 1.10 volts. That is, this nonspontaneous reaction would have to be carried out in an *electrolytic cell*.

Example 21-4

At standard conditions, will chromium(III) ions, Cr^{3+}, oxidize metallic copper to copper(II) ions, Cu^{2+}, or will Cu^{2+} oxidize metallic chromium to Cr^{3+} ions?

Plan

We refer to the table of standard reduction potentials and choose the two appropriate half-reactions. The copper half-reaction has the more positive reduction potential, so we write it first. Then we write the chromium half-reaction as an oxidation, balance the electron transfer, and add the two half-reactions and their potentials.

Solution

			E^0
$3(Cu^{2+} + 2e^- \longrightarrow Cu)$		(reduction)	$+0.337$ V
$2(Cr \longrightarrow Cr^{3+} + 3e^-)$		(oxidation)	$+0.74$ V
$2Cr + 3Cu^{2+} \longrightarrow 2Cr^{3+} + 3Cu$		$E^0_{cell} = +1.08$ V	

E^0_{cell} is positive, so we know this is the spontaneous reaction.

Cu^{2+} ions spontaneously oxidize metallic Cr to Cr^{3+} ions and are reduced to metallic Cu.

EOC 50a

21-17 Electrode Potentials for Other Half-Reactions

It is possible to construct half-cells in which both oxidized and reduced species are in solution as ions in contact with inert electrodes. For example, the standard iron(III) ion/iron(II) ion half-cell contains 1.0 M concentrations of the two ions. It involves the following half-reaction:

$$Fe^{3+} + e^- \longrightarrow Fe^{2+} \qquad E^0 = +0.771 \text{ V}$$

The standard dichromate ($Cr_2O_7{}^{2-}$) ion/chromium(III) ion half-cell consists of a 1.0 M concentration of each of the two ions in contact with an inert electrode. The balanced half-reaction in acidic solution (1.0 M H^+) is

$$Cr_2O_7{}^{2-} + 14H^+ + 6e^- \longrightarrow 2Cr^{3+} + 7H_2O \qquad E^0 = +1.33 \text{ V}$$

Standard electrode potentials for some other reactions are given in Table 21-3 and in Appendix J. These potentials can be used like those of the electromotive series.

Platinum metal is often used as the inert electrode material. These two standard half-cells could be shown in shorthand notation as

Pt/Fe^{3+}(1.0 M), Fe^{2+}(1.0 M)   and

Pt/$Cr_2O_7{}^{2-}$(1.0 M), Cr^{3+}(1.0 M)

Table 21-3
Standard Electrode Potentials for Selected Half-Cells

Reduction Half-Reaction		Standard Electrode Potential E^0 (volts)
$Zn(OH)_4^{2-} + 2e^-$	$\longrightarrow Zn + 4OH^-$	-1.22
$Fe(OH)_2 + 2e^-$	$\longrightarrow Fe + 2OH^-$	-0.877
$2H_2O + 2e^-$	$\longrightarrow H_2 + 2OH^-$	-0.8277
$PbSO_4 + 2e^-$	$\longrightarrow Pb + SO_4^{2-}$	-0.356
$NO_3^- + H_2O + 2e^-$	$\longrightarrow NO_2^- + 2OH^-$	$+0.01$
$Sn^{4+} + 2e^-$	$\longrightarrow Sn^{2+}$	$+0.15$
$AgCl + e^-$	$\longrightarrow Ag + Cl^-$	$+0.222$
$Hg_2Cl_2 + 2e^-$	$\longrightarrow 2Hg + 2Cl^-$	$+0.27$
$O_2 + 2H_2O + 4e^-$	$\longrightarrow 4OH^-$	$+0.40$
$NiO_2 + 2H_2O + 2e^-$	$\longrightarrow Ni(OH)_2 + 2OH^-$	$+0.49$
$H_3AsO_4 + 2H^+ + 2e^-$	$\longrightarrow H_3AsO_3 + H_2O$	$+0.58$
$Fe^{3+} + e^-$	$\longrightarrow Fe^{2+}$	$+0.771$
$ClO^- + H_2O + 2e^-$	$\longrightarrow Cl^- + 2OH^-$	$+0.89$
$NO_3^- + 4H^+ + 3e^-$	$\longrightarrow NO + 2H_2O$	$+0.96$
$Cr_2O_7^{2-} + 14H^+ + 6e^-$	$\longrightarrow 2Cr^{3+} + 7H_2O$	$+1.33$
$MnO_4^- + 8H^+ + 5e^-$	$\longrightarrow Mn^{2+} + 4H_2O$	$+1.51$
$PbO_2 + SO_4^{2-} + 4H^+ + 2e^-$	$\longrightarrow PbSO_4 + 2H_2O$	$+1.685$

Example 21-5

At standard conditions, will tin(IV) ions, Sn^{4+}, oxidize gaseous nitrogen oxide, NO, to nitrate ions, NO_3^-, in acidic solution, or will NO_3^- oxidize Sn^{2+} to Sn^{4+}?

Plan

The NO_3^-/NO reduction half-reaction has the more positive E^0 value, so we write the Sn^{4+}/Sn^{2+} half-reaction as an oxidation. We balance the electron transfer and add the two half-reactions to obtain the equation for the *spontaneous* reaction. Then we add the half-reaction potentials to obtain the overall cell potential.

Solution

$$
\begin{array}{lr}
 & E^0 \\
2(NO_3^- + 4H^+ + 3e^- \longrightarrow NO + 2H_2O) & +0.96 \text{ V} \\
3(Sn^{2+} \longrightarrow Sn^{4+} + 2e^-) & -0.15 \text{ V} \\
\hline
2NO_3^- + 8H^+ + 3Sn^{2+} \longrightarrow 2NO + 4H_2O + 3Sn^{4+} \quad E^0_{cell} = +0.81 \text{ V}
\end{array}
$$

Because E^0_{cell} is positive for this reaction,

nitrate ions spontaneously oxidize tin(II) ions to tin(IV) ions and are reduced to nitrogen oxide in acidic solution,   when all species are present at unit activities.

EOC 51, 52

Now that we know how to use standard reduction potentials, let us use them to explain the reaction that occurs in the electrolysis of aqueous NaCl. The first two electrolytic cells we considered involved *molten* NaCl and *aqueous* NaCl (Sections 21-3 and 21-4). There was no doubt that in molten NaCl metallic Na would be produced by reduction of Na^+, and gaseous Cl_2

would be produced by oxidation of Cl^-. But we found that in aqueous NaCl, H_2O, rather than Na^+, was reduced. This is consistent with the less negative reduction potential of H_2O, compared with Na^+.

$$\begin{array}{ll} & \underline{E^0} \\ 2H_2O + 2e^- \longrightarrow H_2 + 2OH^- & -0.828 \text{ V} \\ Na^+ + e^- \longrightarrow Na & -2.714 \text{ V} \end{array}$$

The more easily reduced species, H_2O, is reduced.

Electrode potentials measure only the relative *thermodynamic* likelihood for various half-reactions. In practice, kinetic factors can complicate matters. For instance, sometimes the electrode process is limited by the rate of diffusion of dissolved species to or from the electrode surface. At some cathodes, the rate of electron transfer from the electrode to a reactant is the rate-limiting step, and a higher voltage (called *overvoltage*) must be developed to accomplish the reduction. As a result of these factors, a half-reaction that is *thermodynamically* more favorable than some other process still might not occur at a significant rate. In the electrolysis of NaCl(aq), Cl^- is oxidized to Cl_2 gas (-1.360 V), instead of H_2O being oxidized to form O_2 gas (-1.229 V), because of the overvoltage of O_2 on Pt, the inert electrode.

21-18 Corrosion

Ordinary **corrosion** is the redox process by which metals are oxidized by oxygen, O_2, in the presence of moisture. There are other kinds, but this is the most common. The problem of corrosion and its prevention are of both theoretical and practical interest. Corrosion is responsible for the loss of billions of dollars annually in metal products. The mechanism of corrosion has been studied extensively. It is now known that the oxidation of metals occurs most readily at points of strain (where the metals are most "active"). Thus, a steel nail, which is mostly iron (Section 22-7), first corrodes at the tip and head (Figure 21-12). A bent nail corrodes most readily at the bend.

A point of strain in a steel object acts as an anode where the iron is oxidized to iron(II) ions, and pits are formed (Figure 21-13):

$$Fe \longrightarrow Fe^{2+} + 2e^- \quad \text{(oxidation, anode)}$$

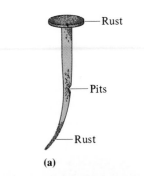

(a)

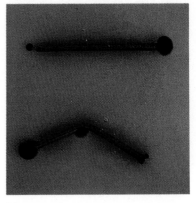

(b)

Figure 21-12
(a) A bent nail corrodes at points of strain and "active" metal atoms. (b) Two nails were placed in an agar gel that contained phenolphthalein and potassium ferricyanide, $K_3[Fe(CN)_6]$. As the nails corroded they produced Fe^{2+} ions at each end and at the bend. Fe^{2+} ions react with $[Fe(CN)_6]^{3-}$ ions to form $Fe_3[Fe(CN)_6]_2$, an intensely blue-colored compound. The rest of each nail is the cathode, at which water is reduced to H_2 and OH^- ions. The OH^- ions turn phenolphthalein pink.

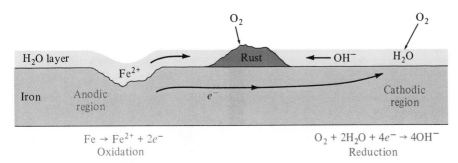

$$Fe \rightarrow Fe^{2+} + 2e^- \qquad O_2 + 2H_2O + 4e^- \rightarrow 4OH^-$$
Oxidation · Reduction

Overall process: $2Fe(s) + \frac{3}{2}O_2(aq) + xH_2O(\ell) \rightarrow Fe_2O_3 \cdot xH_2O(s)$

Figure 21-13
The corrosion of iron. Pitting appears at the anodic region, where iron metal is oxidized to Fe^{2+}. Rust appears at the cathodic region.

Rust is a serious economic problem.

The electrons produced then flow through the nail to areas exposed to O_2. These act as cathodes where oxygen is reduced to hydroxide ions, OH^-:

$$O_2 + 2H_2O + 4e^- \longrightarrow 4OH^- \quad \text{(reduction, cathode)}$$

At the same time, the Fe^{2+} ions migrate through the moisture to the surface. The overall reaction is obtained by balancing the electron transfer and adding the two half-reactions.

$$
\begin{array}{lll}
2(Fe \longrightarrow Fe^{2+} + 2e^-) & \text{(oxidation, anode)} \\
\underline{O_2 + 2H_2O + 4e^- \longrightarrow 4OH^-} & \text{(reduction, cathode)} \\
2Fe + O_2 + 2H_2O \longrightarrow 2Fe^{2+} + 4OH^- & \text{(net reaction)}
\end{array}
$$

The Fe^{2+} ions can migrate from the anode through the solution toward the cathode region, where they combine with OH^- ions to form iron(II) hydroxide. Iron is further oxidized by O_2 to the $+3$ oxidation state. The material we call rust is a complex hydrated form of iron(III) oxides and hydroxides with variable water composition; it can be represented as $Fe_2O_3 \cdot xH_2O$. The overall reaction for the rusting of iron is

$$2Fe(s) + \tfrac{3}{2}O_2(aq) + xH_2O(\ell) \longrightarrow Fe_2O_3 \cdot xH_2O(s)$$

21-19 Corrosion Protection

There are several methods for protecting metals against corrosion. The most widely used are

1. Plating the metal with a thin layer of a less easily oxidized metal
2. Connecting the metal directly to a "sacrificial anode," a piece of another metal that is more active and therefore preferentially oxidized
3. Allowing a protective film, such as a metal oxide, to form naturally on the surface of the metal
4. Galvanizing, or coating steel with zinc, a more active metal
5. Applying a protective coating such as paint

The thin layer of tin on tin-plated steel cans is less easily oxidized than iron, and it protects the steel underneath from corrosion. It is deposited either by dipping the can into molten tin or by electroplating. Copper is also less active than iron (see Table 21-2). It is sometimes deposited by electroplating to protect metals when food is not involved. Whenever the layer of tin or copper is breached, the iron beneath it corrodes even more rapidly

Compare the potentials for the *oxidation* half-reactions (the reverse of tabulated reduction half-reactions) to see which metal is more easily oxidized:

	$E^0_{\text{oxidation}}$
$Fe \longrightarrow Fe^{2+} + 2e^-$	$+0.44$ V
$Sn \longrightarrow Sn^{2+} + 2e^-$	$+0.14$ V
$Cu \longrightarrow Cu^{2+} + 2e^-$	-0.337 V

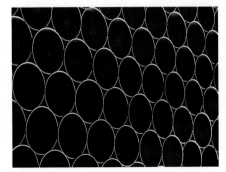

Two methods of protection against corrosion. (Left) Steel cans coated with a polymer to prevent oxidation. (Right) Electrogalvanizing.

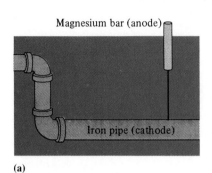

Magnesium bar (anode)

Iron pipe (cathode)

(a)

BRITISH RESOURCE
LONDON

(b)

Figure 21-14
(a) Cathodic protection of buried iron pipe. A magnesium or zinc bar is oxidized instead of the iron. The "sacrificial" anode eventually must be replaced. (b) Cathodic protection of a ship's hull. The small yellow horizontal strips are blocks of titanium (coated with platinum) that are attached to the ship's hull. The hull is steel (mostly iron). When the ship is in salt water, the titanium blocks become the anode, and the hull the cathode, in a voltaic cell. Because oxidation always occurs at the anode, the ship's hull (the cathode) is protected from oxidation (corrosion).

than it would without the coating, because of the adverse electrochemical cell that is set up.

Figure 21-14a shows an iron pipe connected to a strip of magnesium, a more active metal, to protect the iron from oxidation. The magnesium is preferentially oxidized. It is called a "sacrificial anode." Similar methods are used to protect bridges and ships' hulls from corrosion. Other active metals, such as zinc, are also used as sacrificial anodes.

Aluminum, a very active metal, reacts rapidly with O_2 from the air to form a surface layer of aluminum oxide, Al_2O_3, that is so thin that it is transparent. This very tough, hard substance is inert toward oxygen, water, and most other corrosive agents in the environment. In this way, objects made of aluminum form their own protective layers and need not be treated further to inhibit corrosion.

Compare the potentials for the *oxidation* half-reactions to see which metal is more easily oxidized:

	$E^0_{oxidation}$
Fe ⟶ Fe^{2+} + $2e^-$	+0.44 V
Zn ⟶ Zn^{2+} + $2e^-$	+0.763 V
Mg ⟶ Mg^{2+} + $2e^-$	+2.37 V

Acid rain endangers structural aluminum by dissolving this Al_2O_3 coating.

Effect of Concentrations (or Partial Pressures) on Electrode Potentials

21-20 The Nernst Equation

Standard electrode potentials, designated E^0, refer to standard state conditions. These standard state conditions are one-molar solutions for ions, 1 atmosphere pressure for gases, and all solids and liquids in their standard states at 25°C. (Remember that we refer to *thermodynamic* standard state conditions, and not standard temperature and pressure as in gas law calculations.) As any of the standard cells described earlier operates, and concentrations or pressures of reactants change, the observed cell voltage drops. Similarly, cells constructed with solution concentrations different from one molar, or gas pressures different from 1 atmosphere, cause the corresponding potentials to deviate from standard electrode potentials.

The **Nernst equation** is used to calculate electrode potentials and cell potentials for concentrations and partial pressures other than standard state values:

Developed by Walther Nernst (1864–1941).

In this equation, the expression following the minus sign represents how much the *nonstandard* conditions cause the electrode potential to deviate from its standard value, E^0. The Nernst equation is normally presented in terms of base-10 logarithms.

$$E = E^0 - \frac{2.303\,RT}{nF} \log Q$$

where

E = potential under the **nonstandard** conditions

E^0 = **standard** potential

R = gas constant, 8.314 J/mol · K

T = absolute temperature, in K

n = number of moles of electrons transferred in the reaction

F = faraday, 96,487 C/mol e^- × 1 J/(V · C)

    = 96,487 J/V · mol e^-

Q = reaction quotient

The reaction quotient, Q, was introduced in Section 17-4. It involves a ratio of concentrations or pressures of products to those of reactants, each raised to the power indicated by the coefficient in the balanced equation. The Q expression that is used in the Nernst equation is the thermodynamic reaction quotient; it can include *both* concentrations and pressures. Substituting these values into the Nernst equation at 25°C gives

At 25°C, the value of $\frac{2.303\,RT}{F}$ is 0.0592; at any other temperature, this term must be recalculated. Can you show that this term has the units V · mol?

$$E = E^0 - \frac{0.0592}{n} \log Q \qquad (Note: \text{in terms of base-10 log})$$

In general, half-reactions for standard reduction potentials are written

$$x\,\text{Ox} + ne^- \longrightarrow y\,\text{Red}$$

"Red" refers to the reduced species, and "Ox" to the oxidized species. The Nernst equation for any *cathode* half-cell (*reduction* half-reaction) is

$$E = E^0 - \frac{0.0592}{n} \log \frac{[\text{Red}]^y}{[\text{Ox}]^x} \qquad \text{(reduction half-reaction)}$$

For the familiar half-reaction involving metallic zinc and zinc ions,

$$\text{Zn}^{2+} + 2e^- \rightleftharpoons \text{Zn} \qquad E^0 = -0.763\text{ V}$$

the corresponding Nernst equation is

Metallic Zn is a pure solid, so its concentration does not appear in Q.

$$E = E^0 - \frac{0.0592}{2} \log \frac{1}{[\text{Zn}^{2+}]} \qquad \text{(for reduction)}$$

We substitute the E^0 value into the equation to obtain

$$E = -0.763\text{ V} - \frac{0.0592}{2} \log \frac{1}{[\text{Zn}^{2+}]}$$

Example 21-6

Calculate the (reduction) potential for the $\text{Fe}^{3+}/\text{Fe}^{2+}$ electrode when the concentration of Fe^{2+} is five times that of Fe^{3+}.

Plan

The Nernst equation lets us calculate potentials for concentrations other than one molar. The tabulation of standard reduction potentials gives us the value of E^0 for the reduction half-reaction. We use the balanced half-reaction and the given concentration ratio to calculate the value of Q. Then we substitute this into the Nernst equation with n equal to the number of moles of electrons involved in the half-reaction.

Solution

The reduction half-reaction is

$$Fe^{3+} + e^- \longrightarrow Fe^{2+} \qquad E^0 = +0.771 \text{ V}$$

We are told that the concentration of Fe^{2+} is five times that of Fe^{3+}, or $[Fe^{2+}] = 5[Fe^{3+}]$. Calculating the value of Q,

$$Q = \frac{[Red]^y}{[Ox]^x} = \frac{[Fe^{2+}]}{[Fe^{3+}]} = \frac{5[Fe^{3+}]}{[Fe^{3+}]} = 5$$

The balanced half-reaction shows one mole of electrons, or $n = 1$. Putting values into the Nernst equation,

$$E = E^0 - \frac{0.0592}{n} \log Q = +0.771 - \frac{0.0592}{1} \log 5 = (+0.771 - 0.041) \text{ V}$$

$$\boxed{E = +0.730 \text{ V}}$$

Example 21-7

Calculate E for the Fe^{3+}/Fe^{2+} electrode when the Fe^{3+} concentration is five times the Fe^{2+} concentration.

Plan

The approach is the same as for Example 21-6.

Solution

$$Q = \frac{[Red]^y}{[Ox]^x} = \frac{[Fe^{2+}]}{[Fe^{3+}]} = \frac{[Fe^{2+}]}{5[Fe^{2+}]} = \frac{1}{5} = 0.200$$

$$E = +0.771 - \frac{0.0592}{1} \log (0.200) = +0.771 - 0.0592(-0.699)$$

$$= (+0.771 + 0.041) \text{ V}$$

$$\boxed{= +0.812 \text{ V}} \qquad \text{(for reduction)}$$

We see that the correction factor, +0.041 volt, differs from that in Example 21-6 only in sign.

Example 21-8

Calculate the potential of the chlorine–chloride ion, Cl_2/Cl^-, electrode when the partial pressure of Cl_2 is 10.0 atm and $[Cl^-] = 1.00 \times 10^{-3} \ M$.

Plan

The approach is the same as for the preceding two examples. Now the Nernst equation involves the molar concentration of a substance in solution (Cl^-) and the pressure of a gaseous component (Cl_2); for this half-reaction, $n = 2$.

Solution

The half-reaction and standard reduction potential are

$$Cl_2 + 2e^- \longrightarrow 2Cl^- \qquad E^0 = +1.360 \text{ V}$$

The appropriate Nernst equation is

The reaction in Examples 21-9 and 21-10 is thermodynamically favored (spontaneous) under the stated conditions, with a potential of $+0.42$ volt *when the cell starts operation*. As the cell discharges and current flows, the product concentrations, $[Mn^{2+}]$ and $[Fe^{3+}]$, increase. At the same time, reactant concentrations, $[MnO_4^-]$, $[H^+]$, and $[Fe^{2+}]$, decrease. This increases $\log Q$, so the correction factor becomes more negative. Thus, the overall E_{cell} *decreases* (the reaction becomes less favorable). Eventually the cell potential approaches zero (equilibrium), and the cell "runs down."

Concentration Cells

As we have seen, different concentrations of ions in a half-cell result in different half-cell potentials. We can use this idea to construct a **concentration cell**, in which both half-cells are composed of the same species, but in different ion concentrations. Suppose we set up such a cell using the Cu^{2+}/Cu half-cell that we introduced in Section 21-10. We put copper electrodes into two aqueous solutions, one that is $0.10\ M\ CuSO_4$ and another that is $1.00\ M\ CuSO_4$. To complete the cell construction, we connect the two electrodes with a wire and join the two solutions with a salt bridge as usual (Figure 21-15). Now the relevant reduction half-reaction in either half-cell is

$$Cu^{2+} + 2e^- \longrightarrow Cu \qquad E^0 = +0.337\ \text{V}$$

The Nernst equation lets us evaluate the reduction potential for the relevant half-reaction in each half-cell:

In the Half-Cell with $[Cu^{2+}] = 1.00\ M$	In the Half-Cell with $[Cu^{2+}] = 0.10\ M$
The concentration is *standard*, so the reduction potential of this half-cell is equal to E^0:	$E = E^0 - \dfrac{0.0592}{n} \log \dfrac{1}{[Cu^{2+}]}$
$Cu^{2+}(1.00\ M) + 2e^- \longrightarrow Cu \qquad E = E^0 = +0.337\ \text{V}$	$= +0.337 - \dfrac{0.0592}{2} \log \dfrac{1}{0.10} = +0.307\ \text{V}$
	$Cu^{2+}(0.10\ M) + 2e^- \longrightarrow Cu \qquad E = +0.307\ \text{V}$

The more concentrated half-cell has the higher potential, so its half-reaction proceeds as written, i.e., as a reduction. In the more dilute half-cell, Cu is oxidized to Cu^{2+}:

$$
\begin{array}{lll}
 & & E \\
\hline
Cu^{2+}(1.00\ M) + 2e^- \longrightarrow Cu & & +0.337\ \text{V}_{(\text{reduction, cathode})} \\
Cu \longrightarrow Cu^{2+}(0.10\ M) + 2e^- & & -0.307\ \text{V}_{(\text{oxidation, anode})} \\
\hline
Cu^{2+}(1.00\ M) \longrightarrow Cu^{2+}(0.10\ M) & E_{cell} = 0.030\ \text{V}_{(\text{overall rxn})} \\
\end{array}
$$

As the reaction proceeds, $[Cu^{2+}]$ decreases in the more concentrated half-cell and increases in the more dilute half-cell until the two concentrations are equal; at that point $E_{cell} = 0$ and equilibrium has been reached. This equilibrium $[Cu^{2+}]$ is the same concentration that would have been formed had we simply mixed the two solutions directly to obtain a solution of intermediate concentration.

The overall cell potential could be calculated more directly by applying the Nernst equation to the overall cell reaction. In any concentration cell, the spontaneous reaction is always from more concentrated solution to more dilute solution. We must first find E^0, the standard cell potential *at standard*

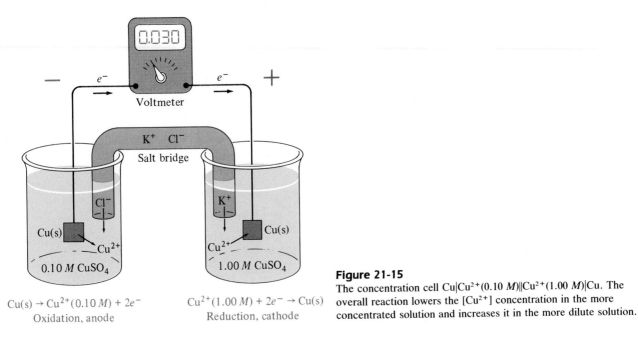

$Cu(s) \rightarrow Cu^{2+}(0.10\,M) + 2e^-$ $\qquad$ $Cu^{2+}(1.00\,M) + 2e^- \rightarrow Cu(s)$
Oxidation, anode $\qquad\qquad\qquad$ Reduction, cathode

Figure 21-15
The concentration cell $Cu|Cu^{2+}(0.10\,M)\|Cu^{2+}(1.00\,M)|Cu$. The overall reaction lowers the $[Cu^{2+}]$ concentration in the more concentrated solution and increases it in the more dilute solution.

concentrations; because the same electrode and the same type of ions are involved in both half-cells, this E^0 is always zero. Thus,

$$E_{cell} = E^0 - \frac{0.0592}{n} \log \frac{[\text{dilute solution}]}{[\text{concentrated solution}]}$$

$$= 0 - \frac{0.0592}{2} \log \frac{0.10}{1.00} = 0.30\ V$$

Exercises 76 and 77 at the end of the chapter involve calculations for concentration cells.

Using Electrochemical Cells to Determine Concentrations

Another way we can apply the ideas of this section is to *measure* the voltage of a cell and then use the Nernst equation to solve for an unknown concentration. The following example illustrates such an application.

Example 21-11

We construct an electrochemical cell at 25°C as follows. One half-cell is a standard Zn^{2+}/Zn cell, i.e., a strip of zinc immersed in a $1.00\ M\ Zn^{2+}$ solution; the other is a *nonstandard* hydrogen electrode in which a platinum electrode is immersed in a solution of *unknown* hydrogen ion concentration with gaseous hydrogen bubbling through it at a pressure of 1.000 atm. This is similar to the zinc–hydrogen cell that we discussed in Section 21-13, except that the hydrogen concentration is not (necessarily) $1.00\ M$. The observed cell voltage is 0.522 V. (a) Calculate the value of the reaction quotient Q. (b) Calculate $[H^+]$ in the second half-cell. (c) Determine the pH of the solution in the second half-cell.

Plan

We saw in Section 21-13 that the zinc–hydrogen cell operated with oxidation at the zinc electrode and reduction at the hydrogen electrode, with a *standard* cell potential of 0.763 V.

$$\text{overall:} \quad Zn + 2H^+ \longrightarrow Zn^{2+} + H_2 \qquad E^0_{cell} = 0.763 \text{ V}$$

(a) We rearrange the Nernst equation to solve for the reaction quotient, Q, from the measured cell voltage and $n = 2$. (b) We substitute concentrations and partial pressures in the expression for Q. Then we can solve for the only unknown, $[H^+]$. (c) The pH can be determined from the $[H^+]$ determined in part (a).

Solution

(a)

$$E_{cell} = E^0_{cell} - \frac{0.0592}{n} \log Q$$

Substituting and solving for Q,

$$0.522 \text{ V} = 0.763 \text{ V} - \frac{0.0592}{2} \log Q$$

$$\frac{0.0592 \text{ V}}{2} \log Q = (0.763 - 0.522) \text{ V} = 0.241 \text{ V}$$

$$\log Q = \frac{(2)(0.241 \text{ V})}{0.0592 \text{ V}} = 8.14$$

$$Q = 10^{8.14} = \boxed{1.4 \times 10^8}$$

(b) We write the expression for Q from the balanced overall equation, and solve for $[H^+]$.

$$Q = \frac{[Zn^{2+}]P_{H_2}}{[H^+]^2}$$

$$[H^+]^2 = \frac{[Zn^{2+}]P_{H_2}}{Q} = \frac{(1.00)(1.00)}{1.4 \times 10^8} = 7.1 \times 10^{-9}$$

$$[H^+] = \boxed{8.4 \times 10^{-5} \text{ M}}$$

(c) $$\text{pH} = -\log [H^+] = -\log (8.4 \times 10^{-5}) = \boxed{4.08}$$

EOC 73

Electrochemical procedures that use the principles illustrated here provide a convenient method for making many concentration measurements.

21-21 The Relationship of E^0_{cell} to ΔG^0 and K

In Section 17-11 we studied the relationship between the standard Gibbs free energy change, ΔG^0, and the thermodynamic equilibrium constant, K.

$$\Delta G^0 = -2.303RT \log K \qquad \text{or} \qquad \Delta G^0 = -RT \ln K$$

There is also a simple relationship between ΔG^0 and the standard cell potential, E^0_{cell}, for a redox reaction (reactants and products in standard states).

$$\Delta G^0 = -nFE^0_{cell}$$

ΔG^0 can be thought of as the *negative of the maximum electrical work* that can be obtained from a redox reaction. In this equation, n is the number of moles of electrons involved in the overall process, and F is the faraday, 96,487 J/V · mol e^-.

Combining these relationships for ΔG^0 gives the relationship between E^0_{cell} values and equilibrum constants:

$$\underbrace{-nFE^0_{cell}}_{\Delta G^0} = \underbrace{-2.303RT \log K}_{\Delta G^0} \qquad \text{or} \qquad \underbrace{-nFE^0_{cell}}_{\Delta G^0} = \underbrace{-RT \ln K}_{\Delta G^0}$$

After multiplying by -1, we can rearrange the equations:

	Using $\log_{10}$	Using $\ln$
	$nFE^0_{cell} = 2.303RT \log K$	$nFE^0_{cell} = RT \ln K$
Rearranging for E^0_{cell}	$E^0_{cell} = \dfrac{2.303RT \log K}{nF}$	$E^0_{cell} = \dfrac{RT \ln K}{nF}$
Rearranging for K	$\log K = \dfrac{nFE^0_{cell}}{2.303RT}$	$\ln K = \dfrac{nFE^0_{cell}}{RT}$

If any one of the three quantities ΔG^0, K, and E^0_{cell} is known, the other two can be calculated using these relationships. It is usually much easier to determine K for a redox reaction from electrochemical measurements than by measuring equilibrium concentrations directly, as described in Chapter 17. Keep in mind the following for all redox reactions *under standard state conditions*.

Forward Reaction	ΔG^0	K	E^0_{cell}	
spontaneous	$-$	>1	$+$	
at equilibrium	0	1	0	(all substances at *standard state conditions*)
nonspontaneous	$+$	<1	$-$	

Example 21-12

Calculate the standard Gibbs free energy change, ΔG^0, in J/mol at 25°C for the following reaction from standard electrode potentials:

$$3Sn^{4+} + 2Cr \longrightarrow 3Sn^{2+} + 2Cr^{3+}$$

Plan

We evaluate the standard cell potential as before. Then we apply the relationship $\Delta G^0 = -nFE^0_{cell}$.

Solution

The standard reduction potential for the Sn^{4+}/Sn^{2+} couple is $+0.15$ volt; that for the Cr^{3+}/Cr couple is -0.74 volt. The equation for the reaction shows Cr being oxidized to Cr^{3+}, so the sign of the E^0 value for the Cr^{3+}/Cr couple is reversed. The overall reaction, the sum of the two half-reactions, has a cell potential equal to the sum of the two half-reaction potentials:

		E^0
$3(Sn^{4+} + 2e^- \longrightarrow Sn^{2+})$		$+0.15$ V
$2(Cr \longrightarrow Cr^{3+} + 3e^-)$		$-(-0.74$ V$)$
$3Sn^{4+} + 2Cr \longrightarrow 3Sn^{2+} + 2Cr^{3+}$	$E^0_{cell} =$	$+0.89$ V

The positive value of E^0_{cell} indicates that the forward reaction is spontaneous.

Recall from Chapter 15 that ΔG^0 can be expressed in joules per *mole of reaction*, that is, per mole of occurrences for the *entire* reaction *as written*. Here we are asking for the number of joules of free energy change that corresponds to the reaction of 2 moles of chromium with 3 moles of tin(IV) to give 3 moles of tin(II) ions and 2 moles of chromium(III) ions.

$$\Delta G^0 = -nFE^0_{cell} = -(6 \text{ mol } e^-)\left(96,500 \frac{J}{V \cdot \text{mol } e^-}\right)(+0.89 \text{ V})$$

$$\Delta G^0 = \boxed{-5.2 \times 10^5 \text{ J/mol}} \quad \text{or} \quad \boxed{-5.2 \times 10^2 \text{ kJ/mol}}$$

EOC 83, 90

Example 21-13

Calculate the value of the equilibrium constant, K, for the reaction in Example 21-12 at 25°C (a) by relating it to ΔG^0 and (b) by relating it to E^0_{cell}.

Plan

We substitute the values of ΔG^0 and E^0_{cell} from Example 21-12 into the following relationships and solve for K.

$$\Delta G^0 = -2.303RT \log K \quad \text{and} \quad \log K = \frac{nFE^0_{cell}}{2.303RT}$$

Solution

(a)
$$\Delta G^0 = -2.303RT \log K$$

$$-5.2 \times 10^5 \text{ J/mol} = (-2.303)\left(8.314 \frac{J}{\text{mol} \cdot K}\right)(298 \text{ K}) \log K$$

$$\log K = \frac{-5.2 \times 10^5}{(-2.303)(8.314)(298)} = +91$$

$$K = \boxed{1 \times 10^{91}}$$

The very large value of *K* is also consistent with the fact that the forward reaction is spontaneous. This tells us nothing about the speed with which the reaction would occur.

(b)
$$\log K = \frac{nFE^0_{cell}}{2.303RT}$$

Substituting the known values gives a numerical expression for K:

$$\log K = \frac{(6)\left(96,500 \frac{1}{V \cdot \text{mol}}\right)(+0.89 \text{ V})}{(2.303)\left(8.314 \frac{J}{\text{mol} \cdot K}\right)(298 \text{ K})} = +90$$

$$K = \boxed{1 \times 10^{90}}$$

This result agrees with that in (a) within roundoff error. Both 1×10^{91} and 1×10^{90} are so very large that the difference between the two is insignificant. The reaction goes essentially to completion. At equilibrium, very little Sn^{4+} is present.

$$K = \boxed{\frac{[Cr^{3+}]^2[Sn^{2+}]^3}{[Sn^{4+}]^3} = 1 \times 10^{91}}$$

Example 21-14

Calculate the value of the equilibrium constant, K, at 25°C for the reaction:

$$2Cu + PtCl_6^{2-} \longrightarrow 2Cu^+ + PtCl_4^{2-} + 2Cl^-$$

Plan

This example combines the ideas we used in Examples 21-12 and 21-13.

Solution

First we find the appropriate half-reactions. Cu is oxidized to Cu^+, so we write the Cu^+/Cu couple as an oxidation and reverse the sign of its tabulated E^0 value. We balance the electron transfer and then add the half-reactions. The resulting E^0_{cell} value can be used to calculate the equilibrium constant, K, for the reaction *as written*.

As the problem is stated, we must keep the equation as written. Therefore, we must accept either a positive or a negative value of E^0_{cell}. A negative value of E^0_{cell} would lead to $K < 1$.

$$
\begin{array}{lll}
2(Cu & \longrightarrow \quad Cu^+ + e^-) & -(+0.521 \text{ V}) \\
PtCl_6^{2-} + 2e^- & \longrightarrow \quad PtCl_4^{2-} + 2Cl^- & +0.68 \text{ V} \\
\hline
2Cu + PtCl_6^{2-} & \longrightarrow \quad 2Cu^+ + PtCl_4^{2-} + 2Cl^- & E^0_{cell} = +0.16 \text{ V}
\end{array}
$$

We then calculate K. Let's use the ln relationship this time:

$$
\ln K = \frac{nFE^0_{cell}}{RT} = \frac{(2)(96.5 \times 10^3 \text{ J/V} \cdot \text{mol})(+0.16 \text{ V})}{(8.314 \text{ J/mol} \cdot \text{K})(298 \text{ K})} = 12.4
$$

$$
K = e^{12.4} = \boxed{2.4 \times 10^5}
$$

At equilibrium, $\quad K = \dfrac{[Cu^+]^2[PtCl_4^{2-}][Cl^-]^2}{[PtCl_6^{2-}]} = 2.4 \times 10^5$.

The forward reaction is spontaneous, and the equilibrium lies far to the right.

EOC 84, 86

Under nonstandard conditions, the relationship between the Gibbs free energy change, ΔG, and the cell potential, E_{cell}, is similar to that at standard conditions.

$$
\Delta G = -nFE_{cell} \qquad \text{(at nonstandard conditions)}
$$

From this relationship, we can see that when E_{cell} reaches zero, ΔG is zero. From Section 17-11, we know that this is the condition of *equilibrium*. At that time, the concentrations are *not* zero; they must have values that satisfy the equilibrium constant, K.

Remember the distinction between ΔG^0 and ΔG.

$$
\Delta G^0 = -RT \ln K = -nFE^0_{cell}
$$

but

$$
\Delta G = -RT \ln Q = -nFE_{cell}
$$

Example 21-15

Calculate ΔG at 25°C for the reaction in Example 21-14 at the concentrations indicated below.

$$[PtCl_6^{2-}] = 1.00 \times 10^{-2} \, M \qquad [Cu^+] = 1.00 \times 10^{-3} \, M$$

$$[Cl^-] = 1.00 \times 10^{-3} \, M \qquad [PtCl_4^{2-}] = 2.00 \times 10^{-5} \, M$$

Plan

The Gibbs free energy change, ΔG, is related to E_{cell} (not E^0_{cell}) by the relationship

$$
\Delta G = -nFE_{cell}
$$

We evaluate E_{cell} by the method shown in Example 21-10.

Solution

From the solution to Example 21-14, E^0_{cell} for this reaction—i.e., at standard conditions—is +0.16 V. For the overall cell reaction, $n = 2$ moles of electrons transferred. We use the Nernst equation to evaluate E_{cell}.

Here we use the Nernst equation to find E_{cell} for the overall reaction, as in Example 21-10. Alternatively, we could find the potential for each half-reaction and then add these to obtain E_{cell}, as in Example 21-9.

$$E_{cell} = E_{cell}^0 - \frac{0.0592}{n} \log \frac{[Cu^+]^2[PtCl_4^{2-}][Cl^-]^2}{[PtCl_6^{2-}]}$$

$$= +0.16 \text{ V} - \frac{0.0592}{2} \log \frac{(1.00 \times 10^{-3})^2(2.00 + 10^{-5})(1.00 \times 10^{-3})^2}{(1.00 \times 10^{-2})}$$

$$= +0.16 \text{ V} - \frac{0.0592}{2} \log (2.00 \times 10^{-15})$$

$$= +0.16 \text{ V} - \frac{0.0592}{2}(-14.699) = \boxed{0.60 \text{ V}}$$

We saw (Example 21-14) that the reaction

$$2Cu + PtCl_6^{2-} \longrightarrow 2Cu^+ + PtCl_4^{2-} + 2Cl^-$$

is spontaneous when all ions are present in 1.0 M concentrations ($E_{cell}^0 = +0.16$ V). We see that E_{cell} is even more positive (+0.60 V) under the stated conditions. The Gibbs free energy change can now be calculated.

$$\Delta G = -nFE_{cell} = -(2)\left(\frac{96.5 \times 10^3 \text{ J}}{\text{V} \cdot \text{mol}}\right)(0.60 \text{ V}) = \boxed{-1.2 \times 10^5 \text{ J/mol}}$$

EOC 89

Primary Voltaic Cells

As any voltaic cell produces current (**discharges**), chemicals are consumed. **Primary voltaic cells** cannot be "recharged." Once the chemicals have been consumed, further chemical action is not possible. The electrolytes and/or electrodes cannot be regenerated by reversing the current flow through the cell using an external direct current source. The most familiar examples of primary voltaic cells are the ordinary "dry" cells that are used as energy sources in flashlights and other small appliances.

21-22 The Dry Cell (Leclanché Cell)

The dry cell was patented by Georges Leclanché in 1866 (Figure 21-16). The container of a dry cell, made of zinc, also serves as one of the electrodes. The other electrode is a carbon rod in the center of the cell. The zinc container is lined with porous paper to separate it from the other materials of the cell. The rest of the cell is filled with a moist mixture (the cell is *not* really dry) of ammonium chloride (NH_4Cl), manganese(IV) oxide (MnO_2), zinc chloride ($ZnCl_2$), and a porous, inert filler. Dry cells are sealed to keep the moisture from evaporating. As the cell operates (the electrodes must be connected externally), the metallic Zn is oxidized to Zn^{2+}, and the liberated electrons flow along the container to the external circuit. Thus, the zinc electrode is the anode (negative electrode):

$$Zn \longrightarrow Zn^{2+} + 2e^- \qquad \text{(oxidation, anode)}$$

The carbon rod is the cathode, at which ammonium ions are reduced:

$$2NH_4^+ + 2e^- \longrightarrow 2NH_3 + H_2 \qquad \text{(reduction, cathode)}$$

Figure 21-16

(a) Some commercial dry cells. The black cell is an alkaline cell and the others are Leclanché cells. (b) The Leclanché cell is a dry cell that generates a potential difference of about 1.6 volts.

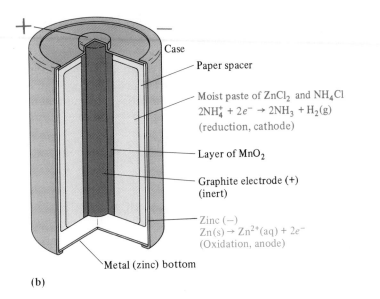

Case

Paper spacer

Moist paste of $ZnCl_2$ and NH_4Cl
$2NH_4^+ + 2e^- \rightarrow 2NH_3 + H_2(g)$
(reduction, cathode)

Layer of MnO_2

Graphite electrode (+)
(inert)

Zinc (−)
$Zn(s) \rightarrow Zn^{2+}(aq) + 2e^-$
(Oxidation, anode)

Metal (zinc) bottom

(b)

Addition of the half-reactions gives the overall cell reaction:

$$Zn + 2NH_4^+ \longrightarrow Zn^{2+} + 2NH_3 + H_2 \qquad E_{cell} = 1.6 \text{ V}$$

As H_2 is formed, it is oxidized by MnO_2 in the cell. This prevents collection of H_2 gas on the cathode, which would stop the reaction.

$$H_2 + 2MnO_2 \longrightarrow 2MnO(OH)$$

The buildup of reaction products at an electrode is called *polarization* of the electrode.

The ammonia produced at the cathode combines with zinc ions and forms a soluble compound containing the complex ions, $[Zn(NH_3)_4]^{2+}$.

$$Zn^{2+} + 4NH_3 \longrightarrow [Zn(NH_3)_4]^{2+}$$

This reaction prevents polarization due to the buildup of ammonia, and it prevents the concentration of Zn^{2+} from increasing substantially, which would decrease the cell potential.

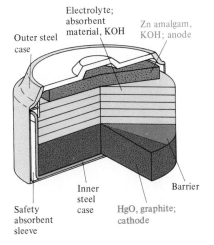

Outer steel case

Electrolyte; absorbent material, KOH

Zn amalgam, KOH; anode

Safety absorbent sleeve

Inner steel case

HgO, graphite; cathode

Barrier

The mercury battery of the type frequently used in watches, calculators, and hearing aids is a primary cell. Although mercury in the water supply is known to cause health problems, no conclusive evidence has been found that the disposal of household batteries contributes to such problems. Nevertheless, manufacturers are working to decrease the amount of mercury in batteries. Between 1980 and 1990, the amount of mercury in Eveready alkaline batteries decreased by a factor of 32, from 0.8% to 0.025%; nevertheless, the life of such batteries increased by 50 percent over that decade.

Alkaline dry cells are similar to ordinary dry cells except that (1) the electrolyte is basic (alkaline) because it contains KOH, and (2) the interior surface of the Zn container is rough; this gives a larger surface area. Alkaline cells have a longer shelf life than ordinary dry cells, and they stand up better under heavy use. The voltage of an alkaline cell is about 1.5 volts. During discharge, the alkaline dry cell reactions are

(anode) $\qquad\qquad\qquad$ $Zn(s) + 2OH^-(aq) \longrightarrow Zn(OH)_2(s) + 2e^-$
(cathode) $2MnO_2(s) + 2H_2O(\ell) + 2e^- \longrightarrow 2MnO(OH)(s) + 2OH^-(aq)$
(overall) $Zn(s) + 2MnO_2(s) + 2H_2O(\ell) \longrightarrow Zn(OH)_2(s) + 2MnO(OH)(s)$

Secondary Voltaic Cells

In **secondary voltaic cells**, or *reversible cells*, the original reactants can be regenerated. This is done by passing a direct current through the cell in the direction opposite to the discharge current flow. This process is referred to as *charging*, or recharging, a cell or battery. The most common example of a secondary voltaic cell is the lead storage battery, used in most automobiles.

21-23 The Lead Storage Battery

The lead storage battery is depicted in Figure 21-17. One group of lead plates bears compressed spongy lead. These alternate with a group of lead plates bearing lead(IV) oxide, PbO_2. The electrodes are immersed in a solution of about 40% sulfuric acid. When the cell discharges, the spongy lead is oxidized to lead ions, and the lead plates accumulate a negative charge.

$$Pb \longrightarrow Pb^{2+} + 2e^- \qquad \text{(oxidation)}$$

The lead ions then combine with sulfate ions from the sulfuric acid to form insoluble lead(II) sulfate. This begins to coat the lead electrode.

$$Pb^{2+} + SO_4^{2-} \longrightarrow PbSO_4(s) \qquad \text{(precipitation)}$$

Thus, the net process at the anode *during discharge* is

$$Pb + SO_4^{2-} \longrightarrow PbSO_4(s) + 2e^- \qquad \text{(anode during discharge)}$$

Figure 21-17
(a) A schematic representation of one cell of a lead storage battery. The reactions shown are those taking place during the *discharge* of the cell. Alternate lead grids are packed with spongy lead and lead(IV) oxide. The grids are immersed in a solution of sulfuric acid, which serves as the electrolyte. To provide a large reacting surface, each cell contains several connected grids of each type as in (b); for clarity, only one of each is shown in (a). Such a cell generates a voltage of about 2 volts. Six of these cells are connected together in series, so that their voltages add, to make a 12-volt battery.

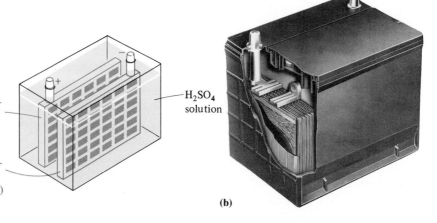

During *discharge*:

PbO₂ plate (cathode)
$PbO_2(s) + 4H^+(aq) + 2e^- \rightarrow Pb^{2+}(aq) + 2H_2O$
$Pb^{2+}(aq) + SO_4^{2-}(aq) \rightarrow PbSO_4(s)$

Pb plate (anode)
$Pb(s) \rightarrow Pb^{2+}(aq) + 2e^-$
$Pb^{2+}(aq) + SO_4^{2-}(aq) \rightarrow PbSO_4(s)$

(a)

H_2SO_4 solution

(b)

The electrons travel through the external circuit and re-enter the cell at the PbO_2 electrode, which is the cathode during discharge. Here, in the presence of hydrogen ions, the lead(IV) oxide is reduced to lead(II) ions, Pb^{2+}. These ions also combine with SO_4^{2-} ions from the H_2SO_4 to form an insoluble $PbSO_4$ coating on the lead(IV) oxide electrode.

$$
\begin{aligned}
PbO_2 + 4H^+ + 2e^- &\longrightarrow Pb^{2+} + 2H_2O \qquad \text{(reduction)} \\
Pb^{2+} + SO_4^{2-} &\longrightarrow PbSO_4(s) \qquad \text{(precipitation)} \\
\hline
PbO_2 + 4H^+ + SO_4^{2-} + 2e^- &\longrightarrow PbSO_4(s) + 2H_2O \quad \text{(cathode during discharge)}
\end{aligned}
$$

The net cell reaction and its standard potential during discharge are obtained by adding the net anode and cathode half-reactions and their tabulated potentials. The tabulated E^0 value for the anode half-reaction is reversed in sign because it occurs as oxidation during discharge.

$$
\begin{array}{lr}
 & E^0 \\
\hline
Pb + SO_4^{2-} \longrightarrow PbSO_4(s) + 2e^- & -(-0.356 \text{ V}) \\
PbO_2 + 4H^+ + SO_4^{2-} + 2e^- \longrightarrow PbSO_4(s) + 2H_2O & +1.685 \text{ V} \\
\hline
Pb + PbO_2 + \underbrace{4H^+ + 2SO_4^{2-}}_{2H_2SO_4} \longrightarrow 2PbSO_4(s) + 2H_2O & E^0_{cell} = +2.041 \text{ V}
\end{array}
$$

One cell creates a potential of about 2 volts. Automobile 12-volt batteries have six cells connected in series. The potential declines only slightly during use, because solid reagents are being consumed. As the cell is used, some H_2SO_4 is consumed, lowering its concentration.

When a potential slightly greater than the potential the battery can generate is imposed across the electrodes, the current flow can be reversed. The battery can then be recharged by reversal of all reactions. The alternator or generator applies this potential when the engine is in operation. The reactions that occur in a lead storage battery are summarized as follows:

$$
Pb + PbO_2 + 2[2H^+ + SO_4^{2-}] \underset{\text{charge}}{\overset{\text{discharge}}{\rightleftharpoons}} 2PbSO_4(s) + 2H_2O
$$

During many repeated charge–discharge cycles, some of the $PbSO_4$ falls to the bottom of the container and the H_2SO_4 concentration remains correspondingly low. Eventually the battery cannot be recharged fully. It can be traded in for a new one, and the lead can be recovered and reused to make new batteries.

The decrease in the concentration of sulfuric acid provides an easy method for measuring the degree of discharge, because the density of the solution decreases accordingly. We simply measure the density of the solution with a hydrometer.

A *generator* supplies direct current (dc). An *alternator* supplies alternating current (ac), so a rectifier (an electronic device) is used to convert this to direct current for the battery.

This is one of the oldest and most successful examples of recycling.

21-24 The Nickel–Cadmium (Nicad) Cell

The nickel–cadmium (nicad) cell has gained widespread popularity because it can be recharged. It thus has a much longer useful life than ordinary (Leclanché) dry cells. Nicad batteries are used in electronic wristwatches, calculators, and photographic equipment.

The anode is cadmium, and the cathode is nickel(IV) oxide. The electrolytic solution is basic. The "discharge" reactions that occur in a nicad battery are

$$
\begin{array}{ll}
\text{(anode)} & Cd(s) + 2OH^-(aq) \longrightarrow Cd(OH)_2(s) + 2e^- \\
\text{(cathode)} & NiO_2(s) + 2H_2O(\ell) + 2e^- \longrightarrow Ni(OH)_2(s) + 2OH^-(aq) \\
\hline
\text{(overall)} & Cd(s) + NiO_2(s) + 2H_2O(\ell) \longrightarrow Cd(OH)_2(s) + Ni(OH)_2(s)
\end{array}
$$

To see why a nicad battery produces a constant voltage, write the Nernst equation for its reaction. Look at Q.

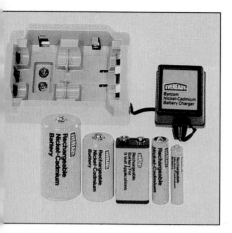

Rechargeable nicad batteries are used to operate many electrical devices.

The solid reaction product at each electrode adheres to the electrode surface. Hence, a nicad battery can be recharged by an external source of electricity; that is, the electrode reactions can be reversed. Because no gases are produced by the reactions in a nicad battery, the unit can be sealed. The voltage of a nicad cell is about 1.4 volts, slightly less than that of a Leclanché cell.

21-25 The Hydrogen–Oxygen Fuel Cell

Fuel cells are voltaic cells in which the reactants are continuously supplied to the cell. The hydrogen–oxygen fuel cell (Figure 21-18) already has many applications. It is used in spacecraft to supplement the energy obtained from solar cells. Liquid H_2 is carried on board as a propellant. The boiled-off H_2 vapor that ordinarily would be lost is used in a fuel cell to generate electrical power.

Hydrogen (the fuel) is supplied to the anode compartment. Oxygen is fed into the cathode compartment. The diffusion rates of the gases into the cell are carefully regulated for maximum efficiency. Oxygen is reduced at the cathode, which consists of porous carbon impregnated with finely divided Pt or Pd catalyst.

$$O_2 + 2H_2O + 4e^- \xrightarrow{\text{catalyst}} 4OH^- \qquad \text{(cathode)}$$

The OH^- ions migrate through the electrolyte, an aqueous solution of a base, to the anode. The anode is also porous carbon containing a small amount of catalyst (Pt, Ag, or CoO). Here H_2 is oxidized to H_2O:

$$H_2 + 2OH^- \longrightarrow 2H_2O + 2e^- \qquad \text{(anode)}$$

The net reaction is obtained from the two half-reactions:

$$
\begin{array}{llll}
O_2 + 2H_2O + 4e^- & \longrightarrow & 4OH^- & \text{(cathode)} \\
2(H_2 + 2OH^- & \longrightarrow & 2H_2O + 2e^-) & \text{(anode)} \\
\hline
2H_2 + O_2 & \longrightarrow & 2H_2O & \text{(net cell reaction)}
\end{array}
$$

The net reaction is the same as the burning of H_2 in O_2 to form H_2O, but combustion does not actually occur. Most of the chemical energy from the formation of H—O bonds is converted directly into electrical energy, rather than into heat energy as in combustion.

The efficiency of energy conversion of the fuel cell operation is 60 to 70% of the theoretical maximum (based on ΔG). This represents about twice the efficiency that can be realized from burning hydrogen in a heat engine coupled to a generator.

When the H_2/O_2 fuel cell is used aboard manned spacecraft, it is operated at a high enough temperature that the water evaporates at the same rate as it is produced. The vapor is then condensed, and the pure water is used for drinking.

Current research is aimed at modifying the design of fuel cells to lower their cost. Better catalysts would speed the reactions to allow more rapid generation of electricity and produce more power per unit volume. The H_2/O_2 cell is nonpolluting; the only substance released is H_2O. Catalysts have been developed that allow sunlight to decompose water into hydrogen and oxygen. These might be used to operate fuel cells, permitting the utilization of solar energy.

Fuel cells have also been constructed using fuels other than hydrogen, such as methane or methanol. Biomedical researchers envision the possibility of using tiny fuel cells to operate heart pacemakers. The disadvantage of other power supplies for pacemakers, which are primary voltaic cells, is

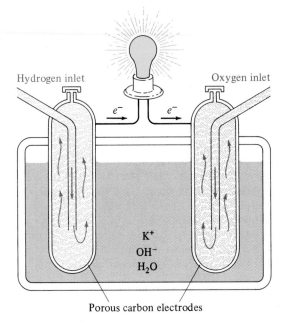

Figure 21-18
Schematic drawing of a hydrogen–oxygen fuel cell.

A hydrogen–oxygen fuel cell that is used in spacecraft.

that their reactants are eventually consumed so that they require periodic surgical replacement. As long as the fuel and oxidizer are supplied, a fuel cell can—in theory, at least—operate forever. It is hoped that, eventually, tiny pacemaker fuel cells can be operated via the oxidation of blood sugar (the fuel) by the body's oxygen at a metal electrode implanted just below the skin.

Key Terms

Activity series   See *Electromotive series*.

Alkaline cell   A dry cell in which the electrolyte contains KOH.

Ampere   Unit of electrical current; 1 ampere equals 1 coulomb per second.

Anode   An electrode at which oxidation occurs.

Cathode   An electrode at which reduction occurs.

Cathodic protection   Protection of a metal (making it a cathode) against corrosion by attaching it to a sacrificial anode of a more easily oxidized metal.

Cell potential   Potential difference, E_{cell}, between oxidation and reduction half-cells under *nonstandard* conditions.

Concentration cell   A voltaic cell in which the two half-cells are composed of the same species but contain different ion concentrations.

Corrosion   Oxidation of metals in the presence of air and moisture.

Coulomb   Unit of electrical charge.

Downs cell   An electrolytic cell for the commercial electrolysis of molten sodium chloride.

Dry cells   Ordinary batteries (voltaic cells) for flashlights, radios, and so on; many are Leclanché cells.

Electrochemistry   The study of the chemical changes produced by electrical current and the production of electricity by chemical reactions.

Electrode potentials   Potentials, E, of half-reactions as reductions versus the standard hydrogen electrode.

Electrodes   Surfaces upon which oxidation and reduction half-reactions occur in electrochemical cells.

Electrolysis   The process that occurs in electrolytic cells.

Electrolytic cell   An electrochemical cell in which electrical energy causes nonspontaneous redox reactions to occur.

Electrolytic conduction   Conduction of electric current by ions through a solution or pure liquid.

Electromotive series   The relative order of tendencies for elements and their simple ions to act as oxidizing or reducing agents; also called the activity series.

Electroplating   Plating a metal onto a (cathodic) surface by electrolysis.

Faraday   An amount of charge equal to 96,487 coulombs; corresponds to the charge on one mole of electrons, 6.022×10^{23} electrons.

Faraday's Law of Electrolysis   One equivalent weight of a substance is produced at each electrode during the passage of one mole of electrons (96,487 coulombs of charge) through an electrolytic cell.

Fuel cell   A voltaic cell in which the reactants (usually gases) are supplied continuously.

Galvanic cell   See *Voltaic cell*.

Half-cell   The compartment in a voltaic cell in which the oxidation or reduction half-reaction occurs.

Hydrogen–oxygen fuel cell   A fuel cell in which hydrogen is the fuel (reducing agent) and oxygen is the oxidizing agent.

Hydrometer   A device used to measure the densities of liquids and solutions.

Lead storage battery   A secondary voltaic cell that is used in most automobiles.

Leclanché cell   A common type of dry cell.

Metallic conduction   Conduction of electrical current through a metal or along a metallic surface.

Nernst equation   An equation that corrects standard electrode potentials for nonstandard conditions.

Nickel–cadmium cell (nicad battery)   A dry cell in which the anode is Cd, the cathode is NiO_2, and the electrolyte is basic.

Polarization of an electrode   Buildup of a product of oxidation or reduction at an electrode, preventing further reaction.

Primary voltaic cell   A voltaic cell that cannot be recharged;

no further chemical reaction is possible once the reactants are consumed.

Sacrificial anode   A more active metal that is attached to a less active metal to protect the less active metal cathode against corrosion.

Salt bridge   A U-shaped tube containing electrolyte, which connects two half-cells of a voltaic cell.

Secondary voltaic cell   A voltaic cell that can be recharged; the original reactants can be regenerated by reversing the direction of current flow.

Standard cell potential   The potential difference, E^0_{cell}, between standard reduction and oxidation half-cells.

Standard electrochemical conditions   1 M concentration for dissolved ions, 1 atm partial pressure for gases, and pure solids and liquids.

Standard electrode potential   By convention, the potential (E^0) of a half-reaction as a reduction relative to the standard hydrogen electrode, when all species are present at unit activity.

Standard electrode   A half-cell in which the oxidized and reduced forms of a species are present at unit activity: 1 M solutions of dissolved ions, 1 atm partial pressure of gases, and pure solids and liquids.

Standard hydrogen electrode (SHE)   An electrode consisting of a platinum electrode that is immersed in a 1 M H^+ solution and that has H_2 gas bubbled over it at 1 atmosphere pressure; defined as the reference electrode, with a potential of *exactly* 0.0000 . . . volt.

Voltage   Potential difference between two electrodes; a measure of the chemical potential for a redox reaction to occur.

Voltaic cell   An electrochemical cell in which spontaneous chemical reactions produce electricity; also called a galvanic cell.

Exercises

Redox Review and General Concepts

1. (a) Define oxidation and reduction in terms of electron gain or loss. (b) What is the relationship between the number of electrons gained and lost in a redox reaction? (c) Why do all electrochemical cells involve redox reactions?

2. Define and illustrate (a) oxidizing agent and (b) reducing agent.

3. For each of the following unbalanced equations, (1) write the half-reactions for oxidation and for reduction, and (2) balance the overall equation using the half-reaction method.
   (a) $I_2 + H_2S \longrightarrow I^- + S + H^+$
   (b) $Cu + Br_2 + OH^- \longrightarrow Cu_2O + Br^- + H_2O$
   (c) $Al + H^+ + SO_4^{2-} \longrightarrow Al^{3+} + H_2O + SO_2$

4. For each of the following unbalanced equations, (1) write the half-reactions for oxidation and reduction, and (2) balance the overall equation using the half-reaction method.
   (a) $S^{2-} + Cl_2 + OH^- \longrightarrow SO_4^{2-} + Cl^- + H_2O$
   (b) $MnO_4^- + IO_3^- + H_2O \longrightarrow MnO_2 + IO_4^- + OH^-$
   (c) $H_2S + Cr_2O_7^{2-} + H^+ \longrightarrow Cr^{3+} + H_2O + S$

5. Compare and contrast ionic conduction and metallic conduction. (b) What is an electrode? (c) What is an inert electrode?

6. Support or refute each of the following statements: (a) In any electrochemical cell the positive electrode is the one toward which the electrons flow through the wire. (b) The cathode is the negative electrode in any electrochemical cell.

Electrolytic Cells—General Concepts

7. (a) Solids such as potassium bromide, KBr, and sodium nitrate, $NaNO_3$, do not conduct electrical current even though they are ionic. Why? Can these substances be electrolyzed as solids? (b) Support or refute the statement that the Gibbs free energy change, ΔG, is negative for any electrolysis reaction.

8. (a) Metallic magnesium cannot be obtained by electrolysis of aqueous magnesium chloride, $MgCl_2$. Why? (b) There are no sodium ions in the overall cell reaction for the electrolysis of aqueous sodium chloride. Why?

9. Consider the electrolysis of molten aluminum oxide, Al_2O_3, dissolved in cryolite, Na_3AlF_6, with inert electrodes. This is the Hall process for commercial production of aluminum (Section 22-6). The following experimental observations can be made when current is supplied:
   (i) Silvery metallic aluminum is produced at one electrode.
   (ii) Oxygen, O_2, bubbles off at the other electrode.
   Diagram the cell, indicating the anode, the cathode, the positive and negative electrodes, the half-reaction occurring at each electrode, the overall cell reaction, and the direction of electron flow through the wire.

10. Do the same as in Exercise 9 for the electrolysis of molten calcium chloride with inert electrodes. The observations are
    (i) Bubbles of pale green chlorine gas, Cl_2, are produced at one electrode.
    (ii) Silvery-white molten metallic calcium is produced at the other electrode.

11. Do the same as in Exercise 9 for the electrolysis of aqueous potassium sulfate, K_2SO_4. The observations are
    (i) Bubbles of gaseous hydrogen are produced at one electrode, and the solution becomes more basic around that electrode.
    (ii) Bubbles of gaseous oxygen are produced at the other electrode, and the solution becomes more acidic around that electrode.

12. Do the same as in Exercise 9 for the electrolysis of an aqueous solution of copper(II) bromide, $CuBr_2$. The observations are
    (i) One electrode becomes coated with copper metal, and the color of the solution around that electrode fades.
    (ii) Around the other electrode, the solution turns brown, as bromine is formed and dissolves in water.

13. (a) Write the equation for the half-reaction when H_2O is reduced in an electrochemical cell. (b) Write the equation for the half-reaction when H_2O is oxidized in an electrochemical cell.

Faraday's Law

14. What are (a) a coulomb, (b) electrical current, (c) an ampere, and (d) a faraday?

15. Calculate the number of electrons that have a total charge of 1 coulomb.

16. Calculate the charge, in coulombs, on a single electron.

17. What mass of zinc ions is reduced by one mole of electrons?

18. How many moles of electrons would be required to reduce 0.100 g of Eu^{3+} to Eu metal?

19. For each of the following cations, calculate (i) the number of faradays required to produce 1.00 mol of free metal and (ii) the number of coulombs required to produce 1.00 g of free metal. (a) Cd^{2+}, (b) Al^{3+}, (c) K^+.

20. For each of the following cations, calculate (i) the number of faradays required to produce 1.00 mol of free metal and (ii) the number of coulombs required to produce 1.00 g of free metal. (a) Co^{3+}, (b) Hg^{2+}, (c) Hg_2^{2+}.

21. The cells in an automobile battery were charged at a steady current of 5.0 A for exactly 5 hr. What masses of Pb and PbO_2 were formed in each cell? The overall reaction is

$$2PbSO_4(s) + 2H_2O(\ell) \longrightarrow$$
$$Pb(s) + PbO_2(s) + 2H_2SO_4(aq)$$

*22. The chemical equation for the electrolysis of a fairly concentrated brine solution is

$$2NaCl(aq) + 2H_2O(\ell) \longrightarrow$$
$$Cl_2(g) + H_2(g) + 2NaOH(aq)$$

What volume of gaseous chlorine would be generated at 745 torr and 85°C if the process were 75% efficient and if a current of 1.5 A flowed for 5.0 hr?

23. Calculate the current required to deposit 0.50 g of elemental platinum from a solution containing $[PtCl_6]^{2-}$ ions within a period of 5.0 hr.

24. How much time would be required to plate an iron platter with 5.0 g of silver, using a solution containing $[Ag(CN)_2]^-$ ions and a current of 1.5 A?

25. What mass of platinum could be plated onto a ring from the electrolysis of a platinum(II) salt with a 0.250-A current for 90.0 s?

26. What mass of silver could be plated onto a spoon from electrolysis of silver nitrate with a 2.00-A current for 25.0 min?

*27. We pass enough current through a solution to plate out *one* mole of nickel metal from a solution of $NiSO_4$. In other electrolysis cells, this same current plates out *two* moles of silver from $AgNO_3$ solution but liberates only *one-half* mole of O_2 gas. Explain these observations.

*28. A current is passed through 500 mL of a solution of CaI_2. The following electrode reactions occur:

anode: $2I^- \longrightarrow I_2 + 2e^-$
cathode: $2H_2O + 2e^- \longrightarrow H_2 + 2OH^-$

After some time, analysis of the solution shows that 53.5 mmol of I_2 has been formed. (a) How many faradays of charge have passed through the solution? (b) How many coulombs? (c) What volume of dry H_2 at STP has been formed? (d) What is the pH of the solution?

29. In a copper-refining plant, a current of 8.00×10^3 A passes through a solution containing Cu^{2+}. What mass, in kilograms, of copper is deposited in 24 hr?

30. The total charge of electricity required to plate out 15.54 g of a metal from a solution of its dipositive ions is 14,475 C. What is the metal?

31. What is the charge on an ion of tin if 7.42 g of metallic tin is plated out by the passage of 24,125 C through a solution containing the ion?

*32. Three electrolytic cells are connected in series; that is, the same current passes through all three, one after another. In the first cell, 1.00 g of Cd is oxidized to Cd^{2+}; in the second, Ag^+ is reduced to Ag; in the third, Fe^{2+} is oxidized to Fe^{3+}. (a) Find the number of faradays passed through the circuit. (b) What mass of Ag is deposited at the cathode in the second cell? (c) What mass of $Fe(NO_3)_3$ could be recovered from the solution in the third cell?

Voltaic Cells—General Concepts

33. (a) What kind of energy is converted into electrical energy in voltaic cells? (b) What kind of process takes place at the cathode of a voltaic cell? At the anode? How does this compare with electrolytic cells?

34. (a) Why must the solutions in a voltaic cell be kept separate and not allowed to mix? (b) What are the functions of a salt bridge?

35. A voltaic cell containing a standard Fe^{3+}/Fe^{2+} electrode and a standard Ga^{3+}/Ga electrode is constructed, and the circuit is closed. Without consulting the table of standard reduction potentials, diagram and completely describe the cell from the following experimental observations. (i) The mass of the gallium electrode decreases, and the gallium ion concentration increases around that electrode. (ii) The ferrous ion, Fe^{2+}, concentration increases in the other electrode solution.

36. Repeat Exercise 35 for a voltaic cell containing standard Co^{2+}/Co and Au^{3+}/Au electrodes. The observations are: (i) Metallic gold plates out on one electrode, and the gold ion concentration decreases around that electrode. (ii) The mass of the cobalt electrode decreases, and the cobalt(II) ion concentration increases around that electrode.

37. What does voltage measure? How does it vary with time in a primary voltaic cell? Why?

*38. In Section 4-6 we learned how to predict from the activity series (Table 4-11) which metals replace which others from aqueous solutions. From that table, we

predict that aluminum will displace silver. The equation for this process is

$$Al(s) + 3Ag^+(aq) \longrightarrow Al^{3+}(aq) + 3Ag(s)$$

Suppose we set up a voltaic cell based on this reaction. (a) What half-reaction would represent the reduction in this cell? (b) What half-reaction would represent the oxidation? (c) Which metal would be the anode? (d) Which metal would be the cathode? (e) Diagram this cell.

39. In a voltaic cell made with metal electrodes, is the more active metal more likely to be the anode or the cathode? Explain.

40. When metallic copper is placed into aqueous silver nitrate, a spontaneous redox reaction occurs. No electricity is produced. Why?

Standard Cell Potentials

41. (a) What are standard electrochemical conditions? (b) Why are we permitted to assign arbitrarily an electrode potential of exactly 0 V to the standard hydrogen electrode?

42. What does the sign of the standard reduction potential of a half-reaction indicate? What does the magnitude indicate?

43. (a) What is the electromotive series? What information does it contain? (b) How was the electromotive series constructed? In general, how do ionization energies and electron affinities of elements vary with their positions in the electromotive series?

44. Standard reduction potentials are 2.9 V for $F_2(g)/F^-$, 0.8 V for $Ag^+/Ag(s)$, 0.5 V for $Cu^+/Cu(s)$, 0.3 V for $Cu^{2+}/Cu(s)$, -0.4 V for $Fe^{2+}/Fe(s)$, -2.7 V for $Na^+/Na(s)$, and -2.9 V for $K^+/K(s)$. (a) Arrange the oxidizing agents in order of increasing strength. (b) Which of these oxidizing agents will oxidize Cu under standard state conditions?

45. Standard reduction potentials are 1.455 V for the $PbO_2(s)/Pb(s)$ couple, 1.360 V for $Cl_2(g)/Cl^-$, 3.06 V for $F_2(g)/HF(aq)$, and 1.77 V for $H_2O_2(aq)/H_2O(\ell)$. Under standard state conditions, (a) which is the strongest oxidizing agent, (b) which oxidizing agent(s) could oxidize lead to lead(IV) oxide, and (c) which oxidizing agent(s) could oxidize fluoride ion in an acidic solution?

46. Arrange the following less commonly encountered metals in an activity series from the most active to the least active: radium [$Ra^{2+}/Ra(s)$, $E^0 = -2.9$ V], rhodium [$Rh^{3+}/Rh(s)$, $E^0 = 0.80$ V], europium [$Eu^{2+}/Eu(s)$, $E^0 = -3.4$ V]. How do these metals compare in reducing ability with the active metal lithium [$Li^+/Li(s)$, $E^0 = -3.0$ V], with hydrogen, and with gold [$Au^{3+}/Au(s)$, $E^0 = 1.5$ V], which is a noble metal and one of the least active of the metals?

47. Arrange the following metals in an activity series from

the most active to the least active: nobelium [$No^{3+}/No(s)$, $E^0 = -2.5$ V], cobalt [$Co^{2+}/Co(s)$, $E^0 = -0.28$ V], gallium [$Ga^{3+}/Ga(s)$, $E^0 = -0.53$ V], thallium [$Tl^+/Tl(s)$, $E^0 = -0.34$ V], polonium [$Po^{2+}/Po(s)$, $E^0 = 0.65$ V].

48. Diagram the following cells. For each cell, write the balanced equation for the reaction that occurs spontaneously, and calculate the cell potential. Indicate the direction of electron flow, the anode, the cathode, and the polarity (+ or −) of each electrode. In each case, assume that the circuit is completed by a wire and a salt bridge.

 (a) A strip of magnesium is immersed in a solution that is 1.0 M in Mg^{2+}, and a strip of silver is immersed in a solution that is 1.0 M in Ag^+.

 (b) A strip of chromium is immersed in a solution that is 1.0 M in Cr^{3+}, and a strip of silver is immersed in a solution that is 1.0 M in Ag^+.

49. Repeat Exercise 48 for the following cells.

 (a) A strip of copper is immersed in a solution that is 1.0 M in Cu^{2+}, and a strip of gold is immersed in a solution that is 1.0 M in Au^{3+}.

 (b) A strip of aluminum is immersed in a solution that is 1.0 M in Al^{3+}, and a strip of copper is immersed in a solution that is 1.0 M in Cu^{2+}.

In answering Exercises 50–64, justify your answer by appropriate calculations. Assume that each reaction occurs at standard electrochemical conditions.

50. (a) Will Fe^{3+} oxidize Sn^{2+} to Sn^{4+} in acidic solution? (b) Will dichromate ions oxidize fluoride ions to free fluorine in acidic solution?

51. (a) Will permanganate ions oxidize arsenous acid, H_3AsO_3, to arsenic acid, H_3AsO_4, in acid solution? (b) Will permanganate ions oxidize hydrogen peroxide, H_2O_2, to free oxygen, O_2, in acidic solution?

52. (a) Will dichromate ions oxidize Mn^{2+} to MnO_4^- in acidic solution? (b) Will sulfate ions oxidize arsenous acid, H_3AsO_3, to arsenic acid, H_3AsO_4, in acid solution?

53. Calculate the standard cell potential, E^0_{cell}, for the cell described in Exercise 35.

54. Calculate the standard cell potential, E^0_{cell}, for the cell described in Exercise 36.

55. (a) Write the equation for the oxidation of $Zn(s)$ by $Cl_2(g)$. (b) Calculate the potential of this reaction under standard state conditions. (c) Is this a spontaneous reaction?

56. An electrochemical cell was needed in which hydrogen and oxygen would react to form water. (a) Using the following standard reduction potentials for the couples given, determine which combination of half-reactions gives the maximum output potential: $E^0 = -0.828$ V for $H_2O(\ell)/H_2(g)$, OH^-; $E^0 = 0.0000$ V for $H^+/H_2(g)$; $E^0 = 1.229$ V for $O_2(g)$, $H^+/H_2O(\ell)$; and $E^0 = $

0.401 V for $O_2(g)$, $H_2O(\ell)/OH^-$. (b) Write the balanced equation for the overall reaction in (a).

57. For each of the following cells, (i) write the net reaction in the direction consistent with the way the cell is written; (ii) write the half-reactions for the anode and cathode processes; (iii) find the standard cell potential, E^0_{cell}, at 25°C; and (iv) tell whether the standard cell reaction actually occurs as given or in the reverse direction. (a) $Pb|Pb^{2+}||Ag^+|Ag$, (b) $Hg|Hg^{2+}||Sn^{2+}|Sn$.

58. Repeat Exercise 57 for the following cells: (a) $Cd|Cd^{2+}||Ag^+|Ag$, (b) $Cr|Cr^{3+}||Fe^{2+}|Fe$.

59. Consult a table of standard reduction potentials and determine which of the following reactions are spontaneous under standard electrochemical conditions.

 (a) $Mn(s) + 2H^+(aq) \rightarrow H_2(s) + Mn^{2+}(aq)$

 (b) $2Al^{3+}(aq) + 3H_2(g) \rightarrow 2Al(s) + 6H^+(aq)$

 (c) $H_2(g) \rightarrow H^+(aq) + H^-(aq)$

 (d) $Cl_2(g) + 2Br^-(aq) \rightarrow Br_2(\ell) + 2Cl^-(aq)$

60. Which of the following reactions are spontaneous in voltaic cells under standard conditions?

 (a) $Si(s) + 6OH^-(aq) + 2Cu^{2+}(aq) \rightarrow$
 $$SiO_3^{2-}(aq) + 3H_2O(\ell) + 2Cu(s)$$

 (b) $Zn(s) + 4CN^-(aq) + Ag_2CrO_4(s) \rightarrow$
 $$Zn(CN)_4^{2-}(aq) + 2Ag(s) + CrO_4^{2-}(aq)$$

 (c) $MnO_2(s) + 4H^+(aq) + Sr(s) \rightarrow$
 $$Mn^{2+}(aq) + 2H_2O(\ell) + Sr^{2+}(aq)$$

 (d) $Cl_2(g) + 2H_2O(\ell) + ZnS(s) \rightarrow$
 $$2HClO(aq) + H_2S(aq) + Zn(s)$$

61. Which of each pair is the stronger oxidizing agent? (a) Cd^{2+} or Al^{3+}, (b) Sn^{2+} or Sn^{4+}, (c) H^+ or Cr^{2+}, (d) Cl_2 or Br_2, (e) MnO_4^- in acidic solution or MnO_4^- in basic solution, (f) F_2 or Pb^{2+}.

62. Which of each pair is the stronger reducing agent? (a) Ag or H_2, (b) Pt or Co^{2+}, (c) Hg or Au, (d) Cl^- in acidic solution or Cl^- in basic solution, (e) Ce^{3+} or Cu, (f) Rb or H_2.

*63. The standard reduction potential for Cu^+ to $Cu(s)$ is 0.521 V, and for Cu^{2+} to $Cu(s)$ it is 0.337 V. Calculate E^0 for the Cu^{2+}/Cu^+ couple.

*64. The element ytterbium forms both 2+ and 3+ cations in aqueous solution. $E^0 = -2.797$ V for $Yb^{2+}/Yb(s)$, and −2.267 V for $Yb^{3+}/Yb(s)$. What is the standard state reduction potential for the Yb^{3+}/Yb^{2+} couple?

*65. Describe the process of corrosion. How can corrosion of an easily oxidizable metal be prevented if the metal must be exposed to the weather?

Concentration Effects; Nernst Equation

*66. How is the Nernst equation of value in electrochemistry? How would the Nernst equation be modified if we wished to use natural logarithms, ln? What is the value of the constant in the following equation at 25°C?

$$E = E^0 - \frac{\text{constant}}{n} \ln Q$$

67. Identify all of the terms in the Nernst equation. What part of the Nernst equation represents the correction factor for nonstandard electrochemical conditions?

68. By putting the appropriate values into the Nernst equation, show that it predicts that the voltage of a standard half-cell is equal to E^0. Use the Zn^{2+}/Zn reduction half-cell as an illustration.

69. Calculate the potential associated with the following half-reaction when the concentration of the cobalt(II) ion is 1.0×10^{-4} M.

$$Co(s) \longrightarrow Co^{2+} + 2e^-$$

70. Calculate the reduction potential for hydrogen ion in a system having a perchloric acid concentration of 1.00×10^{-4} M and a hydrogen pressure of 2.00 atm. (Recall that $HClO_4$ is a strong acid in aqueous solution.)

71. The standard reduction potentials for the $H^+/H_2(g)$ and $O_2(g)$, $H^+/H_2O(\ell)$ couples are 0.0000 V and 1.229 V, respectively.
    (a) Write the half-reactions and the overall reaction, and calculate E^0 for the reaction

    $$2H_2(g) + O_2(g) \longrightarrow 2H_2O(\ell)$$

    (b) Calculate E for the cell when the pressure of H_2 is 5.00 atm and that of O_2 is 0.90 atm.

72. Consider the cell represented by the notation

    $$Zn(s)|ZnCl_2(aq)||Cl_2(g, 1\ atm);\ Cl^-(aq)|C$$

    Calculate (a) E^0 and (b) E for the cell when the concentration of the $ZnCl_2$ is 0.10 mol/L.

73. What is the concentration of Ag^+ in a half-cell if the reduction potential of the Ag^+/Ag couple is observed to be 0.35 V?

74. What must be the pressure of fluorine gas to produce a reduction potential of 2.65 V in a solution that contains 0.10 M F^-?

75. What is the cell potential for each of the following reactions under the specified conditions, at 25°C?
    (a) $Zn(s) + 2H^+(1.0 \times 10^{-3}\ M) \longrightarrow$
    $$Zn^{2+}(3.0\ M) + H_2(g)\ (5.0\ atm)$$
    (b) $Cu(s) + 2Ag^+(1.0 \times 10^{-2}\ M) \longrightarrow$
    $$Cu^{2+}(5.0 \times 10^{-2}\ M) + 2Ag(s)$$

*76. We construct a cell in which identical copper electrodes are placed in two solutions. Solution A contains 0.80 M Cu^{2+}. Solution B contains Cu^{2+} at some concentration known to be lower than that in solution A. The potential of the cell is observed to be 0.045 V. What is $[Cu^{2+}]$ in solution B?

*77. Find the potential of the cell in which identical iron electrodes are placed into solutions of $FeSO_4$ of concentration 1.0 mol/L and 0.10 mol/L.

*78. We construct a standard copper–zinc cell, close the circuit, and allow the cell to operate. At some later time, the cell voltage reaches zero, and the cell is "run down." (a) What will be the ratio of $[Zn^{2+}]$ to $[Cu^{2+}]$ at that time? (b) What will be the concentrations?

*79. Repeat Exercise 78 for a standard copper–nickel cell.

Relationships Among ΔG^0, E^0_{cell}, and K

80. How are the signs and magnitudes of E^0_{cell}, ΔG^0, and K related for a particular reaction? Why is the equilibrium constant K related only to E^0_{cell} and not to E_{cell}?

81. In light of your answer to Exercise 80, how do you explain the fact that ΔG^0 for a redox reaction *does* depend on the number of electrons transferred, according to $\Delta G^0 = -nFE^0_{cell}$?

82. Calculate E^0_{cell} from the tabulated standard reduction potentials for each of the following reactions in aqueous solution. Then calculate ΔG^0 and K at 25°C from E^0_{cell}. Which reactions are spontaneous as written?
    (a) $Sn^{4+} + 2Fe^{2+} \longrightarrow Sn^{2+} + 2Fe^{3+}$
    (b) $2Cu^+ \longrightarrow Cu^{2+} + Cu(s)$
    (c) $3Zn(s) + 2MnO_4^- + 4H_2O \longrightarrow$
    $$2MnO_2(s) + 3Zn(OH)_2(s) + 2OH^-$$

83. Calculate ΔG^0 (overall) and ΔG^0 per mole of metal for each of the following reactions from E^0 values.
    (a) Zinc dissolves in dilute hydrochloric acid to produce a solution that contains Zn^{2+}, and hydrogen gas is evolved.
    (b) Chromium dissolves in dilute hydrochloric acid to produce a solution that contains Cr^{3+}, and hydrogen gas is evolved.
    (c) Silver dissolves in dilute nitric acid to form a solution that contains Ag^+, and NO is liberated as a gas.
    (d) Lead dissolves in dilute nitric acid to form a solution that contains Pb^{2+}, and NO is liberated as a gas.

84. Use tabulated reduction potentials to calculate the value of the equilibrium constant for the reaction

    $$2K(s) + 2H_2O(\ell) \rightleftharpoons 2K^+ + 2OH^- + H_2(g)$$

85. Use tabulated reduction potentials to calculate the equilibrium constant for the reaction

    $$2Br^- + Cl_2(g) \rightleftharpoons Br_2(aq) + 2Cl^-$$

86. Using the following half-reactions and E^0 data at 25°C:

    $$PbSO_4(s) + 2e^- \longrightarrow Pb(s) + SO_4^{2-} \quad E^0 = -0.356\ V$$
    $$PbI_2(s) + 2e^- \longrightarrow Pb(s) + 2I^- \quad E^0 = -0.365\ V$$

    calculate the equilibrium constant for the reaction

    $$PbSO_4(s) + 2I^- \rightleftharpoons PbI_2(s) + SO_4^{2-}$$

87. (a) Given the following E^0 values at 25°C, calculate K_{sp} for cadmium sulfide, CdS.

    $$Cd^{2+}(aq) + 2e^- \longrightarrow Cd(s) \quad E^0 = -0.403\ V$$
    $$CdS(s) + 2e^- \longrightarrow Cd(s) + S^{2-}(aq) \quad E^0 = -1.21\ V$$

    (b) Evaluate ΔG^0 at 25°C for the process

    $$CdS(s) \rightleftharpoons Cd^{2+}(aq) + S^{2-}(aq)$$

88. Refer to tabulated reduction potentials. (a) Calculate K_{sp} for AgBr(s). (b) Calculate ΔG^0 for the reaction

$$AgBr(s) \rightleftharpoons Ag^+(aq) + Br^-(aq)$$

89. Calculate the free energy change, ΔG, for each reaction of Exercise 75 under the stated conditions, at 25°C.

90. Calculate ΔG^0 for the half-reaction

$$\tfrac{1}{2}H_2O_2(aq) + H^+ + e^- \longrightarrow H_2O(\ell)$$

given that $E^0 = 1.77$ V for the $H_2O_2(aq)/H_2O(\ell)$ couple.

Practical Aspects of Electrochemistry

91. Distinguish among (a) primary voltaic cells, (b) secondary voltaic cells, and (c) fuel cells.

92. Sketch and describe the operation of (a) the Leclanché dry cell, (b) the lead storage battery, and (c) the hydrogen–oxygen fuel cell.

93. Why is the dry cell designed so that Zn and MnO_2 do not come into contact? What reaction might occur if they were in contact? How would this reaction affect the usefulness of the cell?

*94. People sometimes try to recharge dry cells, with limited success. (a) What reaction would you expect at the zinc electrode of a Leclanché cell in an attempt to recharge it? (b) What difficulties would arise from the attempt?

95. Briefly describe how a storage cell operates.

96. How does a fuel cell differ from a dry cell or a storage cell?

97. Does the physical size of a commercial cell govern the potential that it will deliver? What does the size affect?

Mixed Exercises

98. Consider the electrochemical cell represented by $Zn(s)|Zn^{2+}\|Fe^{3+}|Fe(s)$. (a) Write the ion–electron equations for the half-reactions and the overall cell equation. (b) The standard reduction potentials for $Zn^{2+}/Zn(s)$ and $Fe^{3+}/Fe(s)$ are -0.763 V and -0.036 V, respectively, at 25°C. Determine the standard potential for the reaction. (c) Determine E for the cell when the concentration of Fe^{3+} is 10.0 mol/L and that of Zn^{2+} is 1.00×10^{-3} mol/L. (d) If 150 mA is to be drawn from this cell for a period of 15.0 min, what is the minimum mass for the zinc electrode?

99. A sample of Al_2O_3 dissolved in a molten fluoride bath is electrolyzed using a current of 1.00 A. (a) What is the rate of production of Al in grams per hour? (b) The oxygen liberated at the positive carbon electrode reacts with the carbon to form CO_2. What mass of CO_2 is produced per hour?

*100. The "life" of a certain voltaic cell is limited by the amount of Cu^{2+} in solution available to be reduced. If the cell contains 25 mL of 0.175 M $CuSO_4$, what is the maximum amount of electrical charge this cell could generate?

*101. Under standard state conditions, the following reaction is not spontaneous:

$$Br^- + 2MnO_4^- + H_2O(\ell) \longrightarrow$$
$$BrO_3^- + 2MnO_2(s) + 2OH^- \qquad E^0 = -0.022 \text{ V}$$

The reaction conditions are adjusted so that $E = 0.100$ V by making $[Br^-] = [MnO_4^-] = 1.5$ mol/L and $[BrO_3^-] = 0.5$ mol/L. (a) What is the concentration of hydroxide ions in this cell? (b) What is the pH of the solution in the cell?

102. How many coulombs of electricity would be required to reduce the iron in 36.0 g of potassium hexacyanoferrate(II), $K_4[Fe(CN)_6]$, to metallic iron?

103. A magnesium bar weighing 5.0 kg is attached to a buried iron pipe to protect the pipe from corrosion. An average current of 0.030 A flows between the bar and the pipe. (a) What reaction occurs at the surface of the bar? of the pipe? In which direction do electrons flow? (b) How many years will it take for the Mg bar to be entirely consumed (1 year = 3.16×10^7 s)? (c) What reaction(s) will occur if the bar is not replaced after the time calculated in part (b)?

104. The production of uranium metal from purified uranium dioxide ore consists of the following steps:

$$UO_2(s) + 4HF(g) \longrightarrow UF_4(s) + 2H_2O(\ell)$$
$$UF_4(s) + 2Mg(s) \xrightarrow{\Delta} U(s) + 2MgF_2(s)$$

What is the oxidation number of U in (a) UO_2, (b) UF_4, and (c) U? Identify (d) the reducing agent and (e) the substance reduced. (f) What current could the second reaction produce if 1.00 g of UF_4 reacted each minute? (g) What volume of HF(g) at 25°C and 10.0 atm would be required to produce 1.00 g of U? (h) Would 1.00 g of Mg be enough to produce 1.00 g of U?

105. Which of each pair is the stronger oxidizing agent? (a) H^+ or Cl_2, (b) Ni^{2+} or Se in contact with acidic solution, (c) $Cr_2O_7^{2-}$ or Br_2 (acidic solution).

106. (a) Describe the process of electroplating. (b) Sketch and label an apparatus that a jeweler might use for electroplating silver onto jewelry. (c) A jeweler purchases highly purified silver to use as the anode in an electroplating operation. Is this a wise purchase? Why?

107. The same quantity of electrical charge that deposited 0.583 g of silver was passed through a solution of a gold salt, and 0.355 g of gold was deposited. What is the oxidation state of gold in this salt?

108. Show by calculation that $E^0 = -1.662$ V for the reduction of Al^{3+} to Al(s), regardless of whether the equation for the reaction is written

(i) $\tfrac{1}{3}Al^{3+} + e^- \longrightarrow \tfrac{1}{3}Al(s)$ $\qquad \Delta G^0 = 160.4$ kJ/mol

or

(ii) $Al^{3+} + 3e^- \longrightarrow Al(s)$ $\qquad \Delta G^0 = 481.2$ kJ/mol

Objectives

As you study this chapter, you should learn

☐ About the roles of some trace elements in nutrition
☐ About major sources of metals
☐ About some pretreatment techniques for ores
☐ About some reduction processes that produce free metals
☐ About some techniques for refining and purifying metals
☐ The specific metallurgies of five metals: magnesium, aluminum, iron, copper, and gold
☐ The basics of conservation of metals

A cluster of 21 minerals that are present in ores, plus two samples of native ores (free metals, Ag and Bi). These are identified below.

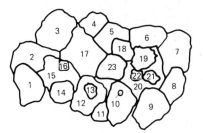

1. Bornite (iridescent)–COPPER
2. Dolomite (pink)–MAGNESIUM
3. Molybdenite (gray)–MOLYBDENUM
4. Skutterudite (gray)–COBALT, NICKEL
5. Zincite (mottled red)–ZINC
6. Chromite (gray)–CHROMIUM
7. Stibnite (*top right*, gray)–ANTIMONY
8. Gummite (yellow)–URANIUM
9. Cassiterite (rust, *bottom right*)–TIN
10. Vanadinite crystal on Goethite (red crystal)–VANADIUM
11. Cinnabar (red)–MERCURY
12. Galena (gray)–LEAD
13. Monazite (white)–RARE EARTHS: cerium, lanthanum, neodymium, thorium
14. Bauxite (gold)–ALUMINUM
15. Strontianite (white, spiny)–STRONTIUM
16. Cobaltite (gray cube)–COBALT
17. Pyrite (gold)–IRON
18. Columbinite (tan, gray stripe)–NIOBIUM, TANTALUM
19. Native BISMUTH (shiny)
20. Rhodochrosite (pink)–MANGANESE
21. Rutile (shiny twin crystal)–TITANIUM
22. Native SILVER (filigree on quartz)
23. Pyrolusite (black, powdery)–MANGANESE

Metals

Metals are widely used for structural purposes in buildings, trains, railroads, ships, aircraft, and cars. They also serve as conductors of heat and electricity. Medical and nutritional research during recent decades has provided much insight into important biological functions of metals. The metals Na, K, Ca, and Mg, as well as some nonmetals (C, H, O, N, P, and S), are present in the human body in substantial quantities. In this chapter, we shall study the occurrence of metals and examine processes for obtaining metals from their ores.

22-1 Occurrence of the Metals

In our study of periodicity, we learned that metallic character increases toward the left and toward the bottom of the periodic table (Section 4-1) and that oxides of most metals are basic (Section 6-8, part 2). The oxides of some metals (and metalloids) are amphoteric (Section 10-8). In Section 13-17 we described metallic bonding and related the effectiveness of metallic bonding to the characteristic properties of metals.

The properties of metals influence the kinds of ores in which they are found, and the metallurgical processes used to extract them from their ores. Metals with negative standard reduction potentials (active metals) are found in nature in the combined state. Those with positive reduction potentials, the less active metals, may occur in the uncombined free state as **native ores**.

You may wish to review some of these sections now.

(text continued on p. 837)

Chemistry in Use. . .
Metals and Life

In 1681 an English physician, Thomas Sedenham, soaked "iron and steel filings" in cold Rhenish wine. He used the resulting solution to treat patients suffering from chlorosis, an iron-deficiency anemia. Iron was the first trace element shown to be essential in the human diet. Around 1850 the French chemist Boussinguault demonstrated that certain salt deposits cured goiter. Those salt deposits contained iodine compounds. Iodine, an essential trace element, is a nonmetal.

In recent years, several additional trace elements have been found to be essential to human nutrition: Cu, Mn, Zn, Co, Mo, Se, and Cr. Six other trace elements have been shown to be essential for proper nutrition of various animals: Sn, V, Ni, F, Si, and As. Essential trace elements are divided into three groups (see table). (One major problem in investigating dietary trace elements is measuring the extremely small amounts of these elements that are present in foods. As an example, the vanadium content of fresh peas is usually less than 4.0×10^{-10} gram per gram of fresh peas. Based on this figure, only 1.0 gram of V is contained in 2700 tons of fresh peas.)

But many of the elements that are essential for nutrition may also be harmful or fatal in larger amounts. Arsenic is a well-known poison. In water meant for human consumption, the maximum allowable limits (according to the Federal Water Pollution Control Administration) for some of the elements in the table are as follows: Zn, 5.0 ppm (parts per million); Cu, 1.0 ppm; Fe, 0.3 ppm; Cr, 0.05 ppm; and As, 0.05 ppm.

A dietary shortage of **iron** is the most common trace element deficiency. Anemia, characterized by a low concentration of hemoglobin in the blood or by a low volume of packed red blood cells, is the usual symptom. The recommended dietary allowance for women ages 23 to 50 years is 80% higher than that for men in the same age group because of the iron lost in menstrual bleeding.

Iodine (a nonmetal) is needed to prevent goiter due to iodine deficiency, which accounts for approximately 96% of incidences of the disease. Iodine is present in two thyroid hormones, thyroxine and triiodothyronine, which increase the metabolic rate and oxygen consumption of cells.

Zinc is present in at least 90 enzymes and in the hormone insulin. Zinc is involved in the functioning of the pituitary and adrenal glands, and in the pancreas and gonads. It plays an important role in the growth processes, including protein synthesis and cell division. Research at the University of Wisconsin in 1936 demonstrated that zinc is essential to animal growth. Human need was established in 1964 by research at Wayne State University. Meat and other animal products are the main dietary sources of zinc for humans.

Copper is essential in the body's oxidation processes. It is a component of several oxidative enzymes. Current theories suggest that a copper deficiency causes anemia because copper is needed for the absorption and mobilization of the iron that is required to make hemoglobin. Human need for copper was established in 1928. Nuts, liver, and shellfish are important sources. Copper serves the same oxygen-carrying function in the fluids of certain marine animals that iron serves in higher animals.

Manganese was shown in 1931 to be essential to animals. One of its functions is the activation of certain enzymes. **Cobalt** is present in vitamin B_{12}, which prevents pernicious anemia. Its need by humans was established in 1935. **Molybdenum** was shown in 1953 to be essential to animals, and human need has also been established. It is a component of several enzymes and is involved in protein formation.

The nonmetal **selenium** is involved in enzyme formation and fat

Many dietary supplements include essential trace elements.

Dietary Trace Elements Shown to Be Essential in Humans and Animals

Element	Date Discovered to Be Essential Dietary Trace Element in Animals	Dietary Sources	Functions in Humans and/or Animals
Group 1 (need in humans established and quantified)			
iron	17th century	meat, liver, fish, poultry, beans, peas, raisins, prunes	component of hemoglobin and myoglobin; oxidative enzymes
iodine*	1850	iodized table salt, shellfish, kelp	needed to make the thyroid hormones thyroxine and triiodothyronine; prevents iodine-deficiency goiter
zinc	1934	meat, liver, eggs, shellfish	present in at least 90 enzymes and in the hormone insulin
Group 2 (need in humans established but not yet quantified)			
copper	1928	nuts, liver, shellfish	component of oxidative enzymes; involved in the absorption and mobilization of iron needed for making hemoglobin
manganese	1931	nuts, fruits, vegetables, whole-grain cereals	involved in formation of enzymes, bone
cobalt	1935	meat, dairy products	component of vitamin B_{12}
molybdenum	1953	organ meats, green leafy vegetables, legumes	involved in formation of enzymes, proteins
selenium*	1957	meat, seafood	involved in enzyme formation, fat metabolism
chromium	1959	meat, beer, unrefined wheat flour	required for glucose metabolism
Group 3 (need established in animals but not yet in humans)			
tin	1970	many†	essential for normal growth of rats
vanadium	1971	many†	needed for optimum growth of chicks and rats
fluorine*	1972	fluoridated water	essential for normal growth of rats
silicon*	1972	many†	needed for growth and bone development in chicks; needed for growth of rats
nickel	1974	many†	needed for normal growth and for formation of red blood cells in rats; needed by chicks and swine
arsenic*	1975	many†	required for normal growth of rats, goats, minipigs; shortage of element can impair reproductive ability of goats, minipigs

* Nonmetals.
† Present in trace quantities in many foods and in the environment.

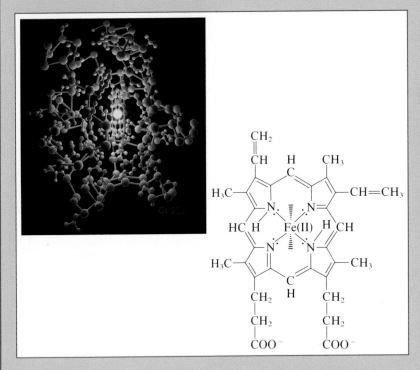

Cytochromes are proteins that are found in all types of organisms, typically in membranes. They take part in electron transport—for instance, in the metabolic process in animals and in the photosynthetic process in green plants. A representation of the amino-acid-chain structure of one of the cytochromes is shown at the left. The yellow region in the center of the model represents a heme group, which contains an iron atom (the white sphere in the center of the heme group). The electron transport carried out by this protein involves alternate reduction of this iron to Fe(II) and oxidation to Fe(III). Hemoglobin, which carries O_2 in the bloodstream, and myoglobin, which stores O_2 in muscle tissue, are other heme proteins. The structure of the heme group of hemoglobin, myoglobin, and one class of cytochromes is shown at the right. The heme groups of other cytochromes differ in the substitution pattern of groups on the ring structure.

metabolism. Animal need for selenium was shown in 1957.

Chromium is essential to glucose metabolism. Chromium deficiency has been observed among children with severe protein deficiency in developing countries, and among some elderly individuals in the United States.

During the period from 1970 to 1975, the trace elements Sn, V, Ni, F, Si, and As were shown to be essential to certain animals. Human need for them has not been demonstrated. Many scientists believe that these trace elements plus several others are probably essential to humans. The metals Cd, Pb, and Li and the nonmetals Br and B may also be essential trace elements.

Two factors have aided in the discovery of the roles of many trace elements. One is the availability of two highly sensitive analytical techniques, activation analysis and electrothermal atomic absorption spectroscopy, which allow detection of these elements in concentrations of only a few parts per billion. The other is the use of special isolation chambers that allow study of animals under carefully controlled conditions, free of unwanted contaminants. The diets fed to animals and their air supply must be carefully purified to keep out even *traces* of unwanted elements, and metals cannot be used in their cages.

New research on the role of trace elements in human nutrition raises interesting questions. For many years, some have suggested that drinking a small quantity of boiled seawater daily supplies essential "trace minerals" that have been leached from the soil by eons of rainfall and may no longer be in our normal foods and drinking water. In view of the recently established importance of several trace elements that are contained in seawater, those who advocate drinking small amounts of seawater may be on the right track.

Let us cite one additional bit of evidence on the importance of trace elements in nutrition. Arthur J. Stattelman of the Poultry Disease Research Center at the University of Georgia has demonstrated that, when the drinking water of chickens contains 10% seawater, the structure of their feathers is greatly improved. These feathers are up to three times more effective as heat insulators than those of chickens fed the same diet but no seawater. The trace element or elements responsible for improved feather quality have not been identified. How is this significant? The cost of producing chickens, as well as other animals used for food, includes the amount of feed required to maintain body temperature at the desired level—not just the amount of feed required for body growth. Chickens with feathers that are more effective heat insulators require less feed to maintain body temperature, and so their production costs are lower.

Table 22-1
Common Classes of Ores

Anion	Examples and Names of Minerals
none (native ores)	Au, Ag, Pt, Os, Ir, Ru, Rh, Pd, As, Sb, Bi, Cu
oxide	hematite, Fe_2O_3; magnetite, Fe_3O_4; bauxite, Al_2O_3; cassiterite, SnO_2; periclase, MgO; silica, SiO_2
sulfide	chalcopyrite, $CuFeS_2$; chalcocite, Cu_2S; sphalerite, ZnS; galena, PbS; iron pyrites, FeS_2; cinnabar, HgS
chloride	rock salt, NaCl; sylvite, KCl; carnallite, $KCl \cdot MgCl_2$
carbonate	limestone, $CaCO_3$; magnesite, $MgCO_3$; dolomite, $MgCO_3 \cdot CaCO_3$
sulfate	gypsum, $CaSO_4 \cdot 2H_2O$; epsom salts, $MgSO_4 \cdot 7H_2O$; barite, $BaSO_4$
silicate	beryl, $Be_3Al_2Si_6O_{18}$; kaolinite, $Al_2(Si_2O_8)(OH)_4$; spodumene, $LiAl(SiO_3)_2$

The most widespread minerals are silicates. But extraction of metals from silicates is very difficult. Metals are obtained from silicate minerals only when there is no other more economical alternative.

(text continued from p. 833)
Examples of the latter are Cu, Ag, Au, and the less abundant Pt, Os, Ir, Ru, Rh, and Pd. Cu, Ag, and Au are also found in the combined state.

Many ''insoluble'' compounds of the metals are found in the earth's crust. Solids that contain these compounds are the **ores** from which metals are extracted. Ores contain **minerals**, comparatively pure compounds of the metals of interest, mixed with relatively large amounts of **gangue**—sand, soil, clay, rock, and other material. Soluble compounds are found dissolved in the sea or in salt beds in areas where large bodies of water have evaporated. Metal ores can be classified by the anions with which the metal ions are combined (Table 22-1 and Figure 22-1).

Pronounce "gangue" as one syllable with a soft final g.

Metallurgy

Metallurgy is the commercial extraction of metals from their ores and the preparation of metals for use. It usually includes several steps: (1) mining the ore, (2) pretreatment of the ore, (3) reduction of the ore to the free metal, (4) refining or purifying the metal, and (5) alloying, if necessary.

Figure 22-1
Major natural sources of the elements. The soluble halide salts are found in oceans, salt lakes, brine wells, and solid deposits. Most helium is obtained from wells in the United States and the U.S.S.R. Most of the other noble gases are obtained from air.

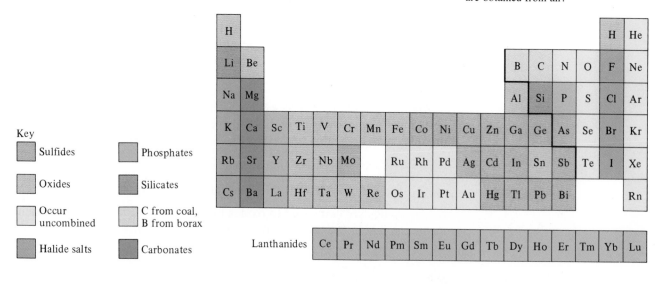

Key
- Sulfides
- Oxides
- Occur uncombined
- Halide salts
- Phosphates
- Silicates
- C from coal, B from borax
- Carbonates

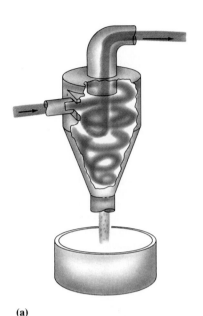

(a)

Figure 22-2

(a) The cyclone separator enriches metal ores. Crushed ore is blown in at high velocity. Centrifugal force takes the heavier particles, with higher percentages of metal, to the wall of the separator. These particles spiral down to the collection bin at the bottom. Lighter particles, not as rich in the metal, move into the center. They are carried out the top in the air stream. (b) A flotation process for enrichment of copper sulfide ore. The relatively light sulfide particles are suspended in the water–oil–detergent mixture and collected as a froth. The denser material sinks to the bottom of the container.

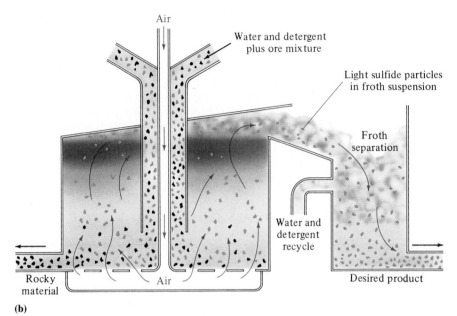

(b)

22-2 Pretreatment of Ores

After being mined, many ores must be concentrated by removal of most of the gangue. Most sulfides have relatively high densities and are more dense than gangue. After pulverization, the lighter gangue particles are removed by a variety of methods. One involves blowing the lighter particles away using a cyclone separator (Figure 22-2). The lighter particles can be sifted out through layers of vibrating wire mesh or on inclined vibration tables.

The **flotation** method is particularly applicable to sulfides, carbonates, and silicates, which are not "wet" by water or else can be made water-repellent by treatment. Their surfaces are easily covered by layers of oil or other flotation agents. A stream of air is blown through a swirled suspension of such an ore in water and oil (or other agent). Bubbles form on the oil on the mineral particles and cause them to rise to the surface. The bubbles are prevented from breaking and escaping by a layer of oil and emulsifying agent. A frothy ore concentrate forms at the surface. By varying the relative amounts of oil and water, the types of oil additive, the air pressure, and so on, it is even possible to separate one metal sulfide, carbonate, or silicate from another (Figure 22-2b).

Another pretreatment process involves chemical modification. This converts metal compounds to more easily reduced forms. Carbonates and hydroxides may be heated to drive off CO_2 and H_2O, respectively.

$$CaCO_3(s) \xrightarrow{\Delta} CaO(s) + CO_2(g)$$

$$Mg(OH)_2(s) \xrightarrow{\Delta} MgO(s) + H_2O(g)$$

Some sulfides are converted to oxides by **roasting**, i.e., heating below their melting points in the presence of oxygen from air. For example,

$$2ZnS(s) + 3O_2(g) \xrightarrow{\Delta} 2ZnO(s) + 2SO_2(g)$$

Iron pyrite, FeS_2 (also known as "fool's gold"), at left, and galena, PbS.

Roasting sulfide ores causes air pollution. Enormous quantities of SO_2 escape into the atmosphere (Section 6-8, part 4), where it causes great environmental damage (Figure 22-3). Federal regulations now require limitation of the amount of SO_2 that escapes with stack gases and fuel gases. Now most of the SO_2 is trapped and used in the manufacture of sulfuric acid (Section 25-11).

Some of these environmental problems are discussed in the Chemistry in Use essay "Acid Rain" in Chapter 6.

22-3 Reduction to the Free Metals

The method used for reduction, or smelting, of metal ores to the free metals depends on how strongly the metal ions are bonded to anions. When bonding is stronger, more energy is required to reduce the metals. This makes reduction more expensive. The most active metals usually have the strongest bonding.

The least reactive metals occur in the free state and thus require no reduction. Examples include Au, Ag, and Pt. This is why gold and silver have been used as free metals since prehistoric times. The less active metals, such as Hg, can be obtained directly from their sulfide ores by roasting. This reduces metal ions to the free metals by oxidation of the sulfide ions.

$$HgS(s) + O_2(g) \xrightarrow{\Delta} SO_2(g) + Hg(g)$$
cinnabar   from air   obtained as vapor;
                                        later condensed

Figure 22-3
Fir trees on Mount Mitchell in North Carolina, killed by acid rain.

Roasting more active metal sulfides produces metal oxides, but no free metals:

$$2NiS(s) + 3O_2(g) \xrightarrow{\Delta} 2NiO(s) + 2SO_2(g)$$

The resulting metal oxides are then reduced to free metals with coke or CO. If C must be avoided, another reducing agent, such as H_2, Fe, or Al, is used.

Coke is impure carbon.

$$SnO_2(s) + 2C(s) \xrightarrow{\Delta} Sn(\ell) + 2CO(g)$$

$$WO_3(s) + 3H_2(g) \xrightarrow{\Delta} W(s) + 3H_2O(g)$$

The very active metals, such as Al and Na, are reduced electrochemically, usually from their anhydrous molten salts. If H_2O is present, it is reduced preferentially. Tables 22-2 and 22-3 summarize reduction processes for some metal ions.

Rhodochrosite, $MnCO_3$, is pink.

Rutile contains TiO_2.

Ultrapure silicon is produced by zone refining.

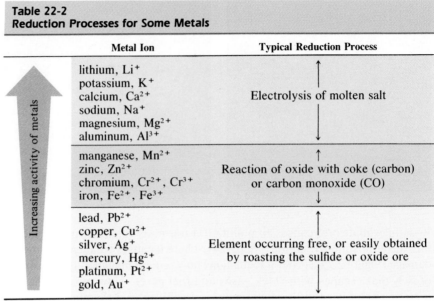

Table 22-2
Reduction Processes for Some Metals

	Metal Ion	Typical Reduction Process
Increasing activity of metals ↑	lithium, Li^+ potassium, K^+ calcium, Ca^{2+} sodium, Na^+ magnesium, Mg^{2+} aluminum, Al^{3+}	Electrolysis of molten salt
	manganese, Mn^{2+} zinc, Zn^{2+} chromium, Cr^{2+}, Cr^{3+} iron, Fe^{2+}, Fe^{3+}	Reaction of oxide with coke (carbon) or carbon monoxide (CO)
	lead, Pb^{2+} copper, Cu^{2+} silver, Ag^+ mercury, Hg^{2+} platinum, Pt^{2+} gold, Au^+	Element occurring free, or easily obtained by roasting the sulfide or oxide ore

22-4 Refining or Purification of Metals

Metals obtained from reduction processes are almost always impure. Further purification (refining) is usually required. This can be accomplished by distillation if the metal is more volatile than its impurities, as in the case of mercury. Among the metals purified electrolytically are Cu, Ag, Au, and Al (Section 21-8). The impure metal is the anode, and a small sample of the pure metal is the cathode. Both are immersed in a solution of the desired metal ion.

Zone refining is often used when extremely pure metals are desired for such applications as solar cells and semiconductors (Section 13-17). An induction heater surrounds a bar of the impure solid and passes slowly from one end to the other (Figure 22-4). As it passes, it melts portions of the bar, which slowly recrystallize as the heating element moves away. The impurity does not fit into the crystal as easily as the element of interest, so most of

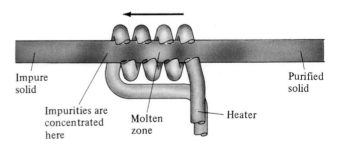

Impure solid

Impurities are concentrated here

Molten zone

Heater

Purified solid

Figure 22-4
A representation of a zone-refining apparatus.

Table 22-3
Some Specific Reduction Processes

Metal	Compound (ore)	Reduction Process	Comments
mercury	HgS (cinnabar)	Roast reduction; heating of ore in air $HgS + O_2 \xrightarrow{\Delta} Hg + SO_2$	
copper	sulfides such as Cu_2S (chalcocite)	Blowing of oxygen through purified molten Cu_2S: $Cu_2S + O_2 \xrightarrow{\Delta} 2Cu + SO_2$	Preliminary ore concentration and purification steps required to remove FeS impurities
zinc	ZnS (sphalerite)	Conversion to oxide and reduction with carbon: $2ZnS + 3O_2 \xrightarrow{\Delta} 2ZnO + 2SO_2$ $ZnO + C \xrightarrow{\Delta} Zn + CO$	Process also used for the production of lead from galena, PbS
iron	Fe_2O_3 (hematite)	Reduction with carbon monoxide: $2C\ (coke) + O_2 \xrightarrow{\Delta} 2CO$ $Fe_2O_3 + 3CO \xrightarrow{\Delta} 2Fe + 3CO_2$	
titanium	TiO_2 (rutile)	Conversion of oxide to halide salt and reduction with an active metal: $TiO_2 + 2Cl_2 + 2C \xrightarrow{\Delta} TiCl_4 + 2CO$ $TiCl_4 + 2Mg \xrightarrow{\Delta} Ti + 2MgCl_2$	Also used for the reduction of UF_4 obtained from UO_2, pitchblende
tungsten	$FeWO_4$ (wolframite)	Reduction with hydrogen: $WO_3 + 3H_2 \xrightarrow{\Delta} W + 3H_2O$	Used also for molybdenum
aluminum	$Al_2O_3 \cdot nH_2O$ (bauxite)	Electrolytic reduction (electrolysis) in molten cryolite, $Na_3[AlF_6]$, at 1000°C: $2Al_2O_3 \xrightarrow{\Delta} 4Al + 3O_2$	
sodium	NaCl (seawater)	Electrolysis of molten chlorides: $2NaCl \xrightarrow{\Delta} 2Na + Cl_2$	Also for calcium, magnesium, and other active metals in Groups IA and IIA

it is carried along in the molten portion until it reaches the end. Repeated passes of the heating element produce a bar of high purity.

The end containing the impurities can be sliced off and recycled.

After purification, many metals are alloyed, or mixed with other elements, to change their physical and chemical characteristics. In some cases, certain impurities are allowed to remain during refining because their presence improves the properties of the metal. For example, a small amount of carbon in iron greatly enhances its hardness. Examples of alloys include brass, bronze, duralumin, and stainless steel.

Mixtures of metals frequently have properties that are more desirable for a particular purpose than are those of a free metal. In such cases, metals are alloyed.

Metallurgies of Specific Metals

The metallurgies of Mg, Al, Fe, Cu, and Au will be discussed as specific examples.

Mg and Al are active metals, Fe and Cu are moderately active, and Au is relatively inactive.

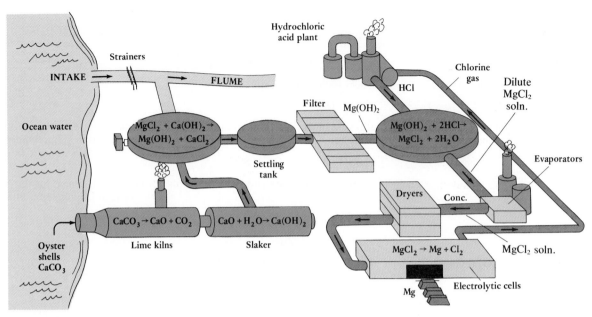

Figure 22-5
Schematic diagram of an industrial plant for the production of magnesium from the Mg^{2+} ions in seawater.

A bed of limestone ($CaCO_3$), Verde River, Arizona.

22-5 Magnesium

Magnesium occurs widely in carbonate ores, but most Mg comes from salt brines and from the sea (Figure 22-5). Seawater is 0.13% Mg by mass. Because of its low density (1.74 g/cm^3), Mg is used in lightweight structural alloys for such things as ladders and aircraft parts.

Magnesium ions are precipitated as $Mg(OH)_2$ by addition of $Ca(OH)_2$ (slaked lime) to sea water. The slaked lime is obtained by crushing oyster shells ($CaCO_3$), heating them to produce lime (CaO), and then adding a limited amount of water (slaking).

$$CaCO_3(s) \xrightarrow{\Delta} CaO(s) + CO_2(g) \qquad \text{(lime production)}$$

$$CaO(s) + H_2O(\ell) \longrightarrow Ca(OH)_2(s) \qquad \text{(slaking lime)}$$

$$Ca(OH)_2(s) + Mg^{2+}(aq) \longrightarrow Ca^{2+}(aq) + Mg(OH)_2(s) \quad \text{(precipitation)}$$

The last reaction occurs because K_{sp} for $Mg(OH)_2$, 1.5×10^{-11}, is much less than that for $Ca(OH)_2$, 7.9×10^{-6}. The milky-white suspension of $Mg(OH)_2$ is filtered, and the solid $Mg(OH)_2$ is then neutralized with HCl to produce $MgCl_2$ solution. Evaporation of the H_2O leaves solid $MgCl_2$, which is then melted and electrolyzed (Figure 22-6) under an inert atmosphere to produce molten Mg and gaseous Cl_2. The products are separated as they are formed, to prevent recombination:

$$Mg(OH)_2(s) + 2[H^+(aq) + Cl^-(aq)] \longrightarrow [Mg^{2+}(aq) + 2Cl^-(aq)] + 2H_2O$$

$$\xrightarrow[\text{then melt solid}]{\text{evaporate solution,}} MgCl_2(\ell) \xrightarrow{\text{electrolysis}} Mg(\ell) + Cl_2(g)$$

Magnesium is cast into ingots or alloyed with other light metals. If no HCl is available, the by-product Cl_2 is allowed to react with CH_4, from natural gas, and with air. This produces more HCl for neutralization of $Mg(OH)_2$.

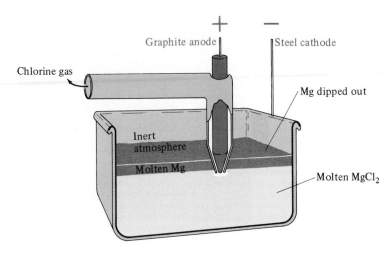

Figure 22-6
A cell for electrolyzing molten $MgCl_2$. The magnesium metal is formed on the steel cathode and rises to the top, where it is dipped off periodically. Chlorine gas is formed around the graphite anode and is piped off.

22-6 Aluminum

Aluminum is the most commercially important nonferrous metal. Its chemistry and uses will be discussed in Section 23-8. Aluminum is obtained from bauxite, or hydrated aluminum oxide, $Al_2O_3 \cdot xH_2O$. Aluminum ions can be reduced to Al by electrolysis only in the absence of H_2O. First the crushed bauxite is purified by dissolving it in a concentrated solution of NaOH to form soluble $Na[Al(OH)_4]$. Then $Al(OH)_3 \cdot xH_2O$ is precipitated from the filtered solution by blowing in carbon dioxide to neutralize the unreacted NaOH and one OH^- ion per formula unit of $Na[Al(OH)_4]$. Heating the hydrated product dehydrates it to Al_2O_3.

> Recall that Al_2O_3 is amphoteric. Impurities such as oxides of iron, which are not amphoteric, are left behind in the crude ore.

$$Al_2O_3(s) + 2NaOH(aq) + 3H_2O(\ell) \longrightarrow 2Na[Al(OH)_4](aq)$$

> For clarity, we have not shown waters of hydration.

$$2Na[Al(OH)_4](aq) + CO_2(aq) \longrightarrow 2Al(OH)_3(s) + Na_2CO_3(aq) + H_2O(\ell)$$

$$2Al(OH)_3(s) \xrightarrow{\Delta} Al_2O_3(s) + 3H_2O(g)$$

The melting point of Al_2O_3 is 2045°C; electrolysis of pure molten Al_2O_3 would have to be carried out at or above this temperature, with great expense. However, it can be done at a much lower temperature when Al_2O_3 is mixed with much lower-melting cryolite, a mixture of NaF and AlF_3 often represented as $Na_3[AlF_6]$. The molten mixture can be electrolyzed at 1000°C with carbon electrodes. The cell used industrially for this process, called the **Hall–Heroult process**, is shown in Figure 22-7.

> A mixture of compounds typically has a lower melting point than any of the pure compounds (Chapter 14).

The inner surface of the cell is coated with carbon or carbonized iron, which functions as the cathode at which aluminum ions are reduced to the free metal. The graphite anode is oxidized to CO_2 gas and must be replaced frequently. This is one of the chief costs of aluminum production.

cathode	$4(Al^{3+} + 3e^- \longrightarrow Al(\ell))$
anode	$3(C(s) + 2O^{2-} \longrightarrow CO_2(g) + 4e^-)$
net reaction	$4Al^{3+} + 3C(s) + 6O^{2-} \longrightarrow 4Al(\ell) + 3CO_2(g)$

Molten aluminum is more dense than molten cryolite, and it collects in the bottom of the cell until it is drawn off and cooled to a solid.

Recently, a more economical approach, the Alcoa chlorine process, has been developed on a modest commercial scale. The anhydrous bauxite is

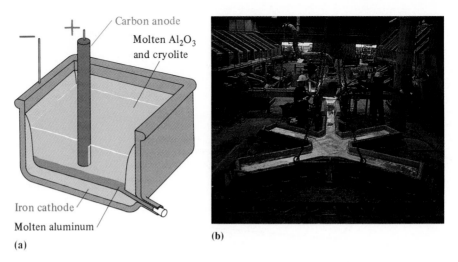

Figure 22-7
(a) Schematic drawing of a cell for producing aluminum by electrolysis of a melt of Al_2O_3 in $Na_3[AlF_6]$. The molten aluminum collects in the container, which acts as the cathode. (b) Casting molten aluminum. Electrolytic cells used in the Hall–Heroult process appear in the background.

Carbon anode

Molten Al_2O_3 and cryolite

Iron cathode

Molten aluminum

(a)

(b)

first converted to $AlCl_3$ by reaction with Cl_2 in the presence of carbon. The $AlCl_3$ is then melted and electrolyzed to give aluminum, and the recovered chlorine is reused in the first step.

$$2Al_2O_3(s) + 3C(\text{coke}) + 6Cl_2(g) \longrightarrow 4AlCl_3(s) + 3CO_2(g)$$

$$2AlCl_3(\ell) \longrightarrow 2Al(\ell) + 3Cl_2(g)$$

This process uses only about 30% as much electrical energy as the Hall–Heroult process.

The use of large amounts of electrical energy in electrolysis makes production of aluminum from ores an expensive metallurgy. Methods for recycling used Al use less than 10% of the energy required to make new metal from bauxite by the Hall–Heroult process. Processing of recycled Al now accounts for more than 25% of the production of this metal. This helps to keep down the cost of aluminum.

Iron ore is scooped in an open-pit mine.

Samples of some ores of iron.

22-7 Iron

The most desirable iron ores contain hematite, Fe_2O_3, or magnetite, Fe_3O_4. As the available supplies of these high-grade ores have dwindled, taconite, which is magnetite in very hard silica rock, has become an important source of iron. The oxide is reduced in blast furnaces (Figure 22-8) by carbon monoxide. Coke mixed with limestone ($CaCO_3$) and crushed ore is admitted at the top of the furnace as the "charge." A blast of hot air from the bottom burns the coke to carbon monoxide with the evolution of more heat:

$$2C(s) + O_2(g) \xrightarrow{\Delta} 2CO(g) + \text{heat}$$

Most of the oxide is reduced to molten iron by carbon monoxide, although some is reduced directly by coke. Several stepwise reductions occur (Figure 22-8), but the overall reactions for Fe_2O_3 can be summarized as follows:

$$Fe_2O_3(s) + 3CO(g) \xrightarrow{\Delta} 2Fe(\ell) + 3CO_2(g) + \text{heat}$$

$$Fe_2O_3(s) + 3C(s) \xrightarrow{\Delta} 2Fe(\ell) + 3CO(g) + \text{heat}$$

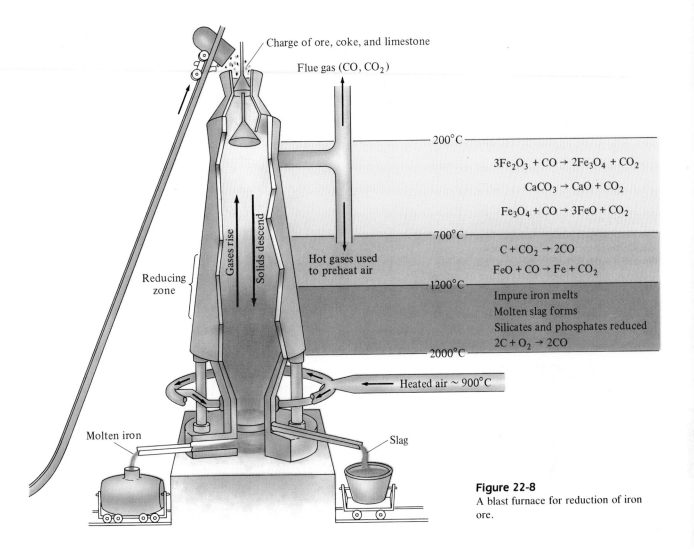

Charge of ore, coke, and limestone

Flue gas (CO, CO$_2$)

200°C

$$3Fe_2O_3 + CO \rightarrow 2Fe_3O_4 + CO_2$$

$$CaCO_3 \rightarrow CaO + CO_2$$

$$Fe_3O_4 + CO \rightarrow 3FeO + CO_2$$

700°C

$$C + CO_2 \rightarrow 2CO$$

$$FeO + CO \rightarrow Fe + CO_2$$

1200°C

Impure iron melts
Molten slag forms
Silicates and phosphates reduced
$$2C + O_2 \rightarrow 2CO$$

2000°C

Hot gases used
to preheat air

Reducing
zone

Gases rise

Solids descend

Heated air ~ 900°C

Molten iron

Slag

Figure 22-8
A blast furnace for reduction of iron
ore.

Much of the CO$_2$ reacts with excess coke to produce more CO to reduce
the next incoming charge.

$$CO_2(g) + C(s) \xrightarrow{\Delta} 2CO(g)$$

The limestone, called a **flux**, reacts with the silica gangue in the ore to
form a molten **slag** of calcium silicate.

$$\underset{\text{limestone}}{CaCO_3(s)} \xrightarrow{\Delta} CaO(s) + CO_2(g)$$

$$CaO(s) + \underset{\text{gangue}}{SiO_2(s)} \xrightarrow{\Delta} \underset{\text{slag}}{CaSiO_3(\ell)}$$

Reaction of a metal oxide (basic) with
a nonmetal oxide (acidic) forms a salt.

The slag is less dense than molten iron; it floats on the surface of the iron
and protects it from atmospheric oxidation. Both are drawn off periodically.
Some of the slag is subsequently used in the manufacture of cement.

The iron obtained from the blast furnace contains carbon, among other
things. It is called **pig iron**. If it is remelted, run into molds, and cooled, it

Considerable amounts of slag are also
used to neutralize acidic soil. If there
were no use for the slag, its disposal
would be a serious economic and
environmental problem.

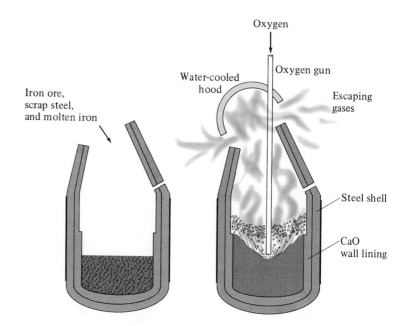

Figure 22-9

The basic oxygen process furnace. Much of the steel manufactured today is refined by blowing oxygen through a furnace that is charged with scrap and molten iron from a blast furnace. After the refined iron is withdrawn into a ladle, alloying elements are added to give the desired steel. The steel industry is one of the nation's largest consumers of oxygen.

Molten steel is poured from a basic oxygen furnace.

It is now profitable to mine ores containing as little as 0.25% copper. The increased use of fiber optics in place of copper in communications cables may help to lessen the demand for this metal. The use of superconducting materials in electricity transmission lines could eventually provide enormous savings.

becomes **cast iron**. This is brittle because it contains much iron carbide, Fe_3C. If all the carbon is removed, nearly pure iron can be produced. It is silvery in appearance, quite soft and of little use. If *some* of the carbon is removed and other metals such as Mn, Cr, Ni, W, Mo, and V are added, the mixture becomes stronger and is known as **steel**. There are many types of steel, containing alloyed metals and other elements in various controlled proportions. Stainless steels show high tensile strength and excellent resistance to corrosion. The most common kind contains 14 to 18% chromium and 7 to 9% nickel.

Pig iron can also be converted to steel by burning out most of the carbon with O_2 in a basic oxygen furnace (Figure 22-9). Oxygen is blown through a heat-resistant tube inserted below the surface of the *molten* iron. Carbon burns to CO, which subsequently escapes and burns to CO_2.

22-8 Copper

Copper is so widely used, especially in its alloys such as bronze (Cu and Sn) and brass (Cu and Zn), that it is becoming very scarce. The U.S. Bureau of Mines estimates that the known worldwide reserves of copper ore will be exhausted during the first half of the next century (Figure 22-10). The two main classes of copper ores are the mixed sulfides of copper and iron—such as chalcopyrite, $CuFeS_2$—and the basic carbonates, such as azurite, $Cu_3(CO_3)_2(OH)_2$, and malachite, $Cu_2CO_3(OH)_2$.

Let us consider $CuFeS_2$ (or $CuS \cdot FeS$). The copper compound is separated from gangue by flotation (Figures 22-2b and 22-11) and then roasted to remove volatile impurities. Enough air is used to convert iron(II) sulfide, but not copper(II) sulfide, to the oxide:

$$2CuFeS_2(s) + 3O_2(g) \xrightarrow{\Delta} 2FeO(s) + 2CuS(s) + 2SO_2(g)$$

Figure 22-11
A copper ore being enriched by
flotation.

Figure 22-10
An open-pit copper mine near Bagdad, Arizona.

The roasted ore is then mixed with sand (SiO_2), crushed limestone ($CaCO_3$),
and some unroasted ore that contains copper(II) sulfide in a reverberatory
furnace at 1100°C. CuS is reduced to Cu_2S, which melts. The limestone and
silica form a molten calcium silicate glass. This dissolves iron(II) oxide to
form a slag less dense than the molten copper(I) sulfide, on which it floats.

$$CaCO_3(s) + SiO_2(s) \xrightarrow{\Delta} CaSiO_3(\ell) + CO_2(g)$$

$$CaSiO_3(\ell) + FeO(s) + SiO_2(s) \xrightarrow{\Delta} CaSiO_3 \cdot FeSiO_3(\ell)$$

The slag is periodically drained off. The molten copper(I) sulfide is drawn
off into a Bessemer converter. There it is again heated and treated with air.
This oxidizes sulfide ions to SO_2 and reduces copper(I) ions to metallic
copper. The overall process is

$$Cu_2S(\ell) + O_2(g) \xrightarrow{\Delta} 2Cu(\ell) + SO_2(g)$$

The impure copper is refined electrolytically (Section 21-8).

A sample that contains two copper-
bearing materials. The blue mineral is
azurite, $Cu_3(CO_3)_2(OH)_2$ or $2CuCO_3 \cdot$
$Cu(OH)_2$. The green mineral is
malachite, $Cu_2CO_3(OH)_2$ or $CuCO_3 \cdot$
$Cu(OH)_2$.

Chemistry in Use. . .
Metal Clusters

One of the frontiers of research in chemistry and physics today is in the study of microscopic particles of metals. Particles so small that they contain only a few atoms are referred to as metal atom clusters or simply *metal clusters*. Metal clusters can be indicated with chemical formulas just like conventional molecules. For example, the clusters Ni_3, Fe_5, and Ag_{17} contain 3, 5, and 17 atoms of nickel, iron, and silver, respectively, but no other elements. It is not yet clear whether clusters of metal atoms represent small pieces of solid metal, or whether they are so different that they have a whole family of unusual properties all of their own, like a new class of molecules. Until recently it was pointless to ask such questions because no one could make metal particles this small. Now, however, laboratories around the world can make clusters out of any metal (or other element) in the periodic table. Systematic studies are under way to measure the physical and chemical properties of these unusual species and to determine what their applications might be.

The studies that eventually made cluster research possible were motivated by the need for new solid materials for the chemical and microelectronics industries. In the chemical industry, small metal particles are used as *catalysts* (Section 16-11). When a catalytic metal powder is added to a mixture of chemicals that is about to react, the reactant molecules stick to the metal surface, and the reaction processes can be modified or controlled. Reactions can be made to occur more rapidly, to produce different products, or to minimize unwanted byproducts. The activity of a catalyst increases as its particle size decreases. Because they are so small, metal clusters may represent the ultimate metal catalysts. Catalysts are especially important in the petroleum industry; they are used to convert the various components of crude oil into useful products. If catalysis were better understood or could be improved by the use of metal clusters, synthetic chemicals and/or energy would be far less expensive.

In the microelectronics industry, a major goal is to make integrated circuits as small as physically possible. New production techniques are needed to produce "nanoscale" transistors, wires, resistors, and so on, which means that these devices are as small as molecules. They can be made in the right sizes and shapes by "plating," or "depositing," metal clusters onto a supporting surface. However, many questions remain unanswered about such small samples of materials. For example, do microscopically thin wires conduct electricity as easily as larger ones? How small can a piece of semiconductor become without ceasing to function? When electrical current flows, does the heating cause small particles to melt or to move on their support? Investigation of the fundamental properties of metal clusters will provide answers for these and many other technical questions.

Clusters are most often produced by a new technique known as *laser vaporization* (Figure a). In this method, a solid piece of the metal to be studied is mounted in a special holder inside a vacuum chamber. A high-powered pulse of laser light, usually from a Nd:YAG laser, is focused with a lens onto the surface of the metal in much the same way that concentrated sunlight can be focused with a magnifying glass to ignite a piece of paper. However, the laser light is so powerful that it generates a temperature of about 10,000 K, which is enough to vaporize a small amount of the metal. The result is an extremely hot vapor composed of metal atoms. A burst of helium gas at room temperature is squirted through the sample holder to cool the metal vapor. Helium does not react, even with hot metal atoms, but through thousands of collisions it cools the metal vapor to near room temperature. As they cool, metal atoms recombine and condense to form solid material again. It is this recombination of atoms that produces the clusters. If allowed to, the recombining metal vapor would plate out again on the surface from which it came. However, the flowing helium gas pushes the small metal particles out into the vacuum chamber in the form of an expanding gaseous spray. The spray, known as a *molecular beam*, is where cluster experiments are conducted. For example, if the cluster molecular beam is sprayed into a mass spectrometer, the masses, and therefore the sizes, of the clusters formed can be measured. Figure (b) shows a mass spectrum of silver clusters produced by this kind of experiment. The various molecules formed with two, three, four, . . . atoms are called *dimers, trimers, tetramers,* and so on. If the cluster spray is intersected by another laser beam, the absorption spectrum of the clusters can be measured to determine their structures (overall shapes, bond lengths, bond angles). Other experiments with lasers and mass spectrometers measure the energy required to ionize clusters of different sizes or to break the chemical bonds that hold cluster atoms together.

Although clusters of virtually any metal can be formed and studied, these molecules are not "stable" in the usual sense of the word. Aggregated atoms in a cluster, not as stable as more conventional molecules, are still far more stable than the

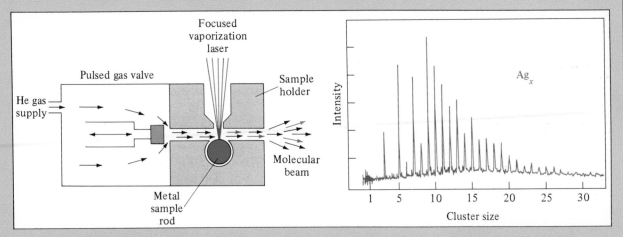

(a) Schematic diagram of the laser vaporization cluster source.

(b) Mass spectrum of silver clusters produced by laser ionization. As shown, not all clusters are produced in equal abundance.

same atoms separated, and so clusters do not spontaneously fly apart. But it is usually not possible to draw simple bonding schemes such as Lewis dot formulas for these kinds of molecules. One exception occurs for dimers of alkali metals, in which each of the atoms has a single s electron in its valence shell, just like a hydrogen atom. Like diatomic hydrogen, H_2, the molecules Li_2, Na_2, K_2, and so on are held together by a two-electron covalent single bond. The same is true for the coinage metal dimers, Ag_2, Cu_2, and Au_2, which have a filled d electron shell not involved in the bonding and a single s electron in the valence shell that forms the covalent bond.

Large particles of sodium, silver, and copper are well-known conductors of electricity, but interestingly the diatomic particles cannot conduct because they contain no free electrons. Transition metals have more complex bonding schemes. For example, a chromium atom has five d electrons and one s electron in its valence shell, and all of these are used in Cr_2, which has a *sextuple* bond! Iron has six d electrons and two s electrons, but Fe_2 has only a single bond. The "unused" electrons in Fe_2 give it "unsatisfied" bonding capacity, and it is therefore an extremely reactive

molecule. When clusters larger than dimers are considered, the chemical bonding becomes too complex for simple description. Some of the electrons are used to hold the atoms together in chemical bonds, and some are delocalized over the surface and volume of the cluster as it begins to "look like" a piece of solid metal. Even with the best computer models available today, it is impossible to predict the chemical bonding scheme in a cluster of about 10 atoms.

Closely related to the chemical bonding in any cluster is its geometric structure. The atoms of solid metals usually are arranged in orderly crystal structures—often close-packed hexagonal or cubic networks in which each atom has 12 nearest neighbors. If these structural patterns were drawn for clusters, they would have sharp edges and flat faces. In reality, effects such as surface tension combine with chemical bonding forces to give clusters a smoother exterior. One especially stable arrangement of cluster atoms that occurs frequently is the 13-atom *icosahedron* shown in Figure (c). This structure resembles close-packed structures in that it has one atom surrounded by 12 nearest neighbors, but its exterior is essentially spherical. Spherical and elliptical structures

are commonplace in cluster structures. It is interesting to note that 12 of the 13 atoms in this icosahedral structure are on the surface of the "particle." Even in a cluster with 100 atoms, about 80 are on the surface. This great surface area, where other molecules can stick and react, makes clusters all the more interesting as potential catalysts. One twist is that chemical bonds in clusters are often weak, making the bonding network easy to disrupt. When this begins to occur, cluster bonds may break and re-form rapidly as though the particle were a liquid droplet instead of a solid! This behavior, which is like melting in larger particles, requires a characteristic temperature that depends on the size of the cluster and the

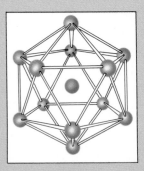

(c) The 13-atom icosahedron structure of many metal clusters.

stability of the bonding arrangement. If they are heated enough, as occurs with laser excitation, clusters may "evaporate" and lose atoms.

As you can see, many of the simple concepts used in chemistry can be applied to clusters, but because of their small size, the concepts must be modified. Full understanding of cluster properties is only now beginning to emerge. Research is still complicated by the conditions under which clusters are studied. Molecular beam laser vaporization can synthesize virtually any kind of cluster, making it possible to study molecules never before known, but it does not make them in large quantities. In the region of experimentation, there are only about 10^8 clusters per cm^3 (about 10^{-16} moles!). At these concentrations, even the most sensitive measurement techniques available are sometimes not good enough. When clusters can be produced in greater quantities or new techniques to measure them can be developed, more of their unusual properties will be revealed. The strange world of cluster research may then become even more fascinating than it already is.

Professor Michael A. Duncan
The University of Georgia

22-9 Gold

Mining low-gold ore in an open-pit mine.

Because of environmental concerns about mercury toxicity, the cyanide process is increasingly preferred. This is not to suggest that mercury is more toxic than cyanide. The problems due to mercury are greater in that it persists in the environment for a long time, and mercury poisoning is cumulative.

Gold is an inactive metal, so it occurs mostly in the native state. It is sometimes found as gold telluride. Because of its high density, metallic gold can be concentrated by panning. In this operation, gold-bearing sand and gravel are gently swirled with water in a pan. The lighter particles spill over the edge, and the denser nuggets of gold remain. Gold is concentrated by sifting crushed gravel in a stream of water on a slightly inclined shaking table that contains several low barriers. These impede the descent of the heavier gold particles but allow the lighter particles to pass over. The gold is then alloyed with mercury and removed. The mercury is distilled away, leaving behind the pure gold.

Gold is also recovered from the anode sludge from electrolytic purification of copper (Section 21-8). Gold is so rare that it is also obtained from very low-grade ores by the cyanide process. Air is bubbled through an agitated slurry of the ore mixed with a solution of NaCN. This causes slow oxidation of the metal and the formation of a soluble complex compound:

$$4Au(s) + 8CN^-(aq) + O_2(g) + 2H_2O(\ell) \longrightarrow$$
$$4[Au(CN)_2]^-(aq) + 4OH^-(aq)$$

After filtration, free gold can then be regenerated by reduction of $[Au(CN)_2]^-$ with zinc or by electrolytic reduction.

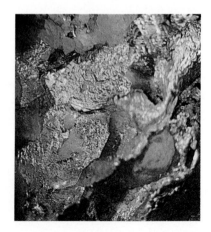

Native gold.

Key Terms

Alloying   Mixing of a metal with other substances (usually other metals) to modify its properties.

Charge   A sample of crushed ore as it is admitted to a furnace for smelting.

Flotation   A method by which hydrophobic (water-repelling) particles of an ore are separated from hydrophilic (water-attracting) particles in a metallurgical pretreatment process.

Flux   A substance added to react with the charge, or a product of its reduction, in metallurgy; usually added to lower a melting point.

Gangue   Sand, rock, and other impurities surrounding the mineral of interest in an ore.

Metallurgy   The overall processes by which metals are extracted from ores.

Native state   An uncombined or free state in nature; used in reference to occurrence of elements.

Ore   A natural deposit containing a mineral of an element to be extracted.

Refining   Purifying of a substance.

Roasting   Heating a compound below its melting point in the presence of air.

Slag   Unwanted material produced during smelting.

Smelting   Chemical reduction of a substance at high temperature in metallurgy.

Zone refining   A method of purifying a bar of metal by passing it through an induction heater; this causes impurities to move along in the melted portion.

Exercises

General Concepts

1. List the chemical and physical properties that we usually associate with metals.

2. Define the term "metallurgy." What does the study of metallurgy include?

3. What kinds of metals are most apt to occur in the uncombined (native) state in nature?

4. List the six anions (and their formulas) that are most often combined with metals in ores. Give at least one example of an ore of each kind. What anion is the most commonly encountered?

5. How does an ore differ from a mineral? Name the three general categories of procedures needed to produce pure metals from ores. Describe the purpose of each.

6. Briefly describe one method by which gangue can be separated from the desired mineral during the concentration of an ore.

7. Give the five general steps involved in extracting a metal from its ore and converting the metal to a useful form. Briefly describe the importance of each.

8. Describe the flotation method of ore pretreatment. Are any chemical changes involved?

9. What kinds of ores are roasted? What kinds of compounds are converted to oxides by roasting? What kinds are converted directly to the free metals?

10. Of the following compounds, which would you expect to require electrolysis to obtain the free metals: KCl, $Cr_2(SO_4)_3$, Fe_2O_3, Al_2O_3, Ag_2S, $MgSO_4$? Why?

11. At which electrode is the free metal produced in the electrolysis of a metal compound? Why?

*12. Write the equation that describes the electrolysis of a brine solution to form NaOH, Cl_2, and H_2. What mass of each substance will be produced in an electrolysis cell for each mole of electrons passed through the cell? Assume 100% efficiency.

13. The following equations represent reactions used in some important metallurgical processes:
   (a) $Fe_3O_4(s) + CO(g) \longrightarrow Fe(\ell) + CO_2(g)$
   (b) $MgCO_3(s) + SiO_2(s) \longrightarrow MgSiO_3(\ell) + CO_2(g)$
   (c) $Au(s) + CN^- + H_2O(\ell) + O_2(g) \longrightarrow$
   $[Au(CN)_2]^- + OH^-$
   Balance the equations. Which one(s) represent reduction to a free metal?

14. Repeat Exercise 13 for
   (a) Al_2O_3 (cryolite solution) $\xrightarrow{\text{electrolysis}}$ $Al(\ell) + O_2(g)$
   (b) $PbSO_4(s) + PbS(s) \longrightarrow Pb(\ell) + SO_2(g)$
   (c) $TaCl_5(g) + Mg(\ell) \longrightarrow Ta(s) + MgCl_2(\ell)$

15. Suggest a method of obtaining manganese from an ore containing manganese(III) oxide, Mn_2O_3. On what basis do you make the suggestion?

16. What is the purpose of utilizing the basic oxygen furnace after the blast furnace in the production of iron?

17. Describe the metallurgy of (a) copper and (b) magnesium.

18. Describe the metallurgy of (a) iron and (b) gold.

19. Briefly describe the Hall–Heroult process for the commercial preparation of aluminum.

20. Name some common minerals that contain iron. Write the chemical formula for the iron compound in each. What is the oxidation number of iron in each substance?

21. During the operation of a blast furnace, coke reacts with the oxygen in air to produce carbon monoxide, which, in turn, serves as the reducing agent for the iron ore. Assuming the formula of the iron ore to be Fe_2O_3,

calculate the mass of air needed for each ton of iron produced. Assume air to be 21% O_2 by mass and assume that the process is 91% efficient.

22. The reaction

$$FeO(s) + CO(g) \longrightarrow Fe(s) + CO_2(g)$$

takes place in the blast furnace at a temperature of 800 K. (a) Calculate ΔH^0_{800} for this reaction, using $\Delta H^0_{f, 800}$ = -268 kJ/mol for FeO, -111 kJ/mol for CO, and -394 kJ/mol for CO_2. Is this a favorable enthalpy change? (b) Calculate ΔG^0_{800} for this reaction, using $\Delta G^0_{f, 800}$ = -219 kJ/mol for FeO, -182 kJ/mol for CO, and -396 kJ/mol for CO_2. Is this a favorable free energy change? (c) Using your values of ΔH^0_{800} and ΔG^0_{800}, calculate ΔS^0_{800}.

23. What is steel? How does the hardness of iron compare with that of steel?

24. Describe and illustrate the electrolytic refining of Cu.

25. Name the undesirable gaseous product formed during the roasting of copper and other sulfide ores. Why is it undesirable?

26. The following reactions take place during the extraction of copper from copper ore:
(a) $2Cu_2S(\ell) + 3O_2(g) \longrightarrow 2Cu_2O(\ell) + 2SO_2(g)$
(b) $2Cu_2O(\ell) + Cu_2S(\ell) \longrightarrow 6Cu(\ell) + SO_2(g)$
Identify the oxidizing and reducing agents. Show that each equation is correctly balanced by demonstrating that the increase and decrease in oxidation numbers are equal.

*27. Assuming complete recovery of metal, which of the following ores would yield the greater quantity of copper on a mass basis? (a) an ore containing 3.80 mass % azurite, $Cu(OH)_2 \cdot 2CuCO_3$, or (b) an ore containing 4.85 mass % chalcopyrite, $CuFeS_2$.

*28. What mass of copper could be electroplated from a solution of $CuSO_4$, using an electrical current of 3.00 A flowing for 5.00 hr? (Assume 100% efficiency.)

*29. (a) Calculate the weight, in pounds, of sulfur dioxide produced in the roasting of 1 ton of chalcocite ore containing 9.2% Cu_2S, 0.77% Ag_2S, and no other source of sulfur. (b) What weight of sulfuric acid can be prepared from the SO_2 generated, assuming 93% of it can be recovered from stack gases, and 87% of that recovered can be converted to sulfuric acid? (c) How many pounds of pure copper can be obtained, assuming 75% efficient extraction and purification? (d) How many pounds of silver can be produced, assuming 82% of it can be extracted and purified?

*30. Forty-five pounds of Al_2O_3, obtained from bauxite, is mixed with cryolite and electrolyzed. How long would a 0.800-A current have to be passed to convert all the Al^{3+} (from Al_2O_3) to aluminum metal? What volume of oxygen, collected at 835 torr and 155°C, would be produced in the same period of time?

*31. Calculate the percentage of iron in hematite ore containing 58.6% Fe_2O_3 by mass. How many pounds of iron would be contained in 1 ton of the ore?

*32. Find the standard molar enthalpies of formation of Al_2O_3, Fe_2O_3, and HgS in Appendix K. Are the values in line with what might be predicted, in view of the methods by which the metal ions are reduced in extractive metallurgy?

*33. Using data from Appendix K, calculate ΔG^0_{298} for the following reactions:
(a) $Al_2O_3(s) \longrightarrow 2Al(s) + \frac{3}{2}O_2(g)$
(b) $Fe_2O_3(s) \longrightarrow 2Fe(s) + \frac{3}{2}O_2(g)$
(c) $HgS(s) \longrightarrow Hg(\ell) + S(s)$
Are any of the reactions spontaneous? Are the ΔG^0_{298} values in line with what would be predicted based on the relative activities of the metal ions involved? Do increases in temperature favor these reactions?

*34. Seawater contains 0.13 mass % Mg^{2+}. What mass of seawater would have to be processed to yield 1.0 ton of the metal if the recovery process were 77% efficient?

23 The Representative Metals

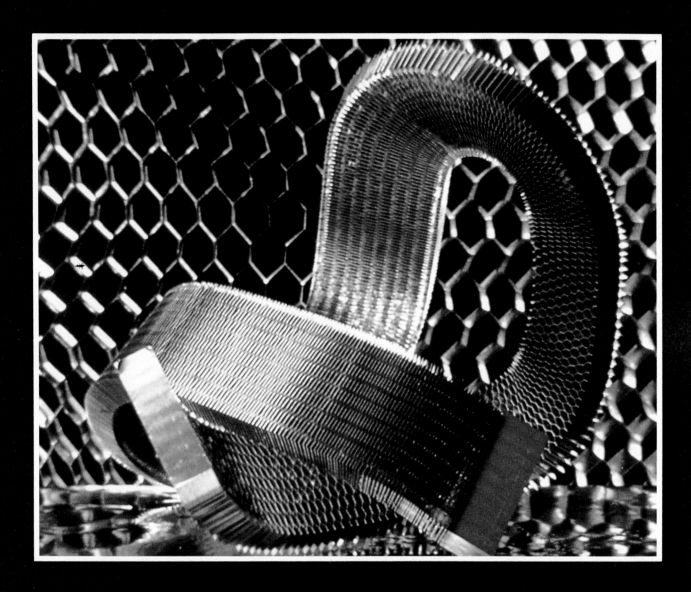

This aluminum honeycomb material is made by bonding aluminum foil sheets to form hexagonal cells. It is used to make sandwich construction panels that have a very high strength-to-weight ratio.

Outline

Objectives

As you study this chapter you should learn

☐ About the properties and occurrence of the Group IA metals
☐ Some important reactions of the Group IA metals
☐ Some important uses of the Group IA metals and their compounds
☐ About the properties and occurrence of the Group IIA metals
☐ Some important reactions of the Group IIA metals

☐ Some important uses of the Group IIA metals and their compounds
☐ About the post-transition metals
☐ Some important periodic trends in the properties of Group IIIA and IVA metals and some of their compounds
☐ About aluminum and a few of its important compounds
☐ About tin, lead, and bismuth and a few of their important compounds

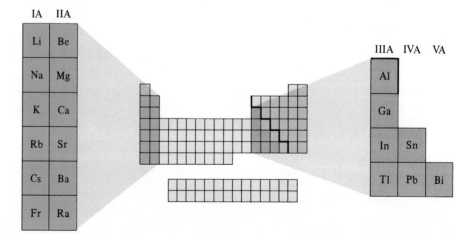

n this chapter, we shall discuss the **representative metals**. The representative elements are those in the A groups of the periodic table. They have valence electrons in their outermost s and p atomic orbitals. Metallic character increases from top to bottom within groups and from right to left within periods. All the elements in Groups IA (except H) and IIA are metals. The heavier members of Groups IIIA, IVA, and VA are called **post-transition metals**.

Hydrogen is included in IA in the periodic table, but it is *not* a metal.

The Alkali Metals (Group IA)

23-1 Group IA Metals: Properties and Occurrence

The alkali metals are not found free in nature, because they are so easily oxidized. They are most economically produced by electrolysis of their molten salts. Sodium (2.6% abundance by mass) and potassium (2.4% abundance) are very common in the earth's crust. The other IA metals are quite rare. Francium consists only of short-lived radioactive isotopes formed by alpha-particle emission from actinium (Section 30-4). Both potassium and cesium also have natural radioisotopes. Potassium-40 is important in the potassium–argon radioactive decay method of dating ancient objects (Section 30-11). The properties of the alkali metals vary regularly as the group is descended (Table 23-1).

See the discussion of electrolysis of sodium chloride in Section 21-3.

The free metals, except lithium, are soft, silvery corrosive metals that can be cut with a knife; lithium is harder. Cesium is slightly golden and melts in the hand (wrapped in plastic because it is so corrosive). The relatively low melting and boiling points of the alkali metals result from their fairly weak bonding forces. Each atom can furnish only one electron for metallic bonding (Section 13-17). Because their outer electrons are so loosely held, the metals are excellent electrical and thermal conductors. They ionize when irradiated with low-energy light (the photoelectric effect). These effects become more pronounced with increasing atomic size. Cesium is used in photoelectric cells.

Alkali metals are good conductors of electricity.

Table 23-1
Properties of the Group IA Metals

Property	Li	Na	K	Rb	Cs	Fr
Outer electrons	$2s^1$	$3s^1$	$4s^1$	$5s^1$	$6s^1$	$7s^1$
Melting point (°C)	186	97.8	63.6	38.9	28.5	27
Boiling point (°C)	1347	904	774	688	678	677
Density (g/cm³)	0.534	0.971	0.862	1.53	1.87	—
Atomic radius (Å)	1.52	1.86	2.31	2.44	2.62	—
Ionic radius, M^+ (Å)	0.60	0.95	1.33	1.48	1.69	—
Electronegativity	1.0	1.0	0.9	0.9	0.8	0.8
E^0 (volts): $M^+(aq) + e^- \longrightarrow M(s)$	−3.05	−2.71	−2.93	−2.93	−2.92	—
Ionization energies (kJ/mol)						
$\quad M(g) \longrightarrow M^+(g) + e^-$	520	496	419	403	376	—
$\quad M^+(g) \longrightarrow M^{2+}(g) + e^-$	7298	4562	3051	2632	2420	—
$\Delta H^0_{hydration}$ (kJ/mol): $M^+(g) + xH_2O \longrightarrow M^+(aq)$	−544	−435	−351	−293	−264	—

The ionization energies of the IA metals show that the single electron in the outer shell is very easily removed. In all alkali metal compounds the metals exhibit the +1 oxidation state. Virtually all are ionic. The extremely high second ionization energies show that removal of an electron from a filled shell is impossible by chemical means.

We might expect the standard reduction potentials of the metal ions to reflect the same trends as the first ionization energies. However, the magnitude of the standard reduction potential of Li, −3.05 volts, is unexpectedly large. The first ionization energy is the amount of energy absorbed when a mole of *gaseous* atoms ionize. The standard reduction potential, E^0, indicates the ease with which a mole of *aqueous* ions is reduced to the metal (Section 21-15). Thus, hydration energies must also be considered (Section 14-2). Because the Li^+ ion is so small, its charge density (ratio of charge to size) is very high. Therefore, it exerts a stronger attraction for polar H_2O molecules than do the other IA ions. These H_2O molecules must be stripped off during the reduction process. Thus, $\Delta H^0_{hydration}$ for the Li^+ ion and E^0 for the Li^+/Li couple are very negative (Table 23-1).

Polarization of an anion refers to distortion of its electron cloud. The ability of a cation to polarize an anion increases with increasing charge density (ratio of charge/size) of the cation.

The high charge density of Li^+ ion accounts for its ability to polarize large anions. This gives a higher degree of covalent character in Li compounds than in other corresponding alkali metal compounds. For example, LiCl is soluble in ethyl alcohol, a less polar solvent than water; NaCl is not. Salts of the alkali metals with small anions are very soluble in water, but salts with large and complex anions (such as silicates and aluminosilicates) are not very soluble.

23-2 Reactions of the Group IA Metals

Many of the reactions of the alkali metals are summarized in Table 23-2. All are characterized by the loss of one electron per metal atom. These metals are very strong reducing agents. Reactions of the alkali metals with H_2 and O_2 were discussed in Sections 6-7 and 6-8, parts 2; reactions with the halogens in Section 7-2; and reactions with water in Section 4-6, part 2.

The high reactivities of the alkali metals are illustrated by their vigorous reactions with water. Lithium reacts readily; sodium reacts so vigorously that it may ignite; and potassium, rubidium, and cesium burst into flames

Table 23-2
Some Reactions of the Group IA Metals

Reaction	Remarks
$4M + O_2 \longrightarrow 2M_2O$	Limited O_2
$4Li + O_2 \longrightarrow 2Li_2O$	Excess O_2 (lithium oxide)
$2Na + O_2 \longrightarrow Na_2O_2$	(sodium peroxide)
$M + O_2 \longrightarrow MO_2$	M = K, Rb, Cs; excess O_2 (superoxides)
$2M + H_2 \longrightarrow 2MH$	Molten metals
$6Li + N_2 \longrightarrow 2Li_3N$	At high temperature
$2M + X_2 \longrightarrow 2MX$	X = halogen (Group VIIA)
$2M + S \longrightarrow M_2S$	Also with Se, Te of Group VIA
$12M + P_4 \longrightarrow 4M_3P$	Also with As, Sb of Group VA
$2M + 2H_2O \longrightarrow 2MOH + H_2$	K, Rb, and Cs react explosively
$2M + 2NH_3 \longrightarrow 2MNH_2 + H_2$	With $NH_3(\ell)$ in presence of catalyst; with $NH_3(g)$ at high temperature (solutions also contain M^+ + solvated e^-)

when dropped into water. The large amounts of heat evolved provide the activation energy to ignite the evolved hydrogen. The elements also react with water vapor in the air or with moisture from the skin.

$$2K + 2H_2O \longrightarrow 2[K^+ + OH^-] + H_2 \qquad \Delta H^0 = -390.8 \text{ kJ/mol rxn}$$

Alkali metals are stored under anhydrous nonpolar liquids such as mineral oil.

As is often true for elements of the second period, Li differs in many ways from the other members of its family. Its ionic charge density and electronegativity are close to those of Mg, so Li compounds resemble those of Mg in some ways. This illustrates the **diagonal similarities** that exist between elements in successive groups near the top of the periodic table.

IA	IIA	IIIA	IVA
Li	Be	B	C
Na	Mg	Al	Si

Lithium is the only IA metal that combines with N_2 to form a nitride, Li_3N. Magnesium readily forms magnesium nitride, Mg_3N_2. Both metals readily combine with carbon to form carbides, whereas the other alkali metals do not react readily with carbon. The solubilities of Li compounds are closer to those of Mg compounds than to those of other IA compounds. The fluorides, phosphates, and carbonates of both Li and Mg are only slightly soluble, but their chlorides, bromides, and iodides are very soluble. Both Li and Mg form normal oxides, Li_2O and MgO, when burned in air at 1 atmosphere pressure. The other alkali metals form peroxides or superoxides.

The IA metal oxides are basic. They react with water to form strong soluble bases.

$$Na_2O(s) + H_2O \longrightarrow 2[Na^+ + OH^-]$$

$$K_2O(s) + H_2O \longrightarrow 2[K^+ + OH^-]$$

The IA cations are derived from strong soluble bases, so they do not hydrolyze.

Sodium reacts vigorously with water.

$$2Na(s) + 2H_2O \longrightarrow$$
$$2[Na^+(aq) + OH(aq)] + H_2(g)$$

Phenolphthalein was added to the water. As NaOH forms, the solution turns pink.

23-3 Uses of Group IA Metals and Their Compounds

Sodium, Na

Sodium is by far the most widely used alkali metal because it is so abundant. Its salts are essential for life. The metal itself is used as a reducing agent in the manufacture of drugs and dyes and in the metallurgy of metals such as titanium and zirconium.

$$TiCl_4(g) + 4Na(\ell) \xrightarrow{\Delta} 4NaCl(s) + Ti(s)$$

Sodium (in the form of sodium–lead alloys) was used in the production of leaded gasoline. The metal is proposed as a heat-transfer liquid in the design of the breeder reactors of newer nuclear power plants. Highway lamps often incorporate Na arcs, which produce a bright yellow glow. A few examples of the uses of sodium compounds are: NaOH, called caustic soda, lye, or soda lye (used for production of rayon, cleansers, textiles, soap, paper, petroleum products); Na_2CO_3, called soda or soda ash, and $Na_2CO_3 \cdot 10H_2O$, called washing soda (also used as a substitute for NaOH when a weaker

The yellowish glow of some highway lamps is due to a sodium arc. Mercury lamps give a bluish glow.

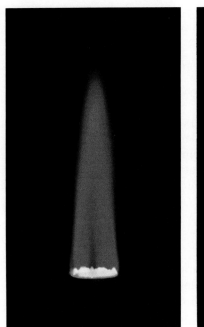
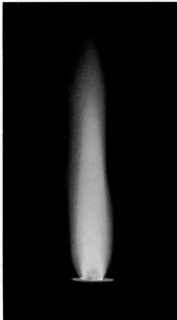
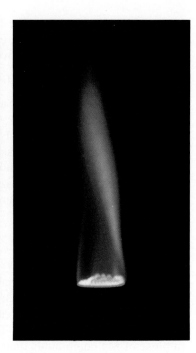

Spacings of energy levels are different for different alkali metals. The salts of the alkali metals impart characteristic colors to flames: lithium (red), sodium (yellow), and potassium (violet).

base is acceptable); $NaHCO_3$, called baking soda or bicarbonate of soda (used for baking and other household uses); $NaCl$ (used as table salt and as the source of all other compounds of Na and Cl); $NaNO_3$, called Chile saltpeter (a nitrogen fertilizer); Na_2SO_4, called salt cake, a by-product of HCl manufacture (used for production of brown wrapping paper and corrugated boxes); and NaH (used for synthesis of $NaBH_4$, which is used to recover silver and mercury from waste water).

Lithium, Li

Metallic lithium has the highest heat capacity of any element. It is used as a heat transfer medium in experimental nuclear reactors. Lithium is used as a reducing agent in the synthesis of many organic compounds. Extremely lightweight lithium–aluminum alloys are being produced for use in aircraft construction. Lithium compounds are used in some lightweight dry cells and storage batteries because they have very long lives, even in extreme temperatures. They are also used in some glasses and ceramics. LiCl and LiBr are very hygroscopic and are used in industrial drying processes and air-conditioning. Lithium compounds are used for the treatment of some types of mental disorders (mainly manic depression).

The highly corrosive nature of both lithium and sodium is a major drawback to applications of the pure metals.

Potassium, K

Like salts of Na (and probably Li), those of potassium are essential for life. KNO_3, commonly known as niter or saltpeter, is used as a potassium and nitrogen fertilizer. Most other major industrial uses for K can be satisfied with the more abundant and cheaper Na.

There are very few practical uses for the rare metals rubidium, cesium, and francium. Cesium is used in some photoelectric cells (Section 5-10).

Table 23-3
Properties of the Group IIA Metals

Property	Be	Mg	Ca	Sr	Ba	Ra
Outer electrons	$2s^2$	$3s^2$	$4s^2$	$5s^2$	$6s^2$	$7s^2$
Melting point (°C)	1283	649	839	770	725	700
Boiling point (°C)	2484	1105	1484	1384	1640	1140
Density (g/cm³)	1.85	1.74	1.55	2.60	3.51	5
Atomic radius (Å)	1.11	1.60	1.97	2.15	2.17	2.20
Ionic radius, M^{2+} (Å)	0.31	0.65	0.99	1.13	1.35	—
Electronegativity	1.5	1.2	1.0	1.0	1.0	1.0
E^0 (volts): $M^{2+}(aq) + 2e^- \longrightarrow 2M(s)$	−1.85	−2.37	−2.87	−2.89	−2.90	−2.92
Ionization energies (kJ/mol)						
$\quad M(g) \longrightarrow M^+(g) + e^-$	899	738	590	549	503	509
$\quad M^+(g) \longrightarrow M^{2+}(g) + e^-$	1757	1451	1145	1064	965	(979)
$\Delta H^0_{\text{hydration}}$ (kJ/mol): $M^{2+}(g) \longrightarrow M^{2+}(aq)$	—	−1925	−1650	−1485	−1276	—

The Alkaline Earth Metals (Group IIA)

23-4 Group IIA Metals: Properties and Occurrence

The alkaline earth metals are all silvery white, malleable, ductile, and somewhat harder than their neighbors in Group IA. Activity increases from top to bottom within the group, with Ca, Sr, and Ba being considered quite active. Each has two electrons in its highest occupied energy level. Both electrons are lost in ionic compound formation, though not as easily as the outer electron of an alkali metal. Compare the ionization energies in Tables 23-1 and 23-3. While most IIA compounds are ionic, those of Be exhibit a great deal of covalent character. This is due to the extremely high charge density of Be^{2+}. Compounds of beryllium therefore resemble those of aluminum in Group IIIA. The IIA elements exhibit the +2 oxidation state in all their compounds. The tendency to form 2+ ions increases from Be to Ra.

The alkaline earth metals show a wider range of chemical properties than the alkali metals. The IIA metals are not quite as reactive as the IA metals, but they are much too reactive to occur free in nature. They are obtained by electrolysis of their molten chlorides. To increase the electrical conductivity of molten anhydrous $BeCl_2$, which is covalent and polymeric, small amounts of NaCl are added to the melt.

Calcium and magnesium are quite abundant in the earth's crust, especially as carbonates and sulfates. Beryllium, strontium, and barium are less abundant. All known radium isotopes are radioactive and are extremely rare.

23-5 Reactions of the Group IIA Metals

Table 23-4 summarizes some reactions of the alkaline earth metals, which, except for stoichiometry, are quite similar to the corresponding reactions of the alkali metals. Reactions with hydrogen and oxygen were discussed in Sections 6-7 and 6-8, parts 2.

Except for Be, all the alkaline earth metals are oxidized to oxides in air. The IIA oxides (except BeO) are basic and react with water to give hy-

In Section 8-5 we found that gaseous $BeCl_2$ is linear. However, the Be atoms in $BeCl_2$ molecules act as Lewis acids. In the solid state, they accept shares in electron pairs from Cl atoms in other molecules, to form polymers:

Limestone is mainly calcium carbonate.

Table 23-4
Some Reactions of the Group IIA Metals

Reaction	Remarks
$2M + O_2 \longrightarrow 2MO$	Very exothermic (except Be)
$Ba + O_2 \longrightarrow BaO_2$	Almost exclusively
$M + H_2 \longrightarrow MH_2$	M = Ca, Sr, Ba at high temperatures
$3M + N_2 \longrightarrow M_3N_2$	At high temperatures
$6M + P_4 \longrightarrow 2M_3P_2$	At high temperatures
$M + X_2 \longrightarrow MX_2$	X = halogen (Group VIIA)
$M + S \longrightarrow MS$	Also with Se, Te of Group VIA
$M + 2H_2O \longrightarrow M(OH)_2 + H_2$	M = Ca, Sr, Ba at room temperature; Mg gives MgO at high temperatures
$M + 2NH_3 \longrightarrow M(NH_2)_2 + H_2$	M = Ca, Sr, Ba in $NH_3(\ell)$ in presence of catalyst; $NH_3(g)$ with heat
$3M + 2NH_3(g) \longrightarrow M_3N_2 + 3H_2$	At high temperatures
$Be + 2OH^- + 2H_2O \longrightarrow Be(OH)_4^{2-} + H_2$	Only with Be

Amphoterism is the ability of a substance to react with both acids and bases.

droxides. Beryllium hydroxide, $Be(OH)_2$, is quite insoluble in water and is amphoteric. Magnesium hydroxide, $Mg(OH)_2$, is only slightly soluble in water. The hydroxides of Ca, Sr, and Ba are strong soluble bases.

Beryllium is at the top of Group IIA. Its oxide is amphoteric, whereas oxides of the heavier members are basic. Metallic character increases from top to bottom within a group and from right to left across a period. This results in increasing basicity and decreasing acidity of the oxides in the same directions, as shown in the following table.

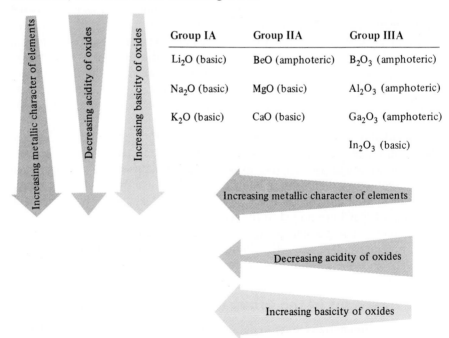

	Group IA	Group IIA	Group IIIA
	Li_2O (basic)	BeO (amphoteric)	B_2O_3 (amphoteric)
	Na_2O (basic)	MgO (basic)	Al_2O_3 (amphoteric)
	K_2O (basic)	CaO (basic)	Ga_2O_3 (amphoteric)
			In_2O_3 (basic)

Calcium, Sr, and Ba react with water at 25°C to form hydroxides and H_2 (see Table 23-4). Magnesium reacts with steam to produce MgO and H_2. Beryllium does not react with pure water even at red heat. It dissolves in alkali as it reacts to form the complex ion, $[Be(OH)_4]^{2-}$, and H_2.

Most simple compounds of Be are covalent, but many are soluble in water. These give solutions that conduct electricity, because of the high hydration energy of Be^{2+}. The tetrahydrated $[Be(OH_2)_4]^{2+}$ ion is formed.

$$BeCl_2(s) + 4H_2O(\ell) \longrightarrow [Be(OH_2)_4]^{2+}(aq) + 2Cl^-(aq) + heat$$

Group IIA compounds are generally less soluble in water than corresponding IA compounds, but many are quite soluble. All form hydrated ions. Because of the strong attraction of Be^{2+} for water's electrons, the Be—O bonds in $[Be(OH_2)_4]^{2+}$ are relatively strong. Thus, the O—H bonds are relatively weak (Section 19-5), and so the ion hydrolyzes to produce acidic solutions.

$$[Be(OH_2)_4]^{2+}(aq) + H_2O(\ell) \longrightarrow$$

$$[Be(OH_2)_3(OH)]^+(aq) + H_3O^+(aq) \qquad K_a = 1.0 \times 10^{-5}$$

The other IIA ions are hexahydrated in most solid compounds. The larger hydrated magnesium ions hydrolyze only very slightly.

$$[Mg(OH_2)_6]^{2+}(aq) + H_2O(\ell) \longrightarrow$$

$$[Mg(OH_2)_5(OH)]^+(aq) + H_3O^+(aq) \qquad K_a = 3.0 \times 10^{-12}$$

Hydrated Ca^{2+}, Sr^{2+}, and Ba^{2+} ions are cations of strong soluble bases and do not hydrolyze.

23-6 Uses of Group IIA Metals and Their Compounds

Calcium, Ca

Calcium and its compounds are widely used commercially. The element is used as a reducing agent in the metallurgy of uranium, thorium, and other metals. It is also used as a scavenger to remove dissolved impurities such as oxygen, sulfur, and carbon in molten metals, and to remove residual gases in vacuum tubes. It is a component of many alloys.

Heating limestone produces *quicklime*, CaO, which can then be treated with water to form *slaked lime*, $Ca(OH)_2$, a cheap base for which industry finds many uses. When slaked lime is mixed with sand and exposed to the CO_2 of the air, it hardens to form mortar and, with a binder, lime plaster for coating walls and ceilings. Careful heating of gypsum, $CaSO_4 \cdot 2H_2O$, produces plaster of Paris, $2CaSO_4 \cdot H_2O$.

Calcium carbonate and calcium phosphate occur in seashells and animal bones.

Magnesium, Mg

Metallic magnesium burns in air with such a brilliant white light that it is used in photographic flash accessories, fireworks, and incendiary bombs. It is very lightweight and is currently used in many alloys for building materials. Like aluminum, it forms an impervious coating of oxide that protects it from further oxidation. Given its inexhaustible supply in the oceans, it is likely that many more structural uses will be found for it as the reserves of iron ores dwindle.

The metal is used as a reagent for many important organic syntheses. When magnesite, $MgCO_3$, is thermally decomposed, *magnesia*, MgO, is produced. Magnesia, an excellent heat insulator, is used in making furnaces, ovens, and crucibles. It can be converted to $Mg(OH)_2$ by reaction with

A laboratory X-ray tube (left) and a close-up view of one of its windows (right). The windows are made of beryllium metal.

aqueous ammonium salts (it does not slake readily). A milky-white aqueous suspension of finely divided $Mg(OH)_2$, called milk of magnesia, is used as a stomach antacid and as a laxative. Anhydrous $MgSO_4$ and $Mg(ClO_4)_2$ are used as drying agents.

Beryllium, Be

Because of its rarity, beryllium has only a few practical uses. It occurs mainly as beryl, $Be_3Al_2Si_6O_{18}$, a gemstone which, with appropriate impurities, may be aquamarine (blue) or emerald (green). The metal itself wasn't readily available for industrial use until 1957. Its very low density and high strength are the basis for its primary use as a structural material. It is also alloyed with copper for use in electrical contacts, springs, and nonsparking tools. Because it is transparent to X-rays, "windows" for X-ray tubes are constructed of beryllium. Beryllium compounds are quite toxic.

Strontium, Sr

Strontium salts are used in fireworks and flares, which show the characteristic red glow of strontium in a flame. Strontium chloride is used in some toothpastes for persons with sensitive teeth. The metal itself has no practical uses.

Barium, Ba

Barium is a constituent of alloys that are used for spark plugs because of the ease with which it emits electrons when heated. It is used as a degassing agent for vacuum tubes. A slurry of finely divided barium sulfate from barite, $BaSO_4$, is used to coat the gastrointestinal tract in preparation for X-ray photographs because it absorbs X-rays so well. It is so insoluble that it is not poisonous; all soluble barium salts are very toxic. A combination of ZnS and $BaSO_4$ (both white) forms a very insoluble, bright-white paint pigment called lithopone.

Table 23-5
Properties of the Group IIIA Elements

Property	B	Al	Ga	In	Tl
Outer electrons	$2s^2 2p^1$	$3s^2 3p^1$	$4s^2 4p^1$	$5s^2 5p^1$	$6s^2 6p^1$
Physical state (25°C, 1 atm)	solid	solid	solid	solid	solid
Melting point (°C)	2300	660	29.8	156.6	303.5
Boiling point (°C)	2550	2367	2403	2080	1457
Density (g/cm³)	2.34	2.70	5.91	7.31	11.85
Atomic radius (Å)	0.88	1.43	1.22	1.62	1.71
Ionic radius, M^{3+} (Å)	(0.20)*	0.50	0.62	0.81	0.95
Electronegativity	2.0	1.5	1.7	1.6	1.6
E^0 (volts): $M^{3+}(aq) + 3e^- \longrightarrow M(s)$	(−0.90)*	−1.66	−0.53	−0.34	0.916
Oxidation states	−3 to +3	+3	+1, +3	+1, +3	+1, +3
Ionization energies (kJ/mol)					
$\quad M(g) \longrightarrow M^+(g) + e^-$	801	578	576	556	586
$\quad M^+(g) \longrightarrow M^{2+}(g) + e^-$	2427	1817	1971	1813	1961
$\quad M^{2+}(g) \longrightarrow M^{3+}(g) + e^-$	3660	2745	2952	2692	2867
$\Delta H^0_{hydration}$ (kJ/mol): $M^{3+}(g) + xH_2O \longrightarrow M^{3+}(aq)$	—	−4750	−4703	−4159	−4117

*For the *covalent* +3 oxidation state.

The Post-Transition Metals

The metals below the stepwise division of the periodic table in Groups IIIA through VA are the **post-transition metals**. These include aluminum, gallium, indium, and thallium from Group IIIA; tin and lead from Group IVA; and bismuth from Group VA. Strictly speaking, the elements *along* the stepwise division are metalloids. However, we shall discuss aluminum here because so many of its properties are characteristic of metals. Aluminum is the only post-transition metal that is considered very reactive.

There are no true metals in Groups VIA, VIIA, and 0.

The properties of Bi (VA) are listed with those of the other Group VA elements in Table 26-1.

23-7 Periodic Trends—Groups IIIA and IVA

The properties of the elements in Groups IIIA (Table 23-5) and IVA (Table 23-6) vary less regularly down the groups than those of the IA and IIA metals. The Group IIIA elements are all solids. Boron, at the top of the

Table 23-6
Properties of the Group IVA Elements

Property	C	Si	Ge	Sn	Pb
Outer electrons	$2s^2 2p^2$	$3s^2 3p^2$	$4s^2 4p^2$	$5s^2 5p^2$	$6s^2 6p^2$
Physical state (25°C, 1 atm)	solid	solid	solid	solid	solid
Melting point (°C)	3570	1410	937	232	328
Boiling point (°C)	sublimes	2355	2830	2270	1740
Density (g/cm³)	2.25 (graphite)	2.33	5.35	7.30 (white tin)	11.35
Atomic radius (Å)	0.77	1.17	1.22	1.40	1.75
Ionic radius, M^{2+} (Å)	—	—	0.73	0.93	1.21
Electronegativity	2.5	1.8	1.9	1.8	1.7
E^0 (volts): $M^{2+}(aq) + 2e^- \longrightarrow M(s)$	(+0.39)*	(+0.10)*	−0.3	−0.14	−0.126
Common oxidation states	±2, ±4	±4	+2, +4	+2, +4	+2, +4

*For the *covalent* +2 oxidation state.

Gallium metal melts below body temperature.

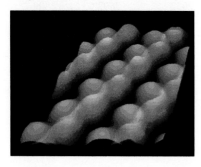

An image of the surface of gallium arsenide, GaAs, produced by a scanning tunneling microscope. This substance is used in microwave generators, lasers, light-emitting diodes, and many other electronic devices.

Compare the radii and densities of these elements with those of the IA and IIA metals in the same rows.

As is generally true, for each pair of compounds, covalent character is greater for the higher (more polarizing) oxidation state of the metal.

group, is a nonmetal. Its melting point, 2300°C, is very high because it crystallizes as a covalent solid. The other elements, aluminum through thallium, form metallic crystals and have considerably lower melting points.

Gallium, Ga

Gallium is unusual in that it melts when held in the hand. It has the largest liquid-state temperature range of any element (29.8°C to 2403°C). It is used in transistors and high-temperature thermometers. Gallium-67 was one of the first artificially produced isotopes to be used in medicine. It concentrates in inflamed areas and in certain melanomas.

Indium, In

Indium is a soft, bluish metal that is used in some alloys with silver and lead to make good heat conductors. Most indium is used in electronics.

Thallium, Tl

Thallium is a soft, heavy metal that resembles lead. It is quite toxic and has no important practical uses.

The elements at the top of Group IVA—C, Si, and Ge—have relatively high melting points (especially C) because they form covalent crystals. The metals, tin and lead, have lower melting points.

The atomic radii do not increase regularly as Groups IIIA and IVA are descended. The atomic radius of Ga, 1.22 Å, is *less* than that of Al, 1.43 Å, which is directly above Ga. The transition elements are located between calcium (IIA) and gallium (IIIA), strontium (IIA) and indium (IIIA), and barium (IIA) and thallium (IIIA). The increase in nuclear charge that accompanies filling of the $(n - 1)d$ subshell results in the contraction of the size of the atoms. This contraction is caused by the stronger attraction of the more highly charged nuclei for the outer electrons. This causes the radii of Ga, In, and Tl to be smaller than would be predicted from the radii of B and Al. In Group IVA, the atomic radii of Ge, Sn, and Pb are contracted for the same reasons. Atomic radii (Figure 6-1), and therefore atomic volumes, strongly influence other properties. For example, the densities of Ga, In, and Tl from Group IIIA and Ge, Sn, and Pb from IVA are much higher than those of the elements above them, due to their smaller atomic radii and atomic volumes.

The smaller atomic volumes of the heavier post-transition elements also result in their valence electrons being held quite strongly. This reduces the reactivities and the metallic character of these elements. Thus, several of the oxides and hydroxides of these metals are amphoteric.

The Group IIIA elements have the ns^2np^1 outer electronic configuration. Aluminum shows only the +3 oxidation state in its compounds. The heavier metals (Ga, In, Tl) can lose or share either the single p valence electron or the p and both s electrons to exhibit the +1 or +3 oxidation state, respectively. The IVA elements have the ns^2np^2 configuration. The IVA metals can lose or share the two p electrons to exhibit the +2 oxidation state. They may also share all four valence electrons to produce the +4 oxidation state. In general, then, the post-transition metals can exhibit oxidation states of $(g - 2)+$ and $g+$ where g = periodic group number. As examples, TlCl and

$TlCl_3$ both exist, as do $SnCl_2$ and $SnCl_4$. The stability of the lower state increases as the groups are descended. This is called the **inert *s*-pair effect** because the two *s* electrons remain nonionized, or unshared, for the $(g - 2)+$ oxidation state. To illustrate, $AlCl_3$ exists but not $AlCl$; $TlCl_3$ is less stable than $TlCl$. Likewise, $GeCl_4$ is more stable than $GeCl_2$, but $PbCl_4$ is less stable than $PbCl_2$. In aqueous solution, the $+3$ compounds of the IIIA metals are stabilized (relative to $+1$) by the very high heats of hydration of the $3+$ ions.

Bi, from Group VA, exhibits the +3 and +5 oxidation states, but the +5 state is quite rare.

Ga, In, and Tl are all quite rare.

23-8 Aluminum

Aluminum is the most reactive of the post-transition metals. It is the most abundant metal in the earth's crust (7.5%) and the third most abundant element. Aluminum is inexpensive compared to most other metals. It is soft and can be readily extruded into wires or rolled, pressed, or cast into shapes.

Because of its relatively low density, aluminum is often used as a lightweight structural metal. It is often alloyed with Mg and some Cu and Si to increase its strength. Many buildings are sheathed in aluminum, which resists corrosion by forming an oxide coating.

Pure aluminum conducts about two thirds as much electrical current per unit volume as copper, but it is only one third as dense. (The density of Al is 2.70 g/cm^3; that of Cu is 8.92 g/cm^3.) As a result, a mass of aluminum can conduct twice as much current as the same mass of copper. Aluminum is now used in electrical transmission lines and has been used in wiring in homes. However, the latter use has been implicated as a fire hazard due to the heat that can be generated during high current flow at the junction of the aluminum wire and fixtures of other metals.

Aluminum is a strong reducing agent:

$$Al^{3+}(aq) + 3e^- \longrightarrow Al(s) \qquad E^0 = -1.66 \text{ V}$$

Aluminum is quite reactive, but a thin, transparent film of Al_2O_3 forms when Al comes into contact with air. This protects it from further oxidation. For this reason it is even passive toward nitric acid, HNO_3, a strong oxidizing agent. When the oxide coating is sanded off, Al reacts vigorously with HNO_3.

$$Al(s) + 4HNO_3(aq) \longrightarrow Al(NO_3)_3(aq) + NO(g) + 2H_2O(\ell)$$

The very negative enthalpy of formation of aluminum oxide makes Al a very strong reducing agent for other metal oxides. The **thermite reaction** is a spectacular example (Figure 23-1). It generates enough heat to produce molten iron for welding steel.

$$2Al(s) + Fe_2O_3(s) \longrightarrow 2Fe(s) + Al_2O_3(s) \qquad \Delta H^0 = -852 \text{ kJ/mol}$$

Anhydrous Al_2O_3 occurs naturally as the extremely hard, high-melting mineral *corundum*, which has a network structure. It is colorless when pure, but becomes colored when transition metal ions replace a few Al^{3+} ions in the crystal. *Sapphire* is usually blue and contains some iron and titanium. *Ruby* is red due to the presence of small amounts of chromium.

Aluminum forms a series of double salts known as *alums*. Common **alums** are hydrated sulfates of the general formula $M^+M^{3+}(SO_4)_2 \cdot 12H_2O$. The most common alum is $KAl(SO_4)_2 \cdot 12H_2O$. The cations Li^+, Na^+, and NH_4^+ can substitute for K^+, and Cr^{3+} and Fe^{3+} can substitute for Al^{3+}, to form

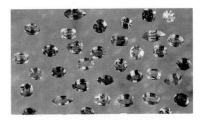

Small amounts of different transition metal ions give different colors to sapphire, which is mostly aluminum oxide.

(a)

(b)

(c)

Figure 23-1
The thermite reaction. A mixture of Fe_2O_3 and aluminum powder was placed in a clay pot with a piece of magnesium ribbon as a fuse. (a) The reaction was initiated by lighting the magnesium fuse. (b) So much heat was produced by the reaction that the iron melted as it was produced. (c) The molten iron dropped out of the clay pot and burned through a sheet of iron that was placed under the pot.

Most oven cleaners contain NaOH (lye). Can you see why they should not be used to clean aluminum pots and pans?

Recall that Al^{3+}(aq) is $[Al(OH_2)_6]^{3+}$

A **dimer** is a molecule formed by the combination of two identical smaller molecules. Compare this with BCl_3 and BF_3 (Section 8-6), which are trigonal planar molecules with sp^2 hybridization at the B atom.

a large variety of alums. All the alums form the same kind of crystals. These large, octahedral crystals are easily grown in the laboratory. Alums are used in large quantities in the dye industry and for sizing in the paper industry. Sizing helps paper to retain its shape and to become water-repellent.

Like its hydroxide, metallic aluminum is amphoteric. Freshly cleaned aluminum dissolves in nonoxidizing acids to form aqueous aluminum salts and H_2.

$$2Al(s) + 6H^+(aq) \longrightarrow 2Al^{3+}(aq) + 3H_2(g)$$

In solutions of strong soluble bases, it forms aqueous aluminate salts—or more properly, tetrahydroxoaluminate salts—and hydrogen.

$$2Al(s) + 2OH^-(aq) + 6H_2O(\ell) \longrightarrow 2[Al(OH)_4]^-(aq) + 3H_2(g)$$
$$\text{aluminate ion}$$

Most anhydrous compounds of aluminum have considerable covalent character, even Al_2O_3. Anhydrous AlF_3 is the only common exception. However, most aluminum salts ionize (dissociate) in aqueous solution. They crystallize from such solutions as hydrated ionic salts because of the high hydration energy of Al^{3+} ion (Table 23-5). For example, $AlCl_3$ crystallizes from aqueous solution as $AlCl_3 \cdot 6H_2O$, which is $[Al(OH_2)_6]Cl_3$.

By contrast, anhydrous $AlCl_3$, like $BeCl_2$, is a three-dimensional polymeric solid in which Cl atoms form bridges between adjacent Al atoms by coordinate covalent bonds. In the vapor phase it exists primarily as the **dimer** of $AlCl_3$, i.e., Al_2Cl_6. In both the solid and vapor phases, each Al atom is tetrahedrally surrounded by four Cl atoms and is sp^3 hybridized. In this way, the Al atom attains an octet of electrons in its valence shell.

Al_2Cl_6

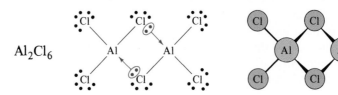

As discussed in Section 19-5, solutions of aluminum salts are acidic because the small, highly charged $Al^{3+}(aq)$ hydrolyzes extensively ($K_a = 1.2 \times 10^{-5}$).

Some antiperspirants contain Al^{3+} salts. The H_3O^+ produced in such hydrolysis reactions kills bacteria that live in perspiration. Look at the label of your deodorant to see if it contains "aluminum chlorhydrate."

23-9 Tin, Lead, and Bismuth

Tin, Sn

Tin is obtained from cassiterite, SnO_2. It is used in alloys such as solder (with lead) and bronze (with copper). Its widest use is in pewter and as tin plate, which is produced by dipping sheet iron or steel into molten tin or by electroplating the tin onto the sheet. Tin exists in three allotropic forms: gray tin, malleable tin, and brittle tin. Malleable tin is most common. It is silver-white and resistant to air oxidation.

Lead, Pb

Lead (Group IVA) is bluish-white and malleable, but is more dense than tin. It is most often obtained from galena, PbS. It is used as a protective absorber of X-rays, in battery plates, and in alloys such as solder. Like most heavy metals, lead is toxic. Its use in gasoline as tetraethyllead, $Pb(C_2H_5)_4$, an antiknock additive, is decreasing because it is now prohibited in new cars. Lead "poisons" the catalysts in catalytic converters. $Pb_3(OH)_2(CO_3)_2$ was formerly used as a pigment in white paints. Pb_3O_4 is still used in red corrosion-resistant *outdoor* paints.

Heavy metals tend to accumulate in the body, where they often inhibit essential enzyme-catalyzed reactions.

The Group IVA metals, tin and lead, exhibit the +2 and +4 oxidation states. All four oxides, SnO, SnO_2, PbO, and PbO_2, are amphoteric. PbO_2 is thermally unstable and decomposes to PbO. Tin(II) oxide, SnO, and lead(II) oxide, PbO, are more basic and have greater ionic character than tin(IV) oxide, SnO_2, and lead(IV) oxide, PbO_2. This is consistent with the general trend that oxides and hydroxides of an element in a higher oxidation state are more acidic and covalent than those of the same element in lower oxidation states.

Both Sn^{2+} and Pb^{2+} ions from *salts* hydrolyze to produce acidic solutions. Aqueous solutions of Sn^{2+} salts are acidified to inhibit hydrolysis and subsequent precipitation of various basic salts (LeChatelier's Principle).

$$[Sn(OH_2)_6]^{2+} + H_2O \rightleftharpoons [Sn(OH_2)_5(OH)]^+ + H_3O^+ \qquad K_a \approx 10^{-2}$$

$$[Pb(OH_2)_6]^{2+} + H_2O \rightleftharpoons [Pb(OH_2)_5(OH)]^+ + H_3O^+ \qquad K_a \approx 10^{-8}$$

Lead(IV) compounds are very strong oxidizing agents:

$$Pb^{4+}(aq) + 2e^- \longrightarrow Pb^{2+}(aq) \qquad E^0 = +1.8 \text{ V}$$

Bismuth, Bi

Bismuth (Group VA) is a dense metal with a yellowish tinge. It is sometimes found in the uncombined form in nature. It occurs also as bismuth(III) oxide, Bi_2O_3, and as bismuth(III) sulfide, Bi_2S_3. It is used in many low-melting alloys to take advantage of its unusual ability to expand upon freezing, thus giving sharp impressions of mold-cast objects. Low-melting Bi alloys are used in fire alarms, automatic sprinkler systems, and electrical fuses. Bismuth subcarbonate, $(BiO)_2CO_3$, bismuth subsalicylate, $(BiO)(C_6H_5O_3)$, and

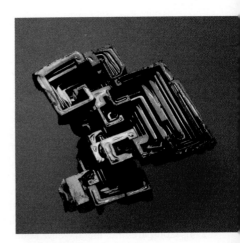

Bismuth metal.

bismuth subnitrate, $(BiO)NO_3$, are used medically in the treatment of stomach and skin disorders.

Bismuth exists primarily in the $+3$ oxidation state, and only rarely in the $+5$ oxidation state. The common oxide, Bi_2O_3, is distinctly basic. It dissolves in acids to produce bismuth(III) salts but does not dissolve in base.

$$Bi_2O_3(s) + 6H^+(aq) \longrightarrow 2Bi^{3+}(aq) + 3H_2O$$

Solutions of bismuth(III) salts hydrolyze so extensively that a precipitate of a basic salt of bismuth forms. For solutions of $Bi(NO_3)_3$, the reaction is

$$Bi^{3+}(aq) + 2H_2O + NO_3^-(aq) \longrightarrow Bi(OH)_2NO_3(s) + 2H^+(aq)$$

$$Bi(OH)_2NO_3(s) \longrightarrow BiONO_3(s) + H_2O$$

Other Bi^{3+} salts react similarly. Hydrolysis can be prevented by acidification.

Bismuth(V) in sodium bismuthate, $NaBiO_3$, is a powerful oxidizing agent. It is used to oxidize Mn^{2+} to permanganate ion, MnO_4^-, in qualitative analysis.

$$NaBiO_3(s) + 6H^+(aq) + 2e^- \longrightarrow$$
$$Bi^{3+}(aq) + Na^+(aq) + 3H_2O \qquad E^0 = {\sim}1.6 \text{ V}$$

Key Terms

Alkali metals   Group IA metals.

Alkaline earth metals   Group IIA metals.

Alums   Hydrated sulfates of the general formula $M^+M^{3+}(SO_4)_2 \cdot 12H_2O$.

Catenation   The bonding together of atoms of the same element to form chains.

Diagonal similarities   Chemical similarities of elements of Period 2 to elements of Period 3 one group to the right; especially evident toward the left of the periodic table.

Dimer   A molecule formed by the combination of two smaller identical molecules.

Double salt   A solid consisting of two co-crystallized salts.

Inert s-pair effect   The tendency of the two outermost s electrons to remain nonionized or unshared in compounds; characteristic of the post-transition metals.

Post-transition metals   Representative metals in the "p block."

Representative metals   Metals in the A groups in the periodic table; their outermost electrons are in s and p orbitals.

Exercises

General Concepts

1. How do the acidities or basicities of metal oxides vary with oxidation number of the same metal?
2. Discuss the general differences in electronic configurations of (a) representative elements, (b) d-transition metals, and (c) f-transition metals.
3. Compare the extents to which the properties (in general) of successive elements across the periodic table differ for (a) representative elements, (b) d-transition metals, and (c) f-transition metals. Explain.
4. Compare the metals and nonmetals with respect to (a) number of outer shell electrons, (b) electronegativities, (c) standard reduction potentials, and (d) ionization energies.
5. How do the physical properties of metals differ from those of nonmetals?
6. Draw electronic configurations ($\underline{\uparrow\downarrow}$ notation) for (a) Mg, (b) Mg^{2+}, (c) Na, (d) Na^+, (e) Sn, (f) Sn^{2+}, and (g) Sn^{4+}.
7. Draw electronic configurations ($\underline{\uparrow\downarrow}$ notation) for (a) K, (b) K^+, (c) Sr, (d) Sr^{2+}, (e) Al, (f) Al^{3+}, and (g) Ga^{3+}.
8. Are the elements in Groups IA and IIA found in the free state in nature? What are the primary sources for these elements? How are the metals obtained?

9. Write the general outer electron configurations for atoms of the IA and IIA metals. What oxidation state(s) would you predict for these elements? What types of bonding would you expect in most of the compounds of these elements? Why?

Alkali and Alkaline Earth Metals

10. Compare the alkali metals with the alkaline earth metals with respect to (a) atomic radii, (b) densities, (c) first ionization energies, and (d) second ionization energies. Explain the comparisons.

11. Summarize the chemical and physical properties of the alkali metals.

12. Summarize the chemical and physical properties of the alkaline earth metals.

13. Describe some uses for (a) lithium and its compounds and (b) sodium and its compounds.

14. Where do the metals of Groups IA and IIA fall with respect to H_2 in the activity series? What does this tell us about their reactivities with water and acids?

15. Write chemical equations describing the reactions of O_2 with each of the alkali and alkaline earth metals. Account for differences within each family.

16. Describe some uses for (a) calcium and its compounds and (b) magnesium and its compounds.

17. Write general equations for reactions of alkali metals with (a) hydrogen, (b) sulfur, and (c) ammonia. Represent the metal as M.

18. Write general equations for reactions of alkali metals with (a) water, (b) phosphorus, and (c) halogens. Represent the metal as M and the halogen as X.

19. Write general equations for reactions of alkaline earth metals with (a) hydrogen, (b) sulfur, and (c) ammonia. Represent the metal as M.

20. Write general equations for reactions of alkaline earth metals with (a) water, (b) phosphorus, and (c) chlorine. Represent the metal as M.

21. What is meant by the term "diagonal relationships?"

22. Give some illustrations of diagonal relationships in the periodic table, and explain each.

23. What is hydration energy? How does it vary for cations of the alkali metals?

24. How does hydration energy vary for cations of the alkaline earth metals?

25. How do the standard reduction potentials of the alkali metal cations vary? Why?

26. How do the standard reduction potentials of the alkaline earth metal cations vary? Why?

27. Why are the standard reduction potentials of lithium and beryllium out of line with respect to group trends?

*28. Calculate ΔH^0 values at 25°C for the reactions of 1 mol of each of the following metals with stoichiometric quantities of water to form metal hydroxides and hydrogen. (a) Li, (b) K, and (c) Ca. Rationalize the differences in these values.

*29. Calculate ΔH^0, ΔS^0, and ΔG^0 for the reaction of 1 mol of Na with water to form aqueous NaOH and hydrogen.

*30. Calculate ΔH^0, ΔS^0, and ΔG^0 for the reaction of 1 mol of Rb with water to form aqueous RbOH and hydrogen. Compare the spontaneity of this reaction with that in Exercise 29.

*31. (a) Calculate the solubility of $Mg(OH)_2$ in a solution that is buffered at pH 5.0. (b) Calculate the solubility of $Mg(OH)_2$ in a solution that is buffered at pH 8.0. (c) How could you make the buffer solutions in parts (a) and (b)?

*32. When CO_2 gas is bubbled through a solution of a Group IIA metal hydroxide, a slightly soluble metal carbonate precipitates.

$$M(OH)_2(aq) + CO_2(g) \longrightarrow MCO_3(s) + H_2O(\ell)$$

(a) Suppose we bubble CO_2 through a solution that is 0.10 M in Ba^{2+} ions and 0.10 M in Sr^{2+} ions. Show by calculation which will precipitate first, $BaCO_3$ or $SrCO_3$. (b) When the second salt of part (a) begins to precipitate, what fraction of the first ion is left in solution? Do you think that this fractional precipitation is a practical method for separation of Ba^{2+} and Sr^{2+} ions from solution?

*33. The standard heat of formation is -426.73 kJ/mol for NaOH(s), -469.23 kJ/mol for NaOH(aq, 6 M), and -469.10 kJ/mol for NaOH(aq, 0.1 M). (a) Calculate the heat of solution to form a 6 M NaOH solution from solid NaOH and water. Comment on your answer. (b) Calculate the heat of dilution to prepare 0.1 M NaOH from 6 M NaOH.

*34. The percentage of limestone in a sample of unknown composition can be determined by dissolving the sample in strong acid, precipitating the Ca^{2+} ions as calcium oxalate (CaC_2O_4), and titrating the $C_2O_4^{2-}$ ions by using MnO_4^- ions in an acidic solution. Write the chemical equations describing this process.

35. What type of solution is formed by dissolving the oxides of Groups IA and IIA in water? How does BeO differ from the others in this behavior? Account for differences in behavior.

Post-Transition Metals

36. Repeat Exercise 9 for the post-transition metals. Explain why some post-transition metals may have more than one positive oxidation state.

37. Describe the "inert-pair effect" associated with the post-transition metals. What is its cause?

38. What are the most likely oxidation states for post-transition metals in the various groups?

39. Write equations illustrating the amphoterism of two post-transition metal hydroxides.

40. Describe some uses of aluminum.

41. What do we mean when we say that a cation is highly polarizing? What do we mean when we say that an anion or molecule is easily polarized?

*42. Why are M^{3+} ions highly polarizing? Does high polarizing power of the metal ion lead to high covalent or ionic character of its bonds? Explain. For a given oxidation state, are the lighter or heavier members of a family more polarizing? Why?

*43. (a) Compare the strengths of Al, Ga, and In as reducing agents. (b) Compare the strengths of Al^{3+}, Ga^{3+}, and In^{3+} as oxidizing agents.

*44. (a) Compare the strengths of Sn and Pb as reducing agents. (b) Compare the strengths of Sn^{2+} and Pb^{2+} as oxidizing agents. (c) Compare the strengths of Sn^{2+} and Pb^{2+} as reducing agents. (d) Compare the strengths of Sn^{4+} and Pb^{4+} as oxidizing agents.

*45. Calculate the pH of a 0.020 M aqueous solution of lead(II) nitrate. Assume that K_a for $[Pb(OH_2)_6]^{2+}$ is 1.0×10^{-8}.

*46. Calculate the pH of a 0.020 M aqueous solution of tin(II) chloride. Assume that K_a for $[Sn(OH_2)_6]^{2+}$ is 1.0×10^{-2}.

47. Which hydrolyzes to the greater extent, Sn^{2+} or Pb^{2+}? Why? How can hydrolysis and precipitation of $Sn(OH)_2$ be prevented in solutions of tin(II) salts? Relate this to LeChatelier's Principle.

*48. A solution is 0.010 M in Pb^{2+} and 0.010 M in Sn^{2+} ions. We add a concentrated solution of KI to this solution. Assume that the KI solution is sufficiently concentrated that the volume change is negligible. Neglect the effects of hydrolysis. (a) Show by calculation which salt begins to precipitate first, PbI_2 or SnI_2. (b) When the second salt of part (a) begins to precipitate, what fraction of the first cation is left in solution? (c) Is fractional precipitation of iodide salts a practical method for separation of Pb^{2+} and Sn^{2+} ions from solution? Why?

*49. Lead is known to crystallize in the cubic system with a unit cell length of 4.95 Å. The density of lead is 11.29 g/cm^3. Is the unit cell primitive, body-centered, or face-centered?

*50. A voltaic cell consists of a tin electrode dipping into a 1 M $Sn(NO_3)_2$ solution and a lead electrode dipping into a 1 M $Pb(NO_3)_2$ solution. The half-cells are connected by a $NaNO_3$ salt bridge. Which electrode is the anode? The standard state reduction potentials are -0.136 V for Sn^{2+} and -0.126 V for Pb^{2+}. What voltage will the cell generate? How can the cell voltage be increased?

White phosphorus reacts with excess chlorine to form phosphorus pentachloride, PCl_5, a white solid.

Outline

Objectives

As you study this chapter, you should learn

☐ About sources, isolation, and uses of the noble gases
☐ About the physical and chemical properties of the noble gases
☐ About compounds of xenon
☐ About the properties of the halogens
☐ Where the halogens occur, how they are produced, and some important uses
☐ Some important reactions of the halogens

☐ About the interhalogens
☐ About the hydrogen halides
☐ How the hydrogen halides are prepared
☐ The properties of aqueous solutions of hydrogen halides (hydrohalic acids)
☐ About halides of some other elements
☐ About the oxyacids of the halogens

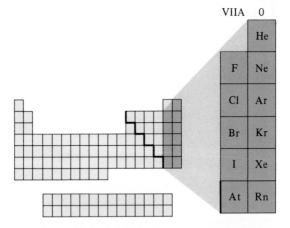

O nly about 20% of the elements are classified as nonmetals. With the exception of H, they are in the upper right-hand corner of the periodic table. In this chapter we shall consider the chemistry and properties of the noble gases (Group 0) and the halogens (Group VIIA). These two groups best illustrate group trends and individuality of elements within groups of nonmetals.

The Noble Gases (Group 0)

24-1 Occurrence, Isolation, and Uses

The noble gases are very low-boiling gases. Except for radon, they can be isolated by fractional distillation of liquefied air. Radon is collected from the radioactive disintegration of radium salts. Table 24-1 gives the percentage of each noble gas in the atmosphere.

Helium is produced in the United States from some natural gas fields. This source was discovered in 1905 by H. P. Cady and D. F. McFarland at the University of Kansas, when they were asked to analyze a nonflammable component of natural gas from a Kansas gas well. Uses of the noble gases are summarized in Table 24-2.

24-2 Physical Properties

The noble gases are colorless, tasteless, and odorless, and all have extremely low melting and boiling points. In the liquid and solid states, the only forces of attraction among the atoms are very weak London or van der Waals forces. Polarizability and interatomic interactions increase with increasing atomic size, and so melting and boiling points increase with increasing atomic number. The attractive forces among He atoms are so small that He remains liquid at 1 atmosphere pressure even at a temperature of 0.001 K. Table 24-3 lists some of the physical properties of the noble gases.

24-3 Chemical Properties

Until the early 1960s, chemists believed that the Group 0 elements would not combine chemically with any elements. In 1962, Neil Bartlett and his research group at the University of British Columbia were studying the powerful oxidizing agent PtF_6. They accidentally prepared and identified $O_2^+PtF_6^-$ by reaction of oxygen with PtF_6. Bartlett reasoned that xenon also should be oxidized by PtF_6 because the first ionization energy of O_2

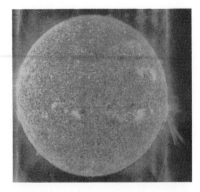

The element He was first discovered in 1868 by analysis of the spectrum of light from the sun. Its name is from the Greek: *helios*, meaning "sun."

Radon is continually produced in small amounts in the uranium radioactive decay sequence (Section 30-10). Radon gas is so unreactive that it eventually escapes from the soil. Recently, measurable concentrations of radon, a radioactive gas, have been observed in basements of many dwellings.

A pressure of about 26 atmospheres is required to solidify He at 0.001 K.

The noble gases are often called the rare gases. They were formerly called the "inert gases" because it was incorrectly thought that they could not enter into chemical combination.

Table 24-1					
Percentages (by volume) of Noble Gases in the Atmosphere					
He	**Ne**	**Ar**	**Kr**	**Xe**	**Rn**
0.0005%	0.0015%	0.94%	0.00011%	0.000009%	~0%

Table 24-2
Uses of the Noble Gases

Noble Gas	Use	Useful Properties or Reasons
helium	1. Filling of observation balloons and other lighter-than-air craft	Nonflammable; 93% of lifting power of flammable hydrogen
	2. He/O_2 mixtures, rather than N_2/O_2, for deep-sea breathing	Low solubility in blood; prevents nitrogen narcosis and "bends"
	3. Diluent for gaseous anesthetics	Nonflammable, nonreactive
	4. He/O_2 mixtures for respiratory patients	Low density, flows easily through restricted passages
	5. Heat transfer medium for nuclear reactors	Transfers heat readily; does not become radioactive; chemically inert
	6. Industrial applications, such as inert atmosphere for welding easily oxidized metals	Chemically inert
	7. Liquid He used to maintain very low temperatures in research (cryogenics)	Extremely low boiling point
neon	Neon signs	Even at low Ne pressure, moderate electric current causes bright orange-red glow; color can be modified by colored glass or mixing with Ar or Hg vapor
argon	1. Inert atmosphere for welding	Chemically inert
	2. Filling incandescent light bulbs	Inert; inhibits vaporization of W and blackening of bulbs
krypton	Airport runway and approach lights	Gives longer life to incandescent lights than Ar, but more expensive
xenon	Xe and Kr mixture in high-intensity, short-exposure photographic flash tubes	Both have fast response to electric current
radon	Radiotherapy of cancerous tissues	Radioactive

(1.31 × 10^3 kJ/mol) is slightly larger than that of xenon (1.17 × 10^3 kJ/mol). He obtained a red crystalline solid initially believed to be $Xe^+PtF_6^-$ but now known to be a more complex compound.

24-4 Xenon Compounds

Since Bartlett's discovery, many other noble gas compounds have been made. Most are compounds of Xe, and the best characterized compounds are xenon fluorides. Oxygen compounds are also well known. Reaction of

Oxygen is second only to fluorine in electronegativity.

Table 24-3
Physical Properties of Noble Gases

Property	He	Ne	Ar	Kr	Xe	Ra
Atomic number	2	10	18	36	54	86
Outer shell e^-	$1s^2$	$2s^22p^6$	$3s^23p^6$	$4s^24p^6$	$5s^25p^6$	$6s^26p^6$
Atomic radius (Å)	0.5	0.70	0.94	1.09	1.30	1.4
Melting point (°C, 1 atm)	−272.2*	−248.6	−189.3	−157	−112	−71
Boiling point (°C, 1 atm)	−268.9	−245.9	−185.6	−152.3	−107.1	−61.8
Density (g/L at STP)	0.18	0.90	1.78	3.75	5.90	9.73
First ionization energy (kJ/mol)	2372	2081	1521	1351	1170	1037

*At 26 atm.

Table 24-4
Xenon Fluorides

Compound	Preparation (Molar ratio $Xe:F_2$)	Reaction Conditions	e^- Pairs Around Xe	Hybridization at Xe	Geometry
XeF_2	1:1–3	400°C or irradiation or elec. discharge	5	sp^3d	
XeF_4	1:5	Same as for XeF_2	6	sp^3d^2	
XeF_6	1:20	300°C and 60 atm or elec. discharge	7	sp^3d^3 (?)	Exact geometry undetermined

Xe with F_2, an extremely strong oxidizing agent, in different stoichiometric ratios produces xenon difluoride, XeF_2; xenon tetrafluoride, XeF_4; and xenon hexafluoride, XeF_6, all colorless crystals (Table 24-4). All involve very electronegative elements.

All the xenon fluorides are formed in exothermic reactions. They are reasonably stable, with Xe—F bond energies of about 125 kJ/mol of bonds. For comparison, strong bond energies range from about 170 to 500 kJ/mol, whereas bond energies of hydrogen bonds are typically less than 40 kJ/mol.

Crystals of the noble gas compound xenon tetrafluoride, XeF_4.

The Halogens (Group VIIA)

The elements of Group VIIA are known as **halogens** (Greek, "salt formers"). The term "**halides**" is used to describe their binary compounds. The heaviest halogen, astatine, is an artificially produced element of which only short-lived radioactive isotopes are known.

24-5 Properties of the Halogens

The elemental halogens exist as diatomic molecules containing single covalent bonds. Properties of the halogens show obvious trends (Table 24-5). Their high electronegativities indicate that they attract electrons strongly. Most binary compounds that contain a metal and a halogen are ionic.

Chlorides occur in salt beds. Twenty-two billion pounds of chlorine were produced in the United States in 1990.

Bromine is a dark red liquid.

Iodine reacts with starch (as in this potato) to form a deep-blue complex substance.

More recently available is an aqueous solution of an iodine complex of polyvinylpyrrolidone, or "povidone." It does not sting when applied to open wounds.

Chlorine

Chlorine (Greek *chloros*, "green") occurs in abundance in $NaCl$, KCl, $MgCl_2$, and $CaCl_2$ in salt water and in salt beds. It is also present as HCl in gastric juices. The toxic, yellowish-green gas is prepared commercially by electrolysis of concentrated aqueous $NaCl$, in which industrially important H_2 and caustic soda ($NaOH$) are also produced (Section 21-4). A laboratory method for the preparation of Cl_2 is the oxidation of HCl with manganese(IV) oxide, MnO_2:

$$4HCl(aq) + MnO_2(s) \longrightarrow MnCl_2(aq) + Cl_2(g) + 2H_2O$$

Chlorine is used to produce many commercially important products. Tremendous amounts of it are used in extractive metallurgy and in chlorinating hydrocarbons to produce a variety of compounds (such as polyvinyl chloride, a plastic). Chlorine is present as Cl_2, $NaClO$, $Ca(ClO)_2$, or $Ca(ClO)Cl$ in household bleaches as well as in bleaches for wood pulp and textiles. Under carefully controlled conditions, Cl_2 is used to kill bacteria in public water supplies.

Bromine

Bromine (Greek *bromos*, "stench") is less abundant than fluorine and chlorine. In the elemental form it is a dense, freely flowing, corrosive, dark-red liquid with a brownish-red vapor at 25°C. It occurs mainly in $NaBr$, KBr, $MgBr_2$, and $CaBr_2$ in salt water, underground salt brines, and salt beds. The major commercial source for bromine is deep brine wells in Arkansas that contain up to 5000 parts per million (0.5%) of bromide. The heated brine is treated with Cl_2, which gives a mixture of Br_2, Cl_2, and salt solution.

$$2\ Br^-(aq) + Cl_2(g) \longrightarrow Br_2(\ell) + 2\ Cl^-(aq)$$

The aqueous layer is poured off, leaving crude Br_2. Bromine is then purified by fractional distillation.

Bromine is used in the production of silver bromide for light-sensitive eyeglasses and photographic film; in the production of sodium bromide, a mild sedative; and in methyl bromide, CH_3Br, a soil fumigant.

Iodine

Iodine (Greek *iodos*, "purple") is a violet-black crystalline solid with a metallic luster. It exists in equilibrium with a violet vapor at 25°C. The element can be obtained from dried seaweed or shellfish or from $NaIO_3$ impurities in Chilean nitrate ($NaNO_3$) deposits. It is contained in the growth-regulating hormone thyroxine, produced by the thyroid gland (Chapter 22). "Iodized" table salt is about 0.02% KI, which helps prevent goiter, a condition in which the thyroid enlarges. Iodine is also used as an antiseptic and germicide in the form of tincture of iodine, a solution in alcohol. Silver iodide is used in "cloud seeding" to cause rain.

The preparation of iodine involves reduction of iodate ion from $NaIO_3$ with sodium hydrogen sulfite, $NaHSO_3$.

$$2IO_3^- + 5HSO_3^- \longrightarrow 3HSO_4^- + 2SO_4^{2-} + H_2O + I_2(s)$$

Iodine is then purified by sublimation. Elemental chlorine or bromine also can be used to displace iodine from iodide salts in aqueous solutions.

$$2I^-(aq) + Cl_2(g) \longrightarrow I_2(s) + 2Cl^-(aq)$$

Table 24-6
Some Common Reactions of the Free Halogens

General Reaction	Remarks
$nX_2 + 2M \longrightarrow 2MX_n$	All X_2 with most metals (most vigorous reaction with F_2 and Group IA metals)
$X_2 + nX_2' \longrightarrow 2XX_n'$	Formation of interhalogens ($n = 1, 3, 5,$ or 7); X is larger than X'
$X_2 + H_2 \longrightarrow 2HX$	
$3X_2 + 2P \longrightarrow 2PX_3$	With all X_2, and with As, Sb, Bi replacing P
$5X_2 + 2P \longrightarrow 2PX_5$	Not with I_2; also Sb $\longrightarrow$ SbF$_5$, SbCl$_5$; As $\longrightarrow$ AsF$_5$; Bi $\longrightarrow$ BiF$_5$
$X_2 + H_2S \longrightarrow S + 2HX$	With all X_2
$X_2' + 2X^- \longrightarrow 2X' + X_2$	$F_2 \longrightarrow Cl_2$, Br$_2$, I$_2$
	$Cl_2 \longrightarrow Br_2$, I$_2$
	$Br_2 \longrightarrow I_2$
$X_2 + C_nH_y \longrightarrow C_nH_{y-1}X + HX$	Halogenation; substitution of many saturated hydrocarbons with Cl_2, Br$_2$
$X_2 + C_nH_{2n} \longrightarrow C_nH_{2n}X_2$	Halogenation of hydrocarbon double bonds with Cl_2, Br$_2$

Iron and chlorine react to form iron(III) chloride, $FeCl_3$.

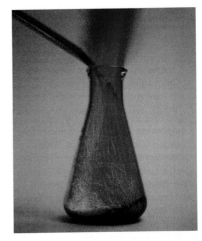

Bromine reacts with powdered antimony so vigorously that the flask vibrates.

24-7 Reactions of the Free Halogens

The free halogens react with most other elements and many compounds. For example, all the Group IA metals react with all the halogens to form simple binary ionic compounds (Section 7-2).

The most vigorous reactions are those of F_2, which usually oxidizes other species to their highest possible oxidation states. Iodine is only a mild oxidizing agent (I^- is a mild reducing agent) and usually does not oxidize substances to high oxidation states. Consider the following reactions of halogens with two metals that exhibit variable oxidation numbers.

With Fe	With Cu
$2Fe + 3F_2 \longrightarrow 2\overset{+3}{Fe}F_3$ (only)	
$2Fe + 3Cl_2$ (excess) $\longrightarrow 2\overset{+3}{Fe}Cl_3$	$Cu + X_2 \longrightarrow \overset{+2}{Cu}X_2$ (X = F, Cl, Br)
$Fe + Cl_2$ (lim. amt.) $\longrightarrow \overset{+2}{Fe}Cl_2$	
$Fe + I_2 \longrightarrow \overset{+2}{Fe}I_2$ (only)	$2Cu + I_2 \longrightarrow 2\overset{+1}{Cu}I$ (only)
$Fe^{3+} + I^- \longrightarrow Fe^{2+} + \frac{1}{2}I_2$	$Cu^{2+} + 2I^- \longrightarrow \overset{+1}{Cu}I + \frac{1}{2}I_2$

Table 24-6 summarizes some of the reactions of the free halogens.

24-8 Interhalogens

The interhalogens are a class of compounds, XX_n', where X and X' are different halogens. X is the larger of the two, and n is 1, 3, 5, or 7. Several interhalogens are known (Table 24-7). Because smaller atoms are grouped around larger atoms, the maximum number of X' atoms increases as the ratio of the radii of X to X' increases. The X—X' bond energies increase as the differences in electronegativities of X and X' increase. Their chemical

Concentrated H_2SO_4 oxidizes HBr to Br_2 and HI to I_2. So H_3PO_4 is used to prepare HBr and HI from their salts because it is a nonoxidizing acid.

$$NaBr(s) + H_3PO_4(\ell) \longrightarrow NaH_2PO_4(s) + HBr(g) \qquad bp = -66.8°C$$

Most *nonmetal halides* hydrolyze to produce hydrogen halides and an acid or oxide of the nonmetal.

$$BCl_3(\ell) + 3H_2O(\ell) \longrightarrow H_3BO_3(aq) + 3HCl(g)$$

$$SiCl_4(\ell) + 2H_2O(\ell) \longrightarrow SiO_2(s) + 4HCl(g)$$

Hydrogen bromide and hydrogen iodide are often prepared by hydrolysis of PBr_3 and PI_3, respectively. They are then isolated by distillation.

$$PX_3 + 3H_2O \longrightarrow H_3PO_3(s) + 3HX(g) \qquad X = Cl, Br, I$$

The phosphorus trihalide is usually formed at the time of reaction by mixing red phosphorus with the free halogen.

A commercially important class of reactions is the *halogenation of saturated hydrocarbons*. Hydrogen halides are by-products of such reactions. A general reaction of this type is

$$C_nH_y + X_2 \longrightarrow C_nH_{y-1}X + HX \qquad X = F, Cl, Br, I$$

However, fluorination of hydrocarbons with F_2 is so explosive that such reactions are not performed. Halogenation of hydrocarbons with I_2 is very slow.

24-11 Aqueous Solutions of Hydrogen Halides (Hydrohalic Acids)

All hydrogen halides react with H_2O to produce *hydrohalic acids* that ionize:

$$H:\ddot{O}: + H:\ddot{X}: \rightleftharpoons H:\overset{+}{\underset{H}{\ddot{O}}}:H + :\ddot{X}:^-$$

The reaction is essentially complete for dilute HCl(aq), HBr(aq), and HI(aq) (Section 10-5). Dilute HF(aq) is a weak acid ($K_a = 7.2 \times 10^{-4}$). In concentrated solutions, more acidic dimeric $(HF)_2$ units are present (Figure 24-2). They ionize as follows.

$$(HF)_2(aq) + H_2O(\ell) \rightleftharpoons H_3O^+(aq) + HF_2^-(aq) \qquad K \approx 5$$

The order of increasing acid strengths of the hydrohalic acids is the same as that of the anhydrous hydrogen halides: HF(aq) $\ll$ HCl(aq) < HBr(aq) < HI(aq) (Section 10-5).

The only acid used in industry to a greater extent than HCl is H_2SO_4. Hydrochloric acid is used in the production of metal chlorides, dyes, and many other commercially important products. It is also used on a large scale to dissolve metal oxide coatings from iron and steel prior to galvanizing or enameling.

Hydrofluoric acid is used in the production of fluorine-containing compounds and for etching glass. The acid reacts with silicates, such as calcium silicate, $CaSiO_3$, in the glass to produce a very volatile and thermodynamically stable compound, silicon tetrafluoride, SiF_4.

$$CaSiO_3(s) + 6HF(aq) \longrightarrow CaF_2(s) + SiF_4(g) + 3H_2O(\ell)$$

Oxidation numbers remain constant in these reactions.

Phosphorus trichloride, PCl_3, is dropped into distilled water containing the indicator methyl orange (yellow at pH = 7). As HCl and H_3PO_3 are formed, the indicator turns red, its color in acidic solutions.

Dilute solution

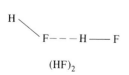

$(HF)_2$

Concentrated solution

Figure 24-2
Hydrogen bonding (dashed lines) in dilute and concentrated aqueous solutions of hydrofluoric acid.

Hydrobromic and hydroiodic acids are less important. They are used to make Br-containing and I-containing organic compounds and in chemical research.

24-12 Halides of Other Elements

The halogens combine with most other elements to form halides. They range from high-melting, water-soluble, ionic electrolytes (metal halides) such as NaCl, KBr, and $CaCl_2$ to volatile, covalent nonelectrolytes (nonmetal halides) such as PCl_3 and SF_6.

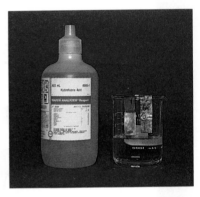

HF cleans oxide coatings from metals.

Halides (increasing nonmetallic character of other elements) →

| **Metal halide**; ionic lattice; metal ions slightly hydrolyzed if at all; X^- not hydrolyzed except F^- Examples: NaCl, MgF_2, $NiBr_2$ | ⇒ intermediate properties ⇒ | **Nonmetal halide** or complex halide; molecular (covalent) compounds, volatile; completely hydrolyzed Examples: BF_3, PCl_5, SF_4, $AsCl_3$ |

24-13 The Oxyacids (Ternary Acids) of the Halogens

Table 24-9 lists the known oxyacids of the halogens, their sodium salts, and some trends in properties. Only three oxyacids, $HClO_4$, HIO_3, and H_5IO_6, have been isolated in anhydrous form. The others are known only in aqueous solution. In all these acids the H is bonded through an O.

The Lewis dot formulas and structures of the chlorine oxyanions are shown in Figure 24-3. The corresponding oxyanions of bromine and iodine have similar structures.

Table 24-9
Oxyacids of the Halogens and Their Salts

Oxidation State	Acid	Name of Acid	Thermal Stability and Acid Strength	Oxidizing Power of Acid	Sodium Salt	Name of Salt	Thermal Stability	Oxidizing Power and Hydrolysis of Anion	Nature of Halogen
+1	HXO	hypohalous acid			NaXO	sodium hypohalite			X = F*, Cl, Br, I
+3	HXO_2	halous acid	Increase	Increases	$NaXO_2$	sodium halite	Increases	Increase	X = Cl, Br (?)
+5	HXO_3	halic acid			$NaXO_3$	sodium halate			X = Cl, Br, I
+7	HXO_4	perhalic acid			$NaXO_4$	sodium perhalate			X = Cl, Br, I
+7	H_5XO_6	paraperhalic acid			several types	sodium paraperhalates			X = I only

*The oxidation state of F is −1 in HOF.

Neither perfluoric acid nor perfluorate ion is known. The other *perhalic acids* and *perhalates* are known. Anhydrous perchloric acid, $HClO_4$, a colorless, oily liquid, distills under reduced pressure (10 to 20 torr) after reaction of a perchlorate salt with nonvolatile, concentrated sulfuric acid.

$$NaClO_4 + H_2SO_4 \longrightarrow NaHSO_4 + HClO_4 \quad \text{(perchloric acid distills)}$$

Hot, concentrated perchloric acid can explode in the presence of reducing agents, especially organic reducing agents. Hot, concentrated $HClO_4$ is a very strong oxidizing agent, but cold, dilute perchloric acid is only a weak oxidizing agent. Even fairly strong reducing agents such as zinc do not reduce ClO_4^- ions in cold dilute $HClO_4$; H_3O^+ ions are reduced to H_2.

$$Zn + 2HClO_4 \text{ (cold, dilute)} \longrightarrow Zn(ClO_4)_2 + H_2$$

Perchloric acid is the strongest of all common acids with respect to ionization, exceeding even HNO_3 and HCl. It is used commercially as a substitute for sulfuric acid, under conditions in which H_2SO_4 could be reduced to SO_2.

Treatment of $HClO_4$ with a strong dehydrating agent such as P_4O_{10} produces the explosive anhydride, dichlorine heptoxide, Cl_2O_7. The oxidation state of Cl is +7 in both compounds.

Key Terms

Halogens   Group VIIA elements; F, Cl, Br, I, and At.

Chain initiation step   The first step in a chain reaction; produces reactive species (such as radicals) that then propagate the reaction.

Chain propagation step   An intermediate step in a chain reaction; in such a step one or more reactive species is consumed and the same or another reactive species is produced.

Chain reaction   A reaction in which reactive species, such as radicals, are produced in more than one step. Con-sists of an initiation step, one or more propagation steps, and one or more termination steps.

Chain termination step   The combination of two radicals, which removes the reactive species that propagate the chain reaction.

Noble gases   Group 0 elements; He, Ne, Ar, Kr, Xe, and Rn.

Radical   An atom or group of atoms that contains one or more unpaired electrons (usually very reactive species).

Exercises

The Noble Gases

1. Why are the noble gases so unreactive?
2. Why were the noble gases among the last elements to be discovered?
3. List some of the uses of the noble gases and reasons for the uses.
4. Arrange the noble gases in order of increasing (a) atomic radii, (b) melting points, (c) boiling points, (d) densities, and (e) first ionization energies.
5. Explain the order of increasing melting and boiling points of the noble gases in terms of polarizabilities of the atoms and forces of attraction between them.
6. What gave Neil Bartlett the idea that compounds of xenon could be synthesized? Which noble gases are known to form compounds? With which elements are the noble gas atoms bonded?
7. Describe the bonding and geometry in XeF_2, XeF_4, and XeF_6.

8. Compare the bond energies of Xe—F bonds with other typical bonds and with hydrogen bonds.
9. Suggest a reason why neon is not likely to form covalent compounds.
*10. How many grams of xenon oxide tetrafluoride, $XeOF_4$, and how many liters of HF at STP could be prepared, assuming complete reaction of 6.50 g of xenon tetrafluoride, XeF_4, with a stoichiometric quantity of water according to the equation below?

$$6XeF_4(s) + 8H_2O(\ell) \longrightarrow$$
$$2XeOF_4(\ell) + 4Xe(g) + 16HF(g) + 3O_2(g)$$

*11. Argon crystallizes at $-235°C$ in a face-centered cubic unit cell with $a = 5.43$ Å. Determine the apparent radius of an argon atom in the solid.
*12. Xenon(VI) fluoride can be produced by the combination of xenon(IV) fluoride with fluorine. Write a chemical

equation for this reaction. What mass of XeF_6 could be produced from 3.62 g of XeF_4?

*13. Xenon hexafluoride reacts rapidly with the SiO_2 in glass or quartz containers to form $XeOF_4(\ell)$ and $SiF_4(g)$. What will be the pressure of SiF_4 in a 1.00-L container at 25°C after 1.00 g of XeF_6 decomposes?

*14. The standard enthalpies of formation are -402 kJ/mol for $XeF_6(s)$ and -261.5 kJ/mol for $XeF_4(s)$. Calculate ΔH^0_{rxn} at 25°C for the preparation of XeF_6 from XeF_4 and $F_2(g)$.

The Halogens

15. Write the electron configuration for each of the atomic halogens. Draw the Lewis symbol for a halogen atom, X. What is the usual oxidation state of the halogens in binary compounds with metals, semiconducting elements, and most nonmetals?

16. Draw the Lewis structure of a halogen molecule, X_2. Describe the bonding in the molecule. What is the trend of bond length and strength going down the family from F_2 to I_2?

17. What types of intermolecular forces are found in molecular halogens? What is the trend in these forces going down the group from F_2 to I_2? Describe the physical state of each molecular halogen at room temperature and pressure.

18. List the halogens in order of increasing (a) atomic radii, (b) ionic radii, (c) electronegativities, (d) melting points, (e) boiling points, and (f) standard reduction potentials.

19. What is the order of increasing X—X (halogen) bond energies? Suggest why the F—F bond energy is less than the Cl—Cl bond energy.

20. Although astatine has been detected in various minerals, most properties of this element have been observed from artificially produced samples or predicted from periodic relationships. Using Table 24-5 predict or estimate the (a) color, (b) melting point, (c) ionic radius, (d) bond energy of this element.

21. Give examples of halogen compounds in which a halogen exhibits the +1, +3, +5, and +7 oxidation states. Are these compounds ionic or covalent?

22. Fluorine cannot be prepared by electrolysis of aqueous solutions of fluoride salts. Why?

23. Write the equations describing the half-reactions and net reaction for the electrolysis of molten KF/HF mixtures. At which electrodes are the products formed? What is the purpose of the HF?

24. Carl O. Christe's preparation of F_2 did not involve direct chemical oxidation. Explain this statement.

*25. A reaction mixture contained 100 g of K_2MnF_6 and 174 g of SbF_5. Fluorine was produced in 38.3% yield. How many grams of F_2 were produced? What volume is this at STP?

26. Write an equation for a common small-scale laboratory preparation of Cl_2 from hydrochloric acid.

27. Discuss the chemistry of the extraction of bromine from water from brine wells.

28. Give two practical uses for compounds of each of the halogens.

29. Write equations describing general reactions of the free halogens, X_2, with (a) Group IA (alkali) metals, (b) Group IIA (alkaline earth) metals, and (c) Group IIIA metals. Represent the metals as M.

30. Write balanced equations for any reactions that occur in aqueous mixtures of (a) NaI and Cl_2, (b) NaCl and Br_2, (c) NaI and Br_2, (d) NaBr and Cl_2, (e) NaF and I_2.

*31. An aqueous solution contains either NaBr or a mixture of NaBr and NaI. Using only aqueous solutions of I_2, Br_2, and Cl_2 and a small amount of CH_2Cl_2, describe how you might determine what is in the unknown solution.

32. Write equations illustrating the tendency of F^- to stabilize high oxidation states of cations and the tendency of I^- to stabilize low oxidation states. Why is this the case?

33. Why are the free halogens more soluble in water than most nonpolar molecules?

34. Draw structures and give hybridizations at the central atom for BrF_3, ClF_5, IBr, and IF_7. Which of these molecules are polar?

35. Ions of the interhalogens (Exercise 34) also exist. An example is the ICl_4^- ion in $KICl_4$. Sketch the ion and indicate the hybridization at I in an ICl_4^- ion.

36. Distinguish between hydrogen bromide and hydrobromic acid.

37. What is the order of decreasing melting and boiling points of the hydrogen halides? Why is the HF "out of line"?

38. Give a reaction illustrating each of the four general methods for preparation of hydrogen halides.

39. Compare the order of acid strengths of the hydrogen halides with that of the hydrohalic acids.

40. Describe the effect of hydrofluoric acid on glass.

41. Compare the general properties of metal halides with those of nonmetal halides.

42. What is the acid anhydride of perchloric acid?

43. Write the equation for the dehydration of $HClO_4$ with tetraphosphorus decoxide.

44. Name the following compounds: (a) $KBrO_3$, (b) KBrO, (c) $NaClO_4$, (d) $NaClO_2$, (e) HBrO, (f) $HBrO_3$, (g) HIO_3, (h) $HClO_4$.

45. Draw the Lewis formulas and structures of the four ternary acids of chlorine.

46. Write equations describing reactions by which the following compounds can be prepared: (a) hypohalous acids of Cl, Br, and I (in solution with hydrohalic acids), (b) hypohalite salts, (c) chlorous acid, (d) halate salts, (e) a perchlorate salt, (f) perchloric acid

47. What is the order of increasing acid strength of the ternary chlorine acids? Explain the order.

48. Choose the strongest acid from each group: (a) HClO, HBrO, HIO; (b) HClO, HClO$_2$, HClO$_3$, HClO$_4$; (d) HIO, HBrO$_3$, HClO$_4$.

49. Write balanced equations for the following chemical reactions: (a) preparation of Cl$_2$ from aqueous KCl by electrolysis, (b) any reaction of Br$_2$ as an oxidizing agent, (c) disproportionation of KClO when it is heated, (d) displacement of one halogen by another in aqueous solution, (e) formation of an iodine chloride from the elements.

50. Based on your knowledge of halogen chemistry, suggest a reaction for preparing (a) KBrO$_3$(aq) from Br$_2$(ℓ); (b) KBrO$_3$(aq) from KBrO(aq); (c) BrF$_5$(g) from Br$_2$(ℓ) (d) Br$_2$(aq) from Br$^-$. Write a balanced ionic equation (or equations) for each of the preparations.

51. Calculate the surface area ($4\pi r^2$), in square Ångstroms (Å^2), for all the halogen atoms and for the halide ions.

52. Calculate the ratio of rates of diffusion of each of the halogen molecules to that of the fluorine molecule.

53. Determine the mass of KClO$_3$ theoretically obtained by the reaction of 35.0 L of Cl$_2$(g), measured at 25°C and 1.00 atm, with hot KOH(aq). What mass of KClO would be theoretically obtainable by the reaction of the same quantity of Cl$_2$(g) with cold KOH(aq)?

*54. A 0.350-A current is applied to molten sodium chloride (see Section 21-3). The Cl$_2$, collected at 26°C and 774 torr pressure, occupies a volume of 35,700 L. How many moles, grams, and pounds of NaCl must have been electrolyzed? How many pounds of metallic sodium were produced? How long did the cell operate?

55. Assuming that "iodized salt" contains 0.02% potassium iodide and 99.98% sodium chloride, calculate the percentage of the total mass of a sample of "iodized salt" that is due to iodine alone.

*56. Use the Nernst equation and the reduction potentials in Appendix J to predict the ratio of [Cl$^-$] to [Br$^-$] at which liquid bromine could liberate gaseous chlorine at 1 atm pressure from a solution of Cl$^-$.

*57. What molarity of hypochlorous acid is produced if 30.0 g of Cl$_2$ is bubbled through 500 mL of water, and 16.7 g of Cl$_2$ escapes unreacted? What molarity of hydrochloric acid will have been produced?

*58. Sea water is a major commercial source of bromine. The average concentration of bromine in sea water is 75 ppm, calculated as Br$^-$ ion. Calculate (a) the volume of seawater in cubic feet that is required to produce one ton of liquid bromine and (b) the volume of chlorine gas, in liters, measured at STP, required to displace all of this bromine. The density of seawater is 64.5 lb/ft^3.

*59. After a 20.65 g sample of argon was bubbled through liquid bromine at 25°C, the combined argon-bromine gas mixture at 734 torr weighed 57.46 g. Using Dalton's law of partial pressures, calculate the vapor pressure of bromine. Ar is not soluble in Br$_2$(ℓ).

*60. A 10.00 g sample of KClO$_3$ is heated. Part of it undergoes the reaction

$$2KClO_3 \longrightarrow 2KCl + 3O_2$$

and part of it undergoes the reaction

$$4KClO_3 \longrightarrow 3KClO_4 + KCl$$

If the residue weighs 7.00 g, what proportion of the KClO$_3$ reacted according to each equation?

61. Identify which of the following are redox reactions. Identify the oxidizing and reducing agent in each of the redox reactions.
(a) $BrO_3^- + 5Br^- + 6H^+ \longrightarrow 3Br_2(\ell) + 3H_2O(\ell)$
(b) $I^- + Br_2(aq) \longrightarrow IBr(aq) + Br^-$
(c) $NaHSO_4(s) + NaCl(s) \xrightarrow{\Delta} Na_2SO_4(s) + HCl(g)$
(d) $2KClO_3(\ell) \xrightarrow{\Delta} 2KCl(s) + 3O_2(g)$

62. Repeat Exercise 61 for:
(a) $4NaClO_3(s) \xrightarrow{\Delta} 3NaClO_4(s) + NaCl(s)$
(b) $3Br_2(\ell) + 6OH^- \longrightarrow BrO_3^- + 5Br^- + 3H_2O(\ell)$
(c) $2NaF(s) + H_2SO_4(conc) \xrightarrow{\Delta} Na_2SO_4(s) + 2HF(g)$
(d) $2Cl_2(g) + HgO(s) + H_2O(\ell) \longrightarrow$
$$HgCl_2(aq) + 2HOCl(aq)$$

The rose on the left is in an atmosphere of sulfur dioxide, SO_2. Gaseous SO_2 and aqueous solutions of HSO_3^- and SO_3^{2-} ions are used as bleaching agents. A similar process is used to bleach wood pulp before it is converted to paper.

Outline

Objectives

As you study this chapter, you should learn

☐ About the occurrence, properties, and uses of the Group VIA nonmetals sulfur, selenium, and tellurium

☐ Some important reactions of these Group VIA nonmetals

☐ About hydrides of the Group VIA nonmetals

☐ About aqueous solutions of the Group VIA hydrides

☐ Some important reactions of the Group VIA hydrides

☐ About some important metal sulfides and their analogs

☐ About the halides of Group VIA nonmetals

☐ About the important Group VIA dioxides: SO_2, SeO_2, and TeO_2

☐ About the important Group VIA trioxides: SO_3, SeO_3, and TeO_3

☐ About some oxyacids of S, Se, and Te

☐ Methods of preparation of, and chemical properties of, aqueous solutions of H_2SO_3, H_2SeO_3, and H_2TeO_3

☐ Methods of preparation of, and chemical properties of, aqueous solutions of H_2SO_4, H_2SeO_4, and H_6TeO_6

☐ About thiosulfuric acid and the thiosulfates

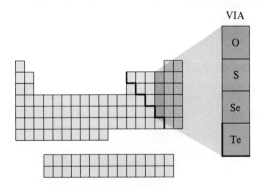

W e shall now discuss the chemistry of the heavier Group VIA elements. The chemistry of oxygen and its compounds was described in Section 6-8. The VIA elements are less electronegative than the halogens. Oxygen and sulfur are clearly nonmetallic, but selenium is less so. Tellurium is usually classified as a metalloid and forms metal-like crystals. Its chemistry is mostly that of a nonmetal. Polonium is a metal. All 29 isotopes of polonium are radioactive.

Irregularities in the properties of elements within a given family increase toward the middle of the periodic table. There are larger differences in the properties of the Group VIA elements than in the properties of the halogens. The properties of elements in the *second period* usually differ significantly from those of other elements in their families, because second-period elements have no low-energy d orbitals. So, the properties of oxygen are not very similar to those of the other Group VIA elements (Table 25-1).

The outer electronic configuration of the VIA elements is ns^2np^4. An atom of each element may gain or share two electrons as it forms compounds. Each forms a covalent compound of the type H_2E in which the VIA element (E) exhibits the oxidation number -2. The maximum number of atoms with which O can bond (coordination number) is four, but S, Se, Te, and probably Po can bond covalently to as many as six other atoms. This is due to the availability of vacant d orbitals in the outer shell of each of the VIA elements except O. One or more of the d orbitals can accommodate additional electrons to form up to six bonds.

The d orbitals do not occur until the third energy level.

Sulfur, Selenium, and Tellurium

25-1 Occurrence, Properties, and Uses of the Elements

Sulfur

Sulfur makes up about 0.05% of the earth's crust. It was one of the elements known to the ancients. It was used by the Egyptians as a yellow coloring, and it was burned in some religious ceremonies because of the unusual odor

Table 25-1
Some Properties of Group VIA Elements

Property	O	S	Se	Te	Po
Physical state (1 atm, 25°C)	Gas	Solid	Solid	Solid	Solid
Color	Colorless (very pale blue)	Yellow	Red-gray to black	Brass-colored, metallic luster	—
Outermost electrons	$2s^22p^4$	$3s^23p^4$	$4s^24p^4$	$5s^25p^4$	$6s^26p^4$
Melting point (1 atm, °C)	-218	112	217	450	254
Boiling point (1 atm, °C)	-183	444	685	990	962
Electronegativity	3.5	2.5	2.4	2.1	1.9
First ionization energy (kJ/mol)	1314	1000	941	869	812
Atomic radius (Å)	0.66	1.04	1.17	1.37	1.4
Ionic (2−) radius (Å)	1.40	1.84	1.98	2.21	—
Common oxidation states	usually -2	$-2, +2, +4, +6$	$-2, +2, +4, +6$	$-2, +2, +4, +6$	$-2, +6$

Native sulfur. Elemental sulfur is deposited at the edges of some hot springs and geysers. This formation surrounds Emerald Lake in Yosemite National Park.

it produced; it is the "brimstone" of the Bible. Alchemists tried to incorporate its "yellowness" into other substances to produce gold.

Sulfur occurs as the free element—predominantly S_8 molecules—and as metal sulfides such as galena, PbS; iron pyrite, FeS_2; and cinnabar, HgS. To a lesser extent, it occurs as metal sulfates such as barite, $BaSO_4$, and gypsum, $CaSO_4 \cdot 2H_2O$, and in volcanic gases as H_2S and SO_2.

Sulfur is found in much naturally occurring organic matter, e.g., petroleum and coal. Its presence in fossil fuels causes environmental and health problems because many sulfur-containing compounds undergo combustion to produce sulfur dioxide (Section 25-8), an air pollutant.

Nearly half of the sulfur used in the United States is recovered from natural gas and oil. Hydrogen sulfide is oxidized to sulfur in the Claus furnace.

$$8H_2S(g) + 4O_2(g) \longrightarrow S_8(\ell) + 8H_2O(g)$$

Elemental sulfur is mined along the U.S. Gulf Coast by the **Frasch process**, or "hot water" process (Figure 25-1). Most of it is used in the production of sulfuric acid, H_2SO_4, the most important of all industrial chemicals. Sulfur is also a component of black gunpowder and is used in the vulcanization of rubber and in the synthesis of many important sulfur-containing organic compounds.

In each of the three physical states, elemental sulfur exists in many forms. The two most stable forms of sulfur, the rhombic (mp 112°C) and monoclinic (mp 119°C) crystalline modifications, consist of S_8 molecules. These are puckered rings containing eight sulfur atoms (Figure 2-3) and all S—S single bonds. The energy of the S—S single bond (213 kJ/mol) is much greater than that of the O—O single bond (138 kJ/mol). This accounts for S atoms bonding to other S atoms to form rings. The rings are arranged differently in the rhombic and monoclinic forms. Below about 150°C, liquid sulfur consists mainly of S_8 molecules. Above 150°C, it becomes increasingly viscous and darkens as the S_8 rings break apart into chains that interlock with each other through S—S bonds. The viscosity reaches a maximum at 180°C, at which point sulfur is dark brown. Above 180°C, the liquid thins as the chains are broken down into smaller chains. At 444°C, sulfur boils to give a vapor containing S_8, S_6, S_4, and S_2 molecules.

Selenium

Like sulfur, selenium exists in a number of allotropic forms, but only two common crystalline modifications are well characterized. These are the gray, metal-like hexagonal form and the red, nonmetallic, monoclinic form. Solid selenium contains mainly Se_8 molecules, but the vapor contains Se_8, Se_6, and Se_2 molecules. There appears to be only one liquid form.

Selenium is quite rare (9×10^{-6} % of the earth's crust). It occurs mainly as an impurity in sulfur, sulfide, and sulfate deposits. It is obtained from the flue dusts that result from roasting sulfide ores and from the "anode mud" formed in the electrolytic refining of copper. It is used as a red coloring in glass. The metal-like form of selenium has an electrical conductivity that is very light-sensitive, so it is used in photocopy machines and in solar cells.

Selenium is often incorporated into glass in the form of Na_2Se. Subsequent heating causes particles of red colloidal selenium to form.

Tellurium

Tellurium is even less abundant (2×10^{-7} % of the earth's crust) than selenium. It occurs mainly in sulfide ores, especially with copper sulfide, and as the tellurides of gold and silver. It, too, is obtained from the "anode mud" from refining of copper. The element forms brass-colored, shiny, hexagonal crystals having low electrical conductivity. It is added to some metals, particularly lead, to increase electrical resistance and improve resistance to heat, corrosion, mechanical shock, and wear. It is also used to color glass red, blue, or brown. This metalloid is a semiconductor.

25-2 Reactions of Group VIA Nonmetals

Some reactions of the Group VIA elements are summarized in Table 25-2.

25-3 Hydrides of VIA Elements

All the VIA elements form covalent compounds of the type H_2E (E = O, S, Se, Te, Po) in which the VIA element is in the -2 oxidation state. H_2O is a liquid that is essential for animal and plant life. H_2S, H_2Se, and H_2Te are colorless, noxious, poisonous gases. They are even more toxic than HCN. Egg protein contains sulfur, and its decomposition forms H_2S, giving off the odor of rotten eggs. H_2Se and H_2Te smell even worse. Their odors are usually ample warning of the presence of these poisonous gases. Small amounts of H_2S cause headaches and nausea. (If these symptoms occur while you are using this common laboratory reagent, you should leave the laboratory and get fresh air immediately.) Exposure to larger amounts can cause fainting and heart or lung failure. H_2S poses an additional problem. After long enough exposure, it severely impairs the sense of smell so that one is no longer aware of its presence. Some properties of the VIA hydrides are summarized in Table 25-3.

Both the melting point and boiling point of water are very much higher than expected by comparison with those of the heavier hydrides (Table 25-3). This is a consequence of hydrogen bonding in ice and liquid water (Section 13-2) caused by the strongly dipolar nature of water molecules. The electronegativity differences between H and the other VIA elements are much

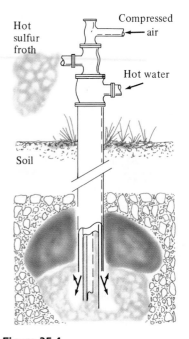

Figure 25-1
The Frasch process for mining sulfur. Three concentric pipes are used. Water at about 170°C and a pressure of 100 lb/in² (7 kg/cm²) is forced down the outermost pipe to melt the sulfur. Hot compressed air is pumped down the innermost pipe. It mixes with the molten sulfur to form a froth, which rises through the third pipe.

Table 25-2
Some Reactions of the VIA Elements (E)

General Equation	Remarks
$xE + yM \longrightarrow M_yE_x$	With many metals
$zE + M_xE_y \longrightarrow M_xE_{y+z}$	Especially with S, Se
$E + H_2 \longrightarrow H_2E$	Decreasingly in the series O_2, S, Se, Te
$E + 3F_2 \longrightarrow EF_6$	With S, Se, Te, and excess F_2
$2E + Cl_2 \longrightarrow E_2Cl_2$	With S, Se (Te gives $TeCl_2$); also with Br_2
$E_2Cl_2 + Cl_2 \longrightarrow 2ECl_2$	With S, Se; also with Br_2
$E + 2Cl_2 \longrightarrow ECl_4$	With S, Se, Te, and excess Cl_2; also with Br_2
$E + O_2 \longrightarrow EO_2$	With S (with Se, use $O_2 + NO_2$)
$3E + 4HNO_3 \longrightarrow 3EO_2 + 2H_2O + 4NO$	With S, Se, Te
$3E + 6OH^- \longrightarrow EO_3^{2-} + 2E^{2-} + 3H_2O$	With S

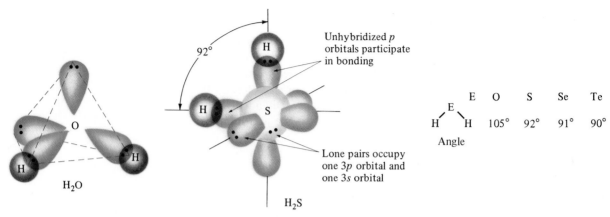

	E	O	S	Se	Te
H—E—H Angle		105°	92°	91°	90°

Figure 25-2
The Group VIA hydrides. The structures of H_2Se and H_2Te are similar to that of H_2S, but the bond angles are slightly smaller because Se and Te atoms are larger than the S atom.

Pumice is a porous solid resulting from solidification of volcanic lava.

smaller than those between H and O, so no H-bonding occurs in H_2S, H_2Se, or H_2Te. The angular, polar water molecule is described by assuming sp^3 hybridization at O (Section 8-9). The bond angles and polarities decrease as the group is descended. The heavier members, H_2S, H_2Se, and H_2Te, have bond angles of nearly 90° (Figure 25-2). They are best described by assuming p^2 bonding at the VIA atom.

Water and hydrogen sulfide are produced from their elements in exothermic reactions, whereas the other hydrides are produced by endothermic reactions. H_2O and H_2S are stable at 25°C, but H_2Se and H_2Te slowly decompose to the elements. H_2S can be prepared from the gaseous elements at high temperatures on pumice chips.

$$8H_2 + S_8 \xrightarrow{600°C} 8H_2S$$

It also can be prepared by reaction between (nonoxidizing) protonic acids and metal sulfides such as iron(II) sulfide.

$$FeS + 2H_2SO_4 \longrightarrow H_2S + Fe(HSO_4)_2$$

H_2S is often prepared in the laboratory by hydrolysis of thioacetamide, a sulfur-containing organic compound. Heating speeds the reaction.

$$CH_3-\overset{\overset{\displaystyle S}{\|}}{C}-N\overset{\displaystyle H}{\underset{\displaystyle H}{\big\langle}} + 2H_2O \xrightarrow{\Delta} CH_3-\overset{\overset{\displaystyle O}{\|}}{C}-O^- + NH_4^+ + H_2S$$

ammonium acetate
NH_4CH_3COO

thioacetamide

Table 25-3
Some Properties of the Group VIA Hydrides

Property	H_2O	H_2S	H_2Se	H_2Te
Melting point (°C)	0.00	−85.60	−60.4	−51
Heat of fusion (kJ/mol)	6.01	2.38	—	—
Boiling point (°C)	100	−60.75	−41.5	−1.8
Heat of vaporization (kJ/mol)	40.7	18.7	19.9	24
Density at boiling point (g/mL)	0.958	0.993	2.004	2.650
Heat of formation at 25°C (kJ/mol)	−286	−20.6	29.7	135
Free energy of formation at 25°C (kJ/mol)	−237	−33.6	15.9	130

25-4 Aqueous Solutions of VIA Hydrides

Aqueous solutions of hydrogen sulfide, selenide, and telluride are acidic, with acid strength increasing as the group is descended: $H_2S < H_2Se < H_2Te$. The same trend was observed for increasing acidity of the hydrogen halides. The acid ionization constants are

H_2S is a stronger acid than H_2O. The solubility of H_2S in water is approximately 0.10 mol/L at 25°C.

	H₂S	**H₂Se**	**H₂Te**
$H_2E \rightleftharpoons H^+ + HE^-$ $\quad K_1$:	1.0×10^{-7}	1.9×10^{-4}	2.3×10^{-3}
$HE^- \rightleftharpoons H^+ + E^{2-}$ $\quad K_2$:	1.3×10^{-13}	$\sim 10^{-11}$	$\sim 1.6 \times 10^{-11}$

25-5 Reactions of VIA Hydrides

The heavier Group VIA hydrides are reducing agents; their reducing strength increases from H_2S to H_2Te. When H_2S acts as a reducing agent it is usually oxidized to elemental sulfur, but strong oxidizing agents may oxidize sulfur to the +4 or +6 oxidation state. Gaseous H_2S reduces moist oxygen only slowly, but solutions of H_2S become clouded with elemental sulfur after a few days, due to dissolved oxygen.

$$2H_2S + O_2 \xrightarrow{\Delta} 2H_2O + 2S$$

Hydrogen telluride solutions undergo a similar reaction in a few minutes.

When heated, H_2S burns in the air to produce SO_2 and steam.

$$2H_2S + 3O_2 \xrightarrow{\Delta} 2SO_2 + 2H_2O$$

When excess H_2S is present, elemental sulfur is produced by the reaction of SO_2 with H_2S. The same reaction occurs rapidly when both gases are bubbled into water.

$$2H_2S + SO_2 \longrightarrow 3S + 2H_2O$$

Many metals displace hydrogen from hydrogen sulfide. Passing H_2S over the surface of magnesium (free of MgO) results in the formation of magnesium sulfide and hydrogen.

$$Mg + H_2S \xrightarrow{\Delta} MgS + H_2$$

The H_2S formed in the decomposition of the sulfur-containing protein of eggs tarnishes silverware by a reaction involving reduction of O_2, oxidation of Ag, and formation of black Ag_2S. In this reaction the H_2S is neither reduced nor oxidized, but apparently aids the oxidation of Ag.

$$4Ag + 2H_2S + O_2 \longrightarrow 2Ag_2S + 2H_2O$$

Tarnished silver is coated with black Ag_2S.

25-6 Metal Sulfides and Analogs

Two classes of binary metal salts of the heavier VIA elements exist: (1) those containing sulfide (S^{2-}), selenide (Se^{2-}), or telluride (Te^{2-}) ions; and (2) those containing hydrosulfide (HS^-), hydroselenide (HSe^-), or hydrotelluride (HTe^-) ions. These salts are the S, Se, and Te analogs of the oxides, O^{2-}, and hydroxides, OH^-, respectively.

Metal sulfides exhibit a wide range of water solubility, and metal ions in aqueous solution are often analytically separated from each other (as in qualitative analysis) with H_2S. Sulfides of some small, highly charged cations hydrolyze completely to form insoluble hydroxides and hydrogen sulfide.

$$Al_2S_3 + 6H_2O \longrightarrow 2Al(OH)_3 + 3H_2S$$

25-7 Halides of VIA Elements

Several halides of the type E_2X_2, EX_2, EX_4, and EX_6 (E = S, Se, Te; X = F, Cl, Br, I) have been prepared. They are listed in Table 25-4.

The large I atom combines only with tellurium, the largest of the three VIA elements. The EX_6 species all contain the small F atoms. Six of the larger halogen atoms cannot fit around the VIA element. This is called **steric hindrance**. The EX_6 species are all octahedral with sp^3d^2 hybridization at the central atom (Section 8-12).

The halides of lower oxidation states—E_2X_2, EX_2, and EX_4—are very reactive. For example, sulfur tetrafluoride reacts with water readily.

$$SF_4(g) + 2H_2O(\ell) \longrightarrow 4HF(g) + SO_2(g)$$

The central atom in each EX_4 species is sp^3d hybridized and has one lone pair of electrons. These nonbonding electrons can be attacked by electron-seeking species, such as H_2O, to initiate a reaction.

However, SF_6 and SeF_6 are stable compounds that are inert toward H_2O. This is an example of *kinetic* stability. The ΔG^0 value for the reaction of SF_6 with H_2O is negative, but the reaction does *not* occur within a reasonable time.

$$SF_6(g) + 3H_2O(\ell) \longrightarrow 6HF(g) + SO_3(g) \qquad \Delta G^0 = -193 \text{ kJ/mol}$$

In contrast, TeF_6 is somewhat less stable and decomposes in H_2O to form telluric acid, H_6TeO_6, within a day. Because of the kinetic stability of gaseous SF_6, it is sometimes inhaled with oxygen for X-ray examinations of the lungs. It could not be used in this way if the SF_6 reacted with H_2O in the lungs to produce HF and SO_3, as might be expected.

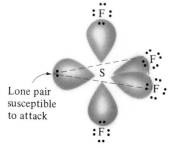

Lone pair susceptible to attack

SF_4 is sp^3d hybridized at S

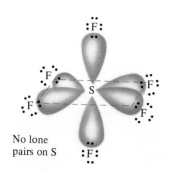

No lone pairs on S

SF_6 is sp^3d^2 hybridized at S

Table 25-4 Group VIA Halides					
Oxidation State	**< +2**	**+2**	**+4**	**+5**	**+6**
sulfur	S_2F_2	$(SF_2)^*$	SF_4	S_2F_{10}	SF_6
	S_2Cl_2	SCl_2	SCl_4		$SClF_5$
	S_2Br_2				$SBrF_5$
	S_nCl_2				
	S_nBr_2				
selenium	Se_2Cl_2	$(SeCl_2)$	SeF_4		SeF_6
	Se_2Br_2	$(SeBr_2)$	$SeCl_4$		
			$SeBr_4$		
tellurium	Te_3Cl_2	$TeCl_2$	TeF_4	Te_2F_{10}	TeF_6
	Te_2Br	$TeBr_2$	$TeCl_4$		
	TeI		$TeBr_4$		
	Te_xI		TeI_4		

*Compounds in parentheses have not been prepared in pure form.

The relative chemical stability of the EX_6 species is due in part to the absence of nonbonding valence electrons around the central atom. Each VIA element has six outer shell electrons, and all six are bonded to fluorine atoms.

Group VIA Oxides

Although others exist, the most important VIA oxides are the dioxides, which are acid anhydrides of H_2SO_3, H_2SeO_3, and H_2TeO_3; and the trioxides, which are anhydrides of H_2SO_4, H_2SeO_4, and H_6TeO_6.

+4 Oxidation State	
Formula	Name
H_2SO_3	sulfurous acid
H_2SeO_3	selenous acid
H_2TeO_3	tellurous acid

+6 Oxidation State	
Formula	Name
H_2SO_4	sulfuric acid
H_2SeO_4	selenic acid
H_6TeO_6	telluric acid

25-8 Group VIA Dioxides: SO_2, SeO_2, and TeO_2

The tendency toward metallic character as a group is descended is illustrated by the stable forms of the dioxides of S, Se, and Te. The very nonmetallic sulfur is covalently bonded to oxygen in SO_2, as is selenium in polymeric SeO_2, but the more metallic tellurium forms ionic TeO_2.

sulfur dioxide (molecular gas)

selenium dioxide (polymeric solid)

tellurium dioxide (ionic crystals)

Double bonds are easily formed between unlike atoms such as S and O, which have *small* electronegativity differences and *large* electronegativity sums. SO_2 and SeO_2 are quite soluble in H_2O, but TeO_2 is only slightly soluble because of its high crystal lattice energy.

SO_2

Sulfur dioxide is a colorless, poisonous, corrosive gas with a very irritating odor. Even in small quantities, it causes coughing and nose, throat, and lung irritation. It is an angular molecule with trigonal planar electronic geometry, sp^2 hybridization at the S atom, and resonance stabilization. Both sulfur–oxygen bonds are of equal length, 1.43 Å, and intermediate in length between typical single and double bonds. The resonance structures are

$119.5°$

Sulfur dioxide is produced in reactions such as the combustion of sulfur-containing fossil fuels and the roasting of sulfide ores.

$$2ZnS + 3O_2 \longrightarrow 2ZnO + 2SO_2$$

SO_2 is a waste product of these operations. In the past, it was released into the atmosphere along with some SO_3 produced by its reaction with O_2. However, efforts are now under way to trap SO_2 and SO_3 and use them to make H_2SO_4. Some coal contains up to 5% sulfur, so both SO_2 and SO_3 are present in the flue gases when coal is burned. No way has been found to remove all the SO_2 from flue gases of power plants. One way of removing most of the SO_2 involves the injection of limestone, $CaCO_3$, into the com-

Sphalerite is ZnS, the mineral in zinc ores. Zinc is an important metal obtained from sulfide ores.

If the SO_2 and SO_3 are allowed to escape into the atmosphere, they cause highly acidic rain.

Volcanoes are natural sources of SO_2 and H_2S.

bustion zone of the furnace. Here $CaCO_3$ decomposes to lime, CaO. This then combines with SO_2 to form calcium sulfite ($CaSO_3$), an ionic solid, which is collected.

$$CaCO_3 \xrightarrow{\Delta} CaO + CO_2 \qquad \text{followed by} \qquad CaO + SO_2 \longrightarrow CaSO_3$$

This process is called scrubbing. A disadvantage of it is the formation of huge quantities of solid waste ($CaSO_3$, unreacted CaO, and by-products).

Catalytic oxidation is now used by the smelting industry to convert SO_2 into SO_3. This is then dissolved in water to make solutions of H_2SO_4 (up to 80% by mass). The gases containing SO_2 are passed through a series of condensers containing catalysts to speed up the reaction. In some cases the impure H_2SO_4 can be used in other operations in the same plant. In other cases it is sold.

Sulfur dioxide is also an important industrial, commercial, and research chemical. It can be prepared in the laboratory by treating sodium hydrogen sulfite with a nonvolatile acid such as sulfuric or phosphoric acid.

$$2NaHSO_3 + H_2SO_4 \longrightarrow Na_2SO_4 + 2SO_2 + 2H_2O$$

It can also be prepared by the action of very weak reducing agents on hot concentrated sulfuric acid. The reaction of metallic copper is an example.

$$Cu + 2H_2SO_4 \longrightarrow CuSO_4 + SO_2 + 2H_2O$$

Large amounts of SO_2 and H_2S are released during volcanic eruptions.

SeO_2 and TeO_2

Selenium and tellurium dioxides can be formed by burning the elements.

$$Se + O_2 \xrightarrow{\Delta} SeO_2 \qquad\qquad Te + O_2 \xrightarrow{\Delta} TeO_2$$

Concentrated HNO_3 dissolves Se and Te to give solutions that contain selenous and tellurous acids, respectively.

$$4\,HNO_3 + Se \longrightarrow H_2SeO_3 + 4\,NO_2 + H_2O$$
$$4\,HNO_3 + Te \longrightarrow H_2TeO_3 + 4\,NO_2 + H_2O$$

When these solutions are evaporated to dryness by heating, the acids decompose to form dioxides.

$$H_2SeO_3 \xrightarrow{\Delta} SeO_2 + H_2O \qquad\qquad H_2TeO_3 \xrightarrow{\Delta} TeO_2 + H_2O$$

25-9 Group VIA Trioxides: SO_3, SeO_3, and TeO_3

SO_3

Sulfur trioxide is a liquid that boils at 44.8°C. It is the anhydride of H_2SO_4. It is formed by the reaction of SO_2 with O_2. The reaction is very exothermic, but ordinarily very slow. It is catalyzed commercially in the **contact process** by spongy Pt, SiO_2, or vanadium(V) oxide, V_2O_5, at high temperatures (400 to 700°C).

$$2SO_2(g) + O_2(g) \xrightleftharpoons{\text{catalyst}} 2SO_3(g)$$

$$\Delta H^0 = -197.6 \text{ kJ/mol}, \Delta S^0 = -188 \text{ J/K} \cdot \text{mol}$$

The high temperature favors SO_2 and O_2 but allows equilibrium to be reached much more rapidly, so it is economically advantageous. The SO_3 is then removed from the gaseous reaction mixture by dissolving it in concentrated H_2SO_4 (95% H_2SO_4 by mass) to produce polysulfuric acids—mainly pyrosulfuric acid, $H_2S_2O_7$. This is called oleum, or fuming sulfuric acid. Addition of fuming sulfuric acid to water produces commercial H_2SO_4.

$$SO_3 + H_2SO_4 \longrightarrow H_2S_2O_7 \quad \text{then} \quad H_2S_2O_7 + H_2O \longrightarrow 2H_2SO_4$$

In the presence of certain catalysts, sulfur dioxide in polluted air reacts rapidly with O_2 to form SO_3. Particulate matter, or suspended microparticles, such as NH_4NO_3 and elemental S, act as efficient catalysts.

Both SO_3 and S_3O_9 molecules are found in all three physical states.

$$3SO_3 \rightleftharpoons S_3O_9$$

SO_3 molecules are **trimers**—large molecules formed by the combination of three small molecules. The SO_3 molecule is trigonal planar, containing sp^2 hybridized sulfur. It is represented by the following resonance formulas:

$$:\!\overset{\displaystyle ..}{\underset{\displaystyle ..}{O}}\!: \quad \overset{\displaystyle .}{\underset{\displaystyle}{O}} \quad :\!\overset{\displaystyle ..}{\underset{\displaystyle ..}{O}}\!: \quad O$$

resonance formulas of SO_3, showing the trigonal planar structure with a $120°$ bond angle.

The S_3O_9 molecule is a puckered ring that can be thought of as three distorted SO_4 tetrahedra (sp^3 hybridization at S) joined by common corners.

Sulfur trioxide decomposes into SO_2 and O_2 at temperatures above 900°C. It acts as an oxidizing agent in many reactions. For example, it converts S_8 to SO_2, SCl_2 to $SOCl_2$ and SO_2Cl_2, P_4 to P_4O_{10}, and PCl_3 to $POCl_3$. It acts as a Lewis acid in reactions with Lewis bases such as magnesium oxide.

$$Mg^{2+} + :\!\overset{..}{\underset{..}{O}}\!:^{2-} + \quad \overset{:\overset{..}{O}:}{\underset{\underset{O}{}{S}}{}} \longrightarrow Mg^{2+} + \left[:\!\overset{..}{\underset{..}{O}}\!:\!S\!:\!\overset{..}{\underset{..}{O}}\!: \right]^{2-}$$

magnesium oxide   sulfur trioxide   magnesium sulfate
(a Lewis base)   (a Lewis acid)

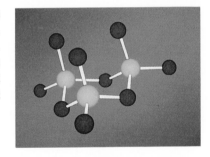

S_3O_9

SeO_3 and TeO_3

Both SeO_3 and TeO_3 are stronger oxidizing agents than SO_3. They are the anhydrides of selenic and telluric acids, respectively. SeO_3 is prepared by the reaction of potassium selenate, K_2SeO_4, and SO_3.

$$K_2SeO_4 + SO_3 \longrightarrow K_2SO_4 + SeO_3$$

Dehydration of telluric acid, H_6TeO_6, at 300 to 600°C produces water-insoluble TeO_3, which decomposes to Te_2O_5 and O_2 above 400°C.

$$H_6TeO_6 \overset{\Delta}{\longrightarrow} TeO_3 + 3H_2O$$

The prefix *pyro* means "heat" or "fire." Pyrosulfuric acid may also be obtained by heating concentrated sulfuric acid, which results in the elimination of one molecule of water per two molecules of sulfuric acid.

$$2H_2SO_4 \longrightarrow H_2S_2O_7 + H_2O$$

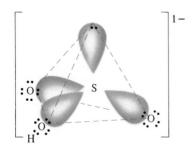

Hydrogen sulfite ion

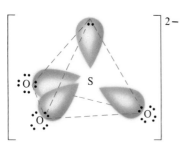

Sulfite ion

Hygroscopic substances readily absorb water from the air.

Oxyacids of Sulfur, Selenium, and Tellurium

25-10 Sulfurous, Selenous, and Tellurous Acids

H_2SO_3

Sulfur dioxide readily dissolves in water to produce solutions of sulfurous acid, H_2SO_3. The acid has not been isolated in anhydrous form.

$$H_2O + SO_2 \rightleftharpoons H_2SO_3$$

The acid ionizes in two steps in water:

$$H_2SO_3 \rightleftharpoons H^+ + HSO_3^- \qquad K_1 = 1.2 \times 10^{-2}$$

$$HSO_3^- \rightleftharpoons H^+ + SO_3^{2-} \qquad K_2 = 6.2 \times 10^{-8}$$

When excess SO_2 is bubbled into aqueous NaOH, sodium hydrogen sulfite, $NaHSO_3$, is produced. This acid salt can be neutralized with additional NaOH or Na_2CO_3 to produce sodium sulfite.

$$NaOH + H_2SO_3 \longrightarrow NaHSO_3 + H_2O$$

$$NaOH + NaHSO_3 \longrightarrow Na_2SO_3 + H_2O$$

The sulfite ion is pyramidal and has tetrahedral electronic geometry.

H_2SeO_3 and H_2TeO_3

Selenous acid, H_2SeO_3, can be prepared by dissolving its anhydride, SeO_2, in water.

$$H_2O + SeO_2 \longrightarrow H_2SeO_3$$

The resulting selenous acid is the only one of the *-ous* acids of sulfur, selenium, and tellurium that can be obtained as a solid. H_2SeO_3 is a white hygroscopic solid. It is a diprotic acid only slightly weaker than H_2SO_3. It ionizes in two steps with $K_1 = 2.7 \times 10^{-3}$ and $K_2 = 2.5 \times 10^{-7}$. Neutralization of H_2SeO_3 produces hydrogen selenite and selenite salts.

Tellurous acid, H_2TeO_3, cannot be produced by dissolving TeO_2 in water because TeO_2 is so insoluble. Tellurites are prepared by dissolving TeO_2 in alkaline solutions. Adding acids to solutions of tellurite salts produces tellurous acid, which has K_a values $K_1 = 2 \times 10^{-3}$ and $K_2 = 1 \times 10^{-8}$.

In view of the weakly metallic character of tellurium, it is not surprising that tellurous acid can also ionize slightly as a base in water.

$$H_2TeO_3 \rightleftharpoons TeO(OH)^+ + OH^- \qquad K \approx 10^{-12}$$

25-11 Sulfuric, Selenic, and Telluric Acids

H_2SO_4

More than 40 million tons of sulfuric acid are produced annually worldwide. The contact process is used for the commercial production of most sulfuric acid. The solution sold commercially as "concentrated sulfuric acid" is 96 to 98% H_2SO_4 by mass, and is about 18 molar H_2SO_4.

Pure H_2SO_4 is a colorless, oily liquid that freezes at 10.4°C and boils at 290 to 317°C while partially decomposing to SO_3 and water. There is some hydrogen bonding in solid and liquid H_2SO_4.

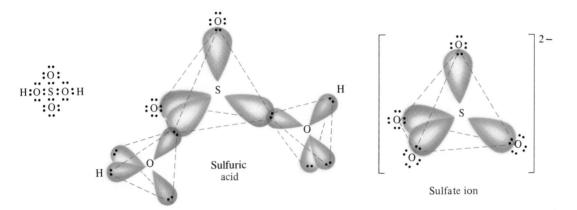

Sulfuric acid

Sulfate ion

Tremendous amounts of heat are evolved when concentrated sulfuric acid is diluted. This illustrates the strong affinity of H_2SO_4 for water. H_2SO_4 is often used as a dehydrating agent. *Dilutions should always be performed by adding the acid to water to avoid spattering the acid.*

Sulfuric acid is a strong acid with respect to the first step of its ionization in water. The second ionization occurs to a lesser extent (Example 18-22).

$$H_2SO_4 \rightleftharpoons H^+ + HSO_4^- \qquad K_1 = \text{very large}$$

$$HSO_4^- \rightleftharpoons H^+ + SO_4^{2-} \qquad K_2 = 1.2 \times 10^{-2}$$

Pouring concentrated H_2SO_4 into an equal volume of H_2O liberates a lot of heat, enough to raise the temperature of the resulting solution from room temperature to 121°C.

The increasing acidity of rain is due in large part to the presence of H_2SO_4 and some H_2SO_3, which are formed by reaction of SO_3 and SO_2 with moisture. The oxides are produced during combustion of sulfur-containing fossil fuels and smelting operations. Even in unpolluted air, rain is *slightly* acidic because of the presence of atmospheric CO_2, which reacts with moisture to produce carbonic acid, H_2CO_3. However, rain in some areas has been found to have a pH of about 2, due mainly to the presence of H_2SO_4. As a reference point, the pH of vinegar is about 2.9. Refer to the Chemistry in Use essay on acid rain in Chapter 6.

Sulfuric acid reacts with structural materials such as limestone and marble (both $CaCO_3$).

$$H_2SO_4 + CaCO_3 \longrightarrow CaSO_4 + CO_2 + H_2O$$

The calcium sulfate then washes away in the rain (see Figure 25-3). H_2SO_4 also breaks down structures composed of metals such as steel (which contains Fe) and Al (which is always coated with an oxide surface when exposed to air).

$$H_2SO_4 + Fe \longrightarrow FeSO_4 + H_2$$

$$3H_2SO_4 + Al_2O_3 \longrightarrow Al_2(SO_4)_3 + 3H_2O$$

Sulfuric acid reacts with many organic compounds, including those in plants and human flesh. The lungs are very susceptible to irritation.

Figure 25-3
Acid rain attacks many structural materials, as demonstrated by this partly dissolved stone statue.

Some of the important properties of H_2SO_4 are summarized below.

1. It is a very strong acid that is corrosive to living and structural materials.
2. It has a high boiling point and can be used to produce volatile acids such as HCl and HCN.
3. It is a mild oxidizing agent. It cannot be used to prepare HBr or HI. Hot concentrated H_2SO_4 dissolves Cu.
4. It is a strong dehydrating agent that is used in organic reactions to remove H_2O and drive reactions toward products, and as a drying agent in desiccators.

H_2SeO_4

Selenic acid is prepared by oxidation of SeO_2 (or H_2SeO_3) with 30% H_2O_2. Selenium trioxide is not used directly because of its instability. H_2SeO_4 can be obtained in the solid form as colorless hygroscopic crystals. Selenates are prepared by neutralization of selenic acid solutions or by oxidation of selenite salts. The acid is about as strong as H_2SO_4. It ionizes in two steps.

$$H_2SeO_4 \rightleftharpoons H^+ + HSeO_4^- \qquad K_1 = \text{very large}$$

$$HSeO_4^- \rightleftharpoons H^+ + SeO_4^{2-} \qquad K_2 = 1.15 \times 10^{-2}$$

H_2SeO_4 and its anions have the same geometries as H_2SO_4 and its anions. They are even stronger oxidizing agents. Interestingly, H_2SeO_4 is the only pure acid that dissolves gold.

H_6TeO_6

H$_6$TeO$_6$ has the stoichiometry $H_2TeO_4 \cdot 2H_2O$. Compare with H_5IO_6 (Section 24-13).

Telluric acid, H_6TeO_6 or $Te(OH)_6$, is obtained by oxidation of TeO_2 with 30% hydrogen peroxide, H_2O_2; or aqueous chloric acid, $HClO_3$; or a solution of nitric acid, HNO_3, and potassium permanganate, $KMnO_4$. Apparently because of its large size, tellurium is able to form a six-coordinate acid; it does so by undergoing sp^3d^2 hybridization. The Te in H_6TeO_6 is octahedrally bonded. Salts of the types $MTeO(OH)_5$ and $M_2TeO_2(OH)_4$ are known, as well as Na_6TeO_6 and Ag_6TeO_6.

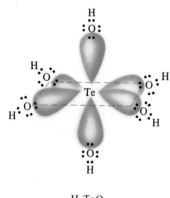

H_6TeO_6

25-12 Thiosulfuric Acid and Thiosulfates

The prefix *thio-* designates the replacement of an oxygen atom with a sulfur atom. Thus, thiosulfuric acid is $H_2S_2O_3$ and the thiosulfate ion is $S_2O_3^{2-}$. Not surprisingly, the acid is prepared by reacting SO_3 with H_2S (rather than with H_2O, which would produce H_2SO_4) at low temperatures in ether.

$$H_2S + SO_3 \xrightarrow[\text{ether}]{\text{cold}} H_2S_2O_3 \qquad \text{(thiosulfuric acid exists only in solution)}$$

The acid is unstable and disproportionates to sulfurous acid and sulfur.

Thiosulfate salts are produced by boiling aqueous sulfite salts with sulfur.

$$Na_2SO_3 + S \xrightarrow{\Delta} Na_2S_2O_3 \qquad \text{sodium thiosulfate}$$

A solution of sodium thiosulfate pentahydrate, $Na_2S_2O_3 \cdot H_2O$, is known as photographer's "fixer." It dissolves unexposed AgBr in photographic film to form soluble $Na_3[Ag(S_2O_3)_2]$, which contains the complex ion $[Ag(S_2O_3)_2]^{3-}$.

The structures of thiosulfuric acid and the thiosulfate ion are analogous to those of sulfuric acid and the sulfate ion.

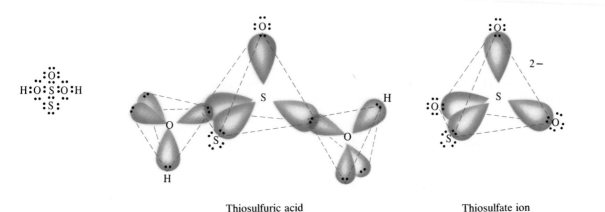

Thiosulfuric acid   Thiosulfate ion

Key Terms

Contact process An industrial process by which sulfur trioxide and sulfuric acid are produced from sulfur dioxide.

Frasch process A method by which elemental sulfur is mined or extracted. Sulfur is melted with superheated water (at 170°C under high pressure) and forced to the surface of the earth as a slurry.

Particulate matter Finely divided solid particles suspended in polluted air.

Polymer A large molecule consisting of chains or rings of linked monomer units, usually characterized by high melting and boiling points.

Exercises

1. Write abbreviated electronic configurations for atomic oxygen, selenium, and polonium.
2. Write out the electronic configurations of oxide, sulfide, and selenide ions.
3. Characterize the Group VIA elements with respect to color and physical state under normal conditions.
4. The Group VIA elements, except oxygen, can exhibit oxidation states ranging from -2 to $+6$, but not -3 or $+7$. Why?
5. List and explain the order of increasing melting and boiling points of the Group VIA elements.
6. Is the order of decreasing first ionization energies of the Group VIA elements consistent with the order of increasing metallic character?
7. Sulfur, selenium, and tellurium are all capable of forming six-coordinate compounds such as SF_6. Give two reasons why oxygen cannot be the central atom in such six-coordinate molecules.
8. For the following species, draw (1) diagrams that show the hybridization of atomic orbitals and (2) three-dimensional structures that show all hybridized orbitals

and outermost electrons. (3) Determine the oxidation state of the Group VIA element (other than oxygen) in each species. (a) H_2S, (b) SF_6, (c) SF_4, (d) SO_2, (e) SO_3.
9. Repeat Exercise 8 for (a) SeF_6, (b) SO_3^{2-}, (c) SO_4^{2-}, (d) HSO_4^-, (e) thiosulfate ion, $S_2O_3^{2-}$ (one S is central atom).
10. Write equations for the reactions of (a) S, Se, and Te with excess F_2; (b) O_2, S, Se, and Te with H_2; (c) S, Se, and Te with O_2.
11. Write equations for the reactions of (a) S and Te with HNO_3; (b) S and Se with excess Cl_2; (c) S and Se with Na, Ca, and Al.
12. What is the order of increasing melting and boiling points and heats of vaporization of the Group VIA hydrides, H_2O, H_2S, H_2Se, and H_2Te? What are their physical states under normal conditions? Why is H_2O out of line with the others?
13. Use Table 25-1 to predict or estimate (a) the color of Po and (b) the ionic radius of Po^{2-}.
14. Use Table 25-3 to estimate the following properties of H_2Po: (a) melting point, (b) boiling point, (c) density.

15. Describe three general methods for the preparation of hydrogen sulfide.

16. Discuss the acidity of the aqueous Group VIA hydrides, including the relative values of acid ionization constants. What is primarily responsible for the order of increasing acidities in this series?

17. Write equations for the reactions of (a) aqueous H_2S with O_2; (b) gaseous H_2S with O_2 in the presence of heat; (c) excess gaseous H_2S with O_2 in the presence of heat; (d) gaseous H_2S with aluminum metal; (e) aqueous H_2S with NaOH.

18. Why is sulfur tetrafluoride a potent poison, whereas SF_6 is not? Why is SF_4 so reactive, whereas SF_6 is not?

19. Compare the structures of the dioxides of sulfur, selenium, tellurium, and polonium. How do they relate to the metallic or nonmetallic character of these elements?

20. Write equations for the following reactions:
    (a) the preparation of sulfur dioxide by the reaction of copper with hot concentrated sulfuric acid
    (b) the reaction of sulfur dioxide with oxygen
    (c) the reaction of sulfur dioxide with water

21. Draw the structure of the S_3O_9 molecule. How are the sulfur atoms hybridized? With what species is this molecule always in equilibrium?

22. Draw a Lewis dash representation of pyrosulfuric acid, $H_2S_2O_7$. Write an equation to show how it is prepared.

23. What are the acid anhydrides of sulfuric acid, selenic acid, and telluric acid?

24. Write equations for reactions of
    (a) NaOH with sulfuric acid (1:1 mole ratio)
    (b) NaOH with sulfuric acid (2:1 mole ratio)
    (c) NaOH with sulfurous acid (1:1 mole ratio)
    (d) NaOH with sulfurous acid (2:1 mole ratio)
    (e) NaOH with selenic acid, H_2SeO_4 (1:1 mole ratio)
    (f) NaOH with selenic acid (2:1 mole ratio)
    (g) NaOH with tellurium dioxide (1:1 mole ratio)
    (h) NaOH with tellurium dioxide (2:1 mole ratio)

25. Which metal sulfides are considered to be water soluble? Will solutions containing the soluble sulfides be alkaline, neutral, or acidic?

26. When zinc metal is added to dilute H_2SO_4, H_2 is generated; when zinc metal is added to concentrated H_2SO_4, H_2S is formed. Write chemical equations for these reactions and explain why there is a difference in the reactions.

*27. How much sulfur dioxide is produced from complete combustion of 1 ton of coal containing 6.23% sulfur?

*28. A sterling silver serving piece contains 137 g of silver. If 0.144 g of silver sulfide (tarnish) forms by reaction of the silver with H_2S from the decomposition of eggs, how much silver must react? What percentage of the silver tarnishes?

*29. Calculate the concentrations of H^+, HSO_3^- and SO_3^{2-} ions present in 0.050 M sulfurous acid, H_2SO_3, solution.

*30. Calculate the concentration of OH^- and the pH in 0.10 M aqueous sodium sulfite, Na_2SO_3.

31. Write reaction equations for the preparation of (a) thiosulfuric acid and (b) potassium thiosulfate.

32. The enthalpy change for dissolving sulfur in six moles of CS_2 at 25°C

$$S(s) + 6CS_2(\ell) \longrightarrow S(in\ CS_2)$$

is 1695 J/mol for rhombic sulfur and 1360 J/mol for monoclinic sulfur. Calculate ΔH^0 for the phase transformation

$$S(rhombic) \longrightarrow S(monoclinic)$$

Which form of sulfur would you predict to be more stable at this temperature?

33. At 502 torr and 750°C, 1.00 L of sulfur vapor is found to weigh 0.5350 g. What is the molecular formula of the major component of the vapor?

34. What mass of H_2SO_4 could be produced in the process given below if 1.00 kg of FeS_2 is used? The unbalanced equations for the process are

$$FeS_2(s) + O_2(g) \longrightarrow Fe_2O_3(s) + SO_2(g)$$
$$SO_2(g) + O_2(g) \longrightarrow SO_3(g)$$
$$SO_3(g) + H_2SO_4(\ell) \longrightarrow H_2S_2O_7(\ell)$$
$$H_2S_2O_7(\ell) + H_2O(\ell) \longrightarrow H_2SO_4(aq)$$

Assume complete reactions.

35. Common copper ores in western United States contain the mineral chalcopyrite, $CuFeS_2$. Assuming that an average commercially useful ore contains 0.263 mass % Cu and that all the sulfur ultimately appears in the smelter stack gases as SO_2, calculate the mass of sulfur dioxide generated by the conversion of 1.00 kg of the ore.

*36. A gaseous mixture at some temperature in a 1.00 L vessel originally contained 1.00 mol SO_2 and 5.00 mol O_2. Once equilibrium conditions were attained, 78.3% of the SO_2 had been converted to SO_3. What is the value of the equilibrium constant (K_c) for this reaction at this temperature?

Nitric acid, HNO₃, reacts with protein-containing materials, staining them yellow. A feather contains protein. Perhaps you have spilled nitric acid on your skin and seen it turn yellow.

Outline

Objectives

As you study this chapter, you should learn

☐ About the occurrence, properties, and importance of nitrogen
☐ About the nitrogen cycle and its importance
☐ About how nitrogen is obtained and the important oxidation states of nitrogen
☐ About some compounds of nitrogen and hydrogen
   ammonia
   the amines
   hydrazine
☐ About the oxides of nitrogen

☐ About some oxyacids of nitrogen and their salts
☐ About phosphorus and arsenic
   occurrence, production, and uses of phosphorus
   arsenic, occurrence and uses
   allotropes of phosphorus and arsenic
   halides of phosphorus and arsenic
   oxides and acids of phosphorus and arsenic

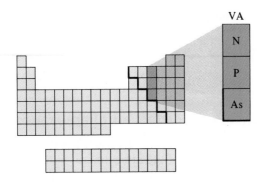

n the nitrogen family, nitrogen and phosphorus are nonmetals, arsenic is predominantly nonmetallic, antimony is more metallic, and bismuth is definitely metallic. Properties of the Group VA elements are listed in Table 26-1.

Oxidation states of the VA elements range from -3 to $+5$. Odd-numbered oxidation states are favored. The VA elements form very few monatomic ions. Ions with a charge of $3-$ occur for N and P, as in Mg_3N_2 and Ca_3P_2. Tripositive cations probably exist for antimony and bismuth in such compounds as antimony(III) sulfate, $Sb_2(SO_4)_3$, and bismuth(III) perchlorate pentahydrate, $Bi(ClO_4)_3 \cdot 5H_2O$. In aqueous solution these are extensively hydrolyzed to SbO^+ or $SbOX(s)$ and BiO^+ or $BiOX(s)$, where X is a univalent anion. These hydrolyzed solutions are strongly acidic (Section 19-5).

All of the Group VA elements show the -3 oxidation state in covalent compounds such as NH_3, PH_3, and AsH_3. The $+5$ oxidation state is found only in covalent compounds such as phosphorus pentafluoride, PF_5, and phosphoric acid, H_3PO_4, and in polyatomic ions such as NO_3^-, PO_4^{3-} and BiO_3^-. N and P show many oxidation states in their compounds, but for As, Sb, and Bi, the common ones are $+3$ and $+5$.

Each Group VA element exhibits the $+3$ oxidation state in one of its oxides: N_2O_3, P_4O_6, As_4O_6, Sb_4O_6, and Bi_2O_3. The first two are acid anhydrides of nitrous acid, HNO_2, and phosphorous acid, H_3PO_3; both are weak acids. As_4O_6 and Sb_4O_6 are amphoteric; As_4O_6 has more acidic character and Sb_4O_6 more basic character. Neither oxide dissolves in water to any

Table 26-1
Properties of the Group VA Elements

Property	N	P	As	Sb	Bi
Physical state (1 atm, 25°C)	Gas	Solid	Solid	Solid	Solid
Color	Colorless	Red, white, black	Yellow, gray	Yellow, gray	Gray
Outermost electrons	$2s^22p^3$	$3s^23p^3$	$4s^24p^3$	$5s^25p^3$	$6s^26p^3$
Melting point (°C)	-210	44 (white)	814 (gray)	631 (gray)	271
Boiling point (°C)	-196	280 (white)	sublimes 613	1750	1560
Atomic radius (Å)	0.70	1.10	1.21	1.41	1.46
Electronegativity	3.0	2.1	2.1	1.9	1.8
First ionization energy (kJ/mol)	1402	1012	947	834	703
Oxidation states	-3 to $+5$	-3 to $+5$	-3 to $+5$	-3 to $+5$	-3 to $+5$

significant extent. The trend of increasing oxide basicity as the group is descended shows increasingly metallic character of the elements.

Nitrogen

26-1 Occurrence, Properties, and Importance

Nitrogen, N_2, is a colorless, odorless, tasteless gas that makes up about 75% by mass and 78% by volume of the atmosphere. Nitrogen compounds form only a minor portion of the earth's crust, but all living matter contains nitrogen. The primary natural inorganic deposits of nitrogen are very localized. They consist mostly of KNO_3 and $NaNO_3$. Most sodium nitrate is mined in Chile.

The extreme abundance of N_2 in the atmosphere and the low relative abundance of nitrogen compounds elsewhere are due to the chemical inertness of N_2 molecules. This results from the very high bond energy of the $N\equiv N$ bond (946 kJ/mol). The Lewis dot formula shows a triple bond and no unpaired electrons, consistent with the experimentally observed diamagnetism of N_2.

Every protein contains nitrogen in each of its fundamental amino acid units.

$:N:::N:$

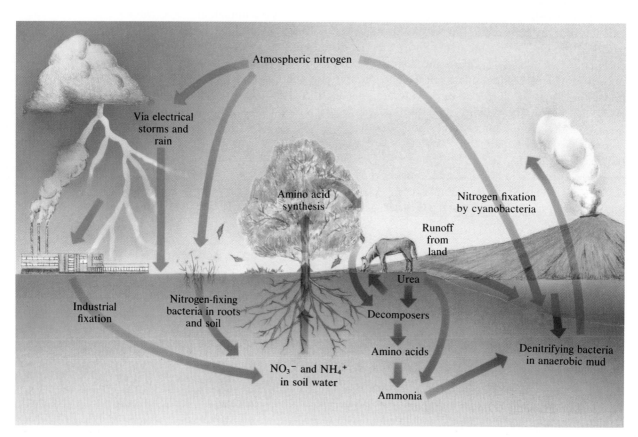

A schematic representation of the nitrogen cycle.

26-2 The Nitrogen Cycle

Although N_2 molecules are unreactive, nature provides mechanisms by which N atoms are incorporated into proteins, nucleic acids, and other nitrogenous compounds. The **nitrogen cycle** is the complex series of reactions by which nitrogen is slowly but continually recycled in the atmosphere, lithosphere (earth), and hydrosphere (water).

When N_2 and O_2 molecules collide in the atmosphere (our nitrogen reservoir) near a bolt of lightning, they can absorb enough electrical energy to produce molecules of NO. An NO molecule is quite reactive because it contains one unpaired electron. NO reacts readily with O_2 to form nitrogen dioxide, NO_2. Most NO_2 dissolves in rainwater and falls to the earth's surface. Bacterial enzymes reduce the nitrogen in a series of reactions in which amino acids and proteins are produced. These are then used by plants, eaten by animals, and metabolized. Then they are excreted as nitrogenous compounds such as urea, $(NH_2)_2CO$, and ammonium salts such as $NaNH_4HPO_4$. These products can also be enzymatically converted to ammonia, NH_3, and amino acids.

Root nodules on soybeans.

Nitrogen is converted directly into NH_3 in another way. Members of the class of plants called legumes (including soybeans, alfalfa, and clover) have nodules on their roots. Within the nodules live bacteria that produce an enzyme called nitrogenase. These bacteria extract N_2 directly from air trapped in the soil and convert it into NH_3. The ability of nitrogenase to catalyze such a conversion, called **nitrogen fixation**, at usual temperatures and pressures with very high efficiency is a marvel to scientists. They must resort to very extreme and costly conditions to produce NH_3 from nitrogen and hydrogen (the Haber process, Section 17-6). We hope to learn through research how the enzymatic conversion is carried out, so that we can accomplish the transformation much more efficiently.

Ammonia is the source of nitrogen in many fertilizers. Unfortunately, nature does not produce NH_3 and related plant nutrient compounds rapidly enough to provide an adequate food supply for the world's growing population. Commercial synthetic fertilizers have helped to lessen this problem, but at great cost for the energy that is required to produce them.

26-3 Production and Oxidation States of Nitrogen

Nitrogen is sold as compressed gas in cylinders. The boiling point of N_2 is $-195.8°C$ ($-320°F$). N_2 is obtained by fractional distillation of liquid air. Tank nitrogen contains traces of O_2, Ar, and H_2O vapor. Further purification is accomplished by passing the gas over hot copper, which reacts with O_2 to form CuO, and then drying it with tetraphosphorus decoxide, P_4O_{10}, a very effective dehydrating agent. The remaining Ar is so inert that its presence usually causes no problems.

Small amounts of very pure, argon-free N_2 are produced by heating pure azides of the Group IA or IIA metals. The reaction produces the metal and pure N_2.

Great advances have been made in cattle breeding in recent decades. Semen from superior bulls can be collected and stored in liquid nitrogen for 30 years or more.

$$2NaN_3(s) \xrightarrow{300°C} 2Na(\ell) + 3N_2(g)$$

sodium azide

Table 26-2
Oxidation States of Nitrogen and Examples

−3	−2	−1	0	+1	+2	+3	+4	+5
NH_3 ammonia	N_2H_4 hydrazine	NH_2OH hydroxylamine	N_2 nitrogen	N_2O dinitrogen oxide	NO nitrogen oxide	N_2O_3 dinitrogen trioxide	NO_2 nitrogen dioxide	N_2O_5 dinitrogen pentoxide
NH_4^+ ammonium ion		NH_2Cl chloramine		$H_2N_2O_2$ hyponitrous acid		HNO_2 nitrous acid	N_2O_4 dinitrogen tetroxide	HNO_3 nitric acid
NH_2^- amide ion						NO_2^- nitrite ion		NO_3^- nitrate ion

No other element exhibits more oxidation states than nitrogen does (Table 26-2).

Nitrogen Compounds with Hydrogen

26-4 Ammonia and Ammonium Salts

34 billion pounds of NH_3 were produced in the United States in 1990. This is approximately 136 pounds per person.

Ammonia is produced commercially by the **Haber process** (Section 17-6).

$$N_2(g) + 3H_2(g) \xrightarrow[\text{high T, P}]{\text{Fe, Fe oxides,}} 2NH_3(g) \qquad \Delta H° = -92.2 \text{ kJ/mol rxn}$$

Ammonia can be prepared in the laboratory by treating aqueous NH_3 or ammonium salts with excess strong base. Excess OH^- favors the reaction to the right. NH_3 is evolved as a gas.

$$NH_4^+ + OH^- \underset{\text{excess}}{\overset{\text{shift}}{\rightleftharpoons}} NH_3 + H_2O$$

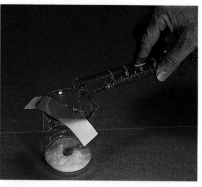

$NH_4Cl(s) + NaOH(aq) \longrightarrow$
$\qquad\qquad NH_3(g) + H_2O$

The NH_3 gas that is formed changes bromthymol blue indicator from yellow to blue.

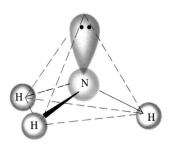

Ammonia, NH_3, mp −77.7°C; bp −33.4°C; polar, pyramidal molecule; tetrahedral (sp^3) electronic geometry.

Ammonia is a colorless gas with a characteristic pungent odor. The molecules are pyramidal and quite polar (Section 8-8). NH_3 is very soluble in water because it forms hydrogen bonds with water. Saturated aqueous NH_3 is 15 M. The lone pair of electrons on the N atom allows NH_3 to function as a base in its reaction with H_2O (Section 10-3).

Household ammonia is a dilute aqueous solution containing about 5% NH_3 and some detergent.

$$NH_3(aq) + H_2O(\ell) \rightleftharpoons NH_4^+(aq) + OH^-(aq) \qquad K_b = 1.8 \times 10^{-5}$$

Ammonia acts as a Lewis base (Section 10-10) when it reacts with metal ions, such as aqueous Co^{2+} or Cu^{2+}, to form complex ions (Sections 20-6 and 29-2), such as $[Co(NH_3)_6]^{2+}$ or $[Cu(NH_3)_4]^{2+}$.

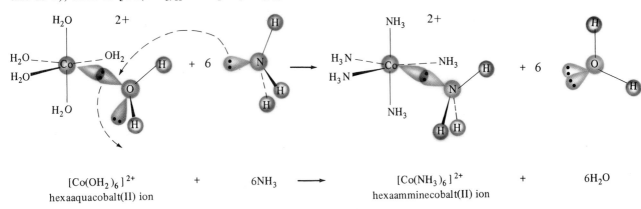

$$[Co(OH_2)_6]^{2+} \qquad + \qquad 6NH_3 \qquad \longrightarrow \qquad [Co(NH_3)_6]^{2+} \qquad + \qquad 6H_2O$$
$$\text{hexaaquacobalt(II) ion} \qquad\qquad\qquad\qquad\qquad \text{hexaamminecobalt(II) ion}$$

Ammonia acts as a Lewis base in its reactions with other strong electron acceptors to form "addition compounds." Its reaction with boron trifluoride, BF_3, is typical.

$$BF_3(g) + {:}NH_3(g) \longrightarrow F_3B{:}NH_3(s)$$

When heated in oxygen, ammonia burns to produce N_2 and water.

$$4NH_3(g) + 3O_2(g) \xrightarrow{\Delta} 2N_2(g) + 6H_2O(g) \quad \Delta H^0 = -1.27 \times 10^3 \text{ kJ/mol rxn}$$

But in the presence of red-hot Pt, NO rather than N_2 is produced:

$$4NH_3(g) + 5O_2(g) \xrightarrow[\text{Pt}]{\Delta} 4NO(g) + 6H_2O(g) \qquad \Delta H^0 = -904 \text{ kJ/mol rxn}$$

This is an important reaction in the Ostwald process for making HNO_3.

Liquid ammonia (bp $-33.4°C$) is used as a solvent for some chemical reactions. It is hydrogen bonded, just as H_2O is, but NH_3 is a much more basic solvent. Its weak *autoionization* produces the ammonium ion, NH_4^+, and the amide ion, NH_2^-. This is similar to H_2O, which ionizes to produce some H_3O^+ and OH^- ions.

$$NH_3(\ell) + NH_3(\ell) \rightleftharpoons NH_4^+ + NH_2^- \qquad K = 10^{-30}$$
$$\text{base}_2 \qquad \text{acid}_1 \qquad\quad \text{acid}_2 \qquad \text{base}_1$$

Solid ammonium nitrate is a powerful oxidizing agent. Here we show its reaction with zinc powder.

Many ammonium salts are known. Most are very soluble in water. They can be prepared by reactions of ammonia with acids.

$$NH_3 + HCN \longrightarrow [NH_4^+ + CN^-] \qquad \text{ammonium cyanide}$$

$$NH_3 + [H^+ + NO_3^-] \longrightarrow [NH_4^+ + NO_3^-] \qquad \text{ammonium nitrate}$$

Ammonium nitrate is used as a source of nitrogen in fertilizers because of its high nitrogen content (35% by mass).

When heated, some ammonium salts that contain oxidizing anions decompose rapidly (and sometimes explosively to produce large volumes of gases). Ammonium nitrate and ammonium dichromate are examples.

$$2NH_4NO_3(s) \xrightarrow{\Delta} 2N_2(g) + 4H_2O(g) + O_2(g) + heat$$

$$(NH_4)_2Cr_2O_7(s) \xrightarrow{\Delta} N_2(g) + Cr_2O_3(s) + 4H_2O(g) + heat$$

26-5 Amines

Amines are organic compounds that are structurally related to ammonia. They are derived from NH_3 by the replacement of one or more hydrogens with organic groups (Section 31-11). All involve sp^3-hybridized N. All are bases because of the lone pair of electrons.

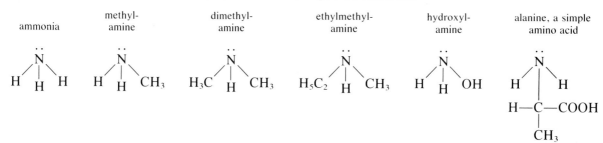

26-6 Hydrazine

Hydrazine, N_2H_4, a weak base, can be considered a derivative of ammonia, related by the replacement of one H by an $-NH_2$ group. It is a colorless, fuming liquid in the anhydrous form. It can be stored even though it is thermodynamically unstable with respect to decomposition ($\Delta G_f^0 = +149.2$ kJ/mol). Its unsymmetrical structure is consistent with its dipole moment of 1.35 D. Hydrazine can be thought of as a nitrogen structural analog of hydrogen peroxide, H_2O_2. It ionizes in two steps.

$$N_2H_4 + H_2O \rightleftharpoons N_2H_5^+ + OH^- \qquad K_1 = 8.5 \times 10^{-7}$$

$$N_2H_5^+ + H_2O \rightleftharpoons N_2H_6^{2+} + OH^- \qquad K_2 = 8.9 \times 10^{-15}$$

Hydrazine is a powerful reducing agent in basic solution, usually being oxidized to N_2. It is used with H_2O_2 and other related compounds for rocket fuels. It is used to make some drugs, industrial chemicals, and dyes.

Hydrazine, H_2NNH_2, mp 2°C, bp 113.5°C.

Nitrogen Oxides

Nitrogen forms several oxides, in which it exhibits positive oxidation states of 1 to 5 (Table 26-2). All have positive free energies of formation, owing to the high dissociation energy of N_2 and O_2 molecules. All are gases except N_2O_5, a solid that melts at 30.0°C. There is evidence for the existence of two very unstable oxides, NO_3 and N_2O_6.

26-7 Dinitrogen Oxide (+1 Oxidation State)

Molten ammonium nitrate undergoes auto-oxidation–reduction (decomposition) at 170 to 260°C to produce dinitrogen oxide, also called nitrous oxide. At higher temperatures, explosions occur, producing N_2, O_2, and H_2O.

$$\overset{(-3)}{N}H_4\overset{(+5)}{N}O_3(s) \xrightarrow{\Delta} \overset{(+1)}{N_2O}(g) + 2H_2O(g) + heat$$

Dinitrogen oxide supports combustion because it produces O_2 when heated.

$$2N_2O(g) \xrightarrow{\Delta} 2N_2(g) + O_2(g) + heat$$

The molecule is linear but unsymmetrical, with a dipole moment of 0.17 D. It is thought to have at least two important contributing resonance structures.

$$:\!\ddot{N}::N::\ddot{O}: \longleftrightarrow :N:::N:\ddot{O}:$$

 1.126 Å N—N—O 1.186 Å

Some dentists use N_2O for its mild anesthetic properties. It is also known as laughing gas because of its side effects.

Dinitrogen oxide, or nitrous oxide, N_2O, mp −90.8°C, bp −88.8°C.

26-8 Nitrogen Oxide (+2 Oxidation State)

The first step of the Ostwald process (Section 26-14) for producing HNO_3 from NH_3 is used for the commercial preparation of nitrogen oxide, NO.

$$4NH_3 + 5O_2 \xrightarrow[\Delta]{catalyst} 4NO + 6H_2O$$

NO is not produced in nature under usual conditions. It is formed by direct reaction of N_2 and O_2 in electrical storms.

NO is a colorless gas that condenses at −152°C to a blue liquid. Gaseous NO is paramagnetic and contains one unpaired electron per molecule.

$$:\!N::\ddot{O}: \longleftrightarrow :N::\ddot{O}:$$

Its unpaired electron makes nitric oxide very reactive. Molecules that contain unpaired electrons are called **radicals**. Surprisingly, in the gas phase NO is predominantly monomeric. In the solid phase it dimerizes to form diamagnetic, polar N_2O_2.

NO reacts with O_2 to form NO_2, a brown, corrosive gas.

$$2NO(g) + O_2(g) \longrightarrow 2NO_2(g)$$

The incidence of skin cancers depends on exposure to ultraviolet (UV) radiation (Section 6-8). In the stratosphere, ozone absorbs much of the harmful UV radiation from the sun before it reaches the surface of the earth. There is concern that NO emitted in the exhaust of supersonic transports (SSTs), which fly in the stratosphere, can catalyze the decomposition of O_3. The natural radiation-induced breakdown of O_3 forms oxygen atoms.

$$\begin{array}{l} NO + O_3 \longrightarrow NO_2 + O_2 \\ NO_2 + O \longrightarrow NO + O_2 \\ \hline O_3 + O \longrightarrow 2O_2 \quad \text{(net reaction)} \end{array}$$

Nitrogen oxide, or nitric oxide, NO, mp −163.6°C, bp −151.8°C, bond distance (1.15 Å) intermediate between $N\equiv O$ (1.06 Å) and $N=O$ (1.20 Å).

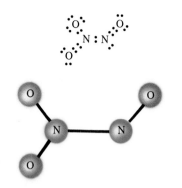

Dinitrogen trioxide, N_2O_3, mp $-102°C$, bp $3.5°C$.

NO is regenerated in the second step, so it causes a chain reaction in which one NO molecule is responsible for the destruction of many O_3 molecules.

26-9 Dinitrogen Trioxide (+3 Oxidation State)

Dinitrogen trioxide, N_2O_3, is the extremely unstable acid anhydride of nitrous acid, HNO_2. It can be isolated in pure form only as a solid (mp $-102°C$). It is prepared as a blue liquid in equilibrium with its reactants by the mixing of equimolar amounts of nitrogen oxide and nitrogen dioxide at $-20°C$.

$$NO + NO_2 \rightleftharpoons N_2O_3$$

Upon boiling ($3.5°C$), it decomposes completely to a mixture of colorless NO and brown NO_2. Its structure is shown in the margin.

26-10 Nitrogen Dioxide and Dinitrogen Tetroxide (+4 Oxidation State)

Nitrogen dioxide is formed by reaction of NO with O_2. It is prepared in the laboratory by heating heavy metal nitrates.

$$2Pb(NO_3)_2(s) \xrightarrow{\Delta} 2PbO(s) + 4NO_2(g) + O_2(g)$$

The temperature dependence of this equilibrium is shown in the photo on page 674.

Each NO_2 molecule contains one unpaired electron. NO_2 readily dimerizes to form colorless, diamagnetic dinitrogen tetroxide, N_2O_4. The formation of N_2O_4 is favored at low temperatures.

$$2NO_2(g) \rightleftharpoons N_2O_4(g) \qquad \Delta H^0 = -57.2 \text{ kJ/mol rxn}$$
$$\text{brown} \qquad \text{colorless}$$

The NO_2 molecule is angular. It is represented by resonance structures.

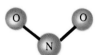

Nitrogen dioxide, NO_2, mp $-11.20°C$, bp $21.2°C$, one unpaired electron, bond length 1.197 Å; brown gas.

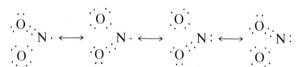

nitrogen dioxide

dinitrogen tetroxide

26-11 Dinitrogen Pentoxide (+5 Oxidation State)

Dinitrogen pentoxide, N_2O_5, is the acid anhydride of nitric acid.

$$N_2O_5(s) + H_2O(\ell) \longrightarrow 2HNO_3(aq)$$

This is called a dehydration because it is equivalent to two occurrences of

$$2HNO_3 \longrightarrow N_2O_5 + H_2O$$

$$H_2O + \tfrac{1}{2}P_4O_{10} \longrightarrow 2HPO_3$$

This is not the actual mechanism of the overall reaction.

N_2O_5 is prepared by dehydrating nitric acid with tetraphosphorus decoxide, P_4O_{10}, a very powerful dehydrating agent.

$$4HNO_3 + P_4O_{10} \xrightarrow{\text{low T}} 4HPO_3 + 2N_2O_5$$

The white solid, N_2O_5, is removed by vacuum sublimation. It is stable below $0°C$, but it decomposes when heated to room temperature.

$$2N_2O_5 \xrightarrow{\Delta} 4NO_2 + O_2 + heat$$

In the gaseous state it consists of N_2O_5 molecules. X-ray diffraction evidence shows that in the solid state it is nitronium nitrate, $NO_2{}^+NO_3{}^-$. Dinitrogen pentoxide is a very strong oxidizing agent.

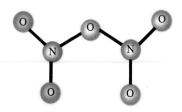

Dinitrogen pentoxide, N_2O_5 (gaseous), mp 30°C, bp 47°C.

one resonance form of gaseous N_2O_5 solid "N_2O_5" is $NO_2{}^+NO_3{}^-$, nitronium nitrate

26-12 Nitrogen Oxides and Photochemical Smog

Nitrogen oxides are produced in the atmosphere by natural processes (Section 26-2). Human activities contribute only about 10% of all the oxides of nitrogen (collectively referred to as NO_x) in the atmosphere, but the human contribution occurs mostly in urban areas, where the oxides may be present in concentrations a hundred times greater than in rural areas.

Just as NO is produced naturally by the reaction of N_2 and O_2 in electrical storms, it is also produced by the same reaction at the high temperatures of internal combustion engines and furnaces.

$$N_2(g) + O_2(g) \rightleftharpoons 2NO(g) \qquad \Delta H^0 = 180 \text{ kJ/mol rxn}$$

At ordinary temperatures the reaction does not occur to a significant extent. Because it is endothermic, it is favored by high temperatures. Even in internal combustion engines and furnaces, the equilibrium still lies far to the left, so only small amounts of NO are produced and released into the atmosphere. However, even very small concentrations of nitrogen oxides cause serious problems.

The NO radical reacts with O_2 to produce NO_2 radicals. Both NO and NO_2 are quite reactive, and they do considerable damage to plants and animals. NO_2 reacts with H_2O in the air to produce corrosive droplets of HNO_3 and more NO.

$$\overset{+4}{3NO_2} + H_2O \longrightarrow \overset{+2}{NO} + 2\overset{+5}{HNO_3} \qquad \text{(nitric acid)}$$

The HNO_3 may be washed out of the air by rainwater, or it may react with traces of NH_3 in the air to form solid NH_4NO_3, a particulate pollutant.

$$HNO_3 + NH_3 \longrightarrow NH_4NO_3$$

This situation occurs in all urban areas, but the problem is worse in warm, dry climates, which are conducive to light-induced (photochemical) reactions. Here ultraviolet (UV) radiation from the sun produces damaging oxidants. The brownish hazes that often hang over such cities as Los Angeles, Denver, and Mexico City are due to the presence of brown NO_2. Problems begin in the morning rush hour as NO is exhausted into the air. The NO combines with O_2 to form NO_2. Then, as the sun rises higher in the sky, NO_2 absorbs UV radiation and breaks down into NO and oxygen radicals:

$$NO_2 \xrightarrow{\text{uv}} NO + O$$

The extremely reactive O radicals combine with O_2 to produce O_3 (ozone):

$$O + O_2 \longrightarrow O_3$$

Obviously, O_3 in the upper atmosphere is not a problem. It is very beneficial!

Ozone is a powerful oxidizing agent that damages rubber, plastic materials, and all plant and animal life. It also reacts with hydrocarbons from automobile exhaust and evaporated gasoline to form secondary organic pollutants such as aldehydes and ketones (Section 31-14). The **peroxyacyl nitrates** (PANs), perhaps the worst of the secondary pollutants, are especially damaging photochemical oxidants that are very irritating to the eyes and throat. Catalytic converters in automobile exhaust systems (Section 16-11) reduce emissions of oxides of nitrogen.

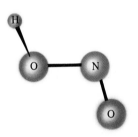

R = hydrocarbon chain or ring

Some Oxyacids of Nitrogen and Their Salts

Nitrogen also forms hyponitrous acid, $H_2N_2O_2$, in which N is in the +1 oxidation state, as well as hyponitrite salts such as $Na_2N_2O_2$.

The main oxyacids of nitrogen are nitrous acid, HNO_2, and nitric acid, HNO_3.

26-13 Nitrous Acid (+3 Oxidation State)

Although nitrous acid, HNO_2, is unstable and cannot be isolated in pure form, N_2O_3 can be considered its anhydride. The reaction of N_2O_3 with aqueous alkali produces nitrite salts.

$$N_2O_3 + 2[Na^+ + OH^-] \longrightarrow 2[Na^+ + NO_2^-] + H_2O$$

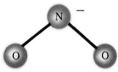

Nitrous acid, HNO_2.

The acid is prepared as a pale blue solution when H_2SO_4 reacts with cold aqueous sodium nitrite. Nitrous acid is a weak acid ($K_a = 4.5 \times 10^{-4}$). It acts as an oxidizing agent toward strong reducing agents and as a reducing agent toward very strong oxidizing agents.

Dot formulas for nitrous acid and the nitrite ion follow, and their structures are shown in the margin.

Nitrite ion, NO_2^-.

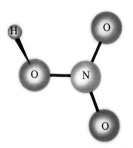

Nitric acid, HNO_3, mp $-42°C$; bp $83°C$; bond lengths N—O (terminal) 1.22 Å, N—O (central) 1.41 Å.

26-14 Nitric Acid (+5 Oxidation State)

Pure nitric acid, HNO_3, is a colorless liquid that boils at 83°C. Light or heat causes it to decompose into NO_2, O_2, and H_2O. The presence of the NO_2 in partially decomposed aqueous HNO_3 causes its yellow or brown tinge. Studies on the vapor phase indicate that the structure of nitric acid is

HNO_3 is commercially prepared by the **Ostwald process**. At high temperatures, NH_3 is catalytically converted to NO, which is cooled and then air-oxidized to NO_2. Nitrogen dioxide reacts with H_2O to produce HNO_3 and some NO. The NO produced in the third step is then recycled into the second step. More than 15 billion pounds of HNO_3 were produced in the United States in 1990.

$$4NH_3(g) + 5O_2(g) \xrightarrow[1000°C]{Pt} 4NO(g) + 6H_2O(g)$$

$$2NO(g) + O_2(g) \xrightarrow{cool} 2NO_2(g)$$

$$3NO_2(g) + H_2O(\ell) \longrightarrow 2[H^+ + NO_3^-] + \boxed{NO(g)}$$

recycle

Copper (left beaker) and zinc (right beaker) react with concentrated nitric acid.

Nitric acid is very soluble in water (~16 mol/L). It is a strong acid and a strong oxidizing agent.

Action of Nitric Acid on Metals

The oxidizing power of HNO_3 enables it to dissolve many metals that do not dissolve in nonoxidizing acids. Copper does not dissolve in HCl (a strong, nonoxidizing acid). But it does dissolve readily in HNO_3. Dissolving Cu in concentrated HNO_3 produces NO_2.

Acids whose anions do not undergo reduction easily are called nonoxidizing acids.

$$Cu + 4[H^+ + NO_3^-] \longrightarrow [Cu^{2+} + 2NO_3^-] + 2NO_2 + 2H_2O$$

In dilute nitric acid (~3 M), the major reduction product is nitrogen oxide.

$$3Cu + 8[H^+ + NO_3^-] \longrightarrow 3[Cu^{2+} + 2NO_3^-] + 2NO + 4H_2O$$

Both reactions probably occur in both cases, but the second predominates in dilute solution and the first in concentrated nitric acid.

When active metals such as zinc are dissolved in nitric acid, several reduction products result; their relative amounts depend on the acid concentration. In general, the higher the acid concentration, the higher the oxidation state of nitrogen in the reduction product.

The counter-ion for Zn^{2+} in all these reactions is NO_3^-.

$$Zn + 2NO_3^- + 4H^+ \longrightarrow Zn^{2+} + 2NO_2 + 2H_2O$$
$$3Zn + 2NO_3^- + 8H^+ \longrightarrow 3Zn^{2+} + 2NO + 4H_2O$$
$$4Zn + 2NO_3^- + 10H^+ \longrightarrow 4Zn^{2+} + N_2O + 5H_2O$$
$$5Zn + 2NO_3^- + 12H^+ \longrightarrow 5Zn^{2+} + N_2 + 6H_2O$$
$$4Zn + NO_3^- + 10H^+ \longrightarrow 4Zn^{2+} + NH_4^+ + 3H_2O$$

Increasing HNO_3 conc. ↑   Increasing ox. no. of N ↑

Action of Nitric Acid on Nonmetals

Many nonmetals disolve in concentrated HNO_3 to form oxides or oxyacids of the nonmetal. Phosphorus and sulfur are oxidized to H_3PO_4 and H_2SO_4, respectively, by hot concentrated HNO_3.

$$P_4 + 20[H^+ + NO_3^-] \longrightarrow 4H_3PO_4 + 4H_2O + 20NO_2$$

$$S + 6[H^+ + NO_3^-] \longrightarrow [2H^+ + SO_4^{2-}] + 2H_2O + 6NO_2$$

For simplicity, S rather than S_8, represents sulfur.

Nitrates

The nitrate ion, NO_3^-, is a planar, resonance-stabilized ion.

$$\left[\begin{array}{c} :\ddot{O}: \\ N \\ :\ddot{O} \quad \ddot{O}: \end{array} \right]^- \longleftrightarrow \left[\begin{array}{c} O \\ N \\ :\ddot{O} \quad \ddot{O}: \end{array} \right]^- \longleftrightarrow \left[\begin{array}{c} :\ddot{O}: \\ N \\ :\ddot{O} \quad \ddot{O}: \end{array} \right]^-$$

Neutralization of HNO_3 by metal hydroxides, carbonates, or oxides produces water-soluble nitrate salts. Some examples are

$$[Ca^{2+} + 2OH^-] + 2[H^+ + NO_3^-] \longrightarrow [Ca^{2+} + 2NO_3^-] + 2H_2O$$

$$CaCO_3 + 2[H^+ + NO_3^-] \longrightarrow [Ca^{2+} + 2NO_3^-] + CO_2 + H_2O$$

$$CaO + 2[H^+ + NO_3^-] \longrightarrow [Ca^{2+} + 2NO_3^-] + H_2O$$

When heated, alkali metal nitrates decompose into nitrites and O_2.

$$2KNO_3(s) \xrightarrow{\Delta} 2KNO_2(s) + O_2(g)$$

Heavy metal nitrates decompose into heavy metal oxides, NO_2 and O_2.

$$2Pb(NO_3)_2(s) \xrightarrow{\Delta} 2PbO(s) + 4NO_2(g) + O_2(g)$$

26-15 $NaNO_2$ and $NaNO_3$ as Food Additives

The brown color of "old" meat is the result of oxidation of blood, and is objectionable to many consumers. Nitrites and nitrates are added to food to retard this oxidation and also to prevent growth of botulism bacteria. Nitrate ion, NO_3^-, is reduced to NO_2^- ion, which is then converted to NO. This in turn reacts with the brown oxidized form of the heme in blood. This reaction keeps meat red longer. However, controversy has arisen concerning the possibility that nitrites combine with amines under the acidic conditions in the stomach to produce carcinogenic *nitrosoamines*.

$$\underbrace{\begin{array}{c} R' \\ \diagdown \\ N \\ \diagup \\ R \end{array}}_{\substack{\text{amine} \\ \text{group}}} \!\! - \!\! \underbrace{N\!\!=\!\!O}_{\substack{\text{nitroso} \\ \text{group}}} \qquad \text{(R and R' = organic groups)}$$

Phosphorus and Arsenic

26-16 Phosphorus: Occurrence, Production, and Uses

Phosphorus is the only element of Group VA that is always combined in nature. Phosphorus is present in all living organisms—as organophosphates and in calcium phosphates such as hydroxyapatite, $Ca_5(PO_4)_3(OH)$, and fluoroapatite, $Ca_5(PO_4)_3F$, in bones and teeth. It also occurs in these and related compounds in phosphate minerals, which are mined mostly in Florida and North Africa.

The tips of "strike anywhere" matches contain tetraphosphorus trisulfide and red phosphorus. Friction converts kinetic energy into heat which initiates a spontaneous reaction.

$$P_4S_3(s) + 8O_2 \longrightarrow P_4O_{10}(s) + 3SO_2(g)$$

Industrially, the element is obtained from phosphate minerals by heating them to 1200 to 1500°C in an electric arc furnace with sand (SiO_2) and coke.

$$2Ca_3(PO_4)_2 + 6SiO_2 + 10C \xrightarrow{\Delta} 6CaSiO_3 + 10CO + P_4$$

calcium phosphate (phosphate rock)   calcium silicate (slag)

Vaporized phosphorus is condensed to a white solid (mp = 44.2°C, bp = 280.3°C) under H_2O to prevent oxidation. Even when kept under H_2O, white phosphorus slowly converts to the more stable red phosphorus allotrope (mp = 597°C; sublimes at 431°C). Red phosphorus and tetraphosphorus trisulfide, P_4S_3, are used in matches. They do not burn spontaneously, yet they ignite easily when heated by friction. Both white and red phosphorus are insoluble in water.

The largest use of phosphorus is in fertilizers. Phosphorus is an essential nutrient, and nature's phosphorus cycle is very slow owing to the low solubility of most natural phosphates. Therefore, phosphate fertilizers are essential. To increase the solubility of the natural phosphates, they are treated with H_2SO_4 to produce "superphosphate of lime," a mixture of two salts. This solid is pulverized and applied as a powder.

This reaction represents the biggest single use of sulfuric acid, the industrial chemical produced in largest quantity.

$$Ca_3(PO_4)_2 + 2H_2SO_4 + 4H_2O \xrightarrow{\text{evaporate}} [Ca(H_2PO_4)_2 + 2(CaSO_4 \cdot 2H_2O)]$$

phosphate rock   calcium dihydrogen phosphate   calcium sulfate dihydrate

superphosphate of lime

26-17 Arsenic

Small amounts of As occur in free form, but most is found in the form of yellow arsenic(III) sulfide, As_2S_3. Its natural occurrence in the +3 oxidation state shows the greater metallic character of arsenic compared to N and P. It also occurs as arsenopyrite, FeAsS. Roasting As_2S_3 ore produces As_4O_6.

$$2As_2S_3 + 9O_2 \xrightarrow{\Delta} As_4O_6 + 6SO_2$$

The oxide can be reduced with coke to produce elemental arsenic, As_4.

$$As_4O_6 + 6C \longrightarrow 6CO + As_4$$

In the absence of air, arsenopyrite decomposes thermally to iron(II) sulfide and elemental arsenic.

$$4FeAsS \xrightarrow[\text{inert atm.}]{\Delta} 4FeS + As_4$$

The mineral orpiment, As_2S_3.

All arsenic compounds are poisonous, and most of their former uses are now banned. Although 0.1-gram doses of arsenic are lethal to humans, minute traces actually stimulate production of red blood cells.

26-18 Allotropes of Phosphorus and Arsenic

Both phosphorus and arsenic exist in several allotropic forms. There are only two important forms of each. One consists of tetrahedral P_4 or As_4 molecules (white phosphorus and yellow arsenic). These allotropes are more

Red phosphorus (left) and white phosphorus (right).

A model of the P_4 molecule (white phosphorus). Yellow arsenic contains analogous As_4 molecules.

volatile, less dense, more soluble in organic solvents, more chemically active, and more toxic than the other allotropes. Red phosphorus crystallizes in a polymeric lattice consisting of bonded tetrahedra in which one bond in each P_4 tetrahedron has been broken and replaced by a bond *between* tetrahedra. Gray arsenic consists of puckered six-membered rings in layers.

White phosphorus is sometimes called yellow phosphorus because, as its surface layer converts to the more stable red form, it appears to turn yellow and eventually red. At very high temperatures, phosphorus and arsenic vapors are thought to contain P_2 and As_2 molecules (compare N_2) in addition to P_4 and As_4 molecules.

26-19 Phosphine and Arsine

Phosphorus and arsenic exhibit the -3 oxidation state in phosphine, PH_3, and arsine, AsH_3. Both are pyramidal molecules with one lone pair of electrons on the central atom, like NH_3. Both are colorless, toxic, ill-smelling gases.

Phosphine is prepared by hydrolysis of calcium phosphide, Ca_3P_2, in a reaction analogous to the production of NH_3 from calcium nitride:

$$Ca_3P_2 + 6H_2O \longrightarrow 3[Ca^{2+} + 2OH^-] + 2PH_3$$

Phosphine is much less stable than NH_3. It reacts with moist air to give clouds of H_3PO_4.

$$PH_3(g) + 2O_2(g) \longrightarrow H_3PO_4(s)$$

Phosphine is a weaker base than NH_3 and forms some phosphonium salts (similar to ammonium salts) such as PH_4I. However, such salts are unstable and decompose in water.

Diphosphine, P_2H_4 or $H_2P—PH_2$ (the analog of hydrazine, N_2H_4), occurs as a by-product of the hydrolytic preparation of PH_3 from Ca_3P_2. It is a colorless liquid that spontaneously ignites in air; it also disproportionates in light to PH_3 and other complex hydrides of greater phosphorus content.

Arsine is even less stable and less basic than phosphine. The reaction of zinc with arsenic compounds such as arsenous acid, H_3AsO_3, in hydrochloric acid solution yields arsine.

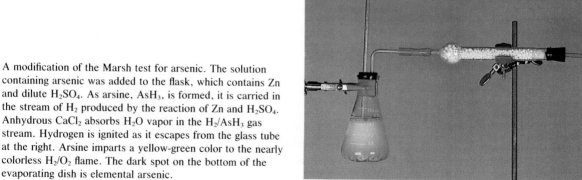

A modification of the Marsh test for arsenic. The solution containing arsenic was added to the flask, which contains Zn and dilute H_2SO_4. As arsine, AsH_3, is formed, it is carried in the stream of H_2 produced by the reaction of Zn and H_2SO_4. Anhydrous $CaCl_2$ absorbs H_2O vapor in the H_2/AsH_3 gas stream. Hydrogen is ignited as it escapes from the glass tube at the right. Arsine imparts a yellow-green color to the nearly colorless H_2/O_2 flame. The dark spot on the bottom of the evaporating dish is elemental arsenic.

$$3Zn + H_3AsO_3 + 6[H^+ + Cl^-] \longrightarrow 3[Zn^{2+} + 2Cl^-] + 3H_2O + AsH_3$$

This reaction is the basis for the Marsh test, which has been used in criminology laboratories to detect the presence of As. When AsH_3, H_2O vapor, and H_2 are passed through a heated glass tube, arsine decomposes. A mirror of metal-like arsenic is deposited inside the tube.

26-20 Phosphorus and Arsenic Halides

The most common and important phosphorus and arsenic halides are the trihalides and pentahalides, in which the phosphorus or arsenic exists in the +3 or +5 oxidation state. All eight possible trihalides (PX_3 and AsX_3 where X = F, Cl, Br, I) are known, as well as the pentahalides PF_5, PCl_5, PBr_5, and AsF_5. Mixed halides such as PF_3Cl_2 also exist.

All the trihalides can be prepared by direct union of the elements in stoichiometric ratios.

$$\left.\begin{array}{l} P_4 + 6X_2 \longrightarrow 4PX_3 \\ As_4 + 6X_2 \longrightarrow 4AsX_3 \end{array}\right\} \quad (X = F, Cl, Br, I)$$

This method is used for the preparation of PCl_3, PBr_3, and PI_3. PF_3 is usually made from AsF_3 by reaction with PCl_3. The mixture is then separated by fractional distillation. The direct reaction of P_4 with F_2 can be explosive.

$$\underset{\text{bp } 62.8°C}{AsF_3} + \underset{\text{bp } 76.1°C}{PCl_3} \longrightarrow \underset{\text{bp } 130°C}{AsCl_3} + \underset{\text{bp } -101.2°C}{PF_3}$$

Indirect methods can also be used for preparing other trihalides. For example, AsF_3 is usually prepared by distilling it from the reaction mixture of calcium fluoride, tetraarsenic hexoxide, and excess concentrated H_2SO_4.

$$6CaF_2 + As_4O_6 + 6H_2SO_4 \longrightarrow 6CaSO_4 + 6H_2O + 4AsF_3$$

The excess H_2SO_4 acts as a dehydrating agent. This removes the H_2O so that it does not react with AsF_3.

The pentahalides can be prepared by reaction of the elements with excess halogen, or of a trihalide with a halogen.

$$P_4 + 10X_2 \longrightarrow 4PX_5 \quad (X = F, Cl, Br)$$

$$As_4 + 10F_2 \longrightarrow 4AsF_5$$

$$PF_3 + F_2 \longrightarrow PF_5$$

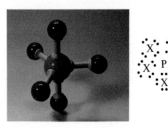

Phosphorus pentahalides, PX_5 (X = F, Cl, Br).

All the pentahalides and phosphorus trihalides hydrolyze rapidly and completely, the arsenic trihalides less so.

$$\overset{+5}{P}X_5 + 4H_2O \longrightarrow \underset{\substack{\text{phosphoric acid}}}{\overset{+5}{H_3PO_4}} + \underset{\substack{\text{hydrohalic acid}}}{5HX}$$

$$\overset{+3}{P}X_3 + 3H_2O \longrightarrow \underset{\substack{\text{phosphorus acid}}}{\overset{+3}{H_3PO_3}} + 3HX$$

The covalent trihalides are all pyramidal molecules with tetrahedral electronic geometry. The pentahalides are trigonal bipyramidal in the gaseous state, but X-ray evidence indicates that solid phosphorus pentachloride is ionic and contains $[PCl_4]^+$ tetrahedra and $[PCl_6]^-$ octahedra. Solid phosphorus pentabromide is apparently $[PBr_4]^+Br^-$.

Phosphorus trihalides, PX_3 (X = F, Cl, Br, I).

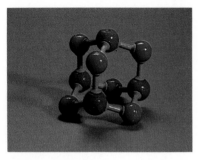

Tetraphosphorus hexoxide, P_4O_6, mp 23.8°C, bp 175.4°C.

Table 26-3
Important Phosphorus Oxyacids

Oxidation State	Formula	Name	No. of Acidic H	Compounds Prepared
+1	H_3PO_2	Hypophosphorous acid	1	Acid, salts
+3	H_3PO_3	(ortho)phosphorous acid	2	Acid, salts
+5	$(HPO_3)_n$	Metaphosphoric acids	n	Salts (n = 3, 4)
+5	$H_5P_3O_{10}$	Triphosphoric acid	3	Salts
+5	$H_4P_2O_7$	Pyrophosphoric acid	4	Acid, salts
+5	H_3PO_4	(ortho)phosphoric acid	3	Acid, salts

26-21 Oxides and Acids of Phosphorus and Arsenic

Several oxides are known. The most important ones are P_4O_6, P_4O_{10}, As_4O_6, and As_2O_5. P_4O_6 can be prepared by heating phosphorus under low pressure of oxygen, whereas high oxygen pressures produce P_4O_{10}. Both are acid anhydrides. As mentioned earlier, P_4O_{10} is a very strong dehydrating agent and will remove H_2O from such compounds as HNO_3. The structure of P_4O_6 is visualized by inserting an oxygen atom between each pair of phosphorus atoms in a P_4 tetrahedron. If an additional oxygen atom is bonded to each phosphorus, the P_4O_{10} structure results (see margin).

Tetraphosphorus decoxide, P_4O_{10}, sublimes at 358°C.

As_4O_6, a white solid, is prepared by roasting As_2S_3. As_2O_5 is prepared by thermal dehydration of arsenic acid, H_3AsO_4.

The oxyacids of phosphorus may contain acidic hydrogens bonded to oxygen,

$$\diagdown\!\!-P-O-H + H_2O \rightleftharpoons \diagdown\!\!-P-O^- + H_3O^+$$

or nonacidic hydrogen atoms bonded directly to phosphorus, $-P-H$. Many such oxyacids are known. The names and formulas of the most important ones are given in Table 26-3. We shall describe phosphorous and phosphoric acids.

Phosphorous Acid (+3 Oxidation State)

Tetraphosphorus hexoxide is the anhydride of (ortho)phosphorous acid, H_3PO_3:

$$P_4O_6 + 6H_2O \xrightarrow{\text{cold}} 4H_3PO_3$$

However, H_3PO_3 is more easily prepared by hydrolysis of PCl_3:

$$PCl_3 + 3H_2O \longrightarrow H_3PO_3 + 3HCl$$

The acid is only diprotic (K_1 = 1.6 × 10^{-2}; K_2 = 7 × 10^{-7}) because one hydrogen is directly bonded to phosphorus and does not ionize (see margin). Its salts are strong reducing agents in basic solution. Phosphite salts containing either $H_2PO_3^-$ or HPO_3^{2-} ions are prepared by neutralizing H_3PO_3 with the appropriate amount of base.

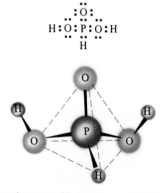

Phosphorous acid, H_3PO_3: mp 73.6°C, bp 200°C (decomposes).

Phosphoric Acid (+5 Oxidation State)

Orthophosphoric acid, or simply phosphoric acid, H_3PO_4, is the most common of the phosphoric acids. Its anhydride is P_4O_{10}.

$$P_4O_{10} + 6H_2O \longrightarrow 4H_3PO_4$$

Pure phosphoric acid is a stable, colorless solid. It is a weak triprotic acid (see Section 18-9). Salts containing each of the three anions can be prepared by stepwise neutralization of H_3PO_4. Salts of the dihydrogen phosphate and hydrogen phosphate ions decompose when heated. For example,

$$3NaH_2PO_4(s) \xrightarrow{\Delta} Na_3(PO_3)_3(s) + 3H_2O(\ell)$$

sodium   sodium
dihydrogen phosphate   trimetaphosphate

The trimetaphosphate ion is a puckered six-membered ring. Each phosphorus atom is sp^3 hybridized and at the center of a distorted tetrahedron, just as in the simple phosphate ion, PO_4^{3-}.

Salts containing trimetaphosphate and similar ions have been used in detergents as "builders." They have the ability to complex with (sequester) ions such as Fe^{3+}, Mg^{2+}, and Ca^{2+}, which are responsible for water hardness. However, when they are so used, these "soluble complex phosphate" ions in waste water flow into rivers and streams in high concentrations. The phosphorus, a nutrient, is in a form that is very easily assimilated by plants and algae. This causes them to grow wildly and reach an undesirable condition of overgrowth called **eutrophication**. When the vegetation dies, its decomposition products are unsightly and have disagreeable odors. The rotting vegetation also uses up the dissolved oxygen supply in the water so that it can no longer support fish and other aquatic animal life. Detergents also cause foaming in flowing waters. For these reasons, the use of phosphate detergents is now discouraged.

Organic polyphosphates are very important biological compounds that serve as reservoirs of chemical energy. The greater the number of connected phosphate groups in an organic polyphosphate group, the more "endothermic" the compound is. The energy derived from food is used to make polyphosphate compounds. When such phosphate–phosphate bonds are later broken, energy is released for muscular and other activity. Two important examples in metabolism are adenosine triphosphate (ATP) and adenosine diphosphate (ADP). ATP is hydrolyzed enzymatically to produce ADP and release energy. If "R" represents the adenosine part of the molecule, the reaction is

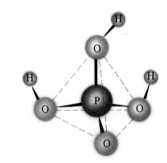

Phosphoric acid, H_3PO_4, mp 42.35°C.

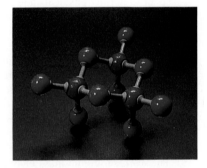

Trimetaphosphate ion, $P_3O_9^{3-}$. This is analogous to the structure of S_3O_9 (Section 25-9).

$$R-O-\overset{\displaystyle O}{\underset{\displaystyle OH}{\overset{|}{\underset{|}{P}}}}-O-\overset{\displaystyle O}{\underset{\displaystyle OH}{\overset{|}{\underset{|}{P}}}}-O-\overset{\displaystyle O}{\underset{\displaystyle OH}{\overset{|}{\underset{|}{P}}}}-OH + H_2O \longrightarrow R-O-\overset{\displaystyle O}{\underset{\displaystyle OH}{\overset{|}{\underset{|}{P}}}}-O-\overset{\displaystyle O}{\underset{\displaystyle OH}{\overset{|}{\underset{|}{P}}}}-OH + HO-\overset{\displaystyle O}{\underset{\displaystyle OH}{\overset{|}{\underset{|}{P}}}}-OH + \text{energy}$$

ATP   ADP   phosphoric acid

Arsenic Oxyacids

The arsenic oxyacids and salts are quite similar to the corresponding phosphorus acids and salts. The acid anhydride of arsenic acid, H_3AsO_4, is As_2O_5, and the anhydride of arsenous acid, H_3AsO_3, is As_4O_6. The oxyacids of arsenic are somewhat weaker than the corresponding phosphorus-containing acids, reflecting the more metallic nature of arsenic.

Key Terms

Haber process A process for the catalyzed industrial production of ammonia from N_2 and H_2 at high temperature and pressure.

Nitrogenases A class of enzymes found in bacteria within root nodules in some plants. They catalyze the reactions by which N_2 molecules from the air are converted to ammonia.

Nitrogen cycle The complex series of reactions by which nitrogen is slowly but continually recycled in the atmosphere, lithosphere, and hydrosphere.

Ostwald process A process for the industrial production of nitrogen oxide and nitric acid from ammonia and oxygen.

PANs Abbreviation for peroxyacyl nitrates, photochemical oxidants in smog.

Photochemical oxidants Photochemically produced oxidizing agents capable of causing damage to plants and animals.

Photochemical smog A brownish smog occurring in urban areas that receive large amounts of sunlight; caused by photochemical (light-induced) reactions among nitrogen oxides, hydrocarbons, and other components of polluted air that produce photochemical oxidants.

Exercises

1. Characterize each of the Group VA elements with respect to normal physical state and color.

2. Write out complete electron configurations for the atoms of the Group VA elements; nitride ion, N^{3-}; and phosphide ion, P^{3-}.

3. Is bismuth classified as a metal or nonmetal? Why? What causes BiO^+ ions to be formed in aqueous bismuth(III) salt solutions such as $Bi(NO_3)_3$? Write an equation for the reaction that occurs. (Section 23-9.)

4. Phosphorus forms two chlorides: PCl_3 and PCl_5. Classify PCl_3 and PCl_5 as ionic or covalent. Represent their three-dimensional structures.

5. The average atomic mass of N is 14.0067 amu. There are two isotopes which contribute to this average: $^{14}_{7}N$ (14.00307 amu) and $^{15}_{7}N$ (15.00011 amu). Calculate the percentage of $^{15}_{7}N$ atoms in a sample of naturally occurring nitrogen.

6. The $N{\equiv}N$ bond energy is 946 kJ/mol and the $N{-}N$ bond energy is 159 kJ/mol. Predict whether four gaseous nitrogen atoms would form two gaseous nitrogen molecules or a gaseous tetrahedral molecule similar to P_4, basing your prediction on the amount of energy released as the molecules are formed. Repeat the calculations for phosphorus using 485 kJ/mol for $P{\equiv}P$ and 213 kJ/mol for $P{-}P$.

7. Compare and contrast the properties of (a) N_2 and P_4, (b) HNO_3 and H_3PO_4, (c) N_2O_3 and P_4O_6.

8. Suggest why corresponding phosphorus and arsenic compounds resemble each other more closely than they do the corresponding nitrogen compounds.

9. List natural sources of nitrogen, phosphorus, and arsenic and at least two uses for each of the first two.

10. Describe the natural nitrogen cycle.

11. Discuss the effects of temperature, pressure, and catalysts on the Haber process for the production of ammonia. (You may wish to consult Section 17-6.)

12. Determine the oxidation states of nitrogen in the following molecules: (a) N_2, (b) NO, (c) N_2O_4, (d) HNO_3, (e) HNO_2.

13. Determine the oxidation states of nitrogen in the following species: (a) NO_3^-, (b) NO_2^-, (c) N_2H_4, (d) NH_3, (e) NH_2^-.

14. Calculate the bond energy of the nitrogen–nitrogen triple bond from data in Appendix K. How is this high value related to the general reactivity of N_2?

15. Draw three-dimensional structures showing all outer shell electrons, describe molecular and ionic geometries, and indicate hybridization (except for N^{3-}) at the central element, for the following species: (a) N_2; (b) N^{3-}; (c) NH_3; (d) NH_4^+; (e) NH_2^-, amide ion; (f) N_2H_4.

16. Draw three-dimensional structures showing all outer shell electrons, describe molecular and ionic geometries, and indicate hybridization at the central element for the following species: (a) NH_2Br, bromamine; (b) HN_3, hydrazoic acid; (c) N_2O_2; (d) $NO_2^+NO_3^-$, solid nitronium nitrate; (e) HNO_3; (f) NO_2^-.

17. Draw three-dimensional structures showing all outer shell electrons for the following species: (a) P_4, (b) P_4O_{10}, (c) As_4O_6, (d) H_3PO_4, (e) AsO_4^{3-}.

18. Write formula unit equations for the following:
    (a) thermal decomposition of potassium azide, KN_3
    (b) reaction of gaseous ammonia with gaseous HCl
    (c) reaction of aqueous ammonia with aqueous HCl
    (d) thermal decomposition of ammonium nitrate at temperatures above 260°C
    (e) reaction of ammonia with oxygen in the presence of red hot platinum catalyst
    (f) thermal decomposition of nitrous oxide (dinitrogen oxide), N_2O
    (g) reaction of NO_2 with water

19. Write formula unit equations for the following:

(a) preparation of "superphosphate of lime"
(b) reaction of phosphorus with limited Cl_2
(c) reaction of phosphorus with excess Cl_2
(d) preparation of phosphorous acid, H_3PO_3
(e) preparation of phosphoric acid

20. Write two equations illustrating the ability of ammonia to function as a Lewis base.

21. In liquid ammonia, would sodium amide, $NaNH_2$, be acidic, basic, or neutral? Would ammonium chloride, NH_4Cl, be acidic, basic, or neutral? Why?

22. Which of the following molecules have a nonzero dipole moment—i.e., are polar molecules? (a) NH_3, (b) NH_2Cl, (c) NO, (d) N_2H_4, (e) HNO_3, (f) PH_3, (g) As_4O_6.

23. Describe with equations the Ostwald process for the production of nitrogen oxide, NO, and nitric acid.

24. Why is NO so reactive?

25. Draw a Lewis formula for NO_2. Would you predict that it is very reactive? How about N_2O_4 (dimerized NO_2)?

26. At room temperature, a sample of NO_2 gas is brown. Explain why this sample loses its color as it is cooled.

27. Some nitrogen-containing compounds used in explosives are ammonium nitrate (NH_4NO_3), hydrazine (N_2H_4), sodium azide (NaN_3), and nitroglycerin ($C_3H_5N_3O_9$). What common characteristic do these substances share? How does this relate to their use?

28. What are the acid anhydrides of (a) nitric acid, HNO_3; (b) nitrous acid, HNO_2; (c) phosphoric acid, H_3PO_4; (d) phosphorous acid, H_3PO_3?

29. Discuss the problem of NO_x emissions with respect to air pollution. Use equations to illustrate the important reactions.

30. Write equations for reactions of nitric acid with two metals and two nonmetals.

31. Why is nitric acid called an oxidizing acid? How does the concentration of nitric acid affect the oxidation states of nitrogen-containing products of its reactions with metals?

32. Discuss the use of sodium nitrite as a meat preservative.

33. Calcium phosphate (phosphate rock) is not applied directly as a phosphorus fertilizer. Why?

34. Using the standard enthalpy of formation of $P_4S_3(s)$ as -154 kJ/mol and other values in Appendix K, calculate the standard enthalpy of combustion of P_4S_3 (the reaction that takes place when a match tip ignites).

*35. Calculate the concentrations of ammonium ions and of amide ions, NH_2^-, in a 7.0×10^{-6} M solution of ammonium chloride in liquid ammonia. Assume that NH_4Cl is completely dissociated. K for the ionization of liquid ammonia is 1×10^{-30}.

36. Using data in Appendix K, calculate the bond energy of each of the species in the following reaction:

$$N_2(g) + O_2(g) \rightleftharpoons 2NO(g)$$

Is it surprising that nitrogen and oxygen do not react significantly at room temperature? Under what conditions do they react in the atmosphere?

*37. The equation for a reaction that is used to produce hydrazine, N_2H_4, is

$$NH_3(aq) + NH_2Cl + NaOH(aq) \longrightarrow$$
$$N_2H_4 + NaCl(aq) + H_2O$$

The reactants are mixed at low temperature and heated to 80–90°C in the presence of a gelatin catalyst. How much hydrazine could be prepared from the reaction of 100 mL of 0.50 M aqueous ammonia with 3.00 g of chloroamine, NH_2Cl, and an excess of sodium hydroxide, assuming 68% yield with respect to the limiting reagent?

*38. What would be the pH of a buffer solution prepared using equal volumes of 0.20 M HNO_2 and 0.20 M KNO_2?

*39. What will be the molarity of 450 mL of a phosphoric acid solution produced by the action of water on 18.0 g of tetraphosphorus decoxide, P_4O_{10}, assuming complete reaction?

*40. Calculate the percentages of phosphorus in P_4O_{10} and in H_3PO_4.

*41. If a household detergent contained 25% by mass sodium trimetaphosphate, $Na_3(PO_3)_3$ or $Na_3P_3O_9$, what percentage of phosphorus would it contain? How many grams of phosphorus would be contained in 10.0 g of the detergent?

*42. The detonator-induced thermal explosion of ammonium nitrate yields $N_2(g)$, $H_2O(g)$, and $O_2(g)$. Calculate the total volume of gas, measured at 1.00 atm and 827°C, theoretically released in the explosive decomposition of 1.00 kg of $NH_4NO_3(s)$.

*43. The following series of balanced equations shows the reactions used to convert pure hydroxyapatite, $Ca_5(PO_4)_3(OH)$, to sodium tripolyphosphate hexahydrate, $Na_5P_3O_{10} \cdot 6H_2O$, a builder for detergents.

$$Ca_5(PO_4)_3(OH) + 5H_2SO_4 \longrightarrow 5CaSO_4 + H_2O + 3H_3PO_4$$
$$3H_3PO_4 \longrightarrow H_5P_3O_{10} + 2H_2O$$
$$H_5P_3O_{10} + 5NaOH \longrightarrow Na_5P_3O_{10} + 5H_2O$$
$$Na_5P_3O_{10} + 6H_2O \longrightarrow Na_5P_3O_{10} \cdot 6H_2O$$

Assuming no loss of phosphorus in any of the reactions, calculate the mass of hydroxyapatite required to produce 10.0 kg of $Na_5P_3O_{10} \cdot 6H_2O$.

*44. Commercial concentrated HNO_3 contains 69.5 mass % HNO_3 and has a density of 1.42 g/mL. What is the molarity of this solution? What volume of the concentrated acid should you use to prepare 10.0 L of dilute HNO_3 solution with a concentration of 6.00 M?

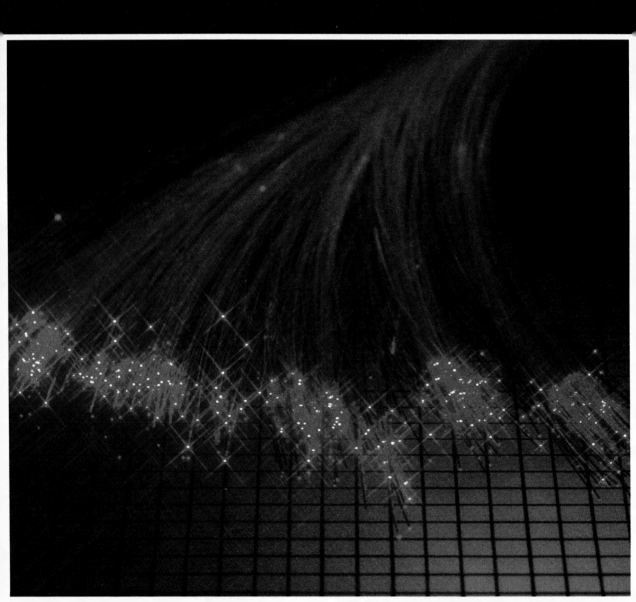

Outline

Objectives

As you study this chapter, you should learn

☐ About carbon and some of its
   important compounds
     occurrence of carbon
     oxides of carbon
     carbonic acid and the
      carbonates
     carbon tetrahalides
☐ About silicon and some of its
   important compounds
     production and uses of silicon
     chemical properties of silicon
     silicon halides

     silicon dioxide (silica)
     silicates and silicic acids
     natural silicates
     silicones
☐ About boron and some of its
   important compounds
     boric acids and boric oxide
     elemental boron
     three-center two-electron
      bonds
     boron hydrides (boranes)
     boron halides

The specialized chemical composition of some glass fibers allows them to conduct light for miles with no significant loss.

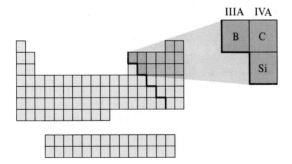

We shall now examine the remaining nonmetals: carbon, silicon, and boron. The general properties of these elements were given in Tables 23-5 and 23-6. Carbon is a nonmetal; silicon and boron are better classified as metalloids. Consistent with their relatively low electronegativities, neither Si nor B forms simple negative ions. Carbon forms the carbide anion, C_2^{2-}, in salt-like

Calcite is crystalline $CaCO_3$.

compounds with the least electronegative metals. There is a high degree of covalent character in carbides. All borides and silicides are covalent.

Some chemical properties of Si (Group IVA) are closer to those of B (IIIA) than to those of C (IVA). The concept of *diagonal relationships* in chemical properties was introduced in Section 23-2. Elements within the same family usually form compounds with similar stoichiometries—for example, CO_2 and SiO_2; B_2O_3 and Al_2O_3. However, the properties of the compounds of the elements of the second period are often closer to those of the corresponding compounds of the third-period element one group to the right. CO_2 is a gas; SiO_2 and B_2O_3 are solids. Silicon and boron halides (except SiF_4) hydrolyze readily, whereas halides of carbon are very inert toward water.

Carbon

Carbon ($1s^2 2s^2 2p^2$) has little tendency to form simple ions. The energy required to remove four outer shell electrons (the sum of the first four ionization energies) is prohibitively large for the existence of C^{4+}. In nearly all of its compounds, carbon is covalently bonded.

As a consequence of the high bond energy of C—C single bonds (347 kJ/mol) and of the C—H bond (414 kJ/mol), carbon has a strong tendency to form hydrocarbon chains and rings. The self-linkage of an element to form chains or rings is called **catenation**. Doubly bonded and triply bonded carbon atoms are also common.

The term "hydrocarbon" refers to a molecule (or a portion of a molecule) consisting of only C and H.

Hydrocarbons are in the domain of *organic chemistry*, the chemistry of most carbon-containing compounds. Carbon is the only element whose compounds constitute an entire branch of chemistry. The term "organic chemistry" arises from the fact that many compounds essential for plant and animal life contain carbon. The distinction between organic carbon compounds and inorganic carbon compounds is arbitrary. Here, we shall consider some of the compounds usually classified as inorganic compounds. Most contain only one carbon atom per formula unit and no C—H bonds. Chapters 31 and 32 introduce organic chemistry.

27-1 Occurrence of Carbon

Carbon makes up only about 0.08% of the combined lithosphere, hydrosphere, and atmosphere. It occurs in the crust of the earth mainly in coal and petroleum and in the form of calcium carbonate or magnesium carbonate rocks (calcite, limestone, dolomite, marble, and chalk). There are some natural deposits of elemental carbon in the form of diamond and graphite (Figure 27-1). Carbon dioxide occurs in the atmosphere, which acts as a CO_2 reservoir for photosynthesis in plants.

Carbon dioxide is incorporated by plants and bacteria into carbohydrates. These are ingested by animals and converted into other biological substances. After plants and animals die, they decompose to hydrocarbons and many other organic substances. These may eventually become coal, peat, natural gas, petroleum, and other **fossil fuels**. Both respiration and the complete combustion of fossil fuels produce CO_2 and H_2O to complete the carbon cycle.

Figure 27-1
Diamond.

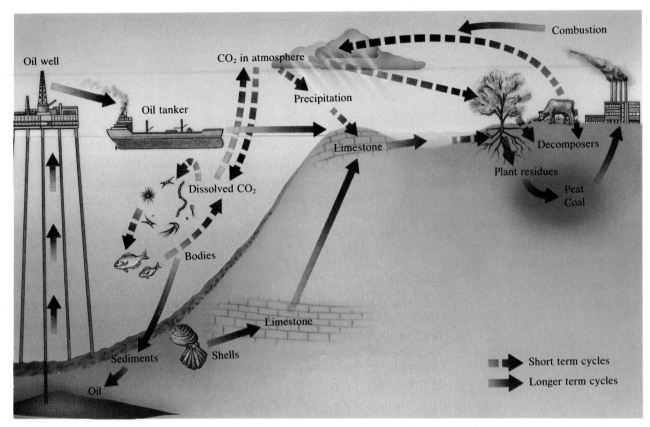

A simplified representation of the carbon cycle. The movement of carbon through the processes designated by dashed arrows is considerably more rapid than through those indicated by solid arrows.

The first prehistoric uses of iron and steel stemmed from the discovery of charcoal fires, which are hotter than wood fires.

Allotropes of Carbon

Carbon exists in two major crystalline forms, graphite and diamond (Section 13-16). A great deal of excitement has been generated by the recent discovery of a new allotropic form of carbon, known as the **fullerenes**. These exist as large symmetrical molecules of which the best characterized is C_{60} (August, 1991). Some of the exciting developments in the chemistry of this new form of carbon (the fullerenes) are discussed in the Chemistry in Use essay in Chapter 31. The so-called amorphous forms, such as charcoal and carbon black, are really arrangements of graphite crystallites. Carbon black is obtained by burning natural gas in a deficiency of air. Charcoal is the residue from the *destructive distillation* of wood. (Wood is heated in the absence of air to drive off volatile substances. Carbonaceous residues are formed at too low a temperature to permit thorough crystallization of the carbon.)

Graphite

At ordinary temperatures and pressures, graphite is the most stable allotrope of carbon. It is soft, black, and slippery, and its density (2.25 g/cm³) is less than that of diamond (3.51 g/cm³). Its properties result from its structure, which consists of sp^2 hybridized carbon atoms arranged in planar layers of six-membered rings (Figure 13-31c). Each carbon atom is at the center of a trigonal planar arrangement. Layers easily slide over each other because only weak van der Waals forces hold one layer to another. For this reason

Charcoal is often used to absorb impurities. The finely divided charcoal in one flask of gasoline has absorbed much of the color that was originally present.

graphite is used as a lubricant, as an additive for motor oil, and in pencil "lead." Certain compounds such as CO_2, H_2O, and NH_3 are able to diffuse between the layers and function as cushions, improving the lubricating ability of graphite. Only three of carbon's four electrons are involved in sigma bonding. The fourth electron of each is an extended pi-overlap system. These electrons are quite mobile, accounting for the ability of graphite to conduct electricity.

Diamond

The structural basis for the heat conductivity of diamond is not well understood. This property is the basis for a common test applied by jewelers to authenticate diamonds.

Diamond is one of the hardest substances known. It is colorless, a poor conductor of electricity, and more dense than graphite. It consists of a network of tetrahedral sp^3 hybridized C atoms, each separated from its neighbors by only 1.54 Å. This structure (Figure 13-31) involves very strong bonds with no mobile electrons. This accounts for diamond's hardness, its poor electrical conductivity, and its very high melting point (~3750°C).

Coal

Coal is a fossil material that has an irregular, graphite-like framework in which some of the C atoms are bonded to H. It contains mineral impurities that remain behind as an ash, as well as other elements such as sulfur that are vaporized as oxides, when the coal is burned. Currently there is renewed interest in research aimed at hydrogenation and liquefaction or gasification of coal.

Coke

Recall that destructive distillation is heating in the absence of air. The discovery of coke was another major advance in metallurgy.

When coal or petroleum is destructively distilled to release valuable volatile hydrocarbons, the residue (impure carbon) is called **coke**. Coke is used to reduce metal oxides in metallurgy. It can be converted to graphite for electrodes by the Acheson process. A large electric current is passed through a pressed rod of coke for several days to heat it sufficiently to recrystallize the carbon.

27-2 Reactions of Carbon

With Oxygen

Carbon is unreactive at 25°C, but it reacts with many substances when heated. It reacts with O_2 to produce either CO or CO_2 (Section 6-8, part 2).

With Metal Oxides

Carbon reduces the oxide of each of the less reactive metals to produce the metal and carbon monoxide (Section 22-7).

$$Fe_2O_3(s) + 3C(s) \xrightarrow{\Delta} 2Fe(\ell) + 3CO(g)$$

In such reactions with metal oxides, carbides of the metals often are produced. All are very hard, high-melting solids. Examples are beryllium carbide, BeC_2, and aluminum carbide, Al_4C_3. In some carbides, carbon behaves as if it were C^{4-} or, more often, C_2^{2-} ($:C:::C:^{2-}$), as in Na_2C_2 and CaC_2. Members of the latter class are also called acetylides; e.g., Na_2C_2 is sodium acetylide.

Synthetic diamonds are used in many cutting and grinding operations.

These release acetylene, HC≡CH (Section 31-4), when treated with water or dilute acid.

With Nonmetal Oxides

Carbon reacts with oxides of some nonmetals to produce very hard covalent carbides. Silicon carbide (carborundum) is one of the hardest substances known. It is produced by reduction of silica (SiO_2, sand) with coke in an electric furnace.

$$SiO_2(\ell) + 3C(s) \xrightarrow{3500°C} SiC(s) + 2CO(g)$$

Silicon carbide is used as the abrasive in sandpaper and grinding wheels. It has the same structure as diamond, except that Si and C atoms alternate in the diamond lattice.

Some grinding and cutting materials made of carborundum (silicon carbide, SiC).

With Steam

Carbon reacts with steam at high temperature in an endothermic reaction to produce "water gas," an equimolar mixture of CO and H_2.

$$C + H_2O \xrightarrow{red\ heat} CO + H_2$$

Both products burn in air, and the mixture is an important industrial fuel.

Water gas is also called "synthesis gas" because it is the starting material for the synthesis of some important chemicals. The Eastman process for making acetic acid is one example.

27-3 Oxides of Carbon

Carbon Monoxide

Carbon monoxide is a colorless, odorless, toxic gas that is formed by burning carbon or a hydrocarbon in a deficiency of oxygen. Because its carbon–oxygen bond length is intermediate between typical double and triple bond lengths, two resonance forms are thought to contribute to its structure.

Oxides with the formulas C_3O_2, C_5O_2, and $C_{12}O_9$ are known, but they are of little importance.

$$:C:::O: \longleftrightarrow :C::\overset{\cdot\cdot}{\underset{\cdot}{O}}$$

CO is produced commercially by the "water gas" reaction, just described, or by incomplete combustion of hydrocarbons, or by reduction of CO_2 by hot carbon. Pure CO is obtained by dehydrating formic acid with concentrated H_2SO_4.

$$\overset{\displaystyle O}{\underset{\displaystyle \|}{} } \\ H{-}C{-}O{-}H + H_2SO_4 \longrightarrow CO + H_2SO_4 \cdot H_2O$$

Carbon monoxide may be considered the anhydride of formic acid, because it dissolves in aqueous bases to yield salts of formic acid, called formates.

$$CO + [Na^+ + OH^-] \xrightarrow[6\ atm]{200°C} [H{-}\overset{O}{\overset{\|}{C}}{-}O^- + Na^+] \quad \text{sodium formate}$$

CO disproportionates to C and CO_2 in a reversible reaction.

$$2CO(g) \rightleftharpoons C(s) + CO_2(g) \qquad \Delta H^0 = -172 \text{ kJ/mol}$$

Carbon monoxide is important industrially as the starting material for the syntheses of many important organic substances. A few examples are

$$CO + Cl_2 \xrightarrow[\text{catalyst}]{\text{uv}} Cl-\underset{\underset{\|}{O}}{C}-Cl$$ phosgene
or carbonyl chloride

$$CO + 2H_2 \xrightarrow[\Delta,\ \text{pressure}]{\text{catalyst}} H-\underset{\underset{|}{H}}{\overset{\overset{H}{|}}{C}}-O-H$$ methyl alcohol
or methanol

$$CO + 3H_2 \xrightarrow[250°C]{\text{catalyst}} H_2O + CH_4$$ methane

Carbon monoxide is formed when any hydrocarbon burns in limited O_2. It is present in all automobile exhaust. An insidious poison, it binds strongly to iron in the oxygen-carrier protein, hemoglobin, in red blood corpuscles. This decreases the ability of blood to carry O_2. The hemoglobin–CO complex is bright red. Because CO is odorless and tasteless, victims are often unaware of the danger until too late.

Many other transition metals react with CO to form coordination complexes (Chapter 29). The carbon donates two electrons into vacant d orbitals of the metal. (In many such carbonyl complexes the metal is assigned an oxidation number of zero.) Because they often occur in the same ores, cobalt and nickel are treated with CO to form carbonyls, which can then be separated easily.

Nickel tetracarbonyl is much more toxic than carbon monoxide, and dicobalt octacarbonyl is somewhat less toxic.

$$Ni(s) + 4CO(g) \longrightarrow Ni(CO)_4(\ell) \qquad \text{nickel tetracarbonyl}$$

$$2Co(s) + 8CO(g) \longrightarrow Co_2(CO)_8(s) \qquad \text{dicobalt octacarbonyl}$$

Production of Ni metal by this method is the Mond process.

Nickel tetracarbonyl is more volatile than dicobalt octacarbonyl, so it can be distilled out of the reaction mixture. Both decompose to the free metal and CO at high temperatures, so the CO can be recycled.

Carbon Dioxide

Carbon dioxide is a colorless, odorless gas. It is the product of complete combustion of elemental carbon and of hydrocarbons. The combustion of gasoline is illustrated by oxidation of octane, C_8H_{18}, one of its components.

$$2C_8H_{18}(\ell) + 25O_2(g) \longrightarrow 16CO_2(g) + 18H_2O(g) + \text{heat}$$

Carbon dioxide is produced commercially as a by-product of the fermentation of sugar or starch for production of ethyl alcohol and alcoholic beverages.

$$\underset{\substack{\text{glucose} \\ \text{(a simple sugar)}}}{C_6H_{12}O_6(s)} \xrightarrow{\text{yeast}} \underset{\substack{\text{ethyl alcohol} \\ \text{or ethanol}}}{2C_2H_5OH(\ell)} + 2CO_2(g)$$

Photosynthesis in plants involves the conversion of carbon dioxide and water into carbohydrates, such as glucose and other sugars.

$$6CO_2 + 6H_2O \xrightarrow{\text{uv}} C_6H_{12}O_6 + 6O_2$$

The CO_2 molecule is linear, with an sp hybridized carbon atom and two carbon–oxygen double bonds.

$$\ddot{O}::C::\ddot{O}$$

Sinkholes result from the collapse of limestone caves.

Carbon dioxide is a minor, but very important, component of the atmosphere. It is present in a concentration of about 325 parts per million. Atmospheric CO_2 tends to achieve, or remain in, equilibrium with the CO_2 dissolved in natural waters, that contained in the earth's crust (such as limestone, $CaCO_3$), and that participating in photosynthesis.

27-4 Carbonic Acid and the Carbonates

Carbon dioxide is moderately soluble in water, and a saturated aqueous solution is about 0.033 M. Its solubility is increased, in accord with LeChatelier's Principle and Henry's Law, by increasing the partial pressure of CO_2. Carbonated beverages contain dissolved CO_2 under pressure. About 0.4% of the dissolved CO_2 reacts with water to form carbonic acid, H_2CO_3.

$$CO_2(g) + H_2O(\ell) \rightleftharpoons H_2CO_3(aq)$$

This reaction is easily reversed by heating. The structure of carbonic acid is

Stalactites and stalagmites are formed by the crystallization of carbonates that were dissolved in ground water.

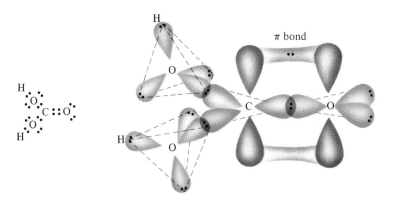

In this drawing the hybrid orbitals on oxygen are shown in red.

This weak diprotic acid has not been isolated in pure form. When neutralized, it can form both hydrogen carbonate and carbonate salts. Carbonate salts can also be formed when basic solids absorb gaseous CO_2.

$$2NaOH(s) + CO_2(g) \longrightarrow Na_2CO_3 \cdot H_2O(s)$$

Alkali metal carbonates are soluble in water. Most other carbonates are quite insoluble.

All natural waters contain dissolved CO_2, so deposits of $CaCO_3$ and $MgCO_3$ are abundant in areas formerly covered by bodies of water. Limestone caves are formed when ground water containing dissolved CO_2 (H_2CO_3) dissolves $CaCO_3$ and forms soluble $Ca(HCO_3)_2$.

$$CaCO_3(s) + H_2CO_3(aq) \longrightarrow [Ca^{2+} + 2HCO_3^-] \quad \text{calcium bicarbonate}$$

As water evaporates and the reaction is reversed, the carbonates are redeposited as stalactites and stalagmites, composed of $CaCO_3$.

$$[Ca^{2+} + 2HCO_3^-] \longrightarrow CaCO_3(s) + CO_2(g) + H_2O$$

Hydrogen carbonate salts are more soluble than the corresponding carbonates. Hydrogen carbonate ion, HCO_3^-, is often called bicarbonate ion.

When heated, metal carbonates decompose to the oxides and CO_2. The alkali metal carbonates require high temperatures for decomposition.

$$Li_2CO_3(s) \xrightarrow{1310°C} Li_2O(s) + CO_2(g)$$

$$PbCO_3(s) \xrightarrow{315°C} PbO(s) + CO_2(g)$$

The CO_3^{2-} ion hydrolyzes, so solutions of carbonates are basic.

$$CO_3^{2-} + H_2O \rightleftharpoons HCO_3^- + OH^- \qquad K_b = 2.1 \times 10^{-4}$$

Solutions of hydrogen carbonate salts are much less basic because HCO_3^- is a weaker base than CO_3^{2-}.

$$HCO_3^- + H_2O \rightleftharpoons H_2CO_3 + OH^- \qquad K_b = 2.4 \times 10^{-8}$$

The carbonate ion is symmetrical and planar, and is represented by three resonance forms. The carbon is sp^2 hybridized, and all bond angles are 120°.

27-5 Carbon Tetrahalides

Carbon forms four tetrahalides: CF_4 is a gas, CCl_4 is a liquid, and CBr_4 and CI_4 are solids. Only CCl_4 is commercially important. It is prepared by the reaction of carbon disulfide (a liquid with bonding similar to that in CO_2) with Cl_2. I_2 or $SbCl_5$ is used as a catalyst. The "carbon tet" is separated from the mixture by fractional distillation.

$$CS_2(\ell) + 3Cl_2(g) \longrightarrow CCl_4(\ell) + S_2Cl_2(\ell)$$

Carbon tetrachloride formerly was used as a dry cleaning agent because it is an excellent nonpolar solvent for dissolving fats, greases, and oils. It was also used to fight electrical fires because it is so dense (1.58 g/mL), does not conduct electricity, and does not support combustion. Stamp collectors use it in detecting printers' watermarks. Because of its toxicity to the liver and high vapor pressure, its use in dry cleaning and in fire extinguishers is now illegal. Used on fires, it produces phosgene, $COCl_2$, which can severely damage the lungs.

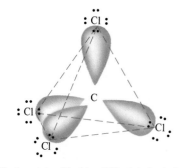

Carbon tetrachloride, CCl_4, tetrahedral electronic geometry, sp^3 hybridization; tetrahedral molecular geometry; mp −23°C; bp 76.5°C; toxic (liver damage).

Silicon

Silicon is a shiny, blue-gray, high-melting, brittle metalloid. It looks like a metal, but it is chemically more like a nonmetal. It is second only to oxygen in abundance in the earth's crust, about 87% of which is composed of silica (SiO_2) and its derivatives, the silicate minerals. The crust is 26% Si, compared to 49.5% O. Silicon does not occur free in nature. Pure silicon crystallizes with a diamond-type structure, but the Si atoms are less closely packed than C atoms. Its density is 2.4 g/cm^3 compared to 3.51 g/cm^3 for diamond.

27-6 Production and Uses of Silicon

Elemental silicon is usually prepared by the high-temperature reduction of silica (sand) with coke. Excess SiO_2 prevents the formation of silicon carbide.

$$SiO_2(s, excess) + 2C(s) \xrightarrow{\Delta} Si(s) + 2CO(g)$$

Reduction of a mixture of silicon and iron oxides with coke produces an alloy of iron and silicon known as *ferrosilicon*. It is used in the production of acid-resistant steel alloys, such as "duriron," and in the "deoxidation" of steel. Aluminum alloys for aircraft are strengthened with silicon.

Elemental silicon is used to make silicone polymers (Section 27-12). Its semiconducting properties (Section 13-17) are used in transistors and solar cells, for which ultrapure silicon is required. Ultrapure silicon is prepared by reducing a tetrahalide in the vapor phase with an active metal. The tetrahalide first must be carefully distilled to remove impurities of boron, aluminum, and arsenic halides.

$$SiCl_4(g) + 4Na(\ell) \longrightarrow 4NaCl(s) + Si(s)$$

The NaCl is then dissolved in hot water, leaving behind quite pure Si, which is melted and cast into bars. To attain less than one part per billion impurities, the Si must be further purified by zone refining (Section 22-4).

Ultrapure silicon. This wafer, 5 inches in diameter, was cut from a rod of silicon that was purified by zone refining.

27-7 Chemical Properties of Silicon

The biggest chemical differences between silicon and carbon are that (1) silicon does not form stable double bonds, (2) it does not form very stable Si—Si bonds unless the silicon atoms are bonded to very electronegative elements, and (3) it has vacant $3d$ orbitals in its valence shell into which it can accept electrons from donor atoms. The Si—O single bond is the strongest of all silicon bonds and accounts for the stability and prominence of silica and the silicates. Silicon forms very few species in which it is bound to more than four atoms. One such species that does exist is the octahedral hexafluorosilicate ion, $[SiF_6]^{2-}$, (sp^3d^2 hybridization at Si).

Si is much larger than C. As a result, Si—Si bonds are too long to permit the effective *pi* bonding that is necessary for multiple bonds.

Elemental silicon is quite unreactive. It does not react with solutions of most acids. A mixture of HNO_3 (an oxidizing acid) and HF dissolves SiO_2 as it forms, to produce hexafluorosilicic acid.

$$Si(s) + 4[H^+ + NO_3^-] + 6HF(aq) \longrightarrow$$
$$[2H^+ + SiF_6^{2-}] + 4NO_2(g) + 4H_2O(\ell)$$
$$\text{hexafluorosilicic acid}$$

Silicon dissolves in solutions of strong bases to produce silicate salts:

$$Si(s) + 2[K^+ + OH^-] + H_2O(\ell) \longrightarrow [2K^+ + SiO_3^{2-}] + 2H_2(g)$$
$$\text{potassium silicate}$$

27-8 Silicon Halides

All four tetrahalides exist. SiF_4 is a gas, $SiCl_4$ and $SiBr_4$ are volatile liquids, and SiI_4 is a solid. Silicon reacts directly with the halogens to produce tetrahalides.

Pure silicon is used in solar cells to collect energy from the sun.

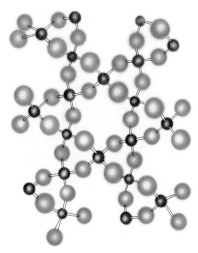

Figure 27-2
Representation of a *part* of a SiO_2 crystal. In SiO_2, each Si atom (gray) is bonded tetrahedrally to four O atoms (red). Each O atom is bonded to two Si atoms.

Natural quartz crystals, SiO_2.

Opal is hydrated silica, $SiO_2 \cdot xH_2O$, that is softer and less dense than quartz. Typically opal shows a marked iridescent play of colors.

$$Si + 2X_2 \longrightarrow SiX_4 \qquad X = F \text{ (explosively), Cl, Br, I}$$

Silicon tetrafluoride can be prepared by reaction of concentrated sulfuric acid with a mixture of silica and calcium fluoride.

$$SiO_2(s) + 2H_2SO_4(aq) + 2CaF_2(s) \longrightarrow 2CaSO_4(s) + 2H_2O(\ell) + SiF_4(g)$$

Most silicon halides are easily hydrolyzed, in contrast to carbon halides (which do not react with water). Although both kinds of reactions are thermodynamically spontaneous, only the latter type actually occurs.

$$CCl_4(\ell) + 2H_2O(\ell) \longrightarrow CO_2(aq) + 4HCl(aq) \qquad \Delta G^0_{298} = -224 \text{ kJ/mol}$$

$$SiCl_4(\ell) + 4H_2O(\ell) \longrightarrow H_4SiO_4(s) + 4HCl(aq) \qquad \Delta G^0_{298} = -277 \text{ kJ/mol}$$

The distinguishing factor is a kinetic one. There are vacant *d* orbitals in the outer shell of Si that can accept an electron pair from an attacking water molecule. The C atom has no such orbitals. As a result, the two reactions proceed by different pathways. That of CCl_4 has a much higher activation energy and is so slow that it is not observable. The availability of outer shell *d* orbitals on a central element often makes it more reactive than it would be otherwise. This situation is similar to the one described in Section 25-7 for *reactive* SF_4 and *inert* SF_6.

27-9 Silicon Dioxide (Silica)

Silicon dioxide exists in two familiar forms in nature: quartz, small chips of which occur in sand; and flint (Latin *silex*), an uncrystallized, amorphous type of silica. Silica is properly represented as $(SiO_2)_n$ because it is a polymeric solid of SiO_4 tetrahedra sharing all oxygens among surrounding tetrahedra (Figures 13-31 and 27-2). For comparison, solid carbon dioxide (Dry Ice) consists of discrete $O{=}C{=}O$ molecules, as does gaseous CO_2.

Some gems and semiprecious stones such as amethyst, opal, agate, and jasper are crystals of quartz with colored impurities. When silica is melted and rapidly cooled, it supercools. That is, it cools to a temperature below its normal melting (freezing) point as it forms an amorphous glass called "fused silica." Because it does not absorb visible or ultraviolet radiation, fused silica is used in optical instruments and sun lamps. It is resistant to most chemical attacks and expands or contracts only very slightly with changing temperature, so it is used to make laboratory glassware. It is, however, rapidly attacked by hydrofluoric acid to form volatile SiF_4.

$$SiO_2(s) + 4HF(aq) \longrightarrow 2H_2O(\ell) + SiF_4(g)$$

27-10 Silicates and Silicic Acids

Silica reacts slowly with strong bases to form various soluble metal silicates.

$$SiO_2(s) + 2[Na^+ + OH^-] \longrightarrow [2Na^+ + SiO_3{}^{2-}] + H_2O$$
sodium metasilicate

$$SiO_2(s) + 4[Na^+ + OH^-] \longrightarrow Na_4SiO_4(aq) + 2H_2O$$
sodium orthosilicate

$$2SiO_2(s) + 6[Na^+ + OH^-] \longrightarrow Na_6Si_2O_7(aq) + 3H_2O$$
sodium pyrosilicate

The parent acids of these salts are metasilicic acid, H_2SiO_3; orthosilicic acid, H_4SiO_4; and pyrosilicic acid or disilicic acid, $H_6Si_2O_7$. They are unstable and cannot be isolated in anhydrous form because they revert to silica. Dehydration of any silicic acid produces SiO_2. Silica in a spongy form with 5% water content is called *silica gel*. Because of its tremendous surface area, it adsorbs water and some vapors in relatively large quantities. It is used as a drying agent.

27-11 Natural Silicates

Most of the crust of the earth is made up of silica and silicates. The natural silicates comprise a large variety of compounds. The structures of all these are based on SiO_4 tetrahedra, with metal ions occupying spaces between the tetrahedra. The extreme stability of the silicates is due to the donation of extra electrons from O into vacant $3d$ orbitals of Si. In many common minerals, called aluminosilicates, Al atoms replace some Si atoms with very little structural change. Because an Al atom has one less positive charge in its nucleus than Si does, it is also necessary to introduce a univalent ion, such as K^+ or Na^+. The major classes of silicates and their structures are summarized in Table 27-1.

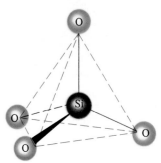

stands for:

Each equilateral triangle in Table 27-1 represents the tetrahedral SiO_4 unit.

Table 27-1
Structures of Silicate Minerals

Structural Arrangement		Si:O Atoms	Anion Unit	Examples
Independent tetrahedra	△	1:4	$(SiO_4)^{4-}$	forsterite, Mg_2SiO_4 olivine $(Mg, Fe)_2SiO_4$ fayalite, Fe_2SiO_4
Two tetrahedra sharing one oxygen	⋈	$1:3\frac{1}{2}$	$(Si_2O_7)^{6-}$	akermanite, $Ca_2MgSi_2O_7$
Closed rings of tetrahedra, each sharing two oxygens		1:3	$(Si_3O_9)^{6-}$, $(Si_4O_{12})^{8-}$, $(Si_6O_{18})^{12-}$	beryl, $Al_2Be_3Si_6O_{18}$
Continuous single chains of tetrahedra, each sharing two oxygens		1:3	$(SiO_3)^{2-}$	pyroxenes: enstatite, $MgSiO_3$; diopside, $CaMg(SiO_3)_2$
Continuous double chains of tetrahedra, sharing two or three oxygens		$1:2\frac{3}{4}$	$(Si_4O_{11})^{6-}$	amphiboles: anthophyllite, $Mg_7(Si_4O_{11})_2(OH)_2$
Continuous sheets of tetrahedra, each sharing three oxygens		$1:2\frac{1}{2}$	$(Si_4O_{10})^{4-}$	talc, $Mg_3Si_4O_{10}(OH)_2$ kaolinite, $Al_4Si_4O_{10}(OH)_8$
Continuous framework of tetrahedra, each sharing all four oxygens	See Figure 27-2	1:2	(SiO_2)	quartz, SiO_2 albite, $NaAlSi_3O_8$

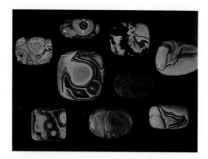

Chalcedony is a form of quartz that consists of extremely small, fibrous crystals. Agate, shown here, is chalcedony that exhibits bands of differing color or texture.

"Asbestos" refers to a group of impure magnesium silicate minerals. As you can see, asbestos is a fibrous material. When inhaled these fibers are highly toxic and carcinogenic.

This sample of mica is lighted from below so that layers are clearly visible.

The physical characteristics of the silicates are often suggested by the arrangement of the SiO_4 tetrahedra. A single-chain silicate, diopside $[CaMg(SiO_3)_2]_n$, and a double-chain silicate, asbestos $[Ca_2Mg_5(Si_4O_{11})_2(OH_2)]_n$, occur as fibrous or needle-like crystals. Talc, $[Mg_3Si_4O_{10}(OH)_2]_n$, a silicate with a sheet-like structure, is flaky. Micas are sheet-like aluminosilicates with about one of every four Si atoms replaced by Al. Muscovite mica is $[KAl_2(AlSi_3O_{10})(OH)_2]_n$. Micas occur in thin sheets that are easily peeled away from each other.

Clays

The clay minerals are silicates and aluminosilicates with sheet-like structures. They result from the weathering of granite and other rocks. The layers have enormous "inner surfaces" that can absorb large amounts of H_2O. Clay mixtures often occur as minute platelets with very large total surface areas. When wet, the clays are easily shaped. When heated to high temperatures, they lose H_2O; when fired in a furnace, they become very rigid. If clays are heated with feldspar, $[KAlSi_3O_8]_n$, and silica, the result is a mixture of crystallites bound by a rigid, glass-like matrix. This is called earthenware, porcelain, or china. Clays are also used to make cement, ceramics, bricks, flowerpots, and other materials.

Glass

Fused sodium silicate, Na_2SiO_3, and calcium silicate, $CaSiO_3$, are the major components of the glass used in such things as drinking glasses, bottles, and window panes. **Glass** is a hard, brittle material that has no fixed composition or regular structure. Because it has no regular structure, it does not break evenly along crystal planes, but breaks to form rounded surfaces and jagged edges. The basic ingredients are produced by heating a mixture of Na_2CO_3 and $CaCO_3$ with sand until it melts, at about 700°C.

$$[CaCO_3 + SiO_2](\ell) \xrightarrow{\Delta} CaSiO_3(\ell) + CO_2(g)$$

$$[Na_2CO_3 + SiO_2](\ell) \xrightarrow{\Delta} Na_2SiO_3(\ell) + CO_2(g)$$

The resulting "soda–lime" glass is clear and colorless (if all CO_2 bubbles escape and if the amounts of reactants are carefully controlled). Other ingredients may be added to produce certain desired characteristics. Soda–lime glass (soft glass) has a low softening temperature and expands and contracts considerably with changing temperature. Pyrex, a type of borosilicate glass, has some of the Na and Ca replaced by boron and has a higher percentage of silica. It is much more resistant to chemical attack and is able to withstand

Table 27-2
Some Substances Used to Color Glass

Substance	Color	Substance	Color
calcium fluoride	milky white	iron(II) compounds	green
copper(I) oxide	red, green, or blue	manganese(IV) oxide	violet
cobalt(II) oxide	blue	tin(IV) oxide	opaque
uranium compounds	yellow, green	colloidal selenium	red

greater thermal and mechanical stress. Flint glasses and lead glasses contain potassium and lead, respectively. They have high refractive indices that make them useful for decorative glassware, prisms, and lenses. A variety of substances may be added in small proportions to give color (Table 27-2).

27-12 Silicones

Silicones are polymeric organosilicon compounds. They contain individual or cross-linked Si—O chains or rings in which some of the oxygens of SiO_4 tetrahedra are replaced by such groups as hydroxyl, —OH; methyl, CH_3—; ethyl, C_2H_5—; or phenyl, C_6H_5—. They possess some of the properties of hydrocarbons and some of the properties of silicon–oxygen compounds. Most are very resistant to chemical attack and thermal decomposition. The linear, high-molecular-weight polymers are very rubbery. Silicone rubber is not attacked by ozone as ordinary rubber is. Liquid silicones are very fluid even at low temperatures, yet retain viscous fluidity at very high temperatures at which hydrocarbon oils decompose. The properties of a particular silicone depend on the substituents and the degree of cross-linking and ring formation. The structures of two typical silicones are illustrated in Figure 27-3.

The toys "Silly Putty" and "Superball" contain flexible, rubbery silicones extended with oil. Other silicones are used as lubricants, caulking materials, and waterproof films. Because it does not crack at low temperatures, a silicone is used for refrigerator gaskets. Some silicones are used in cosmetic surgery because of their chemical inertness and gel-like firmness and pliability.

Boron

Boron is very rare, constituting only 0.0003% of the earth's crust. Much of it is localized in dry lake beds in the southwestern United States, in the form of borax, $Na_2B_4O_5(OH)_4 \cdot 8H_2O$, and kernite, $Na_2B_4O_5(OH)_4 \cdot 2H_2O$, hydrated sodium salts of tetraboric acid. Borax is used mainly as a cleansing agent in laundry detergents and as a mild alkali. It is used in soldering and welding as a flux, because it melts at low temperatures and dissolves metaloxide films.

Fine china. The cup is made of bone china, which contains bone meal. The phosphate ions in bone meal were converted to pyrophosphate ions, $P_2O_7^{4-}$, when the china was fired. Pyrophosphate ions combined with iron(III) ions make the china white. The darker color of the saucer is due to the presence of iron(III) ions without pyrophosphate ions.

Glass may be given a variety of colors by the addition of various substances.

Two silicones. A very useful household product and a toy.

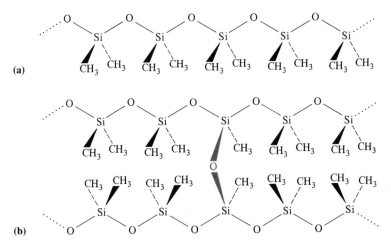

Figure 27-3
Silicones. Portions of (a) an unbranched silicone and (b) a cross-linked silicone.

27-13 Boric Acids and Boric Oxide

Borax reacts with sulfuric acid to produce boric acid, $B(OH)_3$ [or H_3BO_3].

$$Na_2B_4O_5(OH)_4 \cdot 8H_2O(s) + [2H^+ + SO_4^{2-}] \longrightarrow$$
$$[2Na^+ + SO_4^{2-}] + 5H_2O + 4H_3BO_3(aq)$$

Boric acid itself is used as a mildly acidic antiseptic. It is a very weak acid. Boric acid increases $[H^+]$ not by giving up H^+, but by accepting OH^- from H_2O.

Boric acid acts as a *Lewis* acid, not as a Bronsted–Lowry acid. In $B(OH)_3$, boron has a vacant valence shell orbital.

$$B(OH)_3 + H_2O \rightleftharpoons B(OH)_4^- + H^+ \qquad K = 7.3 \times 10^{-10}$$

or

$$H_3BO_3 + H_2O \rightleftharpoons H_4BO_4^- + H^+ \qquad K = 7.3 \times 10^{-10}$$

The solid acid melts and loses water at high temperatures, producing a colorless, transparent melt of boric oxide, B_2O_3.

$$2H_3BO_3(\ell) \xrightarrow{\Delta} B_2O_3(\ell) + 3H_2O(g)$$

Rapid cooling produces a glass-like solid. Boric oxide is used in the fabrication of the sturdy borosilicate glasses (Section 27-11).

Boric oxide is the anhydride of three common boric acids, two of which are obtained from boric acid (or, more properly, orthoboric acid) by dehydration.

$$4H_3BO_3(s) \xrightarrow[-4H_2O]{\Delta} 4HBO_2(s) \xrightarrow[-H_2O]{\Delta} H_2B_4O_7(s)$$

orthoboric acid   metaboric acid   tetraboric acid (pyroboric acid)

A number of borate salts containing anions of these acids are known.

27-14 Elemental Boron

Boron is obtained in low purity in a dark-brown, amorphous form by the high-temperature reduction of boric oxide with magnesium. The magnesium oxide is then dissolved away by hydrochloric acid, leaving the boron behind.

$$B_2O_3(s) + 3Mg(s) \xrightarrow{\Delta} 3MgO(s) + 2B(s)$$

Boron can be obtained in higher purity by reduction of gaseous boron tribromide or trichloride with hydrogen over a hot tungsten or tantalum filament.

$$2BBr_3(g) + 3H_2(g) \xrightarrow[1000-2500°C]{W} 2B(s) + 6HBr(g)$$

Boron exists in several allotropic modifications. α-Rhombohedral boron is the simplest in structure. All are large, complex polyhedral clusters of B atoms. They are very hard and are semiconductors. The α-rhombohedral form consists of icosahedral B_{12} units. One B atom occupies each vertex (Figure 27-4).

27-15 Three-Center, Two-Electron Bonds

Three-center, two-electron bonds, as well as ordinary two-center bonds, are present in all modifications of elemental boron. In these bonds, B atoms from three adjacent icosahedra share two electrons. These are delocalized

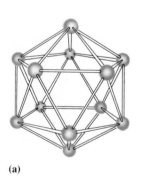

(a)

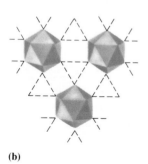

(b)

Figure 27-4
(a) An icosahedron is a regular polyhedron with 20 triangular faces and 12 vertices. Boron and many of its compounds are based on this arrangement, with B atoms at the vertices. (b) Linking of icosahedra of boron atoms in rhombohedral boron.

over a molecular orbital resulting from the overlap of three pseudo-*sp* hybrid orbitals on each B atom. These pseudo-*sp* hybrids have more *p* character than *s* character. A three-center, two-electron bond can be shown as follows:

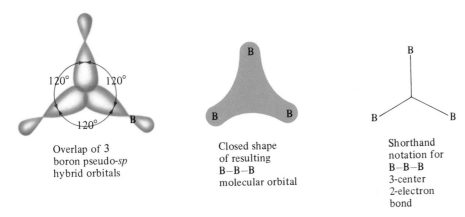

| Overlap of 3 boron pseudo-*sp* hybrid orbitals | Closed shape of resulting B–B–B molecular orbital | Shorthand notation for B–B–B 3-center 2-electron bond |

27-16 Boron Hydrides (Boranes)

Boron forms a number of unusual binary compounds with hydrogen called **boron hydrides**, or **boranes**. They fit into two series, B_nH_{n+4} and the less stable B_nH_{n+6}. Examples are given in Table 27-3, and typical structures are shown in Figure 27-5. A special system of naming is usually applied, in which the prefix indicates the number of B atoms and the parenthetical number gives the number of H atoms. The simplest borane is diborane, B_2H_6. BH_3 exists only as a short-lived reaction intermediate.

The higher boranes are generally prepared by controlled heating of diborane or other boranes.

$$2B_2H_6(g) \xrightarrow{120°C} B_4H_{10}(g) + H_2(g)$$

$$5B_2H_6(g) \xrightarrow{200-240°C} 2B_5H_9(g) + 6H_2(g)$$

$$2B_9H_{15}(s) \xrightarrow{> -30°C} B_8H_{12}(s) + B_{10}H_{14}(s) + 2H_2(g)$$

Decomposition of magnesium boride, Mg_3B_2, by acids also produces a number of boranes, but mainly B_2H_6.

Boron trichloride, BCl_3, or boron trifluoride, BF_3, reacts with lithium aluminum hydride, $LiAlH_4$, in ether solution to produce diborane.

$$4BCl_3(g) + 3LiAlH_4(s) \xrightarrow{ether} 2B_2H_6(g) + 3LiCl(s) + 3AlCl_3(s)$$

Table 27-3
Typical Boron Hydrides

B_nH_{n+4}	Name	Phase (25°C)	B_nH_{n+6}	Name	Phase (25°C)
B_2H_6	diborane	gas	B_4H_{10}	tetraborane(10)	gas
B_4H_8	tetraborane(8)	gas	B_5H_{11}	pentaborane(11)	liquid
B_5H_9	pentaborane(9)	liquid	B_6H_{12}	hexaborane(12)	liquid
B_8H_{12}	octaborane(12)	solid	$B_{10}H_{16}$	decaborane(16)	solid

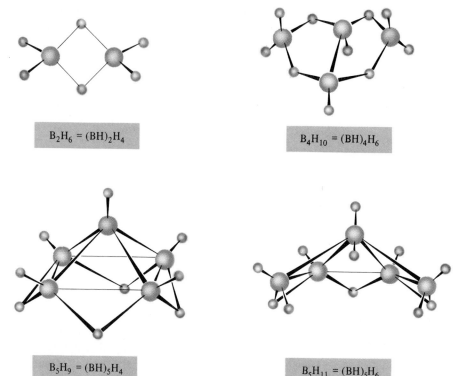

$B_2H_6 = (BH)_2H_4$

$B_4H_{10} = (BH)_4H_6$

Figure 27-5
The structures of some boranes, with their numbering conventions and some interatomic distances. The orange circles represent B atoms, and the white ones represent H. The thin lines connecting B atoms do not show bonds, but only define the polyhedral framework.

$B_5H_9 = (BH)_5H_4$

$B_5H_{11} = (BH)_5H_6$

Many boranes have structures based on fragments of icosahedra, as shown in Figure 27-5. Each borane contains three-center, two-electron bonds. Some are B $\diagdown$ B bonds and others are B $\diagup$ $\diagdown$ B bonds involving "hydrogen bridges," as will be discussed for diborane.

The bonding in diborane, B_2H_6, is not well explained by valence bond theory. Each of the six H atoms contributes one electron, and each of the two B atoms contributes three electrons, to give a total of 12 bonding electrons. If two electrons are allotted to each of the four terminal B—H (two-center) bonds, only four electrons remain to form the other three bonds. It is thought that each B $\diagup$ $\diagdown^H$ B bridge is held together by a three-center, two-electron bond. This results from the overlap of *approximately sp³* hybrid orbitals on *two* B atoms with the $1s$ orbital of the bridging H atoms. Two electrons occupy each of these three-center molecular orbitals:

Here, hybrid orbitals are shown in red.

The boranes are volatile compounds that decompose at high temperatures to B and H. All hydrolyze, but the ease of hydrolysis varies. Halogens, hydrogen halides, or boron trihalides react with boranes to yield halogenated borane derivatives. Diborane is a colorless, poisonous gas with an unpleasant odor. All boranes are easily oxidized to B_2O_3 and H_2O. Most ignite spontaneously in air.

The boranes can also react to produce borohydride anions. The simplest of these, BH_4^-, is tetrahedral. Some syntheses of metal borohydrides are

$$4NaH + B(OCH_3)_3 \xrightarrow{250°C} NaBH_4 + 3NaOCH_3$$

$$2LiH(s) + B_2H_6(g) \xrightarrow{ether} 2LiBH_4(s)$$

$$AlCl_3 + 3NaBH_4 \xrightarrow{\Delta} Al(BH_4)_3 + 3NaCl$$

Sodium borohydride, $NaBH_4$, is important in both inorganic and organic chemistry as a reducing agent and a source of hydride ions, H^-.

As a reducing agent, $NaBH_4$ is less vigorous than $LiAlH_4$.

27-17 Boron Halides

All four boron trihalides are known. All are covalent, planar, nonpolar molecules with 120° X—B—X angles, and all involve sp^2 hybridization (Section 8-6). Boron trifluoride is a gas, BCl_3 and BBr_3 are liquids, and BI_3 is a solid. By comparison, more metallic aluminum, which is just below boron in Group IIIA, forms ionic AlF_3, but the other aluminum trihalides are primarily covalent. All BX_3 molecules hydrolyze to produce boric acid and the corresponding hydrogen halide.

$$BX_3 + 3H_2O \longrightarrow H_3BO_3 + 3HX \qquad X = F, Cl, Br, I$$

The boron trihalides form mixed halides by exchange reactions such as

$$BBr_3 + BCl_3 \rightleftharpoons BBr_2Cl + BBrCl_2$$

The boron halides are Lewis acids, because each boron atom has only six electrons in its outer shell in these compounds (Section 10-10).

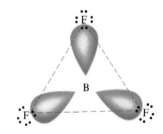

Boron trifluoride, BF_3 (gas phase): trigonal planar electronic geometry, sp^2 hybridization on B; trigonal planar molecular geometry; mp $-127°C$; bp $-100°C$.

Key Terms

Boranes   See *Boron hydrides*.

Boron hydrides   Binary compounds of boron and hydrogen.

Catenation   Bonding of atoms of the same element into chains or rings.

Clay   A class of silicate and aluminosilicate minerals with sheet-like structures that have enormous surface areas, which can absorb large amounts of water.

Coke   An impure form of carbon obtained by destructive distillation of coal or petroleum.

Fossil fuels   Coal, petroleum, natural gas, peat, and oil shale—substances consisting largely of hydrocarbons, derived from the decay of organic materials under geological conditions of high pressure and temperature.

Fullerene   A newly discovered allotropic form of carbon that consists of highly symmetrical molecules, e.g., C_{60}.

Glass   A hard, brittle material that has no fixed composition or regular structure.

Silicones   Polymeric organosilicon compounds; they contain individual or cross-linked Si—O chains or rings in which some oxygens of SiO_4 tetrahedra are replaced by other groups.

Three-center, two-electron bond   A bond holding three atoms together but involving only two electrons; common in boron compounds.

Exercises

Carbon

1. Why are most compounds of carbon covalent?
2. Compare the properties of CO_2 and SiO_2; B_2O_3 and SiO_2. How do you explain the similarities and differences?
3. Give several natural sources of carbon. Summarize the natural carbon cycle.
4. Sketch the structures of graphite and diamond. Describe their properties and relate them to their structures.
*5. What fundamental chemical modification is involved in coal liquefaction and gasification?
6. What is coke? Why is it so important industrially? How can it be converted to graphite?
7. Discuss the possible ways in which the carbon atom can form covalent bonds. Draw Lewis structures for each of the possibilities and predict the angles between the bonds around the carbon atom.
8. Draw Lewis formulas and structures for each of the following species. Indicate the hybridization at each C atom. (a) CO_2, (b) H_2CO_3, (c) CO_3^{2-}.
9. Draw the Lewis formula and structure for each of the following species. Indicate the hybridization at each C atom. (a) CCl_4, (b) HCN, (c) C_2H_2.
10. Write equations for the reactions of carbon at high temperatures with (a) limited O_2, (b) excess O_2, (c) nickel(II) oxide, NiO, (d) silicon dioxide, SiO_2, (e) steam.
11. Write equations for reactions to show how
   (a) carbon monoxide can be obtained from formic acid, HCO_2H.
   (b) sodium formate can be produced from CO.
   (c) carbon monoxide disproportionates.
12. Write equations for reactions to show how
   (a) methanol can be synthesized from CO.
   (b) methane can be synthesized from CO.
   (c) dicobalt octacarbonyl can be prepared.
13. Write the electron configurations of the C and Si atoms. What would you predict for the formulas of the compounds formed between these elements and fluorine? Although silicon can form the $[SiF_6]^{2-}$ ion, the analogous ion for carbon, $[CF_6]^{2-}$, is unknown. Why?
14. (a) Write Lewis structures for N_2, CN^-, CO, and C_2^{2-}. Identify the hybridization of each atom in these structures. (b) Assume that the available molecular orbitals of CN^- and CO are like those of N_2 and C_2^{2-}, respectively (Figure 9-5). Write the molecular orbital electron configurations for these four species.
15. Pieces of dry ice were mixed with 100 mL of 5.00 M NaOH. Write the chemical equation for the reaction to form sodium carbonate. What mass of sodium carbonate would be formed from 3.45 g of CO_2 reacting with the NaOH solution?

*16. A 7.716-g mixture containing only lithium carbonate and lithium oxide was heated to drive off all the carbon dioxide from the carbonate. A total of 1.817 L of dry CO_2 was collected at 747 torr and 32°C. What mass of lithium carbonate was contained in the original sample? What percentages of lithium carbonate and lithium oxide made up the sample?
*17. How many liters of methanol (a liquid) at 20°C can be produced from the complete reaction of 421 ft³ of carbon monoxide, measured at 22.3 atm pressure and 50°C, with an excess of hydrogen? The specific gravity of methanol is 0.791 at 20°C.
18. The largest diamond ever discovered was the Cullinan diamond, which weighed 3106 carats. Calculate the volume of this stone in cm³. The density of diamond is 3.51 g/cm³. One carat is equal to 200 mg.

Silicon

19. What is the fundamental building unit in all silicon-containing substances in the crust of the Earth? How is this building block modified to form the various types of natural silicates?
20. How is silicon obtained from natural sources? How is it purified for use in semiconductors? How does the process of zone refining work?
21. Why are pure silicon and germanium good semiconducting elements?
22. Write equations for the following reactions:
   (a) silicon with aqueous sodium hydroxide
   (b) silicon with bromine
   (c) silicon tetrachloride with water
   (d) carbon tetrachloride with water
23. How does hydrofluoric acid attack glass? Illustrate for $CaSiO_3$.
24. Describe the production of glass. How is glass colored? Give three examples of colored glasses and their colors.
25. What is a general characteristic of most silicones? How do the structures of silicones differ from quartz (SiO_2)?
26. What are compounds that contain only silicon and halogens called? What is the general formula for this series of compounds?
*27. What is the total mass of silicon in the crust of the Earth? Assume that the radius of the Earth is 6400 km, the crust is 50 km thick, the density of the crust is 3.5 g/cm³, and 25.7 mass % of the crust is silicon.
*28. Using the data of Appendix K, calculate the ΔG^0 values at 25°C for each of the reactions below, X = F, Cl. Use standard states for each species.

$$Si + 2X_2 \longrightarrow SiX_4$$

Which reaction has the more negative ΔG^0 value at 25°C?

Boron

29. How can borax be converted to boric acid, H_3BO_3?

30. Show how metaboric acid, HBO_2; tetraboric acid, $H_2B_4O_7$, and boric oxide, B_2O_3, can be prepared by successive dehydration of boric acid.

31. Write equations for the following reactions:
    (a) a preparation of elemental boron
    (b) the preparation of B_5H_9 from diborane
    (c) the preparation of B_4O_{10} from diborane
    (d) the preparation of diborane from BCl_3
    (e) boron trifluoride with water

32. Write equations for two reactions illustrating the ability of BCl_3 to act as a Lewis acid.

33. Is it surprising that boron participates in three-center, two-electron bonds? Why? Have we encountered other examples of these?

34. Describe and illustrate the bonding in two kinds of three-center, two-electron bonds.

35. What elements make up the class of compounds known as the "boranes"? What is the formula of the simplest known borane? Draw a diagram of the molecular structure for this compound.

36. What volume of H_2 measured at STP is required to produce 1.00 g of elemental boron by reduction of BBr_3?

*37. How much borax must react with excess sulfuric acid to produce 280 lb of boric acid, assuming 89.7% yield based on borax?

*38. Calculate the concentrations of all species and the pH in 0.10 M boric acid, $B(OH)_3$, a triprotic acid. $K_1 = 7.3 \times 10^{-10}$, $K_2 = 1.8 \times 10^{-13}$, $K_3 = 1.6 \times 10^{-14}$.

*39. How many grams of diborane, B_2H_6, can be produced from the 88.6% efficient reaction of 23.6 g of boron trichloride with an excess of lithium aluminum hydride in ether solution?

40. A pure sample of a gaseous boron hydride is found to be 78.4% boron by mass. A 0.2438-g sample occupies 131.7 mL at STP. What is the molecular formula of the compound?

41. A student prepared some diborane using the reaction

$$4BF_3(g) + 3LiAlH_4(s) \xrightarrow{\text{ether}} 2B_2H_6(g) + 3LiF(s) + 3AlF_3(s)$$

(a) Assuming 100% yield, what mass of B_2H_6 could be produced from the reaction of 5.0 g of BF_3 with 10.0 g of $LiAlH_4$? (b) The student used ether that had not been carefully dried and lost some diborane to the following reaction:

$$B_2H_6(g) + 6H_2O(\ell) \longrightarrow 2H_3BO_3(\text{ether}) + 6H_2(g)$$

How much of the diborane would react with 0.015 g of water?

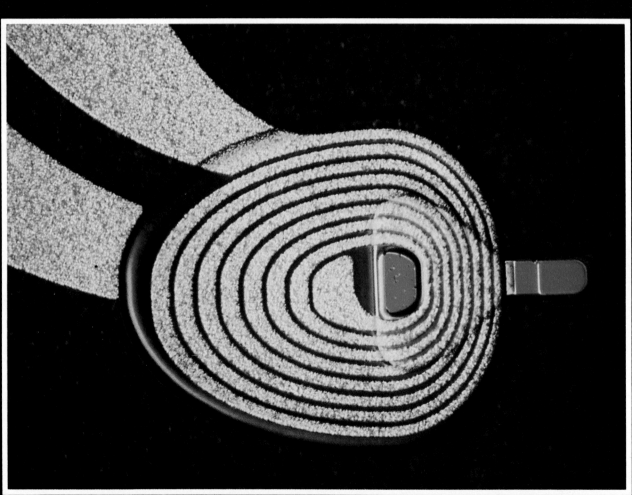

Outline

Objectives

As you study this chapter, you should learn

☐ About the *d*-transition metals and some of their typical compounds

☐ How these metals are classified into subgroups

☐ The origin of color in compounds of *d*-transition metals

☐ About trends in physical properties of *d*-transition metals

☐ About the oxides and hydroxides of two important *d*-transition metals, manganese and chromium

☐ About the role of some *d*-transition metals and their compounds as catalysts

☐ About the *f*-transition metals (the rare earths)

Many transition metals are used in modern technology. The computer disk recording head shown here is made by depositing a thin film of copper (yellow) on a silicon ceramic material over a magnetic core made of a nickel-iron alloy. This "read-write" head is the size of a period at the end of this sentence. To write information to a computer disk, a current is passed through the copper coil to induce a magnetic field in the core. This, in turn, permanently magnetizes a small region of iron oxide on the disk. At a later time, the same head is used to sense which regions on the disk have been magnetized, thus "reading" the stored information.

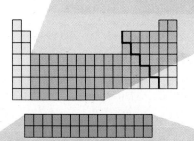

IIIB	IVB	VB	VIB	VIIB	VIIIB			IB	IIB
Sc	Ti	V	Cr	Mn	Fe	Co	Ni	Cu	Zn
Y	Zr	Nb	Mo	Tc	Ru	Rh	Pd	Ag	Cd
La	Hf	Ta	W	Re	Os	Ir	Pt	Au	Hg
Ac	Unq	Unp	Unh	Uns					

Ce	Pr	Nd	Pm	Sm	Eu	Gd	Tb	Dy	Ho	Er	Tm	Yb	Lu
Th	Pa	U		Pu	Am	Cm	Bk	Cf	Es	Fm	Md	No	Lr

Oxides of most nonmetals are acidic, and oxides of most metals are basic (except those having high oxidation states).

The term "transition elements" was coined to denote elements in the middle of the periodic table. They provide a transition between the "base formers" on the left and the "acid formers" on the right. The term applies to both the d- and f-transition elements (d and f atomic orbitals are being filled across this part of the periodic table). All are metals. We commonly use the term "transition metals" to refer to the d-transition metals. The f-transition elements are usually called the rare earths or inner transition elements.

The *d*-Transition Metals

28-1 General Properties

The d-transition metals are located between Groups IIA and IIIA in the periodic table. Strictly speaking, a d-transition metal must have a partially filled set of d orbitals. Zinc, cadmium, and mercury (Group IIB) and their cations have completely filled sets of d orbitals. They are not really d-transition metals. However, they are often discussed with d-transition metals because their properties are similar to those of the d-transition metals. All of the other elements in this region have partially filled sets of d-orbitals, except the IB elements and palladium, which have completely filled sets. Some of the cations of these latter elements have partially filled sets of d orbitals.

Some properties of $3d$-transition metals are listed in Table 28-1. The following are properties of transition elements.

1. All are metals.
2. Most are harder and more brittle and have higher melting points, boiling points, and heats of vaporization than nontransition metals.
3. Their ions and their compounds are usually colored.
4. They form many complex ions (Chapter 29).
5. With few exceptions, they exhibit multiple oxidation states.
6. Many of them are paramagnetic, as are many of their compounds.
7. Many of the metals and their compounds are effective catalysts.

Table 28-1
Properties of Metals in the First Transition Series

Properties	Sc	Ti	V	Cr	Mn	Fe	Co	Ni	Cu	Zn
Melting point (°C)	1541	1660	1890	1850	1244	1535	1495	1453	1083	420
Boiling point (°C)	2831	3287	3380	2672	1962	2750	2870	2732	2567	907
Density (g/cm³)	2.99	4.54	6.11	7.18	7.21	7.87	8.9	8.91	8.96	7.13
Atomic radius (Å)	1.62	1.47	1.34	1.25	1.29	1.26	1.25	1.24	1.28	1.34
Ionic radius, M^{2+} (Å)	—	0.94	0.88	0.89	0.80	0.74	0.72	0.69	0.70	0.74
Electronegativity	1.3	1.4	1.5	1.6	1.6	1.7	1.8	1.8	1.8	1.6
E^0 (V) for $M^{2+}(aq) + 2e^- \rightarrow M(s)$	−2.08*	−1.63	−1.2	−0.91	−1.18	−0.44	−0.28	−0.25	+0.34	−0.76
IE (kJ/mol) first	631	658	650	653	717	759	758	737	745	906
second	1235	1310	1414	1592	1509	1561	1646	1753	1958	1733

* For $Sc^{3+}(aq) + 3e^- \rightarrow Sc(s)$.

Table 28-2
Ground State Electronic Configurations of d-Transition Metals

Period 4		Period 5		Period 6	
$_{21}$Sc	$[Ar]3d^14s^2$	$_{39}$Y	$[Kr]4d^15s^2$	$_{57}$La	$[Xe]5d^16s^2$
$_{22}$Ti	$[Ar]3d^24s^2$	$_{40}$Zr	$[Kr]4d^25s^2$	$_{72}$Hf	$[Xe]4f^{14}5d^26s^2$
$_{23}$V	$[Ar]3d^34s^2$	$_{41}$Nb	$[Kr]4d^45s^1$	$_{73}$Ta	$[Xe]4f^{14}5d^36s^2$
$_{24}$Cr	$[Ar]3d^54s^1$	$_{42}$Mo	$[Kr]4d^55s^1$	$_{74}$W	$[Xe]4f^{14}5d^46s^2$
$_{25}$Mn	$[Ar]3d^54s^2$	$_{43}$Tc	$[Kr]4d^55s^2$	$_{75}$Re	$[Xe]4f^{14}5d^56s^2$
$_{26}$Fe	$[Ar]3d^64s^2$	$_{44}$Ru	$[Kr]4d^75s^1$	$_{76}$Os	$[Xe]4f^{14}5d^66s^2$
$_{27}$Co	$[Ar]3d^74s^2$	$_{45}$Rh	$[Kr]4d^85s^1$	$_{77}$Ir	$[Xe]4f^{14}5d^76s^2$
$_{28}$Ni	$[Ar]3d^84s^2$	$_{46}$Pd	$[Kr]4d^{10}$	$_{78}$Pt	$[Xe]4f^{14}5d^96s^1$
$_{29}$Cu	$[Ar]3d^{10}4s^1$	$_{47}$Ag	$[Kr]4d^{10}5s^1$	$_{79}$Au	$[Xe]4f^{14}5d^{10}6s^1$
$_{30}$Zn	$[Ar]3d^{10}4s^2$	$_{48}$Cd	$[Kr]4d^{10}5s^2$	$_{80}$Hg	$[Xe]4f^{14}5d^{10}6s^2$

Several of the apparent irregularities in these electron configurations can be explained by the special stability of half-filled and filled sets of d orbitals (Section 5-16).

28-2 Electronic Configurations and Oxidation States

Some of the valence electrons of the d-transition metals are in d orbitals one energy level below the highest occupied level. The properties of these metals vary less dramatically for consecutive elements than do those of representative elements, whose valence electrons are all in the highest occupied energy level. Electronic configurations of the three d-transition series are given in Table 28-2. Properties of these metals can be roughly correlated with either the total number of d electrons or the number of unpaired electrons.

Most transition metals exhibit more than one nonzero oxidation state. The *maximum* oxidation state is given by a metal's group number, but this is often not its most stable oxidation state (Table 28-3).

Table 28-3
Nonzero Oxidation States of the 3d-Transition Metals*

IIIB	IVB	VB	VIB	VIIB	VIIIB			IB	IIB
Sc	Ti	V	Cr	Mn	Fe	Co	Ni	Cu	Zn
		+1 r.	+1 r.	+1 r.		+1 r.	+1 r.	+1 r.	
	+2 r.	+2 r.	+2 r.	+2	+2 r.	+2	+2	+2	+2
+3	+3 r.	+3 r.	+3	+3 o.	+3	+3 o.	+3 o.	+3 o.	
	+4	+4	+4 o.	+4 o.	+4 o.	+4 o.	+4 o.		
		+5 o.	+5 o.	+5 o.	+5 o.				
			+6 o.	+6 o.	+6 o.				
				+7 o.					

* Most common oxidation states on color screens. o. = oxidizing agent; r. = reducing agent.

In the "building" of atoms by the Aufbau Principle, the outer *s* orbitals are filled before the inner *d* orbitals (Section 5-16).

The outer *s* electrons lie outside the *d* electrons and are *always* the first ones lost in ionization. In the first transition series, only scandium and zinc exhibit just one nonzero oxidation state. Scandium loses its two 4*s* electrons and its only 3*d* electron to form Sc^{3+}. Zinc loses its two 4*s* electrons to form Zn^{2+}.

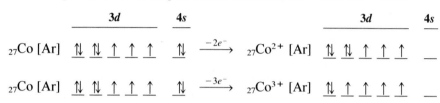

All of the other 3*d*-transition metals exhibit at least two oxidation states in their compounds. For example, cobalt can form Co^{2+} and Co^{3+} ions.

Pentaamminechlorocobalt(III) chloride, $[Co(NH_3)_5(Cl)]Cl_2$, is a compound that contains cobalt in the +3 oxidation state (left). Hexaaquacobalt(II) chloride, $[Co(OH_2)_6]Cl_2$, contains cobalt in the +2 oxidation state (right).

The most common oxidation states of the 3*d*-transition elements are +2, +3, and +4, with +5 and +6 coming next. The elements in the middle of each series exhibit more oxidation states than those to the left and right. As one moves down a group, higher oxidation states become more stable and more common (opposite to the trend for representative elements). This is because the *d* electrons are more effectively shielded from the nucleus as the group is descended and are therefore more easily ionized or more readily available for sharing. For example, cobalt commonly exhibits the +2 and +3 oxidation states. Rh and Ir are just below Co. Their common oxidation states are +3 and +4. The +4 state is slightly more stable for Ir than for the lighter Rh.

28-3 Classification into Subgroups

The transition metals and Zn, Cd, and Hg are subdivided into eight groups designated by Roman numerals (I to VIII) followed by "B." The Roman numeral *usually* designates the maximum oxidation number exhibited by members of the group. This does not mean that simple ions with these charges exist; no simple ions of these elements possess a charge greater than 3+. The elements in corresponding A and B groups form many compounds of similar stoichiometry (Table 28-4). However, their chemical properties are usually dissimilar.

Group VIIIB consists of three columns of three metals each, which have no counterparts among the representative elements. Each horizontal row in VIIIB is called a **triad** and is named after the best-known metal of the row. The three rows are the *iron*, *palladium*, and *platinum triads*.

There are differences between the A and B groups, but the transition metals show many of the same trends as the representative elements. The following trends of the transition metals also apply to representative elements. For corresponding compounds of metals from a B group in the same oxidation state, covalent character usually decreases and ionic character

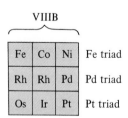

VIIIB

Fe	Co	Ni	Fe triad
Rh	Rh	Pd	Pd triad
Os	Ir	Pt	Pt triad

Table 28-4
Typical Compounds and Ions of the A and B Groups

IA	IIA	IIIA	IVA	VA	VIA	VIIA
NaCl	$MgBr_2$	$Al(NO_3)_3$	CCl_4	$POCl_3$	SO_4^{2-}	Cl_2O_7
KNO_3	$CaCl_2$	$Ga(OH)_3$	PbO_2	PO_4^{3-}	$H_2S_2O_7$	$HClO_4$
IB	**IIB**	**IIIB**	**IVB**	**VB**	**VIB**	**VIIB**
CuCl	$ZnBr_2$	$Sc(NO_3)_3$	$TiCl_4$	$VOCl_3$	CrO_4^{2-}	Mn_2O_7
$AgNO_3$	$CdCl_2$	$Y(OH)_3$	ZrO_2	VO_4^{3-}	$H_2Cr_2O_7$	$HMnO_4$

increases as the group is descended. We observe increasing electrical conductivity of aqueous solutions and increasing melting and boiling points for the heavier compounds. Consider the metal(V) oxides of Group VB.

Oxide	Melting Point	
V_2O_5	690°C	
Nb_2O_5	1460°C	Increasing ionic character
Ta_2O_5	1800°C	↓

For compounds containing the *same elements* in different proportions, the one containing the metal in the lower oxidation state is usually more ionic. Titanium(II) chloride, $TiCl_2$, and $TiCl_3$ are ionic solids, whereas titanium(IV) chloride, $TiCl_4$, is a molecular liquid.

The oxides and hydroxides of lower oxidation states of a given transition metal are basic. Those containing intermediate oxidation states tend to be amphoteric, and those containing high oxidation states tend to be acidic. This will be illustrated for the oxides and hydroxides of Mn and Cr in Sections 28-6 and 28-7.

28-4 Color

Most transition metal compounds are colored, a characteristic that distinguishes them from most compounds of the representative elements. What causes the color? In transition metal *compounds*, the *d* orbitals in any one energy level of the metals are not degenerate. No longer do all have the same energy, as they do in isolated atoms. They are often split into sets of orbitals separated by energies that correspond to wavelengths of light in the visible region (4×10^{-5} cm to 7×10^{-5} cm). The absorption of visible light

Rhodonite is manganese(II) silicate ($MnSiO_3$), which is pink, mixed with calcium silicate ($CaSiO_3$). The black veins are oxides of manganese. Rhodonite is often cut into cabochons, goblets, vases, and other decorative objects.

Table 28-5
Colors of Aqueous Solutions of Nitrate Salts of Some Representative Metal Ions and Some Transition Metal Ions

Representative Metal Ion	Color of Aq. Solution	Transition Metal Ion	Color of Aq. Solution
Na^+	Colorless	Cr^{3+}	Deep blue
Ca^{2+}	Colorless	Mn^{2+}	Pale pink
Mg^{2+}	Colorless	Fe^{2+}	Pale green
Al^{3+}	Colorless	Fe^{3+}	Orchid
Sn^{2+}	Colorless	Co^{2+}	Pink
Sn^{4+}	Colorless	Ni^{2+}	Green
Pb^{2+}	Colorless	Cu^{2+}	Blue

causes electronic transitions between orbitals in these sets. Our eyes are sensitive to these wavelengths and can distinguish among them well. Table 28-5 compares the colors of transition metal nitrates in aqueous solution with those of representative metal nitrates. The colors of transition metal compounds will be discussed in Section 29-10.

28-5 Trends in Physical Properties

Properties of transition metals vary with position in the periodic table somewhat more irregularly than do those of representative elements. This is due to many factors, and a detailed discussion is beyond the level of the text. We will state some generalizations that are usually valid.

Zinc, cadmium, and mercury show properties that are generally inconsistent with the trends. The reasons are related to the unusual stability of electronic configurations with only completely filled sets of orbitals. Each has a pseudo–noble gas electronic configuration; i.e., each has a noble gas core with completely filled sets of d orbitals and s orbitals. Copper, palladium, silver, and gold also have filled sets of d orbitals, and they behave in the same way.

The noble gases exhibit properties that are out of line with the general periodic trends of the representative elements (Chapter 6).

Atomic Radii

The atomic radii of the d-transition (and IIB) metals are given in Table 28-6 and are plotted against periodic position in Figure 28-1. Atomic radii of d-transition metals generally decrease from left to right across a period. This

Table 28-6
Atomic Radii of the d-Transition Metals (Å)

Period 4 (3d)	Sc 1.62	Ti 1.47	V 1.34	Cr 1.25	Mn 1.29	Fe 1.26	Co 1.25	Ni 1.24	Cu 1.28	Zn 1.34
Period 5 (4d)	Y 1.80	Zr 1.60	Nb 1.46	Mo 1.39	Tc 1.36	Ru 1.34	Rh 1.34	Pd 1.37	Ag 1.44	Cd 1.54
Period 6 (5d)	La 1.87	Hf 1.58	Ta 1.46	W 1.39	Re 1.37	Os 1.35	Ir 1.36	Pt 1.38	Au 1.44	Hg 1.57

is because their nuclear charges increase. But these decreases are not as consistent as those for the representative elements. Radii of elements near the end of the transition series increase. This is because the increases in effective nuclear charge are outweighed by greater repulsions among d electrons in a nearly filled or filled set of orbitals.

The radii of representative elements increase substantially as a group is descended and electrons occupy energy levels farther from the nucleus. Likewise, the transition metals of the fifth period have larger radii than those of the fourth period. But the transition metals of the sixth period have nearly the same radii as the metals above them. This phenomenon is known as the **lanthanide contraction**, following lanthanum (atomic number 57). The f electrons shield outer electrons less effectively than do s or p electrons. The insertion of 14 lanthanides, with 14 poorly shielding f electrons two shells inside the outermost shell, results in higher effective nuclear charges being felt by the outermost electrons. These higher effective nuclear charges pull the outer electrons closer to the nuclei, so the radii are smaller than we might expect.

The lanthanide contraction has important effects on the properties of the sixth-period transition and post-transition elements, as we saw in Section 23-7.

Density

Densities increase (Figure 28-2, Table 28-1) as radii decrease (Figure 28-1, Table 28-6) within a given period. They also increase as a group is descended. The metals of Period 6 have higher atomic masses than those of Period 5,

Osmium and iridium, with densities of 22.6 and 22.4 g/cm³, respectively, are the densest elements known.

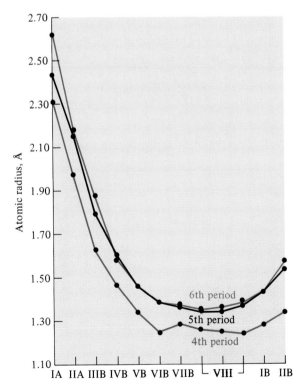

Figure 28-1
Variations in the atomic radii of the alkali, alkaline earth, and d-transition metals of the fourth, fifth, and sixth periods.

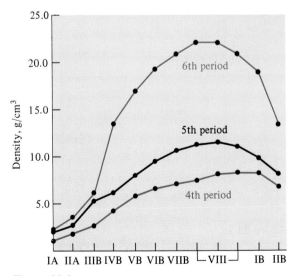

Figure 28-2
Variations in the densities of the alkali, alkaline earth, and d-transition metals of the fourth, fifth, and sixth periods.

but nearly the same atomic volume because of the lanthanide contraction. Thus, the sixth-period transition metals are very dense.

Magnetism

Many transition metals and ions have one or more unpaired electrons and are paramagnetic (Section 5-16). Figure 28-3 shows the good agreement between the number of unpaired electrons (theory) and the degree of paramagnetism (experiment). Magnetic measurements are also important in explaining the colors and bonding of transition metal compounds (Section 29-10).

The metals of the iron triad (Fe, Co, and Ni) are the only *free* elements that exhibit **ferromagnetism**. This property is much stronger than paramagnetism; it allows a substance to become permanently magnetized when placed in a magnetic field. This happens as randomly oriented electron spins align themselves with an applied field. To exhibit ferromagnetism, the atoms must be within the proper range of sizes so that unpaired electrons on adjacent atoms can interact cooperatively with each other, but not to the extent that they pair. Experimental evidence suggests that in ferromagnets, atoms cluster together into **domains** that contain large numbers of atoms in fairly small volumes. The atoms within each domain interact cooperatively with each other.

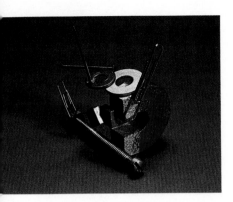

Iron displays ferromagnetism.

Some *alloys* that do not contain one of these metals also exhibit ferromagnetism.

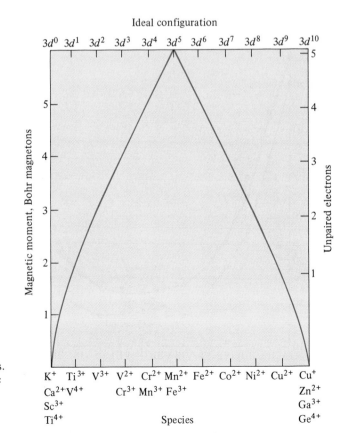

Figure 28-3

Variations in the number of unpaired electrons and the magnetic properties of metal ions in the first transition series and a few representative (A group) metals. This plot shows *idealized* values. Experimental values for many ions cover a wide range. Magnetic moments are a measure of paramagnetism and are related to the number of unpaired electrons. Magnetic moments are expressed in Bohr magnetons.

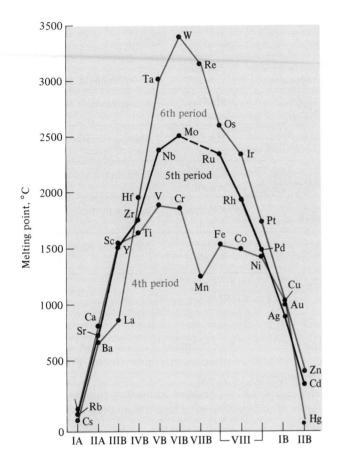

Figure 28-4

Variations in the melting points of the alkali, alkaline earth, and d-transition metals of the fourth, fifth, and sixth periods.

Melting Points

Melting points and boiling points of transition metals are roughly related to the number of unpaired electrons. The strength of metallic bonding increases with the availability of electrons to participate in the delocalized bonding (Section 13-17). Note the similarity of the shapes of the plots in Figures 28-3 and 28-4. Because the alkali and alkaline earth metals have only one or two outer electrons, their melting points are relatively low compared to other solids. The increasing availability of d electrons, especially unpaired electrons, causes an increase in melting points of transition metals from the beginning toward the middle of each series. As electron pairing increases toward the right of each series, melting points decrease.

Although most transition metals are hard, with fairly high melting points, the elements of Group IIB [Zn, Cd, and Hg (a liquid)] are soft with low melting points. This is because they have pseudo–noble gas electron configurations and no unpaired electrons.

28-6 Manganese Oxides and Hydroxides

We shall describe the variations in *chemical* properties of the oxides and hydroxides of manganese and chromium as the oxidation state of the metal increases. These variations are typical of the oxides and hydroxides of many transition metals.

Table 28-7
Some Compounds of Manganese

Ox. State	Oxide	bp (°C) of Oxide*	Hydroxide	Acidic/Basic Character	Related Salt	Name of Salt
+2	MnO green	1785	$Mn(OH)_2$	basic	$MnSO_4$ reddish	manganese(II) sulfate
+3	Mn_2O_3 brown or black	d > 940	$Mn(OH)_3$	weakly basic	$Mn_2(SO_4)_3$ green	manganese(III) sulfate
+4	MnO_2 gray to black	d > 535	H_2MnO_3 $[MnO(OH)_2]$	amphoteric	$CaMnO_3$† brown	calcium manganite
+6	MnO_3 (?) reddish	d	H_2MnO_4 $[MnO_2(OH)_2]$	acidic	K_2MnO_4 green	potassium manganate
+7	Mn_2O_7 green-brown	d > 55	$HMnO_4$	strongly acidic	$KMnO_4$ purple	potassium permanganate

* The symbol ''d'' indicates decomposition above the specified temperature.
† Simplified representation; manganite salts usually exist in condensed forms such as $CaO \cdot 2MnO_2 = CaMn_2O_5$ and
$CaO \cdot 5MnO_2 = CaMn_5O_{11}$.

Solid $MnCl_2 \cdot 4H_2O$ and a concentrated solution of $MnCl_2$ (left). Solid $KMnO_4$ and a dilute solution of $KMnO_4$ (right). The cellulose in a piece of paper towel has reduced $KMnO_4$ to potassium manganate, K_2MnO_4, which is green (center). Some K_2MnO_4 has been reduced to MnO_2, which is brown, under the photographer's hot lights.

Manganese exhibits the +2, +3, +4, +6, and +7 oxidation states in its simple compounds. The +2 oxidation state is the most stable. Potassium permanganate, in which manganese is in the +7 oxidation state, is stable; it is a very strong oxidizing agent and a common laboratory reagent. Table 28-7 shows some compounds of manganese in different oxidation states, and some properties of each.

Just as the oxyacids (nonmetal hydroxides) increase in acidity as the oxidation number of the central nonmetal increases, so do the transition metal hydroxides. Compare, for example, the ''hydroxides'' of chlorine of Group VIIA (Chapter 24) with those of manganese (Group VIIB) in Table 28-8. Permanganic and perchloric acids are both very strong acids and very strong oxidizing agents. Permanganic acid, $HMnO_4$, exists only in solution. It can be neutralized with bases to produce salts such as potassium permanganate, $KMnO_4$, which are very strong oxidizing agents.

$$MnO_4^- + 8H^+ + 5e^- \longrightarrow Mn^{2+} + 4H_2O \qquad E^0 = +1.51 \text{ V}$$

Manganous acid, H_2MnO_3, and manganic acid, H_2MnO_4, are also unstable, although their salts can be isolated. Manganese(II) and manganese(III) hydroxides are relatively stable. $Mn(OH)_2$ is air-oxidized to manganese(III) oxyhydroxide, $MnO(OH)$. Manganese(VII) oxide, Mn_2O_7, a dark-brown, explosive liquid, is the anhydride of $HMnO_4$. (Likewise, dichlorine heptoxide, Cl_2O_7, a colorless, explosive liquid, is the anhydride of $HClO_4$.)

The properties of several of the compounds of manganese and chlorine in high oxidation states are very much alike. However, the chemical properties are usually not as similar for other compounds of corresponding A and B group elements. Manganese is a metal and chlorine is a nonmetal, but the trends in acidity and basicity as a function of oxidation state are similar.

The other Group VIIB elements form many compounds with formulas similar to those of manganese. Technetium consists only of radioisotopes,

Table 28-8
Variation in Aciditiy of Oxyacids (hydroxides) with Oxidation States

			Oxidation State			
+1	+2	+3	+4	+5	+6	+7
ClOH [HClO] hypochlorous acid		ClO(OH) [HClO$_2$] chlorous acid		ClO$_2$(OH) [HClO$_3$] chloric acid		ClO$_3$(OH) [HClO$_4$] perchloric acid
	Mn(OH)$_2$ manganese(II) hydroxide	Mn(OH)$_3$ manganese(III) hydroxide	MnO(OH)$_2$ [H$_2$MnO$_3$] manganous acid		MnO$_2$(OH)$_2$ [H$_2$MnO$_4$] manganic acid	MnO$_3$(OH) [HMnO$_4$] permanganic acid

Increasing covalent character; increasing acid strength

but its chemistry has been studied. The heptoxides, Mn_2O_7, Tc_2O_7, and Re_2O_7, react with water to give strongly acidic solutions.

$$Mn_2O_7(\ell) + 3H_2O \longrightarrow 2[H_3O^+ + MnO_4^-] \qquad \text{permanganic acid}$$

$$Tc_2O_7(s) + 3H_2O \longrightarrow 2[H_3O^+ + TcO_4^-] \qquad \text{pertechnetic acid}$$

$$Re_2O_7(s) + 3H_2O \longrightarrow 2[H_3O^+ + ReO_4^-] \qquad \text{perrhenic acid}$$

Whereas Mn_2O_7 is dangerously explosive, Tc_2O_7 and Re_2O_7 are stable enough to be sublimed. This is evidence of the increasing stability of the higher oxidation states as a transition metal group is descended.

28-7 Chromium Oxides and Hydroxides

Typical of the metals near the middle of a transition series, chromium shows several oxidation states. The most common are +2, +3, and +6 (Table 28-9).

Oxidation–Reduction

The most stable oxidation state of Cr is +3. Solutions of blue chromium(II) salts are air-oxidized to chromium(III). Cr^{2+} is a strong reducing agent.

$$Cr^{3+} + e^- \longrightarrow Cr^{2+} \qquad E^0 = -0.41 \text{ V}$$

Chromium(VI) species are oxidizing agents. Basic solutions containing chromate ions, CrO_4^{2-}, are weakly oxidizing. Acidification produces the dichromate ion, $Cr_2O_7^{2-}$, and chromium(VI) oxide, both powerful oxidizing agents.

$$Cr_2O_7^{2-} + 14H^+ + 6e^- \longrightarrow 2Cr^{3+} + 7H_2O \qquad E^0 = +1.33 \text{ V}$$

Aqueous solutions of some compounds that contain chromium. Left to right: chromium(II) chloride ($CrCl_2$) is blue; chromium(III) chloride ($CrCl_3$) is green; potassium chromate (K_2CrO_4) is yellow; potassium dichromate ($K_2Cr_2O_7$) is orange.

Table 28-9
Some Compounds of Chromium

Ox. State	Oxide	Hydroxide	Name	Acidic/Basic	Related Salt	Name
+2	CrO black	$Cr(OH)_2$	chromium(II) hydroxide	basic	$CrCl_2$ anhydr. colorless aq. lt. blue	chromium(II) chloride
+3	Cr_2O_3 green	$Cr(OH)_3$	chromium(III) hydroxide	amphoteric	$CrCl_3$ anhydr. violet aq. green	chromium(III) chloride
					$KCrO_2$ green	potassium chromite
+6	CrO_3 dk. red	H_2CrO_4 or $[CrO_2(OH)_2]$	chromic acid	weakly acidic	K_2CrO_4 yellow	potassium chromate
		$H_2Cr_2O_7$ or $[Cr_2O_5(OH)_2]$	dichromic acid	acidic	$K_2Cr_2O_7$ orange	potassium dichromate

Chromate–Dichromate Equilibrium

Red chromium(VI) oxide, CrO_3, is the acid anhydride of two acids: chromic acid, H_2CrO_4, and dichromic acid, $H_2Cr_2O_7$. Neither acid has been isolated in pure form, although chromate and dichromate salts are common. CrO_3 reacts with H_2O to produce strongly acidic solutions containing hydrogen ions and (predominantly) dichromate ions.

$$2CrO_3 + H_2O \longrightarrow [2H^+ + Cr_2O_7{}^{2-}] \qquad \text{dichromic acid (red-orange)}$$

From such solutions orange dichromate salts can be crystallized after adding a stoichiometric amount of base. Addition of excess base produces yellow solutions from which only yellow chromate salts can be obtained. The two anions exist in solution in a pH-dependent equilibrium:

$$\overset{+6}{2CrO_4{}^{2-}} + 2H^+ \rightleftharpoons \overset{+6}{Cr_2O_7{}^{2-}} + H_2O \qquad K_c = \frac{[Cr_2O_7{}^{2-}]}{[CrO_4{}^{2-}]^2[H^+]^2} = 4.2 \times 10^{14}$$

$$\underset{\text{yellow}}{\phantom{2CrO_4{}^{2-}}} \qquad\qquad \underset{\text{orange}}{\phantom{Cr_2O_7{}^{2-}}}$$

Adding a strong acid to a solution that contains $CrO_4{}^{2-}/Cr_2O_7{}^{2-}$ ions favors the reaction to the right and increases $[Cr_2O_7{}^{2-}]$. Adding a base favors the reaction to the left and increases $[CrO_4{}^{2-}]$.

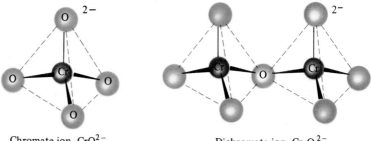

Chromate ion, CrO_4^{2-} Dichromate ion, $Cr_2O_7^{2-}$

Dichromate ions and dimanganese heptoxide are isoelectronic. They also have the same *geometry* as dichlorine heptoxide.

$$\left[\begin{array}{c} :\!\ddot{O}:\quad :\!\ddot{O}: \\ :\!\ddot{O}:\!Cr\!:\!\ddot{O}:\!Cr\!:\!\ddot{O}: \\ :\!\ddot{O}:\quad :\!\ddot{O}: \end{array}\right]^{2-} \qquad \begin{array}{c} :\!\ddot{O}:\quad :\!\ddot{O}: \\ :\!\ddot{O}:\!Mn\!:\!\ddot{O}:\!Mn\!:\!\ddot{O}: \\ :\!\ddot{O}:\quad :\!\ddot{O}: \end{array} \qquad \begin{array}{c} :\!\ddot{O}:\quad :\!\ddot{O}: \\ :\!\ddot{O}:\!Cl\!:\!\ddot{O}:\!Cl\!:\!\ddot{O}: \\ :\!\ddot{O}:\quad :\!\ddot{O}: \end{array}$$

$$Cr_2O_7{}^{2-} \qquad\qquad Mn_2O_7 \qquad\qquad Cl_2O_7$$

Each of these three species involves two tetrahedra sharing a corner.

Dehydration of chromate or dichromate salts with concentrated H_2SO_4 produces CrO_3. Chromium(IV) oxide is a strong oxidizing agent. A powerful "cleaning solution" used for removing greasy stains and coatings from laboratory glassware is made by adding concentrated H_2SO_4 to a concentrated solution of $K_2Cr_2O_7$. The active ingredients are CrO_3, an oxidizing agent, and H_2SO_4, an excellent solvent.

This cleaning solution is very dangerous because it is a strong oxidizing agent and is carcinogenic.

Chromium(III) hydroxide is amphoteric (Section 10-8).

$$Cr(OH)_3(s) + 3H^+ \longrightarrow Cr^{3+} + 3H_2O \qquad\qquad \text{(rxn. with acids)}$$

$$Cr(OH)_3(s) + OH^- \longrightarrow Cr(OH)_4^- \quad \text{or} \quad CrO_2^- \cdot 2H_2O \qquad \text{(rxn. with bases)}$$

Molybdenum and tungsten are just below chromium in Group VIIB. Both MoO_3 and WO_3 are more thermally stable, less acidic, and weaker oxidizing agents than CrO_3.

28-8 Transition Metals as Catalysts

Transition metals and their compounds function as effective catalysts in many reactions, both homogeneous and heterogeneous. Vacant d orbitals in many transition metal ions can accept electrons from reactants to form intermediates. These subsequently decompose to form products. The unreactive transition metals such as Pt, Pd, and Au are often used as finely divided solids to provide surfaces on which heterogeneous reactions can occur. The following reactions are typical of those catalyzed by transition metals and their compounds.

Homogeneous reactions occur in only one phase. Heterogeneous reactions involve more than one phase.

1. Haber process for the production of ammonia (Sections 17-6 and 26-4).

$$N_2 + 3H_2 \xrightarrow[\text{high T, P}]{\text{Fe, Fe oxides}} 2NH_3$$

2. Contact process for the production of sulfur trioxide in the manufacture of sulfuric acid (Section 25-11).

$$2SO_2 + O_2 \xrightarrow[400°C]{V_2O_5} 2SO_3$$

3. Bromination of benzene (Section 32-4).

$$\underset{\text{benzene}}{C_6H_6} + Br_2 \xrightarrow{FeBr_3} \underset{\text{bromobenzene}}{C_6H_5Br} + HBr$$

4. Hydrogenation of olefins, unsaturated hydrocarbons (Section 32-5).

$$\underset{\text{an olefin}}{RCH{=}CH_2} + H_2 \xrightarrow{Pt} \underset{\text{an alkane}}{RCH_2CH_3} \qquad R = \text{organic groups}$$

Transition metal ions are also present in the active sites of many important biological catalysts called enzymes.

Table 28-10
Periodic Table with *f*-Transition Metals (the Lanthanides and Actinides) Highlighted

	IA																		VIIA	0
		IIA												IIIA	IVA	VA	VIA			
1	1 H																		1 H	2 He
2	3 Li	4 Be												5 B	6 C	7 N	8 O		9 F	10 Ne
3	11 Na	12 Mg	IIIB		IVB	VB	VIB	VIIB	——VIII——			IB	IIB	13 Al	14 Si	15 P	16 S		17 Cl	18 Ar
4	19 K	20 Ca	21 Sc		22 Ti	23 V	24 Cr	25 Mn	26 Fe	27 Co	28 Ni	29 Cu	30 Zn	31 Ga	32 Ge	33 As	34 Se		35 Br	36 Kr
5	37 Rb	38 Sr	39 Y		40 Zr	41 Nb	42 Mo	43 Tc*	44 Ru	45 Rh	46 Pd	47 Ag	48 Cd	49 In	50 Sn	51 Sb	52 Te		53 I	54 Xe
6	55 Cs	56 Ba	57 La	58 Ce ▶ 71 Lu	72 Hf	73 Ta	74 W	75 Re	76 Os	77 Ir	78 Pt	79 Au	80 Hg	81 Tl	82 Pb	83 Bi	84 Po		85 At	86 Rn
7	87 Fr	88 Ra	89 Ac	90 Th ▶ 103 Lr	104 Unq*	105 Unp*	106 Unh*	107 Uns*	108	109	110	111	112	113	114	115	116		117	118

Periods (left vertical label) / Groups (top label)

	58 Ce	59 Pr	60 Nd	61 Pm*	62 Sm	63 Eu	64 Gd	65 Tb	66 Dy	67 Ho	68 Er	69 Tm	70 Yb	71 Lu
6 LANTHANIDE SERIES	58 Ce	59 Pr	60 Nd	61 Pm*	62 Sm	63 Eu	64 Gd	65 Tb	66 Dy	67 Ho	68 Er	69 Tm	70 Yb	71 Lu
7 ACTINIDE SERIES	90 Th	91 Pa	92 U	93 Np*	94 Pu*	95 Am*	96 Cm*	97 Bk*	98 Cf*	99 Es*	100 Fm*	101 Md*	102 No*	103 Lr*

*Artificially produced elements.

The Rare Earths (*f*-Transition Metals)

All the rare earth elements are metals in which an *f* sublevel *two* shells inside the outermost occupied shell is being filled (atomic numbers 58–71 and 90–103; see Table 28-10). Chemical properties of elements are governed largely by the electrons in their outermost shells. Each rare earth element has two *s* electrons in its outermost shell and either eight or nine electrons (two *s*, six *p*, and zero or one *d*) in the next shell inward (Table 28-11). This accounts for the chemical similarity of these elements. Most form M^{3+} ions by loss of the two outer *s* electrons and either the *d* electron one shell inward or one of the *f* electrons two shells inward. The *d*-transition metals show a strong tendency to form coordination compounds; the *f*-transition metals show markedly less tendency in this direction.

These elements also show other oxidation states.

The lanthanides are not as rare as the name "rare earths" implies. Mixtures of lanthanides and actinides occur in minerals such as monazite and gadolinite. However, these elements are difficult to obtain in pure form because of their chemical similarities. Historically, the lanthanides and some actinides were tediously separated from each other after extraction from their minerals. Hundreds of fractional recrystallizations of mixtures of their

Table 28-11
Electronic Configurations of Lanthanides and Formulas of Their Chlorides

Atomic Number	Name	Symbol	Electronic Configuration		Formula of Chloride	Ox. State of Lanthanide
57	lanthanum*	La	[Xe]	$5d^1 6s^2$	$LaCl_3$	+3
58	cerium	Ce	[Xe]	$4f^1 5d^1 6s^2$	$CeCl_3$, $CeCl_4$	+3, +4
59	praseodymium	Pr	[Xe]	$4f^3 6s^2$	$PrCl_3$, $PrCl_4$	+3, +4
60	neodymium	Nd	[Xe]	$4f^4 6s^2$	$NdCl_3$	+3
61	promethium	Pm	[Xe]	$4f^5 6s^2$	$PmCl_3$	+3
62	samarium	Sm	[Xe]	$4f^6 6s^2$	$SmCl_2$, $SmCl_3$	+2, +3
63	europium	Eu	[Xe]	$4f^7 6s^2$	$EuCl_2$, $EuCl_3$	+2, +3
64	gadolinium	Gd	[Xe]	$4f^7 5d^1 6s^2$	$GdCl_3$	+3
65	terbium	Tb	[Xe]	$4f^9 6s^2$	$TbCl_3$, $TbCl_4$	+3, +4
66	dysprosium	Dy	[Xe]	$4f^{10} 6s^2$	$DyCl_3$	+3
67	holmium	Ho	[Xe]	$4f^{11} 6s^2$	$HoCl_3$	+3
68	erbium	Er	[Xe]	$4f^{12} 6s^2$	$ErCl_3$	+3
69	thulium	Tm	[Xe]	$4f^{13} 6s^2$	$TmCl_3$	+3
70	ytterbium	Yb	[Xe]	$4f^{14} 6s^2$	$YbCl_2$, $YbCl_3$	+2, +3
71	lutetium	Lu	[Xe]	$4f^{14} 5d^1 6s^2$	$LuCl_3$	+3

* Lanthanum is properly a *d*-transition metal.

compounds were required. Now they are separated more efficiently by chromatography (Section 1-6). This separation relies on differences in attraction of similar compounds in solution for ionic sites on insoluble resins.

The actinides are all radioactive. Most do not occur naturally and have been prepared only since 1940, by nuclear reactions. Uranium (number 92) is the best-known naturally occurring actinide. It has been known for more than 200 years. Otto Hahn and Lise Meitner discovered nuclear fission in uranium in 1939. To extract uranium, its oxide ores are treated with nitric acid. The nitrate that is formed is decomposed to pure oxide, which can be reduced to the metal with calcium. To enrich uranium in its fissionable ^{235}U isotope for use in nuclear power plants (Section 30-15), the oxide is converted to UF_4 with HF. The UF_4 is oxidized to the volatile covalent UF_6 by fluorine. The vapor of UF_6, which contains both $^{238}UF_6$ and $^{235}UF_6$, is then subjected to repeated diffusion through porous barriers to concentrate the $^{235}UF_6$ (Graham's Law). Gas centrifuges are now used for the concentrating process.

Several of the rare earths have some commercial importance. Praseodymium and neodymium are used in sunglasses and glassblowers' goggles. Carefully controlled mixtures of rare earths are used in the fluorescent screens of color televisions. Thorium nitrate has been used for more than a century in gas mantles for lanterns and street lamps.

Key Terms

Domain A cluster of atoms in a ferromagnetic substance, all of which align in the same direction in the presence of an external magnetic field.

Ferromagnetism The ability of a substance to become permanently magnetized by exposure to an external magnetic field.

Lanthanide contraction A decrease in the radii of the elements following the lanthanides compared to what would be expected if there were no *f*-transition metals.

Triad A horizontal row of three elements in Group VIIIB.

Exercises

1. How are the d-transition metals distinguished from other elements?

2. What are the general properties of the d-transition metals?

3. Why are trends in variations of properties of successive d-transition metals less regular than trends among successive representative elements?

4. Write out the electronic configurations for the following species: (a) V, (b) Fe, (c) Cu, (d) Zn, (e) Fe^{3+}, (f) Ni^{2+}, (g) Ag, (h) Ag^+.

5. Why do copper and chromium atoms have "unexpected" electronic configurations?

6. Why are most transition metal compounds colored?

7. Write the general outer shell electron configuration for atoms of the f-transition elements. Which electrons are primarily involved in chemical bonding?

8. What is the most common oxidation state for the lanthanides? Write the general formulas of the oxides, nitrides, and halides.

9. Discuss the similarities and differences among elements of corresponding A and B groups of the periodic table— IIA and IIB, for example.

10. What are the three triads of the d-transition metals?

11. Why do we say that zinc, cadmium, and mercury have pseudo–noble gas configurations?

12. What is the lanthanide contraction? Which transition elements does it affect? How does it affect radii and densities?

13. Copper exists in the +1, +2, and +3 oxidation states. Which is the most stable? Which is a good oxidizing agent and which is a good reducing agent?

14. For a given transition metal in different oxidation states, how does the acidic character of its oxides increase? How do ionic and covalent character vary? Characterize a series of metal oxides as examples.

15. For different transition metals in the same oxidation state in the same group (vertical column) of the periodic table, how do covalent character and acidic character of their oxides vary? Why? Cite evidence for the trends.

16. Chromium(VI) oxide is the acid anhydride of which two acids? Write their formulas. What is the oxidation state of the chromium in these acids?

17. From its general reactivity and its position in the electromotive series, would you expect manganese to occur in nature as the sulfide or the oxide? Name the primary ore of manganese.

18. What are the most common oxidation states of manganese? What uses are made of MnO_2 and MnO_4^-?

19. How is the number of unpaired electrons per atom or ion experimentally determined?

20. What is ferromagnetism? What are domains? Which pure elements are ferromagnetic?

21. What single characteristic appears to be the most important in determining melting points of the d-transition metals?

22. How do you explain what has happened in the following chemical changes?
(a) $MnO_4^{2-} + H^+ \longrightarrow$
                 purple solution + dark brown solid
(b) $MnO_4^{2-} + O_3(g) + H^+ \longrightarrow$ purple solution + gas

23. What is the pH of a 0.015 M solution of permanganic acid? $K_a = 2.0 \times 10^3$ for $HMnO_4$.

24. Compare the structures and properties of Mn_2O_7, $Cr_2O_7^{2-}$, and Cl_2O_7.

25. Give four examples of reactions in which d-transition metals act as catalysts. What common characteristic of the d-transition metals makes them effective as catalysts?

*26. How many grams of Co_3O_4 (a mixed oxide, $CoO \cdot Co_2O_3$) must react with excess aluminum to produce 175 g of metallic cobalt, assuming 69.3% yield?

$$3Co_3O_4 + 8Al \xrightarrow{\Delta} 9Co + 4Al_2O_3$$

*27. What is the ratio of $[Cr_2O_7^{2-}]$ to $[CrO_4^{2-}]$ at 25°C in a solution prepared by dissolving 1.0×10^{-3} mol of sodium chromate, Na_2CrO_4, in enough of an aqueous solution buffered at pH = 2.00 to produce 200 mL of solution?

*28. Answer Exercise 27 for pH = 12.00.

29. Which would you expect to be the stronger oxidizing agent, Co^{3+} or Cr^{3+}? Why?

30. Classify the following substances as acidic or basic: MnO, Mn_2O_3, Mn_2O_7. Explain.

31. Arrange the substances of Exercise 30 in order of increasing covalent character.

Mixed Exercises

*32. Calculate ΔG^0 for the oxidation of ferrous ion to ferric ion by O_2 in acidic solution.

$$4Fe^{2+}(aq) + 4H^+(aq) + O_2(g) \longrightarrow$$
$$4Fe^{3+}(aq) + 2H_2O(\ell)$$

33. Without consulting a table of electrode potentials, indicate which of the following substances should be strong oxidizing agents: Cr, $Co(OH)_2$, $NiO(OH)$, Cu_2O, CrO_3. Justify your choices.

*34. Scandium is quite an active metal. What volume of hydrogen, measured at STP, is produced by the reaction of 5.75 g of scandium with excess hydrochloric acid?

35. Which solution would have a lower pH: 0.1 M $FeCl_3$ or 0.1 M $FeCl_2$? Why? (Table 19-2 may be helpful.)

36. Calculate the pH of a 0.100 M $Fe(NO_3)_2$ solution. $K_a = 3.0 \times 10^{-10}$ for Fe^{2+} at 25°C.

37. $2CrO_4^{2-} + 2H^+ \rightleftharpoons Cr_2O_7^{2-} + H_2O(\ell)$

To what pH must a 0.100 M Na_2CrO_4 solution be adjusted so that the concentrations of CrO_4^{2-} and $Cr_2O_7^{2-}$ are equal?

Compounds that contain the same transition metal in two different oxidation states are intensely colored. The coordination compound Prussian blue, $Fe_4[Fe(CN)_6]_3$, has been used as a dye. It contains Fe^{3+} ions and $[Fe(CN)_6]^{4-}$ ions. The $[Fe(CN)_6]^{4-}$ complex ion contains Fe^{2+}.

Objectives

As you study this chapter, you should learn

☐ To recognize coordination compounds
☐ The metals that form soluble ammine complexes in aqueous solutions and the formulas for common ammine complexes
☐ About the terminology that describes coordination compounds
☐ The rules for naming coordination compounds
☐ To recognize common structures of coordination compounds

☐ About structural isomers
☐ About stereoisomers
☐ To apply the concepts of valence bond theory to coordination compounds
☐ About the crystal field theory and its advantages
☐ About the origin of color in complex species
☐ About the spectrochemical series
☐ About crystal field stabilization energy (CFSE)

Coordination compounds are found in many places on the earth's surface. Every living system includes many coordination compounds. They are also important components of everyday products as varied as cleaning materials, medicines, inks, and paints. A list of important coordination compounds appears to be endless because new ones are discovered every year.

29-1 Coordination Compounds

In Section 10-10 we discussed Lewis acid–base reactions. A *base* makes available a share in an electron pair, and an *acid* accepts a share in an electron pair, to form a **coordinate covalent bond**. Such bonds are often represented by arrows (shown in red here).

Covalent bonds in which the shared electron pair is provided by one atom are called *coordinate covalent bonds*.

ammonia,   Boron trichloride,
a Lewis base   a Lewis acid

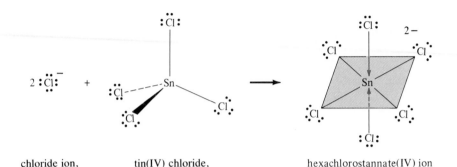

chloride ion, a Lewis base	tin(IV) chloride, a Lewis acid	hexachlorostannate(IV) ion

The arrows do not imply that two Sn—Cl bonds are different from the others. Once formed, all the Sn—Cl bonds in the $[SnCl_6]^{2-}$ ion are alike.

Most d-transition metal ions have vacant d orbitals that can accept shares in electron pairs. Many act as Lewis acids by forming coordinate covalent bonds in **coordination compounds** (**coordination complexes**, or **complex ions**). Complexes of transition metal ions or molecules include $[Cr(OH_2)_6]^{3+}$, $[Co(NH_3)_6]^{3+}$, $[Ni(CN)_4]^{2-}$, $[Fe(CO)_5]$, and $[Ag(NH_3)_2]^+$. Many complexes are very stable, as indicated by their low dissociation constants, K_d (Section 20-6 and Appendix I).

We now understand from molecular orbital theory that all substances have some vacant orbitals—they are *potential* Lewis acids. Most substances have lone pairs of electrons—they are *potential* Lewis bases.

Many important biological substances are coordination compounds. Hemoglobin and chlorophyll are two examples (Figure 29-1). Hemoglobin is a protein that carries O_2 in blood. It contains iron(II) ions bound to large porphyrin rings. The transport of oxygen by hemoglobin involves the coordination and subsequent release of O_2 molecules by the Fe(II) ions. Chlorophyll is necessary for photosynthesis in plants. It contains magnesium ions bound to porphyrin rings. Vitamin B-12 is a large complex of cobalt. Coordination compounds have many practical applications in such areas as water treatment, soil and plant treatment, protection of metal surfaces, analysis of trace amounts of metals, electroplating, and textile dyeing.

Bonding in transition metal complexes was not understood until the pioneering research of Alfred Werner, a Swiss chemist of the 1890s and early

Figure 29-1

(a) A model of a hemoglobin molecule (MW = 64,500 amu). Individual atoms are not shown. The four heme groups in a hemoglobin molecule are represented by disks. Each heme group contains one Fe^{2+} ion and porphyrin rings. A single red blood cell contains more than 265 million hemoglobin molecules and more than 1 billion Fe^{2+} ions. (b) The structure of chlorophyll a, which also contains a porphyrin ring with a Mg^{2+} ion at its center. Chlorophyll is necessary for photosynthesis. The porphyrin ring is the part of the molecule that absorbs light. The structure of chlorophyll b is slightly different.

Porphyrin ring system + iron = heme group (disks at left)

Chlorophyll a

(a)

(b)

Table 29-1
Interpretation of Experimental Data by Werner

Formula	Moles AgCl Precipitated per Formula Unit	Number of Ions per Formula Unit (based on conductance)	True Formula	Ions/Formula Unit	
$PtCl_4 \cdot 6NH_3$	4	5	$[Pt(NH_3)_6]Cl_4$	$[Pt(NH_3)_6]^{4+}$	4 Cl^-
$PtCl_4 \cdot 5NH_3$	3	4	$[Pt(NH_3)_5Cl]Cl_3$	$[Pt(NH_3)_5Cl]^{3+}$	3 Cl^-
$PtCl_4 \cdot 4NH_3$	2	3	$[Pt(NH_3)_4Cl_2]Cl_2$	$[Pt(NH_3)_4Cl_2]^{2+}$	2 Cl^-
$PtCl_4 \cdot 3NH_3$	1	2	$[Pt(NH_3)_3Cl_3]Cl$	$[Pt(NH_3)_3Cl_3]^+$	Cl^-
$PtCl_4 \cdot 2NH_3$	0	0	$[Pt(NH_3)_2Cl_4]$	no ions	

1900s. He won the Nobel prize in chemistry in 1913. Great advances have been made since in the field of coordination chemistry, but Werner's work remains the most important contribution by a single researcher.

Prior to Werner's work, the formulas of transition metal complexes were written with dots, $CrCl_3 \cdot 6H_2O$, $AgCl \cdot 2NH_3$, just like double salts such as iron(II) ammonium sulfate hexahydrate, $FeSO_4 \cdot (NH_4)_2SO_4 \cdot 6H_2O$. The properties of solutions of double salts are the properties expected for solutions made by mixing the individual salts. However, a solution of $AgCl \cdot 2NH_3$, or more properly $[Ag(NH_3)_2]Cl$, behaves differently from either a solution of (very insoluble) silver chloride or a solution of ammonia. The dots have been called "dots of ignorance," because they signified that the mode of bonding was unknown. Table 29-1 summarizes the types of experiments Werner performed and interpreted to lay the foundations for modern coordination theory.

Werner isolated platinum(IV) compounds with the formulas that appear in the first column of Table 29-1. He added excess $AgNO_3$ to solutions of carefully weighed amounts of the five salts. The precipitated AgCl was collected by filtration, dried, and weighed. He determined the number of moles of AgCl produced. This told him the number of Cl^- ions precipitated per formula unit. The results are in the second column. Werner reasoned that the precipitated Cl^- ions must be free (uncoordinated), whereas the unprecipitated Cl^- ions must be directly bound to Pt so they could not be precipitated by Ag^+ ions. He also measured the conductances of solutions of these compounds of known concentrations. By comparing these with data on solutions of simple electrolytes, he found the number of ions per formula unit. The results are in the third column. Piecing the evidence together, he concluded that the correct formulas are the ones listed in the last two columns. The NH_3 and Cl^- within the brackets are bonded by coordinate covalent bonds to the Lewis acid, Pt(IV) ion. The charge on a complex is the sum of its constituent charges.

Double salts are ionic solids resulting from the cocrystallization into a single structure of two salts from the same solution. In the example given, the solid is produced from an aqueous solution of iron(II) sulfate, $FeSO_4$, and ammonium sulfate, $(NH_4)_2SO_4$.

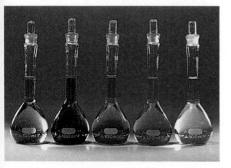

Compounds of the transition metals are often colored, whereas those of A group metals are usually colorless. Aqueous solutions of some nitrate salts (left to right): $Fe(NO_3)_3$, $Co(NO_3)_2$, $Ni(NO_3)_2$, $Cu(NO_3)_2$, and $Zn(NO_3)_2$.

The conductance of a solution of an electrolyte is a measure of its ability to conduct electricity. It is related to the number of and the charges on ions in solution.

29-2 The Ammine Complexes

The **ammine complexes** contain NH_3 molecules bonded to metal ions. Because the ammine complexes are important compounds, we will describe them briefly.

Most metal hydroxides are insoluble in water, and so aqueous NH_3 reacts with nearly all metal ions to form insoluble metal hydroxides, or hydrated

Table 29-2
Common Metal Ions That Form Soluble Complexes with an Excess of Aqueous Ammonia[a]

Metal Ion	Insoluble Hydroxide Formed by Limited Aq. NH_3	Complex Ion Formed by Excess Aq. NH_3
Co^{2+}	$Co(OH)_2$	$[Co(NH_3)_6]^{2+}$
Co^{3+}	$Co(OH)_3$	$[Co(NH_3)_6]^{3+}$
Ni^{2+}	$Ni(OH)_2$	$[Ni(NH_3)_6]^{2+}$
Cu^+	$CuOH \longrightarrow \frac{1}{2}Cu_2O$[b]	$[Cu(NH_3)_2]^+$
Cu^{2+}	$Cu(OH)_2$	$[Cu(NH_3)_4]^{2+}$
Ag^+	$AgOH \longrightarrow \frac{1}{2}Ag_2O$[b]	$[Ag(NH_3)_2]^+$
Zn^{2+}	$Zn(OH)_2$	$[Zn(NH_3)_4]^{2+}$
Cd^{2+}	$Cd(OH)_2$	$[Cd(NH_3)_4]^{2+}$
Hg^{2+}	$Hg(OH)_2$	$[Hg(NH_3)_4]^{2+}$

[a] The ions of Rh, Ir, Pd, Pt, and Au show similar behavior.
[b] CuOH and AgOH are unstable and decompose to the corresponding oxides.

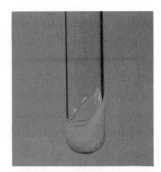

oxides. The exceptions are the cations of the strong soluble bases (Group IA cations and the heavier members of Group IIA—Ca^{2+}, Sr^{2+}, and Ba^{2+}).

$$Cu^{2+} + 2NH_3 + 2H_2O \longrightarrow Cu(OH)_2(s) + 2NH_4^+$$

$$Cr^{3+} + 3NH_3 + 3H_2O \longrightarrow Cr(OH)_3(s) + 3NH_4^+$$

In general terms, we can represent this reaction as

$$M^{n+} + nNH_3 + nH_2O \longrightarrow M(OH)_n(s) + nNH_4^+$$

where M^{n+} represents all of the common metal ions *except* those of the IA metals and the heavier IIA metals.

The hydroxides of some metals and some metalloids are amphoteric (Section 10-8). Aqueous NH_3 is a weak base ($K_b = 1.8 \times 10^{-5}$), so the $[OH^-]$ is too low to dissolve amphoteric hydroxides to form hydroxo complexes.

However, several metal hydroxides do dissolve in an excess of aqueous NH_3 to form ammine complexes. For example, the hydroxides of copper and cobalt are readily soluble in an excess of aqueous ammonia solution:

$$Cu(OH)_2(s) + 4NH_3 \rightleftharpoons [Cu(NH_3)_4]^{2+} + 2OH^-$$

$$Co(OH)_2(s) + 6NH_3 \rightleftharpoons [Co(NH_3)_6]^{2+} + 2OH^-$$

$Cu(OH)_2$ (light blue) dissolves in excess aqueous NH_3 to form $[Cu(NH_3)_4]^{2+}$ ions (deep blue).

Interestingly, all metal hydroxides that exhibit this behavior are derived from the 12 metals of the cobalt, nickel, copper, and zinc families. All the common cations of these metals except Hg_2^{2+} (which disproportionates) form soluble complexes in the presence of excess aqueous ammonia (Table 29-2).

29-3 Important Terms

Let us define and illustrate a few terms. The Lewis bases in coordination compounds may be molecules, anions, or (rarely) cations. They are called **ligands** (Latin *ligare*, "to bind"). The **donor atoms** of the ligands are the atoms that donate shares in electron pairs to metals. In some cases it is not possible to identify donor atoms, because the bonding electrons are not localized on specific atoms. Some small organic molecules such as ethylene, $H_2C=CH_2$, bond to a transition metal through the electrons in their double bonds. Examples of typical simple ligands are listed in Table 29-3.

$Co(OH)_2$ (a blue compound that turns gray quickly) dissolves in excess aqueous NH_3 to form $[Co(NH_3)_6]^{2+}$ ions (yellow-orange).

Table 29-3
Typical Simple Ligands with Their Donor Atoms Shaded

Molecule	Name	Name as Ligand	Ion	Name	Name as Ligand
$:NH_3$	ammonia	ammine	$:\ddot{C}l:^-$	chloride	chloro
$:\ddot{O}H_2$	water	aqua	$:\ddot{F}:^-$	fluoride	fluoro
$:C\equiv O:$	carbon monoxide	carbonyl	$:C\equiv N:^-$	cyanide	cyano[a]
$:PH_3$	phosphine	phosphine	$:\ddot{O}H^-$	hydroxide	hydroxo
$:\ddot{N}=\ddot{O}^-$	nitrogen oxide	nitrosyl	$:N\begin{smallmatrix}\ddot{O}.^-\\\ddot{O}:\end{smallmatrix}$	nitrite	nitro[b]

[a] Nitrogen atoms can also function as donor atoms.
[b] Oxygen atoms can also function as donor atoms, in which case the ligand name is "nitrito."

Ligands that can bond to a metal through only one donor atom at a time are **unidentate** (Latin *dent*, "tooth"). Ligands that can bond simultaneously through more than one donor atom are **polydentate**. Polydentate ligands that bond through two, three, four, five, or six donor atoms are called *bidentate*, *tridentate*, *quadridentate*, *quinquedentate*, and *sexidentate*, respectively. Complexes that consist of a metal atom or ion and polydentate ligands are called **chelate complexes** (Greek *chele*, "claw").

The **coordination number** of a metal atom or ion in a complex is the number of donor atoms to which it is coordinated, not necessarily the number of

Table 29-4
Some Ligands and Coordination Spheres (complexes)

Ligand(s)	Classification	Coordination Sphere	Oxidation Number of M	Coordination Number of M
NH₃ ammine	unidentate	$[Co(NH_3)_6]^{3+}$ hexaamminecobalt(III)	+3	6
H₂N—CH₂—CH₂—NH₂ (or N⌢N) ethylenediamine (en)	bidentate	 $[Co(en)_3]^{3+}$ tris(ethylenediamine)- cobalt(III) ion	+3	6

Table 29-4
(continued)

Ligand(s)	Classification	Coordination Sphere	Oxidation Number of M	Coordination Number of M
Br^- bromo $H_2N-CH_2-CH_2-NH_2$ ethylenediamine (en)	unidentate bidentate	[Cu(en) Br$_2$] dibromoethylenediamine-copper(II) ion	+2	4
$H_2N-(CH_2)_2-N-(CH_2)_2-NH_2$ (or N⌣N⌣N) diethylenetriamine (dien)	tridentate	[Fe(dien)$_2$]$^{3+}$ bis(diethylenetriamine)-iron(III) ion	+3	6
ethylenediaminetetraacetato (edta)	sexidentate	[Co(edta)]$^-$ (ethylenediaminetetraacetato)-cobaltate(III) ion	+3	6

ligands. The **coordination sphere** includes the metal or metal ion and its ligands, but no uncoordinated counter-ions. For example, the coordination sphere of hexaamminecobalt(III) chloride, $[Co(NH_3)_6]Cl_3$, is the hexaamminecobalt(III) ion, $[Co(NH_3)_6]^{3+}$. These terms are illustrated in Table 29-4.

29-4 Nomenclature

Many thousands of coordination compounds are known, and many new ones are discovered each year. The International Union of Pure and Applied Chemistry (IUPAC) has adopted a set of rules for naming them. The rules are based on those originally devised by Werner.

1. Cations are named before anions.
2. In naming the coordination sphere, ligands are named in alphabetical order. The prefixes di = 2, tri = 3, tetra = 4, penta = 5, hexa = 6, and so on, specify the number of each kind of *simple* ligand. For example, in dichloro, the "di" indicates that two Cl^- act as ligands. For complicated ligands (usually chelating agents), other prefixes are used: bis = 2, tris = 3, tetrakis = 4, pentakis = 5, and hexakis = 6. The names of complicated ligands are enclosed in parentheses. These prefixes are not used in alphabetizing. When a prefix denotes the number of substituents on a single ligand, as in dimethylamine, $NH(CH_3)_2$, it *is* used to alphabetize ligands.
3. The names of anionic ligands end in the suffix -o. Examples are F^-, fluoro; OH^-, hydroxo; O^{2-}, oxo; S^{2-}, sulfido; CO_3^{2-}, carbonato; CN^-, cyano; SO_4^{2-}, sulfato; NO_3^-, nitrato; $S_2O_3^{2-}$, thiosulfato.
4. The names of neutral ligands are usually unchanged. Four important exceptions are NH_3, ammine; H_2O, aqua; CO, carbonyl; and NO, nitrosyl.
5. Some metals exhibit variable oxidation states. The oxidation number of such a metal is designated by a Roman numeral in parentheses following the name of the complex ion or molecule.
6. The suffix "-ate" at the end of the name of the complex signifies that it is an anion. If the complex is neutral or cationic, no suffix is used. The English stem is usually used for the metal, but where the naming of an anion is awkward, the Latin stem is substituted. For example, "ferrate" is used rather than "ironate," and "plumbate" rather than "leadate" (Table 29-5).

Several examples are given to illustrate the rules.

$K_2[Cu(CN)_4]$	potassium tetracyanocuprate(II)
$[Ag(NH_3)_2]Cl$	diamminesilver(I) chloride
$[Cr(OH_2)_6](NO_3)_3$	hexaaquachromium(III) nitrate
$[Co(en)_2Br_2]Cl$	dibromobis(ethylenediamine)cobalt(III) chloride
$[Ni(CO)_4]$	tetracarbonylnickel(0)
$[Pt(NH_3)_4][PtCl_6]$	tetraammineplatinum(II) hexachloroplatinate(IV)
$[Cu(NH_3)_2(en)]Br_2$	diammine(ethylenediamine)copper(II) bromide
$Na[Al(OH)_4]$	sodium tetrahydroxoaluminate
$Na_2[Sn(OH)_6]$	sodium hexahydroxostannate(IV)
$[Co(en)_3](NO_3)_3$	tris(ethylenediamine)cobalt(III) nitrate
$K_4[Ni(CN)_2(ox)_2]$	potassium dicyanobis(oxalato)nickelate(II)
$[Co(NH_3)_4(OH_2)Cl]Cl_2$	tetraammineaquachlorocobalt(III) chloride

Table 29-5
Names for Some Metals in Complex Anions

Metal	Name* of Metal in Complex Anions
aluminum	aluminate
antimony	antimonate
chromium	chromate
cobalt	cobaltate
copper	*cupr*ate
gold	*aur*ate
iron	*ferr*ate
lead	*plumb*ate
manganese	manganate
nickel	nickelate
platinum	platinate
silver	*argent*ate
tin	*stann*ate
zinc	zincate

* Stems derived from Latin names for metals are shown in italics.

The term "ammine" (two m's) signifies the presence of ammonia as a ligand. It is different from the term "amine" (one m), which describes a class of organic compounds (Section 31-11) that are derived from ammonia.

Water is written OH_2 rather than H_2O to emphasize that oxygen is the donor atom.

The oxidation state of aluminum is not given because it is always +3.

The abbreviation "ox" represents the oxalate ion $(COO)_2^{2-}$ or $C_2O_4^{2-}$.

Some coordination compounds. Starting at the top left and moving clockwise: [Cr(CO)$_6$] (white), CO is the ligand; K$_3$\{Fe[(COO)$_2$]$_3$\} (green), (COO)$_2{}^{2-}$ (oxalate ion) is the ligand; [Co(H$_2$N—CH$_2$—CH$_2$—NH$_2$)$_3$]I$_3$ (yellow-orange), ethylenediamine is the ligand; [Co(NH$_3$)$_5$(OH$_2$)]Cl$_3$ (red), NH$_3$ and H$_2$O are ligands; K$_3$[Fe(CN)$_6$] (red-orange), CN$^-$ is the ligand. A drop of water fell on this sample.

29-5 Structures

The structures of coordination compounds are governed largely by the coordination number of the metal. Most can be predicted by VSEPR theory (Chapter 8). Lone pairs of electrons in d orbitals usually have only small influences on geometry because they are not in the outer shell. Table 29-6 summarizes the geometries and hybridizations for common coordination numbers.

Transition metal complexes with coordination numbers as high as 7, 8, and 9 are known, but they are very rare. For coordination number 5, the trigonal bipyramidal structure is more common than the square pyramid. The energies associated with these structures are very close. Both tetrahedral and square planar geometries are quite common for complexes with coordination number 4. The tabulated geometries are ideal geometries. Actual structures are sometimes distorted, especially if the ligands are not all the same. The distortions are due to compensations for the unequal electric fields generated by the different ligands.

Isomerism in Coordination Compounds

Isomers are substances that have the same number and kinds of atoms arranged differently. *Because their structures are different, isomers have different properties.* Here we shall restrict our discussion of isomerism to that caused by different arrangements of ligands about metal ions.

Isomers can be broadly classed into two major classes: structural isomers and stereoisomers. Each can be further subdivided as follows.

Structural Isomers	Stereoisomers
1. ionization isomers	**1.** geometric (positional) isomers
2. hydrate isomers	**2.** optical isomers
3. coordination isomers	
4. linkage isomers	

The term "isomers" comes from the Greek word meaning "equal weights."

Stereoisomers are more important. Distinctions between simple stereoisomers involve only one coordination sphere, and the same ligands and donor atoms. Differences between **structural isomers** involve either more than one coordination sphere or different donor atoms on the same ligand. They

Table 29-6
Geometries and Hybridizations for Various Coordination Numbers

Coordination Number	Geometry	Hybridization	Examples
2	linear	sp	$[Ag(NH_3)_2]^+$ $[Cu(CN)_2]^-$
4	tetrahedral	sp^3	$[Zn(CN)_4]^{2-}$ $[Cd(NH_3)_4]^{2+}$
4	square planar	dsp^2 or sp^2d	$[Cu(OH_2)_4]^{2+}$ $[Pt(NH_3)_2Cl_2]$
5	trigonal bipyramidal	dsp^3	$[Fe(CO)_5]$ $[CuCl_5]^{3-}$
5	square pyramidal	d^2sp^2	$[NiBr_3\{P(C_2H_5)_3\}_2]$ $[RuCl_3\{P(C_6H_5)_3\}_2]$
6	octahedral	d^2sp^3 or sp^3d^2	$[Fe(CN)_6]^{4-}$ $[Fe(OH_2)_6]^{2+}$

contain *different atom-to-atom bonding sequences*. Before considering stereoisomers, we shall describe the four types of structural isomers.

29-6 Structural Isomers

Ionization (Ion-Ion Exchange) Isomers

These isomers result from the interchange of ions inside and outside the coordination sphere. For example, red-violet $[Co(NH_3)_5Br]SO_4$ and red $[Co(NH_3)_5SO_4]Br$ are ionization isomers.

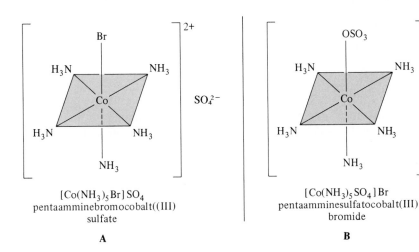

Isomers such as those shown here may *or may not* exist in the same solution in equilibrium. Such isomers are formed by *different* reactions.

$[Co(NH_3)_5Br]SO_4$
pentaamminebromocobalt((III)
sulfate

A

$[Co(NH_3)_5SO_4]Br$
pentaamminesulfatocobalt(III)
bromide

B

In structure A, the SO_4^{2-} ion is free and is not bound to the cobalt(III) ion. A solution of A reacts with a solution of barium, $BaCl_2$, to precipitate $BaSO_4$, but does not react with $AgNO_3$. In structure B, the SO_4^{2-} ion is bound to the cobalt(III) ion and so it does not react with $BaCl_2$ in aqueous solution. However, the Br^- ion is free, and a solution of B reacts with $AgNO_3$ to precipitate AgBr. **Equimolar** solutions of A and B also have different electrical conductivities. The sulfate solution, A, conducts electric current better because its ions have 2+ and 2− charges rather than 1+ and 1−. Other examples of this type of isomerism include

$[Pt(NH_3)_4Cl_2]Br_2$ and $[Pt(NH_3)_4Br_2]Cl_2$
$[Pt(NH_3)_4SO_4](OH)_2$ and $[Pt(NH_3)_4(OH)_2]SO_4$
$[Co(NH_3)_5NO_2]SO_4$ and $[Co(NH_3)_5SO_4]NO_2$
$[Cr(NH_3)_5SO_4]Br$ and $[Cr(NH_3)_5Br]SO_4$

Hydrate Isomers

Hydration isomerism and ionization isomerism are quite similar. In some crystalline complexes, water can be *inside* and *outside* the coordination sphere. For example, when treated with excess $AgNO_3(aq)$, solutions of the following three hydrate isomers yield three, two, and one mole of AgCl precipitate, respectively, per mole of complex (top of page 974).

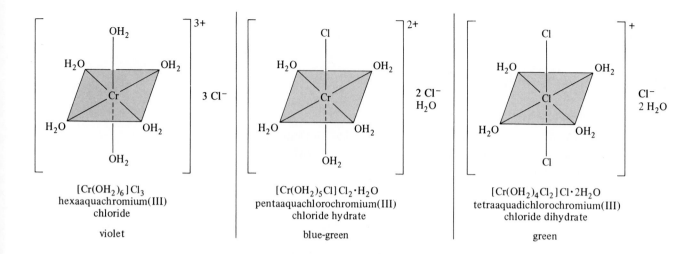

$[Cr(OH_2)_6]Cl_3$
hexaaquachromium(III)
chloride

violet

$[Cr(OH_2)_5Cl]Cl_2 \cdot H_2O$
pentaaquachlorochromium(III)
chloride hydrate

blue-green

$[Cr(OH_2)_4Cl_2]Cl \cdot 2H_2O$
tetraaquadichlorochromium(III)
chloride dihydrate

green

Coordination Isomers

Coordination isomerism can occur in compounds containing both complex cations and complex anions. Such isomers involve exchange of ligands between cation and anion, i.e., between coordination spheres.

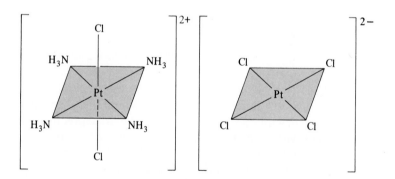

$[Pt(NH_3)_4Cl_2][PtCl_4]$
tetraamminedichloroplatinum(IV)
tetrachloroplatinate(II)

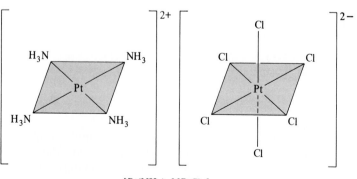

$[Pt(NH_3)_4][PtCl_6]$
tetraammineplatinum(II)
hexachloroplatinate(IV)

Linkage Isomers

Certain ligands can bind to metal ions in more than one way. Examples of such ligands are cyano, $-CN^-$, and isocyano, $-NC^-$; nitro, $-NO_2^-$, and nitrito, $-ONO^-$. The donor atoms are on the left in these representations. Examples of linkage isomers follow.

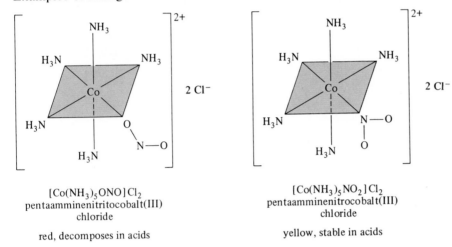

$[Co(NH_3)_5ONO]Cl_2$
pentaamminenitritocobalt(III)
chloride

red, decomposes in acids

$[Co(NH_3)_5NO_2]Cl_2$
pentaamminenitrocobalt(III)
chloride

yellow, stable in acids

29-7 Stereoisomers

Compounds that contain the same atoms and the same atom-to-atom bonding sequences, but that differ only in the spatial arrangements of the atoms relative to the central atom, are **stereoisomers**. Complexes with only *simple* ligands can exist as stereoisomers only *if* they have coordination number 4 or greater. The most common coordination numbers among coordination complexes are 4 and 6, and so they will be used to illustrate stereoisomerism.

A complex with coordination number 2 or 3 that contains only *simple ligands* can have only one spatial arrangement. All apparent "isomers" are equivalent to the complex turned around. Try building models to see this.

Geometrical Isomers

In geometrical isomers, or positional isomers, of *coordination compounds* the same ligands are arranged in different orders within the coordination sphere. Stereoisomers that are not optical isomers are **geometrical isomers**. *Cis–trans* isomerism is one kind of geometric isomerism. *Cis* means "adjacent to" and *trans* means "on the opposite side of." *Cis-* and *trans*-diamminedichloroplatinum(II) are shown below.

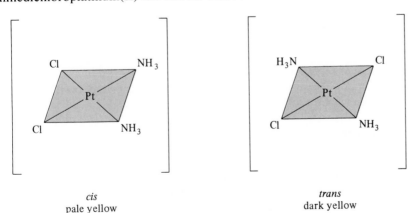

cis
pale yellow

trans
dark yellow

The *cis* isomer has been used in chemotherapy. The *trans* isomer has no such activity.

In the *cis* isomer, the chloro groups are closer to each other (on the same side of the square) than they are in the *trans* isomer. The ammine groups are also closer together in the *cis* complex.

Tetrahedral complexes of the type ML_4 do not exhibit geometrical isomerism. All four ligands are equidistant from each other.

Several types of isomerism are possible for octahedral complexes. For example, complexes of the type $[MA_2B_2C_2]$ can exist in several isomeric forms. Consider as an example $[Cr(OH_2)_2(NH_3)_2Br_2]^+$. First, the members of all three pairs of like ligands may be either *trans* to each other or *cis* to each other.

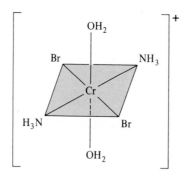

trans-diammine-*trans*-diaqua-*trans*-dibromochromium(III) ion

A

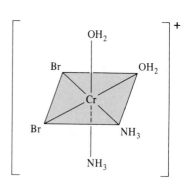

cis-diammine-*cis*-diaqua-*cis*-dibromochromium(III) ion

B

Then, members of one of the pairs may be *trans* to each other, but the members of the other two pairs are *cis*.

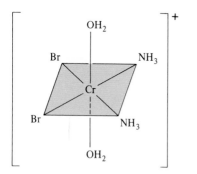

cis-diammine-*trans*-diaqua-*cis*-dibromochromium(III) ion

C

trans-diammine-*cis*-diaqua-*cis*-dibromochromium(III) ion

D

cis-diammine-*cis*-diaqua-*trans*-dibromochromium(III) ion

E

Further interchange of the positions of the ligands produces no new geometric isomers. However, one of the five geometrical isomers (B) can exist in two distinct forms called *optical isomers*.

See whether you can discover why there is no *trans-trans-cis* isomer.

Optical Isomers

The *cis*-diammine-*cis*-diaqua-*cis*-dibromochromium(III) geometrical isomer (B) exists in two forms that bear the same relationship to each other as left and right hands. They are *nonsuperimposable* mirror images of each other and are called **optical isomers** or **enantiomers**.

An object that is not superimposable with its mirror image is said to be *chiral*.

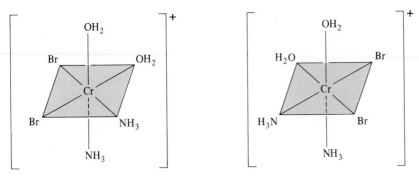

Optical isomers of *cis*-diammine-*cis*-diaqua-*cis*-dibromochromium(III) ion

The dichlorobis(ethylenediamine)cobalt-(III) ion, $[Co(en)_2Cl_2]^+$, exists as a pair of *cis–trans* isomers. Ethylenediamine is represented as N N.

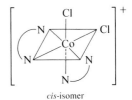

cis-isomer

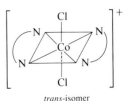

trans-isomer

Cis-dichlorobis(ethylenediamine)cobalt-(III) perchlorate, $[Co(en)_2Cl_2]ClO_4$, is purple.
Trans-dichlorobis(ethylenediamine)cobalt(III) chloride, $[Co(en)_2Cl_2]Cl$, is green.

Optical isomers interact with polarized light in different ways. Separate equimolar solutions of the two rotate a plane of polarized light (see Figures 29-2 and 29-3) by equal amounts but in opposite directions. One solution is **dextrorotatory** (rotates to the *right*) and the other is **levorotatory** (rotates to the *left*). Optical isomers are called *dextro* and *levo* isomers. The phenomenon by which a plane of polarized light is rotated is called **optical activity**. It can be measured with a device called a polarimeter (Figure 29-3) or with more sophisticated instruments. A single solution containing equal amounts of the two isomers is a **racemic mixture**. This solution does not rotate a plane of polarized light. The equal and opposite effects of the two isomers exactly cancel. To exhibit optical activity, the *dextro* and *levo* isomers (sometimes designated as delta, Δ, and lambda, Λ, isomers) must be separated from each other. This is done by one of a number of chemical or physical processes

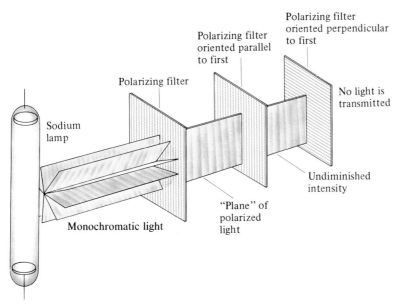

Figure 29-2
Light from a lamp or from the sun consists of electromagnetic waves that vibrate in all directions perpendicular to the direction of travel. Polarizing filters absorb all waves except those that vibrate in a single plane. The third polarizing filter, with a plane of polarization at right angles to the first, absorbs the polarized light completely.

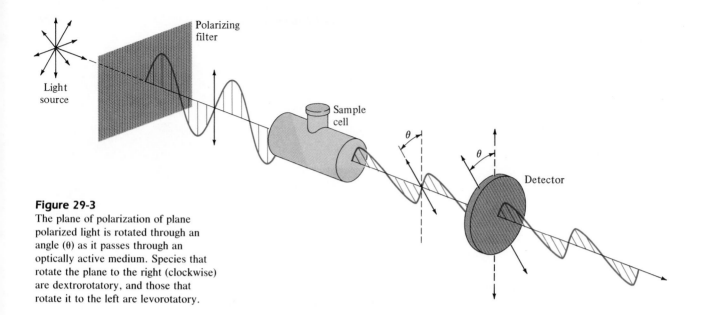

Figure 29-3
The plane of polarization of plane polarized light is rotated through an angle (θ) as it passes through an optically active medium. Species that rotate the plane to the right (clockwise) are dextrorotatory, and those that rotate it to the left are levorotatory.

broadly called **optical resolution**. Another pair of optical isomers is shown below. Both contain ethylenediamine, a bidentate ligand.

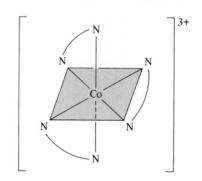

Λ-*tris*(ethylenediamine)cobalt(III)ion

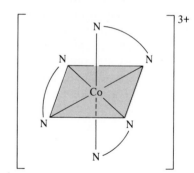

Δ-*tris*(ethylenediamine)cobalt(III) ion

Bonding in Coordination Compounds

Bonding theories for coordination compounds should be able to account for structural features, colors, and magnetic properties. The earliest accepted theory was the **valence bond theory** (Chapter 8). It can account for structural and magnetic properties. It offers no explanation for the wide range of colors of coordination compounds. However, it has the advantage of being a simple description of bonding that uses the classical picture of the chemical bond. The **crystal field theory** gives quite satisfactory explanations of color as well as of structure and magnetic properties for many coordination compounds.

29-8 Valence Bond Theory

An assumption of the valence bond theory applied to coordination compounds is that the bonds between ligands and metal are entirely covalent. This is not really true. All bonds have some degree of both covalent and

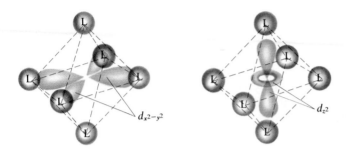

Figure 29-4

Orientation of $d_{x^2-y^2}$ and d_{z^2} orbitals relative to the ligands in an octahedral complex.

ionic character. Because there are so many complex species in which the coordination number of the metal ion is 6, we shall focus our attention on some of them.

Recall the shapes of the *d* orbitals.

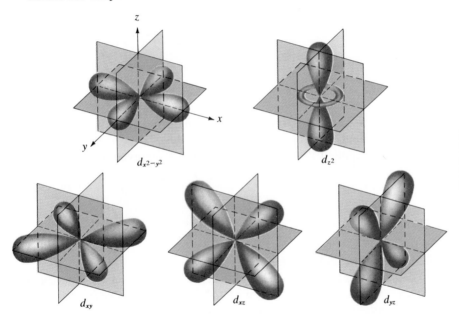

Spatial orientation of *d* orbitals. Note that the lobes of the $d_{x^2-y^2}$ and d_{z^2} orbitals lie along the axes, whereas the lobes of the others lie along diagonals between the axes.

The lobes of the $d_{x^2-y^2}$ and d_{z^2} orbitals are directed along the x, y, and z axes. The lobes of the d_{xy}, d_{yz}, and d_{xz} orbitals lie between the axes. The six ligand donor atoms in an octahedral complex are at the corners of an octahedron, two along each of the three axes. Valence bond theory postulates that the donor atoms must donate electrons into a set of octahedrally hybridized d^2sp^3 or sp^3d^2 metal orbitals. Thus, the two metal *d* orbitals used in the hybridization must be the $d_{x^2-y^2}$ and d_{z^2} orbitals, because they are the only ones directed along the x, y, and z axes (Figure 29-4).

Let us study some complexes to see how the valence bond theory explains their properties. Hexaaquairon(III) perchlorate, $[Fe(OH_2)_6](ClO_4)_3$, is known to be paramagnetic, with a magnetic moment corresponding to five unpaired electrons per iron atom. The valence bond description of the bonding in the $[Fe(OH_2)_6]^{3+}$ ion follows. Atomic iron has the electronic configuration

A magnetic moment indicates the extent of paramagnetism of a molecule or polyatomic ion.

	3*d*					4*s*
$_{26}$Fe [Ar]	⇅	↑	↑	↑	↑	⇅

The Fe^{3+} ion is formed by loss of the $4s$ electrons and one of the $3d$ electrons.

$$Fe^{3+} \; [Ar] \quad \underset{3d}{\uparrow \; \uparrow \; \uparrow \; \uparrow \; \uparrow} \quad \underset{4s}{\underline{}}$$

This is in agreement with Hund's Rule.

To account for the experimentally observed fact that $[Fe(OH_2)_6]^{3+}$ has five unpaired electrons, each $3d$ orbital is assumed to have one unpaired electron in the complex. The vacant $4d_{x^2-y^2}$ and $4d_{z^2}$ orbitals are hybridized with the vacant $4s$ and $4p$ orbitals.

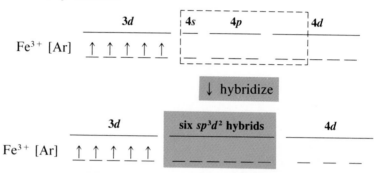

Each of the six H_2O ligands now donates one electron pair into one of the six sp^3d^2 orbitals, forming six coordinate covalent bonds. The electrons originally on the oxygen atoms are represented here as "$\times$" rather than "$\uparrow$," even though electrons are indistinguishable. These are the only *bonding* electrons; none of the $3d$ electrons of iron is involved in bonding.

$$[Fe(OH_2)_6]^{3+} \; [Ar] \quad \underset{3d}{\uparrow \; \uparrow \; \uparrow \; \uparrow \; \uparrow} \quad \underset{sp^3d^2}{\times\times \; \times\times \; \times\times \; \times\times \; \times\times \; \times\times} \quad \underset{4d}{\underline{} \; \underline{} \; \underline{}}$$

A set of *outer d* orbitals is used in hybridization, and so $[Fe(OH_2)_6]^{3+}$ is called an **outer orbital complex**.

In deciding whether the hybridization at the central metal ion of octahedral complexes is sp^3d^2 (outer orbital) or d^2sp^3 (inner orbital), *we must know the results of magnetic measurements*. These indicate the number of unpaired electrons.

A convenient way to describe d-transition metal ions is to indicate the number of nonbonding electrons in d orbitals.

The hexacyanoferrate(III) ion (the ferricyanide ion), $[Fe(CN)_6]^{3-}$, also involves Fe^{3+} (a d^5 ion). But its magnetic moment indicates only one unpaired electron per iron atom. The $3d_{x^2-y^2}$ and $3d_{z^2}$ orbitals (rather than those in the $4d$ shell) are involved in d^2sp^3 hybridization. This forces pairing of all but one of the nonbonding $3d$ electrons of Fe^{3+}.

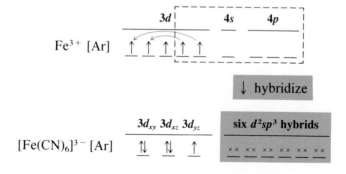

Only inner *d* orbitals are used in hybridization, and so $[Fe(CN)_6]^{3-}$ is called an **inner orbital complex**. As in the previous case, none of the original 3*d* electrons of the Fe^{3+} ion is involved in bonding.

To account for the single unpaired electron in $[Co(NH_3)_6]^{2+}$ ions (Co^{2+} is a d^7 ion) by valence bond theory, we postulate the *promotion* of a 3*d* electron to a 5*s* orbital as d^2sp^3 hybridization occurs.

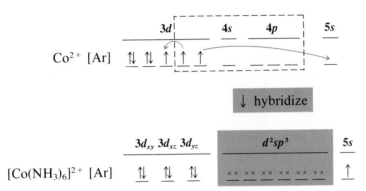

This bonding for $[Co(NH_3)_6]^{2+}$ is consistent with the fact that $[Co(NH_3)_6]^{2+}$ is easily oxidized to $[Co(NH_3)_6]^{3+}$, an extremely stable complex ion.

The electronic configurations and hybridizations of some octahedral complexes are shown in Table 29-7.

29-9 Crystal Field Theory

Hans Bethe and J. H. van Vleck developed the crystal field theory between 1919 and the early 1930s. It was not widely used until the 1950s. In its pure form it assumes that the bonds between ligand and metal ion are completely ionic. Both ligand and metal ion are treated as infinitesimally small, non-polarizable point charges. Valence bond theory assumes complete covalence. Modern ligand field theory is an outgrowth of crystal field theory. It attributes partial covalent character and partial ionic character to bonds.

In a metal ion surrounded by other atoms, the *d* orbitals are at higher energy than they are in an isolated metal ion. If the surrounding electrons were uniformly distributed about the metal ion, the energies of *all* five *d* orbitals would increase by the same amount (a *spherical crystal field*). Because the ligands approach the metal ion from different directions, they affect different *d* orbitals in different ways. Here we illustrate the application of these ideas to complexes with coordination number 6 (*octahedral crystal field*).

The $d_{x^2-y^2}$ and d_{z^2} orbitals are directed along a set of mutually perpendicular *x*, *y*, and *z* axes (see p. 979). As a group, these orbitals are called the e_g orbitals. The d_{xy}, d_{yz}, and d_{xz} orbitals, collectively called the t_{2g} orbitals, lie between the axes. The ligand donor atoms approach the metal ion along the axes to form octahedral complexes. Crystal field theory proposes that the approach of the six donor atoms (point charges) along the axes sets up an electric field (the crystal field). Electrons on the ligands repel electrons in e_g orbitals on the metal ion more strongly than they repel those in t_{2g} orbitals (Figure 29-5). This removes the degeneracy of the set of *d* orbitals

Adding a solution containing Fe^{3+} ions to a solution of potassium hexacyanoferrate(II), $K_4[Fe(CN)_6]$, forms a blue precipitate,

$K\overset{+3}{Fe}[\overset{+2}{Fe}(CN)_6]$. This compound is called **Turnbull's blue**. Hexacyanoferrate(III) ions react with Fe^{2+} ions to form a blue precipitate known as **Prussian blue**. Prussian blue and Turnbull's blue are thought to be

the same compound, $K\overset{+3}{Fe}[\overset{+2}{Fe}(CN)_6]$. They were used in making blueprints.

Recall that degenerate orbitals are orbitals of equal energy.

Table 29-7
Bonding and Hybridization in Some Octahedral Complexes

Metal Ion	Outer Electron Configuration	Complex Ion	Type	Outer Electron Configuration
Cr^{3+} (d^3)	3d 4s 4p 4d	$[Cr(NH_3)_6]^{3+}$	inner orbital (paramagnetic)	3d — d^2sp^3 — 4d
Mn^{2+} (d^5)	3d 4s 4p 4d	$[Mn(CN)_6]^{4-}$	inner orbital (paramagnetic)	3d — d^2sp^3 — 4d
Fe^{2+} (d^6)	3d 4s 4p 4d	$[Fe(CN)_6]^{4-}$	inner orbital (diamagnetic)	3d — d^2sp^3 — 4d
Co^{3+} (d^6)	3d 4s 4p 4d	$[Co(NH_3)_6]^{3+}$	inner orbital (diamagnetic)	3d — d^2sp^3 — 4d
		$[CoF_6]^{3-}$	outer orbital (paramagnetic)	3d — sp^3d^2 — 4d
Co^{2+} (d^7)	3d 4s 4p 4d	$[Co(OH_2)_6]^{2+}$	outer orbital (paramagnetic)	3d — sp^3d^2 — 4d
Ni^{2+} (d^8)	3d 4s 4p 4d	$[Ni(OH_2)_6]^{2+}$	outer orbital (paramagnetic)	3d — sp^3d^2 — 4d

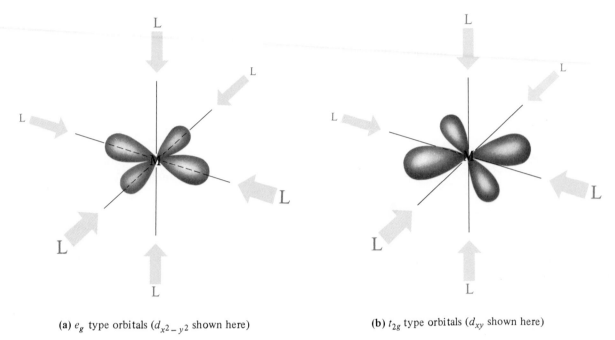

(a) e_g type orbitals ($d_{x^2-y^2}$ shown here)

(b) t_{2g} type orbitals (d_{xy} shown here)

Figure 29-5
Effects of the approach of ligands on the energies of d orbitals on the metal ion. In an octahedral complex, the ligands (L) approach the metal ion (M) along the x-, y-, and z-axes, as indicated by the blue arrows. (a) The orbitals of the e_g type—$d_{x^2-y^2}$ (shown here) and d_{z^2}—point directly toward the incoming ligands, so electrons in these orbitals are strongly repelled. (b) The orbitals of the t_{2g} type—d_{xy} (shown here), d_{xz} and d_{yz}—do not point toward the incoming ligands, so electrons in these orbitals are less strongly repelled.

and splits them into two sets, the e_g set at higher energy and the t_{2g} set at lower energy.

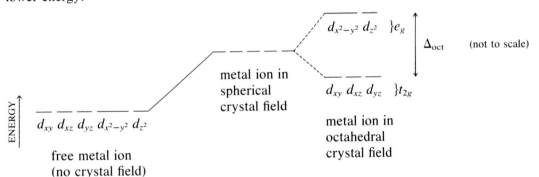

The energy separation between the two sets is called $\Delta_{octahedral}$, or Δ_{oct}. It is proportional to the *crystal field strength* of the ligands—that is, how strongly the ligand electrons repel the electrons on the metal ion.

The d electrons on a metal ion occupy the t_{2g} set in preference to the higher-energy e_g set. Electrons that occupy the e_g orbitals are strongly repelled by the relatively close approach of ligands. The occupancy of these orbitals tends to destabilize octahedral complexes.

Let us now describe the hexafluorocobaltate(III) ion, $[CoF_6]^{3-}$, and the hexaamminecobalt(III) ion, $[Co(NH_3)_6]^{3+}$. Both contain the d^6 Co^{3+} ion;

Δ_{oct} is sometimes called 10*Dq*. Its typical values are between 100 and 400 kJ/mol.

they have already been treated in terms of valence bond theory (Table 29-7). $[CoF_6]^{3-}$ is a paramagnetic outer orbital complex; $[Co(NH_3)_6]^{3+}$ is a diamagnetic inner orbital complex. We will focus our attention on the d electrons.

The Co^{3+} ion has six electrons (four unpaired) in its $3d$ orbitals.

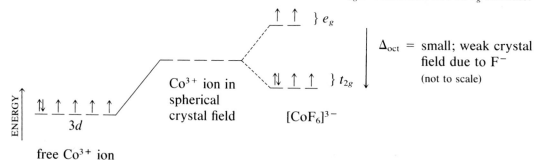

This is a high spin complex.

Its magnetic moment indicates that $[CoF_6]^{3-}$ also has four unpaired electrons per ion. So there must be four electrons in t_{2g} orbitals and two in e_g orbitals.

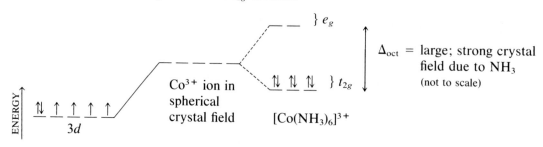

This is a low spin complex.

On the other hand, $[Co(NH_3)_6]^{3+}$ is diamagnetic, so all six d electrons must be paired in the t_{2g} orbitals.

The difference in configurations between $[CoF_6]^{3-}$ and $[Co(NH_3)_6]^{3+}$ is due to the relative magnitudes of the crystal field splitting, Δ_{oct}, caused by the different crystal field strengths of F^- and NH_3. The NH_3 molecule donates electrons into vacant metal orbitals more readily than the F^- ion does. As a result, the crystal field splitting generated by the close approach of six NH_3 molecules (strong field ligands) to the metal ion is greater than that generated by the approach of six F^- ions (weak field ligands).

$$\Delta_{oct} \text{ for } [Co(NH_3)_6]^{3+} > \Delta_{oct} \text{ for } [CoF_6]^{3-}$$

The crystal field splitting for $[CoF_6]^{3-}$ is very small. This means that an energetically more favorable situation results if two electrons remain unpaired in the antibonding e_g orbitals. (Hund's Rule requires that electrons singly occupy a set of degenerate orbitals before pairing.) This avoids the expenditure of energy that is necessary to pair electrons by bringing two negatively charged particles into the same region of space. If the approach of a set of ligands removes the d orbital degeneracy but causes a Δ_{oct} less than the electron pairing energy, P, then the electrons will singly occupy

Table 29-8
High and Low Spin Octahedral Configurations

d^n	Examples	High Spin	Low Spin	d^n	Examples	High Spin	Low Spin
d^1	Ti^{3+}	e_g: _ _ ; t_{2g}: ↑ _ _	same as high spin	d^6	Fe^{2+}, Ru^{2+}, Pd^{4+}, Rh^{3+}, Co^{3+}	e_g: ↑ ↑ ; t_{2g}: ⥮ ↑ ↑	e_g: _ _ ; t_{2g}: ⥮ ⥮ ⥮
d^2	Ti^{2+}, V^{3+}, Zr^{2+}	e_g: _ _ ; t_{2g}: ↑ ↑ _	same as high spin	d^7	Co^{2+}, Rh^{2+}	e_g: ↑ ↑ ; t_{2g}: ⥮ ⥮ ↑	e_g: ↑ _ ; t_{2g}: ⥮ ⥮ ⥮
d^3	V^{2+}, Cr^{3+}	e_g: _ _ ; t_{2g}: ↑ ↑ ↑	same as high spin	d^8	Ni^{2+}, Pt^{2+}, Au^{3+}	e_g: ↑ ↑ ; t_{2g}: ⥮ ⥮ ⥮	same as high spin
d^4	Mn^{3+}, Re^{3+}	e_g: ↑ _ ; t_{2g}: ↑ ↑ ↑	e_g: _ _ ; t_{2g}: ⥮ ↑ ↑	d^9	Cu^{2+}	e_g: ⥮ ↑ ; t_{2g}: ⥮ ⥮ ⥮	same as high spin
d^5	Mn^{2+}, Fe^{3+}, Ru^{3+}	e_g: ↑ ↑ ; t_{2g}: ↑ ↑ ↑	e_g: _ _ ; t_{2g}: ⥮ ⥮ ↑	d^{10}	Zn^{2+}, Ag^+, Hg^{2+}	e_g: ⥮ ⥮ ; t_{2g}: ⥮ ⥮ ⥮	same as high spin

the resulting nondegenerate orbitals. After all d orbitals are half-filled, additional electrons will pair with electrons in the t_{2g} set. This is the case for $[CoF_6]^{3-}$, which is called a **high spin complex**.

For $[CoF_6]^{3-}$: F^- is weak field ligand so $\Delta_{oct} < P$; thus, high spin complex

electron pairing energy

A high spin complex in crystal field terminology corresponds to an outer orbital complex in valence bond theory.

In contrast, the $[Co(NH_3)_6]^{3+}$ ion is a **low spin complex**. Δ_{oct} generated by the strong field ligand, NH_3, is greater than the electron pairing energy. So electrons pair in t_{2g} orbitals before any occupy the antibonding e_g orbitals.

$[Co(NH_3)_6]^{3+}$: NH_3 is strong field ligand so $\Delta_{oct} > P$; thus, low spin complex

A low spin complex corresponds to an inner orbital complex.

Low spin configurations exist only for octahedral complexes having metal ions with d^4, d^5, d^6, and d^7 configurations. For d^1–d^3 and d^8–d^{10} ions, only one possibility exists. In these cases, the configuration is designated as high spin. All d^n possibilities are shown in Table 29-8.

29-10 Color and the Spectrochemical Series

A substance appears colored because it absorbs light that corresponds to one or more of the wavelengths in the visible region of the electromagnetic spectrum (4000 to 7000 Å) and transmits or reflects the other wavelengths. Our eyes are detectors for light in the visible region, and so each wavelength in this region appears as a different color. A combination of all wavelengths

The electron pairing energy is larger than the crystal field splitting energy.

$[Co(OH_2)_6]^{2+}$ ions are pink (bottom). A limited amount of aqueous ammonia produces $Co(OH)_2$, a blue compound that quickly turns gray (middle). $Co(OH)_2$ dissolves in excess aqueous ammonia to form $[Co(NH_3)_6]^{2+}$ ions, which are orange-yellow (top). Because $[Co(NH_3)_6]^{2+}$ ions have single electrons in their $5s$ orbitals (p. 981), they are easily oxidized to $[Co(NH_3)_6]^{3+}$ ions.

Table 29-9
Complementary Colors

Wavelength Absorbed (Å)	Spectral Color (color absorbed)	Complementary Color (color observed)
4100	violet	lemon yellow
4300	indigo	yellow
4800	blue	orange
5000	blue-green	red
5300	green	purple
5600	lemon yellow	violet
5800	yellow	indigo
6100	orange	blue
6800	red	blue-green

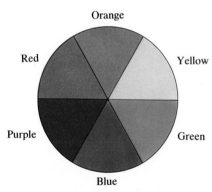

A color wheel shows colors and their complementary colors. For example, green is the complementary color of red. The data in Table 29-9 are given for specific wavelengths. Broad bands of wavelengths are shown in this color wheel.

in the visible region is called ''white light''; sunlight is an example. The absence of all wavelengths in the visible region is blackness.

In Table 29-9 we show the relationships among colors absorbed and colors transmitted or reflected in the visible region. The first column displays the wavelengths absorbed. The **spectral color** is the color associated with the wavelengths of light absorbed by the sample. When certain visible wavelengths are absorbed from incoming ''white'' light, the light *not absorbed* remains visible to us as transmitted or reflected light. For instance, a sample that absorbs orange light appears blue. The **complementary color** is the color associated with the wavelengths that are not absorbed by the sample. The complementary color is seen when the spectral color is removed from white light.

One transition of a high spin octahedral Co(III) complex is depicted as follows.

Ground State		Excited State	Energy of Light Absorbed
e_g ↑ ↑	$\xrightarrow[\text{of light}]{\text{Absorption}}$	↑↓ ↑ - - - - - - - -	
t_{2g} ↑↓ ↑ ↑		↑ ↑ ↑ - - - - - - -	$\Delta E = h\nu$ depends on Δ_{oct}

Planck's constant is
$h = 6.63 \times 10^{-34}$ J·s.

The colors of complex compounds that contain a given metal depend on the ligands. The yellow compound at the left is a salt that contains $[Co(NH_3)_6]^{3+}$ ions. In the next three compounds, left to right, one NH_3 ligand in $[Co(NH_3)_6]^{3+}$ has been replaced by NCS^- (orange), H_2O (red), and Cl^- (purple). The green compound at the right is a salt that contains $[Co(NH_3)_4Cl_2]^+$ ions.

The frequency (ν), and therefore the wavelength and color, of the light absorbed are related to Δ_{oct}.* This, in turn, depends upon the crystal field

---

* The numerical relationship between Δ_{oct} and the wavelength, λ, of the absorbed light is found by combining the expressions $E = h\nu$ and $\nu = c/\lambda$, where c is the speed of light. For *one* mole of a complex,

$$\Delta_{\text{oct}} = EN_A = \frac{hcN_A}{\lambda} \qquad \text{where } N_A \text{ is Avogadro's number}$$

Table 29-10
Colors of Some Chromium(III) Complexes

$[Cr(OH_2)_4Br_2]Br$	green	$[Cr(CON_2H_4)_6][SiF_6]_3$	green
$[Cr(OH_2)_6]Br_3$	bluish gray	$[Cr(NH_3)_5Cl]Cl_2$	purple
$[Cr(OH_2)_4Cl_2]Cl$	green	$[Cr(NH_3)_4Cl_2]Cl$	violet
$[Cr(OH_2)_6]Cl_3$	violet	$[Cr(NH_3)_6]Cl_3$	yellow

Different anions often cause compounds containing the same complex cation to have different colors.

strength of the ligands. So the colors and visible absorption spectra of transition metal complexes, as well as their magnetic properties, yield information about the strengths of the ligand–metal interactions.

By interpreting the visible spectra of many complexes, it is possible to arrange common ligands in order of increasing crystal field strengths.

$$I^- < Br^- < Cl^- < F^- < OH^- < H_2O < (COO)_2^{2-} < NH_3 < en < NO_2^- < CN^-$$

Increasing crystal field strength

Such an arrangement is called a **spectrochemical series**. Strong field ligands, such as CN^-, usually produce low spin complexes, where possible, and large crystal field splittings. Weak field ligands, such as Cl^-, usually produce high spin complexes and small crystal field splittings. Low spin complexes usually absorb higher-energy (shorter-wavelength) light than do high spin complexes. The colors of several six-coordinate Cr(III) complexes are listed in Table 29-10.

In $[Cr(NH_3)_6]Cl_3$, the Cr(III) is bonded to six ammonia ligands, which produce a relatively high value of Δ_{oct}. This causes the $[Cr(NH_3)_6]^{3+}$ ion to absorb relatively high-energy visible light in the blue and violet regions. Thus, we see yellow-orange, the complementary color.

Water is a weaker field ligand than ammonia, and therefore Δ_{oct} is less for $[Cr(OH_2)_6]^{3+}$ than for $[Cr(NH_3)_6]^{3+}$. As a result, $[Cr(OH_2)_6]Br_3$ absorbs lower-energy (longer-wavelength) light. This causes the reflected and transmitted light to be higher-energy bluish gray, the color that describes $[Cr(OH_2)_6]Br_3$.

We see the light that is transmitted (passes through the sample) or that is reflected by the sample.

29-11 Crystal Field Stabilization Energy

Electrons in the t_{2g} orbitals of an octahedral complex are lower in energy, while electrons in the e_g orbitals are higher in energy, than they would be in d orbitals of a metal ion in a spherical field. The lower the total energy of a system, the more stable it is. The **crystal field stabilization energy** (CFSE) of a complex is a measure of the net energy of stabilization (compared to the ion in a spherical field) of a metal ion's electrons. Crystal field stabilization energy is a major factor that contributes to the stability of complex ions with certain electronic configurations ($d^4 - d^7$ ions). Here we consider the CFSEs of some octahedral complexes. For a given set of six ligands, the t_{2g} and e_g orbitals are split by a certain amount of energy, Δ_{oct}. Each t_{2g} orbital is $\frac{2}{5}\Delta_{oct}$ *below* the energy of the set of degenerate d orbitals in a spherical (homogeneous) field. Each e_g orbital is $\frac{3}{5}\Delta_{oct}$ *above* the energy of the unsplit d orbitals.

Colors of some copper(II) compounds. Left to right: $CuBr_2$, $CuSO_4 \cdot 5H_2O$, $CuCl_2 \cdot 2H_2O$, $(CuOH)_2CO_3$.

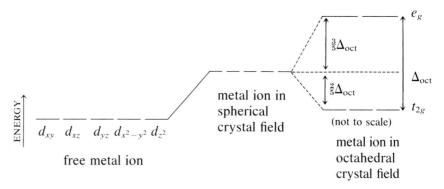

There are three t_{2g} orbitals and two e_g orbitals. The total energy of the sets of t_{2g} and e_g orbitals is the same as the energy of the set of degenerate d orbitals in a spherical field. That is, the change in total energy of a set of d orbitals is zero as a result of the approach of the six ligands along the x, y, and z axes.

$$\Delta E = \text{CFSE} = 3 \left(-\tfrac{2}{5}\Delta_{oct}\right) + 2 \left(\tfrac{3}{5}\Delta_{oct}\right) = 0$$
$$\qquad\qquad\qquad (t_{2g}) \qquad\qquad (e_g)$$

Table 29-8 shows the electronic configurations of some high spin and low spin octahedral complexes of d-transition metals. An energy change of $-\tfrac{2}{5}\Delta_{oct}$ (stabilization) is assigned to each t_{2g} electron, and an energy change of $+\tfrac{3}{5}\Delta_{oct}$ (destabilization) is assigned to each e_g electron. The CFSE is the sum of the energies of all the electrons in the metal atom or ion. Table 29-11 shows crystal field stabilization energies for the kinds of octahedral complexes described in Table 29-8. You should study these two tables together.

For example, for a high spin d^7 complex such as $[Co(OH_2)_6]^{2+}$, the CFSE is $5(-\tfrac{2}{5}\Delta_{oct}) + 2(+\tfrac{3}{5}\Delta_{oct}) = -\tfrac{4}{5}\Delta_{oct}$. So the amount of *energy released* by the octahedral splitting of the occupied orbitals in forming the complex ion is $\tfrac{4}{5}\Delta_{oct}$. Configurations with the most negative CFSEs in Table 29-11 are the ones for which many stable octahedral complexes are known. No configuration can produce a CFSE greater than zero. That is, no d-transition metal ions should be less stable in an octahedral ligand environment than in a spherical crystal field.

Most magnetic and spectral properties of complex species of the d-transition elements can be accounted for by the crystal field theory. This is strong evidence for the general validity of the theory. Consider the heats (enthalpies) of hydration for the series of $2+$ ions shown in Figure 29-6. All these ions form octahedral, weak field complexes with H_2O, $[M(OH_2)_6]^{2+}$.

Table 29-11
Crystal Field Stabilization Energies for Octahedral d^n Complexes

d^n	High Spin	Low Spin	d^n	High Spin	Low Spin
d^0	0	same as high spin			
d^1	$-\tfrac{2}{5}\,\Delta_{oct}$	same as high spin	d^6	$-\tfrac{2}{5}\,\Delta_{oct}$	$-\tfrac{12}{5}\,\Delta_{oct}$
d^2	$-\tfrac{4}{5}\,\Delta_{oct}$	same as high spin	d^7	$-\tfrac{4}{5}\,\Delta_{oct}$	$-\tfrac{9}{5}\,\Delta_{oct}$
d^3	$-\tfrac{6}{5}\,\Delta_{oct}$	same as high spin	d^8	$-\tfrac{6}{5}\,\Delta_{oct}$	same as high spin
d^4	$-\tfrac{3}{5}\,\Delta_{oct}$	$-\tfrac{8}{5}\,\Delta_{oct}$	d^9	$-\tfrac{3}{5}\,\Delta_{oct}$	same as high spin
d^5	0	$-\tfrac{10}{5}\,\Delta_{oct}$	d^{10}	0	same as high spin

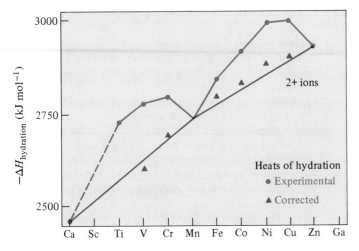

Figure 29-6

Heats of hydration for 2+ transition metal ions. When values of CFSE from spectroscopic Δ_{oct}'s are subtracted from experimental values of $\Delta H_{hydration}$ (●), a plot of the "corrected" values of $\Delta H_{hydrated}$ (▲) versus atomic number is very nearly a straight line.

$\Delta H_{hydration}$ is the amount of energy absorbed when one mole of gaseous ions forms hydrated ions.

$$M^{2+}(g) + 6H_2O \longrightarrow [M(OH_2)_6]^{2+}(aq)$$

The CFSE is zero for weak field (high spin) octahedral d^0, d^5, and d^{10} complexes (Table 29-11). Experimental values for Ca^{2+} (d^0), Mn^{2+} (d^5), and Zn^{2+} (d^{10}) show no CFSE.

> You should be able to apply the reasoning we have outlined to show that CFSE is zero for these complexes.

Key Terms

Ammine complexes Complex species that contain ammonia molecules bonded to metal ions.

Chelate A ligand that utilizes two or more donor atoms in bonding to metals.

cis–trans **isomerism** A type of geometrical isomerism related to the angles between like ligands.

Complementary color The color associated with the wavelengths of light that are not absorbed—i.e., the color transmitted or reflected.

Coordinate covalent bond A covalent bond in which both shared electrons are donated by the same atom; a bond between a Lewis base and a Lewis acid.

Coordination compound or complex A compound containing coordinate covalent bonds.

Coordination isomers Isomers involving exchange of ligands between a complex cation and a complex anion of the same compound.

Coordination number The number of donor atoms coordinated to a metal.

Coordination sphere The metal ion and its coordinated ligands, but not any uncoordinated counter-ions.

Crystal field stabilization energy The net energy of stabilization gained by a metal ion's nonbonding d electrons as a result of complex formation.

Crystal field theory A theory of bonding in transition metal complexes in which ligands and metal ions are treated

as point charges; a purely ionic model. Ligand point charges represent the crystal (electrical) field perturbing the metal's d orbitals that contain nonbonding electrons.

Δ_{oct} The energy separation between e_g and t_{2g} sets of metal d orbitals caused by octahedral complexation of ligands; sometimes called $10Dq$.

Dextrorotatory Describes an optically active substance that rotates the plane of plane polarized light clockwise; also called dextro.

Donor atom A ligand atom whose electrons are shared with a Lewis acid.

e_g **orbitals** A set of $d_{x^2-y^2}$ and d_{z^2} orbitals; those d orbitals within a set with lobes directed along the x, y, and z axes.

Enantiomers Optical isomers.

Geometrical isomers Stereoisomers that are not mirror images of each other; also known as position isomers.

High spin complex The crystal field designation for an outer orbital complex; all t_{2g} and e_g orbitals are singly occupied before any pairing occurs.

Hydrate isomers Isomers of crystalline complexes that differ in terms of the presence of water inside or outside the coordination sphere.

Inner orbital complex The valence bond designation for a complex in which the metal ion utilizes d orbitals one shell inside the outermost occupied shell in its hybridization.

Ionization isomers   Isomers that result from interchange of ions inside and outside the coordination sphere.

Isomers   Different substances that have the same formula.

Levorotatory   Refers to an optically active substance that rotates the plane of plane polarized light counterclockwise; also called levo.

Ligand   A Lewis base in a coordination compound.

Linkage isomers   Isomers in which a particular ligand bonds to a metal ion through different donor atoms.

Low spin complex   The crystal field designation for an inner orbital complex; contains electrons paired in t_{2g} orbitals before e_g orbitals are occupied in octahedral complexes.

Optical activity   The rotation of plane polarized light by one of a pair of optical isomers.

Optical isomers   Stereoisomers that differ only by being nonsuperimposable mirror images of each other, like left and right hands; also called enantiomers.

Outer orbital complex   The valence bond designation for a complex in which the metal ion utilizes d orbitals in the outermost (occupied) shell in hybridization.

Pairing energy   The energy required to pair two electrons in the same orbital.

Plane polarized light   Light waves in which all the electric vectors are oscillating in one plane.

Polarimeter   A device used to measure optical activity.

Polydentate   Describes ligands with more than one donor atom.

Racemic mixture   An equimolar mixture of dextro and levo optical isomers that is, therefore, optically inactive.

Spectral color   The color associated with the wavelengths of light that are absorbed.

Spectrochemical series   An arrangement of ligands in order of increasing ligand field strength.

Square planar complex   A complex in which the metal is in the center of a square plane, with a ligand donor atom at each of the four corners.

Stereoisomers   Isomers that differ only in the way in which atoms are oriented in space; they consist of geometrical and optical isomers.

Strong field ligand   A ligand that exerts a strong crystal or ligand electrical field and generally forms low spin complexes with metal ions when possible.

Structural isomers   (Applied to coordination compounds.) Isomers whose differences involve more than a single coordination sphere or else different donor atoms; they include ionization isomers, hydrate isomers, coordination isomers, and linkage isomers.

t_{2g} orbitals   A set of d_{xy}, d_{yz}, and d_{xz} orbitals; those d orbitals within a set with lobes bisecting the x, y, and z axes.

Weak field ligand   A ligand that exerts a weak crystal or ligand field and generally forms high spin complexes with metals.

Exercises

Basic Concepts

1. What property of transition metals allows them to form coordination compounds easily?

2. Suggest more appropriate designations for $NiSO_4 \cdot 6H_2O$, $Cu(NO_3)_2 \cdot 4NH_3$, and $Ni(NO_3)_2 \cdot 6NH_3$.

3. What are the two constituents of a complex? What type of chemical bonding occurs between these constituents?

4. Define the term "coordination number" for the central atom or ion in a complex. What values of the coordination numbers for metal ions are most common?

5. Describe the experiments of Alfred Werner on the compounds of the general formula $PtCl_4 \cdot nNH_3$ where $n = 2, 3, 4, 5, 6$. What was his interpretation of these experiments?

6. For each of the compounds of Exercise 5, write formulas indicating the species within the coordination sphere. Also indicate the charges on the complex ions.

7. Distinguish among the terms ligands, donor atoms, and chelates.

8. Identify the ligands and give the coordination number and the oxidation number for the central atom or ion in each of the following: (a) $[Co(NH_3)_2(NO_2)_4]^-$, (b) $[Cr(NH_3)_5Cl]Cl_2$, (c) $K_4[Fe(CN)_6]$, (d) $[Pd(NH_3)_4]^{2+}$.

9. Repeat Exercise 8 for (a) $Na[Au(CN)_2]$, (b) $[Ag(NH_3)_2]^+$, (c) $[Pt(NH_3)_2Cl_4]$, (d) $[Co(en)_3]^{3+}$.

10. What is the term given to the phenomenon of ring formation by a ligand in a complex? Describe a specific example.

11. Write a structural formula showing the ring(s) formed by a bidentate ligand such as the ethylenediamine with a metal ion such as Fe^{3+}. How many atoms are in each ring? The formula for this complex ion is $[Fe(en)_3]^{3+}$.

Ammine Complexes

12. Which of the following insoluble metal hydroxides will dissolve in an excess of aqueous ammonia? (a) $Zn(OH)_2$, (b) $Cr(OH)_3$, (c) $Fe(OH)_2$, (d) $Ni(OH)_2$, (e) $Cd(OH)_2$.

13. Write net ionic equations for the reactions in Exercise 12.

14. Write net ionic equations for reactions of solutions of the following transition metal salts in water with a *limited amount* of aqueous ammonia: (It is not necessary to show the ions as hydrated.) (a) $CuCl_2$, (b) $Zn(NO_3)_2$, (c) $Fe(NO_3)_3$, (d) $Hg(NO_3)_2$, (e) $MnCl_3$.

15. Write *net ionic* equations for the reactions of the insoluble products of Exercise 14 with an *excess* of aqueous ammonia, if a reaction occurs.

Naming Coordination Compounds

16. Give systematic names for the following compounds.

  (a) $[Ni(CO)_4]$ (d) $[Cr(en)_3](NO_3)_3$

  (b) $Na_2[Co(OH_2)_2(OH)_4]$ (e) $[Pt(NH_3)_4(NO_2)_2]\,(NO_3)_2$

  (c) $[Ag(NH_3)_2]Br$ (f) $K_2[Cu(CN)_4]$

17. Name the following substances:

  (a) $Na[Au(CN)_2]$ (f) $K[Cr(NH_3)_2(OH)_2Cl_2]$

  (b) $[Pt(NH_3)_4]Cl_2$ (g) $[Ni(NH_3)_4(H_2O)_2](NO_3)_2$

  (c) $[CoCl_6]^{3-}$ (h) $Na[Al(H_2O)_2(OH)_4]$

  (d) $[Co(H_2O)_6]^{3+}$ (i) $[Co(NH_3)_4Cl_2][Cr(C_2O_4)_2]$

  (e) $Na_2[Pt(CN)_4]$

18. Write formulas for the following:

  (a) diamminedichlorozinc,

  (b) tin(IV) hexacyanoferrate(II),

  (c) tetracyanoplatinate(II) ion,

  (d) potassium hexacyanochromate(III),

  (e) tetraammineplatinum(II) ion,

  (f) hexaamminenickel(II) bromide,

  (g) tetraamminecopper(II) pentacyanohydroxoferrate(III).

19. Write formulas for the following compounds:

  (a) *trans*-diamminedinitroplatinum(II),

  (b) rubidium tetracyanozincate,

  (c) triaqua-*cis*-dibromochlorochromium(III),

  (d) pentacarbonyliron(0),

  (e) sodium pentacyanocobaltate(II),

  (f) hexamineruthenium(III) tetrachloronickelate(II).

Structures of Coordination Compounds

20. Write formulas and provide names for three complex cations in each of the following categories.

  (a) cations coordinated to only unidentate ligands,

  (b) cations coordinated to only bidentate ligands,

  (c) cations coordinated to two bidentate and two unidentate ligands,

  (d) cations coordinated to one tridentate ligand, one bidentate ligand, and one unidentate ligand,

  (e) cations coordinated to one tridentate ligand and three unidentate ligands.

21. Provide formulas and names for three complex anions that fit each description in Exercise 20.

22. How many geometrical isomers can be formed by complexes that are (a) octahedral MA_2B_4 and (b) octahedral MA_3B_3? Name any geometrical isomers that can exist. Is it possible for any of these isomers to show optical activity? Explain.

23. Write the structural formulas for (a) two isomers of $[Pt(NH_3)_2Cl_2]$, (b) four isomers (including linkage isomers) of $[Co(NH_3)_3(NO_2)_3]$, (c) two isomers (including ionization isomers) of $[Pt(NH_3)_3Br]Cl$.

24. Determine the number and types of isomers that would be possible for each of the following complexes.

  (a) tetraamminediaquachromium(III) ion

  (b) triamminetriaquachromium(III) ion

  (c) tris(ethylenediamine) chromium(III) ion

  (d) dichlorobis(ethylenediamine)platinum(IV) chloride

  (e) diamminedibromodichlorochromate(III) ion

*25. Indicate whether the complexes in each pair are identical or are isomers.

  (a)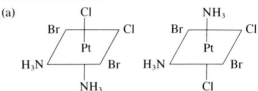

  (b)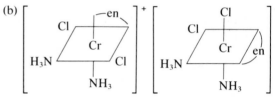

  (c)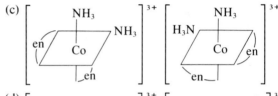

  (d)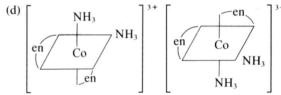

26. Distinguish between structural isomers and stereoisomers.

27. Distinguish between an optically active complex and a racemic mixture.

28. Write the formula for a potential ionization isomer of each of the following compounds. Name each one.

  (a) $[Cr(NH_3)_4I_2]Br$ (c) $[Fe(NH_3)_5CN]SO_4$

  (b) $[Ni(en)_2(NO_2)_2]Cl_2$

29. Write the formula for a potential hydrate isomer of each of the following compounds. Name each one. (a) $[Cu(OH_2)_4]Cl_2$, (b) $[Ni(OH_2)_5Br]Br \cdot H_2O$.

30. Write the formula for a potential coordination isomer of each of the following compounds. Name each one. (a) $[Co(NH_3)_6][Cr(CN)_6]$, (b) $[Ni(en)_3][Cu(CN)_4]$.

31. Write the formula for a potential linkage isomer of each of the following compounds. Name each one. (a) $[Co(en)_2(NO_2)_2]Cl$, (b) $[Cr(NH_3)_5(CN)](CN)_2$.

Valence Bond Theory

32. Describe the hybridization at the metal ion and sketch the structure of the complex part of each of the following. It is not necessary to distinguish between d^2sp^3 and sp^3d^2 hybridization.

  (a) $[Ag(NH_3)_2]Cl$ (d) $[Co(NH_3)_6]_2(SO_4)_3$

  (b) $[Fe(en)_3]PO_4$ (e) $[Pt(NH_3)_4]Cl_2$

  (c) $[Co(NH_3)_6]SO_4$ (f) $(NH_4)_2[PtCl_4]$

33. Repeat Exercise 32 for the following:
    (a) $K_2[PdCl_6]$ (d) $[Co(en)_3]Cl_3$
    (b) $(NH_4)_2[PtCl_6]$ (e) $[Cr(en)_2(NH_3)_2](NO_3)_2$
    (c) $[Co(en)_3]Cl_2$ (f) $[Co(NH_3)_4Cl_2]Cl$

34. Why are the d_{z^2} and $d_{x^2-y^2}$ orbitals utilized in d^2sp^3 and sp^3d^2 hybridization rather than two of the d_{xy}, d_{yz}, and d_{xz} orbitals?

35. On the basis of the spectrochemical series, determine whether each of the following complexes is inner orbital or outer orbital, and diamagnetic or paramagnetic.
    (a) $[Cu(OH_2)_6]^{2+}$ (d) $[Cr(NH_3)_6]^{3+}$ (g) $[Fe(OH_2)_6]^{3+}$
    (b) $[MnF_6]^{3-}$ (e) $[CrCl_4Br_2]^{3-}$ (h) $[Fe(NO_2)_6]^{3-}$
    (c) $[Co(CN)_6]^{3-}$ (f) $[Co(en)_3]^{3+}$

36. Draw diagrams showing outer electron configurations and hybridizations at the metal ions for each of the complexes in Exercise 35.

37. Using the valence bond theory, describe the bonding in a Co^{3+} octahedral complex with ligands that can (a) occupy only the outer orbitals on the Co^{3+} and (b) occupy inner orbitals on the Co^{3+}. What types of hybridization of the atomic orbitals on Co^{3+} are proposed for each?

38. How many unpaired electrons would you predict there to be in each of the following: (a) $[Fe(CN)_6]^{3-}$, (b) $[Fe(OH_2)_6]^{3+}$, (c) $[Mn(OH_2)_6]^{2+}$, (d) $[Co(NH_3)_6]^{3+}$?

39. Consider the compound having the formula $[Co(NH_3)_5(H_2O)]^{3+}[Co(NO_2)_6]^{3-}$. In terms of valence bond theory, describe the bonding in each ion. Would you expect this substance to be paramagnetic or diamagnetic?

Crystal Field Theory

40. Describe clearly what Δ_{oct} is. How is Δ_{oct} actually measured experimentally? How is it related to the spectrochemical series?

41. Δ_{oct} has been referred to as 10 Dq in the past. Suggest a reason for calling Δ_{oct} 10 units Dq, or 10 units anything else, for that matter.

42. On the basis of the spectrochemical series, determine whether the complexes of Exercise 35 are low spin or high spin complexes.

43. Write out the electron distribution in t_{2g} and e_g orbitals for the following in an octahedral field.

Metal Ions	Ligand Field Strength
V^{2+}	weak
Mn^{2+}	strong
Mn^{2+}	weak
Ni^{2+}	weak
Cu^{2+}	weak
Fe^{3+}	strong
Cu^+	weak
Ru^{3+}	strong

44. Write formulas for two complex ions that would fit into each of the categories of Exercise 43. Name the complex ions you list.

*45. Describe the relationship among Δ_{oct}, the electron pairing energy, and whether or not a complex is high spin or low spin. Illustrate the relationship with Fe^{2+} in strong and weak octahedral fields.

*46. What is crystal field stabilization energy? Given the following spectrophotometrically measured values of Δ_{oct}, calculate the CFSE's for the ions. Remember to determine first whether the complex is high spin or low spin.

Complex Ion	Δ_{oct}*
(a) $[Co(NH_3)_6]^{3+}$	22,900 cm^{-1}
(b) $[Ti(OH_2)_6]^{2+}$	20,300 cm^{-1}
(c) $[Cr(OH_2)_6]^{3+}$	17,600 cm^{-1}
(d) $[Co(CN)_6]^{3-}$	33,500 cm^{-1}
(e) $[Co(OH_2)_6]^{2+}$	10,000 cm^{-1}
(f) $[Cr(en)_3]^{3+}$	21,900 cm^{-1}
(g) $[Cu(OH_2)_6]^{2+}$	13,000 cm^{-1}
(h) $[V(OH_2)_6]^{3+}$	18,000 cm^{-1}

*The cm^{-1} is an energy unit; 1 cm^{-1} = 11.96 J/mol.

47. Determine the electronic distribution in (a) $[Co(CN)_6]^{3-}$, a low-spin complex ion, and (b) $[CoF_6]^{3-}$, a high-spin complex ion. Express the crystal field stabilization energy for each complex.

48. Determine the crystal field stabilization energy for
    (a) $[Mn(H_2O)_6]^{3+}$, a high-spin complex
    (b) $[Mn(CN)_6]^{3-}$, a low-spin complex

Mixed Exercises

49. The yellow complex compound $K_3[Rh(C_2O_4)_3]$ can be prepared from the wine-red complex compound $K_3[RhCl_6]$ by boiling a concentrated aqueous solution of $K_3[RhCl_6]$ and $K_2C_2O_4$ for two hours and then evaporating the solution until the product crystallizes.

$$K_3[RhCl_6](aq) + 3K_2C_2O_4(aq) \xrightarrow{\Delta}$$
$$K_3[Rh(C_2O_4)_3](s) + 6KCl(aq)$$

What is the theoretical yield of the oxalato complex if 1.00 g of the chloro complex is heated with 5.75 g of $K_2C_2O_4$? In an experiment, the actual yield was 0.83 g. What is the percent yield?

50. Consider the formation of the triiodoargentate(I) ion.

$$Ag^+ + 3I^- \longrightarrow [AgI_3]^{2-}$$

Would you expect an increase or decrease in the entropy of the system as the complex is formed? The standard state absolute entropy at 25°C is 72.68 J/K mol for Ag^+, 111.3 J/K mol for I^-, and 253.1 J/K mol for $[AgI_3]^{2-}$. Calculate $\Delta S°$ for the reaction and confirm your prediction.

51. Molecular iodine reacts with I^- to form a complex ion.

$$I_2(aq) + I^- \rightleftharpoons [I_3]^-$$

Calculate the equilibrium constant for this reaction given the following data at 25°C:

$$I_2(aq) + 2e^- \longrightarrow 2I^- \quad E° = 0.535 \text{ V}$$
$$[I_3]^- + 2e^- \longrightarrow 3I^- \quad E° = 0.5338 \text{ V}$$

*52. Calculate the pH of a solution prepared by dissolving 0.15 mol of tetraamminecopper(II) chloride, $[Cu(NH_3)_4]Cl_2$, in 1.0 L of solution. Ignore hydrolysis of Cu^{2+}.

*53. Use the following standard reduction potential data to answer the questions.

$Co^{3+} + e^-$	$\rightleftharpoons Co^{2+}$	$E° = 1.808$ V
$Co(OH)_3(s) + e^-$	$\rightleftharpoons Co(OH)_2(s) + OH^-(aq)$	$E° = 0.17$ V
$[Co(NH_3)_6]^{3+} + e^-$	$\rightleftharpoons [Co(NH_3)_6]^{2+}$	$E° = 0.108$ V
$[Co(CN)_6]^{3-} + e^-$	$\rightleftharpoons [Co(CN)_5]^{3-} + CN^-$	$E° = -0.83$ V
$O_2(g) + 4H^+(10^{-7} \text{ M}) + 4e^-$	$\rightleftharpoons 2H_2O(\ell)$	$E° = 0.815$ V
$2H_2O(\ell) + 2e^-$	$\rightleftharpoons H_2(g) + 2OH^-$	$E° = -0.828$ V

Which cobalt(III) species among those listed would oxidize water? Which cobalt(II) species among those listed would be oxidized by water? Explain your answers.

*54. Calculate (a) the molar solubility of $Zn(OH)_2$ in pure water, (b) the molar solubility of $Zn(OH)_2$ in 0.30 M NaOH solution, and (c) the concentration of $[Zn(OH)_4]^{2-}$ ions in the solution of (b).

Nuclear fusion provides the energy of our sun and other stars. Development of controlled fusion as a practical source of energy requires methods to initiate and contain the fusion process. Here a very powerful laser beam has initiated a fusion reaction in a 1-mm target capsule that contained deuterium and tritium. In a 0.5-picosecond burst, 10^{13} neutrons were produced by the reaction $^{2}_{1}H + ^{3}_{1}H \rightarrow ^{4}_{2}He + ^{1}_{0}n$.

Objectives

As you study this chapter, you should learn

☐ About the makeup of the nucleus
☐ About relationships between neutron–proton ratio and nuclear stability
☐ About the band of stability
☐ To calculate mass deficiency and nuclear binding energy
☐ About common types of radiation emitted when nuclei undergo radioactive decay
☐ To write and balance equations that describe nuclear reactions
☐ About different kinds of nuclear reactions undergone by nuclei, depending on their positions relative to the band of stability
☐ About methods for detecting radiation

☐ To understand half-lives of radioactive elements
☐ To carry out the calculations associated with radioactive decay
☐ About disintegration series
☐ About some uses of radionuclides, including the use of radioactive elements for dating objects
☐ About nuclear reactions that are induced by bombardment of nuclei with particles
☐ About nuclear fission and some of its applications, including the atomic bomb and nuclear reactors
☐ About nuclear fusion and some prospects for and barriers to its use for production of energy

Chemical properties are determined by electronic distributions and are only indirectly influenced by atomic nuclei. Up to now, we have discussed ordinary chemical reactions, so we have focused attention on electronic configurations. Nuclear reactions involve changes in the composition of nuclei. These extraordinary processes are often accompanied by the release of tremendous amounts of energy and by transmutations of elements. Some differences between nuclear reactions and ordinary chemical reactions follow.

Nuclear Reaction	Ordinary Chemical Reaction
1. Elements may be converted from one to another.	1. No new elements can be produced.
2. Particles within the nucleus are involved.	2. Usually only outermost electrons participate.
3. Tremendous amounts of energy are released or absorbed.	3. Relatively small amounts of energy are released or absorbed.
4. Rate of reaction is not influenced by external factors.	4. Rate of reaction depends on factors such as concentration, temperature, catalyst, and pressure.

Marie Curie (1867–1934) is the only person to have been honored with Nobel prizes in both physics and chemistry. In 1903, Pierre and Marie Curie and Henri Becquerel shared the prize in physics for the discovery of natural radioactivity. Marie Curie also received the 1911 Nobel prize in chemistry for her discovery of radium and polonium and the compounds of radium. She named polonium for her native Poland (her maiden name was Sklodowska). Marie's daughter, Irene Joliot-Curie, and *her* husband, Frederic Joliot, received the 1935 Nobel prize in chemistry for the first synthesis of a new radioactive element.

Medieval alchemists spent years trying to convert other heavy metals into gold without success. Years of failure and the acceptance of Dalton's atomic theory early in the 19th-century convinced scientists that elements are not interconvertible. Then, in 1896, Henri Becquerel discovered "radioactive rays" (**natural radioactivity**) coming from a uranium compound. Ernest Rutherford's study of these rays showed that atoms of one element may indeed be converted into atoms of other elements by spontaneous nuclear disintegrations. Many years later it was shown that nuclear reactions initiated by bombardment of nuclei with accelerated subatomic particles or other nuclei can also transform one element into another—accompanied by the release of radiation (**induced radioactivity**).

Becquerel's discovery led other researchers, including Marie and Pierre Curie, to discover and study new radioactive elements. Many radioactive isotopes, or **radioisotopes**, now have important medical, agricultural, and industrial uses.

Nuclear fission is the splitting of a heavy nucleus into lighter nuclei. **Nuclear fusion** is the combination of light nuclei to produce a heavier nucleus. Huge amounts of energy are released when these processes occur. They could satisfy a large portion of our future energy demands. Current research is aimed at surmounting the technological problems associated with safe and efficient use of nuclear fission reactors and with the development of controlled fusion reactors.

30-1 The Nucleus

In Chapter 5 we described the principal subatomic particles (Table 30-1). Recall that the neutrons and protons together constitute the nucleus, with the electrons occupying essentially empty space around the nucleus. The nucleus is only a minute fraction of the total volume of an atom, yet nearly all the mass of an atom resides in the nucleus. Thus, nuclei are extremely dense. It has been shown experimentally that nuclei of all elements have approximately the same density, 2.44×10^{14} g/cm^3.

If enough nuclei could be gathered together to occupy one cubic centimeter, the total weight would be about 250 million tons!

Table 30-1
Fundamental Particles of Matter

Particle	Rest Mass	Charge
Electron (e^-)	0.00054858 amu	1–
Proton (p or p^+)	1.0073 amu	1+
Neutron (n or n^0)	1.0087 amu	none

From an electrostatic point of view, it is amazing that positively charged protons (and uncharged neutrons) can be packed so closely together. Yet nonradioactive nuclei do not spontaneously decompose, so they must be stable. In the early 20th century, when Rutherford postulated the nuclear model of the atom, scientists were puzzled by such a situation. Physicists have since detected many very short-lived subatomic particles (in addition to protons, neutrons, and electrons) as products of nuclear reactions. Well over a hundred have been identified by now. Their functions are not entirely understood, but it is now thought that they help to overcome the proton–proton repulsions and to bind nuclear particles (**nucleons**) together. The attractive forces among nucleons appear to be important over only extremely small distances, about 10^{-13} cm.

30-2 Neutron–Proton Ratio and Nuclear Stability

The term "**nuclide**" is used to refer to different atomic forms of all elements. The term "isotope" applies only to different forms of the same element. Most naturally occurring nuclides have even numbers of protons and even numbers of neutrons; 157 nuclides fall into this category. Nuclides with odd numbers of both are least common (there are only five), and those with odd–even combinations are intermediate in abundance (Table 30-2). Furthermore, nuclides with certain "magic numbers" of protons and neutrons are especially stable. Nuclides with a number of protons *or* a number of neutrons *or* a sum of the two equal to 2, 8, 20, 28, 50, 82, or 126 have unusual stability. Examples are $^{4}_{2}\text{He}$, $^{16}_{8}\text{O}$, $^{42}_{20}\text{Ca}$, $^{88}_{38}\text{Sr}$, and $^{208}_{82}\text{Pb}$. This suggests an energy-level (shell) model for the nucleus similar to the shell model of electron configurations.

Figure 30-1 is a plot of the number of neutrons (N) versus number of protons (Z) for the stable nuclides. For low atomic numbers, the most stable nuclides have equal numbers of protons and neutrons ($N = Z$). Above atomic number 20, the most stable nuclides have more neutrons than protons. Careful examination reveals an approximately stepwise shape to the plot, due to the stability of nuclides with even numbers of nucleons.

In Section 5-6 we learned about the nuclide symbol for an element,

$$^{A}_{Z}E$$

where E is the chemical symbol for the element, Z is its atomic number, and A is its mass number.

30-3 Nuclear Stability and Binding Energy

Experimentally, we observe that the mass of an atom is always *less* than the sum of the masses of its constituent particles. We now know why this *mass deficiency* occurs. We also know that the mass deficiency is in the nucleus of the atom and has nothing to do with the electrons. However, because tables of masses of isotopes include the electrons, we shall also include them.

Table 30-2
Abundance of Naturally Occurring Nuclides

Number of protons	even	even	odd	odd
Number of neutrons	even	odd	even	odd
Number of such nuclides	157	52	50	5

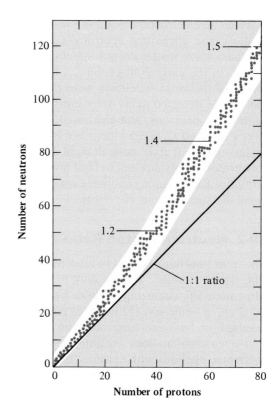

Figure 30-1
A plot of the number of neutrons versus the number of protons in stable nuclei. As atomic number increases, the N/Z ratio of the stable nuclei increases. The stable nuclei are located in an area known as the band of stability. Most radioactive nuclei occur outside this band.

Do you remember how to find the numbers of protons, neutrons, and electrons in a specified atom? Review Section 5-6.

The **mass deficiency**, Δm, for a nucleus is the difference between the sum of the masses of electrons, protons, and neutrons in the atom (calculated mass) and the actual measured mass of the atom:

$$\Delta m = (\text{sum of masses of all } e^-, p^+, \text{ and } n^0) - (\text{actual mass of atom})$$

For most naturally occurring isotopes, the mass deficiency is only about 0.15% or less of the calculated mass of an atom.

Example 30-1
Calculate the mass deficiency for chlorine-35 atoms. The actual mass of a chlorine-35 atom is 34.9689 amu.

Plan
We first find the numbers of protons, electrons, and neutrons in one atom. Then we determine the "calculated" mass as the sum of the rest masses of these particles. The mass deficiency is the actual mass subtracted from the calculated mass. This deficiency is commonly expressed either as mass per atom or as mass per mole of atoms.

Solution
One atom of $_{17}^{35}\text{Cl}$ contains 17 protons, 17 electrons, and $(35 - 17) = 18$ neutrons. First we sum the masses of these particles.

protons:	17×1.0073 amu	$= 17.124$ amu	(masses from Table 30-1)
electrons:	17×0.00054858 amu $=$	0.0093 amu	
neutrons:	18×1.0087 amu	$= 18.157$ amu	

$$\text{sum} = 35.290 \text{ amu} \longleftarrow \text{Calculated mass}$$

Then we subtract the actual mass from the "calculated" mass to obtain Δm.

$$\Delta m = 35.290 \text{ amu} - 34.9689 \text{ amu} = \boxed{0.321 \text{ amu}} = \text{mass deficiency}$$

We have calculated the mass deficiency in amu/atom. Recall (Section 5-8) that 1 gram is 6.022×10^{23} amu. We can show that a number expressed in amu/atom is equal to the same number of g/mol of atoms.

$$\frac{? \text{ g}}{\text{mol}} = \frac{0.321 \text{ amu}}{\text{atom}} \times \frac{1 \text{ g}}{6.022 \times 10^{23} \text{ amu}} \times \frac{6.022 \times 10^{23} \text{ atoms}}{1 \text{ mol } ^{35}\text{Cl atoms}}$$

$$= \boxed{0.321 \text{ g/mol of } ^{35}\text{Cl atoms}} \longleftarrow \text{(mass } deficiency \text{ in a mole of Cl atoms)}$$

EOC 8a, 9a

What has happened to the mass represented by the mass deficiency? In 1905 Einstein set forth the Theory of Relativity. He stated that matter and energy are equivalent. An obvious corollary is that matter can be transformed into energy and energy into matter. The transformation of matter into energy occurs in the sun and other stars. It happened on earth when controlled nuclear fission was achieved in 1939 (Section 30-13). The reverse transformation, energy into matter, has not yet been accomplished on a large scale. Einstein's equation, which we encountered in Chapter 1, is $E = mc^2$. E represents the amount of energy released, m the mass of matter transformed into energy, and c the speed of light in a vacuum, 2.997925×10^8 m/s (usually rounded off to 3.00×10^8 m/s).

A mass deficiency represents the amount of matter that would be converted into energy and released if the nucleus were formed from initially separate protons and neutrons. This energy is the **nuclear binding energy**, **BE**. Thus, we could rewrite the Einstein relationship as

$$BE = (\Delta m)c^2$$

Specifically, if 1 mole of ^{35}Cl nuclei were to be formed from 17 moles of protons and 18 moles of neutrons, the resulting mole of nuclei would weigh 0.321 gram less than the original collection of protons and neutrons (Example 30-1).

Stated differently, the nuclear binding energy for ^{35}Cl is the amount of energy that would be required to separate 1 mole of ^{35}Cl nuclei into 17 moles of protons and 18 moles of neutrons. This has never been done. Let's use the value of Δm for ^{35}Cl atoms to calculate their nuclear binding energy.

Example 30-2

Calculate the nuclear binding energy of ^{35}Cl in joules per mole of Cl atoms. 1 joule (J) = 1 kg $\cdot$ m^2/s^2.

Plan

The mass deficiency that we calculated in Example 30-1 is related to the binding energy by the Einstein equation.

Solution

We use the Einstein equation to obtain the nuclear binding energy. The mass deficiency is 0.321 g/mol = 3.21×10^{-4} kg/mol.

$$BE = (\Delta m)c^2 = (3.21 \times 10^{-4} \text{ kg/mol})(3.00 \times 10^8 \text{ m/s})^2 = 2.89 \times 10^{13} \frac{\text{kg} \cdot \text{m}^2/\text{s}^2}{\text{mol}}$$

$$= \boxed{2.89 \times 10^{13} \text{ J/mol of } ^{35}\text{Cl atoms}}$$

EOC 8b, 9b, 10

The nuclear binding energy of a mole of ^{35}Cl nuclei, 2.89×10^{13} J/mol, is an enormous amount of energy—enough to heat 6.9×10^7 kg (~76,000 tons) of water from 0°C to 100°C!

Nuclear binding energies are frequently expressed in kJ/g of nuclei. Figure 30-2 is a plot of average binding energy per gram of nuclei versus mass number. It shows that nuclear binding energies (per gram) increase rapidly with increasing mass number, reach a maximum around mass number 50, and then decrease slowly. The nuclei with the highest binding energies (mass numbers 40 to 150) are the most stable. Large amounts of energy would be required to separate these nuclei into their component neutrons and protons. Even though these nuclei are the most stable ones, *all* nuclei are stable with respect to complete decomposition into protons and neutrons because all (except ^1H) nuclei have mass deficiencies. In other words, the energy equivalent of the loss of mass represents an associative force that is present in all nuclei except ^1H. It must be overcome to separate the nuclei completely into particles.

> Some unstable radioactive nuclei do emit a single proton, a single neutron, or other subatomic particles as they decay in the direction of greater stability. None decomposes entirely into elementary particles.

30-4 Radioactive Decay

Nuclei whose neutron-to-proton ratios lie outside the stable region undergo spontaneous radioactive decay by emitting one or more particles and/or electromagnetic rays. The type of decay that occurs usually depends on whether the nucleus is above, below, or to the right of the band of stability (Figure 30-1). Common types of radiation emitted in decay processes are summarized in Table 30-3.

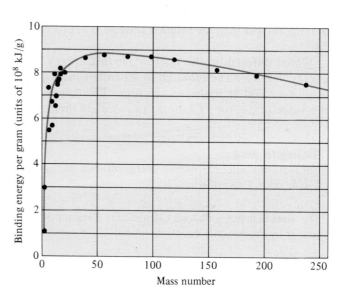

Figure 30-2
Plot of binding energy per gram versus mass number. Very light and very heavy nuclei are relatively unstable.

Table 30-3
Common Types of Radioactive Emissions

Type and Symbol[a]	Identity	Mass (amu)	Charge	Velocity	Penetration
Beta (β^-, $_{-1}^{0}\beta$, $_{-1}^{0}e$)	Electron	0.00055	1$-$	≤90% speed of light	Low to moderate, depending on energy
Positron[b] ($_{+1}^{0}\beta$, $_{+1}^{0}e$)	Positively charged electron	0.00055	1$+$	≤90% speed of light	Low to moderate, depending on energy
Alpha (α, $_{2}^{4}\alpha$, $_{2}^{4}\text{He}$)	Helium nucleus	4.0026	2$+$	≤10% speed of light	Low
Proton ($_{1}^{1}p$, $_{1}^{1}\text{H}$)	Proton, hydrogen nucleus	1.0073	1$+$	≤10% speed of light	Low to moderate, depending on energy
Neutron ($_{0}^{1}n$)	Neutron	1.0087	0	≤10% speed of light	Very high
Gamma ($_{0}^{0}\gamma$) ray	High-energy electro-magnetic radiation such as X-rays	0	0	Speed of light	High

[a] The number at the upper left of the symbol is the number of nucleons, and the number at the lower left is the number of positive charges.
[b] On the average, a positron exists for only about a nanosecond (1×10^{-9} second) before colliding with an electron and being converted into the corresponding amount of energy.

The particles can be emitted at different kinetic energies. These are equal to the energy equivalent of the mass loss of the products relative to reactants (Section 30-3) minus the energy associated with subsequently emitted gamma rays (electromagnetic radiation). Radioactive decay often leaves a nucleus in an excited (high-energy) state. Then the decay is followed by gamma ray emission. The energy of the gamma ray ($h\nu$) is equal to the energy difference between the ground and excited nuclear states. This is like the emission of lower-energy electromagnetic radiation that occurs as an atom in its excited electronic state returns to its ground state (Section 5-11). Studies of gamma ray energies strongly suggest that nuclear energy levels are quantized just

Recall that the energy of electromagnetic radiation is $E = h\nu$, where h is Planck's constant and ν is the frequency.

A technician cleans lead glass blocks that form part of the giant OPAL particle detector at CERN, the European center for particle physics near Geneva, Switzerland.

Robotics technology is used to manipulate highly radioactive samples safely.

as are electronic energy levels. This adds further support for a shell model for the nucleus.

$$\text{(excited nucleus)} \longrightarrow {}_{Z}^{M}E^* \longrightarrow {}_{Z}^{M}E + {}_{0}^{0}\gamma$$

The penetrating abilities of the particles and rays are proportional to their energies. Beta particles and positrons are about 100 times more penetrating than the heavier and slower-moving alpha particles. They can be stopped by a $\frac{1}{8}$-inch-thick (0.3 cm) aluminum plate. They can burn skin severely but cannot reach internal organs. Alpha particles have low penetrating ability and cannot damage or penetrate skin. However, they can damage sensitive internal tissue if inhaled. The high-energy gamma rays have great penetrating power and severely damage both skin and internal organs. They travel at the speed of light and can be stopped by thick layers of concrete or lead.

30-5 Nuclei above the Band of Stability

Nuclei in this region have too high a ratio of neutrons to protons. They undergo decays that *decrease* the ratio, such as **beta emission** or, less commonly, **neutron emission**. A beta particle is an electron ejected *from the nucleus* when a neutron is converted into a proton:

$$_{0}^{1}n \longrightarrow {}_{1}^{1}p + {}_{-1}^{0}\beta$$

Beta emission results in a simultaneous increase by one in the number of protons and decrease by one in the number of neutrons. Examples of beta particle emission are

$$_{88}^{228}\text{Ra} \longrightarrow {}_{89}^{228}\text{Ac} + {}_{-1}^{0}\beta \qquad \text{and} \qquad {}_{6}^{14}\text{C} \longrightarrow {}_{7}^{14}\text{N} + {}_{-1}^{0}\beta$$

Neutron emission decreases the number of neutrons by one without changing the atomic number. A lighter isotope is formed.

$$_{53}^{137}\text{I} \longrightarrow {}_{53}^{136}\text{I} + {}_{0}^{1}n \qquad \text{and} \qquad {}_{7}^{17}\text{N} \longrightarrow {}_{7}^{16}\text{N} + {}_{0}^{1}n$$

In the balanced equation for a nuclear reaction,
1. The sums of left-hand superscripts (mass numbers) must be the same on both sides, *and*
2. The sums of left-hand subscripts (atomic numbers) must be the same on both sides.

> In all equations for nuclear reactions,
> sum of mass numbers of reactants = sum of mass numbers of products
> sum of atomic numbers of reactants = sum of atomic numbers of products

30-6 Nuclei below the Band of Stability

These nuclei can *increase* their neutron-to-proton ratios by undergoing **positron emission** or **electron capture** (*K* capture).

A positron has the mass of an electron but a positive charge. Positrons are emitted when protons are converted to neutrons:

$$_{1}^{1}p \longrightarrow {}_{0}^{1}n + {}_{+1}^{0}\beta$$

Thus, positron emission results in a *decrease* by one in atomic number and an *increase* by one in the number of neutrons, with *no change* in mass number.

$$_{19}^{38}\text{K} \longrightarrow {}_{18}^{38}\text{Ar} + {}_{+1}^{0}\beta \qquad \text{and} \qquad {}_{8}^{15}\text{O} \longrightarrow {}_{7}^{15}\text{N} + {}_{+1}^{0}\beta$$

The same effect can be accomplished by electron capture (*K* capture), in which an electron from the *K* shell ($n = 1$) is captured by the nucleus.

$$^{106}_{47}Ag + ^{0}_{-1}e \longrightarrow ^{106}_{46}Pd \quad \text{and} \quad ^{37}_{18}Ar + ^{0}_{-1}e \longrightarrow ^{37}_{17}Cl$$

Some nuclides, e.g., $^{22}_{11}Na$, undergo both electron capture and positron emission.

$$^{22}_{11}Na + ^{0}_{-1}e \longrightarrow ^{22}_{10}Ne \ (3\%) \quad \text{and} \quad ^{22}_{11}Na \longrightarrow ^{22}_{10}Ne + ^{0}_{+1}\beta \ (97\%)$$

Some nuclei, especially heavier ones, undergo **alpha emission**. Alpha particles are helium nuclei, $^{4}_{2}He$—two protons and two neutrons. Alpha emission also results in an increase of the neutron-to-proton ratio. An example is the alpha emission of lead-204.

$$^{204}_{82}Pb \longrightarrow ^{200}_{80}Hg + ^{4}_{2}\alpha$$

Electron capture by the nucleus differs from an atom gaining an electron to form an ion.

30-7 Nuclei with Atomic Number Greater Than 82

All nuclides with atomic number greater than 82 are beyond the band of stability and are radioactive. Many of these decay by alpha emission.

$$^{226}_{88}Ra \longrightarrow ^{222}_{86}Rn + ^{4}_{2}\alpha \quad \text{and} \quad ^{210}_{84}Po \longrightarrow ^{206}_{82}Pb + ^{4}_{2}\alpha$$

α-Particles carry a double positive charge, but charge is usually not shown in nuclear reactions.

The decay of radium-226 was originally reported in 1902 by Rutherford and Soddy. It was the first transmutation of an element ever observed. A few heavy nuclides also decay by beta emission, positron emission, and electron capture.

Some isotopes of uranium ($Z = 92$) and elements of higher atomic number, the **transuranium elements**, also decay by nuclear fission. In this process, a heavy nuclide splits into nuclides of intermediate mass and neutrons.

$$^{252}_{98}Cf \longrightarrow ^{142}_{56}Ba + ^{106}_{42}Mo + 4\,^{1}_{0}n$$

30-8 Detection of Radiations

Photographic Detection

Emanations from radioactive substances affect photographic plates just as ordinary visible light does. Becquerel's discovery of radioactivity resulted from the unexpected exposure of such a plate, wrapped in black paper, by a nearby enclosed sample of a uranium-containing compound, potassium uranyl sulfate. After a photographic plate has been developed and fixed, the intensity of the exposed spot is related to the amount of radiation that struck the plate. Quantitative detection of radiation by this method is difficult and tedious.

Detection by Fluorescence

Fluorescent substances can absorb high-energy radiation such as gamma rays and subsequently emit visible light. As the radiation is absorbed, the absorbing atoms jump to excited electronic states. The excited electrons return to their ground states through a series of transitions, some of which emit visible light. This method may be used for the quantitative detection of radiation, using an instrument called a **scintillation counter**.

Cloud Chambers

The original cloud chamber was devised by C. T. R. Wilson in 1911. A chamber contains air saturated with vapor. Particles emitted from a radioactive substance ionize air molecules in the chamber. Cooling the chamber

Figure 30-3
A cloud chamber. The emitter is glued onto a pin stuck into a stopper that is mounted on the chamber wall. The chamber has some volatile liquid in the bottom and rests on Dry Ice. The cool air near the bottom becomes supersaturated with vapor. When an emission speeds through this vapor, ions are produced. These ions serve as "seeds" about which the vapor condenses, forming tiny droplets, or fog.

Dry Ice

causes droplets of liquid to condense on these ions. The paths of the particles can be followed by observing the fog-like tracks produced. The tracks may be photographed and studied in detail. Figures 30-3 and 30-4 show a cloud chamber and a cloud chamber photograph, respectively.

Gas Ionization Counters

The Geiger counter can detect only β and γ radiation. The α-particles cannot penetrate the walls of the tube.

A common gas ionization counter is the **Geiger–Müller counter** (Figure 30-5). Radiation enters the tube through a thin window. Windows of different stopping powers can be used to admit only radiation of certain penetrating powers.

collision

Figure 30-4
A historic cloud chamber photograph of alpha tracks in nitrogen gas. The forked track was shown to be due to a speeding proton (going off to the left) and an isotope of oxygen (going off to the right). It is assumed that the alpha particle struck the nucleus of a nitrogen atom at the point where the track forks.

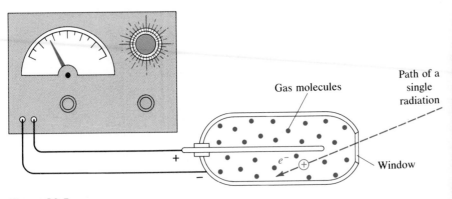

Figure 30-5
The principle of operation of a gas ionization counter. The center wire is positively charged, and the shell of the tube is negatively charged. When radiation enters through the window, it ionizes one or more gas atoms. The electrons are attracted to the central wire, and the positive ions are drawn to the shell. This constitutes a pulse of electric current, which is amplified and displayed on the meter or other readout.

A sample of carnotite, a uranium ore, shown with a Geiger counter.

30-9 Rates of Decay and Half-Life

Radionuclides have different stabilities and decay at different rates. Some decay nearly completely in fractions of a second and others only after millions of years. The rates of all radioactive decays are independent of temperature and obey *first-order kinetics*. In Section 16-4 we saw that the rate of a first-order process is proportional only to the concentration of one substance:

$$\text{rate of decay} = k[A]$$

The integrated rate equation for a first-order process is

$$\log \left(\frac{A_0}{A} \right) = \frac{akt}{2.303} \quad \text{or} \quad \ln \left(\frac{A_0}{A} \right) = akt$$

Here A represents the amount of decaying radionuclide of interest remaining at some time t, and A_0 is the amount present at the beginning of the observation. The k is the rate constant, which is different for each radionuclide. Each atom decays independently of the others, so the stoichiometric coefficient a is *always* 1 for radioactive decay. Therefore, we can drop it from the calculations in this chapter. If N represents the number of disintegrations per unit time, a similar relationship holds:

$$\log \left(\frac{N_0}{N} \right) = \frac{kt}{2.303} \quad \text{or} \quad \ln \left(\frac{N_0}{N} \right) = kt$$

The half-life, $t_{1/2}$, of a reaction is the amount of time required for half of the original sample to react. For a first-order process, $t_{1/2}$ is given by the equation

$$t_{1/2} = \frac{\ln 2}{k} = \frac{0.693}{k}$$

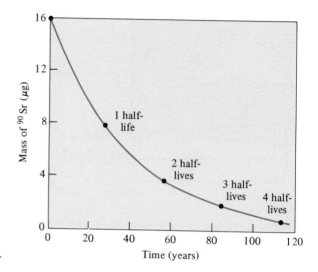

Figure 30-6
The decay of a 16-μg sample of $^{90}_{38}$Sr.

In 1963, a treaty was signed by the United States, the Soviet Union, and the United Kingdom prohibiting the further testing of nuclear weapons in the atmosphere. Since then, strontium-90 has been disappearing from the air, water, and soil according to the curve in Figure 30-6. So the treaty has largely accomplished its aim up to the present (1991).

The isotope strontium-90 was introduced into the atmosphere by the atmospheric testing of nuclear weapons. Because of the chemical similarity of strontium to calcium, it now occurs with Ca in measurable quantities in milk, bones, and teeth as a result of its presence in food and water supplies. It is a radionuclide that undergoes beta emission with a half-life of 28 years. It may cause leukemia, bone cancer, and other, related disorders. If we begin with a 16-μg sample of $^{90}_{38}$Sr, 8 μg will remain after one half-life of 28 years. After 56 years, 4 μg will remain; after 84 years, 2 μg; and so on (Figure 30-6).

Similar plots for other radionuclides all show the same shape of **exponential decay curve**. About ten half-lives (280 years for $^{90}_{38}$Sr) must pass for radionuclides to lose 99.9% of their radioactivity.

Example 30-3

The "cobalt treatments" used in medicine to arrest certain types of cancer rely on the ability of gamma rays to destroy cancerous tissues. Cobalt-60 decays with the emission of beta particles and gamma rays, with a half-life of 5.27 years.

$$^{60}_{27}\text{Co} \longrightarrow {}^{60}_{28}\text{Ni} + {}_{-1}^{0}\beta + {}_{0}^{0}\gamma$$

Gamma rays destroy both cancerous and normal cells, so the beams of gamma rays must be directed as nearly as possible toward only cancerous tissue.

How much of a 3.42-μg sample of cobalt-60 remains after 30.0 years?

Plan

We determine the value of the specific rate constant, k, from the given half-life. This value is then used in the first-order integrated rate equation to calculate the amount of cobalt-60 remaining after the specified time.

Solution

We first determine the value of the specific rate constant.

$$t_{1/2} = \frac{0.693}{k} \quad \text{so} \quad k = \frac{0.693}{t_{1/2}} = \frac{0.693}{5.27 \text{ yr}} = 0.131 \text{ yr}^{-1}$$

This value can now be used to determine the ratio of A_0 to A after 30.0 years:

$$\log\left(\frac{A_0}{A}\right) = \frac{kt}{2.303} = \frac{0.131 \text{ yr}^{-1} (30.0 \text{ yr})}{2.303} = 1.71$$

Taking the inverse log of both sides, $\frac{A_0}{A} = 10^{1.71} = 51$.

$A_0 = 3.42$ μg, so:

$$A = \frac{A_0}{51} = \frac{3.42 \text{ μg}}{51} = \boxed{0.067 \text{ μg } {}^{60}_{27}\text{Co}} \text{ remain after 30.0 years.}$$

EOC 41, 42

30-10 Disintegration Series

Many radionuclides cannot attain nuclear stability by only one nuclear reaction. Instead, they decay in a series of disintegrations. A few such series are known to occur in nature. Two begin with isotopes of uranium, ^{238}U and ^{235}U, and one begins with ^{232}Th. All end with a stable isotope of lead ($Z = 82$). Table 30-4 outlines in detail the ^{238}U, ^{235}U, and ^{232}Th disintegration series, showing half-lives. For any particular decay step, the decaying nuclide is called the **mother** nuclide, and the product nuclide is the **daughter**.

Uranium-238 decays by alpha emission to thorium-234 in the first step of one series. Thorium-234 subsequently emits a beta particle to produce protactinium-234 in the second step. The series can be summarized as shown in Table 30-4a. The *net* reaction for the ^{238}U series is

$$^{238}_{92}\text{U} \longrightarrow {}^{206}_{82}\text{Pb} + 8\,{}^{4}_{2}\text{He} + 6\,{}_{-1}^{0}\beta$$

"Branchings" are possible at various points in the chain. That is, two successive decays may be replaced by alternate decays, but they always result in the same final product. There are also decay series of varying lengths starting with some of the artificially produced radionuclides (Section 30-12).

30-11 Uses of Radionuclides

Radionuclides have practical uses because they decay at known rates or, in some cases, because they emit radiation continuously.

Radioactive Dating

The ages of articles of organic origin can be estimated by **radiocarbon dating**. The radioisotope carbon-14 is produced continuously in the upper atmosphere as nitrogen atoms capture cosmic-ray neutrons:

$$^{14}_{7}\text{N} + {}^{1}_{0}n \longrightarrow {}^{14}_{6}\text{C} + {}^{1}_{1}\text{H}$$

The carbon-14 atoms react with oxygen molecules to form ^{14}CO$_2$. This process continually supplies the atmosphere with radioactive ^{14}CO$_2$, which is removed from the atmosphere by photosynthesis. The intensity of cosmic rays is related to the sun's activity. As long as this remains constant, the amount of ^{14}CO$_2$ in the atmosphere remains constant. ^{14}CO$_2$ is incorporated into living organisms just as ordinary ^{12}CO$_2$ is, so a certain fraction of all carbon atoms in living substances is carbon-14. This decays with a half-life of 5730 years.

$$^{14}_{6}\text{C} \longrightarrow {}^{14}_{7}\text{N} + {}_{-1}^{0}\beta$$

After death, the plant no longer carries out photosynthesis, so it no longer takes up ^{14}CO$_2$. Other organisms that consume plants for food stop doing

Radiocarbon dating is used to estimate the ages of articles found in archaeological studies.

In recent decades, atmospheric testing of nuclear warheads has also caused fluctuations in the natural abundance of ^{14}C.

Table 30-4
Emissions and Half-Lives of Members of Natural Radioactive Series*

(a) ^{238}U Series	(b) ^{235}U series	(c) ^{232}Th Series

(a) ^{238}U Series

$^{238}_{92}U \rightarrow \alpha$

4.51×10^9 y

$^{234}_{90}Th \rightarrow \beta$
24.1 d

$^{234}_{91}Pa \rightarrow \beta$
6.75 h

$^{234}_{92}U \rightarrow \alpha$
2.47×10^5 y

$^{230}_{90}Th \rightarrow \alpha$
8.0×10^4 y

$^{226}_{88}Ra \rightarrow \alpha$
1.60×10^3 y

$^{222}_{86}Rn \rightarrow \alpha$
3.82 d

$\beta \leftarrow {}^{218}_{84}Po \rightarrow \alpha$
0.04%   3.05 m

$\alpha \leftarrow {}^{218}_{85}At$ $^{214}_{82}Pb \rightarrow \beta$
2 s   26.8 m

$\beta \leftarrow {}^{214}_{83}Bi \rightarrow \alpha$
99.96%   19.7 m

$\alpha \leftarrow {}^{214}_{84}Po$ $^{210}_{81}Tl \rightarrow \beta$
1.6×10^{-4} s   1.32 m

$^{210}_{82}Pb \rightarrow \beta$
20.4 y

$\beta \leftarrow {}^{210}_{83}Bi \rightarrow \alpha$
~100%   5.01 d

$\alpha \leftarrow {}^{210}_{84}Po$ $^{206}_{81}Tl \rightarrow \beta$
138 d   4.19 m

$^{206}_{82}Pb$

(b) ^{235}U series

$^{235}_{92}U \rightarrow \alpha$

7.1×10^8 y

$^{231}_{90}Th \rightarrow \beta$
25.5 h

$^{231}_{91}Pa \rightarrow \alpha$
3.25×10^4 y

$\beta \leftarrow {}^{227}_{89}Ac \rightarrow \alpha$
98.8%   21.6 y

$\alpha \leftarrow {}^{227}_{90}Th$ $^{223}_{87}Fr \rightarrow \beta$
18.2 d   22 m

$^{223}_{88}Ra \rightarrow \alpha$
11.4 d

$^{219}_{86}Rn \rightarrow \alpha$
4.00 s

$\beta \leftarrow {}^{215}_{84}Po \rightarrow \alpha$
5×10^{-4}% 1.78×10^{-3} s

$\alpha \leftarrow {}^{215}_{85}At$ $^{211}_{82}Pb \rightarrow \beta$
10^{-4} s   36.1 m

$\beta \leftarrow {}^{211}_{83}Bi \rightarrow \alpha$
99.7%   2.16 m

$\alpha \leftarrow {}^{211}_{84}Po$ $^{207}_{81}Tl \rightarrow \beta$
0.52 s   4.79 m

$^{207}_{82}Pb$

(c) ^{232}Th Series

$^{232}_{90}Th \rightarrow \alpha$

1.41×10^{10} y

$^{228}_{88}Ra \rightarrow \beta$
6.7 y

$^{228}_{89}Ac \rightarrow \beta$
6.13 h

$^{228}_{90}Th \rightarrow \alpha$
1.91 y

$^{224}_{88}Ra \rightarrow \alpha$
3.64 d

$^{220}_{86}Rn \rightarrow \alpha$
55.3 s

$\beta \leftarrow {}^{216}_{84}Po \rightarrow \alpha$
0.014%   0.14 s

$\alpha \leftarrow {}^{216}_{85}At$ $^{212}_{82}Pb \rightarrow \beta$
3×10^{-4} s   10.6 h

$\beta \leftarrow {}^{212}_{83}Bi \rightarrow \alpha$
66.3%   60.6 m

$\alpha \leftarrow {}^{212}_{84}Po$ $^{208}_{81}Tl \rightarrow \beta$
3.0×10^{-7} s   3.10 m

$^{208}_{82}Pb$

* Abbreviations are y, year; d, day; m, minute; and s, second. Less prevalent decay branches are shown in blue.

so at death. The emissions from the ^{14}C in dead tissue then decrease with the passage of time. The activity per gram of carbon is a measure of the length of time elapsed since death. Comparison of ages of ancient trees calculated from ^{14}C activity with those determined by counting rings indicates that cosmic ray intensity has varied somewhat throughout history. The calculated ages can be corrected for these variations. The carbon-14 technique is useful only for dating objects less than 50,000 years old. Older objects have too little activity to be dated accurately.

The **potassium–argon** and **uranium–lead** methods are used for dating older objects. Potassium-40 decays to argon-40 with a half-life of 1.3 billion years.

$$^{40}_{19}K + \ _{-1}^{\ 0}e \longrightarrow \ ^{40}_{18}Ar$$

Because of its long half-life, potassium-40 can be used to date objects up to 1 million years old by determination of the ratio of $^{40}_{19}K$ to $^{40}_{18}Ar$ in the sample. The uranium–lead method is based on the natural uranium-238 decay series, which ends with the production of stable lead-206. This method is used for dating uranium-containing minerals several billion years old. All the ^{206}Pb in such minerals is *assumed* to have come from ^{238}U. Because of the very long half life of $^{238}_{92}U$, 4.5 billion years, the amounts of intermediate nuclei can be neglected. A meteorite that was 4.6 billion years old fell in Mexico in 1969. Results of $^{238}U/^{206}Pb$ studies indicate that our solar system was formed several billion years ago.

Gaseous argon is easily lost from minerals. Therefore, measurement based on the $^{40}K/^{40}Ar$ method may not be as reliable as desired.

Example 30-4

A piece of wood taken from a cave dwelling in New Mexico is found to have a carbon-14 activity (per gram of carbon) only 0.636 times that of wood cut today. Estimate the age of the wood. The half-life of carbon-14 is 5730 years.

Plan

As we did in Example 30-3, we determine the specific rate constant k from the known half-life. The time required to reach the present fraction of the original activity is then calculated from the first-order decay equation.

Solution

First we find the first-order specific rate constant for ^{14}C:

$$t_{1/2} = \frac{0.693}{k}$$

$$k = \frac{0.693}{t_{1/2}} = \frac{0.693}{5730 \ yr} = 1.21 \times 10^{-4} \ yr^{-1}$$

The present ^{14}C activity, N (disintegrations per unit time), is 0.636 times the original activity, N_0:

$$N = 0.636 \ N_0$$

We substitute into the first-order decay equation:

$$\ln\left(\frac{N_0}{N}\right) = kt$$

$$\ln\left(\frac{N_0}{0.636 \ N_0}\right) = (1.21 \times 10^{-4} \ yr^{-1})t$$

We cancel N_0 and solve for t.

$$\ln\left(\frac{1}{0.636}\right) = (1.21 \times 10^{-4} \ yr^{-1})t$$

$$0.452 = (1.21 \times 10^{-4} \ yr^{-1})t \quad \text{or} \quad t = \boxed{3.74 \times 10^3 \ yr \ (\text{or } 3740 \ yr)}$$

EOC 46

Example 30-5

A sample of uranium ore is found to contain 4.64 mg of ^{238}U and 1.22 mg of ^{206}Pb. Estimate the age of the ore. The half-life of ^{238}U is 4.51×10^9 years.

Plan

The original mass of ^{238}U is equal to the mass of ^{238}U remaining plus the mass of ^{238}U that decayed to produce the present mass of ^{206}Pb. We obtain the specific rate constant, k, from the known half-life. Then we use the ratio of original ^{238}U to remaining ^{238}U to calculate the time elapsed, with the aid of the first-order integrated rate equation.

Solution

First we calculate the amount of ^{238}U that must have decayed to produce 1.22 mg of ^{206}Pb, using the isotopic masses.

$$\underline{?} \text{ mg } ^{238}\text{U} = 1.22 \text{ mg } ^{206}\text{Pb} \times \frac{238 \text{ mg } ^{238}\text{U}}{206 \text{ mg } ^{206}\text{Pb}} = 1.41 \text{ mg } ^{238}\text{U}$$

Thus, the sample originally contained 4.64 mg + 1.41 mg = 6.05 mg of ^{238}U. We next evaluate the specific rate (disintegration) constant, k.

$$t_{1/2} = \frac{0.693}{k} \qquad \text{so} \qquad k = \frac{0.693}{t_{1/2}} = \frac{0.693}{4.51 \times 10^9 \text{ yr}} = 1.54 \times 10^{-10} \text{ yr}^{-1}$$

Now we calculate the age of the sample, t.

$$\log\left(\frac{A_0}{A}\right) = \frac{kt}{2.303}$$

$$\log\left(\frac{6.05 \text{ mg}}{4.64 \text{ mg}}\right) = \frac{(1.54 \times 10^{-10} \text{ yr}^{-1})t}{2.303}$$

$$\log 1.30 = (6.69 \times 10^{-11} \text{ yr}^{-1})t$$

$$\frac{0.114}{6.69 \times 10^{-11} \text{ yr}^{-1}} = t \qquad t = 1.70 \times 10^9 \text{ years}$$

The ore is approximately 1.7 billion years old.

Medical Uses of Radionuclides

The use of cobalt radiation treatments for cancerous tumors was described in Example 30-3. Several other nuclides are used as **radioactive tracers** in medicine. Radioisotopes of an element have the same chemical properties as stable isotopes of the same element, so they can be used to "label" an element in compounds. A radiation detector can be used to follow the path of the element throughout the body. Salt solutions containing ^{24}Na can be injected into the bloodstream to follow the flow of blood and locate obstructions in the circulatory system. Thallium-201 tends to concentrate in healthy heart tissue, whereas technetium-99 concentrates in abnormal heart tissue. The two can be used together to survey damage from heart disease. Iodine-131 concentrates in the thyroid gland, liver, and certain parts of the brain. It is used to monitor goiter and other thyroid problems, as well as liver and brain tumors. The energy produced by the decay of plutonium-238 is converted into electrical energy in heart pacemakers. The relatively long half-life of the isotope allows the device to be used for ten years before replacement.

Research Applications for Radionuclides

The pathways of chemical reactions can be investigated using radioactive tracers. When radioactive $^{35}S^{2-}$ ions are added to a saturated solution of cobalt sulfide in equilibrium with solid cobalt sulfide, the solid becomes radioactive. This shows that sulfide ion exchange occurs between solid and solution in the solubility equilibrium:

$$CoS(s) \rightleftharpoons Co^{2+}(aq) + S^{2-}(aq) \qquad K_{sp} = 8.7 \times 10^{-23}$$

Photosynthesis is the process by which the carbon atoms in CO_2 are incorporated into glucose, $C_6H_{12}O_6$, in green plants:

$$6CO_2 + 6H_2O \xrightarrow[\text{chlorophyll}]{\text{sunlight}} C_6H_{12}O_6 + 6O_2$$

The process is more complex than the net equation implies; it actually occurs in many steps and produces a number of intermediate products. By using labeled $^{14}CO_2$, we can identify the intermediate molecules. They contain the radioactive ^{14}C atoms.

Agricultural Uses of Radionuclides

The pesticide DDT is toxic to humans and animals repeatedly exposed to it. DDT persists in the environment a long time. It concentrates in fatty tissues. The DDT once used to control the screw-worm fly was replaced by a radiologic technique. The irradiation of the male flies with gamma rays alters their reproductive cells, sterilizing them. When great numbers of sterilized males are released in an infested area, they mate with females, who, of course, produce no offspring. This results in the reduction and eventual disappearance of the population.

The procedure works because the female flies mate only once. In an area highly populated with sterile males, the probability of a "productive" mating is very small.

Labeled fertilizers can also be used to study nutrient uptake by plants and to study the growth of crops. Gamma irradiation of some foods allows them to be stored for longer periods. For example, it retards the sprouting of potatoes and onions.

Irradiation with gamma rays from radioactive isotopes has kept the strawberries at the right fresh for 15 days, while those at the left are moldy. Such irradiation kills mold spores, but does no damage to the food.

Industrial Uses of Radionuclides

There are many applications of radiochemistry in industry and engineering. Two will be mentioned here. When great precision is required in the manufacture of strips or sheets of metal of definite thicknesses, the penetrating powers of various kinds of radioactive emissions are utilized. The thickness of the metal is correlated with the intensity of radiation passing through it. The flow of a liquid or gas through a pipeline can be monitored by injecting a sample containing a radioactive substance. Leaks in pipelines can also be detected in this way.

30-12 Artificial Transmutations of Elements

The first artificially induced nuclear reaction was carried out by Rutherford in 1915. He bombarded nitrogen-14 with alpha particles to produce an isotope of oxygen and a proton.

$$^{14}_{7}N + {}^{4}_{2}He \longrightarrow {}^{1}_{1}H + {}^{17}_{8}O$$

Such reactions are often indicated in abbreviated form, with the bombarding particle and emitted subsidiary particles shown parenthetically between the mother and daughter nuclei.

$$^{14}_{7}N\ ({}^{4}_{2}\alpha,\ {}^{1}_{1}p)\ {}^{17}_{8}O$$

Several thousand artificially induced reactions have been carried out with bombarding particles such as neutrons, protons, deuterons ($^{2}_{1}H$), alpha particles, and other small nuclei.

Bombardment with Positive Ions

A problem arises with the use of positively charged nuclei as projectiles. For a nuclear reaction to occur, the bombarding nuclei must actually collide with the target nuclei, which are also positively charged. Collisions cannot occur unless the projectiles have sufficient kinetic energy to overcome coulombic repulsion. The required kinetic energies increase with increasing atomic numbers of the target and of the bombarding particle.

Particle accelerators called **cyclotrons** (atom smashers) and **linear accelerators** have overcome the problem of repulsion. A cyclotron (Figure 30-7) consists of two hollow, D-shaped electrodes called "dees." Both dees are in an evacuated enclosure between the poles of an electromagnet. The particles to be accelerated are introduced at the center in the gap between the dees. The dees are connected to a source of high-frequency alternating current that keeps them oppositely charged. The positively charged particles are attracted toward the negative dee. The magnetic field causes the path of the charged particles to curve 180 degrees to return to the space between the dees. Then the charges are reversed on the dees, so the particles are repelled by the first dee (now positive) and attracted to the second. This repeated process is synchronized with the motion of the particles. They accelerate along a spiral path and eventually emerge through an exit hole oriented so that the beam hits the target atoms (Figure 30-8).

In a linear accelerator, the particles are accelerated through a series of tubes within an evacuated chamber (Figure 30-9). The odd-numbered tubes are at first negatively charged and the even ones positively charged. A

The first cyclotron was constructed by E. O. Lawrence and M. S. Livingston at the University of California in 1930.

The path of the particle is initially circular because of the interaction of the particle's charge with the electromagnet's field. As the particle gains energy, the radius of the path increases, and the particle spirals outward.

The first linear accelerator was built in 1928 by a German physicist, Rolf Wideroe.

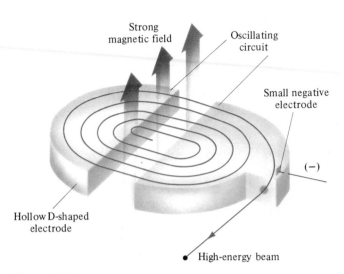

Figure 30-7
Schematic representation of a cyclotron.

Strong
magnetic field

Oscillating
circuit

Small negative
electrode

(−)

Hollow D-shaped
electrode

High-energy beam

Figure 30-8
A beam of protons (bright area) from a cyclotron at the Argonne National Laboratory. Nuclear reactions take place when protons and other atomic particles strike the nuclei of atoms.

positively charged particle is attracted toward the first tube. As it passes through that tube, the charges on the tubes are reversed so that the particle is repelled out of the first tube (now positive) and toward the second (negative) tube. As the particle nears the end of the second tube, the charges are again reversed. As this process is repeated, the particle is accelerated to very high velocities. The polarity is changed at constant frequency, so subsequent tubes are longer to accommodate the increased distance traveled by the accelerating particle per unit time. The bombardment target is located outside the last tube. If the initial polarities are reversed, negatively charged particles can also be accelerated. The longest linear accelerator, completed in 1966 at Stanford University, is about 2 miles long. It is capable of accelerating electrons to energies of nearly 20 GeV.

Many nuclear reactions have been induced by such bombardment techniques. At the time of development of particle accelerators, there were a few gaps among the first 92 elements in the periodic table. Particle accelerators were used between 1937 and 1941 to synthesize three of the four

One gigaelectron volt (ĞeV) = 1 × 10^9 eV = 1.60 × 10^{-10} J. This is sometimes called 1 billion electron volts (BeV) in the United States.

Figure 30-9
Diagram of an early type of linear accelerator. An alpha emitter is placed in the container at the left. Only those alpha particles that happen to be emitted in line with the series of accelerating tubes can escape.

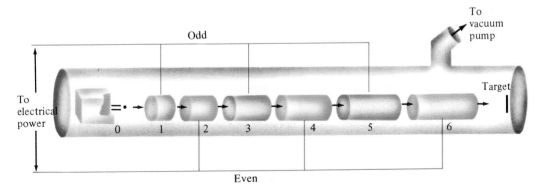

Odd

To
vacuum
pump

Target

To
electrical
power

0   1   2   3   4   5   6

Even

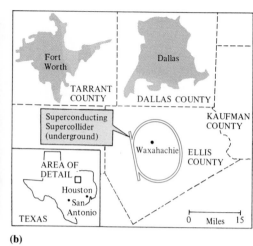

(a)

(b)

(a) An aerial view of the particle accelerator dedicated in 1978 at the Fermi National Accelerator Laboratory (Fermilab), near Batavia, Illinois. This proton accelerator, 4 miles in circumference, accelerates protons to energies of 1 trillion electron volts. (b) Construction of a vastly larger accelerator has begun near Waxahatchie, Texas. This accelerator, known as the superconducting supercollider, or SSC, will be 53 miles in circumference, big enough to fit the District of Columbia inside. Scientists will probe the nature of the nucleus by observing collisions with energies up to 40 trillion electron volts.

"missing" elements: numbers 43 (technetium), 85 (astatine), and 87 (francium).

$$^{96}_{42}\text{Mo} + ^{2}_{1}\text{H} \longrightarrow ^{97}_{43}\text{Tc} + ^{1}_{0}n$$

$$^{209}_{83}\text{Bi} + ^{4}_{2}\text{He} \longrightarrow ^{210}_{85}\text{At} + 3\,^{1}_{0}n$$

$$^{230}_{90}\text{Th} + ^{1}_{1}\text{H} \longrightarrow ^{223}_{87}\text{Fr} + 2\,^{4}_{2}\text{He}$$

Many hitherto unknown, unstable, artificial isotopes of known elements have also been synthesized so that their nuclear structures and behavior could be studied.

Neutron Bombardment

Neutrons bear no charge, so they are not repelled by nuclei as positively charged projectiles are. They do not need to be accelerated to produce bombardment reactions. Neutron beams can be generated in several ways. A frequently used method involves bombardment of beryllium-9 with alpha particles.

$$^{9}_{4}\text{Be} + ^{4}_{2}\text{He} \longrightarrow ^{12}_{6}\text{C} + ^{1}_{0}n$$

Nuclear reactors (Section 30-15) are also used as neutron sources. Neutrons ejected in nuclear reactions usually possess high kinetic energies and are called **fast neutrons**. When they are used as projectiles they cause reactions, such as (n, p) or (n, α) reactions, in which subsidiary particles are ejected. The fourth "missing" element, number 61 (promethium), was synthesized by fast neutron bombardment of neodymium-142.

Fast neutrons move so rapidly that they are likely to pass right through a target nucleus without reacting. Hence, the probability of a reaction is low, even though the neutrons may be very energetic.

$$^{142}_{60}\text{Nd} + ^{1}_{0}n \longrightarrow ^{143}_{61}\text{Pm} + _{-1}^{0}\beta$$

Slow neutrons ("thermal" neutrons) are produced when fast neutrons collide with **moderators** such as hydrogen, deuterium, oxygen, or the carbon atoms in paraffin. These neutrons are more likely to be captured by target nuclei. Bombardments with slow neutrons can cause neutron-capture (n, γ) reactions.

$$^{200}_{80}\text{Hg} + ^{1}_{0}n \longrightarrow ^{201}_{80}\text{Hg} + ^{0}_{0}\gamma$$

Slow neutron bombardment also produces the ^{3}H isotope (tritium):

$$^6_3Li + ^1_0n \longrightarrow ^3_1H + ^4_2He \quad (n, \alpha) \text{ reaction}$$

E. M. McMillan discovered the first transuranium element, neptunium, in 1940 by bombarding uranium-238 with slow neutrons.

$$^{238}_{92}U + ^1_0n \longrightarrow ^{239}_{92}U + ^0_0\gamma$$

$$^{239}_{92}U \longrightarrow ^{239}_{93}Np + ^{\,0}_{-1}\beta$$

Several additional elements have been prepared by neutron bombardment or by bombardment of the nuclei so produced with positively charged particles. Some examples are

$$\left.\begin{array}{l}^{238}_{92}U + ^1_0n \longrightarrow ^{239}_{92}U + ^0_0\gamma \\[4pt] ^{239}_{92}U \longrightarrow ^{239}_{93}Np + ^{\,0}_{-1}\beta \\[4pt] ^{239}_{93}Np \longrightarrow ^{239}_{94}Pu + ^{\,0}_{-1}\beta \end{array}\right\} \quad \text{plutonium}$$

$$^{239}_{94}Pu + ^4_2He \longrightarrow ^{242}_{96}Cm + ^1_0n \qquad \text{curium}$$

$$^{246}_{96}Cm + ^{12}_6C \longrightarrow ^{254}_{102}No + 4\,^1_0n \qquad \text{nobelium}$$

$$^{252}_{98}Cf + ^{10}_5B \longrightarrow ^{257}_{103}Lr + 5\,^1_0n \qquad \text{lawrencium}$$

30-13 Nuclear Fission and Fusion

Isotopes of some elements with atomic numbers above 80 are capable of undergoing fission in which they split into nuclei of intermediate masses and emit one or more neutrons. Some fissions are spontaneous; others require that the activation energy be supplied by bombardment. A given nucleus can split in many different ways, liberating enormous amounts of energy. Some of the possible fissions that can result from bombardment of fissionable uranium-235 with fast neutrons follow. The uranium-236 is a short-lived intermediate.

$$^{235}_{92}U + ^1_0n \longrightarrow [^{236}_{92}U] \begin{cases} ^{160}_{62}Sm + ^{72}_{30}Zn + 4\,^1_0n + \text{energy} \\[3pt] ^{146}_{57}La + ^{87}_{35}Br + 3\,^1_0n + \text{energy} \\[3pt] ^{140}_{56}Ba + ^{93}_{36}Kr + 3\,^1_0n + \text{energy} \\[3pt] ^{144}_{55}Cs + ^{90}_{37}Rb + 2\,^1_0n + \text{energy} \\[3pt] ^{144}_{54}Xe + ^{90}_{38}Sr + 2\,^1_0n + \text{energy} \end{cases}$$

Recall that the binding energy is the amount of energy that must be supplied to the nucleus to break it apart into subatomic particles. Figure 30-10 is a plot of binding energy per nucleon versus mass number. It shows that atoms of intermediate mass number have the highest binding energies per nucleon; therefore, they are the most stable. Thus, fission is an energetically favorable process for heavy atoms, because atoms with intermediate masses and greater binding energies per nucleon are formed.

Which isotopes of which elements undergo fission? Experiments with particle accelerators have shown that every element with an atomic number of 80 or more has one or more isotopes capable of undergoing fission, pro-

The term "nucleon" refers to a nuclear particle, either a neutron or a proton.

$1 \text{ MeV} = 1.60 \times 10^{-13} \text{ J}$

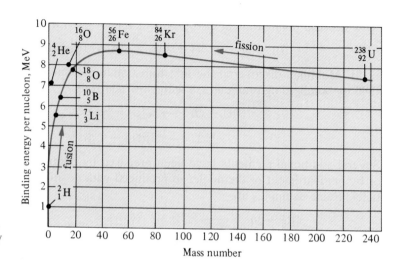

Figure 30-10
Variation in nuclear binding energy with atomic mass.
The most stable nucleus is $^{56}_{26}$Fe, with a binding energy
of 8.80 MeV per nucleon.

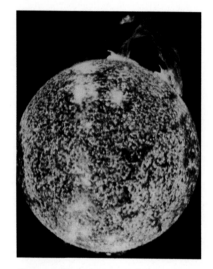

Our sun, like other stars, is a giant
nuclear fusion reactor. It supplies
energy to the earth from a distance of
93,000,000 miles.

The deuteron and triton are the
nuclei of two isotopes of hydrogen,
called deuterium and tritium. Deuterium occurs naturally in water. When
the D_2O is purified as "heavy water," it
can be used for several types of
chemical analysis.

vided they are bombarded at the right energy. Nuclei with atomic numbers
between 89 and 98 fission spontaneously with long half-lives of 10^4 to 10^{17}
years. Nuclei with atomic numbers of 98 or more fission spontaneously with
shorter half-lives of a few milliseconds to 60.5 days. One of the *natural* decay
modes of the transuranium elements is via spontaneous fission. In fact, all
known nuclides with *mass numbers* greater than 250 do this because they
are too big to be stable. Most nuclides with mass numbers between 225 and
250 do not undergo fission spontaneously (except for a few with extremely
long half-lives). They can be induced to undergo fission when bombarded
with particles of relatively low kinetic energies. Particles that can supply
the required activation energy include neutrons, protons, alpha particles,
and fast electrons. For nuclei lighter than mass 225, the activation energy
required to induce fission rises very rapidly.

In Section 30-2 we discussed the stability of nuclei with even numbers of
protons and even numbers of neutrons. We should not be surprised to learn
that both ^{233}U and ^{235}U can be excited to fissionable states by slow neutrons
much more easily than ^{238}U, because they are less stable. It is so difficult
to cause fission in ^{238}U that this isotope is said to be "nonfissionable."

Fusion, the joining of light nuclei to form heavier nuclei, is favorable for
the very light atoms. In both fission and fusion, the energy liberated is
equivalent to the loss of mass that accompanies the reactions. Much greater
amounts of energy per unit mass of *reacting atoms* are produced in fusion
than in fission.

Spectroscopic evidence indicates that the sun is a tremendous fusion
reactor consisting of 73% H, 26% He, and 1% other elements. Its major
fusion reaction is thought to involve the combination of a deuteron, $^2_1$H, and
a triton, $^3_1$H, at tremendously high temperatures to form a helium nucleus
and a neutron.

$$^2_1\text{H} + {}^3_1\text{H} \longrightarrow {}^4_2\text{He} + {}^1_0n + \text{energy}$$

Thus, solar energy is actually a form of fusion energy. This is the only
unlimited kind of energy available to us. Our fossil fuels are just "leftover"
fusion energy.

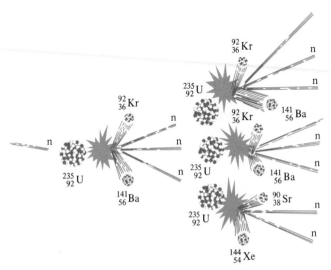

Figure 30-11
A self-propagating nuclear chain reaction. A stray neutron induces a single fission, liberating more neutrons. Each of them induces another fission, each of which is accompanied by release of two or three neutrons. The chain continues to branch in this way, finally resulting in an explosive rate of fission.

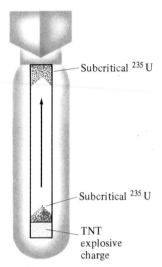

Figure 30-12
One design used in atomic bombs. A conventional explosive is used to bring two subcritical masses together to form a supercritical mass.

30-14 The Atomic Bomb (Fission)

Typically, two or three neutrons are produced per fission reaction. These neutrons can collide with other fissionable atoms to sustain and expand the process. If sufficient fissionable material, the **critical mass**, is contained in a small enough volume, an explosive chain reaction can result. If too few fissionable atoms are present, most of the neutrons escape and no chain reaction occurs. Figure 30-11 depicts a fission chain reaction.

One type of atomic bomb (Figure 30-12) contains two subcritical portions of fissionable material. One portion is driven into the other to form a supercritical mass by an ordinary chemical explosive such as trinitrotoluene (TNT). A nuclear fission explosion results. Tremendous amounts of heat energy are released, as well as many radionuclides whose effects are devastating to life and the environment. The radioactive dust and debris are called **fallout**.

30-15 Nuclear Reactors (Fission)

Controlled fission reactions in nuclear reactors are of great use and even greater potential. The fuel elements of a nuclear reactor have neither the composition nor the extremely compact arrangement of the critical mass of a bomb. Thus, no possibility of nuclear explosion exists. However, various dangers are associated with nuclear energy generation. The possibility of "meltdown" will be discussed with respect to cooling systems in light water reactors. Proper shielding precautions must be taken to ensure that the radionuclides produced are always contained within vessels from which neither they nor their radiations can escape. Long-lived radionuclides from

The launching of the nuclear submarine *Hyman G. Rickover* into the Thames River in Connecticut (August 27, 1983).

spent fuel must be stored underground in heavy, shock-resistant containers until they have decayed to the point that they are no longer biologically harmful. As examples, strontium-90 ($t_{1/2}$ = 28 years) and plutonium-239 ($t_{1/2}$ = 24,000 years) must be stored for 280 years and 240,000 years, respectively, before they lose 99.9% of their activities. Critics of nuclear energy contend that the containers could corrode over such long periods, or burst as a result of earth tremors, and that transportation and reprocessing accidents could cause environmental contamination with radionuclides. They claim that river water used for cooling is returned to the rivers with too much heat (thermal pollution), thus disrupting marine life. (It should be noted, though, that fossil fuel electric power plants cause the same thermal pollution, for the same amount of electricity generated.) The potential for theft also exists. Plutonium-239, a fissionable material, could be stolen from reprocessing plants and used to construct atomic weapons.

Proponents of the development of nuclear energy argue that the advantages far outweigh the risks. Nuclear energy plants do not pollute the air with oxides of sulfur, nitrogen, carbon, and particulate matter, as fossil fuel electric power plants do. The big advantage of nuclear fuels is the enormous amount of energy liberated per unit mass of fuel. At present, nuclear reactors provide about 17% of the electrical energy consumed in the United States. In some parts of Europe, where natural resources of fossil fuels are scarcer, the utilization of nuclear energy is higher. For instance, in France and Belgium, more than 65% of electrical energy is produced from nuclear reactors. With rapidly declining fossil fuel reserves, it appears likely that nuclear energy and solar energy will become increasingly important. However, intensifying opposition to nuclear power may mean that further growth in energy production using nuclear power in the United States must await technological developments to overcome the remaining hazards.

Light Water Reactors

Most commercial nuclear power plants in the United States are "light water" reactors, moderated and cooled by ordinary water. Figure 30-13 is a schematic diagram of a light water reactor plant. The reactor core at the left replaces the furnace in which coal, oil, or natural gas is burned in a fossil fuel plant. Such a fission reactor consists of five main components: (1) fuel, (2) moderator, (3) control rods, (4) cooling system, and (5) shielding.

Fuel   Rods of U_3O_8 enriched in uranium-235 serve as the fuel. Unfortunately, uranium ores contain only about 0.7% $^{235}_{92}U$. Most of the rest is nonfissionable $^{238}_{92}U$. The enrichment is done in processing and reprocessing plants by separating $^{235}UF_6$ from $^{238}UF_6$, prepared from the ore as described in Chapter 28. Separation by diffusion is based on Graham's Law, which relates rate of diffusion to the molecular weights of gases (Section 12-13). Another separation procedure uses the ultra-centrifuge.

A potentially more efficient method of enrichment would involve the use of sophisticated tunable lasers to ionize $^{235}_{92}U$ selectively and not $^{238}_{92}U$. The ionized $^{235}_{92}U$ could then be made to react with negative ions to form another compound, easily separated from the mixture. For this method to work, we must construct lasers capable of producing radiation monochromatic enough to excite one isotope and not the other—a difficult challenge.

The water that comes near the nuclear core flows in a closed system and is not released to the environment.

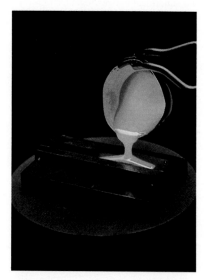

Nuclear waste may take centuries to decompose, so we cannot afford to take risks in its disposal. Suggested approaches include casting it into ceramics, as shown here, to eliminate the possibility of the waste dissolving in ground water. The encapsulated waste could then be deposited in underground salt domes. Located in geologically stable areas, such salt domes have held petroleum and compressed natural gas trapped for millions of years. The political problems of nuclear waste disposal are at least as challenging as the technological ones.

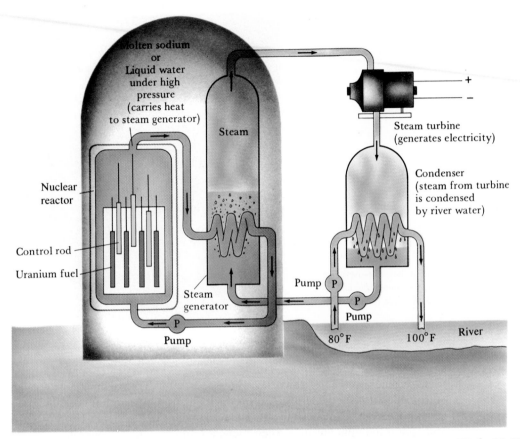

Figure 30-13
A schematic diagram of a light water reactor plant.

Moderator   The most efficient fission reactions occur with slow neutrons. Thus, the fast neutrons ejected during fission must be slowed by collisions with atoms of comparable mass that do not absorb them, called **moderators**. The most commonly used moderator is ordinary water, although graphite is sometimes used. The most efficient moderator is helium, which slows neutrons but does not absorb them all. The next most efficient is "heavy water" (deuterium oxide, $^2_1H_2O$ or $^2_1D_2O$). This is so expensive that it has been used chiefly in research reactors. A Canadian-designed power reactor that uses heavy water is more neutron-efficient than light water reactors.

Control Rods   Cadmium and boron are good neutron absorbers:

$$^{10}_{5}B + ^1_0n \longrightarrow ^7_3Li + ^4_2\alpha$$

The rate of a fission reaction is controlled by the use of movable control rods, usually made of cadmium or boron steel. They are automatically inserted in or removed from spaces between the fuel rods. The more neutrons absorbed by the control rods, the fewer fissions occur and the less heat is produced. Hence, the heat output is governed by the control system that operates the rods.

Cooling System   Two cooling systems are needed. First, the moderator itself serves as a coolant for the reactor. It transfers fission-generated heat to a

Uranium is deposited on the negative electrode in the electrorefining phase of fuel reprocessing. The crystalline mass is about 97% LiCl and KCl. The remaining 3% uranium chloride is responsible for the amethyst color.

steam generator. This converts water to steam. The steam then goes to turbines that drive generators to produce electricity. Another coolant (river water, seawater, or recirculated water) condenses the steam from the turbine, and the condensate is then recycled into the steam generator.

The danger of meltdown arises if a reactor is shut down quickly. The disintegration of radioactive fission products still goes on at a furious rate, fast enough to overheat the fuel elements and to melt them. So it is not enough to shut down the fission reaction. Efficient cooling must be continued until the short-lived isotopes are gone and the heat from their disintegration is dissipated. Only then can the circulation of cooling water be stopped.

The 1979 accident at Three Mile Island, near Harrisburg, Pennsylvania, was due to stopping the water pumps too soon *and* the inoperability of the emergency pumps. A combination of mechanical malfunctions, errors, and carelessness produced the overheating that damaged the fuel assembly. It did not and *could not explode*, although melting of the core material did occur. The 1986 accident at Chernobyl, in the USSR, was far more serious. The effects of that disaster will continue for decades. The Chernobyl accident is discussed in the Chemistry in Use essay in this chapter.

The neutrons are the worst problem of radiation. The human body contains a high percentage of H_2O, which absorbs neutrons very efficiently. A new weapon, the neutron bomb, produces massive amounts of neutrons and so is effective against people, but it does not produce the long-lasting radiation of the fission atomic bomb.

Shielding   It is essential that people and the surrounding countryside be adequately shielded from possible exposure to radioactive nuclides. The entire reactor is enclosed in a steel containment vessel. This is housed in a thick-walled concrete building. The operating personnel are further protected by a so-called biological shield, a thick layer of organic material made of compressed wood fibers. This absorbs the neutrons and beta and gamma rays that would otherwise be absorbed in the human body.

Chemistry in Use. . .
The Chernobyl Legacy

On April 26, 1986, the world's worst nuclear reactor accident occurred at the V. I. Lenin nuclear power plant near the town of Chernobyl in the Soviet Ukraine. A poorly planned and poorly executed experiment on the reactor's safety system literally blew up in the operators' faces. The experiment was planned to coincide with the plant's intermediate maintenance shutdown of reactor number 4, an RBMK 1000 model reactor.

Ironically, this experiment was designed to improve the safety of the reactor. Nuclear reactors generate electricity, but typically that electricity is not used directly to run the reactor's computers, pumps, and other equipment. Instead, the electricity is sent to a central power grid that then returns a portion of it for the reactor's power needs. If the power line returning from the central power grid were incapacitated, the reactor would be without electrical power for a while. The pumps that cool the reactor and the computers that control the reactor would be shut down, and the reactor would be out of human control. Under such dangerous conditions, the reactor would generate enough heat to melt the fuel rods and the zircalloy metal that contains the fuel. The melted material would then fall to the bottom of the reactor and continue to generate large amounts of heat. This situation is called *reactor meltdown* and should be avoided.

Most reactors have large banks of batteries to run the systems for 10 to 15 seconds after power from the central grid is interrupted. Within these 15 seconds, diesel-fueled electrical generators must start up and then come on line to run the pumps and computers. The Chernobyl experiment was an attempt to stretch the time allowed for the diesel-powered generators to start up from 15 to 45 seconds. The idea was to use the reactor's electrical turbogenerators, which would still be spinning at high velocities and generating electricity, to run the pumps for the additional 30 seconds.

As planned, the reactor operators lowered the reactor power level from its normal 3200 megawatts (MW) of heat energy to 1000 MW over 24 hours. Once this power level was reached, they deviated from their experimental plans. At about 11:00 PM on April 25, as the operators lowered the reactor power to the desired 1000-MW range, they made a mistake and allowed the power level to drop to 30 MW. Struggling for the next 2 hours, they could raise the power level to only 200 MW, which was still far below their 1000-MW goal. By this time they had removed nearly all of the 211 control rods and allowed the re-actor to become abnormally cool. The operators had also ignored several warning signals from the reactor systems and bypassed some of the safety systems. Shortly after 1:00 AM on April 26, the decision was made to begin the experiment. The steam from the reactor was shut off and the turbogenerators, which were now providing the sole power to the reactor pumps, began to slow down and generate less electricity. The flow rate of the cooling water to the reactor decreased, and a design flaw in the reactor then began to raise the reactor's power level dramatically. In 2.5 seconds the power level rose from 200 MW to 3800 MW, then in another 1.5 seconds the power leapt to 120 times the normal operating level! This dramatic power increase was sufficient to melt the fuel. An experiment designed to stave off a meltdown had initiated one.

The tremendous increase in heat generated a large steam bubble that blew a 5000-ton biological shield off the reactor's top. Then a second explosion tore a hole in the roof of the

reactor building, exposing the reactor core to the atmosphere. Over the next two weeks the burning graphite reactor core releasd between 50 and 180 megacuries (MCi) of radiation. The most hazardous radioactive material released was in the form of ^{131}I, $t_{1/2} = 8.02$ days, and ^{137}Cs, $t_{1/2} = 30.2$ years. The amounts of these two radionuclides released at Chernobyl were similar to those released by all of the atmospheric nuclear weapons tests between 1945 and 1980.

Atmospheric winds containing this material distributed most of the radioactivity over eastern Europe, smaller amounts over western Europe, and a light sprinkling over the rest of the Northern Hemisphere. Areas in which it was raining when radioactivity passed over received larger amounts of activity. Radiation levels were lethal in the immediate vicinity of the Chernobyl reactor. Fire fighters called in to extinguish fires from the explosion began to notice the effects of radiation poisoning soon after their arrival. Twenty-seven of them would die from their exposure over the next few months. A total of 31 people died as a direct result of the accident. Of the small army of people needed to entomb the reactor, 203 were hospitalized for radiation sickness.

The environmental effects of the accident were so severe that the 135,000 people living within a 30-mile radius of the plant were moved. Many of the trees, cars, and animals and much of the surface soil have been removed and buried. Areas outside the 30-mile zone also received enough radiation to cause problems. For instance, the Laplanders of northern Finland can no longer harvest reindeer for food because the reindeer eat moss that is laden with ^{137}Cs. Forested areas in the Ukraine that were downwind from the explosion now have ^{137}Cs incorporated in the leaves of plants. If fires occurred in these forests, they would redistribute the radionuclide to the atmosphere and cause "fallout from Chernobyl" again. Because of the relatively long half-life of ^{137}Cs, fire prevention in these forests will be a vital concern for a century.

The human toll from Chernobyl, beyond the 31 known dead, will be difficult to assess. Estimates are that during the next 50 years between 28,000 and 45,000 people will develop cancers caused by Chernobyl's radiation. Most of these excess cancers will occur in eastern Europe. Throughout the Northern Hemisphere the increase in the mortality rate is expected to be about 0.03%. Although the expected increase in cancer deaths is a small percentage, it does represent tens of thousands of people who will suffer the ultimate penalty of the Chernobyl legacy.

For more details on the Chernobyl reactor accident, see Charles H. Atwood, "Chernobyl—What Happened?" *Journal of Chemical Educaiton,* volume 65, number 12 (December 1988), 1037–1041.

Charles H. Atwood
Mercer University

Breeder Reactors (Fission)

It is predicted that our limited supply of uranium-235 will last only another 50 years. However, nonfissionable uranium-238 is about 100 times more plentiful and can be converted into fissionable plutonium-239. This takes place to some extent in light water reactors. Fissionable uranium-233 can also be produced by neutron bombardment of thorium-232:

$$^{238}_{92}U + ^{1}_{0}n \longrightarrow ^{239}_{94}Pu + 2\ ^{0}_{-1}\beta \qquad \text{and} \qquad ^{232}_{90}Th + ^{1}_{0}n \longrightarrow ^{233}_{92}U + 2\ ^{0}_{-1}\beta$$

The Soviet Union has placed especially high priority on the development and use of breeder reactors.

It is possible to build reactors, called **breeder reactors**, that not only generate large quantities of heat from fission, but also generate more fuel than they use. In such reactors, neutrons are absorbed in a thorium or uranium "blanket" to cause the above reactions to occur. However, several difficulties are associated with the design of the breeder reactor. This type of reactor uses fast neutrons, so no moderator is needed, but control is more difficult. It also must operate at higher temperatures than light water reactors; thus, water cannot be used as a coolant. Liquid sodium, which is not a neutron moderator, is used instead. Sodium is very reactive and has a tendency to attack the walls of its container. Heat must be transferred very

An additional problem is that plutonium-239 is one of the most toxic substances known, so any release would be disastrous.

efficiently because plutonium-239 melts at the relatively low temperature of 640°C.

30-16 Nuclear Fusion (Thermonuclear) Energy

Fusion reactions are accompanied by even greater energy production per unit mass of reacting atoms than are fission reactions. However, they can be initiated only by extremely high temperatures. The fusion of $_1^2H$ and $_1^3H$ occurs at the lowest temperature of any fusion reaction known, but even this is 40,000,000 K! Such temperatures exist in the sun and other stars, but they are nearly impossible to achieve and contain on earth. **Thermonuclear** bombs (called fusion bombs or hydrogen bombs) of incredible energy have been detonated in tests but, thankfully, never in war. In them the necessary activation energy is supplied by the explosion of a fission bomb.

The explosion of a thermonuclear (hydrogen) bomb releases tremendous amounts of energy.

It is hoped that fusion reactions can be harnessed for generation of energy for domestic power. Because of the tremendously high temperatures required, no currently known structural material can confine these reactions. At such high temperatures all molecules dissociate and most atoms ionize, resulting in the formation of a new state of matter called a **plasma**. A very-high-temperature plasma is so hot that it melts and decomposes anything it touches, including structural components of a reactor. The technological innovation required to build a workable fusion reactor probably represents the greatest challenge ever faced by the scientific and engineering community.

Plasmas have been called the fourth state of matter.

Recent attempts at the containment of lower-temperature plasmas by external magnetic fields have been successful, and they encourage our hopes. However, fusion as a practical energy source lies far in the future at best. The biggest advantages of its use would be that (1) the deuterium fuel can be found in virtually inexhaustible supply in the oceans; and (2) fusion reactions would produce only radionuclides of very short half-life, primarily tritium ($t_{1/2}$ = 12.3 years), so there would be no long-term waste disposal problem. If controlled fusion could be brought about, it could liberate us from dependence on uranium and fossil fuels.

The plasma in a fusion reactor must not touch the walls of its vacuum vessel, which would be vaporized. In the Tokamak fusion test reactor, the plasma is contained within a magnetic field shaped like a doughnut. The magnetic field is generated by D-shaped coils around the vacuum vessel.

Key Terms

Alpha particle (α)   A helium nucleus.

Artificial transmutation   An artificially induced nuclear reaction caused by bombardment of a nucleus with subatomic particles or small nuclei.

Band of stability   A band containing nonradioactive nuclides in a plot of number of neutrons versus atomic number.

Beta particle (β)   An electron emitted from the nucleus when a neutron decays to a proton and an electron.

Binding energy (nuclear binding energy)   The energy equivalent ($E = mc^2$) of the mass deficiency of an atom.

Breeder reactor   A nuclear reactor that produces more fissionable nuclear fuel than it consumes.

Chain reaction   A reaction that, once initiated, sustains itself and expands.

Cloud chamber   A device for observing the paths of speeding particles as vapor molecules condense on them to form fog-like tracks.

Control rods   Rods of materials such as cadmium or boron steel that act as neutron absorbers (not merely moderators), used in nuclear reactors to control neutron fluxes and therefore rates of fission.

Critical mass   The minimum mass of a particular fissionable nuclide, in a given volume, that is required to sustain a nuclear chain reaction.

Cyclotron   A device for accelerating charged particles along a spiral path.

Daughter nuclide   A nuclide that is produced in a nuclear decay.

Fast neutron   A neutron ejected at high kinetic energy in a nuclear reaction.

Fluorescence   Absorption of high-energy radiation by a substance and the subsequent emission of visible light.

Gamma ray (γ)   High-energy electromagnetic radiation.

Half-life of a radionuclide   The time required for half of a given sample to undergo radioactive decay.

Heavy water   Water containing deuterium, a heavy isotope of hydrogen, ^2_1H.

K capture   Absorption of a K shell ($n = 1$) electron by a proton as it is converted to a neutron.

Linear accelerator   A device used for accelerating charged particles along a straight line path.

Mass deficiency   The amount of matter that would be converted into energy if an atom were formed from constituent particles.

Moderator   A substance such as hydrogen, deuterium, oxygen, or paraffin capable of slowing fast neutrons upon collision.

Mother nuclide   A nuclide that undergoes nuclear decay.

Nuclear binding energy   The energy equivalent of the mass deficiency; energy released in the formation of an atom from subatomic particles.

Nuclear fission   The process in which a heavy nucleus splits into nuclei of intermediate masses and one or more protons are emitted.

Nuclear fusion   The combination of light nuclei to produce a heavier nucleus.

Nuclear reaction   A reaction involving a change in the composition of a nucleus; it can evolve or absorb an extraordinarily large amount of energy.

Nuclear reactor   A system in which controlled nuclear fission reactions generate heat energy on a large scale. The heat energy is subsequently converted into electrical energy.

Nucleons   Particles comprising the nucleus; protons and neutrons.

Nuclides   Different atomic forms of all elements (in contrast to isotopes, which are different atomic forms of a single element).

Plasma   A physical state of matter that exists at extremely high temperatures, in which all molecules are dissociated and most atoms are ionized.

Positron   A nuclear particle with the mass of an electron but opposite charge.

Radiation   High-energy particles or rays emitted in nuclear decay processes.

Radioactive dating   A method of dating ancient objects by determining the ratio of amounts of mother and daughter nuclides present in an object and relating the ratio to the object's age via half-life calculations.

Radioactive tracer   A small amount of radioisotope that replaces a nonradioactive isotope of the element in a compound whose path (for example, in the body) or whose decomposition products are to be monitored by detection of radioactivity; also called a radioactive label.

Radioactivity   The spontaneous disintegration of atomic nuclei.

Radioisotope   A radioactive isotope of an element.

Radionuclide   A radioactive nuclide.

Scintillation counter   A device used for the quantitative detection of radiation.

Slow neutron   A fast neutron slowed by collision with a moderator.

Thermonuclear energy   Energy from nuclear fusion reactions.

Transuranium elements   The elements with atomic numbers greater than 92 (uranium); none occurs naturally and all must be prepared by nuclear bombardment of other elements.

Exercises

Nuclear Stability and Radioactivity

1. How do nuclear reactions differ from ordinary chemical reactions?

2. What is the equation that relates the equivalence of matter and energy? What does each term in this equation represent?

3. What is mass deficiency? What is binding energy? How are the two related?

4. What are nucleons? What is the relationship between the number of protons and the atomic number? What is the relationship among the number of protons, the number of neutrons, and the mass number?

5. Define the term "binding energy per nucleon." How can this quantity be used to compare the stabilities of nuclei?

6. Describe the general shape of the plot of binding energy per nucleon against mass number.

7. (a) Briefly describe a plot of the number of neutrons against the atomic number (for the stable nuclides). Interpret the observation that the plot shows a band with a somewhat step-like shape. (b) Describe what is meant by "magic numbers" of nucleons.

8. The actual mass of a ^{64}Zn atom is 63.9291 amu. (a) Calculate the mass deficiency in amu/atom and in g/mol for this isotope. (b) What is the nuclear binding energy in kJ/mol for this isotope?

9. The actual mass of a ^{108}Pd atom is 107.90389 amu. (a) Calculate the mass deficiency in amu/atom and in g/mol for this isotope. (b) What is the nuclear binding energy in kJ/mol for this isotope?

10. Calculate the following for $^{63}_{29}$Cu (actual mass = 62.9298 amu):
   (a) mass deficiency in amu/atom
   (b) mass deficiency in g/mol
   (c) binding energy in ergs/atom
   (d) binding energy in J/atom
   (e) binding energy in kJ/mol

11. Calculate the nuclear binding energy in kJ/mol for each of the following: (a) $^{14}_{7}$N, (b) $^{56}_{26}$Fe, (c) $^{130}_{52}$Te. Their respective atomic masses are 14.00307 amu, 55.9349 amu, and 129.9067 amu. Which of these nuclides has the greatest binding energy per nucleon?

12. Repeat Exercise 11 for (a) $^{20}_{10}$Ne, (b) $^{59}_{27}$Co, and (c) $^{106}_{46}$Pd. The atomic masses are 19.99244 amu, 58.9332 amu, and 105.9032 amu, respectively.

13. Describe how (a) nuclear fission and (b) nuclear fusion generate more stable nuclei.

14. Compare the behaviors of α, β, and γ radiation (a) in an electrical field, (b) in a magnetic field, and (c) with various shielding materials, such as a piece of paper and concrete. What is the composition of each type of radiation?

15. Name some radionuclides that have medical uses, and give the uses.

16. Name and describe four methods for detection of radiation.

17. Describe how radionuclides can be used in (a) research, (b) agriculture, and (c) industry.

18. Compare the general penetrating abilities of α-particles, β-particles, γ-rays, and neutrons. Why are α-particles that are absorbed internally by the body particularly dangerous?

Nuclear Reactions

19. Consider a radioactive nuclide with a neutron/proton ratio that is larger than those for the stable isotopes of that element. What mode(s) of decay might be expected for this nuclide, and why?

20. Repeat Exercise 19 for a nuclide with a neutron/proton radio that is smaller than those for the stable isotopes.

21. Calculate the neutron/proton ratio for each of the following radioactive nuclides and predict how each of the nuclides might decay: (a) $^{13}_{5}$B (stable mass numbers for B are 10 and 11), (b) $^{81}_{38}$Sr (stable mass numbers for Sr are between 84 and 88), (c) $^{213}_{82}$Pb (stable mass numbers for Pb are between 204 and 208).

22. Repeat Exercise 21 for (a) $^{193}_{79}$Au (stable mass number for Au is 197), (b) $^{184}_{75}$Re (stable mass numbers for Re are 185 and 187), and (c) $^{142}_{59}$Pr (stable mass number for Pr is 141).

23. Write the symbols for the daughter nuclei in the following radioactive decays (β refers to an e^-):

   (a) $^{237}_{92}$U $\xrightarrow{-\beta}$ (d) ^{224}Ra $\xrightarrow{-\alpha}$

   (b) ^{13}C $\xrightarrow{-n}$ (e) ^{18}F $\xrightarrow{-p}$

   (c) ^{11}B $\xrightarrow{-\gamma}$ (f) $^{40}_{19}$K $\xrightarrow{+\beta}$

24. Predict the kind of decays you would expect for the following radionuclides:
   (a) $^{60}_{27}$Co   (n/p ratio too high)
   (b) $^{20}_{11}$Na   (n/p ratio too low)
   (c) $^{224}_{88}$Rn
   (d) $^{64}_{29}$Cu
   (e) $^{238}_{92}$U
   (f) $^{11}_{6}$C

25. What are nuclear bombardment reactions? Explain the shorthand notation used to describe bombardment reactions.

26. Fill in the missing symbols in the following nuclear bombardment reactions:
   (a) $^{23}_{11}$Na + ? $\longrightarrow$ $^{23}_{12}$Mg + $^{1}_{0}n$
   (b) $^{96}_{42}$Mo + $^{4}_{2}$He $\longrightarrow$ $^{100}_{43}$Tc + ?
   (c) $^{232}_{90}$Th + ? $\longrightarrow$ $^{240}_{96}$Cm + 4 $^{1}_{0}n$
   (d) ? + $^{1}_{1}$H $\longrightarrow$ $^{29}_{14}$Si + $^{0}_{0}\gamma$

(e) $^{209}_{83}Bi + ? \longrightarrow ^{210}_{84}Po + ^{1}_{0}n$

(f) $^{238}_{92}U + ^{16}_{8}O \longrightarrow ? + 5 \, ^{1}_{0}n$

27. Write the symbols for the daughter nuclei in the following nuclear bombardment reactions:

(a) $^{60}_{28}Ni \, (n, p)$

(b) $^{98}_{42}Mo \, (^{1}_{0}n, \beta)$

(c) $^{35}_{17}Cl \, (p, \alpha)$

(d) $^{20}_{10}Ne \, (\alpha, \gamma)$

(e) $^{15}_{7}N \, (p, \alpha)$

(f) $^{10}_{5}B \, (n, \alpha)$

28. Write the nuclear equation for each of the following bombardment processes:

(a) $^{14}_{7}N \, (\alpha, p) \, ^{17}_{8}O$

(b) $^{106}_{46}Pd \, (n, p) \, ^{106}_{45}Rh$

(c) $^{23}_{11}Na \, (n, \beta^{-})X$. Identify X.

29. Repeat Exercise 28 for the following:

(a) $^{113}_{48}Cd \, (n, \gamma) \, ^{114}_{48}Cd$

(b) $^{6}_{3}Li \, (n, \alpha) \, ^{3}_{1}H$

(c) $^{2}_{1}H \, (\gamma, p)X$. Identify X.

30. Write the shorthand notation for each of the following nuclear reactions:

(a) $^{6}_{3}Li + ^{1}_{0}n \longrightarrow ^{4}_{2}He + ^{3}_{1}H$

(b) $^{31}_{15}P + ^{2}_{1}H \longrightarrow ^{32}_{15}P + ^{1}_{1}H$

(c) $^{238}_{92}U + ^{1}_{0}n \longrightarrow ^{239}_{93}Np + _{-1}^{0}e$

31. Repeat Exercise 30 for the following:

(a) $^{253}_{99}Es + ^{4}_{2}He \longrightarrow ^{256}_{101}Md + ^{1}_{0}n$

(b) $^{27}_{13}Al + ^{1}_{0}n \longrightarrow ^{26}_{13}Al + 2 \, ^{1}_{0}n$

(c) $^{37}_{17}Cl + ^{1}_{1}H \longrightarrow ^{1}_{0}n + ^{37}_{18}Ar$

32. Write the nuclear equations for the following processes: (a) $^{63}_{28}Ni$ undergoing β^{-} emission, (b) two deuterium ions undergoing fusion to give $^{3}_{2}He$ and a neutron, (c) a nuclide being bombarded by a neutron to form $^{7}_{3}Li$ and an α-particle (identify the unknown nuclide), (d) $^{14}_{7}N$ being bombarded by a neutron to form three α-particles and an atom of tritium.

33. Write the nuclear equations for the following processes: (a) $^{228}_{90}Th$ undergoing α decay, (b) $^{110}_{49}In$ undergoing positron emission, (c) $^{127}_{53}I$ being bombarded by a proton to form $^{121}_{54}Xe$ and seven neutrons, (d) tritium and deuterium undergoing fusion to form an α-particle and a neutron, (e) $^{95}_{42}Mo$ being bombarded by a proton to form $^{95}_{43}Tc$ and radiation (identify this radiation).

34. "Radioactinium" is produced in the actinium series from $^{235}_{92}U$ by the successive emission of an α-particle, a β^{-}-particle, an α-particle, and a β^{-}-particle. What are the symbol, atomic number, and mass number for "radioactinium"?

35. An alkaline earth element (Group IIA) is radioactive. It undergoes decay by emitting three α-particles in succession. In what periodic table group is the resulting element found?

36. A nuclide of element unnilquadium, $^{257}_{104}Unq$, is formed by the nuclear reaction of californium-98 and carbon-12, with the emission of four neutrons. This new nuclide rapidly decays by emitting an α-particle. Write the equations for these nuclear reactions, and identify the nuclide that is formed as $^{257}_{104}Unq$ decays.

37. Describe how (a) cyclotrons and (b) linear accelerators work.

Rates of Decay

38. What does the half-life of a radionuclide represent? How do we compare the relative stabilities of radionuclides in terms of half-lives?

39. Why must all radioactive decays be first order?

40. Describe the process by which steady-state (constant) ratios of carbon-14 to (nonradioactive) carbon-12 are attained in living plants and organisms. Describe the method of radiocarbon dating. What factors limit the use of this method?

41. The half-life of $^{19}_{8}O$ is 29 s. What fraction of the isotope originally present would be left after 5.0 s?

42. The half-life of $^{11}_{6}C$ is 20.3 min. How long will it take for 90.0% of a sample to decay? How long will it take for 99.0% of the sample to decay?

43. The activity of a sample of tritium decreased by 5.5% over the period of a year. What is the half-life of $^{3}_{1}H$?

44. A very unstable isotope of beryllium, ^{8}Be, undergoes α emission with a half-life of 0.07 fs. How long does it take for 99.99% of a 1.0-μg sample of ^{8}Be to undergo decay? What would be the volume of helium, measured at STP, generated from the decay of the 1.0-μg sample?

45. The $^{14}_{6}C$ activity of an artifact from the tomb of Hemaka (2930 ± 200 BC) was 8.3 min · g C. The half-life of $^{14}_{6}C$ is 5730 years and the current $^{14}_{6}C$ activity is 15.3/min · g C (that is, 15.3 disintegrations per minute per gram of carbon). How old is the artifact?

46. A piece of wood from a burial site was analyzed using $^{14}_{6}C$ dating and was found to have an activity of 13.4/min · g C. Using the data given in Exercise 45 for $^{14}_{6}C$ dating, determine the age of this piece of wood.

47. Strontium-90 is one of the harmful radionuclides that results from nuclear fission explosions. It decays by beta emission with a half-life of 28 years. How long would it take for 99.99% of a given sample, released in an atmospheric test of an atomic bomb, to disintegrate?

48. Carbon-14 decays by beta emission with a half-life of 5730 years. Assuming a particular object originally contained 7.50 μg of carbon-14 and now contains 0.67 μg of carbon-14, how old is the object?

Fission and Fusion

49. Briefly describe a nuclear fission process. What are the two most important fissionable materials?

50. What is a chain reaction? Why are nuclear fission processes considered chain reactions? What is the critical mass of a fissionable material?

51. Where have continuous nuclear fusion processes been observed? What is the main reaction that occurs in such sources?

52. The reaction that occurred in the first fusion bomb was ^7_3Li (p, α) X. (a) Write the complete equation for the process and identify the product, X. (b) The atomic masses are 1.007825 amu for ^1_1H, 4.00260 amu for α, and 7.01600 amu for ^7_3Li. Find the energy for the reaction, in kJ/mol.

53. Summarize how an atomic bomb works, including how the nuclear explosion is initiated.

54. Discuss the pros and cons of the use of nuclear energy instead of other, more conventional types of energy based on fossil fuels.

55. Describe and illustrate the essential features of a light water fission reactor.

56. How is fissionable uranium-235 separated from nonfissionable uranium-238?

57. Distinguish between moderators and control rods of nuclear reactors.

58. What are the major advantages and disadvantages of fusion as a potential energy source, compared with fission? What is the major technological problem that must be solved to permit development of a fusion reactor?

Mixed Exercises

***59.** Calculate the binding energy, in kJ/mol of nucleons, for the following isotopes: (a) $^{15}_8O$ with a mass of 15.00300 amu, (b) $^{16}_8O$ with a mass of 15.99491 amu, (c) $^{17}_8O$ with a mass of 16.99913 amu, (d) $^{18}_8O$ with a mass of 17.99915 amu, (e) $^{19}_8O$ with a mass of 19.0035 amu. Which of these would you expect to be most stable?

***60.** The first nuclear transformation (discovered by Rutherford) can be represented by the shorthand notation $^{14}_7N$ (α, p) $^{17}_8O$. (a) Write the corresponding nuclear equation for this process. The respective atomic masses are

14.00307 amu for $^{14}_7N$, 4.00260 amu for ^4_2He, 1.007825 amu for ^1_1H, and 16.99913 amu for $^{17}_8O$. (b) Calculate the energy change of this reaction in kJ/mol.

61. A proposed series of reactions (known as the carbon–nitrogen cycle) that could be important in the very hottest region of the interior of the Sun is

$$^{12}C + {}^1H \longrightarrow A + \gamma$$
$$A \longrightarrow B + {}_{+1}^0e$$
$$B + {}^1H \longrightarrow C + \gamma$$
$$C + {}^1H \longrightarrow D + \gamma$$
$$D \longrightarrow E + {}_{+1}^0e$$
$$E + {}^1H \longrightarrow {}^{12}C + F$$

Identify the species labeled A–F.

***62.** Show by calculation which reaction produces the larger amount of energy per atomic mass unit of material reacting.

fission: $^{235}_{92}U + {}^1_0n \longrightarrow {}^{94}_{40}Zr + {}^{140}_{58}Ce + 6 {}_{-1}^0e + 2 {}^1_0n$

fusion: $2 {}^2_1H \longrightarrow {}^3_1H + {}^1_1H$

The atomic masses are 235.0439 amu for $^{235}_{92}U$, 93.9061 amu for $^{94}_{40}Zr$, 139.9053 amu for $^{140}_{58}Ce$, 3.01605 amu for ^3_1H, 1.007825 amu for ^1_1H, 2.0140 amu for ^2_1H.

***63.** The separation of isotopes is commonly done by gaseous diffusion processes. (a) Using Graham's law, calculate the ratio of the rates of effusion for $^1_1H_2(g)$ and $^2_1H_2(g)$. The atomic masses are 1.007825 amu and 2.0140 amu for ^1_1H and ^2_1H, respectively. (b) Repeat the calculation for $^{235}_{92}UF_6(g)$ and $^{238}_{92}UF_6(g)$. The atomic masses of the uranium isotopes are 235.0439 amu and 238.0508 amu, respectively. (c) In which pair of isotopes are the nuclides easier to separate?

31 Organic Chemistry I: Compounds

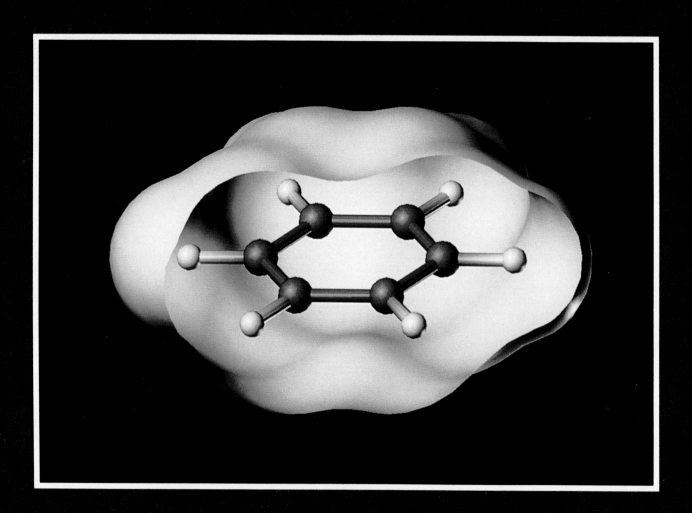

Outline

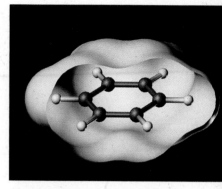

A computer-generated model of a molecule of benzene, C_6H_6. A ball-and-stick model is shown inside a representation of the molecular surface.

Objectives

As you study this chapter, you should learn

☐ About saturated hydrocarbons (alkanes and cycloalkanes)—their structures and their nomenclature

☐ About unsaturated hydrocarbons (alkenes and alkynes)—their structures and their nomenclature

☐ About the occurrence of hydrocarbons in petroleum and natural gas

☐ To visualize structural and geometrical isomerism for hydrocarbons

☐ About some aromatic hydrocarbons—benzene,

condensed aromatics, and substituted aromatic compounds

☐ About some common functional groups, the resulting classes of organic compounds, and how to name the compounds
  halides
  alcohols and phenols
  ethers
  amines
  carboxylic acids and some of
     their derivatives
  aldehydes and ketones

Organic chemistry is the chemistry of compounds that contain C—C or C—H bonds. Why is one entire branch of chemistry devoted to the behavior of the compounds of just one element? The answer is twofold: (1) There are many more compounds that contain carbon than there are compounds that do not, and (2) the molecules containing carbon can be so much larger and more complex.

Originally the term "organic" was used to describe compounds of plant or animal origin. In 1828, Friedrich Wöhler synthesized urea by boiling ammonium cyanate with water.

Of the Group IVA elements, carbon is a nonmetal, silicon and germanium are metalloids, and tin and lead are metallic elements.

Urea, H_2N—CO—NH_2, is the principal end product of metabolism of nitrogen-containing compounds in mammals. It is eliminated in the urine. An adult man excretes about 30 grams of urea in 24 hours.

$$\underset{\substack{\text{ammonium cyanate}\\ \text{(an inorganic compound)}}}{NH_4OCN} \xrightarrow[\text{boil}]{H_2O} \underset{\substack{\text{urea}\\ \text{(an organic compound)}}}{H_2N-\overset{\overset{\textstyle O}{\|}}{C}-NH_2}$$

This disproved the "vital force" theory, which held that organic compounds could be made only by living things. Today many organic compounds are manufactured from inorganic materials.

We encounter organic chemistry in every aspect of our lives. All life is based on a complex interrelationship of thousands of organic substances—from simple compounds such as sugars, amino acids, and fats to vastly more complex ones, such as the enzymes that catalyze life's chemical reactions and the huge DNA molecules that carry genetic information from one generation to the next. The food we eat (including many additives); the clothes we wear; the plastics and polymers that are everywhere; our life-saving medicines; the paper on which we write; our fuels; many of our poisons, pesticides, carcinogens, dyes, soaps and detergents—all involve organic chemistry.

We normally think of petroleum and natural gas as fuel sources, but most synthetic organic materials are also derived from these two sources. More than half of the top 50 commercial chemicals are organic compounds derived in this way. Indeed, petroleum and natural gas may one day be more valuable as raw materials for organic synthesis than as fuel sources. If so, we may greatly regret our delay in developing alternative energy sources while burning up vast amounts of our petroleum and natural gas deposits as fuels.

A carbon atom has four electrons in its outermost shell, with ground state configuration $1s^2 2s^2 2p^2$. The C atom can attain a stable configuration by forming four covalent bonds. As we saw in Chapter 8, each C atom can form single, double, or triple bonds by utilizing various hybridizations. The bonding of carbon is summarized in Table 31-1, using examples that we saw in Chapters 7 and 8. Carbon is unique among the elements in the extent to which it bonds to itself and in the diversity of compounds that are formed. The ability of an element to bond to itself is known as **catenation** ("chain making"). Carbon atoms catenate to form long chains, branched chains, and rings that may also have chains attached to them. Millions of such compounds are known.

Although millions of organic compounds are known, the elements they contain are very few—C and H; often N, O, S, or a halogen; and sometimes another element. The great number and variety of organic compounds are

Table 31-1
Hybridization of Carbon in Covalent Bond Formation

Hybridization and Resulting Geometry	Orbitals Used by C Atom	Bonds Formed by C Atom	Example	
sp^3, tetrahedral	four sp^3 hybrids	four σ bonds	methane	H—C—H with H above and below
sp^2, trigonal planar	three sp^2 hybrids, one p orbital	three σ bonds, one π bond	ethylene	C=C
sp, linear	two sp hybrids, two p orbitals	two σ bonds, two π bonds	acetylene	H—C≡C—H

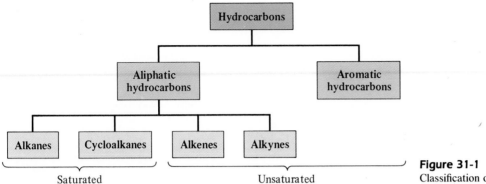

Figure 31-1
Classification of hydrocarbons.

a result of the many different arrangements of atoms, or *structures*, that are possible. The chemical and physical properties of organic compounds are related to the structures of their molecules. Thus, the basis for organizing and understanding organic chemistry is structural theory. (Please review Chapters 7 and 8.)

The vast and fascinating topic of organic chemistry could be the subject of many years of study. We shall give only a brief introduction to it. In this chapter we organize compounds of organic chemistry into classes or "families" according to their structural features and learn how to name various types of compounds. Chapter 32 will present a few of the typical reactions undergone by organic substances.

Organic molecules are based on a framework of carbon–carbon and carbon–hydrogen bonds. Many compounds contain *only* the two elements C and H; they are called **hydrocarbons**. Hydrocarbons that contain delocalized rings, such as the benzene ring (Section 9-6), are called **aromatic hydrocarbons**. Those that do not contain such delocalized systems are called **aliphatic hydrocarbons**. Aliphatic hydrocarbons that contain only sigma (σ) bonds (i.e., only single bonds) are called **saturated hydrocarbons**. Those that contain both sigma and pi (π) bonds (i.e., double, triple, or delocalized bonds) are called **unsaturated hydrocarbons**. These classifications are diagrammed in Figure 31-1. The first seven sections of this chapter are devoted to the study of hydrocarbons.

A **functional group** is a special arrangement of atoms within an organic molecule that is responsible for some characteristic chemical behavior of the compound. Different molecules that contain the same functional groups have similar chemical behavior. We shall follow our study of hydrocarbons with a presentation of some important characteristic functional groups and the resulting classes of compounds.

Saturated Hydrocarbons

31-1 Alkanes and Cycloalkanes

The hydrocarbons are among the simplest and most common compounds of carbon. The *saturated hydrocarbons*, or **alkanes**, are compounds in which each carbon atom is bonded to four other atoms. Each H atom is bonded

The term "saturated" comes from early studies in which chemists tried to add hydrogen to various organic substances. Those to which no more hydrogen could be added were called saturated, by analogy with saturated solutions.

$$CH_4 \quad \text{or} \quad H-\underset{\underset{\displaystyle H}{|}}{\overset{\overset{\displaystyle H}{|}}{C}}-H$$

(a)

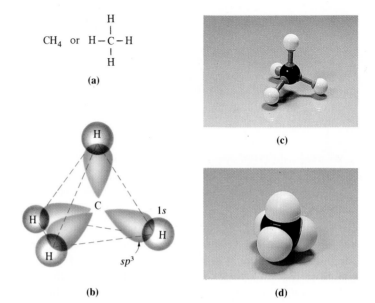

(c)

(b)

(d)

Figure 31-2

Representations of a molecule of methane, CH_4. (a) The condensed and line formulas for methane. (b) The overlap of the four sp^3 carbon orbitals with the s orbitals of four hydrogen atoms forms a tetrahedral molecule. (c) A ball-and-stick model and (d) a space-filling model of methane.

to only one C atom. Saturated hydrocarbons contain only single bonds. Petroleum and natural gas are mostly saturated hydrocarbons.

In Section 8-7 we examined the structure of the simplest alkane, *methane*, CH_4. We saw that methane molecules are tetrahedral, with sp^3 hybridization at carbon (Figure 31-2).

Ethane, C_2H_6, is the next simplest saturated hydrocarbon. Its structure is quite similar to that of methane. Two carbon atoms share a pair of electrons. Each carbon atom also shares an electron pair with each of three hydrogen atoms. Both carbon atoms are sp^3 hybridized (Figure 31-3). One can visualize the formation of a C_2H_6 molecule from two CH_4 molecules by mentally removing one H atom (and its electron) from each CH_4 molecule and then joining the fragments. *Propane*, C_3H_8, is the next member of the family (Figure 31-4). It can be considered the result of removing one H atom from a methane molecule and one from an ethane molecule, and joining the fragments.

Two different compounds have the formula C_4H_{10}, but different structures and hence different properties. Such *isomers* result when two molecules contain the same atoms bonded together in different orders. The structures of these two isomeric *butanes* are shown in Figure 31-5. These two structures correspond to the two ways in which a hydrogen atom can be removed from a propane molecule and replaced by a —CH_3 group. If a —CH_3 replaces

iso = "same"; *mer* = "part." As we saw in Sections 29-5 through 29-7, isomers are substances that have the same numbers and kinds of atoms arranged differently. Isomerism in organic compounds is discussed more systematically in Chapter 32.

Figure 31-3

Models of ethane, C_2H_6. (a) The condensed and line formulas for ethane. (b) A ball-and-stick model and (c) a space-filling model of ethane.

$$CH_3CH_3 \quad \text{or} \quad H-\underset{\underset{\displaystyle H}{|}\,\underset{\displaystyle H}{|}}{\overset{\overset{\displaystyle H}{|}\,\overset{\displaystyle H}{|}}{C-C}}-H$$

(a)

(b)   **(c)**

Figure 31-4
Ball-and-stick and space-filling models of propane, C_3H_8.

an H from either of the end carbon atoms, the result is *normal butane*, abbreviated as *n*-butane. **Normal**, or "straight-chain," hydrocarbons are those in which there is no branching. If the —CH_3 group replaces an H from the central carbon atom of propane, however, the result is 2-methyl-propane, or *isobutane*. This is the simplest *branched-chain hydrocarbon*.

The formulas of the alkanes can be written in general terms as C_nH_{2n+2}, where *n* is the number of carbon atoms per molecule. The first five members of the series are

		CH_4	C_2H_6	C_3H_8	C_4H_{10}	C_5H_{12}
Number of C atoms = *n*	=	1	2	3	4	5
Number of H atoms = 2*n* + 2 =		4	6	8	10	12

The formula of each alkane differs from the next by CH_2, a **methylene group**.

A series of compounds in which each member differs from the next by a specific number and kind of atoms is called a **homologous series**. The properties of members of such a series are closely related. The boiling points of the lighter members of the saturated hydrocarbon series are shown in Figure 31-6. As the molecular weights of the normal hydrocarbons increase, their boiling points also increase regularly. Properties such as boiling point depend on the forces between molecules (Chapter 13). Carbon–carbon and carbon–hydrogen bonds are essentially nonpolar and are arranged tetrahedrally around each C atom. As a result, saturated hydrocarbons are nonpolar molecules, and the only significant intermolecular forces are London forces (Section 13-2). These forces, which are due to induced dipoles, become stronger as the sizes of the molecules and the number of electrons in each

Though somewhat misleading, the term "straight-chain" is widely used. The carbon chains are linear only in the structural formulas that we write. They are actually zigzag, due to the tetrahedral bond angles at each carbon, and sometimes further kinked or twisted. Think of such a chain of carbon atoms as *continuous*. We can trace a single path from one terminal carbon to the other and pass through every other C atom *without backtracking*.

Figure 31-5
Ball-and-stick models of the two isomeric butanes, *n*-butane, $CH_3CH_2CH_2CH_3$, and isobutane, $CH_3CH(CH_3)CH_3$.

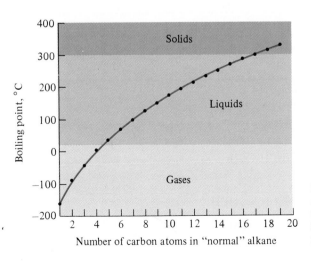

Figure 31-6
A plot of normal boiling point versus the number of carbon atoms in normal, i.e., straight-chain, saturated hydrocarbons.

Table 31-2
Some Normal Hydrocarbons (alkanes)

Molecular Formula	IUPAC Name	Normal bp (°C)	Normal mp (°C)	State at Room Temperature
CH_4	methane	-161	-184	
C_2H_6	ethane	-88	-183	
C_3H_8	propane	-42	-188	gas
C_4H_{10}	n-butane	$+0.6$	-138	
C_5H_{12}	n-pentane	36	-130	
C_6H_{14}	n-hexane	69	-94	
C_7H_{16}	n-heptane	98	-91	
C_8H_{18}	n-octane	126	-57	
C_9H_{20}	n-nonane	150	-54	
$C_{10}H_{22}$	n-decane	174	-30	
$C_{11}H_{24}$	n-undecane	194.5	-25.6	liquid
$C_{12}H_{26}$	n-dodecane	214.5	-9.6	
$C_{13}H_{28}$	n-tridecane	234	-6.2	
$C_{14}H_{30}$	n-tetradecane	252.5	$+5.5$	
$C_{15}H_{32}$	n-pentadecane	270.5	10	
$C_{16}H_{34}$	n-hexadecane	287.5	18	
$C_{17}H_{36}$	n-heptadecane	303	22.5	
$C_{18}H_{38}$	n-octadecane	317	28	
$C_{19}H_{40}$	n-nonadecane	330	32	solid
$C_{20}H_{42}$	n-eicosane	205 (at 15 torr)	36.7	

molecule increase. Thus, trends such as those depicted in Figure 31-6 are due to the increase in effectiveness of London forces.

Some systematic method for naming compounds is necessary. The system in use today is prescribed by the International Union of Pure and Applied Chemistry (IUPAC). The names of the first 20 normal alkanes are listed in Table 31-2. You should become familiar with at least the first ten. The names

Table 31-3
Isomeric Hexanes, C_6H_{14}

IUPAC Name	Formula	Normal bp (°C)	Normal mp (°C)
n-hexane	$CH_3CH_2CH_2CH_2CH_2CH_3$	68.7	-94

2-methylpentane	$CH_3CH_2CH_2CHCH_3$
	              $\mid$
	             CH_3
		60.3	-153.7

3-methylpentane	$CH_3CH_2CHCH_2CH_3$
	         $\mid$
	         CH_3
		63.3	-118

2,2-dimethylbutane	CH_3
	    $\mid$
	 $CH_3CH_2CCH_3$
	    $\mid$
	    CH_3
		49.7	-99.7

2,3-dimethylbutane	$CH_3CH-CHCH_3$
	    $\mid$ $\mid$
	   CH_3 CH_3
		58.0	-128.4

(a) *n*-pentane
bp 36.1°C

(b) methylbutane
bp 27.9°C

(c) dimethylpropane
bp 9.5°C

Figure 31-7
Ball-and-stick models of the three isomeric pentanes. Each contains 5 C atoms and 12 H atoms. The atoms are bonded in a different order in each of the three structural isomers.

of the alkanes starting with pentane have prefixes (from Greek) that give the number of carbon atoms in the molecules. All alkane names have the *-ane* ending.

We have seen that there are two saturated C_4H_{10} hydrocarbons. For the C_5 hydrocarbons, there are three possible arrangements of the atoms. Thus, three different *pentanes* are known.

The number of structural isomers increases rapidly as the number of carbon atoms in saturated hydrocarbons increases. There are five isomeric *hexanes* (Table 31-3). Table 31-4 displays the number of isomers of some saturated hydrocarbons (alkanes). Most of the isomers have not been prepared or isolated. They probably never will be.

As the degree of branching increases for a series of molecules of the same molecular weight, the molecules become more compact (Figure 31-7). A compact molecule can have fewer points of contact with its neighbors than more extended molecules do. As a result, the total induced dipole forces (London forces) are weaker for branched molecules, and the boiling points of such compounds are lower.

Table 31-4
Numbers of Possible Isomers of Alkanes

Formula	Isomers
C_7H_{16}	9
C_8H_{18}	18
C_9H_{20}	35
$C_{10}H_{22}$	75
$C_{11}H_{24}$	159
$C_{12}H_{26}$	355
$C_{13}H_{28}$	802
$C_{14}H_{30}$	1,858
$C_{15}H_{32}$	4,347
$C_{20}H_{42}$	366,319
$C_{25}H_{52}$	36,797,588
$C_{30}H_{62}$	4,111,846,763

Cycloalkanes

The cyclic saturated hydrocarbons, or **cycloalkanes**, have the general formula C_nH_{2n}. The cycloalkanes (and many other ring compounds that we shall encounter later) are often shown in simplified skeletal form, in which each intersection of two lines represents a C atom and we mentally add enough H atoms to give each carbon atom four bonds. The first four cycloalkanes and their simplified representations are

cyclopropane   cyclobutane   cyclopentane   cyclohexane

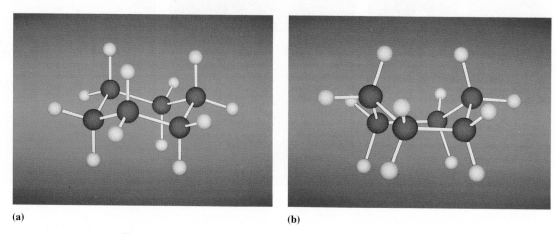

(a)   **(b)**

Figure 31-8
The two stable forms of the cyclohexane ring: (a) chair and (b) boat. The chair form is more stable because the hydrogens (or other substituents) are, on the average, farther from one another than in the boat form.

In some of these structures, the bond angles are somewhat distorted from the ideal tetrahedral angle of 109.5°, the most severe distortions being 60° in cyclopropane and 90° in cyclobutane. As a result, these rings are said to be "strained," and these two compounds are unusually reactive for saturated hydrocarbons. Cyclopentane is quite stable with a nearly flat ring, because the bond angles in a regular pentagon (108°) are near the tetrahedral angle (109.5°).

Cyclohexane can assume two different geometries that are essentially strain-free, but to do this, the ring "puckers," or becomes nonplanar. The two stable arrangements of cyclohexane are called the *chair* and the *boat* forms (Figure 31-8).

31-2 Naming Saturated Hydrocarbons

In this section we introduce the system for naming saturated hydrocarbons. It is important to realize that many compounds (and their names) were so familiar to chemists before the development of the IUPAC system (beginning about 1890) that they continued to be called by their common, or "trivial," names. In this and the next chapter, IUPAC names appear in blue type, and the common alternative names are shown in black type.

The IUPAC naming system is based on the names of the normal hydrocarbons given in Table 31-2 and their higher homologs. To name a branched-chain hydrocarbon, we first find the longest continuous chain of carbon atoms and use the root name that corresponds to that normal hydrocarbon. We then indicate the position and kind of *substituent* attached to the chain. **Alkyl group** substituents attached to the longest chain are thought of as fragments of hydrocarbon molecules obtained by the removal of one hydrogen atom. We give them names related to the parent hydrocarbons from which they are derived, as follows. Other alkyl groups are named similarly (Table 31-5). We use the general symbol R to represent any alkyl group. The cycloalkyl groups derived from the first four cycloalkanes are called cyclopropyl, cyclobutyl, cyclopentyl, and cyclohexyl, respectively.

Table 31-5
Some Alkanes and the Related Alkyl Groups

Parent Hydrocarbon	Alkyl Group, —R	
CH_4, methane	—CH_3, methyl	

$$H—\overset{\displaystyle H}{\underset{\displaystyle H}{C}}—H \quad or \quad CH_4$$

$$—\overset{\displaystyle H}{\underset{\displaystyle H}{C}}—H \quad or \quad —CH_3$$

C_2H_6, ethane	—C_2H_5, ethyl	

$$H—\overset{\displaystyle H}{\underset{\displaystyle H}{C}}—\overset{\displaystyle H}{\underset{\displaystyle H}{C}}—H \quad or \quad CH_3CH_3$$

$$—\overset{\displaystyle H}{\underset{\displaystyle H}{C}}—\overset{\displaystyle H}{\underset{\displaystyle H}{C}}—H \quad or \quad —CH_2CH_3$$

C_3H_8, propane	—C_3H_7, n-propyl	—C_3H_7, isopropyl

$$H—\overset{\displaystyle H}{\underset{\displaystyle H}{C}}—\overset{\displaystyle H}{\underset{\displaystyle H}{C}}—\overset{\displaystyle H}{\underset{\displaystyle H}{C}}—H \quad or \quad CH_3CH_2CH_3$$

$$—\overset{\displaystyle H}{\underset{\displaystyle H}{C}}—\overset{\displaystyle H}{\underset{\displaystyle H}{C}}—\overset{\displaystyle H}{\underset{\displaystyle H}{C}}—H$$

$$H—\overset{\displaystyle H}{\underset{\displaystyle H}{C}}—\overset{\displaystyle H}{C}—\overset{\displaystyle H}{\underset{\displaystyle H}{C}}—H$$

C_4H_{10}, n-butane	—C_4H_9, n-butyl	—C_4H_9, sec-butyl (read as "secondary butyl")

$$H—\overset{\displaystyle H}{\underset{\displaystyle H}{C}}—\overset{\displaystyle H}{\underset{\displaystyle H}{C}}—\overset{\displaystyle H}{\underset{\displaystyle H}{C}}—\overset{\displaystyle H}{\underset{\displaystyle H}{C}}—H \quad or \quad CH_3(CH_2$$

$$—\overset{\displaystyle H}{\underset{\displaystyle H}{C}}—\overset{\displaystyle H}{\underset{\displaystyle H}{C}}—\overset{\displaystyle H}{\underset{\displaystyle H}{C}}—\overset{\displaystyle H}{\underset{\displaystyle H}{C}}—H$$

$$H—\overset{\displaystyle H}{\underset{\displaystyle H}{C}}—\overset{\displaystyle H}{C}—\overset{\displaystyle H}{\underset{\displaystyle H}{C}}—\overset{\displaystyle H}{\underset{\displaystyle H}{C}}—H$$

C_4H_{10}, methylpropane (common name isobutane)	—C_4H_9, t-butyl (read as "tertiary butyl")	—C_4H_9, isobutyl

$$H—\overset{\displaystyle CH_3}{\underset{\displaystyle CH_3}{C}}—CH_3 \quad or \quad CH(CH_3)_3$$

$$—\overset{\displaystyle CH_3}{\underset{\displaystyle CH_3}{C}}—CH_3$$

$$H—\overset{\displaystyle —CH_2}{\underset{\displaystyle CH_3}{C}}—CH_3$$

Summary of IUPAC Rules for Naming Alkanes

1. Find the longest continuous chain of C atoms. Choose the base name that describes the number of C atoms in this chain, with the ending -*ane* (Table 31-2).
2. Number the C atoms in this chain beginning at the end nearer the branching. Always number the longest chain so that the substituents have the lowest possible numbers.
3. Assign the names and position numbers that indicate the substituents.
4. Use the appropriate prefix to indicate the number of each substituent: *di* = 2, *tri* = 3, *tetra* = 4, *penta* = 5, and so on.

There are no hard and fast rules for the order in which substituent names appear. Sometimes we name them in alphabetical order; sometimes they are named in order of increasing complexity (although it can be difficult to define "complexity").

Let us name the following compound.

$$\begin{array}{c}
H \\
| \\
HH-C-HHHHH \\
|||||| \\
H-C-C-C-C-C-C-H \\
|||||| \\
HHHHHH
\end{array}$$

or

$$CH_3CH(CH_3)CH_2CH_2CH_2CH_3$$

We follow Rules 1 and 2 to number the carbon atoms in the longest chain.

$$\begin{array}{c}
CH_3 \\
| \\
CH_3-CH-CH_2-CH_2-CH_2-CH_3 \\
123456
\end{array}$$

The methyl group is attached to the *second* carbon atom in a *six-carbon* chain, and so the compound is named 2-methylhexane.

The following examples further illustrate the rules of nomenclature.

Parentheses are used to conserve space. Formulas written with parentheses must indicate unambiguously the structure of the compound. The parentheses here indicate that the CH₃ group is attached to the C that precedes it.

It is incorrect to name the compound 5-methylhexane, because that violates Rule 2.

Example 31-1

Name the compound represented by the structural formula

$$\begin{array}{c}
CH_3 \\
| \\
CH_3-C-CH_2-CH_2-CH_3 \\
| \\
CH_3
\end{array} \quad or \quad CH_3C(CH_3)_2(CH_2)_2CH_3$$

Plan

We first find the longest carbon chain and number it to give substituents the smallest possible numbers. Then we name the substituents as in Table 31-5 and specify the number of each as indicated in Rule 4 above.

Remember that the (dash) formulas indicate which atoms are bonded to each other. They do *not* show molecular geometry. In this molecule, each C atom is tetrahedrally bonded to four other atoms.

Solution

$$\begin{array}{c}
CH_3 \\
2|345 \\
CH_3-C-CH_2-CH_2-CH_3 \\
| \\
CH_3
\end{array}$$

The longest chain contains five carbons, so this compound is named as a derivative of pentane. There are two methyl groups, both at carbon number 2. The IUPAC name of this compound is   2,2-dimethylpentane.

Example 31-2

Name the compound represented by the structural formula

$$\begin{array}{c}
CH_3 \\
| \\
CH_3-C-CH_2-CH-CH_3 \\
|| \\
CH_3CH_2 \\
| \\
CH_3
\end{array}$$

Plan

The approach is the same as in Example 31-1. We should be aware that the longest continuous carbon chain may not be *written* in a straight line.

Solution

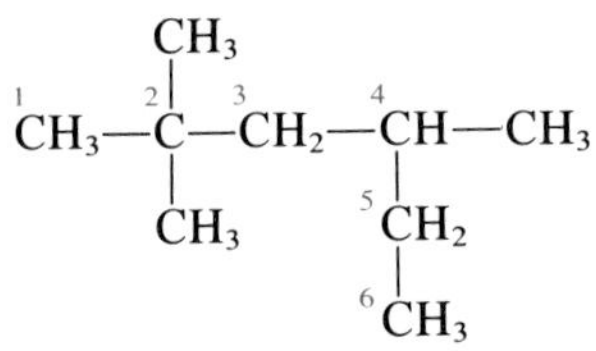

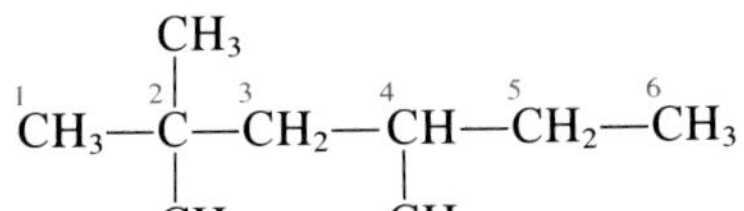

(written to emphasize better the six-C chain)

The longest chain contains six carbons, so this compound is named as a derivative of hexane. There are three methyl substituents—two at carbon number 2 and one at carbon number 4. The IUPAC name of this compund is

2,2,4-trimethylhexane.

EOC 13, 14

Example 31-3

Write the structure for 2,5-dimethyl-4-*tert*-butylheptane.

Plan

The root name "heptane" indicates that there is a seven-carbon chain.

$$C—C—C—C—C—C—C$$

The names and numbers of the substituents tell us where to attach the alkyl groups.

Solution

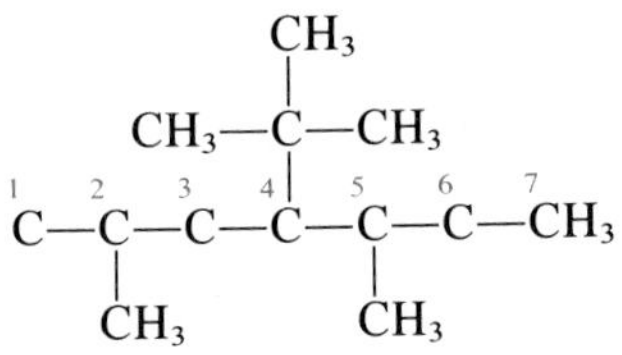

Then we fill in enough hydrogens to saturate each C atom and arrive at the structure

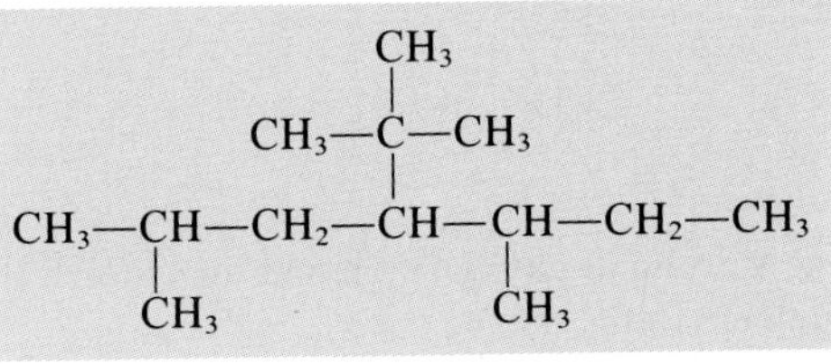

Example 31-4

Write the structure for the compound 2-cyclopropyl-3-ethylpentane.

Plan

The root name "pentane" tells us that the structure is based on a five-carbon chain. We place the substituents at the positions indicated by the numbers in the name.

Solution

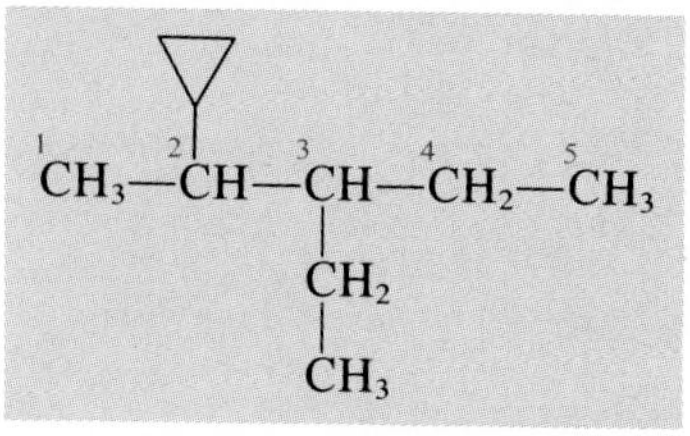

$$CH_3-\overset{2}{CH}-\overset{3}{CH}-\overset{4}{CH_2}-\overset{5}{CH_3}$$
$$\underset{CH_2}{\overset{1}{|}}$$
$$\underset{CH_3}{|}$$

where the symbol  $\bigtriangledown\!\!|$ represents the cyclopropyl group, CH_2-CH_2
$$\underset{|}{\overset{\diagdown CH \diagup}{}}$$

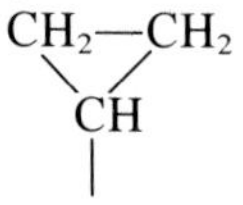

EOC 15, 16

The names of substituted cycloalkanes are derived analogously to those of alkanes. (1) The base name is determined by the number of carbon atoms in the ring, using the same base name as the alkane with the addition of *cyclo* in front. (2) If only one substituent is attached to the ring, no "location number" is required, because all positions in a cycloalkane are equivalent. (3) Two or more functional groups on the ring are identified by location numbers, which should be assigned sequentially to the ring carbons in the order that gives the *smallest sum* of location numbers.

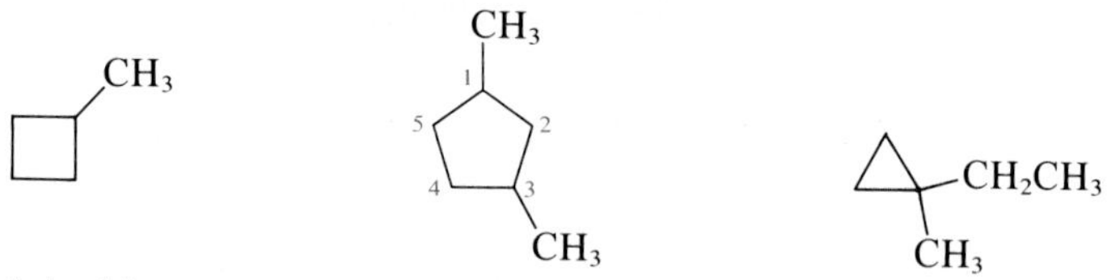

methylcyclobutane

1,3-dimethylcyclopentane
(not 1,4-dimethylcyclopentane)

1-ethyl-1-methylcyclopropane

Example 31-5

Draw the structure for 1,1-dimethyl-3-*sec*-butylcycloheptane.

Plan

The "cycloheptane" base name tells us that the structure contains substituents on a saturated seven-membered ring.

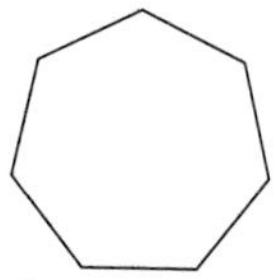

We number this ring starting at any position and then attach the substituents as indicated by the numbers in the name.

Solution

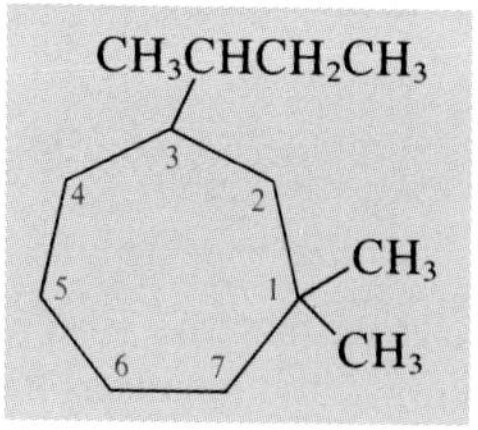

Chemistry in Use. . .
Petroleum

Petroleum, or crude oil, was discovered in the United States (Pennsylvania) in 1859 and in the Middle East (Iran) in 1908. It has been found in many other locations since these initial discoveries, and is now pumped from the ground in many parts of the world. Petroleum consists mainly of hydrocarbons. Small amounts of organic compounds containing nitrogen, sulfur, and oxygen are also present. Each oil field produces petroleum with a particular set of characteristics. Distillation of petroleum produces several fractions, as shown in the table.

Because gasoline is so much in demand, higher hydrocarbons (C_{12} and higher) are "cracked" to increase the amount of gasoline that can be made from a barrel of petroleum. The hydrocarbons are heated, in the absence of air and in the presence of a catalyst, to produce a mixture of smaller alkanes that can be used in gasoline. This process is called *thermal* or *catalytic cracking*.

The **octane number** (rating) of a gasoline indicates how smoothly it burns and how much engine "knock" it produces. (Engine knock is caused by premature detonation of fuel in the combustion chamber.) 2,2,4-Trimethylpentane, isoöctane, has excellent combustion properties and was arbitrarily assigned an octane number of 100. Normal heptane, CH_3—$(CH_2)_5$—CH_3, has very poor combustion properties and was assigned an octane number of zero.

$$CH_3-\underset{\underset{\displaystyle CH_3}{|}}{\overset{\overset{\displaystyle CH_3}{|}}{C}}-CH_2-\underset{}{\overset{\overset{\displaystyle CH_3}{|}}{CH}}-CH_3$$

isoöctane (octane number = 100)

$$CH_3CH_2CH_2CH_2CH_2CH_2CH_3$$

n-heptane (octane number = 0)

Mixtures of these two were prepared and burned in test engines to establish the octane scale. The octane number of such a mixture is the percentage of isoöctane in it. Gasolines burned in standard test engines are assigned octane numbers based on the compression ratio at which they begin to knock. A 90-octane fuel produces the same amount of knock as the 90% isoöctane–10% *n*-heptane mixture. Branched-chain compounds produce less knock than straight-chain compounds. The octane numbers of two isomeric hexanes are

$$CH_3-(CH_2)_4-CH_3$$

n-hexane
octane number = 25

A petroleum refinery tower.

$$(CH_3)_3C-CH_2-CH_3$$

2,2-dimethylbutane
octane number = 92

Depletion of petroleum reserves is having a major impact on the supply, and therefore the cost, of energy. However, we must be aware that petroleum is also the raw material from which most of the carbon-containing substances for a technological society are synthesized. This includes not just fuels, but also pharmaceuticals, cosmetics, synthetic fabrics, plastics, synthetic structural materials, and many more commonly used products.

How long will our supply of petroleum last? Figure (a) displays worldwide consumption since 1900. Note the dramatic increase since 1960. If the current rate of consumption continues, *presently known* reserves of petroleum will be exhausted in about 35 years.

Current technology recovers only about 30% of the petroleum in a field. As our *known* reserves decrease, greatly improved technologies will be required to extract more petroleum from the earth. Such

Petroleum Fractions		
Fraction*	**Principal Composition**	**Distillation Range**
Natural gas	C_1–C_4	Below 20°C
Bottled gas	C_5–C_6	20–60°
Gasoline	C_4–C_{12}	40–200°
Kerosene	C_{10}–C_{16}	175–275°
Fuel oil, diesel oil	C_{15}–C_{20}	250–400°
Lubricating oils	C_{18}–C_{22}	Above 300°
Paraffin	C_{23}–C_{29}	mp 50–60°
Asphalt		Viscous liquid ("bottoms fraction")
Coke		Solid

*Other descriptions and distillation ranges have been used, but all are similar.

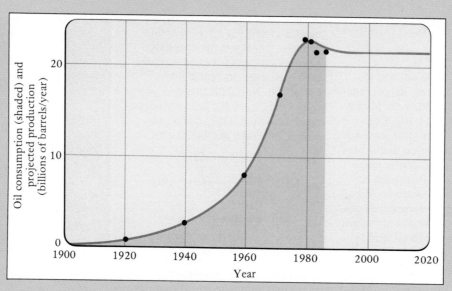

(a) World petroleum use since 1900 and projected use until 2020. Presently known reserves will be exhausted in about 35 years if use continues at the current levels.

technologies will likely be more costly and make petroleum products more expensive.

There are large deposits of *oil* *shale rock* in the United States. However, the cost of recovering petroleum from these deposits is high, and disposal of the waste presents major problems. Improved technologies or much higher oil prices will be required before this source of petroleum is developed.

Unsaturated Hydrocarbons

There are three classes of unsaturated hydrocarbons: (1) the alkenes and their cyclic counterparts, the cycloalkenes; (2) the alkynes and the cycloalkynes; and (3) the aromatic hydrocarbons.

31-3 Alkenes (Olefins)

Recall that the cycloalkanes may also be represented by the general formula C_nH_{2n}.

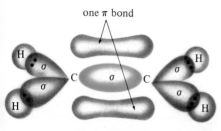

one π bond

Four C—H σ bonds (gray), one C—C σ bond (green), and one C—C π bond (purple) in the planar C_2H_4 molecule.

The simplest **alkenes** contain one carbon–carbon double bond, C=C, per molecule. The general formula for noncyclic alkenes is C_nH_{2n}. The simplest alkene is C_2H_4, which is usually called by its common name, ethylene.

The bonding in ethylene was described in Section 8-13 (see margin). The hybridization (sp^2) and bonding at other double-bonded carbon atoms are similar. Both carbon atoms in C_2H_4 are located at the centers of trigonal planes. Rotation about C=C double bonds does not occur significantly at room temperature. Therefore, compounds that have the general formula (XY)C=C(XY) exist as a pair of *cis–trans* isomers. Figure 31-9 shows the *cis–trans* isomers of 1,2-dichloroethene. The existence of compounds with different arrangements of groups on the opposite sides of a bond with restricted rotation is called **geometrical isomerism**. This *cis–trans* isomerism can occur across double bonds in alkenes and across single bonds in rings.

 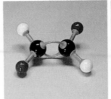 

(a)   *cis*-1,2-dichloroethene   (b)   *trans*-1,2-dichloroethene
m.p. −80.5°C, b.p. 60.3°C   m.p. −50°C, b.p. 47.5°C

Figure 31-9
Two isomers of 1,2-dichloroethene are possible because rotation about the double bond is restricted. This is an example of *geometrical* isomerism. A ball-and-stick model and a space-filling model are shown for each isomer. (a) The *cis* isomer. (b) The *trans* isomer.

Two shared electron pairs draw the atoms closer together than a single electron pair does. Thus, carbon–carbon double bonds are shorter than C—C single bonds, 1.34 Å versus 1.54 Å. The physical properties of the alkenes are similar to those of the alkanes, but their chemical properties are quite different.

The root for the name of each alkene is derived from the alkane having the same number of C atoms as the longest chain containing the double bond. In the trivial (common) system of nomenclature, the suffix *-ylene* is added to the characteristic root. In systematic (IUPAC) nomenclature, the suffix *-ene* is added to the characteristic root.

> ### *Summary of IUPAC Rules for Naming Alkenes and Cycloalkenes*
>
> 1. Locate the C atoms in the *longest* continuous C chain *that contains the double bond.* Use the base name prefix with the ending *-ene.*
> 2. Number the C atoms of this basic chain sequentially, *beginning at the end nearer the double bond.* Insert the number describing the position of the double bond (indicated by its *first* carbon location) before the base name. (This is necessary only for chains of four or more C atoms, because only one position is possible for a double bond in a chain of two or three carbon atoms.)
> 3. In naming alkenes, the double bond takes positional precedence over substituents on the carbon chain. The double bond is assigned the lowest possible number.
> 4. To name compounds with possible geometrical isomers, consider the two largest groups within the carbon chain that contains the double bond—these are indicated as part of the base name. Insert the prefix *cis-* or *trans-* just before the number of the double bond to indicate whether these are on the same or opposite sides, respectively, of the double bond.
> 5. For cycloalkenes, the double bond is assumed to be between C atoms 1 and 2, so no position number is needed to describe it.

The prefix *trans-* means "across" or "on the other side of." As a reminder of this terminology, think of words such as "transatlantic."

Some illustrations of this naming system follow.

$$CH_2{=}CH_2 \qquad CH_3{-}CH{=}CH_2 \qquad \overset{4}{C}H_3\overset{3}{C}H_2{-}\overset{2}{C}H{=}\overset{1}{C}H_2$$

systematic: ethene   propene   1-butene
trivial: (ethylene)   (propylene)   (*n*-butylene)

$$\overset{1}{C}H_3\overset{2}{C}H{=}\overset{3}{C}H\overset{4}{C}H_3 \qquad CH_3{-}\underset{\underset{CH_3}{|}}{C}{=}CH_2$$

2-butene

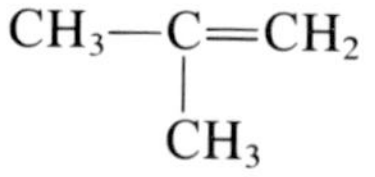

methylpropene
(isobutylene)

The following two names illustrate the application of Rule 3.

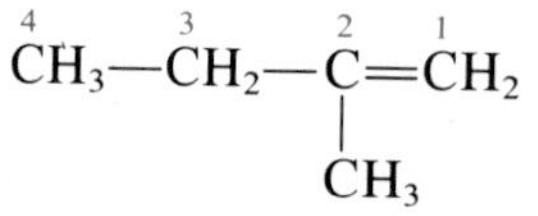

$$\overset{4}{C}H_3{-}\overset{3}{C}H_2{-}\underset{\underset{CH_3}{|}}{\overset{2}{C}}{=}\overset{1}{C}H_2 \qquad \overset{1}{C}H_3{-}\overset{2}{C}H{=}\overset{3}{C}H{-}\underset{\underset{CH_3}{|}}{\overset{4}{C}}H{-}\overset{5}{C}H_3$$

2-methyl-1-butene 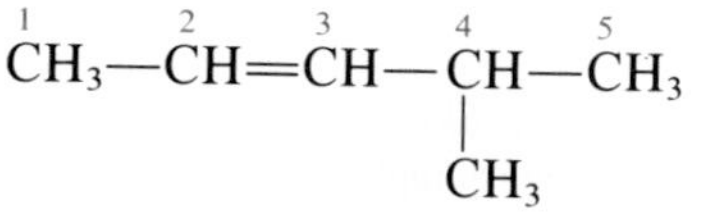   4-methyl-2-pentene

Some alkenes, called **polyenes**, have two or more carbon–carbon double bonds per molecule. The suffixes *-adiene*, *-atriene*, and so on are used to indicate the number of (C=C) double bonds in a molecule.

$$\overset{1}{C}H_2{=}\overset{2}{C}H{-}\overset{3}{C}H{=}\overset{4}{C}H_2 \qquad \overset{4}{C}H_3{-}\overset{3}{C}H{=}\overset{2}{C}{=}\overset{1}{C}H_2$$

1,3-butadiene   1,2-butadiene

1,3-Butadiene and similar molecules that contain *alternating* single and double bonds are described as having **conjugated double bonds**. Such compounds are of special interest because of their polymerization reactions (Section 32-11).

Example 31-6

Name the following two alkenes:

(a)

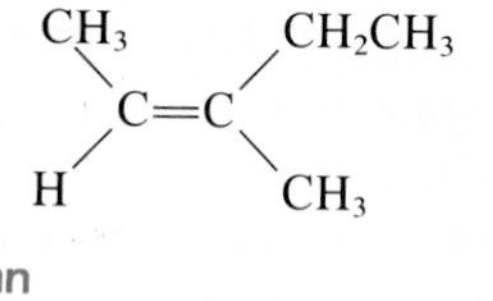

$$\underset{H}{\overset{CH_3}{\diagdown}}C{=}C\underset{CH_3}{\overset{CH_2CH_3}{\diagup}}$$

(b)

$$\underset{H}{\overset{CH_3CH_2}{\diagdown}}C{=}C\underset{CH_2CH_3}{\overset{CH_2CH_3}{\diagup}}$$

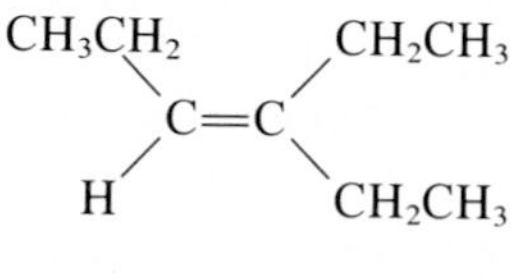

Plan

For each compound, we first find the longest chain that includes the double bond, and then number it beginning at the end nearer the double bond (Rules 1 and 2). Then we specify the identities and positions of substituents in the same way we did for alkanes. In (a), we specify the geometrical isomer by locating the two largest groups in the chain and then describe their relationship using the *cis–trans* terminology.

Solution

(a)

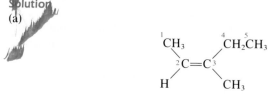

The longest such chain contains five atoms and has a double bond beginning at atom 2; thus the compound is named as a derivative of 2-pentene. Now we must apply Rule 4. The two largest groups in the chain are the terminal —CH_3 (carbon 1) and the —CH_2CH_3 (carbons 4 and 5); these are on the *same side* of the double bond, so we name the compound as a derivative of *cis*-2-pentene. The only substituent is the methyl group at carbon 3. The full name of the compound is

3-methyl-*cis*-2-pentene.

(b) There are two choices for the longest chain, and either one would have an ethyl substituent.

We could number either chain from the other end and still have the double bond starting at carbon 3; we number from the end that gives the carbon bearing the ethyl group the lowest possible position number, 3. Carbon 3 has two equivalent substituents, so geometrical isomerism is not possible, and we do not use the

cis–trans terminology. The name is   3-ethyl-3-hexene.

The cycloalkenes are represented by the general formula C_nH_{2n-2}. Two cycloalkenes and their skeletal representations are

cyclopentene   cyclohexene

Example 31-7
Draw the structure of 3-methylcyclohexene.

Plan
We draw the ring of the specified size with one double bond in the ring. We number the ring so the double bond is between atoms 1 and 2. Then we add the designated substituents at the indicated positions.

Solution
We number the six-membered ring so that the double bond is between atoms 1 and 2.

A methyl group is attached at carbon 3; the correct structure is

Remember that each intersection of two lines represents a carbon atom; there are enough H atoms at each C atom to make a total of four bonds to carbon.

31-4 Alkynes

The **alkynes**, or acetylenic hydrocarbons, contain carbon–carbon triple bonds, —C≡C—. The noncyclic alkynes with one triple bond per molecule have the general formula C_nH_{2n-2}. The bonding in all alkynes is similar to that in acetylene (Section 8-14). Triply bonded carbon atoms are sp hybridized. The triply bonded atoms and their adjacent atoms lie on a straight line (Figure 31-10).

Alkynes are named like the alkenes except that the suffix *-yne* is added to the characteristic root. The first member of the series is commonly called acetylene. Its molecular formula is C_2H_2. It is thermodynamically unstable, decomposing explosively to C(s) and H_2(g) at high pressures. It may be converted into ethene and then to ethane by the addition of hydrogen. These reactions suggest that the formula for acetylene is H—C≡C—H.

Figure 31-10
Models of acetylene, H—C≡C—H.

CH≡CH   CH₃—C≡CH   CH₃—CH₂—C≡CH   CH₃—C≡C—CH₃   CH₃—CH—C≡CH
ethyne   propyne   1-butyne   2-butyne |
(acetylene)   CH₃
                                                                                                    3-methyl-1-butyne

The triple bond takes positional precedence over substituents on the carbon chain. It is assigned the lowest possible number in naming.

Summary of IUPAC Rules for Naming Alkynes

Alkynes are named like the alkenes except for the following two points:

1. The suffix *-yne* is added to the characteristic root.
2. Because the linear arrangement about the triple bond does not lead to geometrical isomerism, the prefixes *cis-* and *trans-* are not used.

Example 31-8
Draw the structure of 5,5-dimethyl-2-heptyne.

Plan

The structure is based on a seven-carbon chain with a triple bond beginning at carbon 2. We add methyl groups at the positions indicated.

Solution

$$\overset{1}{C}-\overset{2}{C}\equiv\overset{3}{C}-\overset{4}{C}-\overset{5}{C}-\overset{6}{C}-\overset{7}{C}$$

There are two methyl groups attached to carbon 5 and sufficient hydrogens to complete the bonding at each C atom.

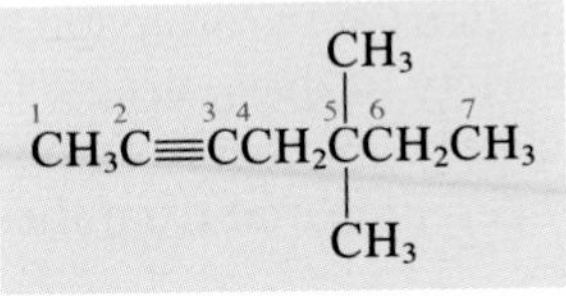

$$CH_3C \equiv CCH_2CCH_2CH_3$$

with methyl groups on C5, numbered 1-7

The carbon chain is numbered to give the multiple bond the lowest possible numbers, even though that results in higher position numbers for the methyl substituents.

Example 31-9

Name the compound

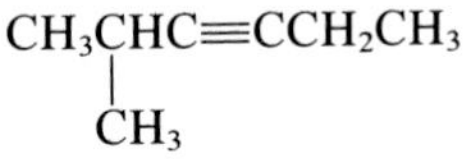

$$CH_3CHC \equiv CCH_2CH_3$$
with CH_3 substituent

Plan

We find the longest chain and number it to give the triple bond the lowest possible number. Then we specify the substituent(s) by name and position number.

Solution

The longest continuous carbon chain that includes the triple bond has six C atoms. There are four ways in which we could choose and number such a chain, and in all four the triple bond would be between C atoms 3 and 4:

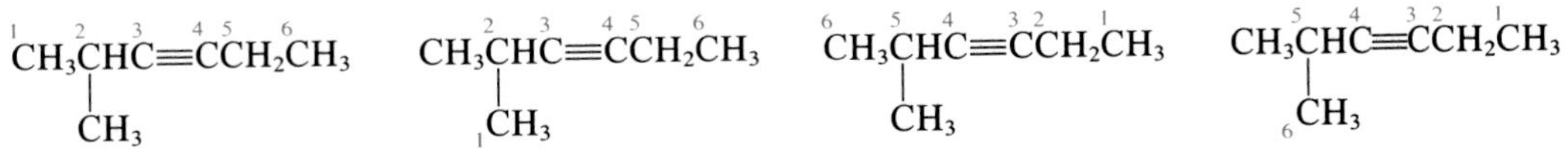

$$CH_3CHC \equiv CCH_2CH_3 \qquad CH_3CHC \equiv CCH_2CH_3 \qquad CH_3CHC \equiv CCH_2CH_3 \qquad CH_3CHC \equiv CCH_2CH_3$$

We want the methyl substituent to have the lowest possible number, so we choose either of the first two possibilities. The name of the compound is

2-methyl-3-hexyne.

EOC 25a, d

Aromatic Hydrocarbons

Originally the word "**aromatic**" was applied to pleasant-smelling substances. The word now describes benzene, its derivatives, and certain other compounds that exhibit similar chemical properties. Some have very foul odors because of substituents on the benzene ring. On the other hand, many fragrant compounds do not contain benzene rings.

Steel production requires large amounts of coke. This is prepared by heating bituminous coal to high temperatures in the absence of air. This process also favors production of *coal gas* and *coal tar*. Because of the enormous amount of coal converted to coke, coal tar is produced in large quantities. It serves as a source of aromatic compounds. For each ton of coal converted to coke, about 5 kg of aromatic compounds are obtained. The 19th-century development of the German coal tar industry greatly stimulated the systematic study of organic chemistry.

The main components of coal gas are hydrogen ($\approx$50%) and methane ($\approx$30%).

Distillation of coal tar produces a variety of aromatic compounds.

Early research on the reactions of the aromatic hydrocarbons led to methods for preparing a great variety of dyes, drugs, flavors and perfumes, and explosives. More recently, large numbers of polymeric materials, such as plastics and fabrics, have been prepared from these compounds.

31-5 Benzene

Benzene is the simplest aromatic hydrocarbon. By studying its reactions, we can learn a great deal about aromatic hydrocarbons. Benzene was discovered in 1825 by Michael Faraday when he fractionally distilled a by-product oil obtained in the manufacture of illuminating gas from whale oil.

Elemental analysis and determination of its molecular weight show that the molecular formula for benzene is C_6H_6. The formula suggests that it is highly unsaturated. But its properties are quite different from those of alkenes and alkynes.

The facts that only one monosubstitution product is obtained in many reactions and that no addition products can be prepared show conclusively that benzene has a *symmetrical ring structure*. Stated differently, every H atom is equivalent to every other H atom, and this is possible only in a symmetrical ring structure (a):

(a)

(skeleton only)

(b)

The debate over the structure and bonding in benzene raged for about 30 years. In 1865, Friedrich Kekulé suggested that the structure of benzene

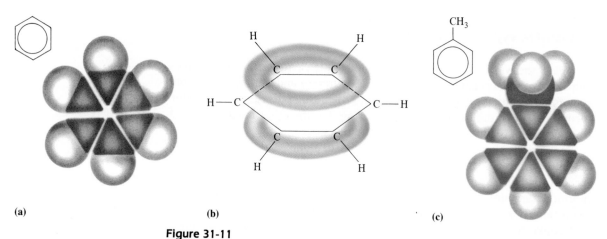

(a)   **(b)**   **(c)**

Figure 31-11
(a) A model of the benzene, C_6H_6, molecule and (b) its electron distribution. (c) A model of toluene, $C_6H_5CH_3$. This is an alkylbenzene, a derivative of benzene in which one H atom has been replaced by a —CH_3 group.

was intermediate between two structures [part (b) page 1048] that we now call resonance structures. We often represent benzene as

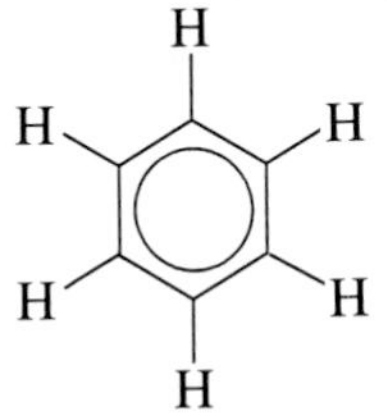

or, more simply

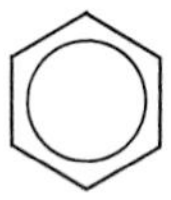

The structure of benzene is described in detail in Section 9-6 in terms of MO theory.

In a benzene molecule, all 12 atoms lie in a plane. This suggests sp^2 hybridization of each carbon. The six sp^2 hybridized C atoms lie in a plane, and the unhybridized p orbitals extend above and below the plane. Side-by-side overlap of the p orbitals forms pi orbitals (see Figure 9-10). The electrons associated with the pi bonds are *delocalized* over the entire benzene ring (Figure 31-11a,b).

31-6 Other Aromatic Hydrocarbons

Benzene molecules bearing alkyl substituents are called **alkylbenzenes**. The simplest of these is methylbenzene (common name, toluene), shown in Figure 31-11c. The xylenes are dimethylbenzenes. Three different compounds (Table 31-6) have the formula $C_6H_4(CH_3)_2$ (see margin). These three xylenes are *structural isomers*. In naming these (as well as other disubstituted benzenes), we use prefixes *ortho-* (abbreviated *o*-), *meta-* (*m*-), or *para-* (*p*-) to refer to relative positions of substituents on the benzene ring. The *ortho*-prefix refers to two substituents located on *adjacent* carbon atoms; e.g., 1,2-dimethylbenzene is *o*-xylene. The *meta*- prefix identifies substituents on C atoms 1 and 3, so 1,3-dimethylbenzene is *m*-xylene. The *para*- prefix refers to substituents on C atoms 1 and 4, so 1,4-dimethylbenzene is *p*-xylene.

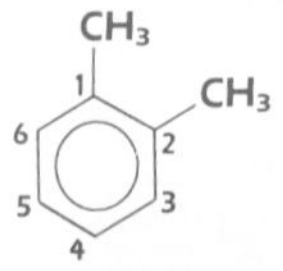

ortho-xylene
bp = 144°C
mp = −27°C

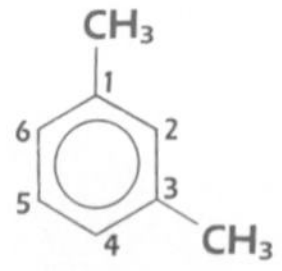

meta-xylene
bp = 139°C
mp = −54°C

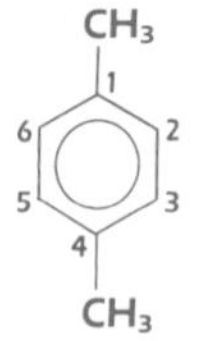

para-xylene
bp = 138°C
mp = 13°C

> *Summary of Rules for Naming Derivatives of Benzene*
>
> 1. If there is only one group on the ring, no number is needed to designate its position.
> 2. If there are two groups on the ring, we use the traditional designations:
>
>   *ortho-* or *o*- for 1,2-disubstitution
>
>   *meta-* or *m*- for 1,3-disubstitution
>
>   *para-* or *p*- for 1,4-disubstitution
>
> 3. If there are three or more groups on the ring, location numbers are assigned so as to give the *minimum sum* of numbers.

Examples are

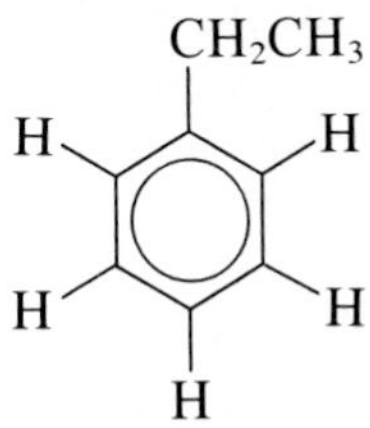

ethylbenzene

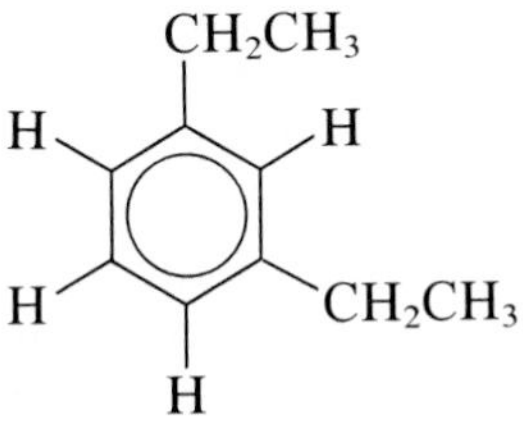

m-diethylbenzene

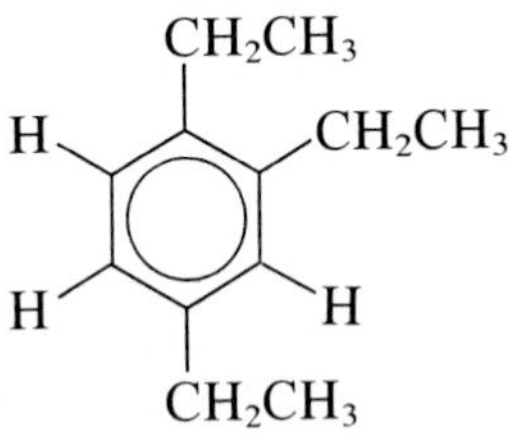

1,2,4-triethylbenzene

Table 31-6
Aromatic Hydrocarbons from Coal Tar

Name	Formula	Normal bp (°C)	Normal mp (°C)	Solubility
benzene	C_6H_6	80	+6	
toluene	$C_6H_5CH_3$	111	−95	
o-xylene	$C_6H_4(CH_3)_2$	144	−27	All
m-xylene	$C_6H_4(CH_3)_2$	139	−54	insoluble
p-xylene	$C_6H_4(CH_3)_2$	138	+13	in
naphthalene	$C_{10}H_8$	218	+80	water
anthracene	$C_{14}H_{10}$	342	+218	
phenanthrene	$C_{14}H_{10}$	340	+101	

When an H atom is removed from a benzene, C_6H_6, molecule, the resulting group, C_6H_5— or ⬡—, is called "phenyl." Sometimes we name mixed alkyl–aromatic hydrocarbons on that basis.

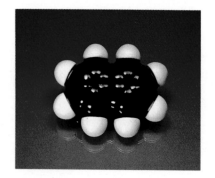

Naphthalene.

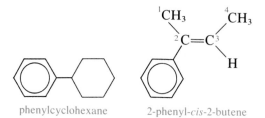

phenylcyclohexane   2-phenyl-*cis*-2-butene

Another class of aromatic hydrocarbons consists of "condensed," or "fused-ring," aromatic systems. The simplest of these are naphthalene, anthracene, and phenanthrene:

naphthalene, $C_{10}H_8$ anthracene, $C_{14}H_{10}$ phenanthrene, $C_{14}H_{10}$

No hydrogen atoms are attached to the carbon atoms that are involved in fusion of aromatic rings, i.e., carbon atoms that are members of two or more aromatic rings.

The traditional name is often used as part of the base name in naming an aromatic hydrocarbon and its derivatives. You should know the names and structures of the fundamental aromatic hydrocarbons discussed thus far: benzene, toluene, the three xylenes, naphthalene, anthracene, and phenanthrene.

Distillation of coal tar provides four volatile fractions as well as the pitch that is used for surfacing roads and in the manufacture of "asphalt" roofing (Figure 31-12). Eight aromatic hydrocarbons are obtained in significant amounts by efficient fractional distillation of the "light oil" fraction (Table 31-6).

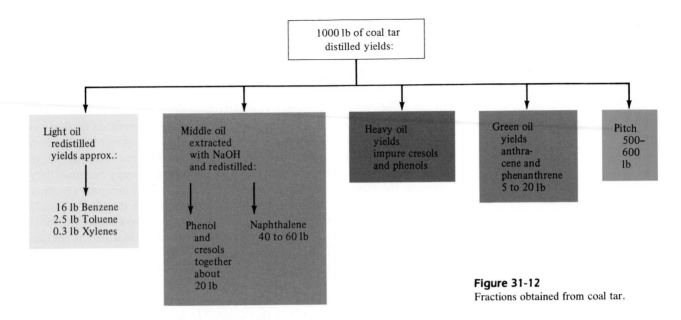

Figure 31-12
Fractions obtained from coal tar.

31-7 Hydrocarbons—A Summary

Figure 31-13 shows a classification of hydrocarbons.

SOME EXAMPLES

Saturated hydrocarbons
Contain only C–C
and
C–H single bonds
(Sections 31–1 and 31–2)

Alkanes, C_nH_{2n+2}
Contain no rings

methane, ethane,
n-hexane,
2-methylhexane,
2,2-dimethylpentane

Cycloalkanes, C_nH_{2n}
Contain rings

cyclopropane,
cyclopentane,
cyclohexane

Aliphatic hydrocarbons
Contain C–C single,
double, and triple bonds,
but no aromatic rings
(Sections 31–1 to 31–4)

Alkenes
Contain one or more
C=C double bonds
(Section 31–3)

Acyclic alkenes, C_nH_{2n}
ethylene, 1-butene,
2-methy-1-butene

Cycloalkenes, C_nH_{2n-2}
cyclopentene,
cyclohexene

Polyenes
Contain more than one
C=C per molecule—
e.g., 1,3-butadiene

Unsaturated hydrocarbons
Contain one or more
C=C double or
C≡C triple bonds
(Sections 31–3 and 31–4)

Alkynes, C_nH_{2n-2}
Contain one or more
C≡C triple bonds
(Section 31–4)

acetylene, 1-butyne,
3-methyl-1-butyne

Hydrocarbons
(contain C, H only)
(Sections 31–1 to 31–6)

Aromatic hydrocarbons
(Sections 31–5 and 31–6)

benzene, naphthalene,
anthracene, toluene,
p-xylene

Figure 31-13
A classification of hydrocarbons.

Functional Groups

The study of organic chemistry is greatly simplified by considering hydrocarbons as parent compounds and describing other compounds as derived from them. In general, an organic molecule consists of a skeleton of carbon atoms with special groups of atoms within or attached to that skeleton. These special groups of atoms are often called **functional groups** because they represent the most common sites of chemical reactivity (function). The only functional groups that are possible in hydrocarbons are double and triple (i.e., pi) bonds. Atoms other than C and H are called **heteroatoms**, the most common being O, N, S, P, and the halogens. Most functional groups contain one or more heteroatoms.

In the next several sections, we shall introduce some common functional groups that contain heteroatoms and learn a little about the resulting classes of compounds. We shall continue to represent hydrocarbon groups with the symbol R—. We commonly use that symbol to represent either an aliphatic (e.g., alkyl) or an aromatic (e.g., an aryl such as phenyl) group. When we specifically mean an aryl group, we shall use the symbol Ar—.

As you study the following sections, you may wish to refer to the summary in Section 31-15.

**Table 31-7
Some Organic Halides**

Formula	Structural Formula	Normal bp (°C)	IUPAC Name	Common Name
CH_3Cl	H—C—Cl (with H above and below)	23.8	chloromethane	methyl chloride
CH_2Cl_2	Cl, H—C—Cl, H	40.2	dichloromethane	methylene chloride
$CHCl_3$	Cl, H—C—Cl, Cl	61	trichloromethane	chloroform
CCl_4	Cl, Cl—C—Cl, Cl	76.8	tetrachloromethane	carbon tetrachloride
$CHCl_2Br$	Cl, H—C—Cl, Br	90	bromodichloromethane	—

31-8 Organic Halides

Almost any hydrogen atom in a hydrocarbon can be replaced by a halogen atom to give a stable compound. Table 31-7 shows some organic halides and their names.

In the IUPAC naming system, the organic halides are named as *halo-* derivatives of the parent hydrocarbons. The prefix *halo-* can be *fluoro-*, *chloro-*, *bromo-*, or *iodo-*. Simple alkyl chlorides are sometimes given common names as alkyl derivatives of the hydrogen halides. For instance, the IUPAC name for CH_3CH_2—Cl is chloroethane; it is commonly called ethyl chloride by analogy to H—Cl, hydrogen chloride.

A carbon atom can be bonded to as many as four halogen atoms, so an enormous number of organic halides can exist. Completely fluorinated compounds are known as **fluorocarbons** or sometimes *perfluorocarbons*. The fluorocarbons are even less reactive than hydrocarbons. Saturated compounds in which all H atoms have been replaced by some combination of Cl and F atoms are called *chlorofluorocarbons* or sometimes **freons**. These compounds have been widely used as refrigerants and as propellants in

Freon is a Du Pont trademark for certain chlorofluorocarbons; other companies' related products are known by other names. Typical freons are trichlorofluoromethane, $CFCl_3$ (called Freon-11), and dichlorodifluoromethane, CF_2Cl_2 (called Freon-12).

Table 31-7 (*continued*)

Formula	Structural Formula	Normal bp (°C)	IUPAC Name	Common Name
$(CH_3)_2CHI$		89.5	2-iodopropane	isopropyl iodide
$CH_3ClC{=}CHCH_2CH_2Cl$		40	2,5-dichloro-*cis*-2-pentene; *trans* isomer is also possible	—
C_5H_7Cl		25	3-chlorocyclopentene is shown; other isomers are also possible	—
C_6H_5I		118	iodobenzene	phenyl iodide
C_6H_4ClBr		204	1-bromo-2-chlorobenzene is shown; other isomers are also possible	*o*-bromochlorobenzene

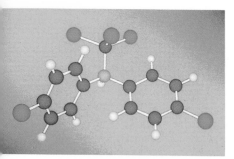

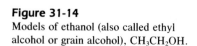

DDT, an organic halide whose full name is 1,1,1-trichloro-2,2-bis(*p*-chlorophenyl)ethane. It was introduced as an insecticide in the 1940s and widely used until the early 1970s. This compound was almost entirely responsible for the virtual eradication of malaria, once the world's most widespread disease, by killing the *Anopheles* mosquito that carried the disease. Unfortunately, this toxic substance stays in the environment for a long time. As a result, it is carried up the food chain and concentrates in the bodies of higher animals. It is especially detrimental in the life cycles of birds. Only about 2% of the DDT in the environment is degraded each year. Even though its use has been banned for many years, it will still take many years before the last traces disappear from the soil.

The simplest phenol is called phenol. The most common member of a class of compounds is often called by the class name. Salt, sugar, alcohol, and phenol are examples.

aerosol cans. However, the release of chlorofluorocarbons into the atmosphere has been shown to be quite damaging to the earth's ozone layer. Since January 1978 the use of chlorofluorocarbons in aerosol cans in the United States has been banned, and efforts to develop both controls for existing chlorofluorocarbons and suitable replacements continue.

31-9 Alcohols and Phenols

Alcohols and phenols contain the hydroxyl group (—O—H) as their functional group. **Alcohols** may be considered to be derived from saturated or unsaturated hydrocarbons by the replacement of at least one H atom by a hydroxyl group. The properties of alcohols result from a hydroxyl group attached to an *aliphatic* carbon atom, —C—O—H. Ethanol (ethyl alcohol) is the most common example (Figure 31-14).

When a hydrogen atom on an aromatic ring is replaced by a hydroxyl group (Figure 31-15), the resulting compound is known as a **phenol**. Such compounds behave more like acids than alcohols. Alternatively, we may view alcohols and phenols as derivatives of water in which one H atom has been replaced by an organic group:

$$H—O—H \qquad H—\overset{\displaystyle H}{\underset{\displaystyle H}{C}}—\overset{\displaystyle H}{\underset{\displaystyle H}{C}}—O—H$$

water   ethanol   phenol

Indeed, this is a better view. The structure of water was discussed in Section 8-9. The hydroxyl group in an alcohol or a phenol is covalently bonded to a carbon atom, but the O—H bond is quite polar. The oxygen atom has two unshared electron pairs, and the C—O—H bond angle is nearly 104.5°.

The presence of a bonded alkyl or aryl group changes the properties of the —OH group. *Alcohols* are so very weakly acidic that they are thought of as neutral compounds. *Phenols* are weakly acidic.

Many properties of alcohols depend on whether the hydroxyl group is attached to a carbon that is bonded to *one*, *two*, or *three* other carbon atoms.

Primary alcohols contain one R group, **secondary alcohols** contain two R groups, and **tertiary alcohols** contain three R groups bonded to the carbon atom to which the —OH group is attached.

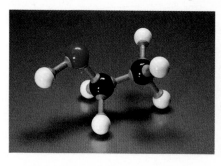

Figure 31-14
Models of ethanol (also called ethyl alcohol or grain alcohol), CH_3CH_2OH.

Representing alkyl groups as R, we can illustrate the three classes of alcohols. The R groups may be the same or different.

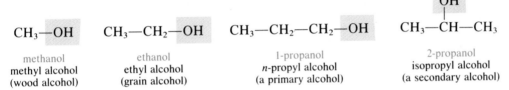

a primary (1°) alcohol   a secondary (2°) alcohol   a tertiary (3°) alcohol

In writing organic structures, we often use primes when we wish to specify that the alkyl groups might be different, e.g., R, R', R''.

Naming Alcohols and Phenols

The systematic name of an alcohol consists of the characteristic stem plus an *-ol* ending. A numeric prefix indicates the position of the —OH group on a chain of three or more carbon atoms.

$$CH_3—OH \qquad CH_3—CH_2—OH \qquad CH_3—CH_2—CH_2—OH \qquad CH_3—\overset{\displaystyle OH}{\underset{\displaystyle |}{CH}}—CH_3$$

methanol
methyl alcohol
(wood alcohol)

ethanol
ethyl alcohol
(grain alcohol)

1-propanol
n-propyl alcohol
(a primary alcohol)

2-propanol
isopropyl alcohol
(a secondary alcohol)

There are four structural isomers of the saturated acyclic four-carbon alcohols with one —OH per molecule.

Acyclic compounds contain no rings.

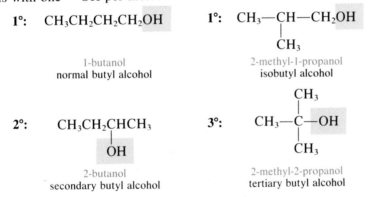

1°: $CH_3CH_2CH_2CH_2OH$

1-butanol
normal butyl alcohol

1°: $CH_3—CH—CH_2OH$
$\qquad\qquad |$
$\qquad\quad CH_3$

2-methyl-1-propanol
isobutyl alcohol

2°: $CH_3CH_2CHCH_3$
$\qquad\qquad\quad |$
$\qquad\qquad OH$

2-butanol
secondary butyl alcohol

3°: $CH_3—\overset{CH_3}{\underset{CH_3}{C}}—OH$

2-methyl-2-propanol
tertiary butyl alcohol

There are eight structural isomers of the analogous five-carbon alcohols. They are often called "amyl" or "pentyl" alcohols. Two examples are

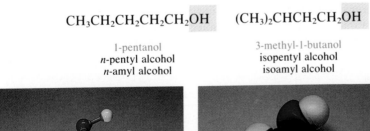

$$CH_3CH_2CH_2CH_2CH_2OH \qquad (CH_3)_2CHCH_2CH_2OH$$

1-pentanol
n-pentyl alcohol
n-amyl alcohol

3-methyl-1-butanol
isopentyl alcohol
isoamyl alcohol

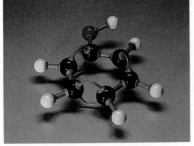

Figure 31-15
Models of phenol, C_6H_5OH.

The **polyhydric alcohols** contain more than one —OH group per molecule. Those containing two OH groups per molecule are called **glycols**. Important examples of polyhydric alcohols include

Polyhydric alcohols are used in permanent antifreeze and in cosmetics.

The *o-*, *m-*, and *p-* notation was introduced in Section 31-6.

CH₂—CH₂	CH₃—CH—CH₂	CH₂—CH—CH₂

1,2-ethanediol
ethylene glycol
(the major ingredient in permanent antifreeze)

1,2-propanediol
propylene glycol

1,2,3-propanetriol
glycerine or glycerol
(a moisturizer in cosmetics)

Phenols are usually referred to by their common names. Examples are

resorcinol   hydroquinone   *o*-cresol   *m*-cresol   *p*-cresol

As you might guess, cresols occur in "creosote," a wood preservative.

Physical Properties of Alcohols and Phenols

The hydroxyl group, —OH, is quite polar, whereas alkyl groups, R, are nonpolar. The properties of alcohols depend on two factors: (1) the number of hydroxyl groups per molecule and (2) the size of the nonpolar portion of the molecule.

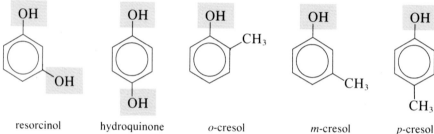

ROH: nonpolar part   polar part

The low-molecular-weight monohydric alcohols are miscible with water in all proportions. Beginning with the four butyl alcohols, solubility in water decreases rapidly with increasing molecular weight. This is because the nonpolar parts of such molecules are much larger than the polar parts. Many

Table 31-8
Physical Properties of Normal Primary Alcohols

Name	Formula	Normal bp (°C)	Solubility in H₂O (g/100 g at 20°C)
methanol	CH_3OH	65	Completely miscible
ethanol	CH_3CH_2OH	78.5	Completely miscible
1-propanol	$CH_3CH_2CH_2OH$	97	Completely miscible
1-butanol	$CH_3CH_2CH_2CH_2OH$	117.7	7.9
1-pentanol	$CH_3CH_2CH_2CH_2CH_2OH$	137.9	2.7
1-hexanol	$CH_3CH_2CH_2CH_2CH_2CH_2OH$	155.8	0.59

polyhydric alcohols are very soluble in water because they contain two or more polar —OH groups per molecule.

Table 31-8 shows that the boiling points of normal primary alcohols increase, and their solubilities in water decrease, with increasing molecular weight. The boiling points of the alcohols are much higher than those of the corresponding alkanes (Table 31-2) because of the hydrogen bonding of the hydroxyl groups.

Most phenols are solids at 25°C. Phenols are only slightly soluble in water unless they contain other functional groups that interact with water.

Some Uses of Alcohols and Phenols

Many alcohols and phenols have considerable commercial importance. Methanol, CH_3OH, was formerly produced by the destructive distillation of wood and is sometimes called wood alcohol. It is now produced in large quantities from carbon monoxide and hydrogen. It is extensively used as a solvent for varnishes and shellacs, as the starting material in the manufacture of formaldehyde (Section 31-14), as a temporary antifreeze (bp = 65°C), and as a fuel additive. It is very toxic and causes permanent blindness when taken internally.

Ethanol, CH_3CH_2OH, also known as ethyl alcohol or grain alcohol, was first prepared by fermentation *a long time ago*. The most ancient written literature refers to beverages that were obviously alcoholic! The syrupy residue from the purification of cane sugar (sucrose) is blackstrap molasses. Its fermentation is one important source of ethanol. The starches in grains, potatoes, and similar foodstuffs can be converted into sugar by malt; this is followed by fermentation to produce ethanol, which is the most important industrial alcohol. Like the other shorter-chain alcohols, ethanol participates in hydrogen bonding and is completely miscible with water.

Fermentation is an enzymatic process carried out by certain kinds of bacteria.

Many simple alcohols are important raw materials in the industrial synthesis of polymers, fibers, explosives, plastics, and pharmaceutical products. Phenols are found in plant products such as flower pigments, tanning agents, and wood. They are widely used in the preparation of plastics and dyes. Dilute aqueous solutions of phenols are used as antiseptics and disinfectants. Polyhydric alcohols are useful for their relatively high boiling points. For instance, glycerine is used as a wetting agent in cosmetic preparations. Ethylene glycol (bp = 197°C), which is completely miscible with water, is used in commercial permanent antifreeze.

H_2C—OH
H_2C—OH
ethylene glycol

31-10 Ethers

When the word "ether" is mentioned, most people think of the well-known anesthetic, diethyl ether. There are many ethers. Their uses range from artificial flavorings to refrigerants and solvents. An **ether** is a compound in which an O atom is bonded to two organic groups:

$$-\overset{|}{\underset{|}{C}}-O-\overset{|}{\underset{|}{C}}-$$

Alcohols are considered derivatives of water in which one H atom has been replaced by an organic group. Ethers may be considered derivatives of water in which both H atoms have been replaced by organic groups.

$$H—O—H \qquad R—O—H \qquad R—O—R'$$

water   alcohol   ether

However, the similarity is only structural because ethers are not very polar and are chemically rather unreactive. (We shall not discuss their reactions here.) In fact, their physical properties are similar to those of the corresponding alkanes; e.g., CH_3OCH_3 is like $CH_3CH_2CH_3$.

Three kinds of ethers are known: (1) aliphatic, (2) aromatic, and (3) mixed. Common names are used for ethers in most cases.

$$H_3C—O—CH_3 \qquad H_3C—O—CH_2CH_3$$

methoxymethane
dimethyl ether
(an aliphatic ether)

methoxyethane
methyl ethyl ether
(an aliphatic ether)

methoxybenzene
methyl phenyl ether
anisole
(a mixed ether)

phenoxybenzene
diphenyl ether
(an aromatic ether)

Models of diethyl ether (top) and methyl phenyl ether (bottom).

Diethyl ether is a very-low-boiling liquid (bp = 35°C). Dimethyl ether is a gas that is used as a refrigerant. The aliphatic ethers of higher molecular weights are liquids, and the aromatic ethers are liquids and solids.

Even ethers of low molecular weight are only slightly soluble in water. Diethyl ether is an excellent solvent for organic compounds. It is widely used to extract organic compounds from plants and other natural sources.

Ethers burn readily, and care must be exercised to avoid fires when ethers are used. At room temperature diethyl ether is oxidized by oxygen in the air to a nonvolatile, explosive peroxide. Thus, ethereal solutions should never be evaporated to dryness, because of the danger of peroxide explosions, unless proper precautionary steps have been taken to destroy all peroxides in advance.

31-11 Amines

The **amines** are derivatives of ammonia in which one or more hydrogen atoms have been replaced by alkyl or aryl groups. Many low-molecular-weight amines are gases or low-boiling liquids (Table 31-9). Amines are basic compounds (Table 18-6; Section 32-7). Their basicity differs, depending on the nature of the organic substituents. The aliphatic amines of low molecular weight are soluble in water. Aliphatic diamines of fairly high molecular

Ammonia acts as a Lewis base because there is one unshared pair of electrons on the N atom (Section 10-10).

Table 31-9
Boiling Points of Some Amines

Name	Formula	Boiling Point (°C)
ammonia	NH_3	−33.4
methylamine	CH_3NH_2	−6.5
dimethylamine	$(CH_3)_2NH$	7.4
trimethylamine	$(CH_3)_3N$	3.5
ethylamine	$CH_3CH_2NH_2$	16.6
aniline	$C_6H_5NH_2$	184
ethylenediamine	$H_2NCH_2CH_2NH_2$	116.5
pyridine	C_5H_5N	115.3

weight are soluble in water because each molecule contains two highly polar —NH_2 groups that form hydrogen bonds with water.

The odors of amines are quite unpleasant; many of the malodorous compounds that are released as fish decay are simple amines. Amines of high molecular weight are nonvolatile, so they have little odor. One of the materials used to manufacture nylon, hexamethylenediamine, is an aliphatic amine. Many aromatic amines are used to prepare organic dyes that are widely used in industrial societies. Amines are also used to produce many medicinal products, including local anesthetics and sulfa drugs.

Amines are widely distributed in nature in the form of amino acids and proteins, which are found in all higher animal forms, and in alkaloids, which are found in most plants. Some of these substances are fundamental building blocks of animal tissue, and minute amounts of others have dramatic physiological effects, both harmful and beneficial. Countless other biologically important substances, including many vitamins, antibiotics, and drugs, contain amino groups, —NR_2 (where R can represent an H, alkyl, or aryl group).

nicotine

strychnine

Structure and Naming of Amines

There are three classes of amines, depending on whether one, two, or three hydrogen atoms have been replaced by organic groups. They are called primary, secondary, and tertiary amines, respectively.

NH_3 RNH_2 R_2NH R_3N

ammonia

methylamine
(a primary amine)

dimethylamine
(a secondary amine)

trimethylamine
(a tertiary amine)

Models of these four molecules are shown in Figure 31-16.

Figure 31-16
Models of (a) ammonia,
(b) methylamine, (c) dimethylamine,
and (d) trimethylamine.

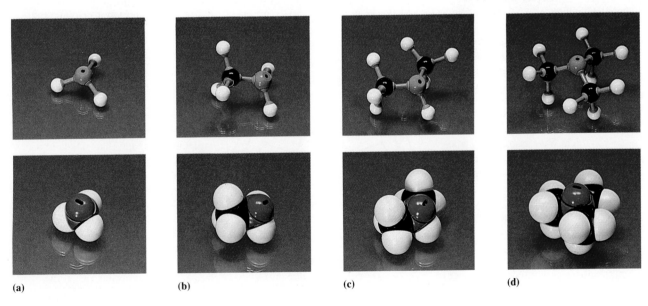

(a)　　　　(b)　　　　(c)　　　　(d)

Chemistry in Use. . .
C₆₀ and the Fullerenes

Some of the most unusual "organic" molecules ever imagined were discovered in the 1980s when molecular beam cluster experiments (see page 848) were used to vaporize solid carbon. A variety of molecules that contained only carbon atoms were produced. In the same way that molecules containing only metal atoms are called *metal clusters,* these species are called *carbon clusters.* Analysis of these carbon molecules of different sizes shows that the molecule containing 60 carbon atoms, C_{60}, is formed far more readily than other sizes and is incredibly stable.

C_{60} was first identified in molecular beam experiments by Professor Richard Smalley and his research team at Rice University. From these early experiments they could only speculate why C_{60} was so stable and so different from the other clusters. Smalley argued that the only structure that could be so stable would be a sphere of carbon formed of interconnected five- and six-membered rings (Figure a). Smalley named the molecule "Buckminsterfullerene" after the architect Buckminster Fuller, who specialized in geodesic dome designs. The structure that he proposed has the same shape as a soccer ball, and the nickname "Buckyball" soon became associated with C_{60}. Unfortunately, because of the small amount of material produced in molecular beam experiments, Smalley could not actually measure the structure to prove that his model was right.

C_{60} might have remained a laboratory curiosity except for an exciting discovery by Donald Huffman (University of Arizona) and Wolfgang Krätschmer (Max Planck Institute for Nuclear Physics, Heidelberg) in 1990. They found that an electrical discharge, or arc, made with graphite electrodes in a helium atmosphere generated carbon soot

with unusual properties. Part of that soot dissolved in ordinary organic solvents such as benzene and toluene; normal soot does not dissolve in these solvents. After filtering away the insoluble material and evaporating the solvent, the scientists isolated a yellowish brown powder. With gram quantities available, traditional analysis using an array of specialized instrumental techniques (nuclear magnetic resonance and infrared spectroscopy, X-ray crystallography) became possible. To everyone's delight (especially Smalley's), the powder contained mostly C_{60}, and it had the structure that he had proposed almost ten years earlier! C_{60} thus became the first cluster of a pure element ever to be isolated and collected in quantities great enough for traditional chemical experiments. These same experiments also produced several larger, less symmetrical, carbon cage molecules (e.g., C_{70} and C_{84}). Taken together, C_{60} and its analogs are now referred to as the *fullerenes.*

The highly symmetrical structure of C_{60} helps to explain many of its unusual properties. It is almost a perfect sphere, with 60 atoms arranged in 20 hexagons and 12 pentagons (32 faces in all). If you compare this structure to Figure (c) on page 849, you can see that the twelve surface metal atoms there and the 12 pentagons in C_{60} are arranged in the same way. Thus, C_{60} also has an *icosahedral* structure.

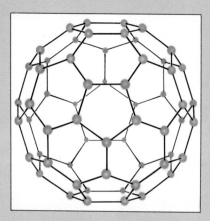

(a) The isosahedral structure of C_{60}.

Each of the 60 carbon atoms is sp^2 hybridized and occupies an identical site in the cluster. Each has two single bonds and one double bond to its neighbors and is located at the juncture of one five-membered ring and two six-membered rings. This structure is very different from those of other forms of carbon. For example, diamond has tetrahedral bonding to four nearest neighbors. Graphite is more similar to C_{60}, with an infinite array of planar six-membered rings. However, an array of six-membered rings cannot be bent into a closed cage because of the strain on the bonds. Substitution of precisely the correct number of five-membered rings, as occurs in C_{60}, relieves this strain.

There are numerous equivalent ways to draw the chemical bonds in C_{60}. Equivalent bonding arrangements, known as *resonance structures,* are also found for molecules such as benzene and give them their "aromatic" character. C_{60} is therefore also an aromatic molecule. Empirical rules have been developed from chemical bonding theory to predict which other molecules similar to C_{60} might be found. In addition to C_{60}, carbon molecules with a multiple of sixty atoms, such as the clusters C_{120}, C_{240}, C_{540}, and C_{960}, are also predicted to be stable aromatic molecules with highly symmetrical open-cage structures. Attempts are underway to isolate and characterize these larger fullerenes.

The spherical cage structure makes C_{60} and the other fullerenes fascinating candidates for all kinds of applications in chemistry and for the preparation of new solid materials. For example, crystals of C_{60} have been prepared in which the molecules arrange themselves in a hexagonal close-packed structure. When alkali metal atoms such as cesium or rubidium are added into the gaps between the balls, the resulting compound is a *superconductor.* The open interior of C_{60} is a cavity about 5 Å wide, large enough to contain other atoms—in particular, metals. Numerous research groups are trying to find the conditions necessary for encapsulating atoms with C_{60} or with other fullerenes that have larger cavities. The unusual shape of the C_{60} molecule is of special interest in situations in which molecular shapes determine chemical activity, as in biological molecules, pharmaceutical drugs, and polymers. To investigate these kinds of applications, other functional groups or reactive organic systems have already been attached to C_{60}. Long chains of the form (C_{60})—R—(C_{60})—R . . . have also been constructed. This peculiar new molecule, C_{60}, and other members of its family are rapidly emerging from the realm of molecular beams into the mainstream of practical chemistry.

Professor Michael A. Duncan
The University of Georgia

The systematic names of amines are based on consideration of the compounds as derivatives of ammonia. Amines of more complex structure are sometimes named as derivatives of the parent hydrocarbon, with the term *amino-* used as a prefix to describe —NH_2.

2-aminobutane or *sec*-butylamine   1-amino-3-ethylcyclohexane

Aniline is the simplest aromatic amine. Many aromatic amines are named as derivatives of aniline.

aniline   3,4,5-tribromoaniline   N,N-dimethylaniline
(primary)

Heterocyclic amines contain nitrogen as a part of the ring, bound to two carbon atoms. Many of these amines are found in coal tar and a variety of natural products. Some aromatic and heterocyclic amines are always called by their common names.

pyridine   pyrrole   quinoline   purine   pyrimidine
(tertiary)   (secondary)   (tertiary)

Genes, the units of chromosomes that carry hereditary characteristics, are essentially long stretches of double helical deoxyribonucleic acid, or DNA. DNA is composed of four fundamental *nucleotide bases*: adenine, guanine, cytosine, and thymine. The first two are modified purines, and the latter two are modified pyrimidines. The sequence of these building blocks

A space-filling model of a portion of the DNA double helical structure.

1061

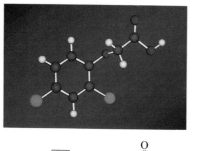

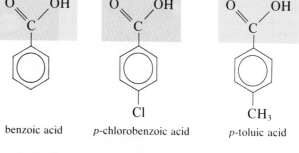

(a)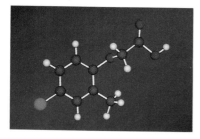

benzoic acid   p-chlorobenzoic acid   p-toluic acid

Many reactions of carboxylic acids involve displacement of the —OH group by another atom or group of atoms. We find it useful to name the non-OH portions of acid molecules because they occur in many compounds. Such compounds are thought of as derivatives of carboxylic acids.

$$R—C—O—H \qquad R—C— \qquad Ar—C—O—H \qquad Ar—C—$$

an aliphatic   an aliphatic   an aromatic   an aromatic
carboxylic acid   acyl group   carboxylic acid   acyl group

Acyl groups are named as derivatives of the parent acid by dropping *-ic acid* and adding *-yl* to the characteristic stem. Some examples are

$$CH_3—C \qquad CH_3CH_2—C \qquad \bigcirc—C$$

acetyl group   propionyl group   benzoyl group

(b)

These two derivatives of phenoxyacetic acids act as herbicides (weed killers) by overstimulating the plant's growth system.

Although many carboxylic acids occur in the free state in nature, many occur as amides or esters (Sections 31-13). Amino acids are substituted carboxylic acids with the general structure

$$R—\overset{H}{\underset{NH_2}{C}}—\overset{O}{C}—OH$$

where R can be either an alkyl or an aryl group. Amino acids are the components of proteins, which make up the muscle and tissue of animals. Many other acids are important in the metabolism and synthesis of fats by enzyme systems. Acetic acid (the acid in vinegar) is the end product in the fermentation of most agricultural products. It is the fundamental unit used by living organisms in the biosynthesis of such widely diverse classes of natural products as long-chain fatty acids, natural rubber, and steroid hormones. It is also a powerful solvent, and an important reagent in the preparation of pharmaceuticals, plastics, artificial fibers, and coatings. Phthalic acid and adipic acid are used in the production of synthetic polymers that are used as fibers (e.g., Dacron and nylon, Section 32-11).

Crystals of glycine viewed under polarized light. Glycine, the simplest amino acid, has the structure shown in the text, with R = H.

31-13 Some Derivatives of Carboxylic Acids

Four important classes of acid derivatives are formed by the replacement of the hydroxyl group by another atom or group of atoms. Each of these derivatives contains an acyl group.

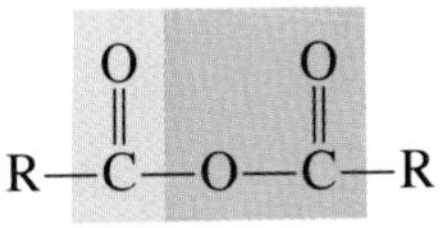

an acid anhydride 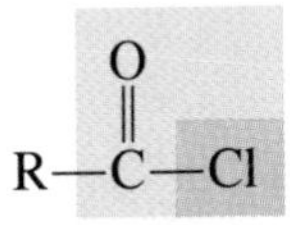   an acyl chloride (an acid chloride)   an ester 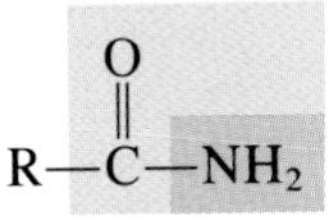   an amide

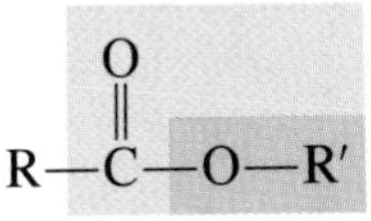

Aromatic compounds of these types (with R = aryl groups) are encountered frequently.

Acid Anhydrides

An acid anhydride can be thought of as the result of removing one molecule of water from two carboxylic acid groups.

The structural relationship between monocarboxylic acids and their anhydrides is

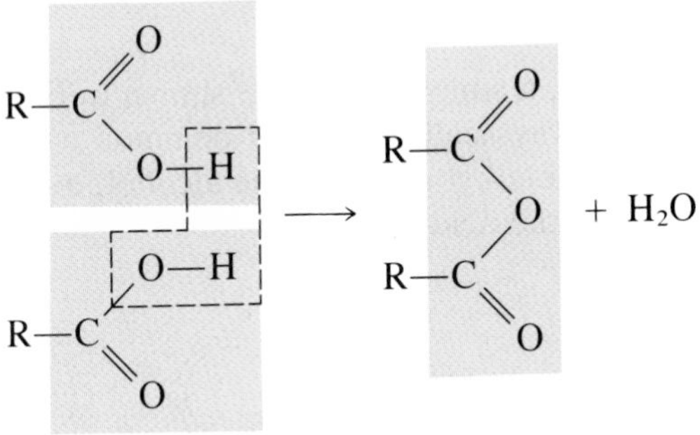

two molecules of a carboxylic acid   one molecule of acid anhydride

Acid anhydrides are usually prepared by indirect methods. This illustration shows the structural relationship that is important for our purposes.

The anhydrides are named by replacing the word "acid" in the name of the parent acid with the word "anhydride." Examples of acids and their anhydrides are

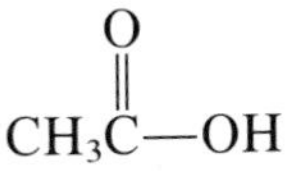

acetic acid 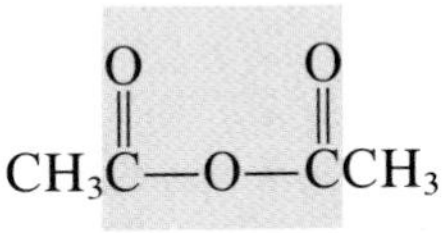   acetic anhydride

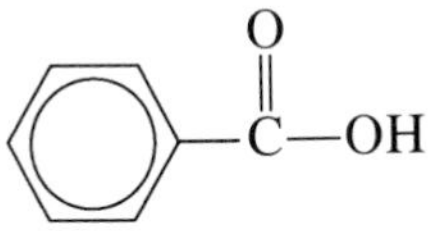

benzoic acid 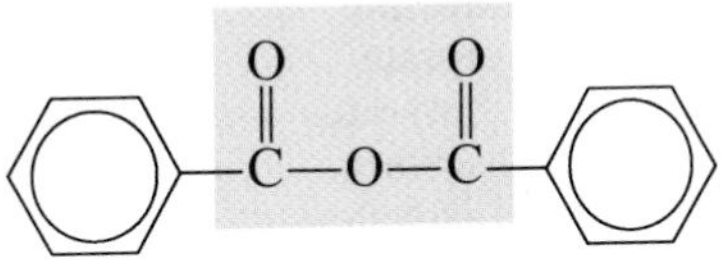   benzoic anhydride

Acyl Halides (Acid Halides)

The **acyl halides**, sometimes called **acid halides**, are structurally related to carboxylic acids by the replacement of the OH group by a halogen, most often Cl. They are usually named by combining the stems of the common names of the carboxylic acids with the suffix *-yl* and then adding the name of the halide ion. Examples are

$$CH_3-\overset{\overset{\displaystyle O}{\|}}{C}-Cl \qquad CH_3CH_2CH_2-\overset{\overset{\displaystyle O}{\|}}{C}-F \qquad \text{(benzene ring)}-\overset{\overset{\displaystyle O}{\|}}{C}-Cl$$

acetyl chloride   butyryl fluoride   benzoyl chloride

Acid halides are very reactive and have not been observed in nature.

As we shall see in Section 32-9, one method of forming esters involves acid-catalyzed reaction of an alcohol with a carboxylic acid.

Esters

Esters can be thought of as the result of removing one molecule of water from a carboxylic acid and an alcohol. Removing a molecule of water from

$$CH_3\overset{\overset{\displaystyle O}{\|}}{C}-OH \qquad \text{and} \qquad HO-CH_2CH_3 \qquad \text{gives} \qquad CH_3\overset{\overset{\displaystyle O}{\|}}{C}-OCH_2CH_3$$

acetic acid   ethyl alcohol   ethyl acetate

Models of ethyl acetate, a simple ester, are shown in Figure 31-18.

Esters are nearly always called by their common names. These consist of, first, the name of the alkyl group in the alcohol, and then the name of the anion derived from the acid.

$$CH_3CH_2\overset{\overset{\displaystyle O}{\|}}{C}-OC(CH_3)_3 \qquad CH_3\overset{\overset{\displaystyle O}{\|}}{C}-O-\text{(benzene ring)} \qquad \text{(benzene ring)}-\overset{\overset{\displaystyle O}{\|}}{C}-OCH_3$$

tert-butyl propionate   phenyl acetate   methyl benzoate

Because of their inability to form hydrogen bonds, esters tend to be liquids with boiling points much lower than those of carboxylic acids of similar molecular weight.

Most simple esters are pleasant-smelling substances. They are responsible for the flavors and fragrances of most fruits and flowers and many of the

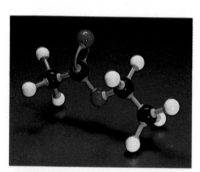

Figure 31-18
Models of ethyl acetate,

$$CH_3-\overset{\overset{\displaystyle O}{\|}}{C}-O-CH_2CH_3,$$ an ester. The

$$CH_3-\overset{\overset{\displaystyle O}{\|}}{C}-$$ fragment is derived from

acetic acid, the parent acid; the $O-CH_2CH_3$ fragment is derived from ethanol, the parent alcohol.

Table 31-12 Some Common Esters		
Ester	**Formula**	**Odor of**
n-butyl acetate	$CH_3COOC_4H_9$	bananas
ethyl butyrate	$C_3H_7COOC_2H_5$	pineapples
n-amyl butyrate	$C_3H_7COOC_5H_{11}$	apricots
n-octyl acetate	$CH_3COOC_8H_{17}$	oranges
isoamyl isovalerate	$C_4H_9COOC_5H_{11}$	apples
methyl salicylate	$C_6H_4(OH)(COOCH_3)$	oil of wintergreen
methyl anthranilate	$C_6H_4(NH_2)(COOCH_3)$	grapes

artificial fruit flavorings that are used in cakes, candies, and ice cream (Table 31-12). Esters of low molecular weight are excellent solvents for nonpolar compounds. Ethyl acetate is an excellent solvent that gives many nail polish removers their characteristic odor.

Fats (solids) and **oils** (liquids) are esters of glycerol and aliphatic acids of high molecular weight. "Fatty acids" are all organic acids that occur in fats and oils (as esters). Fats and oils have the general formula

Apple blossoms.

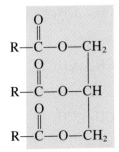

The fatty acid portions, $R-\overset{O}{\underset{\|}{C}}-$, may be saturated or unsaturated. The R's may be the same or different groups.

Fats are solid esters of glycerol and (mostly) saturated acids. Oils are liquid esters that are derived primarily from unsaturated acids and glycerol. The acid portion of a fat usually contains an even number of carbon atoms, often 16 or 18. The acids that occur most frequently in fats and oils are

Most natural fatty acids contain even numbers of carbon atoms because they are synthesized in the body from two-carbon acetyl groups.

butyric	$CH_3CH_2CH_2COOH$
lauric	$CH_3(CH_2)_{10}COOH$
myristic	$CH_3(CH_2)_{12}COOH$
palmitic	$CH_3(CH_2)_{14}COOH$
stearic	$CH_3(CH_2)_{16}COOH$
oleic	$CH_3(CH_2)_7CH=CH(CH_2)_7COOH$
linolenic	$CH_3CH_2CH=CHCH_2CH=CHCH_2CH=CH(CH_2)_7COOH$
ricinoleic	$CH_3(CH_2)_5CHOHCH_2CH=CH(CH_2)_7COOH$

Figure 31-19 is a model of stearic acid, a long-chain saturated fatty acid.

Glycerol is

$$HO-CH_2$$
$$HO-CH$$
$$HO-CH_2$$

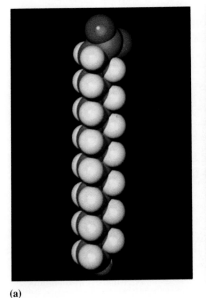

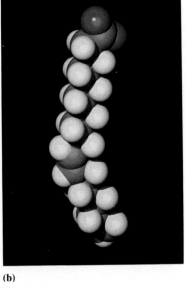

(a)   (b)

Figure 31-19
Models of long-chain fatty acids. The saturated fatty acids (a) are linear and tend to pack, like sticks of wood, to form solid masses in blood vessels, thereby constricting them. The *trans* unsaturated fatty acids have a slight Z-shaped kink in the chain, but are also essentially linear molecules. By contrast, *cis* unsaturated fatty acids (b) are bent and so do not pack as well as linear structures and do not collect in blood vessels as readily. Many natural vegetable fats contain esters of *cis* unsaturated fatty acids or polyunsaturated fatty acids. Health problems associated with saturated fatty acids can be decreased by eating less animal fat, butter, and lard. Problems due to *trans* fatty acids are reduced by avoiding processed vegetable fats.

Naturally occurring fats and oils are mixtures of many different esters. Milk fat, lard, and tallow are familiar, important fats. Soybean oil, cottonseed oil, linseed oil, palm oil, and coconut oil are examples of important oils.

The triesters of glycerol are called glycerides. *Simple glycerides* are esters in which all three R groups are identical. Two examples are

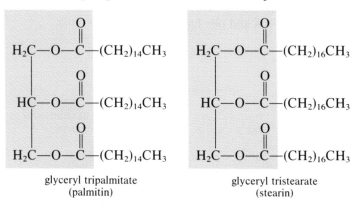

glyceryl tripalmitate
(palmitin)

glyceryl tristearate
(stearin)

Honeybees produce the wax to build their honeycombs.

Glycerides are frequently called by their common names, indicated in parentheses in the examples above. The common name is the characteristic stem for the parent acid plus an *-in* ending.

Waxes are esters of fatty acids and alcohols other than glycerol. Most are derived from long-chain fatty acids and long-chain monohydric alcohols. Both usually contain even numbers of carbon atoms. Beeswax is largely $C_{15}H_{31}COOC_{30}H_{61}$; carnauba wax contains $C_{25}H_{51}COOC_{30}H_{61}$. Both are esters of myricyl alcohol, $C_{30}H_{61}OH$.

Amides

Amides are thought of as derivatives of organic acids and primary or secondary amines. Amides contain the

$$-\overset{\overset{\displaystyle O}{\|}}{C}-N\diagup$$

grouping of atoms. They are named as derivatives of the corresponding carboxylic acids, the suffix *-amide* being substituted for *-ic acid* or *-oic acid* in the name of the parent acid.

$$CH_3\overset{\overset{\displaystyle O}{\|}}{C}-NH_2 \qquad \underset{\text{benzamide}}{\bigcirc\!\!-\overset{\overset{\displaystyle O}{\|}}{C}-NH_2}$$

acetamide   benzamide

The presence of alkyl or aryl substituents attached to nitrogen is designated by prefixing the letter N and the name of the substituent to the name of the unsubstituted amide.

$$CH_3\overset{\overset{\displaystyle O}{\|}}{C}-\overset{\overset{\displaystyle CH_3}{|}}{N}-CH_3 \qquad CH_3\overset{\overset{\displaystyle O}{\|}}{C}-\overset{\overset{\displaystyle H}{|}}{N}-\bigcirc \qquad$$

N,N-dimethylacetamide   N-phenylacetamide
                          (acetanilide)   N,N-diethyl-*m*-toluamide

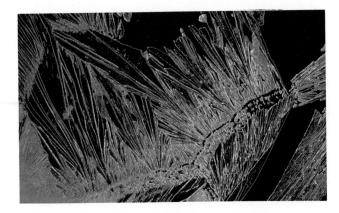

Crystals of acetaminophen (Tylenol) viewed under polarized light. The structure of acetaminophen is

$$CH_3-\overset{\overset{\displaystyle O}{\|}}{C}-\underset{\underset{\displaystyle H}{|}}{N}-\hspace{-0.5em}\bigcirc\hspace{-0.5em}-OH$$

Unsubstituted amides (with the exception of formamide, $HCONH_2$) are crystalline solids at room temperature, with melting and boiling points even higher than those of the carboxylic acids of comparable molecular weight. Dimethylformamide, $HCON(CH_3)_2$, is a good solvent for both polar and nonpolar compounds; it is useful as a reaction medium when such different compounds need to be brought into contact with one another. Acetanilide (sometimes called antifebrin) is the amide of acetic acid and aniline. It is used to treat headaches, neuralgia, and mild fevers. N,N-diethyl-*m*-toluamide, the amide of metatoluic acid and N,N-diethylamine, is the active ingredient in some insect repellents. Proteins are complex amides of high molecular weight. Some synthetic fibers are also polyamides (Section 32-11).

31-14 Aldehydes and Ketones

Aldehydes and ketones contain the carbonyl group, $\diagdown C=O$. In **aldehydes**, at least one H atom is bonded to the carbonyl group. **Ketones** have two alkyl or aryl groups bonded to a carbonyl group. Models of formaldehyde (the simplest aldehyde) and acetone (the simplest ketone) are shown in Figure 31-20.

The simplest aldehyde, formaldehyde,

$$H-\overset{\overset{\displaystyle O}{\|}}{C}-H,$$ has two H atoms and no alkyl or aryl groups. Other aldehydes have one alkyl or aryl group and one H atom bonded to the carbonyl group.

$$\underset{\substack{\text{aliphatic}\\\text{aldehyde}}}{R-\overset{\overset{\displaystyle O}{\|}}{C}-H} \qquad \underset{\substack{\text{aromatic}\\\text{aldehyde}}}{Ar-\overset{\overset{\displaystyle O}{\|}}{C}-H} \qquad \underset{\substack{\text{aliphatic}\\\text{ketone}}}{R-\overset{\overset{\displaystyle O}{\|}}{C}-R} \qquad \underset{\substack{\text{aromatic}\\\text{ketone}}}{Ar-\overset{\overset{\displaystyle O}{\|}}{C}-Ar} \qquad \underset{\substack{\text{mixed}\\\text{ketone}}}{Ar-\overset{\overset{\displaystyle O}{\|}}{C}-R}$$

Aldehydes are usually called by their common names. These are derived from the name of the acid with the same number of C atoms (Table 31-13).

Figure 31-20
(a) Models of formaldehyde, HCHO, the simplest aldehyde. (b) Models of acetone, $CH_3-CO-CH_3$, the simplest ketone.

(a)

(b)

Table 31-13
Properties of Some Simple Aldehydes

Common Name	Formula	Normal bp (°C)
formaldehyde (methanal)	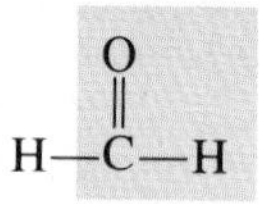 $H-\overset{\overset{\displaystyle O}{\|\|}}{C}-H$	−21
acetaldehyde (ethanal)	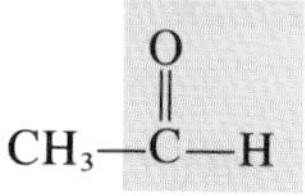 $CH_3-\overset{\overset{\displaystyle O}{\|\|}}{C}-H$	20.2
propionaldehyde (propanal)	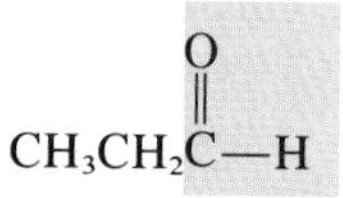 $CH_3CH_2\overset{\overset{\displaystyle O}{\|\|}}{C}-H$	48.8
benzaldehyde	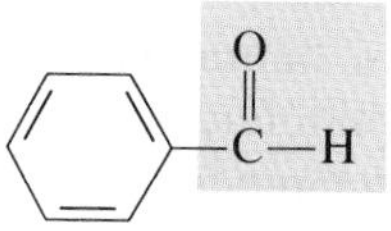 $\langle \bigcirc \rangle-\overset{\overset{\displaystyle O}{\|\|}}{C}-H$	179.5

The systematic (IUPAC) name is derived from the name of the parent hydrocarbon. The suffix *-al* is added to the characteristic stem. The carbonyl group takes positional precedence over other substituents.

Formaldehyde has long been used as a disinfectant and as a preservative for biological specimens (including embalming fluid). Its main use is in the production of certain plastics and in binders for plywood. Many important natural substances are aldehydes and ketones. Examples include sex hormones, some vitamins, camphor, and the flavorings extracted from almonds and cinnamon. Aldehydes contain a carbon–oxygen double bond, so they are very reactive compounds. As a result, they are valuable in organic synthesis, particularly in the construction of carbon chains.

The simplest ketone is called acetone. Other simple, commonly encountered ketones are usually called by their common names. These are derived by naming the alkyl or aryl groups attached to the carbonyl group.

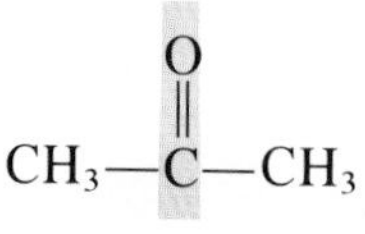

$CH_3-\overset{\overset{\displaystyle O}{\|\|}}{C}-CH_3$

acetone

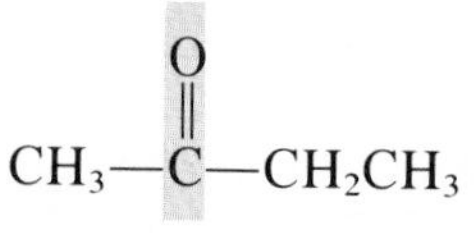

$CH_3-\overset{\overset{\displaystyle O}{\|\|}}{C}-CH_2CH_3$

methyl ethyl ketone

$CH_3CH_2-\overset{\overset{\displaystyle O}{\|\|}}{C}-CH_2CH_3$

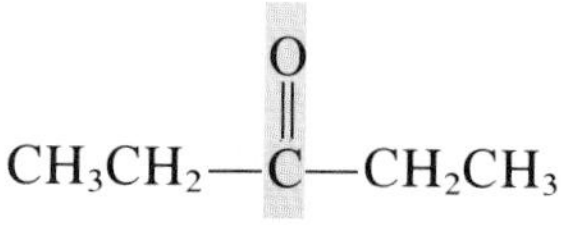

diethyl ketone

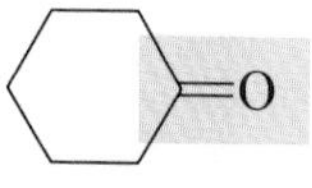

cyclohexanone

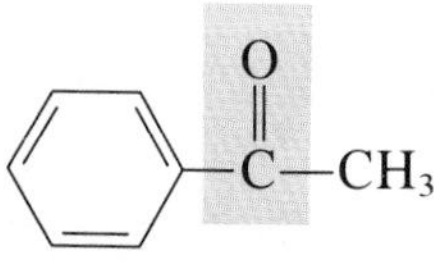

acetophenone
(methyl phenyl ketone)

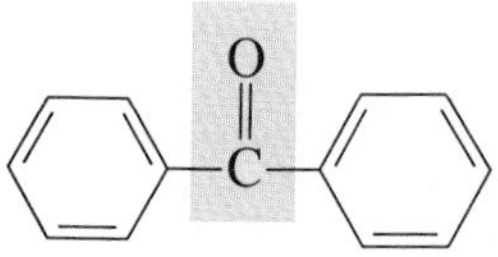

benzophenone
(diphenyl ketone)

When the benzene ring is a substituent, it is called a phenyl group ($-C_6H_5$).

 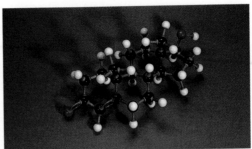

Steroid molecules have similar molecular shapes but different biochemical functions. Progesterone (left), a female sex hormone, and testosterone (right), a male sex hormone. Both are ketones.

The systematic names for ketones are derived from their parent hydrocarbons. The suffix -*one* is added to the characteristic stem.

$$\underset{1}{CH_3}-\underset{2}{\overset{\overset{\displaystyle O}{\|}}{C}}-\underset{3}{CH_2}-\underset{4}{CH_3}$$

2-butanone

$$\underset{1}{CH_3}-\underset{2}{CH_2}-\underset{3}{\overset{\overset{\displaystyle O}{\|}}{C}}-\underset{4}{\overset{\overset{\displaystyle CH_3}{|}}{CH}}-\underset{5}{CH_2}-\underset{6}{CH_3}$$

4-methyl-3-hexanone

The ketones are excellent solvents. Acetone is very useful because it dissolves most organic compounds yet is completely miscible with water. Acetone is widely used as a solvent in the manufacture of lacquers, paint removers, explosives, plastics, drugs, and disinfectants. Some ketones of high molecular weight are used extensively in blending perfumes. Structures of some naturally occurring aldehydes and ketones are

benzaldehyde
(almonds)

cinnamaldehyde
(cinnamon)

vanillin (vanilla)

muscone
(musk deer, used
in perfumes)

testosterone
(male sex hormone)

camphor

31-15 Summary of Functional Groups

Some important functional groups and the corresponding classes of related compounds are summarized in Figure 31-21.

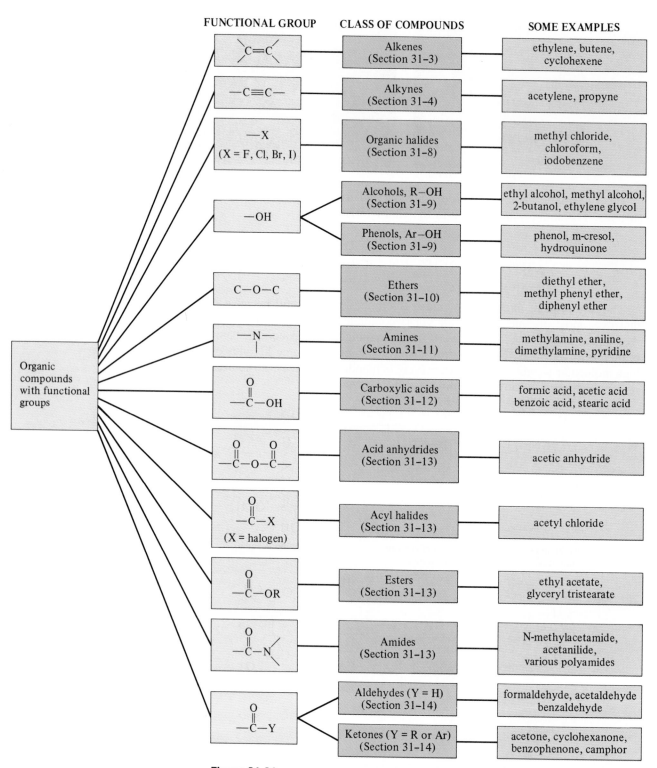

Figure 31-21
Summary of some functional groups and classes of organic compounds.

Key Terms

Acid anhydride A compound produced by dehydration of

$$R-\overset{\overset{\text{O}}{\|}}{C}-O-\overset{\overset{\text{O}}{\|}}{C}-R.$$

a carboxylic acid; general formula is $R-\overset{\overset{\text{O}}{\|}}{C}-O-\overset{\overset{\text{O}}{\|}}{C}-R$.

Acid halide See *Acyl halide*.

Acyl group The group of atoms remaining after removal of an —OH group of a carboxylic acid.

Acyl halide A compound derived from a carboxylic acid by replacing the —OH group with a halogen (X), usually —Cl; general formula is $R-\overset{\overset{\text{O}}{\|}}{C}-X$; also called acid halide.

Alcohol A hydrocarbon derivative containing an —OH group attached to a carbon atom not in an aromatic ring.

Aldehyde A compound in which an alkyl or aryl group and a hydrogen atom are attached to a carbonyl group; general formula is $R-\overset{\overset{\text{O}}{\|}}{C}-H$.

Aliphatic hydrocarbons Hydrocarbons that do not contain aromatic (delocalized) rings.

Alkanes See *Saturated hydrocarbons*.

Alkenes (olefins) Unsaturated hydrocarbons that contain one or more carbon–carbon double bonds.

Alkyl group A group of atoms derived from an alkane by the removal of one hydrogen atom.

Alkylbenzene A compound containing an alkyl group bonded to a benzene ring.

Alkynes Unsaturated hydrocarbons that contain one or more carbon–carbon triple bonds.

Amide A compound containing the $-\overset{\overset{\text{O}}{\|}}{C}-N\!\!\diagup^{\diagdown}$ group.

Amine A compound that can be considered a derivative of ammonia, in which one or more hydrogens are replaced by alkyl or aryl groups.

Amino acid A compound containing both an amino group and a carboxylic acid group.

Amino group The —NH_2 group.

Aromatic hydrocarbons Benzene and its derivatives; contain delocalized rings.

Aryl group The group of atoms remaining after a hydrogen atom is removed from an aromatic system.

Carbonyl group The $-\overset{\overset{\text{O}}{\|}}{C}-$ group.

Carboxylic acid A compound containing a $-\overset{\overset{\text{O}}{\|}}{C}-O-H$ group.

Catenation The ability of an element to bond to itself.

Conformations Structures of a compound that differ by the extent of rotation about a single bond.

Conjugated double bonds Double bonds that are separated from each other by one single bond, as in C=C—C=C.

Cycloalkanes Cyclic saturated hydrocarbons.

Ester A compound of the general formula $R-\overset{\overset{\text{O}}{\|}}{C}-O-R'$ where R and R' may be the same or different, and may be either aliphatic or aromatic.

Ether A compound in which an oxygen atom is bonded to two alkyl or two aryl groups, or one alkyl and one aryl group.

Fat A solid triester of glycerol and (mostly) saturated fatty acids.

Fatty acid An aliphatic acid; many can be obtained from animal fats.

Functional group A group of atoms that represents a potential reaction site in an organic compound.

Geometrical isomers Compounds with different arrangements of groups on the opposite sides of a bond with restricted rotation, such as a double bond or a single bond in a ring; for example, *cis–trans* isomers of certain alkenes.

Glyceride A triester of glycerol.

Heterocyclic amine An amine in which the nitrogen is part of a ring.

Homologous series A series of compounds in which each member differs from the next by a specific number and kind of atoms.

Hydrocarbon A compound that contains only carbon and hydrogen.

Ketone A compound in which a carbonyl group is bound to two alkyl or two aryl groups, or to one alkyl and one aryl group.

Oil A liquid triester of glycerol and unsaturated fatty acids.

Olefins See *Alkenes*.

Organic chemistry The chemistry of substances that contain carbon–hydrogen bonds.

Phenol A hydrocarbon derivative that contains an —OH group bound to an aromatic ring.

Pi bonds A chemical bond formed by the side-to-side overlap of atomic orbitals.

Polyene A compound that contains more than one double bond per molecule.

Polyhydric alcohol An alcohol that contains more than one —OH group.

Primary alcohol An alcohol with no or one R group bonded to the carbon bearing the —OH group.

Primary amine An amine in which one H atom of ammonia has been replaced by an organic group.

Saturated hydrocarbons Hydrocarbons that contain only single bonds. They are also called *alkanes* or *paraffin hydrocarbons*.

Secondary alcohol   An alcohol with two R groups bonded to the carbon bearing the —OH group.

Secondary amine   An amine in which two H atoms of ammonia have been replaced by organic groups.

Sigma bond   A chemical bond formed by the end-to-end overlap of atomic orbitals.

Structural isomers   Compounds that contain the same numbers of the same kinds of atoms in different geometric arrangements.

Tertiary alcohol   An alcohol with three R groups bonded to the carbon bearing the —OH group.

Tertiary amine   An amine in which three H atoms of ammonia have been replaced by organic groups.

Unsaturated hydrocarbons   Hydrocarbons that contain double or triple carbon–carbon bonds.

Exercises

Basic Ideas

1. (a) What is organic chemistry? (b) What was the "vital force" theory? (c) What happened to the "vital force" theory?

2. (a) What is catenation? (b) How is carbon unique among the elements?

3. How many "everyday" uses of organic compounds can you think of? List them.

4. (a) What are the principal sources of organic compounds? (b) Some chemists argue that the ultimate source of all naturally occurring organic compounds is carbon dioxide. Could this be possible? *Hint:* Think about the origins of coal, natural gas, and petroleum.

Aliphatic Hydrocarbons

5. (a) What are hydrocarbons? (b) What are saturated hydrocarbons? (c) What are the alkanes?

6. Describe the bonding in and the geometry of molecules of the following alkanes: (a) methane, (b) ethane, (c) propane, (d) *n*-butane. How are the formulas for these compounds similar? different?

7. (a) What are "normal" hydrocarbons? (b) What are branched-chain hydrocarbons? (c) Cite three examples of each.

8. (a) What is a homologous series? (b) Provide specific examples of compounds that are members of a homologous series. (c) What is a methylene group? (d) How does each member of a homologous series differ from compounds that come before and after it in the series?

9. (a) How do the melting points and boiling points of the normal alkanes vary with molecular weight? (b) Do you expect them to vary in this order? (c) Why or why not?

10. What are structural isomers?

11. (a) What are alkyl groups? (b) Draw structures for and write the names of the first five normal alkyl groups. (c) What is the origin of the names for alkyl groups?

12. What are cycloalkanes? Write the general molecular formula for a cycloalkane. Could a substance with the molecular formula C_5H_{12} be a cycloalkane? Explain your answer.

13. Write the structures for the three isomeric saturated hydrocarbons having the molecular formula C_5H_{12}. Name each by the IUPAC system.

14. Repeat Exercise 13 for the five isomers of C_6H_{14}.

15. Write the structure for 2,2-dimethylpropane.

16. Write the IUPAC name for

17. (a) What are alkenes? (b) What other names are used to describe alkenes?

18. (a) How does the general formula for the alkenes differ from the general formula for the alkanes? (b) Why are the general formulas identical for alkenes and cycloalkanes that contain the same number of carbon atoms?

19. (a) What are cycloalkenes? (b) What is their general formula? (c) Provide three examples.

20. Describe the bonding at each carbon atom in (a) ethene, (b) propene, (c) 1-butene, and (d) 2-butene.

21. (a) What are geometric (structural) isomers? (b) Why is rotation around a double bond not possible at room temperature? (c) What do *cis* and *trans* mean? (d) Draw structures for *cis*- and *trans*-1,2-dichloroethene. How do their melting and boiling points compare?

22. How do carbon–carbon single bond lengths and carbon–carbon double bond lengths compare? Why?

23. (a) What are alkynes? (b) What other name is used to describe them? (c) What is the general formula for alkynes? (d) How does the general formula for alkynes compare with the general formula for cycloalkenes? Why?

24. Describe the bonding and geometry associated with the triple bond of alkynes.

25. Draw the structural formulas of the following compounds: (a) 1-butyne, (b) 2-methylpropene, (c) 2-ethyl-3-methyl-1-butene, (d) 3-methyl-1-butyne.

26. Draw the structural formulas of the following compounds: (a) 3-hexyne, (b) 1,3-pentadiene, (c) 3,3-dimethylcyclobutene, (d) 3,4-diethylhexane.

27. Write the IUPAC names for the following compounds:

(a) $CH_3\overset{\displaystyle CH_3}{\underset{\displaystyle CH_3}{C}}CH_2CH_3$ (c) $CH_3\overset{\displaystyle CH_3}{CH}CHCH_3$

(b) $CH_2{=}C(CH_3)_2$ (d)

28. Repeat Exercise 27 for

(a)   (d) $CH_3\overset{\displaystyle }{C}{=}CHCH_3$
$\qquad\qquad\qquad\qquad\quad \overset{\displaystyle }{\underset{\displaystyle CH_3}{}}$

(b) $CH_3\overset{\displaystyle }{CH}CH_2CH_3$ (e) $CH_3CH_2\overset{\displaystyle }{CH}CH_3$
$\qquad\quad \underset{\displaystyle CH_3}{}$ $\qquad\qquad\qquad\qquad \underset{\displaystyle CH_2CH_3}{}$

(c) $CH_3C{\equiv}CCH_3$ (f) $CH_3\overset{\displaystyle }{CH}CH_2CH_3$
$\qquad\qquad\qquad\qquad\qquad\qquad\quad \underset{\displaystyle CH_2CH_3}{}$

Aromatic Hydrocarbons

29. (a) What are aromatic hydrocarbons? (b) What is the principal source of aromatic hydrocarbons?

30. (a) What is the most common aromatic hydrocarbon? (b) From what source was it first isolated? When?

31. (a) What are resonance structures? (b) Draw resonance structures for benzene. (c) What do we mean when we say that the electrons associated with the π bonds in benzene are delocalized over the entire ring?

32. What is a phenyl group? How many isomeric mono-phenylnaphthalenes are possible?

33. There are three isomeric trimethylbenzenes. Write their structural formulas and name them.

34. How many isomeric dibromobenzenes are possible? What names are used to designate these isomers?

35. Write the structural formulas for the following compounds: (a) *p*-difluorobenzene, (b) *n*-propylbenzene, (c) 1,3,5-tribromobenzene, (d) 1,3-diphenylbutane.

36. Write the IUPAC names for the following compounds:

(a)   (c)

(b)   (d)

Alkyl and Aryl Halides

37. Write the general representation for the formula of an alkyl halide. How does this differ from the representation for the formula of an aryl halide?

38. Name the following halides:

(a)   (c) $CHCl_3$

(b) $CH_3{-}\overset{\displaystyle CH_3}{\underset{\displaystyle }{CH}}{-}CH_2Cl$ (d) $Cl{-}\overset{\displaystyle Cl}{\underset{\displaystyle }{C}}{=}\overset{\displaystyle H}{\underset{\displaystyle }{C}}{-}Cl$

39. Write the structural formulas for the following: (a) 2-chloropentane, (b) 4-bromo-1-butene, (c) 1,2-dichloro-2-fluoropropane, (d) 1,4-dichlorobenzene.

40. Name the following:

(a)   (c)

(b)   (d)

41. The compound 1,2-dibromo-3-chloropropane (DBCP) was used as a pesticide in the 1970s. Recently, agricultural workers have claimed that exposure to DBCP made them sterile. Write the formula for this compound.

Alcohols and Phenols

42. (a) What are alcohols and phenols? (b) How do they differ? (c) Why can alcohols and phenols be viewed as derivatives of hydrocarbons? as derivatives of water?

43. (a) Distinguish among primary, secondary, and tertiary alcohols. (b) Write names and formulas for three alcohols of each type.

44. (a) Draw structural formulas for and write the names of the four (saturated) alcohols that contain four carbon atoms and one —OH group per molecule. (b) Draw structural formulas for and write the names of the eight (saturated) alcohols that contain five carbon atoms and one —OH group per molecule. Which ones may be classified as primary alcohols? secondary alcohols? tertiary alcohols?

45. (a) What are glycols? (b) Draw structures of three examples. (c) Why are glycols more soluble in water than monohydric alcohols that contain the same number of carbon atoms?

46. Refer to Table 31-8 and explain the trends in boiling points and solubilities of alcohols in water.

47. Why are methyl alcohol and ethyl alcohol called wood alcohol and grain alcohol, respectively?

48. Why are most phenols only slightly soluble in water?

49. Write the structural formula for each of the following compounds: (a) 1-butanol, (b) cyclohexanol, (c) 1,4-pentanediol.

50. Name the following compounds:

(a) CH_3CH-CH_2OH
      |
      CH_3

(b) $CH_3-\overset{\overset{\displaystyle CH_3}{|}}{\underset{\underset{\displaystyle CH_3}{|}}{C}}-CH_2-CH_2OH$

(c) $CH_3-CH-CH_2$
            | |
            OH OH

(d) $CH_3-\overset{\overset{\displaystyle CH_3}{|}}{\underset{\underset{\displaystyle CH_3}{|}}{C}}-OH$

51. Which of the following compounds are phenols? Name each compound.

(a) [benzene ring]—CH_2CH_2OH

(b) [cyclohexane ring]—OH

(c) [benzene ring with]—OH and —OH

(d) [fused ring structure]—OH

(e) [benzene ring]—OCH_3

52. Write the structural formulas for the following: (a) p-bromophenol, (b) 4-nitrophenol (the nitro group is —NO_2), (c) m-nitrophenol.

Ethers

53. Distinguish among aliphatic ethers, aromatic ethers, and mixed ethers.

54. Briefly describe the bonding around the oxygen atom in dimethyl ether. What intermolecular forces are found in this ether?

55. What determines whether an ether is "symmetrical" or "unsymmetrical"?

*56. Write the structural formulas for the following: (a) methoxymethane, (b) 1-ethoxypropane, (c) 1,3-dimethoxybutane, (d) ethoxybenzene, (e) methoxycyclobutane.

57. Name the following ethers:

(a) $CH_3-O-CH_2CH_3$

(b) $CH_3-O-CH-CH_3$
                        |
                        CH_3

(c) [benzene ring]—$O-CH_2CH_3$

(d) [cyclohexane ring]—$O-CH_3$

Amines

58. (a) What are amines? (b) Why are amines described as derivatives of ammonia?

59. Write the general representation for the formula for a compound that is (a) a primary amine, (b) a secondary amine, (c) a tertiary amine. Is $(CH_3)_3CNH_2$ a tertiary amine? Give a reason for your answer.

60. Name the following amines:

(a) CH_3-CH_2
              |
              NH
              |
      CH_3-CH_2

(b) O_2N—[benzene ring]—NH_2

(The —NO_2 substituent is called "nitro-.")

(c) [cyclopentane ring]—NH_2

(d) $CH_3CH_2CH_2CH_2-N-CH_2CH_2CH_2CH_3$
                                          |
                              $CH_2CH_2CH_2CH_3$

61. The stench of decaying proteins is due in part to the two compounds whose structures and common names are

$H_2N-CH_2CH_2CH_2CH_2-NH_2$ putrescine
$H_2N-CH_2CH_2CH_2CH_2CH_2-NH_2$ cadaverine

Name these compounds as amino-substituted alkanes.

Carboxylic Acids and Their Derivatives

62. (a) What are carboxylic acids? (b) Draw structural formulas for and write the names of five carboxylic acids.

63. (a) Why are aliphatic carboxylic acids sometimes called fatty acids? Cite two examples.

64. (a) What are acyl chlorides, or acid chlorides? (b) Draw structures for four acid chlorides and name them.

65. (a) What are esters? (b) Draw structures for four esters and write their names.

66. (a) What is the general representation for the formula of an acid anhydride? (b) Draw structures for four acid anhydrides and write their names.

67. Write the structural formulas for the following: (a) 2-methylpropanoic acid, (b) 3-bromobutanoic acid, (c) p-nitrobenzoic acid, (d) potassium benzoate, (e) 2-aminopropanoic acid.

68. List six naturally occurring esters and their sources.

69. (a) What are fats? What are oils? (b) Write the general formulas for fats and oils.

70. (a) What are glycerides? Distinguish between simple glycerides and mixed glycerides. (b) Write names and formulas for three simple glycerides.

71. Write the names and formulas for some acids that occur in fats and oils (as esters).

72. What are waxes?

73. Name the following esters:

(a) $CH_3\overset{\displaystyle O}{\overset{\|}{C}}-OCH_2CH_2CH_3$

(b) $CH_3\overset{\displaystyle O}{\overset{\|}{C}}OCH_3$

74. Name the following esters:

(a)

(b) $CH_3CH_2CH_2\overset{\displaystyle O}{\overset{\|}{C}}OCH_2CH_2CH_2CH_3$

Aldehydes and Ketones

75. (a) Distinguish between aldehydes and ketones. (b) Cite three examples (each) of aliphatic and aromatic aldehydes and ketones by drawing structural formulas and naming the compounds.

76. (a) List several naturally occurring aldehydes and ketones. (b) What are their sources? (c) What are some uses of these compounds?

*__77.__ Name the following compounds:

(a) $CH_3CH_2CH_2CH_2\overset{\displaystyle O}{\overset{\|}{C}}H$

(c) $H-\overset{\displaystyle Br}{\underset{\displaystyle Br}{\overset{\displaystyle |}{\underset{\displaystyle |}{C}}}}-CH_2-\overset{\displaystyle O}{\overset{\|}{C}}H$

(b)

(d)

*__78.__ Write the chemical formulas for the following: (a) 2-methylbutanal, (b) propynal, (c) o-methoxybenzaldehyde, (d) 2-butanone, (e) 1-bromo-2-propanone, (f) 3-hexanone.

Mixed Exercises

*__79.__ Identify the class of organic compounds (ester, ether, ketone, and so on) to which each of the following belongs:

(a)

(b)

(c) $CH_3\overset{\displaystyle O}{\overset{\|}{C}}OC(CH_3)_3$

(d)

(e)

(f) $CH_3\overset{\displaystyle O}{\overset{\|}{C}}CH_2-$

*__80.__ Identify the class of organic compounds (ester, ether, ketone, and so on) to which each of the following belongs:

(a)

(d)

(b)

(c)

(e)

*__81.__ Identify and name the functional groups in each of the following:

(a)

(b) $CH_2{=}CH-\overset{\displaystyle O}{\overset{\|}{C}}-O-CH_2CH_2CH_2CH_3$

(c)

(d)

82. Identify and name each functional group in the following:

(a) $CH_3CH_2-\overset{\overset{\displaystyle O}{\|}}{C}-O-\overset{\overset{\displaystyle O}{\|}}{C}-CH_2CH_3$

(b) $HO-\underset{\underset{\displaystyle OCH_3}{|}}{\bigcirc}-\underset{\underset{\displaystyle OH}{|}}{\overset{\overset{\displaystyle OH}{|}}{C}}HCH_2-\underset{\underset{\displaystyle}{}}{\overset{\overset{\displaystyle H}{|}}{N}}-CH_3$

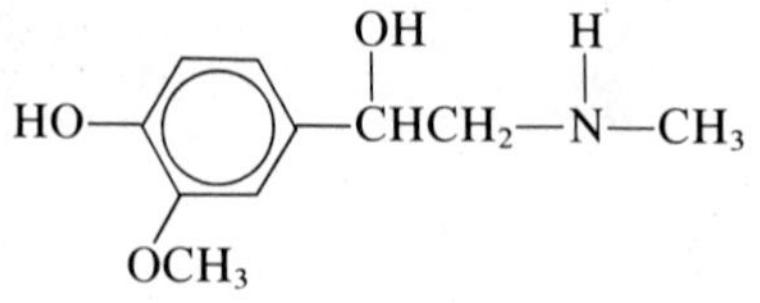

(c) dioxane (also known as 1,4-dioxin)

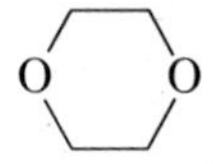

(d) morphine

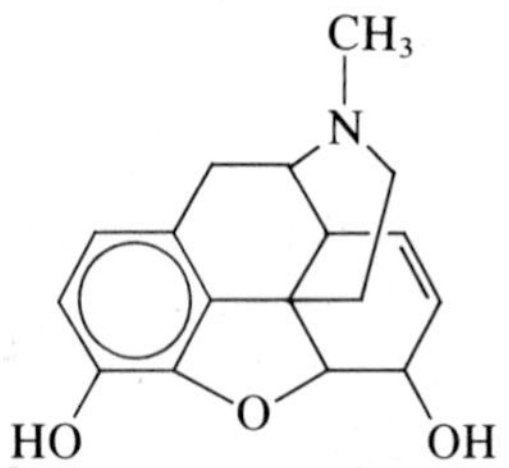

(e) epinephrine (adrenaline)

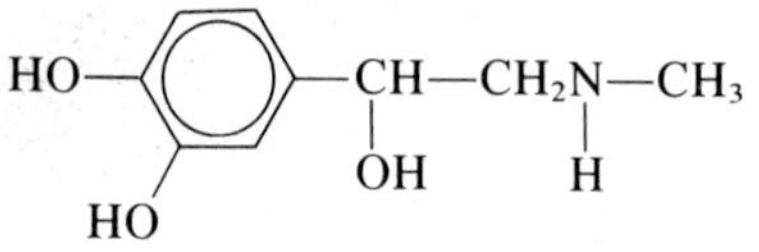

83. Identify and name the functional groups in each of the following:
(a) morpholine

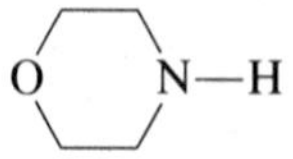

(b) citric acid

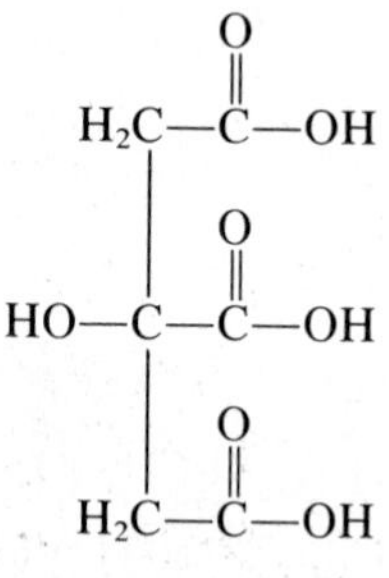

(c) coniine (from the hemlock plant; the poison that Socrates drank)

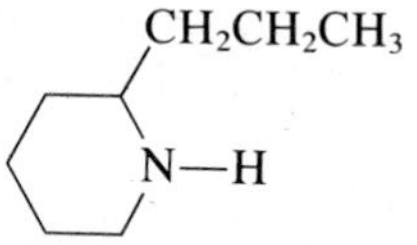

(d) glucose (a simple sugar, also known as dextrose)

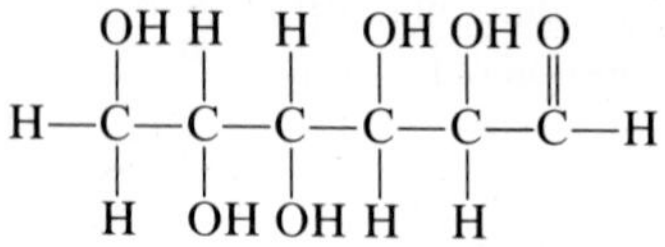

(e) vitamin C (also called ascorbic acid)

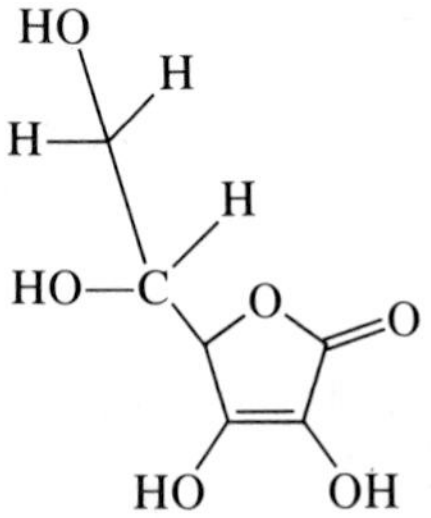

84. Name the following compounds:

(a) $CH_3CH_2CH_2CH_2OH$

(b)

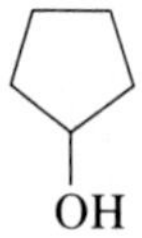

(c) $CH_3-\underset{\underset{\displaystyle NH_2}{|}}{C}H-CH_3$

(d) $CH_3-\underset{\underset{\displaystyle Cl}{|}}{C}=CH_2$

(e) $Br-\bigcirc-Br$

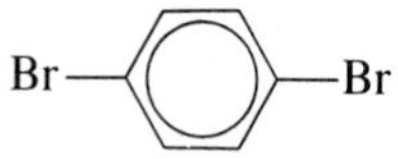

(f) $(CH_3CH_2)_3N$

(g) $\bigcirc-O-\bigcirc$

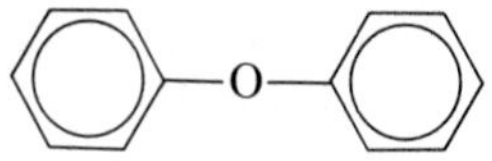

(h)

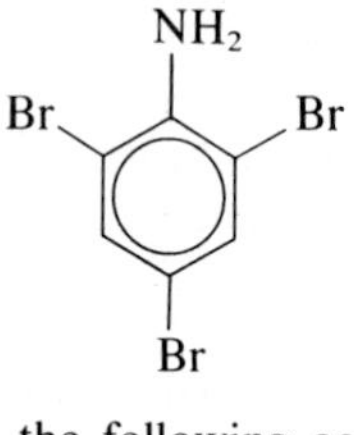

85. Draw the structural formulas for the following compounds: (a) p-bromotoluene, (b) cyclohexanol, (c) 2-methoxy-3-methylbutane, (d) diethylamine, (e) o-chlorophenol, (f) 1,4-butanediol.

86. Name the following compounds:

(a) $CH_3\underset{\underset{\displaystyle CH_3}{|}}{C}HCH_2OH$

(b) $CH_3CH_2CH_2CH_2NH_2$

(c) $CH_3CH_2CH_2CH_2\overset{\overset{\displaystyle O}{\|}}{C}H$

(d)

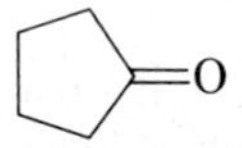

(e) $CH_3CH_2\underset{\underset{\displaystyle OCH_3}{|}}{C}HCH_3$

(f) $CH_3\overset{\overset{\displaystyle H_3C}{}}{\underset{\underset{\displaystyle CH_3}{|}}{C}}-\overset{\overset{\displaystyle O}{\|}}{C}OH$

Organic Chemistry II: Molecular Geometry and Reactions

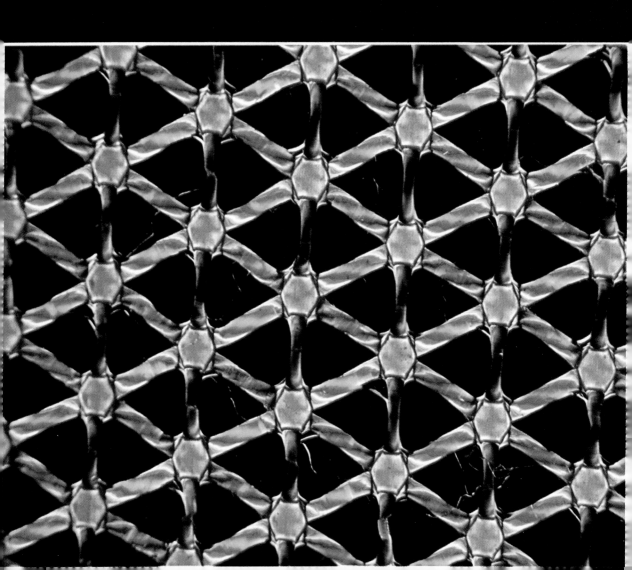

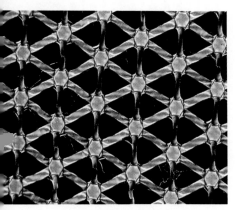

A polyethylene net photographed with polarized light.

Objectives

As you study this chapter, you should learn

☐ About the kinds of isomerism that organic compounds can exhibit—structural, geometrical, and optical

☐ About the conformations that molecules can adopt by rotation about single bonds

☐ To recognize and give examples of the three fundamental classes of organic reactions—substitution, addition, and elimination

☐ About some uses of the fundamental classes of organic reactions and some compounds that are prepared by each class of reaction

☐ About some common organic acids and bases and their relative strengths

☐ About oxidations and reductions of organic molecules, and how to recognize each type of transformation

☐ About the relative ease of oxidation of primary, secondary, and tertiary alcohols and the products that are formed

☐ About oxidation and reduction of alkenes and alkynes, the products that are formed, and the uses for such reactions

☐ About some reactions in which compounds that contain the carbonyl group are reduced, and the products that are formed

☐ About combustion reactions of organic compounds

☐ About some reactions by which esters and amides are formed

☐ About some reactions in which esters are hydrolyzed (saponification reactions)

☐ About some common polymers and the reactions by which they are formed

Geometries of Organic Molecules

As we learned in Chapter 29, the chemical and physical properties of a substance depend strongly on the arrangements, as well as the identities, of its atoms.

Isomers are substances that have the same number and kind of atoms—that is, the same *molecular formula*—but with the atoms arranged differently. *Because their structures are different, isomers have different properties.*

We can describe isomers in another way. Two molecules that cannot be superimposed on one another by any rotation about single bonds are isomers.

Isomers can be broadly divided into two major classes: structural isomers and stereoisomers. In Chapter 29 we discussed isomerism in coordination compounds, and in Chapter 31 we learned about some isomeric organic compounds. We shall now take a more systematic look at some three-dimensional aspects of organic structures—a subject known as **stereochemistry** ("spatial chemistry").

32-1 Structural Isomers

Structural isomers differ in the *order* in which their atoms are bonded together.

In our studies of hydrocarbons in Sections 31-1 through 31-7, we saw some examples of structural isomerism. Recall that there are three structural isomers of pentane.

n-pentane, C_5H_{12} isopentane, C_5H_{12} methylbutane   neopentane, C_5H_{12} dimethylpropane

These three isomers differ in the lengths of their base chains, but not in the functional groups present (i.e., only alkyl groups are present in this case). As a result, they differ somewhat in their melting and boiling points, but differ only very slightly in the reactions that they undergo.

In another type of structural isomerism, the compounds have the same *number* and *kind* of functional groups on the *same base chain* or the *same ring*, but in different positions. We term this *positional isomerism*. Again, such isomers usually have very similar chemical properties, differing mainly in physical properties such as melting and boiling points. The following groups of compounds are examples of positional isomers:

$CH_2{=}CHCH{=}CHCH_2CH_3$ $CH_2{=}CHCH_2CH{=}CHCH_3$ $CH_2{=}CHCH_2CH_2CH{=}CH_2$

1,3-hexadiene, C_6H_{10} 1,4-hexadiene, C_6H_{10} 1,5-hexadiene, C_6H_{10}

1,2-propanediol, $C_3H_8O_2$ 1,3-propanediol, $C_3H_8O_2$

o-dichlorobenzene, $C_6H_4Cl_2$ *m*-dichlorobenzene, $C_6H_4Cl_2$ *p*-dichlorobenzene, $C_6H_4Cl_2$

Sometimes the different order of arrangements of atoms results in different functional groups; such isomers are termed *functional group isomers* or *constitutional isomers*. Some examples of this type of isomerism are

an alcohol and an ether:

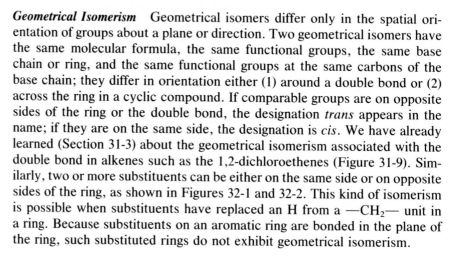

ethanol, C_2H_6O dimethyl ether, C_2H_6O

an aldehyde and a ketone:

butanal, C_4H_8O butan-2-one, C_4H_8O
(propionaldehyde)   (methyl ethyl ketone)

32-2 Stereoisomers

> In **stereoisomers** the atoms are linked together in the same atom-to-atom order, but their arrangements in space are different.

There are two types of stereoisomerism.

Geometrical Isomerism   Geometrical isomers differ only in the spatial orientation of groups about a plane or direction. Two geometrical isomers have the same molecular formula, the same functional groups, the same base chain or ring, and the same functional groups at the same carbons of the base chain; they differ in orientation either (1) around a double bond or (2) across the ring in a cyclic compound. If comparable groups are on opposite sides of the ring or the double bond, the designation *trans* appears in the name; if they are on the same side, the designation is *cis*. We have already learned (Section 31-3) about the geometrical isomerism associated with the double bond in alkenes such as the 1,2-dichloroethenes (Figure 31-9). Similarly, two or more substituents can be either on the same side or on opposite sides of the ring, as shown in Figures 32-1 and 32-2. This kind of isomerism is possible when substituents have replaced an H from a —CH_2— unit in a ring. Because substituents on an aromatic ring are bonded in the plane of the ring, such substituted rings do not exhibit geometrical isomerism.

Optical Isomerism   Some macroscopic objects are mirror images of one another but cannot be superimposed. Your two hands are a familiar example of this; each hand is a nonsuperimposable mirror image of the other (Figure 32-3). "Superimposable" means that if one object is placed over the other, the positions of all parts will match.

An object that is *not* superimposable with its mirror image is said to be **chiral** (from the Greek word *cheir*, meaning "hand"); an object that *is* superimposable with its mirror image is said to be **achiral**. Examples of familiar

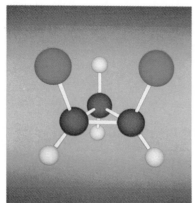

(a)

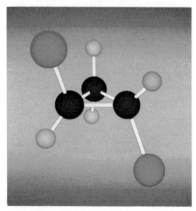

(b)

Figure 32-1
Models of (a) *cis*-dichlorocyclopropane and (b) *trans*-dichlorocyclopropane.

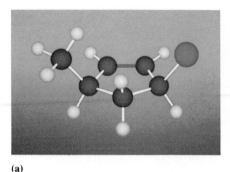
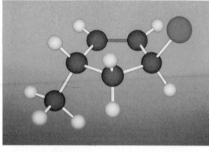

(a)   (b)

Figure 32-2
Models of (a) *cis*-3-chloro-5-methylcyclopentene and (b) *trans*-3-chloro-5-methylcyclopentene. In each model, the shaded bond at the back is the double bond.

objects that are chiral are a screw, a propeller, a foot, an ear, and a spiral staircase; examples of common objects that are achiral are a plain cup with no decoration, a pair of eyeglasses, and a sock.

We speak of a screw or a propeller as being "right-handed" or "left-handed."

> Some molecules can exist in two forms that bear the same relationship to each other as do left and right hands. Such chiral molecules are called **optical isomers** or **enantiomers** of one another.

As an example of this, consider first the two models of chlorobromomethane, CH_2ClBr, shown in Figure 32-4. They are mirror images of one another, and they can be superimposed. Thus, this molecule is *achiral* and is not capable of optical isomerism. Now consider chlorobromoiodomethane, $CHClBrI$ (Figure 32-5). This molecule is not superimposable with its mirror image, so it is *chiral*, and the two forms are said to be optical isomers of one another. Any compound that contains four different groups bonded to the same carbon atom is chiral; i.e., it exhibits optical isomerism. Such a carbon is said to be *asymmetric* (meaning "without symmetry"). Most simple chiral molecules contain at least one asymmetric carbon atom, although there are other ways in which molecular chirality can occur.

Optical isomers of a compound have the same type and number of atoms, connected in the same order, but arranged differently in space. In this way they are like geometrical isomers. However, the two types of isomerism differ in that geometrical isomers have *different* physical and chemical properties, whereas optical isomers have *identical* physical properties (such as

Figure 32-3
Mirror images. Place your left hand in front of a mirror; you will observe that it looks just like your right hand. We say that the two hands are mirror images of one another; each hand is in every way the "reverse" of the other. Now try placing one hand directly over the other; they are not identical. Hence, they are nonsuperimposable mirror images. Each hand is a *chiral* object.

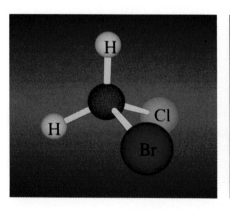
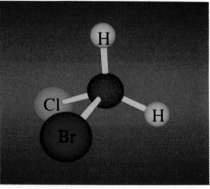

Figure 32-4
Models of two mirror-image forms of chlorobromomethane, CH_2ClBr. The two models are the same (superimposable), so they are achiral. CH_2ClBr does not exhibit optical isomerism.

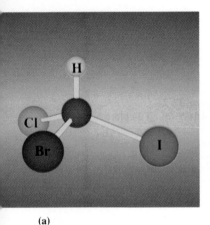

(a)

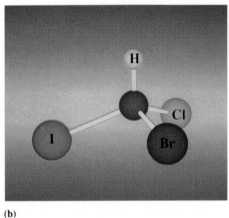

(b)

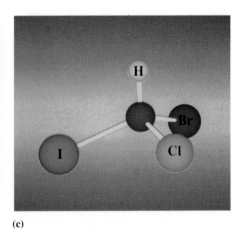

(c)

Figure 32-5

(a, b) Models of the two mirror-image forms of chlorobromoiodomethane, CHClBrI. (c) The same model as in (a), turned so that H and I point the same as in (b); however, the Br and Cl atoms are not in the same positions in (b) and (c). The two models in (a) and (b) cannot be superimposed on one another no matter how we rotate them, so they are chiral. These two forms of CHClBrI represent *different compounds* that are optical isomers of one another.

melting point, boiling point, and density). Optical isomers also undergo the same chemical reactions, except when they interact with other compounds that are themselves chiral. They often exhibit different solubilities in solvents that are composed of chiral molecules.

Optical isomers do differ from each other in one important physical property; they interact with polarized light in different ways. Let us briefly review the main features of this subject from Chapter 29. Separate equimolar solutions of two optical isomers rotate a plane of polarized light (Figures 29-2 and 29-3) by equal amounts but in opposite directions. The solution that rotates the polarized light to the right is **dextrorotatory**, and the other, which rotates it to the left, is **levorotatory**. Optical isomers are called *dextro* (D) and *levo* (L) isomers. The phenomenon in which a plane of polarized light is rotated is called **optical activity**. It can be measured with a polarimeter (Figure 29-3) or with more sophisticated instruments. A **racemic mixture** is a single sample containing equal amounts of the two optical isomers of a compound. Such a solution does not rotate a plane of polarized light, because the equal and opposite effects of the two isomers exactly cancel. To exhibit optical activity, the *dextro* and *levo* isomers must be separated from each other. The chemical or physical processes that accomplish this are broadly called *optical resolution*.

One very important way in which optical isomers differ from one another chemically is in their biological activities. α-Amino acids have the general structure

$$H_2N-\overset{\displaystyle H}{\underset{\displaystyle R}{\overset{\displaystyle |}{\underset{\displaystyle |}{C}}}}-COOH$$

where R represents any of a number of common substituents. The central carbon atom has four different groups bonded to it. All α-amino acids (except glycine, in which R = H) can exist as two optical isomers. Figure 32-6 shows this mirror-image relationship for optical isomers of phenylalanine, in which R = —CH$_2$C$_6$H$_5$. All naturally occurring phenylalanine in living systems is in the L form. In fact, only L-isomers of optically active amino acids are found in proteins.

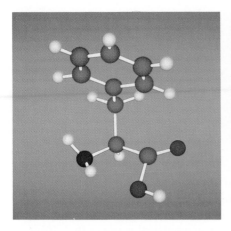
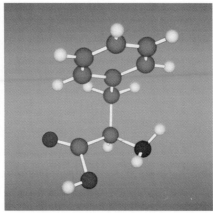

Figure 32-6
Models of the two optical isomers of phenylalanine. The naturally occurring phenylalanine in all living systems is the L form, shown on the left.

32-3 Conformations

A *conformation* is one specific geometry of a molecule. The **conformations** of a compound differ from one another in the *extent of rotation about one or more single bonds*. The C—C single bond length, 1.54 Å, is relatively independent of the structure of the rest of the molecule.

Rotation about single C—C bonds is possible; in fact, at room temperature it occurs rapidly. There might appear to be an infinite number of kinds of ethane molecules, depending on the rotation of one carbon atom with respect to the other. However, at room temperature ethane molecules possess sufficient thermal energy to cause rapid rotation about the single carbon–carbon bond from one conformation to another. Therefore, there is only one kind of ethane molecule, and no isomers are known. The staggered conformation is slightly more stable than the eclipsed conformation (see Figure 32-7). In the staggered conformation there is less repulsive interaction between H atoms on adjacent C atoms.

As we saw in Section 31-3, rotation does not occur around carbon–carbon double bonds at room temperature.

Consider two conformations of *n*-butane (Figure 32-8). Again, staggered conformations are *slightly* more stable than eclipsed ones. At room temperature many conformations are present in a sample of pure *n*-butane.

Take care to distinguish between conformational differences and isomerism. The two forms of *n*-butane shown in Figure 32-8 are *not* isomers of one another. Either form can be converted to the other by rotation about a single bond, which is a very easy process that does not involve breaking any bonds.

Staggered
conformation

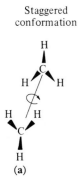

Eclipsed
conformation

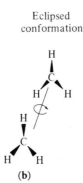

(a)

(b)

Figure 32-7
Two possible conformations of ethane. (a) Staggered. (b) Eclipsed. Rotation of one CH_3 group about the C—C single bond, as shown by the curved arrows, converts one conformation to the other.

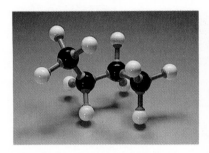
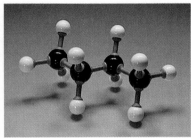

Figure 32-8
Two staggered conformations of *n*-butane, *n*-C_4H_{10}.

By contrast, to convert one isomer to another, at least one chemical bond would have to be broken and then re-formed. This is most obvious with structural isomerism, in which a conversion would change the order of attachment of the atoms. It is also true for geometrical isomers that differ in orientation about a double bond. To convert such a *cis* isomer to a *trans* isomer, it would be necessary to rotate part of the molecule about the double bond. Such a rotation would move the *p* orbitals out of the parallel alignment that is necessary to form the pi component of the double bond (Section 8-13). The breaking of this pi bond is quite costly in terms of energy; it occurs only with the input of energy in the form of heat or light.

We saw in Section 31-1 that cyclohexane exists in two forms, called the *chair* and *boat* forms (Figure 32-9). The chair form is the more stable of the two, because the hydrogens (or other substituents) are, on the average, farther from one another than in the boat form. However, chair and boat cyclohexane *are not* different compounds. Either form is easily converted into the other by rotation around single bonds without breaking any bonds, and the two forms cannot be separated. Thus, they are different *conformations* of cyclohexane.

Looking carefully at the chair conformation of cyclohexane (Figure 32-9a), we see that there are two different kinds of hydrogens on each C atom. One kind lies near the plane generally defined by the ring of carbon atoms. By analogy with the equator of the earth, this kind is called the *equatorial* hydrogens. Hydrogens of the other type are bonded in a direction that is generally perpendicular to the average plane of the ring. Again by analogy with the earth, these are called the *axial* hydrogens. Each carbon in the chair conformation of cyclohexane has one axial and one equatorial hydrogen, and their orientation alternates from each carbon to the next. There are three axial positions on each side of the ring, and these three bonds are

Figure 32-9
The two stable forms of the cyclohexane ring: (a) chair and (b) boat. In the chair conformation of cyclohexane, there are six *axial* positions (the H atoms marked by *a*) and six *equatorial* positions.

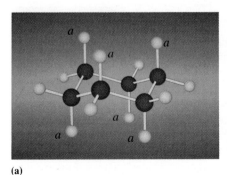

(a)

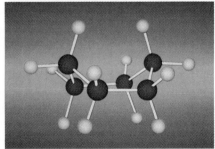

(b)

parallel to one another. By contrast, the equatorial positions point generally *away from* one another. As a result, a monosubstituted cyclohexane often has its substituent in an equatorial position, because the substituent is less crowded there than in an axial position.

Fundamental Classes of Organic Reactions

Organic compounds display very different abilities to react, ranging from the limited reactivity of hydrocarbons and fluorocarbons to the great variety of reactions undergone by the thousands of organic molecules that contain several functional groups. Reactivity depends on structure. We can usually predict the kinds of reactions a compound can undergo by identifying the functional groups it contains. But the electronic and structural features that are *near* a functional group can also affect its reactivity. One of the fascinations of organic chemistry is our ability to ''fine-tune'' both physical and chemical properties by making small changes in structure. The successes of this approach are innumerable, including the development of fuels and their additives or alternatives, the improvement of pharmaceuticals to enhance their effectiveness and minimize their ill effects, and the development of polymers and plastics with an incredible variety of properties and uses.

In the remainder of this chapter we shall present a few of the kinds of reactions that organic compounds undergo. A topic of such vast scope as reactivity of organic compounds can be made manageable only if we organize it. Nearly all organic transformations involve at least one of three fundamental classes of reactions. The following three sections will address these classes. We shall also look at some reaction sequences that combine several reaction steps from more than one of the fundamental classes.

32-4 Substitution Reactions

In a **substitution reaction**, an atom or a group of atoms attached to a carbon atom is removed and another atom or group of atoms takes its place. No change occurs in the degree of saturation at the reactive carbon atom.

The saturated hydrocarbons (alkanes and cycloalkanes) are chemically rather inert materials. For many years they were known as *paraffin* hydrocarbons because they undergo few reactions. They do not react with such powerful oxidizing agents as potassium permanganate and potassium dichromate. However, they do react with the halogens, with oxygen when ignited, and with concentrated nitric acid. As expected, members of a homologous series (Section 31-1) have similar chemical properties. If we study the chemistry of one of these compounds, we can make predictions about the others with a fair degree of certainty.

The saturated hydrocarbons can react without a big disruption of the molecular structure only by *displacement, or substitution of one atom for another*. At room temperature, chlorine and bromine react very slowly with saturated straight-chain hydrocarbons. At higher temperatures, or in the presence of sunlight or other source of ultraviolet light, H atoms in the hydrocarbon can be replaced easily by halogen atoms. These substitution reactions are called **halogenation** reactions. The mechanism of reaction of Cl_2 with CH_4 may be represented as

$$: \overset{..}{\underset{..}{Cl}} : \overset{..}{\underset{..}{Cl}} : \quad \xrightarrow[\text{sunlight}]{\text{heat or}} \quad 2 : \overset{..}{\underset{..}{Cl}} \cdot$$

chlorine molecule   chlorine atoms
(radicals)

A radical is an atom or group of atoms that contains at least one unpaired electron. Some carry an electrical charge, others do not.

$$H : \overset{\overset{\textstyle H}{}}{\underset{\underset{\textstyle H}{}}{C}} : H \; + \; : \overset{..}{\underset{..}{Cl}} \cdot \; \longrightarrow \; H : \overset{\overset{\textstyle H}{}}{\underset{\underset{\textstyle H}{}}{C}} \cdot \; + \; H : \overset{..}{\underset{..}{Cl}} :$$

methane   chlorine   methyl   hydrogen
atom   radical   chloride

$$H : \overset{\overset{\textstyle H}{}}{\underset{\underset{\textstyle H}{}}{C}} \cdot \; + \; : \overset{..}{\underset{..}{Cl}} : \overset{..}{\underset{..}{Cl}} : \; \longrightarrow \; H : \overset{\overset{\textstyle H}{}}{\underset{\underset{\textstyle H}{}}{C}} : \overset{..}{\underset{..}{Cl}} : \; + \; \left(: \overset{..}{\underset{..}{Cl}} \cdot \right) \; \text{and so on}$$

methyl   chlorine   methyl   chlorine
radical   molecule   chloride   atom

The reaction of Cl_2 with CH_4 is called a **free-radical chain reaction**. It is quite similar to the reaction of Cl_2 with H_2 (Section 24-10). Cl_2 molecules absorb light or heat energy and split into very reactive Cl atoms (an *initiation* step). Some Cl atoms attack a methane molecule, removing one of the H atoms to form HCl; the methyl radical, in turn, reacts with Cl_2 molecules to form methyl chloride (*propagation* steps). Eventually Cl atoms may recombine to form Cl_2 molecules. The overall reaction is usually represented as

Note that only one half of the chlorine atoms occur in the organic product. The other half form hydrogen chloride, a commercially valuable compound.

$$\underset{\text{methane}}{H - \overset{\overset{\textstyle H}{|}}{\underset{\underset{\textstyle H}{|}}{C}} - H} \; + \; \underset{\text{chlorine}}{Cl - Cl} \; \xrightarrow[\text{UV}]{\text{heat or}} \; \underset{\substack{\text{chloromethane} \\ \text{(methyl chloride)} \\ \text{bp} = 23.8°C}}{H - \overset{\overset{\textstyle H}{|}}{\underset{\underset{\textstyle H}{|}}{C}} - Cl} \; + \; HCl$$

Many organic reactions produce more than a single product. For example, the chlorination of CH_4 may produce several other products in addition to CH_3Cl, as the following equations show.

$$H - \overset{\overset{\textstyle H}{|}}{\underset{\underset{\textstyle H}{|}}{C}} - Cl \; + \; Cl - Cl \; \longrightarrow \; \underset{\substack{\text{dichloromethane} \\ \text{(methylene chloride)} \\ \text{bp} = 40.2°C}}{Cl - \overset{\overset{\textstyle H}{|}}{\underset{\underset{\textstyle H}{|}}{C}} - Cl} \; + \; HCl$$

$$Cl - \overset{\overset{\textstyle H}{|}}{\underset{\underset{\textstyle H}{|}}{C}} - Cl \; + \; Cl - Cl \; \longrightarrow \; \underset{\substack{\text{trichloromethane} \\ \text{(chloroform)} \\ \text{bp} = 61°C}}{Cl - \overset{\overset{\textstyle H}{|}}{\underset{\underset{\textstyle Cl}{|}}{C}} - Cl} \; + \; HCl$$

$$\underset{\substack{| \\ Cl}}{\overset{\substack{H \\ |}}{Cl-C-Cl}} + Cl-Cl \longrightarrow \underset{\substack{| \\ Cl}}{\overset{\substack{Cl \\ |}}{Cl-C-Cl}} + HCl$$

tetrachloromethane
(carbon tetrachloride)
bp = 76.8°C

The mixture of products formed in an organic reaction can be separated by physical methods such as fractional distillation, which relies on differences in the boiling points of different compounds.

When a hydrocarbon has more than one C atom, its reaction with Cl_2 is more complex. The first step in the chlorination of ethane gives the product that contains one Cl atom per molecule.

$$\underset{\substack{| \ \ | \\ H \ \ H}}{\overset{\substack{H \ \ H \\ | \ \ |}}{H-C-C-H}} + Cl-Cl \xrightarrow[\text{UV}]{\text{heat or}} \underset{\substack{| \ \ | \\ H \ \ H}}{\overset{\substack{H \ \ H \\ | \ \ |}}{H-C-C-Cl}} + HCl$$

ethane   chloroethane
(ethyl chloride)
bp = 13.1°C

Ethyl chloride is widely used by athletic trainers as a spray-on pain killer.

When a second hydrogen atom is replaced, a mixture of the two possible products is obtained.

$$\underset{\substack{| \ \ | \\ H \ \ H}}{\overset{\substack{H \ \ H \\ | \ \ |}}{H-C-C-Cl}} + Cl-Cl \xrightarrow[\text{UV}]{\text{heat or}} \left\{ \begin{array}{c} \underset{\substack{| \ \ | \\ H \ \ H}}{\overset{\substack{H \ \ Cl \\ | \ \ |}}{H-C-C-Cl}} \\ \text{1,1-dichloroethane} \\ \text{bp} = 57°C \\ \\ \underset{\substack{| \ \ | \\ H \ \ H}}{\overset{\substack{H \ \ H \\ | \ \ |}}{Cl-C-C-Cl}} \\ \text{1,2-dichloroethane} \\ \text{bp} = 84°C \end{array} \right\} + HCl$$

The product mixture does not contain equal numbers of moles of the dichloroethanes, so we do not show a stoichiometrically balanced equation. Because reactions of saturated hydrocarbons with chlorine can produce many products, the reactions are not always as useful as might be desired.

The reaction of cyclopentane with Cl_2 produces only one monosubstituted product, chlorocyclopentane (cyclopentyl chloride). Multiple substitution also occurs.

$$\underset{\substack{H_2C-CH_2}}{\overset{\substack{CH_2}}{\underset{H_2C \diagdown \diagup CH_2}{}}} + Cl_2 \xrightarrow[\text{UV}]{\text{heat or}} \underset{\substack{H_2C-CH_2}}{\overset{\substack{Cl \\ | \\ CH}}{\underset{H_2C \diagdown \diagup CH_2}{}}} + HCl$$

cyclopentane   chlorocyclopentane

Substitution is the most common kind of reaction of the aromatic ring. Halogenation, with chlorine or bromine, occurs readily in the presence of iron or anhydrous iron(III) chloride (a Lewis acid) catalyst.

When iron is added as a catalyst, it reacts with chlorine to form iron(III) chloride, which is the true catalyst.

benzene   chlorobenzene

The equation is usually written in condensed form as

Aromatic rings can undergo *nitration,* substitution of the *nitro* group —NO$_2$, in a mixture of concentrated nitric and sulfuric acids at low temperatures.

The H$_2$SO$_4$ is both a catalyst and a dehydrating agent. The H in the product H$_2$O comes from the hydrocarbon; the OH comes from HNO$_3$.

nitric acid   nitrobenzene

The explosive TNT (2,4,6-trinitrotoluene) is manufactured by the nitration of toluene in several steps.

toluene   nitric acid   2,4,6-trinitrotoluene (TNT)

Groups other than hydrogen can be substituted by other atoms or groups of atoms. For example, in many reactions another group replaces the —OH group. These reactions form esters and amides (Sections 31-13, 32-9). Substitution of a halogen for the —OH of a carboxylic acid forms an acyl halide (Section 32-9), a type of compound that is widely used in subsequent synthesis.

The hydrohalic acids react with alcohols in the presence of certain Lewis acid catalysts to form *alkyl halides.* Concentrated hydrochloric acid reacts with primary alcohols very slowly at elevated temperatures, with zinc chloride serving as a catalyst.

$$\text{CH}_3\text{CH}_2\text{OH} \;+\; \text{HCl} \;\xrightarrow[\Delta]{\text{ZnCl}_2}\; \text{CH}_3\text{CH}_2\text{Cl} \;+\; \text{H}_2\text{O}$$

ethanol   chloroethane
                           ethyl chloride

Alcohols react with common inorganic oxyacids to produce **inorganic esters**. For instance, nitric acid reacts with alcohols to produce nitrates by substitution of nitrate, $-ONO_2$, for hydroxyl, $-OH$.

$$CH_3CH_2OH \; + \; HONO_2 \longrightarrow CH_3CH_2-ONO_2 + H_2O$$

ethanol   nitric acid   ethyl nitrate

Simple inorganic esters may be thought of as compounds that contain one or more alkyl groups covalently bonded to the anion of a *ternary* inorganic acid. Unless indicated, the term "ester" refers to organic esters.

The substitution reaction of nitric acid with glycerol produces the explosive nitroglycerine. Alfred Nobel's discovery in 1866 that this very sensitive material could be made into a "safe" explosive by absorbing it into diatomaceous earth or wood meal led to his development of dynamite.

Nobel's brother had been killed and his father permanently crippled in a nitroglycerine explosion in 1864. Nobel willed $9,200,000 to establish a fund for annual prizes in physics, chemistry, medicine, literature, and peace. The prizes were first awarded in 1901.

$$
\begin{array}{l}
H_2C-OH \\
\;\;\;| \\
HC-OH \;\; + \; 3HONO_2 \; \xrightarrow{H_2SO_4} \\
\;\;\;| \\
H_2C-OH
\end{array}
\quad
\begin{array}{l}
H_2C-ONO_2 \\
\;\;\;| \\
HC-ONO_2 \;\; + \; 3H_2O \\
\;\;\;| \\
H_2C-ONO_2
\end{array}
$$

glycerol   glyceryl trinitrate
(nitroglycerine)

Interestingly, nitroglycerine ("nitro") is taken by persons who have heart disease. It acts as a vasodilator (dilates the blood vessels) to decrease arterial tension.

Cold, concentrated H_2SO_4 reacts with alcohols to form **alkyl hydrogen sulfates**. The reaction with lauryl alcohol is an important industrial reaction.

$$CH_3(CH_2)_{10}CH_2-OH \;\; + \; HOSO_3H \longrightarrow CH_3(CH_2)_{10}CH_2-OSO_3H + H_2O$$

1-dodecanol (lauryl alcohol)   sulfuric acid   lauryl hydrogen sulfate

The neutralization reaction of an alkyl hydrogen sulfate with NaOH then produces the sodium salt of the alkyl hydrogen sulfate.

$$CH_3(CH_2)_{10}CH_2-OSO_3H + Na^+OH^- \longrightarrow CH_3(CH_2)_{10}CH_2-OSO_3{}^-Na^+ + H_2O$$

sodium lauryl sulfate (a detergent)

Sodium salts of the alkyl hydrogen sulfates that contain about 12 carbon atoms are excellent detergents. They are also biodegradable. (Soaps and detergents were discussed in Section 14-18.)

Tertiary alcohols are synthesized by substitution reactions in which the $-OH$ group from water replaces a halo group in a tertiary alkyl halide.

$$
\begin{array}{c}
\;\;\;\;CH_3 \\
\;\;\;\;| \\
CH_3-C-Cl \; + \; H_2O \; \xrightarrow[\text{pure } H_2O]{25°C} \\
\;\;\;\;| \\
\;\;\;\;CH_3
\end{array}
\quad
\begin{array}{c}
\;\;\;\;CH_3 \\
\;\;\;\;| \\
CH_3-C-OH \; + \; HCl \\
\;\;\;\;| \\
\;\;\;\;CH_3
\end{array}
$$

tert-butyl chloride   *tert*-butyl alcohol

32-5 Addition Reactions

An **addition reaction** involves an *increase* in the number of groups attached to carbon. The molecule becomes more nearly saturated.

The principal reactions of alkenes and alkynes (and their cyclic analogs) are addition reactions rather than substitution reactions. For example, contrast the reactions of ethane and ethylene with Cl_2.

ethane: $CH_3—CH_3 + Cl_2 \longrightarrow CH_3—CH_2Cl + HCl$ (substitution, slow)

ethylene: $H_2C{=}CH_2 + Cl_2 \longrightarrow$

$$CH_2—CH_2$$
$$\underset{Cl}{|} \quad \underset{Cl}{|}$$

(addition, rapid)

Carbon–carbon double bonds are *reaction sites* and so represent *functional groups*. Most addition reactions involving alkenes and alkynes proceed rapidly at room temperature. By contrast, substitution reactions of the alkanes require catalysts and high temperatures.

Bromine adds readily to the alkenes to give dibromides. The reaction with ethylene is

$$H_2C{=}CH_2 + Br_2 \longrightarrow \underset{\underset{Br}{|} \quad \underset{Br}{|}}{CH_2—CH_2}$$

1,2-dibromoethane
(ethylene dibromide)

The addition of Br_2 to alkenes is used as a simple qualitative test for unsaturation. Bromine, a dark red liquid, is dissolved in a nonpolar solvent. When an alkene is added, the solution becomes colorless as the Br_2 reacts with the alkene to form a colorless compound. This reaction is used to distinguish between alkanes and alkenes.

Hydrogenation is an extremely important addition reaction of the alkenes. Hydrogen adds across double bonds at elevated temperatures, under high pressures, and in the presence of an appropriate catalyst (finely divided Pt, Pd, or Ni).

$$CH_2{=}CH_2 + H_2 \xrightarrow[\text{heat}]{\text{catalyst}} CH_3—CH_3$$

Unsaturated hydrocarbons are converted to saturated hydrocarbons in the manufacture of high-octane gasoline and aviation fuels. Unsaturated vegetable oils can also be converted to solid cooking fats (shortening) by hydrogenation (Figure 32-10).

Figure 32-10
Hydrogenation of the olefinic double bonds in a vegetable oil converts it to a solid fat.

$$
\begin{array}{l}
\overset{\displaystyle O}{\overset{\displaystyle \|}{H_2COC}}(CH_2)_7CH{=}CH(CH_2)_7CH_3 \\[4pt]
\overset{\displaystyle O}{\overset{\displaystyle \|}{HCOC}}(CH_2)_7CH{=}CH(CH_2)_7CH_3 \\[4pt]
\overset{\displaystyle O}{\overset{\displaystyle \|}{H_2COC}}(CH_2)_7CH{=}CH(CH_2)_7CH_3 \\
\qquad\qquad \text{olein (an oil, liquid)}
\end{array}
\xrightarrow[\substack{\text{Ni catalyst}\\\text{heat}}]{3H_2}
\begin{array}{l}
\overset{\displaystyle O}{\overset{\displaystyle \|}{H_2COC}}(CH_2)_{16}CH_3 \\[4pt]
\overset{\displaystyle O}{\overset{\displaystyle \|}{HCOC}}(CH_2)_{16}CH_3 \\[4pt]
\overset{\displaystyle O}{\overset{\displaystyle \|}{H_2COC}}(CH_2)_{16}CH_3 \\
\qquad \text{stearin (a fat, solid)}
\end{array}
$$

The *hydration reaction* (addition of water) is another very important addition reaction of alkenes. It is used commercially for the preparation of a wide variety of alcohols from petroleum by-products. Ethanol, the most important industrial alcohol, is produced industrially by the hydration of ethylene from petroleum, using H_2SO_4 as a catalyst.

$$H_2C\!=\!CH_2 + H_2O \xrightarrow{\ H_2SO_4\ } H_3C\!-\!CH_2OH$$

Think of H_2O as HOH.

This reaction takes place in two steps.

$$H\!-\!C\!=\!C\!-\!H + HOSO_3H \xrightarrow{\ cold\ } H\!-\!\overset{\displaystyle H}{\underset{\displaystyle H}{C}}\!-\!\overset{\displaystyle H}{\underset{\displaystyle H}{C}}\!-\!OSO_3H$$

ethene   sulfuric acid   ethyl hydrogen sulfate

H_2SO_4, the catalyst, is regenerated in the second reaction.

$$H\!-\!\overset{\displaystyle H}{\underset{\displaystyle H}{C}}\!-\!\overset{\displaystyle H}{\underset{\displaystyle H}{C}}\!-\!OSO_3H + H\!-\!OH \longrightarrow H\!-\!\overset{\displaystyle H}{\underset{\displaystyle H}{C}}\!-\!\overset{\displaystyle H}{\underset{\displaystyle H}{C}}\!-\!OH + HOSO_3H$$

steam   ethanol

We see that the first step is an addition reaction in which H—OSO₃H adds across the double bond. This is followed by a substitution reaction in which —OH from water (steam) replaces —OSO₃H from ethyl hydrogen sulfate. These reactions amount to the net addition of water to ethene.

Ethylene glycol (Section 31-9) is an important solvent, industrial chemical, and permanent antifreeze. An important commercial synthesis of ethylene glycol from ethylene involves addition followed by substitution. Ethylene reacts with the hypochlorous acid present in chlorine water, and the product is then hydrolyzed in an aqueous solution of sodium carbonate.

$$CH_2\!=\!CH_2 + \quad HOCl \quad \longrightarrow \quad \underset{\underset{\displaystyle OH\ \ \ \ Cl}{|\quad\ \ |}}{CH_2\!-\!CH_2} \xrightarrow[2Na^+CO_3^{2-}]{H_2O} \underset{\underset{\displaystyle OH\ \ \ OH}{|\quad\ \ |}}{CH_2\!-\!CH_2}$$

hypochlorous acid   ethylene   1,2-ethanediol
(chlorine + water)   chlorohydrin   (ethylene glycol)

Ethylene glycol is completely miscible with water. It is widely used in permanent antifreeze (bp = 197°C).

One of the commercially most important addition reactions of the alkenes forms *polymers*. This reaction will be discussed in Section 32-11.

The alkynes contain two pi bonds, both of which are sources of electrons, and they are more reactive than the alkenes. The most common reaction of the alkynes is addition across the triple bond. The reactions with hydrogen and with bromine are typical.

$$H\!-\!C\!\equiv\!C\!-\!H \xrightarrow{\ H_2\ } \overset{\displaystyle H}{\underset{\displaystyle H}{}}C\!=\!C\overset{\displaystyle H}{\underset{\displaystyle H}{}} \xrightarrow{\ H_2\ } H_3C\!-\!CH_3$$

Each sequence may be thought of as a series of stepwise reactions. Stopping either sequence after the first step is difficult because the product still contains a reactive double bond.

$$H\!-\!C\!\equiv\!C\!-\!H \xrightarrow{\ Br_2\ } \overset{\displaystyle H}{\underset{\displaystyle Br}{}}C\!=\!C\overset{\displaystyle H}{\underset{\displaystyle Br}{}} \xrightarrow{\ Br_2\ } H\!-\!\overset{\displaystyle Br}{\underset{\displaystyle Br}{C}}\!-\!\overset{\displaystyle Br}{\underset{\displaystyle Br}{C}}\!-\!H$$

1,2-dibromoethene   1,1,2,2-tetrabromoethane

Other unsaturated bonds can also undergo addition reactions. Probably the most important example is the carbonyl group, $-\overset{\displaystyle O}{\overset{\|}{C}}-$. Because of the availability of lone pairs of electrons on the oxygen atom, the products can undergo a wide variety of subsequent reactions. For example, HCN adds to the C=O bond of acetone.

$$CH_3-\overset{\displaystyle O}{\overset{\|}{C}}-CH_3 + HCN \xrightarrow{\text{NaOH(aq)}} CH_3-\overset{\displaystyle OH}{\underset{\displaystyle CN}{\overset{|}{\underset{|}{C}}}}-CH_3$$

This is a key early step in the production of the transparent plastic known as Plexiglas or Lucite.

32-6 Elimination Reactions

An **elimination reaction** involves a *decrease* in the number of groups attached to carbon. The degree of unsaturation increases.

Vicinal (or *vic*) dihalides are dihalo compounds in which the halogen atoms are situated on adjacent carbon atoms.

$$CH_3-\underset{\displaystyle Cl}{\overset{|}{CH}}-\underset{\displaystyle Cl}{\overset{|}{CH}}CH_3$$

2,3-dichlorobutane (a *vic* dihalide)

The name *geminal* (or *gem*) dihalide is used for dihalides in which both halogen atoms are attached to the same carbon atom.

$$CH_3-\underset{\displaystyle Cl}{\overset{\displaystyle Cl}{\overset{|}{\underset{|}{C}}}}-CH_2CH_3$$

2,2-dichlorobutane (a *gem* dihalide)

One type of elimination reaction takes place when *vicinal* dihalides are treated with a mixture of zinc dust in acetic acid (or ethanol). The two halogen atoms are eliminated from the adjacent carbons, and the single bond between the two carbons is transformed into a double bond. Such a reaction is called a **dehalogenation** reaction. In this reaction the molecule becomes more *unsaturated*.

$$CH_3-\underset{\displaystyle Br}{\overset{|}{CH}}-\underset{\displaystyle Br}{\overset{|}{CH}}-CH_3 \xrightarrow[\substack{\text{in acetic acid} \\ \text{or ethanol}}]{\text{Zn}}$$

2,3-dibromobutane

$$\begin{matrix} CH_3 \\ \diagdown \\ \quad\quad C=C \\ H \diagup \quad\quad \diagdown CH_3 \\ \\ \textit{trans}\text{-2-butene} \\ \\ + \\ \\ CH_3 \quad\quad CH_3 \\ \diagdown \quad\quad \diagup \\ C=C \\ H \diagup \quad\quad \diagdown H \\ \\ \textit{cis}\text{-2-butene} \end{matrix} \quad + \quad ZnBr_2$$

Another type of elimination reaction can occur for chloro-, bromo- and iodoalkanes and is sometimes called **dehydrohalogenation**. In such a reaction, the halogen, X, from one C atom and a hydrogen from an adjacent C atom are eliminated. The single bond between the carbon atoms is changed to a double bond; again, the molecule becomes *more unsaturated*. The net reaction is the transformation of an alkyl halide (or haloalkane) into an alkene. Dehydrohalogenation reactions are usually catalyzed by strong bases such as sodium hydroxide, NaOH, and sodium ethoxide, CH_3CH_2ONa.

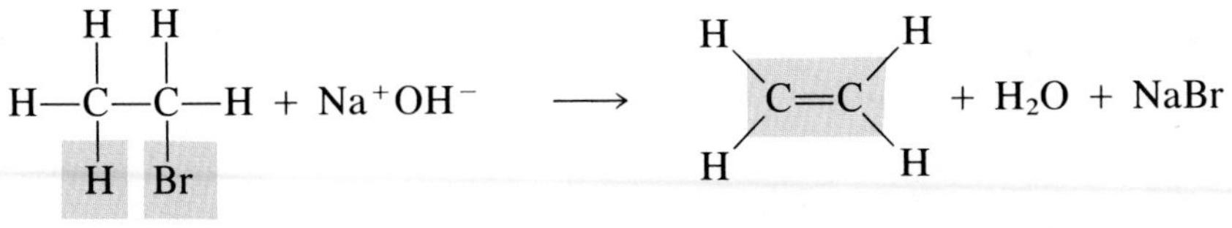

bromoethane   ethene
(ethylene)

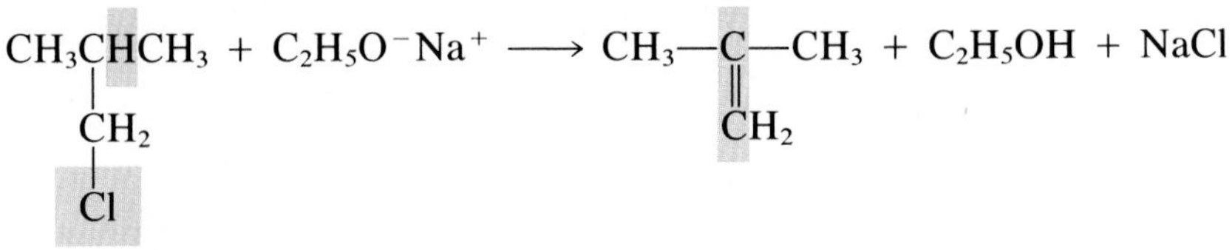

1-chloro-2-   sodium   2-methylpropene   ethanol
methylpropane   ethoxide

Ethanol, C_2H_5OH, is an even weaker acid than water, HOH. So we know that the ethoxide ion, $C_2H_5O^-$ (from C_2H_5ONa), is an even stronger base than OH^- (from NaOH). The strengths of organic acids and bases are discussed in Section 32-7.

A related reaction is **dehydration**, in which an alcohol is converted into an alkene and water by the elimination of —OH and —H from adjacent carbon atoms. The dehydration of an alcohol to form an alkene can be considered the reverse of the hydration of an alkene to form an alcohol (Section 32-5). Dehydration reactions are catalyzed by acids. Dehydration of tertiary alcohols (Section 31-9) is easier than dehydration of secondary alcohols, which in turn is easier than dehydration of primary alcohols. Drastic conditions are required for dehydration of primary alcohols, whereas much gentler conditions can accomplish the change for tertiary alcohols.

3° alcohol:

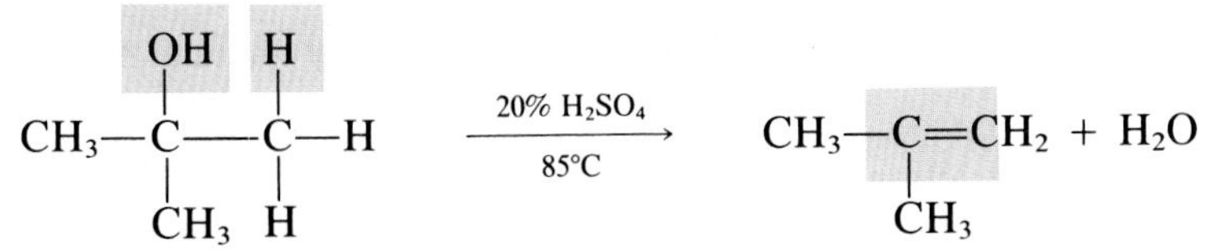

2-methyl-2-propanol   2-methylpropene
(*tert*-butyl alcohol)

2° alcohol:

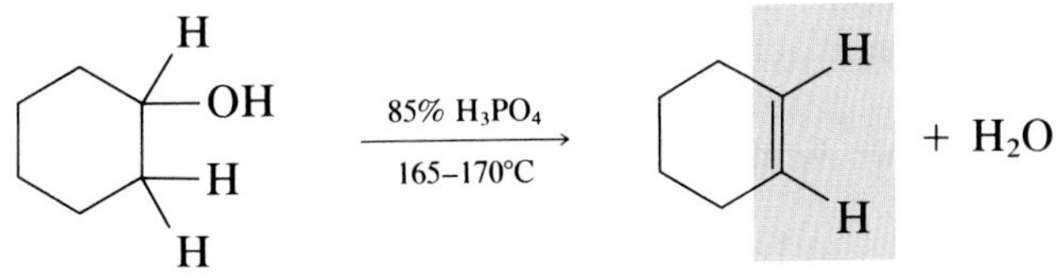

cyclohexanol   cyclohexene

Recall that H_3PO_4 is a weak acid; here it functions as a dehydrating agent.

1° alcohol:

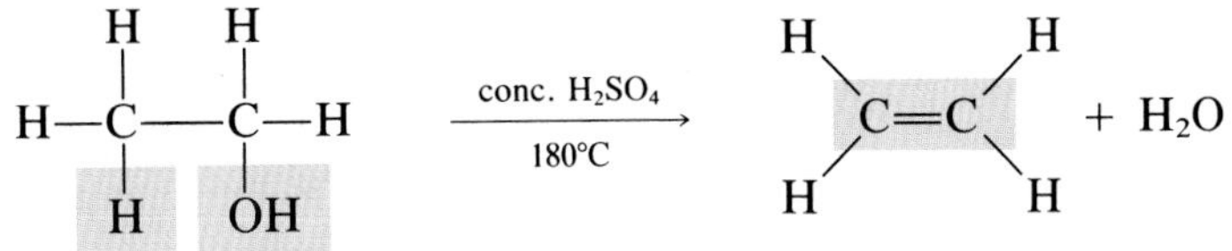

ethanol   ethene
(ethylene)

Concentrated H_2SO_4 is a *very* powerful dehydrating agent.

Such simple elimination reactions are relatively rare. However, elimination reactions frequently occur as individual steps in more complex re-

Concentrated sulfuric acid is an excellent dehydrating agent. Here it removes water from sucrose, a sugar with the formula $C_{12}H_{22}O_{11}$. Dehydration of sucrose produces (mostly) carbon.

action sequences. Many more elimination reactions will be encountered in a course in organic chemistry.

Some Other Organic Reactions

32-7 Reactions of Brønsted–Lowry Acids and Bases

Many organic compounds can act as weak Brønsted–Lowry acids or bases. Their reactions involve the transfer of H^+ ions, or *protons* (Section 10-3). Like similar reactions of inorganic compounds, these acid–base reactions of organic acids and bases are usually fast and reversible. Consequently, we can discuss the acidic or basic properties of organic compounds in terms of equilibrium constants (Section 18-4). Let us look briefly at some common types of organic acids and bases.

In the Brønsted–Lowry description, an *acid* is a *proton donor* and a *base* is a *proton acceptor*. Review the terminology of conjugate acid–base pairs in Section 10-3.

Some Organic Acids

The most important organic acids contain carboxyl groups, $-\overset{\overset{\displaystyle O}{\|}}{C}-O-H$.

They are called carboxylic acids (Section 31-12). They ionize slightly when dissolved in water, as illustrated with acetic acid.

$$CH_3COOH + H_2O \rightleftharpoons CH_3COO^- + H_3O^+ \qquad K_a = \frac{[CH_3COO^-][H_3O^+]}{[CH_3COOH]} = 1.8 \times 10^{-5}$$

$$\text{acid}_1 \qquad\qquad \text{base}_2 \qquad\qquad \text{base}_1 \qquad\qquad \text{acid}_2$$

Sources of some naturally occurring carboxylic acids.

Acetic acid is 1.3% ionized in 0.10 M solution. The acid strengths of the monocarboxylic acids are approximately the same, regardless of the lengths of the chains, with K_a values (in water) in the range 10^{-5} to 10^{-4}. Their acid

Table 32-1
K_a and pK_a Values of Some Carboxylic Acids

Name	Formula	K_a	pK_a
formic acid	HCOOH	1.8×10^{-4}	3.74
acetic acid	CH_3COOH	1.8×10^{-5}	4.74
propionic acid	CH_3CH_2COOH	1.4×10^{-5}	4.85
monochloroacetic acid	$ClCH_2COOH$	1.5×10^{-3}	2.82
dichloroacetic acid	$Cl_2CHCOOH$	5.0×10^{-2}	1.30
trichloroacetic acid	Cl_3CCOOH	2.0×10^{-1}	0.70
benzoic acid	C_6H_5COOH	6.3×10^{-5}	4.20
phenol*	C_6H_5OH	1.3×10^{-10}	9.89
ethanol*	CH_3CH_2OH	$\sim 10^{-18}$	~ 18

* Phenol and ethanol are not carboxylic acids. Phenol is weakly acidic compared to carboxylic acids, whereas ethanol (like other aliphatic alcohols) is far weaker still.

In Section 18-3 we defined

$$pK_a = -\log K_a$$

When K_a goes down by a factor of 10, pK_a goes up by one unit. We see that the weaker an acid, the higher its pK_a value.

Reaction of sodium metal with ethanol gives sodium ethoxide and hydrogen. Ethanol is much less acidic than water, so its reaction with sodium is much less vigorous than the reaction of water with sodium (Section 4-6).

strengths increase dramatically when electronegative substituents are present on the α-carbon atom (K_a values in water range from 10^{-3} to 10^{-1}). Compare acetic acid and the three substituted acetic acids in Table 32-1. There are two main reasons for this increase: (1) The electronegative substituents pull electron density from the carboxylic acid group, and (2) the more electronegative substituents help to stabilize the resulting carboxylate anion by spreading the negative charge over more atoms.

The alcohols are *very weakly acidic* compounds. However, they do not react with strong soluble bases. They are much weaker acids than water (see Table 32-1), but some of their reactions are analogous to those of water.

The very reactive metals react with alcohols to form **alkoxides** with the liberation of hydrogen.

$$2CH_3\text{—}OH + 2Na \longrightarrow H_2 + 2\,[Na^+ + CH_3O^-]$$
$$\text{sodium methoxide}$$
$$\text{(an alkoxide)}$$

$$2CH_3CH_2\text{—}OH + 2Na \longrightarrow H_2 + 2\,[Na^+ + CH_3CH_2O^-]$$
$$\text{sodium ethoxide}$$
$$\text{(an alkoxide)}$$

This is similar to the reaction of water with active metals.

$$2H\text{—}OH + 2Na \longrightarrow$$
$$H_2 + 2[Na^+ + OH^-]$$

The low-molecular-weight alkoxides are strong bases that react with water (hydrolyze) to form the parent alcohol and a strong soluble base.

$$[Na^+ + CH_3CH_2O^-] + H\text{—}OH \longrightarrow CH_3CH_2OH + [Na^+ + OH^-]$$
$$\text{sodium ethoxide} \qquad\qquad\qquad \text{ethanol}$$

Phenols react with metallic sodium to produce **phenoxides**; the reactions are analogous to those of alcohols. Because phenols are more acidic than alcohols, their reactions are more vigorous.

$$2\,\bigcirc\text{—OH} + 2Na \longrightarrow H_2 + 2\left[\bigcirc\text{—O}^- + Na^+\right]$$
$$\text{phenol} \qquad\qquad\qquad\qquad \text{sodium phenoxide}$$

Phenol and its substituted derivatives have K_a values on the order of 10^{-10}, so phenoxide salts are rather strongly basic; $K_b \approx 10^{-4}$.

Carboxylate salts behave in a similar fashion, but to a lesser extent.

$$RCOO^- + HX \rightleftharpoons RCOOH + X^-$$

$$RCOO^- + H_2O \rightleftharpoons RCOOH + OH^- \qquad K_b = \frac{[RCOOH][OH^-]}{[RCOO^-]}$$

$$\underset{base_1}{} \quad \underset{acid_2}{} \qquad \underset{acid_1}{} \quad \underset{base_2}{}$$

Applying the relationship $K_w = K_a K_b$ to the K_a values for carboxylic acids in Table 32-1, we see that typical K_b values for unsubstituted carboxylate salts are on the order of 10^{-9}; stronger acids such as the haloacetic acids give salts that are weaker bases.

In summary, we can rank the base strengths of these common organic bases.

alkoxides > aliphatic amines > phenoxides > carboxylates ≈ aromatic amines ≈ heterocyclic amines

We could describe the oxidation and reduction of organic compounds in terms of changes in oxidation numbers, just as we did for inorganic compounds in Sections 4-7 and 4-8. Formal application of oxidation number rules to organic compounds often leads to fractional oxidation numbers for carbon. For organic species, the descriptions in terms of increase or decrease of oxygen or hydrogen are usually easiest to apply.

32-8 Oxidation–Reduction Reactions in Organic Chemistry

Oxidation of an organic molecule usually corresponds to *increasing* its *oxygen* content or *decreasing* its *hydrogen* content. **Reduction** of an organic molecule usually corresponds to *decreasing* its *oxygen* content or *increasing* its *hydrogen* content.

For example, the oxygen content increases when an alkane is converted to an alcohol, so this process is an oxidation.

This reaction is not easy to carry out, and alcohols are prepared by many simpler methods. We show it here to illustrate the ideas of organic oxidations and reductions.

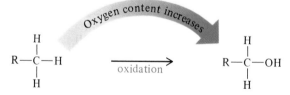

Oxygen content increases

$$R-\underset{\underset{H}{|}}{\overset{\overset{H}{|}}{C}}-H \xrightarrow{\text{oxidation}} R-\underset{\underset{H}{|}}{\overset{\overset{H}{|}}{C}}-OH$$

Converting a primary alcohol to an aldehyde or a secondary alcohol to a ketone is also an oxidation; the hydrogen content decreases.

The notation R′ emphasizes that the two R groups may be the same (R = R′, e.g., in the formation of acetone,

$$CH_3-\overset{\overset{O}{\|}}{C}-CH_3)$$

or different (R ≠ R′, e.g., in the formation of methyl ethyl ketone,

$$CH_3-\overset{\overset{O}{\|}}{C}-CH_2CH_3).$$

Hydrogen content decreases

1° alcohol $R-\underset{\underset{H}{|}}{\overset{\overset{OH}{|}}{C}}-H \xrightarrow{\text{oxidation}} R-\overset{\overset{O}{\|}}{C}-H$ (aldehyde)

2° alcohol $R-\underset{\underset{R'}{|}}{\overset{\overset{OH}{|}}{C}}-H \xrightarrow{\text{oxidation}} R-\overset{\overset{O}{\|}}{C}-R'$ (ketone)

An aldehyde can be oxidized to a carboxylic acid.

In each of these "oxidation" reactions, something else must act as the oxidizing agent (which is reduced). These oxidizing agents are often inorganic species such as dichromate ions, $Cr_2O_7^{2-}$, or permanganate ions, MnO_4^-. The reverse of each of the above reactions is a reduction of the organic molecule. In this reverse reaction the reducing agent (the substance that is oxidized) is often an inorganic compound.

Let us look at a few important types of organic oxidations and reductions.

Oxidation of Alcohols

Aldehydes can be prepared by the oxidation of *primary* alcohols. The reaction mixture is heated to a temperature slightly above the boiling point of the aldehyde so that the aldehyde distills out as soon as it is formed. Potassium dichromate in the presence of dilute sulfuric acid is the common oxidizing agent.

Aldehydes are easily oxidized to carboxylic acids. Therefore, they must be removed from the reaction mixture as soon as they are formed. Aldehydes have lower boiling points than the alcohols from which they are formed, so the removal of aldehydes is easily accomplished.

$$CH_3OH \xrightarrow[\text{dil. } H_2SO_4]{K_2Cr_2O_7} \underset{\substack{\text{methanal} \\ \text{(formaldehyde)} \\ \text{bp} = -21°C}}{H-\overset{\displaystyle O}{\overset{\|}{C}}-H}$$

methanol
bp = 65°C

$$CH_3CH_2CH_2CH_2OH \xrightarrow[\text{dil. } H_2SO_4]{K_2Cr_2O_7} CH_3CH_2CH_2\overset{\displaystyle O}{\overset{\|}{C}}-H$$

1-butanol
bp = 117.5°C

butanal
(butyraldehyde)
bp = 75.7°C

Ketones can be prepared by the oxidation of *secondary* alcohols. Ketones are not as susceptible to oxidation as are aldehydes. Potassium permanganate in alkaline solution may be used as the oxidizing agent.

Ketones are not as easily oxidized as aldehydes, because the further oxidation of a ketone requires the breaking of a carbon–carbon bond. Thus, it is not as important that they be quickly removed from the reaction mixture.

$$CH_3-\overset{\displaystyle OH}{\overset{|}{C}H}-CH_3 \xrightarrow[OH^-]{KMnO_4} CH_3-\overset{\displaystyle O}{\overset{\|}{C}}-CH_3$$

2-propanol
isopropyl alcohol

acetone

These two reactions can also be described as a type of *elimination* reaction (Section 32-6) called *dehydrogenation*. A molecule of hydrogen, H_2, is eliminated to form a C=O double bond.

cyclooctanol   →   cyclooctanone

When the carbon bearing the alcohol —OH group also has a hydrogen attached (a primary or secondary alcohol), the oxidation is easy; when it has no hydrogen attached (a tertiary alcohol), the oxidation is difficult.

$$CH_3—CH_2—\overset{\displaystyle OH}{\underset{|}{CH}}—CH_3 \xrightarrow[OH^-]{KMnO_4} CH_3—CH_2—\overset{\displaystyle O}{\overset{\|}{C}}—CH_3$$

2-butanol

2-butanone
methyl ethyl ketone

Aldehydes and ketones are prepared commercially by a catalytic process that involves passing alcohol vapors and air over a copper gauze or powder catalyst at approximately 300°C. Here the oxidizing agent is O_2.

$$2CH_3OH + O_2 \xrightarrow[300°C]{Cu} 2H—\overset{\displaystyle O}{\overset{\|}{C}}—H + 2H_2O$$

methanol

formaldehyde

Formaldehyde is quite soluble in water; the gaseous compound can be dissolved in water to give a 40% solution.

Acetaldehyde can be prepared by the similar oxidation of ethanol.

$$2CH_3CH_2OH + O_2 \xrightarrow[300°C]{Cu} 2CH_3—\overset{\displaystyle O}{\overset{\|}{C}}—H + 2H_2O$$

Oxidation of tertiary alcohols is difficult because the breaking of a carbon–carbon bond is required. Such oxidations are of little use in synthesis.

Oxidation or Reduction of Alkenes and Alkynes

A qualitative test for carbon–carbon unsaturation involves the oxidation of alkenes or alkynes with cold dilute aqueous solutions of $KMnO_4$. Dilute MnO_4^- solutions are pink. When an unsaturated hydrocarbon is added, the pink permanganate color disappears and MnO_2, a brown solid, is formed. With ethene the reaction is

Frequently, only the reactants and products of interest are shown in organic reactions. The reduction product in this reaction is MnO_2.

$$CH_2{=}CH_2 + MnO_4^- \xrightarrow{H_2O} \overset{\displaystyle OH \quad OH}{\underset{| \quad |}{CH_2—CH_2}} + MnO_2 + \text{other oxidation products}$$

pink

ethylene glycol (brown)

Other alkenes undergo similar oxidation reactions. Hot or concentrated $KMnO_4$ solutions oxidize ethene to carbon dioxide and water. Similarly, the unsat-

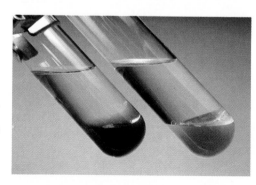

When aqueous $KMnO_4$ is shaken with hexane (left), it retains its pink color. The pink color is removed by reaction with the double bond of 1-hexene (right), and brown solid MnO_2 is formed.

uration in alkynes causes the disappearance of the pink MnO_4^- color and the formation of brown MnO_2.

The hydrogenation reactions by which alkenes are converted to alkanes (Section 32-5) increase the hydrogen content, so those addition reactions can be classed as *reductions*. Similarly, alkynes can be reduced (hydrogenated) to alkenes.

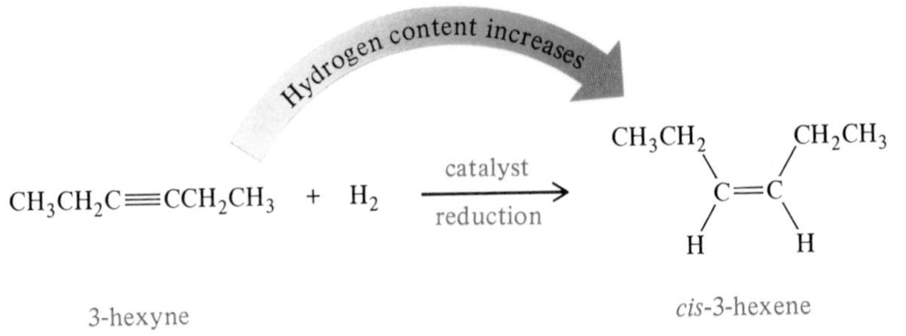

Hydrogen content increases

$$CH_3CH_2C{\equiv}CCH_2CH_3 \ + \ H_2 \ \xrightarrow[\text{reduction}]{\text{catalyst}} \ \underset{H \quad\quad H}{\overset{CH_3CH_2 \quad\quad CH_2CH_3}{C{=}C}}$$

3-hexyne

cis-3-hexene

Reduction of Carbonyl Compounds

Reduction of a variety of compounds that contain the carbonyl group provides synthetic methods to produce primary and secondary alcohols. A common, very powerful reducing agent is lithium aluminum hydride, $LiAlH_4$; other reducing agents include sodium in alcohol and sodium borohydride, $NaBH_4$.

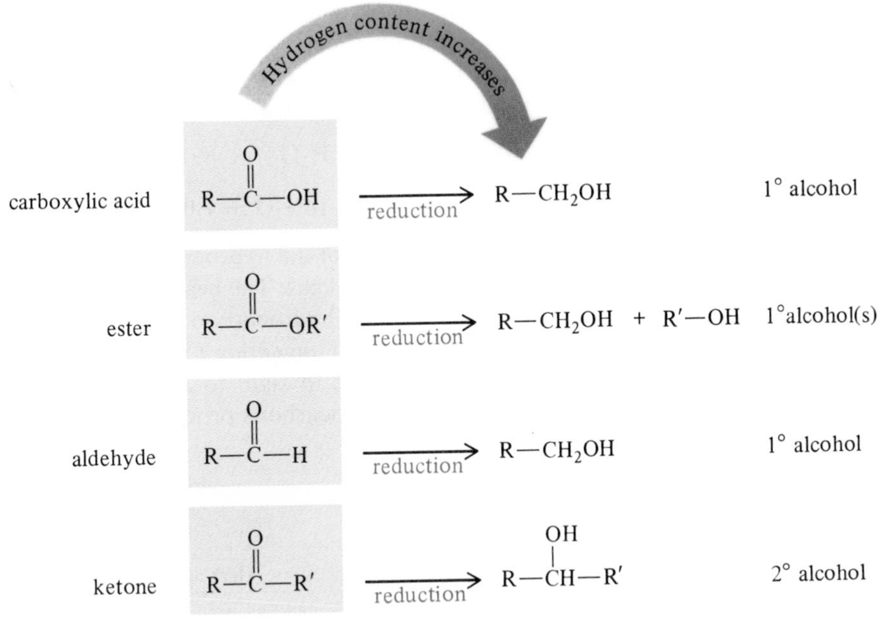

Hydrogen content increases

carboxylic acid $\quad \underset{\text{(}R{-}\overset{\overset{O}{\|}}{C}{-}OH\text{)}}{R{-}\overset{\overset{O}{\|}}{C}{-}OH} \quad \xrightarrow{\text{reduction}} \quad R{-}CH_2OH \qquad 1° \text{ alcohol}$

ester $\quad R{-}\overset{\overset{O}{\|}}{C}{-}OR' \quad \xrightarrow{\text{reduction}} \quad R{-}CH_2OH \ + \ R'{-}OH \quad 1°\text{alcohol(s)}$

aldehyde $\quad R{-}\overset{\overset{O}{\|}}{C}{-}H \quad \xrightarrow{\text{reduction}} \quad R{-}CH_2OH \qquad 1° \text{ alcohol}$

ketone $\quad R{-}\overset{\overset{O}{\|}}{C}{-}R' \quad \xrightarrow{\text{reduction}} \quad R{-}\overset{\overset{OH}{|}}{CH}{-}R' \qquad 2° \text{ alcohol}$

Organic reactions are sometimes written in extremely abbreviated form. This is often the case when a variety of common oxidizing or reducing agents will accomplish the desired conversion.

Oxidation of Aromatic Compounds and Alkylbenzenes

Compounds that are entirely aromatic, such as benzene and the condensed aromatics (Sections 31-5, 31-6), are quite resistant to oxidation by chemical oxidizing agents. The reactions of strong oxidizing agents with alkylbenzenes illustrate the stability of the benzene ring system. Heating toluene with a basic solution of $KMnO_4$ results in a nearly 100% yield of benzoic acid. The ring itself remains intact; only the nonaromatic portion of the molecule is oxidized.

The MnO_2 is removed by filtration of the basic solution. Benzoic acid, an insoluble weak acid, is then precipitated by acidification with a strong inorganic acid.

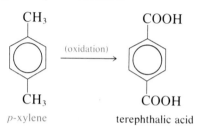

toluene

benzoic acid

Two such alkyl groups on an aromatic ring are oxidized to give a diprotic acid, as the following example illustrates.

p-xylene

terephthalic acid

Terephthalic acid is used to make "polyesters," an important class of polymers (Section 32-11).

Combustion of Organic Compounds

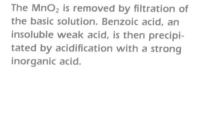

Acetylene is produced by the slow addition of water to calcium carbide.

$CaC_2(s) + 2H_2O(\ell) \longrightarrow$
$\qquad HC{\equiv}CH(g) + Ca(OH)_2(s)$

The light of one kind of headlamp used by miners and cave explorers is given off by the combustion of acetylene.

Recall that ΔH^0 is negative for an exothermic process.

The most extreme oxidation reactions of organic compounds occur when they burn in O_2. Such *combustion reactions* (Section 6-8, part 3) are highly exothermic. When the combustion takes place in excess O_2, the products are CO_2 and H_2O. Examples of alkane combustions are

methane: $\quad CH_4 \quad + 2O_2 \longrightarrow CO_2 \quad + H_2O \quad + 891 \text{ kJ}$

n-octane: $\quad 2C_8H_{18} + 25O_2 \longrightarrow 16CO_2 + 18H_2O + 1.090 \times 10^4 \text{ kJ}$

Such reactions are the bases for important uses of the hydrocarbons as fuels, e.g., gasoline, diesel fuel, heating oil, and natural gas. The **heat of combustion** is the amount of energy *liberated* per mole of hydrocarbon burned. Heats of combustion are assigned positive values *by convention* (Table 32-3) and are therefore equal in magnitude, but opposite in sign, to ΔH^0 values for combustion reactions. The combustion of hydrocarbons produces large vol-

Table 32-3 Heats of Combustion of Some Alkanes			
		Heat of Combustion	
Hydrocarbon		kJ/mol	J/g
methane	CH_4	891	55.7
propane	C_3H_8	2220	50.5
n-pentane	$n\text{-}C_5H_{12}$	3507	48.7
n-octane	$n\text{-}C_8H_{18}$	5450	47.8
n-decane	$n\text{-}C_{10}H_{22}$	6737	47.4
ethanol*	C_2H_5OH	1372	29.8

* Not an alkane; included for comparison only.

umes of gases in addition to large amounts of heat. The rapid formation of these gases at high temperature and pressure drives the pistons or turbine blades in internal combustion engines.

In the absence of sufficient oxygen, partial combustion of hydrocarbons occurs. The products may be carbon monoxide (a very poisonous gas) or carbon (which deposits on spark plugs, in the cylinder head, and on the pistons of automobile engines). Many modern automobile engines now use microcomputer chips and sensors to control the air supply and to optimize the fuel/O_2 ratio. The reactions of methane with insufficient oxygen are

$$2CH_4 + 3O_2 \longrightarrow 2CO + 4H_2O \quad \text{and} \quad CH_4 + O_2 \longrightarrow C + 2H_2O$$

All hydrocarbons undergo similar reactions.

The alkenes, like the alkanes, burn in *excess* oxygen to form carbon dioxide and water in exothermic reactions.

$$CH_2{=}CH_2 + 3O_2 \text{ (excess)} \longrightarrow 2CO_2 + 2H_2O + 1387 \text{ kJ}$$

When an alkene (or any other unsaturated organic compound) is burned in air, a yellow, luminous flame is observed and considerable soot (unburned carbon) is formed. This reaction provides a qualitative test for unsaturation. Saturated hydrocarbons burn in air without forming significant amounts of soot.

Acetylene lamps are charged with calcium carbide. Very slow addition of water produces acetylene, which is burned as it is produced. Acetylene is also used in the oxyacetylene torch for welding and cutting metals. When acetylene is burned with oxygen, the flame reaches temperatures of about 3000°C.

Like other hydrocarbons, the *complete combustion* of aromatic hydrocarbons releases large amounts of energy.

$$2C_6H_6 + 15O_2 \longrightarrow 12CO_2 + 6H_2O + 6548 \text{ kJ}$$

Because they are so unsaturated, aromatic hydrocarbons burn *in air* with a yellow, sooty flame.

32-9 Formation of Carboxylic Acid Derivatives

The carboxylic acid derivatives introduced in Section 31-13 can be formed by *substitution* of another group in place of —OH in the carboxyl group. The acyl halides (acid halides) are usually prepared by treating acids with PCl_3, PCl_5, or $SOCl_2$ (thionyl chloride). In general terms, the reaction of acids with PCl_5 may be represented as

Hexane, C_6H_{14}, an alkane, burns cleanly in air to give CO_2 and H_2O (top). 1-Hexene, C_6H_{12}, an alkene, burns with a flame that contains soot (middle). Burning o-xylene, an aromatic hydrocarbon, produces large amounts of soot (bottom).

The acyl halides are much more reactive than their parent acids. Consequently, they are often used in reactions to introduce an acyl group into another molecule.

When an organic acid is heated with an alcohol, an equilibrium is established with the resulting *ester* and water. The reaction is catalyzed by traces of strong inorganic acids, such as a few drops of concentrated H_2SO_4.

$$CH_3-\overset{\overset{\displaystyle O}{\|}}{C}-OH \;+\; CH_3CH_2-OH \; \underset{}{\overset{H^+, \Delta}{\rightleftharpoons}} \; CH_3-\overset{\overset{\displaystyle O}{\|}}{C}-O-CH_2CH_3 \;+\; H_2O$$

| acetic acid | ethyl alcohol | ethyl acetate, an ester |

In general terms, the reaction of an organic acid and an alcohol may be represented as

Many experiments have shown conclusively that the OH group from the acid and the H from the alcohol are the atoms that form water molecules.

$$R-\overset{\overset{\displaystyle O}{\|}}{C}-(OH) \;+\; R'-O(H) \; \rightleftharpoons \; R-\overset{\overset{\displaystyle O}{\|}}{C}-O-R' \;+\; H_2O$$

| acid | alcohol | ester |

(R and R′ may be the same or different alkyl groups.)

Reactions between acids and alcohols are usually quite slow and require prolonged boiling (refluxing). However, the reactions between most acyl halides and most alcohols occur very rapidly without requiring the presence of an acid catalyst.

A reaction in which a small molecule, such as water or hydrogen chloride, is eliminated and two molecules are joined is often called a *condensation reaction.*

$$CH_3-\overset{\overset{\displaystyle O}{\|}}{C}-(Cl) \;+\; CH_3-CH_2-O(H) \; \longrightarrow \; CH_3-\overset{\overset{\displaystyle O}{\|}}{C}-O-CH_2CH_3 \;+\; HCl$$

| acetyl chloride | ethyl alcohol | ethyl acetate |

Amides are usually *not* prepared by the reaction of an amine with an organic acid. Acyl halides and acid anhydrides react readily with primary and secondary amines to produce amides. The reaction of an acyl halide with a primary or secondary amine produces an amide and a salt of the amine.

In both of these preparations, one half of the amine is converted to an amide and the other half to a salt.

$$2CH_3NH_2 \;+\; CH_3-\overset{\overset{\displaystyle O}{\|}}{C}-Cl \; \longrightarrow \; CH_3-\overset{\overset{\displaystyle O}{\|}}{C}-\underset{\underset{\displaystyle CH_3}{|}}{N}-H \;+\; CH_3NH_3{}^+\,Cl^-$$

| methylamine (a primary amine) | acetyl chloride (an acyl halide) | N-methylacetamide (an amide) | methylammonium chloride (a salt) |

The reaction of an acid anhydride with a primary or a secondary amine produces an amide and a salt of the amine.

$$2(CH_3)_2NH \ + \ CH_3\overset{O}{\overset{\|}{C}}-O-\overset{O}{\overset{\|}{C}}-CH_3 \ \longrightarrow \ CH_3-\overset{O}{\overset{\|}{C}}-\overset{CH_3}{\underset{CH_3}{N}} \ + \ CH_3-\overset{O}{\overset{\|}{C}}-O^- \ ^+H_2N(CH_3)_2$$

| dimethylamine | acetic anhydride | N,N-dimethylacetamide | dimethylammonium |
| (a secondary amine) | (an acid anhydride) | | acetate (a salt) |

32-10 Hydrolysis of Esters

Most esters are not very reactive, so strong reagents are required for their reactions. Esters can be hydrolyzed by refluxing with solutions of strong bases.

$$CH_3-\overset{O}{\overset{\|}{C}}-O-CH_2CH_3 \ + \ Na^+OH^- \ \overset{\Delta}{\longrightarrow} \ CH_3\overset{O}{\overset{\|}{C}}-O^- \ Na^+ \ + \ CH_3CH_2OH$$

ethyl acetate   sodium acetate   ethanol

The hydrolysis of esters in the presence of strong soluble bases is called **saponification** (soap-making). The hydrolysis of fats and oils produces soaps.

In general terms, the hydrolysis of esters may be represented as

$$R-\overset{O}{\overset{\|}{C}}-O-R' \ + \ Na^+OH^- \ \overset{\Delta}{\longrightarrow} \ R-\overset{O}{\overset{\|}{C}}-O^-Na^+ \ + \ R'OH$$

ester   salt of an acid   alcohol

Like other esters, fats and oils (Section 31-13) can be hydrolyzed in strongly basic solution to produce salts of the acids and the alcohol glycerol. The resulting sodium salts of long-chain fatty acids are soaps. In Section 14-18 we described the cleansing action of soaps and detergents.

American pioneers prepared their soaps by boiling animal fat with an alkaline solution obtained from the ashes of hardwood. The resulting soap could be "salted out" by adding sodium chloride, making use of the fact that soap is less soluble in a salt solution than in water.

$$\begin{array}{c}
\overset{\quad\quad\overset{\displaystyle O}{\parallel}}{H_2C-O-C-(CH_2)_{16}CH_3} \\
\overset{\quad\quad\overset{\displaystyle O}{\parallel}}{HC-O-C-(CH_2)_{16}CH_3} \quad + \; 3Na^+\,OH^- \longrightarrow 3CH_3(CH_2)_{16}\overset{\displaystyle O}{\overset{\parallel}{C}}-O^-Na^+ \;+\; \begin{array}{c} H_2C-OH \\ HC-OH \\ H_2C-OH \end{array} \\
\overset{\quad\quad\overset{\displaystyle O}{\parallel}}{H_2C-O-C-(CH_2)_{16}CH_3}
\end{array}$$

glyceryl tristearate   sodium stearate   glycerol
(a fat)   (a soap)

32-11 Polymerization Reactions

The word fragment -mer means "part." Recall that isomers are compounds that are composed of the same (iso) parts (mers). A monomer is a "single part"; a large number of monomers combine to form a polymer, "many parts."

A **polymer** is a large molecule that is an extremely high-molecular-weight chain of small molecules. The small molecules that are linked to form polymers are called **monomers**. Typical polymers consist of hundreds or thousands of monomers and have molecular weights up to thousands or millions of grams per mole.

Polymers are divided into two classes—natural and synthetic. Important biological molecules such as proteins, nucleic acids, and polysaccharides (starches and the cellulose in wood or cotton) are natural polymers. Natural rubber and natural fibers such as silk and wool are also natural polymers. Familiar examples of synthetic polymers include plastics such as polyethylene, Teflon, and Lucite (Plexiglas) and synthetic fibers such as Nylon, Orlon, and Dacron. In this section we shall describe some processes by which polymers are formed from organic compounds.

> **Polymerization** is the combination of many small molecules (monomers) to form large molecules (polymers).

Addition Polymerization

Polymerization is an important reaction of the alkenes. It is an addition reaction (Section 32-5), so polymers formed by this reaction are called **addition polymers**. The formation of polyethylene is an important example. In the presence of appropriate catalysts (a mixture of aluminum trialkyls, R_3Al, and titanium tetrachloride, $TiCl_4$), ethylene polymerizes into chains containing 800 or more carbon atoms.

$$n CH_2{=}CH_2 \xrightarrow{\text{catalyst}} {+}CH_2{-}CH_2{+}_n$$

ethylene   polyethylene

The polymer may be represented as $CH_3(CH_2{-}CH_2)_nCH_3$, where n is approximately 400. Polyethylene is a tough, flexible plastic. It is widely used as an electrical insulator and for the fabrication of such items as unbreakable refrigerator dishes, plastic cups, and squeeze bottles. Polypropylene is made by polymerizing propylene, $CH_3{-}CH{=}CH_2$, in much the same way. Teflon is made by polymerizing tetrafluoroethylene in a similar reaction.

Teflon is a trade name owned by Du Pont, a company that has developed and manufactured many fluorinated polymers.

$$n CF_2{=}CF_2 \xrightarrow[\text{heat}]{\text{catalyst}} {+}CF_2{-}CF_2{+}_n$$

tetrafluoroethylene   "Teflon"

Many cooking utensils with "nonstick" surfaces are coated with a polymer such as Teflon.

The molecular weight of Teflon is about 2×10^6. Approximately 20,000 $CF_2{=}CF_2$ molecules polymerize to form a single giant molecule. Teflon is a very useful polymer. It does *not* react with concentrated acids and bases or with most oxidizing agents, nor does it dissolve in most organic solvents.

Natural rubber is obtained from the sap of the rubber tree, a sticky liquid called latex. Rubber is a polymeric hydrocarbon formed (in the sap) by the combination of about 2000 molecules of 2-methyl-1,3-butadiene, commonly called isoprene. The molecular weight of rubber is about 136,000.

$$2n CH_2{=}\overset{\overset{\textstyle CH_3}{|}}{C}{-}CH{=}CH_2 \longrightarrow \left(CH_2{-}\overset{\overset{\textstyle CH_3}{|}}{C}{=}CH{-}CH_2{-}CH_2{-}\overset{\overset{\textstyle CH_3}{|}}{C}{=}CH{-}CH_2\right)_n$$

isoprene   natural rubber

When natural rubber is warmed, it flows and becomes sticky. To eliminate this problem, **vulcanization** is used. This is a process in which sulfur is added to rubber and the mixture is heated to approximately 140°C. Sulfur atoms combine with some of the double bonds in the linear polymer molecules to form bridges that bond one rubber molecule to another. This cross-linking by sulfur atoms converts the linear polymer into a three-dimensional polymer. Fillers and reinforcing agents are added during the mixing process to increase the durability of rubber and to form colored rubber. Carbon black is the most common reinforcing agent. Zinc oxide, barium sulfate, titanium dioxide, and antimony(V) sulfide are common fillers.

Some synthetic rubbers are superior to natural rubber in some ways. Neoprene is a synthetic elastomer (an elastic polymer) with properties quite similar to those of natural rubber. The basic structural unit is chloroprene, which differs from isoprene in having a chlorine atom rather than a methyl group as a substituent on the 1,3-butadiene chain.

Numerous other polymers are elastic enough to be called by the generic name "rubber."

$$n CH_2{=}CH{-}\overset{\overset{\textstyle Cl}{|}}{C}{=}CH_2 \xrightarrow{\text{polymerization}} \left(CH_2{-}CH{=}\overset{\overset{\textstyle Cl}{|}}{C}{-}CH_2\right)_n$$

chloroprene   neoprene (a synthetic rubber)

Neoprene is less affected by gasoline and oil, and is more elastic than natural rubber. It resists abrasion well and is not swollen or dissolved by hydrocarbons. It is widely used to make hoses for oil and gasoline, electrical insulation, and automobile and refrigerator parts.

If two different monomers are mixed and then polymerized, **copolymers** are formed. Depending on the ratio of the two monomers and the reaction conditions, the order of the units can range from quite regular (e.g., alternating) to completely random. In this way, polymers with a wide variety of properties can be produced. The most important rubber produced in the United States is SBR, a polymer of styrene with butadiene in a 1:3 ratio.

The double bonds in SBR can be cross-linked by vulcanization, as described above for natural rubber. SBR is used primarily for making tires. Other copolymers are used to make car bumpers, body and chassis parts, wire insulation, sporting goods, sealants, and caulking compounds.

Some additional polymers and their uses are listed in Table 32-4.

Condensation Polymerization

Some polymerization reactions are based on *condensation reactions*, in which two molecules combine by splitting out or eliminating a small molecule. For such a polymer to be formed, each monomer must have two functional groups, one on each end. A polymer formed in this way is called a **condensation polymer**. There are many useful condensation polymers, based on a wide variety of bifunctional molecules.

Polyesters (short for "*poly*meric *esters*") are condensation polymers that are formed when *dihydric alcohols* react with *dicarboxylic acids*. An ester linkage is formed at each end of each monomer molecule to build up large molecules. A useful polyester is prepared from ethylene glycol and terephthalic acid:

Dihydric alcohols contain two —OH groups per molecule.

polyethylene terephthalate (Dacron, Mylar)

Table 32-4
Some Important Addition Polymers

Polymer Name (some trade names)	Some Uses	Polymer Production, Tons/Yr in United States	Monomer Formula	Monomer Name (top-50 rank in U.S. chemical production)
polyethylene (Polythene)	electrical insulation; toys and molded objects; bags; squeeze bottles	8 million	$\underset{H}{\overset{H}{>}}C=C\underset{H}{\overset{H}{<}}$	ethylene (4)
polypropylene (Herculon, Vectra)	bottles; films; lab equipment; toys; packaging film; filament for rope, webbing, carpeting; molded auto and appliance parts	2.7 million	$\underset{H}{\overset{H}{>}}C=C\underset{CH_3}{\overset{H}{<}}$	propylene (10)
polyvinyl chloride (PVC)	pipe, siding, gutters; floor tiles; phonograph records	3.5 million	$\underset{H}{\overset{H}{>}}C=C\underset{Cl}{\overset{H}{<}}$	vinyl chloride (20)
polyacrylonitrile (Orlon, Acrilan)	acrylic fibers for carpets, clothing, knitwear	920,000	$\underset{H}{\overset{H}{>}}C=C\underset{CN}{\overset{H}{<}}$	acrylonitrile (40)
polystyrene (Styrene, Styrofoam, Styron)	molded toys, dishes, kitchen equipment; insulating foam, e.g., ice chests; rigid foam packaging	2 million	$\underset{H}{\overset{H}{>}}C=C\overset{H}{<}\!\!\bigcirc$	styrene (21)
polyvinylacetate (PVA)	water-based "latex" paint; adhesives; paper and textile coatings	500,000	$\underset{H}{\overset{H}{>}}C=C\underset{O-C-CH_3}{\overset{H}{<}}$ with $\parallel O$	vinyl acetate (41)
poly(methylmeth-acrylate) (Plexiglas, Lucite)	high-quality transparent objects; water-based paints; contact lenses	450,000	$\underset{H}{\overset{H}{>}}C=C\underset{C-OCH_3}{\overset{H}{<}}$ with $\parallel O$	methyl methacrylate
polytetrafluoro-ethylene (Teflon)	gaskets; pan coatings; electrical insulation; bearings	7,000	$\underset{F}{\overset{F}{>}}C=C\underset{F}{\overset{F}{<}}$	tetrafluoroethylene
polybutadiene	automotive tire tread, hoses, belts; metal can coatings	400,000	$\underset{H}{\overset{H}{>}}C=C\overset{H}{<}$ $C=C\underset{H}{\overset{H}{<}}$	butadiene
ethylene–propylene copolymer	appliance parts; auto hoses, bumpers, body and chassis parts; coated fabrics	150,000	see above	ethylene, propylene
SBR copolymer	tires	1.4 million	see above	styrene, butadiene

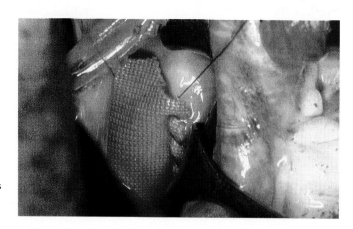

A patch made of Dacron polymer is used to close a defect in a human heart.

More than 2 million tons of this polymer are produced annually in the United States. Dacron, the fiber produced from this polyester, accounts for 50% of all synthetic fibers. It absorbs very little moisture, and its properties are nearly the same whether it is wet or dry. Additionally, it possesses exceptional elastic recovery properties, so it is used to make "permanent-press" fabrics. This polyester can also be made into films of great strength (e.g., Mylar), which can be rolled into sheets 1/30 the thickness of a human hair. Such films can be magnetically coated to make audio and video tapes.

The molecular weight of the polymer varies from about 10,000 to about 25,000. It melts at about 260 to 270°C.

The polymeric amides, **polyamides**, are an especially important class of condensation polymers. **Nylon** is the best known polyamide. It is prepared by heating anhydrous hexamethylenediamine with anhydrous adipic acid, a

Polymers have a wide range of properties and uses. Mylar sheet polymer is used to protect documents, such as this photocopy of a later version of the Declaration of Independence (top). Another kind of polymer coating is used to make shatterproof shields of glass items such as fluorescent light bulbs (bottom).

Nylon is formed at the interface where hexamethylenediamine (in the lower water layer) and adipyl chloride (a derivative of adipic acid, in the upper hexane layer) react. The Nylon can be drawn out and wound on a stirring rod.

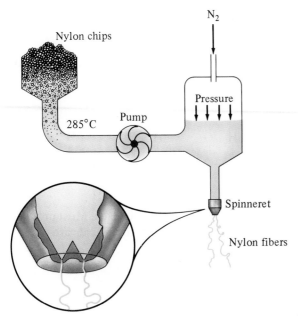

Figure 32-11
Fibers of synthetic polymers are made by extrusion of the molten material through tiny holes, called *spinnerets*. After cooling, Nylon fibers are stretched to about four times their original length to orient the polymer molecules.

dicarboxylic acid. This substance is called Nylon 66 because the parent diamine and dicarboxylic acid each contain six carbon atoms.

$$\left.\begin{array}{c} \underset{\text{adipic acid}}{\text{HO}-\overset{\overset{\text{O}}{\|}}{\text{C}}-(\text{CH}_2)_4-\overset{\overset{\text{O}}{\|}}{\text{C}}-\text{OH}} \\ + \\ \underset{\text{hexamethylenediamine}}{\text{H}_2\text{N}-(\text{CH}_2)_6-\text{NH}_2} \end{array}\right\} \xrightarrow[-\text{H}_2\text{O}]{\text{heat}} -\text{NH}-\left(\overset{\overset{\text{O}}{\|}}{\text{C}}-(\text{CH}_2)_4-\overset{\overset{\text{O}}{\|}}{\text{C}}-\text{NH}-(\text{CH}_2)_6-\text{NH}\right)_n\overset{\overset{\text{O}}{\|}}{\text{C}}-$$

Nylon 66
(a polyamide)

Molten Nylon is drawn into threads (Figure 32-11). After cooling to room temperature, these can be stretched to about four times their original length. The "cold drawing" process orients the polymer molecules so that their long axes are parallel to the fiber axis. At regular intervals there are N—H---O hydrogen bonds that *cross-link* adjacent chains to give strength to the fiber.

Petroleum is the ultimate source of both adipic acid and hexamethylenediamine. We do not mean that these compounds are present in petroleum, only that they are made from it. The same is true for many industrial chemicals. The cost of petroleum is an important factor in our economy because so many different products are derived from petroleum.

Some types of natural condensation polymers play crucial roles in living systems. **Proteins** are polymeric chains of *L-amino acids* (Sections 31-12, 32-1) linked by peptide bonds. A **peptide bond** is formed by the elimination of a molecule of water between the amino group of one amino acid and the carboxylic acid group of another.

peptide bond

$$\underset{\overset{|}{\text{H}}\;\;\overset{|}{\text{R}}}{\text{H}-\overset{\overset{\text{H}}{|}}{\text{N}}-\overset{\overset{|}{|}}{\text{C}}-\overset{\overset{\text{O}}{\|}}{\text{C}}-\text{OH}} + \underset{\overset{|}{\text{H}}\;\;\overset{|}{\text{R}'}}{\text{H}-\overset{\overset{\text{H}}{|}}{\text{N}}-\overset{\overset{|}{|}}{\text{C}}-\overset{\overset{\text{O}}{\|}}{\text{C}}-\text{OH}} \longrightarrow \underset{\overset{|}{\text{H}}\;\;\overset{|}{\text{R}}}{\text{H}-\overset{\overset{\text{H}}{|}}{\text{N}}-\overset{\overset{|}{|}}{\text{C}}-\overset{\overset{\text{O}}{\|}}{\text{C}}}-\underset{\overset{|}{\text{H}}\;\;\overset{|}{\text{R}'}}{\overset{\overset{\text{H}}{|}}{\text{N}}-\overset{\overset{|}{|}}{\text{C}}-\overset{\overset{\text{O}}{\|}}{\text{C}}-\text{OH}} + \text{H}_2\text{O}$$

When this process is carried out repeatedly, a large molecule called a **polypeptide** is formed.

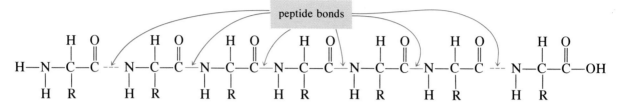

We see that the monomer units of proteins are the amino acids. Most proteins contain 100 to 300 amino acid units. Twenty different amino acids are usually

Table 32-5
Some Functions of Proteins

Example	Function
Enzymes	
amylase	converts starch to glucose
DNA polymerase I	repairs DNA molecule
transaminase	transfers amino group from one amino acid to another
Structural Proteins	
viral coat proteins	outer covering of virus
keratin	hair, nails, horns, hoofs
collagen	tendons, cartilage
Hormones	
insulin, glucagon	regulate glucose metabolism
oxytocin	regulates milk production in female mammals
vasopressin	increases retention of water by kidney
Contractile Proteins	
actin	thin contractile filaments in muscle
myosin	thick filaments in muscle
Storage Proteins	
casein	a nutrient protein in milk
ferritin	stores iron in spleen and egg yolk
Transport Proteins	
hemoglobin	carries O_2 in blood
myoglobin	carries O_2 in muscle
serum albumin	carries fatty acids in blood
cyctochrome c	transfers electrons
Immunological Proteins	
γ-globulins	form complexes with foreign proteins
Toxins	
neurotoxin	blocker of nerve function in cobra venom
ricin	nerve toxin in South American frog (most toxic substance known— 0.000005 g is fatal to humans)

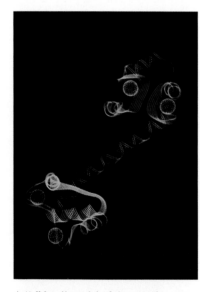

A "ribbon" model of the protein calmodulin. In this type of model, the ribbon represents the polypeptide chain. This protein coordinates with Ca^{2+} ions (white spheres) and aids in transporting them in living systems.

found in proteins. They have in common the general structure indicated on page 1114, but different side chains, (R), distinguish the amino acids from one another.

Proteins make up more than 50% of the dry weight of animals and bacteria. They perform many important functions in living organisms, a few of which are indicated in Table 32-5. Each protein carries out a specific biochemical function. Each is a polypeptide with its own unique *sequence* of amino acids. The amino acid sequence of a protein determines exactly how it folds up in a three-dimensional conformation and how it performs its precise biochemical task.

Key Terms

Achiral Describes an object that *can* be superimposed with its mirror image.

Addition reaction A reaction in which two atoms or groups of atoms are added to a molecule, one on each side of a double or triple bond. The number of groups attached to carbon *increases*, and the molecule becomes more nearly saturated.

Brønsted–Lowry acid A species that acts as a proton donor.

Brønsted–Lowry base A species that acts as a proton acceptor.

Chiral Describes an object that *cannot* be superimposed with its mirror image.

Condensation polymer A polymer that is formed by a condensation reaction.

Condensation reaction A reaction in which a small molecule, such as water or hydrogen chloride, is eliminated and two molecules are joined.

Conformation One specific geometry of a molecule. The conformations of a compound differ from one another only by rotation about single bonds.

Copolymer A polymer formed from two different compounds (monomers).

Dehydration The reaction in which H and OH are eliminated from adjacent carbon atoms, to form water and a more unsaturated bond.

Dehalogenation A reaction in which two halogen atoms are eliminated from a *vicinal* dihalide to form a double bond.

Dehydrogenation Elimination of two hydrogen atoms from a compound to form a double bond.

Dehydrohalogenation An elimination reaction in which a hydrogen halide, HX (X = Cl, Br, I), is eliminated from a haloalkane. A C=C double bond is formed.

Elimination reaction A reaction in which the number of groups attached to carbon *decreases*. The degree of unsaturation in the molecule increases.

Enantiomers See *Optical isomers*.

Esterification The reaction in which an alcohol and a carboxylic acid eliminate water and link to form an ester.

Geometrical isomers Compounds with different arrangements of groups on the opposite sides of a bond with restricted rotation, such as a double bond or a single bond in a ring; for example, *cis–trans* isomers of certain alkenes.

Hydration reaction A reaction in which the elements of water, H and OH, add across a double or triple bond.

Hydrogenation The reaction in which hydrogen adds across a double or triple bond.

Monomers The small molecules from which polymers are formed.

Optical isomers Molecules that are nonsuperimposable mirror images of one another, i.e., that bear the same relationship to each other as do left and right hands; also called *enantiomers*.

Oxidation (as applied to organic compounds) The increase of oxygen content or the decrease of hydrogen content of an organic molecule.

Peptide bond A bond formed by elimination of a molecule of water between the amino group of one amino acid and the carboxylic acid group of another.

Polyamide A polymeric amide.

Polyester A polymeric ester.

Polymerization The combination of many small molecules (monomers) to form large molecules (polymers).

Polymers Large molecules formed by the combination of many small molecules (monomers).

Polypeptide A polymer composed of amino acids linked by peptide bonds.

Protein A naturally occurring polymeric chain of L-amino acids linked together by peptide bonds.

Racemic mixture A single sample containing equal amounts of the two enantiomers (optical isomers) of a compound; does not rotate the plane of polarized light.

Reduction (as applied to organic compounds) The decrease of oxygen content or the increase of hydrogen content of an organic molecule.

Saponification The hydrolysis of esters in the presence of strong soluble bases.

Soap The sodium salt of a long-chain fatty acid.

Stereoisomers Isomers in which the atoms are linked together in the same atom-to-atom order, but with different arrangements in space. See *Geometrical isomers, Optical isomers.*

Structural isomers   Compounds that contain the same number of the same kinds of atoms but that differ in the order in which their atoms are bonded together.

Substitution reaction   A reaction in which an atom or a group of atoms attached to a carbon atom is replaced by another atom or group of atoms. No change occurs in the degree of saturation at the reactive carbon.

Vulcanization   The process in which sulfur is added to rubber and heated to 140°C, to cross-link the linear rubber polymer into a three-dimensional polymer.

Exercises

Geometries of Organic Molecules

1. Name three types of isomerism that are possible in organic molecules, and give an example of each.
2. Which of the following compounds can exist as *cis* and *trans* isomers? Draw them. (a) 1-butene, (b) 2-bromo-1-butene, (c) 2-bromo-2-butene, (d) 1,2-dichlorocyclobutane.
3. Which of the following compounds can exist as *cis* and *trans* isomers? Draw them. (a) 2,3-dimethyl-2-butene, (b) 2,3-dichloro-2-butene, (c) dichlorobenzene, (d) 1,1-dichlorocyclobutane.
4. What are conformations?
5. Distinguish between conformations and isomers.
6. Define optical isomerism. To what do the terms "levorotatory" and "dextrorotatory" refer?
7. Which of the following compounds would exhibit optical isomerism?

(a) CH_3CHCH_3 (c) HO — ... — C — ...
           | |
           Br   Cl

(b) $CH_3CHCH_2CH_3$ (d) $CH_3CH{=}CHCH_3$
           |
           OH

8. Draw three-dimensional representations of the enantiomeric pairs in Exercise 7.
9. Write formulas and names for two alkenes that exhibit geometrical isomerism.
10. Write formulas and names for two substituted cycloalkanes that exhibit geometrical isomerism.
11. Write formulas and names for the isomers of (a) dichlorobenzene, (b) trifluorobenzene, and (c) chlorotoluene. What kind of isomerism is illustrated by each of these sets of compounds?

Reactions of Organic Molecules

12. (a) What is a substitution reaction? (b) What is a halogenation reaction? (c) What is a free-radical chain reaction?
13. (a) Describe the reaction of methane with chlorine in ultraviolet light. (b) Write equations that show formulas for all compounds that can be formed by reaction (a). (c) Write names for all compounds in these equations.
14. (a) Describe the reaction of ethane with chlorine in ultraviolet light. (b) Write equations that show formulas for all compounds that can be formed by reaction (a). (c) Write names for all compounds in these equations.
15. Why are the halogenation reactions of the alkanes of limited value?
16. Why does one-step halogenation of cycloalkanes give a single product, whereas one-step halogenation of normal alkanes gives a multiplicity of products?
17. (a) What are the principal uses of hydrocarbons? (b) Why are hydrocarbons used for these purposes? (c) What does heat of combustion mean? Illustrate.
18. Most reactions of the alkanes that do not disrupt the carbon skeleton are substitution reactions, whereas the alkenes are characterized by addition to the double bond. What does this statement mean?
19. Which of the following compounds could undergo addition reactions? (a) propane, (b) propene, (c) cyclopentene, (d) acetone.
20. Write equations for two reactions in which alkenes undergo addition reactions with halogens. Name all compounds.
21. How can bromination be used to distinguish between alkenes and alkanes?
22. (a) What is hydrogenation? (b) Why is it important? (c) Write equations for two reactions that involve hydrogenation of alkenes. (d) Name all compounds in (c).
23. What is the difference between a vegetable oil and a shortening? Illustrate.
24. (a) What is polymerization? (b) Write equations for three polymerization reactions.
25. Describe two qualitative tests that can be used to distinguish between alkenes and alkanes. (b) Cite some specific examples. (c) What difference in reactivity is the basis for the qualitative distinction between alkanes and alkenes?
26. (a) Why are alkynes more reactive than alkenes? (b) What is the most common kind of reaction that alkynes undergo? (c) Write equations for three such reactions. (d) Name all compounds in part (c).
27. (a) What is the most common kind of reaction that the

benzene ring undergoes? (b) Write equations for the reaction of benzene with chlorine in the presence of an iron catalyst and for the analogous reaction with bromine.

28. Write equations to illustrate both aromatic and aliphatic substitution reactions of toluene using (a) chlorine and (b) bromine.

29. Write the structural formula of the organic compound formed in each of the following reactions:

(a) $CH_3-\underset{}{\bigcirc}-CH_3 \xrightarrow[H_2O, \text{ heat}]{\text{excess KMnO}_4}$

(b) $CH_3CH=CHCH_3 \xrightarrow{HBr}$

(c) $C_6H_5CH=CH_2 \xrightarrow[h\nu]{Br_2}$

30. Write the structural formula for the organic compound or the carbon-containing compound formed in each of the following reactions:

(a) $CH_4 + O_2(\text{excess}) \xrightarrow{\Delta}$

(b) $CH_4 + Cl_2 \xrightarrow{\text{light}}$

(c) $CH_3CH=CH_2 \xrightarrow{H_2SO_4}$

31. Repeat Exercise 30 for

(a) $HC\equiv CH + O_2(\text{excess}) \xrightarrow{\Delta}$

(b) $CH_3Cl + Cl_2 \xrightarrow{\text{light}}$

(c) $CH_3CH=CH_2 \xrightarrow{HCl}$

(d) $\bigcirc \xrightarrow{Br_2}{Fe}$

32. Suppose you have three test tubes. The first contains either hexane or 2-hexene, the second contains either benzene or styrene ($C_6H_5CH=CH_2$), and the third contains either cyclohexene or 2-bromopropane. Describe a simple chemical test that would enable you to determine visually which compound is present in each test tube.

33. (a) What are alkyl hydrogen sulfates? (b) How are they prepared? (c) Can alkyl hydrogen sulfates be classified as acids? Why? (d) What are detergents? (e) What is sodium lauryl sulfate? (f) What is the common use of sodium lauryl sulfate?

34. (a) Write equations for the reaction of nitric acid with the alcohols methanol, ethanol, n-propanol, and n-butanol. (b) Name the inorganic ester formed in each case. (c) What are inorganic esters?

35. (a) What is nitroglycerine? (b) How is it produced? (c) List two important uses for nitroglycerine. Are they similar?

36. Classify each reaction as substitution, addition, or elimination:

(a) $CH_3CH_3 + Cl_2 \xrightarrow{\text{light}} CH_3CH_2Cl + HCl$

(b) $CH_3CH=O + HCN \longrightarrow CH_3\overset{\overset{\displaystyle H}{|}}{\underset{\underset{\displaystyle CN}{|}}{C}}-OH$

(c) $CH_3CH=CH_2 + HOCl \longrightarrow CH_3\underset{\underset{\displaystyle OH}{|}}{C}HCH_2Cl$

37. Classify each reaction as substitution, addition, or elimination:

(a) $CH_3CH_2Br + CN^- \longrightarrow CH_3CH_2CN + Br^-$

(b) $CH_3\underset{\underset{\displaystyle Br}{|}}{C}HCH_2Br + Zn \longrightarrow CH_3CH=CH_2 + ZnBr_2$

(c) $C_6H_6 + H_2SO_4 \longrightarrow C_6H_5SO_3H + H_2O$

38. Why are aqueous solutions of amines basic? Show, with equations, how the dissolution of an amine in water is similar to the dissolution of ammonia in water.

39. Show that the reaction of amines with inorganic acids such as HCl are similar to the reactions of ammonia with inorganic acids.

*40. What are the equilibrium concentrations of the species present in a 0.100 M solution of aniline? $K_b = 4.2 \times 10^{-10}$

$$C_6H_5NH_2(aq) + H_2O(\ell) \rightleftharpoons C_6H_5NH_3^+ + OH^-$$

*41. Which solution would be the more acidic: a 0.10 M solution of aniline hydrochloride, $C_6H_5NH_3Cl$ ($K_b = 4.2 \times 10^{-10}$ for aniline, $C_6H_5NH_2$), or a 0.10 M solution of methylamine hydrochloride, CH_3NH_3Cl ($K_b = 5.0 \times 10^{-4}$ for methylamine, CH_3NH_2)? Justify your choice.

42. Choose the compound that is the stronger acid in each set:

(a) $CH_3CH_2CH_2OH$ or $CH_3-\bigcirc-OH$

(b) CH_3CH_2OH or $\bigcirc-\overset{\overset{\displaystyle O}{\|}}{C}-OH$

(c) $\bigcirc-OH$ or $\bigcirc-\overset{\overset{\displaystyle O}{\|}}{C}-OH$

(d) $\bigcirc-OH$ or $\bigcirc-OH$

43. (a) What are alkoxides? (b) What do we mean when we say that the low-molecular-weight alkoxides are strong bases?

44. (a) Write equations for the reactions of three alcohols with metallic sodium. (b) Name all compounds in these equations. (c) Are these reactions similar to the reaction of metallic sodium with water? How?

45. Which physical property of aldehydes is used to advantage in their production from alcohols?
46. Describe a simple test to distinguish between the two isomers 1-pentene and cyclopentane.
47. How are the terms "oxidation" and "reduction" often used in organic chemistry? Classify the following changes as either oxidation or reduction: (a) CH_4 to CH_3OH, (b) $CH_2=CH_2$ to $CH_3—CH_3$, (c) $CH_3CH_2CH_2OH$ to $CH_3CH_2CH_3$.

    (d) ⬡—CH_3 to   ⬡—$\overset{\displaystyle O}{\overset{\|}{C}}$—OH

48. Classify the following changes as either oxidation or reduction: (a) CH_3OH to CO_2 and H_2O, (b) CH_3CH_2OH to CH_3CHO, (c) CH_3COOH to CH_3CHO, (d) $CH_3CH=CH_2$ to $CH_3CH_2CH_3$.
49. Why is it hazardous to burn a hydrocarbon fuel in an insufficient supply of oxygen?
50. How does the heat of combustion of ethyl alcohol compare with the heats of combustion of low molecular weight saturated hydrocarbons on a per-mole basis and on a per-gram basis?
51. Write equations to illustrate the oxidation of the following aromatic hydrocarbons by potassium permanganate in basic solution. (a) Toluene, (b) ethylbenzene, (c) 1,2-dimethylbenzene.
52. (a) Do you expect aromatic hydrocarbons to produce soot as they burn? Why? (b) Would you expect the flames to be blue or yellow?
53. Describe the preparation of three aldehydes from alcohols and write appropriate equations. Name all reactants and products.
54. Describe the preparation of three ketones from alcohols and write appropriate equations. Name all reactants and products.
55. Write equations for the formation of three different esters, starting with a different acid chloride and a different alcohol in each case. Name all compounds.
56. Write equations for the formation of three different esters, starting with an acid and an alcohol in each case. Name all compounds.
57. (a) What is saponification? (b) Why is this kind of reaction called saponification?
58. Write equations for the hydrolysis of (a) methyl acetate, (b) ethyl formate, (c) n-butyl acetate, and (d) n-octyl acetate. Name all products.

Polymers

59. What is a polymer? What is the term for the smaller molecule that serves as the repeating unit making up a polymer? What are typical molecular weights of polymers?
60. What is necessary if a molecule is to be capable of polymerization? Name three types of molecules that can polymerize.

61. Poly(vinyl alcohol) has a relatively high melting point, 258°C. How would you explain this behavior? A segment of the polymer is

    $$\cdots CH_2CHCH_2CHCH_2CHCH_2CHCH_2CH \cdots$$
    $$\qquad | \qquad\; | \qquad\; | \qquad\; | \qquad\; |$$
    $$\quad OH \quad OH \quad OH \quad OH \quad OH$$

62. What changes could be made in the structures of polymer molecules that would increase the rigidity of the polymer and raise its melting point?
63. Methyl vinyl ketone, $CH_3\overset{\displaystyle O}{\overset{\|}{C}}CH=CH_2$, can be polymerized by addition polymerization. The addition reaction involves only the $C=C$ bond. Write the molecular structure of a four-unit segment of this polymer.
64. (a) What is rubber? (b) What is vulcanization? (c) What is the purpose of vulcanizing rubber? (d) What are fillers and reinforcing agents? (e) What is their purpose?
65. (a) What is an elastomer? (b) Cite a specific example. (c) What are some of the advantages of neoprene compared with natural rubber?
66. What are polyamides? What kind of reaction forms polyamides?
67. (a) What are polyesters? (b) What is Dacron? (c) How is Dacron prepared? (d) What is Mylar? (e) Is it reasonable to assume that a polyester can be made from propylene glycol and terephthalic acid? If so, sketch its structure.
68. Suppose the following diol is used with terephthalic acid to form a polyester. Sketch the structure of the polymer, showing two repeating units.

    $$HOCH_2—⬡—CH_2OH$$

69. (a) What is Nylon? (b) How is it prepared?
70. Common Nylon is called Nylon 66. (a) What does this mean? (b) Write formulas for two other possible Nylons.
*71. A cellulose polymer has a molecular weight of 750,000 g/mol. Estimate the number of units of the monomer, β-glucose ($C_6H_{12}O_6$) in this polymer. This polymerization reaction can be represented as

    $$xC_6H_{12}O_6 \longrightarrow \text{cellulose} + (x-1)H_2O$$

*72. A 2.30-g sample of poly(vinyl alcohol) was dissolved in water to give 101 mL of solution. The osmotic pressure of the solution was 55 torr at 25°C. Estimate the molecular weight of the poly(vinyl alcohol).
73. Describe the structure of a natural amino acid molecule. What kind of isomerism do most amino acids exhibit? Why?
74. How are the amino acid units in a polypeptide joined together? What are the links called?
75. Consider only two amino acids:

$$NH_2-\overset{\overset{\displaystyle H}{|}}{\underset{\underset{\displaystyle R_1}{|}}{C}}-COOH \quad \text{and} \quad NH_2-\overset{\overset{\displaystyle H}{|}}{\underset{\underset{\displaystyle R_2}{|}}{C}}-COOH$$

Write the structural formulas for the dipeptides that could be formed containing one molecule of each amino acid.

*76. How many different dipeptides can be formed from the three amino acids A, B, and C? Write the sequence of amino acids in each. Assume that an amino acid could occur more than once in each dipeptide.

*77. How many different tripeptides can be formed from the three amino acids A, B, and C? Write the sequence of amino acids in each. Assume that an amino acid could occur more than once in each triopeptide.

78. Aspartame (trade name NutraSweet) is a methyl ester of a dipeptide:

$$\text{—CH}_2-\overset{\overset{\displaystyle COOCH_3}{|}}{CH}-NH-\overset{\overset{\displaystyle O}{\|}}{C}-\overset{\overset{\displaystyle CH_2COOH}{|}}{CH}-NH_2$$

Write the structural formulas of the two amino acids that are combined to make aspartame (neglecting optical isomerism).

Mixed Exercises

*79. A laboratory procedure calls for oxidizing 2-propanol to acetone using an acidic solution of $K_2Cr_2O_7$. However, an insufficient amount of $K_2Cr_2O_7$ is on hand, so the laboratory instructor decides to use an acidic solution of $KMnO_4$ instead. What mass of $KMnO_4$ is required to carry out the same amount of oxidation as 1.00 g of $K_2Cr_2O_7$?

*80. The chemical equation for the water gas reaction is

$$C(s) + H_2O(g) \rightleftharpoons CO(g) + H_2(g)$$

At 1000 K, the value of K_p (in atm) for this reaction is 3.2. When we treat carbon with steam and allow the reaction to reach equilibrium, the partial pressure of water vapor is observed to be 15.6 atm. What are the partial pressures of CO and H_2 under these conditions?

*81. (a) In aqueous solution, acetic acid exists mainly in the molecular form ($K_a = 1.8 \times 10^{-5}$). (a) Calculate the freezing point depression for a 0.10 molal aqueous solution of acetic acid, neglecting any ionization of the acid. $K_f = 1.86°C/\text{molal}$ for water. (b) In nonpolar solvents such as benzene, acetic acid exists mainly as dimers

$$CH_3-C\overset{\displaystyle O\cdots H-O}{\underset{\displaystyle O-H\cdots O}{\diagup\diagdown}}C-CH_3$$

as a result of hydrogen bonding. Calculate the freezing point depression for a 0.10 molal solution of acetic acid in benzene. $K_f = 5.12°C/\text{molal}$ for benzene. Assume complete dimer formation.

*82. What is the pH of a 0.10 M solution of sodium benzoate? $K_a = 6.3 \times 10^{-5}$ for benzoic acid, C_6H_5COOH. Would this solution be more or less acidic than a 0.10 M solution of sodium acetate? $K_a = 1.8 \times 10^{-5}$ for acetic acid, CH_3COOH.

83. Identify the major products of each reaction:

(a) [benzene ring with OH and C(=O)—OH substituents] $+ 2NaOH \longrightarrow$

(b) [benzene ring]—$\overset{\overset{\displaystyle O}{\|}}{C}$—OH + CH_3OH $\xrightarrow[\Delta]{H_2SO_4}$

(c) $CH_3CH_2\overset{\overset{\displaystyle O}{\|}}{CH}$ $\xrightarrow[H^+]{MnO_4^-}$

84. Identify the major products of each reaction:

(a) [benzene ring]—CH_2CH_2OH + Na $\longrightarrow$

(b) $CH_3CH_2\overset{\overset{\displaystyle O}{\|}}{C}OCH_3$ $\xrightarrow[\Delta]{KOH(aq)}$

(c) [benzene ring with O$\overset{\overset{\displaystyle O}{\|}}{C}CH_3$ and C(=O)—OH substituents] $\xrightarrow[\Delta]{NaOH(aq)}$

85. What is the molecular structure of the monomer that polymerizes to form the following polymer?

$$\cdots CH_2-\overset{\overset{\displaystyle CH_3}{|}}{\underset{\underset{\displaystyle \bigcirc}{|}}{C}}-CH_2-\overset{\overset{\displaystyle CH_3}{|}}{\underset{\underset{\displaystyle \bigcirc}{|}}{C}}-CH_2-\overset{\overset{\displaystyle CH_3}{|}}{\underset{\underset{\displaystyle \bigcirc}{|}}{C}}-CH_2-\overset{\overset{\displaystyle CH_3}{|}}{\underset{\underset{\displaystyle \bigcirc}{|}}{C}} \cdots$$

*86. (a) What is the osmotic pressure of a 1.00% aqueous solution of sucrose, $C_{12}H_{22}O_{11}$, at 25°C? (b) To what height will the column rise under this pressure? Assume that the density of the solution is 1.00 g/mL.

Appendices

The number to the left of the decimal point in a logarithm is called the *characteristic,* and the number to the right of the decimal point is called the *mantissa.* The characteristic only locates the decimal point of the number, so it is usually not included when counting significant figures. The mantissa has as many significant figures as the number whose log was found.

To obtain the natural logarithm of a number on an electronic calculator, (1) enter the number and (2) press the (ln) (or ln x) button.

$$\ln 4.45 = 1.4929041 = \underline{1.493}$$

$$\ln 1.27 \times 10^3 = 7.1468 \quad = \underline{7.147}$$

Finding Antilogarithms   Sometimes we know the logarithm of a number and must find the number. This is called finding the *antilogarithm* (or *inverse logarithm*). To do this on a calculator, we (1) enter the value of the log, (2) press the (INV) button, and (3) press the (log) button.

On some calculators, the inverse log is found as follows: (1) enter the value of the log, (2) press the 2ndF (second function) button, and (3) press 10^x.

$$\log x = 6.131; \quad \text{so } x = \text{inverse log of } 6.131 = \underline{1.352 \times 10^6}$$

$$\log x = -1.562; \quad \text{so } x = \text{inverse log of } -1.562 = \underline{2.74 \times 10^{-2}}$$

To find the inverse natural logarithm, we (1) enter the value of the ln, (2) press the (INV) button, and (3) press the (ln) or (ln x) button.

On some calculators, the inverse natural logarithm is found as follows: (1) enter the value of the ln, (2) press the 2ndF (second function) button, and (3) press e^x.

$$\ln x = 3.552; \quad \text{so } x = \text{inverse ln of } 3.552 = \underline{3.49 \times 10^1}$$

$$\ln x = -1.248; \quad \text{so } x = \text{inverse ln of } -1.248 = \underline{2.87 \times 10^{-1}}$$

Calculations Involving Logarithms

Because logarithms are exponents, operations involving them follow the same rules as the use of exponents. The following relationships are useful:

$$\log xy = \log x + \log y \qquad \text{or} \qquad \ln xy = \ln x + \ln y$$

$$\log \frac{x}{y} = \log x - \log y \qquad \text{or} \qquad \ln \frac{x}{y} = \ln x - \ln y$$

$$\log y^x = y \log x \qquad \text{or} \qquad \ln y^x = y \ln x$$

$$\log \sqrt[y]{x} = \log x^{1/y} = \frac{1}{y} \log x \qquad \text{or} \qquad \ln \sqrt[y]{x} = \ln x^{1/y} = \frac{1}{y} \ln x$$

A-3 Quadratic Equations

Algebraic expressions of the form

$$ax^2 + bx + c = 0$$

are called **quadratic equations**. Each of the constant terms (a, b, and c) may be either positive or negative. All quadratic equations may be solved by the **quadratic formula**:

$$x = \frac{-b \pm \sqrt{b^2 - 4ac}}{2a}$$

If we wish to solve the quadratic equation $3x^2 - 4x - 8 = 0$, we use $a = 3, b = -4$, and $c = -8$. Substitution of these values into the quadratic formula gives

$$x = \frac{-(-4) \pm \sqrt{(-4)^2 - 4(3)(-8)}}{2(3)} = \frac{4 \pm \sqrt{16 + 96}}{6}$$

$$= \frac{4 \pm \sqrt{112}}{6} = \frac{4 \pm 10.6}{6}$$

The two roots of this quadratic equation are

$$x = 2.4 \quad \text{and} \quad x = -1.1$$

As you construct and solve quadratic equations based on the observed behavior of matter, you must decide which root has physical significance. Examination of the *equation that defines x* always gives clues about possible values for x. In this way you can tell which is extraneous (has no physical significance). Negative roots are often extraneous.

When you have solved a quadratic equation, you should always check the values you obtained by substitution into the original equation. In the above example we obtained $x = 2.4$ and $x = -1.1$. Substitution of these values into the original equation, $3x^2 - 4x - 8 = 0$, shows that both roots are correct. Such substitutions often do not give a perfect check because some round-off error has been introduced.

Appendix B

Electronic Configurations of the Atoms of the Elements

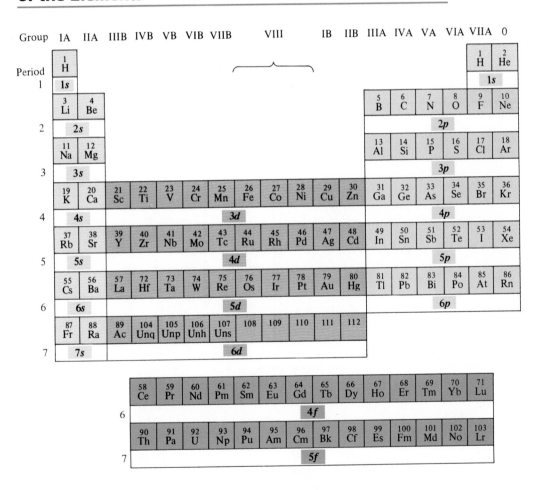

A periodic table colored to show the kinds of atomic orbitals (sublevels) being filled in different parts of the periodic table. The atomic orbitals are given below the symbols of blocks of elements. The electronic structures of the A group and 0 group elements are perfectly regular and can be predicted from their positions in the periodic table, but there are many exceptions in the *d* and *f* blocks. The populations of subshells are given in the table on pages A-10–A-11.

Electronic Configurations of the Atoms of the Elements

Element	Atomic Number	1s	2s	2p	3s	3p	3d	4s	4p	4d	4f	5s
H	1	1										
He	2	2										
Li	3	2	1									
Be	4	2	2									
B	5	2	2	1								
C	6	2	2	2								
N	7	2	2	3								
O	8	2	2	4								
F	9	2	2	5								
Ne	10	2	2	6								
Na	11	Neon core			1							
Mg	12				2							
Al	13				2	1						
Si	14				2	2						
P	15				2	3						
S	16				2	4						
Cl	17				2	5						
Ar	18	2	2	6	2	6						
K	19	Argon core						1				
Ca	20							2				
Sc	21						1	2				
Ti	22						2	2				
V	23						3	2				
Cr	24						5	1				
Mn	25						5	2				
Fe	26						6	2				
Co	27						7	2				
Ni	28						8	2				
Cu	29						10	1				
Zn	30						10	2				
Ga	31						10	2	1			
Ge	32						10	2	2			
As	33						10	2	3			
Se	34						10	2	4			
Br	35						10	2	5			
Kr	36	2	2	6	2	6	10	2	6			
Rb	37	Krypton core										1
Sr	38											2
Y	39									1		2
Zr	40									2		2
Nb	41									4		1
Mo	42									5		1
Tc	43									5		2
Ru	44									7		1
Rh	45									8		1
Pd	46									10		
Ag	47									10		1
Cd	48									10		2

Common Units of Length

$$1 \text{ inch } = 2.54 \text{ centimeters (exactly)}$$

1 mile = 5280 feet = 1.609 kilometers
1 yard = 36 inches = 0.9144 meter
1 meter = 100 centimeters = 39.37 inches = 3.281 feet
    = 1.094 yards
1 kilometer = 1000 meters = 1094 yards = 0.6215 mile
1 Ångstrom = 1.0×10^{-8} centimeter = 0.10 nanometer
    = 1.0×10^{-10} meter = 3.937×10^{-9} inch

Common Units of Volume

$$1 \text{ quart } = 0.9463 \text{ liter}$$
$$1 \text{ liter } = 1.056 \text{ quarts}$$

1 liter = 1 cubic decimeter = 1000 cubic centimeters
    = 0.001 cubic meter
1 milliliter = 1 cubic centimeter = 0.001 liter
    = 1.056×10^{-3} quart
1 cubic foot = 28.316 liters = 29.902 quarts
    = 7.475 gallons

Common Units of Force* and Pressure

1 atmosphere = 760 millimeters of mercury
    = 1.013×10^5 pascals
    = 14.70 pounds per square inch
1 bar = 10^5 pascals
1 torr = 1 millimeter of mercury
1 pascal = 1 kg/m·s^2 = 1 N/m^2

*Force: 1 newton (N) = 1 kg·m/s^2, i.e., the force that, when
applied for 1 second, gives a 1-kilogram mass a velocity of 1
meter per second.

Common Units of Energy

$$1 \text{ joule } = 1 \times 10^7 \text{ ergs}$$

1 thermochemical calorie* = 4.184 joules = 4.184×10^7 ergs
    = 4.129×10^{-2} liter-atmospheres
    = 2.612×10^{19} electron volts
1 erg = 1×10^{-7} joule = 2.3901×10^{-8} calorie
1 electron volt = 1.6022×10^{-19} joule = 1.6022×10^{-12} erg = 96.487 kJ/mol†
1 liter-atmosphere = 24.217 calories = 101.32 joules = 1.0132×10^9 ergs
1 British thermal unit = 1055.06 joules = 1.05506×10^{10} ergs = 252.2 calories

*The amount of heat required to raise the temperature of one gram of water from 14.5°C to
15.5°C.
†Note that the other units are per particle and must be multiplied by 6.022×10^{23} to be
strictly comparable.

Appendix D

Physical Constants

Quantity	Symbol	Traditional Units	SI Units
Acceleration of gravity	g	980.6 cm/s	9.806 m/s
Atomic mass unit (1/12 the mass of ^{12}C atom)	amu or u	1.6606×10^{-24} g	1.6606×10^{-27} kg
Avogadro's number	N	6.022×10^{23} particles/mol	6.022×10^{23} particles/mol
Bohr radius	a_0	0.52918 Å 5.2918×10^{-9} cm	5.2918×10^{-11} m
Boltzmann constant	k	1.3807×10^{-16} erg/K	1.3807×10^{-23} J/K
Charge-to-mass ratio of electron	e/m	1.7588×10^{8} coulomb/g	1.7588×10^{11} C/kg
Electronic charge	e	1.60219×10^{-19} coulomb 4.8033×10^{-10} esu	1.60219×10^{-19} C
Electron rest mass	m_e	9.10952×10^{-28} g 0.00054859 amu	9.10952×10^{-31} kg
Faraday constant	F	96,487 coulombs/eq 23.06 kcal/volt·eq	96,487 C/mol e$^-$ 96,487 J/V·mol e$^-$
Gas constant	R	$0.08206 \dfrac{\text{L} \cdot \text{atm}}{\text{mol} \cdot \text{K}}$ $1.987 \dfrac{\text{cal}}{\text{mol} \cdot \text{K}}$	$8.3145 \dfrac{\text{kPa} \cdot \text{dm}^3}{\text{mol} \cdot \text{K}}$ 8.3145 J/mol·K
Molar volume (STP)	V_m	22.414 L/mol	22.414×10^{-3} m^3/mol 22.414 dm^3/mol
Neutron rest mass	m_n	1.67495×10^{-24} g 1.008665 amu	1.67495×10^{-27} kg
Planck constant	h	6.6262×10^{-27} erg·s	6.6262×10^{-34} J·s
Proton rest mass	m_p	1.6726×10^{-24} g 1.007277 amu	1.6726×10^{-27} kg
Rydberg constant	R_∞	3.289×10^{15} cycles/s 2.1799×10^{-11} erg	1.0974×10^{7} m^{-1} 2.1799×10^{-18} J
Speed of light (in a vacuum)	c	2.9979×10^{10} cm/s (186,281 miles/second)	2.9979×10^{8} m/s

$\pi = 3.1416$
$e = 2.71828$
$\ln X = 2.303 \log X$

$2.303 R = 4.576$ cal/mol·K $= 19.15$ J/mol·K
$2.303 RT$ (at 25°C) $= 1364$ cal/mol $= 5709$ J/mol

Appendix E

Some Physical Constants for Water and a Few Common Substances

Vapor Pressure of Water at Various Temperatures

Temperature (°C)	Vapor Pressure (torr)	Temperature (°C)	Vapor Pressure (torr)	Temperature (°C)	Vapor Pressure (torr)	Temperature (°C)	Vapor Pressure (torr)
−10	2.1	21	18.7	51	97.2	81	369.7
−9	2.3	22	19.8	52	102.1	82	384.9
−8	2.5	23	21.1	53	107.2	83	400.6
−7	2.7	24	22.4	54	112.5	84	416.8
−6	2.9	25	23.8	55	118.0	85	433.6
−5	3.2	26	25.2	56	123.8	86	450.9
−4	3.4	27	26.7	57	129.8	87	468.7
−3	3.7	28	28.3	58	136.1	88	487.1
−2	4.0	29	30.0	59	142.6	89	506.1
−1	4.3	30	31.8	60	149.4	90	525.8
0	4.6	31	33.7	61	156.4	91	546.1
1	4.9	32	35.7	62	163.8	92	567.0
2	5.3	33	37.7	63	171.4	93	588.6
3	5.7	34	39.9	64	179.3	94	610.9
4	6.1	35	42.2	65	187.5	95	633.9
5	6.5	36	44.6	66	196.1	96	657.6
6	7.0	37	47.1	67	205.0	97	682.1
7	7.5	38	49.7	68	214.2	98	707.3
8	8.0	39	52.4	69	223.7	99	733.2
9	8.6	40	55.3	70	233.7	100	760.0
10	9.2	41	58.3	71	243.9	101	787.6
11	9.8	42	61.5	72	254.6	102	815.9
12	10.5	43	64.8	73	265.7	103	845.1
13	11.2	44	68.3	74	277.2	104	875.1
14	12.0	45	71.9	75	289.1	105	906.1
15	12.8	46	75.7	76	301.4	106	937.9
16	13.6	47	79.6	77	314.1	107	970.6
17	14.5	48	83.7	78	327.3	108	1004.4
18	15.5	49	88.0	79	341.0	109	1038.9
19	16.5	50	92.5	80	355.1	110	1074.6
20	17.5						

Specific Heats and Heat Capacities for Some Common Substances

Substance	Specific Heat (J/g · °C)	Molar Heat Capacity (J/mol · °C)
Al(s)	0.900	24.3
Ca(s)	0.653	26.2
Cu(s)	0.385	24.5
Fe(s)	0.444	24.8
Hg(ℓ)	0.138	27.7
H_2O(s), ice	2.09	37.7
H_2O(ℓ), water	4.18	75.3
H_2O(g), steam	2.03	36.4
C_6H_6(ℓ), benzene	1.74	136
C_6H_6(g), benzene	1.04	81.6
C_2H_5OH(ℓ), ethanol	2.46	113
C_2H_5OH(g), ethanol	0.954	420
$(C_2H_5)_2O$(ℓ), diethyl ether	3.74	172
$(C_2H_5)_2O$(g), diethyl ether	2.35	108

Heats of Transformation and Transformation Temperatures of Several Substances

Substance	mp (°C)	Heat of Fusion (J/g)	ΔH_{fus} (kJ/mol)	bp (°C)	Heat of Vaporization (J/g)	ΔH_{vap} (kJ/mol)
Al	658	395	10.6	2467	10520	284
Ca	851	233	9.33	1487	4030	162
Cu	1083	205	13.0	2595	4790	305
H_2O	0.0	334	6.02	100	2260	40.7
Fe	1530	267	14.9	2735	6340	354
Hg	−39	11	23.3	357	292	58.6
CH_4	−182	58.6	0.92	−164	—	—
C_2H_5OH	−117	109	5.02	78.0	855	39.3
C_6H_6	5.48	127	9.92	80.1	395	30.8
$(C_2H_5)_2O$	−116	97.9	7.66	35	351	26.0

Appendix F

Ionization Constants for Weak Acids at 25°C

Acid	Formula and Ionization Equation		K_a
Acetic	CH_3COOH	$\rightleftharpoons H^+ + CH_3COO^-$	1.8×10^{-5}
Arsenic	H_3AsO_4	$\rightleftharpoons H^+ + H_2AsO_4^-$	$2.5 \times 10^{-4} = K_1$
	$H_2AsO_4^-$	$\rightleftharpoons H^+ + HAsO_4^{2-}$	$5.6 \times 10^{-8} = K_2$
	$HAsO_4^{2-}$	$\rightleftharpoons H^+ + AsO_4^{3-}$	$3.0 \times 10^{-13} = K_3$
Arsenous	H_3AsO_3	$\rightleftharpoons H^+ + H_2AsO_3^-$	$6.0 \times 10^{-10} = K_1$
	$H_2AsO_3^-$	$\rightleftharpoons H^+ + HAsO_3^{2-}$	$3.0 \times 10^{-14} = K_2$
Benzoic	C_6H_5COOH	$\rightleftharpoons H^+ + C_6H_5COO^-$	6.3×10^{-5}
Boric*	$B(OH)_3$	$\rightleftharpoons H^+ + BO(OH)_2^-$	$7.3 \times 10^{-10} = K_1$
	$BO(OH)_2^-$	$\rightleftharpoons H^+ + BO_2(OH)^{2-}$	$1.8 \times 10^{-13} = K_2$
	$BO_2(OH)^{2-}$	$\rightleftharpoons H^+ + BO_3^{3-}$	$1.6 \times 10^{-14} = K_3$
Carbonic	H_2CO_3	$\rightleftharpoons H^+ + HCO_3^-$	$4.2 \times 10^{-7} = K_1$
	HCO_3^-	$\rightleftharpoons H^+ + CO_3^{2-}$	$4.8 \times 10^{-11} = K_2$
Citric	$C_3H_5O(COOH)_3$	$\rightleftharpoons H^+ + C_4H_5O_3(COOH)_2^-$	$7.4 \times 10^{-3} = K_1$
	$C_4H_5O_3(COOH)_2^-$	$\rightleftharpoons H^+ + C_5H_5O_5COOH^{2-}$	$1.7 \times 10^{-5} = K_2$
	$C_5H_5O_5COOH^{2-}$	$\rightleftharpoons H^+ + C_6H_5O_7^{3-}$	$7.4 \times 10^{-7} = K_3$
Cyanic	$HOCN$	$\rightleftharpoons H^+ + OCN^-$	3.5×10^{-4}
Formic	$HCOOH$	$\rightleftharpoons H^+ + HCOO^-$	1.8×10^{-4}
Hydrazoic	HN_3	$\rightleftharpoons H^+ + N_3^-$	1.9×10^{-5}
Hydrocyanic	HCN	$\rightleftharpoons H^+ + CN^-$	4.0×10^{-10}
Hydrofluoric	HF	$\rightleftharpoons H^+ + F^-$	7.2×10^{-4}
Hydrogen peroxide	H_2O_2	$\rightleftharpoons H^+ + HO_2^-$	2.4×10^{-12}
Hydrosulfuric	H_2S	$\rightleftharpoons H^+ + HS^-$	$1.0 \times 10^{-7} = K_1$
	HS^-	$\rightleftharpoons H^+ + S^{2-}$	$1.3 \times 10^{-13} = K_2$
Hypobromous	$HOBr$	$\rightleftharpoons H^+ + OBr^-$	2.5×10^{-9}
Hypochlorous	$HOCl$	$\rightleftharpoons H^+ + OCl^-$	3.5×10^{-8}
Nitrous	HNO_2	$\rightleftharpoons H^+ + NO_2^-$	4.5×10^{-4}
Oxalic	$(COOH)_2$	$\rightleftharpoons H^+ + COOCOOH^-$	$5.9 \times 10^{-2} = K_1$
	$COOCOOH^-$	$\rightleftharpoons H^+ + (COO)_2^{2-}$	$6.4 \times 10^{-5} = K_2$
Phenol	HC_6H_5O	$\rightleftharpoons H^+ + C_6H_5O^-$	1.3×10^{-10}
Phosphoric	H_3PO_4	$\rightleftharpoons H^+ + H_2PO_4^-$	$7.5 \times 10^{-3} = K_1$
	$H_2PO_4^-$	$\rightleftharpoons H^+ + HPO_4^{2-}$	$6.2 \times 10^{-8} = K_2$
	HPO_4^{2-}	$\rightleftharpoons H^+ + PO_4^{3-}$	$3.6 \times 10^{-13} = K_3$
Phosphorous	H_3PO_3	$\rightleftharpoons H^+ + H_2PO_3^-$	$1.6 \times 10^{-2} = K_1$
	$H_2PO_3^-$	$\rightleftharpoons H^+ + HPO_3^{2-}$	$7.0 \times 10^{-7} = K_2$
Selenic	H_2SeO_4	$\rightleftharpoons H^+ + HSeO_4^-$	Very large $= K_1$
	$HSeO_4^-$	$\rightleftharpoons H^+ + SeO_4^{2-}$	$1.2 \times 10^{-2} = K_2$
Selenous	H_2SeO_3	$\rightleftharpoons H^+ + HSeO_3^-$	$2.7 \times 10^{-3} = K_1$
	$HSeO_3^-$	$\rightleftharpoons H^+ + SeO_3^{2-}$	$2.5 \times 10^{-7} = K_2$
Sulfuric	H_2SO_4	$\rightleftharpoons H^+ + HSO_4^-$	Very large $= K_1$
	HSO_4^-	$\rightleftharpoons H^+ + SO_4^{2-}$	$1.2 \times 10^{-2} = K_2$
Sulfurous	H_2SO_3	$\rightleftharpoons H^+ + HSO_3^-$	$1.2 \times 10^{-2} = K_1$
	HSO_3^-	$\rightleftharpoons H^+ + SO_3^{2-}$	$6.2 \times 10^{-8} = K_2$
Tellurous	H_2TeO_3	$\rightleftharpoons H^+ + HTeO_3^-$	$2 \times 10^{-3} = K_1$
	$HTeO_3^-$	$\rightleftharpoons H^+ + TeO_3^{2-}$	$1 \times 10^{-8} = K_2$

*Boric acid acts as a Lewis acid in aqueous solution.

Appendix G

Ionization Constants for Weak Bases at 25°C

Base	Formula and Ionization Equation				K_b	
Ammonia	NH_3	$+ H_2O \rightleftharpoons NH_4^+$		$+ OH^-$	1.8×10^{-5}	
Aniline	$C_6H_5NH_2$	$+ H_2O \rightleftharpoons C_6H_5NH_3^+$		$+ OH^-$	4.2×10^{-10}	
Dimethylamine	$(CH_3)_2NH$	$+ H_2O \rightleftharpoons (CH_3)_2NH_2^+$		$+ OH^-$	7.4×10^{-4}	
Ethylenediamine	$(CH_2)_2(NH_2)_2$	$+ H_2O \rightleftharpoons (CH_2)_2(NH_2)_2H^+$		$+ OH^-$	8.5×10^{-5}	$= K_1$
	$(CH_2)_2(NH_2)_2H^+$	$+ H_2O \rightleftharpoons (CH_2)_2(NH_2)_2H_2^{2+}$		$+ OH^-$	2.7×10^{-8}	$= K_2$
Hydrazine	N_2H_4	$+ H_2O \rightleftharpoons N_2H_5^+$		$+ OH^-$	8.5×10^{-7}	$= K_1$
	$N_2H_5^+$	$+ H_2O \rightleftharpoons N_2H_6^{2+}$		$+ OH^-$	8.9×10^{-16}	$= K_2$
Hydroxylamine	NH_2OH	$+ H_2O \rightleftharpoons NH_3OH^+$		$+ OH^-$	6.6×10^{-9}	
Methylamine	CH_3NH_2	$+ H_2O \rightleftharpoons CH_3NH_3^+$		$+ OH^-$	5.0×10^{-4}	
Pyridine	C_5H_5N	$+ H_2O \rightleftharpoons C_5H_5NH^+$		$+ OH^-$	1.5×10^{-9}	
Trimethylamine	$(CH_3)_3N$	$+ H_2O \rightleftharpoons (CH_3)_3NH^+$		$+ OH^-$	7.4×10^{-5}	

Appendix H

Solubility Product Constants for Some Inorganic Compounds at 25°C

Substance	K_{sp}	Substance	K_{sp}
Aluminum compounds		Chromium compounds	
$AlAsO_4$	1.6×10^{-16}	$CrAsO_4$	7.8×10^{-21}
$Al(OH)_3$	1.9×10^{-33}	$Cr(OH)_3$	6.7×10^{-31}
$AlPO_4$	1.3×10^{-20}	$CrPO_4$	2.4×10^{-23}
Antimony compounds		Cobalt compounds	
Sb_2S_3	1.6×10^{-93}	$Co_3(AsO_4)_2$	7.6×10^{-29}
Barium compounds		$CoCO_3$	8.0×10^{-13}
$Ba_3(AsO_4)_2$	1.1×10^{-13}	$Co(OH)_2$	2.5×10^{-16}
$BaCO_3$	8.1×10^{-9}	$CoS\ (\alpha)$	5.9×10^{-21}
$BaC_2O_4 \cdot 2H_2O^*$	1.1×10^{-7}	$CoS\ (\beta)$	8.7×10^{-23}
$BaCrO_4$	2.0×10^{-10}	$Co(OH)_3$	4.0×10^{-45}
BaF_2	1.7×10^{-6}	Co_2S_3	2.6×10^{-124}
$Ba(OH)_2 \cdot 8H_2O^*$	5.0×10^{-3}	Copper compounds	
$Ba_3(PO_4)_2$	1.3×10^{-29}	$CuBr$	5.3×10^{-9}
$BaSeO_4$	2.8×10^{-11}	$CuCl$	1.9×10^{-7}
$BaSO_3$	8.0×10^{-7}	$CuCN$	3.2×10^{-20}
$BaSO_4$	1.1×10^{-10}	$Cu_2O\ (Cu^+ + OH^-)$†	1.0×10^{-14}
Bismuth compounds		CuI	5.1×10^{-12}
$BiOCl$	7.0×10^{-9}	Cu_2S	1.6×10^{-48}
$BiO(OH)$	1.0×10^{-12}	$CuSCN$	1.6×10^{-11}
$Bi(OH)_3$	3.2×10^{-40}	$Cu_3(AsO_4)_2$	7.6×10^{-36}
BiI_3	8.1×10^{-19}	$CuCO_3$	2.5×10^{-10}
$BiPO_4$	1.3×10^{-23}	$Cu_2[Fe(CN)_6]$	1.3×10^{-16}
Bi_2S_3	1.6×10^{-72}	$Cu(OH)_2$	1.6×10^{-19}
Cadmium compounds		CuS	8.7×10^{-36}
$Cd_3(AsO_4)_2$	2.2×10^{-32}	Gold compounds	
$CdCO_3$	2.5×10^{-14}	$AuBr$	5.0×10^{-17}
$Cd(CN)_2$	1.0×10^{-8}	$AuCl$	2.0×10^{-13}
$Cd_2[Fe(CN)_6]$	3.2×10^{-17}	AuI	1.6×10^{-23}
$Cd(OH)_2$	1.2×10^{-14}	$AuBr_3$	4.0×10^{-36}
CdS	3.6×10^{-29}	$AuCl_3$	3.2×10^{-25}
Calcium compounds		$Au(OH)_3$	1.0×10^{-53}
$Ca_3(AsO_4)_2$	6.8×10^{-19}	AuI_3	1.0×10^{-46}
$CaCO_3$	4.8×10^{-9}	Iron compounds	
$CaCrO_4$	7.1×10^{-4}	$FeCO_3$	3.5×10^{-11}
$CaC_2O_4 \cdot H_2O^*$	2.3×10^{-9}	$Fe(OH)_2$	7.9×10^{-15}
CaF_2	3.9×10^{-11}	FeS	4.9×10^{-18}
$Ca(OH)_2$	7.9×10^{-6}	$Fe_4[Fe(CN)_6]_3$	3.0×10^{-41}
$CaHPO_4$	2.7×10^{-7}	$Fe(OH)_3$	6.3×10^{-38}
$Ca(H_2PO_4)_2$	1.0×10^{-3}	Fe_2S_3	1.4×10^{-88}
$Ca_3(PO_4)_2$	1.0×10^{-25}	Lead compounds	
$CaSO_3 \cdot 2H_2O^*$	1.3×10^{-8}	$Pb_3(AsO_4)_2$	4.1×10^{-36}
$CaSO_4 \cdot 2H_2O^*$	2.4×10^{-5}	$PbBr_2$	6.3×10^{-6}

Solubility Product Constants for Some Inorganic Compounds at 25°C (continued)

Substance	K_{sp}	Substance	K_{sp}
Lead compounds (*cont.*)		**Nickel compounds** (*cont.*)	
$PbCO_3$	1.5×10^{-13}	NiS (α)	3.0×10^{-21}
$PbCl_2$	1.7×10^{-5}	NiS (β)	1.0×10^{-26}
$PbCrO_4$	1.8×10^{-14}	NiS (γ)	2.0×10^{-28}
PbF_2	3.7×10^{-8}	**Silver compounds**	
$Pb(OH)_2$	2.8×10^{-16}	Ag_3AsO_4	1.1×10^{-20}
PbI_2	8.7×10^{-9}	$AgBr$	3.3×10^{-13}
$Pb_3(PO_4)_2$	3.0×10^{-44}	Ag_2CO_3	8.1×10^{-12}
$PbSeO_4$	1.5×10^{-7}	$AgCl$	1.8×10^{-10}
$PbSO_4$	1.8×10^{-8}	Ag_2CrO_4	9.0×10^{-12}
PbS	8.4×10^{-28}	$AgCN$	1.2×10^{-16}
Magnesium compounds		$Ag_4[Fe(CN)_6]$	1.6×10^{-41}
$Mg_3(AsO_4)_2$	2.1×10^{-20}	Ag_2O ($Ag^+ + OH^-$)†	2.0×10^{-8}
$MgCO_3 \cdot 3H_2O$*	4.0×10^{-5}	AgI	1.5×10^{-16}
MgC_2O_4	8.6×10^{-5}	Ag_3PO_4	1.3×10^{-20}
MgF_2	6.4×10^{-9}	Ag_2SO_3	1.5×10^{-14}
$Mg(OH)_2$	1.5×10^{-11}	Ag_2SO_4	1.7×10^{-5}
$MgNH_4PO_4$	2.5×10^{-12}	Ag_2S	1.0×10^{-49}
Manganese compounds		$AgSCN$	1.0×10^{-12}
$Mn_3(AsO_4)_2$	1.9×10^{-11}	**Strontium compounds**	
$MnCO_3$	1.8×10^{-11}	$Sr_3(AsO_4)_2$	1.3×10^{-18}
$Mn(OH)_2$	4.6×10^{-14}	$SrCO_3$	9.4×10^{-10}
MnS	5.1×10^{-15}	$SrC_2O_4 \cdot 2H_2O$*	5.6×10^{-8}
$Mn(OH)_3$	$\sim 1.0 \times 10^{-36}$	$SrCrO_4$	3.6×10^{-5}
Mercury compounds		$Sr(OH)_2 \cdot 8H_2O$*	3.2×10^{-4}
Hg_2Br_2	1.3×10^{-22}	$Sr_3(PO_4)_2$	1.0×10^{-31}
Hg_2CO_3	8.9×10^{-17}	$SrSO_3$	4.0×10^{-8}
Hg_2Cl_2	1.1×10^{-18}	$SrSO_4$	2.8×10^{-7}
Hg_2CrO_4	5.0×10^{-9}	**Tin compounds**	
Hg_2I_2	4.5×10^{-29}	$Sn(OH)_2$	2.0×10^{-26}
$Hg_2O \cdot H_2O$		SnI_2	1.0×10^{-4}
($Hg_2^{2+} + 2OH^-$)†	1.6×10^{-23}	SnS	1.0×10^{-28}
Hg_2SO_4	6.8×10^{-7}	$Sn(OH)_4$	1.0×10^{-57}
Hg_2S	5.8×10^{-44}	SnS_2	1.0×10^{-70}
$Hg(CN)_2$	3.0×10^{-23}	**Zinc compounds**	
$Hg(OH)_2$	2.5×10^{-26}	$Zn_3(AsO_4)_2$	1.1×10^{-27}
HgI_2	4.0×10^{-29}	$ZnCO_3$	1.5×10^{-11}
HgS	3.0×10^{-53}	$Zn(CN)_2$	8.0×10^{-12}
Nickel compounds		$Zn_2[Fe(CN)_6]$	4.1×10^{-16}
$Ni_3(AsO_4)_2$	1.9×10^{-26}	$Zn(OH)_2$	4.5×10^{-17}
$NiCO_3$	6.6×10^{-9}	$Zn_3(PO_4)_2$	9.1×10^{-33}
$Ni(CN)_2$	3.0×10^{-23}	ZnS	1.1×10^{-21}
$Ni(OH)_2$	2.8×10^{-16}		

*[H_2O] does not appear in equilibrium constants for equilibria in aqueous solution in general, so it does *not* appear in the K_{sp} expressions for hydrated solids.
†Very small amounts of oxides dissolve in water to give the ions indicated in parentheses. Solid hydroxides are unstable and decompose to oxides as rapidly as they are formed.

Appendix I

Dissociation Constants for Some Complex Ions

Dissociation Equilibrium	K_d
$[AgBr_2]^- \rightleftharpoons Ag^+ + 2Br^-$	7.8×10^{-8}
$[AgCl_2]^- \rightleftharpoons Ag^+ + 2Cl^-$	4.0×10^{-6}
$[Ag(CN)_2]^- \rightleftharpoons Ag^+ + 2CN^-$	1.8×10^{-19}
$[Ag(S_2O_3)_2]^{3-} \rightleftharpoons Ag^+ + 2S_2O_3^{2-}$	5.0×10^{-14}
$[Ag(NH_3)_2]^+ \rightleftharpoons Ag^+ + 2NH_3$	6.3×10^{-8}
$[Ag(en)]^+ \rightleftharpoons Ag^+ + en^*$	1.0×10^{-5}
$[AlF_6]^{3-} \rightleftharpoons Al^{3+} + 6F^-$	2.0×10^{-24}
$[Al(OH)_4]^- \rightleftharpoons Al^{3+} + 4OH^-$	1.3×10^{-34}
$[Au(CN)_2]^- \rightleftharpoons Au^+ + 2CN^-$	5.0×10^{-39}
$[Cd(CN)_4]^{2-} \rightleftharpoons Cd^{2+} + 4CN^-$	7.8×10^{-18}
$[CdCl_4]^{2-} \rightleftharpoons Cd^{2+} + 4Cl^-$	1.0×10^{-4}
$[Cd(NH_3)_4]^{2+} \rightleftharpoons Cd^{2+} + 4NH_3$	1.0×10^{-7}
$[Co(NH_3)_6]^{2+} \rightleftharpoons Co^{2+} + 6NH_3$	1.3×10^{-5}
$[Co(NH_3)_6]^{3+} \rightleftharpoons Co^{3+} + 6NH_3$	2.2×10^{-34}
$[Co(en)_3]^{2+} \rightleftharpoons Co^{2+} + 3en^*$	1.5×10^{-14}
$[Co(en)_3]^{3+} \rightleftharpoons Co^{3+} + 3en^*$	2.0×10^{-49}
$[Cu(CN)_2]^- \rightleftharpoons Cu^+ + 2CN^-$	1.0×10^{-16}
$[CuCl_2]^- \rightleftharpoons Cu^+ + 2Cl^-$	1.0×10^{-5}
$[Cu(NH_3)_2]^+ \rightleftharpoons Cu^+ + 2NH_3$	1.4×10^{-11}
$[Cu(NH_3)_4]^{2+} \rightleftharpoons Cu^{2+} + 4NH_3$	8.5×10^{-13}
$[Fe(CN)_6]^{4-} \rightleftharpoons Fe^{2+} + 6CN^-$	1.3×10^{-37}
$[Fe(CN)_6]^{3-} \rightleftharpoons Fe^{3+} + 6CN^-$	1.3×10^{-44}
$[HgCl_4]^{2-} \rightleftharpoons Hg^{2+} + 4Cl^-$	8.3×10^{-16}
$[Ni(CN)_4]^{2-} \rightleftharpoons Ni^{2+} + 4CN^-$	1.0×10^{-31}
$[Ni(NH_3)_6]^{2+} \rightleftharpoons Ni^{2+} + 6NH_3$	1.8×10^{-9}
$[Zn(OH)_4]^{2-} \rightleftharpoons Zn^{2+} + 4OH^-$	3.5×10^{-16}
$[Zn(NH_3)_4]^{2+} \rightleftharpoons Zn^{2+} + 4NH_3$	3.4×10^{-10}

*The abbreviation "en" represents ethylenediamine, $H_2NCH_2CH_2NH_2$.

Appendix J

Standard Reduction Potentials in Aqueous Solution at 25°C

Acidic Solution	Standard Reduction Potential, E^0 (volts)
$Li^+(aq) + e^- \longrightarrow Li(s)$	-3.045
$K^+(aq) + e^- \longrightarrow K(s)$	-2.925
$Rb^+(aq) + e^- \longrightarrow Rb(s)$	-2.925
$Ba^{2+}(aq) + 2e^- \longrightarrow Ba(s)$	-2.90
$Sr^{2+}(aq) + 2e^- \longrightarrow Sr(s)$	-2.89
$Ca^{2+}(aq) + 2e^- \longrightarrow Ca(s)$	-2.87
$Na^+(aq) + e^- \longrightarrow Na(s)$	-2.714
$Mg^{2+}(aq) + 2e^- \longrightarrow Mg(s)$	-2.37
$H_2(g) + 2e^- \longrightarrow 2H^-(aq)$	-2.25
$Al^{3+}(aq) + 3e^- \longrightarrow Al(s)$	-1.66
$Zr^{4+}(aq) + 4e^- \longrightarrow Zr(s)$	-1.53
$ZnS(s) + 2e^- \longrightarrow Zn(s) + S^{2-}(aq)$	-1.44
$CdS(s) + 2e^- \longrightarrow Cd(s) + S^{2-}(aq)$	-1.21
$V^{2+}(aq) + 2e^- \longrightarrow V(s)$	-1.18
$Mn^{2+}(aq) + 2e^- \longrightarrow Mn(s)$	-1.18
$FeS(s) + 2e^- \longrightarrow Fe(s) + S^{2-}(aq)$	-1.01
$Cr^{2+}(aq) + 2e^- \longrightarrow Cr(s)$	-0.91
$Zn^{2+}(aq) + 2e^- \longrightarrow Zn(s)$	-0.763
$Cr^{3+}(aq) + 3e^- \longrightarrow Cr(s)$	-0.74
$HgS(s) + 2H^+(aq) + 2e^- \longrightarrow Hg(\ell) + H_2S(g)$	-0.72
$Ga^{3+}(aq) + 3e^- \longrightarrow Ga(s)$	-0.53
$2CO_2(g) + 2H^+(aq) + 2e^- \longrightarrow (COOH)_2(aq)$	-0.49
$Fe^{2+}(aq) + 2e^- \longrightarrow Fe(s)$	-0.44
$Cr^{3+}(aq) + e^- \longrightarrow Cr^{2+}(aq)$	-0.41
$Cd^{2+}(aq) + 2e^- \longrightarrow Cd(s)$	-0.403
$Se(s) + 2H^+(aq) + 2e^- \longrightarrow H_2Se(aq)$	-0.40
$PbSO_4(s) + 2e^- \longrightarrow Pb(s) + SO_4^{2-}(aq)$	-0.356
$Tl^+(aq) + e^- \longrightarrow Tl(s)$	-0.34
$Co^{2+}(aq) + 2e^- \longrightarrow Co(s)$	-0.28
$Ni^{2+}(aq) + 2e^- \longrightarrow Ni(s)$	-0.25
$[SnF_6]^{2-}(aq) + 4e^- \longrightarrow Sn(s) + 6F^-(aq)$	-0.25
$AgI(s) + e^- \longrightarrow Ag(s) + I^-(aq)$	-0.15
$Sn^{2+}(aq) + 2e^- \longrightarrow Sn(s)$	-0.14
$Pb^{2+}(aq) + 2e^- \longrightarrow Pb(s)$	-0.126
$N_2O(g) + 6H^+(aq) + H_2O + 4e^- \longrightarrow 2NH_3OH^+(aq)$	-0.05
$2H^+(aq) + 2e^- \longrightarrow H_2(g)$ (reference electrode)	0.000

Standard Reduction Potentials in Aqueous Solution at 25°C (*continued*)

Acidic Solution	Standard Reduction Potential, E^0 (volts)
$AgBr(s) + e^- \longrightarrow Ag(s) + Br^-(aq)$	0.10
$S(s) + 2H^+(aq) + 2e^- \longrightarrow H_2S(aq)$	0.14
$Sn^{4+}(aq) + 2e^- \longrightarrow Sn^{2+}(aq)$	0.15
$Cu^{2+}(aq) + e^- \longrightarrow Cu^+(aq)$	0.153
$SO_4^{2-}(aq) + 4H^+(aq) + 2e^- \longrightarrow H_2SO_3(aq) + H_2O$	0.17
$SO_4^{2-}(aq) + 4H^+(aq) + 2e^- \longrightarrow SO_2(g) + 2H_2O$	0.20
$AgCl(s) + e^- \longrightarrow Ag(s) + Cl^-(aq)$	0.222
$Hg_2Cl_2(s) + 2e^- \longrightarrow 2Hg(\ell) + 2Cl^-(aq)$	0.27
$Cu^{2+}(aq) + 2e^- \longrightarrow Cu(s)$	0.337
$[RhCl_6]^{3-}(aq) + 3e^- \longrightarrow Rh(s) + 6Cl^-(aq)$	0.44
$Cu^+(aq) + e^- \longrightarrow Cu(s)$	0.521
$TeO_2(s) + 4H^+(aq) + 4e^- \longrightarrow Te(s) + 2H_2O$	0.529
$I_2(s) + 2e^- \longrightarrow 2I^-(aq)$	0.535
$H_3AsO_4(aq) + 2H^+(aq) + 2e^- \longrightarrow H_3AsO_3(aq) + H_2O$	0.58
$[PtCl_6]^{2-}(aq) + 2e^- \longrightarrow [PtCl_4]^{2-}(aq) + 2Cl^-(aq)$	0.68
$O_2(g) + 2H^+(aq) + 2e^- \longrightarrow H_2O_2(aq)$	0.682
$[PtCl_4]^{2-}(aq) + 2e^- \longrightarrow Pt(s) + 4Cl^-(aq)$	0.73
$SbCl_6^-(aq) + 2e^- \longrightarrow SbCl_4^-(aq) + 2Cl^-(aq)$	0.75
$Fe^{3+}(aq) + e^- \longrightarrow Fe^{2+}(aq)$	0.771
$Hg_2^{2+}(aq) + 2e^- \longrightarrow 2Hg(\ell)$	0.789
$Ag^+(aq) + e^- \longrightarrow Ag(s)$	0.7994
$Hg^{2+}(aq) + 2e^- \longrightarrow Hg(\ell)$	0.855
$2Hg^{2+}(aq) + 2e^- \longrightarrow Hg_2^{2+}(aq)$	0.920
$NO_3^-(aq) + 3H^+(aq) + 2e^- \longrightarrow HNO_2(aq) + H_2O$	0.94
$NO_3^-(aq) + 4H^+(aq) + 3e^- \longrightarrow NO(g) + 2H_2O$	0.96
$Pd^{2+}(aq) + 2e^- \longrightarrow Pd(s)$	0.987
$AuCl_4^-(aq) + 3e^- \longrightarrow Au(s) + 4Cl^-(aq)$	1.00
$Br_2(\ell) + 2e^- \longrightarrow 2Br^-(aq)$	1.08
$ClO_4^-(aq) + 2H^+(aq) + 2e^- \longrightarrow ClO_3^-(aq) + H_2O$	1.19
$IO_3^-(aq) + 6H^+(aq) + 5e^- \longrightarrow \frac{1}{2}I_2(aq) + 3H_2O$	1.195
$Pt^{2+}(aq) + 2e^- \longrightarrow Pt(s)$	1.2
$O_2(g) + 4H^+(aq) + 4e^- \longrightarrow 2H_2O$	1.229
$MnO_2(s) + 4H^+(aq) + 2e^- \longrightarrow Mn^{2+}(aq) + 2H_2O$	1.23
$N_2H_5^+(aq) + 3H^+(aq) + 2e^- \longrightarrow 2NH_4^+(aq)$	1.24
$Cr_2O_7^{2-}(aq) + 14H^+(aq) + 6e^- \longrightarrow 2Cr^{3+}(aq) + 7H_2O$	1.33
$Cl_2(g) + 2e^- \longrightarrow 2Cl^-(aq)$	1.360
$BrO_3^-(aq) + 6H^+(aq) + 6e^- \longrightarrow Br^-(aq) + 3H_2O$	1.44
$ClO_3^-(aq) + 6H^+(aq) + 5e^- \longrightarrow \frac{1}{2}Cl_2(g) + 3H_2O$	1.47
$Au^{3+}(aq) + 3e^- \longrightarrow Au(s)$	1.50
$MnO_4^-(aq) + 8H^+(aq) + 5e^- \longrightarrow Mn^{2+}(aq) + 4H_2O$	1.51
$NaBiO_3(s) + 6H^+(aq) + 2e^- \longrightarrow Bi^{3+}(aq) + Na^+(aq) + 3H_2O$	~1.6
$Ce^{4+}(aq) + e^- \longrightarrow Ce^{3+}(aq)$	1.61
$2HClO(aq) + 2H^+(aq) + 2e^- \longrightarrow Cl_2(g) + 2H_2O$	1.63
$Au^+(aq) + e^- \longrightarrow Au(s)$	1.68
$PbO_2(s) + SO_4^{2-}(aq) + 4H^+(aq) + 2e^- \longrightarrow PbSO_4(s) + 2H_2O$	1.685
$NiO_2(s) + 4H^+(aq) + 2e^- \longrightarrow Ni^{2+}(aq) + 2H_2O$	1.7
$H_2O_2(aq) + 2H^+(aq) + 2e^- \longrightarrow 2H_2O$	1.77
$Pb^{4+}(aq) + 2e^- \longrightarrow Pb^{2+}(aq)$	1.8
$Co^{3+}(aq) + e^- \longrightarrow Co^{2+}(aq)$	1.82
$F_2(g) + 2e^- \longrightarrow 2F^-(aq)$	2.87

Standard Reduction Potentials in Aqueous Solution
at 25°C (continued)

Basic Solution	Standard Reduction Potential, E^0 (volts)
$SiO_3^{2-}(aq) + 3H_2O + 4e^- \longrightarrow Si(s) + 6OH^-(aq)$	-1.70
$Cr(OH)_3(s) + 3e^- \longrightarrow Cr(s) + 3OH^-(aq)$	-1.30
$[Zn(CN)_4]^{2-}(aq) + 2e^- \longrightarrow Zn(s) + 4CN^-(aq)$	-1.26
$Zn(OH)_2(s) + 2e^- \longrightarrow Zn(s) + 2OH^-(aq)$	-1.245
$[Zn(OH)_4]^{2-}(aq) + 2e^- \longrightarrow Zn(s) + 4OH^-(aq)$	-1.22
$N_2(g) + 4H_2O + 4e^- \longrightarrow N_2H_4(aq) + 4OH^-(aq)$	-1.15
$SO_4^{2-}(aq) + H_2O + 2e^- \longrightarrow SO_3^{2-}(aq) + 2OH^-(aq)$	-0.93
$Fe(OH)_2(s) + 2e^- \longrightarrow Fe(s) + 2OH^-(aq)$	-0.877
$2NO_3^-(aq) + 2H_2O + 2e^- \longrightarrow N_2O_4(g) + 4OH^-(aq)$	-0.85
$2H_2O + 2e^- \longrightarrow H_2(g) + 2OH^-(aq)$	-0.8277
$Fe(OH)_3(s) + e^- \longrightarrow Fe(OH)_2(s) + OH^-(aq)$	-0.56
$S(s) + 2e^- \longrightarrow S^{2-}(aq)$	-0.48
$Cu(OH)_2(s) + 2e^- \longrightarrow Cu(s) + 2OH^-(aq)$	-0.36
$CrO_4^{2-}(aq) + 4H_2O + 3e^- \longrightarrow Cr(OH)_3(s) + 5OH^-(aq)$	-0.12
$MnO_2(s) + 2H_2O + 2e^- \longrightarrow Mn(OH)_2(s) + 2OH^-(aq)$	-0.05
$NO_3^-(aq) + H_2O + 2e^- \longrightarrow NO_2^-(aq) + 2OH^-(aq)$	0.01
$O_2(g) + H_2O + 2e^- \longrightarrow OOH^-(aq) + OH^-(aq)$	0.076
$HgO(s) + H_2O + 2e^- \longrightarrow Hg(\ell) + 2OH^-(aq)$	0.0984
$[Co(NH_3)_6]^{3+}(aq) + e^- \longrightarrow [Co(NH_3)_6]^{2+}(aq)$	0.10
$N_2H_4(aq) + 2H_2O + 2e^- \longrightarrow 2NH_3(aq) + 2OH^-(aq)$	0.10
$2NO_2^-(aq) + 3H_2O + 4e^- \longrightarrow N_2O(g) + 6OH^-(aq)$	0.15
$Ag_2O(s) + H_2O + 2e^- \longrightarrow 2Ag(s) + 2OH^-(aq)$	0.34
$ClO_4^-(aq) + H_2O + 2e^- \longrightarrow ClO_3^-(aq) + 2OH^-(aq)$	0.36
$O_2(g) + 2H_2O + 4e^- \longrightarrow 4OH^-(aq)$	0.40
$Ag_2CrO_4(s) + 2e^- \longrightarrow 2Ag(s) + CrO_4^{2-}(aq)$	0.446
$NiO_2(s) + 2H_2O + 2e^- \longrightarrow Ni(OH)_2(s) + 2OH^-(aq)$	0.49
$MnO_4^-(aq) + e^- \longrightarrow MnO_4^{2-}(aq)$	0.564
$MnO_4^-(aq) + 2H_2O + 3e^- \longrightarrow MnO_2(s) + 4OH^-(aq)$	0.588
$ClO_3^-(aq) + 3H_2O + 6e^- \longrightarrow Cl^-(aq) + 6OH^-(aq)$	0.62
$2NH_2OH(aq) + 2e^- \longrightarrow N_2H_4(aq) + 2OH^-(aq)$	0.74
$OOH^-(aq) + H_2O + 2e^- \longrightarrow 3OH^-(aq)$	0.88
$ClO^-(aq) + H_2O + 2e^- \longrightarrow Cl^-(aq) + 2OH^-(aq)$	0.89

Appendix K

Selected Thermodynamic Values

Species	$\Delta H^0_{f298.15}$ (kJ/mol)	$S^0_{298.15}$ (J/mol·K)	$\Delta G^0_{f298.15}$ (kJ/mol)	Species	$\Delta H^0_{f298.15}$ (kJ/mol)	$S^0_{298.15}$ (J/mol·K)	$\Delta G^0_{f298.15}$ (kJ/mol)
Aluminum				**Cesium**			
Al(s)	0	28.3	0	Cs⁺(aq)	−248	133	−282.0
AlCl₃(s)	−704.2	110.7	−628.9	CsF(aq)	−568.6	123	−558.5
Al₂O₃(s)	−1676	50.92	−1582	**Chlorine**			
Barium				Cl(g)	121.7	165.1	105.7
BaCl₂(s)	−860.1	126	−810.9	Cl⁻(g)	−226	—	—
BaSO₄(s)	−1465	132	−1353	Cl₂(g)	0	223.0	0
Beryllium				HCl(g)	−92.31	186.8	−95.30
Be(s)	0	9.54	0	HCl(aq)	−167.4	55.10	−131.2
Be(OH)₂(s)	−907.1	—	—	**Chromium**			
Bromine				Cr(s)	0	23.8	0
Br(g)	111.8	174.9	82.4	(NH₄)₂Cr₂O₇(s)	−1807	—	—
Br₂(ℓ)	0	152.23	0	**Copper**			
Br₂(g)	30.91	245.4	3.14	Cu(s)	0	33.15	0
BrF₃(g)	−255.6	292.4	−229.5	CuO(s)	−157	42.63	−130
HBr(g)	−36.4	198.59	−53.43	**Fluorine**			
Calcium				F⁻(g)	−322	—	—
Ca(s)	0	41.6	0	F⁻(aq)	−332.6	—	−278.8
Ca(g)	192.6	154.8	158.9	F(g)	78.99	158.6	61.92
Ca²⁺(g)	1920	—	—	F₂(g)	0	202.7	0
CaC₂(s)	−62.8	70.3	−67.8	HF(g)	−271	173.7	−273
CaCO₃(s)	−1207	92.9	−1129	HF(aq)	−320.8	—	−296.8
CaCl₂(s)	−795.0	114	−750.2	**Hydrogen**			
CaF₂(s)	−1215	68.87	−1162	H(g)	218.0	114.6	203.3
CaH₂(s)	−189	42	−150	H₂(g)	0	130.6	0
CaO(s)	−635.5	40	−604.2	H₂O(ℓ)	−285.8	69.91	−237.2
CaS(s)	−482.4	56.5	−477.4	H₂O(g)	−241.8	188.7	−228.6
Ca(OH)₂(s)	−986.6	76.1	−896.8	H₂O₂(ℓ)	−187.8	109.6	−120.4
Ca(OH)₂(aq)	−1002.8	76.15	−867.6	**Iodine**			
CaSO₄(s)	−1433	107	−1320	I(g)	106.6	180.66	70.16
Carbon				I₂(s)	0	116.1	0
C(s, graphite)	0	5.740	0	I₂(g)	62.44	260.6	19.36
C(s, diamond)	1.897	2.38	2.900	ICl(g)	17.78	247.4	−5.52
C(g)	716.7	158.0	671.3	HI(g)	26.5	206.5	1.72
CCl₄(ℓ)	−135.4	216.4	−65.27	**Iron**			
CCl₄(g)	−103	309.7	−60.63	Fe(s)	0	27.3	0
CHCl₃(ℓ)	−134.5	202	−73.72	FeO(s)	−272	—	—
CHCl₃(g)	−103.1	295.6	−70.37	Fe₂O₃(s, hematite)	−824.2	87.40	−742.2
CH₄(g)	−74.81	186.2	−50.75	Fe₃O₄(s, magnetite)	−1118	146	−1015
C₂H₂(g)	226.7	200.8	209.2	FeS₂(s)	−177.5	122.2	−166.7
C₂H₄(g)	52.26	219.5	68.12	Fe(CO)₅(ℓ)	−774.0	338	−705.4
C₂H₆(g)	−84.86	229.5	−32.9	Fe(CO)₅(g)	−733.8	445.2	−697.3
C₃H₈(g)	−103.8	269.9	−23.49	**Lead**			
C₆H₆(ℓ)	49.03	172.8	124.5	Pb(s)	0	64.81	0
C₈H₁₈(ℓ)	−268.8	—	—	PbCl₂(s)	−359.4	136	−314.1
C₂H₅OH(ℓ)	−277.7	161	−174.9	PbO(s, yellow)	−217.3	68.70	−187.9
C₂H₅OH(g)	−235.1	282.6	−168.6	Pb(OH)₂(s)	−515.9	88	−420.9
CO(g)	−110.5	197.6	−137.2	PbS(s)	−100.4	91.2	−98.7
CO₂(g)	−393.5	213.6	−394.4				
CS₂(g)	117.4	237.7	67.15				
COCl₂(g)	−223.0	289.2	−210.5				

Selected Thermodynamic Values (continued)

Species	$\Delta H^0_{f298.15}$ (kJ/mol)	$S^0_{298.15}$ (J/mol·K)	$\Delta G^0_{f298.15}$ (kJ/mol)
Lithium			
Li(s)	0	28.0	0
LiOH(s)	−487.23	50	−443.9
LiOH(aq)	−508.4	4	−451.1
Magnesium			
Mg(s)	0	32.5	0
MgCl₂(s)	−641.8	89.5	−592.3
MgO(s)	−601.8	27	−569.6
Mg(OH)₂(s)	−924.7	63.14	−833.7
MgS(s)	−347	—	—
Mercury			
Hg(ℓ)	0	76.02	0
HgCl₂(s)	−224	146	−179
HgO(s, red)	−90.83	70.29	−58.56
HgS(s, red)	−58.2	82.4	−50.6
Nickel			
Ni(s)	0	30.1	0
Ni(CO)₄(g)	−602.9	410.4	−587.3
NiO(s)	−244	38.6	−216
Nitrogen			
N₂(g)	0	191.5	0
N(g)	472.704	153.19	455.579
NH₃(g)	−46.11	192.3	−16.5
N₂H₄(ℓ)	50.63	121.2	149.2
(NH₄)₃AsO₄(aq)	−1268	—	—
NH₄Cl(s)	−314.4	94.6	−201.5
NH₄Cl(aq)	−300.2	—	—
NH₄I(s)	−201.4	117	−113
NH₄NO₃(s)	−365.6	151.1	−184.0
NO(g)	90.25	210.7	86.57
NO₂(g)	33.2	240.0	51.30
N₂O(g)	82.05	219.7	104.2
N₂O₄(g)	9.16	304.2	97.82
N₂O₅(g)	11	356	115
N₂O₅(s)	−43.1	178	114
NOCl(g)	52.59	264	66.36
HNO₃(ℓ)	−174.1	155.6	−80.79
HNO₃(g)	−135.1	266.2	−74.77
HNO₃(aq)	−206.6	146	−110.5
Oxygen			
O(g)	249.2	161.0	231.8
O₂(g)	0	205.0	0
O₃(g)	143	238.8	163
OF₂(g)	23	246.6	41
Phosphorus			
P(g)	314.6	163.1	278.3
P₄(s, white)	0	177	0
P₄(s, red)	−73.6	91.2	−48.5
PCl₃(g)	−306.4	311.7	−286.3
PCl₅(g)	−398.9	353	−324.6
PH₃(g)	5.4	210.1	13
P₄O₁₀(s)	−2984	228.9	−2698
H₃PO₄(s)	−1281	110.5	−1119
Potassium			
K(s)	0	63.6	0
KCl(s)	−436.5	82.6	−408.8
KClO₃(s)	−391.2	143.1	−289.9
KI(s)	−327.9	106.4	−323.0
KOH(s)	−424.7	78.91	−378.9
KOH(aq)	−481.2	92.0	−439.6

Species	$\Delta H^0_{f298.15}$ (kJ/mol)	$S^0_{298.15}$ (J/mol·K)	$\Delta G^0_{f298.15}$ (kJ/mol)
Rubidium			
Rb(s)	0	76.78	0
RbOH(aq)	−481.16	110.75	−441.24
Silicon			
Si(s)	0	18.8	0
SiBr₄(ℓ)	−457.3	277.8	−443.9
SiC(s)	−65.3	16.6	−62.8
SiCl₄(g)	−657.0	330.6	−617.0
SiH₄(g)	34	204.5	56.9
SiF₄(g)	−1615	282.4	−1573
SiI₄(g)	−132	—	—
SiO₂(s)	−910.9	41.84	−856.7
H₂SiO₃(s)	−1189	134	−1092
Na₂SiO₃(s)	−1079	—	—
H₂SiF₆(aq)	−2331	—	—
Silver			
Ag(s)	0	42.55	0
Sodium			
Na(s)	0	51.0	0
Na(g)	108.7	153.6	78.11
Na⁺(g)	601	—	—
NaBr(s)	−359.9	—	—
NaCl(s)	−411.0	72.38	−384
NaCl(aq)	−407.1	115.5	−393.0
Na₂CO₃(s)	−1131	136	−1048
NaOH(s)	−426.7	—	—
NaOH(aq)	−469.6	49.8	−419.2
Sulfur			
S(s, rhombic)	0	31.8	0
S(g)	278.8	167.8	238.3
S₂Cl₂(g)	−18	331	−31.8
SF₆(g)	−1209	291.7	−1105
H₂S(g)	−20.6	205.7	−33.6
SO₂(g)	−296.8	248.1	−300.2
SO₃(g)	−395.6	256.6	−371.1
SOCl₂(ℓ)	−206	—	—
SO₂Cl₂(ℓ)	−389	—	—
H₂SO₄(ℓ)	−814.0	156.9	−690.1
H₂SO₄(aq)	−907.5	17	−742.0
Tin			
Sn(s, white)	0	51.55	0
Sn(s, grey)	−2.09	44.1	0.13
SnCl₂(s)	−350	—	—
SnCl₄(ℓ)	−511.3	258.6	−440.2
SnCl₄(g)	−471.5	366	−432.2
SnO₂(s)	−580.7	52.3	−519.7
Titanium			
TiCl₄(ℓ)	−804.2	252.3	−737.2
TiCl₄(g)	−763.2	354.8	−726.8
Tungsten			
W(s)	0	32.6	0
WO₃(s)	−842.9	75.90	−764.1
Zinc			
ZnO(s)	−348.3	43.64	−318.3
ZnS(s)	−205.6	57.7	−201.3

Appendix L

When calculations are done by different methods, roundoff errors may give slightly different answers. These differences are larger in calculations with several steps. Usually there is no cause for concern when your answers differ *slightly* from the answers given here.

Answers to Even-Numbered Numerical Exercises

Chapter 1

22. (a) 52,600, (b) 0.00000410, (c) 100., (d) 0.08206, (e) 9346, (f) 0.009346

26. (a) 0.0210 ft^2, (b) 307 in, (c) 49.0 in^3, (d) 6048 m, (e) 3.9×10^4 mi/day

28. (a) kilo-, (b) milli-, (c) mega-, (d) deci-, (e) centi-, (f) deci-, (g) milli-, (h) micro-

30. (a) 1.03×10^{-2} km, (b) 1.03×10^4 m, (c) 2.47×10^5 g, (d) 4.32×10^3 L, (e) 8.59 L, (f) 4.567×10^6 cm^3

32. (a) 1.0×10^2 km/hr, (b) 95 ft/s

34. (a) 24.0 cg, (b) 500 cm, (c) the same, (d) 3.2 m^3

36. (a) 62.7 tons ore, (b) 62.7 kg ore

38. 3.0×10^{-3} kg salt/day   **40.** 27.98 cents/L

42. (a) 0.040 cm^3/drop, 40 μL/drop; (b) 4.2 mm

44. (a) 474 lb, (b) 97.5 kg, (c) 6.10×10^5 cg

46. 4.85×10^7 atoms   **50.** 6.09 g/cm^3

52. 1.65 g/cm^3 **54.** 1.2×10^{14} g/cm^3

56. (a) 25.052 cm^3, (b) 21.073 g, (c) 17.601 g, (d) 17.601 cm^3, (e) 7.451 cm^3, (f) 2.828 g/cm^3

58. (a) 52.9 cm^3, (b) 3.75 cm, (c) 1.48 in

62. (a) $-18°C$, (b) 310.2 K, (c) 77°F, (d) 65.3°F

64. (c) 285.3°R   **66.** 198 J

68. (a) 1.39×10^6 J, (b) 13.0°C   **70.** 97.9°C

74. (b) approx. 80°F, (c) approx. 425 K, (d) 80.33°F, 422.04 K

76. (a) 5.64×10^3 g, (b) 33.4 cm^3

78. 105 mg KCN   **80.** 9.4542×10^{12} km/yr, 5.8746×10^{12} mi/yr

Chapter 2

16. 1.630   **18.** 44.5 g, 534 g, 2.23×10^{24} g

20. (a) 10.811, 10.811; (b) As, 74.922; (c) 26.9815, 26.9815; (d) Cr, 51.9961

22. (a) 70.9054 amu, (b) 18.0152 amu, (c) 183.17 amu, (d) 261.9676 amu

24. (a) 0.988 mol NH_3, (b) 33.2 mol NH_4Br, (c) 0.027 mol PCl_5, (d) 2.265 mol Fe

26. (a) 2.82×10^{-2} mol Cl, (b) 1.661×10^{-24} mol Cl, (c) 9.40×10^{-4} mol Cl

28. (a) 9.1×10^{-8} mol Fe, (b) 6.51×10^{-6} mol CH_4, (c) 76.6 mol O_2, (d) 76.6 mol $Fe(NO_3)_3$

30. (a) 5.1×10^{-6} g Fe, (b) 1.04×10^{-4} g CH_4, (c) 2.45×10^3 g O_2, (d) 1.85×10^4 g $Fe(NO_3)_3$

32. (a) 6.02×10^{23} molecules CO, (b) 6.02×10^{23} molecules N_2, (c) 1.36×10^{23} molecules P_4, (d) 2.72×10^{23} molecules P_2

34. (a) 47.3 g C_6H_5OH, (b) 60.7 g SiO_2, (c) 77.2 g $C_7H_8NSO_3$, (d) 0.0126 g KNO_3

36. 1 mol Na^+, 1 mol Cl^-; 4 mol Na^+, 2 mol SO_4^{2-}; 0.1 mol Ca^{2+}, 0.2 mol NO_3^-; 0.25 mol $(NH_4)_3PO_4$

38. 34.3% Fe, 65.7% Cl

40. Azurite, 55.31% Cu; chalcocite, 79.84% Cu; chalcopyrite, 34.63% Cu; covelite, 66.46% Cu; cuprite, 88.82% Cu; malachite, 57.49% Cu; cuprite, Cu_2O, has the highest copper content on a percent-by-mass basis

42. $C_8H_{11}O_3N$ **44.** C_4H_9 **46.** C_2H_6O **48.** Fe_3O_4

50. (a) C_9H_9N, (b) 131.19 g/mol

52. (a) $C_3H_5O_2$, **(b)** $C_6H_{10}O_4$
54. (a) $C_{17}H_{28}N_4O_7S$, **(b)** $C_{17}H_{28}N_4O_7S$
58. (a) 1.00 g O, **(b)** 1.50 g O
60. 150.8 g Hg   **62.** 79.1 g $KMnO_4$
64. 163 tons Cu_2S **66.** 193 g NaCl
68. (a) 473 g $CuSO_4 \cdot H_2O$, **(b)** 425 g $CuSO_4$
70. (a) 71.41%, **(b)** 51.7%
72. (a) 188 lb $MgCO_3$, **(b)** 587 lb impurities, **(c)** 54.3 lb Mg
74. (a) 3.00 mol CO_2, **(b)** 2.00 mol CO_2, **(c)** 4.00 mol CO_2
76. (a) 1.33 mol O_3, **(b)** 4.00 mol O, **(c)** 64.0 g O_2, **(d)** 42.7 g O_2
78. (a) 64 g C, **(b)** 13.148 g N_2O_5, **(c)** 0.44 g $Al_2(SO_4)_3$, **(d)** 3.6×10^{-9} g HCl
80. (a) 0.526 mol Ag, **(b)** 0.526 mol Ag, **(c)** 2.12×10^{-3} mol Ag, **(d)** 8.74×10^{-4} mol Ag
82. $C_2H_3NBr_2Cl_2$
84. (a) 0.372 mol Pb, **(b)** 77.1 g Pb, **(c)** 2.24×10^{23} atoms Pb
88. (a) 111 g $CuSO_4$, **(b)** 624 g $CuSO_4 \cdot 5H_2O$
90. 0.300 g Cr_2O_3

Chapter 3

8. (b) 300 molecules H_2, **(c)** 200 molecules N_2
10. (b) 264 molecules O_2, **(c)** 198 molecules H_2O
12. (b) 15 mol HCl, **(c)** 7.5 mol H_2O
14. (b) 8.8 mol O_2, **(c)** 7.5 mol H_2O
16. (a) 3 mol O_2/2 mol ZnS, **(b)** 1 mol ZnO/1 mol ZnS, **(c)** 1 mol SO_2/1 mol ZnS
18. (a) 1.50 mol O_2, **(b)** 0.500 mol O_2, **(c)** 0.500 mol O_2, **(d)** 0.500 mol O_2, **(e)** 2.00 mol O_2
20. (a) 1.88 mol O_2, **(b)** 1.50 mol NO, **(c)** 2.25 mol H_2O
22. 112 g Cl_2 **24.** 24 g Fe_3O_4 **26.** 1.75 g F_2
28. 0.651 g Br_2 **30.** 67.8% $CuSO_4$ **32.** 79.7 g NH_3
34. 387 g superphosphate   **36.** 52.45 g K
38. 38.9 g $BaSO_4$
40. 46.4 g Cr present, 33.9 g Cr recovered
42. 49.6 g PCl_5 **44.** 99.2% yield
46. 820 g C_2H_5OBr
48. (a) 220 kg CaO, **(b)** 210 kg CaC_2
50. 69.1 g H_2TeO_3 **52.** 358 g $KClO_3$
54. 1.50 kg pure TiO_2 **56.** 59.8 kg Zn
58. 109 g soln, 9 g NH_4Cl **60.** 69.3 g $(NH_4)_2SO_4$
62. 404 mL soln   **64.** 2.50 M H_3PO_4
66. 0.775 M H_3PO_4
68. (a) 0.175 M $(CH_3)_2CHOH$, **(b)** 1.75×10^{-4} mol $(CH_3)_2CHOH$
70. (a) 8.2 g Na_3PO_4, **(b)** 8.2 g Na_3PO_4
72. (a) 20.0% $CaCl_2$, **(b)** 2.13 M $CaCl_2$
74. 28.7 M HF   **76.** 0.0287 M $BaCl_2$

78. 0.353 L $CuSO_4$ soln   **80.** 0.700 L conc HCl soln
82. 192 mL   **84.** 4.07 M H_2SO_4
86. 8.31×10^{-3} L HNO_3 soln   **88.** 0.00640 M $AlCl_3$ soln
90. 1.63 g AgCl   **94.** 181 mL NaOH soln
96. 0.03616 M H_2SO_4 **98.** 0.106 g $CaCO_3$
100. 24.0 g CS_2

Chapter 4

6. About 0.22 J/g · °C   **8.** About 90°C
68. (a) $+3, +3, +5, +5, +5, +5, +5, +5$; **(b)** $0, -1, +1, +3, +5, +7, +7$; **(c)** $+2, +4, +2, +6, +7, +7$; **(d)** $+2, -2, -1, -\frac{1}{2}$
70. (a) $-3, +3, +5, -\frac{1}{3}, -3$, **(b)** $-1, +1, +5, +7$
76. 12.9 g Zn
94. (a) 0.122 mol O_2, **(b)** 0.147 mol O_2, **(c)** 0.0231 mol O_2

Chapter 5

8. 8.15×10^{-20} C   **12.** 2.3×10^{-14}
14. $^{12}C^+$, 0.0833; $^{12}C^{2+}$, 0.167; $^{13}C^+$, 0.0769, $^{13}C^{2+}$, 0.154; order of increasing ratios is $^{13}C^+ < ^{12}C^+ < ^{13}C^{2+} < ^{12}C^{2+}$
16. (a) 0.021761%, **(b)** 39.958%, **(c)** 60.020%
24. 107.87 amu   **28.** 69.17% ^{63}Cu
32. 73.3 amu; data lost at low-mass end of plot
34. 87.6 amu   **36. (d)** 78.99 amu
40. (a) 3.08×10^{14} s^{-1}, **(b)** 6.10×10^{14} s^{-1}, **(c)** 6.10×10^9 s^{-1}, **(d)** 6.10×10^{18} s^{-1}
42. (a) infrared, **(b)** visible, **(c)** microwave, **(d)** X-rays or gamma rays
44. (a) 4.47×10^{14} s^{-1}, **(b)** 2.96×10^{-19} J/photon, **(c)** red
46. 5.85×10^{-19} J/photon; 3.52×10^5 J/mol or 352 kJ/mol
48. 28 photons   **50.** 9.5×10^{26} photons
54. 320 nm   **58. (d)** 3.20×10^{15} s^{-1}
60. (a) $2 \rightarrow 1$, 121.5 nm; **(b)** $3 \rightarrow 1$, 102.6 nm; **(c)** $4 \rightarrow 1$, 97.24 nm
62. (a) red, **(b)** orange-red, **(c)** blue (indigo), **(d)** red, **(e)** red
64. 192 kJ/mol   **66.** 2.53×10^{18} photons
68. (a) 1.32×10^{-14} m, **(b)** 3.97×10^{-32} m
70. 8×10^{-33} kg

Chapter 6

26. 1.05×10^{15} s^{-1} **36.** 90.06 kJ/g

56. F, 0.710 Å; Cl, 0.990 Å; Cl—F, 1.700 Å
58. K^+, 1.33 Å; Cl^-, 1.82 Å; Li^+, 0.75 Å

Chapter 7

82. Cl, -1; O, -2, Sr, $+2$; K, $+1$; Al, $+3$; Se, -2
84. $+2$, $+4$, $+3$, $+5$, -2, -1
86. (a) $+4$, (b) $+4$, (c) -2, (d) $+5$, (e) $+3$,
(f) $+1$, (g) $+5$, (h) $+6$, (i) $+2$

Chapters 8, 9, 10

None

Chapter 11

4. (a) 0.335 M K_2SO_4, (b) 0.139 M $Al_2(SO_4)_3$,
(c) 0.0716 M $Al_2(SO_4)_3 \cdot 18H_2O$
6. 143 g $(NH_4)_2SO_4$ **8.** 5.292 M H_2SO_4
10. 1.44 M NaCl   **12.** 0.828 M Na_3PO_4
14. 3.00 L NaOH soln, 0.667 L H_3PO_4 soln
16. 0.857 M CH_3COOH **18.** 0.296 L
26. 180 mL NaOH soln   **28.** 0.08964 M NaOH
30. 0.03617 M H_2SO_4 **32.** 25.2% $(COOH)_2 \cdot 2H_2O$
34. 0.106 g $CaCO_3$ **36.** 9.28 mmol HCl
38. 1.17 N H_3PO_4
40. 0.230 M H_3AsO_4, 0.690 N H_3AsO_4
42. 0.0655 M $Ba(OH)_2$ **44.** 0.123 N HCl
46. 0.181 N HCl, 0.181 M HCl
48. 4.0 L $KMnO_4$ soln   **50.** 10 mL $KMnO_4$ soln
52. 0.06900 M I_2, 0.1728 g As_2O_3
54. (a) 0.0167 L HI soln, (b) 0.197 L HI soln,
(c) 0.0519 L HI soln
56. 59.35% Fe
58. (a) 2.05 g $KMnO_4$, (b) 13.0 g NaI,
(c) 1.77 g $K_2Cr_2O_7$
60. 0.200 M $KMnO_4$, 1.00 N $KMnO_4$
62. 0.535 M $FeSO_4$, 0.535 N $FeSO_4$ **64.** 0.1375 M
68. 310 mL   **70.** 27.8 mL

Chapter 12

8. (a) 14.4 psi, (b) 745 mm Hg, (c) 29.3 in Hg,
(d) 9.93 $\times$ 10^4 Pa, (e) 0.980 atm, (f) 33.2 ft H_2O
10. (a) 858 torr, 1.13 atm; (b) 808 torr, 1.06 atm
14. 0.33 atm
16. (a) 2.9 atm, (b) 0.043 L or 43 mL
18. 3400 balloons (to 2 sig. figures)
24. $-177°C$ **26.** (a) 1.43 L, (b) 31 cm
28. At $-78.5°C$, 6.529 L; at $-195.8°C$, 2.59 L, at $-268.9°C$,
0.143 L
32. 4.00 atm   **34.** 305 K or 32°C

38. 2.7×10^{14} molecules CO   **40.** 78 L, 2.4 g/L
42. For $D = 0.178$ g/L, helium (He); for $D = 0.900$ g/L,
neon (Ne)
46. 40.0 atm   **48.** (a) 27 K or $-246°C$, (b) 5.3 g/L
50. 2.0×10^8 L Cl_2 or 7.1×10^6 ft^3 Cl_2, 50 ft (to 1 sig.
figure)
52. 29.6 g/mol, 2%   **54.** 44 g/mol
56. 46.7 g/mol   **60.** 106 atm, 37.2 atm
62. He, 0.259; Ar, 0.358; Xe, 0.383
64. (a) 15.75 atm; (b) 5.25 atm; (c) 5.25 atm, 5.25 atm
66. 385 mL
68. (a) He, 2.78 atm; N_2, 1.78 atm; (b) 4.56 atm;
(c) 0.610
74. 1.17   **80.** 40.6 cm
82. Relative rates of effusion in order H_2:HD:D_2 are
1.4137:1.1545:1.
88. 0.695, 43.2 atm
90. (a) 0.958 atm, (b) 0.937 atm
94. Reaction (b) requires the smallest volume of O_2 per
volume of other gas.
96. 8 atoms of S per molecule
98. 57 g NaN_3 **100.** 4.33 g $KClO_3$
102. 2.95 L NH_3 **104.** 166 g KNO_3 **106.** 17.2% S
108. (a) 2.48 L C_8H_{18}, (b) 39.3 min
110. 4000 L air   **112.** 1.81   **114.** 99.5 g/mol
116. (a) 6130 g H_2O, (b) 6150 mL H_2O
118. 5.10 atm (ideal), 5.02 atm (van der Waals), 2%
120. C_2N_2

Chapter 13

32. 234 torr   **34.** (a) 43.4 kJ/mol
40. For water, 361.8 torr; for heavy water, 338.1 torr
42. (c) 37.1 kJ/mol, (d) 348 K or 75°C
46. 6.6×10^3 J/mol   **48.** 9.7×10^3J
50. 1.338×10^9 J   **52.** 54.1°C
54. 7.73×10^4 J   **56.** 58.6°C
58. (a) 4.38×10^3 J, (b) 26.5 g
86. (a) $a/2$, (b) 6 equidistant neighbors (Cl^- ions),
(c) $\dfrac{a}{\sqrt{2}}$, (d) 6 equidistant neighbors (Na^+ ions)
88. 1.00 g/cm³
90. 208 g/mol (calculated); Pb
92. (a) 8.070×10^{-23} g Mg, (b) 4.616×10^{-23} cm³,
(c) 1.758 g/cm³
94. (a) 8 C atoms/unit cell, (b) 4 nearest neighbors,
(c) $\dfrac{a \sqrt{3}}{4}$, (d) 1.545 Å, (e) 3.517 g/cm³
98. 1.542 Å
106. 0.988 atm (ideal); 0.982 atm (van der Waals); 0.6%
110. 2 atoms/unit cell, body-centered cubic crystal structure
112. 54.7 min
114. (a) solid, 18.36 cm³; liquid, 19.87 cm³; gas, 1.867 $\times$

10^5 cm³ (186.7 L);   **(b)** 13.5 cm³,   **(c)** solid, 0.735; liquid, 0.670; gas, 7.19×10^{-5}

Chapter 14

20. At 25°C, 3.6×10^{-4}; at 50°C, 2.6×10^{-4}; decreases
26. If s = solubility in g solute/100 g H_2O,

solubility as mass % $= \dfrac{s}{s + 100} \times 100$

28. 2.08 m C_6H_5COOH in C_2H_5OH
30. 27.0 g NaCl, 153 g H_2O
32. **(a)** 0.88 m K_2ZrF_6,   **(b)** 0.016
34. Add 14.5 g of NaCl to a 1-liter volumetric flask; add distilled water to dissolve the NaCl(s) and then continue to add water to a total volume of exactly 1 liter.
36. **(a)** 1.75 m $C_6H_{12}O_6$,   **(b)** the same
38. 0.6210 M K_2SO_4, 0.6374 m K_2SO_4, 10.00% K_2SO_4, 0.9886
40. $X_{C_2H_5OH} = 0.439$; $X_{H_2O} = 0.561$
44. **(a)** 12.6 torr,   **(b)** 62.0 torr
46. 23.9 g solute
48. P_{total} = 208 torr; in the vapor, X_{ethane} = 0.947, $X_{propane}$ = 0.053
50. **(a)** 80 torr,   **(b)** 175 torr,   **(c)** 270 torr
56. 1044°C   **58.** fp = -3.72°C, bp = 101.02°C
60. 24.3 g $C_{10}H_8$ **62.** $C_4H_8O_4$
64. **(a)** 70% $C_{10}H_8$, 30% $C_{14}H_{10}$;   **(b)** 80.4°C
68. **(a)** 3,   **(b)** 2,   **(c)** 5,   **(d)** 3
72. 751 torr   **74.** 1%   **76.** $i = 1.79$, 79% ionization
78. **(b)** 244 g/mol
82. 0.286 atm   **84.** 0.062 M
86. $\Delta T_f = 1.91 \times 10^{-5}$ °C; $\Delta T_b = 9.44 \times 10^{-6}$ °C
88. 5.85 atm
90. **(a)** -9.30×10^{-5} °C,   **(b)** 1.22×10^{-3} atm or 0.930 torr,   **(c)** 1000%,   **(d)** 10% error
98. -843.6 kJ/mol
100. **(a)** 60.5 g/mol,   **(b)** 117 g/mol
104. 56% lactose   **108.** 193 g/mol

Chapter 15

12. -366 J
16. **(a)** $+983$ J, work done on the system;   **(b)** 0 J, no work done
20. 220 J/°C
22. **(a)** 2.1×10^3 J,   **(b)** -1.0×10^5 J/mol $Pb(NO_3)_2$
24. -41.8 kJ/g $C_6H_6(\ell)$; -3260 kJ/mol $C_6H_6(\ell)$
30. **(a)** 71.4 kJ evolved,   **(b)** 598 g O_2
40. -583 kJ/mol   **42.** -988.4 kJ/mol rxn
44. -220.1 kJ/mol rxn
46. **(a)** -36.0 kJ/mol rxn,   **(b)** -237 kJ/mol rxn,   **(c)** 25.5 kJ/mol rxn
48. **(a)** 95.1 kJ/mol O_2,   **(b)** 143 kJ/mol O_3,   **(c)** 2.97 J/g O_2,   **(d)** 2.97 kJ/g O_3

50. 96.4 kJ released ($\Delta H°$ = -96.4 kJ)
52. -65.3 kJ/mol SiC(s)
54. **(b)** -890.3 kJ/mol CH_4, -5451.4 kJ/mol $C_8H_{18}(\ell)$, -6829.2 kJ/mol $C_{10}H_{22}(\ell)$ **(c)** $C_{10}H_{22}(\ell)$,   **(d)** -55.50 kJ/g $CH_4(g)$, -47.723 kJ/g $C_8H_{18}(\ell)$, -48.00 kJ/g $C_{10}H_{22}(\ell)$; $CH_4(g)$ produces the most heat per gram of compound
58. -536 kJ/mol rxn   **60.** -302 kJ/mol rxn
62. 192 kJ/mol   **64.** 327.0 kJ/mol
66. 294 kJ/mol   **72.** -649 kJ/mol
74. -2463 kJ/mol
88. **(a)** $+1383$ J/mol · K,   **(b)** -188.0 J/mol · K,   **(c)** -24.9 J/mol · K
92. -4.78 kJ/mol rxn   **94.** -51.89 kJ/mol
96. **(a)** 8.15 kJ/mol rxn,   **(b)** 5.4 kJ/mol rxn,   **(c)** -364.1 kJ/mol rxn
102. **(a)** $\Delta H°_{rxn} = -196.0$ kJ/mol, $\Delta G°_{rxn} = -233.6$ kJ/mol, $\Delta S°_{rxn} = +125.6$ J/mol · K
104. 370 K or 97°C
106. **(a)** spontaneous,   **(b)** rate of reaction is very slow,   **(c)** not at equilibrium at any temperature
110. **(a)** -156 kJ/mol $C_6H_{12}(\ell)$,   **(b)** -165 kJ/mol $C_6H_5OH(\ell)$
112. **(a)** 0.13 J/g · °C, tungsten (W);   **(b)** 0.26 J/g · °C, molybdenum (Mo)
114. $q = +854$ J, $w = -63.1$ J, $\Delta E = 791$ J
116. **(a)** 2.88 kJ/°C,   **(b)** 22.514°C

Chapter 16

10. Rate of disappearance of O_2 = 1.0 M/min; rate of appearance of NO = 0.80 M/min; rate of appearance of H_2O = 1.2 M/min
14. Rate = $(2.5 \times 10^{-2}$ M^{-1} · min^{-1})[B][C]
16. **(a)** Rate of reaction = $(12$ M^{-2} · s^{-1})[NO]²[O₂];   **(b)** Rate = 6.1×10^{-4} M/s;   **(c)** rate of disappearance of NO = 1.2×10^{-3} M/s, rate of disappearance of O_2 = 6.1×10^{-4} M/s, rate of formation of NO_2 = 1.2×10^{-3} M/s
18. Decrease, 8
20. **(a)** Rate = $(1.1 \times 10^2$ M^{-2} · s^{-1})[ClO₂]²[OH⁻],   **(b)** second order with respect to ClO_2, first order with respect to OH⁻, third order overall
22. Rate = $(2.5 \times 10^{-3}$ M^{-2} · s^{-1})[A][B]²
24. Rate = $(20$ M^{-4} · s^{-1}) [A]²[B]³
26. Experiment 2, Rate = 0.10 M/s; experiment 2, Rate = 0.40 M/s
30. 4.91×10^{-2} min or 2.95 s
32. **(a)** 2.72×10^{-3} M,   **(b)** 2.86×10^{-2} M,   **(c)** 171 s or 0.0475 hr,   **(d)** 3.77×10^3 s or 1.05 hr
34. **(a)** 1.8×10^9 s or 57 yr,   **(b)** 57 g NO_2,   **(c)** $[NO_2]_{produced}$ = 1.38 M
36. 1680 s or 28.0 min   **38.** **(b)** 0.55 M
40. **(a)** zero order,   **(b)** 5.6 s

46. 83.8 kJ/mol rxn   **48. (a)** -200 kJ/mol
50. 103 kJ/mol rxn
52. (a) 2.61×10^3 kJ/mol,   **(b)** 5.78×10^{-21} s^{-1},
(c) 704 K
54. 8×10^2 kg CO$_2$/L · hr
56. (a) 0.23 s^{-1},   **(b)** 287 K or 14°C
66. 64 kJ/mol
68. Rate = $(6.2 \times 10^{-4}$ s$^{-1})$[N$_2$O$_5$]
70. (a) 4.5×10^{-6} mol N$_2$O$_5$ remaining;   **(b)** 30 s
72. (a) Rate = k[Hb][CO],   **(b)** 0.280 L/μmol · s,
(c) 0.252 μmol/L · s

Chapter 17

12. Products favored in (b), (c), and (d)
14. 1.1×10^{-5} **16.** 0.13
18. (a) 8.9×10^5,   **(b)** 1.3×10^{-12},   **(c)** 1.6×10^{-24}
20. (a) 7.4×10^{-5} mol H$_2$, 1.8×10^{-3} mol HI;   **(b)** 92
24. 0.12   **26.** 0.424 M CO, 0.576 M CO$_2$
30. $Q > K$, reverse reaction predominates; $Q < K$, forward
reaction predominates
32. (a) false,   **(b)** false,   **(c)** false,   **(d)** false
34. 0.048 M HCN   **36.** [Cl$_2$] = 3.4 M
38. 2.50×10^{-2}
40. [SbCl$_5$] = 7.4×10^{-4} M, [SbCl$_3$] = 7.0×10^{-3} M,
[Cl$_2$] = 2.6×10^{-3} M
42. (a) false,   **(b)** false,   **(c)** true,   **(d)** true,   **(e)** false,
(f) false
44. 63.8% of PCl$_5$ is dissociated
56. (a) 16,   **(b)** 0.44 M
58. (a) 0.13;   **(b)** [A] = 0.13 M, [B] = [C] = 0.12 M;
(c) [A] = 0.70 M, [B] = [C] = 0.30 M
60. (a) 1.8;   **(b)** P_{Cl_2} = 6.4 atm, P_{PCl_3} = 6.4 atm, P_{PCl_5} =
23 atm;   **(c)** 36 atm;   **(d)** 29 atm
62. (a) [N$_2$O$_4$] = 9.8×10^{-3} M, [NO$_2$] = 7.56×10^{-3} M;
(b) [N$_2$O$_4$] = 4.3×10^{-3} M, [NO$_2$] = 4.98×10^{-3} M;
(c) [N$_2$O$_4$] = 0.022 M, [NO$_2$] = 0.0113 M
66. 1.6×10^{-9} **68.** 0.166
70. 7.76   **72.** 6.0×10^{-4} atm
74. (a) 33.2% H$_2$ unreacted,   **(b)** less
78. Experiment 1, 1.81; experiment 2, 1.71; experiment 3,
1.94; average, 1.82
82. 6.9×10^{24}
84. At 400°C, K_p = 8.3×10^{-3}, K_c = 1.5×10^{-4}; at
800°C, K_p = 16, K_c = 0.18
86. (a) -37.9 kJ/mol rxn at 25°C; **(b)** K_p = 4.9×10^{-4};
(c) 68.0 kJ/mol rxn at 800°C
88. (a) 1.1×10^5,   **(b)** 2.2×10^2,   **(c)** 1.0×10^5
92. K_c = 2.72×10^{22}; Q_c = 0.2; reaction favors products
(very strongly, because K_c is so large)
94. (a) 7.9×10^{11},   **(b)** 8.9×10^5,   **(c)** 1.3×10^{-12},
(d) 1.6×10^{-24}
96. 4.32×10^{-3}

Chapter 18

2. (a) [H$^+$] = [Br$^-$] = 0.10 M;   **(b)** [K$^+$] = [OH$^-$] =
0.040 M;   **(c)** [Ca^{2+}] = 0.0020 M, [Cl$^-$] = 0.0040 M
4. (a) [K$^+$] = [OH$^-$] = 0.036 M;   **(b)** [Ba^{2+}] = 0.014 M,
[OH$^-$] = 0.028 M,   **(c)** [Ca^{2+}] = 0.161 M, [NO$_3^-$] =
0.322 M
12. In 0.040 M KOH, [H$_3$O$^+$] = 2.5×10^{-13} M; in
0.020 M Sr(OH)$_2$, [H$_3$O$^+$] = 2.5×10^{-13} M; in 0.014 M
Ba(OH)$_2$, [H$_3$O$^+$] = 3.6×10^{-13} M; in pure water,
[H$_3$O$^+$] = 1.0×10^{-7} M
14. [H$^+$] = 2.38×10^{-9} M
16. (a) 3.2×10^{-17} M,   **(b)** 1.0×10^{-30} M
18. (a) -3.28,   **(b)** 0.62,   **(c)** -11.24,   **(d)** -6.31
20. (a) 5.4×10^{10},   **(b)** 1.9×10^{-11},   **(c)** 0.014,   **(d)** 4.0
22. In 0.040 M KOH, pH = 12.60; in 0.020 M Sr(OH)$_2$,
pH = 12.60; in 0.014 M Ba(OH)$_2$, pH = 12.44; in pure
water, pH = 7.00
24. 10.30   **26.** 6.3×10^{-4} M
28. Each part in order [H$_3$O$^+$], [OH$^-$], pH, pOH:
(a) 2.9×10^{-13} M, 0.035 M, 12.54, 1.46;   **(b)** 0.035 M,
2.9×10^{-13} M, 1.46, 12.54;   **(c)** 1.4×10^{-13} M, 0.070 M,
12.85, 1.15
30. (a) 2.00×10^{-7} M,   **(b)** 6.699
32. (a) 6.182 at 37°C,   **(b)** neutral
38. 2.9×10^{-5} **40.** 7.3×10^{-5}
42. 1.8×10^{-8} M
44. [C$_6$H$_5$COOH] = 0.35 M; [H$_3$O$^+$] = [C$_6$H$_5$COO$^-$] =
4.7×10^{-3} M; [OH$^-$] = 2.1×10^{-12} M
46. 1.55×10^{-3} **48.** 5.8%   **50.** 5.0×10^{-4}
52. (a) [OH$^-$] = 1.3×10^{-3} M, 1.3%, 11.11;   **(b)** [OH$^-$]
= 6.8×10^{-3} M, 6.8%, 11.83
54. For triethylamine, [OH$^-$] = 2.0×10^{-3} M; for trimethylamine, [OH$^-$] = 8.2×10^{-4} M
56. 2.07   **58.** [H$_3$O$^+$] = 2.8×10^{-3} M, pH = 2.55
64. 3.08   **68.** K_a = 6×10^9, pK_a = 8.22
72. 4.5×10^{-4} M
74. (a) 3.44,   **(b)** 4.74
76. (a) 9.08,   **(b)** 9.08
82. [H$_3$O$^+$] = [HClO$_4$] = 0.10 M, pH = 1.00
84. (a) 9.34,   **(b)** 9.26,   **(c)** 1.00
86. [acid] = 6.7×10^{-2} M, [salt] = 0.13 M
88. 0.0015 M C$_2$H$_5$NH$_3^+$
90. 0.19 M CH$_3$CH$_2$COOH
92. 0.26 L NaOH, 0.74 L CH$_3$COOH
94. (a) 0.500 M,   **(b)** 0.250 M,   **(c)** 0.500 M,   **(d)** 1.8×10^{-5} M,   **(e)** 4.74
96. 8.9×10^{-3} M
98. [H$_3$AsO$_4$] = 0.193 M, [H$_3$O$^+$] = 0.0071 M,
[H$_2$AsO$_4^-$] = 0.0071 M, [HAsO$_4^{2-}$] = 5.6×10^{-8} M,
[OH$^-$] = 1.4×10^{-12} M, [AsO$_4^{3-}$] = 2.4×10^{-18};
[H$_3$PO$_4$] = 0.076 M, [H$_3$O$^+$] = 0.024 M, [H$_2$PO$_4^-$] =
0.024 M, [HPO$_4^{2-}$] = 6.2×10^{-8} M, [OH$^-$] = 4.2×10^{-13} M, [PO$_4^{3-}$] = 9.3×10^{-19} M

100. $[H_3O^+] = 1.4 \times 10^{-4}\ M$, $[HCO_3^-] = 1.4 \times 10^{-4}\ M$, $[OH^-] = 7.1 \times 10^{-11}\ M$, $[CO_3^{2-}] = 4.8 \times 10^{-11}\ M$
102. $[H^+] = 5.0 \times 10^{-6}\ M$, pH $= 5.30$
104. (a) 1.07,   (b) $6.4 \times 10^{-5}\ M$
106. 1.00

Chapter 19

12. 1.4×10^{-11}
14. (a) 2.2×10^{-11},   (b) 2.9×10^{-7},   (c) 1.6×10^{-10}
16. (a) 1.2×10^{-3} %,   (b) 0.14%,   (c) 3.3×10^{-3} %
18. 8.08
22. (a) 5.04,   (b) 5.76,   (c) 3.00
30. (a) 2.81, 0.75%;   (b) 5.19, 7.6×10^{-3} %;   (c) 6.17, 4.5×10^{-4} %
34. (a) 0.60,   (b) 0.82,   (c) 1.12,   (d) 1.78,   (e) 7.00, (f) 12.10
36.

Moles NaOH Added	$[H_3O^+]$	$[OH^-]$	pH	pOH
0	$4.2 \times 10^{-4}\ M$	$2.4 \times 10^{-11}\ M$	3.38	10.62
0.00200	$7.2 \times 10^{-5}\ M$	$1.4 \times 10^{-10}\ M$	4.14	9.85
0.00400	$2.7 \times 10^{-5}\ M$	$3.7 \times 10^{-10}\ M$	4.57	9.43
0.00500	$1.8 \times 10^{-5}\ M$	$5.6 \times 10^{-10}\ M$	4.74	9.25
0.00700	$7.7 \times 10^{-6}\ M$	$1.3 \times 10^{-9}\ M$	5.11	8.89
0.00900	$2.0 \times 10^{-6}\ M$	$5.0 \times 10^{-9}\ M$	5.70	8.30
0.00950	$9.0 \times 10^{-7}\ M$	$1.1 \times 10^{-8}\ M$	6.00	8.00
0.0100	$4.2 \times 10^{-9}\ M$	$2.4 \times 10^{-6}\ M$	8.38	5.62
0.0105	$2.0 \times 10^{-11}\ M$	$5.0 \times 10^{-4}\ M$	10.70	3.30
0.0120	$5.0 \times 10^{-12}\ M$	$2.0 \times 10^{-3}\ M$	11.30	2.70
0.0150	$2.0 \times 10^{-12}\ M$	$5.0 \times 10^{-3}\ M$	11.70	2.30

Phenolphthalein is a suitable indicator.
38. (a) 2.47 (usual approximation is not valid for this solution);   (b) 3.14;   (c) 5.46;   (d) 7.70;   (e) 9.92; (g) neutral red, bromthymol blue, and phenolphthalein are appropriate indicators.
44. 8.94, phenolphthalein
46.

Moles NH$_3$ Added	$[H_3O^+]$	$[OH^-]$	pH	pOH
0	$4.2 \times 10^{-4}\ M$	$2.4 \times 10^{-11}\ M$	3.38	10.62
0.00100	$1.6 \times 10^{-4}\ M$	$6.2 \times 10^{-11}\ M$	3.80	10.21
0.00400	$2.7 \times 10^{-5}\ M$	$3.7 \times 10^{-10}\ M$	4.57	9.43
0.00500	$1.8 \times 10^{-5}\ M$	$5.6 \times 10^{-10}\ M$	4.74	9.25
0.00900	$2.0 \times 10^{-6}\ M$	$5.0 \times 10^{-9}\ M$	5.70	8.30
0.00950	$9 \times 10^{-7}\ M$	$1 \times 10^{-8}\ M$	6.0	8.0
0.0100	$(1.0 \times 10^{-7}\ M)$	$(1.0 \times 10^{-7}\ M)$	7.00	7.00
0.0105	$1 \times 10^{-8}\ M$	$9 \times 10^{-7}\ M$	8.0	6.0
0.0130	$1.8 \times 10^{-9}\ M$	$5.4 \times 10^{-6}\ M$	8.73	5.27

No suitable indicator.
48. 8.2, phenolphthalein or bromthymol blue

Chapter 20

8. (a) 3.5×10^{-5},   (b) 7.7×10^{-19},   (c) 6.9×10^{-15}, (d) 8.8×10^{-12}
10.

Compound	Molar Solubility, M	K_{sp} (calculated)
$SrCrO_4$	5.9×10^{-3}	3.5×10^{-5}
BiI_3	1.3×10^{-5}	7.7×10^{-19}
$Fe(OH)_2$	1.2×10^{-5}	6.9×10^{-15}
$Zn(CN)_2$	1.3×10^{-4}	8.8×10^{-12}

(a) $SrCrO_4$,   (b) BiI_3,   (c) $SrCrO_4$,   (d) BiI_3
12. (a) 2.3×10^{-6} mol CuI/L, $[Cu^+] = 2.3 \times 10^{-6}\ M$, $[I^-] = 2.3 \times 10^{-6}\ M$, 4.4×10^{-4} g/L
(b) 6.5×10^{-7} mol $Ba_3(PO_4)_2$/L, $[Ba^{2+}] = 2.0 \times 10^{-6}\ M$, $[PO_4^{3-}] = 1.3 \times 10^{-6}\ M$, 3.9×10^{-4} g/L
(c) 2.1×10^{-3} mol PbF_2/L, $[Pb^{2+}] = 2.1 \times 10^{-3}\ M$, $[F^-] = 4.2 \times 10^{-4}\ M$, 0.51 g/L
(d) 4.7×10^{-6} mol Ag_3PO_4/L, $[Ag^+] = 1.4 \times 10^{-5}\ M$, $[PO_4^{3-}] = 4.7 \times 10^{-6}\ M$, 2.0×10^{-3} g/L
14.

Compound	K_{sp}	Molar Solubility, M	Solubility, g/L
CuI	5.1×10^{-12}	2.3×10^{-6}	4.4×10^{-4}
$Ba_3(PO_4)_2$	1.3×10^{-29}	6.5×10^{-7}	3.9×10^{-4}
PbF_2	3.7×10^{-8}	2.1×10^{-3}	0.51
Ag_3PO_4	1.3×10^{-20}	4.7×10^{-6}	2.0×10^{-3}

(a) PbF_2,   (b) $Ba_3(PO_4)_2$,   (c) PbF_2,   (d) $Ba_3(PO_4)_2$
16. 6.5×10^{-3} mol Ag_2SO_4/L
18. (a) 1.5×10^{-7} mol $Mg(OH)_2$/L,   (b) 1.9×10^{-5} mol $Mg(OH)_2$/L
20. 1.0×10^{-9} mol $BaCrO_4$/L, 3.4×10^{-6} mol Ag_2CrO_4/L, Ag_2CrO_4 is more soluble
22. $Q_{sp} = 1.5 \times 10^{-6}$, $Q_{sp} < K_{sp}$, so no precipitate
24. $Q_{sp} = 9.7 \times 10^{-11}$; $Q_{sp} > K_{sp}$, so precipitate will form
26. (a) (i) $2.0 \times 10^{-11}\ M$, (ii) $1.0 \times 10^{-8}\ M$, (iii) $7.3 \times 10^{-19}\ M$;
(b) (i) $1.7 \times 10^{-7}\ M$, (ii), $1.0 \times 10^{-8}\ M$, (iii) $7.3 \times 10^{-19}\ M$
28. 48%
32. (a) AuCl first, then AgCl, finally CuCl;   (b) $[Au^+] = 1.1 \times 10^{-4}\ M$, 99.89% Au^+ precipitated (ppt'd);
(c) $[Au^+] = 1.1 \times 10^{-7}\ M$, $[Ag^+] = 9.5 \times 10^{-5}\ M$
34. (a) $PbCrO_4$ will ppt first;   (b) $1.8 \times 10^{-13}\ M$;
(c) $1.8 \times 10^{-7}\ M$;   (d) $1.0 \times 10^{-6}\ M$
36. $Q_{sp} < K_{sp}$, so $Mg(OH)_2$ will not ppt; 7.61
38. 1.3×10^3 mol CaF_2/L
40. $Q_{sp} > K_{sp}$, so ppt will form but will not be seen
42. 4.30
44. (a) 9.65,   (b) 2.0×10^{-4} g/100 mL
54. $1.4 \times 10^{-22}\ M$
56. $2.1 \times 10^{-7}\ M$; $Q_{sp} < K_{sp}$, so Ag_2SO_4 will not ppt
58. (a) $1.3 \times 10^{-4}\ M$ in Ag_2CrO_4, $1.4 \times 10^{-5}\ M$ in $BaCrO_4$;   (b) Ag_2CrO_4 has greater molar solubility; $BaCrO_4$ has larger K_{sp}

60. $[Cd^{2+}] = 2.0 \times 10^{-6} M$, $[Pd^{2+}] = 7.7 \times 10^{-16} M$
62. (a) no,   **(b)** 0.44 M
64. $Q_{sp} > K_{sp}$, so ZnS would ppt

Chapter 21

16. 1.602×10^{-19} C/e^- **18.** 1.97×10^{-3} mol e^-
20. (a) 3 faradays, 4.91×10^4 C;   **(b)** 2 faradays, 962 C;
(c) 1 faraday, 481 C
22. 3.0 L Cl_2 **24.** 3.0×10^3 s or 50 min
26. 3.35 g Ag
28. (a) 0.107 faradays,   **(b)** 1.03×10^4 C,   **(c)** 1.20 L H_2,
(d) 13.33
30. Pb **32. (a)** 0.0178 faraday, **(b)** 1.92 g Ag, **(c)** 4.31 g $Fe(NO_3)_3$
44. (a) increasing oxidizing strength: $K^+ < Na^+ < Fe^{2+} < Cu^{2+} < Cu^+ < Ag^+ < F_2$;   **(b)** F_2 and Ag^+ can oxidize Cu
46. Decreasing activity: Eu > Ra > Rh; Eu is more active than Li; Eu and Ra are more active than H; Eu, Ra, and Rh are more active than Au
48. (a) +3.17 V,   **(b)** +1.54 V
50. (a) yes, $E^0_{cell} = +0.62$ V;   **(b)** no, $E^0_{cell} = -1.54$ V
52. (a) no, $E^0_{cell} = -0.18$ V;   **(b)** no, $E^0_{cell} = -0.41$ V for reduction to H_2SO_3 and $E^0_{cell} = -0.38$ V for reduction to SO_2
54. +1.78 V
56. (a) Combining half-reactions $H_2/H_2,OH^-$ and $O_2,H^+/H_2O$ would give the highest voltage, +2.057 V;
(b) $O_2 + 2H_2 + 4H^+ + 4OH^- \rightarrow 6H_2O$
58. (a) (i) $Cd + 2Ag^+ \rightarrow Cd^{2+} + 2Ag$;
    (ii) oxidation: $Cd \rightarrow Cd^{2+} + 2e^-$;
    reduction: $Ag^+ + e^- \rightarrow Ag$;
    (iii) +1.202 V;
    (iv) the standard reaction occurs as written
(b) (i) $2Cr + 3Fe^{2+} \rightarrow 2Cr^{3+} + 3Fe$;
    (ii) oxidation: $Cr \rightarrow Cr^{3+} + 3e^-$;
    reduction: $Fe^{2+} + 2e^- \rightarrow Fe$;
    (iii) +0.30 V;
    (iv) the standard reaction occurs as written
60. (a) $E^0_{cell} = +2.04$ V, spontaneous;   **(b)** $E^0_{cell} = +1.71$ V, spontaneous;   **(c)** $E^0_{cell} = +4.12$ V, spontaneous;
(d) $E^0_{cell} = -3.07$ V, not spontaneous
62. (a) H_2,   **(b)** Pt,   **(c)** Hg,   **(d)** Cl^- in basic solution,
(e) Cu,   **(f)** Rb
64. +0.530 V   **70.** -0.246 V
72. (a) +2.123 V,   **(b)** +2.194 V
74. 3×10^{-10} atm   **76.** 0.024 M
78. (a) 1.5×10^{37};   **(b)** $[Cu^{2+}] = 2 \times 10^{-37} M$, $[Zn^{2+}] = 2.00 M$
82. (a) $E^0_{cell} = -0.62$ V, not spontaneous, $\Delta G^0 = 120$ kJ/mol rxn, $K = 10^{-21}$;

(b) $E^0_{cell} = +0.368$ V, spontaneous, $\Delta G^0 = -35.5$ kJ/mol rxn, $K = 1.6 \times 10^6$;
(c) $E^0_{cell} = +1.833$ V, spontaneous, $\Delta G^0 = -1061$ kJ/mol rxn, $K = 10^{186}$
84. 7.7×10^{70} **86.** 2 (to one sig. figure)
88. (a) 5×10^{-13},   **(b)** +68 kJ/mol rxn
90. -171.4 kJ/mol rxn
98. (a) $3Zn + 2Fe^{3+} \rightarrow 3Zn^{2+} + 2Fe$,   **(b)** +0.727 V,
(c) +0.836 V,   **(d)** 0.0457 g Zn
100. 840 C   **102.** 18,900 C
104. (a) +4,   **(b)** +4,   **(c)** 0,   **(d)** Mg(s),   **(e)** UF_4(s),
(f) 20.5 A,   **(g)** 0.0411 L HF(g),   **(h)** yes
108. -1.662 V (using either half-reaction)

Chapter 22

12. 40.00 g NaOH, 1.01 g H_2, 35.45 g Cl_2
22. (a) -15 kJ/mol rxn, favorable;   **(b)** +5 kJ/mol rxn, unfavorable;   **(c)** -25 J/mol · K
28. 17.8 g Cu
30. (a) 1.45×10^8 s (1680 days or 4.60 yr),   **(b)** 9.59×10^3 L O_2
34. 1.0×10^3 tons seawater

Chapter 23

28. (a) -201.4 kJ/mol Li,   **(b)** -138.9 kJ/mol K,   **(c)** -415.0 kJ/mol Ca. E^0 for Li^+/Li is more negative than E^0 for K^+/K; therefore, ΔH^0 is more negative in (a) than in (b). In reaction (c), 1 mol of H_2 is produced, whereas in (a) and (b) only $\frac{1}{2}$ mol of H_2 is produced; therefore, ΔH^0 is even more negative for (c). All three E^0 values are very negative, so differences among E^0's are less important than the amount of reaction that occurs.
30. $\Delta H^0 = -195.4$ kJ/mol, $\Delta S^0 = 29.4$ J/mol · K, $\Delta G^0 = -204.0$ kJ/mol
32. (a) $SrCO_3$, when $[CO_3^{2-}] > 9.4 \times 10^{-9} M$;   **(b)** 0.12 of Sr^{2+} remains in solution; no
46. 2.00
48. (a) PbI_2, when $[I^-] > 9.3 \times 10^{-4} M$;   **(b)** 8.7×10^{-5} $(8.7 \times 10^{-3} \%)$; yes
50. Tin; 0.010 V; lowering T, decreasing $[Sn^{2+}]$, increasing $[Pb^{2+}]$

Chapter 24

10. 2.33 g $XeOF_2$, 1.88 L HF
12. 4.28 g XeF_6 **14.** -140 kJ/mol rxn
52. $Cl_2/F_2 = 0.732$, $Br_2/F_2 = 0.488$, $I_2/F_2 = 0.387$, $At_2/F_2 = 0.301$

54. 2.96×10^3 mol NaCl, 1.73×10^5 g NaCl, 3.81×10^2 lb NaCl, 150 lb Na, 8.16×10^8 s (25.9 yr)
56. 5×10^4
58. (a) 4.1×10^5 ft³ seawater,   (b) 1.3×10^5 L Cl_2
60. 76.6% by first reaction, 23.4% by second reaction

Chapter 25

28. 0.125 g Ag reacted, 0.0912%
30. $[OH^-] = 1.3 \times 10^{-4}$ M, pH $= 10.11$
32. $+335$ J/mol, rhombic form more stable
34. 1.64×10^3 g "new" H_2SO_4
36. 2.8

Chapter 26

34. -3721 kJ/mol
36. 945.4 kJ/mol N≡N bonds, 498.4 kJ/mol O=O bonds, 631.7 kJ/mol N=O bonds
38. 3.35
40. 43.7% P in P_4O_{10}, 31.6% P in H_3PO_4
42. 3.95×10^3 L of gas
44. 15.7 M HNO_3, 3.82 L

Chapter 27

16. 5.27 g Li_2CO_3, 68.3% Li_2CO_3, 31.7% Li_2O
18. 177 cm³
28. for X = F, -1573 kJ/mol rxn; for X = Cl, -617.0 kJ/mol rxn
36. 3.11 L H_2
38. $[H_3BO_3] = 0.10$ M, $[H_3O^+] = [H_2BO_3^-] = 8.5 \times 10^{-6}$ M; $[HBO_3^{2-}] = 1.8 \times 10^{-13}$ M, $[BO_3^{3-}] = 3.4 \times 10^{-22}$ M, pH $= 5.07$, $[OH^-] = 1.2 \times 10^{-9}$ M
40. B_3H_9

Chapter 28

26. 344 g Co_3O_4
28. $[Cr_2O_7^{2-}]/[CrO_4^{2-}] = 4.2 \times 10^{-12}$
32. -177 kJ/mol
34. 4.29 L H_2
36. 5.26

Chapter 29

46. (a) -657 kJ/mol,   (b) -194 kJ/mol,   (c) -253 kJ/mol,   (d) -962 kJ/mol,   (e) -95.7 kJ/mol,   (f) -314 kJ/mol,   (g) -93.3 kJ/mol,   (h) -172 kJ/mol
48. (a) $-\frac{3}{5}\Delta$,   (b) $-\frac{8}{5}\Delta$
50. Decrease, -153.5 J/mol · K
52. 10.38
54. (a) 2.2×10^{-6} mol $Zn(OH)_2$/L,   (b) 0.15 mol $Zn(OH)_2$/L,   (c) 0.15 M

Chapter 30

8. (a) 0.602 amu/atom, 0.602 g/mol;   (b) 5.42×10^{10} kJ/mol
10. (a) 0.594 amu/atom,   (b) 0.594 g/mol,   (c) 8.88×10^{-4} erg/atom,   (d) 8.88×10^{-11} J/atom,   (e) 5.35×10^{10} kJ/mol
12. (a) 1.55×10^{10} kJ/mol (1.29×10^{-12} J/nucleon);   (b) 5.01×10^{10} kJ/mol (1.42×10^{-12} J/nucleon);   (c) 8.81×10^{10} kJ/mol (1.38×10^{-12} J/nucleon) $^{59}_{27}$Co has highest binding energy/nucleon
22. (a) 1.44, positron emission or electron capture;   (b) 1.45, positron emission or electron capture;   (c) 1.41, β emission
42. 90.0% in 67.5 min, 99.0% in 135 min
44. 0.93 fs, 5.6 μL   **46.** 1100 yr
48. 2.00×10^4 yr   **52.** -1.68×10^9 kJ/mol
60. 1.15×10^8 kJ/mol
62. Fission, 1.40×10^{-3} J/amu; fusion, 1.52×10^{-13} J/amu; fusion

Chapter 31

None

Chapter 32

40. $[OH^-] = 6.5 \times 10^{-6}$ M, $[C_6H_5NH_3^+] = 6.5 \times 10^{-6}$ M, $[C_6H_5NH_2] = 0.100$ M, $[H^+] = 1.5 \times 10^{-9}$ M
72. 7.7×10^3 g/mol
80. $P_{CO} = P_{H_2} = 7.1$ atm
82. 8.60, more acidic
86. (a) 0.714 atm,   (b) 738 cm (24.2 ft)

Illustration and Table Credits

Chapter 1

Chapter opener: NASA; **unnum. figure p. 3:** from Petit Format/Nestle, Photo Researchers, Inc.; **1-1(a):** Charles D. Winters; **1-1(b, c):** James Morgenthaler; **1-2(a–c), 1-4(a–d), 1-6(a, b), 1-10, 1-11(a, b), 1-13, 1-14, 1-18, 1-19(a), unnum. figures pp. 12, 32:** Charles Steele; **unnum. figure p. 15:** © Tony Stone Worldwide; **unnum. figure p. 16:** from Standard Oil of Ohio; **1-15(a):** courtesy of OHAUS; **1-15(b, c):** courtesy of Mettler; **1-16:** courtesy of National Institute of Standards and Technology; **1-17:** James Morgenthaler; **1-19(b):** from Biophoto Associates, N.H.P.A.

Chapter 2

Chapter opener: Charles D. Winters; **2-4(parts 1 and 2), 2-6, 2-8, 2-9, 2-10, unnum. figures pp. 49, 70, 75, 80:** Charles Steele; **2-4(parts 3 and 4), unnum. figure pp. 62, 81:** Charles D. Winters; **2-5:** Philippe Plailly, Science Source/Photo Researchers, Inc.; **unnum. figure p. 79:** © Gemological Institute of America; **unnum. figure p. 82:** William Felger from Grant Heilman.

Chapter 3

Chapter opener: Robert W. Metz; **unnum. figure p. 84:** from Atlanta Gas Light Company; **unnum. figure p. 86:** by permission of the British Library; **3-4(d), unnum. figures pp. 94, 98, 99(bottom), 100, 107, 110, 115:** Charles Steele; **unnum. figure p. 99(center):** © Tom McHugh/Photo Researchers, Inc.; **3-2(a–c), 3-3(a–c), 3-4(a–c):** Charles D. Winters; **unnum. figure p. 106:** James Morgenthaler.

Chapter 4

Chapter opener: J. Morgenthaler; **4-1:** from *Annalen der Chemie und Pharmacie,* VIII, Supplementary Volume for 1872; **4-3(a, b), 4-4, 4-5, unnum. figures pp. 122, 123, 125(middle), 129, 142, 146, 148(top), 150, 151, 153:** Charles Steele; **unnum. figure p. 125(top):** courtesy of Bethlehem Steel; **unnum. figure p. 125(bottom);** from AT&T Bell Laboratories; **unnum. figure p. 126(left):** from J. W. van Spronsen, *The Periodic System of Chemical Elements: A History of the First Hundred Years,* Elsevier, 1969; **unnum. figure p. 126(right):** from Edward G. Mazurs, *Types of Graphic Representation of the Periodic System of Chemical Elements,* 1957; **4-2(a–c), unnum. figures pp. 140, 143:** Charles D. Winters; **unnum. figure p. 136:** Kevin Schafer/Tom Stack & Associates; **unnum. figures pp. 141, 148(bottom), 149, 156:** J. Morgenthaler.

Chapter 5

Chapter opener: courtesy of the California Institute of Technology; **unnum. figure p. 167:** UPI/Bettmann; **unnum. figures pp. 170, 182, 186(top):** The Bettmann Archive; **unnum. figure p. 172:** © Scott Camazine 1991; **unnum. figure p. 173:** Charles D. Winters; **unnum. figure p. 174:** University of Oxford, Museum of History of Science, courtesy of AIP Niels Bohr Library; **5-7:** adapted from D. B. Murphy and V. Rousseau, *Foundations of College Chemistry,* 1st ed., Ronald Press, 1969; **5-10(a):** courtesy of Finnigan Corporation; **5-10(b):** Charles Steele; **unnum. figure p. 184:** courtesy of Canon U.S.A., Inc.; **unnum. figure p. 186(bottom):** from J. M. Pasachoff, *Astronomy: From Earth to the Universe,* 3rd ed., Saunders College Publishing, 1987, © Gary Ladd, 1972; **unnum. figure p. 192(top):** from Philips Electronic Instruments, Inc.; **unnum. figure p. 192(bottom):** Spielman/CNRI/Phototake, NYC; **unnum. figure p. 204:** Runk/Schoenberger from Grant Heilman; **unnum. figure p. 214:** courtesy of James Mauseth; **unnum. figure p. 215(left):** from W. L. Masterton, E. J. Slowinski, C. L. Stanitski, *Chemical Principles,* 6th ed., Saunders College Publishing, 1985; **unnum. figure p. 215(right):** Peter Arnold, Inc./by Dagmar Hailer-Hamann.

Chapter 6

Chapter opener: AT&T Bell Laboratories; **unnum. figure p. 220, 6-8, 6-10:** Charles D. Winters; **unnum. figures pp. 221, 241(top left and right):** Charles Steele; **6-5:** from H. C. Metcalfe, J. E. Williams, J. F. Costka, *Modern Chemistry,* Holt, Rinehart and Winston, 1986 and CHEM Study Film: *Bromine—Element from the Sea;* **6-6:** Grant Heilman from Grant Heilman; **6-7:** from NASA/Goddard Space Flight Center; **6-11:** from Georgia Power Company by Max Fundom; **6-12:** K10234(2) courtesy Department Library Services, American Museum of Natural History; **unnum. figures p. 247:** EPA; **unnum. figures p. 248(top left and right):** Dean and Chapter of Lincoln; **unnum. figure p. 248(bottom left):** Ohio Edison; **6-13:** from National Center for Atmospheric Research/National Science Foundation.

Chapter 7

Chapter opener: J. Weber; **unnum. figures pp. 257(top), 280:** Charles D. Winters; **unnum. figure p. 257(bottom):** from Levin; **unnum. figures pp. 259, 292:** Charles Steele; **unnum. figures p. 274:** Robert C. Simpson/Tom Stack & Associates; **unnum. figure p. 275:** Carolina Biological Supply Company; **unnum. figure

p. 291: National Center for Atmospheric Research/National Science Foundation.

Chapter 8

Chapter opener: © NIH/Science Source/Photo Researchers, Inc.; **unnum. figures pp. 305, 308, 310(right), 314, 316, 318, 323:** Charles Steele; **unnum. figures pp. 310(left), 315:** Charles D. Winters; **unnum. figure p. 328:** courtesy of Bethlehem Steel.

Chapter 9

Chapter opener: Leon Lewandowski.

Chapter 10

Chapter openers: Marna G. Clarke; **unnum. figures pp. 359, 368, 369:** James W. Morgenthaler; **unnum. figure p. 364(top):** Martin Dohrn/Science Photo Library/Photo Researchers, Inc.; **unnum. figure p. 364(bottom):** Charles Steele; **unnum. figures pp. 366, 375:** Charles D. Winters.

Chapter 11

Chapter opener: Brinkmann Instruments; **unnum. figures pp. 382, 383:** James W. Morgenthaler; **11-1(a, b):** Charles D. Winters; **unnum. figures pp. 391, 393:** Charles Steele.

Chapter 12

Chapter opener: © E. R. Degginger; **unnum. figure p. 403:** from Metcalfe and CHEM Study Film: *Molecular Motion;* **unnum. figure p. 405:** Wolfgang Bayer/Bruce Coleman Inc., New York; **12-2(a):** from Highsmith, *Physics and Our World,* Saunders College Publishing, 1975; **12-2(b):** Taylor Scientific Instruments; **12-2(c), 12-15(b), unnum. figure p. 443(bottom):** Charles Steele; **unnum. figure p. 411:** The Granger Collection, New York; **unnum. figures pp. 413, 441:** Charles D. Winters; **unnum. figure p. 418:** National Center for Atmospheric Research/National Science Foundation; **12-8:** Marna G. Clarke; **unnum. figure p. 442:** Robert W. Metz; **unnum. figure p. 443(top):** courtesy of Chrysler; **unnum. figure p. 447:** NASA/Jet Propulsion Lab; **unnum. figure p. 450:** Oak Ridge National Laboratory; **unum. figure p. 451:** courtesy Fisher Scientific.

Chapter 13

Chapter opener: David Parker/Science Photo Library/Photo Researchers, Inc.; **13-1, 13-9, 13-14, 13-31(b, bottom), unnum. figures pp. 462, 477(bottom left and right), 478:** Charles D. Winters; **13-11(c), 13-31(c, bottom), unnum. figures pp. 464(top center), 464(bottom left margin), 471, 477(top right), 496, 498(bottom left and right):** Charles Steele; **unnum. figure p. 464(top right):** from Rainex; **unum. figure p. 464(middle left margin):** Manfred Danegger/Peter Arnold, Inc.; **13-16:** Marna G. Clarke; **unnum. figure p. 479:** from Joesten, *The World of Chemistry,* Program 8, "Chemical Bonds"; **13-18(b):** from Marvin L. Hackert; **13-31(a, bottom):** © Gemological Institute of America; **unnum. figure p. 494:** Chem. Design/Science Photo Library/Photo Researchers, Inc.; **unnum. figure p. 495:** Masato Murakimi/Istec/© 1991 Discover Publications; **unnum. figure p. 500(bottom):** R. E. Davis.

Chapter 14

Chapter opener: Richard Megna, Fundamental Photographs, New York; **14-4, 14-5, 14-6, unnum. figures pp. 515, 519(top left), 520,** 521, 549, 554: Charles Steele; **unnum. figure p. 517:** Kip Peticolas, Fundamental Photographs, New York; **14-14(a), unnum. figures pp. 518(top left), 548:** J. W. Morgenthaler; **unnum. figure p. 518(bottom left):** Bill Wood/Bruce Coleman Limited; **14-20(left), unnum. figures pp. 519(top right), 523:** © Robert W. Metz; **unnum. figure p. 519(bottom left):** Leon Lewandowski; **unnum. figures p. 525:** courtesy of Professor M. R. Willcott, III; **14-14(b):** Union Carbide Industrial Gases, Linde Division; **unnum. figure p. 534:** Bethlehem Steel; **unnum. figure p. 535:** Union Carbide Corporation; **unnum. figure p. 542:** Charles D. Winters; **unnum. figures p. 543:** courtesy of R. F. Baker, University of Southern California Medical School; **unnum. figure p. 544:** from NALCO Chemical Company; **unnum. figure p. 546:** courtesy E. I. Du Pont de Nemours and Company; **unnum. figure p. 547:** SIU School of Medicine/Bruce Coleman Inc., New York; **unnum. figure p. 550:** © Paolo Koch, Photo Researchers, Inc.; **unnum. figure p. 553(top):** © The Stock Market/Michael Furman, 1987; **unnum. figure p. 553(bottom):** Tony Freeman/PhotoEdit; **unnum. figure p. 555:** © The Stock Market/Dick Frank, 1987.

Chapter 15

Chapter opener, unnum. figure p. 605: Atlanta Gas Light Company; **15-2(a, b), unnum. figures pp. 598, 604:** Charles D. Winters; **unnum. figure p. 570:** from Fisher Scientific; **unnum. figure p. 588:** Charles Steele; **unnum. figure p. 592:** NASA; **unnum. figure p. 603:** courtesy of General Motors.

Chapter 16

Chapter opener: © Robert Herko/The Image Bank; **unnum. figures pp. 617(top), 648(top):** Charles D. Winters; **unnum. figures pp. 617(bottom), 634, 641, 650:** Charles Steele; **unnum. figure p. 620:** Leon Lewandowski; **unnum. figures pp. 642, 643, 651:** J. Morgenthaler; **unnum. figure p. 647:** from Harshaw/Filtrol Partnership; **16-17(b):** General Motors.

Chapter 17

Chapter opener: Dresser Industries/Kellogg Company; **unnum. figures pp. 672, 675, 684:** Charles Steele; **unnum. figures p. 647:** Charles D. Winters; **unnum. figure p. 677:** courtesy of M. W. Kellogg Company.

Chapter 18

Chapter opener, 18-3: Marna G. Clarke; **18-4, unnum. figures pp. 703, 709, 728,** Charles Steele; **18-1, unnum. figures pp. 712, 716, 734:** from Beckman Instruments; **18-2:** Charles D. Winters; **unnum. figures pp. 718, 720, 724:** J. Morgenthaler.

Chapter 19

Chapter opener, unnum. figure p. 750: Charles Steele; **unnum. figures pp. 745, 746:** from Beckman Instruments; **unnum. figures p. 752:** Charles D. Winters.

Chapter 20

Chapter opener, unnum. figures pp. 766, 768, 770: Charles Steele; **unnum. figure p. 760(bottom):** M. Timothy O'Keefe/Tom Stack & Associates; **unnum. figure p. 761(margin):** CNRI/Science Photo Library/Photo Researchers, Inc.; **unnum. figures p. 761(bottom):** Brian Parker/Tom Stack & Associates; **unnum. figures pp. 775, 776:** J. Morgenthaler; **unnum. figure p. 777:** Charles D. Winters.

Chapter 21

Chapter opener: Patricia Caufield, Photo Researchers, Inc.; **21-12(b)**, **unnum. figure p. 788:** Charles D. Winters; **21-5(c):** ASARCO, Inc.; **21-6(b)**, **21-16(a)**, Charles Steele; **unnum. figures pp. 796, 797:** J. Morgenthaler; **unnum. figure p. 798:** courtesy of Donald M. West; **Table 21-2:** from G. Charlot, *IUPAC SUPPLE-MENT*, "Selected Constants, Oxidation–Reduction Potentials in Aqueous Solutions," 1971; **unnum. figure p. 808(top):** M. D. Ippolito; **unnum. figure p. 808(bottom left):** courtesy of E. I. Du Pont de Nemours and Company; **unnum. figure p. 808(bottom right):** from Bethlehem Steel; **21-14(b):** from Kotz and Purcell, 2/e; **unnum. figures pp. 821(bottom right), 824,** courtesy of Eveready Battery Company; **21-17(b):** courtesy of Delco Remy Division, General Motors Corporation; **unnum. figure p. 825:** from United Technologies.

Chapter 22

Chapter opener: Paul Silverman, Fundamental Photographs, New York; **unnum. figure p. 834:** Charles Steele; **Table p. 835:** reprinted with permission from *Chemical and Engineering News,* 26 March 1979, © 1979 American Chemical Society; **unnum. figure p. 836:** Illustration copyright by Irving Geis; **unnum. figure p. 839(top left):** Ward's Natural Science Establishment; **unnum. figures pp. 839(top right), 840(bottom):** Charles D. Winters; **22-3:** John Shaw/Tom Stack & Associates; **Table 22-2:** from *Chemistry: A Contemporary Approach,* by G. T. Miller, Jr., © 1976 by Wadsworth Publishing Co., Inc., reprinted by permission; **unnum. figures pp. 840(top), 847:** Brian Parker/Tom Stack & Associates; **unnum. figure p. 840(middle):** Allen B. Smith/Tom Stack & Associates; **Table 22-3:** from R. G. Gymer, *Chemistry in the Natural World,* 1977, D. C. Heath and Co., reprinted by permission of the publisher; **22-10, unnum. figure p. 842:** James Cowlin; **22-7(b):** from the Aluminum Association, Inc.; **unnum. figures pp. 844, 846:** courtesy of Bethlehem Steel; **22-11:** from H. C. Metcalfe, J. E. Williams, J. F. Costka, *Modern Chemistry,* Holt, Rinehart and Winston, 1986; **unnum. figure p. 850(margin):** courtesy of E. I. Du Pont de Nemours and Company; **unnum. figure p. 850(bottom):** © Gemological Institute of America.

Chapter 23

Chapter opener: courtesy American Cyanamid Company; **unnum. figures pp. 855, 857(top), 861(bottom), 862:** Charles Steele; **unnum. figure p. 857(bottom):** Steve Elmore/Tom Stack & Associates; **unnum. figures p. 858:** from Masterton, Slowinski, and Stanitski; **unnum. figure p. 859:** Dick George/Tom Stack & Associates; **unnum. figure p. 861(top):** Kevin Schafer/Tom Stack & Associates; **unnum. figure p. 864(top):** Leon Lewandowski; **unnum. figure p. 864(bottom):** courtesy IBM; **unnum. figure p. 865:** © Gemological Institute of America; **23-1(a–c):** Charles D. Winters; **unnum. figure p. 867:** Paul Silverman, Fundamental Photographs, New York; **Tables 23-1, 23-2, 23-3, 23-5, 23-6:** adapted from Rochow and by permission of the publisher from *Inorganic Chemistry* by Jacob Kleinberg, William J. Argersinger, Jr., and Ernest Griswold (D. C. Heath and Company, 1960); **Table 23-4:** adapted from Kleinberg, Argersinger, and Griswold and from T. Moeller, *Advanced Organic Chemistry* (Wiley, 1952).

Chapter 24

Chapter opener: Charles Steele; **unnum. figure p. 873:** from NASA; **unnum. figure p. 875:** from Argonne National Labora-tory; **unnum. figure p. 878(top):** Kevin Schafer/Tom Stack & Associates; **unnum. figures pp. 878(middle and bottom), 879:** Charles Steele; **unnum. figures pp. 882, 883:** J. Morgenthaler; **Tables 24-1, 24-8:** from T. Moeller; **Tables 24-3, 24-5:** adapted from Moeller and from D. B. Murphy and V. Rousseau, *Foundations of College Chemistry,* Ronald Press, 1969; **Tables 24-6, 24-7, 24-9:** adapted from Moeller and by permission of the publisher from *Inorganic Chemistry* by Kleinberg, Argersinger, and Griswold (D. C. Heath, 1960); **Table 24-8:** adapted from Moeller.

Chapter 25

Chapter opener: Charles D. Winters; **unnum. figure p. 892(top):** Don and Pat Valenti/Tom Stack & Associates; **unnum. figure p. 892(bottom):** J. Morgenthaler; **unnum. figures pp. 895, 901:** Charles Steele; **unnum. figure p. 897:** Brian Parker/Tom Stack & Associates; **unnum. figure p. 898:** William Felger from Grant Heilman; **unnum. figure p. 899:** R. E. Davis; **25-3:** courtesy Field Museum of Natural History; **Table 25-1:** adapted from Rochow; **Table 25-2:** adapted from Moeller; **Table 25-3:** adapted from Kleinberg, Argersinger, and Griswold; **Table 25-4:** from R. Steudel, *Chemistry of the Non-Metals,* English Edition, by F. C. Nachod, J. J. Zuckerman, Waller de Gruyter, Berlin, New York, 1977.

Chapter 26

Chapter opener: Charles D. Winters; **unnum. figures pp. 909(top), 917:** from Metcalfe, Williams, and Castka and Walter O. Scott; **unnum. figure p. 909(bottom):** from American Breeders Service; **unnum. figures pp. 910, 912, 918, 919(top), 920(top), 921, 922, 923(middle):** Charles Steele; **unnum. figures pp. 911, 919(middle and bottom):** Charles D. Winters; **unnum. figure p. 920(bottom):** J. Morgenthaler; **Table 26-1:** adapted from Rochow.

Chapter 27

Chapter opener: Comstock; **unnum. figure p. 928:** Allen B. Smith/Tom Stack & Associates; **27-1, unnum. figure p. 930(bottom):** © Gemological Institute of America; **unnum. figures pp. 929(bottom), 931, 935(top), 936(middle), 938(top):** Charles Steele; **unnum. figure p. 932:** M. Timothy O'Keefe/Tom Stack & Associates; **unnum. figure p. 933:** Dick George/Tom Stack & Associates; **unnum. figure p. 935(bottom):** from Standard Oil Company (Ohio); **unnum. figure p. 936(bottom):** Paul Silverman, Fundamental Photographs, New York; **unnum. figure p. 938(middle):** Particulate Mineralogy Unit, Avondale Research Center, U. S. Bureau of Mines; **unnum. figures pp. 938(bottom), 939(bottom):** Charles D. Winters; **unnum. figures pp. 939(top), 939(middle):** J. Morgenthaler; **Table 27-1:** adapted from B. Mason, *Principles of Geochemistry,* 3/e (John Wiley and Sons, Inc., 1966).

Chapter 28

Chapter opener: courtesy IBM; **unnum. figures pp. 950, 951, 956:** J. Morgenthaler; **unnum. figures pp. 954, 957:** Charles Steele; **Table 28-3:** adapted from T. Moeller; **Table 28-7:** adapted from R. T. Sanderson, *Inorganic Chemistry,* Reinhold, 1967 and N. V. Sedgwick, *The Chemical Elements and Their Compounds,* Vol. II, Oxford, 1950; **Tables 28-8, 28-9:** adapted from Sanderson.

Chapter 29

Chapter opener: Charles D. Winters; **unnum. figures pp. 966, 967, 977, 985, 987:** J. Morgenthaler; **unnum. figure p. 971:** Charles D.

Winters; **unnum. figures p. 981:** Charles Steele; **unnum. figure p. 986(bottom):** from Masterton, Slowinski, and Stanitski.

Chapter 30

Chapter opener: University of California, Lawrence Livermore Laboratory; **unnum. figure p. 996:** AIP Niels Bohr Library, W. F. Meggers Collection; **unnum. figure p. 1001(bottom left):** Philippe Plailly/Science Photo Library/Photo Researchers, Inc.; **unnum. figure p. 1001(bottom right):** from Oak Ridge National Laboratory; **30-4:** from Brescia, Mehlman, Pellegrini, and Stambler, *Chemistry: A Modern Introduction,* 2/e, Saunders College Publishing, 1978; **unnum. figure p. 1005:** Dave Davidson/Tom Stack & Associates; **unnum. figure p. 1007:** Don and Pat Valenti/Tom Stack & Associates; **Table 30-4:** from *Handbook of Chemistry and Physics,* R. C. Weast, Ed., © CRC Press, Inc., 1973, used by permission of CRC Press, Inc., Boca Raton, FL, and from *Lange's Handbook of Chemistry,* J. A. Dean, Ed., McGraw-Hill, New York, 1973; **unnum. figure p. 1011:** from International Atomic Energy Agency; **30-8 and unnum. figure p. 1020:** Argonne National Laboratory; **unnum. figure(a) p. 1014:** Fermilab photo.; **unnum. figure p. 1016:** courtesy NASA; **unnum. figure p. 1017:** courtesy of E. I. Du Pont de Nemours and Company; **unnum. figure p. 1018:** U. S. Dept. of Energy/Science Photo Library/Photo Researchers, Inc.; **unnum. figure p. 1021:** Tass/Sovfoto; **unnum. figure p. 1023(top):** U. S. Navy/Science Photo Researchers, Inc.; **unnum. figure p. 1023(bottom):** courtesy of Princeton Plasma Physics Laboratory.

Chapter 31

Chapter opener: J. Weber, University of Geneva, Switzerland; **31-2(c, d), 31-3(b, c), 31-4, 31-7, 31-9(a, b), 31-10, 31-14, 31-15,** **31-16, 31-17(4 photographs), 31-18, 31-20(a, b), unnum. figures pp. 1050, 1056, 1058, 1071(left, right):** Charles Steele; **31-5, unnum. figure p. 1068:** Charles D. Winters; **31-8(a, b), unnum. figures pp. 1054(top left), 1064(top and middle, a, b):** R. E. Davis; **unnum. figure p. 1041:** courtesy American Petroleum Institute; **unnum. figure p. 1061:** Illustration copyright by Irving Geis; **unnum. figure p. 1064(bottom):** Photomicrograph by © Herb Charles Ohlmeyer/Fran Heyl Associates; **unnum. figure p. 1067:** Biophoto Associates; **31-19:** Leonard Lessin; **unnum. figure p. 1069:** Phillip A. Harrington/Fran Heyl Associates; **31-12:** reprinted with permission of Macmillan Publishing Co. from Conant and Blatt, *The Chemistry of Organic Compounds,* Fifth Edition, © 1959 by Macmillan Publishing Co.; **Table 31-4:** adapted from Fieser and Fieser, *Organic Chemistry,* 2/e, D. C. Heath and Co., Boston, 1950, p. 32; **Table 31-8:** from H. Hart, *Organic Chemistry, A Short Course,* 7/e p. 190, © 1987 by Houghton Mifflin Company, used by permission.

Chapter 32

Chapter opener: Dr. Harold Rose/Science Photo Library/Photo Researchers; **32-1, 32-2, 32-4, 32-5, 32-6, 32-9:** R. E. Davis; **32-3:** Ray Ellis/Photo Researchers, Inc.; **37-7(2 photographs), 32-8, unnum. figures pp. 1102, 1105:** Charles Steele; **32-10:** J. Morgenthaler; **unnum. figures pp. 1096, 1097, 1112(bottom right):** Charles D. Winters; **unnum. figure p. 1104:** Charles D. Winters—photographs, John Reynolds—spelunker; **unnum. figure p. 1107:** North Wind Picture Archives; **unnum. figure p. 1112(top):** courtesy of Drs. James L. Monro and Gerald Shore and the Wolfe Medical Publications, London, England; **unnum. figures p. 1112(bottom left, parts 1 and 2):** courtesy of E. I. Du Pont de Nemours and Company; **unnum. figure p. 1114:** Evans and Sutherland.

Index

Entries in *italics* indicate illustrations; page numbers followed by *t* indicate tables; page numbers followed by *m* indicate marginal notes.

LOCATION OF COMMONLY USED INFORMATION

Atomic and Molecular Properties

Atomic and ionic radii	223
Aufbau order, aufbau principle	202
Bond energies	581–582
Electron affinities	229
Electron configurations	Appendix B
Electronegativity values	232
Electronic, molecular, and ionic geometries	330
Hund's rule	204
Ionization energies	225
Molecular orbital diagrams	341
Pauli exclusion principle	203
Planck's equation	182
Quantum numbers (rules)	195–196

Thermodynamic Properties, Kinetics, Equilibrium, States of Matter

Absolute entropies	Appendix K
Arrhenius equation	641–643
Clausius–Clapeyron equation	471
Colligative properties (equations)	527–528, 532, 534, 541
Enthalpies of formation	576, Appendix K
Free energies of formation	Appendix K
Gas laws	419, 424, 436, 439
Heats of fusion, vaporization	469, 474, Appendix E
Hess' Law	577, 579
LeChatelier's Principle	671–676
Specific heats, heat capacities	Appendix E
van't Hoff equation	689
Vapor pressure of water	Appendix E

Acids, Bases, and Salts

Acid–base indicators	716, 751
Amphoteric hydroxides	367
Common acids and bases	129, 131, 362
Dissociation constants (complex ions), K_d	Appendix I
Ionization constants for weak acids, K_a	707, Appendix F
Ionization constants for weak bases, K_b	715, Appendix G
Names and formulas of common ions	285
Solubility product constants, K_{sp}	Appendix H
Solubility rules	132–133

Electrochemistry

Activity series of the metals	141
Faraday's Law	789–790
Nernst equation	810
Standard reduction potentials, E^0	803, 806, Appendix J

Miscellaneous

Classification of organic compounds	1051, 1072
Lewis formulas (rules)	271–273, 278–279
Naming alkanes	1036–1037
Naming coordination compounds	970
Naming inorganic compounds	282–287
Organic functional groups	1072
Oxidation states (common)	144, 145